STRUCTURAL GEOLOGY
OF ROCKS AND REGIONS

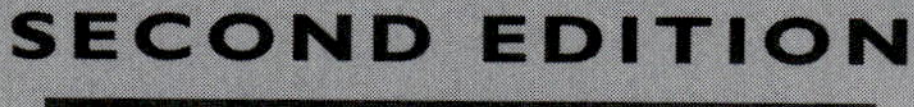

SECOND EDITION

STRUCTURAL GEOLOGY
OF ROCKS AND REGIONS

GEORGE H. DAVIS
The University of Arizona—Tucson

STEPHEN J. REYNOLDS
Arizona State University—Tempe

JOHN WILEY & SONS, INC.

ACQUISITIONS EDITOR Clifford Mills
MARKETING MANAGER Catherine Faduska
PRODUCTION EDITOR Deborah Herbert
DESIGN SUPERVISOR Ann Marie Renzi
MANUFACTURING MANAGER Mark Cirillo
ILLUSTRATION Sandra Rigby

Cover art by David A. Fischer
Cover designed by Ann Marie Renzi
Text designed by David Levy

This book was set in 10 on 12 Times Roman by BI-COMP, Inc. and printed by Courier Stoughton and bound by Courier Westford. The cover was printed by New England Book Component.

Recognizing the importance of preserving what has been written, it is a policy of John Wiley & Sons, Inc. to have books of enduring value published in the United States printed on acid-free paper, and we exert our best efforts to that end.

To order books or for customer service please, call 1(800)-CALL-WILEY (225-5945).

Library of Congress Cataloging-in-Publication Data
Davis, George H. (George Herbert), 1942–
Structural geology of rocks and regions.—2nd ed./George H. Davis, Stephen J. Reynolds.
p. cm.
Includes bibliographical references and indexes.
ISBN 978-0-471-52621-6 (Cloth)
1. Geology, Structural. I. Reynolds, Stephen J. II. Title.
QE601.D3 1996
551.8—dc20 95-40790
CIP

Printed in the United States of America

20

We dedicate this book to our parents. Through their love and generosity, a marvelous life of exploring and teaching, and the desire to continuously learn and grow, were made possible.

George acknowledges his parents, as well as his parents by marriage:

Ruth L. Davis and David H. Davis
Margaret G. Siepert and Albert F. Siepert

And Steve acknowledges his parents:

Elizabeth Ann Reynolds and John R. Reynolds

PREFACE

Our purpose in writing this textbook is to communicate the physical and geometric elegance of geologic structures within the Earth's crust and to describe the ways in which geologic structures reflect the nature and origin of crustal deformation through time. Geologic structures provide part of the basis for recognizing and reconstructing the profound changes that have marked the physical evolution of the Earth's outer layers, as observed from the scale of the plates down to the scale of the microscopic. Understanding the nature and extensiveness of deformational structures in the Earth's crust has both scientific value and practical benefit. But, there is a philosophical value as well. Our perceptions of who we are and where we are in time and space are shaped by facts and interpretations regarding the historical development of the crust of the planet on which we live. Knowing fully the extent to which our planet is dynamic, not static, is a reminder of the lively and special environment we inhabitat.

For example, we have come to understand that our earthly foundations are not fixed. Instead, we live on continent-size plates that are in a continual state of slow motion. The interaction of these plates has played a dominant role in both the formation and deformation of rock bodies in the Earth's crust. Knowledge of present-day plate tectonic processes aids us in interpreting past structural movements. Furthermore, evaluating and interpreting past dynamics may help us predict what the present actions hold for the future.

The focus of this book favors applications in regional tectonics, exploration geology, active tectonics, and geohydrology, but at all times it underscores the importance first of understanding fundamental concepts and principles. Perhaps, what is emphasized most is how structural geologists think about deformed rocks. Once the conceptual framework within which structural geologists operate is grasped, the Earth begins to look different. In fact, natural physical processes and natural physical phenomena, whether geologic or not, never quite look the same again. At least, that has been our experience.

We have arranged this text in three parts. *Fundamentals* (*Part I*) provides essential background for analyzing *Structures* (*Part II*). *Descriptive Analysis* (*Part III*) describes how to function in the field, and what to do with the data that are collected.

Nature of Structural Geology (*Chapter 1*), introduces the beauty, the challenging geometry, and the practical value of structural geology. It also presents the basic philosophical approach to structural geology as used throughout the book. The approach is known as **detailed structural analysis**. The steps are descriptive (the physical and geometric description of geologic structures and deformed rocks), kinematic (evaluation of changes in location, orientation, shape, and size of rocks during deformation), and dynamic (evaluation of the forces and stresses that caused deformation, and appraisal of the strength and behavior of the rock at the time of deformation). These three steps of detailed structural analysis provide the leverage for unraveling and interpreting structures and systems of structures at any scale, from rocks to regions.

Kinematic Analysis (*Chapter 2*), presents the means for evaluating and describing how rocks have changed location, orientation, shape, and size during the course of deformation. *The very presence of structures and systems of structures in rocks in the Earth's crust reflects translation* (*change in the location*), *rotation* (*change in orientation*), *distortion* (*change in shape*), *and dilation* (*change in size*) *of the rocks in which they are found. If no such changes were required, the structures would have no reason to exist.* Kinematic analysis is applied at all scales, from movements within the lattice of an individual crystal to movements of plates.

Dynamic Analysis (*Chapter 3*) probes the origin of deformation in terms of forces, stresses, rock strength, and mechanics (the relationship of stress to strain). Down-to-earth examples give a "feel" for forces and stresses and the math that is used to describe them. Conventional laboratory testing of rock strength is employed as a useful context to think about forces and stresses, and to apply the basic math. Furthermore, laboratory evaluation of the response of rocks to force and stress permits us to see more clearly and quantitatively the influence of variables like lithology, temperature, confining pressure, fluid pressure, and strain rate on resistance to deformation.

Deformation Mechanisms And Microstructures (*Chapter 4*) explores what actually takes place at the microscopic and submicroscopic scales that enables "hard" rock to change shape and size—as if it were soft. In this chapter we see that temperature, pressure, mineralogy, strain rate, and the presence (or absence) of fluids determine which mechanisms will operate during deformation. The processes that permit rocks to deform on the grain and subgrain scales range from brittle microfracturing and grinding to plastic deformation involving the subtle "creep" of crystals. The ease (or difficulty) with which microscopic deformational mechanisms spring into action when rock is stressed controls the level of stress that rock can support.

Part II, *Structures*, presents the chief classes of structures. Structures are described and analyzed according to the methods and principles presented in Part I, "Fundamentals." In addition, the mechanics of formation of each class of structure is examined, using principles of applied physics. The geometry and formation of the structures is viewed from the microscopic to the regional, although the greatest descriptive attention is at the outcrop and map scale.

Within *Part II* are the following chapters: *Joints and Shear Fractures* (*Chapter 5*), *Faults* (*Chapter 6*), *Folds* (*Chapter 7*), *Cleavage, Foliation, and Lineation* (*Chapter 8*), and *Shear Zones and Progressive Deformation* (*Chapter 9*). Part II concludes with *Plate Tectonics* (*Chapter 10*), a chapter that expresses the modern concept that much, if not most, crustal deformation is directly or secondarily linked to processes of seafloor spreading and plate tectonics. In the "plates" chapter we will explore the intriguing geometry and motions of plates on a sphere. We examine the nature and magnitude of forces created by plate movements. And we view the character of active tectonism at or near modern plate boundaries.

In *Concluding Thoughts* following Chapter 10, we emphasize that structures do not exist in isolation, but are integrated within networks of interconnecting structures, at all scales, to accomplish just the right amount of deformational response to the prevailing tectonic stresses. *Concluding Thoughts* is a jumping off point for studying systems of structures at all conceivable scales.

Part III, *Descriptive Analysis*, is subdivided into sections A to N. It is intended to be used as preparation for field work, and as a guide to "reducing" data collected in the field. The sections in Part III are entitled: (A) Geologic Mapping, (B) Mapping Contact Relations, (C) Identifying Primary Structures, (D) Measuring Structural Orientations with a Compass, (E) Preparing Geologic Cross Sections, (F) Preparing Subsurface Contour Maps, (G) Orthographic Projection, (H) Stereographic Projection, (I) Stereographic Evaluation of Rotation, (J) Determining Slip on a Fault Through Orthographic and Stereographic Projection, (K) Stereographic Relationship of Faults To Principal Stress Directions, (L) Methods in Joint Analysis, (M) Some Additional Tips on Fold Analysis, and (N) Studying Shear Zones in the Field.

We have written this book in a way that proceeds from basic physical concepts and methods of analysis to the description and interpretation of structures and systems of structures. Yet we have also made each chapter as self-contained as possible, knowing that some users will want to read or present the material in a different order. Throughout the book, techniques, methods, experiments, and calculations are described in detail, with the aim of inviting active participation and discovery through laboratory and field work.

Those who have come to know the first edition of *Structural Geology of Rocks and Regions* will find some notable changes in this the second edition. (1) The authorship has expanded from one to two. We, Davis and Reynolds, have been close colleagues for years and now are both professors at neighboring universities. (2) The geologist pictured on the cover of the first edition, who had been sitting on the same outcrop for 11 years (!), is now up and walking around again, checking out new relationships. We asked our cover illustrator to step back a bit, to see if there was anyone else working in the area. In fact, there is. And there may be others. (3) Two new chapters have been added, and these are ones that bear importantly on the understanding of modern structural geology: *Deformation Mechanisms and Microstructures* (*Chapter 4*) and *Shear Zones and Progressive Deformation* (*Chapter 9*). (4) Two former chapters have been eliminated (Contacts and Primary Structure), although central concepts of each have been incorporated elsewhere in the text. (5) Nitty-gritty descriptive analysis (formerly Chapter 2) is now positioned as the core of Part III. (6) The organization is more streamlined, allowing the reader to move directly from the fundamentals (kinematic analysis, dynamic analysis, deformation mechanisms) into the major structures themselves (joints, faults, folds, cleavage, foliation, lineation, and shear zones), without an interlude. Moreover, by sweeping most nitty-gritty descriptive operations into Part III, there are fewer roadblocks to smooth, uninterrupted reading. (7) *Plate Tectonics* (*Chapter 10*) comes at the end (although it could be read at the beginning), providing a natural transition into texts and courses in regional tectonics, active tectonics, and plate tectonics. Through it all, we have tried hard to make the second edition as user-friendly as the first.

Structural geology is an essential tool in unraveling the geologic history of any given area or region within the Earth, especially in mountain belts where original arrangements of rocks and geologic contacts are profoundly modified by deformational movements.

When viewed at geological scales of time and spatial dimension, *rocks must be regarded as materials almost without strength, capable of being deformed continuously even under the slightest of pressures*. The geologic mapping of mountain belts has shown this to be true time and time again. The understanding of this paradox, which we try to capture in the expression "*soft as a rock*," yields a transformed view of the strength and behavior of rocks and regions.

Practical applications of structural geology are broad ranging and powerful—traditionally in petroleum geology, exploration geology, and mining engineering, and now in environmental geology, hydrology, geothermal energy development, and civil engineering. The practical value of structural geology derives largely from the fact that movement and trapping of fluids through rock are strongly influenced by fracturing, faulting, and folding. Thus, a firm understanding of principals of structural geology is beneficial when exploring for oil and gas, when investigating for the source of contaminants in a part of the groundwater system, when evaluating the energy potential of a natural geothermal system, when appraising daring but dangerous proposals for the subterranean disposal of radioactive waste, when examining basin sediments as potential groundwater reservoirs, when assessing the fundamental stability of steep slopes underlain by rain-infiltrated fractures in the bedrock, or when planning the excavation of mines or tunnels. Because most ore deposits owe their existence to the movement of mineral-bearing hydrothermal solutions through fractured bedrock, structural geology is a basic tool in the exploration for metals, and in the "development" of an ore body. Finally, the understanding of active tectonic events like earthquakes, volcanic eruptions, and massive landsliding are intellectually demanding activities that require knowledge derived from intimacy with the principals of fracturing and faulting.

Field relations are the primary sources for structural geologists. What value "the field" has for those intent on understanding Earth dynamics! Dr. Howard Lowry, a brilliant scholar in the field of English literature, captured the significance of primary sources. His words have special meaning for geologists:

> By [primary sources] we mean the first-hand things, the authentic ground of facts and ideas, the original wells and springs out of which all the rest either is drawn or flows. . . . Regard for the primary sources makes one forever the enemy of preconceptions, of manipulated data, raw opinion, and guesswork—of all the sleek shortcuts to wisdom in ten easy lessons. . . . Exclusive reliance on second-hand things makes second-hand men and women. It deludes us into thinking we are wiser than we are. . . . Breadth of knowledge, even knowing a little about a lot, has its obvious value. But breadth that perpetually sends down no clean, strong roots in the primary sources—into the deep earth and "the hidden rivers murmuring in the dark" of the rocks—such breadth clarifies very little. It merely puts our bewilderment on a broader basis. It leads us into incredible naivete and gullibility. It makes us too quick to believe all we read. (From *College Talks* by H. F. Lowry, edited by J. R. Blackwood, pp. 86–87. Published with permission of Oxford University Press, New York, copyright © 1969.)

This text on structural geology is based on and directed to the primary sources.

One final note. Both of us like to tell stories to illustrate and to bring to life the material we present. When George tells his stories, he goes right into it: "Going to the freezer for a midnight snack, I pulled out a pizza whose form is portrayed in the geologic map shown in . . . ." When Steve tells his stories, he inserts an (SJR): "When I (SJR) prepared the Geologic Map of Arizona, I gave careful attention to which rocks would be . . . ." In this way you will know who is telling which personal story within a book that "we" have so enjoyed writing together.

George H. Davis and Stephen J. Reynolds

ACKNOWLEDGMENTS

Many people helped enormously in the preparation of this book. Wallace P. Varner (who was the main illustrator for the first edition) and Susie Gillatt generated most of the line drawings and artwork, and patiently took care of alterations and modifications whenever they were required. Furthermore, Wally and Susie worked to assure that all drawings were accounted for on disks, and in the proper formats. Given the number of illustrations, and the complexity of many of them, we are especially grateful for their efforts. Jim Abbott, yet a third illustrator, came into the project when we needed him most to help carry the heavy load of line drawings and artwork, and thus advanced the project. David A. Fischer rendered the artwork for the cover and most of the cartoons, just as he did for the first edition. Bill Keller produced high-quality prints from the negatives and slides that we provided.

Angela Smith assisted us in countless ways, including, but not limited to, tracking down books, journal articles, and illustrations; assembling, copying, and mailing draft versions of the text and illustrations; helping with the securing of permissions; and working on the bibliography. She did her work, as always, cheerfully and effectively. In addition, we wish to express our appreciation for the backup support provided by Lori Roe and Alexei Peristykh, in particular for their substantial assistance in assembling the references section of the book.

Certain colleagues helped us in ways that go far beyond the norm by giving especially careful reading of text, providing and/or giving permission to use exceptional numbers of quality illustrations, and clarifying difficult technical points. These colleagues include Atilla Aydin, Ed Beutner, Steve Boyer, Terry Engelder, Simon Hanmer, Martin Huber, Andy Kronenberg, Steve Marshak, Ken McClay, Win Means, Nick Nickelsen, Cees Passchier, Donal Ragan, John Ramsay, Randy Richardson, Charles Sammis, Carol Simpson, John Suppe, Jan Tullis, Chris Wilson, and Mary Lou Zoback.

Outlines of the organization of this edition and early drafts of several chapters were formally reviewed. The input and observations received from the following reviewers were very beneficial: Edward C. Beutner, Franklin & Marshall College; Wallace A. Bothner, University of New Hampshire; Susan M. Cashman, Humboldt State University; Ibrahim Cemen, Oklahoma State University; Colleen Elliott, Concordia University; Randy Forsythe; State University of New Jersey—Rutgers; John R. Griffin, University of Nebraska, Lincoln; Nancy Lindsley-Griffin, University of Nebraska, Lincoln; Eric Nelson, Colorado School of Mines; George C. Stephens, George Washington University; James L. Whitford-Stark, Sul Ross State University. We continue to be grateful to the reviewers of the first edition, who are Edward C. Beutner, Frederick W. Cropp, Randall D. Forsythe, Donal M. Ragan, and David V. Wiltschko.

We appreciate immensely the high quality of support rendered by John Wiley & Sons, Inc., in transforming manuscript and figures to printed page. We extend our deepest thanks and heartfelt praise for the scrupulous care in step-by-step editing and production to Deborah Herbert, Production Editor. Similarly, we are indebted to Ishaya Monokoff, Illustration Director, and Sandra Ribgy, Illustration Assistant, for their work in assembling and presenting our illustration materials. We thank Ann Renzi, Designer, for a design so well suited to the purpose and personality of the book. We thank Editors Joan Kalkut, Chris Rogers, and Clifford Mills for their guidance, help, and patience throughout the project.

Last but not least, George acknowledges the support of Merrily Davis, and Steve acknowledges the support of Susie Gillatt. As those who have attempted a textbook know, the time investments are heavy and the results slow in coming. We really depended on the support of these two special people in our lives.

G. H. D.
S. J. R.

CONTENTS

STRUCTURAL GEOLOGY
OF ROCKS AND REGIONS

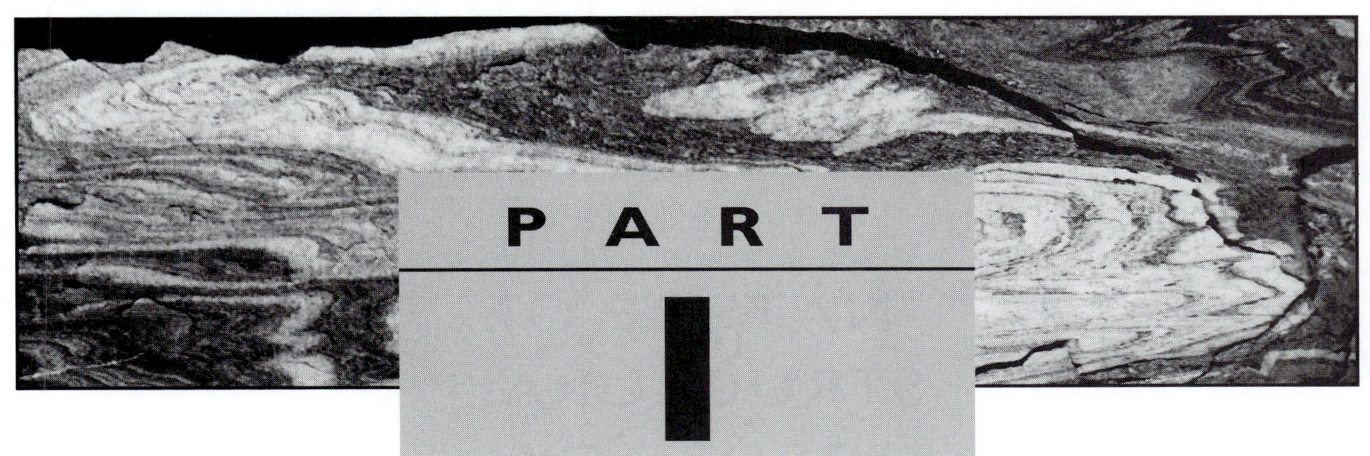

PART I

FUNDAMENTALS

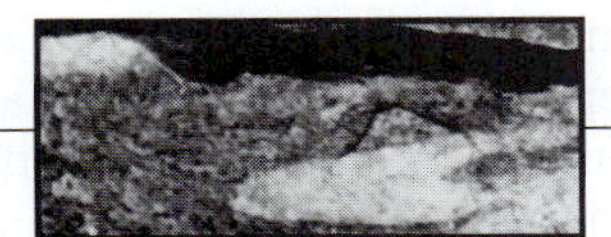

CHAPTER 1

NATURE OF STRUCTURAL GEOLOGY

DEFORMATION OF THE EARTH'S CRUST

The start of any journey into unfamiliar territory is often spurred by dreaming, a kind of dreaming that spawns not lightheadedness but intense curiosity and the setting of goals. Our journey will explore the architecture of the crust of the planet on which we live. We will be concerned primarily with architectural forms that have developed through deformation as a response to forces and stresses.

Deformation is a word that is used in several ways. It refers to the structural changes that take place in the original location, orientation, shape, and volume of a body of rock. It refers to the physical and chemical processes that produce the structural changes. And it refers to the geologic structures that form to accommodate the changes. Any body of rock, no matter how hard, will deform if the conditions are right. This concept emerges in historic photographs of a fence line located at the site of the Hebgen Lake earthquake, a destructive quake that wracked southwestern-most Montana in 1959. As a result of shifts in the ground surface, the fence was forced to shorten. Where shifts were modest, shortening was accommodated by bending (Figure 1.1*A*). But where shortening exceeded the bending limit of the wooden slats, the fence fractured and splintered abruptly (Figure 1.1*B*).

Structural deformation results from stresses (for now, think of "pressures") that exceed rock strength. When strength is exceeded, the rock will fail by brittle (fracture) or ductile (flow) deformation, depending on how the physical environment has affected the ability of the rock to resist the stresses. For example, the ability of a rock to withstand stresses decreases with increasing temperature. Stresses are created in nature in countless ways: The weight of thousands of meters of sediments within a depositional basin creates a vertical stress that generally results in the thinning and compaction of the sediments as they are buried deeper and deeper. The forceful intrusion of magma can "shoulder aside" rocks and

Figure 1.1 (*A*) Buckled fence and (*B*) broken fence, Hebgen Lake earthquake area, Montana. Fences, like rocks, respond in different ways to shortening. (Photograph by J. R. Stacy. Courtesy of United States Geological Survey.)

produce folding and faulting or stretching and thinning of the country rock that is invaded. The cooling of an igneous body, such as a basalt lava flow, causes shrinkage and contraction expressed in columnar jointing (Figure 1.2). The slow, steady "head-on" convergence of plates at plate boundaries produces major fault systems and fold belts, in some cases raising beds to vertical orientations (Figure 1.3). The spreading apart of plates along the oceanic ridges stretches the oceanic crust by faulting, rifting, and the injection of swarms of dikes. Where tectonic plates slide past one another, such as along the San Andreas fault in California, the buildup of stress results in sudden punctuated movement announced by earthquakes. The gravitational collapse of volcanoes above evacuated magma chambers can produce enormous craterlike calderas, like Crater Lake in Oregon (Figure 1.4).

Stresses that cause deformation generally build slowly but persistently, but in some situations incredibly high stresses "just show up." We have in mind Meteor Crater, located in northern Arizona, where asteroid impact created a bull's-eye of deformational destruction (Figure 1.5*A*). The now-

Figure 1.2 Columnar joints formed in basalt, exposed at San Miguel Regla, Hidalgo, Mexico. (Photograph by C. Fries. Courtesy of United States Geological Survey.)

Figure 1.3 Steeply inclined limestone beds of Cretaceous age in the Central Andes east of Lima, Peru. The lake in the foreground is at an elevation of 14,000 ft (over 4 km); rocks in the background reach 18,000 ft (over 5 km). The limestone beds were originally deposited below sea level! (Photograph by G. H. Davis.)

upturned, pervasively fractured and distorted sedimentary rocks never knew what hit them (Figure 1.5*B*), as revealed by the telltale presence of a peculiar microscopic mineral texture aptly named "shocked" quartz (Figure 1.5*C*).

ARCHITECTURE AND STRUCTURES

Jacob Bronowski (1973), in his superb set of essays entitled *The Ascent of Man*, suggests that our conception of science today is a description and exploration of the underlying structures of nature, and he points out that words like "structure," "pattern," "plan," "arrangement," and "architecture" constantly occur in every description that we try to make. He believes:

> The notion of discovering an underlying order in matter is man's basic concept for exploring nature. The architecture of things reveals a structure below the surface, a hidden grain, which when it is laid bare, makes it possible to take natural formations apart . . . . (From *The Ascent of Man* by J. Bronowski, p. 95. Published with permission of Little, Brown and Company, Boston, copyright © 1973.)

Bronowski's remarks apply beautifully to **structural geology**, which can be most succinctly defined as the study of the architecture of the Earth's crust, insofar as it has resulted from deformation (Billings, 1972, p. 2). The expression "architecture of the Earth" is very appropriate because structural geology addresses the form, symmetry, geometry, and certainly the elegance and artistic rendering of the components of the Earth's crust on all scales (Figure 1.6). At the same time, structural geology focuses on the strength and mechanical properties of crustal materials, both at the time of their deformation and now.

Although architecture and structural geology have much in common, the challenges of the architect and the structural geologist are quite different. The architect designs a structure, perhaps a building or a bridge,

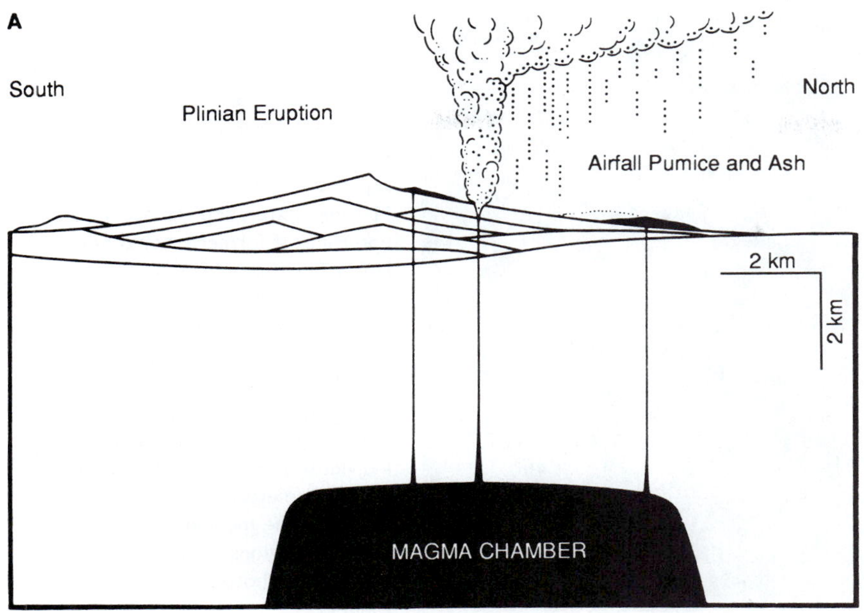

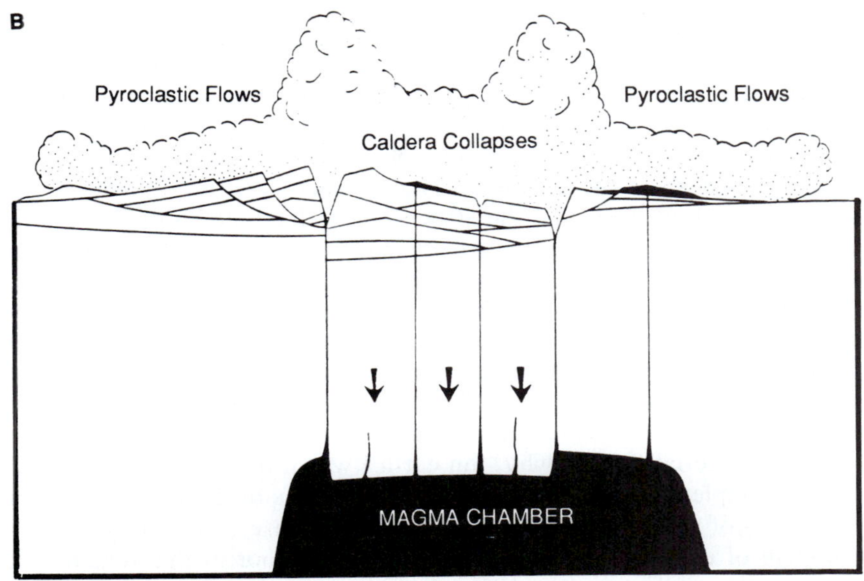

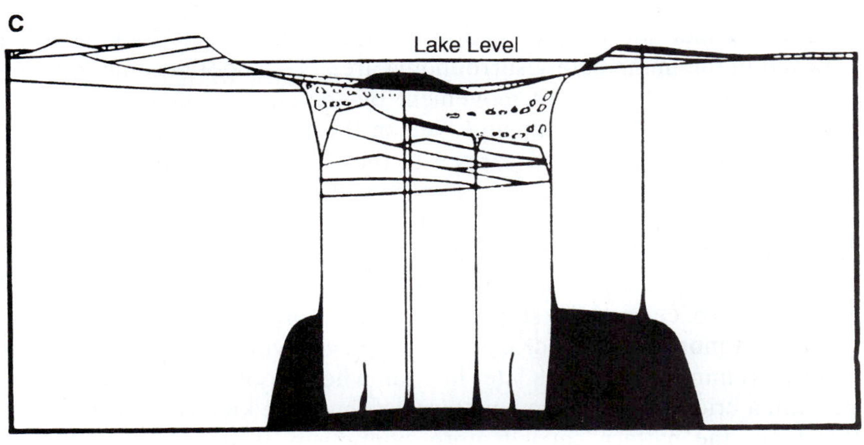

Figure 1.4 (*A*) Eruptive formation of prehistoric volcano, Mount Mazama. (*B*) Caldera collapse of the volcanic edifice into the emptied magma chamber. (*C*) Lake forms within the caldera structure. At the center of the lake minor eruptions build a small cinder cone, part of which (not shown) is Wizard Island. [Drawing by Charles R. Bacon. Courtesy of United States Geological Survey (1988).]

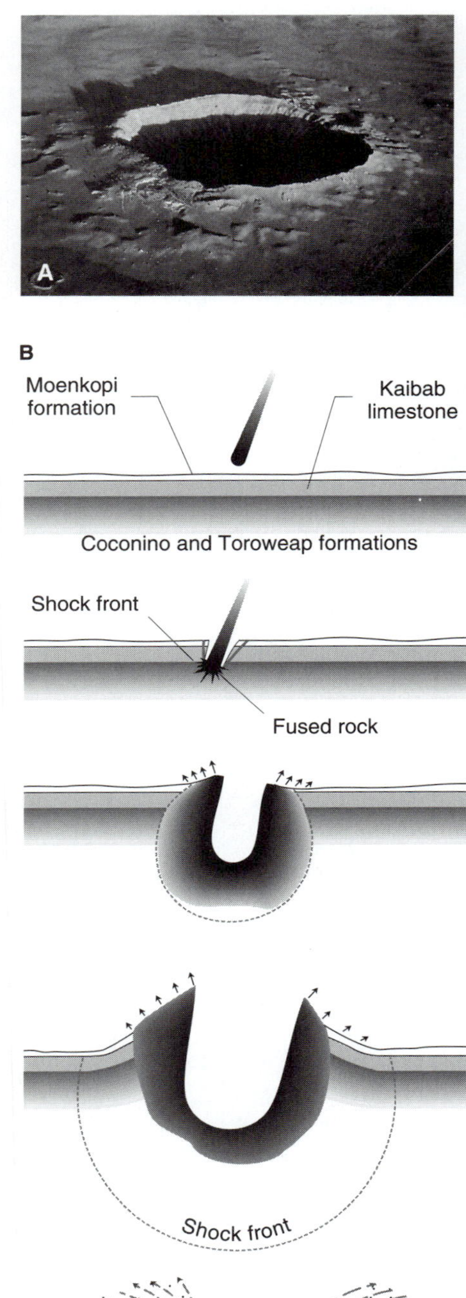

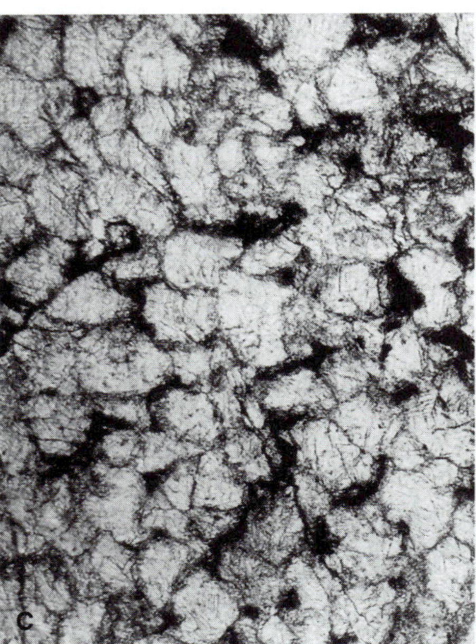

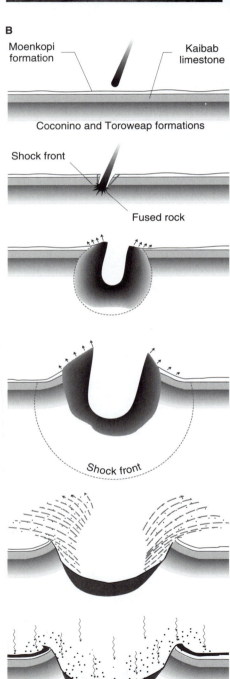

Figure 1.5 (*A*) Oblique aerial photo of Meteor Crater, Arizona. (Photograph and copyright by Peter Kresan). (*B*) Diagrammatic representation of the formation of Meteor Crater, and the accompanying upturning of formerly horizontal strata. [After Shoemaker (1979), fig. 4, p. 11.] (*C*) Shocked quartz collected from Meteor Crater. The rock, derived from the Coconino Sandstone, is 75% quartz, 20% coesite, and 5% glass. The light-colored quartz grains are deformed to fit into a mosaic. Individual quartz grains are close to 0.1 mm in diameter. The black opaque areas and the medium gray areas are the main regions of coesite and stishovite. (Photomicrograph by Susan Kieffer).

giving due attention to function, appearance, geometry, material, size, strength, cost, and other such factors. Then the architect supervises the process of construction daily, or perhaps weekly, making changes where necessary. In the end, the architect may be the only person who is aware of discrepancies between the original plan and the final product.

In contrast, the structural geologist is greeted in nature by what looks like a finished product, like the structural product shown in Figure 1.7, and is challenged to ask a number of questions. What is the structure? What starting materials were used? What is the geometry of the structure? How did the materials change shape during deformation? What was the source of the stresses that caused the deformation? And what was the sequence of steps in construction? Attempts to answer these questions generate even more questions. When was the job done? How long did it take? What were the temperature and pressure conditions? How strong were the materials? And, why ''on earth'' was it done?

The complexity of interpreting natural systems hit home to me in the reflections of a small pool I encountered amid dense underbrush—within the bush of eastern Canada. Rock exposure is poor in this region, and thus clues regarding structural history are meager. Yet the surface waters of this pool were marked by foam patterns that conveyed geological insight. Delicately fashioned, these patterns resembled the layering of rocks deformed under hot, deep conditions—exactly the kind of rocks that are exposed in the immediately surrounding area. This pool became my one-day laboratory. Patterns of movement on the surface of the pool were both complex and ever-changing. Seeing the patterns come and go, my mind shifted to what would happen to these structures when winter set in. Some single pattern would be frozen; one of an infinite number of patterns would be preserved; and yet, that pattern might or might not be representative of the kinds of motion I had watched. I began to realize more fully that every geologic record we examine is but one out of millions of possible frozen records, stop-action points, tiny scenarios from a much longer and more complex drama that we never will know in full.

I was reminded of all this later in Utah when I looked down into a pool beneath a bridge and saw flow patterns of the same kind (Figure 1.8). But this time, the pattern carried more symbolism. The ordered patterns,

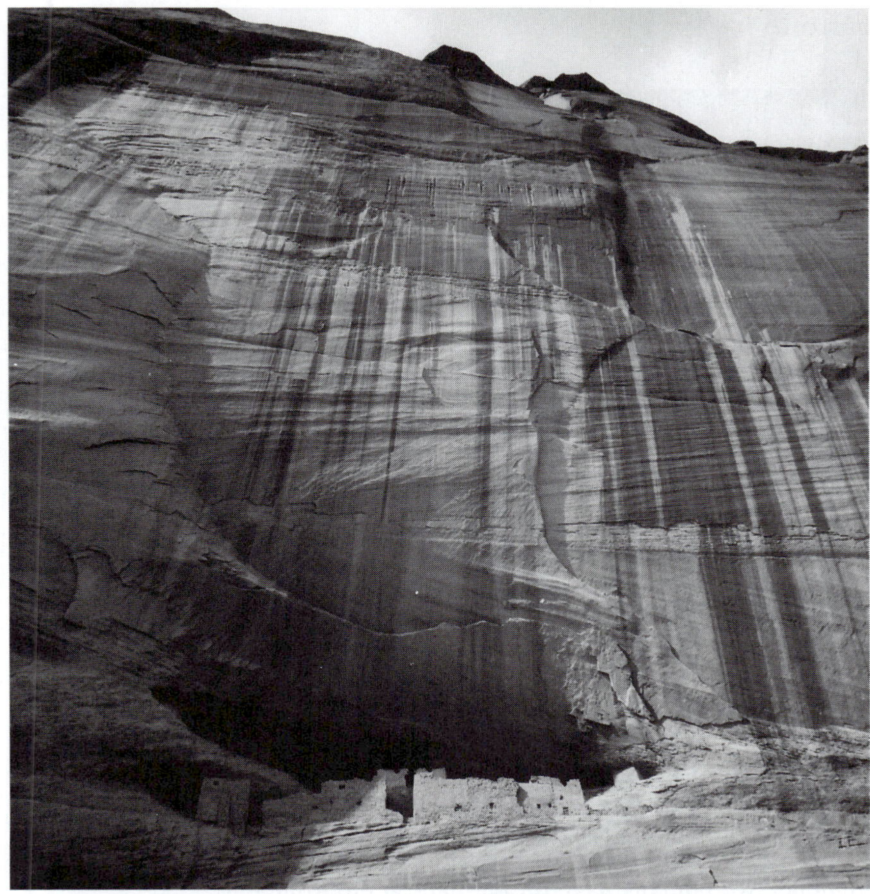

Figure 1.6 White House Ruin in Canyon de Chelly, Arizona, a sublime blend of the architecture of nature and that of the Ancient Ones. (Photograph by G. H. Davis.)

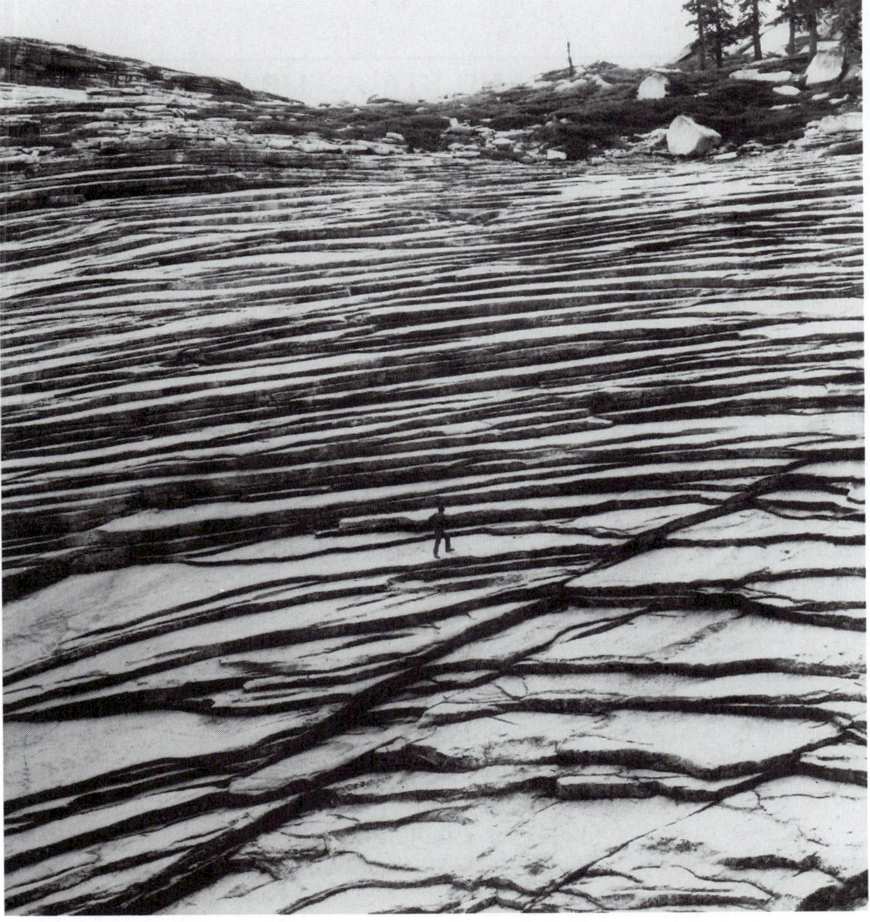

Figure 1.7 Geologist confronting the structure of nature, in this case an exfoliation jointing in granite near Shuteye Peak in the Sierra Nevada. (Photograph by N. K. Huber. Courtesy of United States Geological Survey.)

Figure 1.8 Deformed foam layers in a Rocky Mountain pool. Paper cup (upper right corner) is being blown by the wind from right to left, while the water is being pulled by gravity toward the lower right. (Photograph by G. H. Davis.)

guided ultimately by the flow of water under the compelling tug of gravity, were being modified simultaneously by the competing movement of a paper cup, blown by the wind, superimposing the imprint of its wake. In similar fashion, we see in matters of geologic record the complex, interfering effects of competition among the agents of gravity, heat, and tectonic stress in fashioning architectural form. And we see in the paper cup the influence we humans have on the natural environment.

PLATE TECTONICS AND STRUCTURAL GEOLOGY

Plate tectonics provides an essential backdrop for understanding the significance of structures, especially regional structures. It is the basis for understanding the dynamic circumstances that give rise to deformational movements. Plate interactions create rock-forming environments, which in turn give rise to the fundamental, original properties of regional rock assemblages. Furthermore, plate motions, both during and after the construction of regional rock assemblages, generate the stresses that impart to rocks their chief deformational characteristics.

Plate motions in the past have been responsible for shaping **orogenic belts** (or simply **orogens**), which are long, broad, and generally linear to arcuate belts in the Earth's crust where extreme mechanical deformation and/or thermal activity are concentrated. The Appalachians, Alps, Andes, and Himalayas are examples. Major regional structures abound in orogens, and these reflect systematic **distortion** (i.e., change in shape) of the crust in which the structures are found. Mountain systems are a physiographic expression of orogenic belts, but the presence of mountains is not integral to our view of an orogen. Ancient orogens, still recognizable as sites of regional distortion, are beveled to flatlands in the interior of continents. And of the presently forming orogens, the structurally interesting parts may not lie in the mountains, but instead may be 10, 50, or even 700 km below the Earth's surface. In this perspective, mountains, if they exist at all, are just the roofline of an orogen.

The generally accepted view among geologists today is that orogenic belts evolve through the interference of slowly moving rigid plates com-

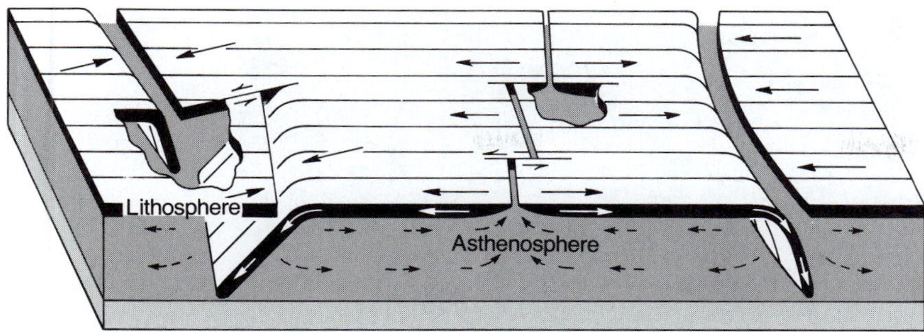

Figure 1.9 The slow, steady, continuous, inevitable movement of the plates, and the shaping of the architecture of the lithosphere. Here we see the convergence, divergence, and strike-slip of lithospheric plates. (From B. Isacks, J. Oliver, and L. R. Sykes, *Journal of Geophysical Research*, v. 73, fig. 1, p. 5857, copyright © 1968 by American Geophysical Union.)

posed of lithosphere (Figure 1.9). Lithosphere is made of continental and/or oceanic crust as well as uppermost mantle material. It can be thought of as the Earth's mechanically competent outer rind, which sluggishly moves on a part of the mantle, known as asthenosphere, that is capable of flowing continuously.

By studying the present configuration of plates, we learn that orogenic belts mainly form along plate margins at or near plate boundaries. It is along plate boundaries that plates interfere. The breadth of an orogen reflects the degree to which plate margins are internally distorted by plate interference. Even though most distortion is concentrated in boundary regions between plates, some regional structures form well within the interior of plates, apparently through transmission of stresses for great distances from plate boundaries.

THE FUNDAMENTAL STRUCTURES

There are three categories of fundamental structures: contacts, primary structures, and secondary structures. **Contacts** are the boundaries that separate one rock body from another. They include normal depositional contacts, unconformities, intrusive contacts, fault contacts, and shear zone contacts (see Part III-B). **Primary structures** are the structures that develop during the formation of a rock body, for example, in sediment before the sediment becomes sedimentary rock, or in lava or magma before it becomes volcanic or intrusive igneous rock (see Part III-C). Primary structures, such as cross-bedding or ripple marks in sandstone, or gas vesicles or ropy texture in basalt, generally reflect the local conditions of the environment within which the rock forms.

Secondary structures are the principal focus of this text. These are structures that form in sedimentary or igneous rocks after lithification, and in metamorphic rocks during or after their formation. The stresses that create secondary structures are commonly relatable to regional deformation. The distinction between primary and secondary structures is sometimes difficult and arbitrary. For example, in regions of active regional deformation, some of the materials that are affected are wet sediments that have not had sufficient time to become lithified. Thus, unlike mud cracks in sediment or pillows in basalt, the "primary" structures within them may have regional tectonic significance.

The fundamental secondary structures in nature are joints and shear fractures; faults; folds; cleavage, foliation, and lineation; and shear zones. These are introduced here and discussed more fully in Part II.

Joints and **shear fractures** are smooth, planar cracks that interrupt the cohesion of the rock and along which there has been *almost* imperceptible

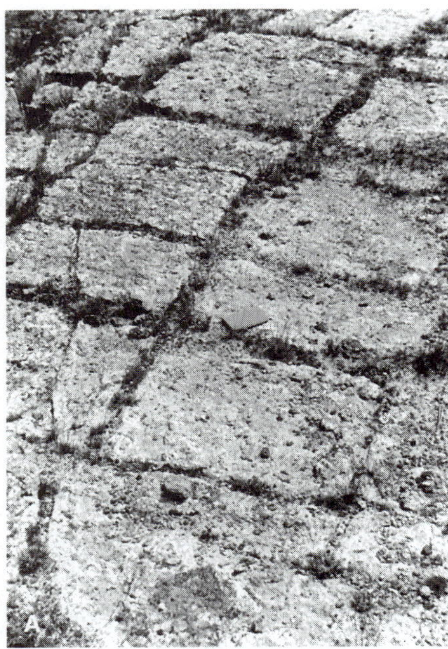

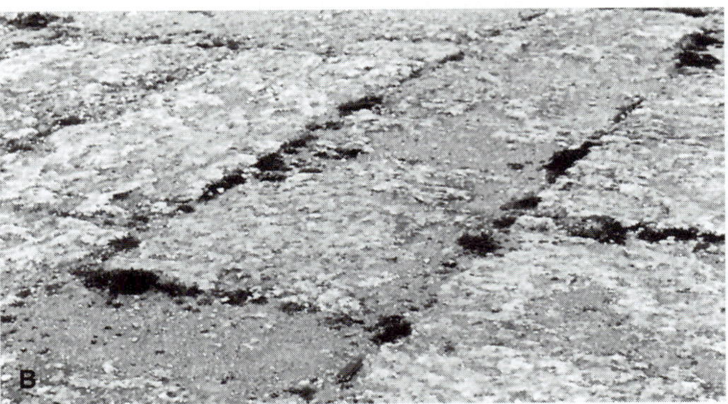

Figure 1.10 The spacing and orientation of jointing is often so regular and systematic that it looks man-made. Compare (*A*) the jointing in limestone in northwestern New Mexico (see notebook for scale) to (*B*) sidewalk-size slabs of limestone carefully placed by Romans in the ancient city of Corinth, Greece. (Photographs by G. H. Davis.)

Figure 1.11 Joint face in the Pennsylvanian Bonaventure Formation along the shore of Chaleur Bay, New Brunswick, Canada. The joint face is marked by an exquisite display of plumose markings. Geologists is Wayne Nesbitt. (Photograph by G. H. Davis.)

movement. They occur in families, or sets, of parallel through-going features. **Joints** form in tension (i.e., a pulling apart) in response to tectonic and thermal stresses that force the rock to extend ever-so-slightly perpendicular to the plane of fracture. Movement parallel to the plane of fracture is negligible. Joints are often remarkable in their consistency of spacing and orientation (Figure 1.10). Figure 1.10*A* is a "low-altitude" aerial photo I took from the top of a 6-ft-high juniper tree in northwestern New Mexico. The pattern looked "man-made," reminding me of the Roman-built plaza in Corinth, Greece (Figure 1.10*B*). Although individual joints commonly are only meters or tens of meters in length and in spacing, they form in sets that can have regional continuity. The tensional nature of joints is revealed in a distinctive decorative ornamentation on the joint surface, known as plumose structure (Figure 1.11), which forms when the tensional crack propagates at velocities sometimes exceeding half the speed of sound.

In contrast to joints, **shear fractures** form by an ever-so-slight sliding or shearing movement parallel to the plane of fracture; there is no tensional opening. Most commonly the products of tectonic loading, shear fractures are especially abundant in rocks that have been folded and faulted. (Joints, in contrast, are found everywhere, even in flat-lying sedimentary rocks that have never experienced tectonic stresses sufficient to cause folding or faulting.) **Slickenlines**, which for now can be thought of as scratches "tooled" into the surfaces during movement (Figure 1.12), testify to origin by shear and disclose the direction of movement. Movement is so small that even when a shear fracture cuts a distinctive marker, like a pebble in a conglomerate or a fossil in a limestone, the offset is rarely discernible to the naked eye. **Transitional tensile fractures**, part joint and part shear fracture, form through an oblique movement that combines tensional opening and shear.

Elevated fluid pressures in rocks can have a profound influence on the ease of formation, and reactivation, of fractures. Movement of fluid along

Figure 1.12 Bob Krantz (at center in sunglasses) and other slickenline aficionados check out the lines on a fault surface that is part of the San Andreas system. Location is the Mecca Hills in southeastern California. (Photograph by G. H. Davis.)

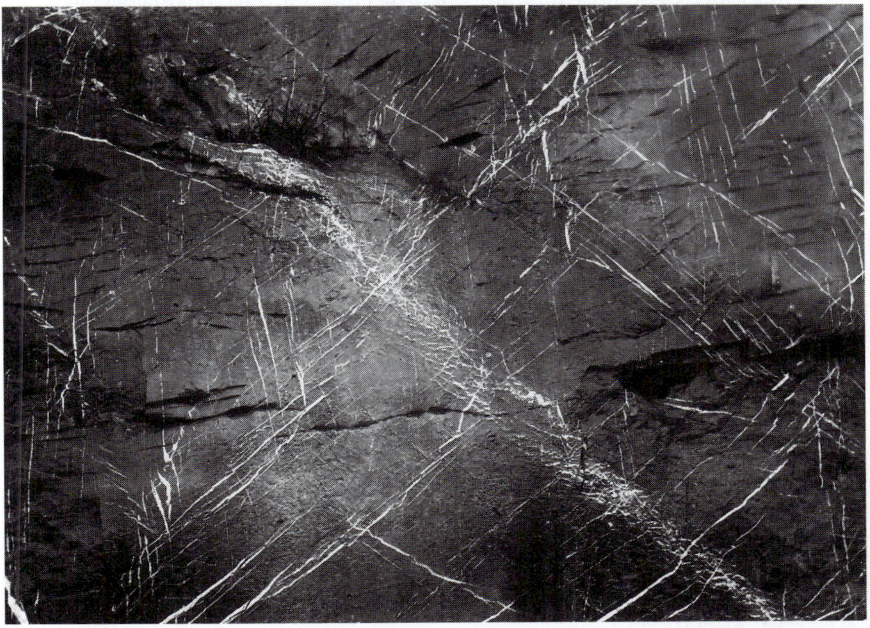

Figure 1.13 Calcite veins in limestone, near Limekiln Point, Highgate Springs, Vermont. (Photograph by C. D. Walcott. Courtesy of United States Geological Survey.)

fractures is commonly recorded in the presence of **veins** (Figure 1.13), composed of minerals that precipitated from solution under favorable conditions of temperature and pressure. The growth directions, and changes in growth directions, of minerals within some veins record the history of opening of the fractures through time. Where the veins contain precious metals, like gold, things become especially interesting.

Faults are discrete fracture surfaces or zones of fractures along which rocks have been offset by movements parallel to the fracture surface(s) (Figure 1.14*A*). The magnitude of the offset can range from centimeters to kilometers. Fault movement can result in a breakdown of rock on either

Figure 1.14 (*A*) Repeated truncation and offset of a white quartz layer by faulting, Tortolita Mountains, Arizona. Note coin (a quarter) used for scale. (Photograph by G. H. Davis.) (*B*) Slickensided, slickenlined fault surface cutting Tertiary volcanic rocks in the Hieroglyphic Mountains, Arizona. (Photograph by S. J. Reynolds.)

side into various products, including clayey gouge; polished, striated, and grooved slickensided surfaces (Figure 1.14*B*); and aggregates of angular fragmented and broken rock known as breccia. During deformation of the Earth's crust, and in response to tectonic loading, rocks are commonly forced to move past one another to achieve a better fit. Certain types of faults permit the crust to be shortened (Figure 1.15). Other types permit parts of the crust to be lengthened (Figure 1.16), or to cause rocks on either side of the fault to shift horizontally. Major faults exert their influence in broad-ranging ways—from creating sites of catastrophic earthquakes to determining locations of boomtown precious-metal camps where mineralization is fault controlled. Because of increasing concern about disasters associated with catastrophic earthquakes, the study of contemporary faulting has emerged as an active arena of structural and seismological analysis.

Figure 1.15 Thrust faulting of Jurassic sedimentary rock at Ketobe knob along the San Rafael swell in central Utah. The thrusts serve to shorten the sequence. [From Neuhauser (1988). Published with permission of Geological Society of America.]

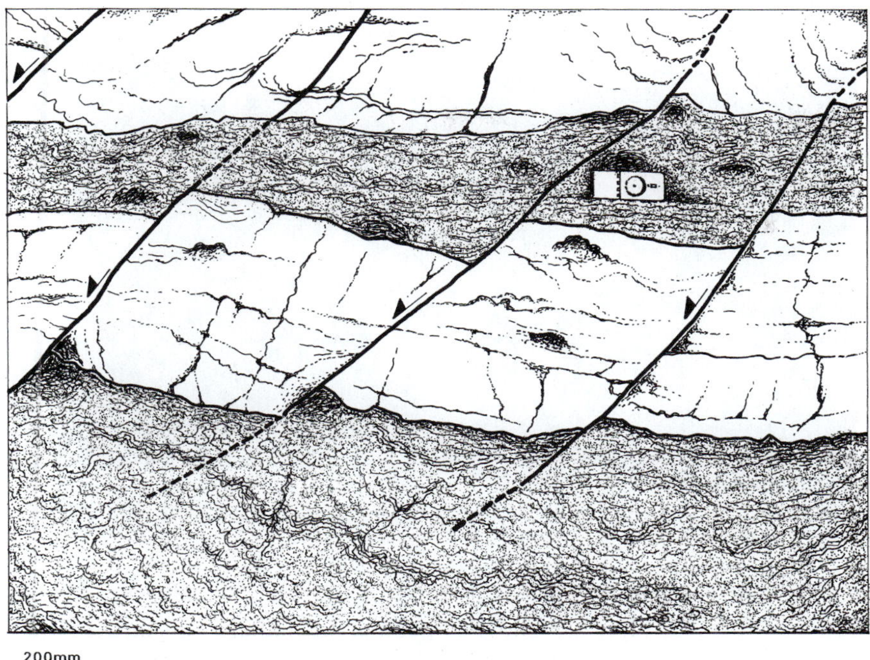

Figure 1.16 Line drawing of normal faults that serve to extend sandstone (white layer) and marls (gray layers) in the Watchet region of the British Isles. [From Peacock and Sanderson (1992). Published with permission of The Geological Society.]

Folds are structures that form when beds and layers are transformed into curved, bent, and crumpled shapes (Figure 1.17). They come in all sizes and shapes, revealing in their internal form something of the conditions under which they developed. Folds can form in many ways. They commonly reflect end-on, viselike buckling and shortening of originally horizontal layers (Figure 1.18). Most folds are intimately associated with faults and shear zones. Under high-temperature conditions, gravitational loading acting on plastically deforming layers can produce cascades of folds that pile up like taffy candy. In addition, folds form in unconsolidated sediments that may slump and flow, as well as in viscous lava flows or dikes as these hot bodies move under the influence of gravity. Folds,

Figure 1.17 Folded dolomite and limestone near Danby, Vermont. (Photograph by A. Keith. Courtesy of United States Geological Survey.)

Figure 1.18 Raplee anticline beautifully exposed along the San Juan River near Mexican Hat, Utah. Regionally horizontal sedimentary rocks abruptly swing to steep dips as a result of compression-induced folding. (Photograph by G. H. Davis.)

however formed, offer a wonderful geometric challenge in three-dimensional visualization and analysis. Details of the folded form of a given layer can commonly be related to the contrasts in stiffness and strength from layer to layer within the sequence of rocks at the time of folding.

Cleavage, **foliation**, and **lineation** are structures that form under conditions of elevated temperature and/or pressure, where mineral grains can change shape, selectively dissolve or precipitate, and recrystallize. These structures are penetrative; that is, they pervade the rock bodies in which they occur so completely that the structures are closely spaced, often at the microscopic scale.

Foliations are very closely spaced parallel planar alignments of features like micas, ribbons of quartz, aligned phenocrysts, shear surfaces, or flattened objects like pebbles or worm burrows (Figure 1.19). A special category of foliation is **cleavage** (Figure 1.20), closely spaced subparallel surfaces with a concentration and regularity that impart a splitting property to rocks. Cleavage forms in response to flattening and shortening and commonly is intimately associated with folding (see Figure 1.20). Some cleavage accommodates nearly imperceptible faulting. But most cleavage reflects discontinuities along which rock has been partially removed by stress-induced dissolution, to permit a shortening of the rock mass above and beyond what could be achieved by folding alone.

Lineations are preferred linear alignments of elements like hornblende needles, mineral aggregates, bundles of tiny folds, or striations and grooves that pervade rock bodies (Figure 1.21). Whether weakly or strongly developed, a lineation may reveal in its geometry and its physical character the direction of shearing or flowing of the rock.

Linear structure is neither delicate enough nor pervasive enough to be penetrative at the microscopic scale; instead, it forms large rodlike and groovelike structures that arise in a variety of ways, including buckling, shearing, and faulting of rocks. Under exceedingly rare circumstances, some linear structure is large enough for children and structural geologists to ride (Figure 1.22).

The presence of cleavage, foliation, and lineation in a rock reflects significant changes in the shape and size of the rock. The movements that produce foliation and lineation pervade the entire mass of rock, just like the internal movements that are generated within a penny when it is

Figure 1.19 (*A*) Photograph of gneissic foliation in sheared granite in the Rincon Mountains, Arizona. Foliation at outcrop scale defined by subparallel alignment of feldspar porphyroclasts (white) in darker matrix of quartz and feldspar. Foliation orientation is accentuated by the presence of white aplite sills dipping gently from upper right to lower left. Knife for scale. (Photograph by G. H. Davis.) (*B*) Foliation defined by alignment of deformed cobbles in the Purgatory Conglomerate, Rhode Island (quarter used for scale). [After Mosher (1981). Copyright © 1981 by The University of Chicago. All rights reserved.]

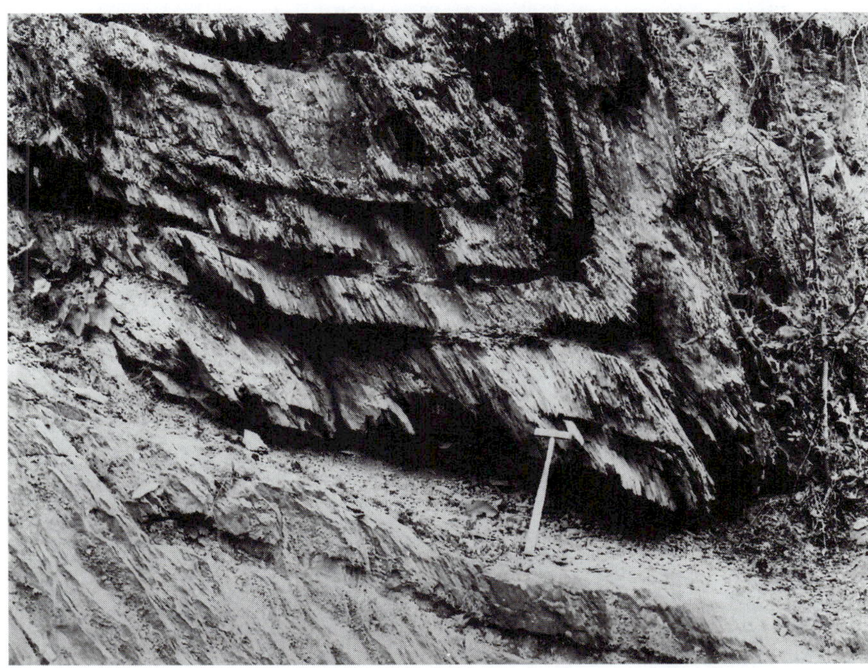

Figure 1.20 Cleavage in banded slate near Walland, Tennessee. (Photograph by A. Keith. Courtesy of United States Geological Survey.)

Figure 1.21 Photograph of outcrop of strongly lineated "mylonitic" gneiss in the South Mountains bordering Phoenix, Arizona. The lineated mylonite was produced by the shearing of a Miocene granite during profound regional extension. The penetrative lineation is parallel to the direction of shearing. (Photograph by S. J. Reynolds.)

Figure 1.22 Linear structure in the "extraordinary striated outcrop" at Saqsaywaman, Peru (see Feininger, 1978). To kids in Cusco this lineated rock is known as *el rodadera* (the rollercoaster). The ride is fast! To slow down, riders spit on their hands and then use the hands as brakes. (Photograph by L. A. Lepry.)

flattened by a moving train. The dramatic change in the shape of the rock or coin can be measured and defined quantitatively, as long as reference markings of known original size and shape are recognizable, like fossils in the deformed rock, or Abe Lincoln's profile in the flattened penny (Figure 1.23).

Shear zones represent the final category of basic geologic structures. Like faults, they accommodate offset, but the offset is distributed across the thickness of a tabular zone that is centimeters, meters, or even kilometers thick. Unlike ordinary fault surfaces, shear zones commonly do not display any discrete physical break. Instead, displacement is achieved without loss of cohesion and continuity, although rocks "caught up" in shear zones undergo extreme changes in shape and orientation (Figure 1.24). The penetrative distributed offset within shear zones can be ex-

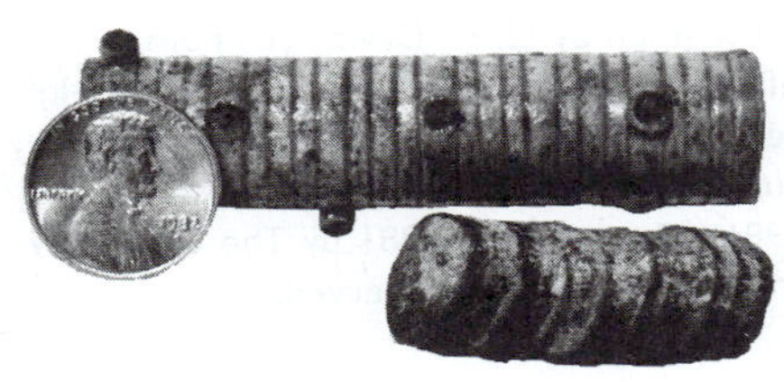

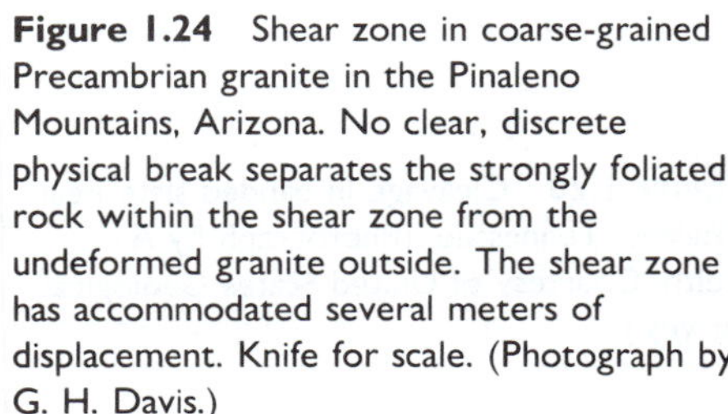

Figure 1.23 Deformed and undeformed crinoid stems. Deformed and undeformed pennies.

Figure 1.24 Shear zone in coarse-grained Precambrian granite in the Pinaleno Mountains, Arizona. No clear, discrete physical break separates the strongly foliated rock within the shear zone from the undeformed granite outside. The shear zone has accommodated several meters of displacement. Knife for scale. (Photograph by G. H. Davis.)

pressed by the presence of pervasive foliation and lineation. Shear zones typically represent the deep roots of faults at levels where elevated temperature permits the crustal rocks to flow.

CONCEPT OF DETAILED STRUCTURAL ANALYSIS

If our work in structural geology is to lead to fresh discoveries and understanding, it cannot be carried out in a casual, generalized, unimaginative way. The complexity of structural systems and processes demands much more. As Howard Lowry states

> Excellence and learning are not commodities to be bought at the corner store. Rather they dwell among rocks hardly accessible, and we must almost wear our hearts out in search of them. (From *College Talks* by H. F. Lowry, edited by J. R. Blackwood, p. 116. Published with permission of Oxford University Press, New York, copyright © 1969.)

In studying basic structures and structural systems, we base our work on a branch of structural geology known as **detailed structural analysis**, with particular emphasis on strain analysis (Sander, 1930; Knopf and Ingerson, 1938; Turner and Weiss, 1963; Ramsay, 1967; Ramsay and Huber, 1983, 1987; Suppe, 1985). Three fundamental interlocking strategies are harnessed in detailed structural analysis: descriptive analysis, kinematic analysis, and dynamic analysis. Each of these looks at geologic structure from a different standpoint.

Descriptive analysis is concerned with recognizing and describing structures and measuring their locations, geometries, and orientations. **Kine-**

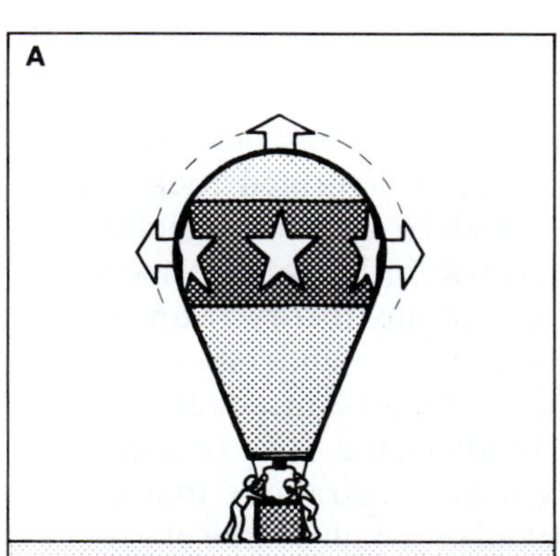

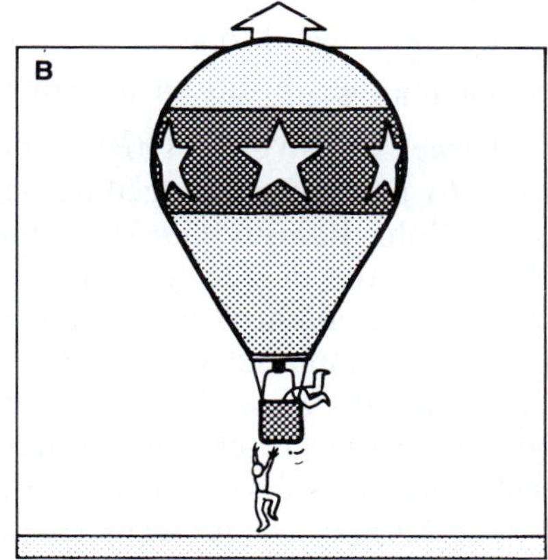

Figure 1.25 The basic movements of (*A*) **dilation** (change in volume), (*B*) **translation** (change in position), (*C*) **rotation** (change in orientation), and (*D*) **distortion** (change in shape). (Artwork by D. A. Fischer.)

matic analysis focuses on interpreting the deformational *movements* responsible for the development of the structures. The basic movements are **translation** (change in position), **rotation** (change in orientation), **distortion** (change in shape), and **dilation** (change in size) (Figure 1.25).

Dynamic analysis interprets deformational movements in terms of forces and stresses responsible for the formation of structures, as well as evaluating the strength of the materials during deformation. Dynamic analysis is generally the most interpretive part of detailed structural analysis, but it derives remarkable power from thoughtful experimental and theoretical studies drawn from principles of mechanics. Both kinematic and dynamic analysis are anchored in descriptive fact—the primary source.

Basic to detailed structural analysis are discovering and describing the degree to which deformed rock bodies are marked by profound geometric order, at all levels of observations. Detailed structural analysis leads us to recognize that the geometry and symmetry of the internal structure of rocks reflect the geometry and symmetry of movements and stresses responsible for the deformation (Sander, 1930).

Knopf and Ingerson (1938), who helped introduce detailed structural analysis, illustrated the main points of the symmetry principle with simple but provocative illustrations. For example, birds on a telephone line commonly face the same direction (Figure 1.26). The geometric order is obvious and easy to document, but puzzling kinematic questions are raised:

- What, if anything, does the geometric alignment tell us about the movement of each bird during landing?
- What do we conclude about the single bird among 50 that faces in the opposite direction?
- Different flight paths? Different destinations?
- And what "forces" caused this preferred alignment?
- Facing upwind to land at a lower air speed?
- Facing upwind to keep the feathers smooth?
- Facing the warmth of the sun?

Figure 1.26 Preferred orientation of birds on a wire. (Photograph by B. J. Young.)

Turner and Weiss (1963) summarized valuable methods that can be used to pick apart the geometric order within deformed rocks, at any scale. John Ramsay (1967), in turn, led modern structural geologists to recognize the degree to which the geometric order in deformed rocks reflects the mathematical requirements of strain theory. Suppe (1985) creatively emphasized the degree to which deeper understanding of the origin and dynamic significance of structures can be discovered through the application of carefully articulated fundamental laws of mechanics. Throughout this book there are recurrent demonstrations that geologic structures and structural systems are marked by a high degree of geometric order; that the geometric order expresses the kinematics of deformation; and that the application of experimental and theoretical principles is helpful in explaining the relation of geometric order and the movements to that which originally caused the deformation in the first place.

DESCRIPTIVE ANALYSIS

The Basis

In descriptive analysis we recognize structures, measure their orientations, and describe, literally inside and out, their physical and geometric components. Descriptive analysis results in facts regarding the physical properties, orientations, and internal configuration of the structures. The basis for descriptive analysis is broad-ranging: direct observation of field relationships, as featured in the cover illustration; examination of rocks deformed experimentally in the laboratory; drilling into the subsurface; geophysical monitoring and probing of the subsurface; exploring the structures of the ocean floor; and studying the stratigraphy and/or petrography of the rocks in which the structures occur.

The foundation for solid descriptive analysis in structural geology is geologic mapping (see Part III-A to C). Geologic mapping reveals the nature of contacts between rock bodies, which in turn discloses the sequential history of major events. Moreover, geologic mapping results in a three-dimensional portrayal of the geometric architecture of the rock systems under investigation, which in turn becomes a basis for interpreting the structural and tectonic history.

During geologic mapping a careful inventory is made of the orientations of contacts, rock units, primary structures, and secondary structures. This is achieved through countless measurements using special compasses equipped to measure not only azimuth, but also inclination (see Part III-D). The three-dimensional measuring challenges are at first a little difficult, for they require "seeing" the structures that need to be measured, understanding whether the structure under consideration is "linear" or "planar," and adroitly if not acrobatically positioning oneself and the compass at the proper angles to achieve the correct readings.

Geologic mapping of contacts in combination with compass measurements of contacts and structures yields a database that can be used to construct geologic cross sections. Geologic cross sections are as important or more important than the maps themselves, for "structure sections" represent the "best" interpretation of the geology as it projects into the subsurface or, up into the sky. The basic steps used in constructing simple geologic cross sections are presented in Part III-E.

Petroleum exploration companies, mining exploration companies, and firms and agencies specializing in hydrogeology commonly need more than standard geologic maps and structure sections to carry out their work. Instead, they set up drilling programs and directly sample the subsurface geology, not only the rocks themselves, but also the depths at which specific rock formations are penetrated. On this basis, a variety of subsurface structural geologic maps can be generated, including structure contour maps and isopach maps. Subsurface contour maps in particular are an incredibly valuable resource in evaluating the structural geology of a region (see Part III-F).

Descriptive structural analysis places a premium on the exacting evaluation of three-dimensional geometric and spatial relationships. Angles between lines, angles between planes, depths to inclined planes, inclinations of intersections between planes—all become terribly important. Tackling the three-dimensional spatial and geometric relationships is probably the most challenging task-oriented requirement in structural geology. We pursue it in this book through two different kinds of tools, two different kinds of projections, with which in the early stages you will develop a love-hate relationship. One is called orthographic projection (see Part III-G). The other is called stereographic projection (see Part III-H,I). We refer to mastering these projections as "sick fun."

Ideally, descriptive analysis is free of interpretation, but when structures are excessively complex or poorly exposed, the mapping and measuring operations must be nudged by interpretation. Also, the observations we make and the questions we ask ourselves while observing are influenced by experience and by framing preliminary interpretations and hypotheses.

The Scale of Things

The specific structural features that can be recognized and described depend importantly on the **scale of observation**. Figure 1.27, which provides a useful nongeologic example of the influence of scale on geologic observation, features two different views of a log pile. As we look at the logs from a distance (Figure 1.27*A*), we are struck by the circular cross-sectional forms of the logs and by differences in diameter. In addition, we see linear alignments of the circular faces of the logs, such as the obvious alignment marked with arrows. In closer view (Figure 1.27*B*), we are struck immediately by the irregularity of the cross-sectional shapes of the logs. The sawed ends are by no means uniformly circular. Furthermore, it becomes apparent that there is internal structure marked by patterns of radial fractures displayed by each of the logs. Looking even more closely, we see that the cross-sectional face of each log is marked by striations produced by sawing.

Application of these woodpile principles to the world of structural geology is easy to imagine. I had some free time between flights at the airport in Oslo, Norway, and decided to look for rocks and structures. As it turned out, the outcrops shown in Figure 1.28 are within easy reach. When these outcrops are viewed from middle distance, such as in Figure 1.28*A*, layering is the dominant descriptive feature. At outcrop scale, however, a pervasive systematic fracturing at a high angle to layering emerges as the most noticeable structure (Figure 1.28*B*).

What we must remember above all is that shifts in scale of observation are absolutely necessary to make a complete inventory of the rock and structural components of any deformed body of rock. Jack Oliver said it best in thinking back to the discovery days of seafloor spreading and plate tectonics: "We learned that our observations must match the scale of the problem we are trying to address."

A number of geologic studies serve as superb models for descriptive analysis. For example, in their classical structural analysis of folded and faulted rocks in the Canadian Rockies, Price and Mountjoy (1970) concisely summarized the results of an enormous amount of data. Their descriptions, plus the geologic cross section they constructed (Figure 1.29), leave little to the imagination. They specifically called attention to

Figure 1.27 What we see in a log pile partly depends on scale of observation. (*A*) Middle-distant view. Arrows point out linear alignment of apparently circular faces of logs. (*B*) Close-up view reveals that the circular faces are not so circular. The close-up view also discloses some interesting fracture patterns. (Photographs by B. J. Young)

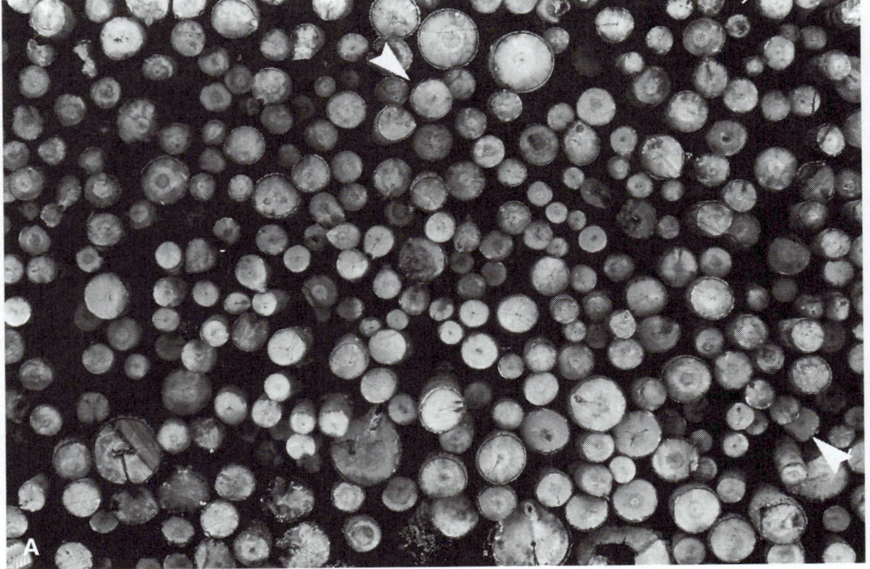

Figure 1.28 What we see in an outcrop partly depends upon scale of observation. (*A*) Middle distant view reveals strongly layered, banded appearance of metasedimentary rocks along the shorefront near the Oslo, Norway, airport. (*B*) The close-up view reveals that a fine, delicate cleavage runs parallel to the layering, and that closely spaced fractures cut across the layering. (Photographs by G. H. Davis.)

types of structure (thrust faults and folds), structural orientations (southwest-dipping faults, upright folds), shapes of structures (concave-upward faults), relation of faults to bedding (faults cut up-section), and relation of the folds to the faults (thrusts die out as folds). The important descriptive statements and phrases are here shown in boldface type to emphasize the extent of descriptive information and the economy of word choice.

> The structure of this part of the Canadian Cordillera is dominated by **thrust faults** which are generally **southwest-dipping** and **concave upward in profile**. The **faults flatten with depth** and have the **upper side displaced relatively northeastward and upward**. They gradually **cut up through the stratigraphic layering northeastward, but commonly follow the layering** over large areas. . . . Many of the faults **bifurcate [split] upwards into numerous splays**, and the total **dis-**

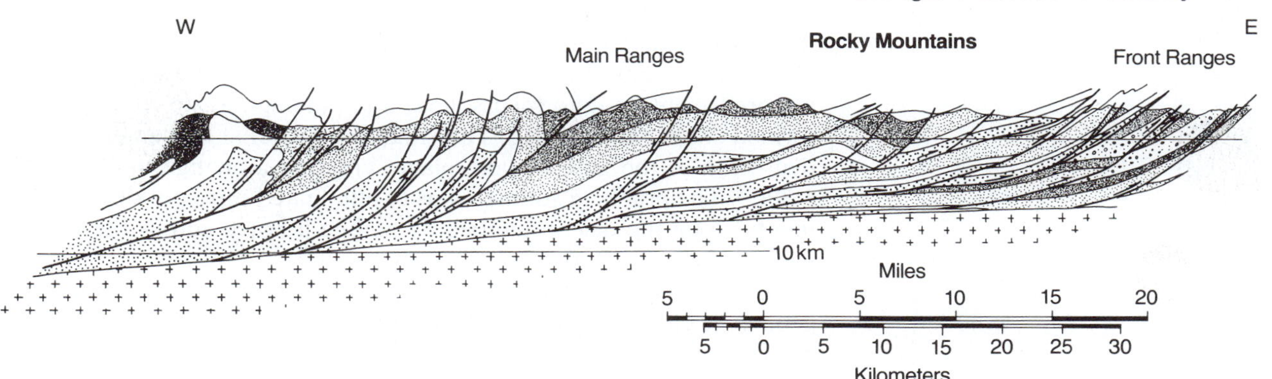

Figure 1.29 Geologic cross section of folded and faulted strata in a part of the Canadian Rockies. [After Price and Mountjoy (1970). Published with permission of Geological Association of Canada.]

placement along them becomes distributed among these splays. Folds are widespread and have developed in conjunction with the thrusting. . . . **Many of the thrust faults themselves are folded** along with the sedimentary layering. . . . **The folds generally are inclined to the northeast or are upright. Many of the thrusts die out as folds.** . . . [From Price and Mountjoy (1970), p. 10. Published with permission of Geological Association of Canada.]

Structural Elements

Learning to recognize and describe the main types of structure is not quite enough. Many different varieties of each type of structure exist in nature. Furthermore, each structure is composed of **structural elements** that in turn must be identified and described, to permit us to carry out a complete descriptive analysis. The variety of structures and structural forms is astonishing.

Structural elements are the physical and geometric components of structures. The **physical elements** are real and tangible, and they have measurable geometry and orientation. The **geometric elements** are imaginary lines and surfaces, invisible but identifiable in the field; they too have measurable geometries and orientations. A troublesome gate I once built along a stretch of fence in my side yard not only may illustrate the approach to analyzing structural elements, but also may clarify how descriptive analysis leads naturally to kinematic analysis.

I first isolated the part of the fence to become the gate by sawing through the two-by-fours forming the horizontal support for the fence line. Then I fastened strap hinges to one side of the gate and to the supporting, adjacent fence line. To the other side of the gate I attached a latch (Figure 1.30*A*). So that the bottom of the gate would clear the top of the adjacent cement driveway, I trimmed the base of the gate with my saw, leaving about 1 in. (25 mm) of clearance. In just a few days, a structural problem had developed. The bottom of what was now a sagging gate no longer cleared the driveway (Figure 1.30*B*). Impulsively, I trimmed 2 in. (50 mm) from the base of the gate. In three more days, the gate was again dragging on the cement. Clearly it was time for descriptive and kinematic analysis.

The gate consists of both physical and geometric elements. The physical elements are two two-by-fours, nine slats, two hinges, and the metal latch. All the elements lie in the same plane. The strap hinges and the two-by-fours are oriented horizontally, the slats vertically. Between the slats are thin, rectangular cracks, or in the language of structural geology, discontinuities. These constitute the only important geometric elements within the gate. Like the slats, they are vertically inclined.

The kinematic solution to the structural problem was discovered by lifting the latch and thus raising the gate off the pavement. The gate did not rotate at all where hinged. Rather, the "uplift" was absorbed within the gate by small translational movements along each of the discontinuities between the slats. In effect, the small translations allowed the gate to be restored from its deformed shape, a parallelogram, to its original undeformed shape, a rectangle. The small translational movements permitted distortion of the gate to take place, creating a change in shape (Figure 1.30*C*). The practical solution was to eliminate future possibility of movements along the discontinuities by nailing a brace across the face of the gate (Figure 1.30*D*).

In matters of geology, too, descriptive analysis takes into consideration geometric elements. A fold, like the one shown in Figure 1.31, may consist of folded layers, the bedding-plane discontinuities between layers (bedding

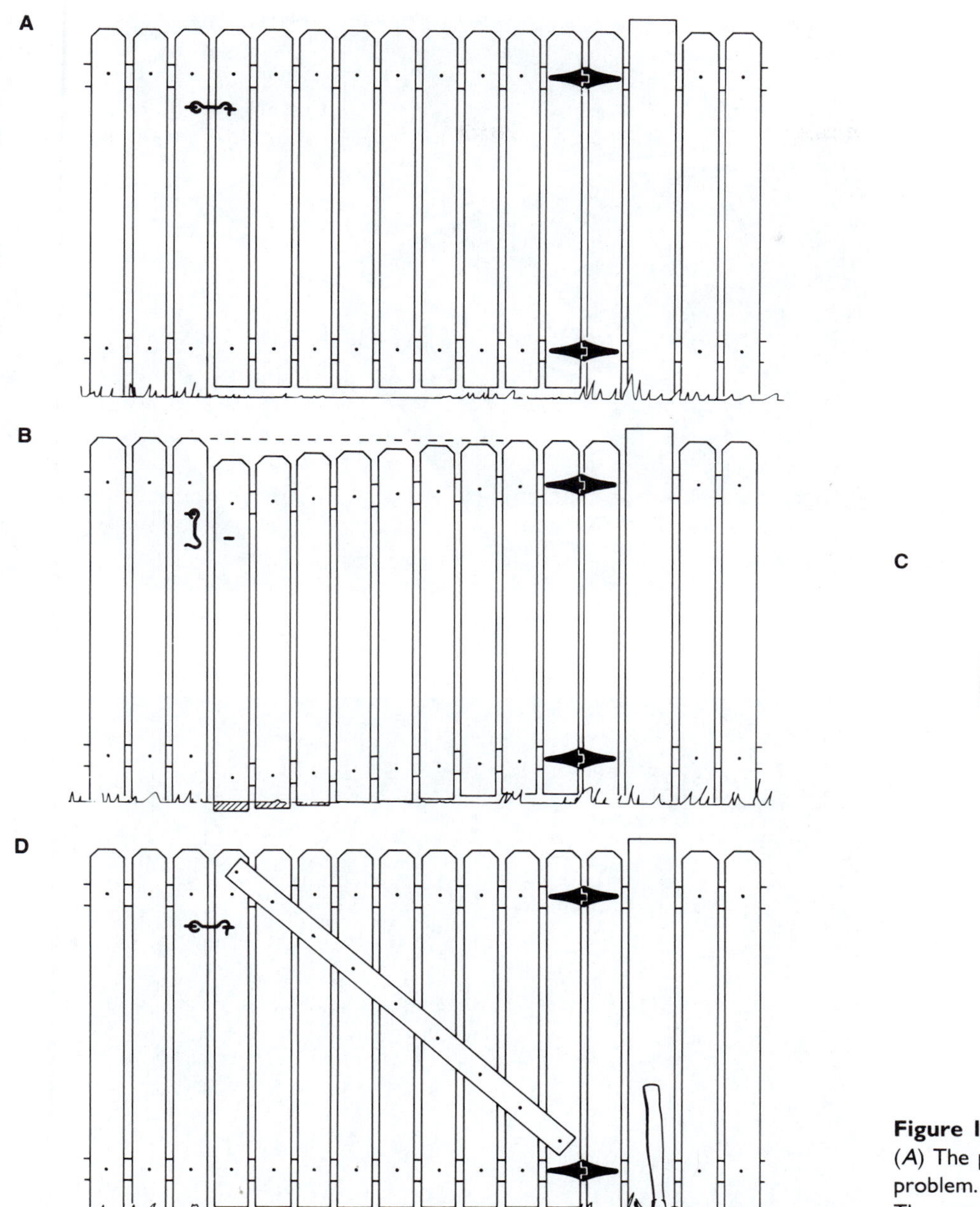

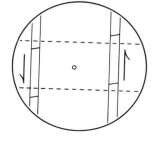

Figure 1.30 Saga of the troublesome gate. (*A*) The perfect gate. (*B*) The structural problem. (*C*) The kinematic solution. (*D*) The remedy.

surfaces), the hinge point (point of maximum curvature of top or bottom of folded layer), and axial surface (passes through hinge lines, from one layer to the next to the next). The **folded layers** are physical and real, composed of the rocks that have been folded. The hinge is also real, fixed in position and contained in real rock. The axial surface and bedding-plane discontinuities are geometric and imaginary. The bedding-plane discontinuities separate each of the folded layers. The axial surface is a convenience for helping to define the orientation and form of the fold. Although the beds themselves reveal the clear expression of the folding, the presence of the discontinuities made the folding possible. Just as the pages of a ream of paper slide past one another when the ream is folded, so the beds slip past one another along the bedding-plane discontinuities to permit the folded forms to develop.

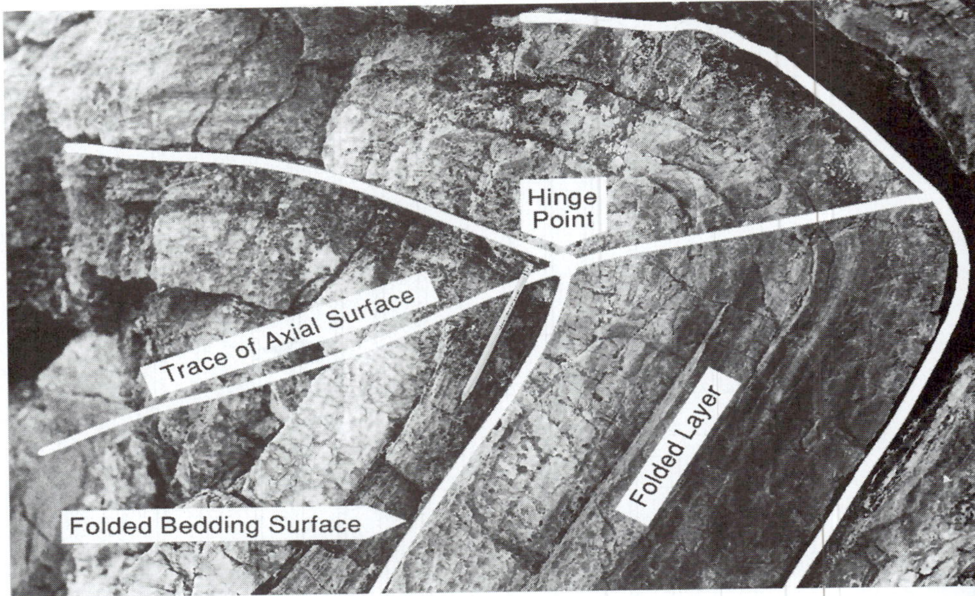

Figure 1.31 Physical and geometric elements of a chevron fold. (Photograph by G. H. Davis.)

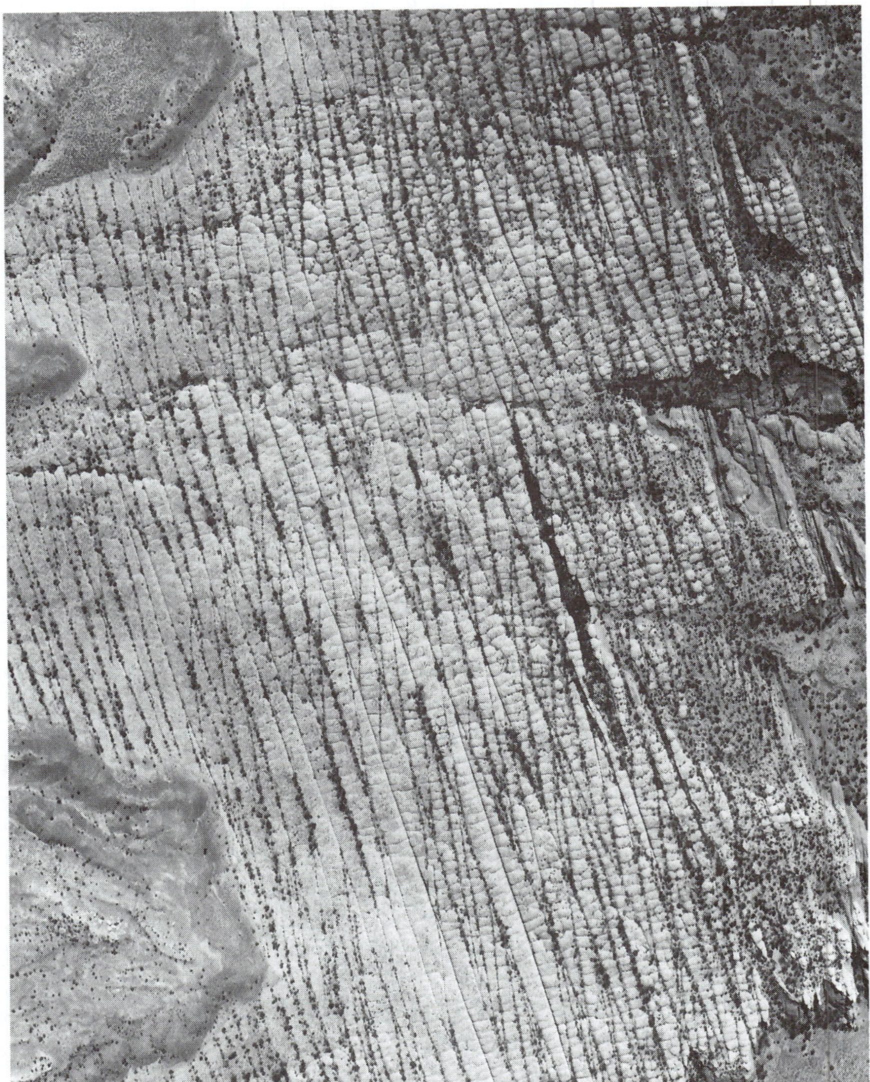

Figure 1.32 Aerial photograph of jointing in the Entrada Sandstone (Jurassic) near the campground at Arches National Park, Utah. Two prominent sets of joints come into clear, crisp resolution when the photo is viewed at a low angle to the plane of the page, parallel to the trend of the joints. (Photograph by R. Dyer.)

Focusing attention on the bedding-plane discontinuities between the rock layers rather than on the rock layers themselves is quite a shift in emphasis. It is like suggesting, perhaps quite correctly, that the key elements of venetian blinds are the openings between adjacent slats, not the slats themselves. Whether the physical or geometric elements are emphasized in structural analysis depends on the problem being addressed. In the heat of the Arizona summer, in our east-facing family room in Tucson, I appreciate the physical elements (the slats) in the morning, but turn to the geometric elements (the openings) in the afternoon.

The degree of geometric order in a deformed rock body is evaluated by measuring the orientations of large numbers of structural elements. The orientations are plotted graphically and evaluated statistically to discern the quality of preferred orientation. Through this process, **sets** of structures or structural elements can be defined. Sets are composed of elements sharing common geometric and/or physical appearance and parallel orientation. For example, the joints shown in Figure 1.32 may be subdivided into three sets on the basis of orientation and spacing. Two or more sets of like structures or structural elements constitute a **system**. All such systems, taken together, plus all structures that do not conveniently arrange themselves into sets, comprise the total **structural system**.

KINEMATIC ANALYSIS

General Approach

Kinematic analysis takes off where descriptive analysis ends. It is about movements. Kinematic analysis deals with recognizing and describing the changes that, during deformation, are brought about by movement of the body as a whole, or by internal movements within the body. When a body, or some part of a body, is forced to change its location or position, it undergoes **translation** (Figure 1.33*A*). When forced to change its orientation, it undergoes **rotation** (Figure 1.33*B*). When forced to change size, it undergoes **dilation** (Figure 1.33*C*). When forced to change shape, it undergoes **distortion** (Figure 1.33*D*).

The overall goal of kinematic analysis is to interpret the combination of translations, rotations, distortions, and dilations that altered the location, orientation, shape, and size of a body of rock. Kinematic analysis is carried out at all scales, from submicroscopic to regional. Ideally, as we "work out the kinematics," we do not even think about the forces or stress that created the deformational movements in the first place, saving this step for dynamic analysis. In this way, we are more likely to establish an objective foundation of fact on which interpretations of cause can proceed.

Evaluating changes in shape and size brought about by deformation is the focus of **strain analysis**. Strain analysis is basic to modern structural analysis. It requires a quantitative evaluation of changes in original size and shape of geologic objects. As a sneak preview, take a look at the deformed worm burrows shown in Figure 1.34*A*. Knowing that these elliptical objects are actually circular in the parts of the rock body that are not deformed, and assuming no change in area during the deformation, we can then determine the direction and amount of shortening and stretching that accompanied deformation. We approach the problem by calculating the area of the ellipse (πab) and setting it equal to the area of the circle (πr^2) (see Figure 1.34*B*). This will yield the radius of the original circular object and give us a basis for comparison.

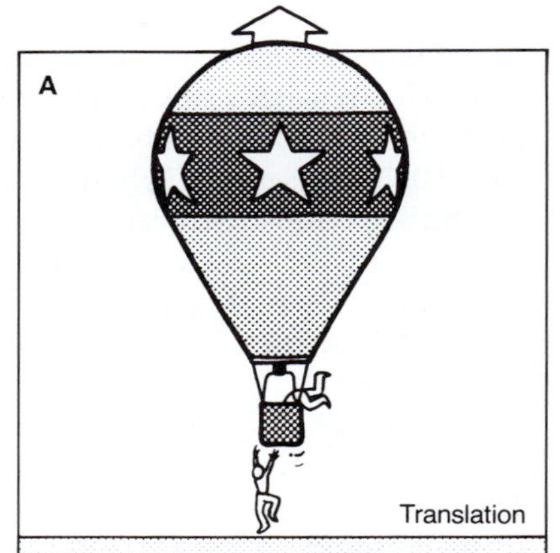

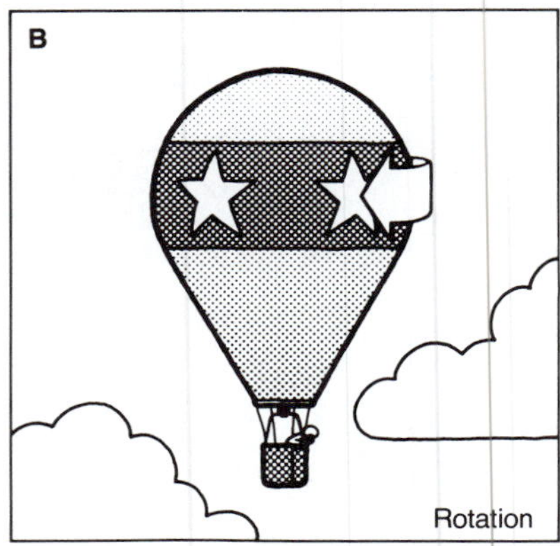

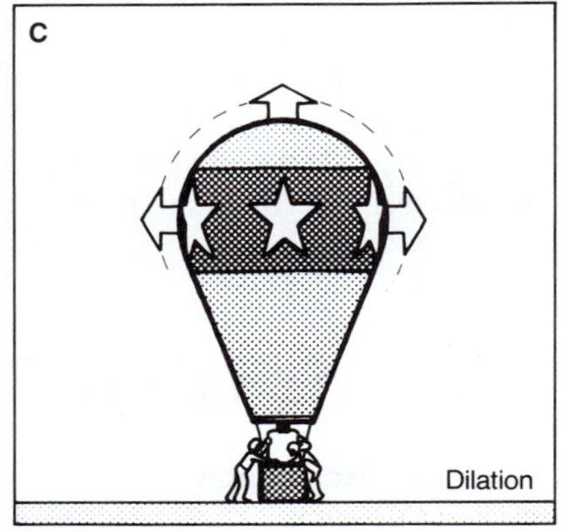

Figure 1.33 (*A*) Translations and (*B*) rotations change location and orientation without necessarily changing size or shape. (*C*) Dilations and (*D*) distortions change size and shape without necessarily changing location and orientation. (Artwork by D. A. Fischer.)

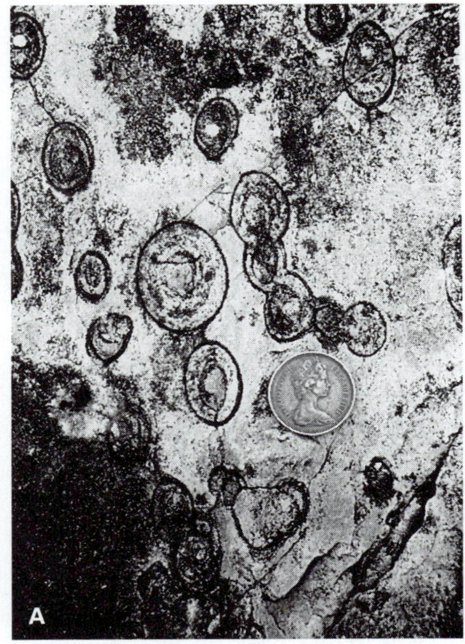

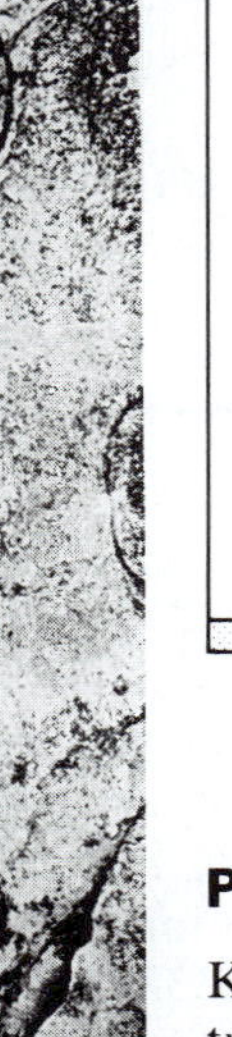

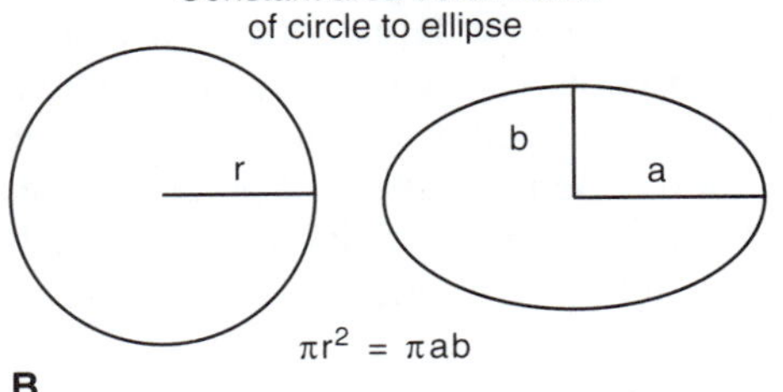

Figure 1.34 (*A*) Photo of originally circular, now deformed worm burrow in Cambrian sandstone along the Moine thrust near Durness, Scotland. (Photograph by G. H. Davis.) (*B*) Simple calculation of amount of strain, based on the assumption that the original circular cross-sectional area of the burrow was the same as the elliptical cross-sectional area now observed.

Penetrative Deformation

Kinematic analysis in general is influenced by the degree to which structural elements are penetrative in the rock body under study (Turner and Weiss, 1963). For structures to be penetrative, they must be spaced so closely with respect to the size of the rock body under consideration that they appear to be everywhere. Figure 1.35 provides a useful image: individual logs within the wooden islands are considered to be penetrative because they are very small and very closely spaced compared to the size of the floating bodies they comprise.

What is learned about structural systems depends partly on the scale at which the structures are penetrative. In Figure 1.36*A*, siltstone and shale are pervasively cut by a closely spaced cleavage. The cleavage is penetrative at the scale of single hand specimens and largely masks the original textures of the sedimentary rock. Only thin relics of rock containing the original primary structure are preserved. Given this situation, numerous data bearing on the nature of cleavage can be drawn from single outcrops of this rock. But clues regarding the nature of the original rock would have to be sought at the microscopic scale. Conversely, primary stratigraphic relationships, if they are to be recognized at all, would have to be explored in a more regional context (Figure 1.36*B*).

In yet another example of this concept (Figure 1.36*C*), two faults, spaced kilometers apart, cut and displace a sequence of shale, limestone, and siltstone. The investigation of individual outcrops and small map areas within this geologic system would yield stratigraphic and petrologic

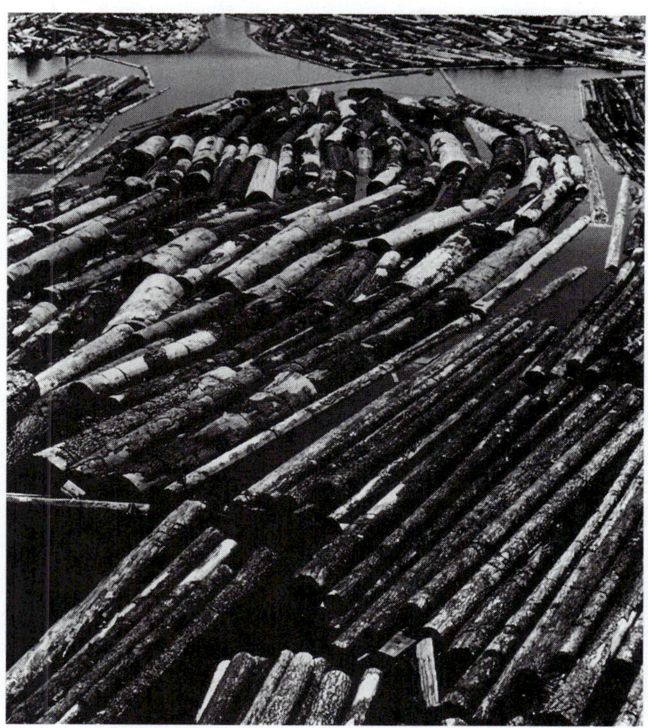

Figure 1.35 Penetrative, parallel logs in a river. (Photograph by B. J. Young.)

information regarding the primary rock assemblage. But only limited structural kinematic insight would be derived from the study of single faults. To really understand this structural system, large regional domains would have to be studied. Seen in a regional view, faults would be considered to be penetrative (Figure 1.36*D*). The structural kinematic evaluation of the system of faults, viewed as a whole, would aid enormously in interpreting the deformational history of the region.

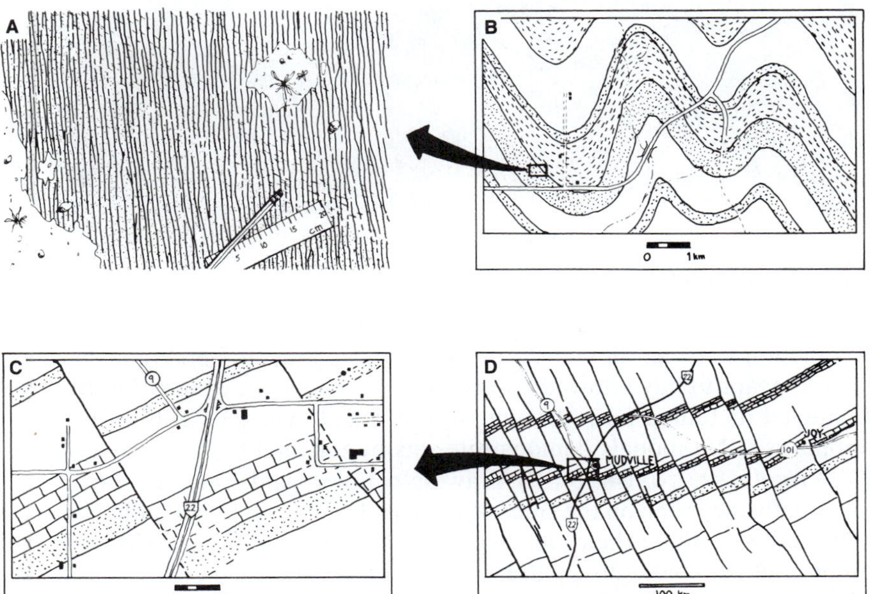

Figure 1.36 The influence of scale on what we see. (*A*) Penetrative cleavage masks original bedding in the outcrop view, (*B*) but not in the regional view. (*C*) At a relatively large scale of observation, the faults appear to be widely spaced. (*D*) At a smaller scale of view, the faults can be considered to be penetrative. (Artwork by R. W. Krantz.)

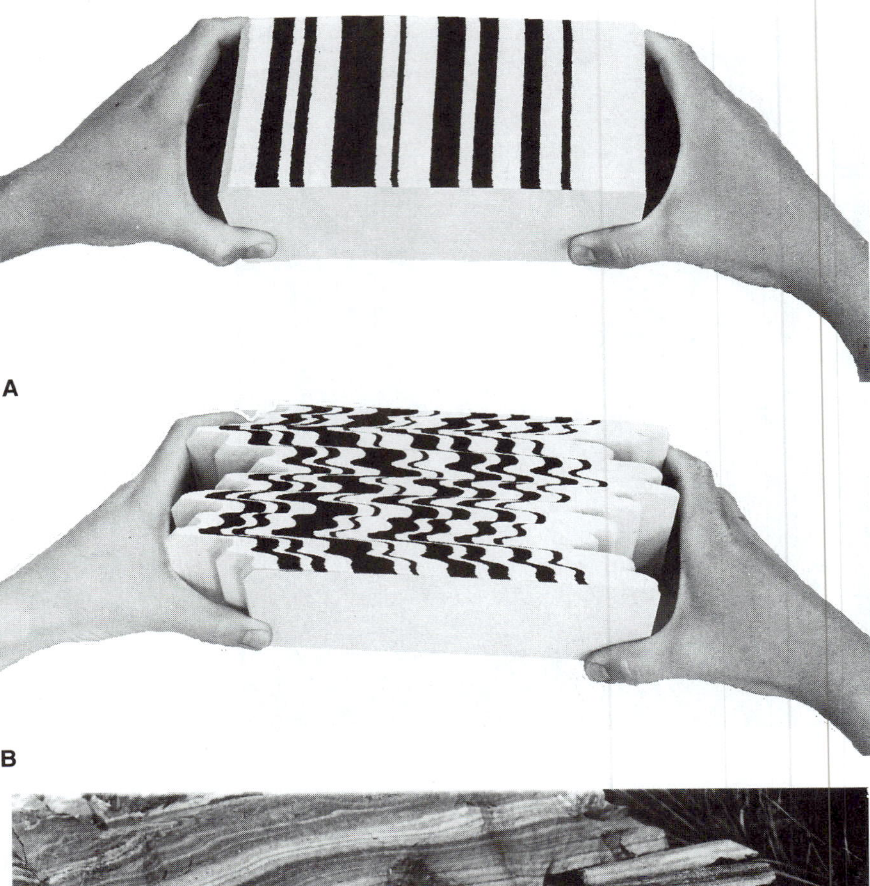

Figure 1.37 O'Driscoll-like experiment carried out by Paige Bausmann. (*A*) Computer card deck with stripes simulating layering. (*B*) Distortion of the deck by penetrative slip achieves "flow" of the layers. (Photograph by G. Kew.) (*C*) Flow folds in metamorphic rock west of the Romback window near Bjornfeld in northern Norway. (Photograph by G. H. Davis.)

Slip, Flow, and Distortion

Kinematic movements are often described in terms of **slip** or **flow**. The distinction between flow and slip is scale dependent. Sometimes a layer that appears to have flowed can be seen upon close examination to have achieved its folded form through myriad small displacements along "microfaults." No models demonstrate this better than those carried out by O'Driscoll (1962, 1964a,b). By drawing parallel bands representing layers of undeformed rock on the sides of card decks, and systematically translating the cards along their close-spaced slip surfaces, O'Driscoll was able to produce dramatic flow patterns (Figure 1.37*A*,*B*), which are not unlike folding seen in metamorphic rocks (Figure 1.37*C*).

Price and Mountjoy understood the scale dependence of slip and flow in their evaluation of the Canadian Rockies.

> All of the individual thrust faults are small when the area within which they occur is compared to the total area of the Rocky Mountains. . . on the appropriate scale of observation, the overall structural pattern within the Rocky Mountains reflects a type of "plastic deformation" or flow involving large-scale strain and translation within a coherent mass of layered sedimentary rock. Slip within the mass was concentrated in a myriad array of discrete, but discontinuous, interleaved, and interlocked shear surfaces, all of which were contained within the mass and are now represented by the various thrust faults. [From Price and Mountjoy (1970), p. 10. Published with permission of Geological Association of Canada.]

The major kinematic lesson that emerges from both experimental studies and field work is this: systematic movements on relatively close-spaced surfaces of slip can produce significant distortion (i.e., strain) of a body of rock.

DYNAMIC ANALYSIS

General Approach

Dynamic analysis interprets forces, stresses, and the mechanics that give rise to structures. For dynamic analysis to be meaningful, it must explain the physical and geometric character of the structures, the kinematics, and the relationship between stress and strain. The major aim of the analysis is to describe the orientation and magnitude of the stress, and the response of the material. This is a challenging step in detailed structural analysis, because significant inferences must be made regarding the environment of deformation (i.e., temperature, pressure, . . .), the strength and physical state of the materials during deformation, the rate at which deformation proceeded, and the boundary conditions (e.g., relationship to plate movements and plate boundaries).

Physical Models

Geological and engineering literature is full of models that are helpful in interpreting the origin of structures. The models are descriptions of the conditions under which geologic structures form. The basis for dynamic analysis is theoretical and experimental research. Most dynamic models are valid in principle, but most structural systems can be explained satisfactorily by more than one dynamic model. Choosing among alternative models is challenging and speculative science.

Think of dynamic models as "recipes" for making structures and structural systems. Recipes describe the amount, the kind, and the composition of the ingredients as well as the conditions to which the ingredients are subjected. If the recipe is valid, the final product is guaranteed, provided the conditions are met. Such recipes can be prepared on the basis of experimental or theoretical studies in which the conditions of deformation are predetermined, measured, or arbitrarily established. Whether a particular model applies to a particular geological system depends largely on interpretation of the conditions under which deformation was achieved. For complex structural systems or those formed approximately 2 billion

Figure 1.38 Some of the ingredients and products of rock-deformation experiments. *Left to right:* a copper jacket; a 1-in. (2.54 cm) core of marble; a 1-in. core of siltstone; a faulted, jacketed core of siltstone; a faulted, slightly barrel-shaped core of marble. (Photograph by R. W. Krantz.)

years ago, interpretation of the conditions of deformation is very difficult. Similar end products may arise from various recipes, just as similar loaves of bread will result from recipes that call for mixing of the dry ingredients before, during, or after the adding of water.

One of the interesting arenas of dynamic modeling is the experimental deformation of small cylindrical cores of rock under conditions of regulated temperature, confining pressure, rate of deformation, and fluid pressure. Empirical observations derived from experimental testing of materials have produced broad-ranging guidelines regarding the conditions under which rocks will fail by faulting, and even the orientations and characteristics of the faults thus formed. Important results of such work are "recipes" for producing faults. One such recipe might read this way (Figure 1.38):

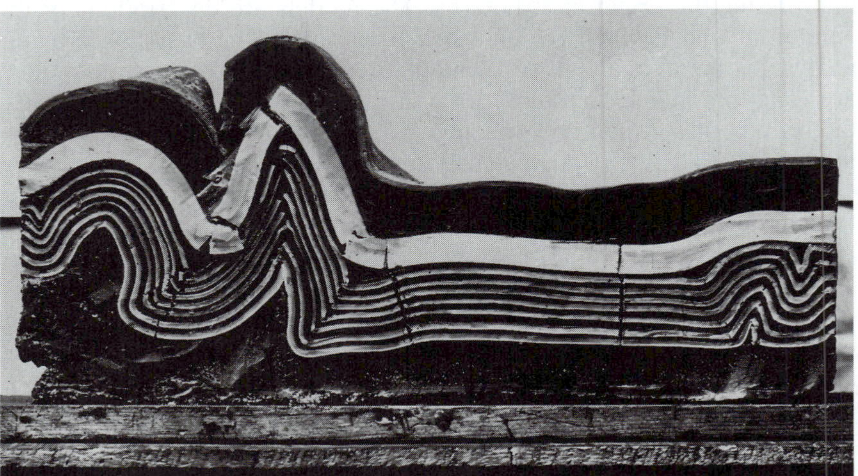

Figure 1.39 An example of one of Bailey Willis' classic experimental deformation models. This one, made of wax, accommodated layer-parallel shortening by folding. Much of the work of Willis, carried out in the nineteenth century, was directed toward understanding mountain building in the Appalachians (see Willis, 1894). (Photographs by J. K. Hillers. Courtesy of United States Geological Survey.)

Use a core-drill device to cut a 1.3-cm diameter, 2.5-cm-long cylinder from a specimen of fine-grained sandstone. Bevel ends of cylinder so that the top and bottom surfaces are planar and parallel. Insert specimen in copper jacket and place in deformation apparatus. Pressurize its environment with 1400 kg/cm^2 confining pressure, thus simulating deep burial. Load the ends of the specimen with a steadily increasing force by mechanically moving a piston into the deformation chamber. Add the force at a rate of 400 kg/min. The specimen should fracture by faulting when the force reaches approximately 3600 kg. Expect one or two faults to form, each inclined at about 28° to the direction of loading (see Figure 1.38).

Each of the major geologic structures has been the subject of decades of dynamic analysis. Faults and fault patterns have been modeled experimentally, not only using real rock materials, but also using soft materials like clay (Figure 1.39). When the sizes and strengths of the soft materials chosen to replicate regional relationships create perfect **scale models** (Hubbert, 1937), the resulting structures can bear an uncanny similarity to the natural structures (Figure 1.40).

Mathematical Models

Theoretical, mathematical analysis of structures has been pursued effectively from the perspective of engineering and fluid mechanics. The equations that ''picture'' the deformation are hardly Kodachromes of outcrop features. They are more like abstract art.

$$L = 2\pi t \sqrt[3]{\frac{\eta}{6\eta_1}} \tag{1.1}$$

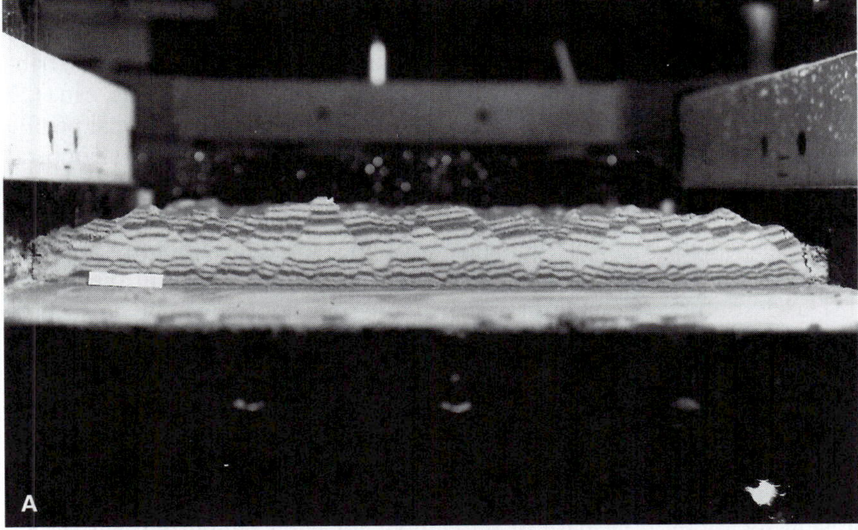

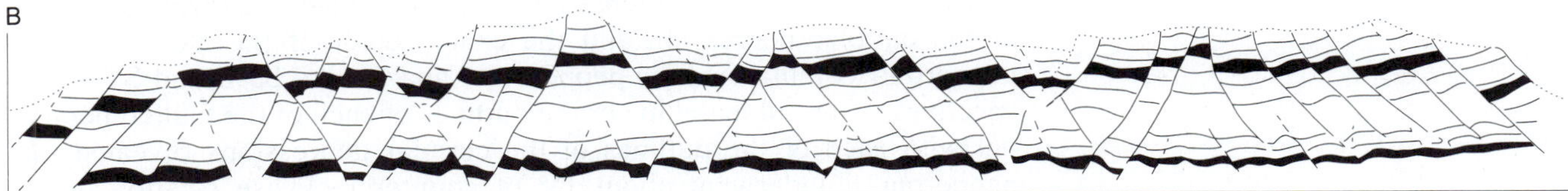

Figure 1.40 Sand models uniformly and horizontally stretched to produce closely spaced normal faulting characteristic of extended regions of the crust. [From Vendeville and others (1987). Published with permission of The Geological Society.]

A

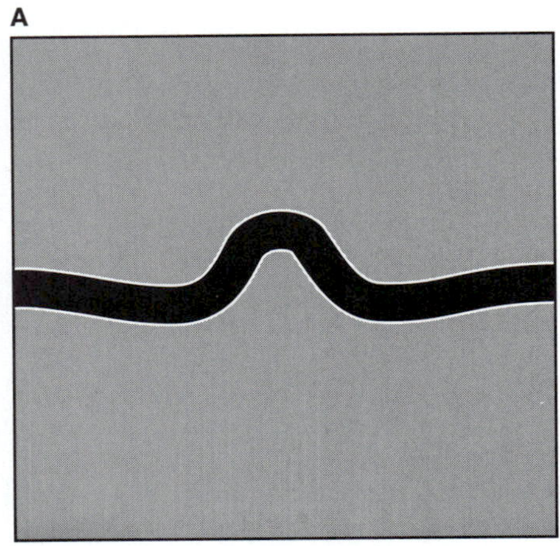

B

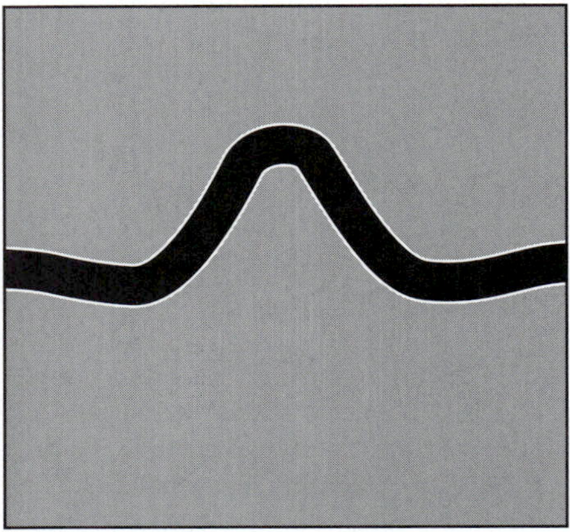

Figure 1.41 Computer simulation images of the buckling of a stiff layer (black) within a soft surrounding matrix (gray). Shortening (at 20%) is the same for each experiment, but contrast in viscosity between stiff layer and soft matrix is different. (*A*) Ductility contrast of 5 to 1. (*B*) Ductility contrast of 100 to 1. [Reprinted with permission from *Journal of Structural Geology*, v. 15, Zhang, Hobbs, and Jessell (1993), Elsevier Science Ltd., Pergamon Imprint, Oxford, England.]

But equations describe dynamic relationships in ways that words and photographs never could. Decoding the equations simply requires knowledge of what the variables represent. When Equation 1.1 is decoded, it reads:

> The wavelength of a fold in a layer of viscosity η and thickness t is equal to the product of (1) 6.2832 times thickness and (2) the cube root of the ratio of the viscosity of the layer to 6 times the viscosity of the rock in which the layer is contained.

The physical expression of this relationship is displayed in Figure 1.41, a computer simulation of a folding event in which a stiff (dark-colored) layer, embedded in a soft matrix, is shortened by 20%. The dramatic differences in form and size are due to significant differences in **ductility contrast** between layer and matrix in the first (Figure 1.41*A*) versus the second experiment (Figure 1.41*B*). The amount of shortening was the same in each experiment.

TWO EXAMPLES OF DETAILED STRUCTURAL ANALYSIS

Detailed Structural Analysis of a Pizza

We close this chapter by presenting two examples to clarify how detailed structural analysis works. Let us start with a true-life but nongeologic example. The example, although a little bizarre, may serve to clarify kinematic analysis and show the bridge that kinematic analysis builds between descriptive analysis and dynamic analysis.

Going to the freezer for a midnight snack, I pulled out a pizza whose form is portrayed in the geologic map shown in Figure 1.42*A*. The pizza, as shown, is arbitrarily oriented with respect to north. In most respects this was a normal pizza. A thin stratum of cheese rested atop tomato sauce and crust. And pepperoni, lightly dusted with cheese, were distributed across the face of the pizza. The diameter of the pizza was 9 in. (23 cm). Topographic relief of the pizza, as revealed in cross-sectional profile (Figure 1.42*A*), was merely 0.5 in. (13 mm). My delight as a structural geologist came upon observing that two pepperoni-sized circular depressions existed near the edge of the pizza in the northeast and northwest quadrants. There was not even a trace of cheese in these depressions, let alone pepperoni. Furthermore, opposite these depressions, in the southeast and southwest quadrants, two pepperoni overlapped in low-angle (overthrust) fault contact. The cross section of Figure 1.42*A* clarifies this relationship.

The physical and geometric properties of the pizza demanded the following kinematic interpretation (Figure 1.42*B*): somehow the pepperoni that had once occupied the circular depressions now devoid of cheese had translated 5.1 and 3.7 in. (12.9 and 9.4 cm), respectively, to their present locations. The path of movement for each pepperoni was determined by measuring the orientation of a line connecting the exact center of the circular depression and the exact center of the closest faulted pepperoni. Orientations for pepperoni 1 and 2 were S5°E and S4°W, respectively. Careful matching of the outlines of each of the faulted pepperoni with each of the outlines of the circular depressions revealed that pepperoni 1 underwent about 15° of counterclockwise rotation during

A

A′

0 1 2 3 4

Scale in Inches

Stratigraphy

Cheese

Pepperoni

Tomato Sauce

Crust

Symbols

Topographic Depression

Thrust Fault

A

1

Structure Profile

A

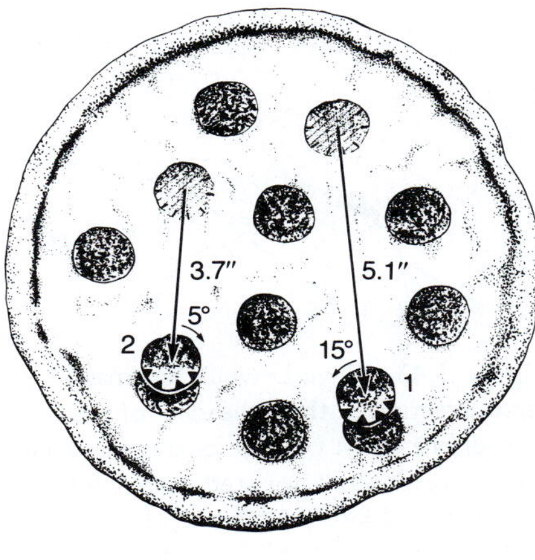

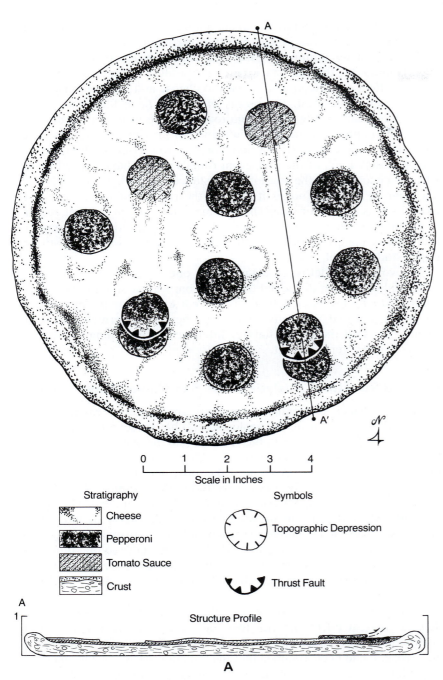

Figure 1.42 (*A*) Geologic map and structure profile of a medium-sized pepperoni pizza. (*B*) Kinematic model of the translation and rotation of the pepperoni.

translation and pepperoni 2 underwent about 5° of clockwise rotation during its translation. No microscopic study of the frozen sauce or cheese was undertaken to evaluate internal distortion due to the translation.

What can be concluded about the dynamics of origin of this structural system? By far, this is the most speculative part of the analysis, but the most enjoyable. (You will note that the language used in dynamic analysis is always cautious.) First, it would *seem* that the force(s) that triggered movement of the pepperoni did not violate the general integrity of the pizza. The box was not crushed in any way, nor was the crust distorted beyond primary kneading and shaping. The probable force causing deformation was gravity. Gravitational forces would not have triggered movement of the two pepperoni, however, unless the pizza had been tilted at some stage in its history.

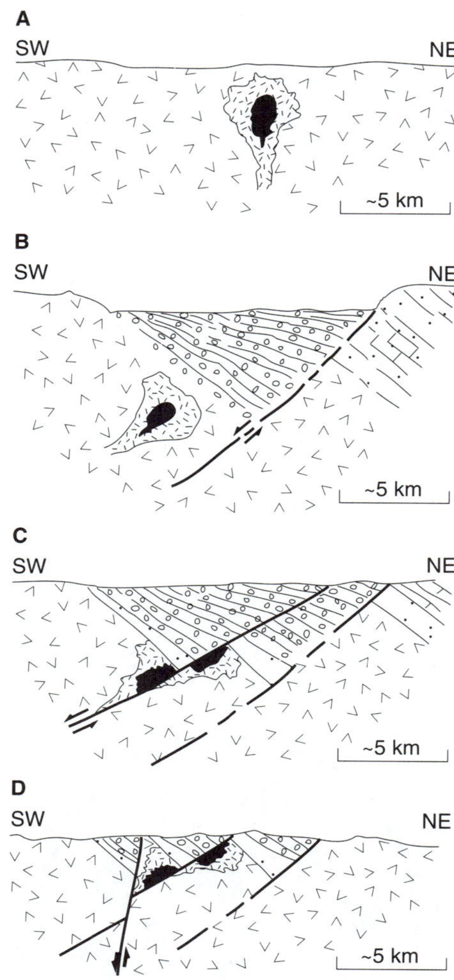

Figure 1.43 Dave Lowell's kinematic reconstruction of the beheading of the San Manuel ore body. (*A*) The quartz monzonite, and the ore body associated with it (black), was emplaced into Precambrian basement rocks approximately 65 Ma ago. Alteration zones (not shown) occur within the igneous body. (*B*) Extensional faulting (approximately 20 Ma ago) made room for deposition of basin fill. The extensional faulting was accompanied by rotation. (*C*) The San Manuel fault formed about 18 Ma ago, and it beheaded the top of the ore body and moved it 2.5 km to the southwest. (*D*) Approximately 10 Ma ago Basin and Range faulting offset the San Manuel fault. (Reproduced from Economic Geology, 1968, Vol. 63, p. 647.)

My working model is that the manufacturer, after preparing the pizzas, chose not to stack the boxed pizzas horizontally in freezer compartments. The manufacturer may have concluded, *perhaps* on the basis of experimentation, that tall stacking of pizza-filled boxes *might have* the adverse affect of flattening cardboard to cheese and tomato sauce before freezing set in. Instead, the pizzas *may have been* filed vertically. *If* stacked vertically while cheese and tomato sauce were yet warm and/or moist, the pepperoni, under the influence of gravitational forces, *might have been* vulnerable to translation along the low-viscosity tomato sauce discontinuity. Each of the pepperoni rounds would have ceased moving when it encountered the frictional resistance of another one. What was not clear to me then, nor is it now, is the rate at which the pepperoni moved—was it rapid or sluggish? The magnitude of the stresses required to initiate movement is also a puzzle. In fact, interpreting the strength of the various materials as a function of temperature would constitute a major study in itself.

My working model is, of course, only one interpretation. Maybe the structural event that dislodged the pepperoni was of an entirely different nature. *Maybe* my interpretation is correct except for timing: after all, the pizza, stacked vertically, may have been undeformed and solidly frozen . . . until the power failure. Interpreting the timing of structural events is often very difficult.

Detailed Structural Analysis of the San Manuel Fault

A beautiful piece of detailed structural analysis was carried out by J. David Lowell, a consulting geologist who has discovered more copper than anyone else in history. In the 1960s he had the opportunity to analyze the structural geologic setting of the San Manuel porphyry copper deposit, located north of Tucson, Arizona. Like most porphyry copper deposits, the San Manuel ore body was associated with a granitic pluton, which had intruded into the upper crust about 65 million years ago. The country rock into which the magma invaded became hydrothermally altered, and concentric zones of distinctive alteration patterns were cylindrically developed around the ore body (Figure 1.43*A*). Lowell deduced that the "cylinders" of alteration, and the granitic pluton itself, were rotated, and not in the vertical orientations that characterized their original emplacement (Figure 1.43*B*).

It had been known for some time that the uppermost part of the ore deposit was cut off by a major low-angle fault. The fault trends north-northwest and is inclined approximately 30° southwestward. Prior to Lowell's work, it had been assumed that the fault developed during an episode of profound crustal shortening, that it displaced the top of the ore body upward and to the northeast by an undetermined distance, and that this part of the ore body was subsequently removed by erosion (Figure 1.43*A*). Lowell determined the exact line of movement (N60°E/S60°W) by measuring grooves and slickenlines on the fault surface. Furthermore, he saw structural evidence along the fault that the upper part of the ore body may have been moved down and to the southwest, which held open the possibility that the displaced part of the ore body might be preserved at depth.

To test this working hypothesis, Lowell examined the geology to the southwest of the San Manuel property and soon learned that some exploration drilling had once been carried out in the area of interest. The drilling was churn drilling, yielding not cores of rock but chips of crushed rock. After persistently inquiring about the whereabouts of the chip samples, he learned that they had been locked behind the metal door of an old explosives storehouse dug into the side of a hill. Behind the door, Dave encountered thousands of bees, which he removed with the assistance of a local beekeeper. The sample bags inside the storehouse had completely deteriorated (aided by packrats!), except for tags showing sample number. He carefully spooned each pile of chip samples into a new sample bag and analyzed the chemistry of the minerals contained. Based on knowledge of the location of each drill hole, and the approximate depths at which the samples had been retrieved during drilling, Lowell reconstructed an alteration pattern that seemed to match that of the San Manuel ore body (Lowell, 1968) (Figure 1.43*C*). Piecing together the alteration patterns above and below the fault, he reconstructed the history (Figure 1.43*B–D*), determining that the top of the ore body had been displaced 4 km southwestward. He predicted the depth where the displaced ore body would lie in the subsurface (Figure 1.43*C*). The very first exploration drill hole proved that his descriptive and kinematic analysis had been carried out perfectly, for the hole penetrated the top of the ore body, now actively mined as the Kalamazoo deposit!

THE TIME FACTOR

Detailed structural analysis feeds naturally into broader geologic synthesis, but not without a grasp of the dimension of time. Analysis of crustal movements in relation to the overall physical evolution of the Earth simply cannot be pursued outside a time reference framework. Interpreting the timing of deformational movements provides a way to recognize discrete periods of deformation. First the relative timing of events is determined, through such means as cross-cutting relationships (e.g., imagine an older fault cut and offset by a younger fault). Then, with the help of age dates and knowledge of the ages of rock formations, the absolute timing may be established, or at least constrained. At a more sophisticated level of interpretation, it may be possible to recognize diachronous events: events, like the movement of a cold front, that steadily migrate across a region through time.

Interpreting timing permits the results of structural investigations to be integrated within a framework of broader geological processes and events. It promotes an understanding that structural deformations are just a small part of a much larger orchestration, the knowledge of which serves to clarify why certain structural relationships developed in the first place.

There is a word of caution: Ted Smiley (1964) has emphasized that all dating is a matter of interpretation, whether it be a date on a structure in an archaeological site, the date of a particular rock formation, or the date of a specific geological event. He stresses that geochronological interpretations can be strengthened greatly by knowing the *exact* physical makeup of the sample to be dated, by determining the *complete* geological history of the rock body from which the sample was removed, and by learning the *precise* association between the material studied and the event whose age is being analyzed. In other words, meaningful age determinations are derived from paying close attention to *the primary sources*. Modern radiometric isotopic determinations are yielding ages for rocks and events that are often beautifully consistent with relative age relationships determined independently on the basis of careful evaluation of the historical geology. Yet, in structurally complex areas, it is not uncommon to face problems of interpreting whether a given radiometric age determi-

ERA	PERIOD		EPOCH	M.Y.B.P.
CENOZOIC	QUATERNARY (Q)		Holocene	.01
			Pleistocene	2
	TERTIARY (T)	NEOGENE	Pliocene	5
			Miocene	25
		PALEOGENE	Oligocene	
			Eocene	55
			Paleocene	67
MESOZOIC	CRETACEOUS (K)		Upper	100
			Lower	140
	JURASSIC (J)		Upper	160
			Middle	180
			Lower	200
	TRIASSIC (TR)		Upper	205
			Middle	215
			Lower	250
PALEOZOIC	PERMIAN (P)		Upper	270
			Lower	290
	PENNSYLVANIAN (IP)			330
	MISSISSIPPIAN (M)			365
	DEVONIAN (D)		Upper	385
			Middle	390
			Lower	405
	SILURIAN (S)		Upper	415
			Lower	425
	ORDOVICIAN (θ)		Upper	460
			Middle	485
			Lower	500
	CAMBRIAN		Upper	515
			Middle	540
			Lower	570
Sorry, not enough room for 90% of Earth's time, ie. PRECAMBRIAN				

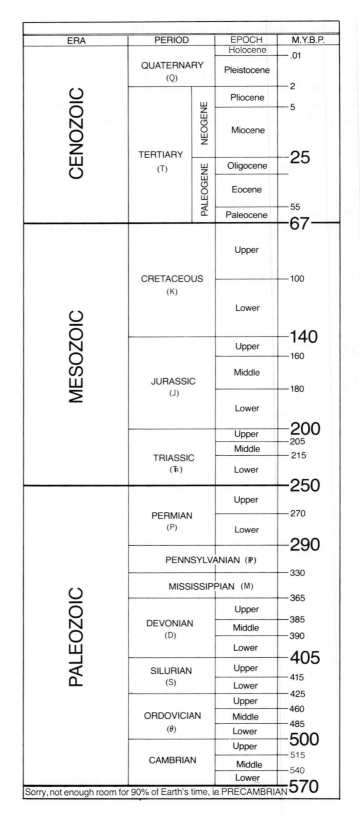

Figure 1.44 Geologic time scale. Although not digital, it keeps great time.

nation reflects an age of crystallization, an age of metamorphism, a time of uplift and cooling, or perhaps nothing geological at all.

The most practical approach for us to gaining familiarity with the timing of events and the ages of rocks within a specific region of interest is to search out and study geologic maps. The explanation of any good geologic map provides the time frame for the local or regional geological column. It is based on understanding of the established stratigraphy, a knowledge of the critical contact relationships, and age determinations for igneous and metamorphic rocks. The column is, in effect, the timepiece to be used as a starting point in investigating the geology of the area or region of interest.

Of course the "alphabet" of time is the **geologic time scale** (Figure 1.44). Learning this time scale places us in the proper framework for contributing, communicating, and understanding. It is first-order business. The day after graduation from high school I entered the office of J. Von Feld, exploration geologist for Consolidation Coal Company. His first statement to me was, "I understand you want to be a geologist." His second statement was in the form of a question: "Have you learned the geologic time scale?"

CHAPTER 2

KINEMATIC ANALYSIS

STRATEGY

Kinematic analysis is the reconstruction of *movements* that take place during the formation and deformation of rocks, *at all scales*. It is carried out without reference to the forces or pressures that cause the deformation. "Tracking" and describing the step-by-step movements is all that counts, without prejudice regarding what generated the movements. Of main interest is the analysis of secondary deformation, like the rotation of rock layers during folding; the displacements along a fault or within a fault system; or the opening up of wall rock as a dike is intruded.

We also apply kinematic analysis to primary movements accompanying the very formation of rocks (see Part III-C), like the wrinkling of the skin of the top of a lava flow to create the ropy "pahoehoe" structures found in basalt (Figure 2.1); the movement of a dune field across a flat landscape to produce the cross-bedding in an aeolian sandstone; or the slumping and landsliding in sands and muds at the edge of a basin to produce "chaotic structure" in a sedimentary sequence.

There are two different strategies for kinematic analysis of deformational structures, and which one is used depends on whether the rocks under consideration have behaved as rigid or as nonrigid bodies during the deformation. The distinction between rigid and nonrigid rests on whether the rock body, at the scale of observation, moves *intact* without a change in shape or size (rigid body deformation), or with a change in shape or size (non–rigid body deformation). The test is whether each point in the body, during the deformation, maintains the same exact location *relative* to neighboring points. If so, it is a rigid body deformation. If not, it is a non–rigid body deformation.

During **rigid body deformation**, rock is translated and/or rotated in such a way that original size and shape are preserved. A schematic example of rigid body translation without rotation is portrayed in Figure 2.2*A*. Block *abde* has shifted its position, but has changed in such a way that its original size and shape are maintained. There is no change in the

Figure 2.1 This landscape of ropy "pahoehoe" lava on the island of Hawaii presents a snapshot of kinematics. The rock looks as if it is still moving. Kinematic analysis would distinguish the timing and sense of motion of lava down the main channel versus the timing and sense of movement of lava up and over the "banks." (Photograph by G. H. Davis.)

configuration of points within the block. Similarly, Figure 2.2*B* portrays rigid body deformation, in this case a rotation. Again, there is no change in the configuration of points within the block.

A geologic example of rigid body deformation is pictured in Figure 2.3*A*. It shows the submerged foundation of a second-century church along the southern edge of the Isthmus of Corinth. Active faulting over the centuries (Figures 2.3*B,C*) has caused the bedrock on which the church rests to move as a rigid body. Along the northern edge of the isthmus, the bedrock has risen out of the sea as a result of faulting (see Figure 2.3*C*). If we had been able to examine closely a given cubic meter of bedrock beneath the church foundation both before and after movement,

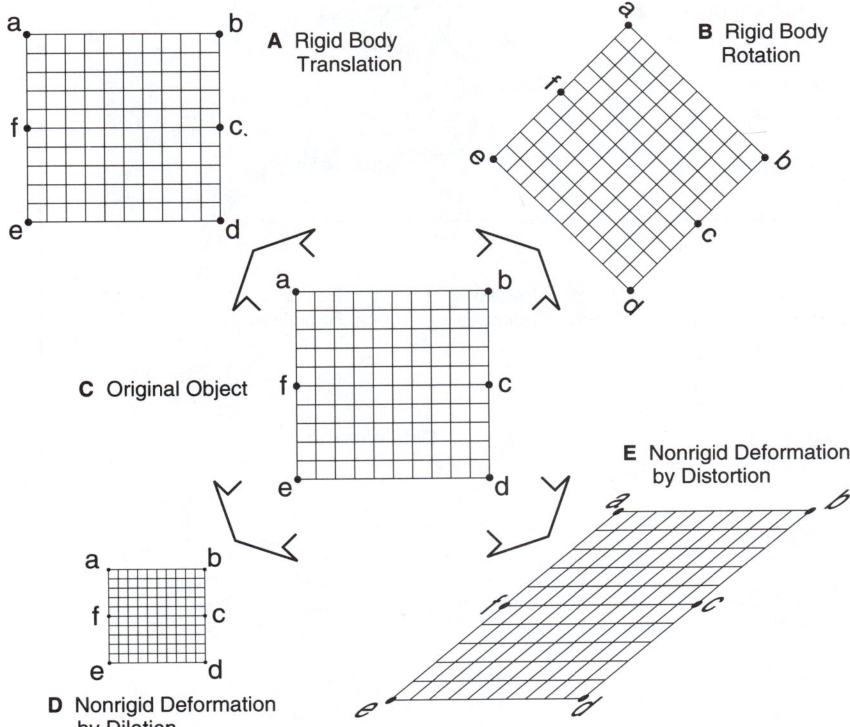

Figure 2.2 Originally undeformed body (*C*) in center of diagram (i.e., square *abde*) is deformed by (*A*) rigid body translation, (*B*) rigid body rotation, (*D*) non–rigid body dilation, and (*E*) non–rigid body distortion.

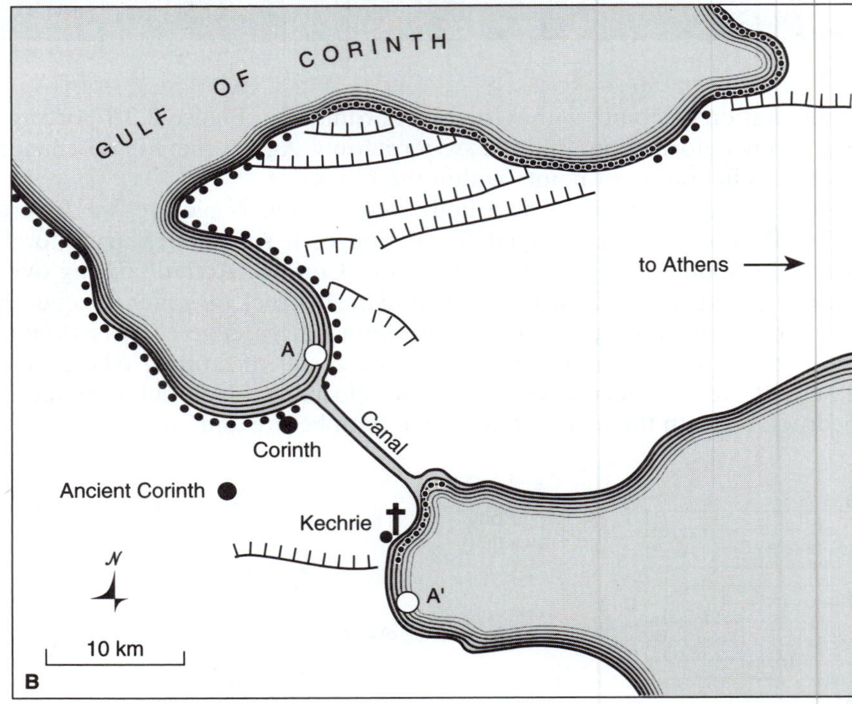

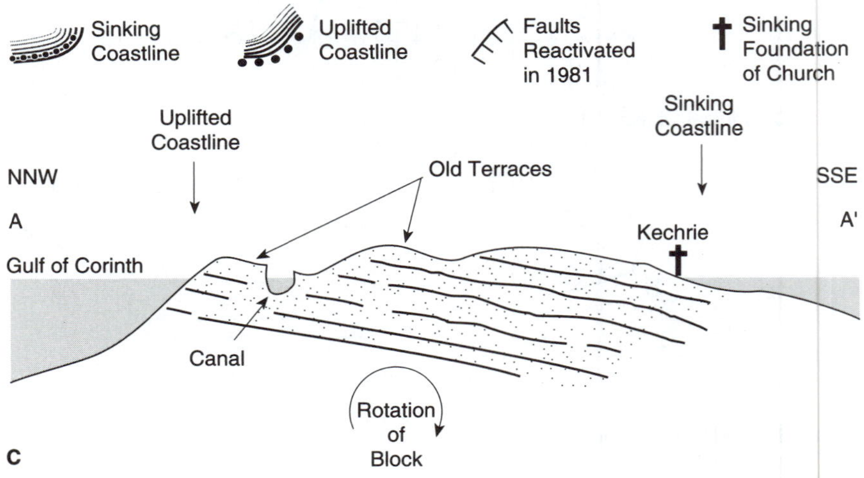

Figure 2.3 (*A*) Submerged foundation of a second century church at Kechrie along the southern edge of the Isthmus of Corinth. Submergence due to tilting of the isthmus by active faulting. Mike Davis for scale. (Photograph by G. H. Davis.) (*B*) Map showing the Kechrie church location in relation to faults that were reactivated during earthquake activity in 1981; the sinking coastline; and the uplifted coastline. [Adapted from Jackson and others (1982).] (*C*) Generalized cross section showing topography, faulting, and tilting along line *A–A′*. [Adapted from Jackson and others (1982), with permission from *Earth and Planetary Sciences Letters*, v. 57, Elsevier Science, Amsterdam.]

we would have seen absolutely no difference in the configuration of points within it. The only observable changes would have been elevation and orientation of the cube.

Strain results from **non–rigid body deformation** when rock undergoes a change in size (**dilation**) and/or a change in shape (**distortion**). Strain expressed as dilation or distortion results from a change in configuration of points within a body. A single body, during a single deformational event, may experience both dilation and distortion.

Pure dilation is a change in size without a change in shape. Figure 2.2*D* pictures a decrease in size (negative dilation); the spacing of points within the original body is cut in half. An example of increase in size (positive dilation) is shown in Figure 2.4, a photomicrograph of a shattered phenocryst of quartz in a fine-grained igneous rock from the Mariana Islands in the western Pacific. The brittle, rigid mass of quartz has responded to deformation by acting like ice breaking up on a frozen sea. Although there is no change in the configuration of points *within* each of the offset pieces, the quartz crystal *as a whole* experienced an increase in area. The increase was accommodated by infilling of the space between the quartz fragments (white) by fine-grained groundmass (black), which flowed into place.

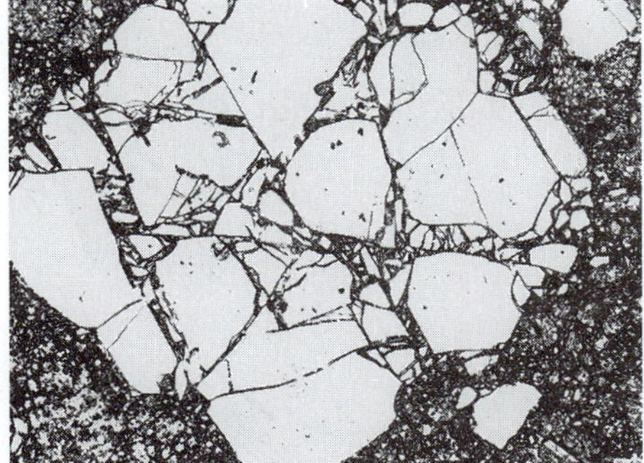

Figure 2.4 Photomicrograph of shattered quartz crystal. The fragmented crystal displays the results of non-rigid body deformation. (Photograph by R. G. Schmidt. Courtesy of United States Geological Survey.)

Significant dilation accompanies such non–rigid structural processes as the shrinkage of mud to produce mud cracks, the compaction of sediments during burial to produce thinning, the cooling of basalt to produce columnar joints, and the dissolving away of impure limestones under directed pressure to produce cleavage. Small volumetric changes are probably common as well, but they are hard to detect because of the difficulty of precisely reconstructing the original size of deformed geologic bodies.

Pure distortion is a change in shape without a change in size. As shown in Figure 2.2*E*, the change in shape of a body from a square to a rhomb is made possible by systematic changes in spacing between points in the body. The distance between points *a* and *e* increases from 10 units to 16 units. Angular relations between alignments of points change as well. For example, angle *aed* is reduced from 90° to 40°.

When non–rigid body deformation results in the systematic distortion of what we normally regard as solid rock, the results can be unusually interesting. The perfect ellipses shown in Figure 2.5 were originally circular sections through volcanic lapilli. Strong flattening transformed the circles into ellipses through non–rigid body distortion. Better than words, the elliptical shapes communicate the full extent of the non–rigid body distortion.

Figure 2.5 Distorted lapilli in tuff from the Lake district of England. [From Oertel (1970). Published with permission of Geological Society of America and the author.]

Non–rigid and rigid body deformation commonly operate together. Fault movements are normally considered rigid body movements (Figure 2.6*A*), but if the faults are very closely spaced at the scale of view, the result is non–rigid body deformation (Figure 2.6*B*).

TRANSLATION

General Concept

During pure translation, a body of rock is displaced in such a way that all points within the body move along parallel paths. Sliding down a steep snow-covered slope on a really stiff cafeteria tray is an example of pure translation, . . . provided you don't "fishtail" or spin out (Figure 2.7). If the tray is perfectly rigid, all the points within it will move exactly the same distance and along the same path.

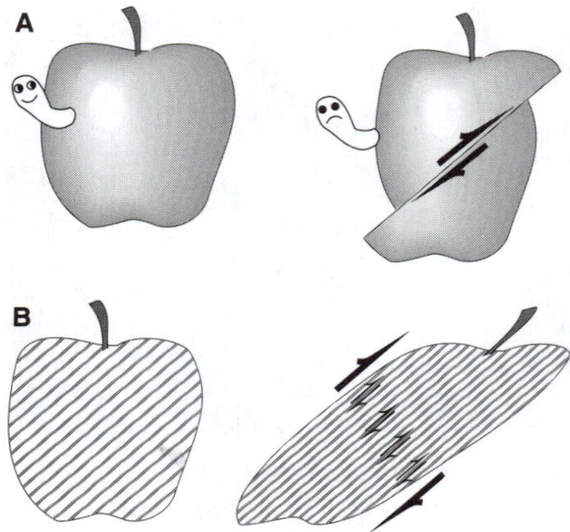

Figure 2.6 (*A*) Rigid versus (*B*) non–rigid deformation of objects by faulting is partly a matter of the closeness of spacing of the structures within the chosen field of view.

Rigid bodies translate past one another along surfaces where the integrity of the rock body, at whatever scale, is interrupted, and along which movement is possible. The interface between tray and snow is the surface of translation in the tray-riding example (Figure 2.7). Surfaces that accommodate translation in the geologic world include faults, joints, and bedding planes. Ground rupture associated with earthquakes may express rigid body translation along fault surfaces (Figure 2.8). Joint surfaces are fracture discontinuities along which rigid rock on either side moved apart, *ever so slightly*. This movement may be helped along by the presence of warm, if not hot, pressurized fluids that force open cracks and precipitate minerals in the form of veins. The **crystal fiber vein** shown in Figure 2.9 formed during opening of the walls of the joint it now occupies. The direction of opening was parallel to the fibers. In yet another example of rigid body movements along surfaces (Figure 2.10) bedding surfaces were activated during **flexural-slip folding**—the beds slipped relative to one another, like pages in a book.

Translation can be considered at extreme scales as well: on the one hand, the world; on the other hand, the crystal lattice. The concept of plate tectonics (Chapter 10) is a view of the entire outer shell of the Earth composed of an array of broad, thin, rigid plates of crust and uppermost mantle (**lithosphere**). The rigid plates translate (and rotate) with respect to one another, moving on hot, ductile, non–rigid mantle material (**asthenosphere**). The plate motions are described mathematically as rigid body translations and rotations. At the opposite end, individual tiny crystals are built in a lattice framework of atoms. We will see in Chapter 4 (''Deformation Mechanisms and Microstructures'') that suitably oriented

Figure 2.7 If the movement is purely rigid body translation, all the points in the body will move exactly the same distance along exactly the same path. If not, ouch! (Artwork by D. A. Fischer.)

Figure 2.8 Kaparelli fault scarp in the eastern Gulf of Corinth region, Greece. The most recent movement along the fault zone took place in a 1981 magnitude 6.4 earthquake. Not all of the scarp shown in the photograph was produced during this earthquake. The height of the 1981 component of scarp formation is expressed in the prominent white stripe of limestone. Mike Davis for scale. (Photograph by G. H. Davis.)

Figure 2.9 Crystal fiber vein cutting melange at Cowhead Point, San Juan Islands, Washington. Note that the crystal fibers are not perfectly perpendicular to the walls of the vein. (Photograph by G. H. Davis.)

layers of atoms within a lattice may translate with respect to other layers when forced to do so.

Displacement Vectors

Rigid body translations can be expressed conveniently in terms of **displacement vectors**, which describe translation in terms of three parameters (Ramsay, 1969): distance of transport, direction of transport, and sense of transport. **Distance of transport** can range from fractions of millimeters to hundreds or thousands of kilometers. **Direction of transport** is expressed as the **trend** and **plunge** (see Part III-D) of the line of movement: we might say, "the movement was along a northeast-southwest line"; or "the movement was along a line plunging 10° N40°E." **Sense of transport** refers to the "polarity" of the movement: we might say, "the movement was *from* southwest to northeast"; or "the movement was *to* the southwest, along a line plunging 10° in a direction S40°W."

These three components of displacement vectors are illustrated in Figure 2.11, a photograph and interpretive diagram of one of the "sliding stones" on Racetrack Playa in California (Sharp and Carey, 1976). The stones translate across slippery wet mud surfaces, propelled, apparently, by the force of the wind (Figure 2.11*A*). Trails left by the stones allow the displacement vector to be constructed (Figure 2.11*B*). The stone slid smoothly along most of its course, but then it encountered a small stone obstacle that caused it to flip over several times and come to rest. This kinematic plan vaguely resembles that which can be interpreted in Figure 2.11*C*, which shows what happened at about midnight on a summer night in 1959 at Hebgen Lake, Montana, just north of Yellowstone. A large-magnitude earthquake hit, triggering major landsliding, rockfalls, and ground displacements. People were killed. One family fleeing the area in their car in darkness sped over a fresh fault scarp whose presence they could not have imagined, or seen. Thankfully, and astonishingly, no one in the car was seriously injured. The car moved (translated) and flipped (rotated) in rigid body ways, and on impact was deformed through non–rigid body distortion. Each member of the family managed to crawl out through the windows to safety.

A

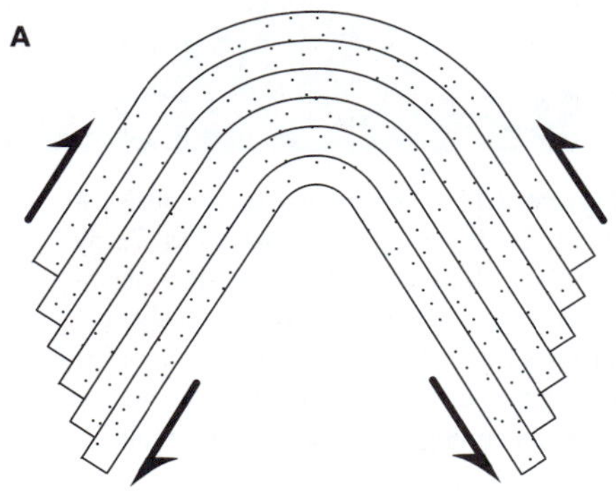

B

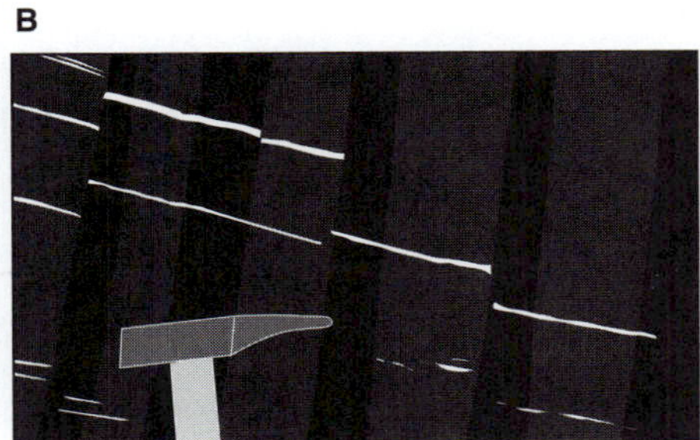

Figure 2.10 (*A*) Flexural-slip folding creates sliding along the layers as if they were pages in a book. (*B*) Slip parallel to the beds of a flexural-slip fold near Cornwall, England, resulted in offset of veins (white) oriented at a high angle to bedding. [Adapted from Ramsay and Huber (1987), Figure 21.9, p. 451, with permission of Harcourt Brace.]

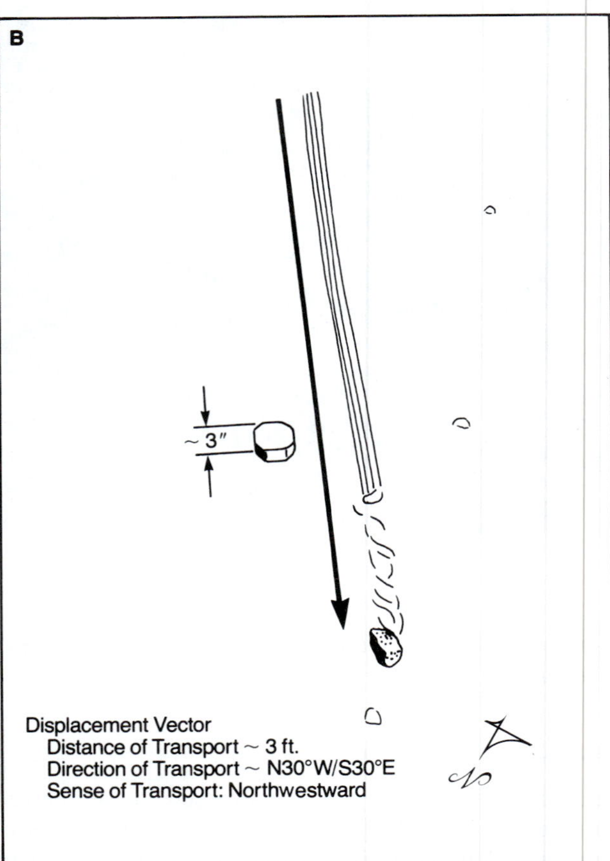

Figure 2.11 (*A*) Sliding stone on Racetrack Playa, California. [From Sharp and Carey (1976). Published with permission of Geological Society of America and the authors.] (*B*) Kinematic description of the sliding stone. (*C*) Overturned car at rest after careening off Hebgen Lake fault scarp in the middle of the night.

It might be useful to illustrate the concept of displacement vector with a river-rafting example. Those who raft the Colorado River, from Lees Ferry to Pearce Ferry, travel 280 river miles (Figure 2.12). The Colorado River follows a circuitous route in covering that distance. The raft trip displacement vector yields a direction of transport of N68°E/S68°W, a sense of transport toward S68°W, and a linear distance of transport of 165 miles. The displacement vector expresses the location of the finishing

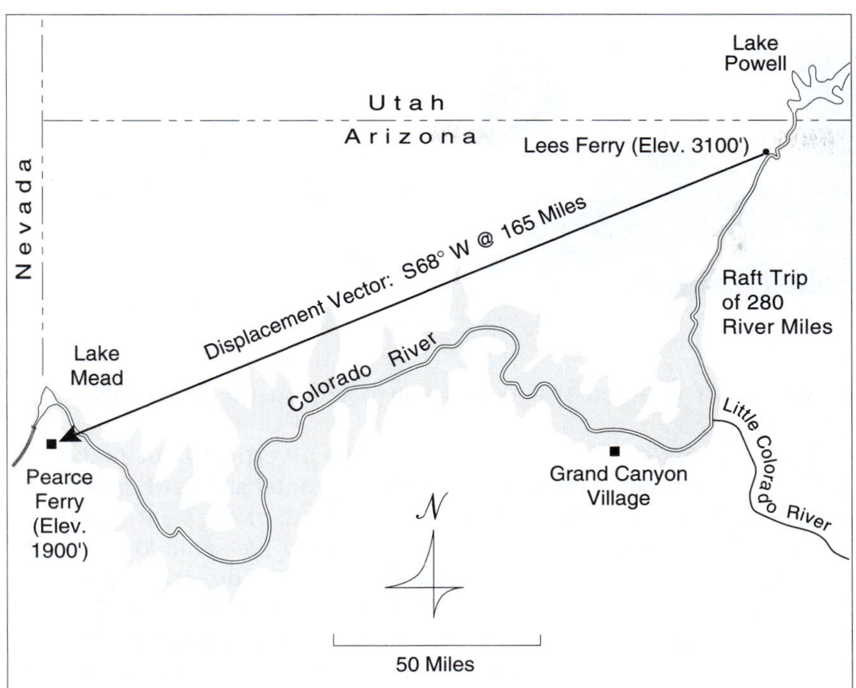

Figure 2.12 Displacement vector describing float-trip translation through the Grand Canyon from Lees Ferry to Pearce Ferry along the Colorado River.

point (the final "takeout") relative to the starting point (the "jumping-off point").

Actually, the raft trip displacement vector is a two-dimensional approximation of the true, three-dimensional solution. The true displacement vector must take into account the difference in elevation between Lees Ferry (3100 ft) and Pearce Ferry (1900 ft) (Figure 2.12), a difference in elevation of about 1200 ft. The true displacement vector is oriented in the same direction as that of the two-dimensional solution, but its plunge is about 0.1° southwest, not horizontal. The sense of transport remains S68°W, but the distance of transport is greater by a tiny fraction of a mile. In this example, the difference between the two- and three-dimensional solutions is negligible, but in most geological problems the third dimension cannot be ignored.

The movement of the Indian Plate provides a plate scale example of the meaning of displacement vectors. India is inferred to have been positioned in deep southerly latitudes during the early Cenozoic Era, and, later, the plate in which India is embedded was translated far to the north (Figure 2.13). The values describing the displacement vector for India's northward ride depend crucially on the accuracy of early Cenozoic reconstructions of plates in the southern hemisphere. Assuming that the reconstruction shown in Figure 2.13 is reasonably accurate, the direction of transport for India is N12°E/S12°W, the sense of transport is north–northeast, and the distance of transport is 7000 km. This is an approximation of the true displacement vector.

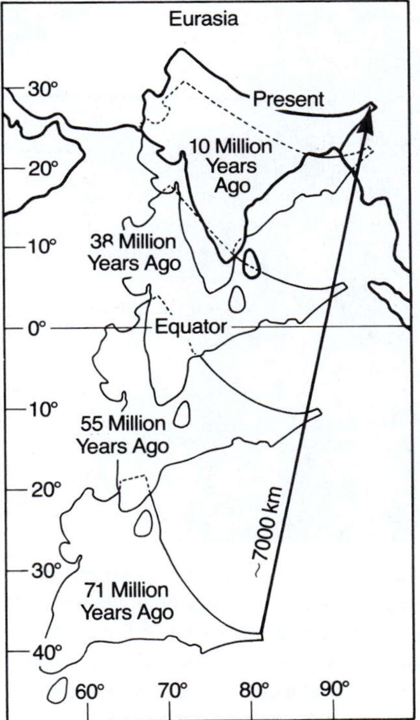

Figure 2.13 Reconstruction of the northward translation of India during the Cenozoic Era. Calculation of the displacement vector depends upon the interpretation of the starting position of India. (From Molnar and Tapponnier, *Science,* v. 189, p. 419–425, copyright © 1977 by American Association for the Advancement of Science.)

Slip on Faults

The concept of a displacement vector as a description of translation can be applied to fault analysis. To construct displacement vectors for faulting, it is necessary to identify two reference points, one on each side of a given fault, that shared a common location in the body before faulting. Translation is described in terms of **slip** (Figure 2.14), *the actual relative*

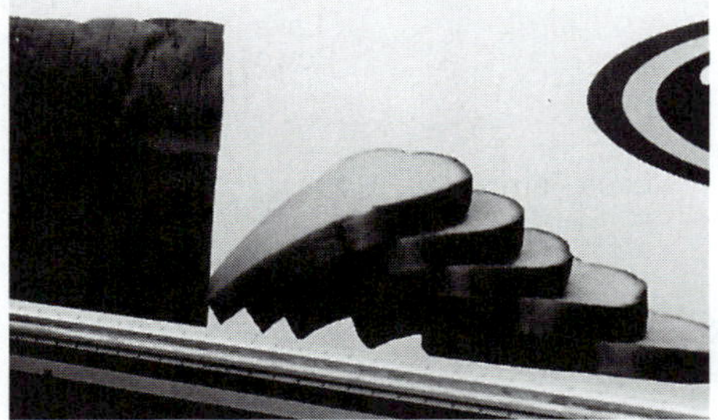

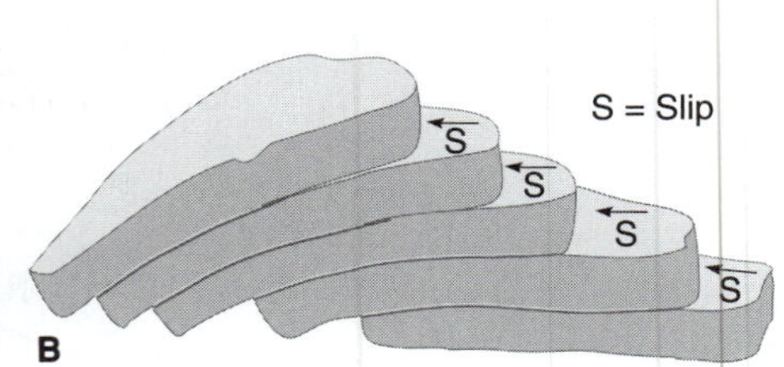

Figure 2.14 (*A*) Photograph of the side of Rainbow Bread Truck, showing bread slices, both before and after deformation. (Photograph, on the move, by G. H. Davis.) (*B*) Tracing of photo that reveals the *slip* on each fault surface.

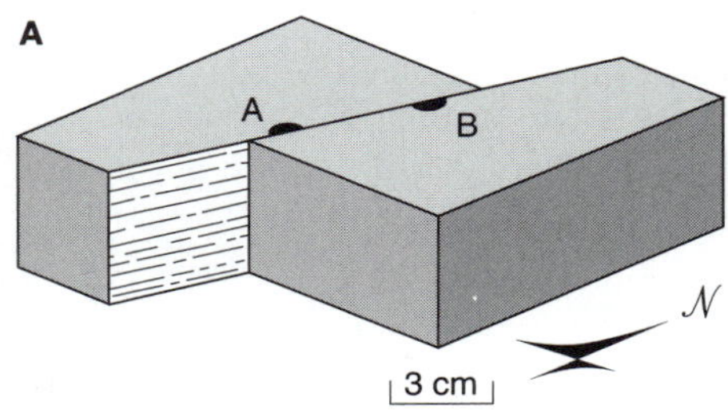

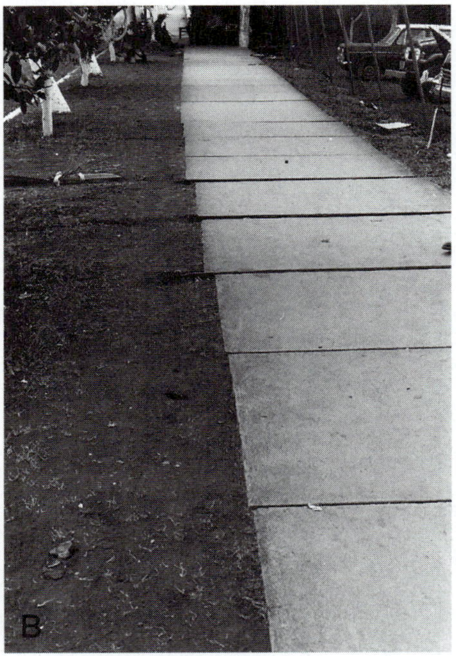

Figure 2.15 (*A*) Schematic portrayal of horizontal, left-handed slip on a vertical fault. The magnitude of translation is 5 cm. (*B*) Real-life examples of horizontal, left-handed slip on vertical faults. The sidewalk panels were shifted by faulting during the earthquake in Managua, Nicaragua, in 1972. Faulting of the sidewalk took advantage of preexisting discontinuities. Maximum translation is about 3 cm. Aggregate displacement across the 12-m wide exposed zone is 28.6 cm. (Photograph by R. D. Brown, Jr. Courtesy of United States Geological Survey.)

displacement between two points that occupied the same location before faulting.

Figure 2.15*A* shows a rigid block cut by a vertical fault that strikes N20°E. Translation on the fault is strictly horizontal and **left-handed**—that is, a faulted marker, when followed to the trace of the fault, is offset to the left. Since points *A* and *B* initially occupied the same location, they provide a means to evaluate slip along the fault. The direction of transport (or direction of slip) is N20°E/S20°W; the sense of transport (sense of slip) of *A* relative to *B* is S20°W; the distance of transport (net slip) is 5 cm. A real-life example of small left-slip movements is pictured in Figure 2.15*B*, sidewalk panels that shifted in left-slip fashion during the Managua earthquake of 1972.

A more general illustration of translation along a fault surface is shown in Figure 2.16*A*. The fault surface strikes N60°E and dips 50°SE. Translation is such that the southeast block of the fault moves down relative to the northwest block. Additionally, the fault accommodates left-handed movement. The actual path of slip during fault translation runs oblique on the fault plane. The displacement vector describing this faulting connects *B* to *A*, two points that occupied the same location before faulting. The orientation of the displacement vector can be determined stereographically (see Part III-H) by plotting both the fault plane and the slip direction, and then interpreting trend and plunge (Figure 2.16*B*). The sense of slip of *B* relative to *A* is 26° N85°E. The magnitude of the total displacement (i.e., the **net slip**), is 5.0 cm. The horizontal component of the slip (**strike–slip** component) is 4.1 cm; and the down-the-dip component of translation (**dip-slip** component) is 2.9 cm (Figure 2.16*C*).

In almost all geological examples, it is impossible to be sure whether a given fault block was stationary or in motion during faulting. All that can be reported is **slip**, the **actual relative displacement** that takes place as a result of translation. The translation portrayed in Figure 2.16*A* could have been achieved in a number of ways: movement of the southeast block downward along a northeastward trend; movement of the northwest block upward along a southwestward trend; movement of the southeast block down and to the northeast accompanied by simultaneous movement of the northwest block up and to the southwest; movement of both blocks downward and northeastward, but such that the southeast block moves a greater distance.

In regions of active faulting, such as the San Andreas fault system in California, there are many natural and man-made reference points that permit fault translation(s) to be measured. Displacement vectors can be calculated on the basis of faulted streams (Figure 2.17), faulted fence lines and roads (see Figure 2.11*C*), and even faulted city streets (Figure 2.18).

Where faults have long been inactive, it is very difficult to locate reference points that can be used to analyze net translation. The "*A*s" and

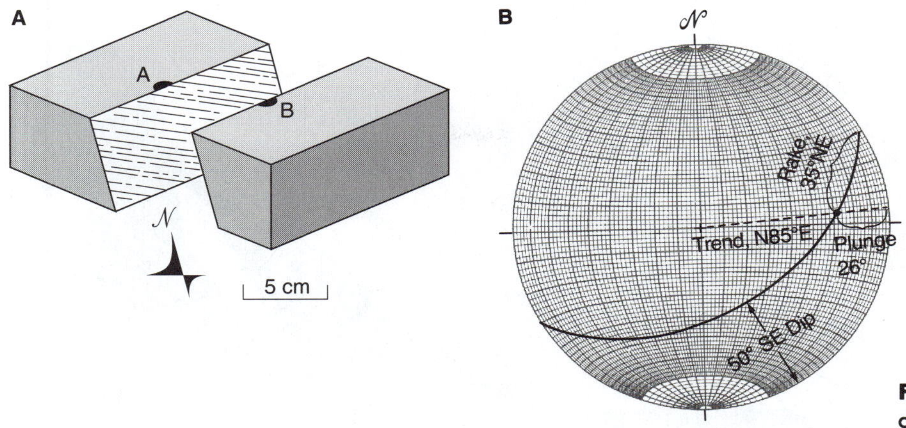

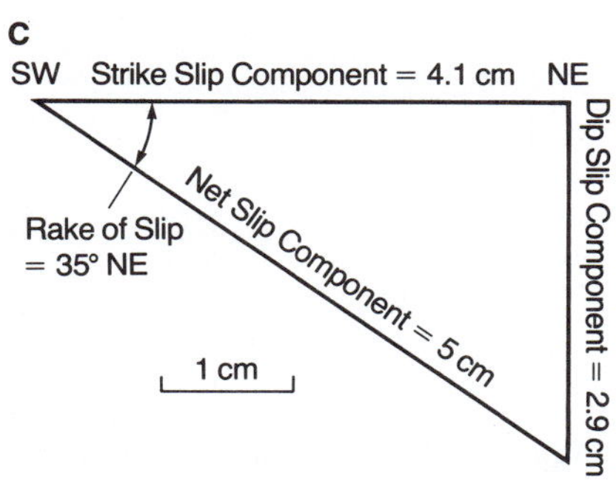

Figure 2.16 (*A*) Left-handed oblique slip (5 cm) on steeply dipping fault. Slickenlines "rake" at an anle of 35° on the fault surface. (*B*) Stereographic determination of the trend and plunge of the displacement vector. (*C*) Net-slip, strike–slip, and dip-slip components of the displacement vector.

"*B*s" of Figures 2.15 and 2.16*A* are difficult to find in nature. However, chance relationships have permitted some elegant kinematic appraisals. Anderson (1973), working in the Lake Mead region of southern Nevada, recognized a Miocene andesitic stratovolcano, the Hamblin–Cleopatra volcano, which was cut in half between 15 and 10 million years ago by the Hamblin Bay fault (Figure 2.19). The northwest part of the volcano was translated approximately 19.3 km relative to the southeast half, in a left-handed sense. Discovering offset geological features like igneous plugs and volcanoes is the rare exception and not the rule. But there are ways to make the most of what we are given. You might want to go to "Descriptive Analysis," (Part III-J) to practice the analysis of translation using orthographic and stereographic projection.

Figure 2.17 Right-handed offset of stream due to movement(s) on the San Andreas fault as exposed in the Carrizo Plains of California. (Photograph by R. E. Wallace. Courtesy of United States Geological Survey.)

Figure 2.18 Faulting (and fissuring) of a street as a result of the Great San Francisco Earthquake of 1906. (Photograph of Bluxom Street, near Sixth Street by G. K. Gilbert. Courtesy of United States Geological Survey.)

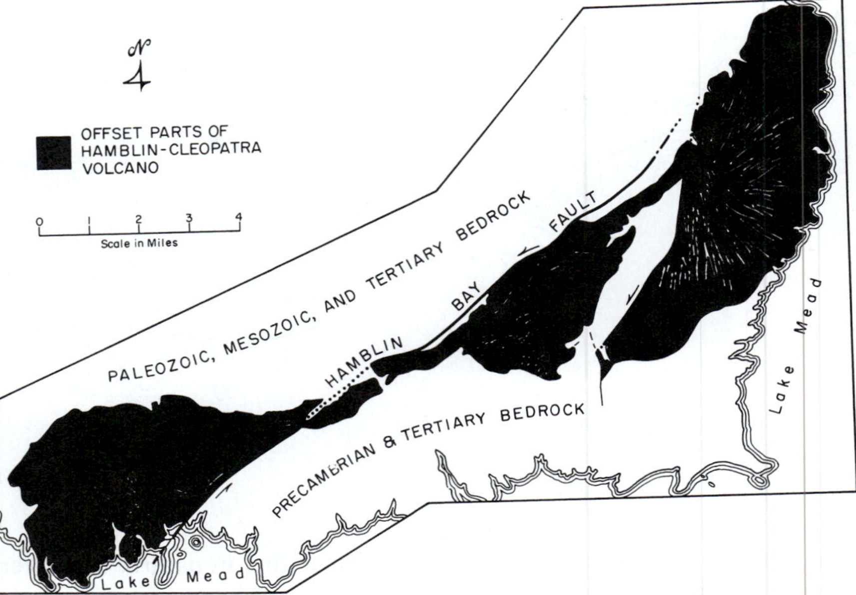

Figure 2.19 The Hamblin–Cleopatra volcano, Miocene in age, was cut in half and rearranged by faulting during the time interval 15 Ma to 10 Ma before present. The offset parts of the volcano, including its once-radial dike swarm, permit the magnitude of the displacement vector for the faulting to be calculated. Total displacement is approximately 12 miles (19 km). [From Anderson (1973). Courtesy of United States Geological Survey.]

ROTATION

General Concept

Rotation is a rigid body operation that changes the configuration of points in a way best described by rotation about some common axis. Amusement parks thrive on rotational operations (Figure 2.20*A*). Axes of rotation may be horizontal (Ferris wheel), vertical (merry-go-round), and inclined (rotor), or all of the above (Tilt-a-Whirl, Roll-o-Plane, the Hammer, and other ghastly rides). The changes in locations of points are described by the orientation of the **axis of rotation** (trend and plunge), the **sense of rotation** (clockwise versus counterclockwise), and the **magnitude of rotation** (measured in degrees). Based on this set of facts, the locations of points before and after rotation can be calculated.

Designating a clockwise versus a counterclockwise sense of rotation depends partly on the direction of view. The operator behind the Ferris wheel sees the cars rotating clockwise about the horizontal center axis (Figure 2.20*B*). The bystander in front, on the other hand, observes counterclockwise motion. (The sense of motion of the hands of a clock would be counterclockwise if a clock with a transparent housing were viewed from behind; my kind of clock!) To avoid ambiguity, a special convention is adopted in structural geology to specify sense of rotation: we describe the sense of rotation while looking *down* the axis of rotation. If the axis is strictly horizontal, we describe not only the sense of motion but also the direction in which the structure is observed. For example, the Ferris wheel is rotating counterclockwise when viewed to the north (Figure 2.20*B*).

Geological Examples

A number of geological examples serve to illustrate why rotation is an important kinematic operation. Originally horizontal layered rocks can become inclined as a result of folding and faulting processes. Consider strata in the middle limb of the Hunter's Point **monocline**, a steplike fold,

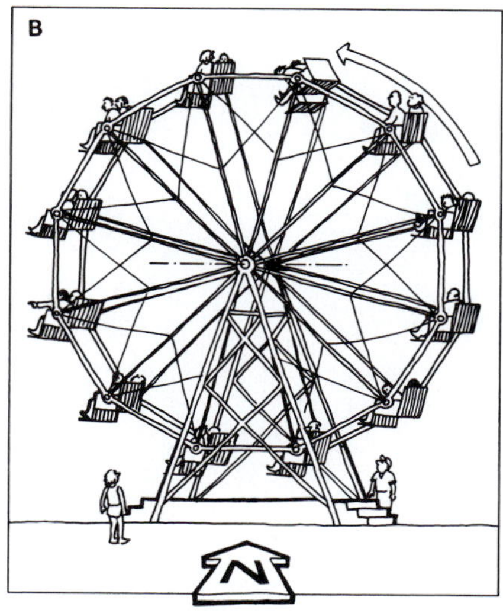

Figure 2.20 (*A*) Amusement parks thrive on rotational operations. Kinematic analysis of rotation includes describing the orientation of the axis of rotation, the sense of rotation, and the magnitude of rotation. (*B*) Sense of rotation depends on the direction of view.

shown in Figure 2.21. The overall form of the fold is revealed in the distant view, where horizontal bedding on the west (left) abruptly bends down to very steep dips (upper right corner). The bold white rock exposures in the foreground show that bedding has been rotated to dips as steep as 90°. The inclined strata can be thought of as having rotated as a nearly rigid body about a horizontal axis. The orientation of the axis of rotation of bedding in the Hunter's Point monocline is 0° N5°W; the sense of rotation of middle limb strata is clockwise when viewed to the north; and the magnitude of rotation is as much as 90°.

Faulting too can result in the rotation of strata. In the kind of faulting known as **listric normal faulting**, the faulted rock moves downward along curved concave-upward faults (Figure 2.22). The geometric consequence of translation of originally horizontal layered strata along curved fault surfaces is the rotation of strata to moderate or steep dips. For the example of listric faulting shown in Figure 2.22, the axis of rotation is parallel to the strike of the fault, the sense of rotation is clockwise when viewed to the north, and the magnitude of rotation ranges from 10° to 90°.

Figure 2.21 Rotation of bedding in the Hunter's Point monocline, northeastern Arizona. Bedding in middle background is horizontal, but it is rotated clockwise to a steep dip in the right background. White outcrops in foreground display near-vertical bedding. Strata in the foreground represent the strike projection of the steeply dipping strata in the fold in the distance. (Photograph by G. H. Davis.)

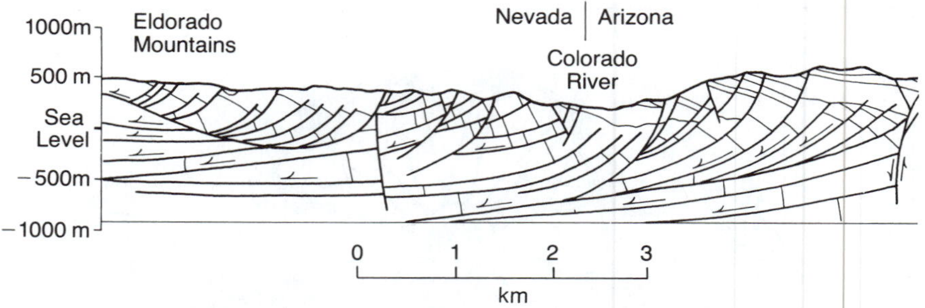

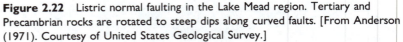

Figure 2.22 Listric normal faulting in the Lake Mead region. Tertiary and Precambrian rocks are rotated to steep dips along curved faults. [From Anderson (1971). Courtesy of United States Geological Survey.]

An example of rotational kinematics at the plate tectonic scale is evident in a comparison of the orientation of the African continent in Late Pennsylvanian time to its orientation today (Figure 2.23). In addition to the translational movement of the continent by some 1000 km from equatorial to northern latitudes, the continent underwent a counterclockwise rotation of some 60° about a vertical axis.

Rotated garnets provide a good example of rotational kinematics at the microscopic scale (Figure 2.24). Some sheared rocks contain garnets that display spiral patterns of inclusions, which were trapped as the garnets rotated and grew during shearing. The kinematic development is crudely analogous to the contamination of pure snow during the ''rotational'' stage of building a snowman. Clockwise versus counterclockwise rotation of garnets is easy to spot where the pattern is strong, but it must be specified with respect to the direction of observation (e.g., ''the garnet rotated clockwise as viewed from south to north''). The amount of rotation can be interpreted by measuring the degree of spiraling.

Tracking the kinematics of rotation is best carried out stereographically. ''Descriptive Analysis'' (Part III-I) offers some additional opportunities in analyzing rotations.

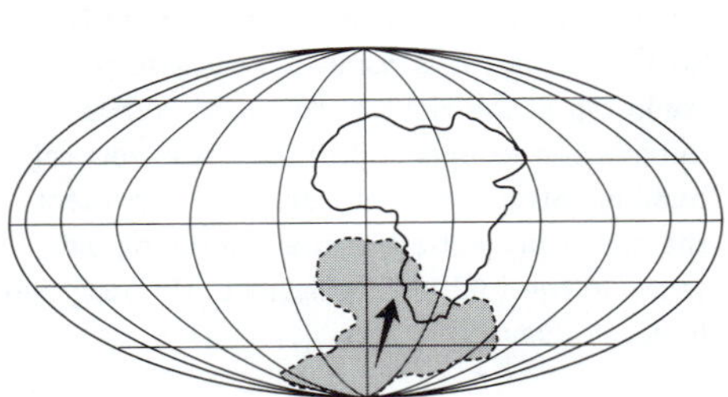

Figure 2.23 Both the position and the orientation of Africa changed radically between the late Pennsylvanian (shaded outline) and the present. Rotation was counterclockwise about a vertical axis. Positioning of Africa based upon Scotese (1991).

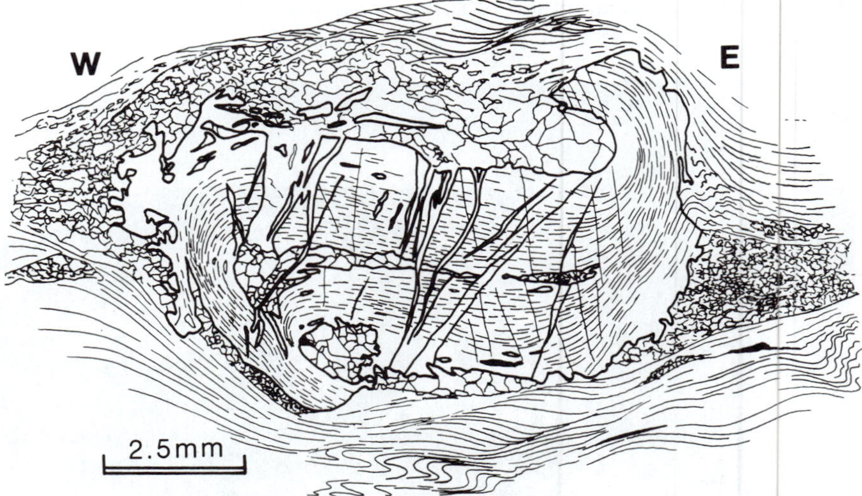

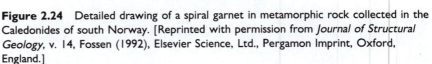

Figure 2.24 Detailed drawing of a spiral garnet in metamorphic rock collected in the Caledonides of south Norway. [Reprinted with permission from *Journal of Structural Geology*, v. 14, Fossen (1992), Elsevier Science, Ltd., Pergamon Imprint, Oxford, England.]

STRAIN

General Concept

Strain produces **dilation**, a change in size, and **distortion**, a change in shape. Points within strained bodies do not retain their original spacing and configuration relative to one another. The original spacing of points within the body is changed.

Where pure dilation takes place without change in shape, internal points of reference spread apart or pack closer together in such a way that line lengths between points become uniformly longer or shorter. Overall shape remains the same. During distortion, the changes in spacing of points in a body are such that the overall shape of the body is altered, with or without a change in size (see Figure 2.2).

As structural geologists we are indebted to John Ramsay, whose text *Folding and Fracturing of Rocks* (1967) underscores the importance of strain analysis and presents the fundamentals and applications of this discipline in a thorough and enlightening way. Although mathematical in nature, strain analysis is fundamentally a geometrical challenge (Ramsay, 1967, p. 50).

The object of strain analysis is a formidable one. We are asked to describe the changes in size and shape that have taken place in a non–rigid body during deformation, and to describe how every line in a body has changed length and relative orientation.

The Ground Rules

We simplify our work in strain analysis by studying the theory in two, not three, dimensions. For more advanced applications, the three-dimensional approach to strain analysis is available in other texts, for example, Ramsay (1967), Jaeger and Cook (1976), Means (1976), Ramsay and Huber (1983), Ragan (1985), and Twiss and Moores (1992).

A customary simplification is to restrict strain analysis to the description of **homogeneous deformation**. It is almost impossible to apply mathematical theory to unwieldy, irregularly deformed, **heterogeneous** structural systems (Ramsay, 1967). Instead, strain analysis focuses on the properties of bodies, or parts of bodies, that have deformed in a regular, uniform manner. The chief constraints that the "homogeneous deformation" clause imposes are:

- Straight lines that exist in the non–rigid body before deformation remain straight after deformation.
- Lines that are parallel in the body before deformation remain parallel after deformation.

For these conditions to hold, the strain must be systematic and uniform across the body that has been deformed. A simple test for homogeneity will become obvious: homogeneous deformation transforms perfect circles into perfect ellipses, and perfect spheres into perfect ellipsoids. Strained rocks that we find in nature typically depart from the rules of homogeneous deformation. However, we deal with this problem by subdividing regions of study into small domains within which strain can be regarded as statistically homogeneous.

The Magic of Strain

The elegance and systematic rendering of Earth structure is a gift of the "rules" of homogeneous strain. The study and application of strain theory result in surprising disclosures regarding the regularity and predictability of changes in lengths and orientations of lines.

A useful way to visualize the two-dimensional properties of strain is through homogeneous distortion of a circle. If a body containing a perfectly circular reference marker is subjected to perfectly homogeneous deformation, the reference circle will be transformed into a perfect ellipse. Furthermore, if the body is subjected to a second homogeneous deformation, the elliptical form of the deformed reference marker will be transformed into yet another perfect ellipse, regardless of the magnitude and orientation of the second deformation. There is only one exception: the special case of a second deformation that is the exact **reciprocal** of the first, thus returning the ellipse to its original circular form. In general, however, no matter how many times an ellipse is distorted homogeneously, it will be repeatedly transformed into a new ellipse.

When one of my students, Chuck Kiven, sprained his ankle badly playing rugby, he did not have the presence of mind to draw a circle on his ankle immediately, to monitor dilation and distortion during swelling; but he had the wisdom at peak swelling to draw a circle on the dilated, distorted ankle region and thus was able to monitor the change from circle to reciprocal ellipse as the swelling went down. In this way he recovered the data he needed to do the calculation.

To explore the geometric elegance of strain, it is useful to keep on hand a deck of cards, the flank of which is embossed with a perfect circle and a perfect ellipse (Figure 2.25*A*). Through shearing of the deck of cards, the outline of the starting circle can be transformed into a perfect ellipse. This is achieved by holding tight to the right end of the card deck with the right hand (Figure 2.25*A*), flexing the deck into the form of a fold (Figure 2.25*B*), grasping tight the left end of the deck with the left hand and then releasing the right hand (Figure 2.25*C*) (Ragan, 1969). It even works left-handed, though the **sense of shear** is opposite. For this type of shear, the orientations of the long and short axes of the ellipse on the cards keep changing.

By repeating the flex-and-shear drill a number of times, any ellipse can be made more and more elongate (Figure 2.25*D–H*). The only limit to flattening and extending an ellipse through shearing is the difficulty of trying to hang onto the deck as it progressively thins and lengthens (Figure 2.25*I*). Nature has no such limit.

Note the outline of the ellipse embossed on the flank of the deck prior to shearing (Figure 2.25*A*). It becomes transformed by shearing into a variety of ellipses, and in one special case into a perfect circle (see Figure 2.25*H*).

The same "magic" holds if we subject a block, on which we have embossed the outlines of a circle and several ellipses, to a flattening (Figure 2.26*A*). As the block is progressively shortened in one direction and lengthened in another, without changing area, the circle transforms to ellipses of greater and greater **aspect ratio** (long axis/short axis) (see Figure 2.26*A–C*). If the progressive deformation is halted just at the right moment (Figure 2.26*B*), a given ellipse might be seen passing through the form of a circle. The orientations of the long and short axes of the strain ellipse remain constant throughout the deformation, in contrast to the steady rotation of the long and short axes in the shear deck example.

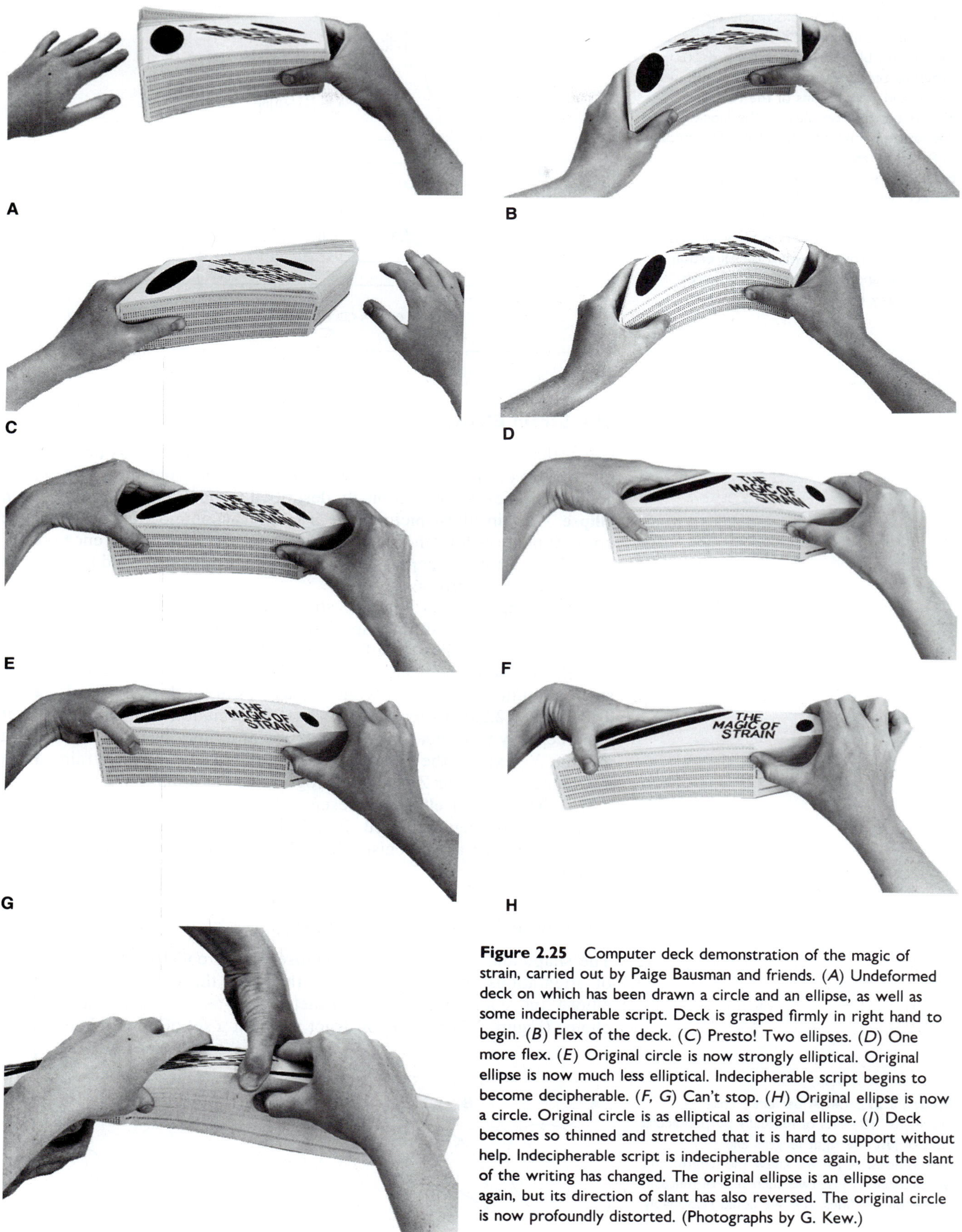

Figure 2.25 Computer deck demonstration of the magic of strain, carried out by Paige Bausman and friends. (*A*) Undeformed deck on which has been drawn a circle and an ellipse, as well as some indecipherable script. Deck is grasped firmly in right hand to begin. (*B*) Flex of the deck. (*C*) Presto! Two ellipses. (*D*) One more flex. (*E*) Original circle is now strongly elliptical. Original ellipse is now much less elliptical. Indecipherable script begins to become decipherable. (*F, G*) Can't stop. (*H*) Original ellipse is now a circle. Original circle is as elliptical as original ellipse. (*I*) Deck becomes so thinned and stretched that it is hard to support without help. Indecipherable script is indecipherable once again, but the slant of the writing has changed. The original ellipse is an ellipse once again, but its direction of slant has also reversed. The original circle is now profoundly distorted. (Photographs by G. Kew.)

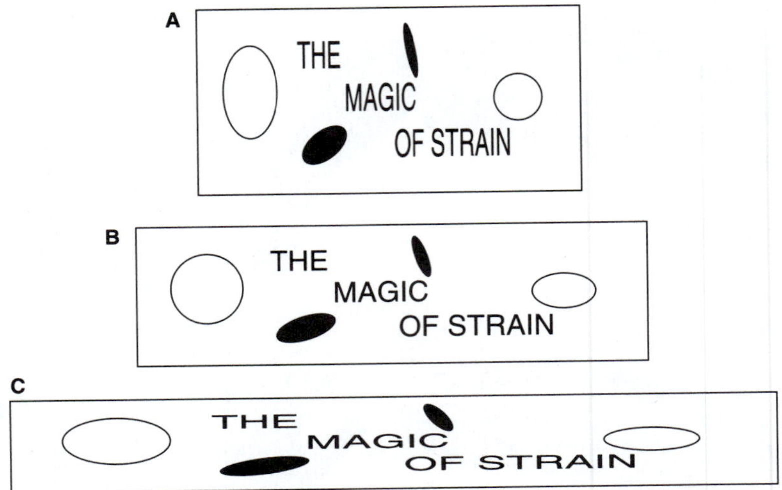

Figure 2.26 (*A*) A deformable block is embossed with a circle, a vertically oriented ellipse, two black ellipses, and words. (*B*) When flattened and extended, changes in the shapes and orientations of the reference objects on the front face of the block record the nature of the internal strain. The amount of flattening and stretching is just enough to transform the original vertical ellipse into a perfect circle. (*C*) With even more flattening and stretching, the white ellipses become tighter and tighter; the two black ellipses continuously rotate toward the direction of stretching; and the letters of the words continuously change font.

The Strain Ellipse

Given the perfection of ellipses derived from homogeneous deformation (distortion) of circles and ellipses, it is no wonder that the strain within geologic bodies is conventionally described through the image of a **strain ellipse**. A strain ellipse pictures the distortion accommodated by a geologic body. It pictures how the shape of an imaginary circular reference object, or perhaps a not-so-imaginary circular geologic object, would be changed as a result of distortion.

We distinguish two kinds of strain ellipse: the instantaneous strain ellipse and the finite strain ellipse. An instantaneous **strain ellipse** is used to portray how a circle is affected by a tiny increment of deformation. The ellipse should be nearly circular because it represents infinitesimally small amounts of strain. It would be as if, during the "flex and shear" drill (Figure 2.25), we created the final deformation through a series of imperceptibly tiny shear movements, one at a time, over the course of a month. We explain the details and use of the instantaneous strain ellipse in Chapter 9, "Shear Zones and Progressive Deformation."

The **finite strain ellipse** represents the total strain experienced by a circle that has been deformed. It is the final result of deformation, which we normally see as geologists. It is the summation of all the incremental components.

Looking at Lines Inside an Ellipse

The shape and orientation of an ellipse can be constructed on the basis of orientations and lengths of lines that pass through the center of the ellipse and connect with the perimeter. Let us take any ellipse of our choice (Figure 2.27*A*) and construct within it a set of perpendicular lines

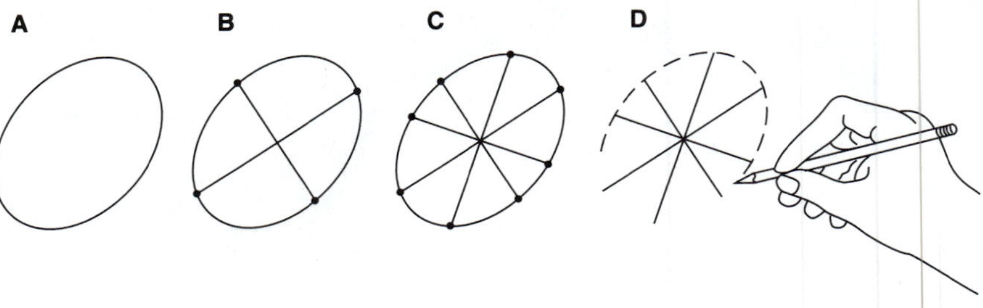

Figure 2.27 (*A*) A perfect ellipse "contains" (*B*) sets of mutually perpendicular lines of just the right lengths and orientations. (*C*) If we knew the lengths and orientations of just two such sets of mutually perpendicular lines, (*D*) we could reasonably draw the ellipse.

that intersect at the "center" of the ellipse (Figure 2.27*B*). Of course, each of these two lines has the perfect length just to touch, but not extend beyond, the perimeter of the ellipse. If we were given the two sets of mutually perpendicular lines (Figure 2.27*C*), we could construct the size and shape and orientation of the ellipse within which they lie (Figure 2.27*D*).

I would love to make a shiny metal mechanical model of a wheel with spokes. This would be no ordinary wheel. It would be deformable into ellipses of any orientation and shape (Figure 2.28). The spokes would "telescope" to shorter and longer lengths as required by the shape of the ellipse. Moreover, the spokes would be hinged to a free-spinning sprocket so that they could rotate to new orientations as the shape of the wheel progressively changed. Playing with this wheel I could see how, through telescoping and rotating, each spoke (i.e., each line in the ellipse) would adapt to the requirements of the new state of strain. With it, I could visualize the "geometric interior" of a finite strain ellipse.

The mastery of strain analysis requires keeping track of changes in the orientations and lengths of lines, and keeping track of the angles between lines. If we learn to do this, we do not have to rely exclusively on the relatively uncommon occurrence of perfect circular or spherical objects being transformed and preserved in the geologic record as perfect ellipses or ellipsoids. Instead, we can describe the state of strain in a deformed body based on a surprisingly small amount of information bearing on changes in lengths in lines and changes in the angles between lines. In fact, **the changes in lengths of lines**, and **the change in angles between lines that were originally perpendicular**, are sufficient to convey the magnitudes and directions of greatest shortening or stretching in the rock body as a whole!

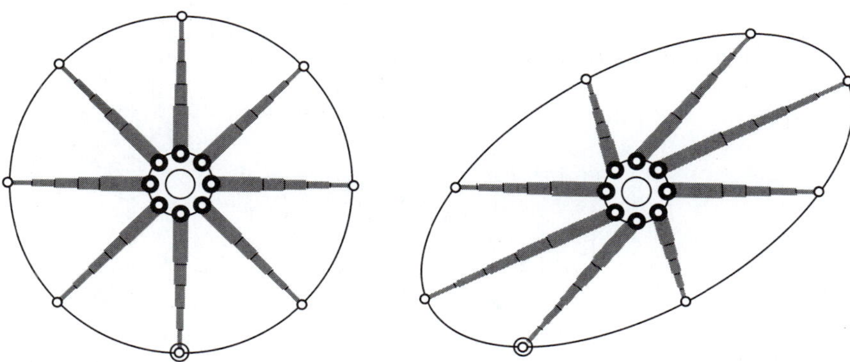

Figure 2.28 The relationship between a strain ellipse and the lengths and orientations of lines within it could be visualized easily with the aid of a nonfrictional retrodeformable elliptocyclometer.

Describing Changes in Lengths of Lines

There are two parameters that permit changes in lengths of lines to be described easily. One is **extension**, symbolized by $\boldsymbol{e}$; the other is **stretch**, symbolized by $\boldsymbol{S}$ (Ramsay, 1967; Means, 1976). Consider line L whose **original length** (l_0) is 5 cm (Figure 2.29*A*). During deformation, the non–rigid body in which L is contained changes shape and/or size such that the line stretches to a **final length** (l_f) of 8 cm (Figure 2.29*B*). The **change in length** (Δl) is 3 cm.

The magnitude of extension (e) in the direction of lengthening is the change in unit length of the line.

$$e = \frac{l_f - l_0}{l_0} \qquad (2.1)$$

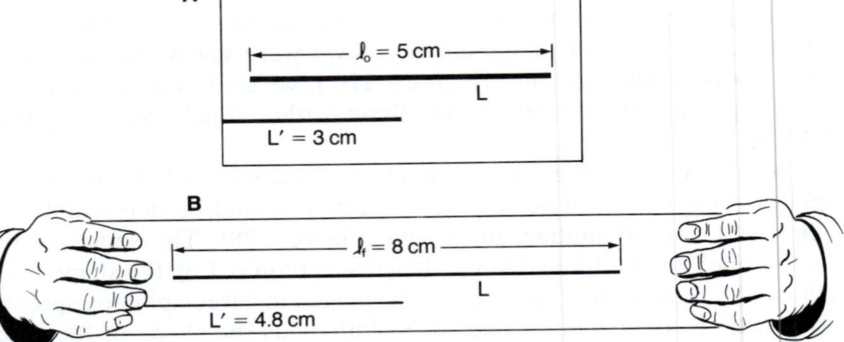

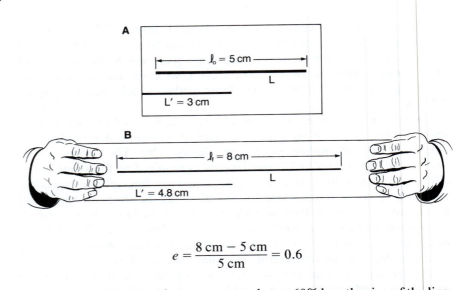

Figure 2.29 (*A*) Lines *L* and *L′* before stretching. (*B*) Lines *L* and *L′* after stretching.

$$e = \frac{8\text{ cm} - 5\text{ cm}}{5\text{ cm}} = 0.6$$

A 0.6 value for extension e corresponds to a 60% lengthening of the line. **Percent lengthening** (or **percent shortening**) is determined by multiplying e by 100%.

A second way to describe the magnitude of the change in length of line L is in terms of the stretch (S). Stretch is equal to final length (l_f) divided by original length (l_0), which is also equal to the value of extension plus one (i.e., $1 + e$). Stretch tells us the final length of a line originally of unit length. A stretch of three means that a line was lengthened 3×. For the example we are considering,

$$S = \frac{8\text{ cm}}{5\text{ cm}} = 1.6$$

Here is how this relationship is derived:

$$e = \frac{l_f - l_0}{l_0}$$

$$e = \frac{l_f}{l_0} - 1$$

$$e + 1 = \frac{l_f}{l_0}$$

$$S = \frac{l_f}{l_0} = 1 + e \qquad (2.2)$$

If line L in Figure 2.29*B* lies within a body that has undergone homogeneous deformation, the values of $e = 0.6$ and $S = 1.6$ must hold for *all* the lines in the body that are parallel to L. Line L' is such a line (see Figure 2.29*A*). If the length of L' before deformation is 3 cm, its length after deformation can be determined by using Equation 2.1:

$$e = \frac{l_f - l_0}{l_0}$$

$$0.6 = \frac{l_f - 3\text{ cm}}{3\text{ cm}}$$

$$l_f = (0.6)(3\text{ cm}) + 3\text{ cm}$$

$$l_f = 4.8\text{ cm (see Figure 2.29}B\text{)}$$

An even simpler way to compute l_f for this line is to multiply l_0 by the stretch (S).

$$l_f = 1.6 \times 3 \text{ cm} = 4.8 \text{ cm (Figure 2.29}B\text{)}$$

Percent lengthening or shortening is determined by multiplying 100% times ($S - 1.0$).

The permissible values for e range from -1 (severe shortening) through zero (no change in length) to $+\infty$ (severe stretching). This range in variation is pictured in cartoon fashion in Figure 2.30. The possible values for S range from 0 (severe shortening) through 1.0 (no change in length) to $+\infty$ (severe stretching) (see Figure 2.30). Figure 2.31 presents quite a number of lines that all started out the same length. Extension (e) and stretch (S) values are given next to each to help us "calibrate."

Change in Length of a Deformed Belemnite Fossil

Extension and stretch are used in reporting numerically the degree of stretching or shortening of geological lines. The stretched belemnite fossil

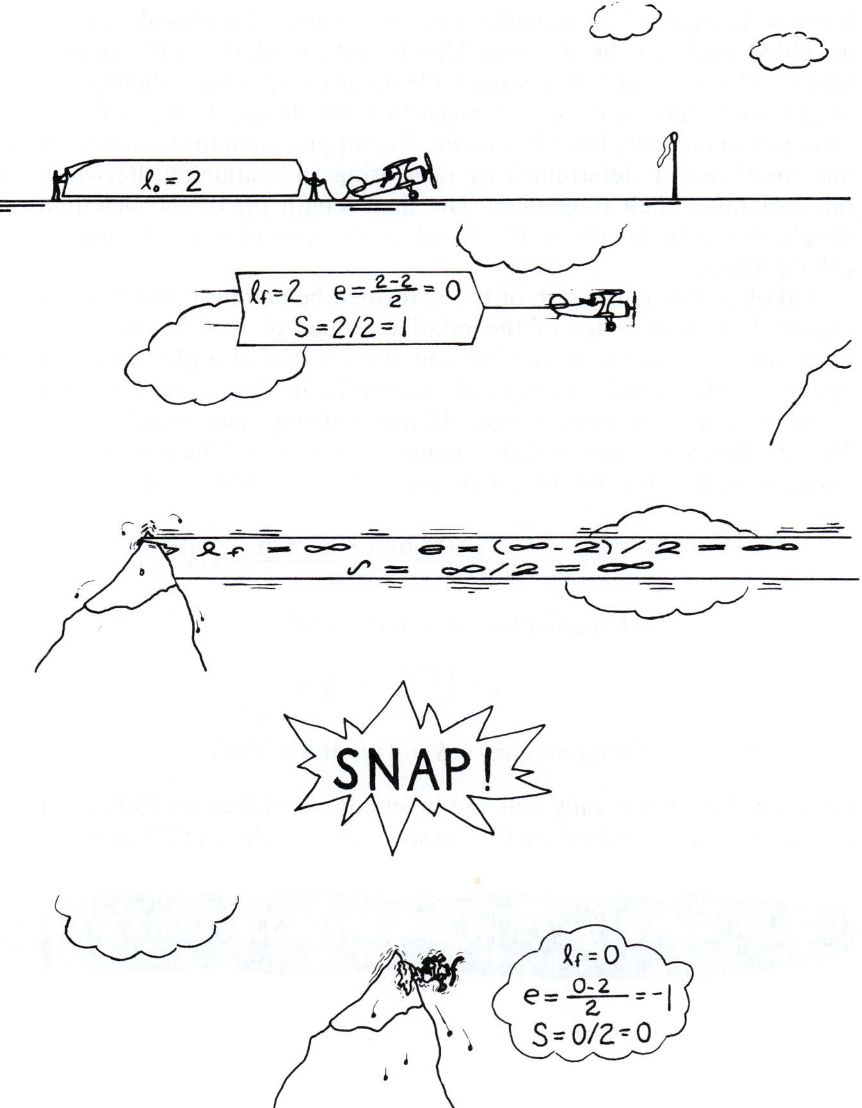

Figure 2.30 The ribbonlike banner that is neither stretched nor shortened has an e value of zero and an S value of 1.0. If the banner is stretched toward infinite length, both its e and S values approach infinity. If the banner is shortened toward zero, its e value approaches -1 and its S value approaches 0. On this particular flight, the airplane to which the banner is attached undergoes non–rigid deformation. (Artwork by R. W. Krantz.)

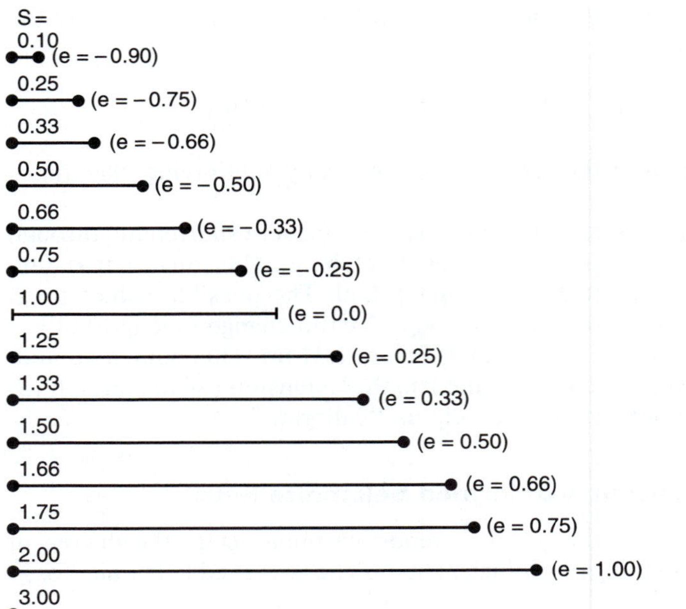

Figure 2.31 Each of the lines in this array started out with a length of 1.0, and with the exception of one line all were either shortened or lengthened. Values of *e* and *S* are given for each of the lines, just to help us "calibrate" the physical significance of stretch and extension values.

featured in Figure 2.32 provides a good example. This fossil, discovered in folded rocks in the western Alps by Albert Heim in the nineteenth century (Milnes, 1979), was stretched into an array of rigid shell fragments of approximately equal size. Spaces that developed during deformation were simultaneously filled by calcite. The original length (l_0) of the belemnite fossil can be determined by measuring and summing the widths of the individual shell fragments. The final length (l_f) of the belemnite is simply the total length of the fossil in its present state, including the calcite filling.

Although the actual size of the deformed belemnite is not revealed in Figure 2.32, knowledge of the relative values of l_0 and l_f is all that is necessary to calculate extension and stretch. Using a photocopy of the belemnite illustration as originally presented in Milnes (1979), we measured an l_0 value of approximately 82 mm and an l_f value of approximately 185 mm. Based on these measurements, we calculated the values of extension and stretch for the direction along which the fossil lies.

$$e = \frac{185\text{ mm} - 82\text{ mm}}{82\text{ mm}} = 1.3$$

$$\%\text{ lengthening} = e \times 100 = 130\%$$

$$S = \frac{185\text{ mm}}{82\text{ mm}} = 2.3$$

$$\%\text{ lengthening} = (S - 1) \times 100 = 130\%$$

Let's not forget: not only was the belemnite stretched by 130%, but the rock in which the belemnite is encased was stretched 130% as well!

Figure 2.32 Stretched belemnite, broken into an array of separated fragments (dark) between which calcite (white) has precipitated. Lengthening is approximately equal to 125%. [From Milnes (1979). Published with permission of Geological Society of America and the author.]

Expressing Changes in Length due to Folding and Faulting

One of the payoffs in constructing geologic cross sections is being able to use them to measure how much stretch or extension is accommodated by the geologic structures. Figure 2.33 presents two sets of geologic cross sections that were used by the geologists who constructed them to calculate percentage shortening and percentage lengthening. In Figure 2.33*A* are cross sections through folds and thrust faults in southeastern Sicily. The sections were prepared by Butler, Grasso, and LaManna (1992). Reference lines *a–b, c–d,* and *e–f* (Figure 2.33*A*) were used by Butler, Grasso, and LaManna (1992) to measure final lengths (l_f). Original lengths (l_0) were determined by measuring the folded, faulted trace lengths of a key limestone bed. Extension values were found to range from −0.25 to −0.40, corresponding to percentage shortening from 25% to 40%, and stretch values of 0.75 to 0.6.

Presented in Figure 2.33*B* are cross sections through normal faults and associated folds in the Virgin River depression, a part of the Basin and Range province in southeastern Nevada and northwestern Arizona. Bohannon and others (1993), who published the cross sections, carefully "restored" the sections as they would have looked before faulting and tilting (Figure 2.33*B*). By comparing the length (l_f) of each cross section of present-day relationships to each restored cross section, they calculated stretch, extension, and percentage lengthening. Stretch values for the two cross sections shown in Figure 2.33*B* are 1.56 and 1.72, respectively, corresponding to extension values of 0.56 and 0.72, and lengthening of 56 and 72%.

Line Length Changes When a Circle Becomes an Ellipse

The elliptical cross section of the worm burrow shown in Figure 2.34*A* was circular before deformation transformed its shape. The deformed worm burrow is a finite strain ellipse that captures the state of strain of the quartzite within which the worm burrow is preserved. The rock was lengthened parallel to the long axis of the finite strain ellipse and contracted parallel to the direction of the short axis. Thus the deformed worm burrow can be used as a strain gauge. The stretch and extension values in these two *special* directions describe the *amount* of lengthening and shortening that the quartzite experienced at this location. How do we go about determining S and e?

Stretch and extension values can be calculated if we know final length (l_f) and original length (l_0) of the long axis and the short axis of the ellipse. Determining the final-length values is easy, for we simply measure them along the two axes. They are 2.6 and 2.2 cm, respectively (see Figure 2.34*B*). But how do we determine what the original diameter of the circular object was before deformation occurred??

We will assume that there was no change in area during the deformation. Having made this assumption, we can say that the final cross-sectional area of the ellipse is equal to the original cross-sectional area of the circle from which the ellipse was derived (Figure 2.34*B*):

$$\text{area of ellipse} = \text{area of circle}$$

$$\pi ab = \pi r^2$$

where a = major semiaxis of ellipse = 1.3 cm
b = minor semiaxis of ellipse = 1.1 cm
r = radius of circle

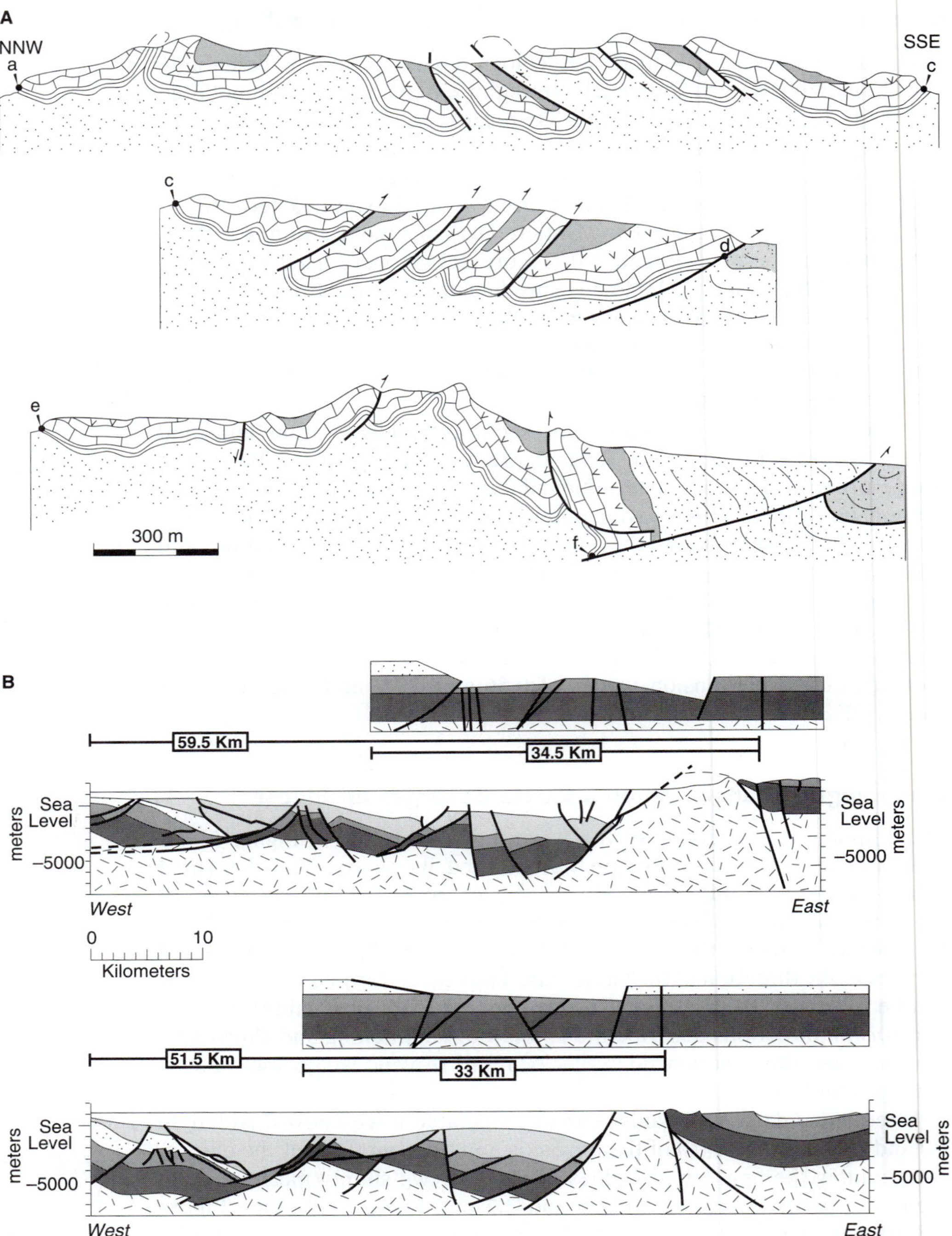

Figure 2.33 (*A*) Cross sections through folded and faulted sedimentary rocks in southeastern Sicily. After constructing the sections, Butler, Grasso, and LaManna (1992) calculated extension and percent shortening values based on changes in line length. [From Butler, Grasso, and LaManna (1992). Published with permission of The Geological Society.] (*B*) Cross sections through faulted and tilted Precambrian, Paleozoic, and Mesozoic rocks in the Virgin River depression region of Nevada and Arizona. After constructing the sections, Bohannon and others (1993) then restored the section to the geometry that existed before faulting and tilting. By comparing section lengths before and after deformation, they were able to calculate extension, stretch, and percent lengthening. [From Bohannon and others (1993). Published with permission of the Geological Society of America.]

Thus

$$r = \sqrt{ab}$$
$$r^2 = ab = 1.4 \text{ cm}^2$$
$$r = \sqrt{ab} = \sqrt{1.4 \text{ cm}^2} = 1.2 \text{ cm}$$

Now we can compute stretch (S) and extension (e). Let us call the long axis A, and set it equal to $2a$ or 2.6 cm (see Figure 2.34A). Let us call the short axis B, and set it equal to $2b$ or 2.2 cm (see Figure 2.34B). Before deformation, lines A and B were the same length and equal to twice the radius of the original circle (i.e., $2r = 2.4$). Therefore,

$$S_A = \frac{l_f}{l_0} = \frac{2a}{2r} = \frac{2.6}{2.4} = 1.1$$

$$S_B = \frac{l_f}{l_0} = \frac{2b}{2r} = \frac{2.2}{2.4} = 0.92$$

$$e_A = \frac{l_f - l_0}{l_0} = \frac{2a - 2r}{2r} = .083$$

$$e_B = \frac{l_f - l_0}{l_0} = \frac{2b - 2r}{2r} = -.083$$

We conclude that lengthening parallel to the long direction of the finite strain ellipse (line A) was approximately 8.3% (i.e., $e_A \times 100\%$), and that shortening parallel to the short axis of the ellipse (line B) was approximately 8.3% (i.e., $e_B \times 100\%$).

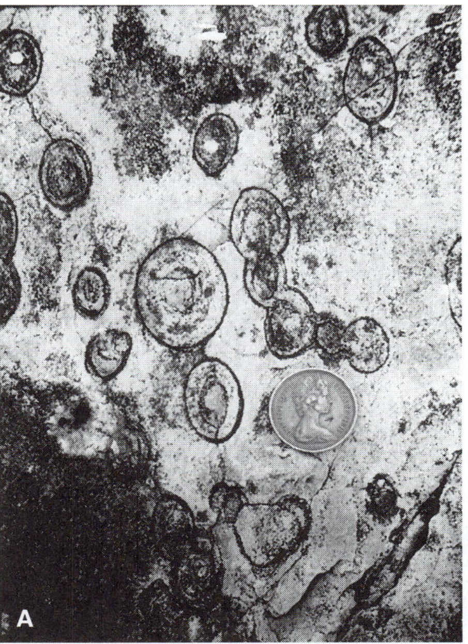

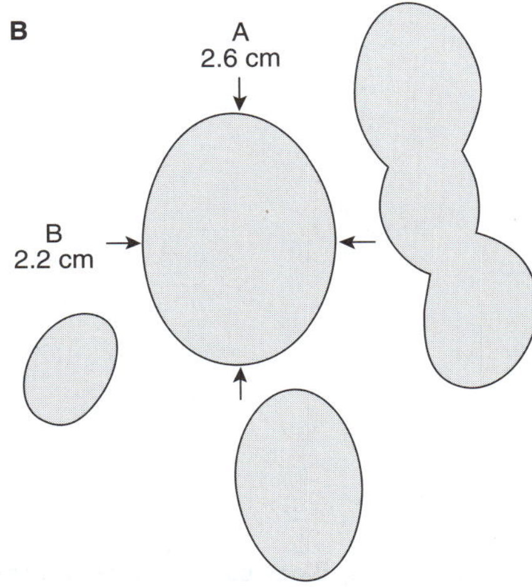

Figure 2.34 (*A*) Deformed worm burrow (genus *Skolithus*) in the Pipe Rock of Cambrian age along the Moine thrust zone of northernmost Scotland. (Photograph by G. H. Davis.) (*B*) Dimensions of the elliptical burrow, measured along the long and short axes.

Angular Shear: Measure of Change in Angles Between Lines

Although strain parameters of extension and stretch effectively describe changes in lengths of lines in deformed bodies, they provide no information regarding changes that take place in the angles between lines. A parameter known as **angular shear**, symbolized by the Greek letter psi (ψ), comes to the rescue. To determine the angular shear along a given line, L, in a strained body, it is essential to identify a line that was originally perpendicular to L. **Angular shear describes the departure of this line from its perpendicular relation with L** (Figure 2.35). The full description requires a sign (**positive equals clockwise; negative equals counterclockwise**) and a magnitude expressed in degrees.

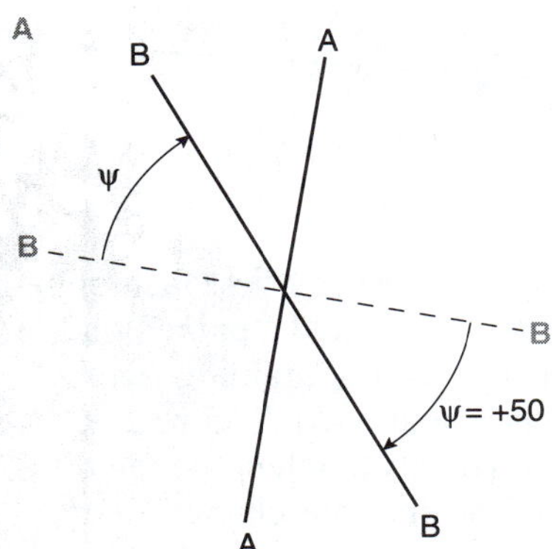

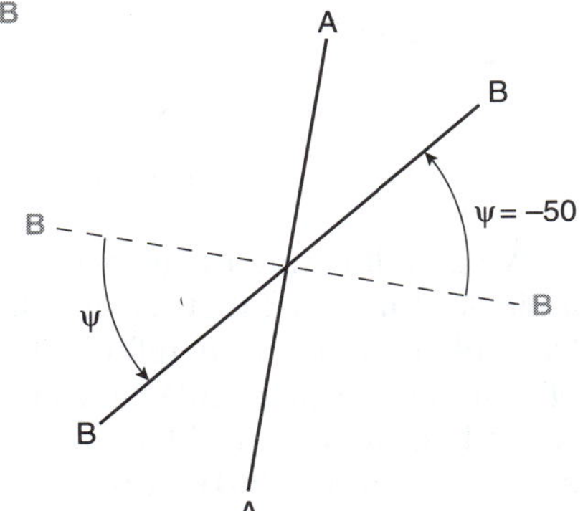

Figure 2.35 Sign conventions for angular shear. (*A*) Determination of the angular shear of line *A* requires identifying a line, in this case *B*, that was originally perpendicular to *A*. The original orientation of line *B* relative to line *A* is shown by the dash line. Angular shear is the shift in angle of $B_{original}$ versus B_{final}. Because the shift is clockwise, the angular shear is positive (+). (*B*) In this example, the angular shear of line *A* is −50°. A counterclockwise shift is denoted by a negative (−) sign.

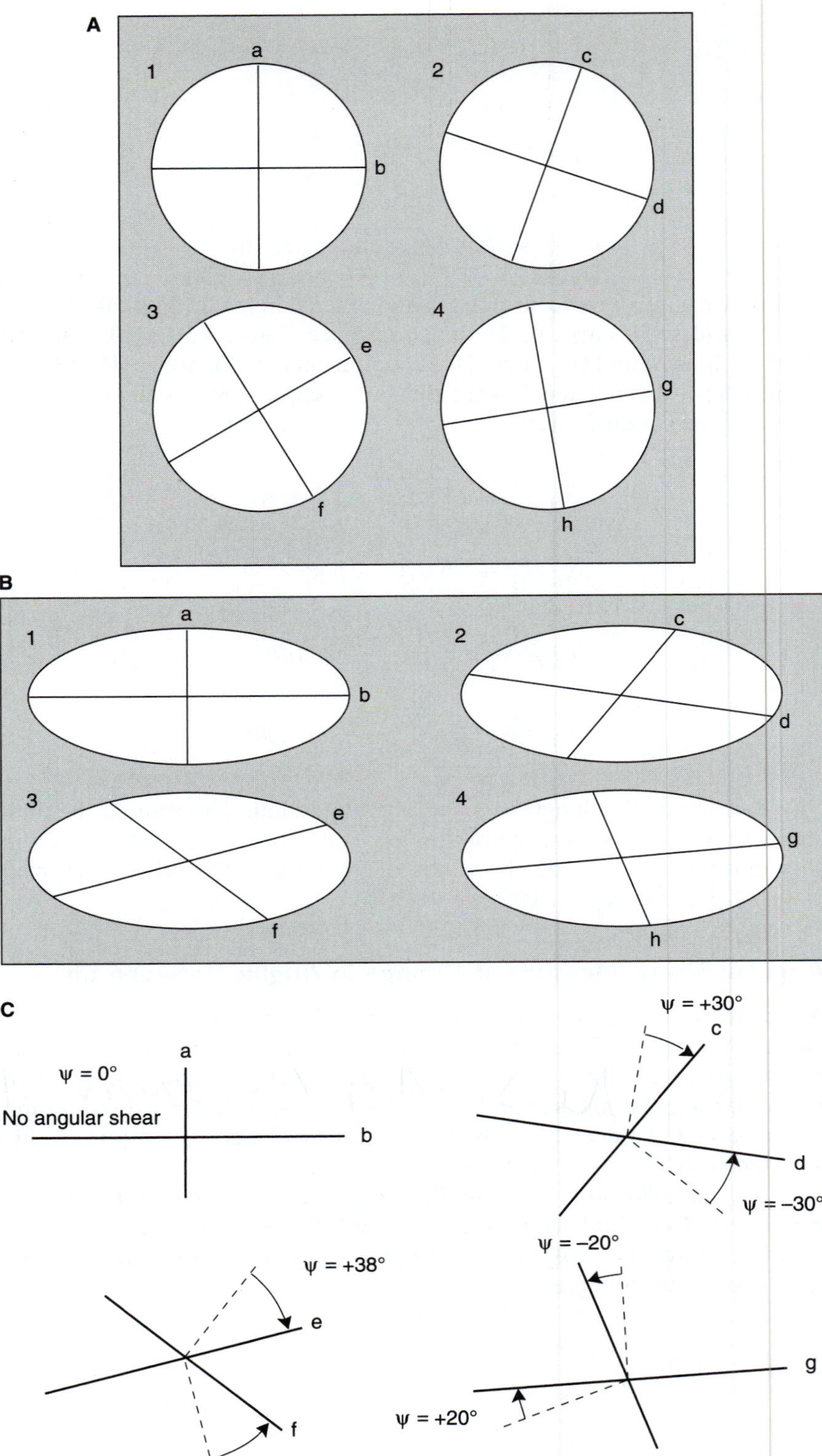

Figure 2.36 (*A*) Block containing reference circles and lines, before deformation. (*B*) Shape of the block after deformation. Original reference circles now are ellipses. The originally mutually perpendicular reference lines have all changed length, and most have changed orientation as well. (*C*) Angular shear along any line can be determined by first identifying a line originally perpendicular to it, and then measuring the angular shift. Remember, clockwise shifts are positive (+); counterclockwise shifts are negative (−).

We can illustrate the measurement of angular shear by first fashioning a block of material that we will deform by flattening, and ''paint'' on the front of the block four reference circles (1–4), each containing two sets of mutually perpendicular lines (*a–b, c–d, e–f,* and *g–h*) (Figure 2.36*A*). After deformation, the block has shortened vertically and lengthened horizontally (Figure 2.36*B*). Furthermore, each of the lines has changed length, six of the lines have changed orientation, and three sets of the lines have moved out of the original right-angle relationship.

It would not be difficult to calculate the extension and stretch values for each of the lines in the flattened block. We already know how to do that.

We can describe the angular shear (ψ) for any given line by identifying a line that was originally perpendicular to it, then measuring the angle through which the perpendicular line moved during deformation. Within ellipse 1, there is no angular shear along line a, nor is there any along line b, for the original perpendicular relationship remains after deformation (Figure 2.36*C*). In ellipse 2, the angular shear along c is $-30°$ and the angular shear along d is $+30°$ (see Figure 2.36*C*). For ellipse 3, the angular shear along e is $+38°$ and the angular shear along f is $-38°$ (see Figure 2.36*C*). Finally, for ellipse 4, the angular shear along g is $-20°$ and the angular shear along h is $+20°$.

If you keep your eyes open, you will see expressions of angular shear (Figure 2.37*A*,*B*). Some fossils contain perpendicular lines in the makeup of their shells. The modification of such originally perpendicular lines by distortion can be used as a means of determining angular shear. The

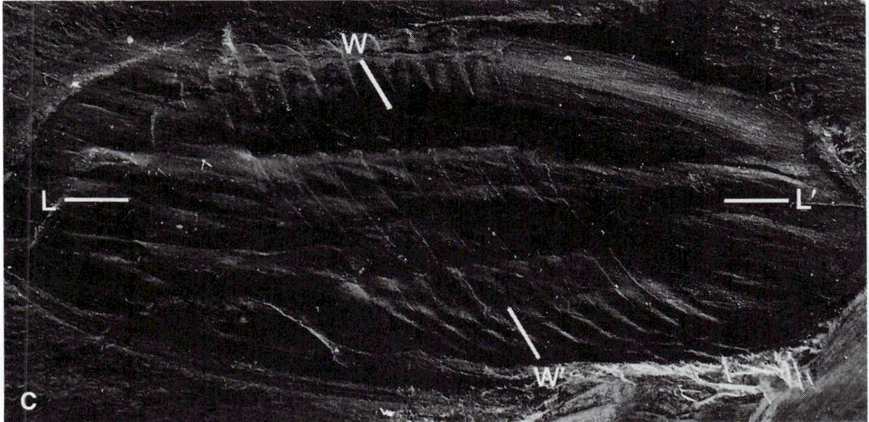

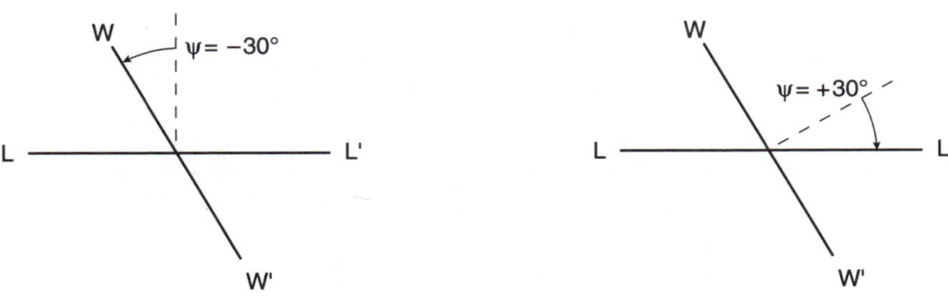

Figure 2.37 (*A*) A fourteenth-century home in Rennes, France (Photograph by G. H. Davis) and (*B*) a twentieth-century barn in Saskatchewan, Canada (Photograph by S. J. Boardman) both have the sheared look. (*C*) Distorted trilobite in Cambrian shale, Caernarvonshire, Wales. Angular shear of rock within which this fossil is found can be determined by measuring the angular relationship between lines *L–L′* and *W–W′*, lines that were perpendicular before the deformation. (From *The Minor Structures of Deformed Rocks: A Photographic Atlas* by L. E. Weiss. Published with permission of Springer-Verlag, New York, copyright © 1972.) Angular shear along *L–L′* is −30°. Angular shear along *W–W′* is +30°.

distorted trilobite featured in Figure 2.37*C* readily lends itself to appraisal of angular shear. Lines parallel to the original length (line L–L') and to the original width (line W–W') of the trilobite are assumed to have been perpendicular before distortion. Now they intersect at 60°. The angular shear along line L–L' is −30° (Figure 2.37*C*). This value is determined by focusing on the line W–W', which was originally perpendicular to L–L', and describing the sense and amount of deflection of that line. In the same fashion, the angular shear along W–W' is found to be +30° (Figure 2.37*C*). In this case, we focus on line L–L', which was originally perpendicular to W–W', and we measure the magnitude and sense of rotation of L–L' with respect to W–W'.

Shear Strain

Let us consider how points on a line move as a response to angular shear. Points *1* to *4* on line A_0 in Figure 2.38*A* are translated by various distances as a result of the rotation of the line on which they reside. Line A_0 is the locus of points *1* to *4*. Line A_f is the locus of the same points in their deformed locations (Figure 2.38*B*). Since angular shear was systematic and deformation was homogeneous, lines A_0 and A_f are straight. Points *1* to *4* move a distance that is directly related to the angular shear and to the distance of each point above the point of intersection with the complementary line. If the distance of each point above the intersection is denoted as y (Figure 2.38*B*), the horizontal distance of translation can be found as follows (Ramsay, 1967):

$$\tan \psi = \frac{\Delta x}{y} \qquad (2.3)$$

$$\Delta x = y \tan \psi$$

Thus **tan ψ** is another way of describing relative shifts in orientations of lines that were originally perpendicular. It is called **shear strain**, symbolized by the Greek letter gamma (γ),

$$\gamma = \tan \psi \qquad (2.4)$$

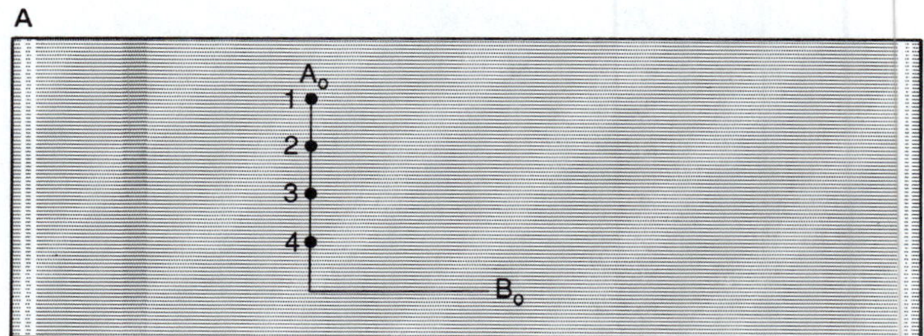

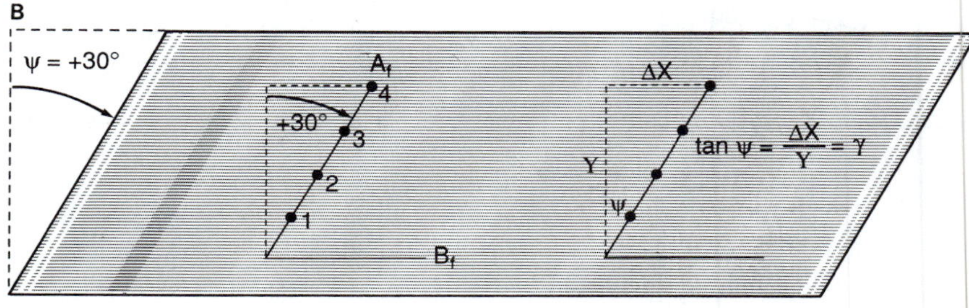

Figure 2.38 Simulation of the shearing of a computer card deck. (*A*) Deck embossed with lines A_0 and B_0 and points 1 to 4 before deformation. (*B*) Configuration of the deck, including the reference lines and points, after shearing.

Shear strain along a line may be positive or negative, depending on the sense of rotation (deflection) of the line originally perpendicular to it. The range of shear strain is zero to infinity. For the example shown in Figure 2.38*B*, the shear strain of line B_f is +tan 30°, or +0.58. The shear strain of line A_f is −tan 30°, or −0.58. For the example of the distorted trilobite shown in Figure 2.37*C*, the shear strain of line *L–L′* is −tan 30°, or −0.58; and the shear strain of line *W–W′* is +tan 30°, or +0.58.

Figure 2.39 presents geometric relationships that may help draw together some of this information. The vertical faces of a square are subjected to progressive angular shear, ranging from $\psi = 0°$ (Figure 2.39*A*) to $\psi = +71.5°$ (Figure 2.39*E*). This corresponds to shear strains ranging from $\gamma = 0.0$ to $\gamma = 3.0$ (Figure 2.39*E*). The dotted vertical reference lines in Figure 2.39 are spaced at a distance that is equal to the length of the edge of the square before deformation, which we will consider to be one unit. Notice that when the top left-hand corner of the square moves horizontally half the distance of the original length of the side of the cube, the shear strain (γ) is +0.5 (Figure 2.39*B*). When it moves a distance of three times the distance of the original length of the side of the cube, the shear strain is +3.0. This is the tangent relationship at work.

Notice as well that the vertical edges of the original square continuously lengthen during this deformation, whereas the horizontal edges of the square do not change in length. As for the edges of the square within the square (Figure 2.39*A*), *DC* and *AB* first shorten then lengthen during the

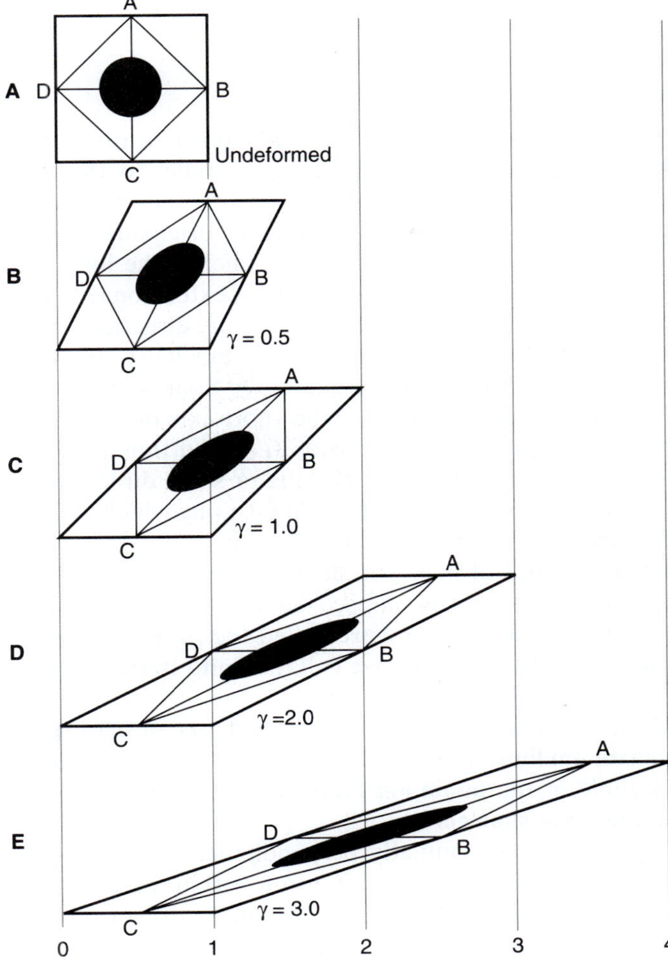

Figure 2.39 One way to get a feel for how the lengths of lines and the angles between lines change during shearing is to deform, using a computer, a fancy geometric object containing sets of mutually perpendicular lines of various orientations. Adding a reference circle to the ornamentation of the originally undeformed object permits changes in the shape and orientation of the strain ellipse to be tracked. The progressive deformation from (*A*) to (*E*) is described in the text. Notice how the long and short axes of the strain ellipse, at each stage of the progressive shearing, continuously change both in length and orientation.

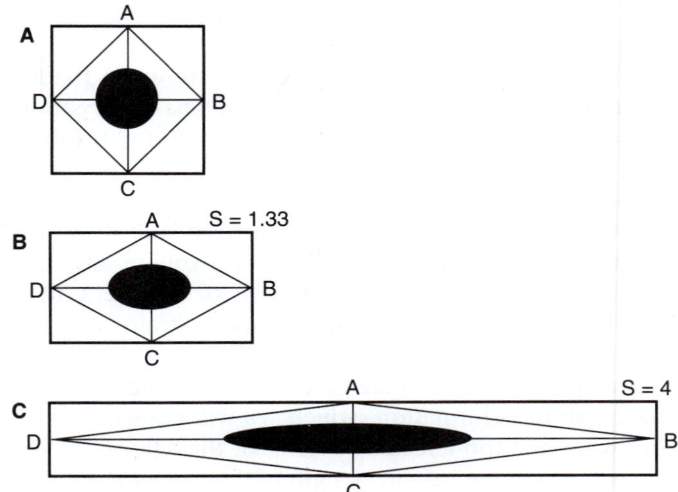

Figure 2.40 In this case, the fancy geometric object is progressively flattened, not sheared, from (*A*) to (*C*). Notice how the long and short axes of the strain ellipse, at each stage of the progressive flattening, change only in length, not in orientation.

progressive deformation (Figures 2.39*A–E*), whereas *DA* and *CB* just keep on lengthening.

A flattening rendition is shown in Figure 2.40. The long and short axes of the ellipse do not rotate as the stretch (*S*) increases from 1.0 to 4.0. Although the edges of the outer square do not change orientation, they do change length. The edges of the inner square change orientation and length; they just keep on lengthening.

The Finite Strain Ellipse

When we look inside an ellipse created by homogeneous deformation of a circle, we make an important discovery. Lines parallel to the long direction of the ellipse are ones along which extension and stretch are greatest. In other words, of all the lines drawn within an originally circular body before deformation, the lines that end up parallel to the long axis of the ellipse are marked by the greatest magnitude of extension (e_1), and thus the greatest stretch (S_1). Lines parallel to the short direction of the ellipse are ones along which extension and stretch are least. In other words, of all the lines drawn within an originally circular body before deformation, the lines that end up parallel to the short axis of the ellipse are marked by the least magnitude of extension (e_3), and thus the least stretch (S_3). Furthermore, lines parallel to the long or short directions of the ellipse are the only lines (i.e., the only directions) along which angular shear (ψ) and shear strain (γ) are zero.

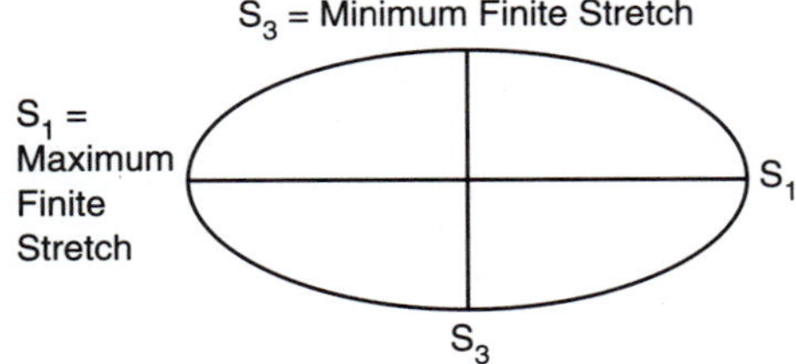

Figure 2.41 The finite strain ellipse and its principal axes. The long axis, S_1, is the direction of maximum finite stretch. The short axis, S_3, is the direction of minimum finite stretch.

Given the unique properties of lines parallel and perpendicular to the long and short directions of the strain ellipse, these directions are given special attention in strain analysis. They are called the **principal axes of the finite strain ellipse**, which for brevity we call the **finite stretching axes** (Figure 2.41).

The long, $\boldsymbol{S_1}$ **axis** of the finite strain ellipse represents the direction and magnitude of the **maximum finite stretch**. The short, $\boldsymbol{S_3}$ **axis** represents the direction and magnitude of the **minimum finite stretch**, which for plane strain is the **maximum finite shortening**.

The finite stretching axes are mutually perpendicular and parallel to the directions of maximum and minimum extension within the deformed body. This is illustrated in Figure 2.42. Lines *A* and *B*, drawn through the center of an undeformed reference circle (Figure 2.42*A*), are the two lines that will become aligned parallel to the finite stretching axes, S_1 and S_3, as a result of deformation (Figure 2.42*C*). Line *A* becomes line *A″*, changing in length from $l_0 = 19.0$ units to $l_f = 31.5$ units. Line *B* becomes line *B″*, changing in length from $l_0 = 19.0$ units to $l_f = 11.5$ units. The stretch value for line *A′* is thus $S_{A'} = S_1 = 1.7$. The stretch value for line *B′* is $S_{B''} = S_3 = 0.6$. If, patiently, we were to proceed to measure the original and deformed lengths of each and every line shown in Figure 2.42, computing the values of stretch for each, we would find that no stretch value exceeds that calculated for line *A″*. And no stretch value is less than that computed for line *B″*. Thus, of all lines in the original starting circle, the one that ends up parallel to the long axis (S_1) of the strain ellipse lengthens the most. And the one that ends up parallel to the short axis (S_3) of the strain ellipse shortens the most.

The axes of the finite strain ellipse are special in yet another way. They are **directions of zero angular shear**. This means that the lines that end up parallel to finite stretching axes must have been perpendicular before deformation as well. Compare Figure 2.42*A* with Figure 2.42*C*. In fact, they are the only perpendicular lines in the distorted body that were perpendicular before distortion. Along the way, however, they departed from orthogonal (Figure 2.42*B*).

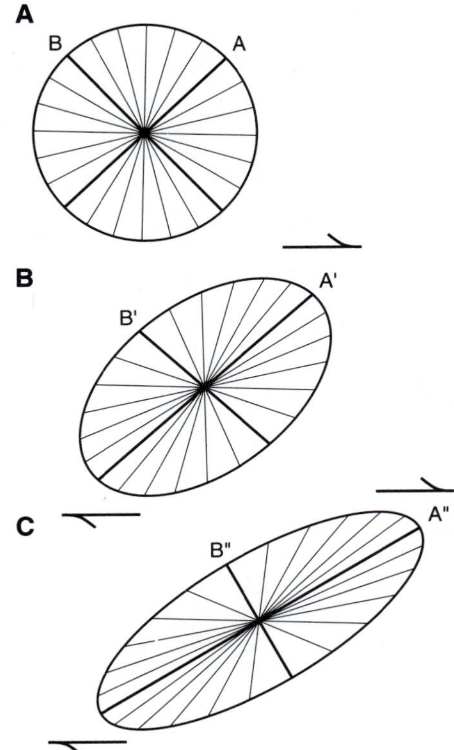

Figure 2.42 (*A*) Undeformed circular body, whose diameter is 19 units, inscribed with lines of common length but different orientations. (*B*) Body is subjected to deformation by shearing. (*C*) After this deformation, almost all the lines have changed in length and orientation. Line *A* ends up parallel to the direction of maximum finite stretch (S_1). Line *B* ends up parallel to the direction of minimum finite stretch (S_3). Of all the lines, *A* lengthened the most and *B* shortened the most. Moreover, of all of the lines drawn in the originally undeformed circular body, only *A* and *B* were perpendicular both before and after deformation. Along the way (e.g., at stage *B*), lines *A* and *B* departed from being mutually perpendicular.

Calibrating the Finite Strain Ellipse

The strain ellipse can be "calibrated" for purposes of evaluating changes in lengths and relative orientations of lines during non–rigid body distortion. If the radius of an original unstrained reference circle is considered to be one unit (Figure 2.43*A*), the lengths of the finite stretching axes (S_1 and S_3) of the strain ellipse can be described in a very convenient manner (Ramsay, 1967). The final length (l_f) of a line, we have learned, is equal to its original length (l_0) multiplied by stretch (*S*). Thus the semiaxis length of S_1 (Figure 2.43*A*) is equal to the radius of the undeformed reference circle ($l_0 = 1.0$) multiplied by the stretch in the S_1 direction. Since the radius of the reference circle is taken to be 1.0, the length of the semiaxis of the strain ellipse in the S_1 direction is simply S_1. Similarly, the length of the semiaxis of the ellipse measured in the S_3 direction is equal to S_3 (Figure 2.43*B*).

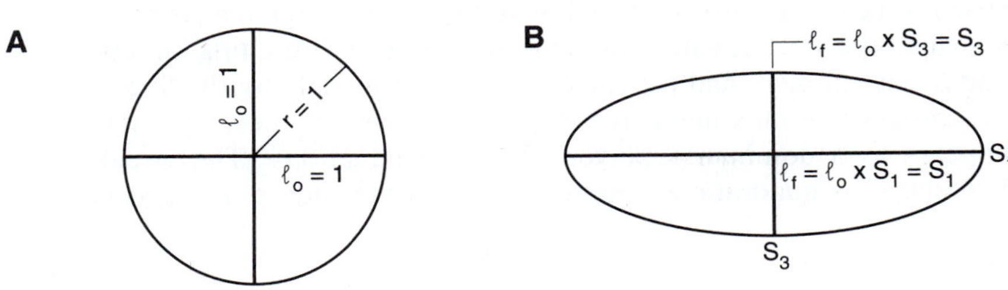

Figure 2.43 (*A*) Original unstrained reference circle. Radius of the circle is 1 unit. (*B*) Strain ellipse resulting from homogeneous deformation of the circle. The lengths of the principal semiaxes of the ellipse are calibrated with respect to stretch (*S*) and original length (l_0).

Evaluating the Strain of Lines in a Body

In carrying out strain analysis, we evaluate changes in lengths and relative orientations of *all* lines in a geologic body, not just the special lines that end up parallel to the finite stretching axes. Figure 2.44 pictures the hypothetical experimental deformation of a clay cake. It shows how a perfectly deformable clay cake ought to distort, given the rules of homogeneous strain. A reference circle with internal reference lines (Figure 2.44*A*) shows the undistorted nature of the clay cake before deformation. The deformed state of the original reference circle is shown in Figure 2.44*B*. Given this special opportunity to view both the starting materials and the finished product, we can describe the strain of lines of any orientation in the model. Let us start out by examining the change in length and the angular shear along line *L* (Figure 2.44*B*).

Before deformation, the length of line *L* was 1.0 unit; in the strained state, *L* is 1.11 units. The stretch and extension values for *L* in the deformed state can be determined as follows:

$$S = \frac{l_f}{l_0} = \frac{1.11}{1.0} = 1.11$$

$$e = \frac{l_f - l_0}{l_0} = \frac{1.11 - 1.00}{1.0} = 0.11$$

In its undeformed state (see Figure 2.44*A*), line *L* made an angle of $\theta = -50°$ with the S_1 axis. In its strained state (see Figure 2.44*B*), it makes an angle of $\theta_d = -26.5°$ with the S_1 axis. This change in relative orientation with respect to the S_1 axis is called the **internal rotation**. The internal rotation (α) for *L* is +23.5°: the + sign always means clockwise change. *This value is not angular shear*, for the angular shear (ψ) along *L* is the deflection from perpendicular of a reference line that was originally normal to *L*.

We can directly measure the angular shear of line *L* by studying its relationship with line *M*, which was perpendicular to *L* before deformation (see Figure 2.44*B*). After deformation, *M* and *L* are no longer perpendicular (see Figure 2.44*B*). The angular shear along *L* is −44.5° (Figure 2.44*C*). The shear strain, $\gamma = \tan \psi$, is −0.98.

Note that the sign convention we use for angular shear and shear strain is **clockwise = positive, counterclockwise = negative** (Figure 2.45*A*). Similarly, when we identify the orientation (θ_d) of a line, *L*, with respect to S_1, we refer to the angle as **positive if measured clockwise from S_1, negative if we measure counterclockwise from S_1** (Figure 2.45).

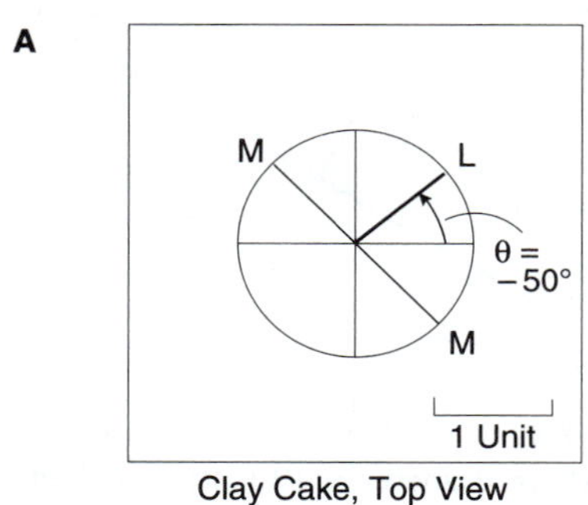

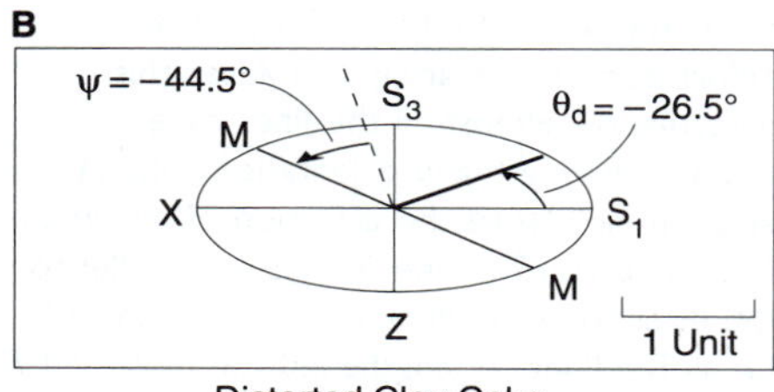

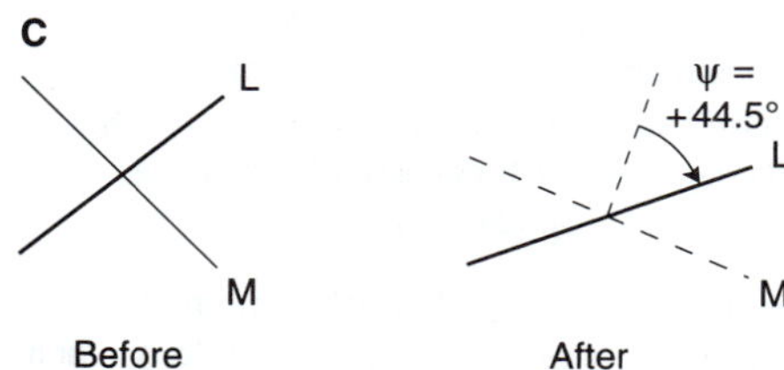

Figure 2.44 Deformation of a hypothetical clay cake that is forced to distort in an ideally homogeneous way. Circle with lines *L* and *M* can be used to monitor the strain. (*A*) Undeformed state. (*B*) Deformed state. (*C*) Lines *L* and *M* before and after deformation. Angular shear of line *M* is equal to +45°.

The Fundamental Strain Equations

There are two fundamental equations that permit the stretch and the shear strain to be determined for any line of any orientation in a strained body, provided the stretch values in the principal finite stretching directions (S_1 and S_3) are known, and provided the angle (θ_d) made by the line (*L*) with the direction of maximum stretch (S_1) is known. A "new" parameter, λ, appears as a dominant "player" in these equations. The Greek letter lambda (λ) is **quadratic elongation**, and it is the square of the stretch:

$$\lambda = S^2 \tag{2.5}$$

The strain equations present quadratic elongation as **reciprocal quadratic elongation** (λ'), where

$$\lambda' = \frac{1}{\lambda} = \frac{1}{S^2}$$

The ratio of shear strain to quadratic elongation (γ/λ) also figures importantly in the strain equations. For a given line in the deformed body, the magnitude of this ratio describes the mix of change in angle versus change in length. Lines along which the value of γ/λ approaches zero are charac-

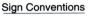

Sign Conventions

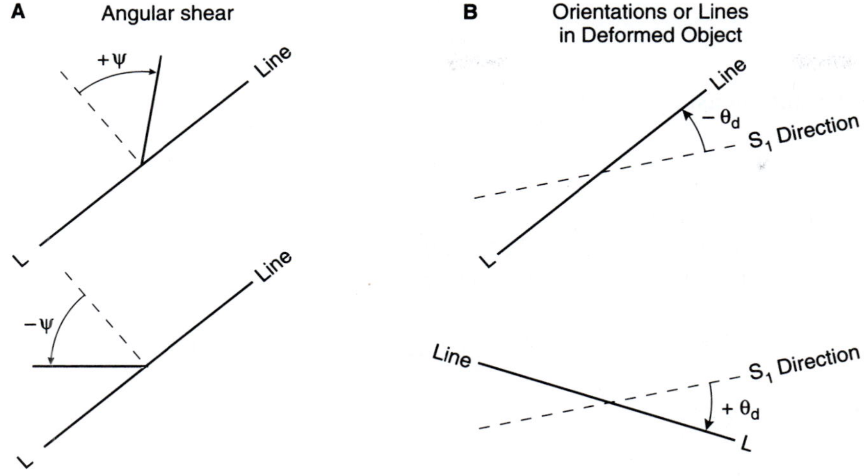

Figure 2.45 Sign conventions are such a pain, but so necessary. Thankfully, we will simply use the convention that in all matters clockwise = positive (+) and counterclockwise = negative (−). (*A*) Positive vs. negative angular shear (ψ) along line *L*. (*B*) Angle θ_d in strain analysis describes the angle between any given line (e.g., *L*), and the direction of maximum finite stretch (S_1). A counterclockwise angle from S_1 is considered to be negative (−); clockwise is positive (+).

terized by change in length with very little shear strain. The γ/λ ratio reaches a maximum value for lines making an angle of 45° to the S_1 direction; shear strain is maximum along lines of this orientation.

The fundamental strain equations (Ramsay, 1967) look like this:

$$\lambda' = \frac{\lambda_3' + \lambda_1'}{2} - \frac{\lambda_3' - \lambda_1'}{2} \cos 2\,\theta_d \qquad (2.6)$$

which also can be written

$$\lambda' = \frac{1}{2}\left(\frac{1}{\lambda_3} + \frac{1}{\lambda_1}\right) - \frac{1}{2}\left(\frac{1}{\lambda_3} - \frac{1}{\lambda_1}\right) \cos \theta_d$$

and

$$\frac{\gamma}{\lambda} = \frac{\lambda_3' - \lambda_1'}{2} \sin 2\,\theta_d \qquad (2.7)$$

which also can be written

$$\frac{\gamma}{\lambda} = \frac{1}{2}\left(\frac{1}{\lambda_3} - \frac{1}{\lambda_1}\right) \sin 2\theta_d$$

where

$$\lambda' = \frac{1}{\lambda},\ \lambda_1' = \frac{1}{\lambda_1},\ \lambda_3' = \frac{1}{\lambda_3}$$

$\lambda = S^2$ along L, which makes an angle of θ_d with S_1

$\lambda_1 =$ greatest quadratic elongation $= S_1^2$

When we see the subscript "1" attached to stretch (S), extension (e), or quadratic elongation (λ), we know we are talking about measurements parallel to the long axis (S_1) of the finite strain ellipse. When we see the subscript "3" attached to stretch, extension, or quadratic elongation, we know we are talking about measurements parallel to the short axis (S_3) of the strain ellipse.

The strain equations can be solved readily with the aid of a calculator if values of λ_1, λ_3, and θ_d are known. To calculate λ and γ for line L in Figure 2.44*B*, the values of λ_1 and λ_3 are computed from stretch values, S_1 and S_3. Since the diameter of the original reference circle is 1.0 unit, the semiaxis lengths of the strain ellipse (see Figure 2.44*B*) are S_1 and S_3, respectively. Since the measured length of S_1 is 1.55 units,

$$\lambda_1 = S_1^2 = (1.55)^2 = 2.40$$

Since the measured length of S_3 is 0.65,

$$\lambda_3 = S_3^2 = (0.65)^2 = 0.42$$

The measured value of θ_d is $-26.5°$.

Using Equations 2.6 and 2.7, the quadratic elongation and shear strain for line *L* are calculated as follows (remember, we are using reciprocals):

$$\lambda' = \frac{1}{2}\left(\frac{1}{0.42} + \frac{1}{2.4}\right) - \frac{1}{2}\left(\frac{1}{0.42} - \frac{1}{2.4}\right) \cos 53° = 0.81$$

$$\lambda = \frac{1}{\lambda'} = 1.2$$

$$S = \sqrt{\lambda} = 1.1$$

Continuing with the solution of the equations:

$$\frac{\gamma}{\lambda} = \frac{1}{2}\left(\frac{1}{0.42} - \frac{1}{2.4}\right) \sin -53° = -0.78$$

$$\gamma = \left(\frac{\gamma}{\lambda}\right) \lambda = (-0.78)(1.2) = -0.94$$

$$\psi = \arctan(-0.94) = -43°$$

Calculating the Variations in Strain

To see the power of the strain equations (Equations 2.6 and 2.7) and the relations they depict, let us try them out in an applied, visual way. Figure 2.46*A* shows the lengths and orientations of a number of lines (lines *a–s*) that are plotted at 10° intervals in the deformed clay cake. The original orientations and lengths of lines *a–s* are shown in Figure 2.46*B*. Figure 2.46*C* extracts lines *a–s* and illustrates how each changes in length and orientation as a result of the deformation.

The images serve to underscore the fundamental properties of homogeneous strain:

First, the finite stretching axes are directions of maximum (S_1) and minimum (S_3) stretch and zero shear strain.

Second, within any body that has undergone a plane strain, there are always **two lines of no finite stretch** along which there has been neither lengthening or shortening; lines oriented in these directions are characterized by stretch values of 1.0.

Third, within any homogeneously distorted body there are two directions marked by maximum shear strain.

Fourth, stretch and shear strain values increase and decrease systematically according to direction in a deformed body; specific values of quadratic elongation and shear strain depend on the magnitudes of S_1, S_3, and θ_d.

The Mohr Strain Diagram

The fundamental strain equations (2.6 and 2.7) seem to have an unduly complicated form, given the reciprocals and the quadratic elongations. But actually the form is quite elegant. Otto Mohr (1882) recognized that

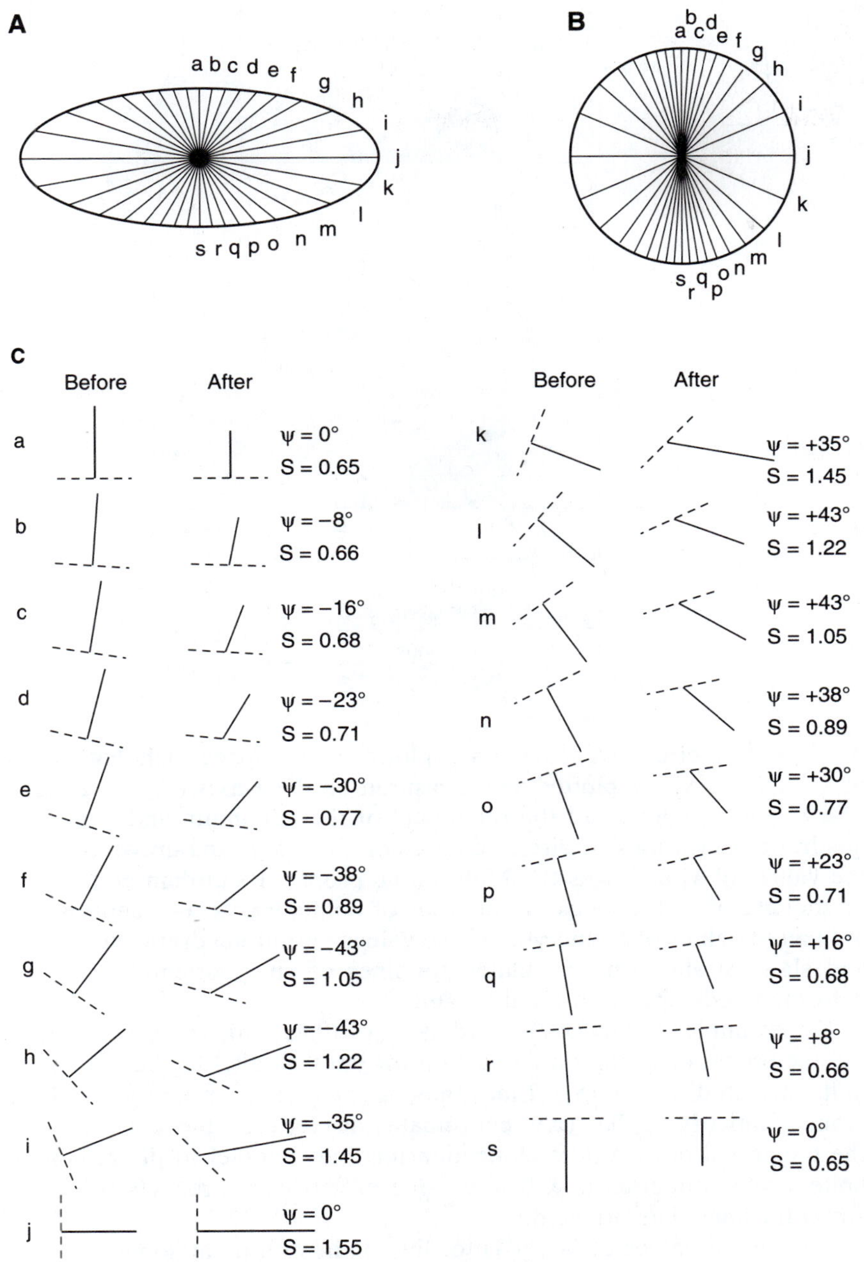

Figure 2.46 (*A*) Clay cake in deformed state, inscribed with lines *a–s* at 10° intervals. (*B*) Orientation of lines *a–s*, as the lines would look if the strain were removed. (*C*) Changes in length and orientation as a result of the distortion of lines *a–s*.

equations written in this way could be represented graphically as a circle. The **Mohr circle strain diagram**, the graphical construction of the strain equations, presents the systematic variations in quadratic elongation and shear strain in a way that is both practical and versatile.

To try it out, let us construct a Mohr diagram to represent the same state of strain as that displayed by the hypothetically deformed clay cake (see Figure 2.44): $\lambda_1 = 2.40$ ($S_1 = 1.55$); $\lambda_3 = 0.42$ ($S_3 = 0.65$). Another rendition of the deformation experiment is presented in Figure 2.47*A*, this one containing a single reference line *A*, which in the deformed state is line A_d. Line A_d makes an angle of $\theta_d = +15°$ (clockwise acute angle!) with the S_1 axis of the finite strain ellipse.

Our goals are to construct a Mohr circle strain diagram on the basis of this information and to determine the values of quadratic elongation (λ) and stretch (S), as well as shear strain (γ) and angular shear (ψ) for line

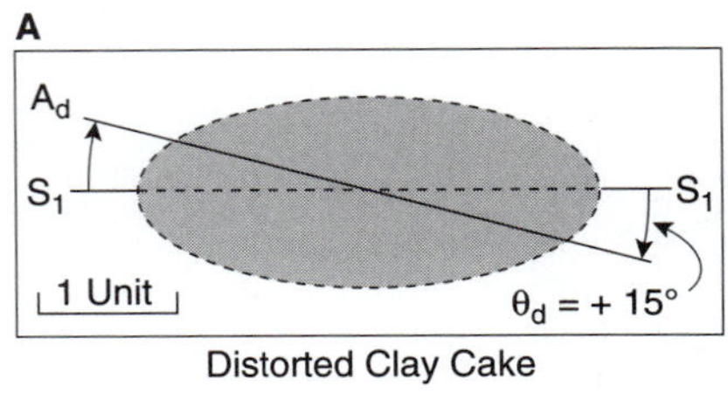

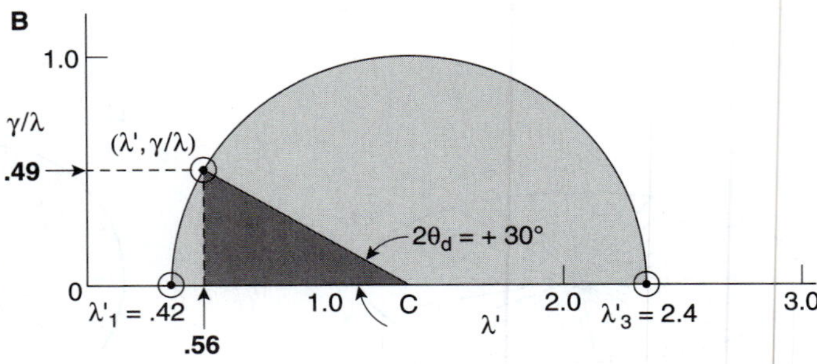

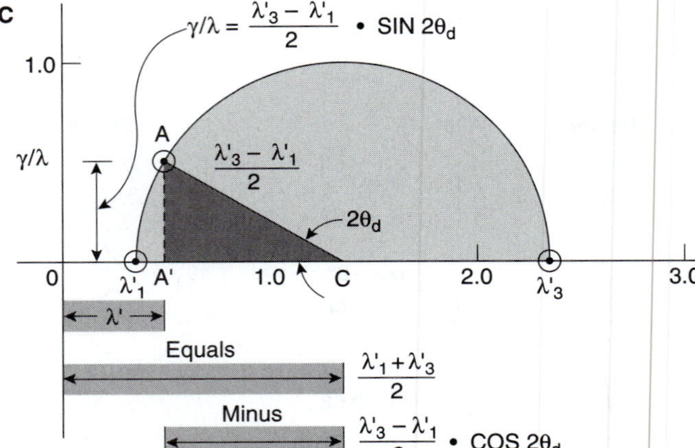

Figure 2.47 (*A*) Distortion of line *A* to line A_d, such that A_d is oriented 15° clockwise (+) from the maximum finite stretch direction (S_1). (*B*) Mohr circle strain diagram showing line A_d within the overall state of strain. (*C*) Mohr circle strain diagram, labeled to show the relation of the geometry of the diagram to the components of the basic strain equations.

A_d. The Mohr circle strain diagram is plotted in x–y space such that values of λ', λ'_1, and λ'_3 are plotted and measured on the x-axis (Figure 2.47*B*). The λ'_1 and λ'_3 values are the reciprocal of the maximum and minimum quadratic elongations. A circle centered on the x-axis and drawn through the values of λ'_1 and λ'_3 is the Mohr circle proper. Its circumference can be thought of as the locus of hundreds of points whose x–y coordinates are paired values of λ' and γ/λ. These values permit quadratic elongation and shear strain to be calculated for lines of *every* orientation in the deformed body shown in Figure 2.46*A*.

For example, λ'_1 has (x,y) coordinates of (0.42,0.0), corresponding to the reciprocal of quadratic elongation for lines parallel to the maximum finite stretch direction (S_1). Lines thus oriented are marked by 0.0 shear strain. Similarly, λ'_3 has (x,y) coordinates of (2.4,0.0), the x value being the reciprocal of quadratic elongation for lines parallel to the minimum finite stretch direction (S_3). The y value of zero again reflects 0.0 shear strain for lines thus oriented.

The paired values of $(\lambda', \gamma/\lambda)$ for line A_d lie at a point somewhere on the perimeter of the Mohr circle. But where? Line A_d lies 15° **clockwise** from the maximum finite stretch direction (S_1) (see Figure 2.47*A*). The location of the point on the Mohr circle representing strain values for this line can be found by plotting a radius **clockwise** from λ'_1 by an angle of $2\theta_d$ or +30° (Figure 2.47*B*). The point where the radius intersects the perimeter of the circle has (x,y) values that correspond to $(\lambda', \gamma/\lambda)$. For line A_d,

$$\frac{\gamma}{\lambda} = 0.49,$$

$$\lambda' = 0.56$$

$$\lambda = \frac{1}{\lambda'} = \frac{1}{0.56} = 1.88$$

$$\gamma = \left(\frac{\gamma}{\lambda}\right)\lambda = (0.49)(1.8) = -0.88 \text{ (negative = counterclockwise)}$$

Since

$$\gamma = \tan \psi$$

$$\psi = \arctan \gamma = 41°$$

A *close-up* view of the Mohr circle strain diagram (Figure 2.47*C*) helps to explain the relation between the geometry of the construction and the elements of the fundamental strain equations. We want to be certain that the λ' value determined for A_d in the diagram really conforms to Equation 2.6, which is

$$\lambda' = \frac{\lambda_3' + \lambda_1'}{2} - \frac{\lambda_3' - \lambda_1'}{2} \cos 2\theta_d$$

The first component of the equation,

$$\frac{\lambda_3' + \lambda_1'}{2}$$

is the x value of the center (C) of the Mohr circle, and this value is equal to the length of OC in Figure 2.47*C*, where O lies at the point q origin. The second component of the equation,

$$\frac{\lambda_3' - \lambda_1'}{2}$$

is the length of the radius of the Mohr circle. Its value is equal to the length of radii, such as line CA (Figure 2.47*C*). The third component of the equation

$$\cos 2\theta_d$$

is equal to the quotient CA'/CA. Substituting into Equation 2.6,

$$\lambda' = \frac{\lambda_3' + \lambda_1'}{2} - \frac{\lambda_3' - \lambda_1'}{2} \cos 2\theta_d$$

$$\lambda' = OC - CA\frac{CA'}{CA}$$

$$\lambda' = OC - CA' \qquad \text{(see Figure 2.47}C\text{)}$$

Examining Equation 2.8 in the same fashion (Figure 2.47*C*),

$$\sin 2\theta_d = \frac{AA'}{CA}$$

$$\frac{\lambda_3 - \lambda_1}{2} \sin 2\theta_d = CA\frac{AA'}{CA} = AA'$$

What if line A_d were to make a counterclockwise acute angle with respect to the direction of maximum finite strain (S_1)? What would this look like on the Mohr circle strain diagram? Line A_d in Figure 2.48*A* makes an angle of $\theta_d = -35°$ with respect to S_1. The state of strain of this

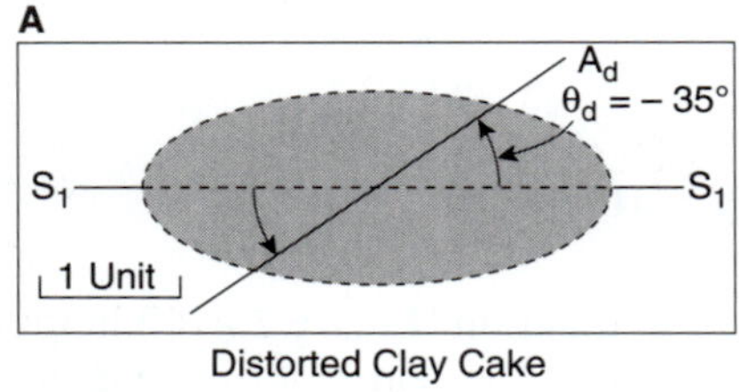

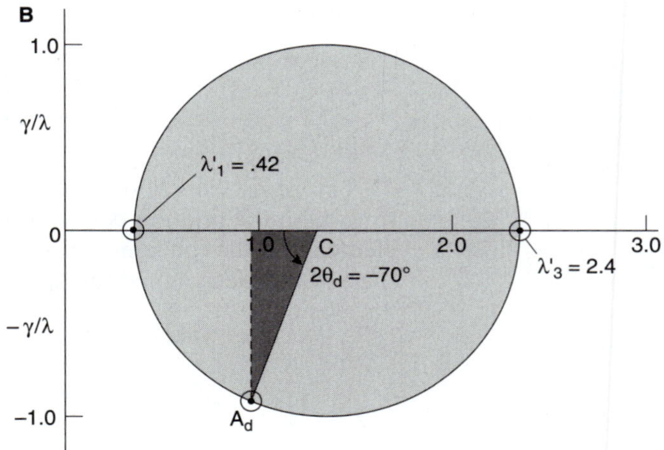

Figure 2.48 (*A*) Distortion of line *A* to line A_d, such that A_d is oriented 35° counterclockwise (−) from the maximum finite stretch direction (S_1). (*B*) Mohr circle strain diagram showing line A_d within the overall state of strain.

line may be found in the Mohr diagram (Figure 2.48*B*) by constructing a radius at an angle of −70° counterclockwise from the *x*-axis.

To gain confidence in the Mohr diagram and its use, feel free to solve for the λ' and γ/λ values for any of the lines shown in Figure 2.46*A*, but this time extract the values directly from the Mohr strain circle. For example, the λ' and γ/λ values for line q ($\theta_d = +70°$) are found by plotting a radius at an angle of $2\theta_d = +140°$ clockwise from the *x*-axis, and then noting the (λ', γ/λ) coordinates of the point where this radius pierces the circumference.

Using Mohr Circle Strain Analysis at the Outcrop Scale

Let us put the Mohr strain diagram into practice. We will start by deforming a slab of limestone containing two brachiopod fossils (Figure 2.49): "fossil E" and "Fossil M." In the undeformed state (Figure 2.49*A*), the hinge line (*H*) and the "back" (*B*) in each fossil are mutually perpendicular. The change in this relationship in the deformed state (Figure 2.49*B*) permits angular shear (ψ) to be determined.

Let us assume that the state of strain in this outcrop had already been analyzed carefully by our friend and colleague, Don Ragan, who gave us the original lengths of the hinge lines of each of the brachiopods. If we had this information, we could compare the hinge line lengths before and after deformation and calculate stretch (*S*) and then quadratic elongation (λ) in the hinge line directions of fossils E and M after deformation. We would find that the calculated values are $\lambda = 0.74$ for fossil E and $\lambda = 1.23$ for fossil M.

The next step is to determine the angular shear values for hinge lines H_E and H_M. We can do this ourselves simply by measuring, with a protractor, the angular relationship between the hinge line (*H*) and the back (*B*) for each of the fossils (Figure 2.49*C*). For fossil E, the angular shear (ψ) of H_E is equal to −30°; and for fossil M, the angular shear (ψ) of H_M is equal to +43°. The angular shear values then can be converted to shear strain values ($\gamma = \tan \psi$).

Finally, we measure the angle made by the hinge line with the maximum finite stretch direction (S_1) (Figure 2.49*D*). The hinge line (H_E) of fossil E makes an angle of $\theta_d = -48°$ with respect to S_1; the hinge line (H_M) of fossil M makes an angle of $\theta_d = +30°$ with respect to S_1.

We now have sufficient information (Table 2.1) to define the complete state of strain in the deformed slab of limestone. To proceed, we construct a Mohr strain diagram (Figure 2.50*A*) in which we first plot the λ' and γ/λ values for the hinge lines of fossils E and M, respectively. The paired

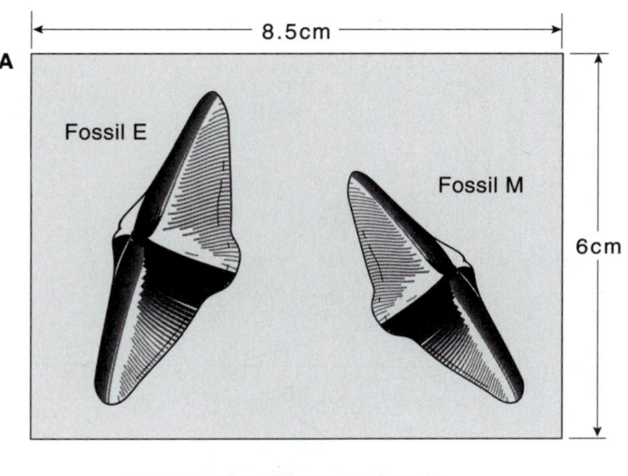

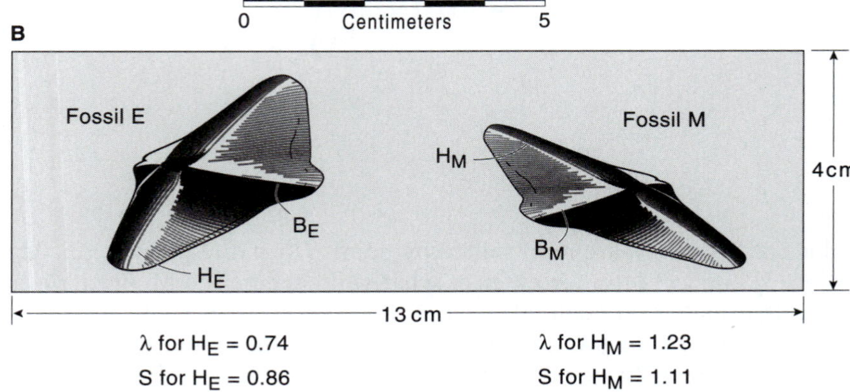

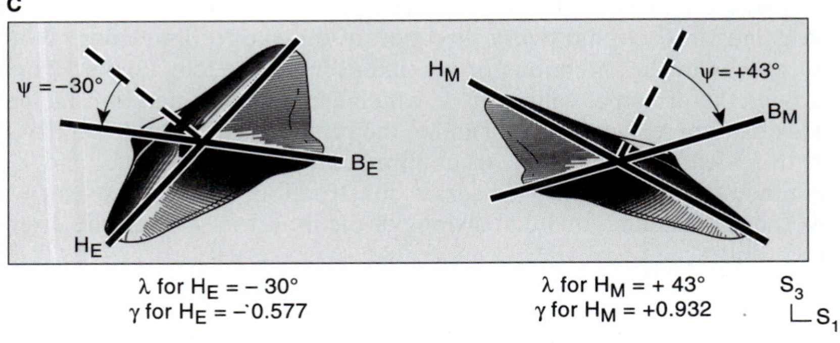

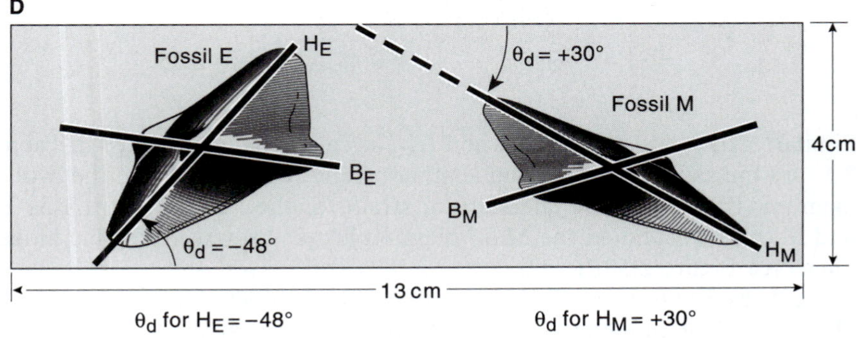

Figure 2.49 (*A*) Slab of undeformed limestone containing two brachiopod fossils, one labeled E, one labeled M. (*B*) Brachiopod fossils E and M in the deformed state. Hinge lines of the fossils are labeled *H*. The "backs" of the fossils are labeled *B*. (*C*) Measurement of the angular shear values for the hinge line (H_E) of fossil E, and for the hinge line (H_M) in fossil M. (*D*) Measurement of the angle between the hinge line (H_E) of fossil E and the direction of maximum finite stretch (S_1); and the angle between the hinge line (H_M) of fossil M and the S_1 direction.

TABLE 2.1
"Strain" data for hinge lines of fossil E and fossil M

Line	S	λ	λ'	ψ	γ	θ_d	γ/λ
H_E	0.86	0.74	1.35	−30°	−0.58	−48°	−0.78
H_M	1.11	1.23	0.81	+43°	0.93	+30°	0.76

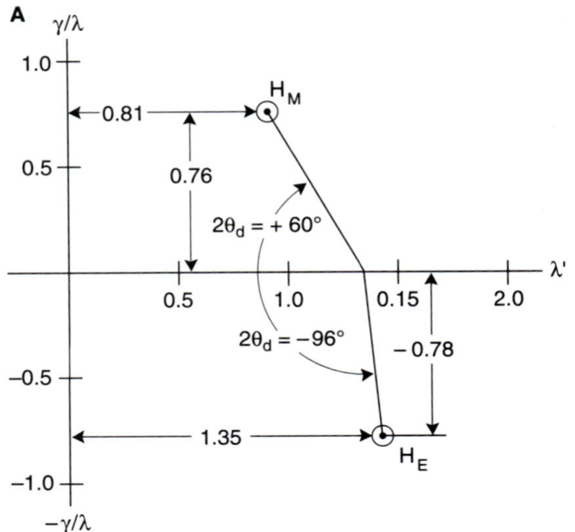

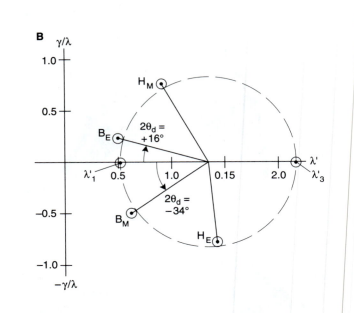

Figure 2.50 Determining the state of strain of the deformed limestone slab using the Mohr strain circle approach. (*A*) A plot of the raw data presented in Table 2.1. (*B*) Fitting a Mohr strain circle to the raw data. The completed diagram discloses the complete state of strain, such that stretch and shear strain can be determined for *any* direction through the slab.

values of λ' and γ/λ are represented by points H_E and H_M (Figure 2.50*A*). Point H_E connects to the center of what will become the Mohr circle by a line that makes an angle of twice θ_d (i.e., $2\theta_d = -96°$) with respect to the λ' axis of the diagram; point H_M connects to the center by a line that makes an angle of $2\theta_d$ equal $+60°$ with respect to the λ' axis (Figure 2.50*A*).

The Mohr strain diagram is completed by simply drawing a best-fit circle (Figure 2.50*B*), which is the locus of the paired values of λ' and γ/λ for lines in each and every direction in the slab of limestone. Note the λ' values in the directions of maximum finite stretch. Points λ_1' and λ_3' provide the principal values of λ', which can be converted to principal values of λ (where λ equals $1/\lambda'$) and to the principal stretches (S_1 and S_3).

In this ideal example we can confirm that the values of λ_1' and λ_3', determined through the Mohr diagram construction, are accurate, for we know the original and final dimensions of the limestone slab in the directions S_1 and S_3:

$$\lambda_1 = S_1^2 = \left(\frac{l_f}{l_o}\right)^2 = \left(\frac{13 \text{ cm}}{8.5 \text{ cm}}\right)^2 = 2.3; \lambda_1' = \frac{1}{\lambda_1} = 0.43$$

$$\lambda_3 = S_3^2 = \left(\frac{l_f}{l_o}\right)^2 = \left(\frac{4 \text{ cm}}{6 \text{ cm}}\right)^2 = 0.44; \lambda_3' = \frac{1}{\lambda_3} = 2.25$$

Similarly, the γ/λ values for H_E and H_M, as calculated and posted in Table 2.1, are the same as the γ/λ values that can be read directly on the Mohr diagram (Figure 2.50*B*). The state of strain for the ''backs'' of fossils E and M can be found on the Mohr diagram just as they were for the hinge lines (see Figure 2.50*B*).

Texts by Ramsay (1967), Means (1976), Ragan (1985), and Ramsay and Huber (1983) contain extensive guidelines on strain-analysis procedures, giving various kinds of starting information. Some of the methods are absolutely ingenious, custom-made for each of the starting conditions that nature, either bountifully or meagerly, has provided. Places like the Moines in northern Scotland are loaded with excellent starting material. White ellipsoidal deformed worm burrows on dark rock give the impression that millions of birds flew over the outcrops after a giant feast. If on the other hand useful starting materials are in scant supply, the quantitative

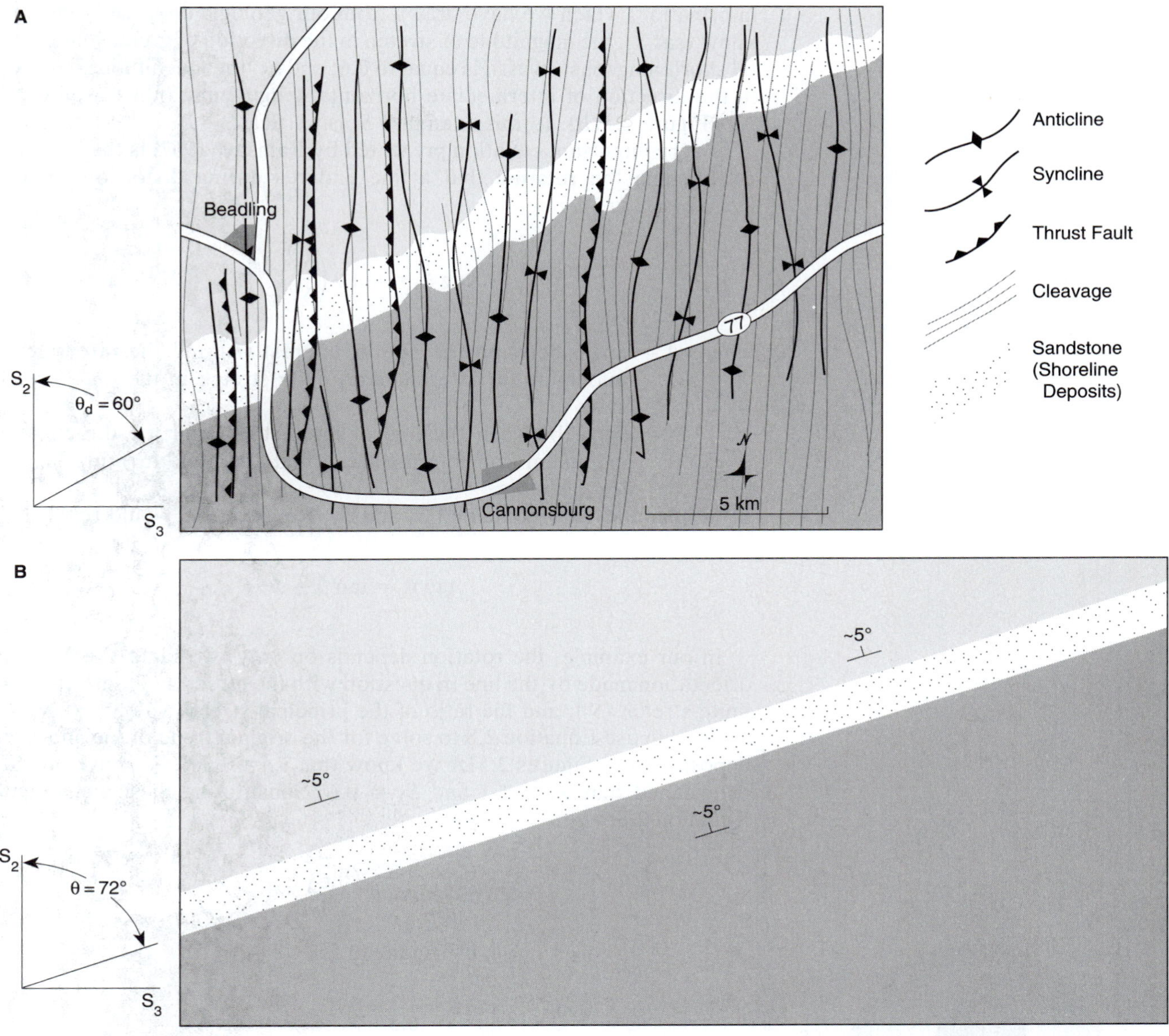

Figure 2.51 (*A*) Map of folded and thrusted sedimentary rocks. The white stippled pattern denotes a sedimentary facies representing an ancient beach separating nonmarine deposits (dark gray) to the south and shallow marine limestones (light gray) to the north. The direction of minimum finite stretch (S_3) is east–west; the direction of intermediate finite stretch (S_2) is north–south; and the direction of maximum finite stretch (S_1) is vertical. (*B*) When the folded and thrusted region is "retrodeformed" to its original state, the beach line is seen to run N72°E, different from its N60°E trend in the deformed state.

strain picture that can be drawn is fragmentary. Commonly the best we can do on the basis of incomplete and/or inadequate data is to prepare partial descriptions of the strain. Describing only part of the strain picture is like drawing only part of a strain ellipse: the orientations of the axes of the strain ellipse might be known, but not shape or size; the shape of the strain ellipse might be known, but not the size or orientation; the size of the strain ellipse might be known, but not the orientation.

Another Example: The Retrodeforming of Regional Strain

Suppose that mapping of an ancient shoreline deposit within a large region of strongly deformed sedimentary rocks revealed the trend line of an ancient beach. We suspect that the present mapped trend of N60°E (Figure 2.51*A*) is a distortion of the original line of beach. We want to determine the original trend of the shoreline deposit. Fortunately we have more information, which we have drawn from our geologic cross sections. We know that S_3, the magnitude of stretch in the direction of greatest horizontal shortening (east–west), is equal to 0.6; and S_2, the magnitude of stretch in the direction of intermediate horizontal lengthening (north–south), is 1.1 (Figure 2.51*B*). In this example, S_1 is vertical.

A very important equation presented by Ramsay (1967) is the basis for establishing the **rotation** that a line endures during distortion (Figure 2.51*B*):

$$\tan \theta_d = \tan \theta \left(\frac{S_3}{S_2}\right) \tag{2.8}$$

where θ = angle between the line of interest (L) and the direction of maximum (or intermediate) finite stretch in the *undeformed* state

θ_d = angle between the line of interest (L) and the direction of maximum (or intermediate) finite stretch (S_1) in the *deformed* state

If the plane of view contained S_1 and S_3, Equation 2.8 would read:

$$\tan \theta_d = \tan \theta \frac{S_3}{S_1}$$

In our example, the rotation depends on only two factors: the initial orientation made by the line in question with the direction of intermediate finite stretch (S_2), and the ratio of the principal stretch values, S_3/S_2.

We can use Equation 2.8 to solve for the original trend of the shoreline deposit. From Figures 2.51*A* we know that $\theta_d = +60°$. We have already determined that $S_2 = 1.1$ and $S_3 = 0.6$. Substituting these values into Equation 2.8,

$$\tan 60° = \tan \theta \left(\frac{0.6}{1.1}\right)$$

$$1.7 = \tan \theta (0.55)$$

$$\tan \theta = \frac{1.7}{0.5} = 3.1$$

$$\theta = \arctan 3.1756 = 72°$$

We conclude (Figure 2.51*B*) that before deformation, the shoreline deposit was originally along a trend 72° east of north.

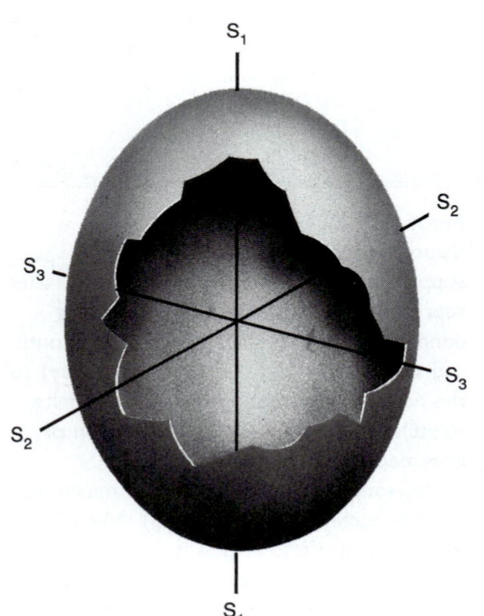

Figure 2.52 The strain ellipsoid: S_1 is the direction of maximum finite stretch, S_2 is the direction of intermediate finite stretch, and S_3 is the direction of minimum finite stretch.

The Finite Strain Ellipsoid and Plane Strain

The most complete strain analyses are three-dimensional. The three-dimensional counterpart of the strain ellipse is called the **strain ellipsoid**. It pictures how the shape of an imaginary *spherical* reference object would be changed as a result of distortion. It is defined by three mutually perpendicular finite stretching axes (Figure 2.52).

If stretching in the S_1 direction is perfectly compensated by shortening in the S_3 direction, and there is no change in length along the S_2 direction, there will be no change in volume of the deformed body. Under such conditions and assuming homogeneous deformation, a perfect sphere of a certain size will be transformed to a perfect ellipsoid of the same volume. This special state of strain is known as **plane strain**. However, it is very common for strain to be truly three-dimensional—that is, the intermediate finite stretch (S_2) is *not* always one.

The long, $\boldsymbol{S_1}$ **axis** represents the direction and magnitude of the **maximum finite stretch**, which for plane strain can be thought of as the direction of maximum finite lengthening. The short, S_3 axis represents the direction and magnitude of the **minimum finite stretch**, which for plane strain is the direction of maximum finite shortening. The intermediate, S_2 axis represents the direction and magnitude of the **intermediate finite stretch**. For plane strain, no finite strain takes place in the S_2 direction, normal to the plane containing the strain ellipse (Figure 2.52).

As you now might imagine, there are equations that describe the quadratic elongation and shear strain of any line of any orientation in a deformed three-dimensional body (Ramsay, 1967; Means, 1976; Ramsay and Huber, 1983).

The Strain Ellipsoid and Its Application

Pioneering applied strain analysis was carried out by Ernst Cloos (1947). He painstakingly analyzed tiny ellipsoids derived from originally near-spherical primary objects (ooids) in Cambrian and Ordovician carbonates exposed in the South Mountain fold in Maryland and Pennsylvania. His study is a classic in modern strain analysis.

The fold that Ernst Cloos examined is a large anticline of the Appalachian Mountain system. The reference materials for his strain analysis, the ooids, are small spherical or slightly ellipsoidal objects commonly found in limestones. The ooids are generally calcareous and have grown concentrically and/or radially outward from centers of nucleation (Figure 2.53*A*). After discovering that the ooids had been deformed into ellipsoids within the South Mountain fold (Figure 2.53*B*), Cloos astutely recognized

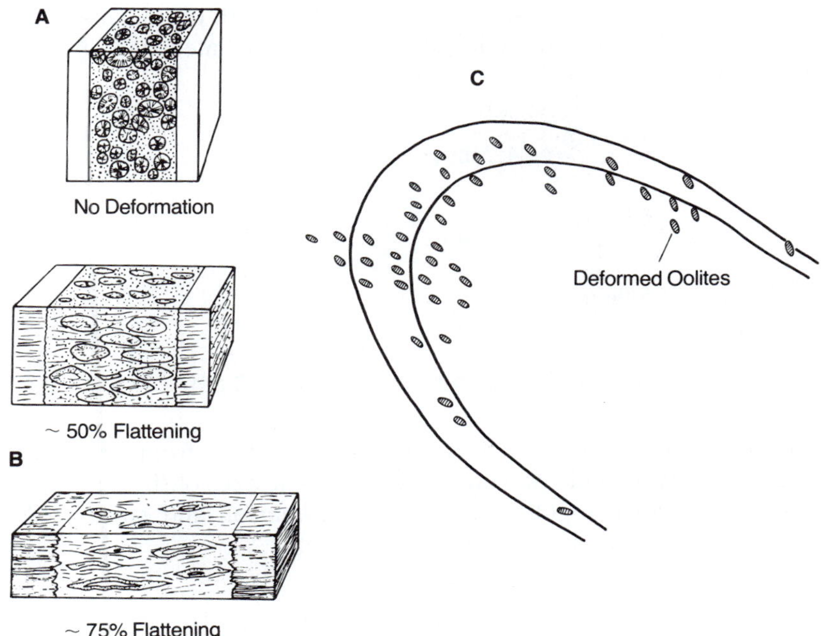

Figure 2.53 (*A*) Undeformed ooids in rocks of South Mountain, Maryland and Pennsylvania. (*B*) Ooids captured in different progressive stages of distortion. (*C*) Schematic digram showing variation in shape and orientation of ooids as a function of position on the South Mountain fold. [From *Structural Geology of Folded Rocks* by E. T. H. Whitten, after Cloos (1947). Originally published by Rand-McNally and Company, Skokie, Illinois, copyright © 1966. Published with permission of John Wiley & Sons, Inc., New York.]

that these primary elements could be harnessed to conduct detailed strain analysis. And detailed it was!

Cloos collected 227 oriented specimens of ooid-bearing limestone. Measurements of all visible structures like cleavage and bedding and jointing were made at each station where the specimens were collected. With the aid of a microscope in the field, Cloos made preliminary measurements and assessments of the strain expressed by the ooids. In the laboratory he prepared 404 oriented thin sections of the ooid-bearing rocks, and these thin sections provided a view of the details of size, shape, and orientation of the strain reference features.

In calculating strain, Cloos assumed that each ooid, before distortion, had been a perfect sphere. He also assumed that distortion of the ooids was accomplished without change in volume. In effect, each perfectly spherical ooid was assumed to have had an original volume defined by

$$V_S = \tfrac{4}{3}\pi r^3 \tag{2.9}$$

where r is the radius.

Each sphere was transformed through distortion into an ellipsoid. The volume of an ellipsoid is defined by

$$V_E = \tfrac{4}{3}\pi abc$$

where a, b, and c are the long, intermediate, and short semiaxes, respectively.

Given the assumption of no change in volume, we write

$$V_S = V_E$$
$$r = \sqrt[3]{abc}$$

Thus by measuring the length of the axes of the deformed and ellipsoidal ooids, Cloos gathered data that allowed the radius of the original sphere to be determined. For example, he reported that at locality 300, the long axis of the ellipsoid ($2a$) measured 8.45 mm, the length of the short axis ($2c$) measured 5.06 mm, and the length of the intermediate axis ($2b$) measured 6.74 mm. Therefore,

$$a = \frac{8.45\text{ mm}}{2} = 4.2\text{ mm}$$

$$b = \frac{6.74\text{ mm}}{2} = 3.4\text{ mm}$$

$$c = \frac{5.06\text{ mm}}{2} = 2.5\text{ mm}$$

$$r = \sqrt[3]{4.2\text{ mm} \times 3.4\text{ mm} \times 2.5\text{ mm}} = 3.3\text{ mm}$$

Knowing the length of the radius of the initial sphere from which the ellipsoid was derived, and knowing the length of the axes of the deformed ellipsoids, he could compute the values of extension and stretch. To calculate stretch in the direction of axis a,

$$S_a = S_1 = \frac{l_f}{l_0} = \frac{4.2}{3.3} = 1.27$$

Figure 2.54 (*A*) Vein of stilpnomelane and quartz. The vein invaded thin-bedded wall rock in the North Hillcrest mine area, Minnesota. The very center of the vein bears a faint line, perhaps a vestige of the former fracture trace that guided the hydrothermal solutions that gave rise to the vein. Spreading apart of the walls was directed at right angles to the centerline of the vein and to the contact of the vein with wall rock. Note that the conspicuous parting (*p*) in wall rock on the left wall of the vein is offset in a way that perfectly matches dilational opening perpendicular to the walls of the vein. (Photograph by R. G. Schmidt. Courtesy of United States Geological Survey.) (*B*) Dan Lynch leans against a dike that ascended through basaltic rock in the Pinacate volcanic field, Sonora, Mexico. The emplacement of a dike accommodates a spreading apart of wall rock along a fracture discontinuity.

To calculate extension (e) in the direction of axis a,

$$e_a = e_1 = \frac{l_f - l_0}{l_0} = \frac{4.2 - 3.3}{3.3} = 0.27$$

Percentage lengthening parallel to line a is

$$0.27 \times 100\% = 27\%$$

By similar calculations, we write

$$S_b = S_2 = 1.03$$

$$e_b = e_2 = 0.03 \text{ (essentially no change)}$$

$$S_c = S_3 = 0.76$$

$$e_c = e_3 = -0.24 \text{ (24\% shortening)}$$

Some of the results of Cloos' strain findings are presented in Figure 2.53*C*, an idealized structure profile of the South Mountain fold showing the variation in orientation of the plane of flattening across the structure.

Dilational Changes

We have assumed in all examples presented so far that the strain has been a pure distortion without change in volume. This assumption has eased our entry into strain theory, but it would be misleading not to underscore the reality of changes in volume during strain. Distortion can be accompanied by either increases or decreases in volume. Easiest to visualize is distortion that opens up the rock, producing extensional cracks that are filled by vein materials like quartz or calcite (Figure 2.54*A*), or by dikes (Figure 2.54*B*). The net effect is an increase in the volume of the distorted rock. Conversely, a rock body may undergo a decrease in volume during

A

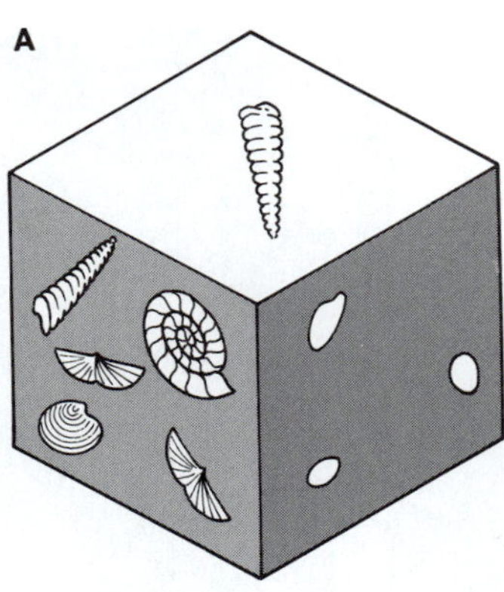

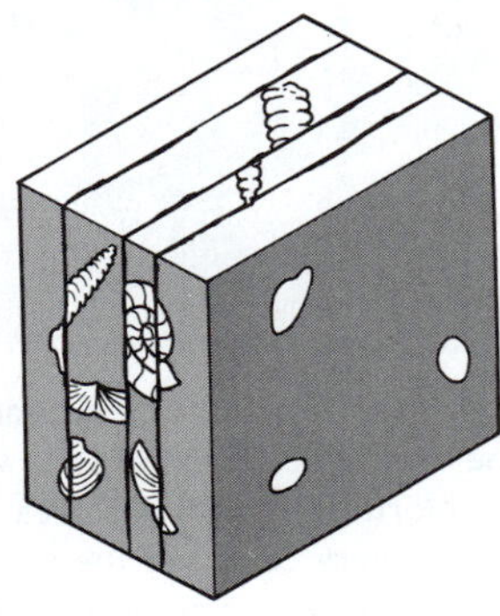

Figure 2.55 (*A*) Negative dilation (i.e., loss of volume) accommodated by pressure-solution along cleavage. (Artwork by R. W. Krantz.) (*B*) These quartz veins in metamorphic rock in northern Norway were originally much longer. They are now short and stubby because of pressure solution removal of material. Notice how the upper limits of each are abruptly truncated. (Photograph by G. H. Davis.)

distortion. One of the stunning discoveries in structural geology is that rock can actually dissolve away in response to compression (Figure 2.55). Part of the rock is removed to facilitate and achieve even more shortening than would be attainable through folding and faulting alone. Material goes into solution and is redeposited in parts of the distorted body that are undergoing an increase in volume. **Stylolites** are structures that represent locations where material is "forced" into solution by directed pressure, leaving behind a black residue of insoluble material (Figure 2.56).

Considered two-dimensionally, increases and decreases in the area of objects can be readily assessed on the basis of two-dimensional strain relationships (Figure 2.57).

$S_1S_3 = 1.0 =$ no area change (Figure 2.57*A*,*B*)

$S_1S_3 < 1.0 =$ area decrease (Figure 2.57*A*,*C*)

$S_1S_3 > 1.0 =$ area increase (Figure 2.57*A*,*D*)

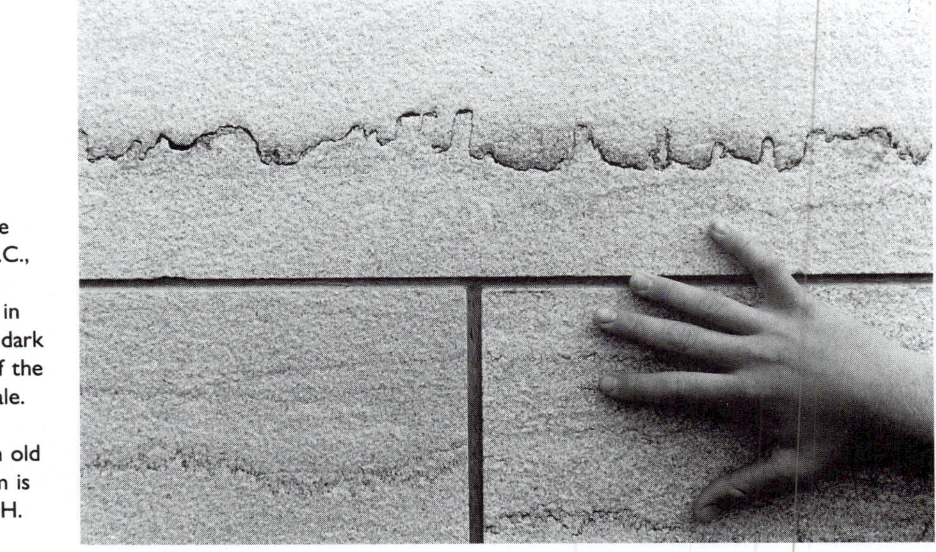

Figure 2.56 While waiting in line at the Air and Space Museum in Washington, D.C., the Davis family focused attention on the well-developed brain-suturelike stylolites in the building stone (limestone). Note the dark insoluble residue that lines the interior of the stylolitic surface. Matt Davis' hand for scale. Normally the best stylolites are seen inside—on limestone and marble steps in old buildings, or in restrooms where the trim is marble or limestone. (Photograph by G. H. Davis.)

The different levels of dilation that may accompany distortions in rocks can be visualized through John Ramsay's **strain field diagram** (see Ramsay, 1967, p. 112). Using this diagram, we find that the physical structural significance of the full range of combinations of stretch values becomes more meaningful (Figure 2.58).

For ease of reference, we have given names to the components of these diagrams. The **field of expansion** is reserved for strain characterized by stretch values (S_1 and S_3) greater than 1.0. It is a field marked by distortion accompanied by an increase in area ($S_1S_3 > 1.0$). Physically the rock stretches in two directions. The **field of contraction**, in contrast, is marked by stretch values less than 1.0. Distortion is accompanied by a decrease in area ($S_1S_3 < 1.0$), and the rock is shortened in two mutually perpendicular directions. In the **field of compensation**, S_1 is greater than 1.0 and S_3 is less than 1.0. Plane strain occurs in this field ($S_1S_3 = 1.0$), but this is a special case in a field where increases or decreases in area are the rule.

The boundary line between the field of contraction and the field of compensation is the **field of linear shortening**, symbolizing strain marked by shortening in one direction, with no strain at right angles to the direction of shortening ($S_1 = 1.0$; $S_3 < 1.0$). The boundary line between the fields of compensation and expansion is the **field of linear stretching**. It describes strain values marked by $S_1 > 1.0$ and $S_3 = 1.0$. It represents deformation marked by extension in one direction only, with no strain at right angles to the direction of extension. Finally, the lines of stretching and shortening intersect at a point, the **field of no strain**, where S_1 and S_3 both are 1.0. We use the strain field diagram to classify structures according to strain significance.

Coaxial and Noncoaxial Strain

In addition to dilational effects, there is yet more complexity to strain. It is the full realization that during distortion the axes of the strain ellipse, which portray the strain at each stage of deformation, are usually not fixed in orientation. Rather, the strain axes commonly rotate! Structural geologists picture this reality through the use of card deck models (e.g., see Ragan, 1969, 1973) or computer models. As we have seen, if a circle is imprinted on the side of a deck of computer cards, and the cards of the deck are homogeneously sheared with respect to one another, the circle

A

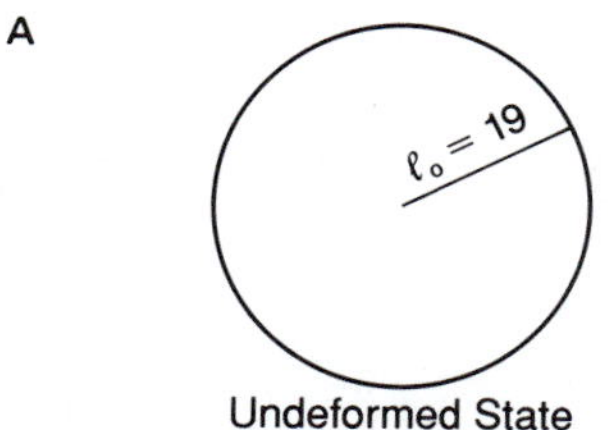

B

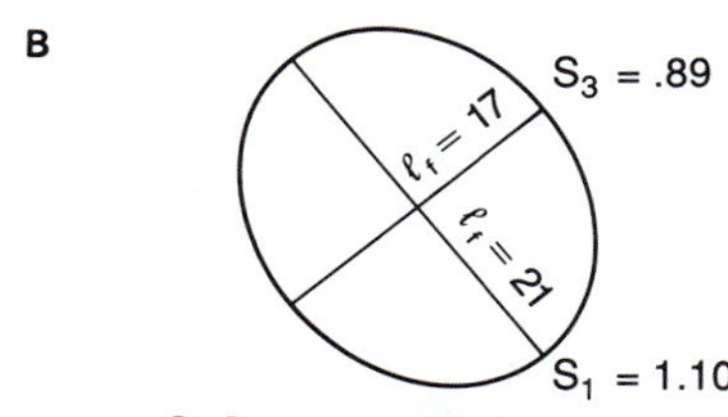

C

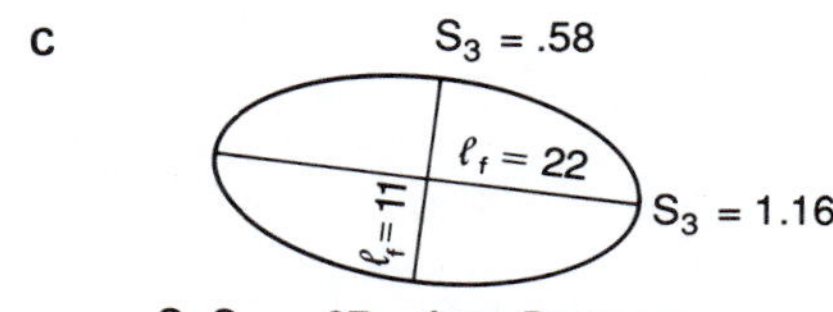

D

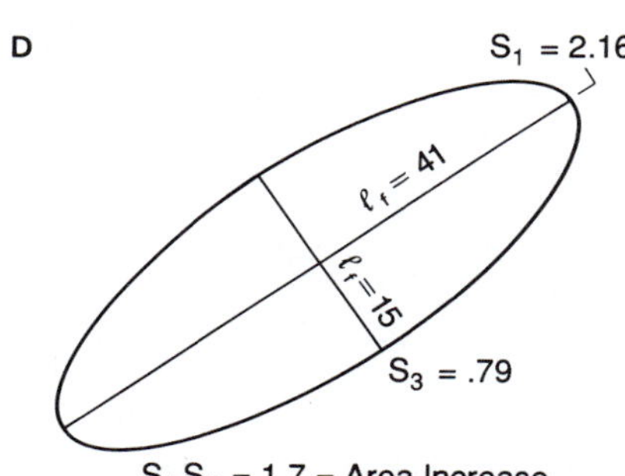

Figure 2.57 Stretch values sensitively reflect changes in area (and volume). (*A*) No distortion, no dilation. (*B*) Distortion without dilation. (*C*) Distortion accompanied by area decrease. (*D*) Distortion accompanied by area increase.

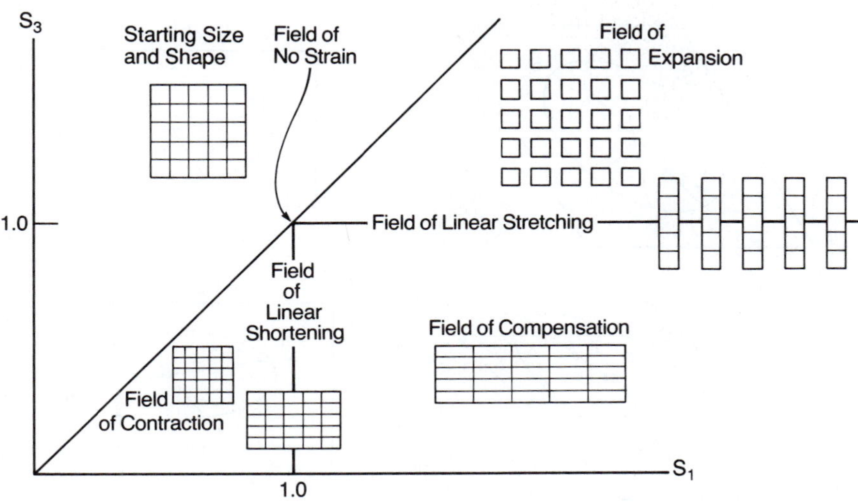

Figure 2.58 Strain field diagram. [Modified from Ramsay (1967).]

will be transformed progressively into an ellipse. If we view the size and orientation of the ellipse at different stages of the deformation (Figure 2.59*A*), we see that the finite stretching axes of the strain ellipse rotate in space. This is an **external rotation**, to be distinguished from **internal rotation**, in which lines rotate *relative* to the finite stretching axes.

As has been emphasized, part of the uniqueness of the finite stretching axes is that these are the only mutually perpendicular lines in the distorted body that were perpendicular before deformation. When the finite stretching axes do not remain fixed in orientation during deformation, the deformation is described as a **noncoaxial strain** (Figure 2.59*A*). But when the finite stretching directions have the same orientation both before and after deformation, the deformation is described as a **coaxial strain** (Figure 2.59*B*).

Pure Shear and Simple Shear

Pure shear and simple shear are types of strain, even though they sound as if they should be the names of processes or movements. They represent two special end members of plane strain. Pure shear is nearly synonymous

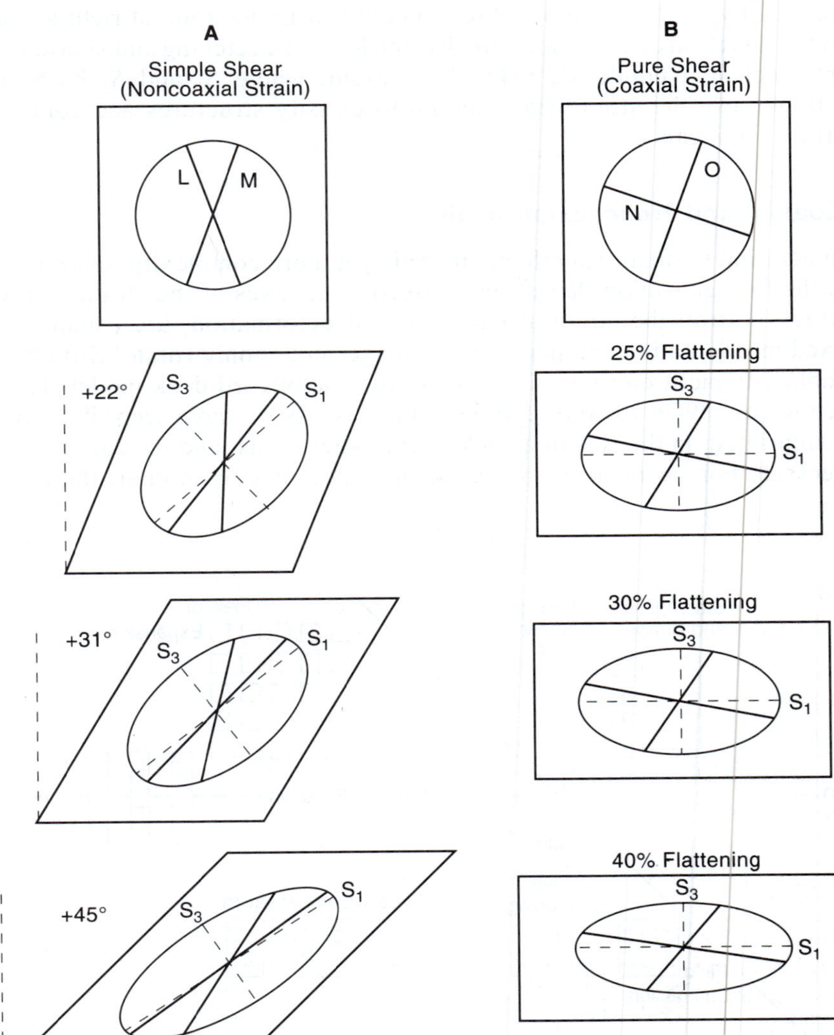

Figure 2.59 (*A*) Progressive deformation of a reference circle and two lines (*L* and *M*) by noncoaxial strain, which in this example is a simple shear. The principal strain axes rotate continuously to different orientations during the deformation. Line *L* at first shortens and then lengthens. Line *M* lengthens throughout the deformation. (*B*) Progressive deformation of a reference circle and two lines (*N* and *O*) by coaxial strain, which in this example is a pure shear. The orientations of the principal strain axes remain fixed throughout the deformation. Line *N* continuously lengthens. Line *O* continuously shortens.

with coaxial strain. Simple shear is nearly synonymous with noncoaxial strain. Both can be illustrated by examining the front face of a cube of material, flattened or sheared in such a way that there is no change in volume, and neither shortened nor stretched in a direction perpendicular to the front face (Figure 2.59).

In **simple shear**, the cube of rock is sheared like a deck of cards (Figure 2.59*A*). The square is converted to a parallelogram. The sides of the parallelogram progressively lengthen as deformation proceeds, but the top and bottom surfaces neither stretch nor shorten. Instead, they maintain their original length, which is the original length of the edge of the cube. The finite stretching axes rotate during the deformation.

In **pure shear**, the cube of rock is shortened in one direction and extended in the perpendicular direction within the plane of strain (Figure 2.59*B*). The square is converted to a rectangle, and the original sides of the square remain parallel after deformation. The finite stretching axes do not rotate.

As structural geologists analyzing strain in outcrop, we are not often able to judge whether deformation was accomplished by pure or simple shear, (i.e., by coaxial or noncoaxial strain). Without information about how the strain accrued, we can describe only the total **finite strain** as we see it. We will see in Chapter 9 (''Shear Zones and Progressive Deformation'') that the key to distinguishing pure and simple shear lies in understanding how incremental strain accrues during **progressive deformation**.

Just a Word on Progressive Deformation

The ways in which incremental strains are added have a profound influence on the physical and geometric nature of structures that develop in a distorted body of rock. Whether a given line (or layer) undergoes stretching or shortening at any instant of time depends on the orientation of the instantaneous strain ellipse with respect to the line (or layer) in question. Whether a given line (or layer) undergoes total finite stretching or shortening depends on the nature of superposition of the incremental strains through time.

We can get a feel for this by observing the changes in lengths of lines L and M in Figure 2.59*A*, as the body in which they reside undergoes progressive shear, and lengths of lines N and O in Figure 2.59*B*, as the body in which they reside undergoes progressive flattening. L shortens at first, and then lengthens, whereas M lengthens continuously; N lengthens continuously, whereas O shortens continuously. We conclude that the superposition of lengthening on shortening does not necessarily mean that there have been two periods of deformation, separated in time and superimposed on each other.

GENERAL SHEAR, AND CHANGING THE SHAPES OF THINGS, IN PRACTICE

General shear represents a complete spectrum of strain intermediate between simple and pure shear. It can consist of everything from a simple shear with only a minor component of pure shear or dilation to mostly pure shear or dilation with only a minor component of simple shear.

We close this chapter with some strain operations that we hope will reassure us all that it is possible to deal with the deformation of whole objects other than circles and lines. The idea is to take an object of any

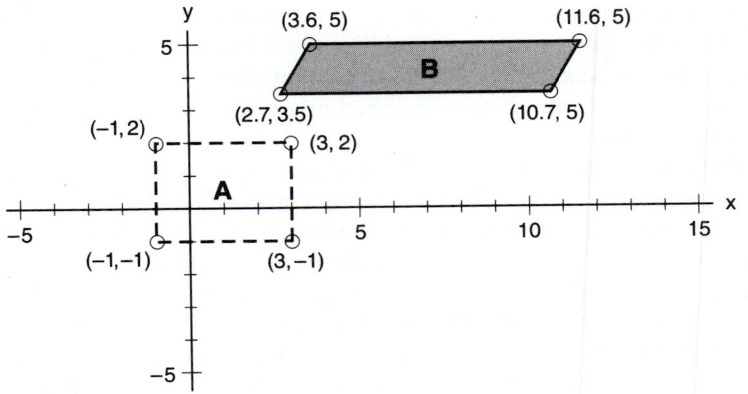

Figure 2.60 We want to be able to define the *x,y* coordinates of the outline of any object, such as rectangle *A*, and then use strain "transformation" equations to calculate the shape, orientation, and position of the object (*B*) in the deformed state. Digitizing tables and computers permit applying transformation equations to complex objects, not just simple shapes.

initial shape and change its shape in ways that conform to the strain parameters with which we are provided.

Let us operate on a rectangular shape, like the one shown as *A* in Figure 2.60. The position and form of the rectangle are "pinned" to (x,y) coordinates on a graph. Our goal is to calculate what the new coordinates (x',y') will be for the deformed rectangle (*B*, Figure 2.60) following translation, rotation, dilation, distortion, some of the above, or all of the above. **Transformation equations** express the final coordinates of points in the body in terms of the original coordinates and the parameters that describe the movements. These parameters are ones we know about already (Figure 2.61):

A: the distance of translation parallel to the *x*-axis

B: the distance of translation parallel to the *y*-axis

ω: the degree of clockwise or counterclockwise rotation about the origin of the coordinate system

S_x: the magnitude of stretch parallel to the *x* axis of the coordinate system

S_y: the magnitude of stretch parallel to the *y* axis of the coordinate system

tan $\psi_x S_y$: the shear strain parallel to the *x*-axis of the coordinate system

tan $\psi_y S_x$: the shear strain parallel to the *y*-axis of the coordinate system

The transformation equations themselves are presented in Figure 2.62. After choosing values for the basic parameters in these equations (Table 2.2), we can simply pick an operation, solve for the new coordinates, and watch the transformation. Solutions are shown in Figure 2.62.

There is one more thing. If, for any transformation, we wish to know whether there has been a change in area, we multiply the values for S_x and S_y, and then subtract the product of multiplying S_y tan ψ_x and S_x tan ψ_y. If the result is 1.0, there is no change in volume. If the result is (e.g.) 0.72, the final volume of the deformed object is 72% of the original. The magic of strain.

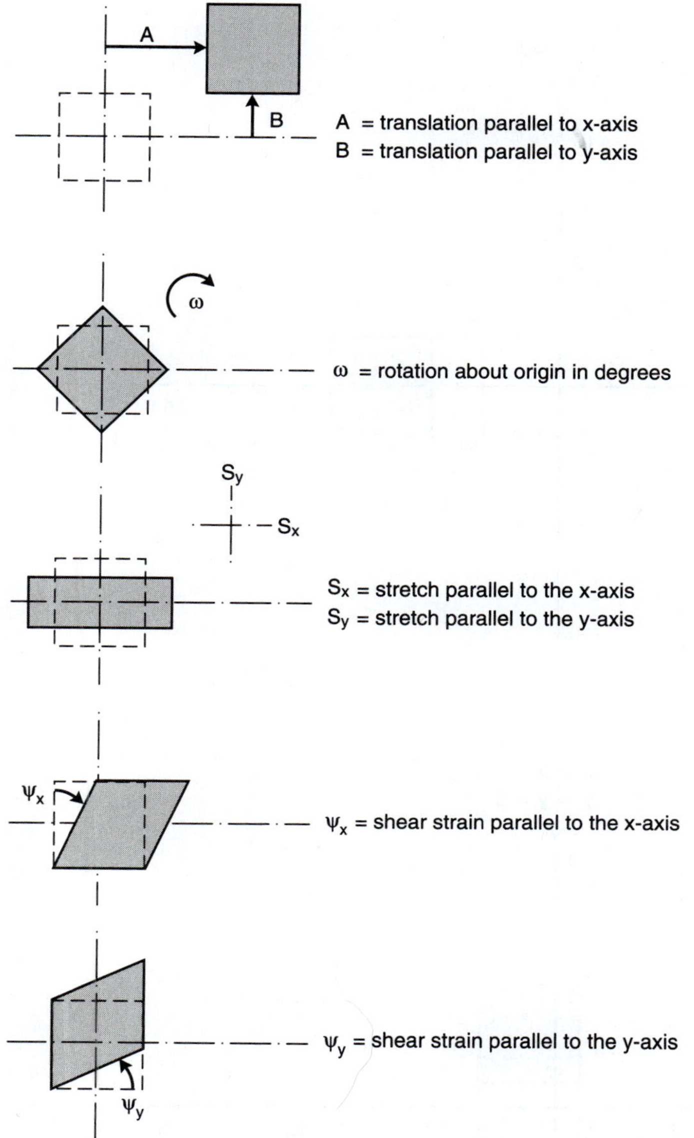

Figure 2.61 The transformation equations are guided by key parameters related to components of translation, rotation, stretch, and shear strain. Here are the conventions.

TABLE 2.2
Values of "transformation parameters" used in Figure 2.62

A	6
B	3
ω	+35°
S_x	3
S_y	0.33
ψ_x	+50°
ψ_y	−20°

A

Translation parallel to x-axis

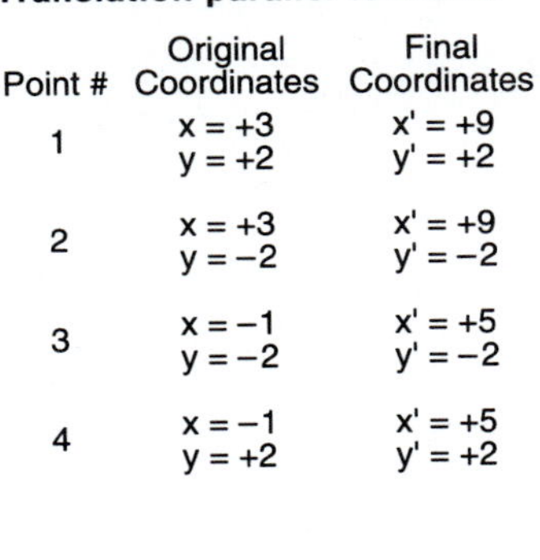

Point #	Original Coordinates	Final Coordinates
1	x = +3 y = +2	x' = +9 y' = +2
2	x = +3 y = −2	x' = +9 y' = −2
3	x = −1 y = −2	x' = +5 y' = −2
4	x = −1 y = +2	x' = +5 y' = +2

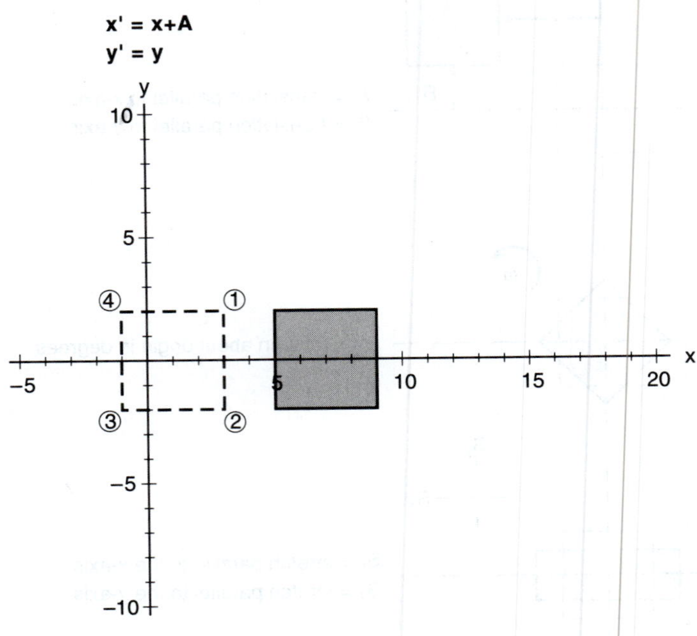

B

Translation parallel to y-axis

Point #	Original Coordinates	Final Coordinates
1	x = +3 y = +2	x' = +3 y' = +5
2	x = +3 y = −2	x' = +3 y' = +1
3	x = −1 y = −2	x' = −1 y' = +1
4	x = −1 y = +2	x' = −1 y' = +5

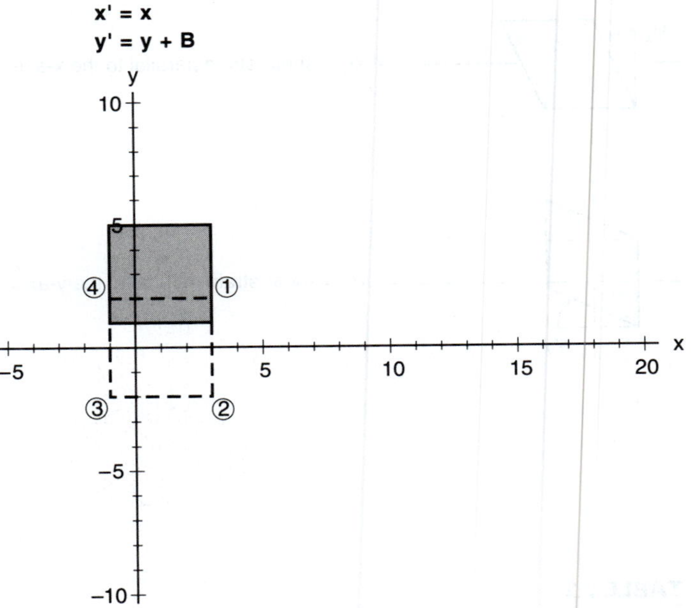

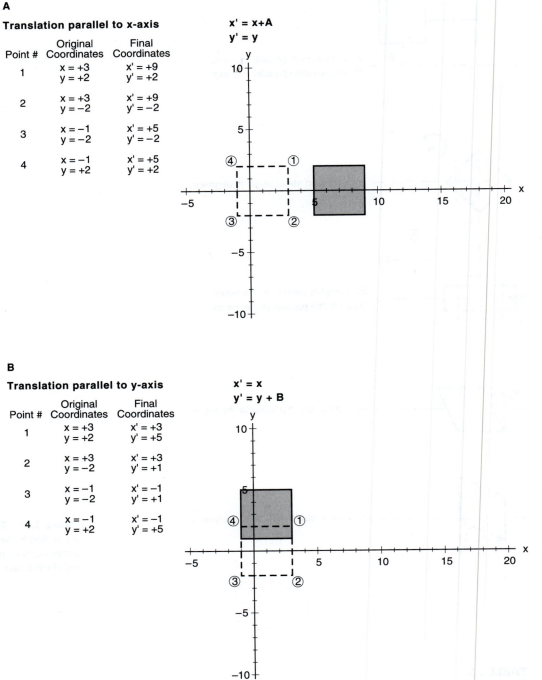

Figure 2.62 Now we apply the transformation equations. Each example begins with equations that describe the final coordinates of a point (x',y') in relation to the original coordinates (x,y) and the relevant parameters. (*A*) Translation parallel to *x*-axis. (*B*) Translation parallel to *y*-axis. (*C*) Clockwise rotation (and counterclockwise as well). (*D*) Extension parallel to the *x*-axis. (*E*) Extension parallel to the *y*-axis. (*F*) Pure shear. (*G*) Uniform dilation. (*H*) Shear parallel to the *x*-axis. (*I*) Shear parallel to the *y*-axis. (*J*) Shear components parallel to both the *x*-axis and the *y*-axis. (*K*) The general transformation, which includes translation, rotation, and shear.

C
Clockwise rotation

Point #	Original Coordinates	Final Coordinates
1	x = +3 y = +2	x' = +3.5 y' = +0.05
2	x = +3 y = −2	x' = +1.4 y' = −3.2
3	x = −1 y = −2	x' = +1.9 y' = −1.1
4	x = −1 y = +2	x' = +0.2 y' = +2.17

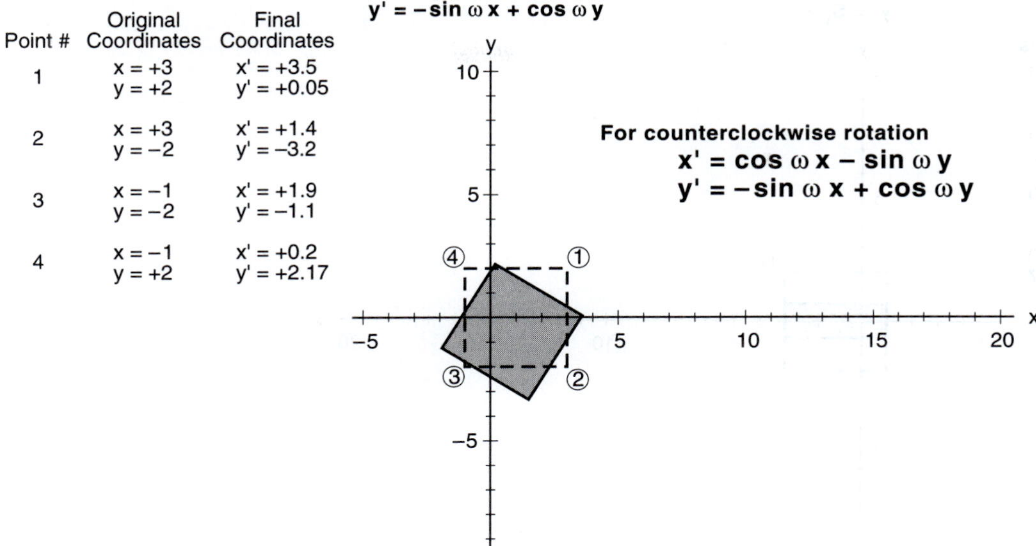

D
Extension parallel to x-axis

Point #	Original Coordinates	Final Coordinates
1	x = +3 y = +2	x' = +9 y' = +2
2	x = +3 y = −2	x' = +9 y' = −2
3	x = −1 y = −2	x' = −3 y' = −2
4	x = −1 y = +2	x' = −3 y' = +2

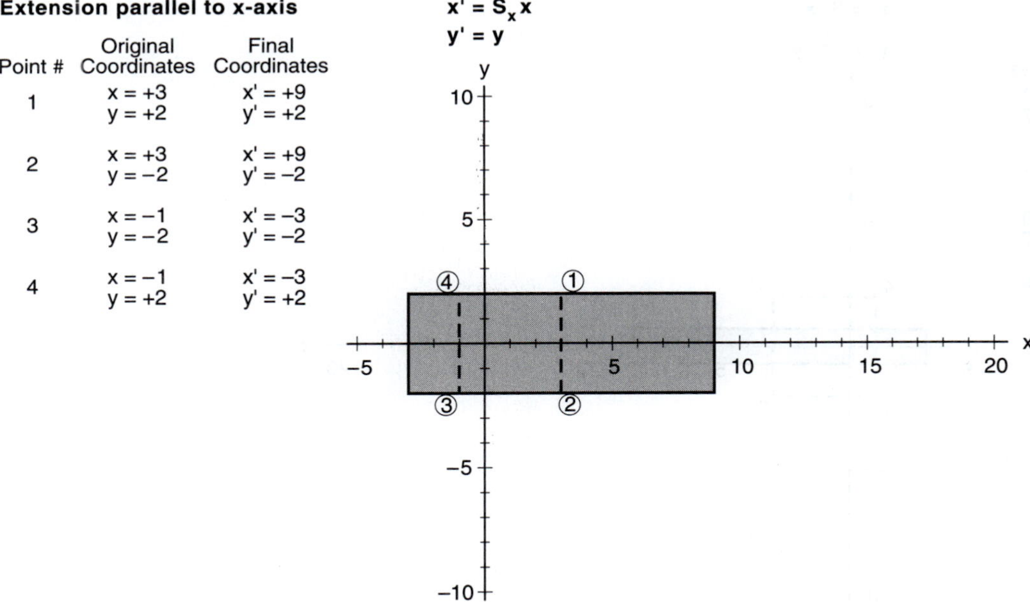

Figure 2.62 (*Continued*)

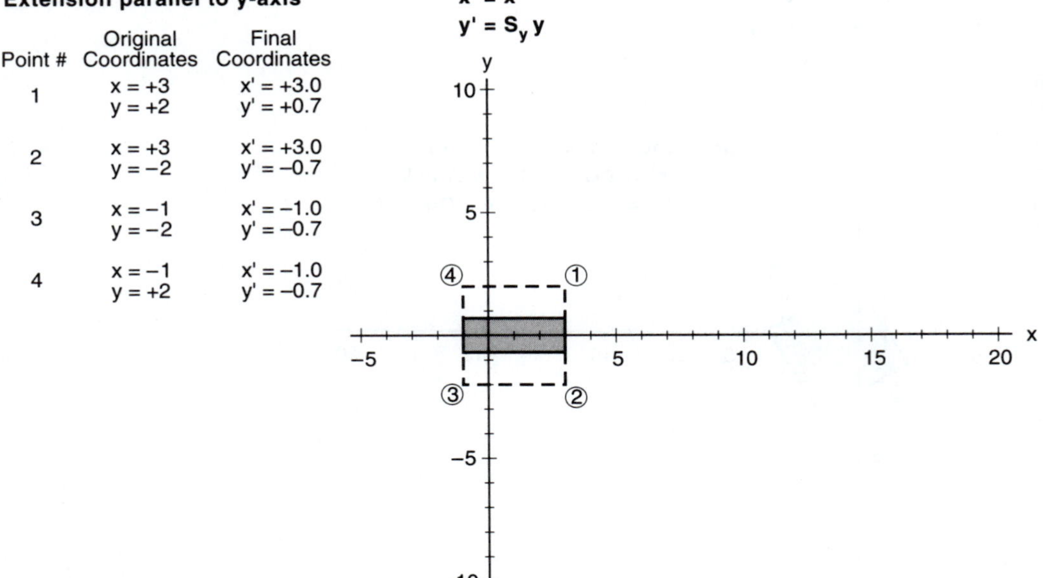

E

Extension parallel to y-axis

$x' = x$

$y' = S_y y$

Point #	Original Coordinates	Final Coordinates
1	x = +3 y = +2	x' = +3.0 y' = +0.7
2	x = +3 y = −2	x' = +3.0 y' = −0.7
3	x = −1 y = −2	x' = −1.0 y' = −0.7
4	x = −1 y = +2	x' = −1.0 y' = −0.7

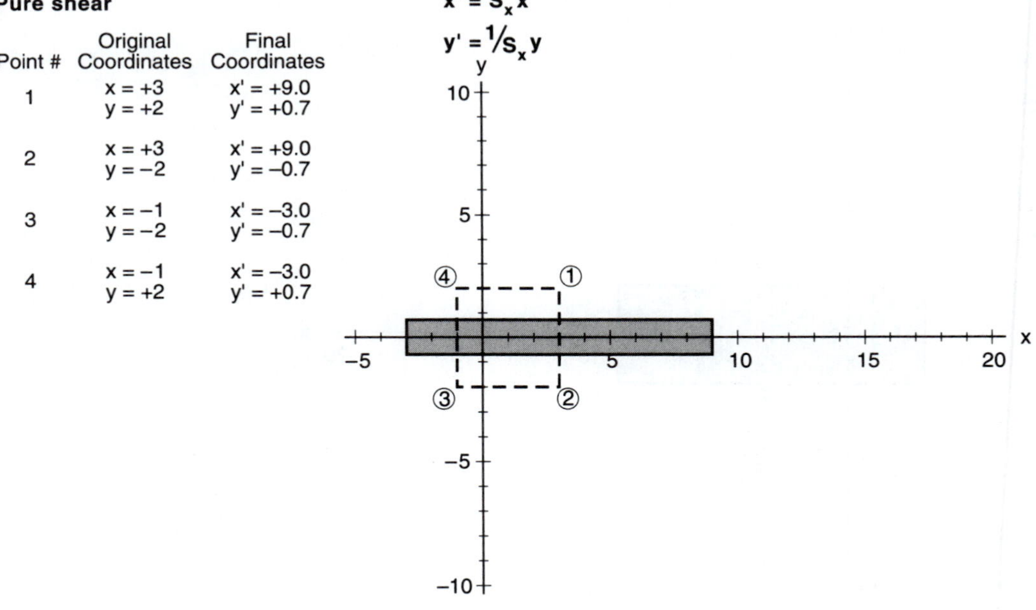

F

Pure shear

$x' = S_x x$

$y' = {}^1\!/_{S_x} y$

Point #	Original Coordinates	Final Coordinates
1	x = +3 y = +2	x' = +9.0 y' = +0.7
2	x = +3 y = −2	x' = +9.0 y' = −0.7
3	x = −1 y = −2	x' = −3.0 y' = −0.7
4	x = −1 y = +2	x' = −3.0 y' = +0.7

Figure 2.62 (*Continued*)

G

Uniform dilation

Point #	Original Coordinates	Final Coordinates
1	x = +3 y = +2	x' = +9 y' = +6
2	x = +3 y = −2	x' = +9 y' = −6
3	x = −1 y = −2	x' = −3 y' = −6
4	x = −1 y = +2	x' = −3 y' = −6

$$x' = S_x\,x$$
$$y' = S_x\,y$$

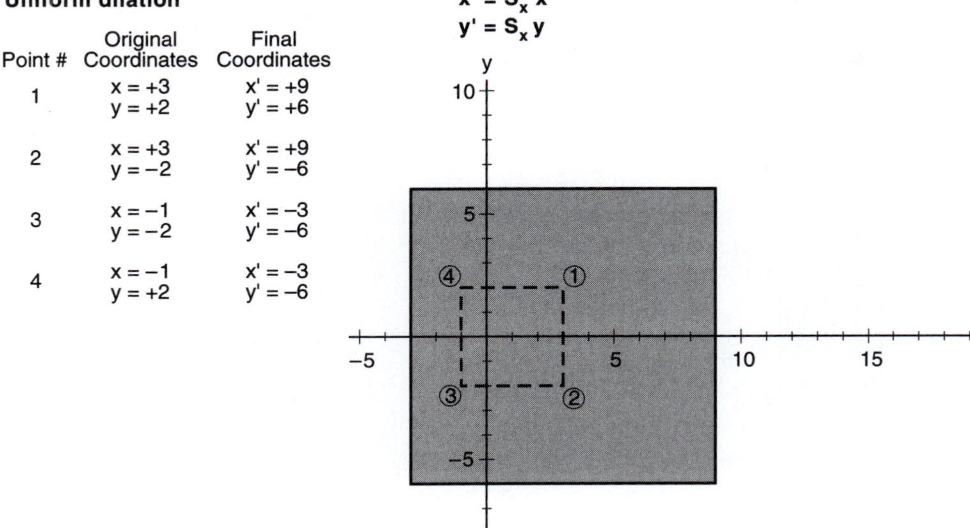

H

Shear parallel to x-axis

Point #	Original Coordinates	Final Coordinates
1	x = +3 y = +2	x' = +5.4 y' = +2.0
2	x = +3 y = −2	x' = +0.6 y' = −2.0
3	x = −1 y = −2	x' = −3.4 y' = −2.0
4	x = −1 y = +2	x' = +1.4 y' = +2.0

$$x' = x + \tan\psi_x\,y$$
$$y' = y$$

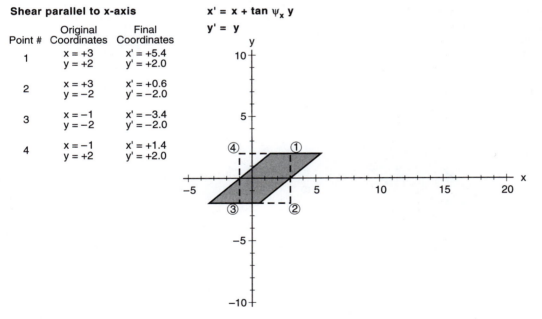

Figure 2.62 (*Continued*)

I

Shear parallel to y-axis

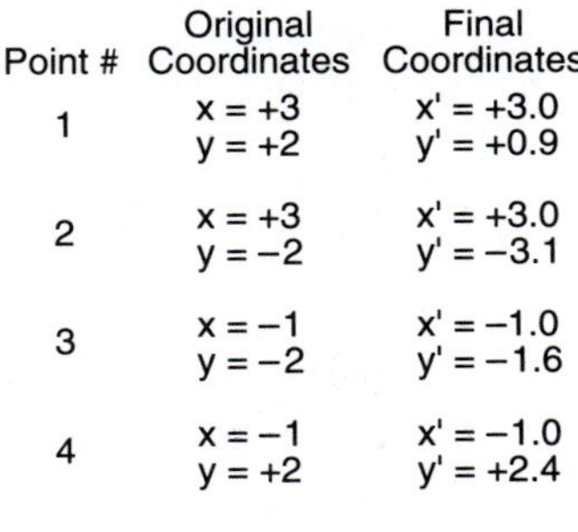

Point #	Original Coordinates	Final Coordinates
1	x = +3 y = +2	x' = +3.0 y' = +0.9
2	x = +3 y = −2	x' = +3.0 y' = −3.1
3	x = −1 y = −2	x' = −1.0 y' = −1.6
4	x = −1 y = +2	x' = −1.0 y' = +2.4

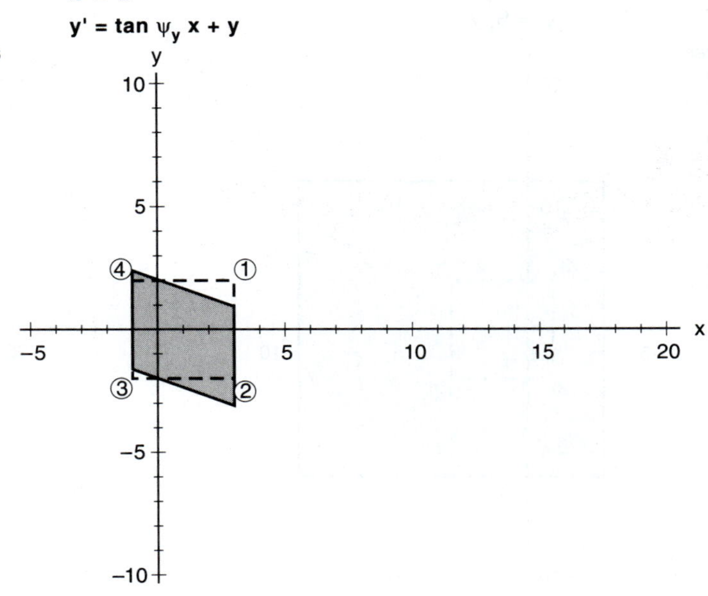

J

Shear parallel to x' and y-axis

Point #	Original Coordinates	Final Coordinates
1	x = +3 y = +2	x' = +5.4 y' = +0.9
2	x = +3 y = −2	x' = +0.6 y' = −3.1
3	x = −1 y = −2	x' = −3.4 y' = −1.6
4	x = −1 y = +2	x' = +1.4 y' = +2.4

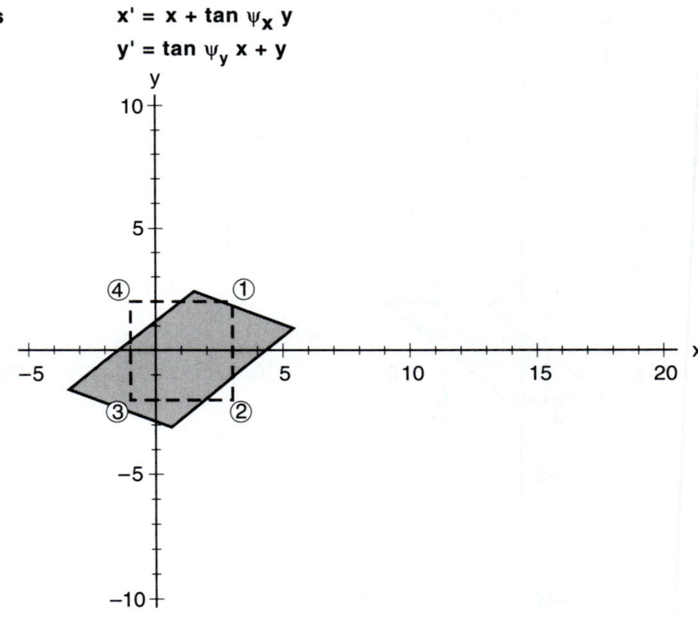

Figure 2.62 (*Continued*)

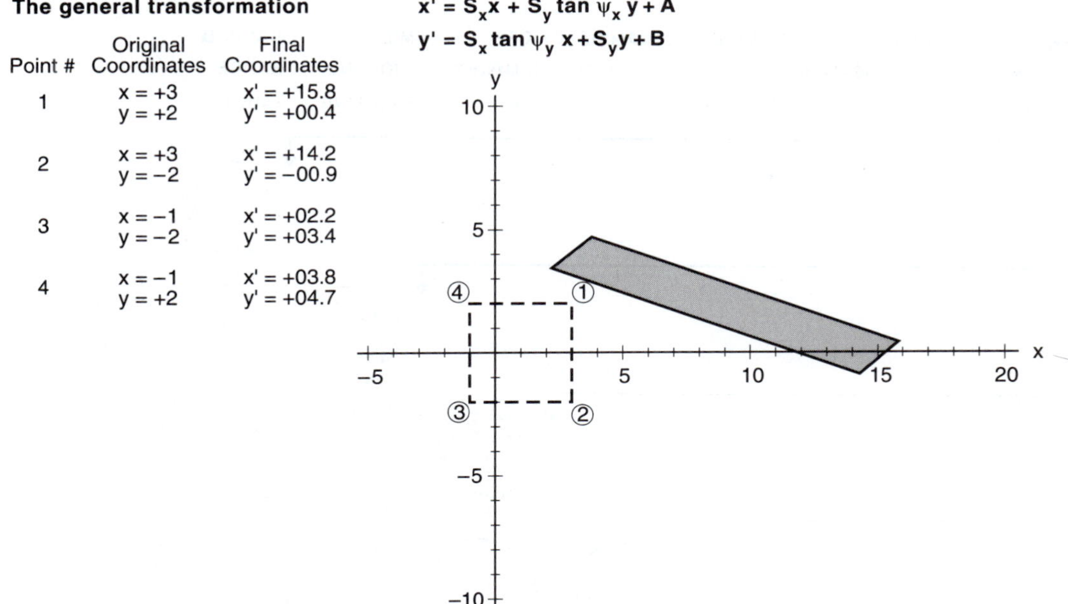

Figure 2.62 (*Continued*)

THE ISSUE OF STRUCTURAL COMPATIBILITY

Kinematics is a three-ring circus when applied to matters of structural geology. Geometrically and mathematically we are required to keep track simultaneously of rigid body translations, rigid body rotations, distortions (coaxial and noncoaxial), and dilations. Sometimes we are required to keep track of superposed deformations (Figure 2.63). Sometimes we just imagine what could happen, for example, ''down under'' (Figure 2.64).

Structures are kinematically linked in systems. The displacements that at one location are accommodated by faulting and internal distortion can be transferred to any number of other structures or combinations of structures: folds, cleavages, stylolites, joints, shear fractures, dike swarms. If structures were *perfectly* exposed, as if the crust were made of glass, we could see the nature of the interrelationships and grasp at once the kinematic beauty of the compatibility among all of the structural ingredients.

But geologic structures are never perfectly exposed—even in the Alps, even in the Colorado Plateau, even in the Mojave Desert. And drilling and tunneling and trenching and seismic profiling are never extensive enough to provide the detail we would like. For this reason we must rely on a backup, namely the concept of **structural compatibility**.

John Ramsay once emphasized to me that the concept of structural compatibility lies at the very heart of structural geology (personal communication, Ascona, Switzerland, 1977). He spoke about it with the passion of a medical researcher talking about the path to a cure. Nothing rivals it in importance. It is a conversation I will never forget.

''Structural compatibility'' is a code expression for assessing whether the kinematic displacements assigned to faults, the geometries postulated for folds, the loss of volume attributed to pressure solution, the extensional strains assigned to joints, the shear strains calculated for shear zones, the distortions calculated from deformed fossils . . . are in harmony for a

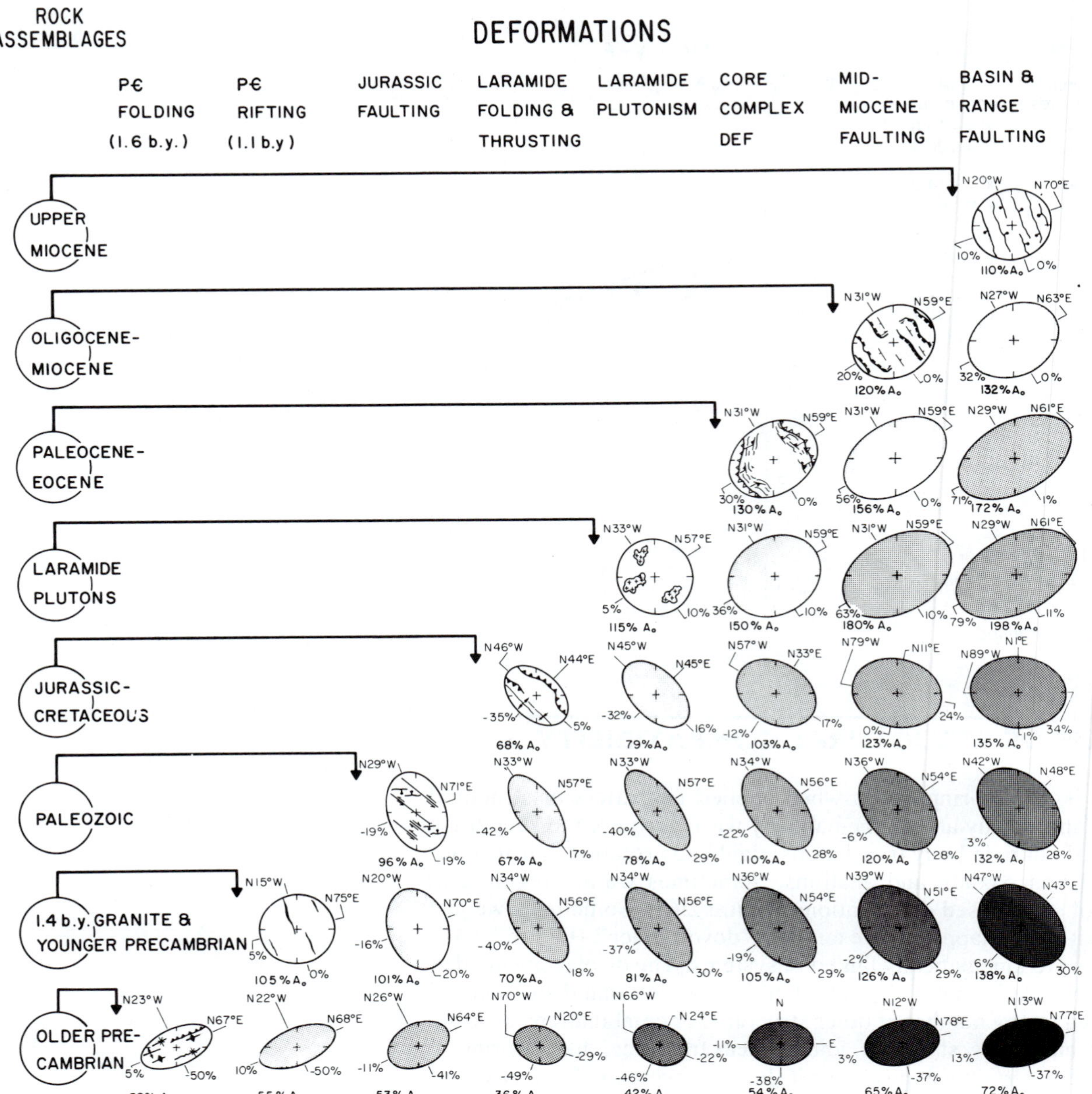

Figure 2.63 Progressive strain diagram portraying regional deformation in southeastern Arizona. Undeformed state of rock assemblages of different ages represented by circles in left column. Regional deformation to which the rock assemblages were subjected are shown at top. Ellipses show state of strain of each assemblage after each deformation. Fine print shows directions of maximum and minimum stretch, percentages of lengthening (+) and shortening (−), and changes in the original surface area (A_0) of each assemblage. [From Davis (1981), Fig. 18, p. 166. Published with permission of Arizona Geological Society.]

given outcrop, for a given study area, for a given region. If everything is "in sync," a "palinspastic reconstruction" of the structural system will result in a perfectly restored geological image, without embarrassing gaps or overlaps. If things are "out of sync," a palinspastic reconstruction of the structural system will be flawed; it will fail the test of being geologically reasonable.

I look back on our "kids" growing up and think of structural compatibility in this way. Tossed into the corner of a bedroom is a pair of 501 Jeans (Figure 2.65*A*). Without touching or moving or lifting-up even a corner of the lump of Levis, I map and measure what I see (Figure 2.65*B*). My map, with its fold symbols for the creases, with its fault symbols for the rips and tears, with its normal fault representation of the unbuttoned fly, becomes the control for cross sections. Based on the cross sections, I

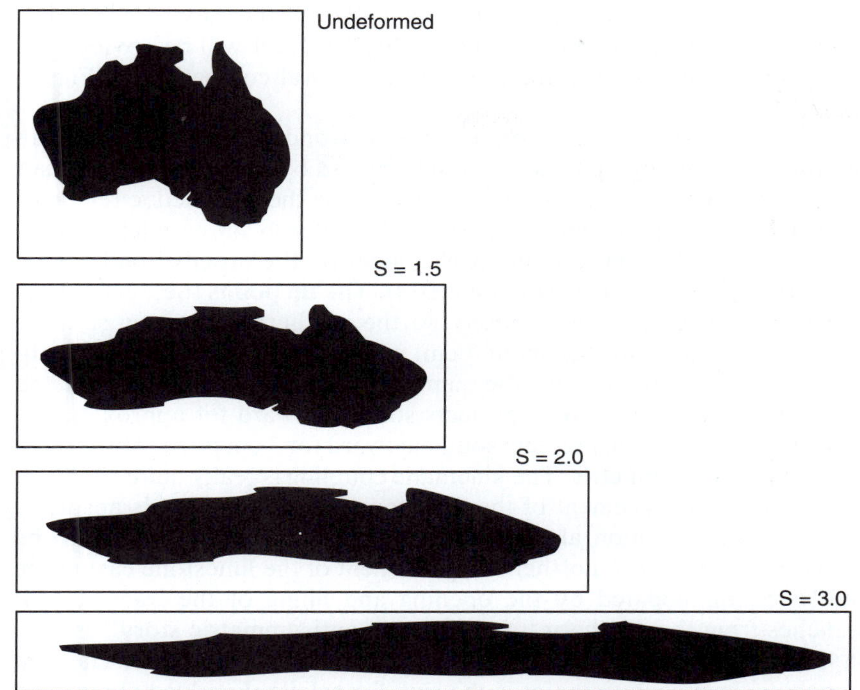

Figure 2.64 A hypothetical progressive deformation "down under."

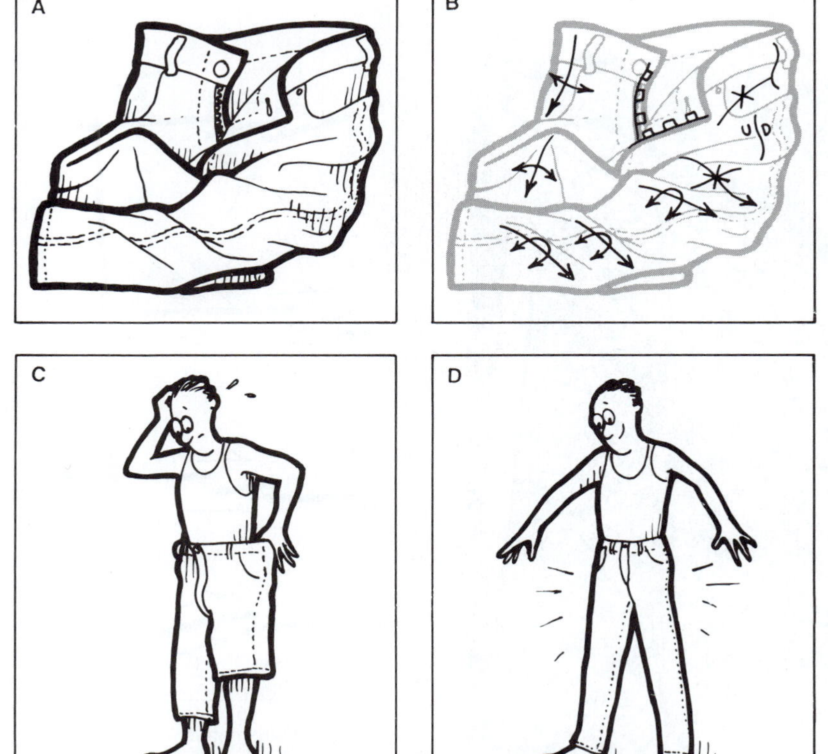

Figure 2.65 (*A*) The structural geologic system, about to be mapped. (*B*) The map. (*C*) Reconstruction based on ineffective cross sections. (*D*) Reconstruction based on effective cross sections. See text for complete explanation. (Artwork by D. A. Fischer.)

can figuratively reconstruct the shape and size of the Levis. When one of the boys next tries on his "virtual reality" 501s, it will be obvious if I have paid any attention to the issues of structural compatibility (Figure 2.65*C,D*).

The concept of structural compatibility is wonderfully illustrated in an outcrop-scale photograph taken and interpreted by Rispoli (1981) (Figure 2.66*A*). The location of the limestone outcrop is the Languedoc region of southern France. It should be a shrine. The outcrop shows a left-handed strike-slip fault with maximum displacement on the order of millimeters and a trace length of 15 cm (Figure 2.66*B*). The **tip points** (i.e., end points of the fault trace) are well exposed. At the northern tip point a stylolite projects northeastward for about 5 cm, and a calcite vein projects northwestward for approximately the same distance (Figure 2.66*A*). At the southern tip point, a stylolite projects southwestward for approximately 5 cm, and a calcite vein projects southeastward for 5 cm or so. The overall geometry is **antisymmetric**. The kinematic coupling is really quite amazing. The northern displacement of the limestone east of the fault is absorbed by pressure dissolution along the stylolitic surface branching from the northern tip point. The northern displacement of the limestone east of the fault is accommodated by the opening and filling of the fracture that branches from the southern tip point. An "antisymmetric story" is read on the opposite side at the same end of the fault (Figure 2.66*A*). The vein branching from each tip point is the **sink** for calcite dissolved along each stylolite. It is quite a system! But there is more. The stylolite teeth, which reflect the direction of shortening, are oriented parallel to the direction of strike-slip, and they progressively decrease in size from the tip point

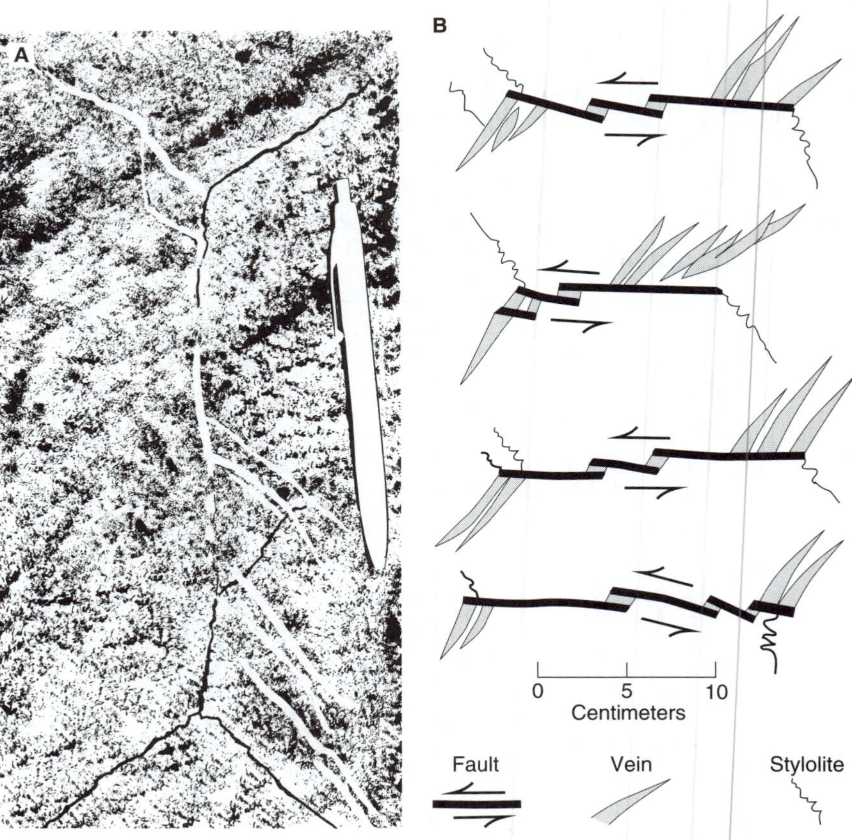

Figure 2.66 (*A*) Remarkable display of the perfect coordination that can exist among structures. Microfault connects an "antisymmetric" coupling of stylolites and veins. This structure was discovered by René Rispoli (see Rispoli, 1981). [Tracing of a photograph by Groshong (1981). Published with permission of Geological Society of America and the author.] (*B*) Map of small faults in limestone, based on work of Rispoli (1981). Veins and stylolitic solution surfaces extend from near the tips of the faults in antisymmetric arrangements. [Adapted from Pollard and Segall (1987).]

to the termination of the stylolitic surface; the crystal-fiber lineation within each calcite vein, which reflects the direction of lengthening, is oriented parallel to the direction of strike-slip, and the width of each vein progressively decreases from tip point to termination.

Rispoli's example is one that underscores the incredible harmony that exists within structural systems. Rispoli identified other examples as well (Figure 2.66*B*). Are these examples unusual? Only in their clarity and elegance. But every structural system possesses a cohesive internal kinematic compatibility. Everything fits. When in the face of reasonably complete information we discover that things do not fit, then we realize we are missing something important.

ON TO DYNAMICS

Translation, rotation, distortion, and dilation are responses of rocks that, under optimum conditions, can be described and kinematically interpreted in considerable detail. But what causes the movements in the first place? And how can we begin to decipher the origins of kinematic movements, especially in ancient systems where the structures formed long ago? Suddenly we enter the arena of dynamic analysis and the interrelationships between stress and strain. Hang on.

CHAPTER 3

DYNAMIC ANALYSIS

CONCEPT OF DYNAMIC ANALYSIS

The goals of dynamic analysis are to interpret the stresses that produce deformation, to describe the nature of the forces from which the stresses are derived, and to evaluate the overall relationships among stress, strain, and rock strength. The description of stress and force is mathematical. Mathematical parameters and equations are necessary to bring forces and stresses to life. In the long run, we would like to be able to reconstruct the stress directions and stress magnitudes that once were active, based on the structures and microstructures they produced. As an aid to our thinking, it helps at times to physically measure the present stress in the crust of the Earth and relate measurements to actively forming geologic structures. The interpretation of ''paleodynamics'' is especially difficult, however, because a given rock will respond to stress in an elastic–brittle manner in one environment, in a plastic manner in another, and in a viscous manner in yet another. Moreover, a rock that responds to stress in a plastic manner in one instant of time may shift to a brittle mode if the rate of application of stress suddenly increases.

Overall, dynamic analysis is all about the interplay between the stresses that tend to deform, and the strengths that tend to resist.

Figure 3.1 Ultimate Frisbee is force in action. (Artwork by D. A. Fischer.)

FORCE

Definition of Force

Translations, rotations, distortions, and dilations are responses of rocks to stresses that are generated by forces. **Force** is classically defined as that which changes, or tends to change, the state of rest or state of motion of a body. The sports world abounds with examples of forces in action (Figure 3.1). Basketballs and Frisbees, footballs and tennis balls, skiers and gymnasts, and bats and bungee cords undergo, or tend to undergo, changes in state of rest or motion when affected by forces. Only a force can cause something to move that had been stationary. Only a force can cause something to change its speed. Only a force can cause something to change its course of travel.

For example, when the body of a studio wrestler is subjected to the force produced by the accelerating mass of a huge opponent, its state of rest or state of motion may be significantly changed (Figure 3.2*A*). The body is usually both translated and rotated. During a "pinning" (Figure 3.2*B*), the body might even undergo slight negative dilation (decrease in volume). In rare instances, the body, or parts of the body, may endure the permanent strain of distortion, involving breaking or stretching or ripping. This does not happen often during studio wrestling because the wrestlers are well aware of the human body's "elastic limit," and they "pull their punches" accordingly. Whether strain is expressed in permanent deformation depends on two factors: the strength of the body and the concentration of the **stress**, which is that which tends to deform a body of material.

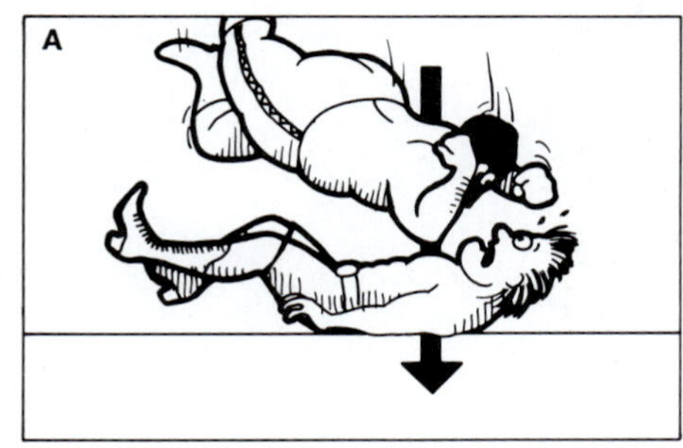

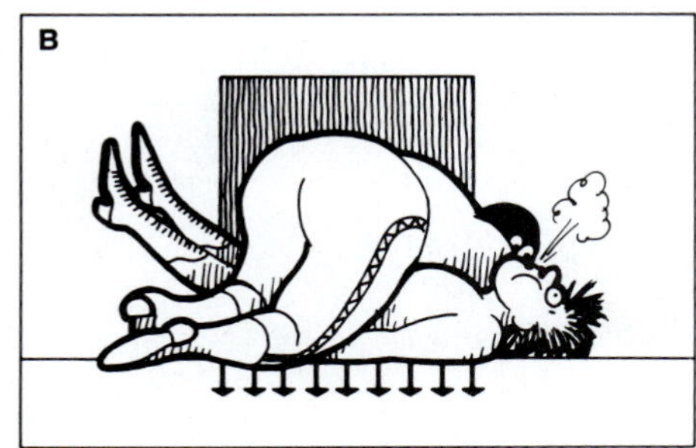

Figure 3.2 Studio wrestling: an exercise in controlling the effects of forces. Mass is a constant for each event. Acceleration is not. (*A*) The force generated by the mass and acceleration of one of the Zolanger brothers (in air) is going to tend to change the state of rest of his opponent. (*B*) The gravitational loading generated by Mr. Zolanger (everybody calls him "Mr.") can be effective even though contact over a large surface area results in low stress concentrations. (Artwork by D. A. Fischer.)

Newton, through his *first law of motion*, described the concept of force in this way: an object at rest will remain at rest and an object in motion will continue in motion with a constant velocity unless it experiences a **net force**, in which case it is caused to accelerate (or deaccelerate). A net force arises when forces are not balanced. In his *second law of motion*, Newton observed that the acceleration of an object is directly proportional to the net force on it, and inversely proportional to the mass of the object. The definition of force is, in fact, based upon **mass (*m*)** and **acceleration (*a*)**:

$$\text{force} = \text{mass} \times \text{acceleration}$$

$$F = ma \tag{3.1}$$

Mass and acceleration are reciprocal.

$$m_1 a_1 = m_2 a_2$$

Thus, if a given force accelerates a 1 kg object by 3 m/s^2, it will accelerate a 2 kg object by 1.5 m/s^2, for:

$$1 \text{ kg} \times 3 \text{ m/s}^2 = 2 \text{ kg} \times 1.5 \text{ m/s}^2$$

Mass and Weight

The **mass (*m*)** of a body is the amount of material the body contains. The mass of a body can be expressed in units of kilograms (International System), grams (cgs system), or slugs (English Standard) (Table 3.1). Mass can be readily calculated if volume and density are known. **Volume (*V*)** is the space occupied by the mass and is expressed commonly in units of cubic centimeters or cubic meters. **Density (*ρ*)** is the measure of the mass of a body per unit volume, and is most commonly expressed in grams per cubic centimeter or kilograms per cubic meter. The relationship among mass, volume, and density is

$$m = \rho V$$

$$\rho = \frac{m}{V}$$

Mass causes a body to have weight in a gravitational field, but mass and weight are not the same. The **weight** of a body of a given mass is the magnitude of the force of gravity acting on the mass, and it varies according to location. The force of gravity acting on the mass of a body on the moon

TABLE 3.1
Units and conversions

Convert	*(Symbol)*	*to*	*Multiply by*
		Length	
centimeter	(cm)	inch	0.39370
centimeter	(cm)	meter	10^{-2}
meter	(m)	feet	3.2808
meter	(m)	centimeter	10^2
kilometer	(km)	mile	0.621371
kilometer	(km)	meter	10^3
mile	(mi)	kilometer	1.6093
feet	(ft)	meter	0.3048
inch	(in.)	centimeter	2.54
micrometer	(μ)	meter	10^{-6}
		Area	
square meter	(m^2)	cm^2	10^4
square meter	(m^2)	ft^2	10.7636
		Volume	
cubic meter	(m^3)	cm^3	10^6
cubic meter	(m^3)	ft^3	35.3134
		Weight	
gram	(g)	pound	0.0022046
pound	(lb)	kilogram	0.453592
kilogram	(kg)	pound	2.20462
		Mass	
kilogram	(kg)	gram	10^3
kilogram	(kg)	slug	68.521×10^{-3}
		Acceleration	
meter/second (sec)	(m/s^2)	ft/s^2	3.2808
		Force	
newton	(N)	kg-m-s^{-2}	1
dyne	(d)	g-cm-s^{-2}	1
newton	(N)	dyne	10^5
dyne	(d)	newton	10^{-5}
newton	(N)	poundal	7.2330
		Pressure, Stress	
pound/sq inch	(psi)	bar	0.0689
bar	(b)	psi	14.5038
bar	(b)	atmosphere	0.98692
bar	(b)	dyne/cm^2	10^6
newton/m^2	(N/m^2)	pascal	1
pascal	(Pa)	newton/m^2	1
kilopascal	(KPa)	Pa	10^3
megapascal	(MPa)	Pa	10^6
gigapascal	(GPa)	Pa	10^9
bar	(b)	Pa	10^5
bar	(b)	megapascal	10^{-1}

will be *less* than the force of gravity acting on the same body on Earth. When we say that we have gained a few pounds, we should actually be saying that we gained a few slugs! Slug is a unit of mass, whereas pound is a unit of weight (Serway, 1990). In the unlikely event that the word "slug" crops up in the context of physics, as opposed to the context of soft gooey garden creatures, we know we are talking about mass. But when we see "kilogram" or "gram," we must examine the context of the measurement to determine whether the reference is to mass or weight.

Units of Force

The **newton** (N) is the basic unit of force in the International System. A newton is the force required to impart an acceleration of one meter per second per second to a body of one kilogram mass. In the cgs system, the basic unit of force is the **dyne**, which is the force required to impart an acceleration of one centimeter per second per second to a body whose mass is one gram. A force of one newton is equivalent to the force of 10^5 dynes.

What Does a Newton Feel Like?

Randy Richardson (University of Arizona) helped us with this one. Let's say we go to the bowling alley in our purple and yellow shirts with flyaway collars. As amateurs we pick up a bowling ball (it weighs 16 lb), dry our fingers, go into our rhythm, extend arm and ball all the way backward during the approach (for an instant the ball is at rest!), then launch the ball forward, and then release. Randy estimates the time (t) that transpires from the ball at rest to release is approximately 0.5 s.

In calling around to the alleys, Randy learned that the velocity of the ball at release for amateurs is on the order of 13 to 15 miles per hour (mph). He then calculated acceleration (a) by dividing velocity (V) by time (t) (converting to metric in the process):

$$V = 14\ \text{mph} = \frac{14\ \text{mi}}{\text{hr}} \times \frac{\text{km}}{0.62\ \text{mi}} \times \frac{\text{hr}}{3600\ \text{s}} \times \frac{10^3\ \text{m}}{\text{km}} = 6.27\,\frac{\text{m}}{\text{s}}$$

$$\frac{V}{t} = \frac{6.27\ \text{m/s}}{0.5\ \text{s}} = 12.54\ \text{m/s}^2$$

To calculate force (F), in newtons, Randy multiplied mass (m) times acceleration (a), again converting to metric, and determined that the constant force needed to accelerate the ball to the point of release is about 90 N:

$$F = \text{ma}$$

$$m = 16\ \text{lb} \times 16\,\frac{\text{oz}}{\text{lb}} \times \frac{28.35\ \text{g}}{\text{oz}} \times \frac{\text{kg}}{10^3\ \text{g}} = 7.26\ \text{kg}$$

$$F = ma = 7.26\ \text{kg} \times 12.54\ \text{m/s}^2$$

or

$$F = 91.1\ \text{kg} \times \text{m/s}^2$$

or

$$F = 91.1\ \text{N}$$

Randy indicates that the biggest uncertainty in the calculation is t, and that next time at the alleys we should all use a stopwatch to time the movement from the back of the swing to release.

Forces as Vectors

Describing the magnitude of a force, in either newtons or dynes, is not sufficient to completely define the force. *Forces are vector quantities,* and therefore the direction in which the force acts also must be specified. The vector property of forces permits them to be added and subtracted using principles of vector algebra, and this in turn makes it possible to evaluate whether forces on a body are in balance.

For example, at the instant of time captured in the "tug-of-war" contest illustrated in Figure 3.3*A*, the team on the left is pulling with a combined force of 240 newtons (240 N), while the team on the right is pulling with a force of 225 newtons (225 N). The net force is 15 N, the advantage going to the team on the left. We all have played the game, and thus we all know that circumstances and conditions can change quickly. For example, at a second instant of time (Figure 3.3*B*), the team on the right digs in on dry ground, while some of the members of the team on the left have slipped down on wet ground, losing both their footing and the force of their legs. Full-body force versus upper-body force turns the advantage to the team on the right, with a net force of 25 N. The team on the right smells victory.

In some instances forces are perfectly balanced, as shown in Figure 3.4*A*. We see the start of yet another team competition, the object of which is to shove a giant ball into the opponent's end zone. We have experienced players here, and each applies his or her force perfectly horizontally toward the center of the ball. At this instant of the game, and probably only at this instant, the forces on the two sides are perfectly balanced, as can be illustrated through vector addition.

For true balancing, the forces must be such that there is no acceleration of the body on which the forces are acting, nor any tendency toward rotation. Imagine how readily the giant ball would spin if all the forces and all the torques were not perfectly in balance (Figure 3.4*B*). **Torque** is the tendency of a force to rotate a body about some axis. It is the product of the force and the perpendicular distance (**moment arm**) from the line of action of the force to the axis of rotation. A force of one newton acting perpendicular to a moment arm one meter long creates a torque of

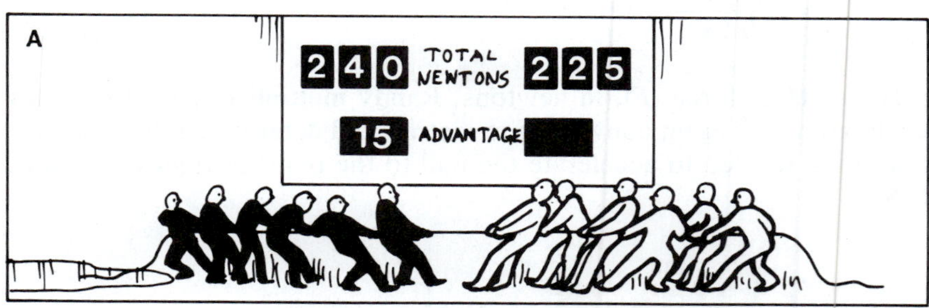

Figure 3.3 (*A*) It is nearly a standoff until (*B*) low coefficient of sliding friction, created by wet grass, swings the advantage to the team on the right. We wish there were more scoreboards calibrated in newtons so that *we* could better calibrate. (Artwork by D. A. Fischer.)

Figure 3.4 Have you played this game? (*A*) Perfectly balanced forces. No torque. (*B*) Enter torque. (*C*) Static equilibrium: forces and torques are perfectly balanced and nothing moves. (*D*) Dynamic equilibrium: Torques are balanced, but forces are not, and the movement is at constant velocity in a straight line. "Get out of the way!" (Artwork by D. A. Fischer.)

one meter-newton (1 m · N). For the instant of time illustrated in Figure 3.4*C*, both the torque and the forces are balanced, and this is why the giant ball is perfectly stationary, in a state of **static equilibrium**.

Static equilibrium requires translational and rotational equilibrium, such that the resultant external force on a body is zero, and the resultant external torques about any chosen axis are zero (Serway, 1990). Static equilibrium is the exception to the rule in the "giant ball" contest. The rule is a condition in which the torques and forces are out of balance and the game is out of control. The paid coaches cannot do the calculations quickly enough to gain an advantage. The game appears to be smooth only when the ball is in a state of **dynamic equilibrium**, moving at a constant velocity in a straight line without spinning (Figure 3.4*D*).

Forces in the Subsurface World

Thinking about forces in terms of fast-moving games is one thing. But to make the transition from the dynamics of motion in the everyday, action world of basketball and car racing, or avalanches and volcanic eruptions, to the steady, subsurface, slow-moving world of structural geology, we need to envision forces acting on and within bodies that are themselves fundamentally at rest in static equilibrium, or in a slow-motion state of dynamic equilibrium. We will most commonly be dealing with situations of a rock body essentially at rest, in which case all the forces tend to be balanced. When net forces are created that make things happen, they trigger accelerations that are generally of two kinds:

- Unimaginably slow acceleration of the larger geologic unit as a whole, such as a major tectonic plate that undergoes an increase in velocity from 6 cm/yr to 7 cm/yr over hundreds or thousands of years.
- Incredibly fast, short-lived accelerations of tiny parts of the larger body, such as a rock adjacent to a fault that shifts one meter during a major earthquake, from one position at rest to another position at rest during the briefest period of time (nanoseconds), achieving huge accelerations during the ride.

In the world of structural geology, forces commonly build slowly, until sooner or later the strength of the entire body, or some part of the body, is overcome, thus triggering internal adjustments involving translation, rotation, distortion, or dilation. Some of these adjustments are recoverable when the forces are removed. Other adjustments result in permanent damage that is nonrecoverable, creating a structural geologic record of the action of forces.

Types of Forces

There are two fundamental classes of forces that affect geologic bodies: body forces and contact forces. **Body forces** act on the mass of a body in a way that depends on the amount of material in the body but is independent of the forces created by adjacent surrounding materials (Means, 1976; Twiss and Moores, 1992). **Contact forces** are pushes or pulls across real or imaginary surfaces of contact, such as a fault between adjacent parts of a rock body (Means, 1976).

The most important body forces are gravitational forces and electromagnetic forces. The force of gravity is ultimately responsible for many geologic actions, such as the downhill flow of lava or glaciers, the fall of rock slides and avalanches, the downslope slumping of wet sediments, the vertical rise of low-density, buoyant magmas and salt domes, the settling of crystals within certain magmas, and plate tectonic forces including "ridge push" and "slab pull."

The body force called gravity can create structural deformation at a scale that is commonly large and visible. In contrast, electromagnetic forces are body forces whose structural geologic presence dwells in the submicroscopic realm. These are among the significant forces that must be overcome to produce deformation (Serway, 1990). Electromagnetic forces hold individual minerals intact, and thus hold rocks together (see Chapter 4). When the effect of contact forces overwhelms these electrostatic forces, the structure of the lattice can be disturbed through translation, rotation, dilation, and distortion on the submicroscopic scale. In this way, hard rocks can be forced to behave as if they were soft.

The loads that create contact forces can arise in any number of ways, some of which are secondary effects of body forces. Suppe (1985) identifies three main mechanisms of loading, each of which produces contact forces: gravitational loading, thermal loading, and displacement loading. **Gravitational loading** is an omnipresent mechanism in which the weight of the rock column generates forces at depth that impinge on each and every physical surface of every size and orientation within the crust. In **thermal loading**, the heating or cooling of rocks creates forces in rocks that, because of confinement, are not able to expand or contract. Imagine the complex pattern of forces generated by heating a confined mass of tightly interlocking rocks and minerals, each of which has a different capacity for thermal expansion or contraction. **Displacement loading** generates forces through large-scale mechanical disturbance of rocks. Examples include the collision of plates, regional bending and arching, and the shouldering aside of country rock by an intruding magma or an impacting asteroid.

Definition of Load

Load is a prominent word in the definition of contact force, and it presents the need to introduce yet a second way of describing the effect of a force. The first way, which we have learned, is to express the effect of a force in terms of how much acceleration the contact force will impart to the mass of a body in a given time. The second way is to describe the effect of the force in terms of how much weight it can support (Price and Cosgrove, 1990). The weight is called the load. Since force equals mass times acceleration, and weight equals mass times acceleration due to gravity, it follows that there should be a close relationship between force and weight. We sometimes miss that connection because the units in which load and weight are expressed are typically reported in shorthand. For example, a load or weight of "10,000 kg" actually means "10,000 kg · m/s^2," which is the same as 10,000 newtons.

Pumping iron in the local gym, the person in Figure 3.5 is demonstrating the relation between load and force. To support the load of the 2.3 kg mass, which weighs 22.7 kg · m/s^2 (50 lb), there is a need to exert a force of 22.7 newtons. Actually lifting the mass (i.e., giving it an upward acceleration) requires a force greater than 22.7 N.

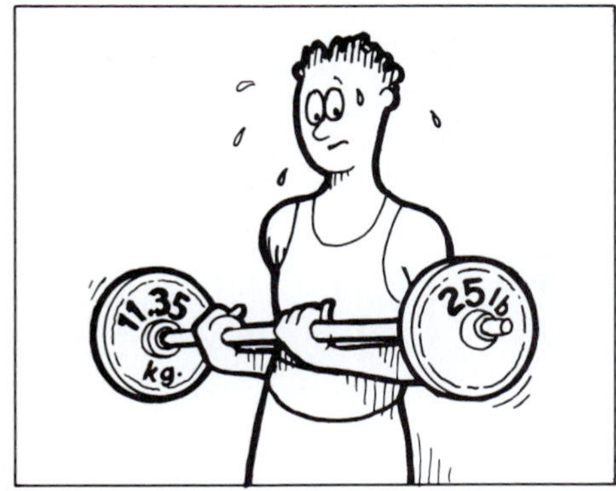

Figure 3.5 The barbell has created a load, measured in kg-m/s^2, which is being supported by a force, measured in newtons. Static equilibrium! (Artwork by D. A. Fischer.)

STRESS

Definition of Stress

Body forces, and contact forces created through loading, are ultimately responsible for geologic deformation. These forces work through the "undercover agent" known as **stress.**

We can think of stress as that which tends to deform a body of material (Jaeger and Cook, 1976). Stress will deform a body if the strength of the body is exceeded. The magnitude of **stress (σ)** is not simply a function of the force (F) from which it was derived, but it relates as well to the **area (A)** on which the force acts:

$$\sigma = \frac{F}{A} \tag{3.2}$$

Nongeologic examples of the relation of stress to force and area abound. Consider the strategy of rescuing a skater who has fallen through the ice in a frozen pond (Figure 3.6). The skater broke through thin ice because the strength of the ice was exceeded by high stress values produced by the concentration of the load of the 170 lb (77 kg) skater onto the thin blades of the skates. Assuming that the 170 lb load was distributed uniformly along just one of the blades, the area of the contact between ice and blade was only about 2 in.^{2} (5.1 cm^2) (Figure 3.6*A*). The stress level produced at the

A

B

C

Figure 3.6 The stress of skating. (Artwork by D. A. Fischer.)

ice–blade interface must have been

$$\sigma = \frac{F}{A} = \frac{170\text{ lb}}{2\text{ in.}^2} = 85\text{ lb/in.}^2 = 6\text{ kg/cm}^2 \text{ (see Figure 3.6}B\text{)}$$

In contrast, the 170 lb rescuer, informed in ways of dynamic analysis, moves in close to the victim while lying on a 10 × 1.5 ft (3 × 0.45 m) plank (Figure 3.6*C*). The total load of rescuer and plank is about 180 lb (82 kg). The area of the plank is 2160 in.^{2} (13,935 cm^2). The load of the rescuer is distributed in such a way that stress concentration at any point under the plank is uniformly small, and well below the breaking strength of the ice.

$$\sigma = \frac{F}{A}$$

$$\sigma = \frac{180\text{ lb}}{2160\text{ in.}^2}$$

$$\sigma = 0.08\text{ lb/in.}^2 = 0.006\text{ kg/cm}^2$$

Units of Stress

There are several ways to express the magnitude of stress. When describing stress levels in the Earth's crust or in specimens subjected to deformation in the laboratory, the preferred unit of measure is the **pascal (Pa):**

> *A stress of one pascal is created by the force of one newton acting on an area of one square meter.*

Because of the small magnitude of a single pascal in comparison to the greater magnitude of stresses in the Earth, we commonly precede the term pascal with the prefix kilo-, mega-, or giga-, where:

one **kilopascal (kPa)** = 1000 pascals (10^3 Pa)
one **megapascal (MPa)** = 1,000,000 pascals (10^6 Pa)
one **gigapascal (GPa)** = 1,000,000,000 pascals (10^9 Pa)

Stress is also commonly expressed in units of load per area, for example **pounds per square inch (psi)** or **kilograms per square centimeter (kg/cm^2)**. Conversions among the common stress units are presented in Table 3.1.

Museum-Piece Stress Calculation

Imagine a large block of granite resting on a marble column (Figure 3.7). The weight of the granite block constitutes a gravitational load that imposes a stress on the marble column. The magnitude of the stress is found by dividing the force created by the load of the granite block by the cross-sectional area of the top of the marble column.

The force created by the granite block is the product of mass (m) times acceleration (g) due to gravity. Mass is volume (V) times density (ρ). In this case:

$$\text{volume } (V) = \text{width } (W) \times \text{breadth } (B) \times \text{height } (H) = 2\text{ m} \times 2\text{ m} \times 2\text{ m}$$

$$V = 8\ m^3$$

$$\text{density } (\rho) = 2.7\text{ g/cm}^3 = 2700\text{ kg/m}^3$$

$$\text{mass } (m) = V\rho = 8\text{ m}^3 \times 2700\text{ kg/m}^3 = 21{,}600\text{ kg}$$

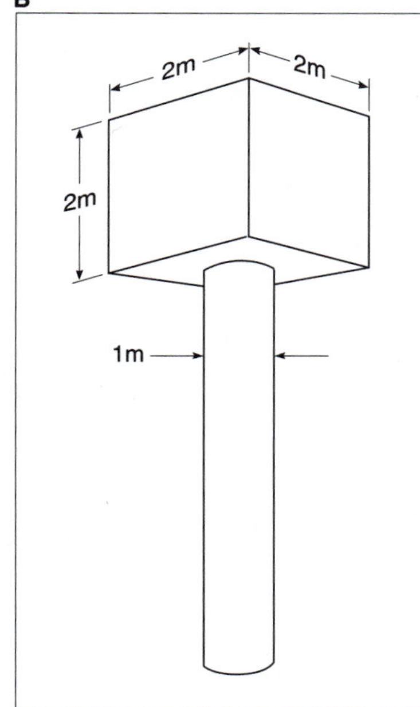

Figure 3.7 (*A*) This piece of sculpture of granite on marble creates an opportunity for calculating the force exerted by the granite block, and the stress generated by it on the top of the marble column. (*B*) Force is determined by measuring the volume of the granite block and multiplying it by its density and by the acceleration due to gravity. The stress is determined by dividing the force by the cross-sectional area of the marble column.

Acceleration due to gravity (g) is given as follows:

$$\text{acceleration } (g) = 9.8 \text{ m/s}^2$$

and force, as just stated, is computed by multiplying mass (m) times gravity (g):

$$\text{force } (F) = \text{mass } (m) \times \text{acceleration } (g) = mg$$

$$F = 21{,}600 \text{ kg} \times 9.8 \text{ m/s}^2 = 211{,}680 \text{ kg} \cdot \text{m/s}^2 = 211{,}680 \text{ N}$$

The stress, represented by the Greek letter sigma (σ), created by the load of the granite block on the marble column is force divided by area:

$$\text{stress } (\sigma) = \frac{\text{force } (F)}{\text{area } (A)}$$

$$\text{area } (A) = \pi r^2 = 3.14 \times (0.5 \text{ m})^2 = 0.79 \text{ m}^2$$

$$\sigma = \frac{F}{A} = \frac{211{,}680 \text{ N}}{0.79 \text{ m}^2} = \frac{267{,}949 \text{ N}}{\text{m}^2} = 267{,}949 \text{ Pa} = 268 \text{ kPa}$$

The stress produced by the granite block tends to deform the marble column. The marble column will permanently deform when its capacity to respond elastically, like a spring, is exceeded.

Calculating Stress Underground

In a manner quite similar to the museum-piece example, we can calculate the stress created by the weight of a *very* large cube of granite in the upper crust (Means, 1976, pp. 112–114). Let us picture a region of the Earth where the upper several kilometers of the crust are entirely composed of granite. Then let us calculate, for a given depth level, the stress created by the load of the granite. We will choose -1000 m as the depth level of interest to us. To set up the calculation, it is helpful to visualize the -1000 m depth level as overlain by a giant cube of granite 1000 m on a side (Figure 3.8). Our goal is to compute the stress level at the base of the block.

The force generated by the weight of the block is determined by multiplying the volume of the block (V = 1000 m × 1000 m × 1000 m) times the density of the granite (ρ = 2700 kg/m^3) times the acceleration due to gravity (g = 9.8 m/s^2) (Figure 3.8):

$$F = 1000\text{ m} \times 1000\text{ m} \times 1000\text{ m} \times 2700\text{ kg/m}^3 \times 9.8\text{ m/s}^2$$

The stress (σ) created by the weight of the block of granite acting on the base of the block is determined by dividing the force (F) by the area (A = 1000 m × 1000 m):

$$\sigma = \frac{F}{A} = \frac{1000\text{ m} \times 1000\text{ m} \times 1000\text{ m} \times 2700\text{ kg/m}^3 \times 9.8\text{ m/s}^2}{1000\text{ m} \times 1000\text{ m}}$$

$$\sigma = 1000\text{ m} \times 2700\text{ kg/m}^3 \times 9.8\text{ m/s}^2 = 26{,}460{,}000\text{ Pa} = 26.5\text{ MPa}$$

This tells us the **lithostatic stress gradient** at depth. Lithostatic stress increases 26.5 MPa/km, which is equivalent to 265 bars or 0.265 kbar/km. There is another way to say it: for each 3.8 km depth, lithostatic stress increases by 1 kbar or 100 MPa.

There is a shortcut to determining the stress at any given depth in the granite. Stress (σ) is simply the product of density (ρ) times gravity (g) times depth (h) (Figure 3.8):

$$\sigma = \rho g h = 2700\text{ kg/m}^3 \times 9.8\text{ m/s}^2 \times 1000\text{ m} = 26.5\text{ MPa}$$

The calculated stress level of 26.5 MPa is very similar to direct "in situ" measurements of stress in deep mines at depth levels in the range of 1000 m. For example, Suppe (1985) discusses the work of Bjorn (1970), who reported the results of direct in situ stress measurements in mines

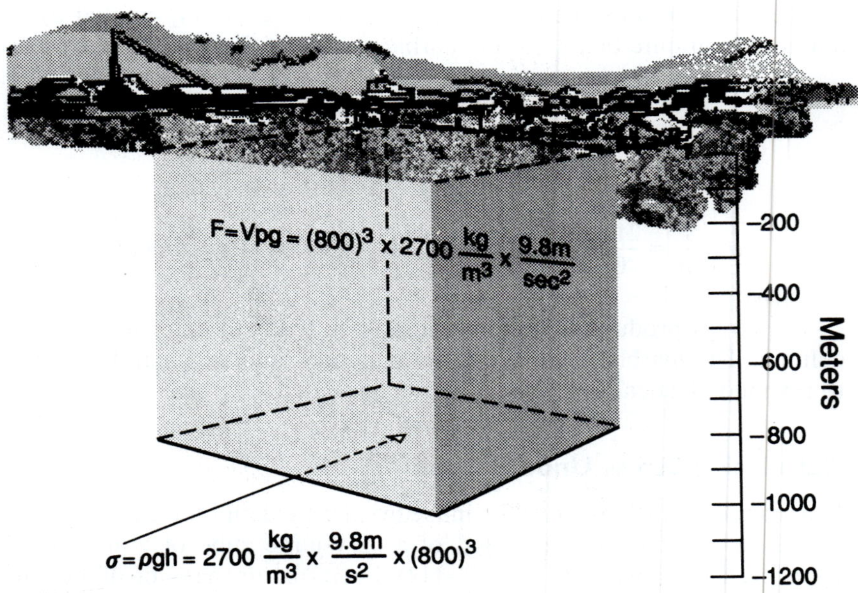

Figure 3.8 We can apply our museum-piece calculation to a huge block of granite nestled in the crust of the Earth.

in Norway, including a measurement of approximately 23 MPa at a depth level of approximately 815 m. This is equivalent to 28 MPa at a depth level of approximately 1 km.

A Fuller Definition of Stress

What we are calling "stress" is more accurately called "traction." Indeed, **traction** is force per unit area acting on a surface. Like force, it is a vector, having both magnitude and orientation. The more accurate and encompassing definition of "stress" elevates stress beyond traction, beyond being a mere vector, to an entity that cannot be described by a single pair of measurements (i.e., magnitude and orientation). "Stress," strictly speaking, refers to the whole collection of stresses (i.e., tractions) acting on each and every plane of every conceivable orientation passing through a single tiny discrete point (*P*) in a body at a given instant of time.

When we try to picture traction, we should envision a single vector arrow stuck in a target at some oblique or perpendicular angle (Figure 3.9). The "target" is the surface on which the traction acts. The exposed arrow length is proportional to the magnitude of the traction. The angle at which the arrow hits the target is the orientation of the traction. When we try to picture stress, we should envision a spherical target hit by countless vector arrows oriented in every conceivable direction. Each different "traction-vector arrow" is stuck in its own uniquely oriented plane within the target (Figure 3.9). The "archer" is uncanny, for the collection of vector arrows comprising any given stress has the shape of a perfect sphere or a perfect ellipsoid. The variation in magnitude of tractions is very orderly.

Throughout this text we will continue to use "stress" when referring to "traction," because this usage is so ingrained in the geological literature. But we will use the expression **"stress tensor"** when referring to the whole collection of tractions at a given point.

Stress and Stress Tensor Analysis

We want to learn how to calculate the stress (i.e., the traction) on individual planes of all possible orientations at a given point (*P*) within a body of rock that is subjected to forces. On the basis of this information, we will describe fully the stress tensor (i.e., the whole collection of tractions) at *P*. If we can define the stress tensor at each and every point within the body, we can fully describe the **stress field**, which is the entire collection of stress tensors. Description of the stress field becomes an integral part of the foundation for evaluating the relationship of stress to strain.

Figure 3.9 A single arrow stuck in a target is like a traction acting on a plane. A traction is usually oblique to the plane on which it acts, though in special cases it may be perpendicular. Tractions are vectors, and the length of the arrow is proportional to magnitude. A large array of arrows (tractions) clustered around a point is like stress. Stress is defined by the field of tractions. The tractions comprising stress are very well organized, giving rise to ellipsoidal geometric images. (Artwork by D. A. Fischer.)

Setting up an Example of Stress Analysis

We now go underground again, at least in our imaginations, to a tiny point (*P*) deep in granite, located 1500 m beneath the surface (Figure 3.10). The region of crust within which point *P* resides is tectonically active and is experiencing a modest east–west horizontal compression. Thus, in addition to the vertical lithostatic compressive stress felt at point *P* due to the load imposed by the overlying granite, there is a small horizontal externally imposed compression of tectonic origin.

Now let us picture within the interior of the granite at point *P* an infinite number of imaginary planes, or real physical planes, of all orientations. The real physical planes include preexisting hairline fractures, cleavage surfaces in minerals, crystal faces, grain boundaries, and veins. Across each one of these planes "there will exist a field of forces equivalent to the loads exerted by material on one side of the plane on the other" (Ragan, 1985, pp. 111–112). Focusing from point to point on any one of these planes, we would discover that the forces are not uniform in either direction or magnitude. Yet, if we were to examine the magnitude and direction of force acting on a smaller and smaller surface area of one of the planes cutting through point *P*, we would discover that the force per unit area (i.e., the stress) would approach a fixed value, . . . fixed in both magnitude and direction.

The Stress Calculations

Let us go through the steps in calculating the stresses at point *P*. Compressional stress operating vertically is 40 MPa, and compressional stress operating horizontally is 20 MPa (Figure 3.11*A*). We will label the vertical stress as σ_z and the horizontal stress as σ_x. They operate parallel to the *z*- and *x*-axes, respectively, of a Cartesian coordinate system about which point *P* is arbitrarily suspended.

Let us first calculate the stress on the plane, passing through point *O*, that makes an angle (θ) of 65° relative to the *z*-axis (Figure 3.11*A*). To calculate the stress (σ) on this plane, consider the plane to be part of a small wedge of rock that is triangular in cross section (Figure 3.11*B*). The dimensions of this wedge of rock are very, very small, on the order of 10^{-12} m^2.

The hypotenuse (*XZ*) of our rock triangle (Figure 3.11*B*) represents the trace of the plane for which we plan to calculate stress. Its length is "*n*" millionths of a meter. The legs of the triangle represent directions (*OX*

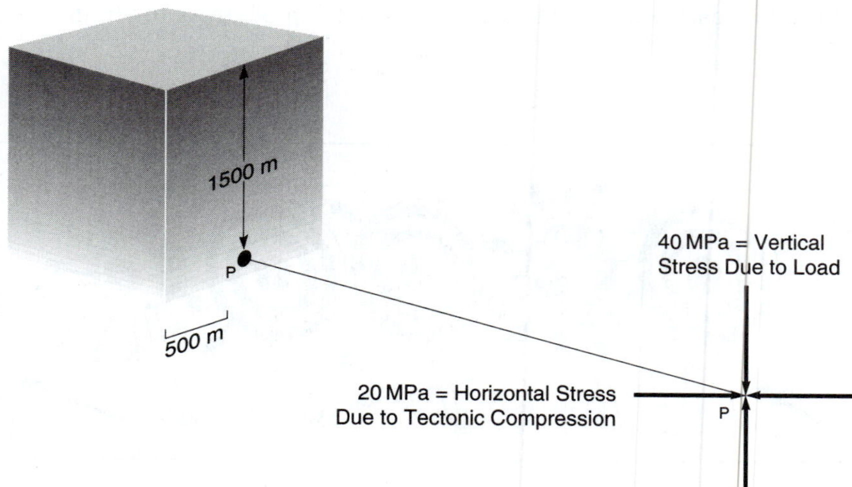

Figure 3.10 We picture a point (*P*) deep down in granite, which for this example we assume to be an ideal elastic body of rock. The stresses acting at *P* are generated by a combination of gravitational loading, and tectonic loading.

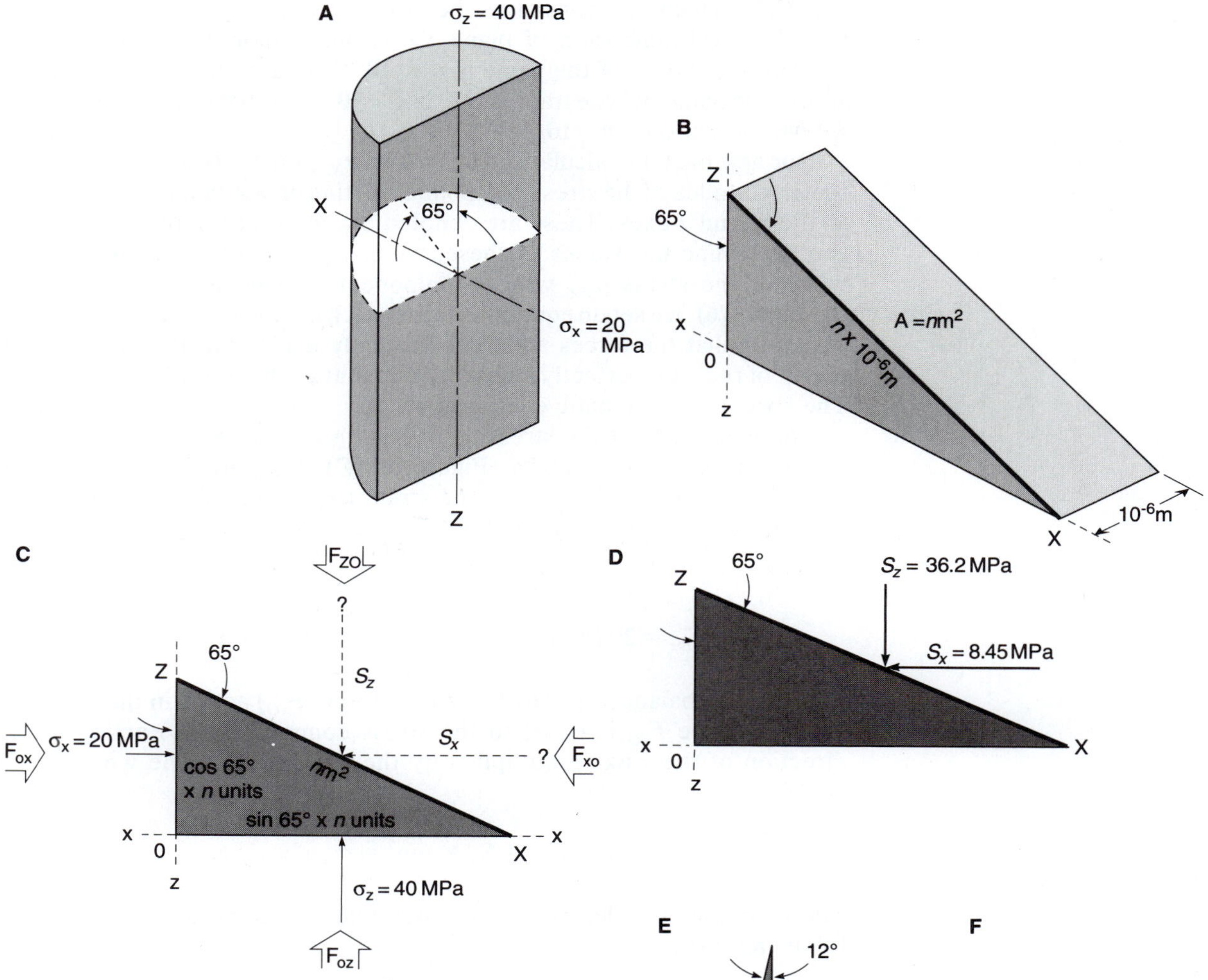

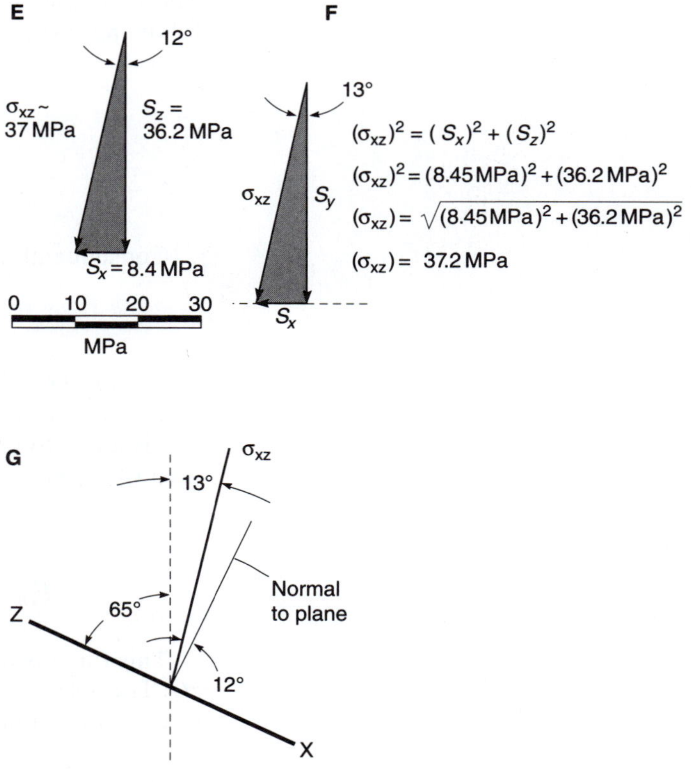

$$(\sigma_{xz})^2 = (S_x)^2 + (S_z)^2$$

$$(\sigma_{xz})^2 = (8.45\,\text{MPa})^2 + (36.2\,\text{MPa})^2$$

$$(\sigma_{xz}) = \sqrt{(8.45\,\text{MPa})^2 + (36.2\,\text{MPa})^2}$$

$$(\sigma_{xz}) = 37.2\,\text{MPa}$$

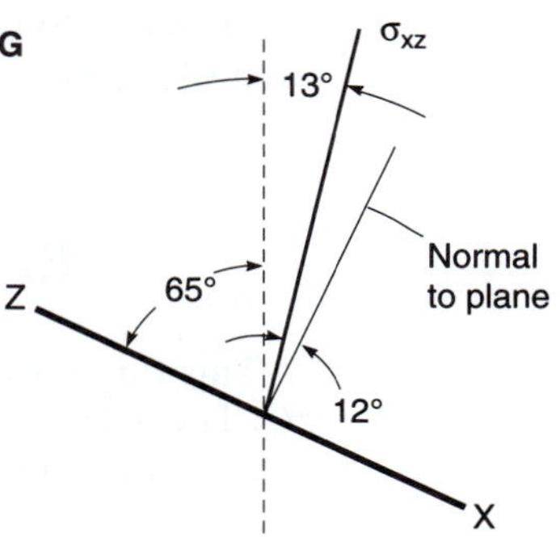

Figure 3.11 Computing stresses on a plane at point *P* deep in granite. (*A*) A look inside the granite at a plane that is inclined 65° to σ_z. (*B*) Block diagram view of the plane and the parcel of rock of which it is a part. (*C*) The balance of forces on the parcel of rock. (*D*) The calculated values of stresses acting parallel to *x* and *z*. (*E*) Graphical determination of the value of the stress vector (σ) acting on the plane. (*F*) Calculation of the stress vector using the Pythagorean theorem. (*G*) Geometric relation of the stress vector to the plane.

and OZ), which are parallel to the stresses σ_x and σ_z. We will assume that the third dimension of plane XZ is one-millionth of a meter long. Therefore the area of the plane is $n \times 10^{-12}$ square meters $\times\ 10^{-12}$. The area of the plane, whose trace is OZ, is $n \times 10^{-12} \cos 65° \text{ m}^2$. For simplicity, we will now drop out "10^{-12}."

Our approach in calculating the stress (σ_{XZ}) on XZ is to first determine the magnitudes of the stress components acting on XZ that operate parallel to the x- and z-axes. These are denoted σ_X and σ_Z (Figure 3.11C). If we can determine the values of these, we can proceed to calculate the full value of the stress σ_{XZ} whose components are σ_X and σ_Z. Following Means (1976), we set up equations from which σ_X and σ_Z can be calculated, assuming that the forces acting horizontally and vertically on this small chunk of rock are perfectly balanced, and that all the torques are balanced. The specimen is in static equilibrium.

The magnitude of the stress σ_x operating on OZ can be calculated by balancing forces horizontally (Figure 3.11C). The force (F_{Ox}), operating in the direction Ox, is equal to the stress (σ_x) acting in the direction Ox multiplied by the area (A_{OZ}) of O_Z.

$$F_{Ox} = \sigma A_{OZ}$$

$$F_{Ox} = 20 \text{ MPa} \times \cos 65° \times n\text{m}^2 = 20n \times \cos 65° \text{ N}$$

This force is balanced by the horizontal force (F_{xO}) acting in the direction xO. The force F_{xO} is equal to the stress component (S_X) acting in the direction of the x-axis, multiplied by the area of the plane whose trace is XZ.

$$F_{xO} = S_X \times n\text{m}^2 = \sigma_X n \text{ N}$$

The equation that describes the horizontal balancing of forces has the following form:

$$F_{Ox} = F_{xO}$$

$$20 \text{ MPa } (\cos 65°)\,(n\text{m}^2) = S_X(n\text{m}^2)$$

$$20 \text{ MPa } (0.42) = S_X$$

Consequently,

$$S_X = 8.4 \text{ MPa}$$

Similarly, the value of σ_Z operating on XY can be calculated by balancing forces vertically (Figure 3.11C). The force (F_{Oz}), operating in the direction Oz, is equal to the stress (σ_z) acting in the direction Oz multiplied by the area (A_{OX}) of the plane whose trace is OX.

$$F_{Oz} = \sigma_z \times A_{OX}$$

$$F_{Oz} = 40 \text{ MPa} \times \sin 65° \times n\text{m}^2 = 40n \times \sin 65° \text{ N}$$

This force is balanced by the vertical force (F_{zO}) acting in the direction zO. The force F_{zO} is equal to the stress component S_Z acting in the direction of the z-axis, multiplied by the area of the plane whose trace is XZ.

$$F_{zO} = S_Z \times n\text{m}^2$$

Therefore, the equation that describes the vertical balancing of forces has the following form:

$$F_{Oz} = F_{zO}$$

$$40 \text{ MPa} (\sin 65°) (nm^2) = S_Z(nm^2)$$

$$40 \text{ MPa} (0.90) = S_Z$$

Consequently,

$$S_Z = 36 \text{ MPa}$$

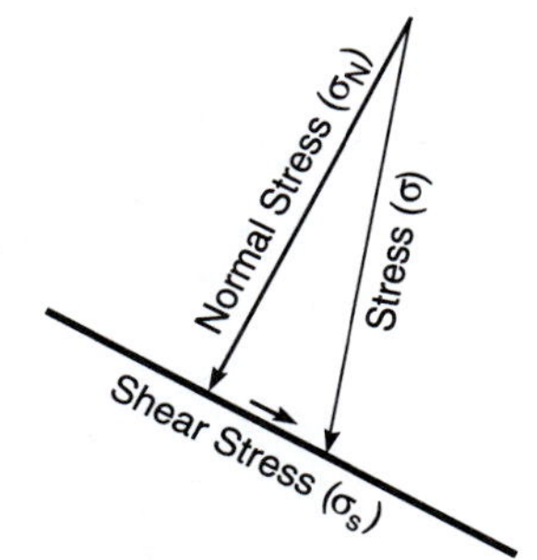

Figure 3.12 A stress (traction) can be resolved into normal (σ_N) and shear stress (σ_S) components.

Having calculated the values of S_X (8.4 MPa) and S_Z (36 MPa) (Figure 3.11*D*), let us determine the stress (σ_{XZ}) for which these are components. We can do this graphically or numerically. Graphically, we map the vectors S_X and S_Z as shown in Figure 3.11*E*. Then we construct the stress (σ_{XZ}) from the tail of S_Z to the tip of S_X. Measuring its length according to the scale of the figure, we find that σ_{XZ} has a value of about 37 MPa. The stress makes an angle θ of approximately 12° with the *z*-axis. It is thus inclined at approximately 78°.

A more accurate way to determine the magnitude and orientation of the stress is through the use of the Pythagorean theorem (Figure 3.11*F*).

$$(\sigma_{XZ})^2 = (S_X)^2 + (S_Z)^2$$

$$(\sigma_{XZ})^2 = (8.4 \text{ MPa})^2 + (36.2 \text{ MPa})^2$$

$$(\sigma_{XZ})^2 = 71 + 1310 = 1381 \text{ MPa}$$

$$\sigma_{XZ} = 37 \text{ MPa}$$

The angle that the stress makes with respect to *z* can be found trigonometrically.

$$\sin \beta = \frac{S_X}{\sigma_{XZ}} = \frac{8.45 \text{ MPa}}{37 \text{ MPa}} = 0.23$$

$$\beta = \arcsin 0.23 = 13°$$

Resolving Normal Stress and Shear Stress

The path to the full description of the stress tensor takes us through steps that include subdividing stress (i.e., traction) into components. As emphasized by Means (1976), stresses are not generally perpendicular to the plane for which they have been calculated. For example, the stress (σ_{XZ}) we just calculated is not orthogonal to plane *XZ*, but instead departs from 90° by 12° (Figure 3.11*G*). Thus, in almost all cases, a given stress can be resolved into two components: one perpendicular to the plane for which the stress has been calculated, the other parallel to the plane.

The component perpendicular to the plane is the **normal stress (σ_N)**, and the component parallel to the plane (Figure 3.12) is the **shear stress (σ_S)**. Normal compressive stress tends to inhibit sliding along a plane; shear stress tends to promote sliding. *Normal stresses are considered to be positive if they are compressive (i.e., directed inward), negative if they are tensile (i.e., directed outward)* (Jaeger and Cook, 1976) (Figure 3.13). Shear stresses are labeled positive or negative on the basis of their sense of shear. *Right-handed shear stresses are here considered to be positive, whereas left-handed shear stresses are negative* (see Figure 3.13).

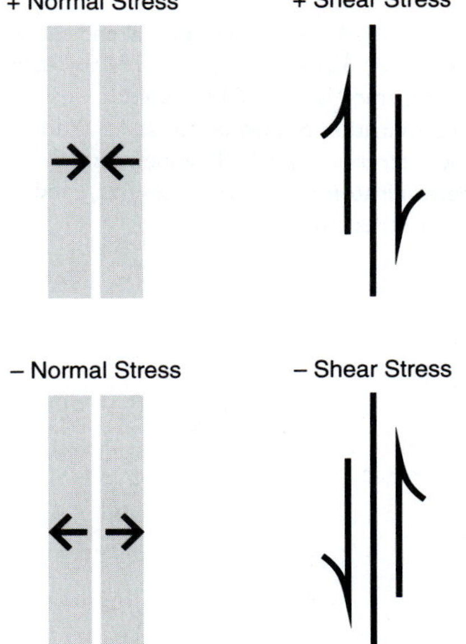

Figure 3.13 Sign conventions are necessary in describing the different kinds of stresses. Here is the convention that we use. Normal stress is positive (+) if compressive, negative (−) if tensile. Shear stress is positive (+) if right-handed, negative (−) if left-handed.

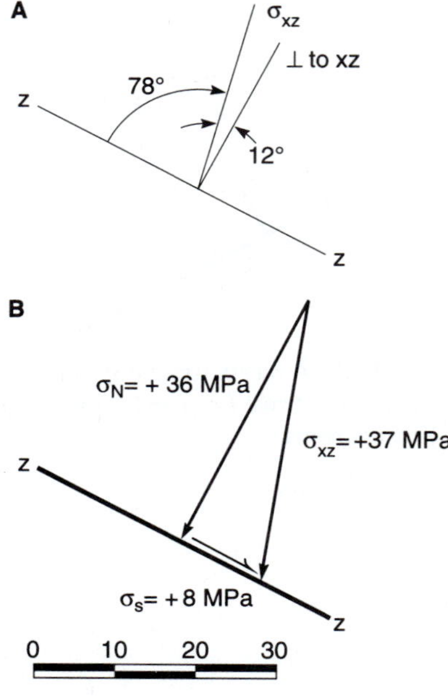

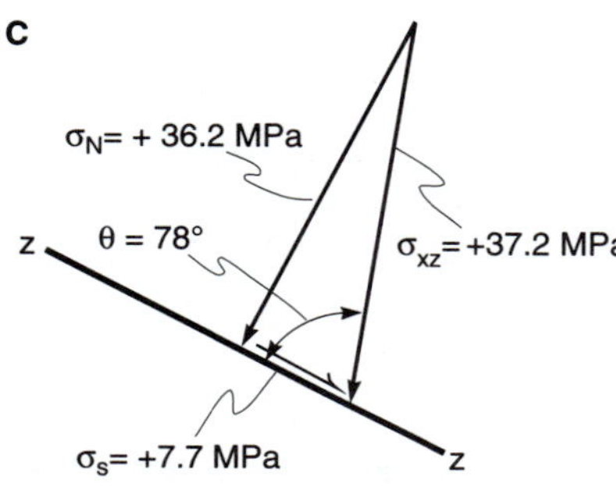

Figure 3.14 Resolving normal and shear stress. (*A*) Stress is inclined +78° to plane whose trace is *XZ*. (*B*) Graphical determination of normal stress (σ_N) and shear stress (σ_S). (*C*) Trigonometric determinaton of normal stress (σ_N) and shear stress (σ_S).

Resolving a stress into normal and shear stress components is actually quite straightforward. It can be done graphically or numerically (Figure 3.14). The graphical construction involves "mapping" the orientation and length of the stress relative to the trace of the plane on which it acts. We start with the orientation of the stress (σ_{XZ}) relative to the plane (*XZ*) on which it acts (Figure 3.14*A*). **The acute angle (θ) between σ_{XZ} and plane *XZ* is approximately +78°. The angle θ is positive when measured clockwise from the plane to the stress direction and negative when measured counterclockwise from the plane to the stress direction.**

Next we show the magnitude of the stress (37.2 MPa) by adding a "map" scale to the diagram (Figure 3.14*B*). Then, from the tail of the arrow representing the stress (σ_{XZ}), the normal stress component (σ_N) is constructed perpendicular to the plane (see Figure 3.14*B*). Its magnitude (+36 MPa) can be read directly using the scale of the drawing. The shear stress component (σ_S) is constructed as a vector drawn from the tip of the normal stress component (σ_N) to the tip of σ_{XZ}, and thus parallel to the plane. Its value (+8 MPa; i.e., right-handed) also can be directly read as well (see Figure 3.14*B*).

The numerical solution for computing the magnitudes of the normal and shear stress components is trigonometric (Figure 3.14*C*). If $\theta = +78°$ is the angle between the plane *XZ* and the stress (σ_{xz}) that acts on it, normal stress (σ_N) can be computed as follows (Figure 3.14*C*):

$$\sigma_N = \sigma_{XZ} \sin\theta = 37\text{ MPa}\,(\sin 78°) = 36\text{ MPa}$$

The magnitude of the shear stress (σ_S) can be determined in a comparable fashion (Figure 3.14*C*):

$$\sigma = \sigma_{XZ} \cos\theta = 37\text{ MPa}\,(\cos 78°) = 7.7\text{ MPa}$$

Computing the Stress Tensor

Stresses, normal stresses, and shear stresses can be calculated for planes of all orientations that pass through point *P*. For example, the magnitude of the stress to a plane that makes an angle (θ) of 25° with the *z*-axis is +24.6 MPa (Figure 3.15*A*). The stress is inclined +55° to the plane and can be resolved into normal stress (σ_N) and shear stress (σ_S) components of +23.6 and +7.6 MPa, respectively (Figure 3.15*B*).

Stresses to planes inclined at $\theta = 0°$ and $\theta = 90°$ are special cases. The plane marked by an orientation of $\theta = 90°$ is horizontal, parallel to *x* (Figure 3.16*A*). The stress calculated for this plane is perpendicular to it and has the same magnitude as σ_z, namely +40 MPa. This stress is purely a normal stress and has no shear stress component. Similarly, the stress to a plane whose orientation is defined by $\theta = 0°$ (in this example, a vertical plane) is a normal stress of 20 MPa (Figure 3.16*B*).

Using a calculator, or better yet a computer spreadsheet that summarizes the mathematics of these calculations, we can determine the orientations and absolute values of stresses, normal stresses, and shear stresses for a host of planes inclined at 5° intervals through point *O*. We have done this for stress conditions of $\sigma_z = 40$ MPa and $\sigma_x = 20$ MPa (Figure 3.17). It is clear that normal stress and shear stress magnitudes vary systematically as a function of orientation: shear stress is zero for planes oriented parallel to the *x*- and *z*-axes ($\theta = 0°$ and 90°, respectively). Shear stress (σ_S) steadily increases from 0.0 MPa to a maximum of 10 MPa as

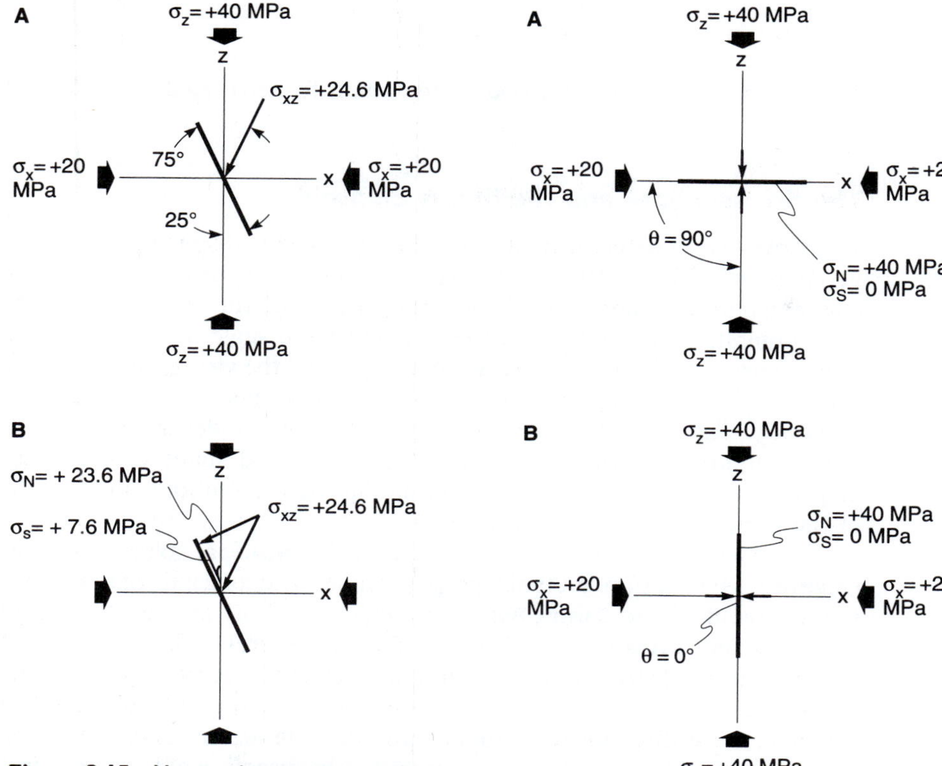

Figure 3.15 Yet another example of (*A*) stress on a plane and (*B*) resolution of the stress into normal and shear stress components.

Figure 3.16 The special case of stresses parallel and perpendicular to planes. Purely normal stresses result. Shear stresses are zero.

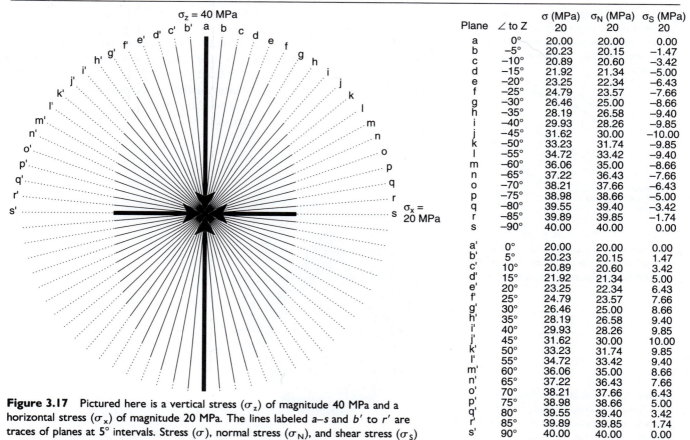

Plane	∠ to Z	σ (MPa) 20	σ_N (MPa) 20	σ_S (MPa) 20
a	0°	20.00	20.00	0.00
b	−5°	20.23	20.15	−1.47
c	−10°	20.89	20.60	−3.42
d	−15°	21.92	21.34	−5.00
e	−20°	23.25	22.34	−6.43
f	−25°	24.79	23.57	−7.66
g	−30°	26.46	25.00	−8.66
h	−35°	28.19	26.58	−9.40
i	−40°	29.93	28.26	−9.85
j	−45°	31.62	30.00	−10.00
k	−50°	33.23	31.74	−9.85
l	−55°	34.72	33.42	−9.40
m	−60°	36.06	35.00	−8.66
n	−65°	37.22	36.43	−7.66
o	−70°	38.21	37.66	−6.43
p	−75°	38.98	38.66	−5.00
q	−80°	39.55	39.40	−3.42
r	−85°	39.89	39.85	−1.74
s	−90°	40.00	40.00	0.00
a'	0°	20.00	20.00	0.00
b'	5°	20.23	20.15	1.47
c'	10°	20.89	20.60	3.42
d'	15°	21.92	21.34	5.00
e'	20°	23.25	22.34	6.43
f'	25°	24.79	23.57	7.66
g'	30°	26.46	25.00	8.66
h'	35°	28.19	26.58	9.40
i'	40°	29.93	28.26	9.85
j'	45°	31.62	30.00	10.00
k'	50°	33.23	31.74	9.85
l'	55°	34.72	33.42	9.40
m'	60°	36.06	35.00	8.66
n'	65°	37.22	36.43	7.66
o'	70°	38.21	37.66	6.43
p'	75°	38.98	38.66	5.00
q'	80°	39.55	39.40	3.42
r'	85°	39.89	39.85	1.74
s'	90°	40.00	40.00	0.00

Figure 3.17 Pictured here is a vertical stress (σ_z) of magnitude 40 MPa and a horizontal stress (σ_x) of magnitude 20 MPa. The lines labeled *a–s* and *b'* to *r'* are traces of planes at 5° intervals. Stress (σ), normal stress (σ_N), and shear stress (σ_S) have been calculated for each of the planes to show the systematic variations.

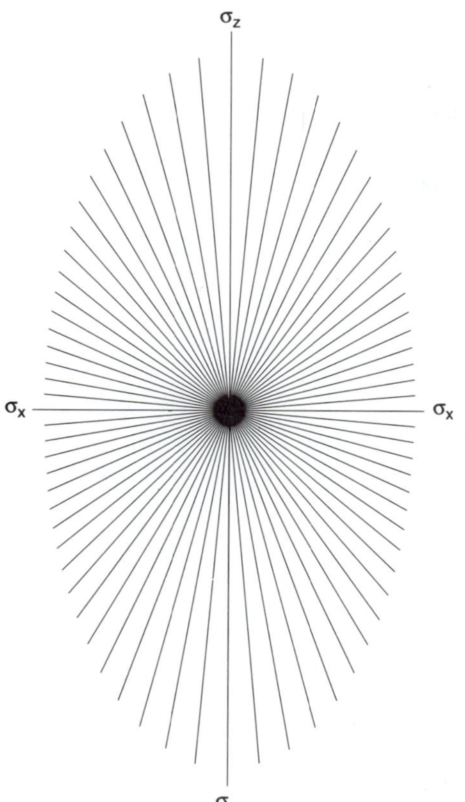

Figure 3.18 Stress ellipse created by arranging all of the stresses (σ) calculated in the preceding figure and arranging them in such a way that their tips meet at a point. Schematic.

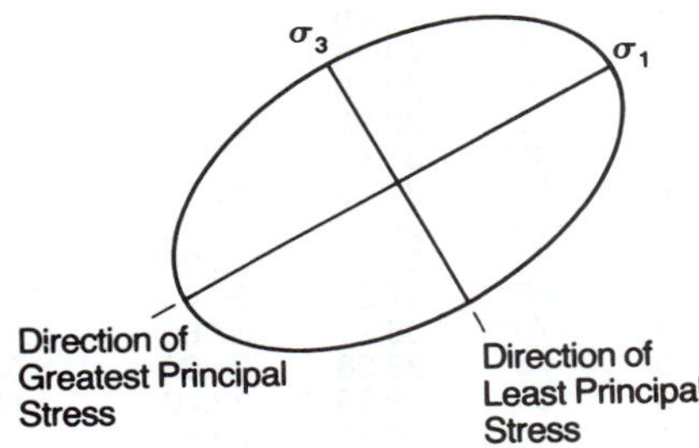

Figure 3.19 The stress ellipse, an image that shows the directions and ratios of the greatest and least principal stresses. It is a two-dimensional portrayal of a stress tensor.

θ increases from 0° to 45°, decreasing again to zero from θ = 45° and θ = 90°.

The Stress Ellipse and the Stress Ellipsoid

The stress data posted in Figure 3.17 are incredibly systematic. This can be appreciated by plotting all the stresses such that their tails meet at a common point, the tiny point O containing the planes for which the stresses were computed. When the stresses are plotted to scale in this fashion, a stress tensor of elliptical form is generated, called the **stress ellipse** (Figure 3.18). The ellipse graphically portrays the stress tensor at a single point in a body. The stress tensor is not a single vector. To describe the stress tensor requires describing the orientation, size, and shape of the stress ellipse. We accomplish this by determining the orientations and lengths of the principal axes of the stress ellipse.

The axes of the stress ellipse are called the **principal stress directions** (Figure 3.19). They are mutually perpendicular. The long axis of the ellipse is the **axis of greatest principal stress (σ_1)**; the short axis is the **axis of least principal stress (σ_3)**. These axes define the stress ellipse. Of all the values of normal stress (σ_N) computed for stresses operating about some point, the maximum normal stress component is parallel to the greatest principal stress direction (σ_1). Indeed this stress is strictly a normal stress. It has no shear stress component. Similarly, the direction of least principal stress (σ_3) is marked by a finite normal stress, by a zero shear stress. The stress operating parallel to σ_3 is the smallest of all the normal stress values. The form (i.e., aspect ratio) of the stress ellipse, of course, changes according to the absolute and relative values of the principal stresses. The ratio of the greatest to least principal stress values will reflect the local stress state.

In *three-dimensional* dynamic analysis, it is the **stress ellipsoid** that provides the description of the stress tensor at a point (Figure 3.20). In addition to the axes of greatest and least principal stress, the stress ellipsoid is characterized by an **axis of intermediate principal stress (σ_2)**, which is oriented perpendicular to the plane of σ_1 and σ_3. Like σ_1 and σ_3, axis σ_2 is a direction of zero shear stress.

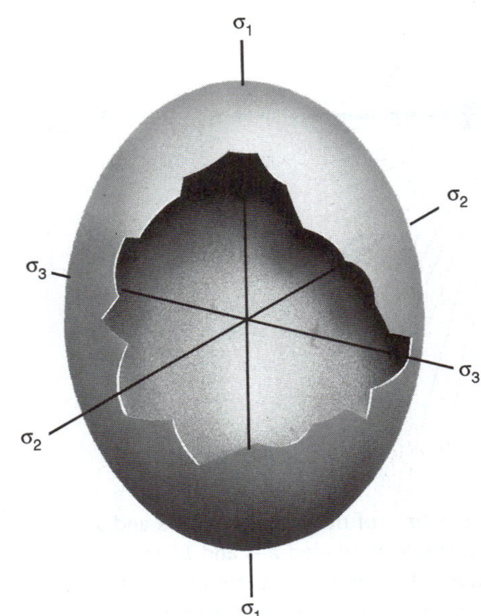

Figure 3.20 The stress ellipsoid, an image that shows the directions and ratios of the greatest, intermediate, and least principal stresses. It is the full three-dimensional portrayal of a stress tensor.

The Special Case of Hydrostatic Stress

A very special stress field, called **hydrostatic**, is characterized by equal stress magnitudes in *all* directions. The next time you go off the high dive, pause at the bottom of the pool for a few seconds to think about hydrostatic stress. You will feel its pressure, a pressure that tends to dilate (negatively in this case) but not distort. When we calculate stresses about a point within a hydrostatic stress field, we confirm that all the stresses have the same value. Moreover, we find that each stress is oriented perpendicular to the plane for which it was calculated. This is the nature of a hydrostatic stress field: no shear stress is generated, and the normal stress components have equal magnitudes in all directions.

By way of example, let us go underground again and envision a location in the shallow crust in granite where the vertical stress and the horizontal stress happen to be exactly equal to one another, namely 12 MPa (Figure 3.21*A*). For this special point (P' in Figure 3.21*A*), let us proceed to determine the stress (σ) on a plane inclined at $\theta = +48°$ (Figure 3.21*A*). By the balancing of forces we find that $S_X = 8.02$ MPa and $S_Z = 8.91$ MPa (Figure 3.21*B*). When we compute the stress (σ) for which S_X and S_Z are components, we find that $\sigma = 12$ MPa and is oriented perpendicular to the plane (Figure 3.21*C*). For this stress, $\sigma_N = 12$ MPa and $\sigma_S = 0$ MPa. When, as before, we calculate stresses for planes at intervals of 5° (Figure 3.22), we discover that each combination of σ_X and σ_Z values always yields a 12 MPa stress and is always oriented exactly perpendicular to the plane for which it is calculated. It is no wonder that hydrostatic stresses cannot distort rocks; there are no shear stresses in hydrostatic stress fields.

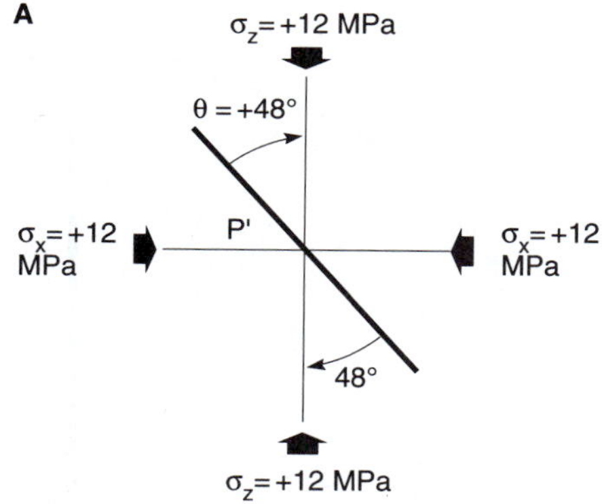

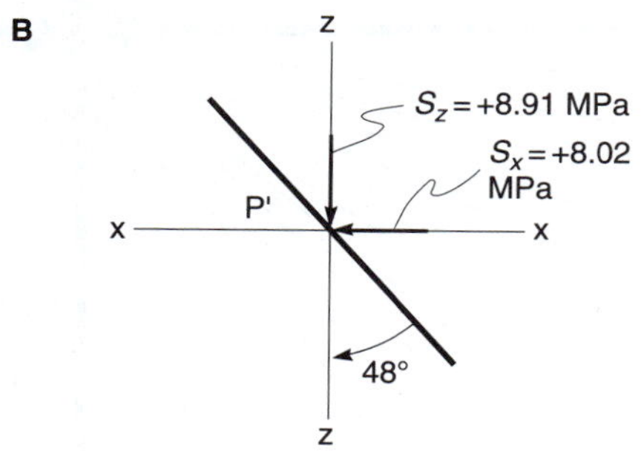

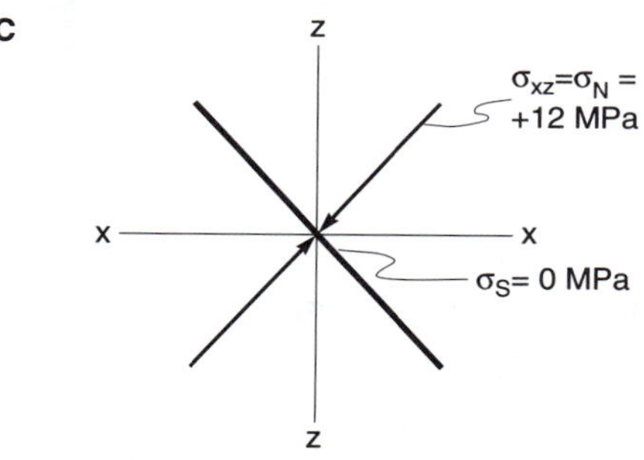

Figure 3.21 Demonstration that no shear stresses exist within a hydrostatic stress field. (*A*) The hydrostatic starting condition. Stress parallel to the *x*-axis and stress parallel to the *z*-axis are the same, 12 MPa. (*B*) Balancing of forces at point *P* results in a S_z value of 8.91 MPa and a S_x value of 8.02 MPa. (*C*) The single stress (σ) for which S_x and S_z are components has a magnitude of—you guessed it!—12 MPa, and it is oriented perfectly perpendicular to the plane. This stress (σ) has no shear stress component.

The Stress Equations

Thankfully, once the magnitudes and orientations of the principal stresses at a point are known, we can quickly calculate the normal stress or shear stress on a plane of any orientation using the **fundamental stress equations** derived in standard engineering and structural geological texts (e.g., Ramsay, 1967; Jaeger and Cook, 1976; Means, 1976). The form of the stress equations is identical to that of strain (Equations 2.6 and 2.7). They are written as follows:

$$\sigma_N = \frac{\sigma_1 + \sigma_3}{2} - \frac{\sigma_1 - \sigma_3}{2}\cos 2\theta \qquad (3.3)$$

$$\sigma_S = \frac{\sigma_1 - \sigma_3}{2}\sin 2\theta \qquad (3.4)$$

In illustrating the use of these equations, let us continue to focus on the stress at *P* at a depth of 1500 m in granitic crust. We determined $\sigma_1 = 40$ MPa and $\sigma_3 = 20$ MPa. Our goal is to calculate σ_N and σ_S for a plane of any arbitrary orientation (e.g., $\theta = +30°$) with respect to σ_1.

$$\sigma_N = \frac{40\text{ MPa} + 20\text{ MPa}}{2} - \frac{40\text{ MPa} - 20\text{ MPa}}{2}\cos 60°$$

$$\sigma_N = 30\text{ MPa} - 10\text{ MPa}\,(0.5000)$$

$$\sigma_N = 25\text{ MPa}$$

A

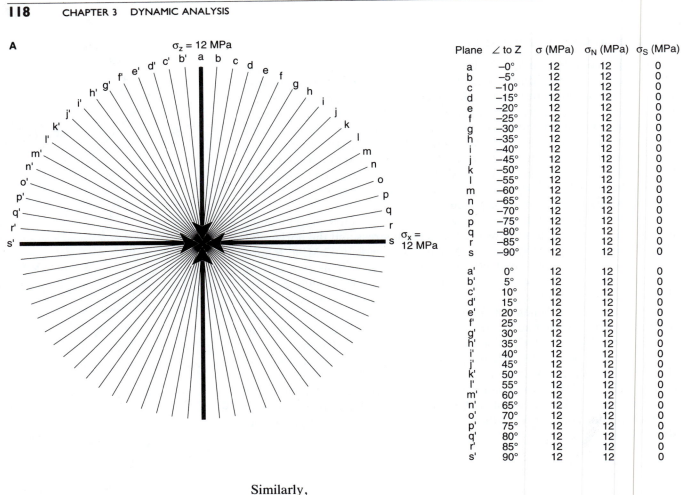

Plane	∠ to Z	σ (MPa)	σ_N (MPa)	σ_S (MPa)
a	−0°	12	12	0
b	−5°	12	12	0
c	−10°	12	12	0
d	−15°	12	12	0
e	−20°	12	12	0
f	−25°	12	12	0
g	−30°	12	12	0
h	−35°	12	12	0
i	−40°	12	12	0
j	−45°	12	12	0
k	−50°	12	12	0
l	−55°	12	12	0
m	−60°	12	12	0
n	−65°	12	12	0
o	−70°	12	12	0
p	−75°	12	12	0
q	−80°	12	12	0
r	−85°	12	12	0
s	−90°	12	12	0
a'	0°	12	12	0
b'	5°	12	12	0
c'	10°	12	12	0
d'	15°	12	12	0
e'	20°	12	12	0
f'	25°	12	12	0
g'	30°	12	12	0
h'	35°	12	12	0
i'	40°	12	12	0
j'	45°	12	12	0
k'	50°	12	12	0
l'	55°	12	12	0
m'	60°	12	12	0
n'	65°	12	12	0
o'	70°	12	12	0
p'	75°	12	12	0
q'	80°	12	12	0
r'	85°	12	12	0
s'	90°	12	12	0

B

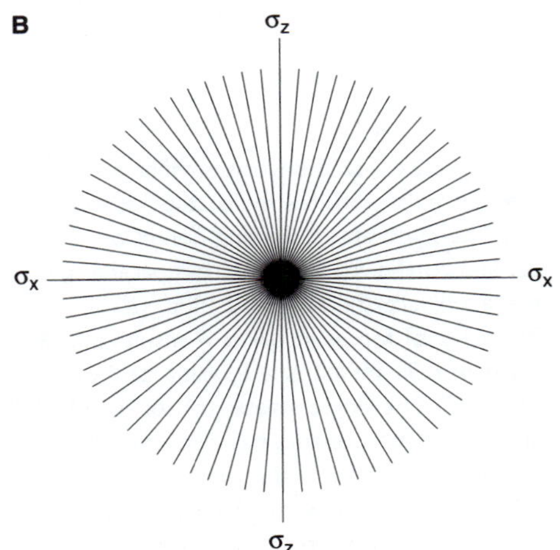

Figure 3.22 (*A*) Pictured here is a vertical stress (σ_z) of magnitude 12 MPa and a horizontal stress (σ_x) of magnitude 12 MPa. The lines labeled *a–s*, and *b'* to *r'* are traces of planes at 5° intervals. Stress (σ), normal stress (σ_N), and shear stress (σ_S) have been calculated for each of the planes. There are no variations. (*B*) Stress "ellipse" created by arranging all of the calculated stresses (σ) and arranging them in such a way that their tips meet at a point. Schematic.

Similarly,

$$\sigma_S = \frac{40\text{ MPa} - 20\text{ MPa}}{2}\sin 60^\circ$$

$$\sigma_S = 10\text{ MPa }(0.8660)$$

$$\sigma_S = 8.6\text{ MPa}$$

For a condition of hydrostatic stress, such as that found at point P' where $\sigma_1 = \sigma_3 = 12$ MPa, we can use the stress equations to calculate σ_N and σ_S for a plane inclined at $\theta = +30^\circ$ to σ_1.

$$\sigma_N = \frac{12\text{ MPa} + 12\text{ MPa}}{2} - \frac{12\text{ MPa} - 12\text{ MPa}}{2}\cos 60^\circ = 12\text{ MPa}$$

$$\sigma_S = \frac{12\text{ MPa} - 12\text{ MPa}}{2}\sin 60^\circ = 0\text{ MPa}$$

The Mohr Stress Diagram

Just as the Mohr strain diagram provides a picture of strain variations within a body (Chapter 2), the Mohr stress diagram gives us a very useful display of the stress equations. The equations describe a circular locus of paired values (σ_N, σ_S) of the normal and shear stresses that operate on planes of any and all orientations within a given body subjected to known values of σ_1 and σ_3. Using the Mohr stress diagram, we can identify a plane of any orientation relative to σ_1 and then read directly

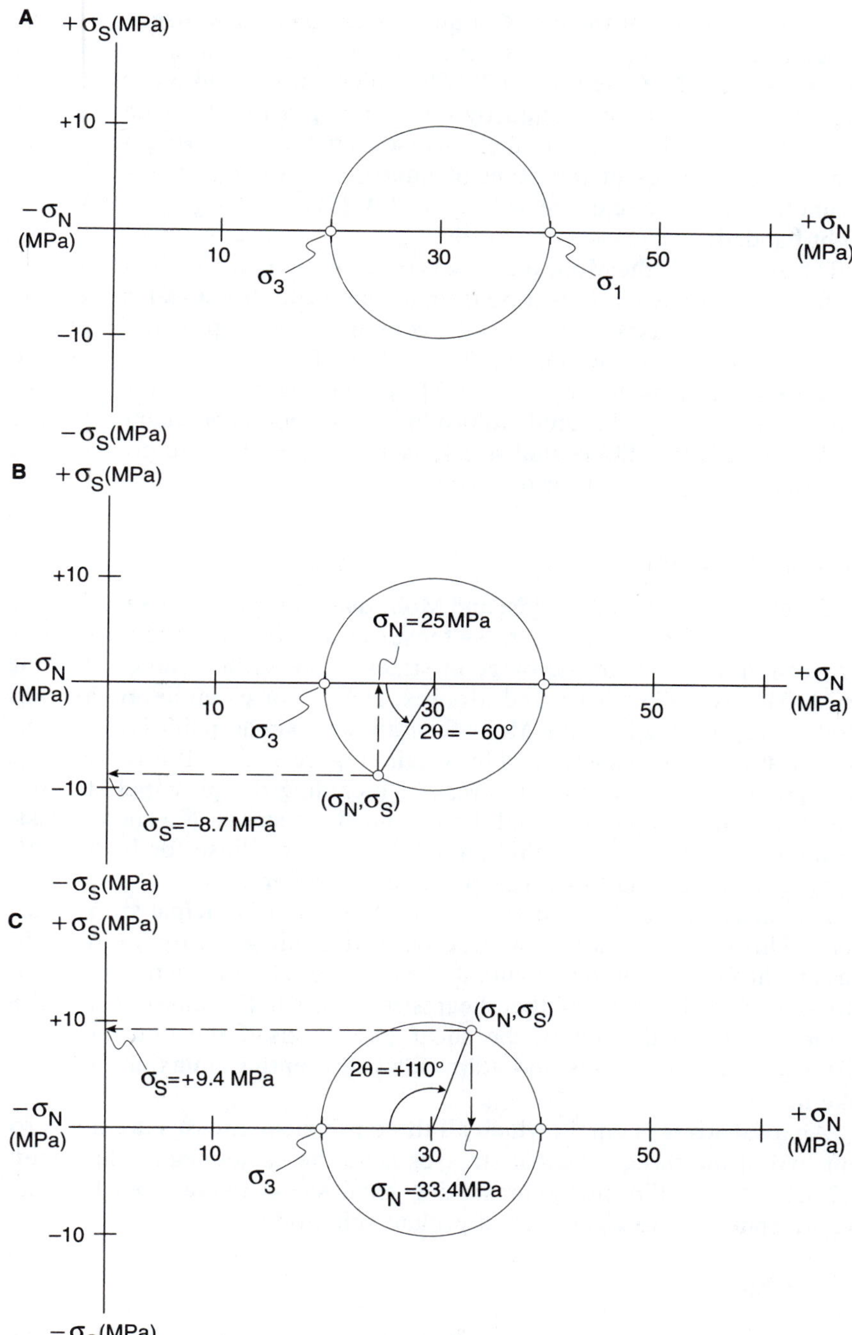

Figure 3.23 Construction of a Mohr stress circle. (*A*) Plot greatest (σ_1) and least (σ_3) principal stresses within *x–y* coordinate system. Normal stresses (σ_N) are plotted along *x*-axis; shear stresses (σ_S) are plotted along *y*-axis. Construct circle that passes through the principal stress values. Center of circle lies along the *x*-axis. (*B*) Determine the normal stress (σ_N) and shear stress (σ_S) values for a plane that makes an angle of $\theta = -30°$ with respect to the direction of greatest principal stress (σ_1). Do this by constructing an angle of $-60°$ (counterclockwise) from the *x*-axis. The intersection of the radius with the circle yields a point whose (*x*,*y*) coordinates are (σ_N, σ_S). (*C*) In this example the (σ_N, σ_S) values are found for a plane that makes an angle of $\theta = +55°$ with the direction of greatest principal stress (σ_1).

from the diagram the values of normal stress (σ_N) and shear stress (σ_S) acting on the plane.

The construction of the Mohr stress circle proceeds as follows. Principal normal stress values (σ_1 and σ_3) are plotted on the *x*-axis of the diagram (Figure 3.23*A*). A circle is drawn through these points such that ($\sigma_1 - \sigma_3$) constitutes the circle's diameter. In the example of stress at *P* where $\sigma_1 = 40$ MPa and $\sigma_3 = 20$ MPa, all the paired values of σ_N and σ_S as listed in Figure 3.17 exist as points on the perimeter of the circle. To define σ_N and σ_S for a specific plane (e.g., a plane oriented at $\theta = -30°$ *counterclockwise* from the plane to σ_1) (Figure 3.23*B*), a radius is constructed on the Mohr stress diagram at an angle of 2θ, or $-60°$, measured *counterclockwise* from the *x*-axis (Figure 3.23*B*). Where this radius intersects the perimeter of the circle, a point is established whose x,y coordinates are the (σ_N,σ_S) values for the plane in question: $\sigma_N = +25$ MPa and $\sigma_S = -8.6$ MPa (left-handed). The values of σ_N and σ_S for a plane oriented +55° to σ_1 are found by constructing a radius at an angle of $2\theta = +110°$, this time measured clockwise from the x-axis (Figure 3.23*C*). The x,y coordinates of the point of intersection of this radius with the perimeter of the circle are $\sigma_N = +33.4$ MPa and $\sigma_S = +9.4$ MPa (right-handed).

The anatomy of the Mohr diagram is revealing (Figure 3.24). The center of the Mohr stress circle is a point that represents the **mean stress,** that is, $(\sigma_1 + \sigma_3)/2$. This is the hydrostatic component of the principal stresses, and it tends to produce dilation. The radius of the circle represents the **deviatoric stress,** that is, $(\sigma_1 - \sigma_3)/2$. The deviatoric stress is the nonhydrostatic component, and it tends to produce distortion. The diameter of the circle is called the **differential stress,** that is, $\sigma_1 - \sigma_3$. The greater it is, the greater the potential for distortion.

Images of Stress

Stress ellipses, stress ellipsoids, and Mohr diagrams provide useful images of the characteristics of the stress tensor at a point, and they provide a way of reminding us of the range of states of stress that can exist in the crust (Means, 1976; Twiss and Moores, 1992). For example, **hydrostatic stress** is represented on the Mohr diagram as a single point lying on the normal stress axis to the right of the origin (Figure 3.25*A*). It is represented as a point because all normal stresses, including the principal stresses, are of the same magnitude, and there is no shear stress. No shear stress, no deviatoric stress, no Mohr[3] circle! The stress ellipse for hydrostatic stress is a circle, and the stress ellipsoid is a sphere.

A **uniaxial stress** is one in which two of the three principal stresses are zero. This condition is represented on a Mohr diagram by a circle that passes through the origin (Figure 3.25*B*). The circle representing uniaxial stress lies to the right of the shear stress axis if the uniaxial stress is compressive, to the left of the shear stress axis if the uniaxial stress is tensile. Stress ellipses and ellipsoids representing uniaxial stress are like needles.

An **axial stress** is one in which all three principal stresses are nonzero, but two of the three principal stresses have the same magnitude (Figure 3.25*C*). Stress ellipsoids representing axial stress range from flattened oblate spheroids to cigar-shaped prolate spheroids.

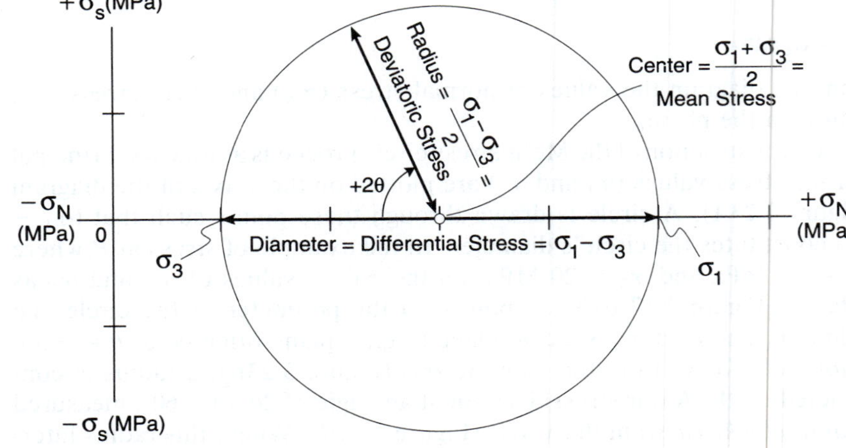

Figure 3.24 The center of the Mohr stress circle represents mean stress, which is the hydrostatic component of the stress field. Mean stress tends to produce dilation. The radius of the Mohr stress circle represents deviatoric stress, which is the nonhydrostatic component of the stress field. Deviatoric stress tends to produce distortion. The diameter of the Mohr stress circle represents differential stress. The larger it is, the greater the potential for distortion.

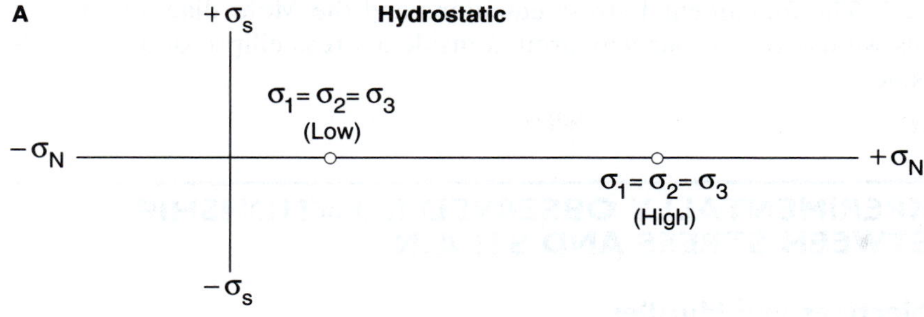

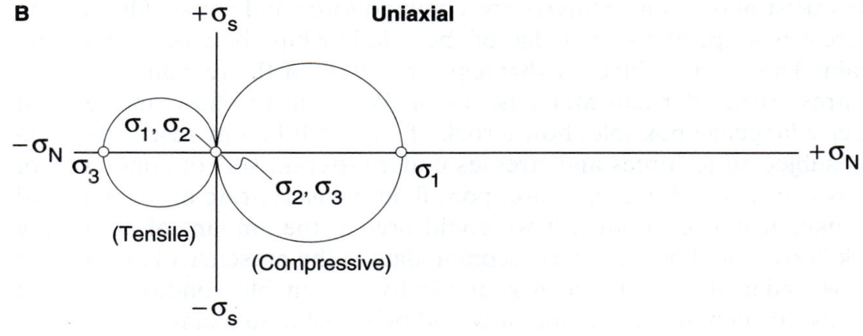

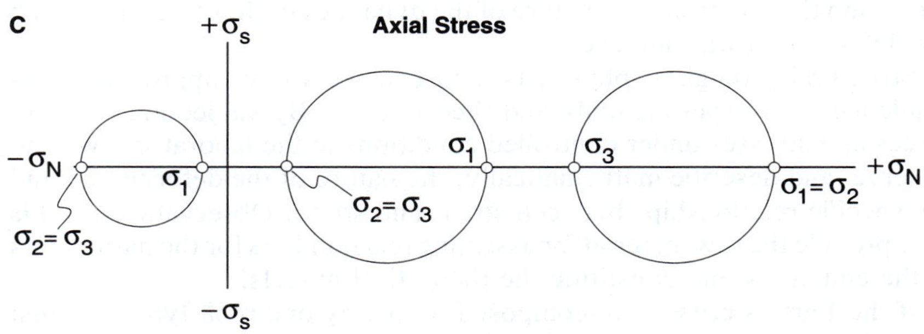

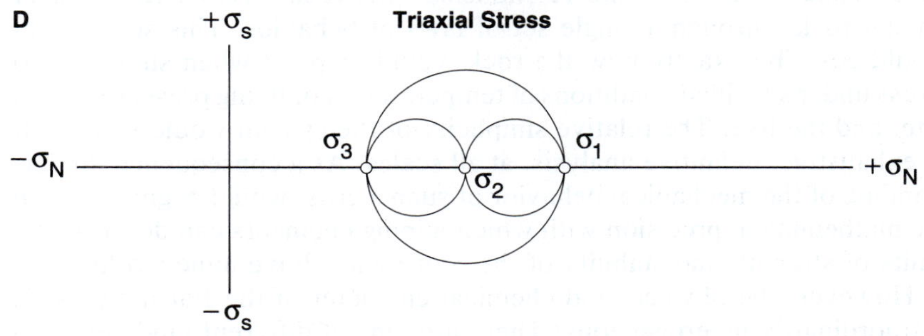

Figure 3.25 Various states of stress can be represented on the Mohr stress diagram: (*A*) hydrostatic stress, (*B*) uniaxial stress, (*C*) axial stress, and (*D*) triaxial stress.

The general case of stress at a point is known as **triaxial stress**. All three principal stresses are of different magnitude, and all are nonzero (Figure 3.25*D*). There are no restrictions regarding whether the principal stresses are tensile or compressional. All can be compressional. All can be tensile. Or there can be some of each.

We need to try to remember that the stress ellipsoids, whatever their sizes and shapes, are stress tensors. The form and orientation of these three-dimensional images represent an elegant way to depict "the whole collection of tractions acting on each and every plane of every conceivable orientation passing through a single point in a body at a given instant of time." The fundamental stress equations and the Mohr diagram are the tools we use to find our way around inside a stress ellipse or a stress ellipsoid.

EXPERIMENTALLY OBSERVED RELATIONSHIPS BETWEEN STRESS AND STRAIN

Objectives and Hurdles

The thrust of dynamic analysis goes beyond force and stress. Of ultimate interest is a specific knowledge of the relationships between stress and strain. This is the subject of **rheology**, the study of the response of rocks to stress (Engelder and Marshak, 1988). We want to know, in the most precise language possible, how a rock of a given lithology responds when it is subjected to forces and stresses under different sets of conditions of temperature, confining pressure, pore fluid pressure, rate of loading, and the like. It would be ideal if we could predict the amount of strain any rock body would be forced to accommodate in the presence of any known stress under any given set of geologically reasonable conditions. If we possessed such understanding, it would be possible to examine the structures in deformed rock, reconstruct the movements that created the structures, and then interpret the nature of the dynamic conditions under which the deformation was achieved.

Structural geologists, physicists, and engineers have approached this challenge both experimentally and theoretically. By subjecting rocks to forces and stresses under controlled conditions in the laboratory, we can observe, and describe mathematically, the nature of the deformation and the specific relationships between stress and strain. Observations of this type provide the raw material for assigning real numbers for the parameters in the equations that constitute the theoretical models.

If the Earth's crust were composed of merely one rock type with just one mineral species (e.g., a pure homogeneous calcite marble), it would be possible to describe the relationships between stress and strain of crustal rocks through a single set of laws of behavior. This set of laws would describe exactly how the rock would respond when subjected to stress under specified conditions of temperature, confining pressure, strain rate, and the like. The relative simplicity of the system would lend itself to exhaustive, definitive analysis, at all scales. As a consequence, understanding of the mechanical behavior of such a crust would begin to match the mathematical precision with which mining engineers can describe the limits of strength and stability of the roof rocks above mine workings.

However, the physical and chemical character of the Earth's crust is extraordinarily heterogeneous! There are tens of different kinds of common rocks, a dozen or more common rock-forming minerals, and hundreds of different kinds of common textures and primary internal structures. Volcanic, plutonic, metamorphic, and sedimentary rock bodies are brought together into intimate, interlinked three-dimensional contact, at all scales, through extrusion, intrusion, deposition, faulting, and shearing. Then, when such heterogeneous bodies of rock are subjected to tectonic stress, each rock and each mineral in each textural configuration responds in its own way to the stress as a function of (1) the physical/chemical characteristics of each rock and mineral and (2) the conditions of deformation. The number of possible combinations of starting materials and deformational conditions is infinite. No single set of laws will do; only general laws that, as they become more refined, may be close approximations.

The Value of Laboratory Deformational Experiments

We find it useful to create in the laboratory the very stresses whose impact we wish to examine. We do this by subjecting rock specimens to controlled loading under known, prescribed conditions in a deformation apparatus. The nature and origin of structural deformation of rocks becomes clearer through images and experiences in experimental deformation. Consider placing an undeformed cylindrical core of limestone into a thick-walled steel pressure vessel, squeezing the rock by hydraulic loading, and then removing a conspicuously deformed specimen bearing structural characteristics identical to those of faulted rocks seen in natural outcrops (Figure 3.26). The experimental process provides a cause-and-effect glimpse of dynamics that few experiences could match. Furthermore, to perform the experiment successfully, we need to learn how to "regulate" force and stress mechanically and mathematically, and how to describe the strain and strength of rocks through the conventional parameters.

Sample Preparation

To prepare for a deformation experiment, a small core of rock is extracted from a rock, like limestone, through the use of a drill press and a diamond drill coring device. The ends of the cylinder of rock are beveled to smooth, planar parallel surfaces on a grinding wheel. If the ends were not ground flat, small irregularities of rock projecting from the ends would cause high stress concentrations, like those focused by the tips of skates on ice or a nail being hammered into wood. After the specimen has been thus prepared, a micrometer is used to measure length (l_0) and diameter (d_0) of the specimen, preferably to three significant figures.

The specimen is then placed in a **jacket** of a thin-walled cylinder of copper or some other material of negligible strength. The jacket serves to seal the rock from whatever fluid (e.g., kerosene) occupies the pressure vessel. Once all this has been done, the jacketed specimen is fitted with an anvil at its base and an upper piston specimen holder at its top (Figure 3.27*A*). Both these components are made of stainless steel and equipped with O-rings to prevent entry of fluids into the jacket from above or below. Then the specimen and its trimmings are screwed into the **pressure vessel** (Figure 3.27*B*), a steel vessel of sufficient wall thickness and inherent strength to resist fracturing under conditions of very high pressure.

During the test, the fluid surrounding the jacketed specimen in the pressure vessel is pressurized to produce a **confining pressure** to simulate burial at depth. The **temperature** of the rock core can be increased to the desired level by increasing the power to the furnace surrounding the

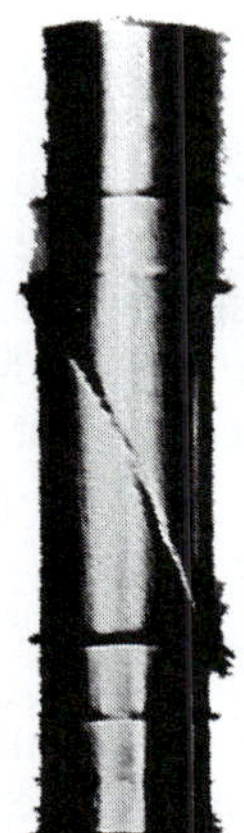

Figure 3.26 Deformation of experimentally deformed cylinders of rocks. Structure is reflected in the thin-walled copper jacket that envelops each rock specimen. *Left to right:* fault in slate, conjugate faults in sandstone, ductile flow in limestone. [From Donath (1970a). Published with permission of National Association of Geology Teachers, Inc.]

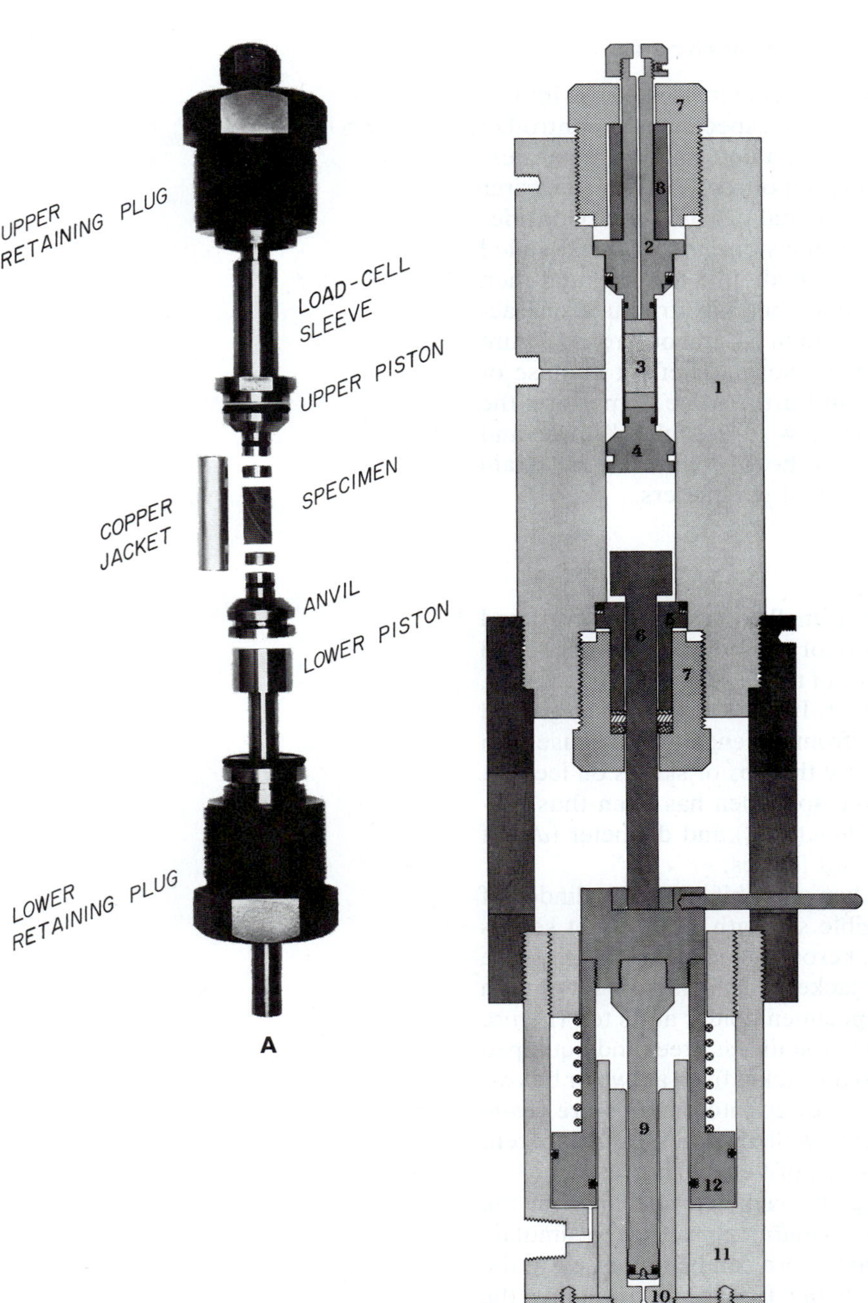

Figure 3.27 (*A*) Internal parts of pressure vessel, showing the relation of the cylindrical rock specimen (and copper jacket) to pistons and anvil. [From Donath (1970a). Published with permission of National Association of Geology Teachers, Inc.] (*B*) Schematic drawing of the vessel–press assembly in the Donath apparatus: 1, pressure vessel; 2, upper piston and seal; 3, specimen; 4, anvil; 5, gland and lower seal; 6, lower piston; 7, retaining plug; 8, load cell; 9, equalization piston; 10, equalization cylinder; 11, ram body; 12, ram piston; 13, collar. [From Donath (1970a). Published with permission of National Association of Geology Teachers, Inc.]

sample in the pressure vessel. **Pore fluid pressure** within the jacketed sample can be elevated as well, and as an independent variable separate from confining pressure. Machining of the pressure vessel is so precise that fluids can be pumped at high pressure directly into the jacketed specimen, again to whatever pressure level is reasonable for the test being conducted.

Types of Tests

There is more than one way to squeeze a rock. In the most common procedure, the cylindrical sample is subjected to **axial compression** (Figure 3.28*A*). Vertical axial compressive stress (σ_1) parallel to the length of the core is taken to higher levels than the horizontal radial compressive stress—that is, the **confining pressure** ($\sigma_2 = \sigma_3$) acting on the body of the rock cylinder. Because the axial stress is the greatest principal stress and the radial stress is the least principal stress, the specimen undergoes length-parallel shortening.

A less common procedure is one in which the core is subjected to **axial extension** (Figure 3.28*B*). In this case, horizontal radial compressive stress, the confining pressure ($\sigma_1 = \sigma_2$) acting on the body of the rock cylinder, is greater than the vertical axial compressive stress (σ_3) parallel to the length of the sample. As a result, the specimen undergoes length-parallel extension.

Axial compression and axial extension tests carried out in this manner are referred to as **triaxial deformation experiments**, although "triaxial" is misleading because most equipment does not permit each of the three principal stresses to be set separately and independently.

In some tests the rock samples are not squeezed at all; rather they are pulled apart (i.e., stretched). Most commonly these are unconfined **tensile strength tests** (Figure 3.28*C*), which aim to determine the smallest amount of stress that will cause a rock to fail in tension. Rocks are much, much weaker in tension than in compression, and, if given a chance to fail in tension, will do so. As a consequence, material scientists, engineers, and structural geologists have a keen interest in the weakness of rocks in tension and have worked to calibrate it by means of a wide variety of tensile strength tests (Price and Cosgrove, 1990).

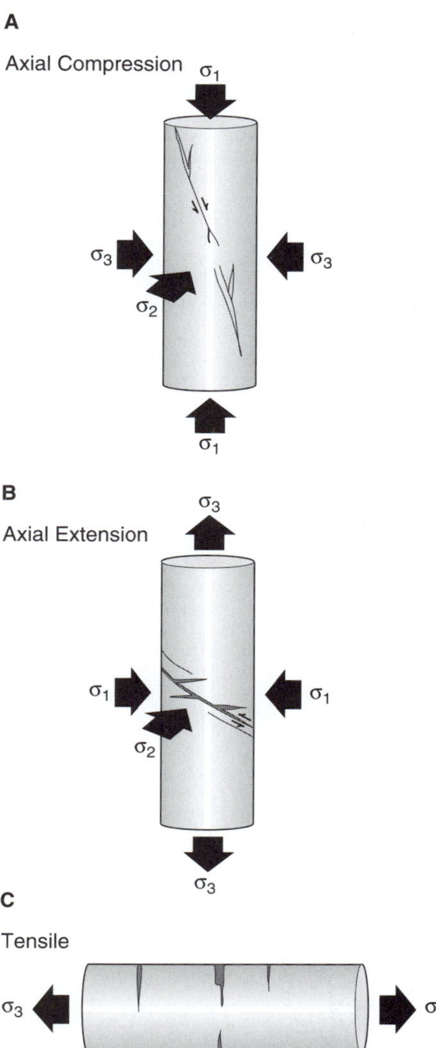

Figure 3.28 Types of deformation experiments: (*A*) Axial compression; (*B*) axial extension; (*C*) tensile.

Pressures, Stresses, and Loads

The basic procedures and techniques of the axial compression test can be understood using a simple though elegant deformation setup known as the **Donath triaxial deformation apparatus** (Figure 3.29). Fred Donath (1970) designed and built this experimental rig in such a way that it is safe, portable, and easy to operate. When I first began to teach, it gave me an appreciation for the experimental approach to rock deformation, as well as a deeper appreciation for the advanced work that scientists like

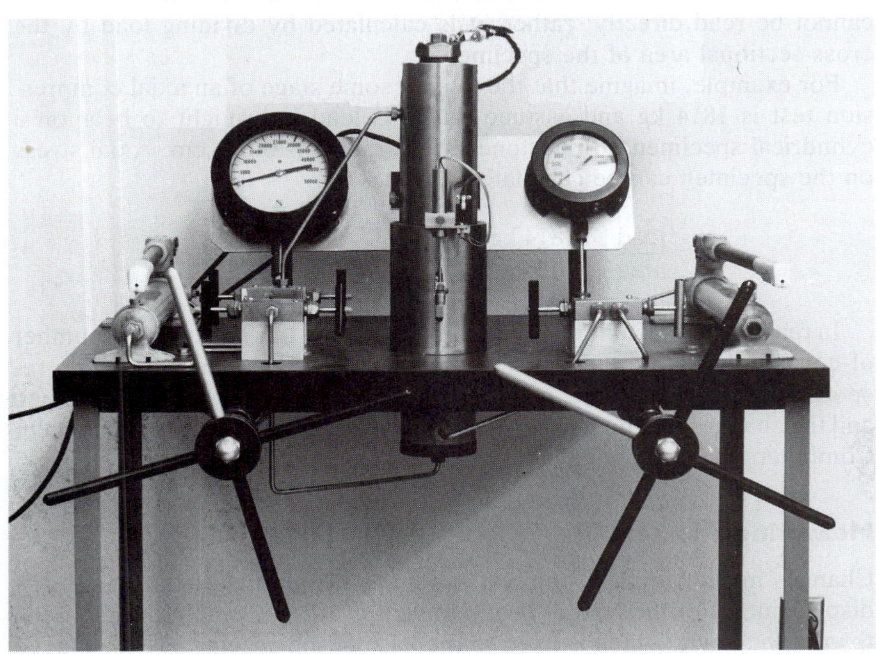

Figure 3.29 The Donath apparatus.

Jan Tullis (Brown University) have been carrying out over the years using sophisticated technologies and imaginative approaches.

The Donath apparatus accepts small cylindrical specimens, with diameters of 0.5 in. (1.3 cm) and lengths of 1 in. (2.5 cm). Burial pressure conditions are imposed on test specimens by pumping kerosene into the specimen chamber to produce an equal, all-sided **confining pressure** on the specimen. In nature, confining pressure at any specified level is derived from the weight of water occupying pore spaces in the overlying column of rock **(hydrostatic stress)** and the weight of the rock column itself **(lithostatic stress).**

In experiments using the Donath apparatus, the level of confining pressure can be read directly from a calibrated gauge. Raising the level of confining pressure by even small amounts results in a small reduction in volume of the specimen. The specimen quite naturally undergoes negative dilation. Hydrostatic stress, even at the highest levels, cannot produce distortion. The equal all-sided compressive stresses simply reduce the size of the specimen uniformly.

When the appropriate level of confining pressure is established, a test specimen can be shortened by applying a vertical **axial load**. The load is applied to the specimen by manually pumping hydraulic jack fluid into the vessel–press assembly. As pressure of the jack fluid increases, a ram forces a piston to advance slowly upward until it butts against the anvil fixed to the base of the specimen (see Figure 3.27*A*). This is a critical moment in the test. When the piston makes contact with the anvil, the stresses are transmitted through the anvil into the specimen. Moreover, the stresses are transmitted through the specimen into the specimen holder and to the steel cap that seals the top of the thick-walled pressure vessel. The steel cap can be thought of as a very strong steel spring, which deforms elastically when subjected to stress.

By knowing the relation of stress magnitude to shortening of the steel cap, it is possible to calibrate the exact amount of load that is being applied by the piston at each stage of the experiment. In practice, electronic signals describing the shortening of the steel cap are transmitted from a load cell in the cap to an X–Y recorder. The signals are "interpreted" by the recording device in terms of load, again directly in pounds for the Donath apparatus. The **axial stress** at any stage during the experiment cannot be read directly; rather, it is calculated by dividing load by the cross-sectional area of the specimen.

For example, imagine that the load at some stage of an axial compression test is 1814 kg and assume that this load is brought to bear on a cylindrical specimen of limestone whose radius is 0.617 cm. Axial stress on the specimen can be calculated as follows:

$$\sigma = \frac{\text{load}}{\text{area}} = \frac{1814\ \text{kg}}{1.19\ \text{cm}^2} = 1524\ \text{kg/cm}^2 = 152\ \text{MPa}$$

In reality, the calculation is more involved than this because of a number of correction factors that must be introduced to compensate for the response of the various steel members that carry the stress to the specimen and the changes that take place in the cross-sectional area of the specimen. Computers make these calculations and corrections easy.

Measuring Shortening

Changes in the length of the test specimen are derived from measuring displacements of the piston. Progressive upward movement of the piston toward the anvil, which is fixed to the base of the specimen, is displayed on the $X-Y$ recorder by pen movement horizontal and to the right (Figure 3.30).

The instant the piston makes contact with the anvil, the pressure gauge that monitors load jumps from zero to some small positive value. As seen on the $X-Y$ recorder (Figure 3.30), this point of contact, known as the **seating position**, is marked by a jump of the pen upward, recording the fact that the piston has encountered resistance. Any further movement of the piston into the specimen chamber records shortening of the specimen. If rocks were perfectly unyielding, the piston could be raised no further, no matter how much force were applied to the anvil. However, rocks are deformable, and thus added increments of load always result in shortening. Shortening is revealed on the $X-Y$ plotter in the form of displacement of the pen to the right of the seating position (Figure 3.30).

Shortening of the core of rock at any stage of an experiment can be described in terms of its extension (e) or stretch (S). Suppose that original specimen length was 2.3 cm and shortening was merely 0.02 cm:

$$e = \frac{\Delta l}{l_o} = \frac{-0.02 \text{ cm}}{2.3 \text{ cm}} = -0.0087 = 0.87\% \text{ shortening}$$

$$S = \frac{l_f}{l_o} = \frac{2.28 \text{ cm}}{2.30 \text{ cm}} = 0.99$$

Notice that "centimeters" cancel out, and both extension (e) and stretch (S) are unitless.

Measuring Strain Rate

The rate at which a rock is shortened or stretched has a very important bearing on how it deforms. Thus we keep track of **strain rate** ($\dot{\varepsilon}$) during

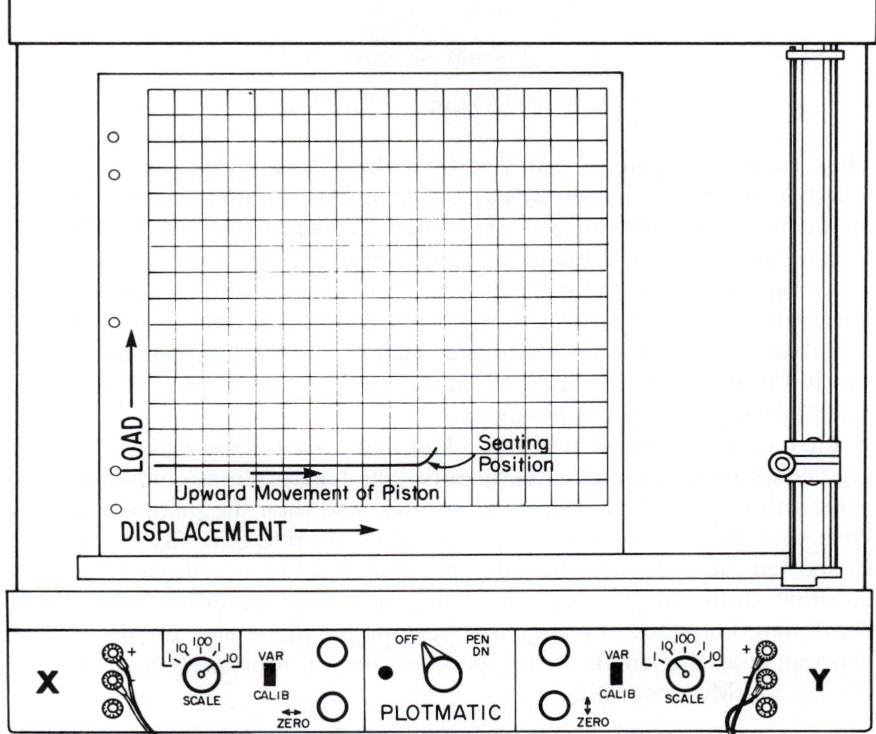

Figure 3.30 An $X-Y$ recorder. The load to which the test specimen is subjected is recorded on the y-axis of the recorder. Displacement of the piston is recorded on the x-axis. Pen path shows upward movement of piston and eventual contact of piston with anvil attached to the base of the specimen. Point of contact is known as the seating position.

deformational experiments. Strain rate is determined by dividing extension (e) by time (t).

$$\text{strain rate} = \dot{\varepsilon} = \frac{e}{t}$$

Strain rate is calibrated in units of reciprocal seconds (s^{-1}). For example, a strain rate of 10^{-5} s^{-1} means that the amount of extension per second is 0.00001. Describing strain rate in these units (s^{-1}) gives the impression, at first, that something is missing, but this is only because extension (e) is unitless.

Let us try an example. Suppose that during a one-hour experiment we shortened a sample of mudstone from its original length of 2.297 cm to a final length of 2.280 cm. What is the strain rate for this experiment?

$$\dot{\varepsilon} = \frac{e}{t} = \frac{\dfrac{2.280 - 2.297}{2.297}}{60 \text{ min} \times 60 \text{ s/min}} = \frac{\dfrac{-0.017}{2.297}}{3600 \text{ s}}$$

$$\dot{\varepsilon} = \frac{-0.0074}{3600 \text{ s}} = \frac{0.0000021}{S} = -2.1 \times 10^{-6} \text{ s}^{-1}$$

Let us try one more example. We are about to subject an undeformed cylindrical core of limestone to shortening. Its length is 3.00 cm. If we were to shorten it at a constant strain rate of -10^{-5}, by how much would we need to reduce the length of the specimen in the first 5 minutes of the experiment?

$$\text{extension } (e) = \frac{\text{length change}}{\text{original length}} = \frac{-\Delta l}{l_o} = \frac{-\Delta l}{3.0 \text{ cm}}$$

$$\text{strain rate } (\dot{\varepsilon}) = \frac{\text{extension (e)}}{\text{second}} = -10^{-5} \text{ s}^{-1}$$

$$\dot{\varepsilon} = \frac{-\Delta l/3 \text{ cm}}{300 \text{ s}} = -10^{-5} \text{ s}^{-1}$$

$$\Delta l = -900 \text{ cm} \cdot \text{s} \times 10^{-5} \text{ s}^{-1}$$

$$\Delta l = -0.009 \text{ cm} = -0.09 \text{ mm}$$

Thus, at a strain rate of -10^{-5} s^{-1}, the limestone specimen of original length 3.00 cm will shorten by approximately one-tenth of one millimeter in the first 5 minutes of the experiment. A strain rate of 10^{-5} seems incredibly slow, but actually it is extremely fast in comparison to the strain rates responsible for the development of mountain systems, which are on the order of 10^{-14} s^{-1}. For sick fun you might want to try to calculate how much a limestone specimen of 3.00 cm original length will shorten in the first 5 minutes of shortening at a strain rate of 10^{-14} s^{-1}. Then again, you might not want to.

Most axial compression tests (and for that matter, axial extension tests) are **"constant strain rate"** experiments. The temperature, confining pressure, and pore fluid pressure are first set, and then the piston, which is flush with the specimen, is moved at a constant rate in a way that serves to shorten the specimen. Strictly speaking, the test is not carried out at constant strain rate because each successive fraction-of-a-centimeter of shortening constitutes a comparatively larger increment of *percentage* shortening as the length of the specimen becomes progressively smaller (Twiss and Moores, 1992).

Some axial compression tests are **"constant stress"** experiments, but once again the name given to the procedure is somewhat misleading because it is the load that is kept constant while the rock sample shortens. Even while load is held constant, axial stress may steadily decrease (slightly) during the experiment as the cross-sectional area of the sample increases (slightly) as a result of flattening (Twiss and Moores, 1992). Constant stress and constant strain rate tests are "driven" by computers in such a way that piston displacement and amount of load are continuously adjusted to maintain the guiding requirements for the test.

A Standard Axial Compression Test

With this background in mind, let us carry out a compressional deformation experiment from start to finish. We will subject a specimen of limestone to a confining pressure of 28 MPa and then load it end-on. The temperature conditions will be cool (i.e., room temperature). The purpose of the test is to "observe" how the limestone might respond to compression-induced shortening at very shallow crustal conditions. We will load the sample rapidly, such that the whole experiment will take only 10 or 20 minutes.

The initial length (l_o) of the specimen is 2.5 cm and the radius (r) is 0.635 cm. When the specimen is subjected to 27.58 MPa confining pressure, the length of the specimen decreases by some tiny amount—approximately 0.025 mm—because of negative dilation. The decrease in radius is negligible. Thus we calculate the cross-sectional area (A) of the specimen on the basis of the initial radius.

$$A = \pi r^2 = 3.1416\,(0.635\text{ cm})^2 = 1.267\text{ cm}^2$$

When load is applied by the piston to the base of the anvil, compressional stresses are transmitted through the rock, and the cylinder of limestone begins to shorten. In the early stage of this compression, pen movement on the X–Y recorder reveals that the load-displacement curve is a straight line, which indicates that load and displacement are directly proportional (Figure 3.31*A*).

The **load–displacement curve** can be transformed into a **stress–strain diagram**, the standard display of experimental tests. Stress–strain diagrams plot differential stress on the y-axis against percentage strain on the x-axis (Figure 3.31*B*).

Differential stress (σ_d), which is plotted on the y-axis of the stress–strain curve, is the difference between the greatest (σ_1) and least (σ_3) stresses:

$$\sigma_d = \sigma_1 - \sigma_3$$

Percentage strain, plotted on the x-axis of the stress–strain curve, is percentage shortening:

$$\%\text{ shortening} = \frac{\Delta l}{l_o} \times 100\%$$

The straight-line nature of the stress–strain diagram indicates that the limestone specimen is behaving elastically, like a spring. The elastic behavior is a dilational spring-action phenomenon, derived from the rock's ability to recover tiny non–rigid body changes in atomic spacings in crystal lattices of its mineral components (Ramsay, 1967). During the elastic stage

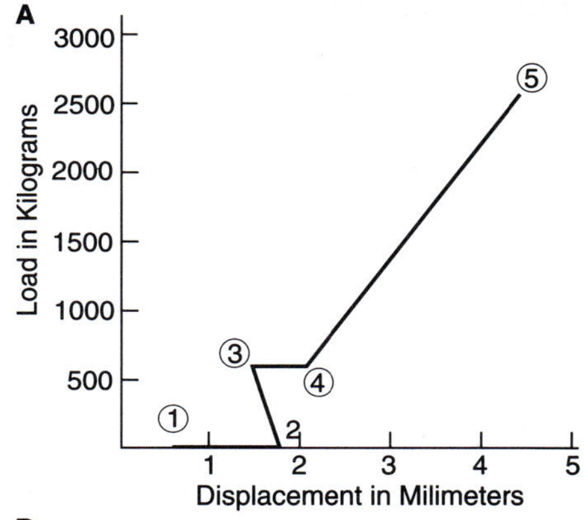

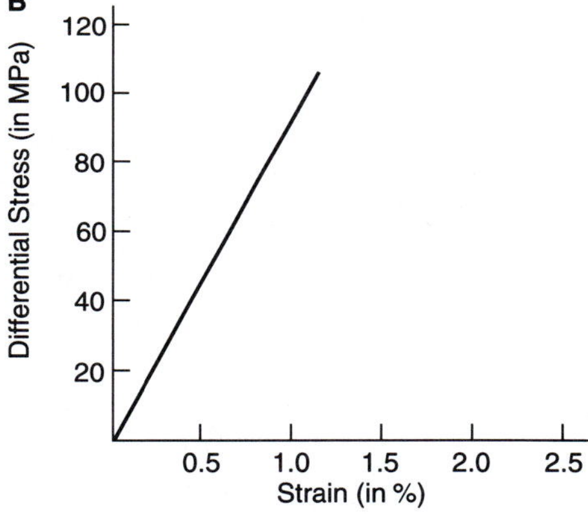

Figure 3.31 (*A*) Load–displacement curve (i.e., the raw data). The steel piston is brought into contact with the anvil, which is in turn in direct contact with specimen (1–2). This is called the "seating" of the specimen. Confining pressure is raised (2–3), and while this takes place the piston is forced away from specimen. The specimen is reseated (3–4). The specimen is loaded, and it responds elastically such that the load–displacement curve during the actual deformation is a straight-line relationship (4–5). (*B*) Transformation of the load–displacement curve to a stress–strain diagram.

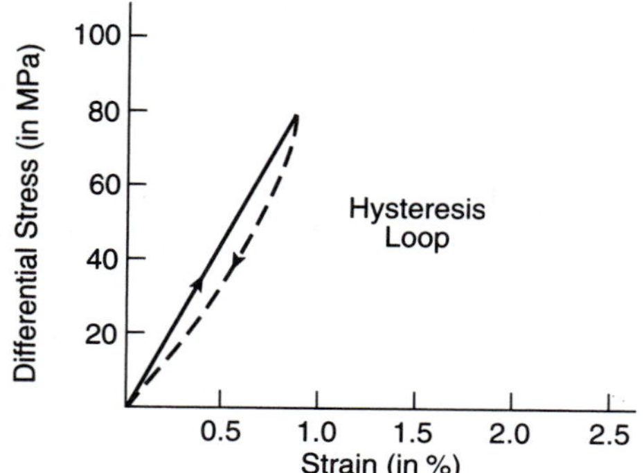

Figure 3.32 When the load is removed from an elastically deforming specimen, the specimen will return to its original length. The return is along a looping path, signifying a time lag in recovery.

of deformation it is possible to remove the load and observe that the limestone cylinder almost instantly springs back to its original length. As load is decreased to 0.0 kg, the pen on the $X-Y$ recorder may not identically retrace the route of its former ascent. Instead, it may loop back to its starting position, reflecting a brief time lag in the **recovery** (elimination) of all the bound-up longitudinal strain (Figure 3.32). Such retardation is known as **hysteresis**, and the loop displayed on the graph is sometimes called a **hysteresis loop**.

If, instead of dropping the level of stress, we continually raise the stress, the limestone may eventually begin to deform plastically. In essence, the range of elastic behavior is surpassed, and nonrecoverable permanent strain begins to accumulate in the rock specimen. A "hook" develops in the load–displacement and stress–strain curve, signifying departure from a straight-line relationship between stress and strain (see Figure 3.33*A*,*B*). **Plastic deformation** produces a permanent change in shape of a solid without failure by rupture.

Anyone who has tried to repair a toy metal Slinky that has been stretched too far knows about the nonrecoverability of plastic deformation. The life expectancy of a Slinky is always short. (Our experience suggests that the half-life of a Slinky is about 4 hours.) In its all-too-brief youth a Slinky can smoothly descend any flight of stairs. Then without warning, it becomes internally entangled. Rescue efforts typically lead to stretching segments of the metal beyond its elastic range, at which point the material becomes permanently plastically deformed. No efforts, however great, can undo the damage.

The onset of plastic deformation during the experiment occurs when the load–displacement curve (or stress–strain curve) departs from its straight-line elastic mode and begins to bend to form a convex-upward curve (Figure 3.33*A*,*B*). The decrease in slope of the curve signifies that proportionally less stress is required to produce a given amount of shortening of the limestone. The point of departure from elastic behavior to plastic behavior is called the **elastic limit**. Its value, known as **yield strength**, is measured in stress. Below its yield strength, a rock behaves as an elastic solid. Above the elastic limit, the rock begins to flow. If limestone behaved perfectly plastically when axially compressed under conditions of 28 MPa confining pressure, it would flow continuously without rupture as long as the axial stress were applied. However, limestone under conditions of low confining pressure is so brittle that it usually ruptures by faulting almost as soon as plastic deformation begins (see Figure 3.33). Indeed, the limestone may experience a **true brittle failure** by sudden fracturing in the straight-line range of the stress–strain curve.

Faulting of limestone under low confining pressure conditions usually is punctuated by a muffled "pop" within the pressure vessel, an event that is almost always marked by a sudden drop in stress level and by spectator cheering. The fault movement shortens the specimen in such a way that axial stress is at least momentarily relieved. In some experiments we have conducted, the axial load drops so fast that the soft-metal needle of the pressure gauge plummets to 0 psi and wraps its tip around the metal peg on which it usually gently rests. However, the drop in pressure is ordinarily modest and is marked on the load–displacement curve and stress–strain curve by a short fishhooklike bend (Figure 33*A*,*B*). If desired, axial load can again be applied, but the level of the load usually does not rise to its former magnitude because even small increments of axial stress create slippage on the fault surface that now cuts the specimen (Figure 3.33*C*).

The test usually is terminated when the limestone breaks. The confining pressure is bled off and so is the axial load. Before the limestone core is removed from the pressure vessel, the net change in length of the specimen is measured on the basis of the seating position of the piston, before and after deformation. For this test, the shortening was merely 0.025 cm. This is nonrecoverable permanent strain. All initial elastic strain is recovered when confining pressure and axial load are removed. Total permanent strain is calculated as follows.

$$e = \frac{\Delta l}{l_o} = \frac{-0.025 \text{ cm}}{2.5 \text{ cm}} = -0.01, = 1\% \text{ shortening}$$

When the specimen is extracted from the pressure vessel, the imprint of the fault trace is clearly visible on the surface of the copper jacket that surrounds the specimen. The angle (θ) made by the trace of the fault with the long axis of the cylinder can be measured with a protractor. For compression tests carried out at low levels of confining pressure, θ is commonly around 25°. Upon removal of the copper jacket, the physical characteristics of the faulted limestone specimen can be examined (Figure 3.34).

Overall, the limestone behaved in **brittle** fashion during this test. Brittle rocks first shorten elastically and then fail by the formation of discrete fractures and faults. Failure may occur abruptly without any plastic deformation. Alternatively, failure by fracturing may follow a modest amount of plastic deformation. For rocks to be considered brittle, the amount of shortening before complete failure by fracturing must be less than 5%.

Compression Test at Higher Confining Pressure

Testing the response of rocks to stress in the laboratory is like eating potato chips. It is impossible to complete one test without starting on another. It is impossible to squeeze rocks without asking, *What if?* The compression test we just carried out provided information regarding the response of limestone to axial compression under conditions of 28 MPa confining pressure, room temperature, and very rapid loading. But *what if* we raise the confining pressure to 103 MPa? How would the limestone then respond?

Under confining pressure conditions of 103 MPa, the yield strength (elastic limit) is found to be much higher than before (curve A, Figure 3.35). Above this elevated yield strength, the limestone begins deforming plastically, as reflected in the bowing of the stress–strain curve.

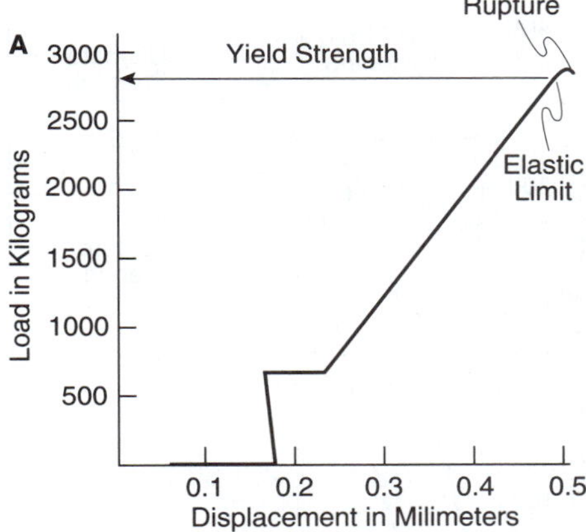

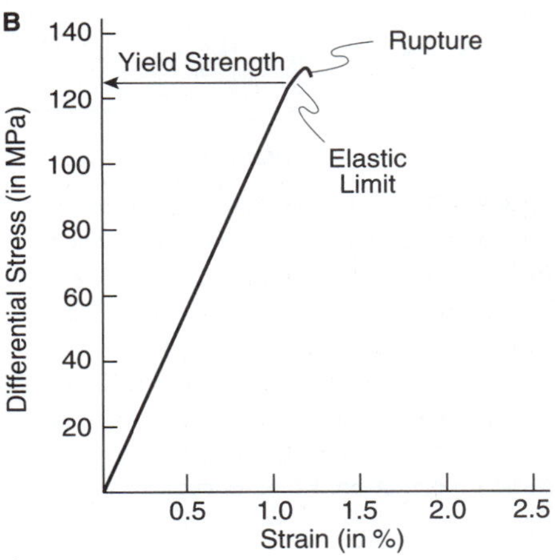

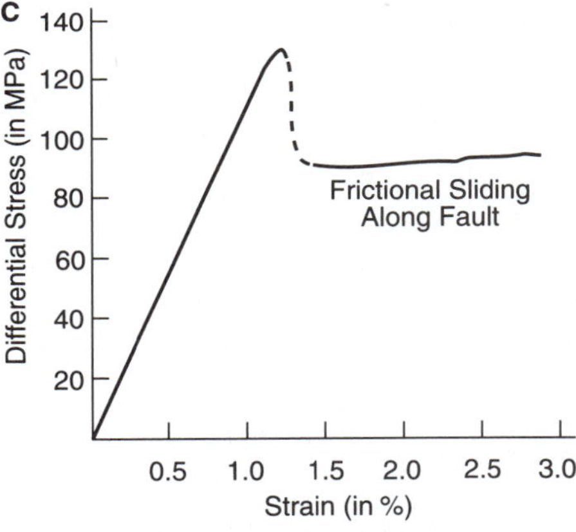

Figure 3.33 Onset of plastic deformation, then rupture, as shown in *(A)* a load–displacement diagram and *(B)* an equivalent stress–strain diagram. *(C)* If we press on with the experiment after rupture, continued shortening of the specimen is achieved by frictional sliding along the fault.

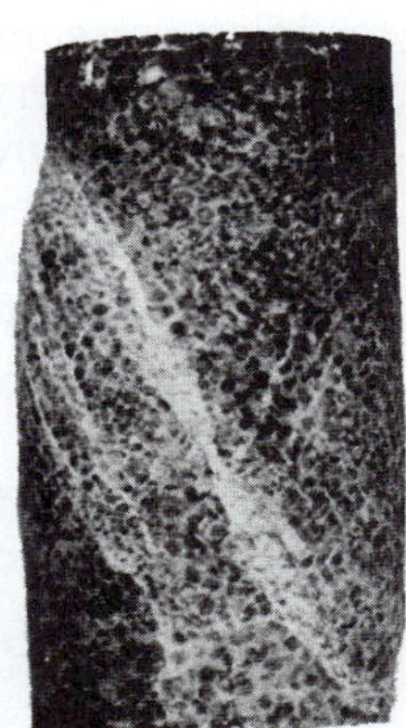

Figure 3.34 Faulted specimens of limestone. [Reprinted by permission, "Some Information Squeezed Out of Rock," by F. A. Donath, *American Scientist,* v. 58, fig. 7, pp. 54–72 (1970b).]

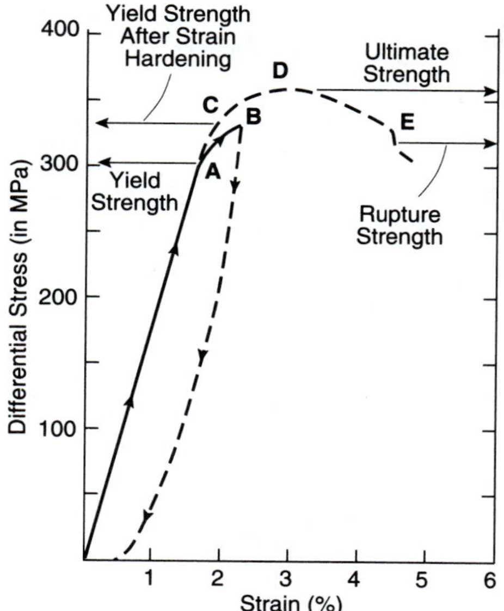

Figure 3.35 Stress–strain diagram for limestone subjected to deformation under confining pressure condition of 103 MPa. Point *A*, onset of plastic deformation. Point *B*, removal of axial load. Note the nonrecoverable plastic deformation. Curve *C*, the now permanently strained specimen is subjected to a second loading. Curve *D*, plastic deformation. Point *E*, rupture.

Before taking the axial stress to higher levels, it is interesting to bleed off all of the axial load shortly after the limestone has begun to deform plastically (curve *B*, Figure 3.35). When this is done, it becomes evident that shortening due to elastic deformation is quickly recovered, but shortening due to plastic deformation is nonrecoverable. Plastic deformation is bound up in the strained rock permanently. When load is reapplied to the specimen, the limestone once again behaves as an elastic solid, but this time the elastic limit is higher than the first yield strength we observed (curve *C*, Figure 3.35). The yield strength of the rock increases because the original fabric of the rock was modified slightly by the plastic deformation. The limestone is said to have undergone **strain hardening**, thus raising its yield strength.

But let us now deal the limestone sample its final blow. As more and more load is applied, the specimen is "pushed" toward failure. The stress–strain curve displays plastic deformation of the limestone in the form of a smooth, gently sloping curve (*D* in Figure 3.35). Unlike the test carried out at 28 MPa confining pressure, the onset of faulting does not follow immediately after a small amount of plastic deformation. Instead, the limestone seems to be able to endure a surprising amount of plastic flow. The limestone becomes so weak that the curve begins to descend to the right, steeply, signifying accelerated plastic deformation; this behavior is called **strain softening** because less stress is required to produce each new increment of strain. Eventually, there is a feeling of impending doom by rupture. **Rupture** by faulting finally occurs, and stress drops abruptly (curve *E*, Figure 3.35).

The stress–strain curve accurately records the details of this deformational experience (Figure 3.35). Above the yield strength the curve is convex upward, recording the gradual dissipation of rock strength during plastic deformation. The zenith of the curve corresponds to the **ultimate strength** of the limestone under the experimental conditions. The stress level at which faulting occurs is the **rupture strength**. Ultimate strength and rupture strength, like yield strength, are measured in terms of stress (e.g., in megapascals).

The results of our series of tests reveal that limestone becomes stronger at higher levels of confining pressure. This holds true regardless of the parameter used to describe the strength: yield strength, ultimate strength, and rupture strength are all higher for the test carried out at 103 MPa than for the test at 28 MPa. Moreover, the limestone undergoes greater plastic deformation at higher levels of confining pressure, assuming that temperature, rate of loading, and other such factors are held constant. The angle (θ) that the trace of the fault makes with the long axis of the core is seen to be greater, about 30°. The specimen of limestone that had faulted under confining pressure conditions of 28 MPa (what depth would that be??) was seen to be very fragile, somewhat powdery along the fault, and broken by a single fracture. In contrast, the specimen subjected to rupture at 103 MPa is more cohesive, slightly barrel shaped, and affected by a relatively wide zone of distributed fault surfaces, one of which accommodated the ultimate rupture (Figure 3.36). Under the higher confining pressure conditions, a greater volume of the rock was affected by the deformation.

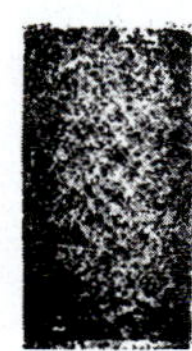
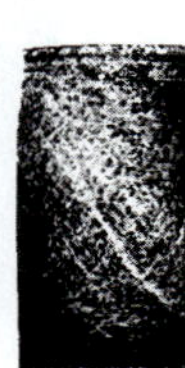

Figure 3.36 Specimens of limestone deformed to approximately the same total strain (about 15%) at different confining pressures. Increased confining pressure causes a transition in deformational mode from brittle to ductile faulting. [Reprinted by permission, "Some Information Squeezed Out of Rock," by F. A. Donath, *American Scientist*, v. 58, pp. 54–72 (1970b).]

TABLE 3.2
Summary of compressional testing of limestone under confining pressure conditions of 27.6, MPa, 103 MPa, and 207 MPa

	Specimen 1	*Specimen 2*	*Specimen 3*
Confining pressure	28 MPa	103 MPa	207 MPa
Differential stress at failure	124 MPa	318 MPa	552 MPa
σ_1 at failure	152 MPa	421 MPa	759 MPa
σ_3 at failure	28 MPa	103 MPa	207 MPa
Angle (θ) between fault and σ_1	25°	30°	33°

In this test, the rock responded in **semibrittle** fashion. Semibrittle rocks deform initially by elastic deformation, and they ultimately fail by the formation of discrete fractures and faults. However, the amount of plastic deformation following elastic deformation and preceding failure is much greater than that for brittle rock. For rock to be considered semibrittle, the amount of shortening before complete failure must be between 5 and 10%.

Still Higher Confining Pressure

When compression of the limestone is carried out at even greater confining pressure, for example, up to the red warning line on the Donath apparatus, at 207 MPa, the limestone responds with even greater strength and ductility. There is still an initial, though limited, elastic deformation, followed by sustained plastic deformation before failure. The limestone responds in a **ductile** fashion. Ductile rocks have nearly continuously curved stress–strain curves. They may initially deform elastically, but if they do, it is seldom long before they begin to respond plastically. They may or may not ultimately fail by the formation of discrete fractures or faults. For rock to be considered "ductile," the ultimate failure, if it does occur, takes place after the rock has shortened by at least 10%.

Table 3.2 presents a summary of the conditions, measurements, and results of compression tests on limestone at confining pressure levels of 28, 103, and 207 MPa. When these data are all plotted on a Mohr stress diagram (Figure 3.37), it becomes even more obvious that proportionately

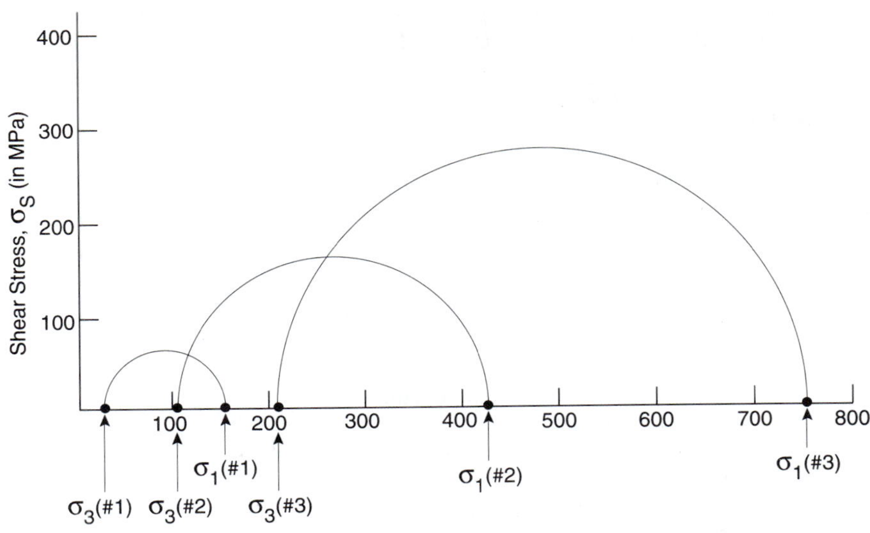

Figure 3.37 Mohr stress diagram summarizing the experimental testing of the limestone under three distinctly different confining pressure conditions.

greater levels of differential stress ($\sigma_1 - \sigma_3$) are necessary to cause rupture when confining pressure (σ_3) is raised. In Chapters 5, "Joints and Shear Fractures," and 6, "Faults," we will see that composite Mohr diagrams of this type become the basis for establishing the *laws* that describe the conditions under which rocks fail by fracturing and faulting.

Strength and Ductility

The mechanical response of rocks to stress is different for different conditions. If we were to examine hundreds of stress–strain diagrams generated during laboratory testing of every rock imaginable under every conceivable set of conditions, we would see significant differences in strength and ductility from graph to graph. One of the chief conclusions that can be drawn from experimental compression tests is that measurements of rock strength and ductility are almost meaningless unless the conditions under which the deformation was achieved are also given. Furthermore, each variable such as temperature or strain rate, has a different effect on the rheology of the rock being tested.

Lithology

All other things being equal, it is possible to arrange the common rock lithologies in order of increasing strength for specific conditions, such as confining pressure, temperature, or rate of loading. Stress–strain diagrams provide the basis for the rankings. The rankings are only approximate, being strongly influenced by the composition, texture, and general condition of the rocks that happened to serve as "representative" test specimens for each lithology. Moreover, the nature and orientation(s) of mechanical heterogeneity—that is, **anisotropy**, resulting from fractures, layers, foliations, and the like—profoundly influence rock strength.

Table 3.3 lists lithologies according to strength based on stress–strain diagrams summarizing compressional tests of rocks at room temperature under low levels of confining pressure. Rocks like salt, anhydrite, shale, and mudstone are seen to be weak (and ductile) compared to rocks of intermediate strength like limestone or calcite-cemented sandstone. Quartzite, granite, and quartz-cemented sandstone are brittle and very strong by comparison.

In a sequence of different lithologies, the rocks that are likely to behave in the most ductile manner when subjected to stress are commonly referred to as **incompetent**. Rocks that are likely to deform in a brittle manner,

TABLE 3.3
General ranking of lithology according to strength, based on tests at room temperature and low confining pressure

Strongest	Quartzite
	Granite
	Quartz-cemented sandstone
	Basalt
	Limestone
	Calcite-cemented sandstone
	Schist
	Marble
	Shale/mudstone
	Anhydrite
Weakest	Salt

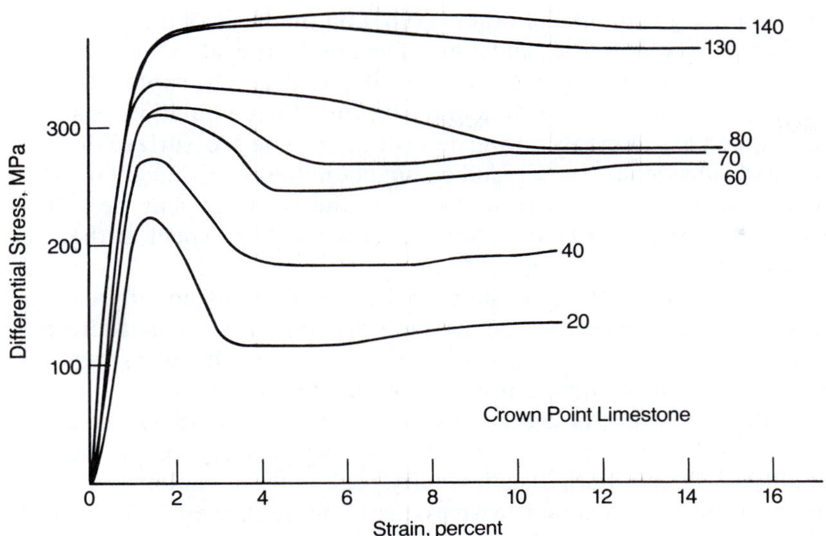

Figure 3.38 Stress–strain diagrams for limestone deformed at a variety of confining pressures. Tests conducted at room temperature. The magnitude of confining pressure (in MPa) for each run is shown next to each curve. Both strength and plasticity increase with greater confining pressure. [Reprinted by permission, "Some Information Squeezed Out of Rock," by F. A. Donath, *American Scientist*, v. 58, pp. 54–72 (1970b).]

with no obvious ductile deformation, are described as **competent**. "Competency" and "incompetency" are relative terms. The ordering of lithologies by competency may change if the conditions of deformation are changed. Changes in confining pressure, temperature, strain rate, and the presence of pore fluid may affect different rocks in different ways.

Confining Pressure and Pore Fluid Pressure

We have seen that increasing the confining pressure on a rock specimen in a compression test has the effect of increasing the strength and ductility of the rock. This has been firmly documented through experimental testing (e.g., see Handin and Hager, 1957). For any given lithology, the yield strength, ultimate strength, rupture strength, and ductility attain greater and greater values with increasing confining pressure. The stress–strain curves shown in Figure 3.38 reveal how the strength and ductility of the Crown Point limestone increase with increasing confining pressure.

The effect of confining pressure can be partially or completely offset by the presence of elevated **pore fluid pressure** in the rock (or test specimen) undergoing deformation. Figure 3.39 shows this nicely by means of graphs

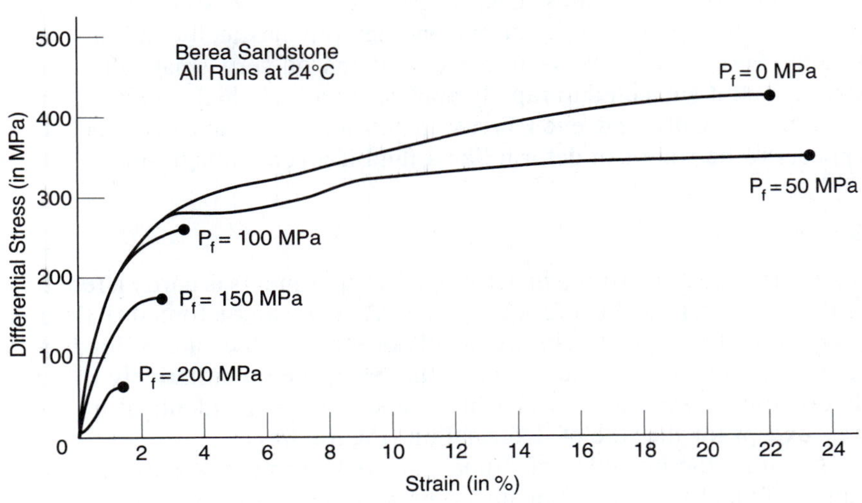

Figure 3.39 Laboratory derived stress–strain curves showing the influence of pore fluid pressure on the strength and ductility of rocks. All tests were carried out at 200 MPa confining pressure. Fluid pressure (P_f) for each run is shown next to each curve. Based on experimental work by Handin and others (1963). (From Marshak/Mitra, BASIC METHODS OF STRUCTURAL GEOLOGY, copyright © 1988, p. 201. Reprinted by permission of Prentice Hall, Upper Saddle River, New Jersey.)

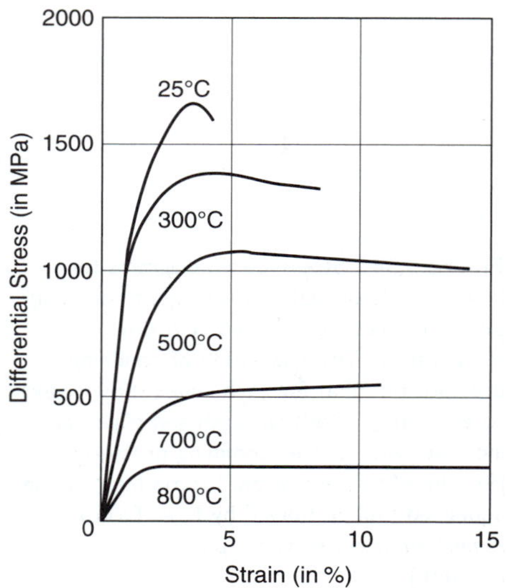

Figure 3.40 Stress–strain diagram for basalt deformed at 5-kbar confining pressure under a variety of temperature conditions. [From Griggs, Turner, and Heard (1960), Geological Society of America.]

constructed by Handin and others (1963) on the basis of a series of compressional tests on Berea sandstone. The graphs reveal that elevated fluid pressure can dramatically decrease ultimate strength, rupture strength, and ductility. In natural sedimentary basins of accumulation, water that is entrapped in sediments during deposition may be pressurized during the course of subsidence, burial, and compaction due to loading by overlying, younger sediments. As a result, a pore fluid pressure may be achieved that exceeds the hydrostatic stress level we would expect at that depth (Hubbert and Rubey, 1959).

Elevated hydrostatic pore fluid pressure conditions counteract the effects of confining pressure on strength and ductility. A measure of the net effect of confining pressure and fluid pressure is **effective stress**: effective stress equals confining pressure minus fluid pressure.

If effective stress is high, a given rock will be relatively strong and ductile; if effective stress is low, the rock will display less strength and ductility. A deeply buried sedimentary rock with anomalously high pore fluid pressure will respond to stress as if the rock were deforming in a relatively low confining pressure environment.

Temperature

An increase in the temperature of a rock generally depresses yield strength, enhances ductility greatly, and lowers ultimate strength. Some rocks are more responsive to the effects of temperature than others. Igneous rocks are less affected by modest increases in temperature than sedimentary rocks (Ramsay, 1967); they are more at home in high-temperature environments. Figure 3.40 presents a typical example of the profound influence of temperature on strength and ductility.

If heated sufficiently, rocks may deform in a plastic or viscous fashion and thus undergo very large permanent strains without ever rupturing or losing cohesion. Viscous materials, in effect, flow when subjected to *any* differential stress, no matter how weak. Unlike materials that deform plastically, truly viscous materials possess no fundamental strength threshold that otherwise would have to be overcome to induce flow. Elevated temperatures promote viscous deformation and cause rocks to flow.

Very seldom do upper crustal rocks behave in ideally viscous fashion. Even though converted to viscous fluids by greatly elevated temperature conditions, most rocks retain the capacity of behaving elastically under relatively rapid strain rates. One of the best examples of a material that possesses this dual capacity of responding both elastically and viscously is Silly Putty, which flows under conditions of very small differential stress. Yet, if subjected to rapidly applied, relatively high stresses, it will behave elastically and even break in tension. In essence, a rock that appears to be solid can deform like a fluid if given enough time.

Strain Rate

Rock strength as measured in deformation experiments is partly a function of the rate at which the stress is applied. A rock can be forced to deform plastically at comparatively low levels of stress if the rate of loading is slow. Heard (1963) helped to verify this principle quantitatively through deformational experiments in which conditions were identical in every way except for the rate of deformation (Figure 3.41).

In testing the influence of "time" on rock strength, we can compare values of yield strength, ultimate strength, and rupture strength as a function of the rate of loading. But normal practice is to carry out a constant

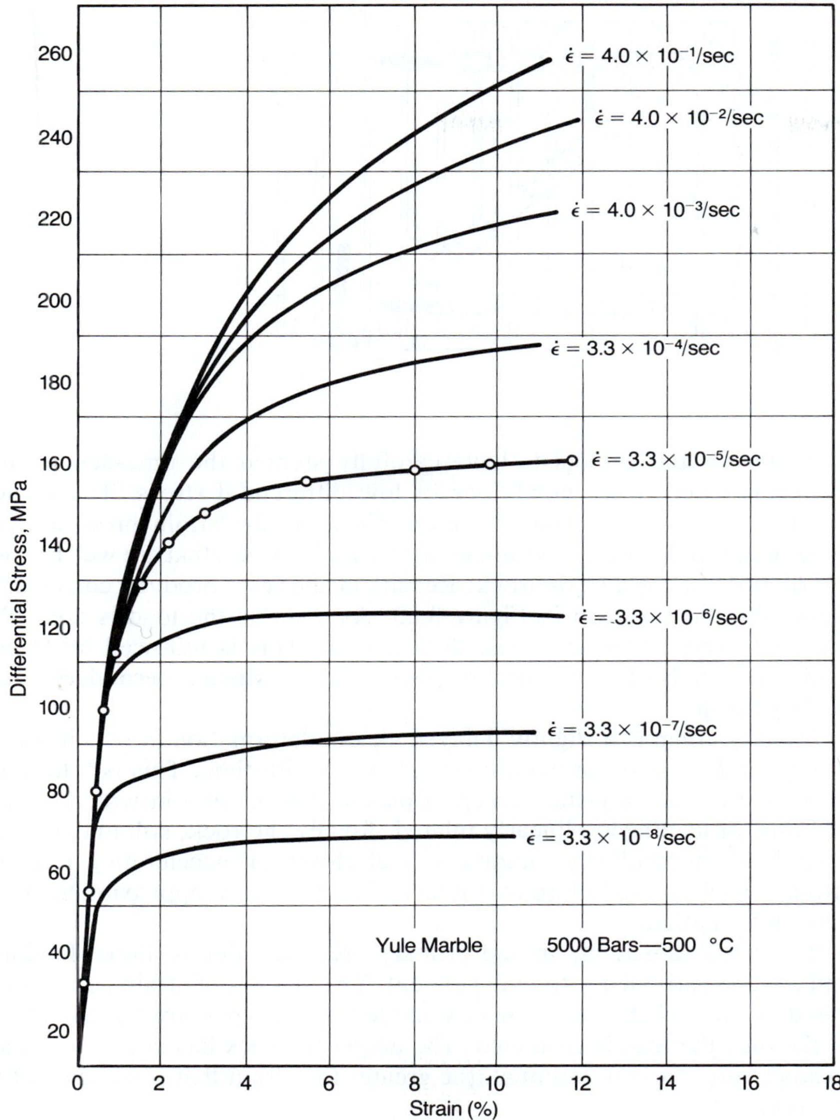

Figure 3.41 Stress–strain diagram for Yule marble deformed at different strain rate conditions. The higher the strain rate, the stronger the rock. [After Heard (1963). Copyright © 1963 by the University of Chicago. All rights reserved.]

strain test, wherein the rate of strain is held constant. When the strain rate is relatively low, the amount of stress required to produce plastic deformation and ultimate failure is smaller than for experiments with higher strain rate.

The observable decrease in rock strength as a function of long-sustained stress is not too surprising. It can be thought of as a fatigue that sets in with time. Stress fractures in athletes are a human manifestation of the same phenomenon. Consider the marathon runner who trained 15–25 km/day for weeks, ran a competitive marathon, and then, after still more hard training, ran a second competitive marathon less than a month later. The fracture of his femur was an expression of the capacity of materials to fail at very low differential stress levels. Such failure happens when stress is permitted to operate, even intermittently, for long periods of time (R. Mark Blew, M.D., personal communication, 1981). Perhaps it is timely to confess that my gate no longer holds, in spite of the nailed brace (see Figure 1.30). Fatigue due to repeated low-stress usage by my boys fetching bikes and basketballs rendered it useless. But look at it now, pinned with 92 bolts (Figure 3.42)!

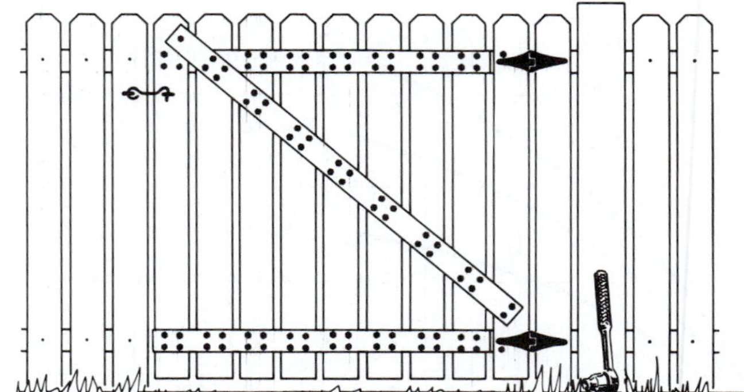

Figure 3.42 Fighting the weakness that comes with time and use. The ultimate (!?) solution to strengthening the gate.

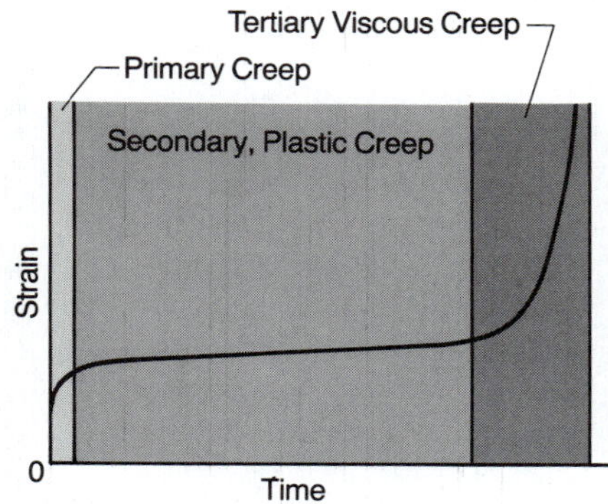

Figure 3.43 Strain versus time diagram, showing the fundamental classes of creep: primary, secondary, and tertiary.

Engineers and geologists have carefully studied the time-dependent strain produced under conditions of low differential stress. The name given to this kind of strain is **creep**. Creep is the strain produced in experiments of long duration under differential stresses that are well below the rupture strength of the rock. Results of the tests produce consistent plots, like those shown in Figure 3.43. As soon as the load is applied, the rock experiences an elastic deformation. This is followed by three distinctive kinds of mechanical response, called **primary**, **secondary**, and **tertiary creep**.

Primary creep is a slightly delayed elastic deformation in which there is a general decrease in the amount of strain with time. This is followed by secondary, steady-state creep, a plastic deformation in which strain and time of loading are linearly related. Finally the rock, still under constant load, dramatically fatigues and discloses an accelerating rate of strain. This leads to failure by rupture. Tertiary creep approximates viscous deformation.

The initial elastic strain and primary creep are due to initial loading and are not particularly time-dependent. The amount of strain accommodated by the rock during secondary and tertiary creep is strictly a function of the time the load is sustained. The longer the rock has been subjected to some tiny stress differential, the greater the strain that is sustained by the rock.

Experimentalists talk about "trading time for temperature" (Jan Tullis, personal communication, November 1992). They are not in a position to routinely run laboratory deformational experiments that last for decades, let alone centuries or millennia. Temperature for time becomes the tradeoff. Long-duration experiments can be simulated by increasing the temperature at which the experiment is carried out. The expectation is that the increased temperature will achieve the level of rock weakening that otherwise would be achieved by lowering the strain rate.

Time

It is natural that geologists (i.e., scientists who have 4.5 billion years to work with) focus on time dependency as one of the most important and perhaps least understood of all the independent variables that influence what we call the strength of rocks. S. W. Carey (1953), a brilliantly creative structural geologist, came to question whether rocks possess *any* fundamental strength. He reasoned that if rocks convert to plastic and viscous substances under conditions of long-duration loading, eventually they must fail by flow and then by rupture, following accelerated tertiary creep. Carey made the scientific community aware of a state of matter that is not encompassed in the terms "solid," "liquid," and "gas." The state is that of a **rheid:**

> A substance whose temperature is below the melting point, and whose deformation by viscous flow during the time of the experiment is at least three orders of magnitude (i.e., 1000×) greater than the elastic deformation under the given conditions. (From "The Rheid Concept in Geotectonics" by S. W. Carey, p. 71. Published with permission of Geological Society of Australia, Inc., copyright © 1953.)

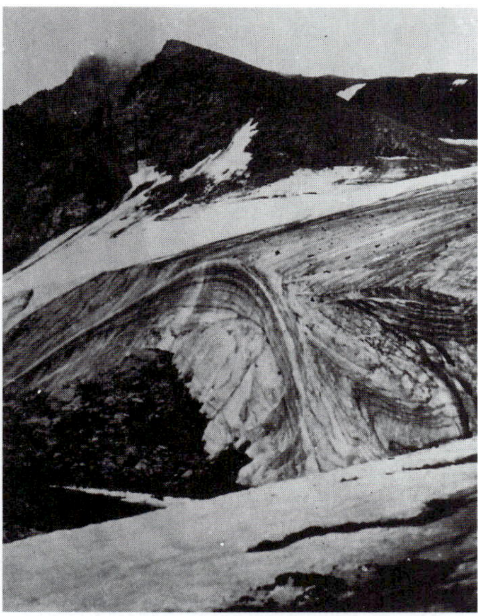

Figure 3.44 Glacial ice, a rheid in action. Ice front of Grasshopper Glacier, Montana, shows contorted laminae caused by differential movement in the ice. (Photograph by T. S. Lovering. Courtesy of United States Geological Survey.)

Under small values of differential stress, ice behaves like a rheid after only two weeks. The ice of a valley glacier "creeping" down a mountain flank expresses a rheid in action (Figure 3.44). Salt, under small values of differential stress, behaves like a rheid after only 10 years. Salt glaciers move down mountain flanks in the Middle East (Figure 3.45), and salt domes ascend like pillars through strata beneath the Gulf of Mexico. Where penetrated by mines in Texas, salt domes display folds and flow patterns that are characteristic of viscous deformation.

Salt and ice behave like rheids, but what about "real rocks"? Marble is a real rock, one we usually think of as rigid. Yet look at the marble bench in Figure 3.46. It is behaving like a rheid after times that are short even by human standards. How is it for the granite tombstone beyond the marble bench (see Figure 3.46)? If granite, below its melting point, were subjected to the force of gravity for 5, or 10, or 15 billion years, would it flow like butter?

Preexisting Weaknesses

In preparing samples for rock deformation experiments, structural geologists are normally meticulous about using pristine, fresh rocks that have no visible cracks or other flaws. It is thought that the presence of a crack or fracture will ruin the experiment. And yet we know that outcrop-scale

Figure 3.45 Kuh-i-Namak salt glacier, Iran. The mountain of salt is 6 miles across and 4300 feet high. The contact between salt and the Cretaceous rocks it penetrates is overturned in places. [Drawing of photograph in Holmes (1964), Figure 171, p. 241.]

Figure 3.46 Marble bench bent downward by its own weight and that of the occasional occupant. The marble rheid is located in cemetery north of Soldiers' Home, Washington, D.C. (Photograph taken in 1925 by W. T. Lee. Courtesy of United States Geological Survey.)

bodies of rock, and regional-scale bodies of rock, are absolutely pervaded by flaws, at all scales: from spaced, through-going faults, fractures, and bedding surfaces; to pervasive joints and veins and stylolites; to webs of microscopic hairline cracks and voids. If we were somehow able to extract a 1 km^3 cube of granite from a very deep quarry at the surface of the Earth and place it on an "Empire State Building" of infinite strength, the cube of granite would probably simply fall apart onto the streets below, as if it were cohesionless.

Larger bodies of rocks are weaker than smaller bodies of rock, in part because of preexisting internal flaws and weaknesses. The influence of the presence of such weaknesses will be greater when the conditions of deformation are fundamentally brittle. When the conditions are ductile, the whole body of rock tends to be affected by the deformation, and the influence of individual preexisting weaknesses like fractures and faults and bedding surfaces will not be as great.

The ease or difficulty with which preexisting fracture surfaces can be activated depends on the combination of friction and normal stress. When normal stresses on preexisting fractures are low, the friction along the fracture surfaces is dependent upon **surface roughness** (Byerlee, 1967, 1978). With higher normal stress, the friction along fracture surfaces becomes independent of rock type (Figure 3.47).

Size

Size alone has a lot to do with strength as well, as dramatically illustrated in the scaled deformation experiments carried out by structural geologists like Vendeville and others (1987), McClay (1989), Merle and Guillier (1989), and Brun and others (1994). When the sizes and strengths of materials chosen to replicate regional relationships, such as all of the Himalayas or the full length of the Andes, are precisely scaled (Hubbert, 1937), the materials chosen are extraordinarily weak. If the materials chosen are too strong, they resist the deforming influence of gravity that

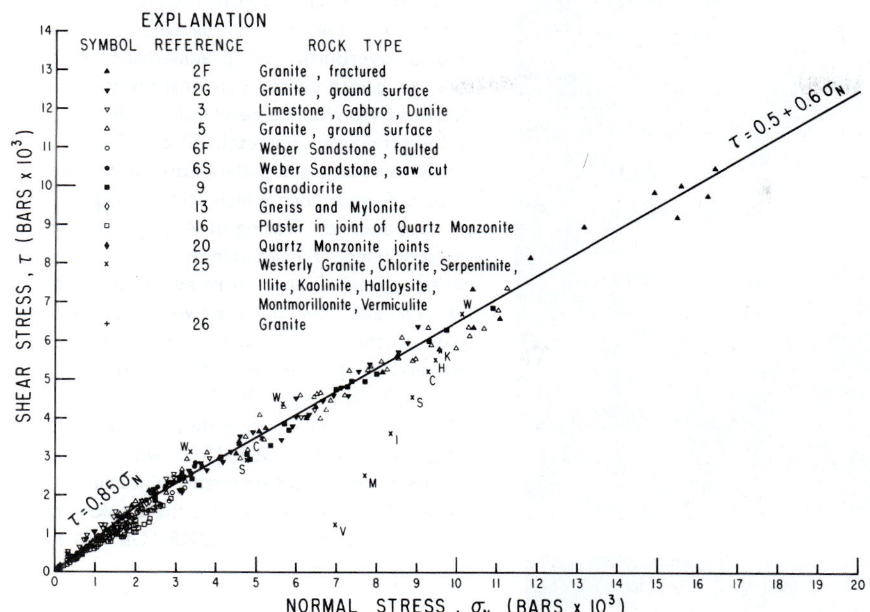

Figure 3.47 Byerlee demonstrated that the frictional properties of almost all rocks are the same. Except for the lowest normal stresses on a fracture surface, there is a direct linear relationship between amount of normal stress (σ_N) acting on the fracture surface versus the amount of shear stress (σ_S) required to overcome friction and initiate movement. [From Byerlee (1978). Permission from *Pure and Applied Geophysics*, v. 116, Byerlee, J. D., Friction of rocks, 1978, Birkhäuser Verlag AG, Basel, Switzerland.]

attends all natural deformation. If the role of gravity is missing from experimental deformation, the structures that are formed will not bear exact similarity to what is found in nature. Ramberg (1967) overcame this problem by increasing the force of gravity in his models . . . by deforming the clays while they whirled at high *G*s in a centrifuge. Ramberg ingeniously filmed the deformation live by mounting a TV camera inside the centrifuge!

Looking for a more versatile, less expensive, and less complicated solution, Cobbold "scaled down" the strength of the materials even further, and in a way that conformed *precisely* to Hubbert's requirements. For the crust of the Earth he selected loose sand, and for the upper mantle directly below it, Silly Putty. For the softer mantle he chose Czechoslovakian honey, which is reknowned for its perfect transparency. Then he fashioned these materials in stratified layers to form meter-scale physical models mechanically equivalent to a regional tract of the Earth's outer crust and upper mantle. The structures resulting from compressional shortening or extensional stretching of such multilayered, scaled models bore an absolutely uncanny similarity to what is found in regional systems of structures (Figure 3.48). Cobbold has added a final touch. To examine the structure of the base of the crust, he simply vacuums up the sand so that the top of the Silly Putty layer comes into view. To examine the structure of the base of the stiffer mantle, he looks up at the base of the Silly Putty layer through the transparent Czechoslovakian honey. To examine internal structure within the deformed upper crust, Cobbold adds some cement to the sand before it is stratified into layers. Then, upon completion of the experimental deformation, he adds a little water to the sand, lets it harden, and saws the deformed, scaled-down version of the upper crust along cross sections of choice.

A

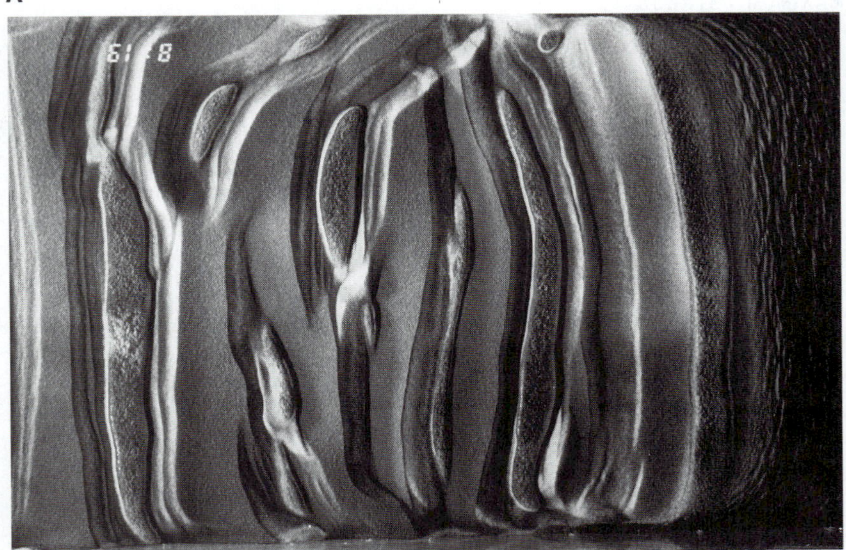

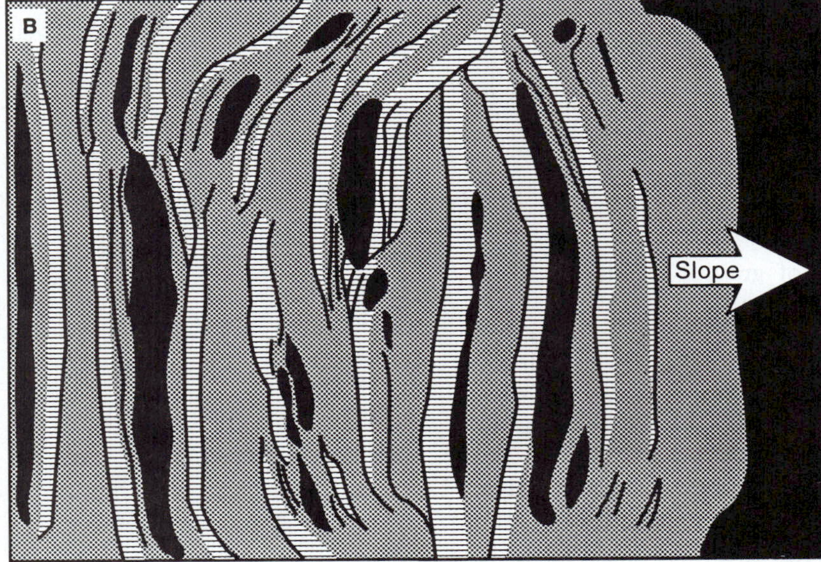

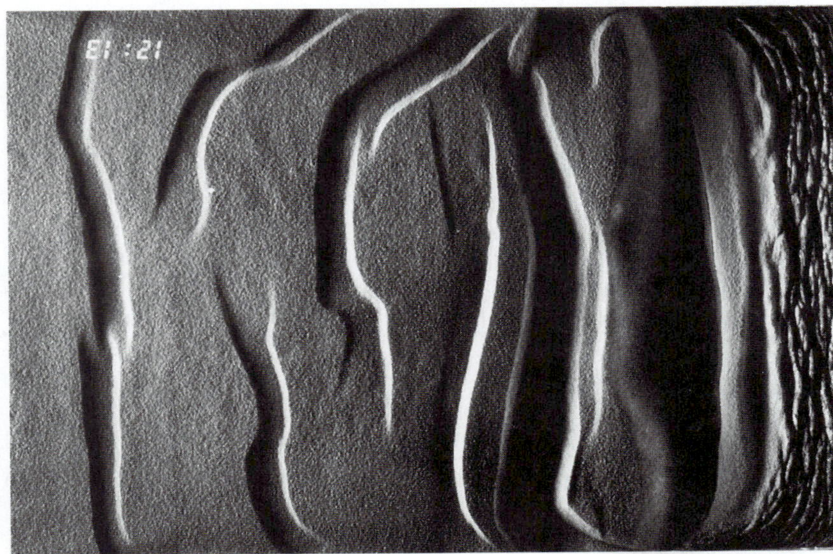

Figure 3.48 Two-stage deformational model consisting of initially tabular viscous silicone (representing salt) overlain by sand representing a tabular overburden of "prekinematic" sandstone deposited before deformation began. (*A*) Overhead view of the result of the silicone and sand extending to a stretch (*S*) of 1.75 by gravity spreading and gliding down a 2° slope. Grabens formed into which silicone diapirs rose and intruded. (*B*) Tracing of *A*. (*C*) Extension was continued, and simultaneously more sand was deposited. Grabens formed, as revealed in this overhead view, but they were not intruded by the silicone. (*D*) Structure section of the final condition. (Reprinted from Marine and Petroleum Geology, v. 9, Vendeville and Jackson, "The rise of diapirs during thinned skinned extension," pp. 331–353, 1992, with kind permission from Butterworth-Heinemann journals, Elsevier Science Ltd, the Boulevard, Langford Lane, Kidlington, OX5 1GB, UK.)

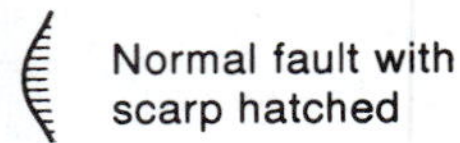

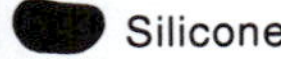

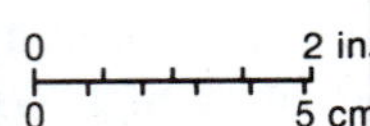

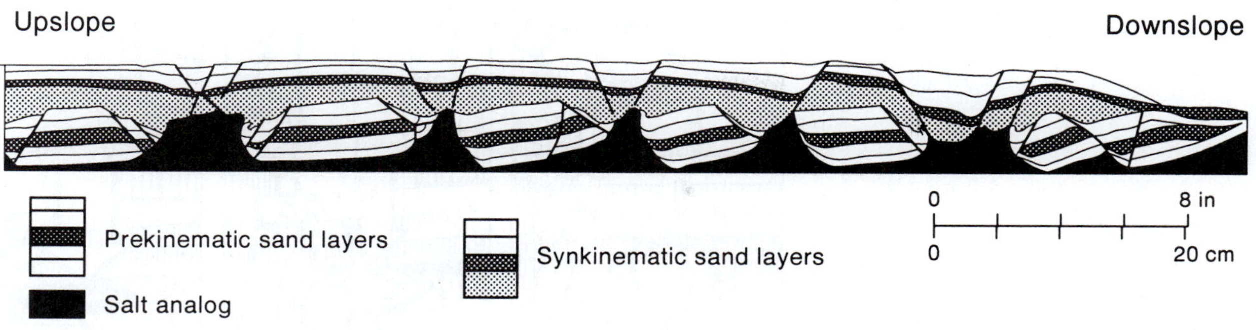

ELASTIC, PLASTIC, AND VISCOUS MODELS OF ROCK BEHAVIOR

The Need for Models of Behavior

To determine whether a test specimen of rock has responded to axial compression in a brittle, semibrittle, or ductile manner, all we need to know is the percentage of shortening before failure by fracturing. Furthermore, if we were to enter the lab after an experiment and view the naked, unjacketed test specimen after deformation, we probably would be able to tell right away if the rock had behaved in a brittle or ductile fashion. Ductile rocks tend to accommodate the shortening (i.e., deform) without loss of cohesion; they do this by distributing the deformation throughout the body, or at the very least within broad zones. Brittle rocks, on the other hand, accommodate shortening with pronounced loss of cohesion along through-going fractures. Instead of being distributed, the deformation is highly concentrated in narrow zones.

As useful as the terms "brittle" and "ductile" may be, they fundamentally emphasize strain. What we need at this point is a set of models that helps us envision the full interplay of stress *and* strain, helps us envision the fundamental ways in which rocks have been found to respond to stress. Indeed, there are three basic models: **elastic**, **plastic**, and **viscous behavior**.

Elastic Behavior

Think of springs under the bed of a flatbed truck. If we load the back of a flatbed truck with solid concrete blocks, the leaf springs of the truck will shorten by an amount that is directly related to the magnitude of the load (Figure 3.49). The bed of the truck gets lower and lower as the blocks are loaded on. But when the blocks are unloaded, the leaf springs recover and so does the bed of the truck. The relationship of the load of the blocks to the displacement of the bed is an equation that describes a straight line . . . and it works perfectly no matter whether you are loading or unloading. It is the same straight-line relationship that we observed during the early, elastic stage of formation of the limestone specimens.

The equation of the straight line describing the proportional relationship of stress to strain for elastic bodies is **Hooke's law** (Figure 3.50):

$$\sigma = Ee \qquad (3.5)$$

Figure 3.49 Elastic behavior is springlike and recoverable. (Artwork by D. A. Fischer.)

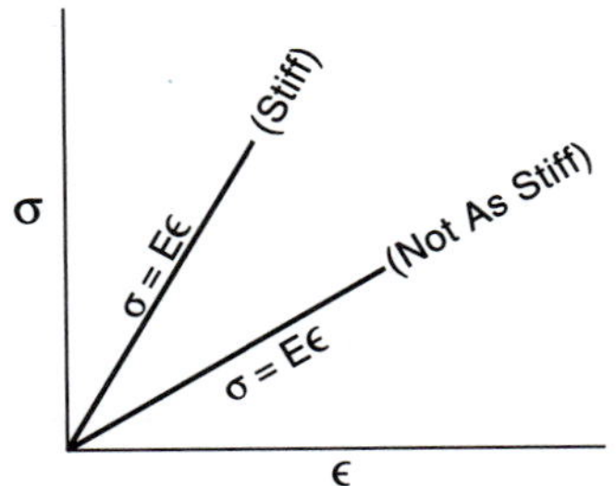

Figure 3.50 Portrayal of Hooke's law: Stress (σ) and strain (ϵ) are directly and linearly related. The constant of proportionality (E) is known as Young's modulus. It is the slope of the line.

where

$$\boldsymbol{E} = \textbf{Young's modulus}$$

$$E = \frac{\sigma}{e} = \frac{\text{stress}}{\text{strain}}$$

The value of ***E***, **Young's modulus**, describes the slope of the straight-line stress–strain curve. Even under the same conditions of deformation, the value of E will vary from rock to rock, reflecting natural differences in the resistance of rock to elastic deformation. Thus the slope of a straight-line stress–strain curve is a measure of the **stiffness** of the rock. Typical values of Young's modulus are presented in Table 3.4: the higher the values, the stiffer the rock. Because extension is unitless, Young's modulus is given in units of stress (e.g., MPa). If stress is compressive, and therefore positive (+), extension will be negative (−), and thus Young's modulus will be negative as well.

In the context of our experimental work, Young's modulus (E) can be thought of as an **elastic modulus** that describes how much stress is required to achieve a given amount of length-parallel elastic shortening of a core of rock. A second elastic modulus, known as **Poisson's ratio** and represented by the Greek letter ν (pronounced nu), describes the degree to which a core of rock bulges as it shortens. Poisson's ratio describes the ratio of lateral strain to longitudinal strain:

$$\nu = \frac{e_{\text{lat}}}{e_{\text{long}}} \tag{3.6}$$

TABLE 3.4
Typical values of Young's modulus (E)

Rock	*E(× 10⁴ MPa)*
Westerly granite	−5.6
Cheshire quartzite	−7.9
Karroo diabase	−8.4
Tennessee marble	−4.8
Witwatersrand shale	−6.8
Solenhofen limestone	−5.3

Poisson's ratio is unitless, for it is a ratio of extensions. Values of Poisson's ratio are presented in Table 3.5.

Poisson's ratio meant very little to me until one day in structural lab my students and I placed a core of granite in a deformation press and squeezed it. The core was 0.76 cm in diameter and about 13 cm long. We compressed it in an unconfined state (i.e., in open air) using a "soil testing" ram. We wired the core with strain gauges to monitor changes in length (longitudinal strain) and width (lateral strain). The wires were connected to an X–Y plotter to permit us to measure tiny magnitudes of elastic strain during the deformation. As we loaded the specimen, the rock began to shorten elastically, but to our surprise the shortening was not compensated by any increase in specimen diameter. The stress did not produce the expected lateral bulging, the "barreling," which we had anticipated. Volume decreased and stress somehow was being stored . . . that is, until the rock exploded with a blast like the sound of a shotgun. The largest rock fragment we found in the lab after the explosion was only pea-sized. We were grateful to have been protected by a Plexiglas shield. I will never forget the significance of very low values of Poisson's ratio. Rock bursts in deep underground mines commonly represent stress releases in rocks of low ν.

The granite would not have exploded if the core of granite had been surrounded by a confining pressure, one so strong that the granite had not been able to extend itself horizontally. To be sure, horizontal pressures would have been generated by the vertical loading, but explosive failure of the specimen would have been prevented. The generation of such horizontal stresses by vertical loading is known as the **Poisson effect**. The magnitude of such horizontal stresses ($\sigma_2 = \sigma_3$) is related not only to the vertical stress (σ_1), but also to Poisson's ratio (ν):

$$\sigma_2 = \sigma_3 = \frac{\nu}{1 - \nu}\sigma_1$$

Since Poisson's ratio for common rocks is approximately 0.25, the magnitude of stress generated by the elastic phenomenon described here as the Poisson effect is approximately one-third of the greatest principal stress.

$$\sigma_3 = \frac{0.25}{1 - 0.25}\sigma_1 = \frac{0.25}{0.75}\sigma_1 = \frac{1}{3}\sigma_1$$

There are two other parameters that describe the elastic relationship between stress and strain (Serway, 1990). One is the **bulk modulus (K)**, which is the resistance that elastic solids offer to changes in their volume. It is determined by dividing the change in hydrostatic pressure by the amount of dilation produced by the change in pressure. The other is the **shear modulus (G)**, which is the resistance that elastic solids offer to the shearing of planes past each other. The shear modulus is determined by dividing shear stress (σ_s) by the shear strain (γ) it produced.

$$K = \text{bulk modulus} = \frac{\Delta_{\text{hydrostatic stress}}}{\Delta_{\text{dilation}}}$$

$$G = \text{shear modulus} = \frac{\sigma_s}{\gamma}$$

TABLE 3.5
Typical values of Poisson's ratio (ν)

Rock Type	
Limestone, fine grained	0.25
Aplite	0.20
Limestone, porous	0.18
Limestone, oolitic	0.18
Limestone, chalcedonic	0.18
Limestone, medium grained	0.17
Limestone, stylolitic	0.11
Granite	0.11
Shale, quartzose	0.08
Graywacke, coarse grained	0.05
Diorite	0.05
Granite, altered	0.04
Graywacke, fine grained	0.04
Shale, calcareous	0.02
Schist, biotite	0.01

An Example of the Significance of Young's Modulus

Serway (1990) presents some wonderful examples of the use of elastic moduli. Here is one that shows the applied power of Hooke's law.

Steel railroad track has an average **coefficient of linear thermal expansion (α)** of 11×10^{-6}. This means that steel track that is 30 m long at 0°C will become 0.013 m (i.e., 1.3 cm) longer when the steel warms to a temperature of 40°C:

$$\alpha = \frac{1}{l_o} \times \frac{\Delta l}{\Delta T} = 11 \times 10^{-6}$$

$$\frac{1}{30\text{ m}} \times \frac{\Delta l}{40°\text{C}} = 11 \times 10^{-6}, \quad \Delta l = 0.013\text{ m}$$

Figure 3.51 Warped trolley tracks in Asbury Park, NJ. It might at first appear as if deformation of the steel was achieved by tectonic compression of New Jersey. However, all of the deformation took place during the course of a single day (date), an exceedingly hot July day. Deformation was due to thermal loading. [Drawing based on photograph in Serway (1990), p. 514.]

The 1.3 cm change in length looks insignificant until we look up Young's modulus for steel ($E = 8.67 \times 10^7$ N/m^2) and calculate first the stress, and then the force, created by the thermal expansion.

$$\sigma = Ee = \frac{F}{A}$$

$$\frac{F}{A} = E \times \frac{\Delta l}{l_o} = 20 \times 10^{10}\ \text{N/m}^2 \times 0.000433 = 8.67 \times 10^7\ \text{N/m}^2$$

Since

$$A = \text{area of steel rail} = 30\ \text{cm}^2$$

$$\text{force } (F) = \sigma \times A = 8.67 \times 10^7\ \text{N/m}^2 \times 0.003\ \text{m}^2 = 2.6 \times 10^5\ \text{N}$$

$$F = 2.6 \times 10^5\ \text{N} = 58{,}500\ \text{lb} = 29\ \text{tons!!}$$

This level of force and stress can exceed even the capacity of steel to respond elastically (Figure 3.51).

If we know Young's modulus (E), Poisson's ratio (ν), and the magnitudes of stresses in three mutually perpendicular directions (x, y, and z), it is straightforward to compute the amount of elastic strain (e) that would take place in directions x, y, and z. Here are the equations. They are marvelous summary statements of the interrelationships stress and elastic strain:

$$e_z = \frac{1}{E}[\sigma_z - \nu(\sigma_x + \sigma_y)]$$

$$e_y = \frac{1}{E}[\sigma_y - \nu(\sigma_z + \sigma_x)]$$

$$e_x = \frac{1}{E}[\sigma_x - \nu(\sigma_z + \sigma_y)]$$

With these equations you could revisit Figure 3.10, whip out your calculator, assume reasonable values for E and ν, and compute in a matter of minutes the percentage of (elastic) shortening vertically, horizontally north–south, and horizontally east–west. Powerful equations!

Plastic Behavior

Think of solid concrete blocks that must be slid from the front to the back of the bed of the flatbed truck (Figure 3.52). The concrete block is not plastic. But the response of the block to our push is analogous to the response of a plastic body. Ideal plastic bodies will not deform until a critical threshold of yield stress is exceeded. As we begin to push on the concrete block, nothing happens until the shear stress along the base of the block is great enough to overcome the resistance to sliding. If the floor of the bed is smooth and polished from years of this kind of activity, the resistance to sliding might be relatively low. If, however, there is leftover roofing tar smeared on the floor of the bed, the resistance to shoving might be relatively high. But once the yield stress has been overcome, and provided we are able to maintain the push at least at the level of yield stress, and assuming that the frictional resistance along the base of the bed does not change, we will be able to slide the block to the back of the truck without stopping. Even if we managed to maintain a push at exactly the level of yield stress, the rate of movement of the block would

Figure 3.52 The forcing of a block to move by building stress to a critical threshold is analogous to plastic deformation. Plastic deformation is non recoverable. (Artwork by D. A. Fischer.)

not be uniform, for stress and strain rate are not linearly connected in the case of plastic bodies.

If we do stop shoving and take a rest, we will not lose ground; the block will not rebound to the front of the bed as if it were spring-loaded. The movement of the block is nonrecoverable. When we decide to move the block a second time, we will need to put our backs into it to build the pressure to the yield stress and beyond.

An ideally plastic body undergoes no strain whatsoever below the yield stress, and this would be represented on a stress–strain curve by a vertical line that terminates at the value of yield stress (Figure 3.53). When the yield stress is achieved, the plastic body will strain as long as the yield stress magnitude is maintained (horizontal line).

"Plastic" rock bodies are not ideal. More typically, an initial elastic response is followed by steady increments of strain, each of which requires a steady, slightly increasing stress. A gentle rise of the stress–strain curve reflects this requirement, and the slope of this line is a measurement of the amount of **strain hardening** taking place within the material. It is as if the sliding friction along the floor of the bed of the truck is steadily increasing from front to back, perhaps because small clots of the roofing tar are scraped off the bed and accumulate along the leading edge of the blocks. Under some circumstances, less and less stress is required to maintain the strain: **strain softening** is taking place inside the rock. It is as if the sliding friction along the floor of the bed of the truck steadily decreased, becoming more and more slippery, from front to back.

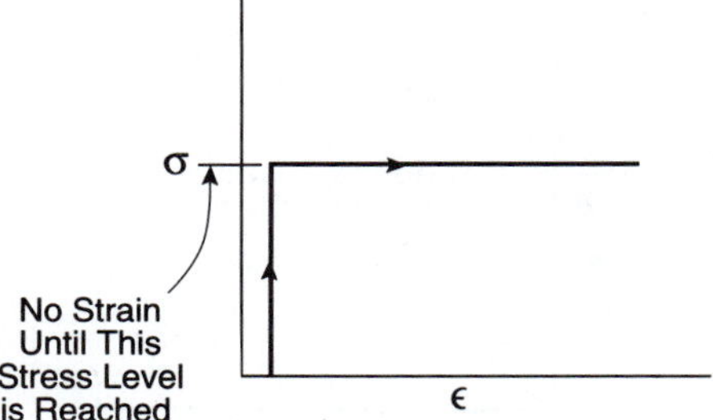

Figure 3.53 Portrayal of ideally plastic deformation. Stress (σ) is raised, but no strain (ϵ) accrues until a critical threshold is exceeded. From that point on, under ideal conditions, deformation continues as long as the stress level is maintained.

Viscous Behavior

Think of shock absorbers underlying the bed of the flatbed truck (Figure 3.54). Viscous bodies are fluids, and the behavior of a viscous body in response to stress is analogous to the inside action of a shock absorber. Stresses parallel to the axis of the shock absorber are generated by loading of the solid concrete blocks into the bed of the truck, sliding and jostling the blocks from the front to the back of the bed, and, once the truck is in motion, hitting bumps along the road. The loading and the bumps cause the piston within the shock to move within the cylinder, which is full of hydraulic fluid. The piston can move under the influence of stress because of opening(s) that permit the hydraulic fluid inside the shock to move from one side of the piston to the other. Even the smallest stress will displace the piston within the shock cylinder. There is absolutely no yield stress

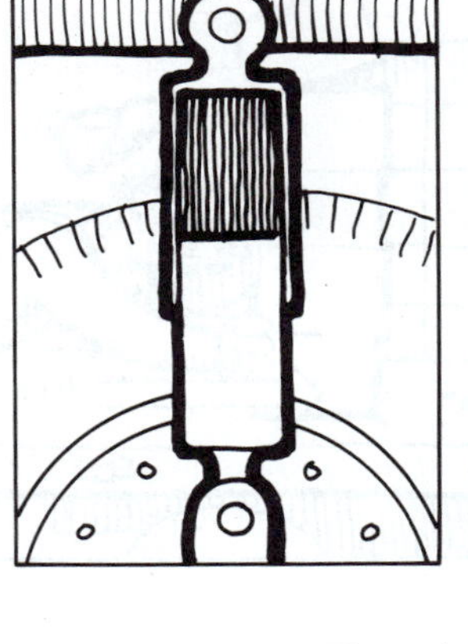
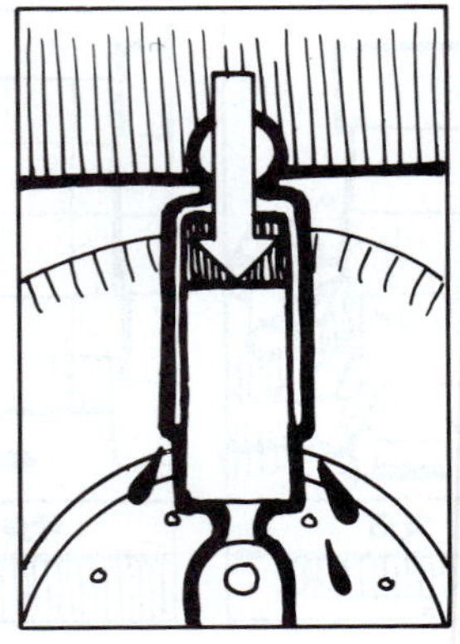

Figure 3.54 The action of shock absorbers is analogous to viscous deformation. Even the tiniest increment of stress will produce flow. Deformation is permanent, nonrecoverable. (Artwork by D. A. Fischer.)

to be overcome. There is no elastic deformation. The piston simply moves at the slightest provocation. If movement of the piston instantly stops, the hydraulic fluid instantly stops deforming. The displacement is not recoverable. And there is not even an elastic recovery, for there is no elastic deformation.

The rate of movement of the piston depends upon the amount of stress and the resistance of the fluid. For an ideal **Newtonian** fluid, there is a straight-line relationship between magnitude of stress and rate of strain (Figure 3.55):

$$\sigma_d = \eta\dot{\epsilon} \tag{3.5}$$

where σ_d = differential stress
η = viscosity
$\dot{\epsilon}$ = strain rate

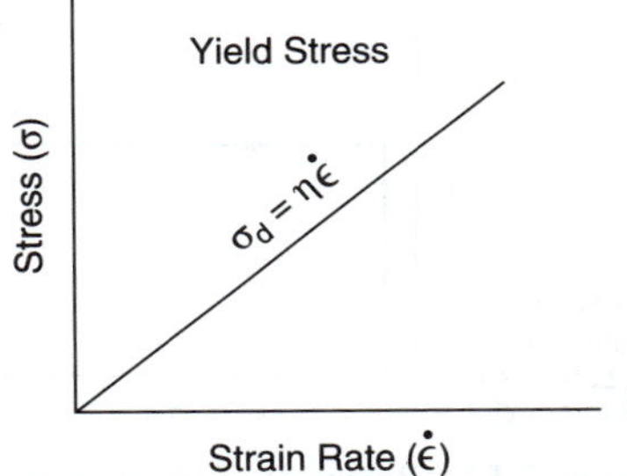

Figure 3.55 Portrayal of ideal viscous behavior on plot of stress (σ) versus strain rate (ϵ).

Viscosity is a measure of resistance to flow (Figure 3.55), just as Young's modulus can be thought of as a measure of resistance to elastic distortion. The greater the viscosity, the greater the internal friction of the fluid. Viscosity is measured in poises: "if a shear stress of one dyne/cm^2 acts on a liquid and gives rise to a strain rate of 1.0 sec^{-1}, the liquid has a viscosity of 1 poise (. . . where 10 poises = 1 Pa sec)" (Price and Cosgrove, 1990, p. 21). Viscosities for common fluids are presented in Table 3.6.

I learned from Jerry, one of the mechanics at Desert Toyota in Tucson, that the opening between hydraulic compartments in some modern shocks is "smart": it becomes smaller when the force on the piston is greater, and expands again when the force goes back to normal load. The auto engineers thus found a way to change the effective viscosity of the fluid without changing fluids. When the hole closes down, the apparent viscosity increases, for it takes more force to "drive" the fluid through the opening.

TABLE 3.6
Viscosities of common fluids

Most viscous	The Earth's mantle	10^{23} poises
	Salt	10^{17}
	Rhyolite lava	10^{9}
	Roofing tar	10^{7}
	Basalt lava	10^{3}
	Corn syrup	10^{2}
	Castor oil	10^{1}
	Heavy machine oil	6
	Olive oil	.8
	Turpentine	.01
Least viscous	Water at 30°C	.008

Figure 3.56 The flatbed truck, while negotiating the bumpy road, utilizes both its springs and shocks and redistributes its load of blocks in an effort to achieve a smoother ride. (See Twiss and Moores, 1992, p. 367). Rocks in any part of the crust of the Earth, in analogous fashion, utilize imaginative combinations of elastic, viscous, and plastic behaviors to relieve stresses and achieve a smoother strain. (Artwork by D. A. Fischer.)

Summary

The Earth, like the flatbed truck we loaded with concrete blocks and drove away, has some parts that are behaving elastically, some that are behaving plastically, and some that are behaving viscously. And all these behaviors are going on simultaneously (Figure 3.56). To make matters even more interesting, the parts work in tandem such that there are combined mechanical responses for every occasion: viscoelastic, elastic–plastic, viscoplastic, firmoviscous (see Ramsay, 1967; Suppe, 1986; Price and Cosgrove, 1990; Twiss and Moores, 1992). Rocks have an appropriate mechanical response for every conceivable situation.

CONCLUSIONS

We commonly believe that the magnitude of the largest stresses that can be generated in the crust is fundamentally due to the forces that create the stresses. However, the really important limitation is the strength of the rocks themselves: the weaker the rocks, the lower the stresses that can be achieved and sustained. In the upper crust, where rock temperatures are not especially high, the strength and ductility of rocks will increase with depth as confining pressure is increased. Relatively high stress levels derived from tectonic stresses and stresses due to loading can be sustained at moderate depths. The rocks will behave as elastic bodies, and when they deform they will deform in a brittle or semibrittle manner. However at deeper levels, where the temperatures are higher, the rocks will be weaker and more ductile, and these properties will cause the rocks to behave as plastic or viscous bodies. The stress levels that can be achieved and sustained are very low.

The empirical behaviors of rocks are now well established. Stress–strain curves are becoming more and more predictable. But what is really happening? What is really happening *inside* the rock that permits it to behave elastically in one situation, semibrittley in another, and plastically or viscously in yet another? We really must go "inside" to understand what is going on. We must move to the scale of individual crystals, and lattices of crystals, to discover how rocks can behave elastically under one set of conditions, and plastically or viscously under some different set of conditions. This is the subject of the next chapter.

CHAPTER 4

DEFORMATION MECHANISMS AND MICROSTRUCTURES

The broad range of structures that we see at the scale of individual outcrops and hand specimens (Figure 4.1*A–C*) draws our attention to the interior of rocks and mineral grains to see what is going on. We go "inside" to explore what processes operate at the microscopic and atomic scales that enable rocks, which seem so hard, to change size and shape as if they were soft. We journey inside to determine what mechanisms are available to give rocks such versatility in deforming elastically *or* plastically *or* viscously, depending on conditions. In short, by going inside for a while and looking around, we will emerge with a better understanding of how deformation is actually achieved.

Processes that permit rocks to deform at the microscopic and atomic scales are called **deformation mechanisms**. The fine-scale structures that are produced are called **microstructures**. Nature has a long list of deformation mechanisms to choose from, and this accounts for the wide variety of observed microstructures and textures. Some deformation mechanisms are brittle, like the splitting of minerals along microscopic cracks or cleavages; others are ductile, like the slippage on hundreds of parallel crystallographic planes or the forced-march migration of atoms from one side of a mineral to another.

We can get a rough idea of some microstructures in the field, through hands-and-knees, close-up peering at outcrop surfaces through a 10× hand lens. But the real work is done through examination of small specimens using a petrographic microscope or an ultra–high magnification transmission or scanning electron microscope.

We can go from microstructures to the deformation mechanisms that produced them by creating microstructures under controlled laboratory conditions, describing the microstructures on the basis of high magnifica-

Figure 4.1 Structures in outcrop reveal the influence of different deformation mechanisms. (*A*) Breccia texture within a Tertiary landslide, Hieroglyphic Mountains, central Arizona. (*B*) Folded pebbles in a deformed carbonate-clast conglomerate, Rincon Mountains, Arizona. (*C*) Gneissic and mylonitic foliations formed by ductile mechanisms, such as dislocation creep, in banded gneiss, Chemiheuvi Mountains, southeastern California. (Photographs by S.J. Reynolds.)

tion microscopy, comparing the microstructures produced in the laboratory to those produced in nature, and applying concepts and interpretations consistent with theoretical analysis of the strength of minerals. In fact, we can deduce the step-by-step evolution of microscopic deformation by means of controlled, start-and-stop experiments.

We begin with a discussion of how crystalline structure and interatomic bonding influence the strength of minerals. We then present the "players," that is, the geologically important deformational mechanisms. Before the chapter ends, we hope to be able to recognize the different types of microstructure when we see them, and to use microstructures as a basis for interpreting the deformation mechanisms and physical conditions that prevailed during deformation.

CRYSTALLINE STRUCTURE AND THE STRENGTH OF SOLIDS

Bonding and the Lattice of Crystals

The **lattice** of a crystal is a three-dimensional repetitive array of atoms, ions, or molecules (Figure 4.2). Both the positioning and the spacing of atoms within the lattice of a crystal are very systematic. The preferred configuration is one that requires minimum energy to maintain. This so-called **equilibrium configuration** reflects the delicate balancing of competing interatomic forces.

The equilibrium configuration of ions of Na^+ and Cl^- in halite is maintained by a dynamic balancing act: the positively charged nuclei of the sodium (Na) and chloride (Cl) atoms repel one another, and the Na^+ and Cl^- ions attract one another because of their opposite charges. The attracting forces tend to hold the atoms together via the process of **bonding**. If the repelling forces exceed the attracting forces, the bonds break and the individual atoms may become more mobile, such as when rocks are heated up and begin to melt.

There are three fundamental types of bonding in minerals: covalent, ionic, and metallic. **Covalent bonds** form when two adjacent atoms share one or more electron. In diamond, adjacent carbon atoms form covalent bonds by sharing electrons with each other (Figure 4.3*A*). **Ionic bonds** form when an atom loses one or more electrons to another atom. Halite (NaCl) is held together by ionic bonds formed when a sodium atom loses a "spare" electron to a nearby, electron-hungry chlorine atom (Figure 4.3*B*). **Metallic bonds** exist where electrons are able to move relatively freely through the material, rather than being strongly attached to any particular atom (Figure 4.3*C*). The mobility of the weakly attached electrons gives metals such as copper their capacity to conduct electricity.

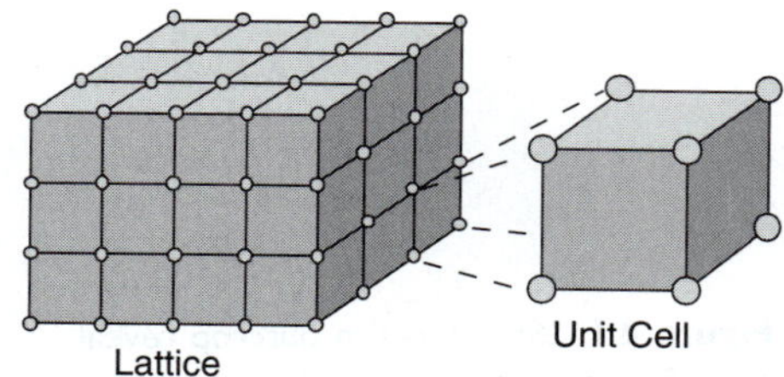

Figure 4.2 Crystalline lattice of a simple solid, showing a unit cell, the smallest known repetitive structure composing a crystal. Spheres represent atoms, ions, or molecules, and connecting lines represent bonds. (Adapted from *Crystalline plasticity and solid state flow in metamorphic rocks: selected topics in geological sciences* by A. Nicolas and J.P. Poirier, copyright © 1976. Published with permission of John Wiley & Sons, Inc., New York.)

Elastic Deformation of a Lattice

The normal, equilibrium spacing of atoms in a crystalline lattice is altered when the lattice is subjected to stress, as created for example, by gravitational loading, thermal loading, or tectonic loading. When an **ideal crystal** (no flaws or defects!) is hydrostatically loaded (equal, all-sided stress) and forced to deform elastically, the distance between atoms is shortened by an amount related to the imposed compressive stress and to the strength of the interatomic forces (Figure 4.4*A,B*) (Nicolas and Poirier, 1976). When the load is removed, the stored potential energy in the compressed

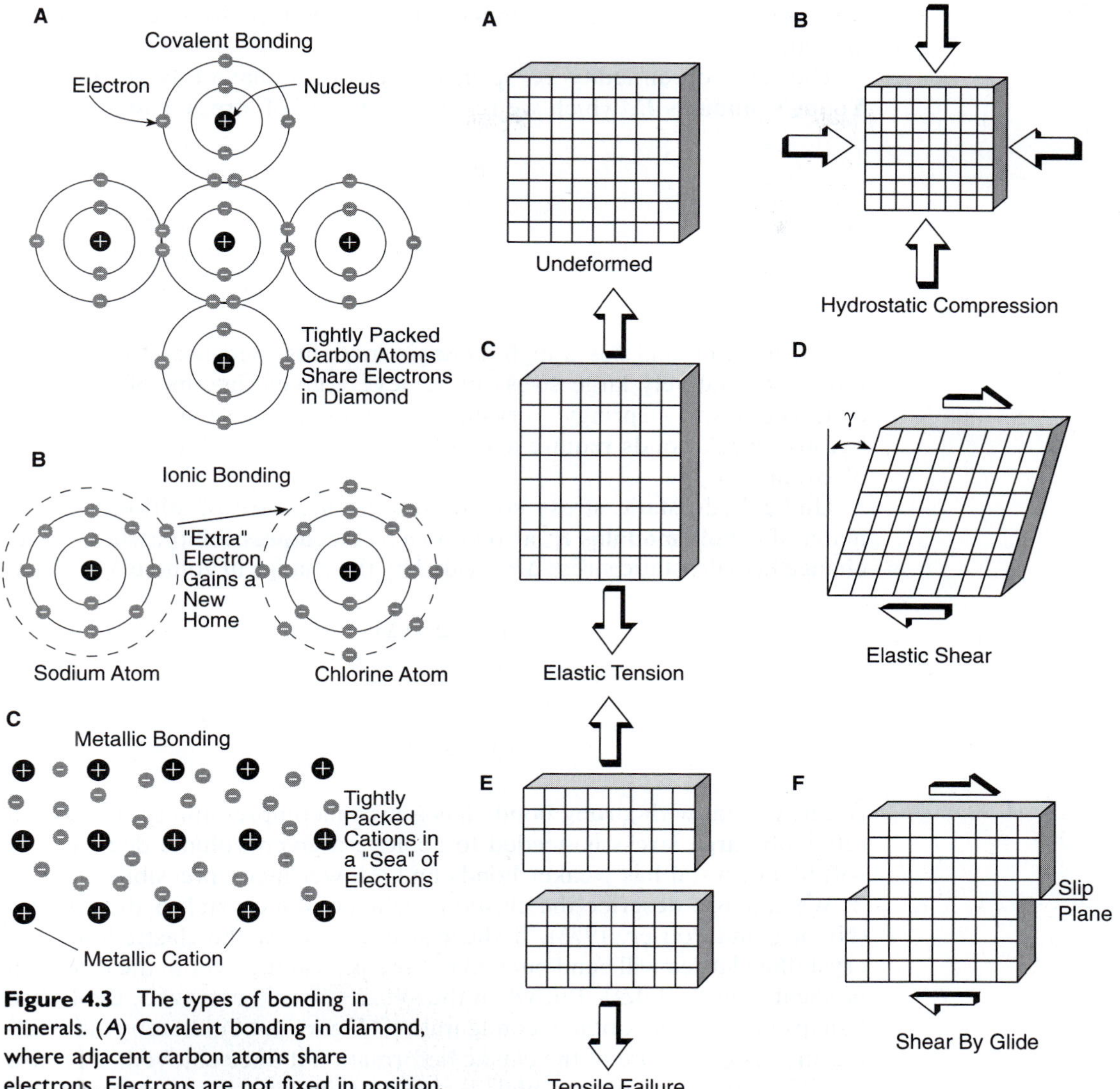

Figure 4.3 The types of bonding in minerals. (*A*) Covalent bonding in diamond, where adjacent carbon atoms share electrons. Electrons are not fixed in position, but instead may reside anywhere within the spherical electron shells. A shared electron may spend equal time around each of the two adjacent nuclei. (*B*) Ionic bonding in halite, where a sodium atom loses a spare electron to a chlorine atom. (*C*) Metallic bonding, where electrons are able to move fairly freely through a three-dimensional framework of positively charged cations.

Figure 4.4 Elastic and inelastic deformation of an ideal crystal. (*A*) Starting configuration. (*B*) Elastic response to an increase in confining stress (hydrostatic compression). The bonds shorten and the crystal decreases slightly in size. The energy is stored as potential energy, as in a compressed spring. (*C*) Elastic response to an imposed tensile stress. The amount of elastic strain shown is greatly exaggerated compared to that which is possible in real geologic materials. (*D*) Elastic response to an imposed shear stress. (*E*) Rupture when tensile stresses exceed the yield strength under tension. (*F*) Shearing via glide or slippage along a crystallographic plane (slip plane). (Adapted from *Crystalline plasticity and solid state flow in metamorphic rocks: selected topics in geological sciences* by A. Nicolas and J. P. Poirier, copyright © 1976. Published with permission of John Wiley & Sons, Inc., New York.)

bonds causes the lattice to rebound to its original equilibrium configuration (Figure 4.4*A*). Because the deformation produced is perfectly elastic, the crystal does not sustain any permanent strain, and thus there is no visible record of the deformation. Similarly, when a crystal is subjected to a small amount of tensile stress, the bonds will stretch by a small amount, permitting the lattice to expand ever so slightly (Figure 4.4*C*). As long as the stretching is less than the tensile elastic limit of the crystal, the lattice will rebound to its original configuration as soon as the stresses are removed.

The ease or difficulty of springlike *elastic* behavior is revealed in **Young's modulus *E***, which we recall is the ratio of stress σ to strain ϵ:

$$E = \sigma/\epsilon$$

or

$$\sigma = E\epsilon$$

A stiff ideal crystal has a high Young's modulus, because strong bonds require a relatively large stress to achieve a given amount of strain. In contrast, a soft "springy" crystal possesses a low Young's modulus, because weak bonds require a smaller stress to achieve a given amount of strain.

Under hydrostatic stress conditions, the measure of stiffness of the material is **bulk modulus *K***, also known as **incompressibility**, which is the change in hydrostatic stress $\Delta\sigma$ divided by the change in volume ΔV. Thus:

$$K = \Delta\sigma/\Delta V$$

or

$$\Delta\sigma = K\Delta V$$

A stiff crystal with strong bonds has a high incompressibility, because a relatively large stress is needed to achieve a given volume decrease. A soft ideal crystal has weaker bonds and a lower incompressibility.

We can also describe the elastic resistance of a crystal to shear stress (Nicolas and Poirier, 1976). If shear stress is below the elastic limit, the crystalline lattice will bend but will not break, leaning over in the direction of shear (Figure 4.4*D*). But when the shear stress is removed, the lattice "snaps back" to its original configuration (Figure 4.4*A*), thereby eliminating any visible record of the elastic deformation it once enjoyed. The ease or difficulty of creating an elastic shear strain in a crystal is expressed via the **shear modulus *G***, which is the ratio of shear stress σ_S to shear strain γ:

$$G = \sigma_s/\gamma$$

or

$$\sigma_S = G\gamma$$

An ideal crystal with a high shear modulus *G* has relatively rigid, inflexible bonds compared to a crystal with a lower shear modulus.

Exceeding the Elastic Limit

If the tensile stress acting on a lattice exceeds the tensile elastic limit of the crystal, a row of bonds will be severed and the crystal will fracture or cleave (Figure 4.4*E*). The fracture may irregularly crosscut the lattice, or it may follow a crystallographic plane of weakness. It all depends on internal structure and the relationship of the direction of stress to the internal structure. The stress required to rupture the bonds in tension is

TABLE 4.1
Experimentally determined compressive and tensile strengths of some rocks and minerals

Mineral or Rock Type	*Compressive Strength* (MPa)	*Tensile Strength* (MPa)
Clay	10	—
Calcite	14	—
Halite	27	—
Shale	30	8
Sandstone	50	10
Limestone	80	10
Basalt	100	10
Granite	160	14
Quartzite	360	—

Note: Compressive and tensile strengths listed are representative values from a range of experimentally determined values. They are derived from uniaxial compression or tension under conditions of no confining pressure. Strengths for rocks, such as shale, vary greatly depending on the orientation of the stresses relative to the planes of weaknesses, such as fissility. Values from Handin (1966) and Middleton and Wilcock (1994).

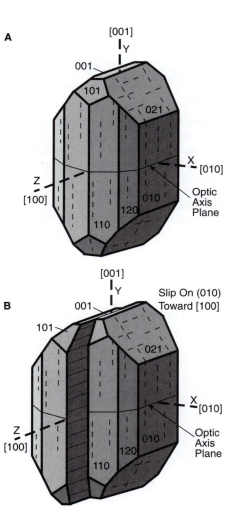

called the **tensile strength** of the crystal. We present the tensile and compressive strengths for common rocks and minerals in Table 4.1.

Similarly, if the magnitude of shear strain exceeds the elastic capacity of the lattice to shift, either a fracture will develop, or the lattice will slip. Slip occurs when atoms on one side of a crystallographic plane suddenly shear with respect to atoms on the opposite side of the plane (Figure 4.4*F*). After slip occurs, the lattice within the undisturbed part of the crystal remains coherent, whereas the crystal is seen to be permanently deformed where the shift occurred. In square-dance style, atoms will have changed partners. The outline of the crystal itself preserves a record of deformation in the form of slight offset of its edge and slight change in shape (Figure 4.4*F*).

Whether a crystal deforms by discrete fracture or by slip along a crystallographic plane depends on whether the stress required for fracture is greater or less than the stress required for slip. If it takes more stress to fracture than to cause slip, the crystal will deform by slip and may behave in a ductile manner. In contrast, if it takes more stress to cause slip than to produce fracture, the crystal will behave brittlely and fracture.

Slip Systems and Crystallographic Control

Where slip along a suitably weak, suitably oriented crystallographic plane is the preferred mode of inelastic deformation, it is necessary to identify the slip plane and the direction of slip, much as in the evaluation of faulting, but at the atomic scale. The crystallographic plane along which slip occurs is the **slip plane**. But it takes more than identification of the slip plane to define the slip *system*. Full description of the **slip system** includes the slip plane and the line of slip within the slip plane (Figure 4.5). The kinematic character of a slip system is analogous to shearing of a microscopic set of playing cards. Each card remains undeformed, but tiny amounts of slip between each adjacent card can dramatically alter the shape of the entire deck.

The crystallographic direction along which a crystal slips is largely controlled by the strength and type of bonds, and by the geometric arrange-

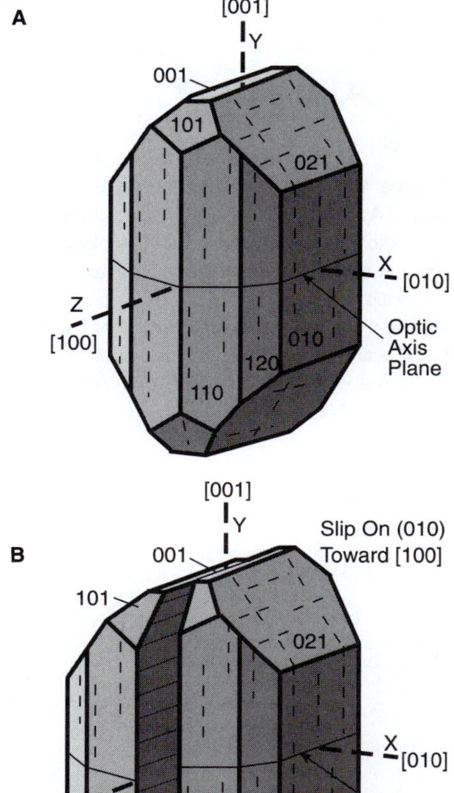

Figure 4.5 A slip system in olivine, where slip occurs along a [100] direction on the (010) plane. (*A*) Intact crystal. (*B*) Offset crystal after slip parallel to the ruled lines on the slip plane. (Part A from *Crystalline plasticity and solid state flow in metamorphic rocks: selected topics in geological sciences* by A. Nicolas and J.P. Poirier, copyright © 1976. Published with permission of John Wiley & Sons, Inc., New York.)

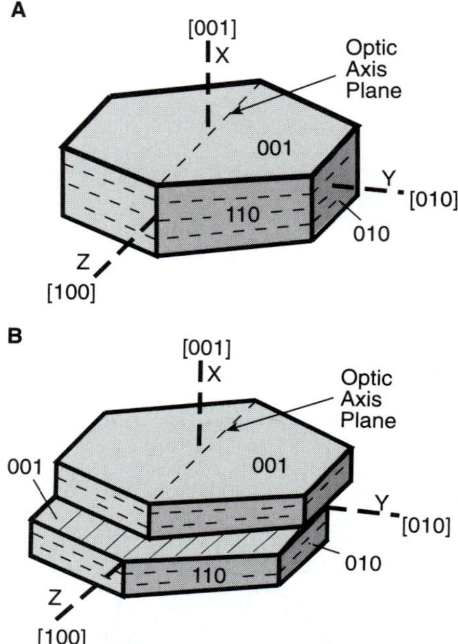

Figure 4.6 Cleavage and slip system in mica. (*A*) Intact crystal. (*B*) Offset crystal after slip along [100] on the (001) plane. (Part A adapted from *Crystalline plasticity and solid state flow in metamorphic rocks: selected topics in geological sciences* by A. Nicolas and J. P. Poirier, copyright © 1976. Published with permission of John Wiley & Sons, Inc., New York.)

ment of atoms and bonds in the lattice (Nicolas and Poirier, 1976). Slip will occur along planes marked by the weakest bonding.

The directions of cleavage in a mineral disclose the crystallographic orientations in the lattice along which bonds are most easily broken. The relative strength of development of cleavage in the various cleavage directions reflects the relative strength of bonding along the various crystallographic directions. We are not surprised to learn that the weakest bonds in muscovite mica are parallel to the perfect (001) cleavage plane in mica (Figure 4.6). In the old days, Russians took advantage of this property and peeled away large thin sheets of muscovite to use as windows.

Some minerals do not possess cleavage. Quartz is one of these. No natural quartz windows. There are no favorable pathways for fractures to cut through quartz. All crystallographic pathways through the lattice offer about the same resistance to splitting. This is why quartz crystals display conchoidal fracture.

Slip Systems and Bonding

The degree of development of cleavage and slip planes is related to the type of bonding (metallic, ionic, or covalent) of minerals (Lawn and Wilshaw, 1975; Nicolas and Poirier, 1976). Take metals, for example. Metallic bonding consists of a lattice of densely packed positive ions (cations) surrounded by a "sea" of very weakly attached electrons. Bonds are relatively weak, permitting metals to slip relatively freely. Slip is easiest in planes and directions with the closest spacing of cations. The metal slips in a direction that minimizes the distance of displacement required to move the lattice from one cation to another along the slip plane.

Imagine trying to cross a stream by jumping on large stones, which by some strange coincidence are arranged like atoms in some metals (Figure 4.7*A*). The easiest way to cross is by jumping to the nearest stone, which is along a diagonal, rather than along a row. Other directions are commonly out of reach. This arrangement of stones, like that of the atoms they represent, can be extended into three dimensions (Figure 4.7*B,C*).

Think of it this way. For many metals, such as copper (Cu), gold (Au), silver (Ag), and aluminum (Al), the shortest *line* of slip between adjacent cations lies within the diagonal (111) plane, where cations are densely spaced (Figure 4.7*C*).

Within ionic crystals such as halite (NaCl), positive and negative ions alternate in the lattice (Figure 4.8), and easiest slip will generally be in a direction that does *not* bring ions of *like* charge into close contact (Nicolas and Poirier, 1976). A simple case features cubic crystals with ionic bonding, (e.g., halite and a number of oxides). The cubic crystal will slip within the (110) plane along the [$\bar{1}$10] diagonal.

Many minerals have complex crystalline structures, much more complex than halite or copper. Many of the common rock-forming silicates, for example, can possess covalent *and* ionic bonds both in the same lattice. The general rule is this: *Complex silicates will fracture or slip along planes that break the weakest bonds or the fewest strong bonds*. Cleavage surfaces and slip surfaces will tend to avoid breaking the bonds between silicon and oxygen atoms, which have a mixed covalent-ionic character and are especially strong (Nicolas and Poirier, 1976). Quartz, which is composed *only* of Si and O, is so strong because slip must break through the strong Si—O bonds. In phyllosilicates (e.g., micas), cleavage and slip are easiest along the (001) plane (Figure 4.6), where weak ionic bonds hold a layer of cations between double sheets of strongly bonded silica

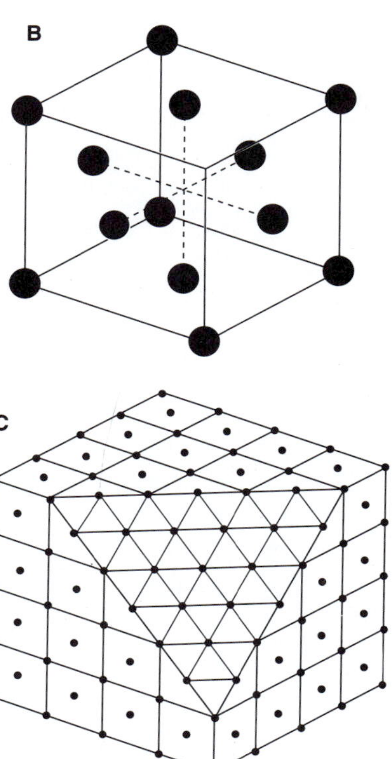

Figure 4.7 Direction of easiest slip. (*A*) Rocks in a stream bed, distributed like cations in a plane within a face-centered metal. To cross the pond with the shortest leaps, you would hop parallel to the diagonal lines. Other routes require more energy and may be impossible. [Artwork by D.A. Fisher.] (*B*) Three-dimensional arrangement of atoms in a face-centered metal. (*C*) Slip plane within a face-centered metal, oriented along the surface with the densest spacing of cations. [Adapted from D. Hull and D.S. Bacon, *Introduction to dislocations* (1984). Published with permission from Butterworth-Heinemann, Ltd.]

tetrahedron. The relative weakness of the ionic bonds accounts for the perfect (001) cleavage of micas. Slip systems for some common minerals are listed in Table 4.2.

Theoretical Strength of Crystals

In addition to controlling the *directions* of fracture and slip, crystallography and bonding profoundly influence the overall **yield strength**—that is, how much elastic deformation a crystal can handle before it yields permanently by either frature or slip. In theory, crystal physicists can predict the yield strength of an ideal crystal simply by knowing the type and strength of bonds, the charge and spacing of atoms, and the crystal symmetry of the mineral (e.g., cubic vs. hexagonal). We list in Table 4.3 some **theoretical yield strengths** for some common minerals.

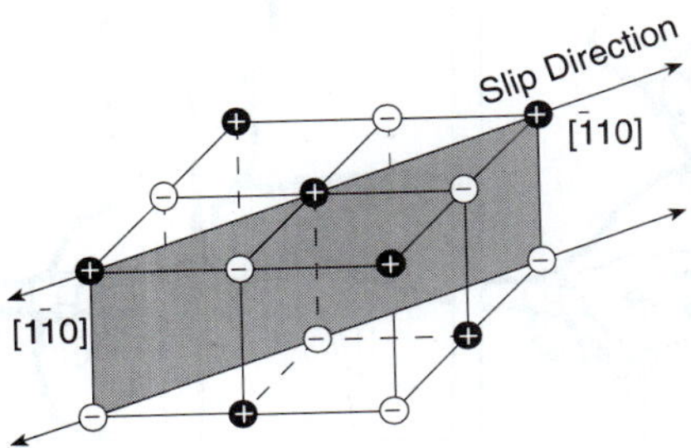

Figure 4.8 Direction of easiest slip in halite, which is along the shortest direction that does not bring ions of like charge into close proximity. (From *Crystalline plasticity and solid state flow in metamorphic rocks: selected topics in geological sciences* by A. Nicolas and J. P. Poirier, copyright © 1976. Published with permission of John Wiley & Sons, Inc., New York.)

TABLE 4.2
Some slip systems for common minerals

Mineral	*Slip System*
Halite	(110)[1$\bar{1}$0]
	(001)[1$\bar{1}$0]
Calcite	(10$\bar{1}$1)[$\bar{1}$012]
	($\bar{2}$021)[1$\bar{1}$02]
Mica	(001)[110]
Quartz	(0001)[11$\bar{2}$0]
	(10$\bar{1}$0)[0001]
	(10$\bar{1}$0)[1$\bar{2}$10]
	(10$\bar{1}$0)[1$\bar{2}$13]
Olivine	(100)[001]
	(110)[001]
	(110)[001]
	(010)[100]

The value in parentheses is the Miller indices of the slip plane, indicating where the plane intersects the crystallographic axes for that mineral. The slip direction within that plane is indicated by the number within brackets. Some slip systems given have symmetrical equivalents. If more than one slip system is given for a mineral, the low-temperature slip systems are listed first.

Sources: Compiled from Hobbs, Means, and Williams, 1976; Nicolas and Poirier, 1976; Nicolas, 1988; and Twiss and Moores, 1992.

TABLE 4.3
Theoretical shear and tensile strengths of some common defect-free solids

Material	*Shear Strength (GPa)*	*Cleavage (tensile) Strength (GPa)*
Metallic Cu	1.2	3.9
NaCl (ionic solid)	2.84	0.43
MgO (simple oxide)	1.6	3.7
Quartz (covalent solid)	4.4	16
Diamond (covalent solid)	121	205

Sources: Lawn and Wilshaw, 1975; Murrell, 1990.

We can see in the numbers presented in Table 4.3 how type of bonding influences yield strength. The predicted theoretical yield strengths for quartz and diamond, which have covalent bonding, are much higher than the predicted yield strength for halite, which has ionic bonding. The theoretical strength of halite, in turn, is higher than the theoretical yield strength for copper, which has metallic bonding.

Minerals with very low yield strengths generally can accommodate simultaneous movement on dozens and dozens of slip planes, permitting a ductile flow of the mineral. At room temperature, the covalent bonds in quartz and diamond are much too strong to permit such ductile behavior to occur. However, copper is a different story. The metallic bonds in copper are so weak that we can make it deform right at the workbench out in the garage, by bending a copper pipe, twisting a copper wire, or buckling a penny in a vise (Figure 4.9).

Bonding can affect strength in yet another way—by influencing the chemical properties of a mineral. Halite and other ionic solids are generally more soluble in hydrous fluids and more chemically reactive than covalent minerals. Soluble minerals may respond to deformation by performing a great disappearing act—by simply dissolving away under the influence of directed stress.

Crystals and Lattices in the Real World

Thus far we have been talking about ideal crystals and ideal crystal lattices. In reality, crystals and lattices are not perfect. They have **defects** and **flaws**, and these affect the real-world strengths of minerals and the ways in which they are able to deform.

The observed tensile strengths of minerals, as measured through deformation experiments under conditions of room temperature, are generally significantly less than the theoretically predicted strengths (Table 4.3). The observed tensile strengths of most minerals and rocks are in fact

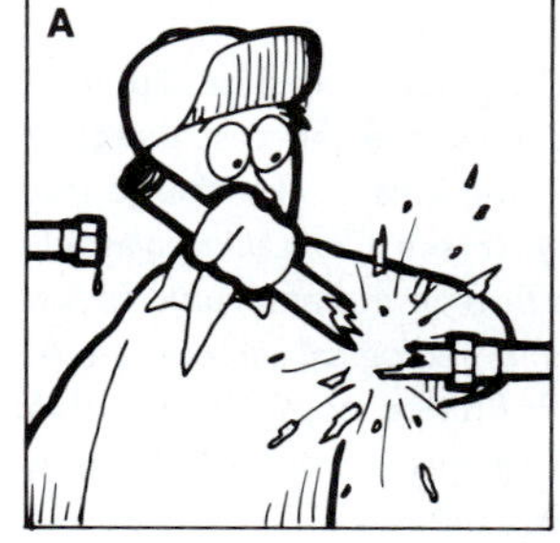

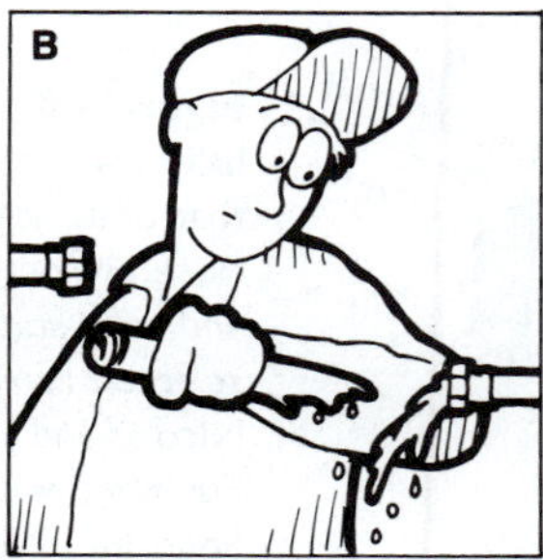

Figure 4.9 Internal atomic/crystal structure determines external material properties. If it is an optimum water connection you want, avoid pipes made of (*A*) quartz, which is too covalent and brittle, and (*B*) halite, which is too ionic and soluble. (*C*) Go for metallic-bonded copper. [Artwork by D. A. Fisher.]

several orders of magnitude lower than the theoretical tensile strength. What could cause minerals to be that much weaker than we would expect??

The culprits are defects and flaws—deviations from the ideal crystalline structure. **Defects** largely govern the strength of rocks and minerals by concentrating stress in ways that permit the material to fracture or slip at stress levels well below the theoretical yield strength. Pole vaulters understand this principle and forever are inspecting their fiberglass "sky-poles" for possible nicks or deep scratches. Climbers do the same with their equipment.

For conditions of brittle failure, microcracks are the most influential defects. They are so small that they can occupy individual grains and crystals at the microscopic and submicroscopic scales. Yet their influence is potent, because the tips of the slitlike cracks can invite the buildup of very high stress concentrations, thus permitting crack propagation.

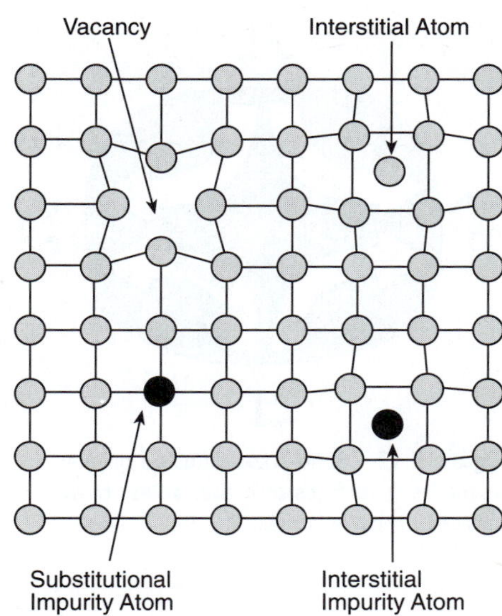

Figure 4.10 Various types of point defects within a crystal.

Defects

There are defects other than microcracks, and their presence can influence the ease or difficulty with which a crystal or grain deforms, under conditions of low and high temperature and low and high pressure. Most defects tend to make it easier for a lattice to deform, but some in fact "harden" the lattice. At the atomic scale, there are three basic varieties of defects: **point defects**, **line defects**, and **planar defects** (Hobbs, Means, and Williams, 1976).

Point Defects

Point defects include **vacancies**, **interstitial atoms**, and **impurities** (Figure 4.10). A vacancy, where an atom is missing from a lattice site that is normally occupied, is like an empty parking space (Figure 4.11). Interstitial atoms are extra atoms that straddle the line between parking places that are already occupied (Figure 4.11); they are extra atoms that rightfully belong in the lattice but are parked illegally between the normal, proper lattice positions. Impurities are unusual vehicles in parking spots normally occupied by another element (Figure 4.11); they are foreign atoms that either are interstitial or occupy a regular space.

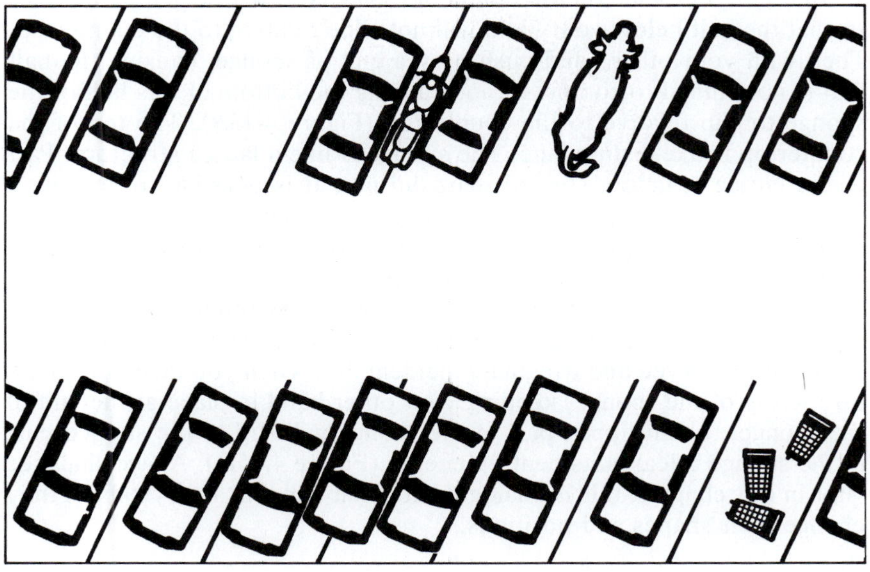

Figure 4.11 Quiz time. Can you find a vacancy, two interstitial atoms, and unexpected impurities in this parking lot analogue of point defects? (Artwork by D. A. Fisher.)

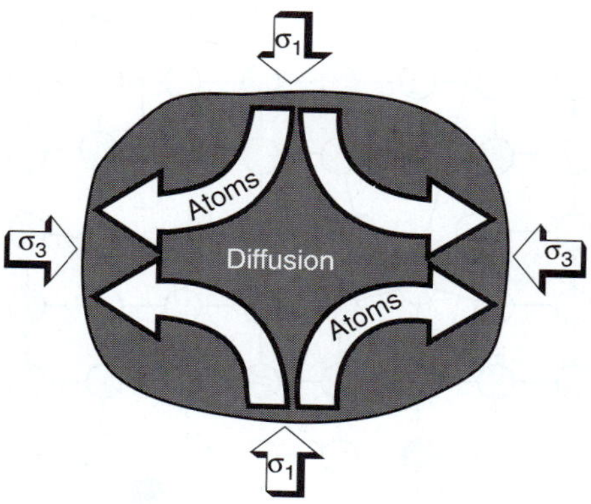

Figure 4.12 Solid-state diffusion drives atoms from surfaces of higher stress toward surfaces of lower stress.

Figure 4.13 Game of VacancyMove, illustrating how vacancies influence the strength of a crystal. (*A*) "Normal" version of VacancyMove. (*B*) "Turbo" version, with three squares removed. Great for short trips. (Artwork by D. A. Fisher.)

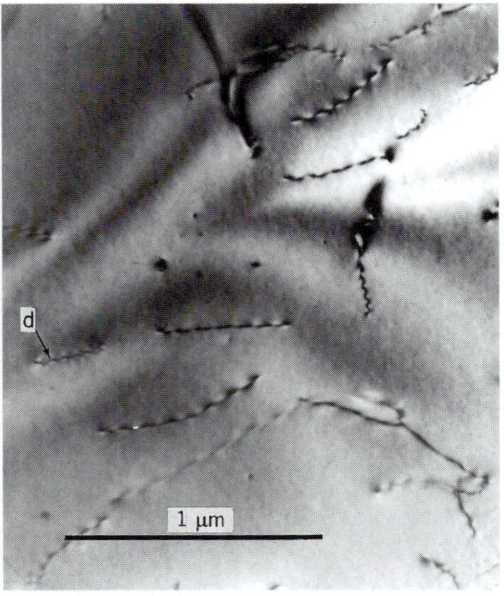

Figure 4.14 Transmitting electron microscope (TEM) image of linear defects, or dislocations. (Photograph courtesy of Andrew Kronenberg.)

The presence of point defects affects the strength of a crystal, and this can be illustrated through a preview of the deformation mechanism known as **solid-state diffusion**. Solid-state diffusion is one way in which a crystal can change its shape and volume in response to directed stress. Solid-state diffusion is achieved by the migration of atoms *away* from surfaces of greatest stress, *toward* surfaces of least stress (Figure 4.12). Vacancies make it so much easier for atoms to move within a lattice during solid-state diffusion. We can envision why this is so by recalling the 15-14-13-12 . . . game that everyone plays in elementary school (Figure 4.13*A*). In Pittsburgh we used to call it "VacancyMove" because it reminded us so much of the role of point defects in crystal lattices. The object of the game is to slide the small plastic squares until the numbers are in the correct reverse order. The game can be difficult for beginners because there is only one "vacancy." We used to pull out two of the squares and play 13-12-11-10 . . . . It's a faster game! Toy manufacturers have now caught on, and they are selling the new improved "Turbo" version (three squares are removed) for people who don't have much time for games (Figure 4.13*B*). After playing "Turbo" once or twice, you will see that a crystal with abundant vacancies will be better able to rearrange itself in response to directed stress.

Interstitial atoms and impurities also affect the capacity of a mineral to deform internally. An impurity or an interstitial atom can cause a mineral to be harder to deform by hindering the movement of "normal" atoms. The hardening reminds me of the time back in Pittsburgh when a rival player glued down one of the squares in my VacancyMove game; it really slowed me down! It made the game *harder*.

Line Defects

Line defects are known as **linear defects** or **dislocations** (Figure 4.14). Line defects can become barriers to the otherwise smooth movement of *planes* of atoms. Dislocations are a microstructure produced by yet another deformation mechanism that we will soon examine. **Dislocations** are difficult to define, and difficult to visualize. But again, as a preview, let's start with an analogy. Sometime when you're doing the dishes, take the kitchen sponge and wet it and wring it out. Then place it on the countertop (Figure 4.15*A*). Put your hand on top of half of it and exert a little gentle pressure so that the half below your hand will not slide relative to the countertop. Then with your other hand push on the end of sponge, causing the half that is not pinned to compress, and causing the bottom of this half of the sponge to slip relative to the countertop (Figure 4-15*B*). The top of the countertop is like a slip plane. The sponge is like a lattice structure. Part of the lattice structure slipped; part did not. It is possible to imagine a line on the countertop, under the sponge, that separates the part of the base of the sponge that moved from the part that did not move (Figure 4-15*C*). This line represents the **dislocation**.

A dislocation can be straight or curved. We have found through countless repetitions of the sponge experiment that the dislocation always seems to be a little curved and irregular. Incidentally, when you take your hand off the top of the sponge, keeping your other hand in place at the end of the sponge, the unslipped part of the sponge will slip as the slipped part of the sponge releases its pent-up energy (Figure 4.15*D*). As we shall see later in this chapter, this is analogous to what happens in crystals as they change their shapes and positions.

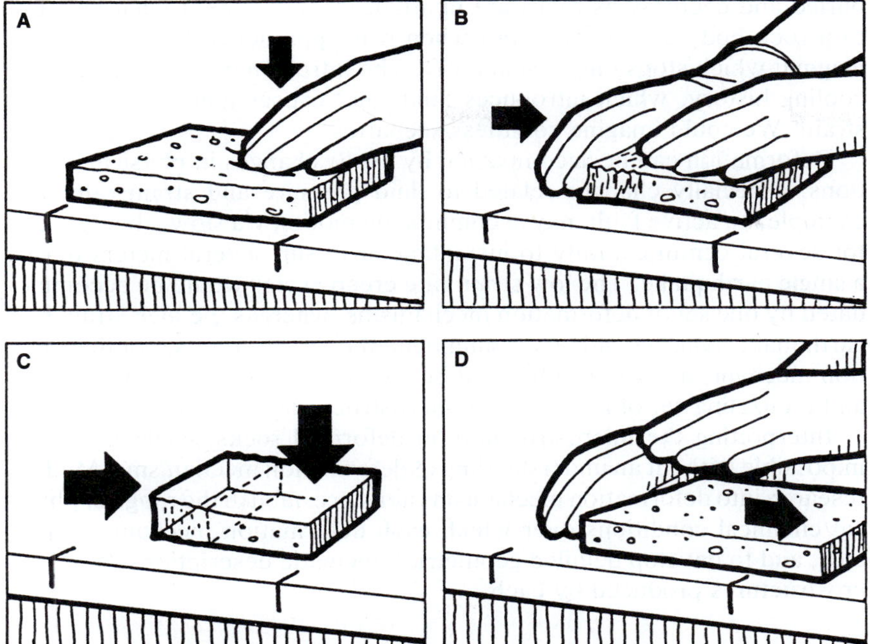

Figure 4.15 Put on your rubber gloves and do the dislocation drill on the kitchen counter. (*A*) Place damp sponge on kitchen countertop, pushing down on one end of sponge, pinning it in place against the countertop. (*B*) Push other end of sponge, compressing the part that is not held down. The countertop is the slip plane and the sponge is like the lattice. (*C*) The line separating slipped versus unslipped parts of sponge is equivalent to a dislocation. (*D*) Remove your hand pinning the sponge and observe the slip of other half of sponge. (Artwork by D. A. Fisher.)

Planar Defects

Planar defects can really interrupt the continuity of the lattice, thereby influencing how a crystal will respond to directed stress. Planar defects of greatest interest to us are **grain boundaries**, **subgrains** (which are slightly misoriented regions within a crystal), **mechanical twins** (like calcite twins or plagioclase twins), and **stacking faults** (where the regular repeating pattern of a lattice is interrupted by the insertion or omission of a partial plane of atoms) (Figure 4.16).

In addition to point defects, dislocations, and planar defects, there are other flaws (e.g., fluid inclusions) at the microscopic and submicroscopic scales. The most important of these are the ubiquitous **microcracks**, which may be contained within a single crystalline grain or may cut several adjacent grains.

DEFORMATION MECHANISMS

When subjected to stress, the crystalline structures and defects within a rock can deform by various mechanisms. The combination of deformation mechanisms that operate within the rock at a given time and place depends on a number of factors. Of these factors, the most important are mineralogy (including grain size), temperature, confining pressure, fluid pressure, differential stress, and strain rate. For a single mineral phase (e.g., calcite) under a set of uniform conditions, one or two deformation mechanisms will typically be enough to accommodate most of the strain and will govern the type of microstructures produced. Most rocks, however, contain more than one mineral phase, and each phase may respond to deformation by a different combination of mechanisms. Furthermore, it is common for deformation to take place under continuously changing conditions. For example, temperature and confining pressure will increase as rocks are

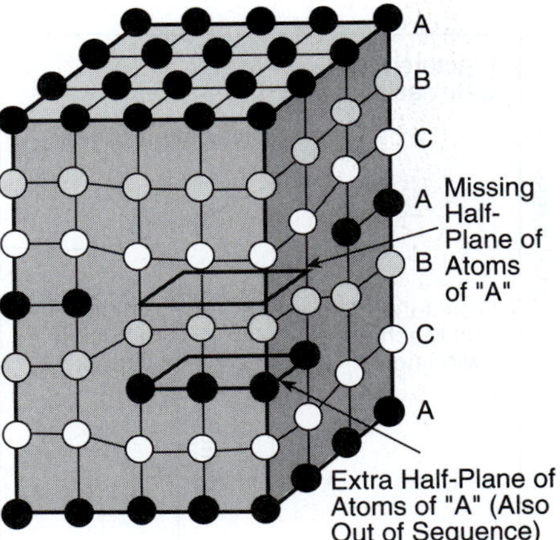

Figure 4.16 Stacking faults, a type of planar defect in a lattice. There are two stacking faults in this example: one produced by an extra half-plane of atoms; the other by a missing half-plane.

buried and decrease as rocks are uplifted. Or think about what happens to a rock body at a given depth when it is approached by an intruding magma, which stops only 100 m away. The intrusion produces a heating/cooling history, which introduces fluids and causes metamorphism and strain. We could imagine countless scenarios.

Deformation can be accompanied by *cyclic* changes in physical conditions, especially changes related to fluid pressure and strain rate. For example, an active fault may accommodate motion via slow, steady creep for several centuries, only to instantaneously slip several meters during a single earthquake. The low strain rate creep cycles would be accommodated by one set of deformation mechanisms, whereas the high strain rate earthquake event would be accommodated by a different set of deformation mechanisms. The result could be a diverse array of microstructures and a puzzling set of crosscutting microstructural relations.

Interpreting the microstructures in deformed rocks would be nearly impossible without an understanding of deformation mechanisms. Modern research into deformation mechanisms aims to establish the range of physical/chemical conditions over which each deformation mechanism operates, and to develop detailed geometric/kinematic descriptions of the microstructures produced by each.

The Main Mechanisms

We find it useful to arrange the main deformation mechanisms into five general categories: (1) microfracturing, cataclasis, and frictional sliding, (2) mechanical twinning and kinking, (3) diffusion creep, (4) dissolution creep, and (5) dislocation creep. These mechanisms are sometimes aided by other important processes, including **recovery** and **recrystallization**, both of which lead to less strained lattices.

We can illustrate the physical conditions under which each mechanism dominates deformation by use of a **deformation map** (Figure 4.17), which generally plots temperature versus differential stress or some other variable. For example, in Figure 4.17 fracture and cataclasis are shown as dominating at low temperatures and high differential stress, whereas several types of diffusion creep occur at higher temperature and lower differential stress.

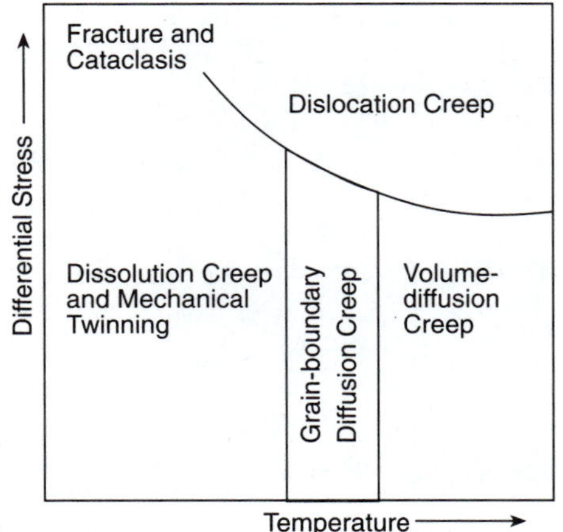

Figure 4.17 Simplified deformation map, showing the general conditions where each deformation mechanism dominates relative to one another.

Microfracturing, **cataclasis**, and **frictional sliding** involve the formation, lengthening, and interconnecting of microcracks; frictional sliding along microcracks and grain boundaries; and the formation and flow of pervasively fractured, brecciated, and comminuted (i.e., ground down) rock and crystal fragments.

Mechanical twinning and **kinking** are less aggressive deformational mechanisms than microfracturing and frictional sliding. Strain is achieved by bending, not breaking, of lattices.

The three kinds of **creep** have different characteristics (Figure 4.17), although all three are mechanisms geared to change the shape and size of crystals in response to directed stress, almost as if the crystals were plastic. **Diffusion creep** changes the shape and size of crystals through the movement of vacancies and atoms within crystals and along grain boundaries. **Dissolution creep** changes the shape and size of crystals through dissolution and reprecipitation of material, aided by fluids along grain boundaries or within pore spaces. Dissolution creep takes from "here" and adds to "there," almost like shaping a snow sculpture. **Dislocation creep**, the fanciest deformation mechanism, operates through intercrystalline slip of the lattice structure.

Microfracturing, Cataclasis, and Frictional Sliding

Microfracturing, cataclasis, and frictional sliding are brittle deformation mechanisms that operate at the grain and subgrain scales. As a response to stress, individual microcracks form, propagate, and link up with other microcracks to form full-blown microfractures and fractures. Individual microcracks can be opened up in tension, or they can accommodate slippage by frictional sliding. In faults and fault zones, which are characterized by the pervasive development of microcracks and fractures, the grain size of a rock may become dramatically reduced by **cataclasis** (crushing!), and the crushed material can move by **cataclastic flow**, as if it were a bunch of ball bearings.

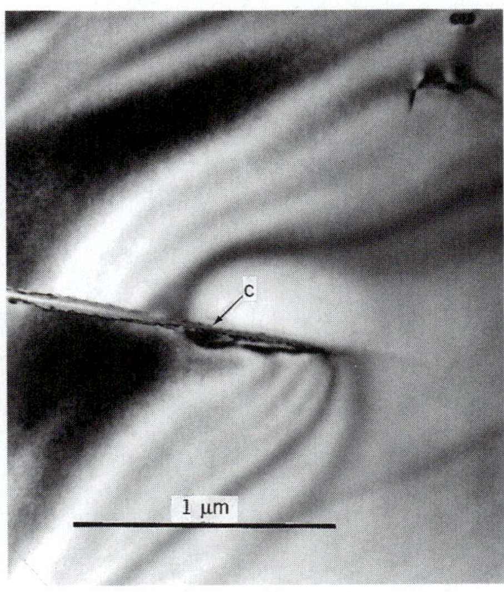

Figure 4.18 TEM bright-field image of microcrack in feldspar. (From A. K. Kronenberg, P. Segall, and G. H. Wolf, *Geophysical Monograph,* v. 56, fig. 15A, p. 33, copyright © 1990 by American Geophysical Union.)

Formation of Microcracks

In low temperature deformation experiments, rocks and minerals accommodate elastic strains of less than 1% before they fracture (Means, 1990). The differential stress at which they fracture is well below predicted theoretical strength. Theoretical predictions do not take into account the presence of **microcracks** (Figure 4.18), microscopic to submicroscopic cracks and surfaces. Like a magnifying glass that can focus normal sunlight to produce fire, microcracks focus ordinary low level stresses to produce fractures. Griffith (1924) worked this out years ago.

Everyday analogues of microcracking abound. A small pebble hit the lower-right corner of the windshield of my new pickup truck, and out of that flaw has grown, day by day, an ever-lengthening fracture; I'm not even driving on rough roads, and the fracture still grows! The way to end the growth and propagation is to eliminate the ''corner'' of the crack where stresses concentrate. I can do this by pulling out my Black & Decker drill, inserting a drill bit designed for brick and concrete, and drilling a small hole in the windshield right through the tip of the crack.

Most microcracks are produced by stress that builds at points of concentration, such as grain boundaries, inclusions, pores, twins, dislocations, and earlier formed microcracks. At the atomic scale, the initiation and growth of microcracks involves the severing of bonds. The breaking of bonds in a crystalline lattice is a little like stretching springs beyond their limit, or even severing the springs (Figure 4.19).

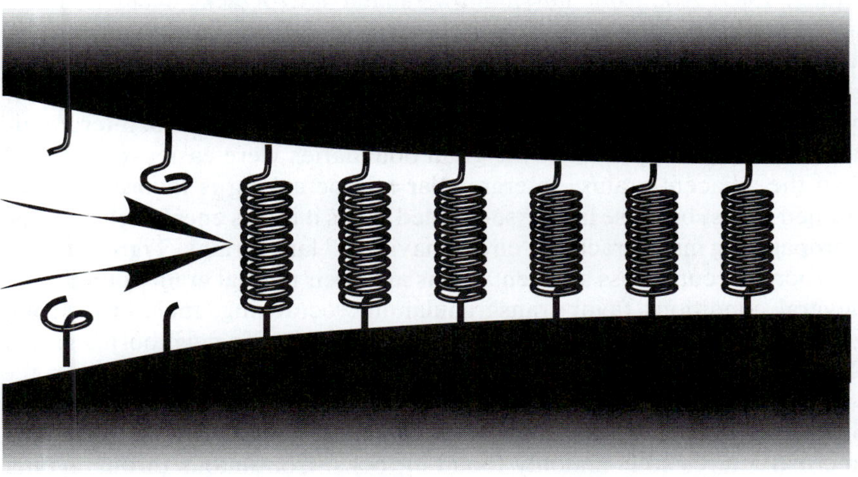

Figure 4.19 Cartoon of how microcracks break across bonds. [Adapted from Lawn and Wilshaw (1975).]

The stresses that cause microcracks may result from tectonic, gravitational, or thermal loading—in other words, from any loading process that causes a geometric misfit or stress concentration. During thermal loading (heating or cooling), neighboring minerals will expand or contract by different amounts if they have different coefficients of thermal expansion. Where the stresses produced are great enough, microcracks will form at the points of concentration.

Some minerals are anisotropic with respect to thermal expansion and contraction; that is, they expand or contract the most along one specific crystallographic direction. Quartz has an especially large and anisotropic coefficient of thermal expansion (Kranz, 1983). When it "indents" a neighboring mineral, such as feldspar, that cannot get out of the way, microcracks form. Differential thermal expansion between adjacent quartz and feldspar grains actually may be the main cause of microcracks in granite. The relative sparsity of microcracks in rocks like diabase may be simply due to the relative absence of quartz as a "hammer" (Nur and Simmons, 1970).

Burial or unroofing of rocks also causes microcracks, not only because of expansion and contraction due to thermal effects, but because of gravitational loading and unloading. As rocks are buried, confining pressure increases and grains increasingly impinge on one another because of compaction. Stress concentrations arise at grain–grain contacts, especially where grains indent or wedge into one another. Preexisting microcracks, particularly those that are low-dipping or horizontal, tend to close during burial. During uplift, microcracks will tend to open up as erosional or tectonic unroofing causes the steady reduction of confining pressure.

Microcracks also form as a result of tectonic loading, especially near preexisting microcracks and grain contacts. During semibrittle or ductile deformation, stress concentrations may develop where propagating twins and dislocations encounter obstacles such as grain boundaries, cavities, or other twins and dislocations. Microcracks form when the motion of the twin or dislocation can no longer accommodate the strain (Lloyd and Knipe, 1992).

Microcracks and Grain-Scale Fractures

Microcracks are commonly subdivided into three types: **intragranular**, **intergranular**, and **transgranular** (Figure 4.20) (Lawn and Wilshaw, 1975; Kranz, 1983; Atkinson, 1987). **Intragranular microcracks** occur within a single grain, most commonly along a cleavage plane. Intragranular fractures form where the fracture strength of the grain is less than that of grain boundaries. **Intergranular microcracks** exploit grain boundaries, propagating around rather than through grains. The presence of intergranular microcracks implies that the grain boundaries were easier to "crack" than the adjacent grains. Intergranular microcracking is favored in fine-grained rocks, because in coarse-grained rocks it is less energy efficient for a propagating microcrack to "circumnavigate" large grains. **Transgranular** microcracks cut across adjacent grains and their mutual grain boundaries. Several conditions favor transgranular microcracking, including strong grain boundaries and similar orientations of cleavage in neighboring grains (Lawn and Wilshaw, 1975; Groshong, 1988).

Microcracks of all three types may be present in a single rock because of the diversity of minerals, crystallographic orientations, textures, and microstructures so commonly found in rocks. Continuous through-going fractures, such as those observed at the scale of a hand specimen or outcrop, generally form by linkage of numerous microcracks, rather than

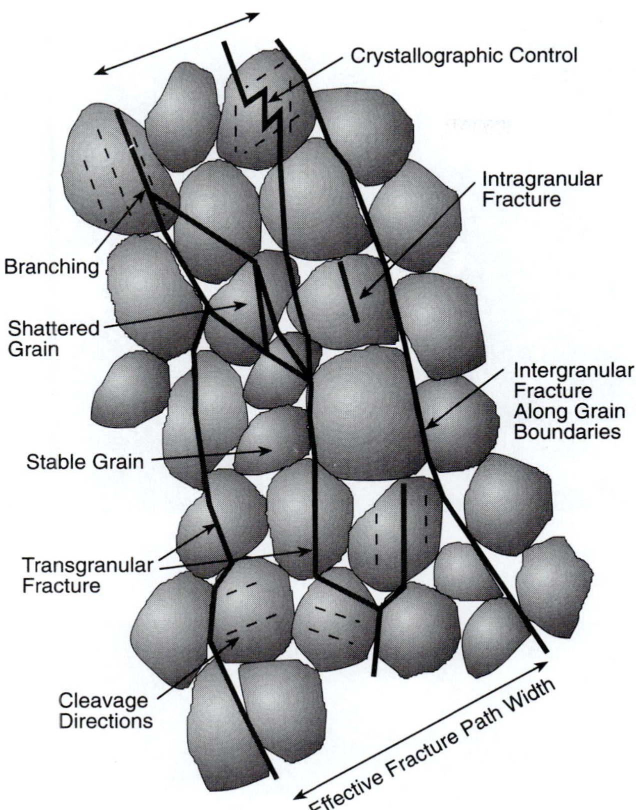

Figure 4.20 Types of microcracks, including intragranular fracture within a grain, intergranular fracture along grain boundaries, and transgranular fractures cutting across several adjacent grains (Modified from *Journal of Structural Geology,* v. 14, G. E. Lloyd and R. J. Knipe, Deformation mechanisms accommodating faulting of quartzite under upper crustal conditions, 1992, Elsevier Science Ltd., Pergamon Imprint, Oxford, England.)

propagation of a single fracture. Only in true tensile conditions does a single microcrack evidently propagate into a large, discrete fracture (Sammis et al., 1986, p. 69).

Cataclasis and Cataclastic Flow

Cataclasis is the pervasive brittle fracturing and granulation of rocks, generally along faults and fault zones. It produces an aggregate of highly fractured grains and rock fragments in a matrix of even smaller, crushed grains (Figures 4.21 and 4.22). Once formed, such crushed aggregates are

Figure 4.21 Brecciated, shattered rock formed by fracturing and cataclasis of granite in the footwall of a Tertiary normal fault, Newberry Mountains, southern Nevada. (Photograph by S. J. Reynolds.)

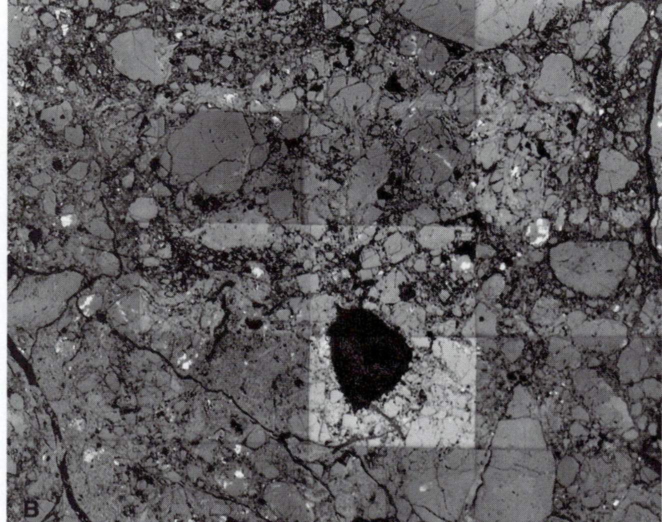

Figure 4.22 Cataclastic rocks at various scales of observation. (*A*) Outcrop of cataclastic rock along Whipple detachment fault, Whipple Mountains, southeastern California. (Photograph by S. J. Reynolds.) (*B*) Scanning electron microscope (SEM) image of fine-grained cataclastic rock. Width is 4.3 mm. (*C*) SEM close-up of part of *B*. Width is 0.2 mm. (Reprinted with permission from *Pure and Applied Geophysics,* v. 125, C. Sammis and others, The kinematics of gouge formation, 1986, Birkhauser Verlag Ag, Basel, Switzerland.)

able to flow by repeated fracturing, frictional sliding, and rigid body rotation of grains and fragments, a process termed **cataclastic flow**. Although flow occurs by brittle processes, it may be homogeneous at the scale of fractions of millimeters or centimeters, to hundreds of meters.

Cataclastic flow is somewhat analogous to **granular flow**, which involves the frictional sliding and rolling of particles in relatively unlithified sediments. In granular flow, most grains remain intact, rather than being repeatedly fractured as in cataclastic flow. Granular flow can occur only where grain boundaries are much weaker than the individual grains and where effective confining pressure is very low, such as on or near the Earth's surface or in zones with high pore fluid pressure; it is most commonly observed in the slumping of unconsolidated sediments. Microcracking and cataclasis clearly involve differential stresses that exceed the rupture strength of the rock, whereas granular flow occurs in rocks with little or no cohesive strength.

Cataclastic rocks are characterized by pervasive cracks and generally sharp, angular grains and fragments (Groshong, 1988). Most cataclastic rocks look remarkably similar at all scales of observation, from the scale

Figure 4.23 Experimentally produced cataclastic textures and fault zone. [Photograph courtesy of Charles Sammis; see Sammis et al. (1986).]

of an outcrop (Figure 4.22*A*) down to a scale that can be resolved only through electron microscopy (Figure 4.22*B*,*C*).

Cataclasis results in a progressive *decrease in grain size* as larger grains are continually broken into smaller ones. It also results in a *decrease in sorting,* as smaller and smaller fragments are produced and intermixed with larger ones. Cataclasis generally causes an *increase in volume,* a process known as **dilatancy** (Brace et al., 1966), as pore space is created between separating fragments. The dominant cause of fracturing in cataclastic rocks is stress concentration, where grains indent or impinge on one another during tectonic loading.

Cataclasis and cataclastic flow have been investigated through detailed examination of natural fault rocks and by experimental investigations of frictional sliding along precut specimens (Tullis and Tullis, 1986). The natural and experimentally produced cataclastic textures are remarkably similar (Figure 4.23), which implies that the experimentally observed behavior is likely to apply, within some limits, to natural fault zones. In laboratory experiments, cataclastic flow becomes progressively more difficult at higher confining pressures because higher confining pressures impede frictional sliding and dilatancy. As a consequence, cataclasis and cataclastic flow are most important in the shallow parts of the crust, where they commonly occur along fault zones.

Mechanical Twinning and Kinking

Mechanical Twinning

Mechanical twinning is a deformation mechanism that produces a bending, rather than a breaking, of a crystalline lattice. In the simplest case, a mechanical twin is formed when the crystalline lattice is bent by shearing parallel to a favorable crystallographic plane (Figure 4.24). The crystalline lattice on one side of the twin plane is sheared by a constant angle of rotation with respect to the lattice on the other side. The two parts of the lattice end up as mirror-image reflections. The amount of bending of the lattice, and thus its angle of rotation, are limited by the crystal structure of the mineral. Mechanical twinning is especially common in calcite and plagioclase feldspar, minerals that possess a suitable crystalline structure for twinning.

A different type of mechanical twin is produced when stresses cause one part of a crystal to rotate about an axis perpendicular to a favorable twin plane. In this type of twinning, the two parts of a crystal on opposite sides of a twin plane are slightly misaligned by a pivoting motion, much like opening a newly dealt hand in a card game. Compared to "normal" twins formed during crystal growth, mechanical twins of both types are more lenticular and more likely to end by a gradual tapering (Groshong, 1988).

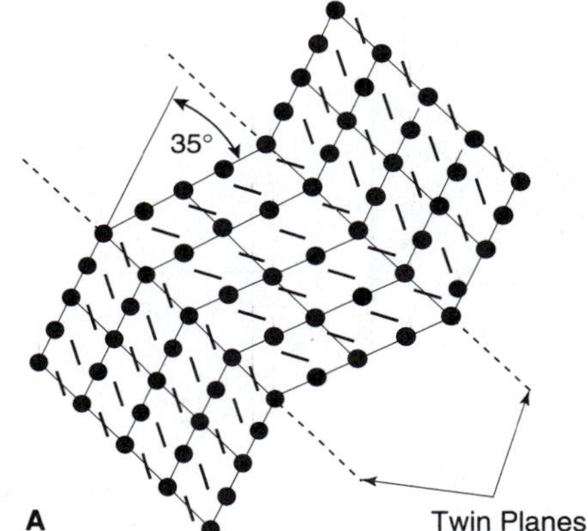

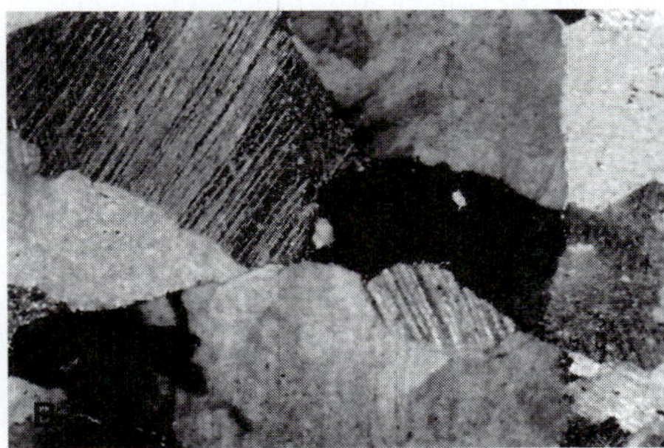

Figure 4.24 Mechanical twins in calcite. (*A*) Formation of a mechanical twin by shearing the calcite lattice parallel to a crystallographic plane. Solid dots represent Ca atoms and the short lines represent the CO_3 groups. [Adapted from Twiss and Moores(1992) and from Hobbs, Means, and Williams (1976).] (*B*) Photomicrograph of mechanical twins in grains of calcite. (Photograph by Richard H. Groshong, Jr.)

Type 1 : Thin, Straight	Type 2 : Thick (>>1μ m) Straight	Type 3 : Thick, Curved	Type 4 : Thick, Patchy
Temp. < 200° C	150 - 300° C	> 200° C	> 250° C

Figure 4.25 Types of twins as a function of temperature. [Reprinted with permission from *Journal of Structural Geology*, v. 15, M. Burkhard, Calcite twins, their geometry, appearance and significance as stress-strain indicators of tectonic regime: a review (1993), Elsevier Science Ltd., Pergamon Imprint, Oxford, England.]

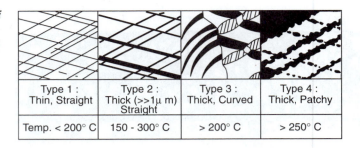

Conditions Favoring Mechanical Twinning

Two conditions are necessary for twinning to occur: (1) there must be a vulnerable twin plane across which shearing or rotation can take place; and (2) the plane must be oriented such that the shear stress along the twin plane is sufficient to distort the lattice. Mechanical twinning is not particularly sensitive to the influence of confining pressure, because twinning does not involve frictional sliding or dilatancy. Moreover, susceptibility to twinning is not greatly influenced by temperature because, unlike deformation mechanisms such as diffusion, mechanical twinning does not involve a temperature-activated process. Twinning *does* require relatively higher differential stresses than some other deformational mechanisms, because it involves energy-intensive bending of the lattice.

Twinning is a comparatively rapid process (Nicolas, 1987), but the amount of shear strain that can be accommodated by twinning is limited by the crystallography of the mineral. The bonds can be rotated only so far before they break. Once the lattice has been bent by the optimum and required angle to form a twin, it cannot accommodate any further strain by twinning. Any additional strain requirements must be accommodated by some other deformation mechanism. At low temperatures, twinning may lead to brittle fracture, in part because a twin, once it has formed, may concentrate stress and cause the crystal to rupture along the twin plane or another favorably oriented cleavage plane (Lloyd and Knipe, 1989).

We can use the geometry and appearance of calcite twins to gauge the *temperature* and *amount* of deformation (Groshong, 1988; Burkhard, 1993). Thin, straight twins represent low temperatures and strains, whereas thick, curved twins form at higher temperatures and strains (Figure 4.25).

A

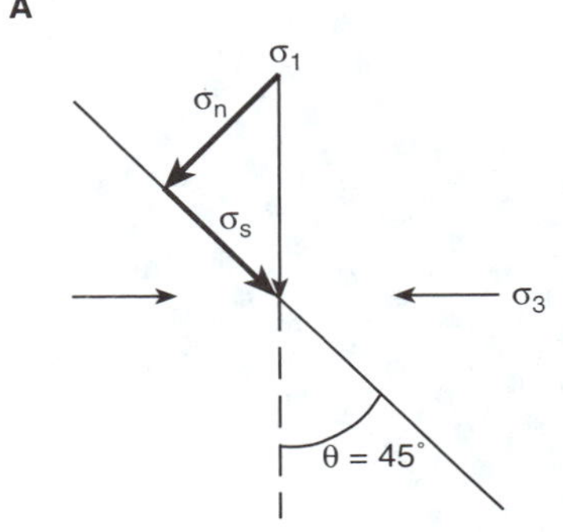

B

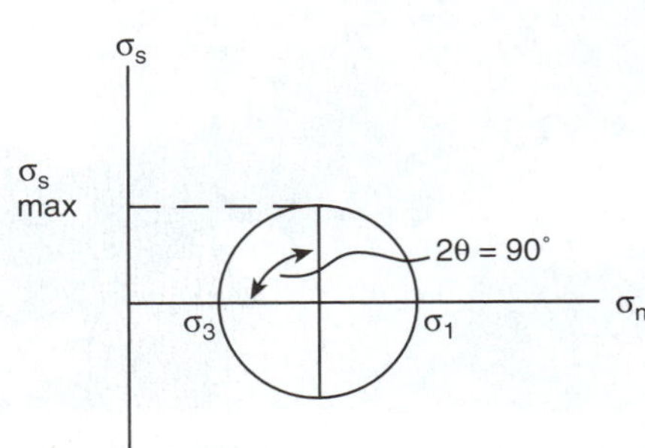

Figure 4.26 The most favorable orientation of a twin plane will be at 45° to σ_1, which is the plane of maximum resolved shear stress. (*A*) σ_1 can be resolved into a normal component perpendicular to and a shear component parallel to the twin plane. (*B*) Mohr circle showing that the maximum resolved shear stress is for surfaces at 45° to σ_1.

Determining Strain and Stress from Mechanical Twins

The orientation of mechanical twin planes in a crystal can be used to determine the orientations of the principal directions of strain and, given a few assumptions, the orientations and magnitudes of the principal stresses (Groshong, 1988; Burkhard, 1993). We are able to reconstruct the stresses, normally a very difficult feat in structural geology, because only those potential twinning planes that are oriented favorably with respect to the imposed stresses will be activated during twinning. The ideal orientation is a twin plane at 45°, because this is the plane of maximum resolved shear stress (Figure 4.26). Various techniques have been devised, but all are applicable only to rocks with small strains (Burkhard, 1993). For calcite, the crystallographic "*c*-axis," whose orientation can be determined microscopically, attempts to rotate into parallelism with σ_1, the direction of greatest principal compressive stress (Figure 4.26). Therefore, the statistically determined preferred orientation of "*c*-axes" of twin planes within the deformed rock tends to reflect the orientation of σ_1.

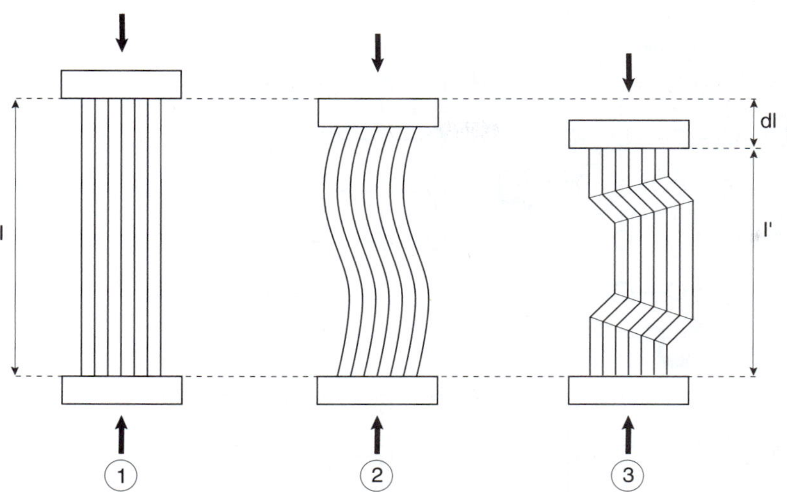

Figure 4.27 Kinking caused by end-on compression of a layered material. Initial length (*l*) is converted to *l*′ with a change in length of *dl*. (From *Principles of rock deformation: petrology and structural geology,* by A. Nicolas, 1987. Reprinted by permission of Kluwer Academic Publishers.)

We might at first think that the use of calcite twins to reconstruct stresses would be limited to sedimentary rock sequences containing calcite–limestone. However Craddock and Pearson (1994) use the technique in imaginative ways, including examining twinning in calcite grains that filled small gas vesicles in billion-year-old basalts in northern Minnesota to reconstruct the ancient stress picture.

Kinking

Kinking, like twinning, involves the bending of a lattice, utilizing planes of weakness (Figure 4.27). Kinking commonly takes place within a discrete band through a crystal. As seen in this section, kink bands usually display a crystallographic orientation or angle of extinction that is different from the rest of the mineral. They show up microscopically as **extinction bands** (Groshong, 1988). Micas and other platy minerals are especially prone to kinking, especially when shortened in a direction parallel to cleavage. End-on compression of confined computer cards can produce similar results (Figure 4.28). Kinks can be pretty tight, because the amount of lattice rotation is *not* limited to a specified angle, as it is in the case of mechanical twinning.

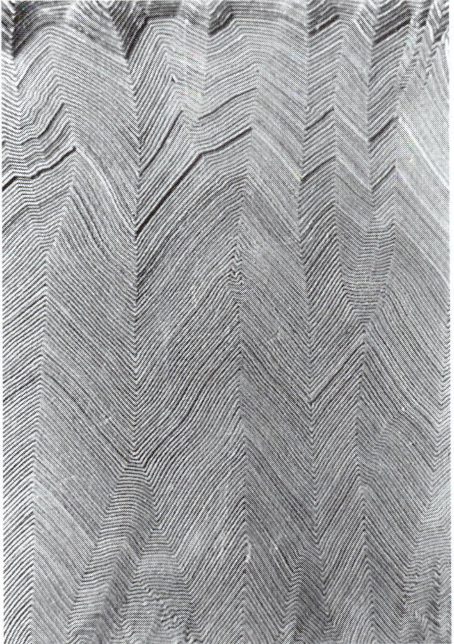

Figure 4.28 Photograph of kink bands in a stack of computer cards experimentally deformed by applying an end-on load and a confining pressure. [Photograph and copyright by Peter Kresan.]

Diffusion Creep

Under many geologic conditions, rocks accommodate deformation by creep, rather than by fracturing, cataclasis, and frictional sliding and rather than by mechanical twinning and kinking. **Creep** is a slow, time-dependent strain; *it occurs at differential stresses well below the rupture strength of the rock*. Creep can be accomplished through several distinctly different deformation mechanisms: **diffusion creep**, **dissolution creep**, and **dislocation creep**, which we refer to loosely as "the three creeps." We begin with the process of diffusion and diffusion creep.

Diffusion

Diffusion involves the movement of atoms through the interior of grains, along grain boundaries, and across pore fluids between grains. In effect, atoms jump "site by site" through a mineral (Figure 4.29). Diffusion is thermally activated. Higher temperatures excite atoms and increase the probability that an atom will migrate. Diffusion is a relatively slow and

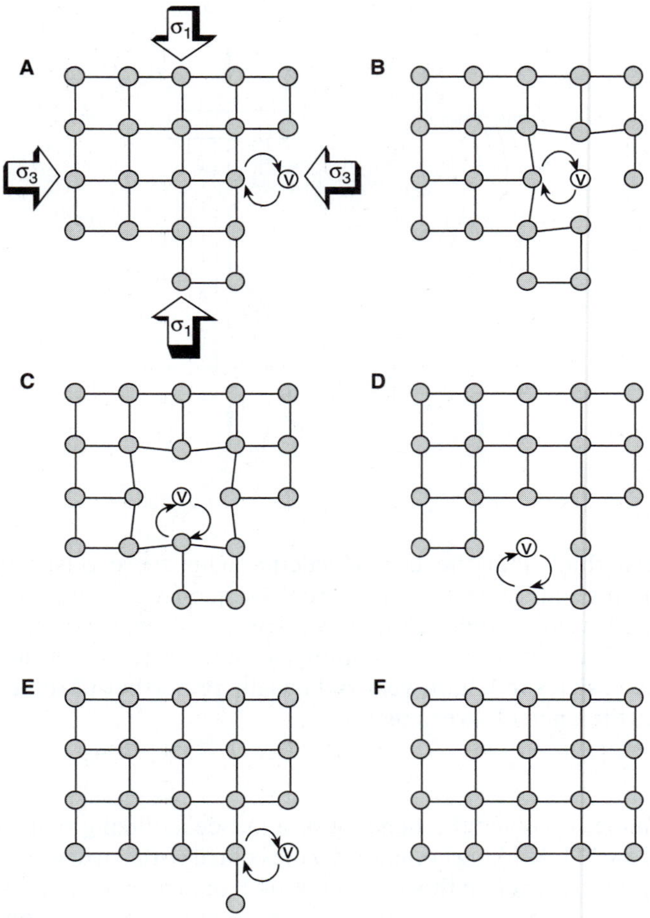

Figure 4.29 Volume-diffusion creep. Vacancy is shown by circle containing a "v". As atoms successively switch places with the vacancy, the vacancy moves through the lattice and the crystal changes shape. The atoms move from surfaces of higher stress to those of lower stress. (*A–D*) A vacancy begins at the edge of the crystal and moves one step at a time to another edge of the crystal. (*E–F*) Another vacancy moves through the crystal in two jumps (one not shown), permitting it to change shape in response to the imposed stress.

inefficient process in a crystalline solid, where atoms are bonded tightly together in more or less fixed positions. It is much faster and much more efficient in a fluid, such as in fluids along grain boundaries, where atoms can move freely.

Because diffusion through a crystalline solid relies on the presence of vacancies and other defects, diffusion is very sluggish in pure, undeformed lattices and is more rapid in lattices with abundant impurities or deformation-related imperfections. Diffusion is most effective in fine-grained rocks, where the distance that the atoms must cover is relatively short.

Volume-Diffusion Creep

At high temperatures, and in the presence of directed stress, diffusion *within* mineral grains can be fast enough to permit the grains to change their shapes. The presence of vacancies (i.e., unoccupied sites) makes this possible. Atoms systematically and sequentially swap positions with vacancies, in a way not unlike a game of checkers. The site hopping changes the shape of the mineral, as atoms and ions move away from sites of high compressive stress to neighboring sites of lower stress, and as vacancies move toward sites of high compressive stress and away from sites of lower stress. This process is known as **volume-diffusion creep**, or **Nabarro–Herring creep**.

When a crystal is subjected to differential stress, vacancies diffuse toward surfaces of high compressive stress, where they are destroyed (Twiss and Moores, 1992). This migration of vacancies from site to site

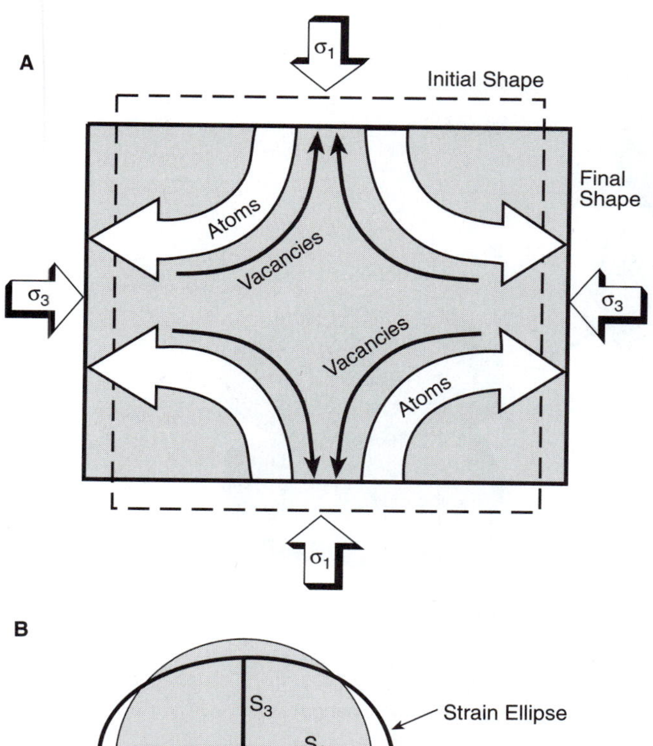

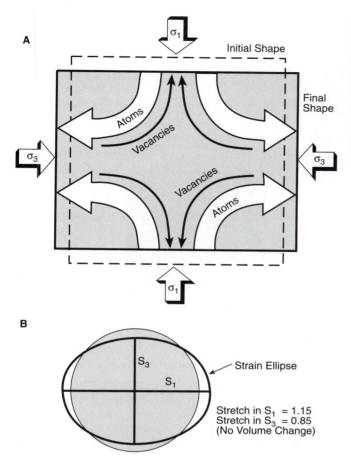

Figure 4.30 Volume-diffusion creep. (*A*) Atoms migrate from surfaces with high compressive stress and toward surfaces of lower compressive stress, whereas vacancies migrate in the opposite direction. (*B*) Finite strain ellipse produced by volume-diffusion creep in *A*. (Part A adapted from *Principles of rock deformation: petrology and structural geology*, by A. Nicolas 1987. Reprinted by permission of Kluwer Academic Publishers.)

across the lattice is matched by a counterflow of atoms in the opposite direction (i.e., from high compressive stress to low compressive stress) (Figure 4.30*A*). The motion of vacancies and atoms results in the selective removal of material from surfaces of high compressive stress and its accumulation on surfaces of low compressive stress. By this process, the crystal changes shape and accommodates strain (Figure 4.30*B*). Diffusion creep is most effective in equant grains, where the diffusive path is short; it becomes progressively more difficult as grains are elongated during deformation. Although volume-diffusion creep is an elegant deformation mechanism, it is so slow that it is difficult to reproduce experimentally in the laboratory with typical rock-forming minerals. Experimental work and deformation theories indicate that volume-diffusion creep is a dominant mechanism only at low differential stresses and at temperatures so high that they approach the melting temperature of the mineral (Figure 4.31).

Grain-Boundary Diffusion Creep

Minerals can also creep by diffusion of material *along grain boundaries*, a process known as **grain-boundary diffusion creep** or **Coble creep** (Figure 4.32). As in the case of volume diffusion, material migrates from surfaces of high compressive stress and accumulates on surfaces of low stress. The migration paths may be longer because the atoms and vacancies move around grains rather than through grains. However, the rate of travel of vacancies and atoms is faster along grain boundaries than directly through a crystal. Therefore, grain-boundary diffusion creep tends to be more

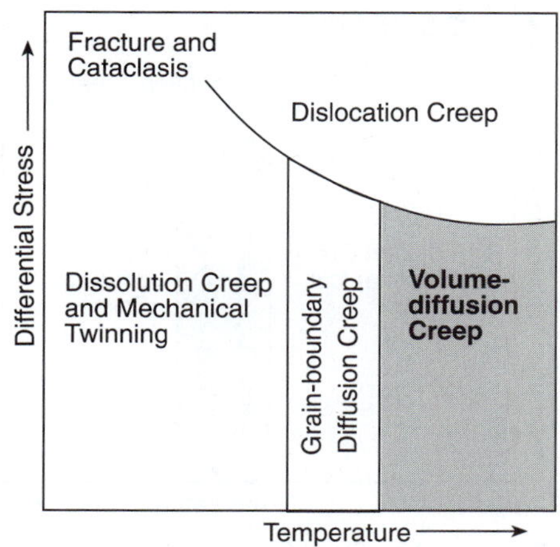

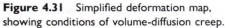

Figure 4.31 Simplified deformation map, showing conditions of volume-diffusion creep.

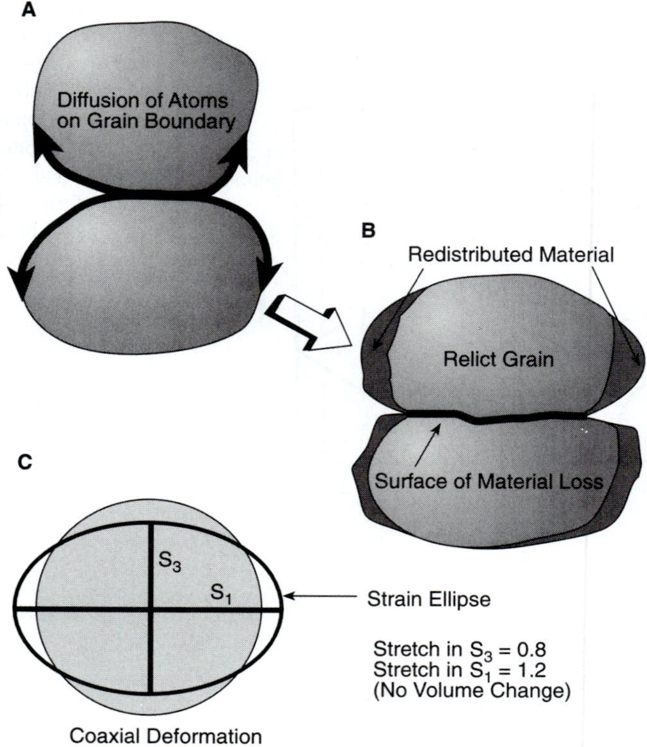

Figure 4.32 Grain-boundary diffusion creep. (*A–B*) Atoms diffuse around the grain boundary from surfaces of high compressive stress to those of low compressive stress. (*C*) Finite strain ellipse for *B*.

efficient than volume-diffusion creep and can occur at lower temperatures (Figure 4.33).

Superplastic Creep

Superplastic creep sounds like a dorky unpopular superhero, but actually it is another diffusion-related creep mechanism. It operates in the experimental deformation of certain metals, and it may also be an important deformation mechanism in rocks (Ashby and Verrall, 1973; Nicolas and Poirier, 1976; Schmidt et al., 1981).

Superplastic creep involves the combination of grain boundary sliding and grain-boundary diffusion (Figure 4.34). Most of the strain that accrues through this mechanism is achieved by the grain-boundary sliding. The role of the grain-boundary diffusion is to permit each of the grains to change shape as it slides so that no voids form between grains. Individual grains remain more or less equant and internally unstrained, even after significant amounts of total strain of the crystalline aggregate. Thus, almost as superheroes, the mineral aggregates can tolerate incredible levels of strain and continue to take more without becoming strain hardened. It is amazing how adaptive and clever minerals can be when they are placed under stress.

Superplastic creep achieves higher strain rates than other types of diffusion creep, probably because grain-boundary sliding is a relatively efficient deformational process and because material must diffuse only a short distance along grain boundaries to accommodate the desired change in shape. The ideal conditions for superplastic creep are fine grain size (a few micrometers), to keep diffusion paths short, and relatively high temperatures (greater than half of the melting temperature), to ensure that diffusion will occur at a sufficiently high rate.

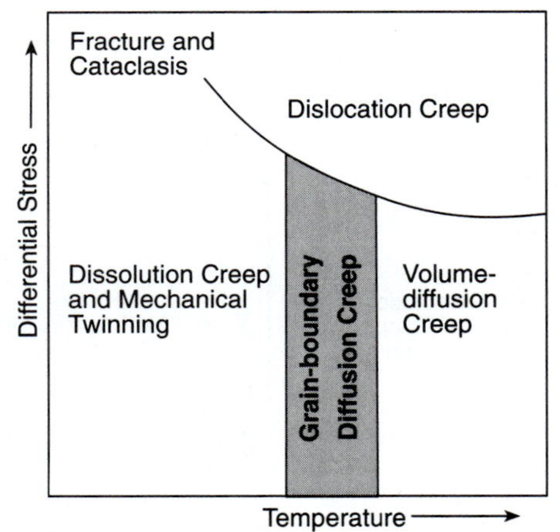

Figure 4.33 Simplified deformation map, showing conditions of grain-boundary diffusion creep.

Dissolution Creep

Dissolution creep, or **pressure solution** as it is often called, involves the selective removal, transport, and reprecipitation of material through fluid films along grain boundaries or pore fluids between grains. The presence of a fluid phase along grain boundaries and in pores between grains greatly increases the efficiency with which material can be removed from sites of high compressive stress and transported to those of lower stress. When subjected to differential stress, grains can change shape by selective dissolution, transport, and reprecipitation of material via the fluid phase (Kerrich, 1978; Groshong, 1988; Knipe, 1989). Dissolution creep differs from the various types of diffusion creep in that it involves transport of material through intergranular fluid.

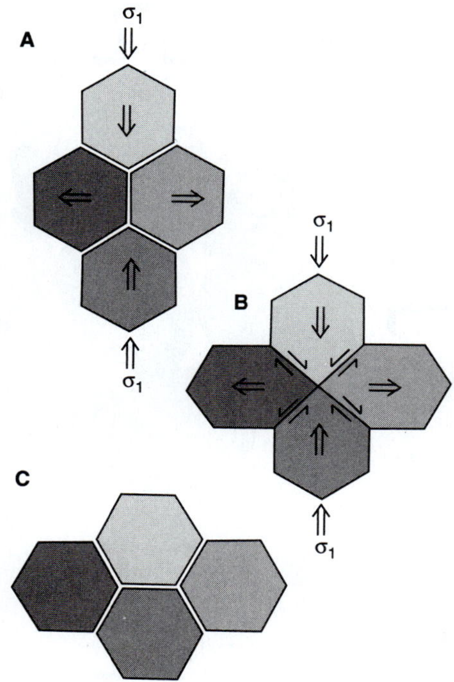

Figure 4.34 Superplastic creep. Shortening occurs via grain-boundary sliding and localized diffusion that allow the grains to change shape (*A,B*) as they slide by one another. (*C*) Grains end up switching neighbors. (Reprinted from *Acta Metallurgia et Materiallia*, v. 21, M. F. Ashby and R. A. Verrall, Diffusion accommodated flow and superplasticity, pp. 149–163, copyright © 1973, with kind permission from Elsevier Science Ltd., The Boulevard, Langford Lane, Kidlington 0X5 1GB, U.K.)

Processes of Dissolution Creep

Dissolution creep depends on three interconnected processes: dissolution at the source, diffusion or migration of the dissolved material along some pathway, and reprecipitation (Figure 4.35). In response to an applied differential stress, grains become preferentially corroded along segments of grain boundaries that are being subjected to high compressive stress. Such segments include those oriented at high angles to the greatest principal compressive stress (σ_1) and those in which rigid grains or objects impinge on one another to concentrate stress (Figure 4.36). Of course, highly soluble minerals will dissolve preferentially to those with lower solubilities. In impure carbonate rocks, for example, calcite typically dissolves more readily than quartz, clays, and iron–manganese oxides. Grains with impurities and those whose crystalline lattices have been somehow damaged, perhaps by the work of other deformation mechanisms, are likewise more susceptible to dissolution than pristine grains.

As solid grains dissolve, they enrich the fluid in their constituents, especially near sites of more rapid dissolution. In contrast, little or no dissolution may occur near grain boundaries normal to the least principal compressive stress (σ_3). The differences in dissolution rates result in **chemical concentration gradients** within the fluid, causing dissolved constituents to diffuse away from the dissolution sites and toward sites of lower com-

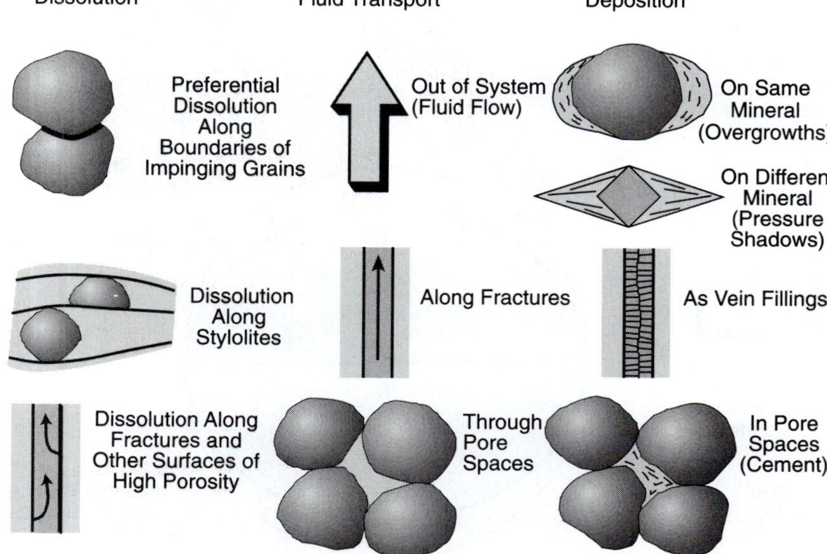

Figure 4.35 Processes of dissolution creep. [Adapted from Nicolas (1987) and Ramsay (1980b).]

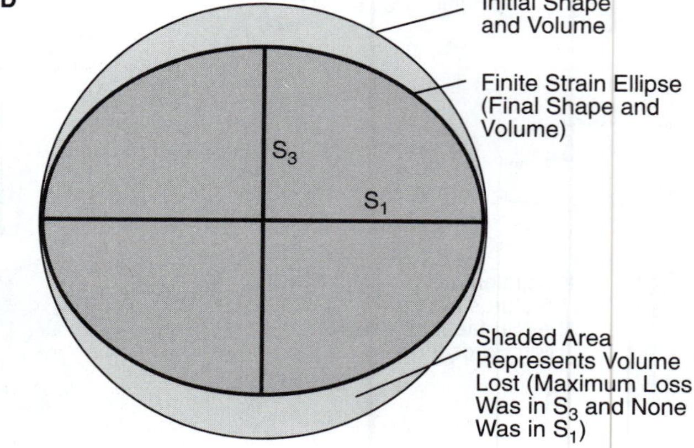

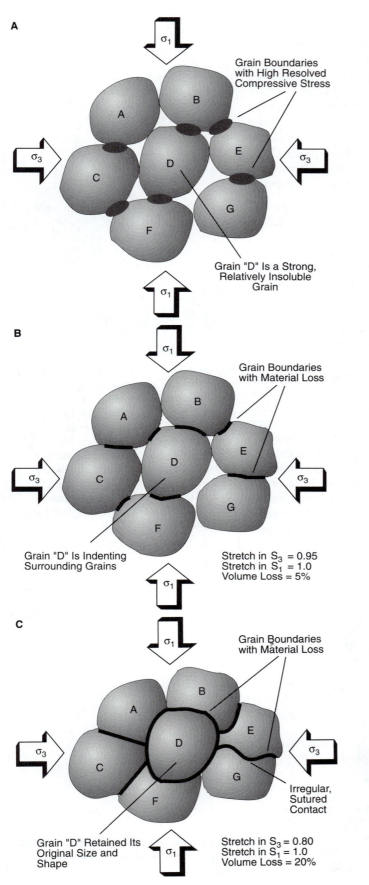

Figure 4.36 Dissolution creep and impinging objects. (*A*) Directed stress is concentrated on grain contacts oriented at a high angle to σ_1. (*B*) Grains preferentially dissolve along these contacts. Note that grain "*D*" is relatively strong and insoluble compared to surrounding grains. (*C*) Final fabric, in which rock has been shortened parallel to σ_1, with the shortening being accommodated by volume loss, rather than lengthening in a perpendicular direction. (*D*) Strain ellipse reflects shortening via volume loss.

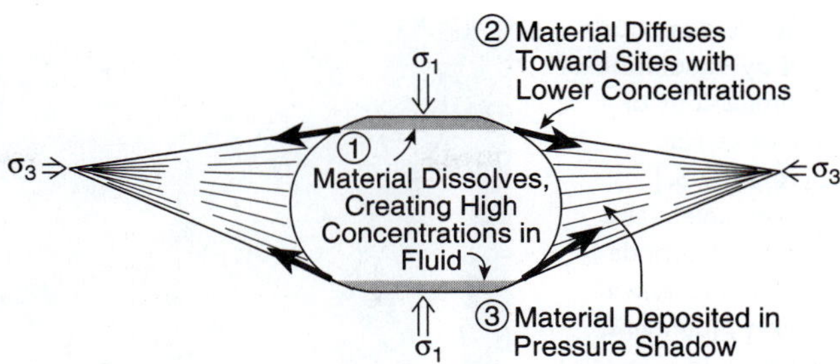

Figure 4.37 Concentration gradients during dissolution creep. Material is dissolved from surfaces with high resolved compressive stress and diffuses to sites of lower compressive stress, such as in pressure shadows.

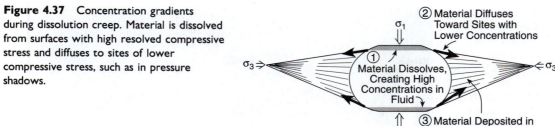

pressive stress (Figure 4.37). Diffusion of the dissolved constituents can occur through the fluid layer adsorbed on grain boundaries or through a static pore fluid. Alternatively, the constituents can be transported by actual movement of their host fluid. Dispersal of the constituents away from dissolution sites, whether by diffusion or fluid flow, is necessary to allow further dissolution to occur.

I learned of a marvelous analog to dissolution creep from my brother the dentist (D. E. Davis, personal communication, 1994). An impacted wisdom tooth may finally decide to emerge, but if it strays off course it may ram into the roots of a neighboring molar. Unprotected by enamel, the root of the molar is vulnerable to pressure solution at the point of contact beween it and the wisdom tooth. If the wisdom tooth maintains its directed tectonic stress, the roots of the molar will eventually disappear, and the molar will no longer be anchored. Dispersal of the dissolved material from the tooth is handled by the bloodstream, the counterpart of which in the world of geology is intergranular fluid in mineral aggregates.

There are telltale clues that reveal that a rock has experienced dissolution creep. Sites of continued dissolution are commonly marked by **stylolites** and the accumulations of less soluble material, such as clays, mica, carbonaceous organic residue, and iron–manganese oxides (Figure 4.38*A*). Material dissolved in the fluid is reprecipitated locally as overgrowths on existing minerals (Figure 4.38*B*), as **fibers** or wedge-shaped **beards** of crystal fibers within pressure shadows (Figure 4.38*C*), and in crystal–fiber veins (Figure 4.38*D*). **Overgrowths** and **pressure shadows** typically form in the protected lee areas next to relatively large, rigid grains, where compressive stresses are low. Fibers within pressure shadows and fibers in crystal–fiber veins grow in the direction of least principal stress (σ_3). This *new* growth, coupled with dissolution along surfaces of high compressive stress, enables the original grains to undergo a shape transformation that reflects the differential stress environment.

Dissolution creep at a given site may ultimately become too inefficient to continue. For example, as grains change shape, the migration paths increase in length, resulting in a drop in the strain rate. Furthermore, dissolution and reprecipitation may result in cementation, compaction, and sealing of cracks and other defects, which in turn leads to strain hardening and the end of creep.

In some circumstances dissolved material may be transported by fluids great distances from the source. For example, dissolution creep is commonly accompanied by prograde metamorphism, which liberates water and other volatiles from minerals through dehydration reactions. The net increase in volume of fluid may drive fluids out of the system, carrying away the dissolved constituents. In such circumstances, the rocks that experienced diffusion creep may actually lose volume as a result of the deformation.

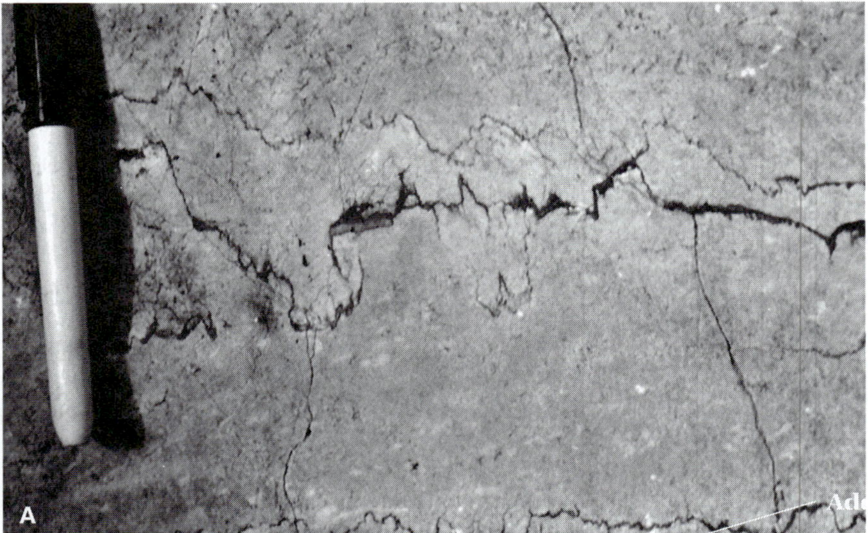

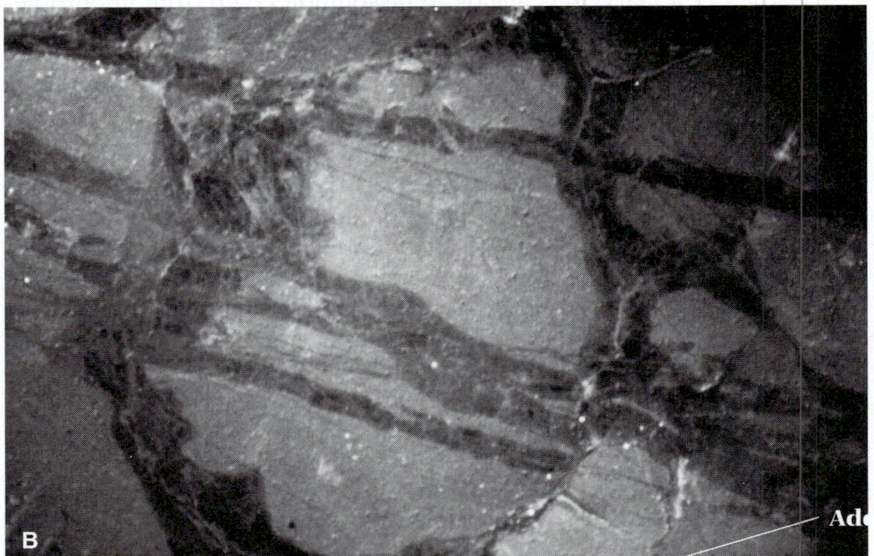

C

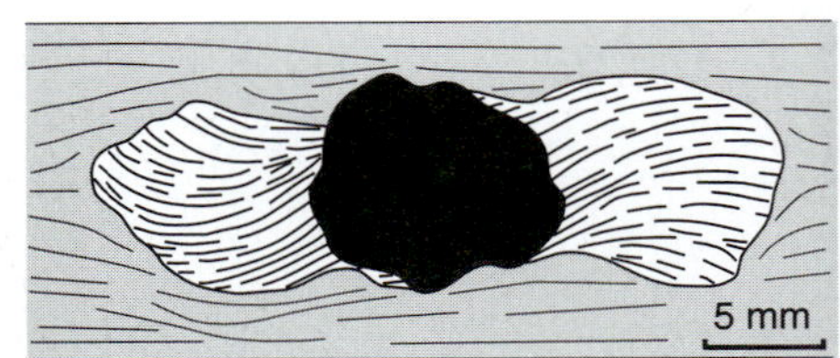

Figure 4.38 Characteristics of dissolution creep. (*A*) Stylolites formed by dissolution of carbonate, leaving dark accumulations of more insoluble material, Papoose Flat, eastern California. [Photograph by S. J. Reynolds.] (*B*) Siliceous microveinlets and overgrowths appear dark in this Cathode Luminescence image of Silurian Tuscarora Sandstone, central Appalachian Mountains, Virginia. [Photograph by Charles M. Onasch.] (*C*) Pressure shadows with fibers around pyrite. [Reprinted with permission from *Journal of Structural Geology*, v. 9, A. Etchecopar and J. Malavielle, Computer Models of pressure shadows; a method for strain measurement and shear-sense determination, (1987), Elsevier Science Ltd., Pergamon Imprint, Oxford, England.] (*D*) Crystal–fiber veins composed of gypsum in Triassic Moenkopi Formation, Lake Mead area, southern Nevada. (Photograph by S. J. Reynolds.)

Conditions Favoring Dissolution Creep

Dissolution creep can take place over a broad range of temperature–pressure conditions, *provided an intergranular fluid is present*. It can even operate efficiently at the relatively low temperatures that accompany diagenesis and low grade metamorphism. For example, horizontal sequences of limestones are typically marked by relatively close-spaced horizontal stylolitic surfaces, which represent sites from which material was removed by dissolution during the gravitational loading that accompanied burial and diagenesis.

A wide range of rock types can accommodate dissolution creep, but this deformation mechanism works best in immature lithologies, such as shales, calcareous shales, and impure carbonate rocks. Fine-grained rocks are especially vulnerable to dissolution creep. Strain rate is inversely proportional to the cube of the grain size (Rutter, 1976); hence, grain-size reduction by dissolution leads to a quickening of the pace of dissolution, a type of strain softening.

Dissolution creep operates at low differential stresses (Figure 4.39), but regrettably at such low strain rates that it has been difficult to produce dissolution creep in the typically short durations of deformation experiments.

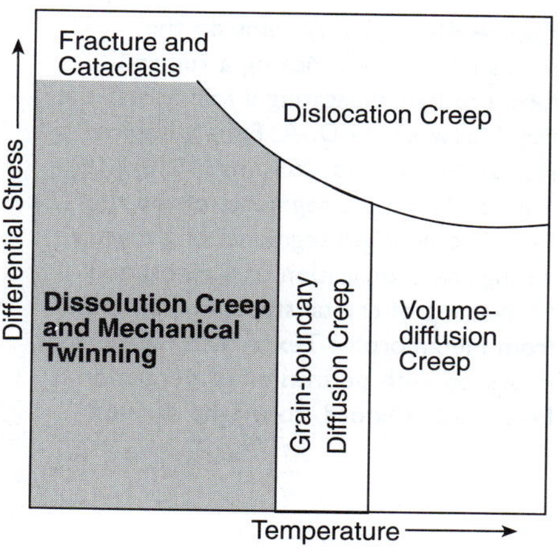

Figure 4.39 Simplified deformation map, showing conditions of dissolution creep.

Dislocation Creep

A mineral grain in a rock can also change shape by dislocation creep. The shape change is achieved by a combination of very localized, temporary distortion of the lattice and shearing of the lattice along a favorable crystallographic plane (the slip plane). If we eliminate for a moment the "distortion" part, the picture would be pretty simple.

For example, imagine a slip plane that cuts through the middle of a crystal (Figure 4.40*A*) and manages to move the entire upper half of the crystal one complete lattice spacing relative to the lower half of the crystal (Figure 4.40*B*). This mechanism is geometrically straightforward but mechanically difficult, because each increment of slip requires that *all* bonds along the slip plane be broken at the same time. Breaking all the bonds at the same time requires more shear stress than is generally available. Dislocation creep is necessary to solve the problem!

Nature's clever way around this problem is to activate only a small part of the slip plane at any one time, thereby breaking fewer bonds at once. This concept can be illustrated by considering the movement of a rug across a hardwood floor (Figure 4.41*A–D*). Picture a carpet so large and heavy and so laden with furniture that a line of people pulling on it from one side cannot budge it. To move (translate) the carpet without removing the furniture, it is necessary to curl the rug into a small anticlinal fold along one edge (this is the temporary, localized distortion), and systematically propagate the fold across the entire length of the rug. This way, the rug can be translated a short distance across the floor. This becomes a simple, though time-consuming job, because only a small part of the entire slip plane is active at any one time. The line that separates the part of the rug that has slipped from that which has not slipped is a **dislocation** (Figure 4.41*E,F*), which in the world of deformation mechanism is the boundary between slipped and unslipped parts of a crystal. (Remember the sponge analogy, Figure 4.15).

In a roundabout way, we have come to the definition. **Dislocation creep** is the production, motion, and destruction of dislocations through crystals and grains, as we will see, accommodated by recovery and recrystallization.

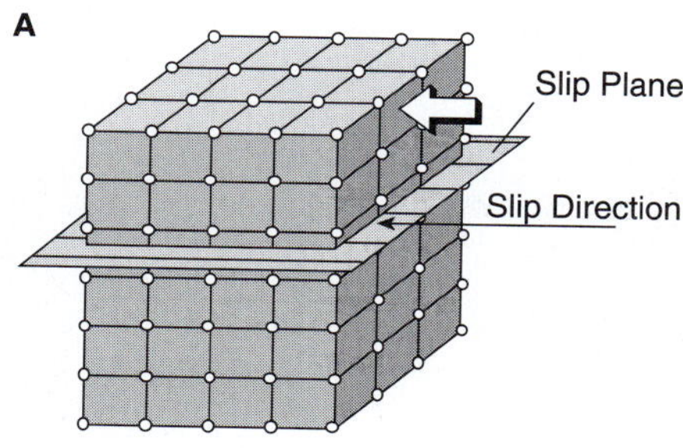

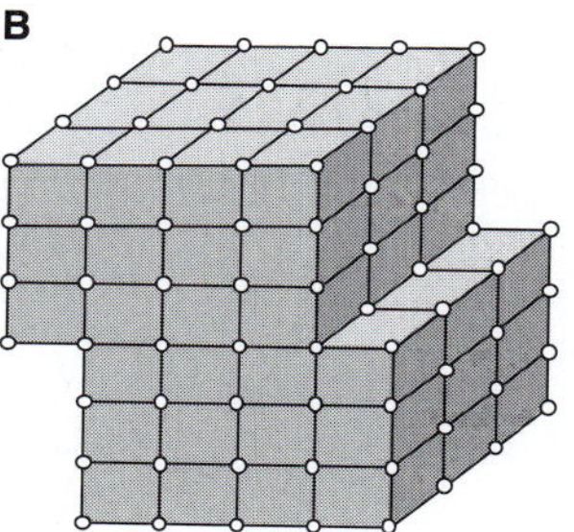

Figure 4.40 Slip of crystal along a slip plane. (*A*) Intact lattice, showing slip plane. (*B*) Offset lattice.

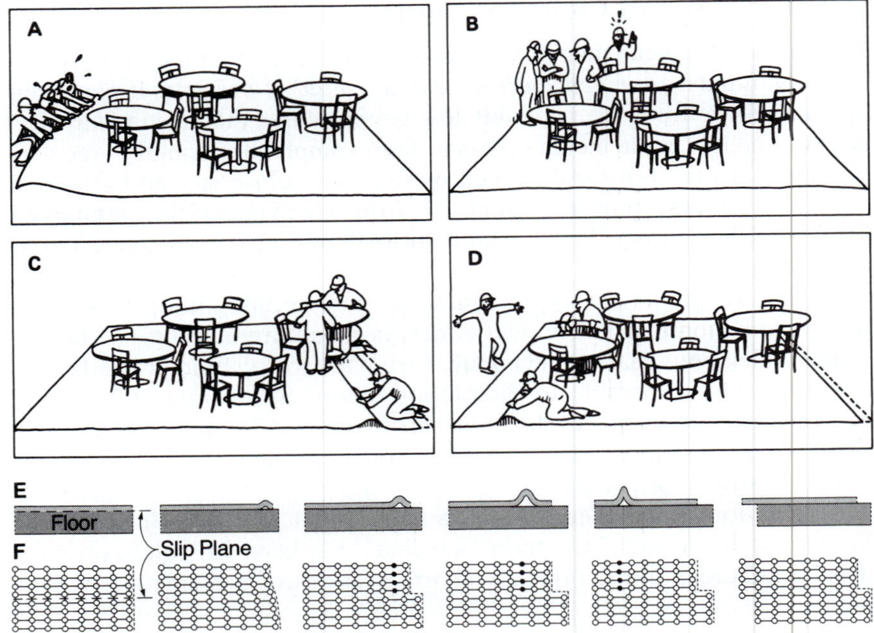

Figure 4.41 Slip systems do the impossible. (*A–D*) Moving a rug and furniture by propagating a fold across the rug. [Artwork by D. A. Fisher. Inspired by Hobbs, Means, and Williams (1976).] (*E*) Rigid and nonrigid segments of the rug. (*F*) Rigid and nonrigid segments of a crystal during the propagation of a dislocation. (Rendering of crystal structure reprinted from Metamorphic Textures by A. H. Spry. Published with permission of Pergamon Press, Ltd., Oxford, copyright © 1969.)

Types of Dislocations

Dislocations in crystals form and propagate in several ways, depending on the orientation of the dislocation relative to the direction of slip. **Edge dislocations**, like the fold in the rug, are oriented *perpendicular to the direction of slip*. The formation and propagation of an edge dislocation in a crystal is illustrated in Figure 4.42. As the perfect crystal lattice is subjected to a shear stress (Figure 4.42*A*,*B*), a small part of the upper right-hand side of the lattice compresses and begins to move with respect to a crystallographic plane below. Compression and distortion of the upper right-hand part of the crystal causes the lattice to snap into a new position, isolating an extra half-plane of atoms in between two others that occupy proper sites (Figure 4.42*C*). The line of intersection of the extra half-plane of atoms and the slip plane is the edge dislocation. It represents the boundary between the part of the lattice that slipped and the parts that did not. With continued stress-induced movement, the extra half-plane shifts to a proper site, forcing the previous occupant to move into an extra half-plane position (Figure 4.42*D*). Thus the edge dislocation migrates. In this way, dislocations are propagated one step at a time through the lattice.

When the extra half-plane reaches the edge of the crystal, it emerges as a small offset and the dislocation ceases to exist (Figure 4.42*E*,*F*). The crystal has changed its shape ever so slightly and, for an instant, is once again defect free—until a new dislocation is produced in the same manner as the first. Through this recurring cycle of formation and propagation of edge dislocations, the crystal changes shape in a *plastic* manner. This process of propagation of a dislocation through a crystal, called **dislocation glide**, occurs on crystallographic planes along which bonds are relatively weak.

Screw dislocations are more difficult to picture. They are oriented *parallel to the direction of slip* (Figure 4.43*A*,*B*). One way to visualize a screw dislocation is to think of a stack of Wheaties cereal boxes on the shelves of the neighborhood grocery store (Figure 4.44). When originally stocked on the shelves, the boxes were neatly aligned in rows. Later, however, someone bumps the shelves, causing one column of boxes to tilt over.

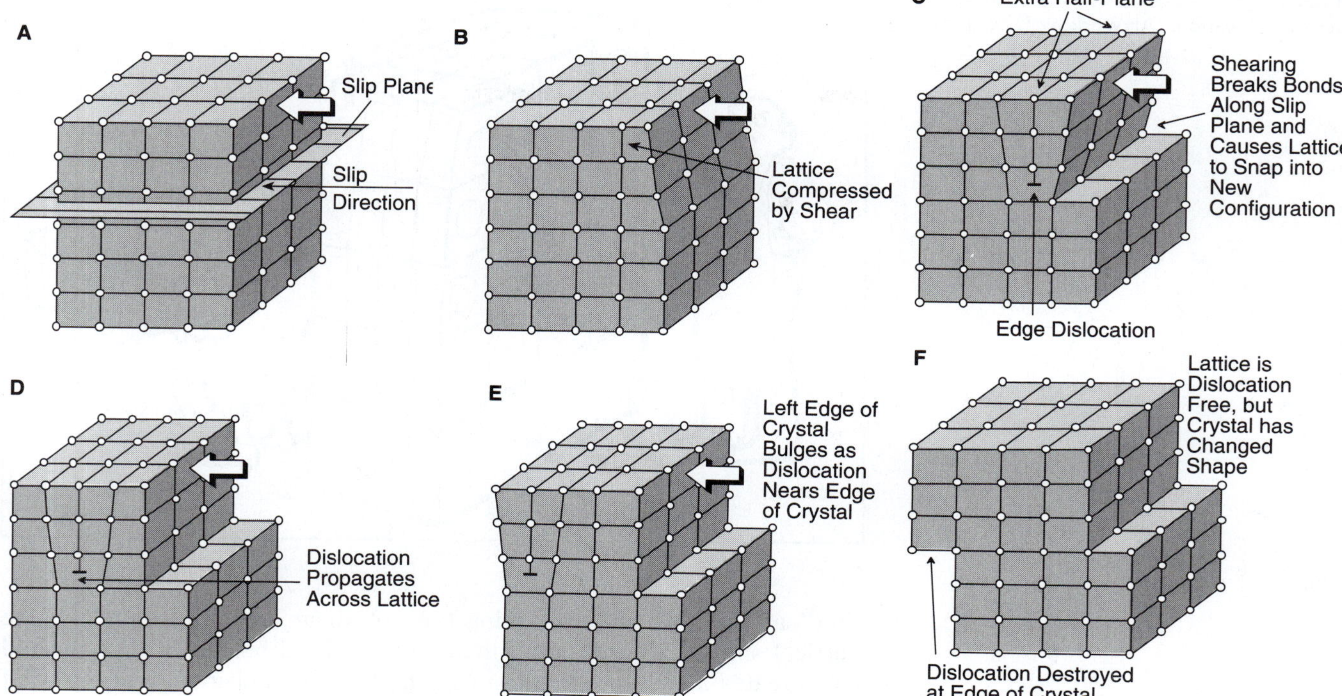

Figure 4.42 Formation and propagation of an edge dislocation. (*A*) Slip system within an undeformed crystal. White arrow indicates imposed stress on upper part of crystal. (*B*) Upper right part of crystalline lattice shortens elastically. (*C*) Stress exceeds yield strength of crystal, and lattice partially slips, forming an isolated extra half-plane of atoms above the slip plane. The bottom of the half-plane is an edge dislocation oriented perpendicular to the direction of slip. (*D* and *E*) Continued deformation causes the dislocation to propagate across the lattice, one step at a time. (*F*) Dislocation propagates to and offsets left side of crystal. The lattice is again defect free, but the crystal has changed shape. [Modified after Hobbs, Means, and Williams (1976).]

Boxes at the base of the tilted column are still essentially aligned with the boxes in the undisturbed column beside it, but boxes at the top of the tilted column have become aligned with the next box back in the undisturbed columns. The tilted boxes have retained their original "lattice position" at the base, but have been sheared backward one "lattice posi-

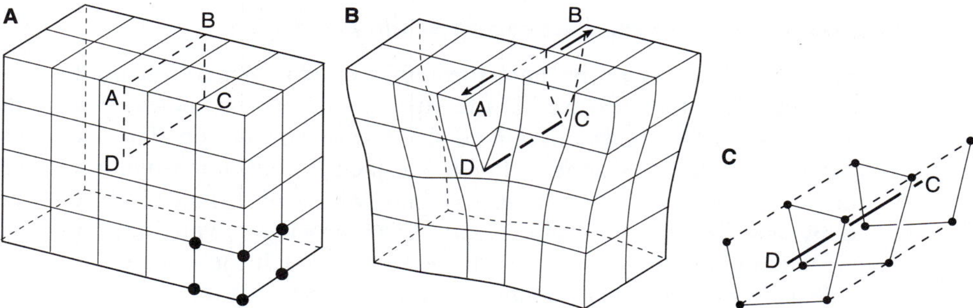

Figure 4.43 Screw dislocation. (*A*) Intact crystal. (*B*) Slip produces a screw dislocation parallel to the direction of slippage. (*C*) Adjacent atoms connect in a helical or spiral pattern, winding around the screw dislocation [Reprinted from D. Hull and D. S. Bacon, *Introduction to dislocations* (1984), p. 18, with permission from Butterworth-Heinemann Ltd.]

Figure 4.44 Grocery store analogy of screw dislocation. (Artwork by D. A. Fisher.)

tion'' at their top. This pivoting type of shear, accommodated by the vertical slip plane separating tilted and nontilted boxes to the right and left, creates a screw dislocation. In the three-dimensional real world of minerals, screw dislocations have extraordinary helical arrangements of atoms, much like a spiral staircase, which wind around the dislocation (Figure 4.43*C*).

Edge and screw dislocations are generally not the straight or systematically curved features that we might envision in an ideal situation. Instead, they curve via a series of offsets, termed kinks and jogs (Figure 4.45). Kinks are offsets of the dislocation within the same slip plane (Figure 4.45*A*), whereas in jogs the dislocation steps from one slip plane to another (Figure 4.45*B*). Through kinks and jogs, an edge dislocation can curve into and become a screw dislocation, and vice versa. Dislocations that consist of both edge and screw segments, and those that are *oblique to the direction of slip* are called **mixed dislocations**. Dislocations with both edge and screw segments may get carried away and form closed **dislocation loops**, which encircle the part of the crystal that has slipped. Dislocation loops may start as a point or very small loop, perhaps centered on a stress-concentrating vacancy or impurity, and progressively grow outward in multiple directions as slip continues.

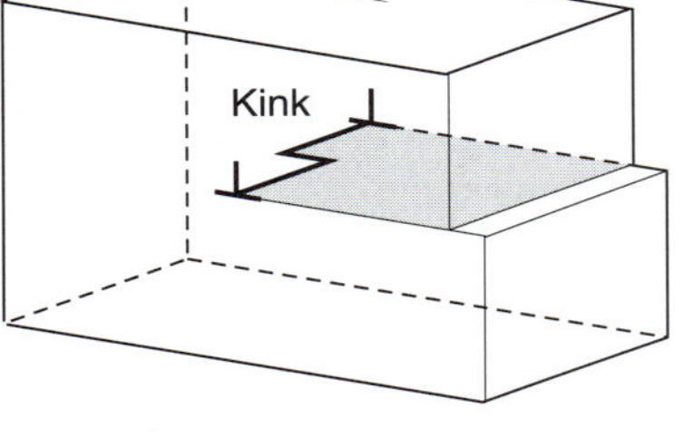

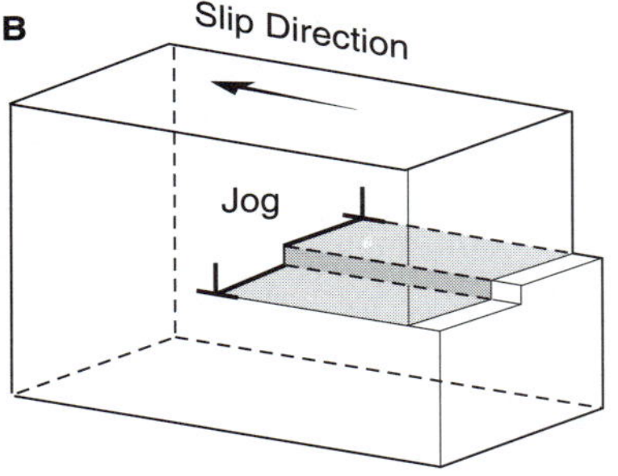

Figure 4.45 Dislocations are commonly marked by (*A*) kinks, which are bends within the slip plane, and (*B*) jogs, in which an edge dislocation jumps from one slip plane to another. [Reprinted from D. Hull and D. S. Bacon, *Introduction to dislocations* (1984), p. 61, with permission from Butterworth-Heinemann Ltd.]

Interactions with the Lattice and Strain Hardening

As a dislocation is propagated through a crystal, it may encounter and interact with other features of the lattice, such as vacancies, interstitial atoms, impurities, and even other dislocations. The motion of a dislocation may be impeded or stopped when the associated extra half-plane encounters an interstitial atom or impurity. This will occur if the energy required to bypass the obstacle or incorporate it into the advancing half-plane exceeds the energy driving the dislocation through the lattice.

Whether one dislocation impedes another depends on the geometry of the interaction. Figure 4.46 shows three possible interactions beween two dislocations, one of which is the bottom of an extra half-plane of atoms, the other, the top of an extra half-plane. Such dislocations are referred to as being of ''opposite sign.'' Both dislocations are utilizing the same slip system, but they are propagating in opposite directions. If the two dislocations are using the same slip plane and they meet coming from

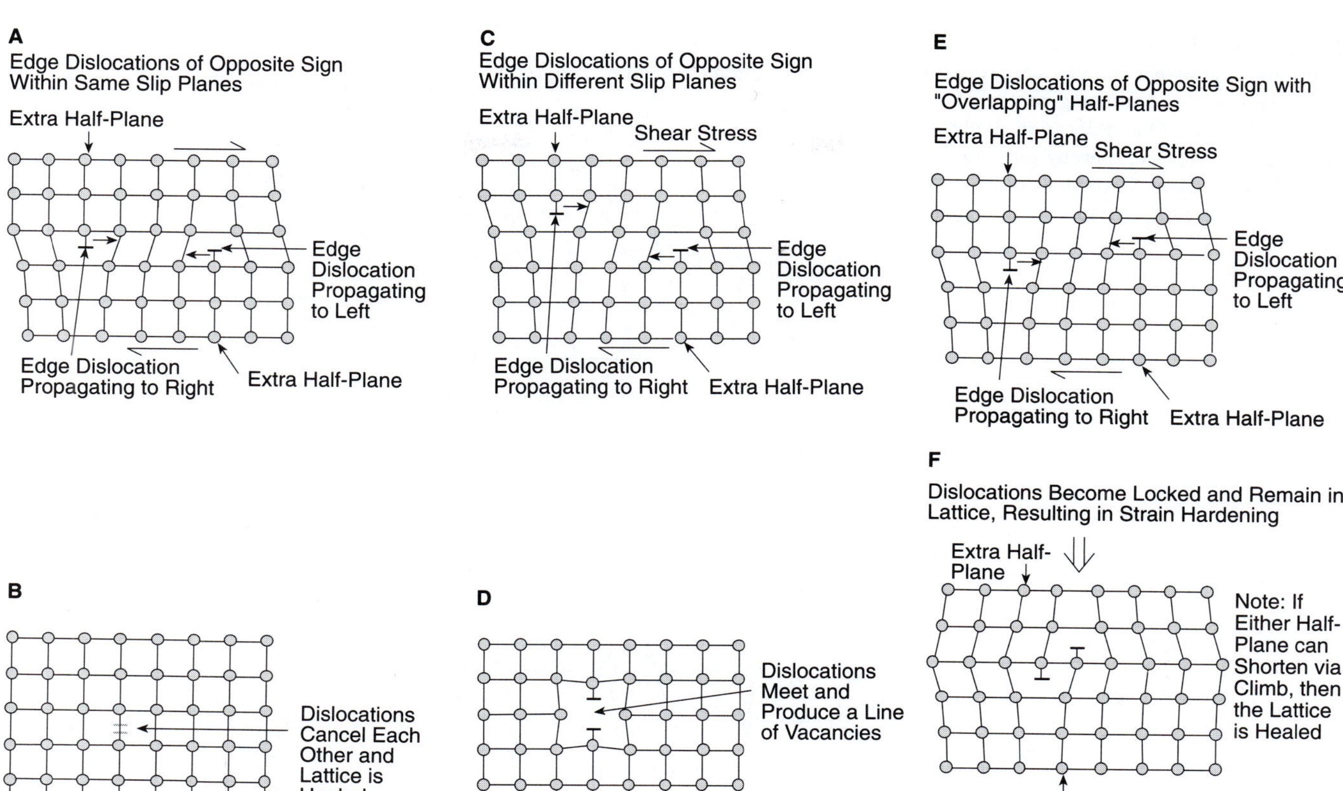

Figure 4.46 Geometry of interactions between edge dislocations within the same slip system. (*A–B*) Dislocations propagating toward each other along the same slip plane will cancel each other and heal the lattice. (*C–D*) Dislocations propagating toward each other along different planes produce a vacancy where they meet. (*E–F*) Dislocations with overlapping half-planes may become locked where they intersect, resulting in strain hardening of the lattice.

different directions, they cancel each other—poof! (Figure 4.46*A,B*). If they are one slip plane apart, they may line up, leaving an intervening vacancy (Figure 4.46*C,D*). If their half-planes overlap, they will likely become locked or "pinned," unable to move further (Figure 4.46*E,F*). Dislocations moving in the same direction may act like cars and trucks driving blind in an afternoon dust storm, piling up on one another when the leading dislocation encounters an obstacle that simply cannot be overwhelmed (Figure 4.47). It takes a tremendous amount of work and effort to clear the way for traffic to resume.

Two dislocations utilizing *different slip systems* may also become pinned where they intersect. **Dislocation tangles** form where a number of dislocations intersect and become pinned in a seemingly intertwined mess. It is the atomic equivalent of a four-way stop sign (Figure 4.48). Tangles and lesser dislocation-pinnings result in a **strain hardening** of the lattice. As a result, dislocation glide becomes increasingly difficult as more and more dislocations are created within a crystal. Continued deformation, therefore, *requires* that dislocations somehow be destroyed or be able to bypass obstacles. Fortunately, nature always seems to have a solution.

Recovery and Recrystallization

Dislocation glide introduces new defects into a crystal. In contrast, recovery and recrystallization cause a *healing* of the crystalline structure by eliminating as many defects as possible, and by depleting leftover stored

Figure 4.47 Dislocation pile up. (*A–C*) Dislocations intersect and pile up, producing strain hardening and stopping continued deformation. (*D*) One dislocation climbs into next higher slip plane, thereby canceling the obstacle dislocation. Distortions of lattice are schematic, and exaggerated.

A

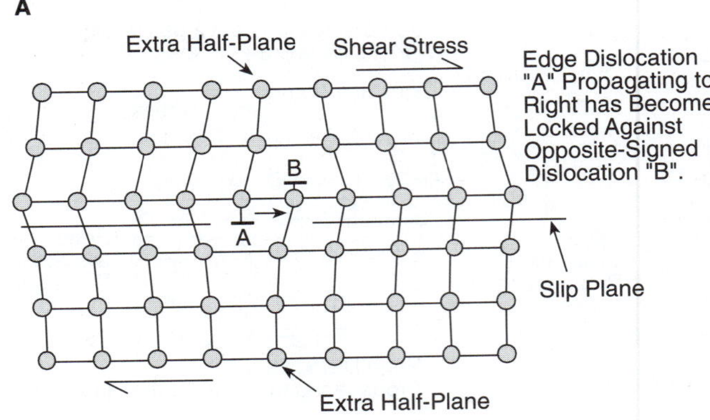

B

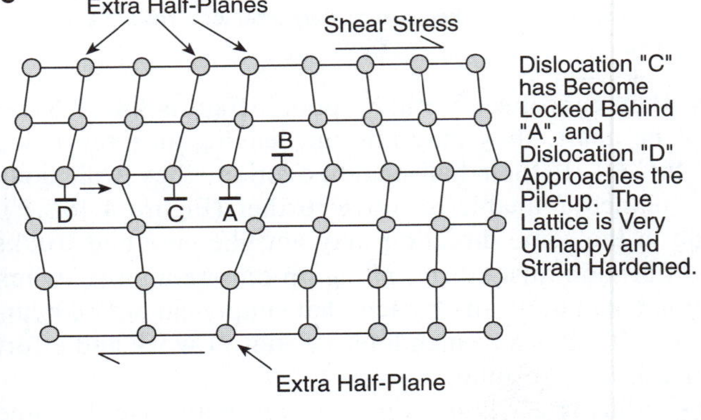

C

Extra Half-Planes

Shear Stress

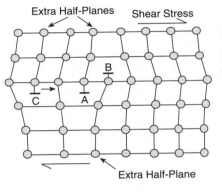

Dislocation "C" has Become Locked Behind "A", and Dislocation "D" Approaches the Pile-up. The Lattice is Very Unhappy and Strain Hardened.

B

D C A

Extra Half-Plane

Figure 4.48 TEM (transmitting electron microscope) image of a mess of tangled dislocations. (Courtesy of Andrew Kronenberg.)

D

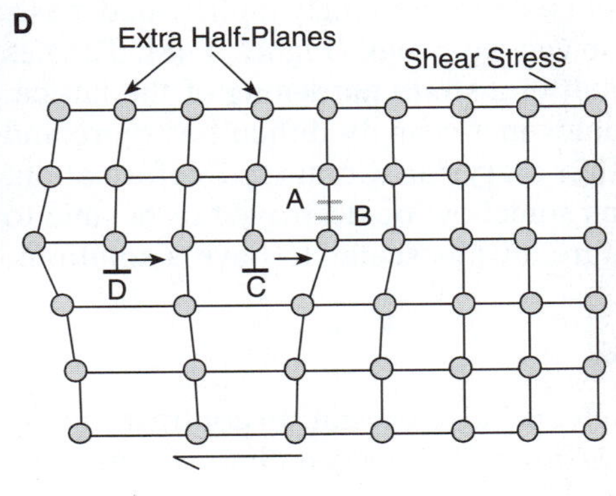

Dislocation "A" has Climbed up Into Next Higher Slip Plane and Cancelled "B", Permitting "C" and "D" to Continue Gliding, and Undoing Some Strain Hardening.

energy in the lattice. **Recovery** promotes healing through the rearrangement and destruction of dislocations. **Recrystallization** and **neomineralization** promote healing by transforming the old "defective" grains into brand-new grains or new configurations of grains.

Ordinary recrystallization takes place within a single grain or within adjacent grains of a common mineralogy (Figure 4.49). Neomineralization, on the other hand, forms new minerals or changes the boundary between two *different* minerals that were in contact (Urai, Means, and Lister, 1986).

Recovery and recrystallization *during* deformation are known as **dynamic recrystallization** (Figure 4.49*B–D*). Recovery and recrystallization *after* deformation is known as **annealing** (Figure 4.49*E,F*). Dynamic re-

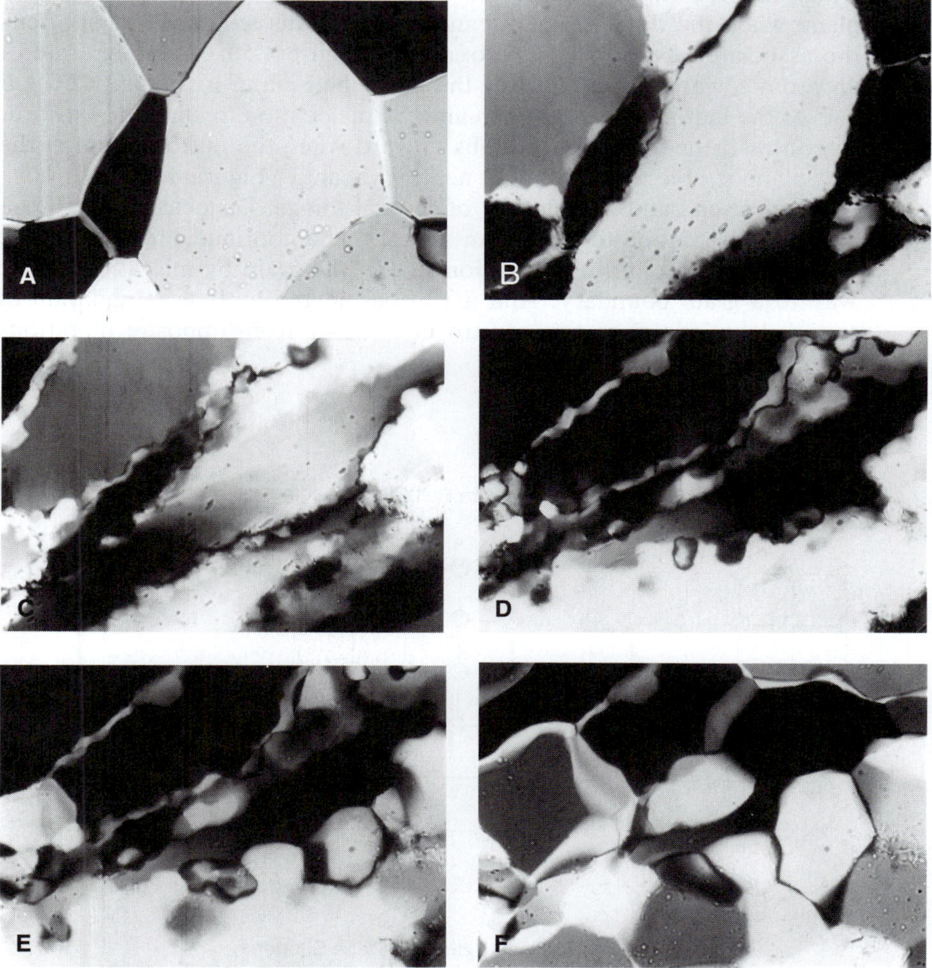

Figure 4.49 Recrystallization produced experimentally by deforming octachloropropane, a synthetic material that deforms ductilely at low temperatures. (*A*) Starting material with undeformed grains bounded by straight boundaries that meet at 120° angles. Bubbles within grains can be used to monitor the amount of strain during subsequent deformation. (*B–D*) Progressive dextral (right-handed) shear, resulting in internal deformation of grains, as recorded by the elliptical shape of bubbles (miniature strain ellipses). Dynamic recrystallization during deformation causes bulges in grain boundaries, some of which become new, small grains in *D*. (*E–F*) Recrystallization and recovery after deformation (annealing) result in growth of grains, the healing of internal defects, and the straightening of grain boundaries. Annealing leads to an appearance in *F* similar to the initial, undeformed material in *A*. (Courtesy of Win Means, SUNY, Albany.)

crystallization counteracts strain hardening and accommodates the continued glide of dislocations, permitting the rock to sustain steady-state flow via dislocation creep. In fact, dislocation creep is possible only where dynamic recrystallization continuously keeps pace with strain hardening. *The rate of dislocation creep is controlled by the rates of recovery and recrystallization.* When dynamic recrystallization and annealing do their work, many of the ordinary telltale signatures of microstructural deformation are eliminated, creating the illusion that the minerals were never deformed in the first place.

Recovery and Rotation Recrystallization

Recovery is achieved principally through the process of **dislocation climb**, which involves the movement of a dislocation to a higher or lower slip plane when the diffusion of vacancies and atoms lengthens or shortens the associated extra half-plane of atoms (Figure 4.50). The migration of an atom toward the bottom of the extra half-plane in Figure 4.50*A*,*B* causes the half-plane to lengthen and the dislocation to climb downward. In contrast, the dislocation climbs upward when the bottom atom in the half-plane is able to jump into a nearby vacancy (Figure 4.50*C*,*D*).

Dislocation climb can do one of several things. Dislocation climb can permit a dislocation to bypass an obstacle and continue slipping (Figure 4.47*D*). It can permit a dislocation to exit the grain by migration all the way to a grain boundary. And it can permit the mutual annihilation of two dislocations migrating toward each other from opposite directions (Figure 4.51).

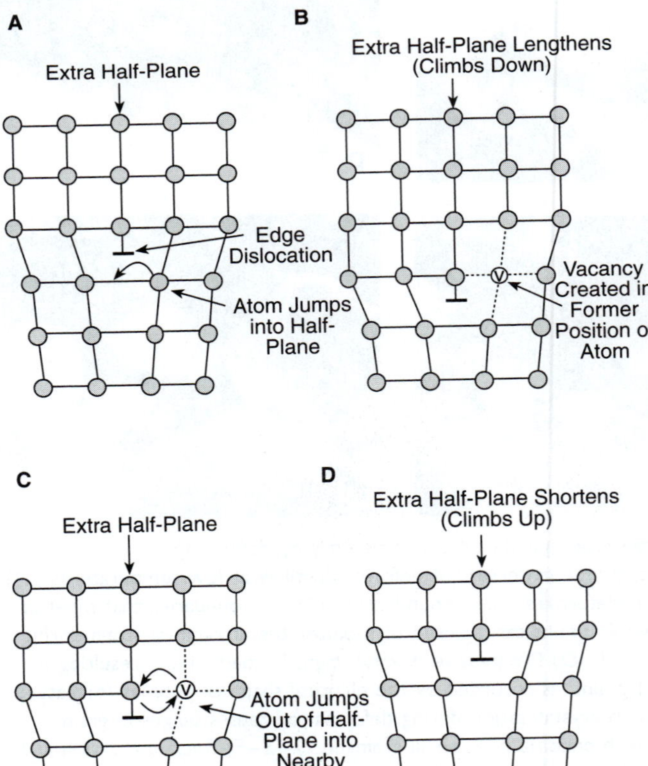

Figure 4.50 Dislocation climb. (*A–B*) Dislocation climbs down as an adjacent atom jumps into the half-plane, lengthening it. (*C–D*) Dislocation climbs up as an atom jumps into an adjacent vacancy, shortening the half-plane. Dislocation is destroyed if it can climb all the way to the grain boundary.

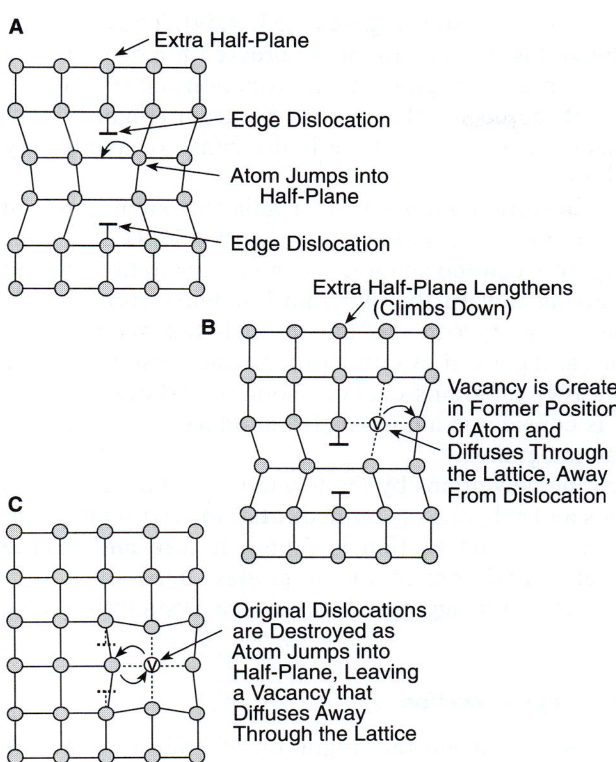

Figure 4.51 Two dislocations climb into and annihilate one another.

Dislocation climb also permits a screw dislocation to bypass an obstacle by changing crystallographic planes via a process called **cross-slip** (Figure 4.52). Cross-slip can result in some complex geometries of edge and screw dislocations, all to get around an obstacle.

Alternatively, dislocation climb may permit adjacent dislocations to arrange themselves into **walls** that form **low-angle boundaries** between

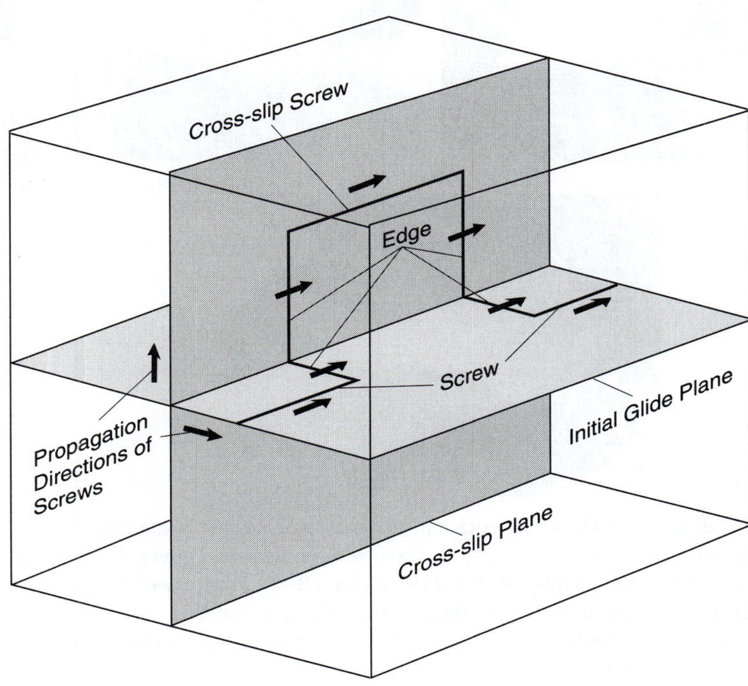

Figure 4.52 Cross-slip, whereby a screw dislocation bypasses an obstacle by moving to a different crystallographic plane. Arrows represent the direction and amount of slip. The slip is equal to one lattice spacing, which you may remember from mineralogy is called the Burgers vector. [Modified after Twiss and Moores (1992).]

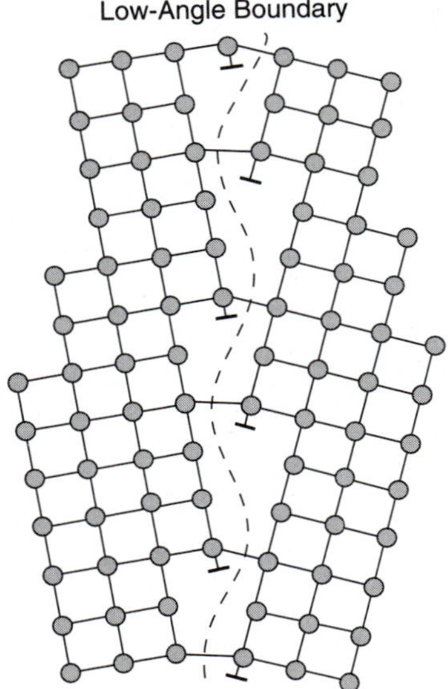

Figure 4.53 Low-angle boundary, where edge dislocations permit a slight misalignment of two adjacent parts (subgrains) of a crystal. [Modified after Twiss and Moores (1992) and Hobbs, Means, and Williams (1976).]

slightly misoriented parts of a crystal (Figures 4.53, 4.54). These misoriented parts, termed **subgrains**, are commonly expressed microscopically in thin section as small areas of a grain with extinction angles slightly different from what is observed for the rest of the grain (Figure 4.55). Such progressive misalignment of the lattice is the cause of undulatory extinction (Figure 4.55*B*).

The crystallographic misorientation between adjacent subgrains can increase during deformation in two main ways (Figure 4.56) (Urai et al., 1986): (1) the boundary between the subgrains can remain stationary and collect dislocations moving toward the subgrain boundary from one or both sides; or (2) the boundary can migrate through the material and accumulate dislocations as it goes. If by either process the crystallographic mismatch between the two subgrains exceeds some threshold, perhaps 10–15°, the boundary is considered a **high-angle boundary** separating distinct grains rather than subgrains.

This process of creating new grains by the formation and accentuation of low-angle boundaries and subgrains is termed **rotation recrystallization**. Evidence for the process in thin section is found in **core and mantle structure** when an intact central core of a grain grades successively outward into subgrains and then into aggregates of new, recrystallized grains (Figure 4.57).

Boundary-Migration Recrystallization

Recrystallization may also occur via the migration of grain boundaries, known as **boundary-migration recrystallization**. On the one hand, a grain may remain stationary and grow at the expense of a neighboring grain, adding to its collection of atoms as it does so (Figure 4.58*A,B*). On the

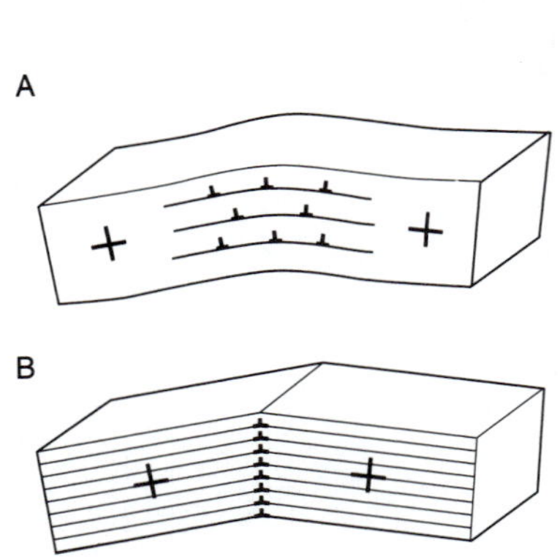

Figure 4.54 Low-angle boundary produced when dislocations become arranged into a wall. (*A*) Widely distributed dislocations of the same sign produce a gentle bending of the lattice, which would be visible in thin section as undulatory extinction. (*B*) Dislocations arranged into a wall separating two slightly misaligned subgrains. [Reprinted from D. Hull and D. S. Bacon, *Introduction to dislocations* (1984), p. 180, with permission from Butterworth-Heinemann Ltd.]

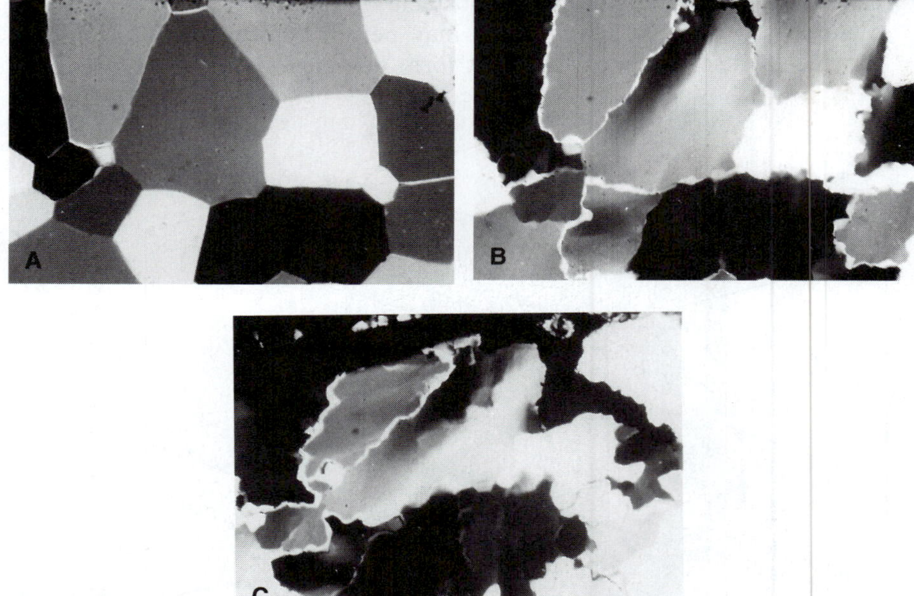

Figure 4.55 Subgrains produced by rotation recrystallization within experimentally deformed octachloropropane. (*A*) Starting material with undeformed grains bounded by straight boundaries that commonly meet at 120° angles. (*B–C*) Progressive dextral (right-handed) shear, resulting in internal deformation of grains, rotation recrystallization, and the development of subgrains and undulatory extinction. (Courtesy of Win Means, SUNY, Albany.)

other hand, a grain may be mobile and migrate laterally through the material during deformation, changing size, shape, *and* position as it does so (Figure 4.58*C*,*D*). If the boundaries of such a grain migrate far enough across the material, a recrystallized grain may end up enclosing a collection of atoms totally different from the one with which it started.

Boundary-migration recrystallization requires that atoms "jump" across the grain boundary from one grain to another, a process that requires energy (Urai et al., 1986). An energy source that can be tapped to accomplish this is the excess energy introduced into the grains during deformation and stored in defects, such as dislocations (Figure 4.58*A*,*B*). A grain that is full of dislocations has a higher energy state than a less deformed grain. The overall energy within a deformed system of minerals is decreased, therefore, if atoms are able to jump ship from a grain full of dislocations to one that is much less internally deformed. This mutiny of atoms causes grain boundaries to migrate as defect-rich grains are consumed by defect-poor ones (Figure 4.58*C*,*D*). This may result in strongly curved, sutured grain boundaries (Figure 4.59).

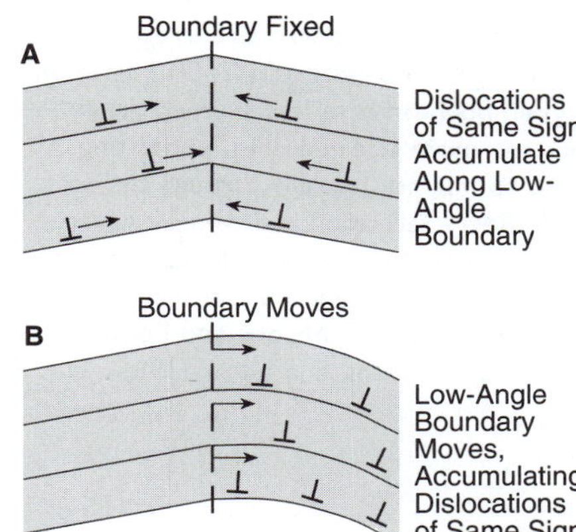

Figure 4.56 Two ways to increase the misorientation of adjacent subgrains. (*A*) Subgrain boundary is fixed and subgrains become more misoriented as dislocations of the same sign propagate into the boundary. (*B*) Subgrain boundary moves, accumulating dislocations of the same sign. (From J. L. Urai, W. D. Means, and G. S. Lister, *Geophysical Monograph,* v. 36, fig. 7, p. 167, copyright © 1986 by American Geophysical Union.)

Figure 4.57 Photomicrograph of experimentally deformed Black Hills Quartzite. (*A*) Shortened 60%, producing flattened original grains with sweeping undulatory extinction, some visible subgrains, and a fringe of small recrystallized grains forming incipient core and mantle structure. Recrystallization in this sample has occurred primarily by subgrain rotation (Hirth and Tullis, 1992). (*B*) Same sample annealed by increasing the temperature after deformation had ceased. Grains are now essentially strain free, as indicated by the even extinction of individual grains and relatively straight grain-boundaries. (Photographs kindly provided by Jan Tullis, Brown University.)

Figure 4.58 Processes of recrystallization via grain boundary migration. (*A–B*) Recrystallization is driven by stored strain energy, leading to more irregular, serrated grain boundaries. (*C–D*) Migration of grain boundaries as less strained grains consume more strained ones, permitting grains to migrate laterally through the rock. (*E–F*) Recrystallization is driven by surface energy, whereby grain boundaries become more straight and tend to intersect at 120° angles. [In part after Nicolas and Poirier (1976); Urai, Means, and Lister (1986).]

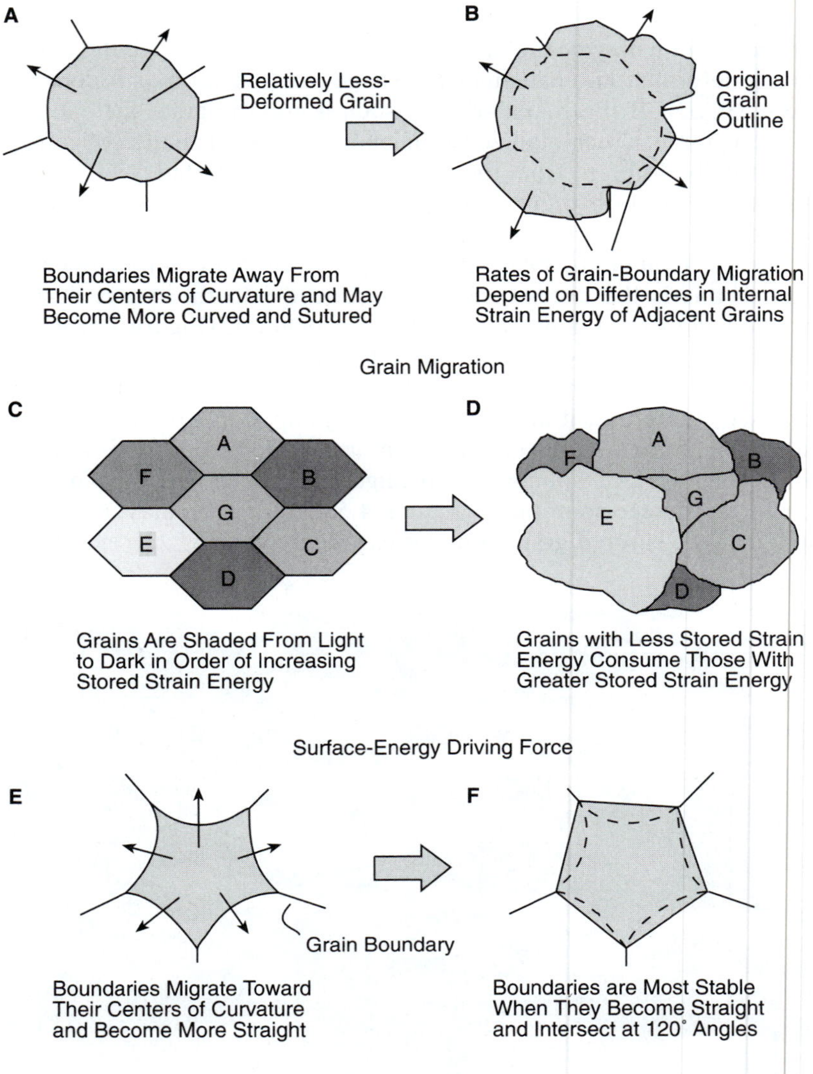

Strain-Energy Driving Force
A
B
Relatively Less-Deformed Grain
Original Grain Outline
Boundaries Migrate Away From Their Centers of Curvature and May Become More Curved and Sutured
Rates of Grain-Boundary Migration Depend on Differences in Internal Strain Energy of Adjacent Grains
Grain Migration
C
D
Grains Are Shaded From Light to Dark in Order of Increasing Stored Strain Energy
Grains with Less Stored Strain Energy Consume Those With Greater Stored Strain Energy
Surface-Energy Driving Force
E
F
Grain Boundary
Boundaries Migrate Toward Their Centers of Curvature and Become More Straight
Boundaries are Most Stable When They Become Straight and Intersect at 120° Angles

Figure 4.59 Photomicrographs of the effects of recrystallization in experimentally deformed octachloropropane. (*A*) Initial sample, containing opaque particles of grinding powder that permit us to track the motion of grain boundaries and to estimate the amount of strain. The sample displays straight boundaries with 120° grain-boundary intersections, such as is common in rocks that have been thermally annealed. (*B–D*) Progressive horizontal shortening results in dynamic recrystallization by grain-boundary migration, leaving serrated grain boundaries. Note that the dark grain at the top center becomes highly strained (undulatory extinction and subgrains) and is almost totally eaten by its less deformed neighbors. [Photographs courtesy of Win Means, SUNY, Albany.]

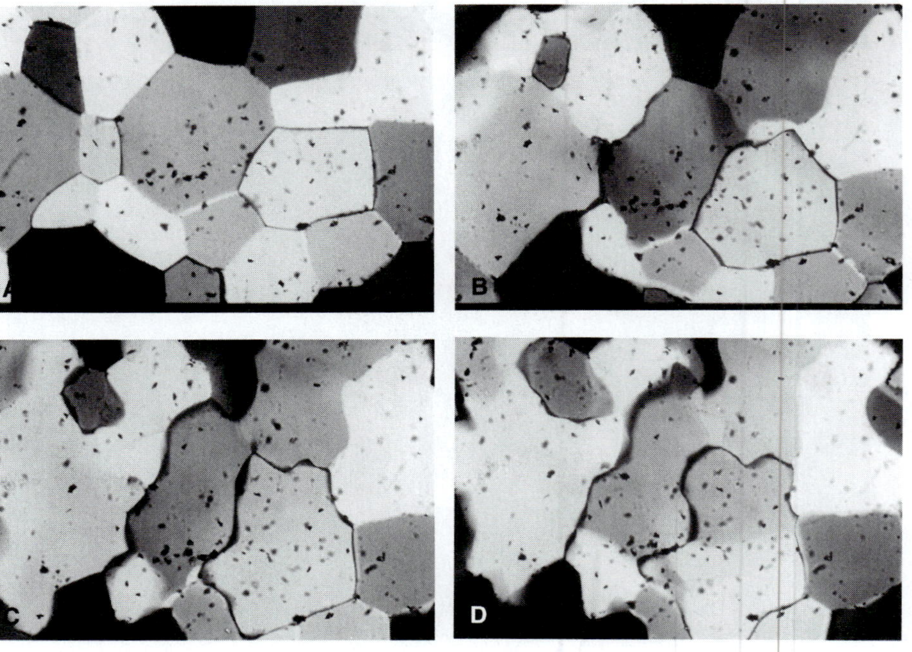

A
B
C
D

Another energy source that strongly influences the shapes and orientations of grain boundaries is the surface energy of grain boundaries. Minimizing grain-boundary energy favors straight or gently curved grain boundaries over strongly curved boundaries, large grains over numerous small grains, and 120° grain boundary intersections that intersect at a point (Figures 4.58*E,F*; 4.59*A*).

Conditions Favoring Recrystallization

Boundary-migration recrystallization is favored by conditions of moderate to high temperatures, under which atoms are vibrating actively and can more easily jump from one grain to another. Migration rates are strongly influenced by variations in the concentration of defects, by the orientation of a grain boundary relative to the crystallographic orientation of the crystal, and by the presence of fluid or impurities along grain boundaries. Grain-boundary migration is probably enhanced by the presence of pore fluid, which helps atoms diffuse more easily from grain to grain. In contrast, grain-boundary migration tends to be impeded by impurities, especially if they accumulate along the migrating grain boundaries.

The interplay of all the factors that influence grain-boundary migration causes the migration rate of grain boundaries to vary significantly between adjacent grains and between different parts of the same grain. This commonly results in very irregular, **serrated grain boundaries** (Figure 4.59), especially during dynamic recrystallization, when defects continually form, accumulate, and are destroyed. Where thermal processes outlast deformation (annealing), surface energy may overwhelm these factors and produce equant grains with geometrically simple boundaries (Figure 4.59*A*).

Rotation recrystallization is favored by differential stresses high enough to generate and propagate dislocations and by temperatures high enough to support dislocation climb, which is a thermally activated process. At moderate temperatures, where climb is slow, rotation recrystallization occurs, but it is outpaced by grain-boundary migration, which then dominates the generation of microstructures. Because rotation and grain-boundary recrystallization are both thermally activated, dislocation creep is most efficient at moderate to high temperatures (Figure 4.60).

Summary

Nature has a vast array of mechanisms that permit rocks to deform. Which deformation mechanism operates depends on a long list of factors, especially mineralogy, grain size, temperature, differential stress, confining pressure, strain rate, the presence or absence of fluids, and fluid pressure. Several deformation mechanisms may be active in a deforming rock mass at the same time, and these may interact in important, but rather complex, ways. One mechanism may dominate the deformation until some strain limitation is reached, whereupon a different mechanism takes over. Dislocation glide may accommodate only a small amount of strain at low temperatures before strain hardening and stress concentration lead to brittle fracture. In other cases, two mechanisms may proceed synchronously and work together to permit more strain, as in the case of dislocation creep accompanied by dynamic recrystallization. Also, two processes may interfere with each other to such an extent that neither one can operate. For example, frictional sliding can cause microcracking and dilation, which lowers the fluid pressure, thereby inhibiting more frictional sliding and microcracking until either the shear stress or fluid pressure builds back up again.

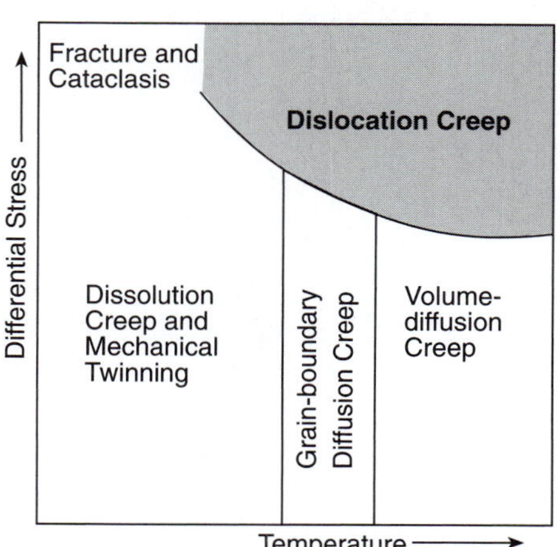

Figure 4.60 Simplified deformation map, showing the conditions of dislocation creep.

DEFORMATION EXPERIMENTS

We will see in Chapter 5 (''Joints and Shear Fractures'') that important insights regarding such brittle deformation mechanisms as microcracking and frictional sliding have been derived through laboratory deformation of rocks and minerals under conditions of low temperature and a wide range of confining pressures. Experimental procedures are exacting, especially when the distribution of microcracks must be monitored and mapped acoustically (microseismically) during the course of an experiment. Even more exacting and technically challenging is the laboratory study of creep, recovery, and recrystallization. To try to generate diffusion creep, dissolution creep, and dislocation creep during short-term experiments, and to factor in the role of recovery and recrystallization, the experiments must be carried out under conditions of very high temperature and confining pressure. Moreover, it is necessary to use very small samples of fine-grained materials. The experimental work carried out in this way has yielded important information on the strength of minerals, the mechanisms by which minerals deform, and the conditions under which each deformation mechanism operates.

The starting materials for such experiments vary, depending on the scientific goals. Typical materials are natural or synthetic single crystals of quartz or calcite or olivine, natural quartzite and novaculite, artificially sintered quartz or olivine aggregates, natural or synthetic quartz–feldspar aggregates (fine-grained granite and aplite), and various types of calcareous rocks and rock salt. To evaluate the conditions favoring each deformation mechanism, the same starting material is deformed under a range of conditions of temperature, confining pressure, differential stress, and strain rate. The mechanical behavior of the material is monitored using the relation between stress, strain, and strain rate throughout the experiment. At the end of the run, the deformed specimen is examined closely, using various microscopic techniques to characterize the microstructures and to evaluate which deformational mechanisms operated during the run under the given set of conditions.

Creep Experiments

Experiments and theories of rock deformation indicate that the three fundamental creep mechanisms (diffusion, dissolution, and dislocation) may occur in a single rock at the same time, but each mechanism will operate at a different rate, depending on temperature, pressure, differential stress, mineralogy, grain size, and availability of fluid. The physical conditions under which each creep mechanism is dominant can be illustrated on **deformation maps** (Figure 4.61), such as those that plot differential stress versus temperature *or* differential stress versus grain size (Ashby, 1972; Rutter, 1976). Strain rate is generally shown on deformation maps by a series of contours that cross from one **deformation field** into another. Deformation maps are constructed partly from theories of deformation and partly from extrapolations of results from the rock deformation experiments.

Deformation maps for creep mechanisms of individual mineral species (Figure 4.61) generally show that dissolution creep is the dominant creep mechanism at low stresses and low temperature, where ''low temperature'' is considered to be less than 40% of the melting temperature of the mineral. Higher stress favors dislocation creep and higher temperature favors diffusion creep (Figure 4.61).

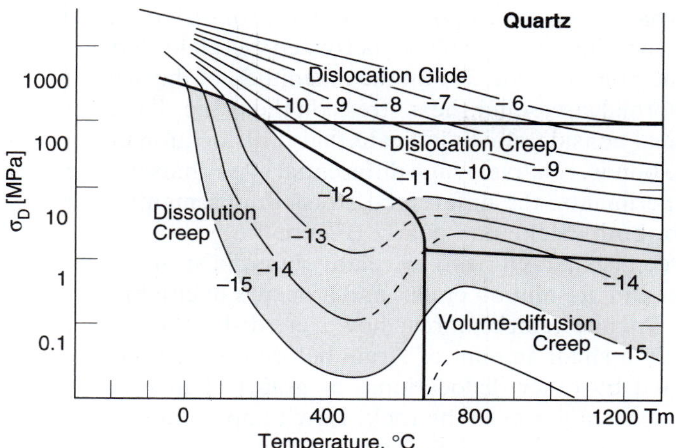

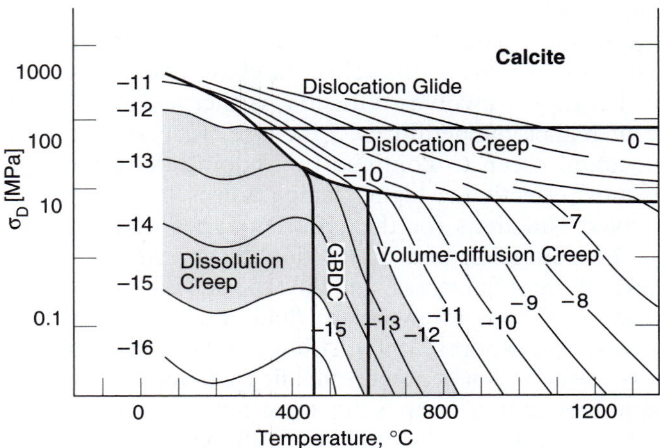

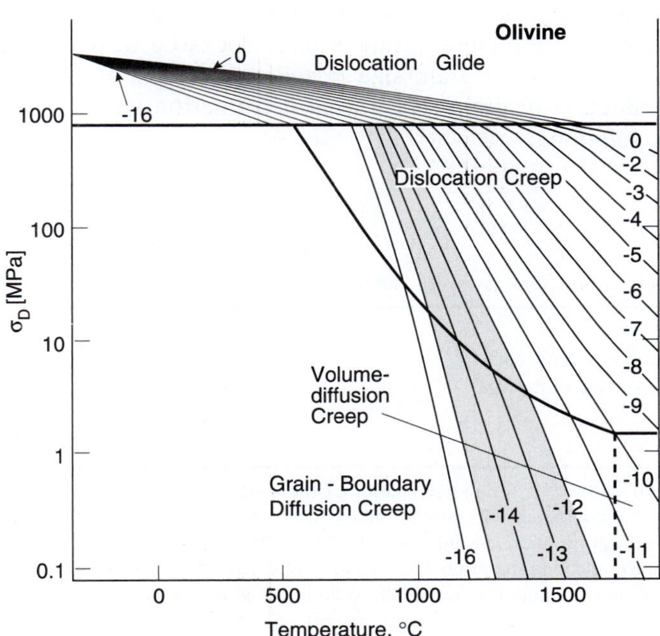

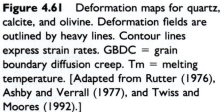

Figure 4.61 Deformation maps for quartz, calcite, and olivine. Deformation fields are outlined by heavy lines. Contour lines express strain rates. GBDC = grain boundary diffusion creep. Tm = melting temperature. [Adapted from Rutter (1976), Ashby and Verrall (1977), and Twiss and Moores (1992).]

Most minerals have several slip systems available for dislocation creep, but whether a given slip system will be activated by glide depends not only on the atomic bonding, but also on the magnitude of the shear stress and the ratio of normal stress and shear stress on the plane. Experimental data and theoretical considerations indicate that if dislocation creep is to break through the bonds, relatively high differential stress must be present. In laboratory experiments, the differential stress requirements approach 10 to 100 MPa for both calcite and quartz (Figure 4.61).

Dislocation creep is characterized by relatively rapid strain rates compared to diffusion and dissolution creep, and it occurs over a broad range of temperatures. At low temperatures, however, dislocation creep can accommodate only a small amount of strain before dislocations become pinned and tangled by other dislocations, or grains change shape and impinge against adjacent grains in the rock. Low-temperature dislocation creep therefore leads to strain hardening and the concentration of stresses at obstacles to glide and at grain boundaries, commonly resulting in fracture (Lloyd and Knipe, 1992).

At higher temperatures or slower strain rates, dislocation creep is possible because dislocations can bypass obstacles via climb and cross slip. Recovery by climb, because it involves diffusion, is thermally activated and most effective at higher temperatures. Likewise, recrystallization, which permits dislocation creep to continue by undoing the effects of strain hardening, is most effective at higher temperatures.

Theoretically derived equations for the different types of creep are listed in Table 4.4. For diffusion and dissolution creep, strain rate is directly proportional to differential stress, and thus the behavior is comparable to that of a linear viscous (Newtonian) fluid. For some models, dislocation creep (Twiss and Moores, 1992), strain rate is proportional to the differential stress raised to some power, yielding a behavior referred to as **power-law creep**. Dislocation creep is therefore capable of generating much higher strain rates for a given increment of differential stress.

For all three varieties of creep (i.e., diffusion, dissolution, and dislocation), strain rate is proportional to temperature because all are "rate limited" by diffusion, which is thermally activated. Diffusion creep and dissolution creep are most effective in fine-grained rocks because strain rate is inversely proportional to grain size raised to a power of 1 to 3. This means that reduction of grain size during deformation will lead to pronounced strain softening.

TABLE 4.4
Theoretically derived equations (flow laws) for different creep processes

Process	Flow law
Volume diffusion creep	$\dot{e} = A_1 \Omega D_v \sigma / kTd^2$
Grain-boundary diffusion creep (dry)	$\dot{e} = A_2 \Omega D_b S \sigma \pi / kTd^3$
Grain-boundary diffusion creep (wet)	$\dot{e} = A_3 \Omega D_f C S \sigma / kTd^3$
Dissolution creep	$\dot{e} = A_4 \Omega I \sigma / kTd$
Dislocation creep	$\dot{e} = (A_5 G D_v b / kT)(\Omega / G)^n$

A_1–A_5 are numerical constants. Ω is the molar volume of the solid. D_v is the coefficient of volume diffusion. D_b is the coefficient of grain boundary (dry) diffusion. D_f is the effective diffusivity of the solid in the grain-boundary fluid. σ is the differential stress. K is Boltzmann's constant. T is temperature and d is the grain size. S is the effective grain-boundary width for diffusion. C is the solubility (mole fraction of the solid in the grain-boundary fluid). I is the velocity of dissolution or growth of the solid (the slowest). G is the shear modulus (shear stress divided by shear strain). b is the Burgers vector, which represents the amount and direction the lattice is displaced across a dislocation.
Source: Knipe, 1989.

Recovery–Recrystallization Experiments

Experimental results by Jan Tullis and others have documented different regimes of deformation, which depend on the relative rates of dislocation production, dislocation climb, and grain boundary migration (Hirth and Tullis, 1992). Each regime is characterized by a distinctive microstructure that largely reflects the operation of different mechanisms of dynamic recrystallization (Figure 4.62).

At relatively high temperatures and slower strain rates, the rate of dislocation climb is fast enough to accommodate dislocation creep via both rotation recrystallization and grain boundary migration (Tullis et al., 1990). The resulting microstructure displays abundant subgrains, recrystallized grains, and irregular, sutured grain boundaries (Figure 4.62*B*). Subgrains and recrystallized grains become larger at successively higher temperatures and begin to more closely resemble those produced by annealing (J. Tullis, 1990b). As a result of easy climb, dislocations have a fairly uniform distribution within grains and are curved, because segments were able to climb or cross-slip out of the slip plane (Figure 4.63*A*).

At moderate temperatures or higher strain rates, rotation recrystallization is not important because dislocation climb, which is limited by the rate of diffusion, is too slow. Instead, dislocation creep is accommodated by grain boundary migration (Tullis et al., 1990). The resulting microstructure lacks subgrains but contains abundant sutured grains and other evidence for grain boundary migration (Figure 4.62*C*). Dislocations are relatively straight, and tangles are present because the dislocations could not climb out of the slip plane (Figure 4.63*B*). Also there are large variations in dislocation density between strongly deformed "old" grains that contain numerous dislocations, and recrystallized "new" grains with relatively few dislocations. Once they have formed, the "new" recrystallized grains concentrate further deformation because they are strain free and weaker than the strain-hardened "old" grains (Tullis et al., 1990).

At even lower temperatures and faster strain rates, dislocation glide accommodates only a small amount of strain before strain hardening prohibits further glide. Dislocations may concentrate stress, leading to fracture. Dislocations, if present, will be straight because climb is very, very slow at low temperatures.

Analogue Experiments

Much of our new understanding of ductile deformation mechanisms and dynamic recrystallization comes from direct observations under the microscope of materials that deform and recrystallize at lower temperatures and higher strain rates than typical rock-forming minerals. These materials, including ice, salt, and strange mineral analogues such as octachloropropane and paradichlorobenzene, behave ductilely and recrystallize readily when deformed near room temperature. Thus, the behavior of individual grains and grain boundaries can be observed through the microscope, essentially in real time, as the material is being deformed.

Wilson (1981, 1986) has comprehensively studied deformation mechanisms in ice as a rock analogue. Ice is an uncanny crystallographic analogue for quartz. Wilson prepares fine-grained ice samples in such a way that crystallographic *c*-axis orientations are randomly dispersed before deformation. Then he deforms the samples, step by step, by pure shear (coaxial strain) and simple shear (noncoaxial strain). His videotapes of the deformation are marvelous, because in watching we suddenly realize that dislocation creep, mechanical twinning, kinking, grain-boundary mi-

Figure 4.62 Microstructures of experimentally deformed aplite. (*A*) Starting material, consisting of fine-grained (0.15 mm) quartz and feldspar with relatively unstrained grains. (*B*) Deformation under conditions of high temperature and slower strain rate permits dislocation climb, resulting in strongly deformed ribbons and grains of quartz with subgrains. Some sutured grain boundaries also record grain-boundary migration. (*C*) Climb is limited during deformation at lower temperature and higher strain rate, resulting in a lack of subgrains. Recrystallization by grain-boundary migration produces some sutured boundaries and other evidence for grain-boundary migration (see also Figure 4.59 and Hirth and Tullis, 1992). [Photographs courtesy of Jan Tullis, Brown University].

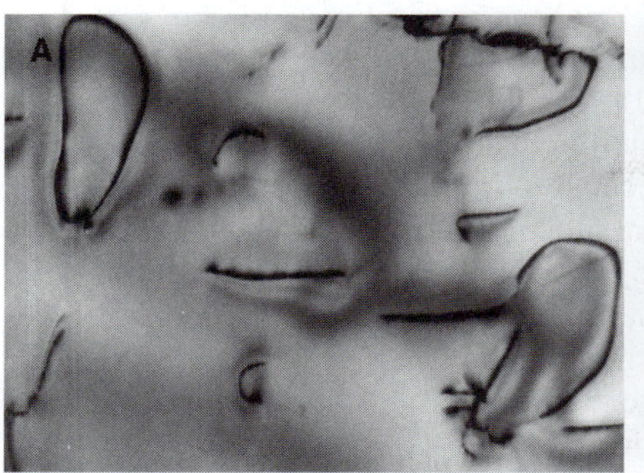

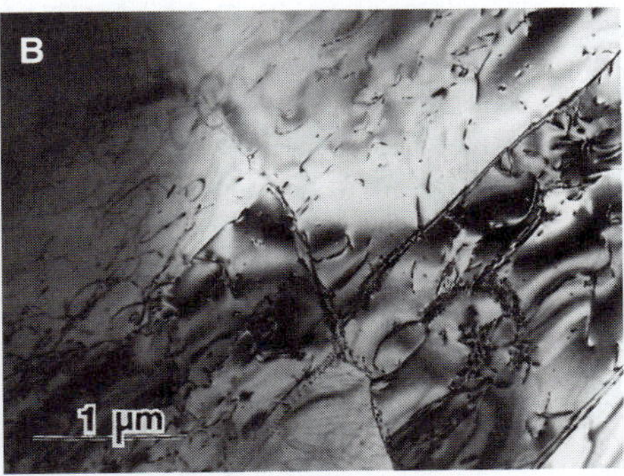

Figure 4.63 TEM (transmitting electron microscope) images of dislocations. (*A*) Curved dislocations and a fairly uniform dislocation density due to mobility of dislocations (climb was easy) due to higher temperature conditions of deformation (see Figure 4.62*B*). (Photograph by Andrew Kronenberg.) (*B*) Straight and tangled dislocations, indicative of only limited dislocation climb. These are similar to those observed in the sample in Figure 4.62*C*. [Photographs courtesy of Jan Tullis, Brown University.] [Reprinted with permission from *Journal of Structural Geology*, Dislocation creep in quartz aggregates, v. 14, G. Hirth and J. Tullis (1992), Elsevier Science Ltd., Pergamon Imprint, Oxford, England.]

gration, production of subgrains, and rotation of grains all are taking place simultaneously in different parts of the microscopic field of view. Obvious deformational microstructures come and go, as ice crystals deform and then experience dynamic recrystallization (Figure 4.64). Wilson tracks the percentge of shortening at each step of the pure shear experiments, and shear strain at each step of the simple shear experiments, allowing us to gain a sense of the appearance of microfabrics as a function of degree of deformation. Furthermore, Wilson shows in real time the stereographic migration of randomly oriented *c*-axes of the ice at the beginning of a given experiment to strongly preferred orientations of *c*-axes at the end of the given experiment. This information instills confidence that microscopic analysis of deformation fabrics can be used to distinguish whether strain has been achieved by coaxial or noncoaxial deformation.

Win Means (1989) pioneered the use of nonrock materials such as octachloropropane (think of a material like mothballs) and paradichloropropane to simulate, through deformation, the geometry and kinematics of dislocation creep and recrystallization. In some experiments he tracks the position of grain boundaries relative to some material reference frame

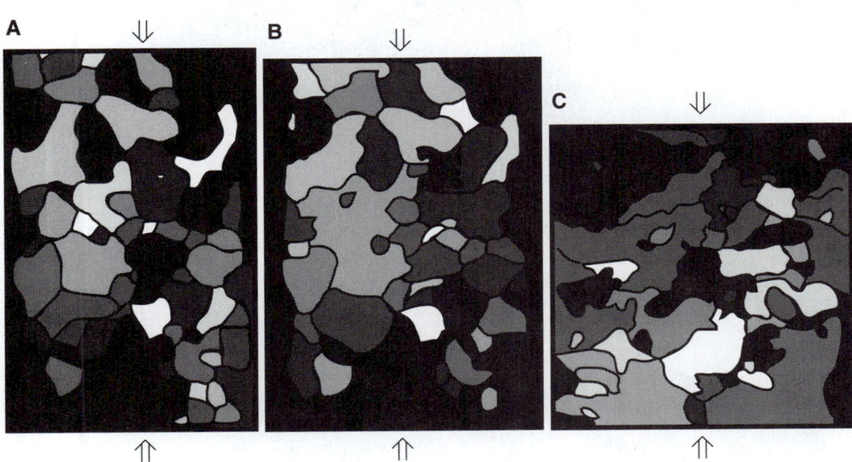

Figure 4.64 Progressive deformation of polycrystalline ice. Shades of gray represent different crystallographic orientations. (*A*) Initial specimen was 19 mm wide with random crystallographic orientations of individual grains. (*B–C*) Progressive vertical shortening parallel to length of page produces serrated grain boundaries, preferential growth of some grains, and a more consistent crystallographic orientation (similar gray appearance of many grains). [Adapted from Wilson (1986), Fig. 2a,b,d.]

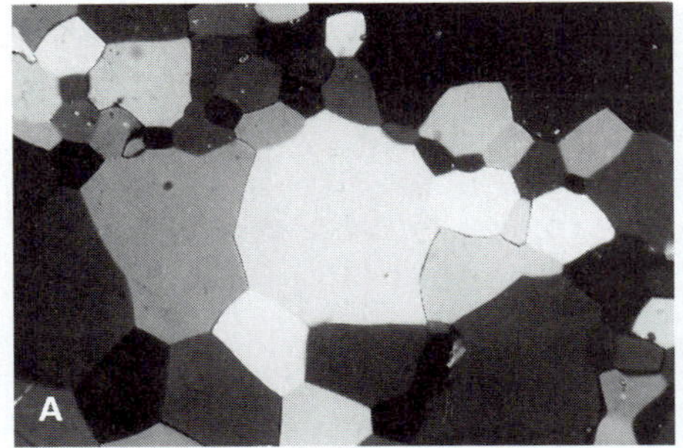

Figure 4.65 Experimental deformation of octachloropropane. (*A*) Initial specimen, showing straight boundaries and internally unstrained grains. (*B–C*) Progressive dextral shear resulting in development of undulatory extinction and subgrains via rotation recrystallization. Some grain-boundary migration is reflected by serrated grain boundaries. (Photographs kindly provided by Win Means, SUNY, Albany.)

by placing small, rigid inclusions, such as iron filings, in the materials before deformation. Figures 4.65 and 4.66 show the results of two such experiments in which the material was subjected to simple shear. In the first experiment, octachloropropane undergoes rotation recrystallization and the formation of subgrains because climb was easy. In the second, paradichlorobenzene displays boundary-migration recrystallization: the grains change size and shape as they migrate through the material. Grains with abundant defects or excess surface energy are consumed by those with lower energy. In these experiments, the material develops various microstructures, including undulatory extinction, subgrains, and sutured grain boundaries, during deformation.

Using Computers to Model Deformation and the Development of Crystallographic Preferred Orientations

The arrival of fast and affordable computers has opened up a new and exciting avenue for exploring how deformation is expressed at the microscopic level. Computer programs that run on personal computers can take the latest theoretical models of deformation and do ''virtual deformation'' of an imaginary aggregate of grains (i.e., a ''virtual rock''). We can observe—in real time—how the geometry and crystallographic preferred orientation of each grain changes during deformation. We can watch how differences in original crystallographic orientation or size cause two adjacent grains to have different behaviors during deformation and different final geometries and orientations.

Some of the most elegant computer models have explored the behavior of quartz and similar minerals (Lister and Hobbs, 1980; Etchecopar and Vasseur, 1987, and Jessell, 1988). These models usually start with an

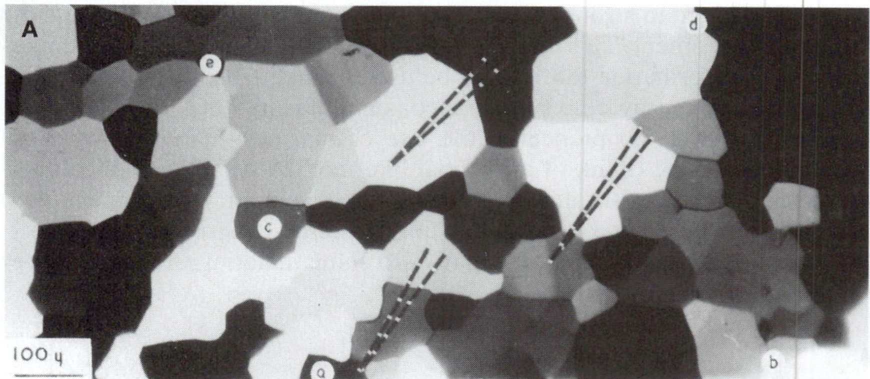

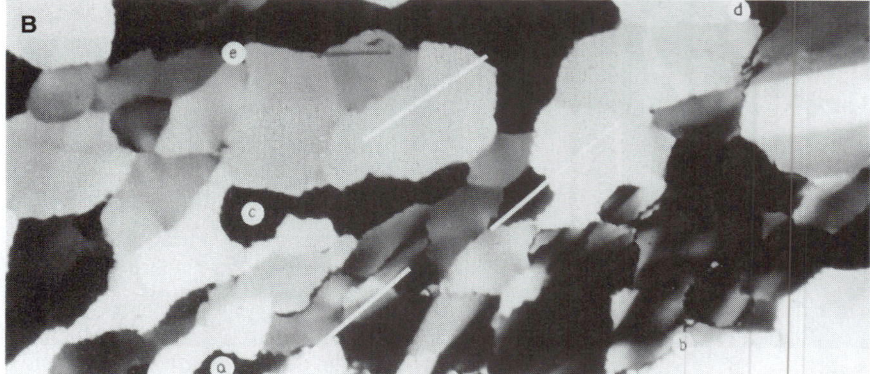

Figure 4.66 Experimental deformation and annealing of octachloropropane. (*A*) Initial specimen, showing straight boundaries and internally unstrained grains. Reference lines and selected grains are labeled.

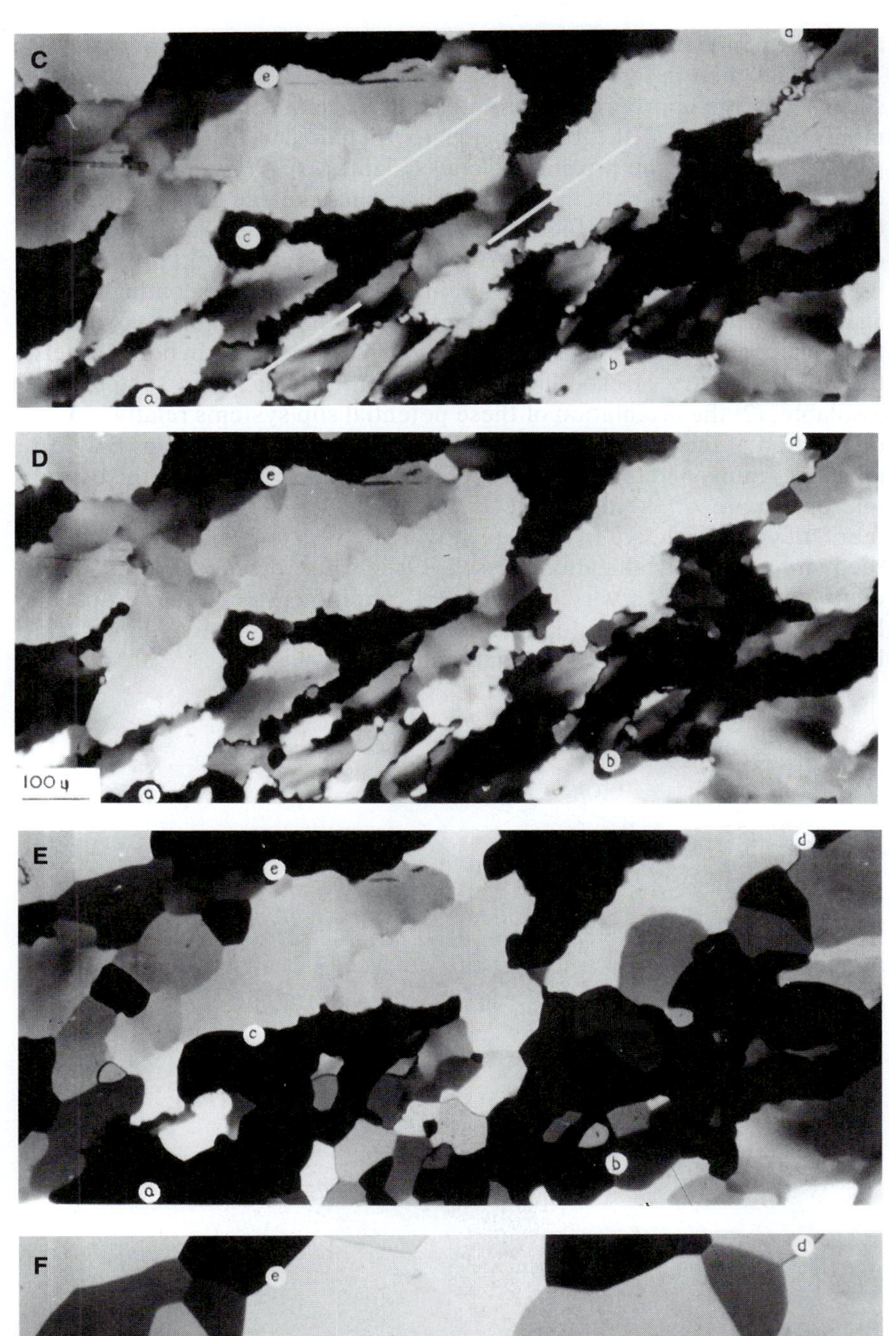

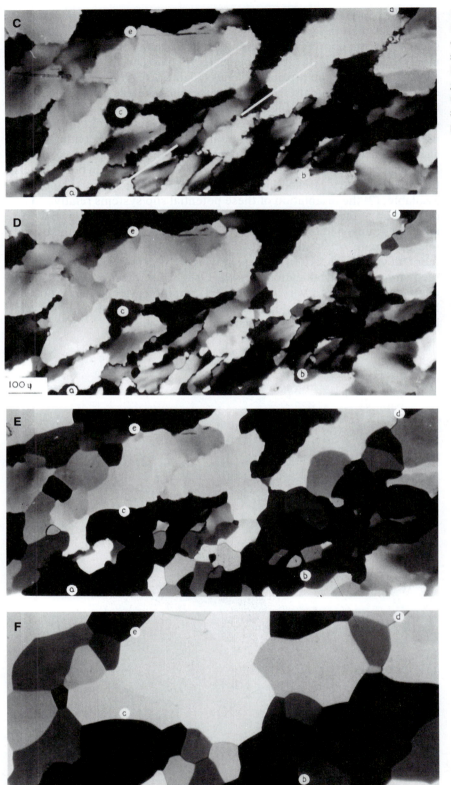

Figure 4.66 (*continued*) (*B–C*) Progressive dextral shear resulting in development of undulatory extinction, subgrains, and some serrated grain boundaries. (*D–F*) Progressive annealing and grain boundary migration, leading to larger undeformed grains with straight boundaries and 120° intersections. [Photographs kindly provided by Win Means, SUNY, Albany.]

aggregate of undeformed grains with random crystallographic orientations. The model then deforms the aggregate, with each grain deforming via shear along the appropriately oriented slip systems. In quartz, slip commonly occurs along the basal plane (perpendicular to the *c*-axis) and parallel to each of the prismatic planes (Figure 4.67*A*). Some models use only a few slip systems while trying to minimize the geometric misfit between adjacent grains (Figure 4.67*B,C*). Other models, based on a theory referred to as **Taylor-Bishop-Hill** after the scientists who proposed it, use more slip systems to let each grain deform without becoming locked against its neighbors. Five independent slip systems are required to permit this to happen, a requirement called the **von Mises criteria**. In both models, the final geometry of each grain depends on (1) which slip systems are available, (2) the orientation of these potential slip systems relative to the orientation of the imposed stresses, (3) the nature of interactions between adjacent grains, and (4) the amount of finite strain. In addition, the overall microstructure of the aggregate is controlled by the rate at which the fabric develops relative to the rate at which it is changed or destroyed by recovery and recrystallization (Jessell, 1988). Thus, the most sophisticated computer models can vary the physical conditions (temperature and the like) in successive computer runs to explore aspects such as how grain-boundary migration affects the final fabric.

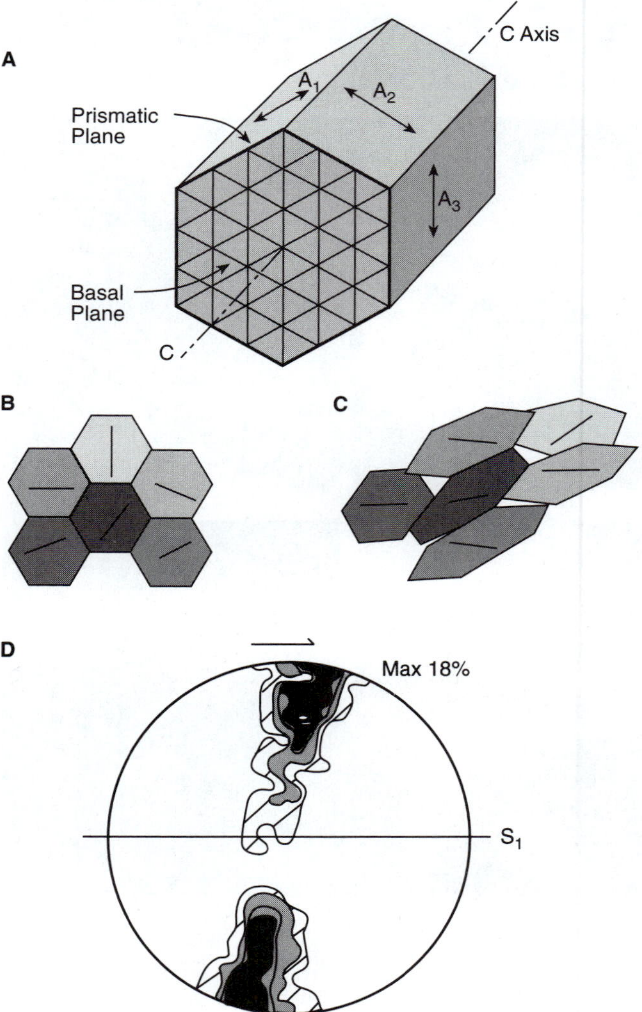

Figure 4.67 Computer simulation of the development of preferred crystallographic orientation in a mineral-like quartz. (*A*) Basal plane and three prismatic slip planes (A_1, A_2, and A_3) in a quartz crystal. (*B*) Initial aggregate of quartz crystals, showing a single slip plane that is permitted to operate for each crystal. (*C*) Final shape of grains for dextral shear and the operation of only one slip plane for each crystal. (*D*) Stereographic distribution of *c*-axes after deformation defines a girdle that leans over in the sense of shear relative to the normal to the foliation (S_1). [Reprinted with permission of *Journal of Structural Geology*, v. 9, A. Etchecopar and G. Vasseur, A 3-D kinematic model of fabric development in polycrystalline aggregates: comparisons with experimental and natural examples (1987), Elsevier Science Ltd., Pergamon Imprint, Oxford, England.]

In addition to displaying the shape of each grain, the programs can calculate stereoplots showing the distribution of the crystallographic axes (Figure 4.67*D*). These plots, called *c*-axis figures for short, largely reflect the amount and kind of strain imposed on the aggregate (e.g., coaxial flattening, noncoaxial simple shear, etc.). Depending on the strain and other factors, a *c*-axis figure may contain a single point maxima, a single girdle of points along a great circle (Figure 4.67*D*), a girdle with deflections (called "doglegs") at the ends, or even two crossing girdles. The power of this method is to compare computer-generated *c*-axis figures for different types of strain against *c*-axis figures derived from measuring, under the microscope, hundreds or thousands of *c*-axes in naturally deformed, real rocks (Law et al., 1986). We can go further and bring in *c*-axis figures measured from some of Jan Tullis' experimentally deformed rocks. It's a great example of how field studies, deformation experiments in the lab, and quantitative theories of deformation as played out on a computer can all complement each other and provide us with much more insight than we would have gained from any single approach.

THE BRITTLE–DUCTILE TRANSITION

The wealth of information that has been generated through the study of deformation mechanisms, when integrated with basic information about changes in the rock composition of the crust with depth, and changes in temperature and pressure with depth, provides a basis for interpreting how the Earth's crust deforms. Deformation is dominated by brittle mechanisms in the upper levels of the crust and by ductile mechanisms at deeper levels. The depth at which deformation switches from dominantly brittle to dominantly ductile mechanisms is called the **brittle–ductile transition**.

Within continental crust, the brittle–ductile transition generally occurs at 10–15 km below the Earth's surface, or within the middle part of the crust. The brittle–ductile transition is not a sharp, discrete break, but instead is gradational across a zone probably several kilometers thick. This gradation is partly due to the presence of different minerals, each of which has its own characteristic mechanical properties and deformational response to a given regime of temperature, pressure, strain rate, and fluid pressure. For this reason, the brittle–ductile transition is probably best depicted as the depth at which brittle and ductile mechanisms contribute equally to the deformation. Ductile mechanisms can occur for some distance above the transition, but they are less important than brittle mechanisms. Likewise, brittle mechanisms can operate below the transition, but they are subordinate to ductile ones; this is especially true for transient (short-duration), high strain rate events, such as earthquakes.

The brittle–ductile transition, in essence, marks the depth at which temperatures are sufficiently high to permit ductile deformation to dominate. In the continental crust, with rocks of dominantly granitic composition, this temperature is probably around 300–350°C, the temperature at which quartz is interpreted to become ductile under wet conditions. The depth at which this temperature is reached will vary according to regional variations in the thermal structure of the crust. In typical continental crust with a geothermal gradient of 20–25°C/km, the depth to the brittle–ductile transition might be approximately 15 km. In regions of recent tectonism, with a geothermal gradient of 50°C/km, the transition might be at 6 km. The transition may move up through the crust during an episode of regional crustal heating, such as during pluton emplacement, and down through the crust as the thermal pulse ends. In a similar manner, initially ductile rocks may be uplifted through the brittle–ductile transition and overprinted by brittle structures. Alternatively, shallow level, initially brittle rocks may become ductile if they are buried to sufficient depths and temperatures.

Using the concept of the brittle–ductile transition, we can model the strength of the entire crust by using a brittle "frictional" failure criterion for the upper crust and a ductile failure criterion for the lower crust (Figure 4.68). For the upper crust, we infer that the strength of the crust increases linearly with depth according to a relation called **Byerlee's law** (see Chapter 5). Byerlee accumulated data on the strength of rocks and noted a linear correlation of strength to confining pressure (depth). This linear relation proves to be very similar to what we would expect of a crust full of faults (i.e., surfaces of frictional sliding), each of which becomes more difficult to move as the confining pressure increases with depth.

To represent the ductile strength of the middle and lower crust, we consider the strength of granitic rocks to be controlled by their weakest abundant mineral—quartz. Accordingly, we use a curved strength envelope derived from the flow laws for wet quartz. The brittle–ductile transition is represented as the depth where the quartz flow law curve intersects the line representing Byerlee's law (Figure 4.68). At this depth, the crust has exactly the same resistance to brittle *and* ductile mechanisms. At shallow depths, the rocks fail brittlely before stresses can build up enough to cause ductile flow. At deeper levels, the rocks deform ductilely before the brittle failure criterion is reached. Note that different levels of the crust are characterized by different strain rates and by different contributions of each deformation mechanism to the total strain (Figure 4.68).

The concept of the brittle–ductile transition as portrayed through the **crustal strength envelope** provides a powerful way to view the entire crust. It can be used to explain why most large earthquakes in the continental

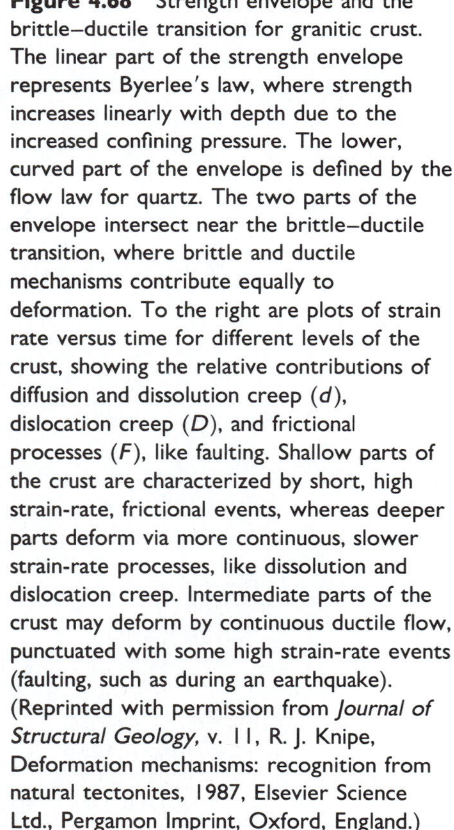

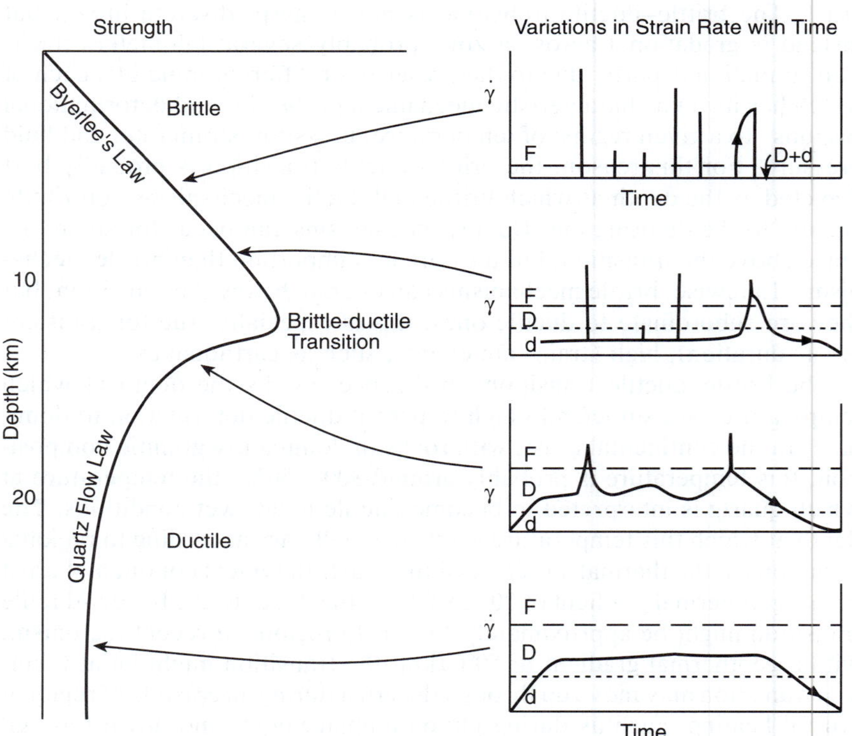

Figure 4.68 Strength envelope and the brittle–ductile transition for granitic crust. The linear part of the strength envelope represents Byerlee's law, where strength increases linearly with depth due to the increased confining pressure. The lower, curved part of the envelope is defined by the flow law for quartz. The two parts of the envelope intersect near the brittle–ductile transition, where brittle and ductile mechanisms contribute equally to deformation. To the right are plots of strain rate versus time for different levels of the crust, showing the relative contributions of diffusion and dissolution creep (*d*), dislocation creep (*D*), and frictional processes (*F*), like faulting. Shallow parts of the crust are characterized by short, high strain-rate, frictional events, whereas deeper parts deform via more continuous, slower strain-rate processes, like dissolution and dislocation creep. Intermediate parts of the crust may deform by continuous ductile flow, punctuated with some high strain-rate events (faulting, such as during an earthquake). (Reprinted with permission from *Journal of Structural Geology*, v. 11, R. J. Knipe, Deformation mechanisms: recognition from natural tectonites, 1987, Elsevier Science Ltd., Pergamon Imprint, Oxford, England.)

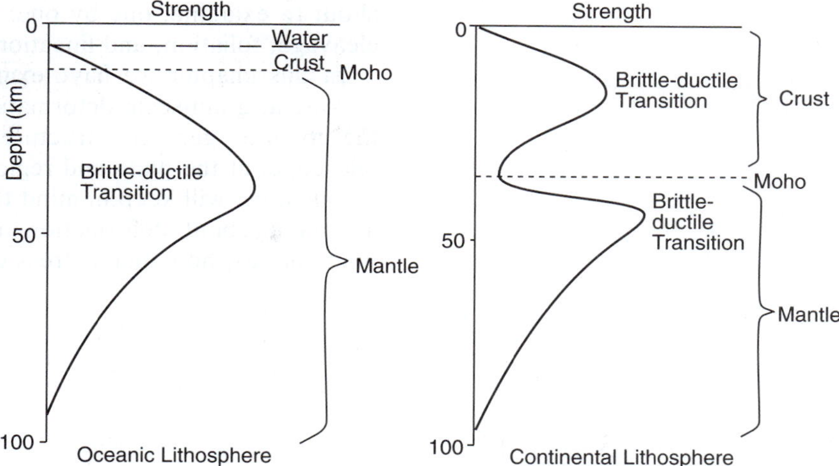

Figure 4.69 Strength envelopes and the brittle–ductile transitions for oceanic and continental lithosphere. (*A*) The strength of oceanic lithosphere increases with depth according to Byerlee's law, until the brittle–ductile transition, where the strength decreases because of the ductile flow of olivine. (*B*) The strength envelope for continental lithosphere has two maximums. Strength within the granitic crust increases downward according to Byerlee's law, until higher temperatures cause the onset of ductile flow of quartz. The strength increases at the Moho, where the quartz-rich crust gives way to the olivine-rich upper mantle. The strength of the underlying mantle part of the lithosphere decreases downward as olivine begins to flow ductilely. (Reprinted with permission from *Nature*, v. 335, Continental tectonics in the aftermath of plate tectonics by P. Molnar, p. 131–137. Copyright 1988, Macmillan Magazines Limited.)

crust initiate at a depth of 10–15 km: earthquakes at this depth have to break the strongest part of the crust. We can also see why the overall strength of the crust would be less in regions of higher heat flow and steeper geothermal gradients. Furthermore, we can envision how rocks may change behavior, *and strength,* as they are uplifted or buried through different levels of the crust.

The concept of strength envelopes can be extended to the entire lithosphere by characterizing the strength of the mantle using ductile flow laws for olivine, the most abundant mineral in the upper mantle. According to this model (Figure 4.69), there is an abrupt increase in strength at the Moho, where the weak, quartz-bearing lower crust is replaced downward by stronger, olivine-bearing mantle. At still greater depths, an increase in temperature and pressure results in ductile flow of olivine and a marked decrease in strength. The transition to underlying weak mantle is the lithosphere–asthenosphere boundary, with the mantle part of the lithosphere being capable of brittle deformation, whereas the asthenosphere would deform ductilely.

A FEW FINAL THOUGHTS

Rocks that seem so hard and rigid prove to be ''soft'' by virtue of weaknesses that are exploitable by deformation mechanisms operating at the microscopic and atomic scales. Minerals and rocks under certain conditions can flow, almost as if they had no strength, . . . almost as if they were not able to withstand even the smallest stress. William Shakespeare, uncanny as always, seemed to understand this when, through Leonato in *Much Ado About Nothing,* he said:

> Being that I *flow* in grief
> The smallest twine may lead me.

Isn't it interesting that the study of deformation mechanisms at the microscopic and atomic scales can prove to be so essential to understanding the mechanical and rheological behavior of the entire lithosphere? In a comparable fashion, knowledge of deformation mechanisms is essential to understanding the fundamental geologic structures, which we are now about to examine, one by one: joints and shear fractures; faults; folds, cleavage, foliation, and lineation; and shear zones.

In this chapter we have examined the Earth through a microscope, looking at grain-scale deformation. After we have made a pass through the common geologic structures, we will look at the Earth through a telescope, at the plate and regional scales. Throughout the journey that remains, we will keep in mind that no matter what scale of structure we are talking about, deformation mechanisms make it all possible, and place limits on just how much stress can build.

STRUCTURES

CHAPTER 5

JOINTS AND SHEAR FRACTURES

DEFINITIONS AND DISTINCTIONS

General Nature Of Joints

Joints are reasonably continuous and through-going planar fractures, commonly on the scale of centimeters to tens or hundreds of meters in length, along which there has been imperceptible "pull-apart" movement more or less perpendicular to the fracture surface (Figure 5.1). Joints are products of brittle failure, and they form when the tensile strength of stressed rock is exceeded. Although the tiny movement that is accommodated by a joint is most commonly an opening perpendicular to the joint surface, the movement can be oblique, marked both by opening and shear movement. In most cases the movement that takes place when a joint forms is nearly microscopic in scale. Thus the amount and direction of movement represented by the existence of a joint cannot be identified in outcrop except where the joint cuts a discrete object (e.g., a pebble, fossil, or inclusion) and the magnitude of offset, although very small, is large enough to be discernible. Joints form to permit minor adjustments to take place as the regional rock bodies within which they are found change in location, orientation, size, and/or shape in response to such actions as burial and compaction; heating and expansion; uplift, cooling, and contraction; and tectonic loading, causing shortening or stretching.

Joints commonly become sites where minerals are precipitated in the form of **veins** (Figure 5.2). The precipitation of minerals "seals" the fracture (Ramsay, 1980b). Mineralizing solutions invade rock bodies along joints and other fractures, precipitating minerals from solution in the open space when the cracks are forced open and the chemical conditions are just

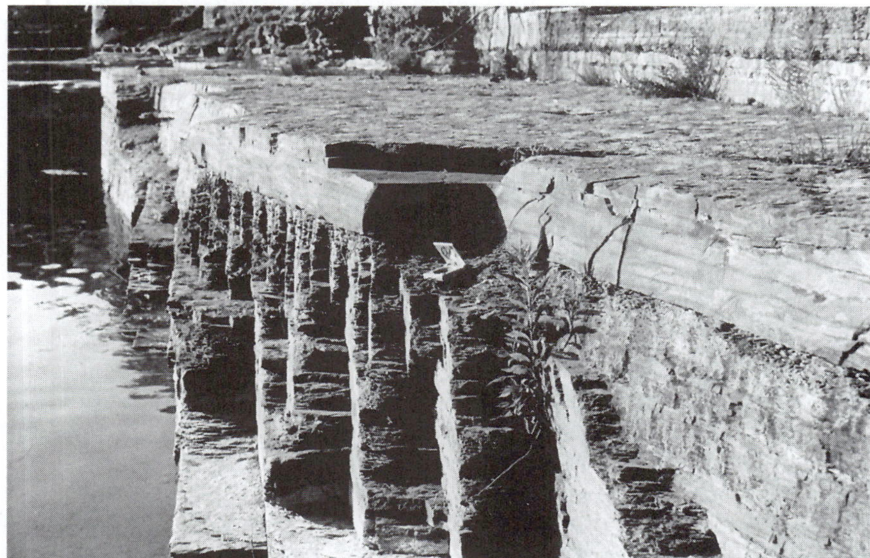

Figure 5.1 Outcrop expression of jointing in siltstone (upper, stiff, tabletop unit) and underlying shales in the upper portion of the Devonian Genessee Group at Taughannock Falls State Park, New York. [Reprinted with permission from *Journal of Structural Geology*, vol. 7, Engelder, T., Loading paths to joint propagation during a tectonic cycle: an example from the Appalachian Plateau 1985), Elsevier Science Ltd., Pergamon Imprint, Oxford, England.]

right. Precipitation is triggered by favorable temperature and/or pressure conditions, and sometimes by the mixing of different fluids that happen to meet at the intersection of fracture-controlled channelways.

Joints are found in all outcrops of rock, and thus they are among the most abundant of geologic structures. The lengths and spacings of joints are related to the size and/or thickness, as well as the stiffness, of the rock body in which the joints occur. Weak, thin units are marked by very closely spaced joint surfaces; stiff, thick units are marked by relatively widely spaced joints. Distances between joints are commonly on the same order of magnitude as the thickness of the rock layer in which the joints are found.

As families of fractures, joints commonly display systematic preferred orientations (Figure 5.3) and often show striking symmetry. The best developed joints are eye-catching **systematic joints** (Figure 5.3*A*), which are planar, parallel, and evenly spaced at distances of centimeters, meters, tens of meters, or even hundreds of meters. A given outcrop area or region of study is typically marked by more than one **set** of joints (Figure 5.3*B*), each with its own distinctive orientation and spacing. Two or more sets

Figure 5.2 Quartz veins in cleaved siltstone in the Whetstone Mountains, Arizona. The veins occupy joints along which dilational opening has occurred. (Photograph by G. H. Davis.)

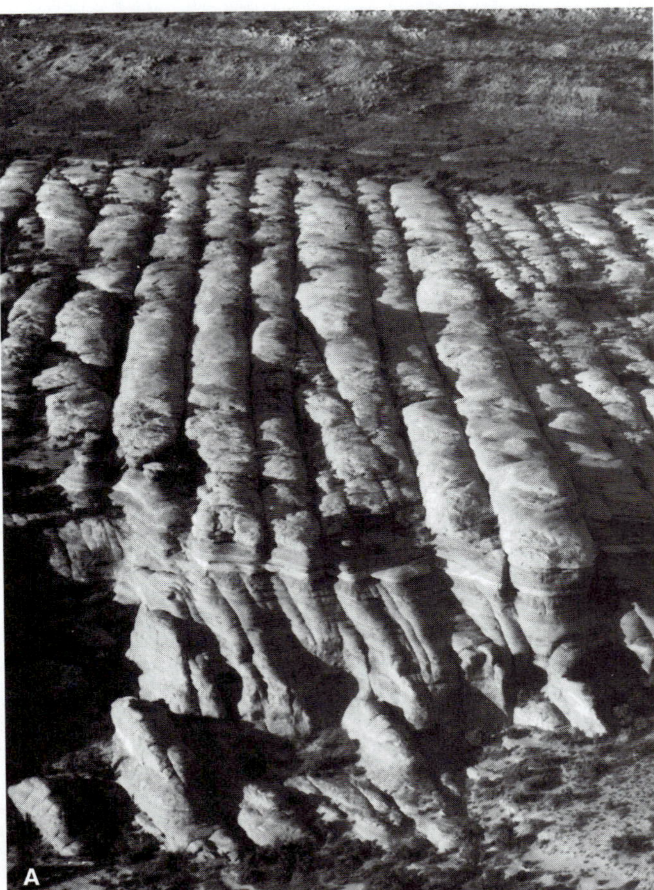

Figure 5.3 (*A*) Aerial photograph of jointing in the Moab Member of the Entrada Sandstone in Arches National Park. A single set of joints consists of individual vertical surfaces spaced from 25 to 50 m. (Photograph by R. Dyer.) (*B*) Aerial view of two sets of joints in sandstone in Canyonlands, Utah. The larger joint-bounded blocks are about 50 m on a side. Eroded aisles are zones of closely spaced joints. (Photograph by G. F. McGill; taken through a hole in the floor of a low-flying small plane.)

constitute a **joint system.** Some joints are so irregular in form, spacing, and orientation that they cannot be readily grouped into distinctive, through-going sets. These are **nonsystematic joints**.

Joints have the effect of subdividing a rock body into myriad faceted chunks and pieces, each of which ''joins'' to neighbors along joints, sometimes architecturally resembling blocks set side by side (Figure 5.4*A*). The joint itself consists of an exceedingly narrow slit, roughly rectangular, elliptical, or circular in overall form, bounded on each side by matching planar surfaces of rock (Figure 5.4*B*) (Pollard and Aydin, 1988). The **joint surfaces** are commonly ornamented by markings that record the propagation of the joint (Figure 5.5).

General Nature of Shear Fractures

Some jointlike fractures are actually **shear fractures**, which are formed in the same way that faults are formed: there is shear parallel to the fracture surface. Yet they are not called faults because offset by shear is generally not visible at the scale of outcrop or hand specimen (Figure 5.6*A*). A shear fracture may be the product of a single event in which the formation

Figure 5.4 Sometimes human architectural products and the natural architecture of jointing bear a similarity to one another. (*A*) Greek structure at Delphi. (*B*) Natural jointing exposed on the north coast of Brittany. (Photographs by G. H. Davis.)

of the fracture surface and the shear along it occur simultaneously. On the other hand, a given shear fracture may represent the result of reactivation and sliding along a joint surface formed much earlier under different stress conditions.

Shear fractures are of the same size and scale as joints, and they too occur in sets of planar parallel fractures. The spacing and the density of shear fractures vary in relation to the mechanical properties of the rocks in which they occur, and in relation to the presence or absence of major structures, like faults and folds. In the early stages of a mapping project, some shear fractures and some joints may be indistinguishable because of the absence of diagnostic ornamentation on featureless fracture surfaces, and the lack of any discernible movement or offset. However, if some of the fractures are shear fractures and not joints, they will eventually be recognized by the presence of **slickenlines** (Figure 5.6*B*), which reflect a shearing movement parallel to the surface as opposed to dilation or opening perpendicular to the surface. The slickenlines on shear fractures

Figure 5.5 (*A*) Close-up of ornamentation details on the face of a joint surface in granitic rock in the Rincon Mountains, Arizona. As we will learn, the joint surface ornamentation features reveal the kinematics of joint propagation. (*B*) Anasazi petroglyph ornamentation on a joint face along the San Juan River between Bluff and Mexican Hat, Utah. Dark patina is desert varnish on the joint face, beneath which is the fresh white sandstone. Human figures are 3 m tall. (Photographs by G. H. Davis.)

are most commonly fine-scale, delicate **ridge-in-groove lineations** (Twiss and Moores, 1992) developed on the adjoining fracture surfaces, or coatings of **crystal fibers** that have grown in the direction of shear displacement. Ultimately it may be seen that the shear fractures occur in sets that are parallel to recognizable faults.

Systems of Joints and Shear Fractures

Joints and shear fractures do not exist in isolation. Rather they are members of enormously large families of fractures with literally millions of members. Every regional rock assemblage, like a granitic batholith or a plateau of sedimentary rock, is pervaded by jointing. It is impossible to try to explain the origin of every joint or shear fracture in an outcrop, let alone every joint or shear fracture within a regional rock assemblage. Instead, we try to explain the origin of dominant sets of joints and shear fractures that can be identified through statistical analysis of the orienta-

Figure 5.6 (*A*) Shear fractures in Cretaceous sandstones in the Tucson Mountains, Arizona. (*B*) Slickenlines on shear fracture cutting Jurassic siltstones in the Tucson Mountains, Arizona. The slickenlines are best expressed in the quartz vein coatings of the fracture surfaces. (Photographs by G. H. Davis.)

tions and physical properties of the joints and shear fractures within a given system.

To begin to discover order among millions upon millions of joints and shear fractures, it is essential to subdivide regional rock assemblages into **structural domains**, each of which may be thought of as containing its own fracture system.

Structural domains are designated on the basis of geographic boundaries, lithologic contacts, structural subdivisions, ages of rock formations, and combinations of these and other factors. The criteria vary according to the scope of the investigation. Nickelsen and Hough (1967), in their analysis of the jointing in the Appalachian Plateau region of Pennsylvania and New York, subdivided domains on the basis of lithology: one domain was restricted to all joints measured in coal, another to all joints measured in sandstone. Rehrig and Heidrick (1972), in their analysis of the joints in late Cretaceous plutons in southern Arizona, established domains of mineralized versus unmineralized plutons, for they wanted to assess the degree to which copper mineralization was associated with preferred fracture orientations. Later, Rehrig and Heidrick (1976) compared joint patterns in late Cretaceous versus Middle Tertiary plutons, attempting to recognize differences in fracture directions that might disclose differences in the regional tectonic stress patterns between the two time periods. To evaluate the influence of joint orientation(s) on the trends of tunnels and shapes of rooms in the Carlsbad Caverns, Jagnow (1979) subdivided the caverns into domains on the basis of cave anatomy. When I carried out joint analysis of a folded massive sulfide ore body in eastern Canada (Davis, 1972), I established three domains—one on each of the limbs and one covering the hinge of the fold.

The substance of detailed joint and shear fracture analysis lies in evaluating the geologic characteristics of the fracture system that occupies each structural domain. There is no single conventional procedure for doing this, but it is useful to approach the analysis of joints and shear fractures by treating them as systems.

Other Fractures

Certain fractures do not qualify as joints, nor do they qualify as conventional shear fractures. Within fault zones, for example, rocks may be pervasively **shattered** along apparently nonsystematic, extremely closely spaced fractures (Figure 5.7). The density of the fracturing is much greater than what is reasonable for ordinary jointing and shear fracturing. Furthermore, the regularity of orientation and spacing of fracture surfaces, so typical of jointing and shear fracturing, is absent. Fractures in such shattered rocks are simply called fractures, not joints. The rocks are described as "shattered," not "jointed." We will have more to say about shattered and broken rock in the next chapter ("Faults").

For ages, quarry workers have taken advantage of fracture-controlled planes of weakness in removing building blocks of granite and limestone from bedrock (Figure 5.8). These fracture weaknesses exert profound control on weathering and erosion, and thus on fashioning landscape. Many scenic attractions owe much of their uniqueness to weathering and erosion of horizontal layers of rock systematically broken by joints. There are great examples of this throughout the world. The walls and buttes in

Figure 5.7 Absolutely pervasively shattered outcrop of granite beneath a fault, in southeastern California. (Photograph by Stephen J. Reynolds.)

Figure 5.8 Joints and joint surfaces exposed in an old rock quarry near Glen Echo, Maryland. (Photograph by G. K. Gilbert. Courtesy of United States Geological Survey.)

systematically jointed rocks in Canyonlands National Park (Utah), and the chimneys and columns of Chiricahua National Monument (Arizona), serve as especially inviting examples (Figure 5.9).

Beyond their scenic value, fractures constitute a structure of indisputable geologic and economic significance. The presence of joints, shear fractures, and faults invites circulation of fluids, including rain and ground-

Figure 5.9 Jointing exerts a major control on landform development. (*A*) Joint-pervaded landscape in Canyonlands National Park, Utah. (From G. E. McGill and A. W. Stromquist, *Journal of Geophysical Research,* v. 4, pp. 4547–4563. Copyright © 1979 by American Geophysical Union.) (*B*) Jointing in flat-lying Miocene ignimbrites, Chiricahua National Monument, Arizona. (Photograph by G. H. Davis.)

water, pollutants and contaminants, hydrothermal mineralizing solutions, geothermal waters, and oil and natural gas.

The reality of the flow of fluids through rock is especially well recorded in caves, and is appreciated by those who love to explore caverns. The shapes and orientations of rooms and passageways in caves are commonly controlled by the selective solution-driven removal of limestone along major joint trends. For example, Left Hand Tunnel of Carlsbad Caverns in New Mexico is elongate east–northeast, parallel to the predominant set of joints that cuts the limestone in that part of the caverns (Jagnow, 1979). Tall, narrow passages are centered on prominent vertical joints that once guided the groundwater circulation.

Explorationists appreciate the benefits of circulation of fluids through fractured rocks. Petroleum geologists evaluate the nature and degree of development of joints as one guide to the reservoir quality of sedimentary formations. In fact, to increase the yield of reservoir rocks in oil and gas fields where production is waning, it is common practice to "crack" the rocks artificially, either through explosives or through the high pressure pumping of fluids into the well(s).

The natural circulation of hydrothermal fluids through joints and other fractures in hot rocks at depth constitutes a potentially significant source of energy, namely geothermal energy. For geothermal systems to be operational, thoroughly jointed rocks are as essential as heat and fluids.

Joints, shear fractures, and faults can serve as sites of deposition of metallic and nonmetallic minerals. In most deposits, a part of the mineralization is localized in and around fractures. The minerals are deposited either through open-space filling of joints or through selective replacement of chemically favorable rocks adjacent to the fracture surfaces along which hydrothermal fluids once circulated. Even where joints and other fractures do not "carry" economically significant levels of mineralization, they may be marked by veinlets and/or alteration assemblages of distinctive silicate and sulfide minerals. Economic geologists use alteration minerals as clues to possible hidden locations of ore deposits (Guilbert and Park, 1986).

Groundwater contamination in countless nations now places increasing importance on the fields of hydrology and geohydrology. Computer models of water moving through sediments and rocks have become very sophisticated. The best models take into consideration the movement of fluids through fractures. The fractures, after all, contribute to the bulk porosity and permeability of the system. The more we know about the character and "connectivity" of fracture systems, the better the mathematical models.

There are other potentially harmful side effects to the ease of circulation of fluids through jointed rock systems. Landsliding, slumping, and other processes of mass wasting on steep hillslopes are enhanced by the saturation of fractured rocks with rainwater and groundwater. The fluid pressures exerted by groundwater in cracks in rocks weaken the level of normal stress on fracture surfaces, thus enhancing the potential for slip. Where fluid-filled joints in rocks beneath a hillslope dip outward toward the free face of the hill, the steady force of gravity, in concert with fluid pressure exerted by water in the cracks, can cause sliding of earth, and houses, outward and downslope (Figure 5.10).

Engineers and consulting geologists address the problem of mass wasting not only in residential areas and municipal construction sites, but also in the designing of open-pit mines. Stable slope angles for a given open-pit mine depend on a number of factors, including fracture orientation and abundance of fractures. The engineering problem demands the judgment and balance required of a tightrope walker (Figure 5.11). The object is to maximize the slope angle of the pit, so that buried parts of the ore

Figure 5.10 Block glide at Point Fermin, near Los Angeles. (Photograph by Spence Air Photos. Published with permission of National Research Council.)

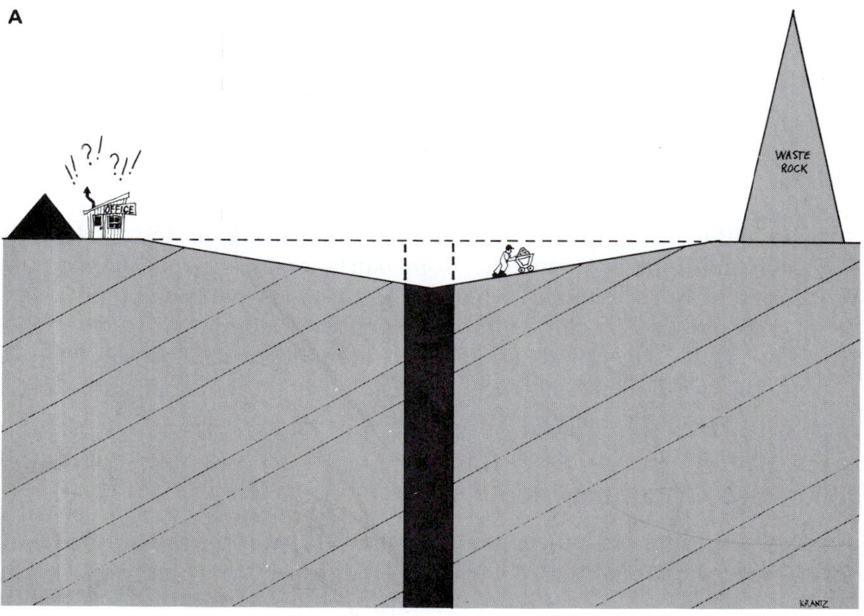

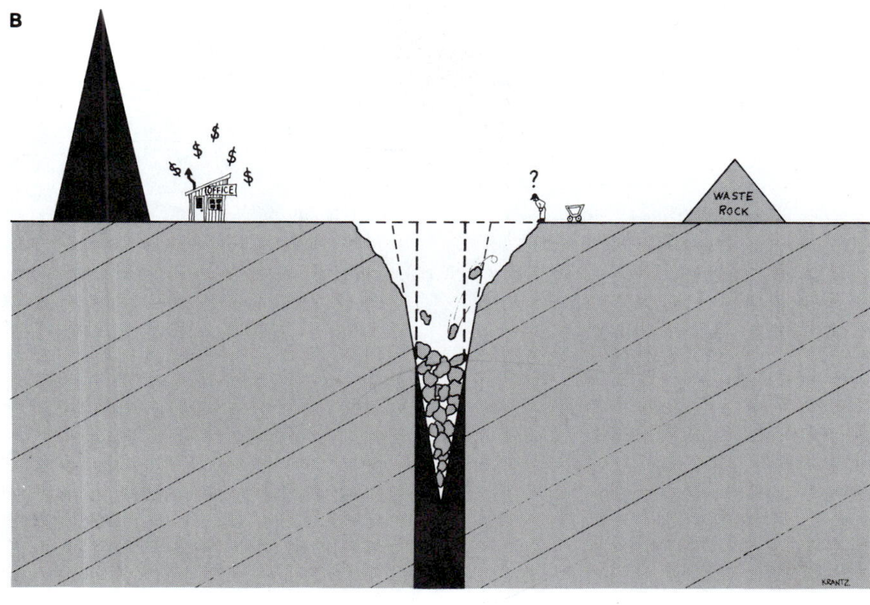

Figure 5.11 Trials and tribulations of pit–slope design. (*A*) Gently dipping pit walls are safe but uneconomical. (*B*) Steeply dipping pit walls are profitable in the short term but risky in the longer view. (Artwork by R. W. Krantz.)

deposit can be uncovered and exploited through removal of the least amount of waste rock overburden. At the same time, the engineers must minimize the risk of slope failure by making certain that the slopes of the pit are not oversteepened, given the geological conditions that exist. Slope failure, when it occurs, results in loss of equipment, disruption of road and/or track systems for hauling ore, and infilling of the open pit by waste rock. Pit slope failure can shut down marginally economic mining operations.

A DETAILED LOOK AT INDIVIDUAL JOINT SURFACES

Shapes of Joint Surfaces

Individual joints are planar to curviplanar surfaces that intersect the tops and flanks of outcrops as lines. Erosion and spalling of rocks along joint surfaces reveal joint faces (Figures 5.12 and 5.13). Joint faces are partial exposures of joint surfaces whose complete two-dimensional form in the plane of the surface is almost never seen in full. Woodworth (1896) concluded long ago, on the basis of a careful inspection of three-dimensional exposures of jointed bedrock, that joint surfaces are elliptical. Now it is generally known that the two-dimensional form of a joint surface is strongly influenced by the three-dimensional shape of the rock body in which the joint is "housed." Joint surfaces in layered sedimentary rocks tend to be rectangular, whereas joints in massive rock bodies, such as igneous plutons, indeed tend to be elliptical (Pollard and Aydin, 1988). In conventional structural studies, no attempt is made to describe or classify the two-dimensional form of the joint shape. The joints are simply treated as planar partial exposures of larger surfaces and left at that.

Joint-Face Ornamentation

An ideal exposure of an ideal joint consists of a smooth, planar **main joint face** bordered by roughly hewn **fringes** (Hodgson, 1961) (Figure 5.14). The

Figure 5.12 Exposure of part of a vertical joint face cutting cross-bedded Navajo Sandstone (Jurassic) in Zion National Park. (Photograph by G. H. Davis.)

Figure 5.13 Exposure of gently dipping joint surface cutting Precambrian rock in northwestern Arizona. Sue Beard is taking strike and dip. (Photograph by G. H. Davis.)

fringes project outward from the main joint face by some small amount. Fringes of joints are the outermost margins of a given joint surface. They are terminations of the main joint face, where the energy required for propagating the joint dissipates. Well-developed and unweathered fringes display a serrated appearance produced by very closely spaced **en echelon joints** that are misaligned with respect to the main face (Roberts, 1961). En echelon joints are marked by three or more relatively short joints that are parallel and overlapping and arranged in a line. En echelon joints in fringes intersect the main joint face at angles of 20°–25° (Figure 5.14).

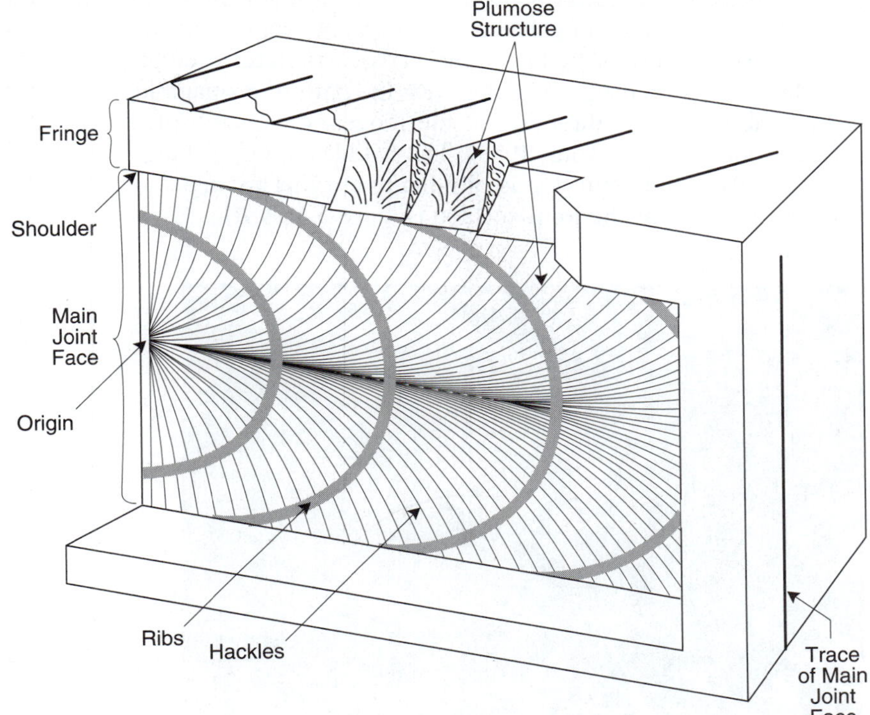

Figure 5.14 This representation of an ideal exposure of a joint face contains all the main "ornamentation" elements. The **origin** marks the site of first movement of the joint. **Hackles** record the direction of propagation. **Ribs** are marked by a slight shift in joint orientation where the velocity of joint propagation slowed down or was arrested. Each rib records the position of the front of the propagating joint at some point in time. The margin of a joint face is marked by an abrupt transition from a smooth planar fracture to roughly hewn **fringes**, where the single joint surface is replaced by quite a number of misaligned en echelon short joint segments. The fringes record the dissipation of the last bit of energy consumed in the joint-forming process. [Modified from Hodgson (1961), *American Journal of Science*, v. 259.]

Although the greatest percentage of exposed joint faces might appear to be featureless, in part because of weathering along joints, a careful, close-up look in favorable lighting reveals that many display subtle to richly expresssed **ornamentation** (Figure 5.15). The ornamentation provides a basis for interpreting the kinematics of development of individual joints. Ornamentation on joint surfaces in stiff units like sandstone or siltstone may be very conspicuous.

The most characteristic ornamentation of joint surfaces includes **origins**, **hackles**, and **ribs** (DeGraff and Aydin, 1987; Pollard and Aydin, 1988) (Figure 5.14). The **origin** of a joint is the site of initial propagation of the joint surface. It is analogous to the focus of an earthquake, the site of first movement, the place where energy is first released to form the break. The origin usually coincides with a mechanical defect, flaw, or irregularity in the rock, like a cavity, fossil, inclusion, or cusp (Figure 5.15) (Helgeson and Aydin, 1991). The origin is identifiable as the spot from which hackles radiate and diverge.

Hackles are linear to systematically curved markings that can occur on joint faces (Figures 5.14 and 5.16). They converge toward the origin of the joint. Hackles are physically composed of tiny ridges and troughs, a microtopographic relief on the main joint face (Roberts, 1961). Hackle patterns on adjacent rock faces that "join" at a common surface are virtually identical. The ridges and troughs on one face nestle perfectly into the troughs and ridges on the adjacent face. The perfect fit indicates that minuscule movements are sufficient to create plumose markings. Hackles record the direction of crack propagation, outward from the origin in the direction of divergence (Figure 5.17). They can be especially well developed on en echelon joint surfaces within the fringe of the joint, and they are the lines of intersection of closely spaced fracture surfaces whose attitudes diverge slightly from the attitude of the main face (Pollard and Aydin, 1988).

The hackles collectively display featherlike **plumose structure**, with the "hackle feathers" radiating from one or more axes (Engelder, 1984). The simplest plume patterns are marked by **straight plume axes** (Figure 5.18*A*,*B*). They are distinguished by a straight, continuous axis that lies parallel to the trace of bedding; the hackles radiate to either side. The straight plume axis itself coincides with the path of propagation of the tip of the fracture in the direction of the longest dimension of the fracture surface (Pollard and Aydin, 1988). The hackles diverge sharply at angles of 30–35° from the central axis, gradually curving to angles of about 75° near the margins of the joint surface (Figure 5.18*A*,*B*).

A

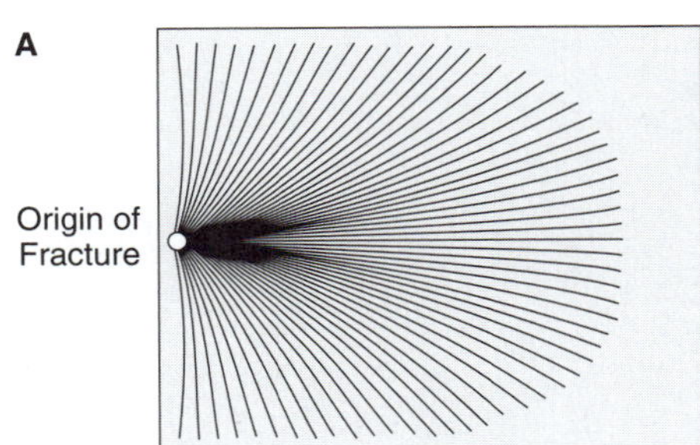

B

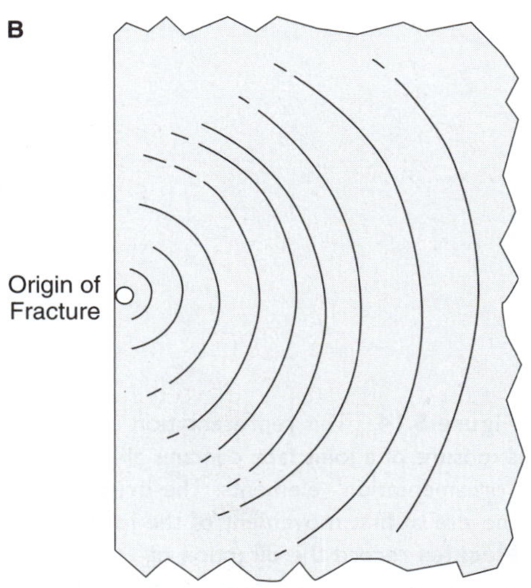

Figure 5.15 (*A*) Diagram of origin of fracture in relation to hackles. (*B*) Diagram of origin of fracture in relation to ribs. (Reprinted with permission from *Tectonophysics*, vol. 12, no. 5, Gash, S. P. J., A study of surface features relating to brittle and semi-brittle fracture, 1971, Elsevier Science B.V., Amsterdam, The Netherlands.) (*C*) Origins commonly coincide with inhomogeneities in the rock, in this case a small cusp (at tip of pencil) along the base of a siltstone layer. [From Pollard and Aydin (1988). Published with permission of the Geological Society of America.]

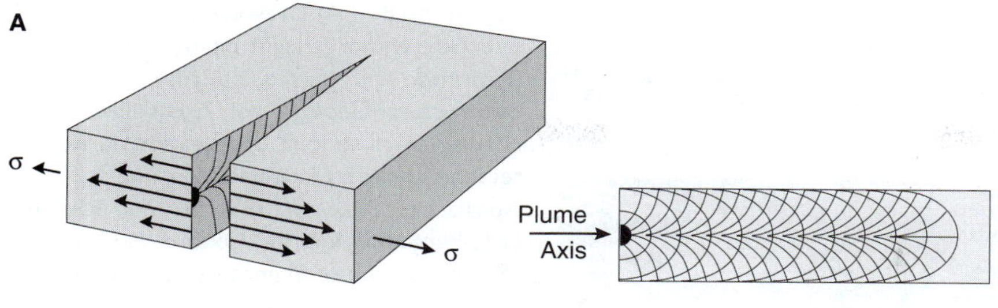

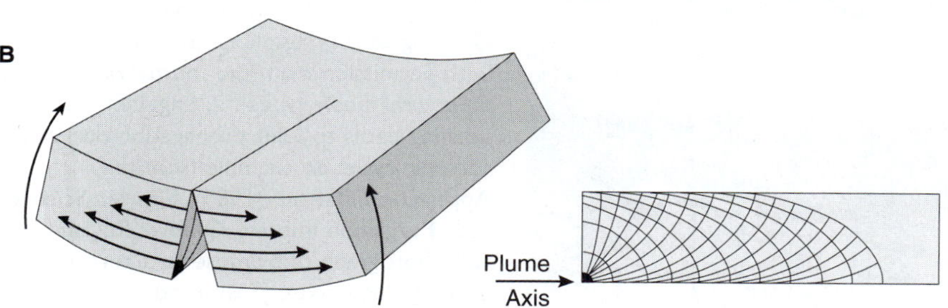

Figure 5.16 (*A*) Hackles produced on joint surface in rock layer broken in tension. Solid black dot is origin. Hackles radiate from origin. Parabolic curves in insert describe propagating joint front, sometimes preserved as ribs. (*B*) Hackles produced on joint surface produced through bending of a rock layer. Solid black dot is origin. Hackles radiate from origin. Solid curves represent propagating joint front. [From DeGraff and Aydin (1987). Published with permission of Geological Society of America.]

Curving plume axes are more complicated (Figure 5.18*C,D*). A single curved axis divides into branches, and then the branches divide again and again, with hackles radiating to either side of the main axis and the secondary branches.

Experimental work has revealed that the presence of plumose markings signals a fracturing achieved by a rapid, near-instantaneous, almost explosive snapping apart of the rock. The velocity of propagation of a joint front averages about half the speed of sound. The direction of propagation of the opening of individual joints is opposite to the direction in which the plumes comprising the plumose structure ''V.'' The plumose markings on a given joint ''V'' or converge toward the origin (see Figure 5.16) (Secor, 1965). Engelder (1987) points out that surrounding the point of origin of a ''joint'' in glass, or in Jell-O (Price and Cosgrove, 1990), there is a mirror-smooth circular or elliptical area that registers the result of a slow, accelerating rupture. When the fracture front picks up speed, it develops the energy to make any necessary orientation corrections and can break off at small angles to the original plane, leaving hackles as a record.

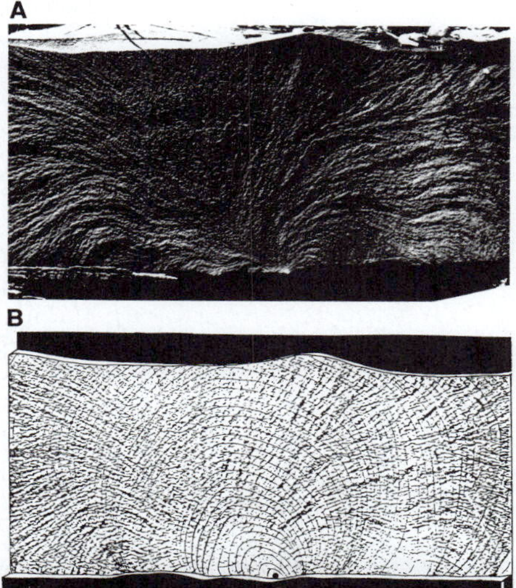

Figure 5.17 Plumose patterns and the reconstruction of joint fronts from the observed hackle patterns. (*A*) Photograph of joint face in siltstone layers. Hackles converge downward to origin. (*B*) Drawing showing origin (black dot) at point of convergence of hackles. The thin, curved black lines are everywhere constructed perpendicular to hackles, and they represent collectively a map of the propagation of the joint front up and out from the origin. [Reprinted with permission from the *Journal of Structural Geology*, vol. 13, Helgeson and Aydin, Characteristics of joint propagation across layer interfaces in sedimentary rocks (1991), Elsevier Science Ltd., Pergamon Imprint, Oxford, England.]

Ribs are also present on most joint faces that display plumose structure (Figures 5.14 and 5.19) (DeGraff and Aydin, 1987; Pollard and Aydin, 1988). Ribs represent positions of the joint front at some past time; they are a ''fossil record'' of a propagating joint front. In some cases they ''map'' the temporary arrest of a propagating joint front; in other cases they record the location at which the velocity of propagation of the joint front diminished. Ribs appear as parabolically curved spaced markings, the same kind of markings that on a much smaller and finer scale distinguish conchoidal fracture in obsidian, or in Plexiglas (Figure 5.20).

At any given point on a joint surface, rib markings are perpendicular to hackles (Ramsay and Huber, 1987). Where ribs cross the plume axis, they are aligned perpendicular to the plume axis (see Figure 5.19). The surfaces or steps of the ribs may be millimeters to centimeters wide. The rib surfaces display orientations in the third dimension that depart slightly from that of the main joint face, and this is what gives ribs their physical

S - type Plume

Plume Axis

A

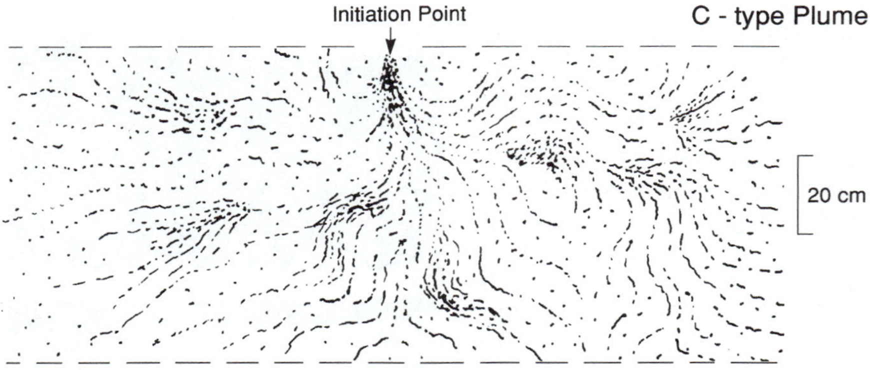

C

Figure 5.18 (*A*) Drawing of plumose structure with a straight plume axis. [Reprinted with permission from the *Journal of Structural Geology,* vol. 7, Engelder, T., Loading paths to joint propagation during a tectonic cycle: an example from the Appalachian Plateau (1985), Elsevier Science Ltd., Pergamon Imprint, Oxford, England.] (*B*) Photograph of plumose structure with straight plume axis. The rock is siltstone within shale of the Devonian Genessee Group, Appalachian Plateau, New York. (Photograph by Terry Engelder.) (*C*) Drawing of curving plume axes. [Reprinted with permission from the *Journal of Structural Geology,* vol. 7, Engelder, T., Loading paths to joint propagation during a tectonic cycle: an example from the Appalachian Plateau, (1985), Elsevier Science Ltd., Pergamon Imprint, Oxford, England.] (*D*) Photograph of plumose structure with curving plume axes. [Reprinted with permission from *Tectonophysics,* vol. 104, Bahat and Engelder, Surface morphology on cross joints of the Appalachian Plateau (1984), Elsevier Science B.V., Amsterdam, The Netherlands.]

Figure 5.19 (*A*) Conchoidal fracture surfaces with well expressed ribs in canyon wall of Navajo Sandstone. Height of face is 8 m. (Photograph by G. H. Davis). (*B*) Large joint face with well developed ribs and hackles in Navajo Sandstone in the Cottonwood Canyon area along the East Kaibab monocline, southern Utah. Height of joint face is 6 m. (Photograph by G. H. Davis.) (*C*) Close-up of ribs in small slab of Straight Cliffs Sandstone (Cretaceous). (Photograph and copyright by Peter Kresan.)

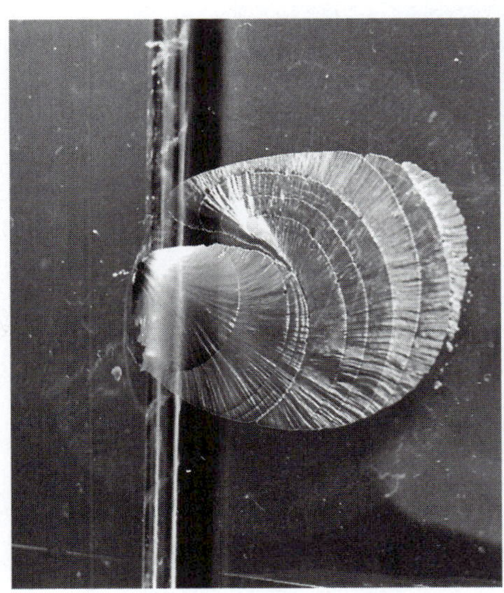

Figure 5.20 Beautiful expression of the ribs and hackles on a joint face produced in a block of Plexiglas. As the block of transparent Plexiglas was being subjected to differential stress, fluid was injected through a tube (at left) to the point of origin for the joint. The joint was thus produced by "hydrofracture" when the tensile strength of the Plexiglas was exceeded. The crack is oriented perpendicular to the direction of least principal stress, which was tensional. Height of the crack is approximately 7 cm. [From Rummel (1987). Fracture mechanics approach to hydraulic fracturing stress measurements. Published with permission from Academic Press.]

expression. In profile view, the change from the main joint face to a rib may be seen to be a smooth curve or a sharp kink. The distances between two ribs on a joint face, when measured along the trace of hackles on the joint face, give a picture of the relative velocities of propagation of the joint front (Pollard and Aydin, 1988).

The presence of plumose structure and ribs tells us that we should not view the formation of a single joint as if it were the product of a single smooth, continuous process. The ornamentation reveals a more complex and discontinuous path of formation, including multiple arrests and branching fronts (Engelder, 1987; Helgeson and Aydin, 1991).

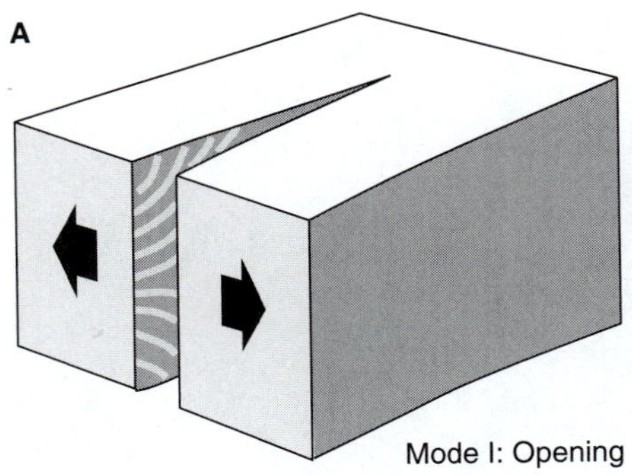

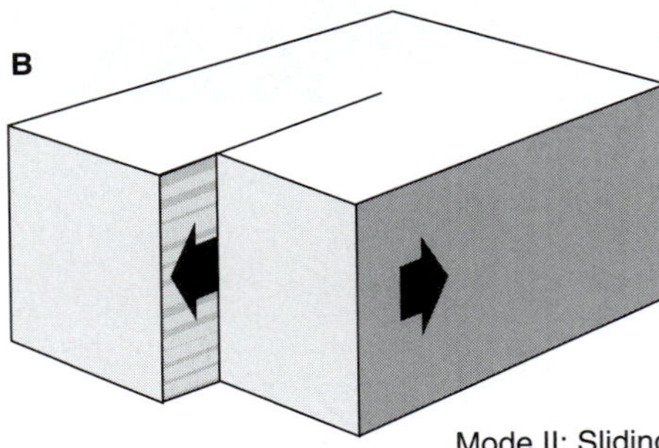

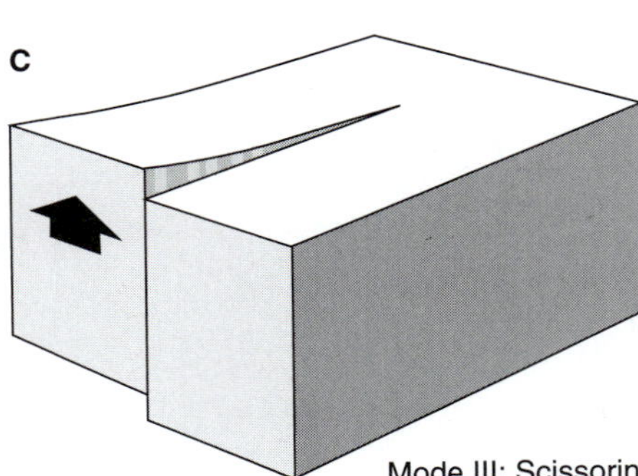

Figure 5.21 The three fundamental fracture modes: (*A*) Mode I opening perpendicular to the walls of the fracture surface. (*B*) Mode II sliding in a direction parallel to the fracture surface and perpendicular to the fracture front. (*C*) Mode III scissoring parallel to the fracture surface and parallel to the fracture front. [Modified from Atkinson (1987), Introduction to fracture mechanics and its geophysical applications.]

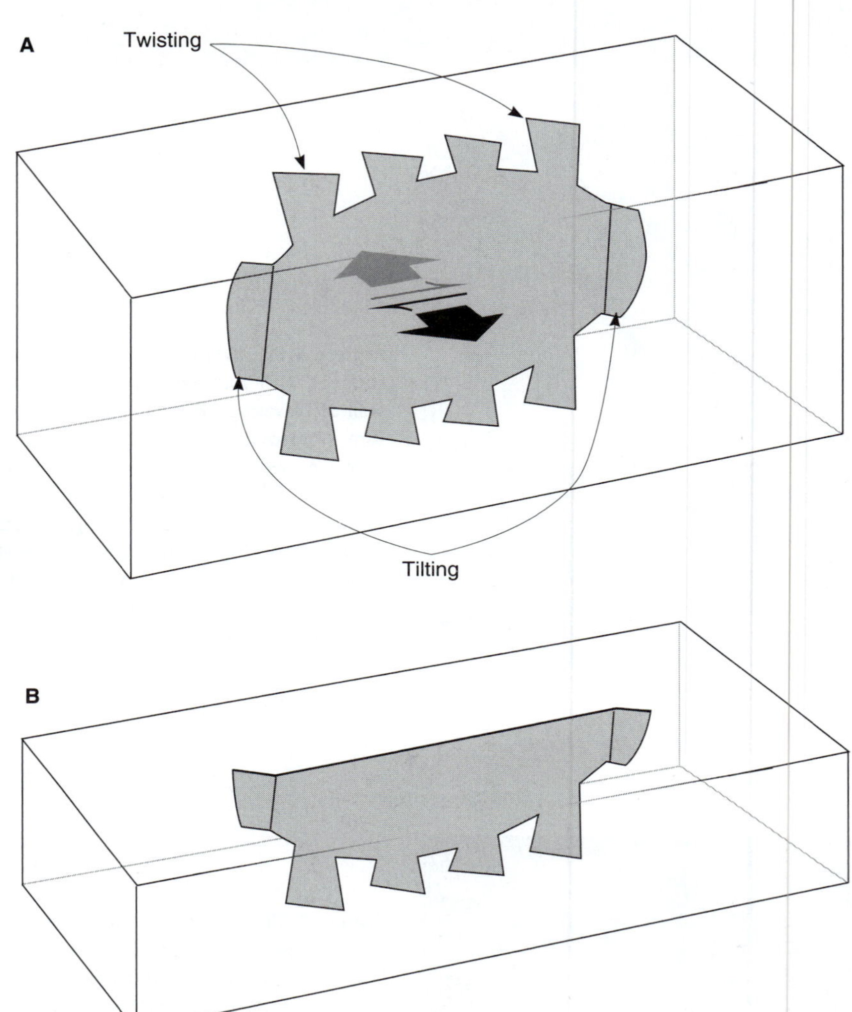

Figure 5.22 (*A*) Here is a joint (gray) in a transparent block of rock. The joint has formed as a result of the combination of mode I opening and mode II sliding (right-handed). Along the fringe of the main joint face are tilted and twisted segments, all misoriented in a clockwise sense due to the right-handed shear component. The axis of tilt is perpendicular to the direction of joint propagation and lies in the plane of the main joint surface. The axis of twist is essentially parallel to the direction of joint propagation and lies in the plane of the main joint surface. The tilts are mode II. The twists are mode III. (*B*) A different level of exposure. [Reprinted with permission from the *Journal of Structural Geology,* vol. 13, Cruikshank, K. M., Zhao, G., and Johnson, A. M., Analysis of minor fractures associated with joints and faulted joints (1991), Elsevier Science Ltd., Pergamon Imprint, Oxford, England.]

Propagation of Individual Fracture Surfaces

There are three fundamental fracture modes that can be used as "end members" to describe *any* combination of movements that occur in the formation and propagation of joints and shear fractures: **opening (mode I)**, **sliding (mode II)**, and **scissors (mode III)** (Atkinson, 1987; Engelder, 1987).

Joints are mode I fractures: extension fractures that open perpendicular to the plane of the joint (Figure 5.21*A*). Mode II and mode III fractures are both shear fractures, marked by movement parallel to the fracture surface. Mode II movements are characterized by a sliding movement parallel to the fracture surface and perpendicular to the fracture front (Figure 5.21*B*). Mode III movements are characterized by a scissors movement parallel to the fracture surface and parallel to the fracture front (Figure 5.21*C*).

As mode I fractures, joints are cracks that try to maintain a perpendicular orientation to the local direction of least principal stress. Planes of such orientation are principal planes, and thus there is no shear stress along them. Yet, as a mode I fracture propagates parallel to its own plane, it may suddenly enter a "region" in which the local stress has shifted, and the crack must make an orientation correction to achieve the most stable orientation (Engelder, 1987).

A joint succeeds in making an orientation correction through a **tilt** or a **twist** (Figure 5.22). Each change in orientation may be described in reference to an axis of rotation. An **axis of tilt** is in the plane of the joint surface and perpendicular to the direction of joint propagation. Ribs are an expression of tilt (Pollard and Aydin, 1988). An **axis of twist** is also in the plane of the joint surface, though parallel to the direction of propagation. A tilt can be achieved without any physical disruption of the joint surface, but a twist cannot, and the surface breaks up into discrete (i.e., "twisted") segments (Engelder, 1987). Hackles are in fact an expression of twist, as are the short, discrete en echelon fractures that form on a joint fringe (Figure 5.22) (Pollard and Aydin, 1988).

A CLOSE LOOK AT PROPAGATION RELATIONSHIPS AMONG JOINT SURFACES

Columnar Joints

We can learn a lot about jointing through examination of columnar joints. Columnar joints are primary volcanic structures produced by a fracturing that accommodates contraction during congealing and shrinking of the flow as it cools (Figure 5.23). Columns tend to be polygonal in the same way that mud cracks resulting from shrinkage of mud are polygonal. Where columnar joints are well developed, the comparison of architecture and structural geology is irresistible and unavoidable (see Figure 5.23).

DeGraff and Aydin (1987), who took a very close look at columnar joints in basalt in Hawaii, discovered kinematic relationships that prove to be valuable to understanding the propagation of joints in general, and the mechanical relationship of one joint surface to another. The joint faces of each column display **bands** that run horizontally, parallel to layering and perpendicular to the column (Figure 5.24). The bands superficially resemble stratification; however, an individual band on the flank of an individual column is in fact an individual joint face. The bands studied by DeGraff and Aydin (1987) ranged in width from 20 to 100 cm (i.e., the width of a column face) and 3–12 cm in height. Each band on each column

Figure 5.23 Architectural display of columnar jointing in the lower colonnade of the prehistoric Mauna flow exposed at the Boiling Pots near Hilo, Hawaii. [From Ryan and Sammis (1978). Published with permission of Geological Society of America and the authors.]

flank is distinctive because of variations in geometry, roughness, orientation, and general texture. Most important, each band possesses its own complete plumose structure and rib markings, revealing that each band is a single fracture "event" (Figure 5.25). Thus, the flank of a column many meters high, which appears at first glance to be a single discrete joint face, is actually composed of tens to hundreds of discrete joint faces. The column grew incrementally.

Cooling of a basalt flow sets up tensile stresses between the part of the flow that is contracting through cooling and the hotter or more fluid part that is not (Pollard and Aydin, 1988). Since a flow cools from the outside in, bands form simultaneously from the top down and from the bottom up; they meet *below* the middle of the flow, because the lower half of a flow stays warmer longer. New cracks propagate at the edges of old cracks, more particularly from origins located at flaws such as voids or large grains (Figure 5.26). In this way, the cracks propagate upward or downward toward the middle of the flow. Thus: "A column face is the net result of many discrete crack events, each of which produces a well-defined segment of the face" (DeGraff and Aydin, 1987, p. 608).

Figure 5.24 Atilla Aydin sits in the shade of columnar joint faces in Snake River Basalt. The faces are marked by parallel, discontinuous bands oriented perpendicular to column axes. The physical expression of the bands is created by textural and geometric variations from band to band. Photograph taken along the Boise River at Lucky Peak Dam in Idaho. [From DeGraff and Aydin (1987). Published with permission of Geological Society of America.]

Connections Between Joints in a Siltstone Sequence

Helgeson and Aydin (1991) studied individual joint surfaces in a sequence of siltstone beds, *separated by thin films of shale*, in the Appalachian Plateau of New York and Pennsylvania. They found that each individual siltstone bed is marked by discrete joint surfaces that do not extend beyond the upper and lower contacts of the bed (Figure 5.27). Each joint surface within each bed is clearly discrete because it possesses its own plumose structure and rib markings. Tracing hackles to points of origin on each face in each bed, Helgeson and Aydin (1991) determined that the joints propagated incrementally downward through the sequence (Figure 5.28). Points of origin were almost always along the upper contacts of the siltstone layers (see Figure 5.27). The vertical alignment of origins reveals the mechanical continuity within the whole sequence.

Patterns of Intersection and Termination of Joints

Detailed observation and mapping of steeply dipping joints in outcrop reveal some common patterns in the ways in which the traces of joints

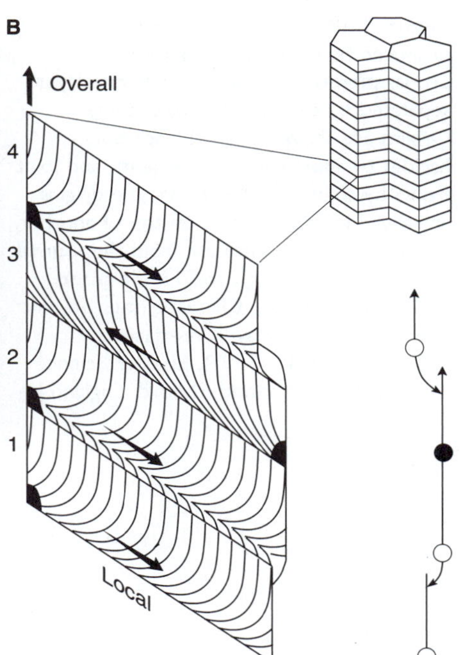

Figure 5.25 Each horizontal band of a column face comes complete with its own origin and hackle pattern, indicating that each band represents an individual crack formed during a discrete fracture event. (*A*) This photograph was taken near a flow base. The white arrows describe the horizontal propagation directions for each band. (Photograph by Atilla Aydin.) (*B*) Diagrammatic representation of the progressive growth of a columnar joint face. The origins occur on the upper edges of preceding segments. The black arrows describe the horizontal propagation for each band. The numbers give the sequence of formation of the bands. [Reprinted with permission from the *Journal of Volcanology and Geothermal Research*, vol. 38, DeGraff, J. M., Long, P. E., and Aydin, A., Use of joint-growth directions and rock textures to infer thermal regimes during solidification of basaltic lava flows (1989), Elsevier Science B.V., Amsterdam, The Netherlands.]

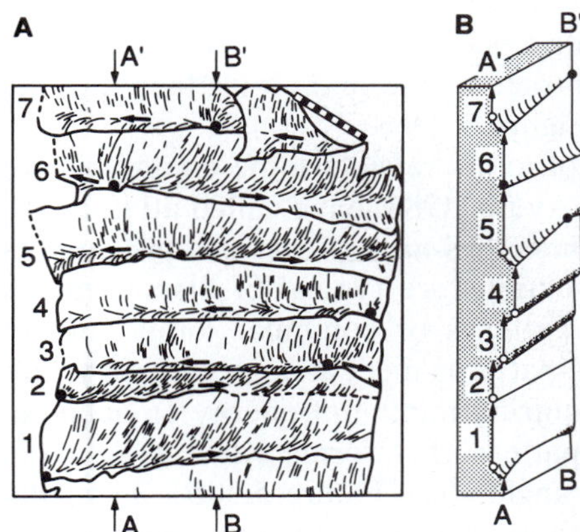

Figure 5.26 Actual data on which interpretation of the progressive development of a columnar joint face is based. (*A*) Line drawing of seven hackle-marked bands on the column face. Origins are noted with solid black dots, and black arrows give horizontal propagation directions with respect to origins. (*B*) Profile showing projected vertical propagation directions. The column face emerges as the product of individual discrete fracture events. [From DeGraff and Aydin (1987). Published with permission of Geological Society of America.]

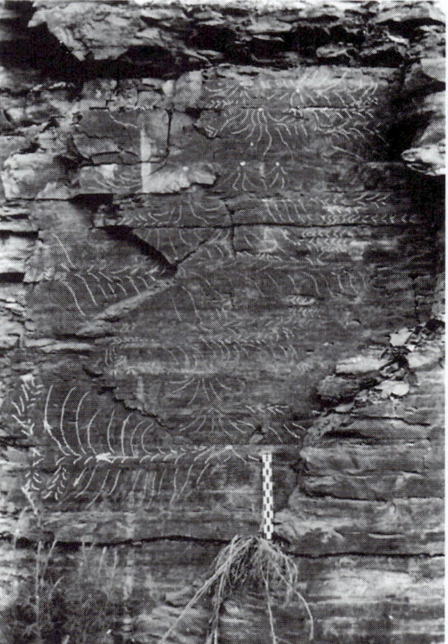

Figure 5.27 Joint faces with hackles in a sequence of siltstones and shales. Some of the hackles have been traced with chalk. [Photograph by Daniel Helgeson. From Pollard and Aydin (1988). Published with permission of Geological Society of America.]

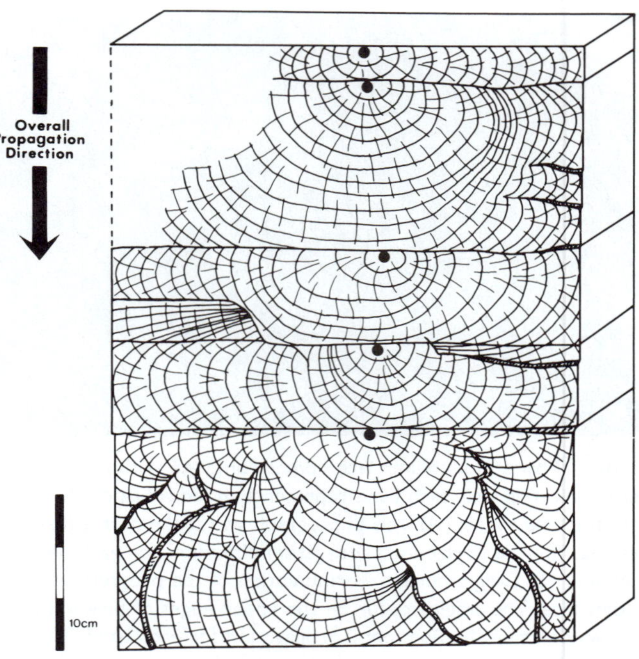

Figure 5.28 A composite joint face produced by systematic *downward* propagation through several siltstone layers separated by ever so thin shale layers. Origins are shown by solid black dots, hackles by short dashlike segments, and joint fronts by solid lines. Note that the origins are aligned vertically, one on top of another. [Reprinted with permission from the *Journal of Structural Geology,* vol. 13, Helgeson, D. E., and Aydin, A., Characteristics of joint propagation across layer interfaces in sedimentary rocks (1991), Elsevier Science Ltd., Pergamon Imprint, Oxford, England.]

intersect and terminate. First of all, the trace expression of any individual ideal joint will depend on the level of exposure relative to the geometry of the whole joint face, including its fringe elements (Figure 5.29*A*). Where the level of exposure intercepts only the fringe of the joint face (Figure 5.29*B*), the trace of the joint might well be a series of en echelon lines at an angle to the main joint. Where the level of exposure intercepts the main part of the joint (Figure 5.29*C*), the trace of the joint will be a single line. Where the level of exposure cuts a deeper level of the joint face, the result may be a combination of the trace of the main joint face and the en echelon fractures of the fringe (Figure 5.29*B*).

The mapping of the intersection of joints reveals some commonly repeated patterns as well. Pollard and Aydin (1988) have called attention to several intersection geometries: **Y-intersections**, **X-intersections**, and **T-intersections** (Figure 5.30). Y-intersections are typical of discontinuous contraction joints, like mud crack patterns or columnar joints (Figure 5.30*A*). The joints that intersect in Y-patterns meet at approximately 120° angles, which is a configuration requiring minimum energy to achieve. X-intersections are the inevitable pattern that forms when systematic continuous joints intersect at acute angles (see Figure 5.30*B*). Typically one joint trace runs continuously without disruption through the intersection, whereas the other is stepped slightly to the right or left, or simply stops. In a system of two sets of vertical faults that intersect at a high angle, there is usually no consistency in which joint set is stepped versus which joint continues without a break through the intersection. T-intersections are common in orthogonal joint systems, where individual joint traces meet at right angles (see Figure 5.30*C*). Typically, the younger joint trace can be followed to its termination at a preexisting joint oriented at right angles.

Joints die out in trace expression in a variety of common ways. The joint trace may abruptly **hook** and stop (Figure 5.31*A,B*), or hook in such a way that forms a T-intersection with an adjacent joint trace that was coming from the other direction (Figure 5.31*C*). A single joint trace, as it dies out, may be replaced by a series of en echelon fractures (Figure

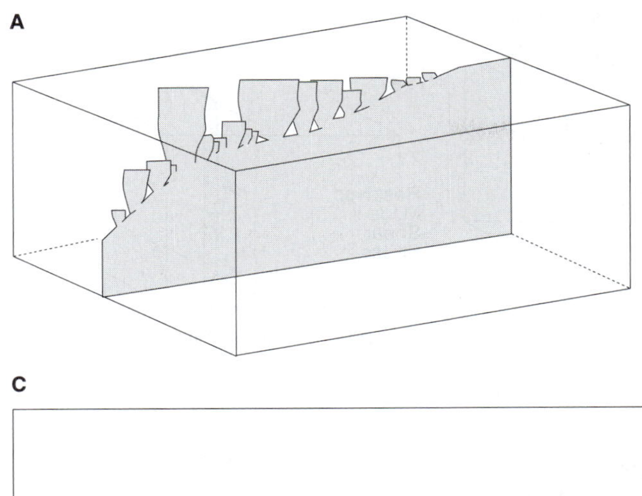

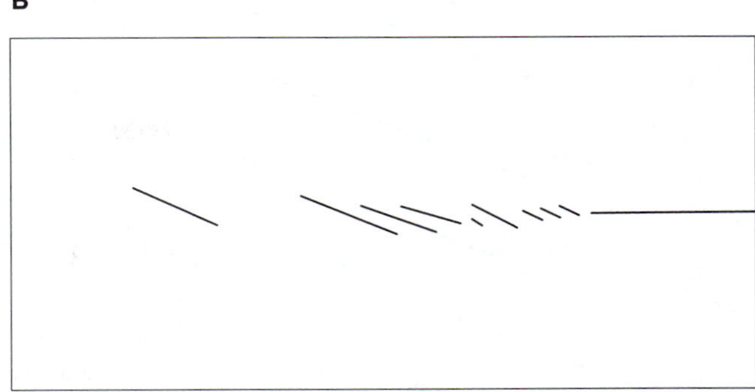

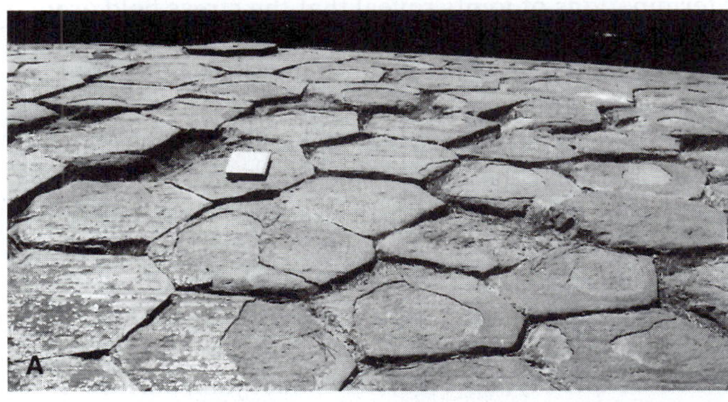

Figure 5.29 (*A*) Block diagram of perfectly transparent rock containing a joint (gray). The joint is composed of a main joint face as well as en echelon cracks along its fringe. (*B*) Map of joint at the surface reveals a set of en echelon fractures that give the appearance that they are not attached in any way to one another, or to the main face whose trace is seen at the right-hand margin of the map. (*C*) Map of joint at the level of the base of the block reveals simply a straight continuous line representing the main joint face. [Reprinted with permission from the *Journal of Structural Geology,* vol. 13, Cruikshank, K. M., Zhao, G., and Johnson, A. M., Analysis of minor fractures associated with joints and faulted joints (1991), Elsevier Science Ltd., Pergamon Imprint, Oxford, England.]

Figure 5.30 Joint intersection patterns. (*A*) Y-intersections of joints in a lava flow. (Reprinted with permission from *Science,* vol. 239, Aydin, A., and DeGraff, J. M., Evolution of polygonal fracture patterns in lava flows. Copyright © 1988 American Association for the Advancement of Science.) (*B*) X-intersections of joints in siltstone exposed at Kimmeridge Bay, United Kingdom. (Photograph by Terry Engelder.) (*C*) T-intersection of joints in quartzite in the Salt River Canyon region, Arizona. (Photograph by G. H. Davis.)

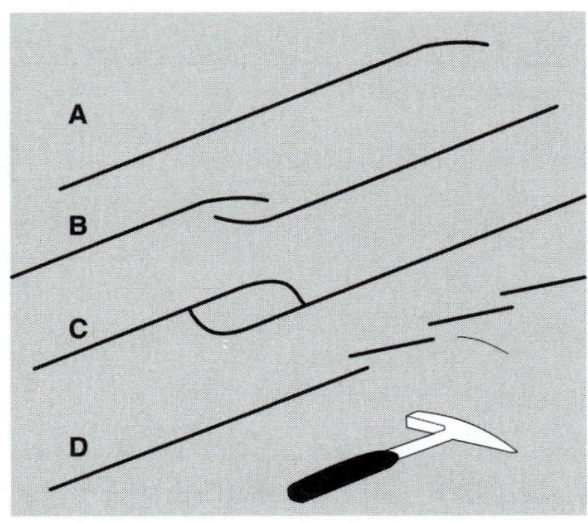

Figure 5.31 Ways in which individual joints die out. (*A*) Joint trace hooks and stops. (*B*) Joint traces hook and stop. (*C*) Joint trace hooks radically to form a T-intersection with adjacent joint. (*D*) Joint trace breaks up into en echelon segments.

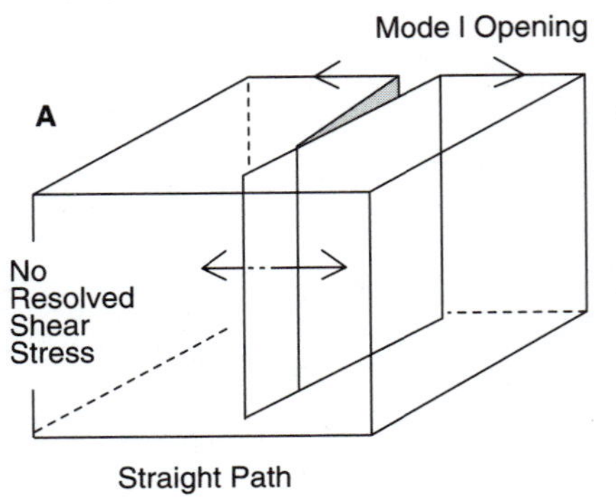

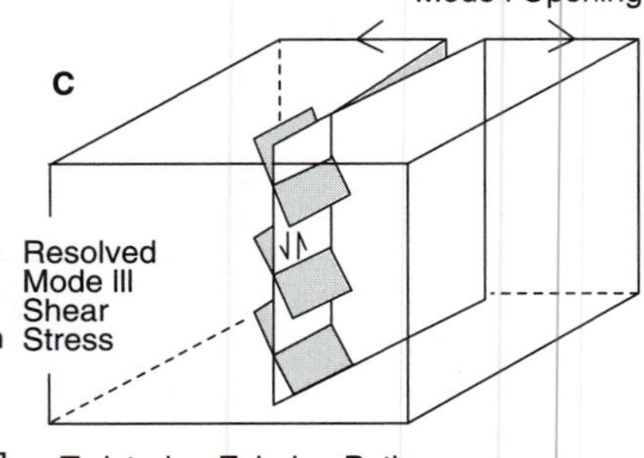

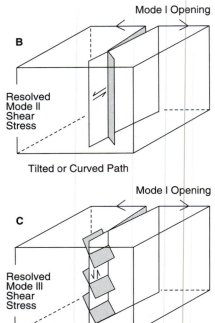

Figure 5.32 The progressive development of hook-shaped terminations and en echelon terminations. (*A*) Plain old mode I joint. (*B*) Mode I joint with mode II action at tip, creating hook shaped termination. (*C*) Mode I joint with mode III action at tip, creating en echelon fractures at termination. [From Pollard and Aydin (1988). Published with permission of Geological Society of America.]

5.31*D*). In many instances a joint trace will simply come to an end without changing trend or breaking up into segments. The manner in which joints intersect and die out can provide information on relative timing of joint set development.

The patterns of terminations of joints reveal that the stress field around one joint can affect the growth and growth direction of a neighboring joint. Pollard and Aydin (1988), and Olson and Pollard (1988, 1989) have explained that as two joints propagate toward each other, each enhances the growth of the close neighbor by imparting an additional increment of tensile stress in the area around the joint front (Figure 5.32). However, as the tips become very close and overlap, shear stresses are induced on neighboring joints. As a consequence, each joint may tilt and curve into the other, producing the characteristic hook-shaped terminations. Or the joints may twist, creating a termination pattern of en echelon joints.

Truncated joints at T-intersections are younger than the joints by which they are truncated (Pollard and Aydin, 1988). The older joint surface along which a younger joint terminates is a free surface, lacking any shear stress component. Therefore, principal stresses are parallel and perpendicular to the joint surface (Suppe, 1985), ''forcing'' a younger joint to be aligned in a perpendicular arrangement.

CREATION OF JOINTS AND SHEAR FRACTURES IN THE LABORATORY

Incentives for Experimental Testing

Insight regarding the conditions under which joints and shear fractures are formed in rocks can be derived from experimental deformation in the laboratory. The fracture strengths of rocks can be measured, and the orientations of fractures with respect to principal stress directions can be observed. Experimental testing of rock strength and empirical observation of the conditions under which rocks fracture form the fundamental basis for exploring the actual mechanics of fracturing. It provides a necessary complement to theoretical analysis of the mechanics of fracturing.

There have been significant economic and practical incentives to carry out extensive testing over the decades. Mining engineering, civil engineering, aeronautical engineering, and materials science are some of the disciplines vitally concerned about the stability of rocks, soils, and hill slopes, as well as manufactured objects like airplane wings and elevator cables. As a result, these professions are constantly pursuing, both experimentally and theoretically, the basic and applied sciences of understanding the mechanisms by which cracks form, and the conditions under which cracks propagate. Furthermore, the research divisions of petroleum companies have carried out exhaustive experimental deformation of rocks to better understand the joint and fracture systems that so importantly govern the large-scale porosity and permeability within petroleum and natural gas reservoirs. The goal of the work has been to develop **laws** describing the conditions under which fractures form. The laws are expressed in basic mathematical equations that describe the critical relationships. Over the decades, and continuing today, there has been a ceaseless effort to refine the laws so that they may "speak" accurately to all situations.

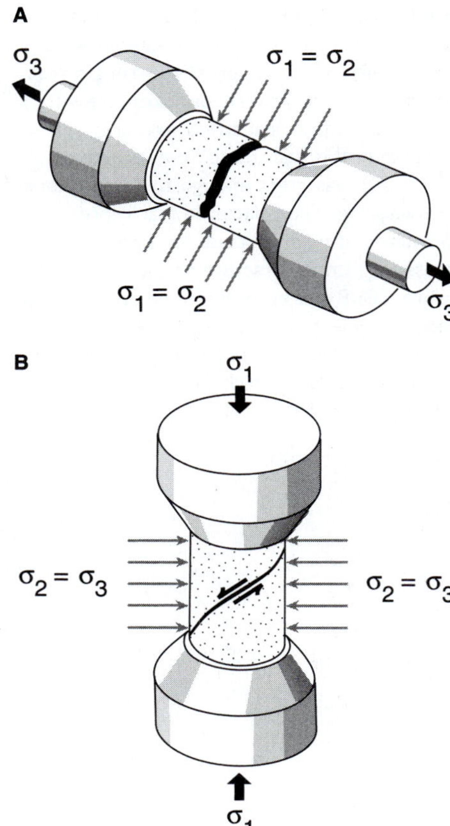

Figure 5.33 The basic setup for (*A*) tensile strength tests and (*B*) compressive strength tests.

Basic Approach in Experimental Testing

For each experimental test, the basic procedure is to hold one principal stress constant, and then progressively increase or decrease the other to create an ever-increasing differential stress. For **tensile strength tests**, an increasing tensile stress (σ_3) is applied to the ends of the specimen while compressive confining pressure ($\sigma_1 = \sigma_2$) perpendicular to the flank of the cylindrical specimen is held constant (Figure 5.33*A*). When the tensile strength of the rock is overcome, the specimen breaks in a mode I tension fracture. For **compressive strength tests**, increasing compressive stress (σ_1) is applied to the ends of the test specimen while compressive confining pressure ($\sigma_3 = \sigma_2$) perpendicular to the flank of the cylindrical specimen is held constant (Figure 5.33*B*). When the compressive strength of the rock is overcome, the specimen breaks by shear failure. For both tensile and compressive strength tests, the magnitudes of σ_1 and σ_3 at failure are calculated and recorded. The specimen is then removed from the pressure vessel, whereupon the orientation of the fracture can be measured relative to the stress field that caused the failure. From this information, we can determine the values of normal stress and shear stress on the fracture at the instant of failure.

As we learned in Chapter 3, the Mohr circle diagram provides a graphical means to describe the values of shear stress and normal stress on any plane within a body subjected to known values of greatest and least principal stress. We can now use the Mohr diagram to "map" the failure values of normal stress and shear stress for each experiment, as well as the conditions of greatest and least principal stress that prevailed at the instant of failure (Figure 5.34*A*). After a series of tests has been carried out on the same lithology under different conditions of confining pressure, a whole family of failure values (or **points of failure**) can be identified. Collectively these define an **envelope of failure** (Figure 5.34*B*), which separates the differential stress conditions under which the rock will remain unfractured and stable, versus the differential stress conditions under which the rock will fail by fracturing.

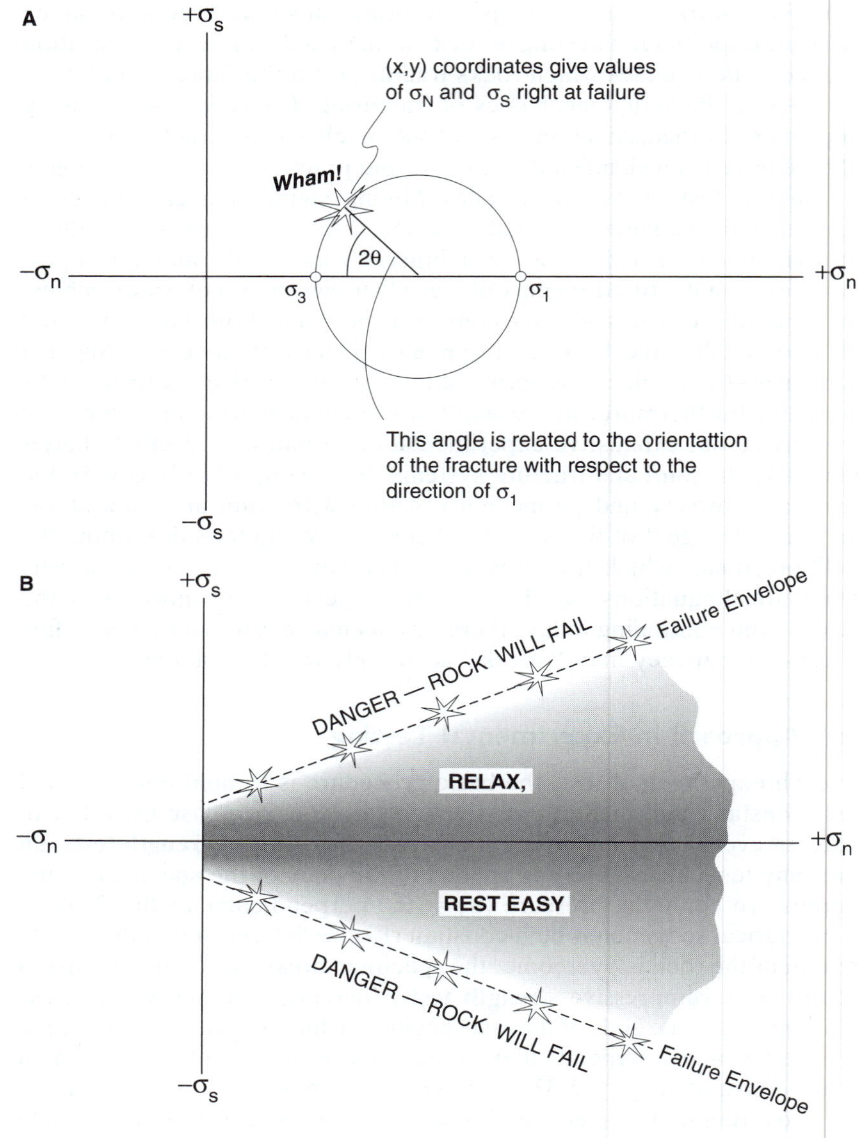

Figure 5.34 (*A*) The Mohr stress diagram can be used to plot the stress conditions of a test, in this case a compressive strength test where both the greatest principal stress (σ_1) and the least principal stress (σ_3) are positive. In this case the differential stress ($\sigma_1 - \sigma_3$) was great enough to cause failure by fracturing. The splattered star represents the values of normal stress (σ_N) and shear stress (σ_S) on the fracture at the instant of failure. (*B*) By carrying out a number of tests under a variety of differential stress conditions, it is possible to define "failure envelopes" within which the rock will not fail, beyond which it will.

Tensile Strength Tests

A variety of tensile strength tests are carried out to evaluate the mechanical response of rocks to tension. These include, but are not limited to, straight pull tests, pull tests with clamped ends (see Figure 5.33*A*), pull tests with cemented ends, and pull tests on "dog-bone"-shaped specimens (Figure 5.35). Price and Cosgrove (1990, p. 30) have emphasized that tensile strength tests are technically tricky, and unlike compressive strength tests, the test results for any given rock are commonly inconsistent from method to method. For example, the tensile strengths measured for the Pennant Sandstone tested by Price, using a wide variety of techniques, ranged from approximately −15 MPa to approximately −65 MPa (Price and Cosgrove, 1990, p. 31). Suppe (1985) reported that the common range of tensile strength for a variety of common rocks is −5 MPa to −20 MPa.

The results of tensile strength tests, when compared with the results of compressive strength tests, have made it obvious that rocks are very weak in tension. The ratios of strength in tension to strength in unconfined compression is commonly 2 : 1, but may exceed 30 : 1 (Price and Cosgrove,

Figure 5.35 (*A*) Some of the materials used in "dog-bone" tensile-strength tests. (*B*) The basic setup for so-called "dog-bone" tensile strength tests carried out informally on the patio. Experimentalists are Matt Davis and Angel.

1990). Everyday experience confirms this. If we want to break a Popsicle stick, we bend it, creating tension in the outer arc of the bend and compression in the inner arc. Weaker in tension, the stick snaps (i.e., fails) along the outer arc.

The sequential stages of a tensile strength experiment can be represented on a Mohr diagram. The state of stress just before tensional stresses are applied can be represented by a single point, indicating that there is no differential stress (i.e., $\sigma_1 = \sigma_3 = 0$) (Figure 5.36). A state of hydrostatic stress exists. As tensile stresses build parallel to the length of the specimen,

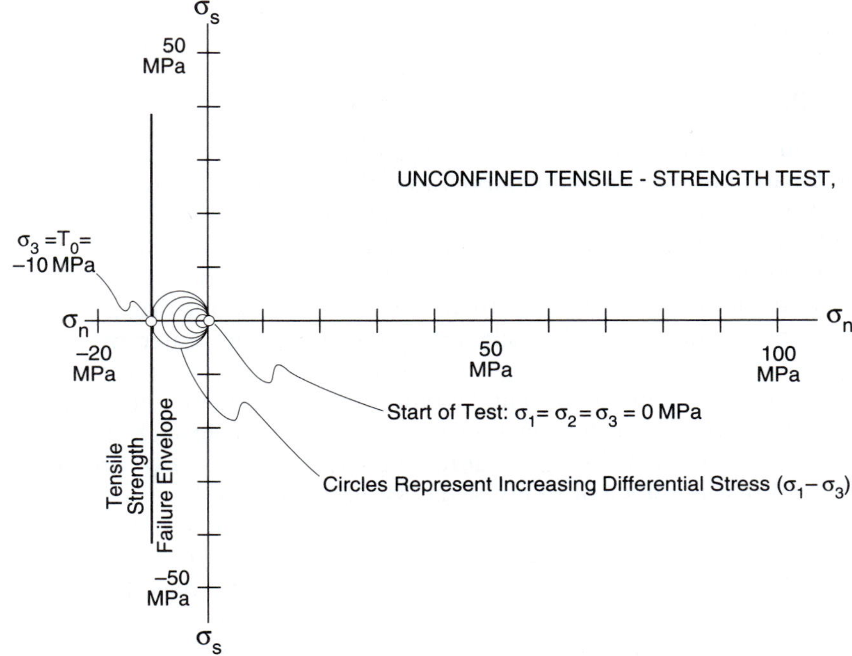

Figure 5.36 Mohr circle representation of a tensile strength test, from start to finish. There is no confinement on the specimen. Tensile stress, parallel to the length of the specimen, is gradually increased until the tensile strength of the specimen is exceeded and it breaks.

a differential stress develops. For a small increment of tensile stress, the differential stress on the Mohr diagram is represented by a circle of very small diameter (Figure 5.36). Stress perpendicular to the axis of the specimen essentially remains the same. It becomes by default the direction and magnitude of greatest principal stress (σ_1). During the test, stress parallel to the axis of the specimen becomes increasingly tensile. This stress is *negative* (tensile), and thus is the least principal stress (σ_3). The buildup of differential stress ($\sigma_1 - \sigma_3$) that takes place during a test is portrayed on the Mohr diagram as a family of circles of increasing diameter (Figure 5.36). When the tensile strength of the rock is exceeded, the rock breaks perpendicular to the direction of tension (Figure 5.35). A mode I joint is formed. The test is run again and again to pin down the average tensile strength for the lithology being examined.

The mode I fracture is parallel to σ_1 and perpendicular to σ_3. In the Mohr diagram, the radius that connects the center of the differential stress circle with the point of failure lies along the x-axis (normal stress axis) and makes an angle of $2\theta = 180°$ with σ_1.

The "law" formulated on the basis of the tensile strength tests is in the following form:

$$\sigma_3 = T_0 \tag{5.1}$$

The equation simply indicates that a rock will fail by mode I fracturing if the magnitude of the least principal stress equals or exceeds the fundamental tensile strength of the rock.

The envelope of failure for a given rock under conditions of unconfined uniaxial tension can be constructed on a Mohr diagram simply by drawing a vertical line through the value of tensile strength (T_0) for the rock (Figure 5.36). The vertical line represents the value of tensile normal stress that must be exceeded before the rock will fail in tension. Tensile stress is negative in the convention we are using. Thus, the magnitudes of tensile stress increase from right to left along the normal stress axis to the left of origin. When the magnitude of tensile stress becomes great enough to touch the envelope, the rock fails by mode I fracturing. The extension fracture breaks perpendicular to the direction in which the specimen was pulled or stretched.

Tensile and Compressive Strength Tests

We will now run a triaxial test in which specimens are subjected to a combination of tension and compression. We do this by applying a small compressive confining pressure to the flanks of a cylindrical specimen of sandstone, and, at the same time, imposing length-parallel tensile stress. On the Mohr diagram, a point representing the constant confining pressure (σ_1) would lie along the normal stress axis to the right of the origin (Figure 5.37). Increasing levels of tensile stress would be represented by points (σ_3) moving to the left of the origin along the normal stress axis. Increasing differential stress, in turn, would be represented by a family of circles of larger and larger diameter, each of which would pass through the point representing σ_1. Ultimately, a **differential stress ($\boldsymbol{\sigma_1 - \sigma_3}$)** would be achieved that is sufficient to break the rock (Figure 5.37).

When tensile and compressive strength tests are carried out for sandstone at confining pressures in the range $\sigma_1 = 3$ to $-5T_0$ (i.e., from three to five times the tensile strength of the sandstone), the collective points of failure map out a parabolic curve that rises steeply from the point on the normal stress axis representing the tensile strength (T_0) of the sand-

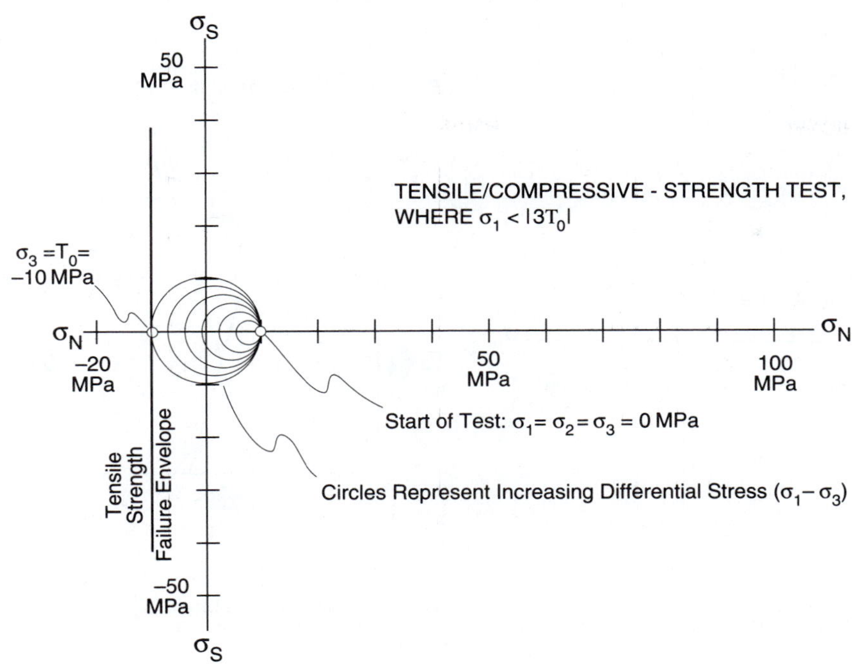

Figure 5.37 Mohr circle representation of a tensile and compressive strength test. The specimen is laterally confined by a compressive stress of 10 MPa, and then is subjected to ever-increasing tensile stress parallel to the length of the specimen. When the tensile stress, which is the magnitude of the least principal stress (σ_3), reaches the tensile strength of the rock, the rock breaks.

stone, then flattens slightly as it passes upward and to the right to barely cross the shear stress axis (Figure 5.38*A*).

Whenever differential stress conditions in a tensile or compressive strength test are such that the Mohr circle representing the specific state of stress touches the parabolic fracture envelope, the rock will fail by fracture. The points that define this **parabolic failure envelope** (Suppe, 1985; Twiss and Moores, 1992, p. 175) have (x,y) coordinates corresponding to the magnitudes of normal stress and shear stress on the fracture at the instant of failure (Figure 5.38*A*).

The parabolic fracture envelope imposes sharp constraints on the types of fracture that can develop. When the confining pressure (σ_1) is less than or equal to $3T_0$, the only type of failure that is possible is mode I tensile failure. However, if the magnitude of confining pressure lies between $3T_0$ and $5T_0$, the rock will fail along fractures that are a combination of extension and shear. This response is known as **transitional tensile behavior** (Suppe, 1985). Examination of the fractured specimen will reveal that the fracture is not strictly mode I, but is a combination of extension (mode I) and shear (mode II) (Figure 5.38*B*). The shearing component will be evident in the form of slight offset and, perhaps, slickenlines. Unlike pure mode I fractures, the fracture that forms will not be oriented parallel to the direction of greatest principal stress (σ_1), but at an angle somewhere between 0° and 30°, depending on the magnitude of confining pressure (Figure 5.38*B*).

A special condition exists when $\sigma_1 = 5T_0$ and σ_3 is just slightly less than the tensile strength of the rock. The Mohr circle representing this state of differential stress intersects the parabolic fracture envelope along the y-axis, at a point representing a normal stress value of 0 MPa and a shear stress value of approximately $2T_0$ (Figure 5.39). This y-intercept for the parabolic fracture envelope represents the fundamental **cohesive strength (σ_0)** of the rock. It represents the strength of the rock when normal stress on the potential plane of fracture equals zero.

The geometry and position of the parabolic fracture envelope can be defined by an equation, and the equation in turn can be thought of as law,

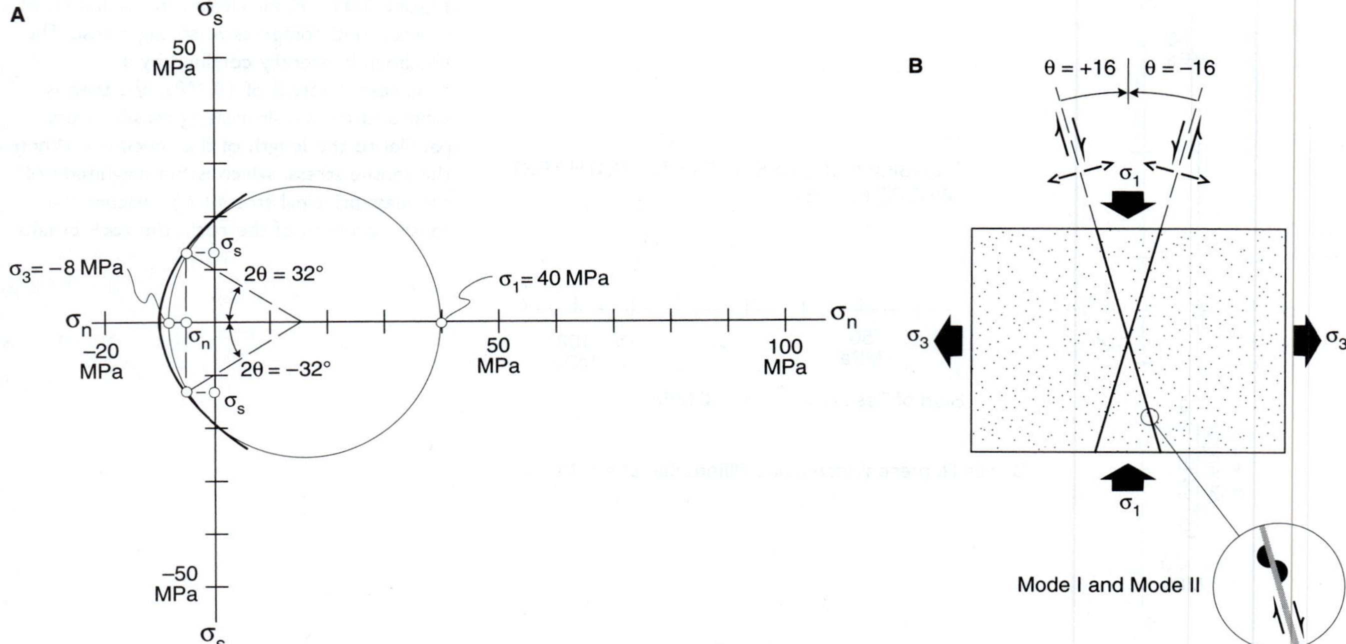

Figure 5.38 (*A*) Mohr circle representation of a tensile and compressive strength test under the special condition where confining pressure has a value greater than 3× tensile strength (T_0) but less than 5× tensile strength (T_0). The failure envelope for these conditions is parabolic (see darkened curve). (*B*) These conjugate fractures formed under conditions of transitional tensile behavior. Note that they are partly tensile (mode I) and partly shear. They intersect at an unusually small angle (in this case, 32°), which is bisected by the direction of greatest principal stress (σ_1).

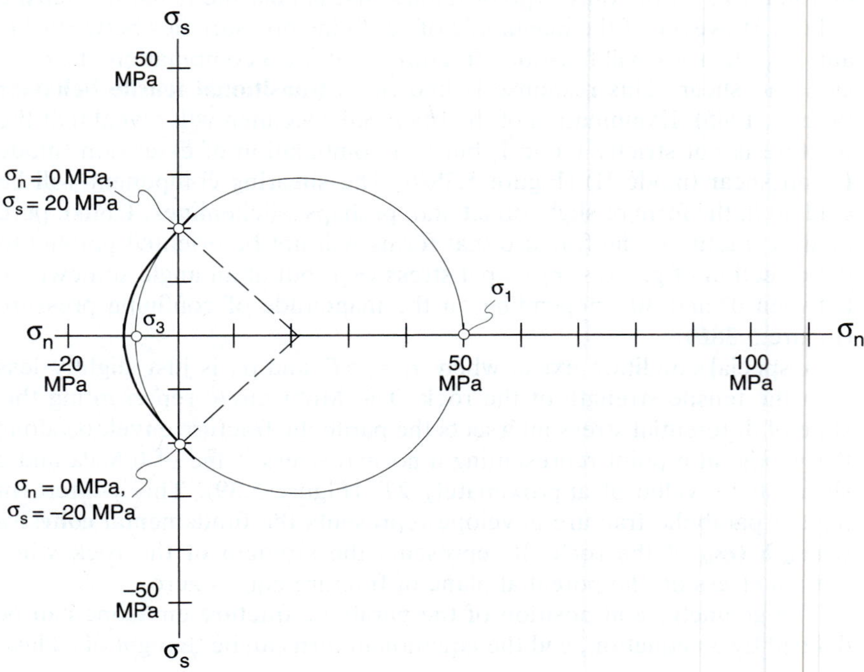

Figure 5.39 Mohr circle representation of a tensile and compressive test under the special condition where confining pressure equals 5× tensile strength (T_0). The normal stress (σ_N) on the fracture at failure is zero. The shear stress (σ_S) is equal approximately to twice the tensile strength (T_0).

the **Griffith law of failure**, which describes transitional tensile behavior (Suppe, 1985). The equation looks like this:

$$\sigma_c = \sqrt{4T_0\sigma_N - 4T_0^2} \tag{5.2}$$

where σ_c = critical shear stress required for faulting

T_0 = tensile strength

σ_N = normal stress

It indicates that the critical shear stress required to break a rock in the transitional tensile field depends on tensile strength T_0 and the magnitude of the normal stress (σ_N) on the fracture. As we shall see, the law is named after A. A. Griffith because he discovered the underlying reason for the behavior. More on this later.

Compressive Strength Tests

To explore the relationship among differential stress, confining pressure, and fracture strength of the sandstone in compression, we apply a compressive axial stress (σ_1) to specimens under different confining pressures (σ_3). In this way we explore the "compression" side of the Mohr diagram, mapping an envelope of failure within which the rock is stable and beyond which the sandstone fractures.

By way of example, let us subject cylindrical cores of sandstone to compressive loading under confining pressure conditions of 10, 20, and 30 MPa (Figure 5.40). The axial stress at the instant of failure is calculated

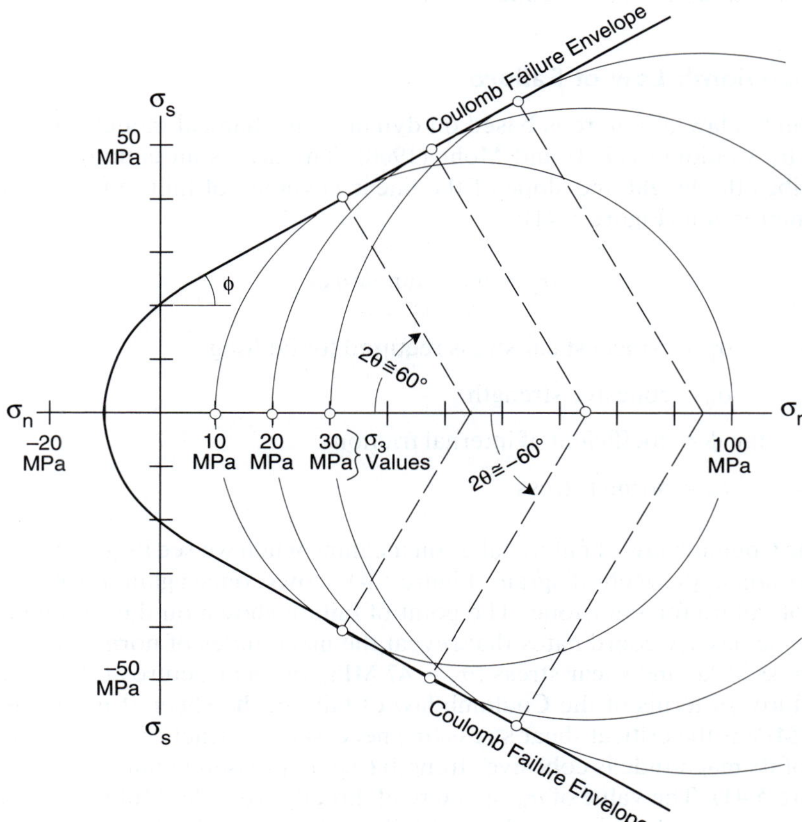

Figure 5.40 Mohr circle representation of a series of compressive strength tests carried out under different conditions of confining pressure (10, 20, 30 MPa). Points of failure lie on a straight line, the Coulomb envelope of failure.

for each of the tests. And after each faulted specimen has been removed from the pressure vessel, the angle (θ), measured from the fault trace to the direction of greatest principal stress (σ_1), is recorded (Figure 5.40). In such compressive strength tests the fractures typically form at angles in the range of 25°–35° to the direction of greatest principal stress. The average is 30°, and the consistency is remarkable. The fractures that form are shear fractures (mode II), marked by offset and by slickenlines in the direction of slip along the surface(s). There is no opening (mode I) component on these fractures.

To proceed from testing to representing the results on a Mohr diagram, the values of σ_1 and σ_3 at failure for each test are first plotted, and circles are drawn through each set of points (Figure 5.40). To each circle is added a radius that makes an angle 2θ with the σ_N axis of the diagram; the 2θ angle is measured positively in clockwise fashion, negatively in counterclockwise fashion. The values of normal stress (σ_N) and shear stress (σ_S) on each fault plane at failure are reflected by the coordinates of the failure points marking the intersection of each circle with its radius (Figure 5.40).

By connecting the failure points for each of the stress circles, a failure envelope for sandstone in the compressive regime is established (Figure 5.40). In this case it is essentially a straight line, climbing from lower left to upper right. It intersects the shear stress axis (y-axis) of the Mohr diagram at small positive values, with actual positioning a function of rock type. The slope and the straightness of the envelope reveal that the compressive strength of the sandstone increases linearly with increasing confining pressure. The actual angle of slope is described as the **angle of internal friction (ϕ)** (Figure 5.40). The envelope itself is called the **Coulomb envelope**, in reference to the Coulomb fracture criterion, a law that captures the specific conditions under which a rock will fail by shear fracturing under compressive stress conditions.

The Coulomb Law of Failure

Coulomb's law of failure is based on dynamic/mechanical models developed by Coulomb (1773) and Mohr (1900). The law is an equation that describes the height and slope of the linear envelope of failure for rocks in compression (Figure 5.41):

$$\sigma_c = \sigma_0 + \tan\phi(\sigma_N) \tag{5.3}$$

where σ_c = critical shear stress required for faulting

σ_0 = cohesive strength

$\tan\phi$ = coefficient of internal friction

σ_N = normal stress

The Coulomb law of failure takes on meaning when we see its geometric expression in the Mohr diagram. Figure 5.41 shows, once again, the envelope of failure for sandstone. The point of failure shown on the Coulomb envelope has x,y coordinates that reveal the magnitudes of normal stress (σ_N = 43 MPa) and shear stress (σ_S = 47 MPa) on the fracture at the time of failure. In terms of the Coulomb law of failure, the shear stress value of 47 MPa is the critical shear stress (σ_c) necessary for fracturing to occur. Part of its magnitude is cohesive strength (σ_0), expressed in units of stress (Figure 5.41). The value of σ_0 can be read directly from the Mohr diagram as the y-intercept of the envelope of failure (Figure 5.41). The rest of σ_c

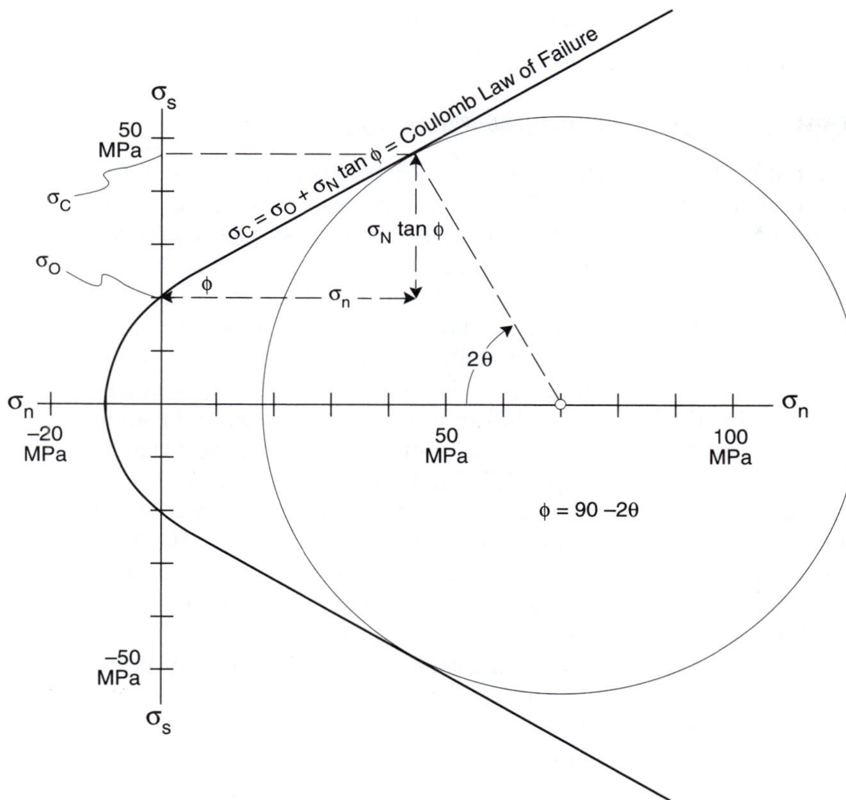

Figure 5.41 According to the Coulomb law of failure, the critical shear stress (σ_c) required to break a rock by shear failure is equal to the cohesive strength of the rock (σ_0) plus another increment of shear equal to the product of normal stress (σ_N) and the coefficient of friction of the rock (tan ϕ). Graphically, the angle of internal friction (ϕ) is simply the slope angle of the Coulomb envelope.

is the stress required to overcome internal frictional resistance to triggering movement on the fracture. This component is labeled tan $\phi(\sigma_N)$ in Figure 5.41, a value expressed in terms of the normal stress (σ_N) acting on the fault plane and the angle of the internal friction (ϕ), which is the slope of the envelope of failure. Thus, as indicated in Figure 5.41,

$$\sigma_c = \sigma_0 + \tan\phi(\sigma_N)$$

We learn from this that the stress level at which shear fractures will form in compression is strongly influenced by ϕ, the angle of internal friction. The angle of internal friction for most rocks lies between 25° and 35°. Consequently, the coefficient of internal friction, tan ϕ, commonly ranges from 0.466 to 0.700. If we can assume that cohesive strength is a very small part of the critical shear stress required for shear fracturing, most shear fractures form when shear stress on the plane of failure reaches a level that is slightly more than 50% of the normal stress acting on the surface. The exact optimum ratio of shear stress to normal stress will vary from rock to rock.

The angle of internal friction (ϕ) determines (θ), the angle between the fracture surface and the direction of greatest principal stress (σ_1). Based on the geometry of the Mohr diagram (Figure 5.41), we have

$$\phi = 90° - 2\theta$$

Thus,

$$2\theta = 90° - \phi$$

$$\theta = \frac{90° - \phi}{2}$$

Since most rocks in nature possess an angle of internal friction of about 30°, the value of θ for most shear fractures is also 30°.

Application of the Coulomb Criterion

Coulomb's law of failure can be used to predict the conditions under which shear fractures will form in compression. Suppose we wanted to know the magnitude of the greatest principal stress that would be required to fracture sandstone under confining pressure conditions of 18 MPa (Figure 5.42). Would 60 MPa be high enough? Not a chance! A stress circle drawn through points on the σ_N axis of the Mohr diagram corresponding to $\sigma_1 = 60$ MPa and $\sigma_3 = 18$ MPa is not large enough to intersect the envelope of failure (Figure 5.42). The stress circle resides entirely within the field of stability. For the differential stress $(\sigma_1 - \sigma_3)$ to be great enough to cause failure, σ_1 must be raised much higher than 60 MPa. Indeed, if σ_1 is raised to the level of 123 MPa, while confining pressure is held constant, the stress circle becomes tangential to the envelope of failure at failure point $\sigma_N = 43$ MPa, $\sigma_S = 47$ MPa (Figure 5.42). On this basis we would surmise that σ_1, when raised to 123 MPa, would cause the specimen to fault. Moreover, the angle (θ) between the fault and σ_1 could be predicted to be approximately $2\theta/2$, or 30° (Figure 5.42).

Predicting the failure point of a rock is some of the fun of rock deformation experiments. Great shouts, screams, and moans, not to mention the clinking of nickels and dimes, have issued forth from our rock deformation lab. Guessing the value of greatest principal stress (σ_1) required to break the very first specimen in a series of tests is almost mindless speculation. But after stress data and θ values for two or three tests have been posted on a Mohr diagram, the predicted stress values for failure begin to cluster within a *very* narrow range. Beads of sweat begin to form as data are plotted and failure envelopes are constructed.

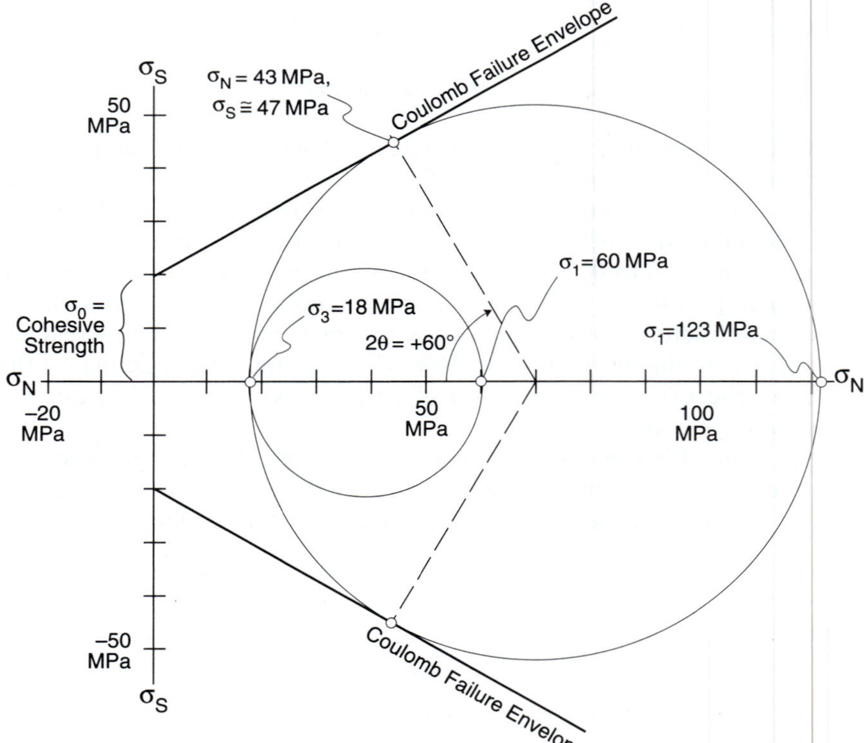

Figure 5.42 Having established the Coulomb failure envelope for this rock, we can predict the conditions under which the rock will fail. If confining pressure is set at 18 MPa, will the raising of greatest principal stress (σ_1) to a level of 60 MPa be enough to cause shear failure? Not a chance! On the other hand, 123 MPa would be just the right amount.

Raising Confining Pressures Even Higher

Were we to subject specimens of sandstone to compressional deformation at high confining pressures, we would discover that Coulomb's criterion for fracturing no longer holds. With increasing confining pressure, rocks behave in a less brittle, more ductile fashion. We have seen this in stress–strain curves, expressed in a departure from a strict linear relationship between stress and strain. In a somewhat analogous manner, we learn through compressive strength tests that the linear (Coulomb) relationship between fracture strength and confining pressure only goes so far, and beyond a certain point, the rock begins to weaken. The expression of this in the world of Mohr diagrams is a change from the straight-line Coulomb envelope to a concave downward envelope of lesser slope (Figure 5.43).

The flattening of the envelope expresses two things: (1) with each increment of increased confining pressure, proportionally less shear stress is required to fracture rock in the brittle–ductile transition, and (2) when rocks in the brittle–ductile transition do fracture, they fail by shear fractures that make an angle θ greater than 30° with respect to the direction of greatest principal stress. In fact, the angle of failure approaches 45° as the rock becomes increasingly ductile.

The law that describes the deformational behavior above the brittle–ductile transition is called the **von Mises criterion** (Ramsay, 1967). When the critical yield stress (which varies from rock to rock) is surpassed, the rock will fail by ductile shear along "planes" of maximum shear stress, oriented at 45° with respect to the direction of greatest principal stress (Figure 5.43).

The Envelopes, All Together

Figure 5.43 is a summary diagram that serves to combine the envelopes for tensile behavior, transitional tensile behavior, Coulomb behavior, and ductile von Mises behavior into a single "grand envelope" for failure

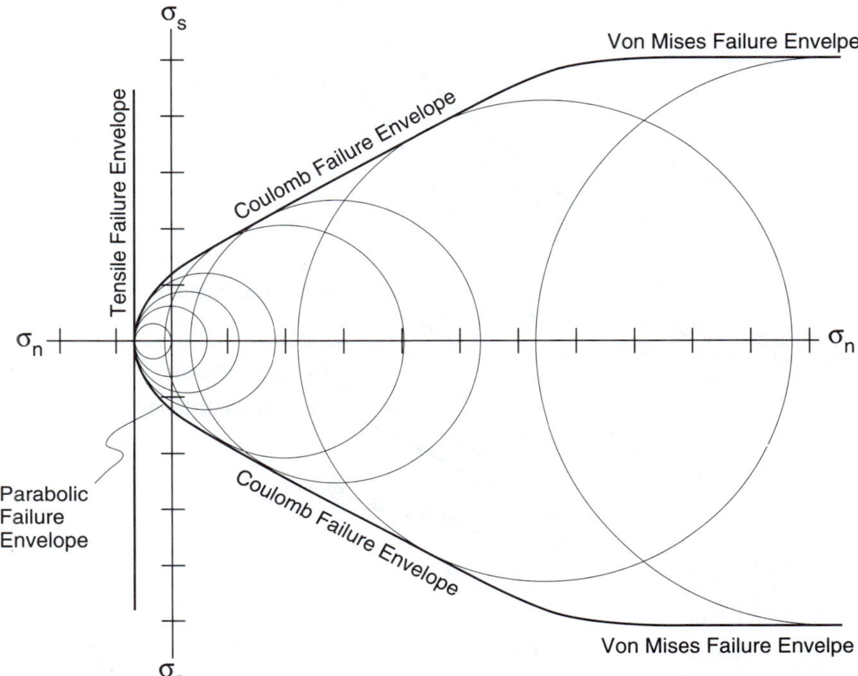

Figure 5.43 For some rocks, if confining pressure (σ_3) is raised high enough, the Coulomb failure criterion no longer holds. Instead, the failure envelope flattens out. A new failure law takes over, known as von Mises.

(Suppe, 1985; Twiss and Moores, 1992). The form of the grand envelope is common for almost all rocks, although the positioning and shape of the envelope will vary according to tensile strength (T_0), cohesive strength (σ_0), and internal friction (ϕ). The values for tensile strength, cohesive strength, and internal friction for common rock types are shown in Figure 5.44*A*. Figures 5.44*B* and *C* illustrate the differences in a failure envelope for a rock of low tensile strength, low cohesive strength, and low internal friction versus a rock of high tensile strength, high cohesive strength, and high internal friction.

Testing of Prefractured Rocks

When an already fractured rock is tested for tensile and compressive strength, it is found to possess neither tensile strength nor cohesive strength. As a result, failure of the specimen simply requires movement on one or more of the preexisting fractures. Since there is no cohesive strength to overcome, the critical level of shear stress required to produce movement on a preexisting fracture is simply related to the frictional resistance to sliding.

Frictional resistance to sliding offered by a fracture surface is related to a number of factors, the most important of which is the normal stress acting on the fracture surface. Friction is directly proportional to the normal force. Resistance to sliding is also related to the character of the surface and to rock type. The magnitude of frictional resistance is less where the fracture surfaces are smooth and planar and greater where the fracture surfaces are rough. Moreover, frictional resistance increases where there are **asperities** (i.e., tiny protuberances, bumps, and other irregularities, along the fracture surface) (Suppe, 1985).

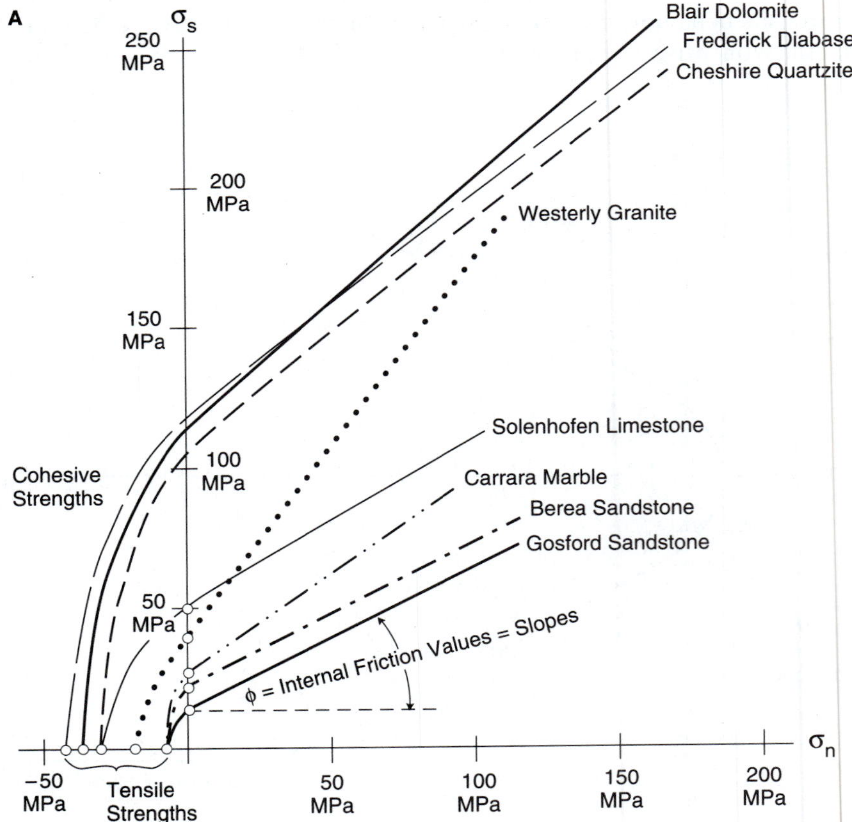

Figure 5.44 (*A*) A Mohr diagram of actual measured values of tensile strength, cohesive strength, and internal friction for a wide range of rock types.

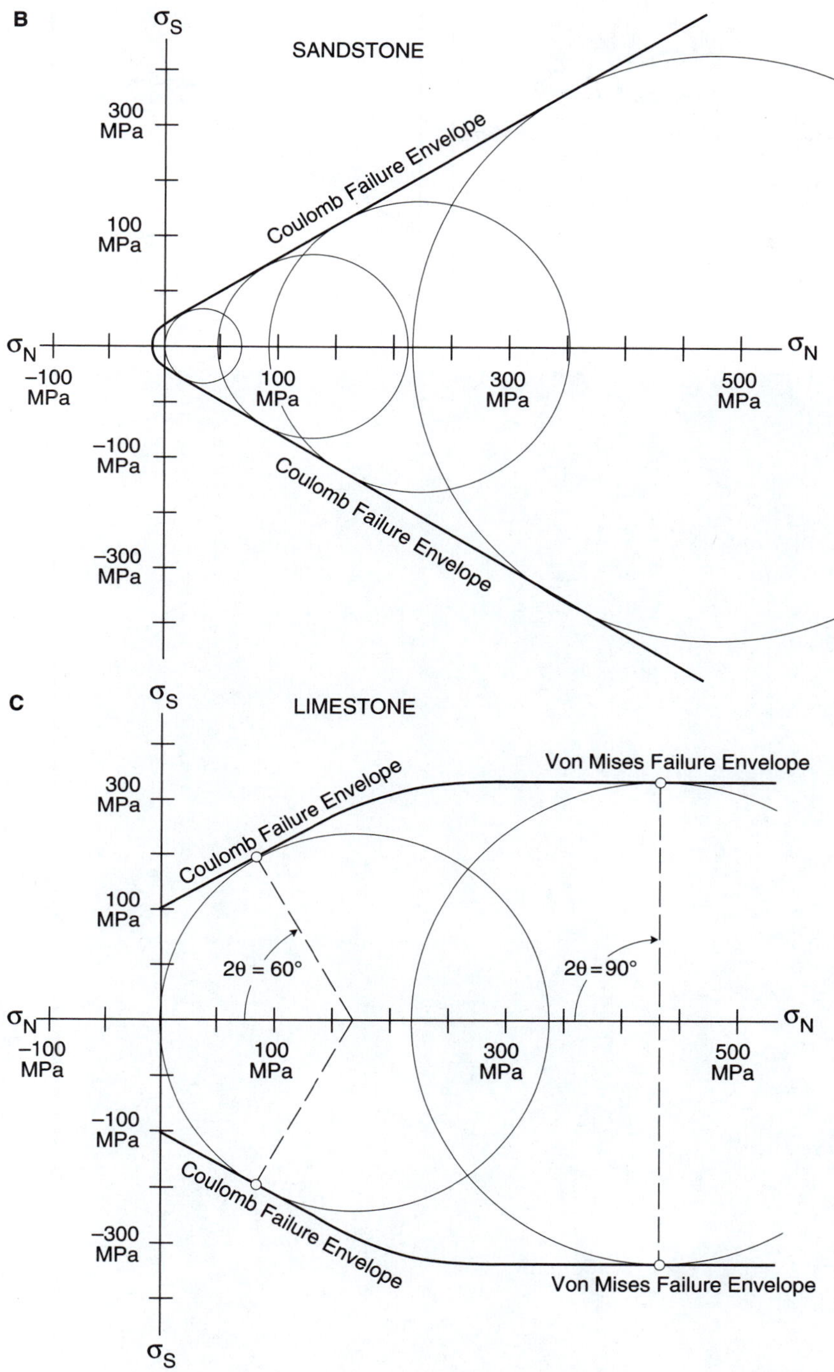

Figure 5.44 (*continued*) (*B*) Schematic portrayal of failure envelope for a rock marked by low tensile strength, low cohesive strength, and low internal friction. (*C*) Schematic portrayal of failure envelope for a rock marked by high tensile strength, high cohesive strength, and high internal friction.

Consider how easy (or how hard) it is to push a sled on a cold winter afternoon (Figure 5.45). The chief determining factors are (1) the smoothness of the interface between the runners and the underlying hard-packed snow, and (2) the weight of the person on the sled. Sleds on ice are faster than sleds on snow or snow mixed with black cinders. The weight of the person on a sled is analogous to the level of normal force acting on the discontinuity between runners and ice. It is easier to give a push to a small child on a sled than to try to budge the overweight dad who has not been able to give up his old Flexible Flyer. Forget budging Dad if his sled is on a cinder patch.

Figure 5.45 The relationship of the dynamics of sliding friction to sled riding. (Artwork by D. A. Fischer.)

A

B

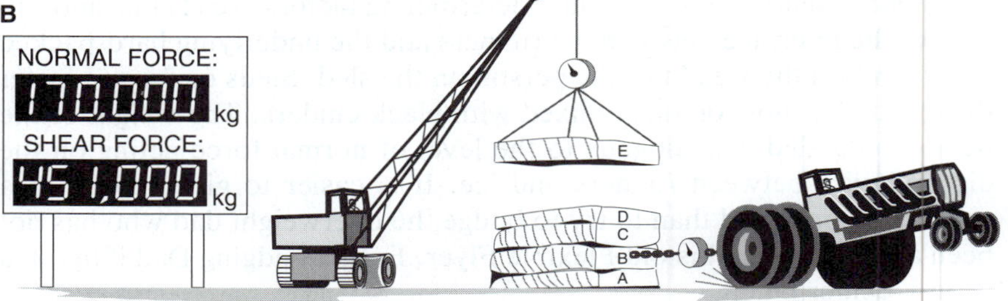

Figure 5.46 (*A*) Close-spaced jointing in Zion National Park has broken the Navajo Sandstone into slabs. (Photograph by G. H. Davis.) In virtual reality we collect slabs *A–E*. (*B*) Tractor pull extravaganza features measuring the relationship between shear force and normal force.

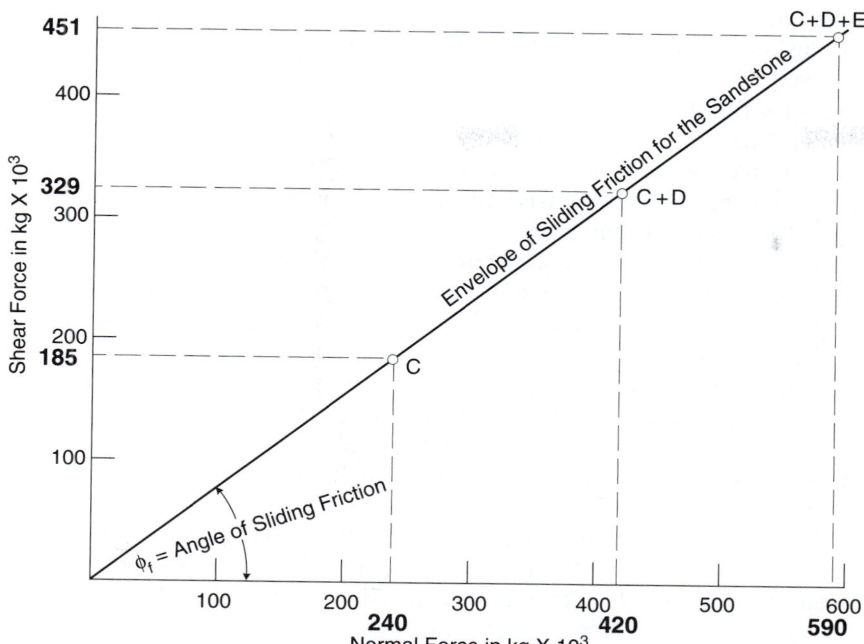

Figure 5.46 (*continued*) (*C*) Mohr-diagram plot shows the results—a straight-line relationship!!

Resistance to sliding on a given fracture in a given rock can be quantitatively measured. Frictional tests are run which yield for each rock type a measure known as the **angle of sliding friction**. They are carried out in a technically sophisticated way, but we can picture the process by carrying out a "thought" experiment in Zion National Park. We start by collecting slabs (*A, B, C, D, E*) from a mountain of rock that is split by a set of natural joints (Figure 5.46*A*). Hauling these out on a flatbed truck, we lay slab *A* on its side, such that the joint surface that originally separated it from slab *B* is on top and horizontal (Figure 5.46*B*). After weighing slab *B* and resting it on slab *A*, we measure how much shear force is required to initiate movement along the joint surface. The values of normal force (weight of slab *B*) and shear force (spring-scale reading) are recorded. We then weigh and add slab *C* to the stack, and once again determine the level of shear force required to trigger movement along the joint surface between slabs *A* and *B*. Then slab *D* is added, and the shear force required to cause movement is measured again. By plotting all three sets of the paired values of normal force and shear force, it is possible to construct a **failure envelope for frictional sliding** (Figure 5.46*C*). The slope of the straight-lined envelope is called the **angle of sliding friction** (ϕ_f) (Figure 5.46*C*). The tangent of the angle of sliding friction is known as the **coefficient of sliding friction** ($\mu_f = \tan \phi_f$) (Figure 5.46*C*), which expresses the ratio of shear stress to normal stress on the fracture surface at the instant that sliding initiates.

The law that describes the conditions under which an existing fracture will move is a modification of the Coulomb law of failure. Since a fractured rock has no cohesion, the term describing cohesive strength is removed from the Coulomb equation. What is left is a simple expression:

$$\sigma_c = \tan \phi_f(\sigma_N) \tag{5.5}$$

This equation, called **Byerlee's law**, says that the critical shear stress necessary to cause reactivation of an existing fracture is equal to the coefficient of sliding friction of the rock multiplied by the normal stress acting on the fracture surface. To the surprise of most, Byerlee (1978) showed that the maximum coefficient of sliding friction ($\mu_f = \tan \phi_f$) is the same for almost all rocks under conditions of moderate to high normal stress (Figure 5.47). For relatively low magnitudes of confining pressure, the average value for the coefficient of sliding friction is 0.85, corresponding to an angle of sliding friction of approximately 40°. For moderate to higher levels of confining pressure, the average angle of sliding friction is approximately 35°. The coefficient of sliding friction becomes smaller as frictional sliding pulverizes mineral grains into clayey **gouge**.

The envelope for frictional sliding lies beneath the Coulomb envelope (Figure 5.48*A*). Rocks will preferentially fail by movement on preexisting fractures, provided the preexisting fractures are suitably oriented. When this occurs, differential stress need not reach the magnitude required to form brand-new fractures. The geologic evidence we observe for repeated movements along joints and other fractures gives a tangible sense of such reactivation.

How can we determine what fracture orientations are suitable for reactivation? We can picture the answer to this question by constructing a Mohr diagram with both the Coulomb and Byerlee (frictional sliding) envelopes for sandstone, and then showing, through circles of increasing diameter, a buildup of differential stress (Figure 5.48*A*). In this example, the slope (ϕ) of the Coulomb envelope is 30°; the slope (ϕ_f) of the Byerlee envelope is 40° (a little higher than normal); and the cohesive strength (σ_0) of the sandstone is 20 MPa. Confining pressure (σ_3) is held constant, while the greatest principal stress (σ_1) is steadily increased. The stress circle representing differential stress eventually achieves a diameter sufficiently large to become tangent to the envelope of frictional sliding (Figure 5.48*A*). The point of intersection of the circle and the envelope represents the shear stress and normal stress magnitudes on a fracture that makes an angle of 25° relative to σ_1 (Figure 5.48*B*). If a fracture of this orientation exists in the rock, it will slip; otherwise the rock will remain completely stable.

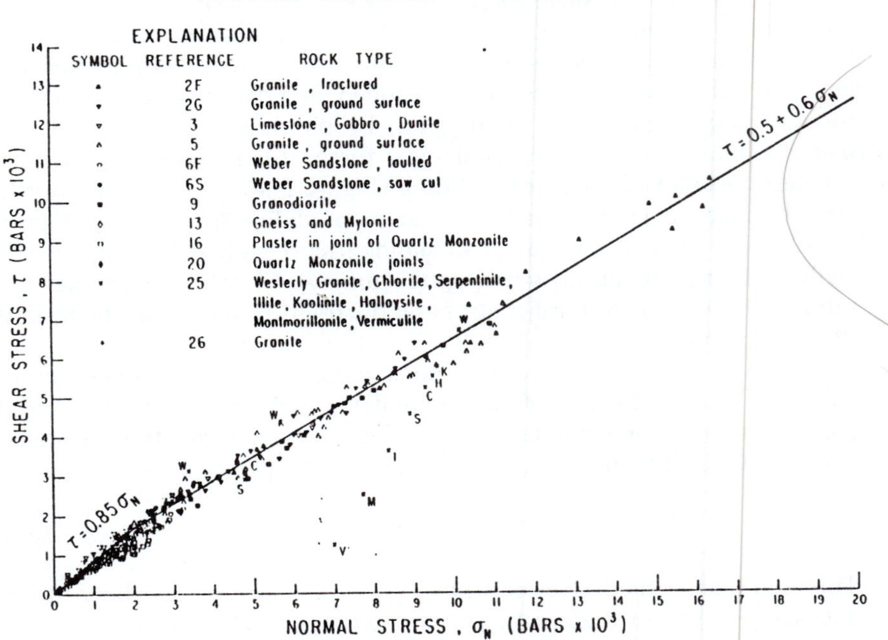

Figure 5.47 Byerlee determined that the sliding friction is the same for almost all rocks under conditions of moderate to high normal stress. The envelope of sliding friction describes the basic relationship and can be used to predict how much shear stress (σ_S) is necessary to cause movement along a preexisting fracture surface subjected to some known normal stress (σ_N). (Reprinted with permission from *Pure and Applied Geophysics*, vol. 116, Byerlee, J. D., Friction of rocks, 1978, Birkhäuser Verlag Ag, Basel, Switzerland.)

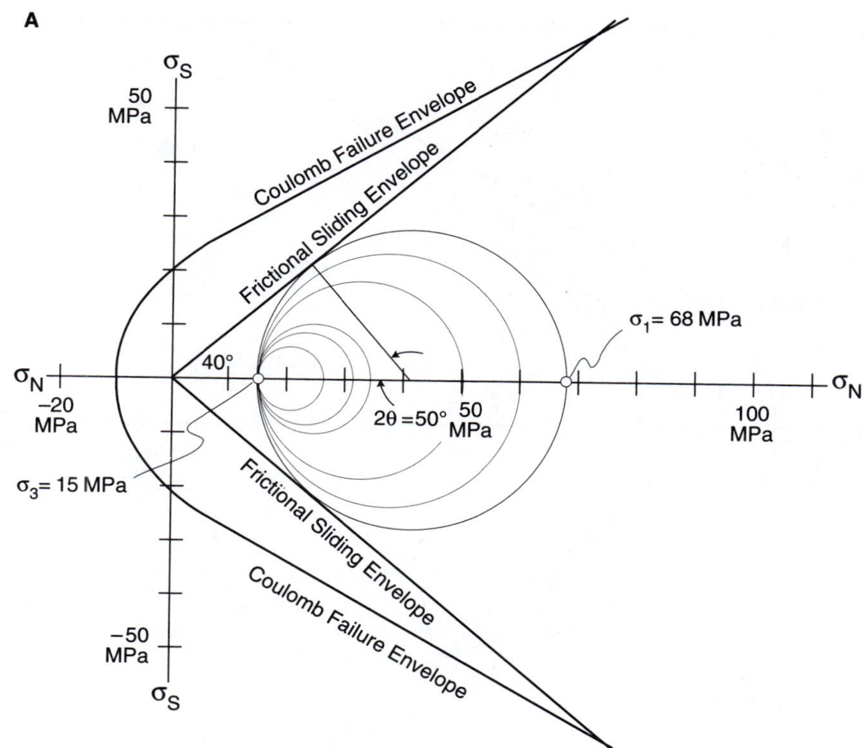

Figure 5.48 Mohr diagrams showing both the Coulomb envelope of failure and an envelope for sliding friction. In this example, the friction is a little higher than normal. (*A*) Confining pressure (σ_3) is held at 15 MPa while greatest principal stress (σ_1) is progressively raised. Eventually the stress difference, as expressed visually through the stress circle, is sufficiently large to "touch" the frictional sliding envelope. (*B*) If a preexisting fracture exists that makes an angle of $\theta = 25°$ with respect to the direction of greatest principal stress (σ_1), the fracture will slip and the rock will fail.

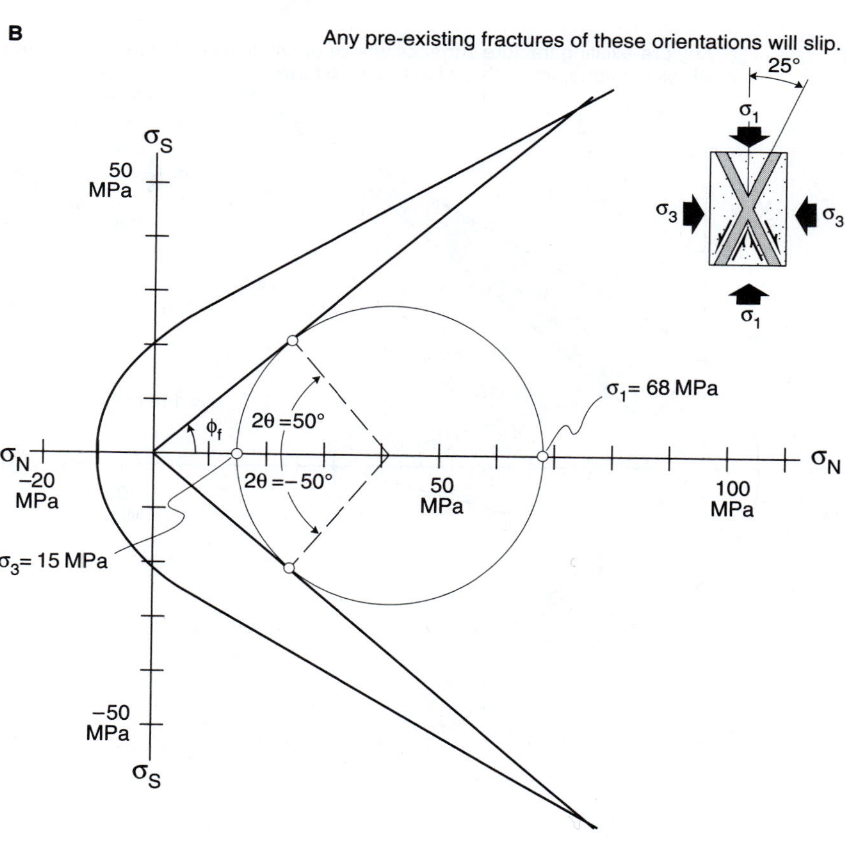

Figure 5.48 (*continued*) (*C*) Assuming that the rock has not failed, the differential stress continues to climb. As it does so, an ever-increasing number of fracture orientations become favorably oriented for slip. For example, before σ_1 reaches 100 MPa, any preexisting fracture in the range 10° to 39° would have been reactivated. (*D*) Assuming that the rock still has not failed, σ_1 is raised further. At 114 MPa, the stress circle intersects the Coulomb failure envelope, at which point the rock will break along a shear fracture oriented 30° to the direction of greatest principal stress. σ_1 could never have climbed that high had there been any preexisting fractures oriented anywhere from 7.5° to 42° to the direction of greatest principal stress (σ_1).

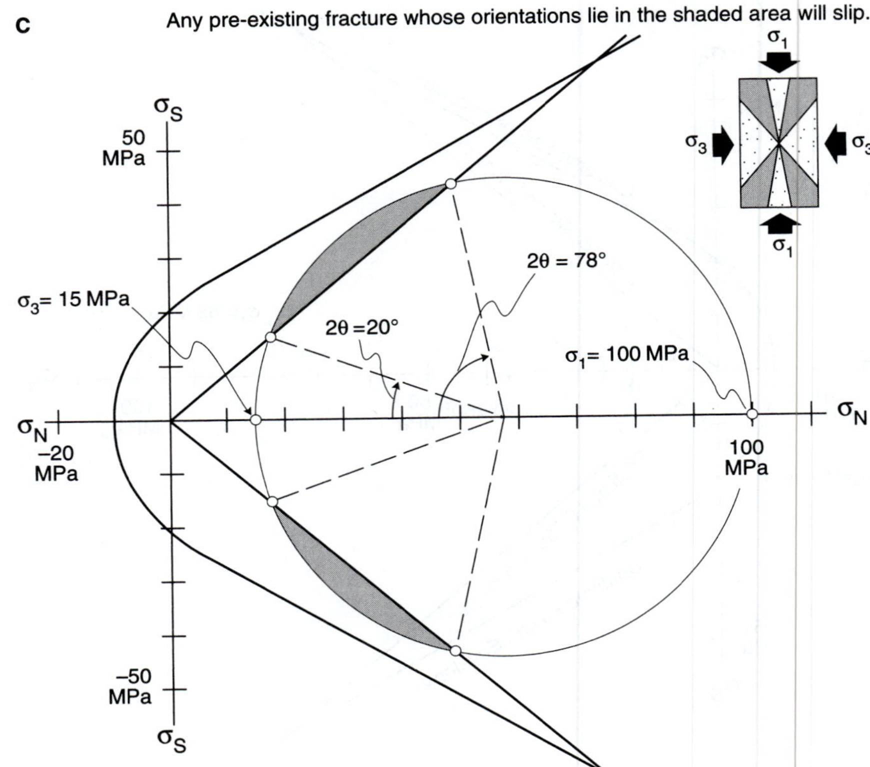

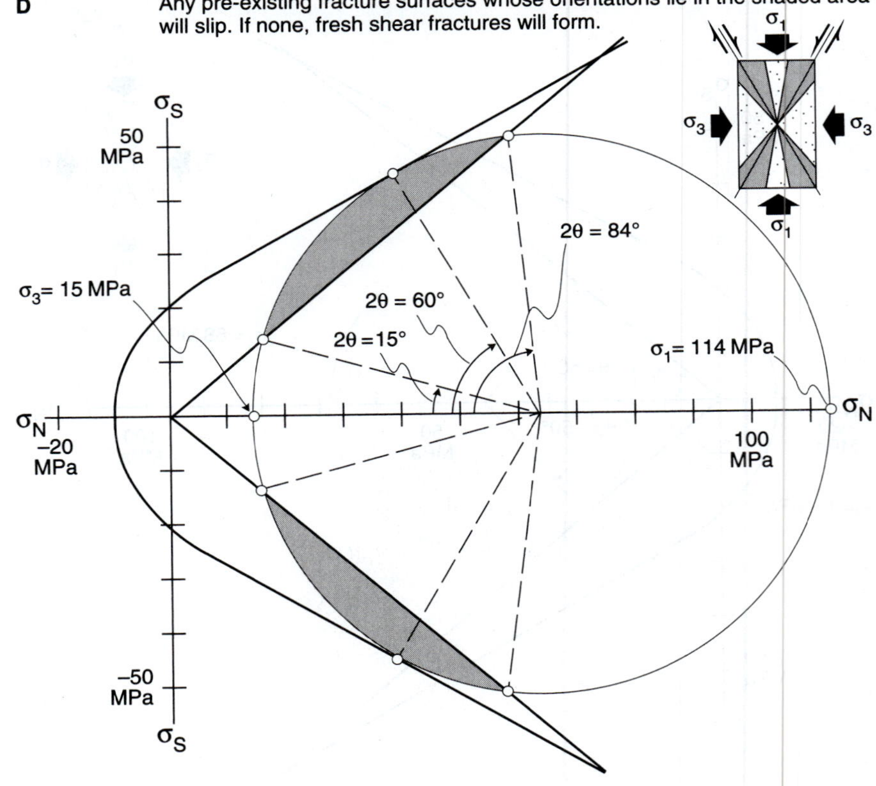

As differential stress increases, the stress circle invades territory across the frictional sliding envelope, which passes through the outer edge of the circle as a chord to the circle (Figure 5.48*C*). Points along the circle between the end points of the chord represent the normal stress and shear stress values on all the fracture orientations that are conducive to reactivation. When the differential stress achieves a level that intercepts the Coulomb fracture envelope (Figure 5.48*D*), new fractures can form at angles of 30° with respect of σ_1. At this level of differential stress, a whole spectrum of fracture orientations, ranging from 7.5° to 42°, is vulnerable to reactivation.

The capacity of preexisting fractures to reactivate and slide in the presence of elevated stress serves as an impediment to stress rising to yet higher levels. In other words, stresses can never build up to values higher than the fractured rock can bear.

THE INFLUENCE OF PORE FLUID PRESSURE

The Classic Paper by Hubbert and Rubey

In a masterpiece article published in 1959, Hubbert and Rubey completely transformed the way in which geologists think about joints, shear fractures, and faults. Specifically, these authors disclosed the dramatic role of elevated pore fluid pressure in the formation and movement of fractures and faults. Hubbert and Rubey were aware that petroleum companies, in drilling for oil, routinely encounter highly elevated fluid pressures in rocks in many sedimentary basins in the world, such as the Gulf of Mexico. Petroleum explorationists had learned through everyday experience that sedimentary rocks in the subsurface are commonly marked by stratigraphically controlled zones of fluid pressure that are so highly elevated that they approach the value of the lithostatic stress produced by the load of the overlying sedimentary cover. In such settings, the fluid pressure essentially supports the weight of all the rock above.

Hubbert and Rubey were able to show that **high pore fluid pressure** (**P_f**) in rocks tends to offset the magnitude of lithostatic normal stress (σ_N) acting on the plane of faulting (or fracturing). They introduced the concept of **effective stress** (σ^*), which is the difference between normal stress and fluid pressure ($\sigma_N - P_f$). They then modified the Coulomb law of failure in a way that makes it clear that reducing the normal stress by fluid pressure also lowers the value of critical shear stress required to produce faulting.

The original Coulomb law of failure (Equation 5.3) looks like this:

$$\sigma_c = \sigma_0 + \tan\phi(\sigma_N)$$

When modified by Hubbert and Rubey (1959), it looks like this:

$$\sigma_c = \sigma_0 + \tan\phi(\sigma_N - P_f) \qquad (5.6)$$

$$\sigma_c = \sigma_0 + \tan\phi(\sigma_N^*)$$

The "simple" change was replacing normal stress with effective stress. But what a difference it makes! It reduces significantly the stress needed for failure. In fact, when effective stress becomes zero, which it does in certain circumstances, one entire term of the Mohr–Coulomb law of failure is eliminated:

$$\tan\phi(\sigma_N - P_f) = \tan\phi(0) = 0$$

What is left is the expression:

$$\sigma_c = \sigma_0$$

When this state exists, a fault or fracture can be formed simply by breaking the cohesion of the rock. And if the rock is already broken along a preexisting fracture, the job is even easier. The importance of effective stress, as emphasized decades ago by Hubbert and Rubey (1959), has been substantiated by exhaustive experimental testing. Time and again it has been demonstrated that rock samples fail by fracturing under significantly smaller loads when pore fluid pressure is added to the sample (Paterson, 1978; Tullis and Tullis, 1986).

Fluid Pressures in the Real World

There is plenty of evidence that high fluid pressures are achieved in the crust of the Earth, especially below several kilometers. Twiss and Moores (1992) point out that above 3 km, the fluid pressure is typically *normal,* equal to **hydrostatic pressure**. Flow is unrestricted. The magnitude of hydrostatic pressure (P_h) is the product of the density of the fluid (ρ_f), gravity (g), and height of the column of groundwater (h):

$$P_h = \rho_f g h \tag{5.7}$$

Deeper than 3 km, fluid pressure generally exceeds hydrostatic pressure because of the effects of compaction. During compaction, pore fluids are forced to occupy less and less space, and fluid pressures mount. Furthermore, because of the effect of the geothermal gradient, **aquathermal pressuring** develops, causing pore fluids to acquire greater pressure as they become hotter. This occurs because water has a higher coefficient of thermal expansion than sediment. Thus pore fluid pressure can build to a much greater value than the calculated extrapolation of hydrostatic pressure based on depth. Moreover, pore fluid pressure can rise to levels approximating, or even exceeding, the **lithostatic pressure** (i.e., the pressure derived from the weight of the entire column of overlying rock). Lithostatic pressure (P_l) is equal to the average density of the column of rock (ρ_r) multiplied by gravity (g) and depth (h):

$$P_l = \rho_r g h \tag{5.8}$$

Hubbert and Rubey introduced an expression for the **fluid pressure ratio (λ)** to describe the ratio of the pore fluid pressure (P_f) to the lithostatic pressure (P_l):

$$\lambda = \frac{P_f}{P_l} = \frac{P_f}{\rho_r g h} \tag{5.9}$$

When the fluid pressure is hydrostatic, and not elevated, λ values typically range from 0.37 to 0.47 (Suppe, 1985). Elevated fluid pressures are referred to as **abnormal fluid pressures**. Values of λ in regions of abnormal fluid pressure typically range from 0.5 to 0.9 (Suppe, 1985).

When pore fluid pressure is high, the magnitude of the second component of the modified Coulomb equation (Equation 5.6) is reduced substantially. And if the rock is already broken along an orientation favorable for reactivation, the need for the first component (a stress magnitude equal to cohesive strength) disappears altogether.

Fluid Pressure and the Development of Joints

The role of abnormal pore fluid pressure can be used to explain the paradoxical fact that tension joints can form at depths of 5 and 10 km and greater (!), within environments where the confining pressures created by overburden are strongly compressional. How can this happen??

Secor (1965) found the answer by thinking carefully about the role of pore fluid pressure. His solution is most graphically illustrated on a Mohr diagram containing a grand envelope of failure. Shown in Figure 5.49*A* is a stress circle that displays the values of greatest and least principal stress (σ_1 and σ_3) that might exist at depth in a region of weak horizontal compressive stress. Both σ_1 and σ_3 are compressive. The differential stress value ($\sigma_1 - \sigma_3$) is not great enough to cause failure by fracture, for the stress circle does not intercept the envelope of failure (see Figure 5.49*A*). If, however, the rocks at this depth possessed high pore fluid pressure, the state of stress would be radically modified. The respective values of σ_1 and σ_3 would *each* be reduced by the exact value of the fluid pressure (P_f): the **greatest principal effective stress (σ_1^*)** would equal $\sigma_1 - P_f$; and the **least principal effective stress (σ_3^*)** would equal $\sigma_3 - P_f$. When the effective stress levels of σ_1 and σ_3 are then plotted (Figure 5.49*A*), the stress circle shifts to the left toward the field of tensional stress. If the magnitude of fluid pressure is great enough, and the differential stress is small enough, the stress circle can be "driven" into collision with the tensile failure envelope (Figure 5.49*A*). As soon as σ_3 achieves the value of tensile strength of the rock, tensional joints form perpendicular to σ_3. Alternatively, if the differential stress is a little higher, the Mohr stress circle will intercept the parabolic failure envelope in the transitional tensional field (Figure 5.49*B*). Transitional tensional fractures then form at low angles to σ_1.

In stress fields of high differential stress, the gradual elevation of fluid pressure will drive the stress circle into collision with the Coulomb failure envelope in the compressional field (Figure 5.49*C*). In this situation, shear fractures form, and they form at lower stress levels than would normally be expected.

When fluid pressure builds sufficiently to trigger fracturing, the fluid pressure may be immediately dissipated, and the original magnitudes of σ_1 and σ_3 may be reestablished. This condition is represented on the Mohr diagram by a "rebounding" of the differential stress circle from left to right (Figure 5.49*D*). Fluids then move from pores in the rock into fractures, where the pressure builds again, and, POW!! The fracture is propagated further. Bahat and Engelder (1984) have suggested that the rhythmic plume patterns, with repeated fans of hackles, reflects fracture driven by fluid pressure (Figures 5.49*E* and *F*). The "drive" suddenly ceases when the crack has propagated a short distance. The fluid pressure falls and tensile stress disappears . . . that is, until fluid pressure builds again for another hit.

Gelatin–Hydrofrac Experiment

Nick Nickelsen (personal communication, July 1993) described to me an experiment that brings to life the role of fluid pressure in jointing. He ran the experiment at Bucknell University for years. It is an experiment originally devised by M. King Hubbert to demonstrate that **hydrofracs** (i.e., mode I joints that develop in the presence of elevated fluid pressures) form perpendicular to σ_3, the direction of least principal stress. It is an amazing experiment!

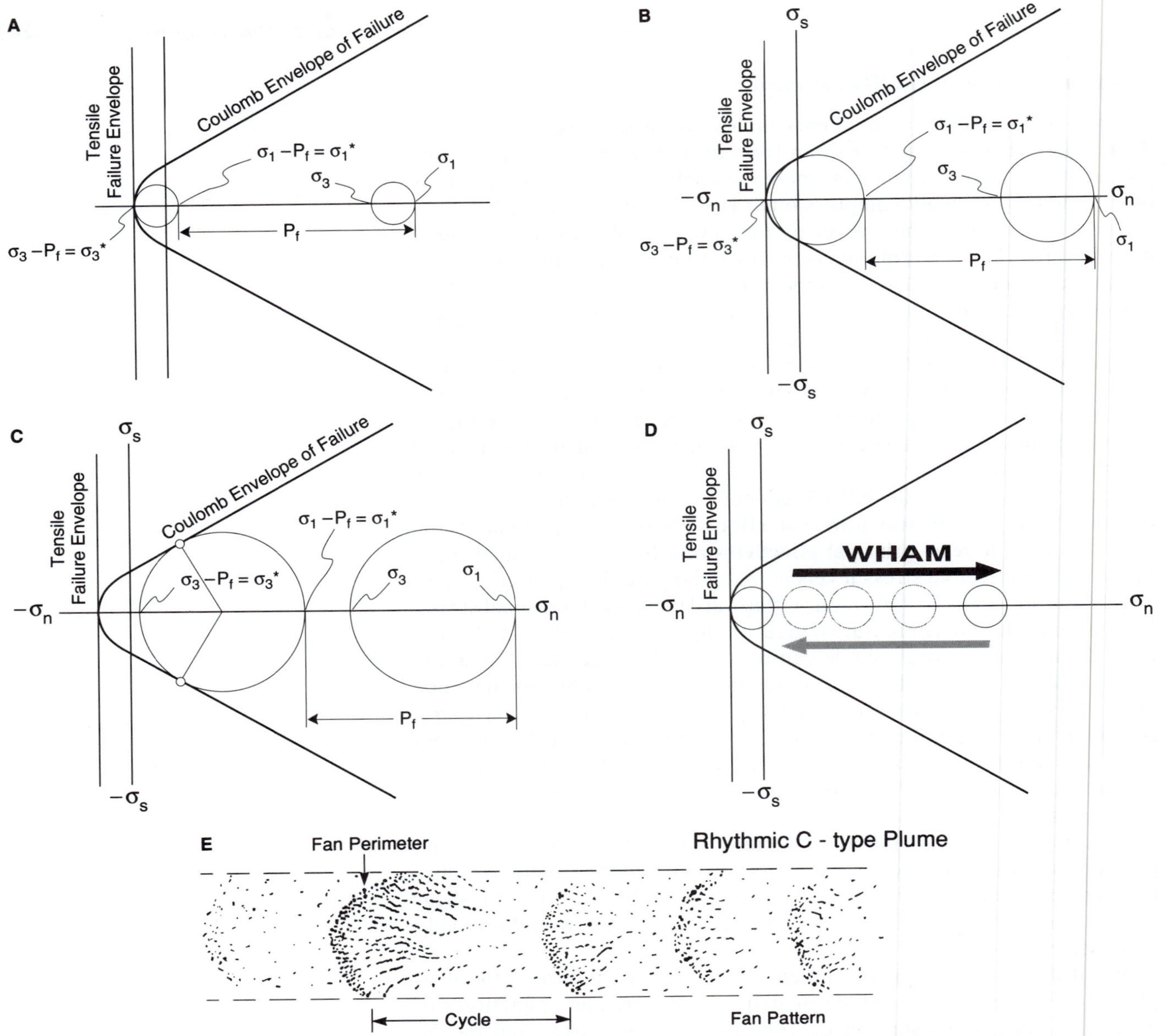

Figure 5.49 Secor's (1965) model of the role of fluid pressure in jointing. (*A*) We start with greatest (σ_1) and least (σ_3) principal stresses being compressive, and with low differential stress ($\sigma_1 - \sigma_3$). When fluid pressure (P_f) is raised, σ_1 and σ_3 are reduced by the same amount (P_f). Because differential stress was very small to begin with, the stress circle passes all the way to the tensile failure envelope, and the rock breaks by mode I failure. (*B*) In this case differential stress is greater, and when fluid pressure is raised, the stress circle crashes into the parabolic failure envelope and the rock breaks through transitional tensile failure. (*C*) If differential stress is yet greater, and fluid pressure is raised, the rock will break according to Coulomb's law of failure. (*D*) Repeated tensile failure in a low differential stress field is achieved by raising the fluid pressure (circle moves left slowly), cracking the rock in tension and releasing the fluid pressure (quick move of the circle to the right!), followed again by raising the fluid pressure. (*E*) Rhythmic plumose ornamentation of joints at Watkins Glen, New York, interpreted by Bahat and Engelder (1984) to be the expression of sequential cracking responding to the rhythms of fluid-pressure buildup and fracture-induced release. [Reprinted with permission from *Tectonophysics,* vol. 104, Bahat, D., and Engelder, T., Surface morphology on cross-joints of the Appalachian Plateau (1984), Elsevier Sciences B.V., Amsterdam, The Netherlands.] (*F*) Rhythmic plumose ornamentation on joint in siltstone within the Devonian Genesee Group, New York. The cyclic nature of this joint is accented by rubbing chalk on the surface of the joint. Stacy Loewy is doing the chalking. (Photograph kindly provided by Terry Engelder.)

Buy some Knox Gelatin at the grocery store, mix up a batch, and pour it into a transparent plastic liter bottle (e.g., mineral water bottle). When the gelatin has set up, apply a clamp (a bent piece of metal will do) to the middle of the bottle to create a stress field (Figure 5.50*A*). The effect of the clamp will be visible distortion of the bottle from a circular to elliptical shape, as viewed in cross section. Now stick a thin glass tube (e.g., a pipette tube), or a thin straw, vertically into the gelatin, such that the bottom of the tube (i.e., the bottom of the drill hole) is at the depth level of the clamp. Next, load a syringe with a slurry of plaster of paris and forcefully inject the plaster of paris down the tube into the gelatin. Immediately a disk-shaped mode I joint (i.e., **hydrofrac**) will form perpendicular to the direction of least principal stress. The location and size of the hydrofrac will be obvious for it will be filled with plaster of paris. Allow the plaster of paris to harden completely, and then cut away the bottle, reach down into the gelatin, and carefully scoop out the filled fracture.

What you will extract is a joint disk, perhaps complete with fringe joints, a perfect image of the three-dimensional geometry of a joint and its relationship to principal stress directions (Figure 5.50*C*).

Fluid Pressure and the Development of Veins

The undeniably important role of fluid pressure in the opening of joints is preserved in certain veins, including **crystal fiber veins**. As the term implies, crystal fiber veins are marked by elongate fiberlike or needlelike aggregates of fine-grained minerals, commonly quartz and calcite (Figure 5.51) (Durney and Ramsay, 1973; Urai, Williams, and Van Roermund, 1991). The fibers may be straight or systematically curved (Figure 5.52). Where straight, they are not necessarily perpendicular to wall rock surfaces. Durney and Ramsay (1973) have shown that the formation of crystal fibers in veins is due to an opening of the walls of the fracture or joint accompanied by simultaneous vein filling, without the actual development of open space. Elevated fluid pressures contribute to the opening of the fractures.

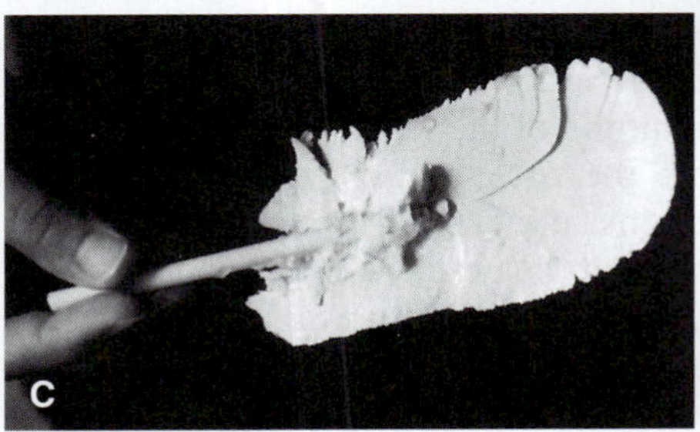

Figure 5.50 The Knox Gelatin "hydrofrac" experiment, as taught to us by Nick Nickelsen. (*A*) Knox Gelatin inside a clamped plastic seltzer bottle. A glass tube, suspended by a piece of cardboard resting on top of the bottle, was inserted into the gelatin while it was solidifying. A fluid slurry of plaster of paris is about to be injected down the tube. (*B*) Injection of the slurry creates a mode I fracture, which fills immediately with the plaster of paris. (*C*) After the plaster of paris dries, the tube and the hardened joint filling can be extracted. [Experiment by Angela Smith. Photographs by Brian Bennon and Brook Riley.]

Figure 5.51 An assortment of crystal fiber veins. Crystal fiber calcite vein in dark limestone (back, right). Although very slightly curved, the fibers are essentially perpendicular to the walls of the vein. Mode I all the way. Crystal fiber quartz vein in sandstone (middle, left). Fibers are nearly perpendicular, but not quite, to the walls of the vein. Calcite crystal fiber veins in siltstone (front, center). [Photograph and copyright © by Peter Kresan.]

A

B

C

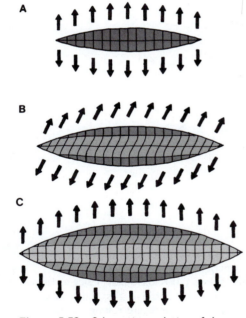

Figure 5.52 Schematic rendering of the development of curved and kinky crystal fiber veins as a response to changes in the direction of fracture opening. (*A*) First increment reflects pure mode I opening. (*B*) Second increment reflects transitional tensile behavior, i.e., opening plus shear. (*C*) Final record "maps out" the kinematic history of the opening of the fracture within which the vein is deposited.

Crystals are precipitated from the fluids during step-by-step incremental dilations of the joint. The final product, the crystal fiber vein, reflects the total movement accomplished by the dilation. In many veins the crystals grow perpendicular to the walls of the joint, reflecting a pure mode I opening (Figure 5.52*A*). But where the opening is accompanied by shear components, the crystal fibers display orientations that are oblique to the vein walls (Figure 5.52*B*). The growth of the crystal fibers is ultrasensitive to changes in the direction of progressive opening of the vein, and thus the final patterns may be strikingly curved (Figure 5.52*C*). During each incremental step, the crystal fibers grow in the direction of differential displacement, which corresponds to the direction of least principal stress. With changing directions of differential displacement, the crystal fibers take on curved forms. The curved forms "map out" the history of incremental movement along the fracture. Where the youngest crystals are added incrementally along the vein walls, the movement history is read from the inside out (**antitaxial veins**) (see Figure 5.53*A*). Where the youngest crystals are added incrementally along the medial line, the center of the vein, the movement history is read from the outside in (**syntaxial veins**) (Figure 5.53*B*).

Some veins are marked by clear evidence of repeated, recurrent fracturing followed in each case by precipitation of new vein mineralization. Such veins are called **crack–seal veins** (Ramsay, 1980). The presence of the fractures and the fibers may or may not be recognizable by the naked eye. For example, Ramsay (1980) described a crystal fiber vein composed of 11 small veinlets, each approximately 45 μm wide, which form what would appear to the naked eye to be a single white vein 0.5 mm thick (Figure 5.54). Upon formation of the initial crack and initial vein, new cracks formed again and again at the vein–rock contact, and after each cracking event they were immediately sealed by a new increment of veining. Such repeated cracking typically leaves a trail of wall rock inclusions in the veins, tiny slivers of rock that were broken off at the time of cracking.

The cracks form through **hydraulic fracturing**. Fluid pressure builds, decreasing effective stress until the magnitude of the least principal stress σ_3 matches the tensile strength (T_0) of the rock and/or vein material. After the crack has formed, and as it opens, mineral fibers are deposited incrementally from the fluids moving through the cracked rock. The crack eventually becomes completely sealed, only to be broken again when fluid pressures build to sufficient magnitude to crack the rock again . . . and again . . . and again.

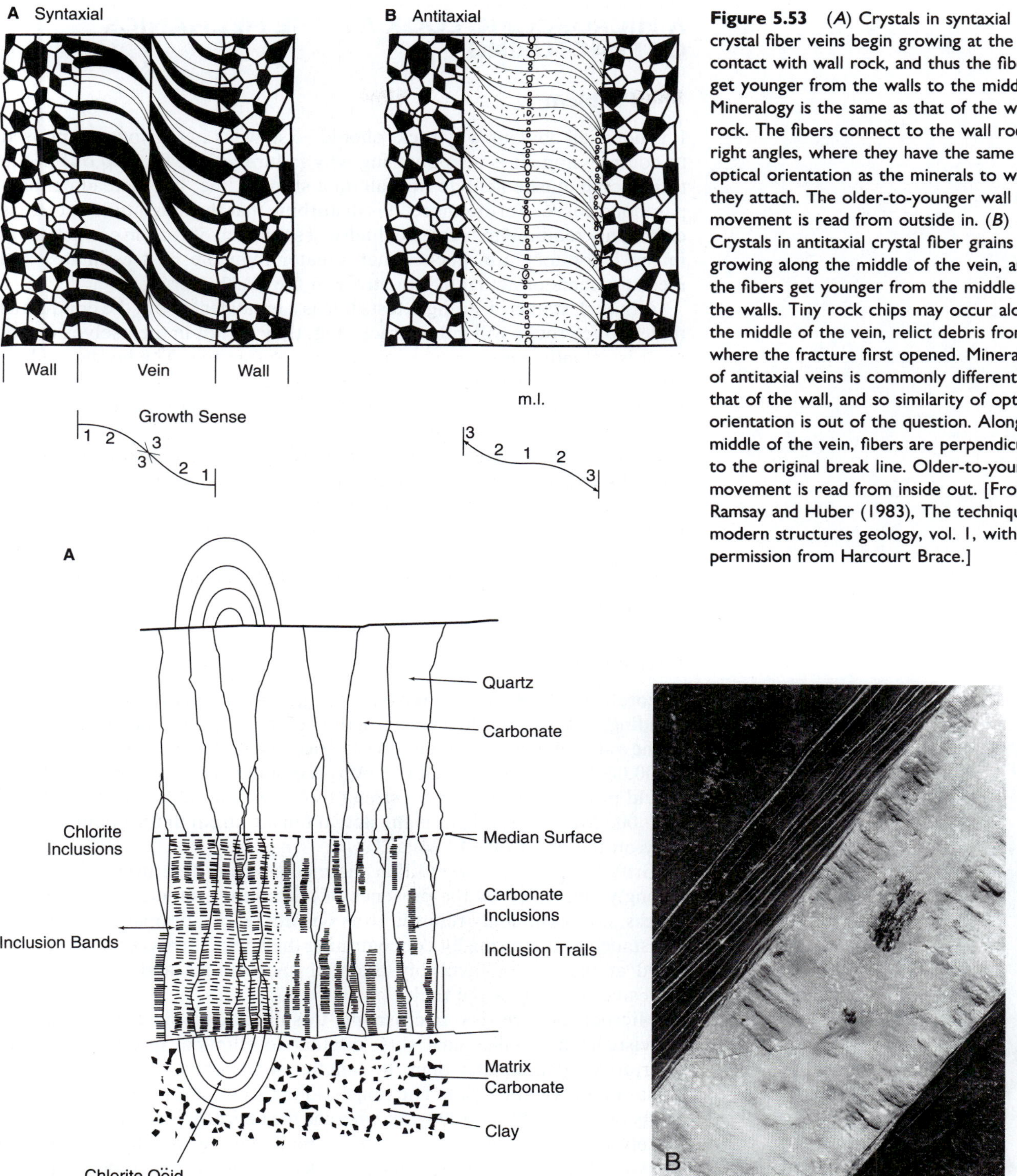

Figure 5.53 (*A*) Crystals in syntaxial crystal fiber veins begin growing at the contact with wall rock, and thus the fibers get younger from the walls to the middle. Mineralogy is the same as that of the wall rock. The fibers connect to the wall rock at right angles, where they have the same optical orientation as the minerals to which they attach. The older-to-younger wall rock movement is read from outside in. (*B*) Crystals in antitaxial crystal fiber grains begin growing along the middle of the vein, and so the fibers get younger from the middle to the walls. Tiny rock chips may occur along the middle of the vein, relict debris from where the fracture first opened. Mineralogy of antitaxial veins is commonly different than that of the wall, and so similarity of optical orientation is out of the question. Along the middle of the vein, fibers are perpendicular to the original break line. Older-to-younger movement is read from inside out. [From Ramsay and Huber (1983), The techniques of modern structures geology, vol. 1, with permission from Harcourt Brace.]

Figure 5.54 (*A*) The crack–seal mechanism created this crystal fiber vein, which as can be seen microscopically is composed of 11 tiny veinlets. Time and again this part of the rock was cracked and sealed. (Reprinted with permission from *Nature,* vol. 284, Ramsay, J., The crack-seal mechanism of rock deformation. Copyright © 1980 Macmillan Magazines Limited.) (*B*) Hand-specimen scale expression of crack–seal mechanism in limestone collected near Burlington, Vermont. Note the ultrathin crack–seal veins adjacent to margin of the large crystal fiber vein of calcite. Width of field of view is 7 cm. (Photograph and copyright © by Peter Kresan.)

A MICROSCOPIC LOOK AT THE MECHANICS OF FRACTURING

The Problem

It is not at all obvious why rocks should be so weak in tension, as opposed to compression. It is also puzzling why failure values for (σ_N,σ_S) in the transitional tensile field do not fall on a straight line, but instead define a parabolic curve. Furthermore, it is disturbing that the magnitudes of stress commonly required to break a cylindrical specimen of rock in compression in the laboratory are much, much greater than the best estimates and measurements of actual in situ stress in the Earth's crust.

Although standard strength tests tell us a great deal about the conditions under which rocks will fail by fracturing, they do not in themselves reveal what is actually happening to cause rocks to fracture. The laws of failure are *descriptions* of what is observed, *not explanations* of the actual fracture mechanics. Consequently we must move to a yet finer scale to determine what may really be happening.

In particular we need to look at **stress concentrators**, imperfections that have the effect of raising the stress at a ''point'' to values that far exceed background levels. A hole would be an example, such as the axial canal of the crinoid stem shown in Figure 5.55. This opening concentrated the background stress sufficiently to cause mechanical twinning of the adjacent calcite. Yet the stress concentrators that concern us the most are at an even finer scale.

Microcracks

Theoretical calculations of tensile strength of rock, based on interatomic bonding, suggest that tensile strength should be approximately one-tenth of the value of Young's modulus (E) (Price and Cosgrove, 1990). Choosing −100,000 MPa as a reasonable Young's modulus for a common rock, we would predict that the tensile strength of rock should be on the order of −10,000 MPa. Why, then, is the actual tensile strength for common rock only on the order of −10 MPa??

Figure 5.55 Slightly deformed crinoid stem, viewed microscopically and in cross section. The circular opening of the crinoid columnal concentrated background stress enough to cause the mechanical twinning of the calcite. (This amazing photomicrograph is courtesy of Terry Engelder.)

Griffith (1924) discovered that the mechanics of fracturing of rocks is strongly controlled by the presence of microscopic cracks. These **microcracks**, now commonly referred to as **Griffith cracks**, dramatically weaken resistance to fracturing by concentrating the stress. Stress concentrations build at the tips of favorably oriented microcracks and overcome the cohesive bonding of the rock.

Microcracks are tiny microscopic-to-submicroscopic cracks that can be visualized as ellipsoidal slits, circular to elliptical in plan, with thin apertures much smaller than the lengths of the cracks (Engelder, 1987; Pollard and Aydin, 1988). The longest dimension of a microcrack is normally between 100 and 1000 μm, whereas the aperture is normally approximately 1 μm wide. Microcracks are so small that they occur wholly within individual grains as **intragranular cracks**, although sometimes they are larger and extend through a number of grains as **transgranular cracks** or along the boundaries between grains as **intergranular cracks** (Richter, 1976; Lloyd and Knipe, 1992). Microcracks do not typically display a preferred orientation; instead, they are generally randomly oriented.

The distribution of microcracks in rocks varies quite a bit. Igneous rocks are marked by a rather uniform distribution, reflecting homogeneous cooling and contraction that affected the whole body. Most rocks, whether igneous or sedimentary or metamorphic, are apt to show an increase in density of microcracking in the vicinity of major local structures, especially faults (Engelder, 1987).

Amplification of Stress at the Tips of Microcracks

Microcracks indeed are amazing stress concentrators. Stresses become amplified by many orders of magnitude at the tips of the cracks. The amplification may create stresses of sufficient magnitude to overcome the strength of the interatomic bonding of the minerals in the rock. If this happens, the microcrack propagates. As a result of this mechanical reality, modest levels of **remote stress** can be amplified to create **local stress** of sufficient magnitude to create a joint. "Remote" is used here in reference to stresses operating from a distance, to be distinguished from "local" stresses that become concentrated at or near the tip of a microcrack (Pollard and Aydin, 1988).

Griffith was able to demonstrate that modest remote tensile stresses are greatly amplified to higher magnitude tensile stresses at the tips of microcracks. Moreover, he showed that *tensile* stresses develop at the tips of microcracks even when the remote stress is compressive (Price and Cosgrove, 1990). Griffith's work led to the discovery that both joints and shear fractures ultimately originate at tensional stress sites at the tips of microcracks.

The amplification of stresses at tips of microcracks depends on a number of factors, including orientation and the length and width of each crack (Suppe, 1985). If the remote stress (σ_3) is tensional and the microcrack is oriented at right angles to the direction of the remote tensional stress, the stress (σ_t) at the tip of a microcrack can be calculated as follows:

$$\sigma_t = \frac{2}{3}\sigma_3\left(\frac{a^2}{b}\right) \tag{5.10}$$

where a is the length of microcrack and b is its aperture

For example, picture a microcrack 100 μm long, having an aperture of 1 μm (Figure 5.56). Assume that it is oriented at right angles to a remote tensional stress (σ_3) whose magnitude is -2 MPa. The local stress (σ_t) that will develop at each tip of the microcrack is determined by Equation 5.10:

$$\sigma_t = \frac{2}{3}(-2\text{ MPa}) \times \frac{100^2}{1}$$

$$\sigma_t = -13{,}000\text{ MPa}$$

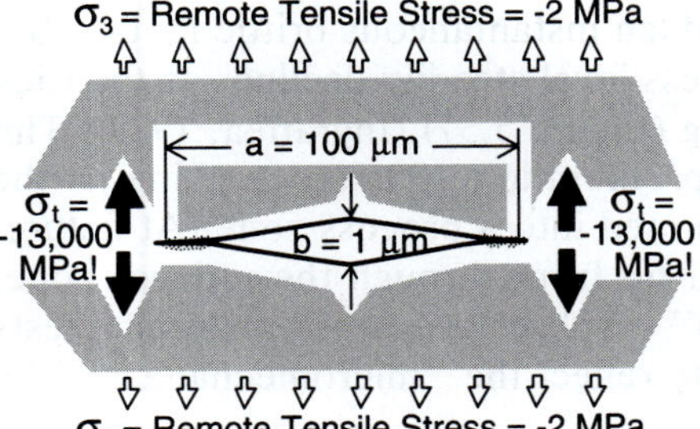

Figure 5.56 Even when "remote" tensile stresses (σ_3) are very, very low, huge tensional tip stresses (σ_t) can form at the tips of microcracks.

This value far exceeds the tensile stress required to fracture the rock beyond the tip of the crack and propagate a joint. Moreover, if the microcrack is marked by elevated pore fluid pressure, even a lower magnitude of remote stress is sufficient to propagate a mode I joint.

Critical Orientation of Microcracks

Tensile stresses are greatest at the tips of microcracks that are most favorably oriented with respect to the principal stresses. Thus within a family of microcracks in a rock body, there will be certain favorably oriented microcracks that will nucleate through-going fractures. The most favorable orientation, (θ), is related to the values of the principal stresses in the following way:

$$\cos 2\theta = \frac{\sigma_1 - \sigma_3}{2(\sigma_1 + \sigma_3)} \tag{5.11}$$

Consistent with this relationship, microcracks oriented perpendicular to the direction of least principal stress (σ_3) are characterized by "tip stresses" larger than those that develop at the tips of microcracks of all other orientations. We know from standard tensile strength tests that a rock will fail by mode I fracturing when $\sigma_3 = T_0$ and $\sigma_1 = 3T_0$. If we plug these values into Equation 5.11 we find that the preferred orientation for microcracks formed in tension is parallel to σ_1 and perpendicular to σ_3:

$$\cos 2\theta = \frac{3T_0 - (-T_0)}{2(3T_0 + (-T_0))} = \frac{4T_0}{4} = 1$$

$$2\theta = 0$$

Stress–Strain Relationships at the Scale of Microcracks

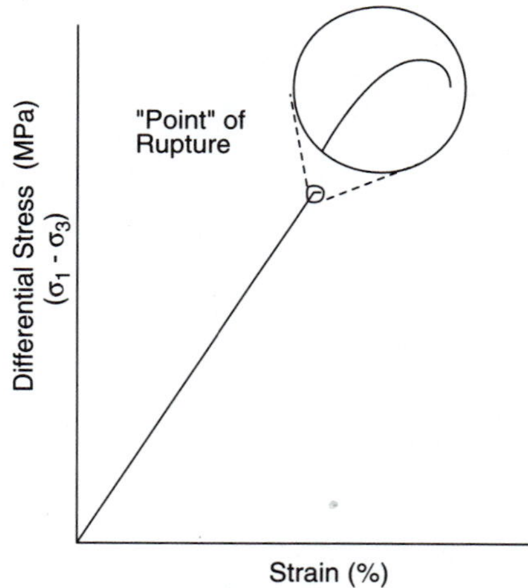

Figure 5.57 The relationship between stress and strain at the tip of a microcrack. This is what is happening at the "point" of rupture. Stress–strain curves can give the impression of quick snap, without plastic deformation. In reality, the failure follows inelastic deformation within the process zone.

At the scale of laboratory testing of cores of rocks, we conclude that jointing is purely an elastic–brittle phenomenon: When the tensile strength of an elastically deforming rock is exceeded, the rock instantaneously breaks in tension. This is not the case at the scale of the tips of microcracks (Ingraffea, 1987). Rock adjacent to the microcrack is stress free. Just beyond the tip of a microcrack, a **process zone** forms as the tip stresses build. Within the process zone, a swarm of microcracking develops as the crack comes closer and closer to propagating. Because of the dense microcracking, the rock behaves inelastically within the process zone, and the propagation of the crack to form a fracture proves to be an elastic–plastic process at this scale of observation. Beyond the process zone, the rock suffers some strain, but the behavior is strictly elastic.

If we were to build a stress–strain curve describing the behavior of rock just beyond the tip of a microcrack, it would display a period of elastic behavior followed *not* by sudden instantaneous brittle failure, but by **strain softening**, in which the stress level steadily declines as the rock moves swiftly to failure by cracking (Figure 5.57) (Ingraffea, 1987). The strain softening would reflect the proliferation of microcracking as the area beyond the tip became transformed into a process zone. At failure, the crack extends from the original tip front through the process zone. Thus the elastic–brittle "quick snap" we observe in tensile–strength tests in the laboratory does not accurately reflect the "micromechanisms" by which the joint comes to life.

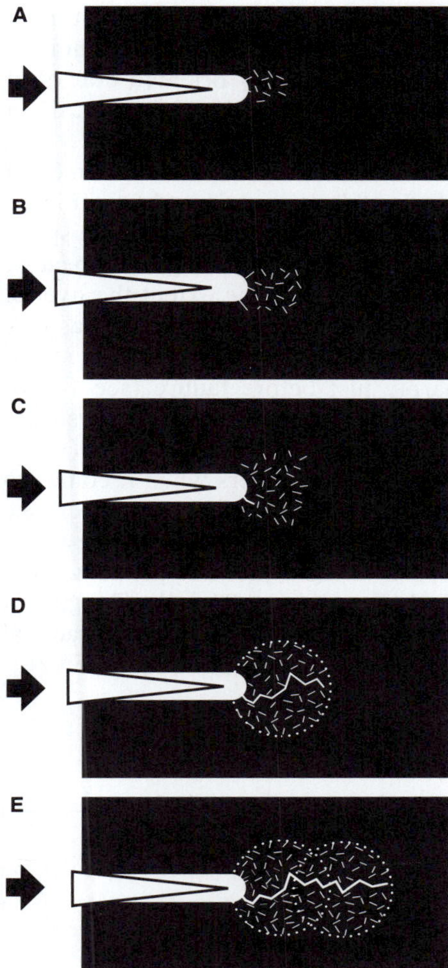

Figure 5.58 The slot experiment. (*A*) A slot is cut into rock, and, figuratively speaking, a wedge is driven into it. (*B–C*) The outward tensional pressures amplify tensional stresses at the tip of the slot. Microcracking proliferates in the process zone. (*D–E*) Eventually, certain favorably oriented microcracks link up in an incipient mode I fracture. [After Atkinson (1987), Introduction to fracture mechanics and its geophysical applications.

Relation of Microcracks to Jointing

As failure stress is approached, the cloud of microcracks developing in the rock becomes more localized and dense; and when shear failure occurs, the surface is really the coalescence of hundreds and hundreds of microcracks (Kranz, 1983). Atkinson (1987) describes a classic experiment analogous to the ripping of a bedsheet, but it is done in rock. A slot is cut in rock, and then outward-directed pressure is placed on the walls of the slot, almost like driving a wedge into the notch, thus placing the tip of the slot in tension (Figure 5.58*A*). At first, isolated microcracks form beyond the slot, and the material behaves elastically. With further loading, a cloud of microcracks forms a process zone and the solid behaves in a nonlinear manner (Figure 5.58*B*,*C*). Finally, a crack forms by the linking up of microcracks (Figure 5.58*D*,*E*). The crack is the incipient joint.

An analogous process takes place when the rock is placed in compression. Scholz (1968a,d) subjected rock samples to compressional deformation and at the same time ''listened in'' to the acoustical expressions of high frequency elastic vibrations emitted by the development and propagation of microcracks leading to shear fracturing. He showed that just before failure, there is a rapid increase in microfracturing frequency (Figure 5.59). Furthermore, Scholz (1968a,d) and Lockner and Byerlee (1977a,c), using special detectors, were able to ''map'' the three-dimensional locations of microcracking within the deforming specimens. Their data show that shear fractures form out of a complex traffic pattern in which thousands of microcracks grow and interfere and interact. Out of the chaos emerges a reasonably discrete ''surface'' marked by the coalescence and linking up of microcracks of various orientations (Figure 5.60). The compressional

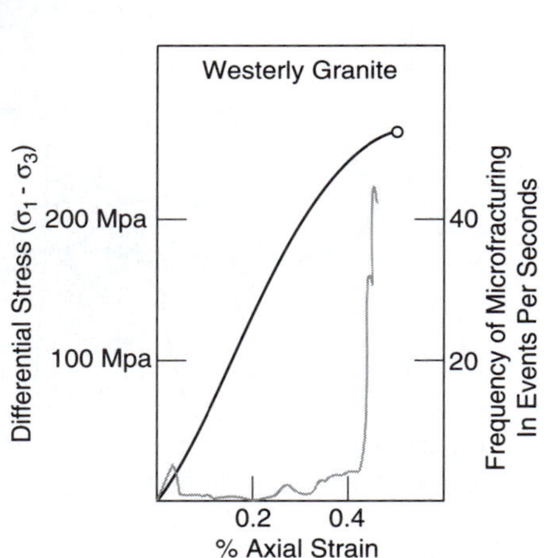

Figure 5.59 Listening in on high frequency elastic vibrations, Scholz (1968b) discovered that just prior to ultimate failure by brittle fracture there is a tremendous increase in microfracturing events. Here are the data: a typical stress–strain curve for the Westerly Granite, upon which is superposed the record of microfracturing events per second. (From C. H. Scholz, *Journal of Geophysical Research,* v. 73, fig. 6, p. 1423, copyright © 1968 by American Geophysical Union.)

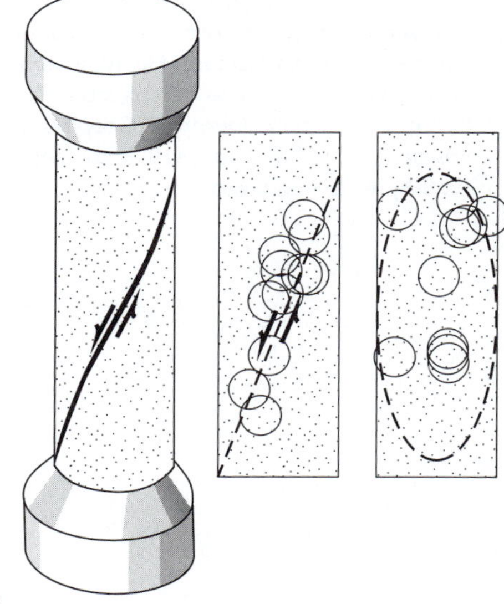

Figure 5.60 Using special acoustical detectors, Scholz (1968a) mapped the sites of prefailure microcracking and confirmed a clear geometric relationship between the distribution of microcracks and the orientation and location of the through-going fracture that forms at failure. (From C. H. Scholz, *Journal of Geophysical Research,* v. 73, fig. 4, p. 1452, copyright © 1968a by American Geophysical Union.)

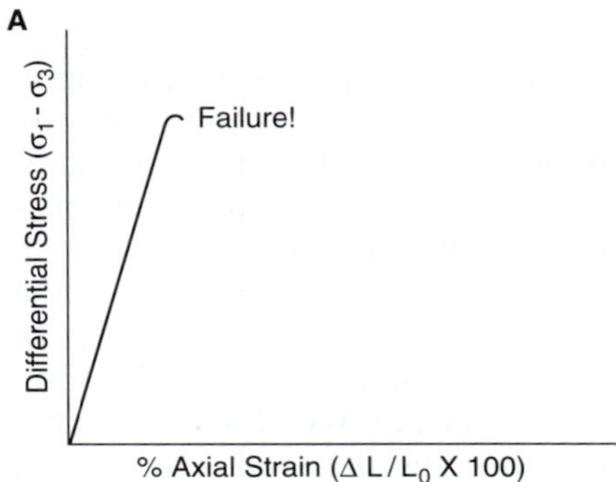

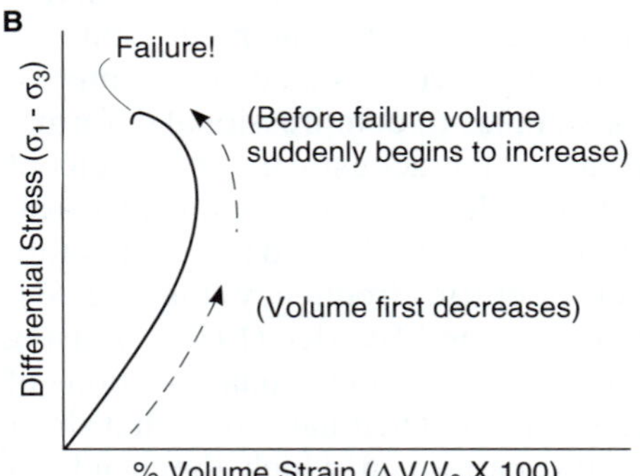

Figure 5.61 (*A*) Normally we just keep track of differential stress and percent strain during a compressive strength test. (*B*) But Scholz (1968b), who was interested in studying dilatancy during deformation, tracked volume change as well. He discovered that just before failure, volume suddenly begins to increase. He attributed such volume increases to microfracturing. (From C. H. Scholz, *Journal of Geophysical Research,* v. 73, fig. 1, p. 1418, copyright © 1968 by American Geophysical Union.)

confining pressure initially closes the microcracks. But with continued loading, crack growth begins and the cracks open up. Mode I opening at the tips of the cracks makes it possible for suitably oriented cracks to accommodate shear. The ultimate shear fracture is, at the microscale, a combination of mode I cracks and shear cracks.

The closing and opening of the microcracks is revealed in experimental tests that track both changes in length and changes in volume (Figure 5.61) (Scholz, 1968d). Specimen volume initially decreases as pore space and microcracks are closed up (Figure 5.61). When crack growth begins, the attendant volume increase tends to offset to some extent the overall specimen volume reduction accompanying shortening. However, as failure is approached, and as clouds of process zones develop, overall specimen dilation occurs, increasing sharply just before failure (see Figure 5.61). Dilation as a precursor to earthquakes figures prominently in strategies aimed at accurate forecasting of destructive earthquakes.

As it turns out, some rocks are *tougher* than others when it comes to propagation of microcracks to form fractures. It is a little like the difference between trying to rip a bedsheet versus trying to rip a pair of Levi's.™ **Fracture toughness** (*K*) varies from rock to rock, just like the toughness of a bedsheet is much different from the toughness of Levi's. Toughness (*K*) is dependent on a whole host of variables, but the essence and practical importance of fracture toughness can be easily imagined (Figure 5.62).

Figure 5.62 Application of fracture toughness (*K*) while flying cross-country. It applies to the wings of the airplane as well.

A SIMULATION OF THE FORMATION OF JOINTS IN NATURE

The Necessary Loading Conditions

The formation of joints requires creating situations in which the magnitude of the effective stress acting parallel to the least principal direction (σ_3^*) attains the tensile strength (T_0) of the rock. Differential stress will be less than $4 \times T_0$. The formation of transitional tensile fractures requires that the effective least principal stress (σ_3^*) is tensile, the effective greatest principal stress (σ_1^*) is compressive, and the differential stress ($\sigma_1^* - \sigma_3^*$) is of sufficient magnitude to intersect the Griffith parabolic fracture envelope, that is, $>4T_0$. The formation of shear fractures requires that the greatest effective principal stresses (σ_1^*) be compressive and the differential stress ($\sigma_1^* - \sigma_3^*$) large enough to intercept the Coulomb envelope.

The geologic processes that can meet some or all of these conditions include burial and uplift, heating and cooling, and tectonic shortening and stretching. These correspond to gravitational loading (and unloading), thermal loading (and unloading), and tectonic loading (and unloading). When we examine the stress conditions that develop in sedimentary rocks during burial and heating followed by uplift and cooling, we learn that jointing is, very simply, inevitable. Moreover, when fluid pressure is high, the rocks do not have a chance. No wonder jointing is so pervasive, even on a regional scale.

Suppe (1985), Engelder (1985), and Twiss and Moores (1992) have laid the groundwork for thinking about what happens mechanically to rock as it is buried and then uplifted. We have built on their approach and have done the calculations, both going down and coming back up again.

Picturing the Geologic Conditions

Let us imagine a thick sequence of sand deposited in a marine nearshore environment at the surface of the Earth. With continued sedimentation, it becomes buried. At a depth of approximately 1.5 km, the sand becomes reasonably lithified, forming sandstone. The sedimentary deposit heats up as it goes down; we are assuming that the geothermal gradient is 25°C/km.

Fluid pressure is an important factor in the environment we are envisioning. The fluid pressure ratio ($\lambda = P_f/P_r$) has a value of 0.4 between 0 and 3 km, but it steps up to 0.7 between 3 and 5 km.

The actual mechanical properties of the sand and sandstone can be described through Young's modulus E and Poisson's ratio υ. Values for sand are different than values for sandstone.

As the sand (or sandstone) is buried and gravitationally loaded, it will tend to expand horizontally because of the Poisson effect. When such expansion is prevented by lateral boundary constraints, horizontal stresses will build even in the absence of horizontal tectonic stresses. Furthermore, as the sand (or sandstone) is heated, it will tend to expand. But if this is impossible because of the confinement, additional internal compressive stresses will build.

Conversely, when the sandstone is uplifted toward the surface, it will tend to contract horizontally because of the Poisson effect and because of the cooling. Fluid pressures, at all times, will operate against the effect of vertical and horizontal compressive stresses.

When the sandstone is fully buried to a depth of 5 km, we will then subject the sandstone to a horizontal compressive tectonic stress of 20 MPa. This horizontal compressive stress will be sufficient to create shear fractures (and other structures). We assume that the deformation will disrupt the sandstone in ways that dissipate the elevated fluid pressure.

After uplift of the sandstone from 5 km to 3 km, we will relax the lateral boundary conditions ever so slightly. From the depth of 3 km to the surface, we will thus permit the sandstone layer to extend (stretch) by a tiny amount, that is, by an extension e of 0.001 per kilometer, or 0.003 in total. In other words, in the final phase of uplift ascent, the sandstone will be permitted to extend itself horizontally by three-tenths of one percent.

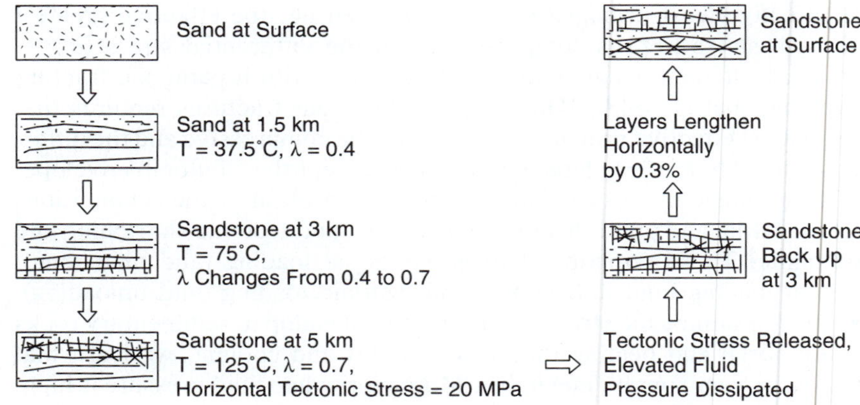

Figure 5.63 Imagining the conditions of burial and uplift, and when and where joints might form.

The array of conditions that we visualize during the burial, the tectonic loading, and the uplift are summarized schematically in Figure 5.63. Now let us see what may happen.

The Calculations

Our goal is to compute the greatest effective principal stress σ_1^* and the least effective principal stress σ_3^* on the basis of the information given. The greatest effective principal stress proves to be vertical in this example ($\sigma_1^* = \sigma_v^*$), whereas the least effective principal stress is consistently horizontal ($\sigma_3^* = \sigma_h^*$). The effective vertical stress is calculated using the now familiar equation:

$$\sigma_v^* = \rho g h - P_f$$

where σ_v^* = effective vertical stress
ρ = density of overlying column
g = gravitational constant
h = height of rock column (i.e., depth)
P_f = fluid pressure

The equation that permits us to calculate horizontal stress is provided by Engelder (1985b) and Twiss and Moores (1992):

$$\sigma_h^* = \left[\left(\frac{v}{1-v}\right)\sigma_v - \left(\frac{E}{1-v}\right)\alpha \Delta T\right] - P_f$$

where σ_h^* = effective horizontal stress
v = Poisson's ratio
σ_v = vertical stress
E = Young's modulus
α = coefficient of thermal expansion
ΔT = change in temperature

Display of the Calculations

Burial

We calculated horizontal and vertical stress at 0.5 km intervals during burial and uplift. We begin with variations in σ_v^* and σ_h^* during burial

(Figure 5.64*A*,*B*). The vertical stress is compressive throughout the burial process, increasing from 0 MPa to 48 MPa from the surface to a depth of 3 km. At the 3 km level, fluid pressure suddenly increases, significantly, and this rise is reflected in a sudden drop of the vertical stress from 48 MPa to 22 MPa (Figure 5.64*C*). Between 3 and 5 km, the vertical effective compressive stress increases steadily to 40 MPa.

Also shown in Figure 5.64*A–C* is the "behavior" of horizontal effective stress, which happens to be the least principal stress. During the burial phase, σ_h^* decreases steadily from 0 MPa to −5 MPa from the surface to 1.5 km (Figure 5.64*A*), where the sand turns to sandstone. The conversion of sand to sandstone imparts a stiffness that permits the effective horizontal stress σ_h^* to rise to a 4 MPa compressive stress at the 1.5 km depth, increasing steadily to 9 MPa between 1.5 and 3 km (Figure 5.64*B*). There, the sudden increase in fluid pressure results in an immediate drop in σ_h^* to −14 MPa (Figure 5.64*C*), and it, theoretically, continues to drop to −25 MPa during burial to 5 km.

Jointing *must* take place during the burial, and this becomes clear in Mohr diagram portrayals of the stress conditions. Notice that the sand is

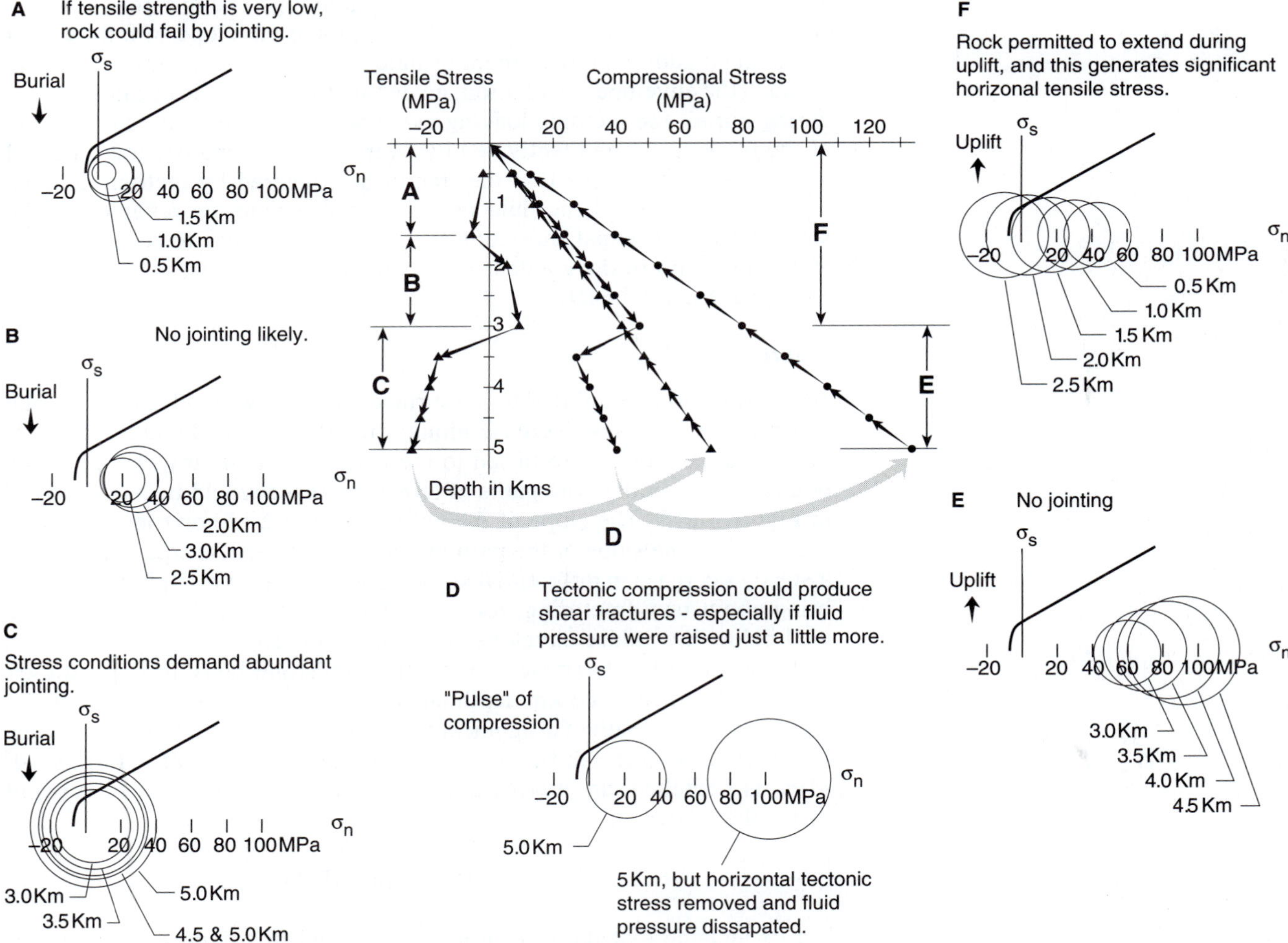

Figure 5.64 Greatest principal stress (σ_1) and least principal stress (σ_3) are calculated and plotted every step along the way. In turn, the stress data are plotted on Mohr diagrams complete with failure envelopes, to try to picture when and where fracturing is most likely to occur.

likely to fail in tension (mode I) at shallow levels when it begins to become firm, because horizontal stresses are tensile, and the tensile strength of the sand is very low (Figure 5.64*A*). In the depth range of 1.5–3 km, however, jointing is not favored; the differential stress circles lie well below the "grand" envelope of failure (see Figure 5.64*B*).

Enter fluid pressure: when the depth level of 3 km is reached, and with it an environment of elevated fluid pressure, the sandstone fails in transitional tensile fracturing (Figure 5.64*C*). The Griffith parabolic failure envelope is punctured. Because of the fracturing, the horizontal tensile stresses calculated for this depth range and shown in Figure 5.63*C* can never be reached, for the stress is relieved by the fracturing.

Tectonic Loading

When, and only when, the sandstone has reached a depth of 5 km, we call upon 20 MPa of tectonic loading to compress the rock horizontally. The values for the calculated horizontal and vertical effective stresses are shown in Figure 5.64*D*. The vertical stress does not change from the level achieved by gravitational loading alone. The horizontal stress increases by 20 MPa. The new state of stress, as pictured in Figure 5.64*D*, is sufficient to crack the rock by shear fracturing and faulting. The differential stress circle lies "dangerously" close to the Coulomb envelope of failure, and just a small additional increment of fluid pressure would cause the circle to touch the envelope. In addition, given that the sandstone became jointed during burial, the tectonic loading event would undoubtedly cause some of the fractures to be reactivated by overcoming sliding friction. We will assume in our example that the fracturing triggered by tectonic loading ultimately caused all the fluid pressure to dissipate, as reflected in the shift of least principal (horizontal) compressive stress to a value of 68 MPa and a shift of the greatest (vertical) compressive stress to a value of 132 MPa (Figure 5.64*D*).

Uplift

As the sandstone is uplifted to the surface, both the vertical and horizontal compressive stresses decrease along straight-line, predictable paths. If the sandstone body were raised to the surface without any change in the boundary conditions, the paths of stress decrease would be as portrayed in Figure 5.64*E*. The slope of the path of decrease of horizontal stress is steeper than the slope of the path of decrease of vertical stress, reflecting a steady decrease in differential stress. Such a "passive" uplift would not bring about new fracturing, for the differential stress circle(s) would lie well below the failure envelope (see Figure 5.64*E*).

If, on the other hand, we permit the sandstone body to expand ever so slightly, starting at 3 km depth and continuing to the surface, fractures will form as the result of a significant buildup in horizontal tensile stresses. If a stiff rock, like sandstone with a Young's modulus of -16.5×10^3 MPa, is stretched, an internal tensile stress will be generated as a result of Hooke's law:

$$\sigma = (-16.5 \times 10^3 \text{ MPa})\epsilon$$

Let us imagine that the sandstone is permitted to extend by a strain (e) of 0.0005 (i.e., stretched by 5/100 of 1%) per *half*-kilometer. This small amount of stretching creates substantial levels of tensile stresses (Figure 5.64*F*). As a result, the values of horizontal stress, in Mohr diagram imagery, slam leftward into tensile territory and penetrate the parabolic envelope and the tensile strength of the sandstone (Figure 5.64*F*) (Suppe, 1985). Joints form effortlessly. Joints will first "break out" in the relatively stiff units (high Young's modulus), where the rate of buildup of tensile stress per unit of extension is high. According to Suppe (1985), the formation of a single joint will relieve the built-up tensile stress within one joint length of the fracture. Thus, joints of this origin tend to be spaced at distances corresponding to the thickness of the unit in which they occur.

INTERPRETATIONS OF REGIONAL JOINTING

Jointing During Uplift

Price (1959), decades ago, recognized the potential of uplift to create jointing. He prepared a simple geometric/strain image to illustrate the concept, emphasizing the lateral expansion that must accompany uplift of a large regional sequence of sediments (Figure 5.65). Wise (1964) recognized the products of uplift-related fracturing in Precambrian rocks in the Wyoming province. He measured 6500 microjoints in mountains of Precambrian basement within parts of enormous fault blocks that were raised thousands of meters in early Tertiary time. Wise interpreted the microcracks to be tension joints that helped to relieve locked-in stresses during uplift. The microjoints, although visible, are hairline and spaced less than 3 mm apart. Wise went on to suggest that the **rift** and **grain** of granites in New England, the weaknesses that quarry workers have long used to split out commercial building blocks (Dale, 1923), may be controlled by microcracks.

Jointing of Several Origins in the Appalachian Plateau

Engelder (1985b) presented a sophisticated analysis of the development of regional jointing in the Appalachian Plateau province. He was able to distinguish the relative ages of four discrete classes of joints, and to interpret them in the context of the tectonic history of the region. **Tectonic joints** and **hydraulic joints** are "cross-fold" joints that formed as mode I tensile joints oriented vertically and generally parallel to the compressional direction responsible for folding (Figure 5.66). Both sets of joints formed under conditions of elevated pore fluid pressure and directed tectonic stress. By sorting the cross-fold joints according to plumose ornamentations, and by noting the relative ages of each class of joints and their distributions in siltstone versus shale, Engelder concluded that the tectonic joints formed when pore fluid pressure became elevated as a result of tectonic squeezing and compression. The hydraulic joints, on the other hand, are thought to have formed when pore fluid pressure became elevated as a result of vertical gravitational loading.

In contrast, two sets of younger joints, **release joints** and **unloading joints**, formed near the surface during uplift and erosion. As the siltstones and shales were brought closer to the surface during uplift and erosion, they cooled and contracted and relaxed elastically, causing stress buildup and jointing. In effect, compressive stress was released along preexisting structural elements (such as cleavage), and the "release joints" formed as near-vertical mode I tension joints perpendicular to the former direction of tectonic compression. Thus they are "strike joints," parallel to the trends of folds (Figure 5.66). The unloading joints are also mode I tensile

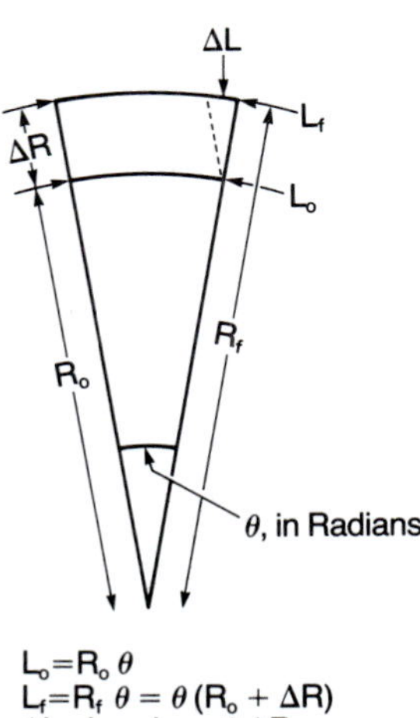

$L_o = R_o\,\theta$
$L_f = R_f\,\theta = \theta(R_o + \Delta R)$
$\Delta L = L_f - L_o = \theta\,\Delta R$

Figure 5.65 Geometric picture of layer-parallel extension accompanying regional uplift. [From Price (1959), *Geological Magazine*, v. 96, Mechanics of jointing in rocks, Reprinted with the permission of Cambridge University Press.]

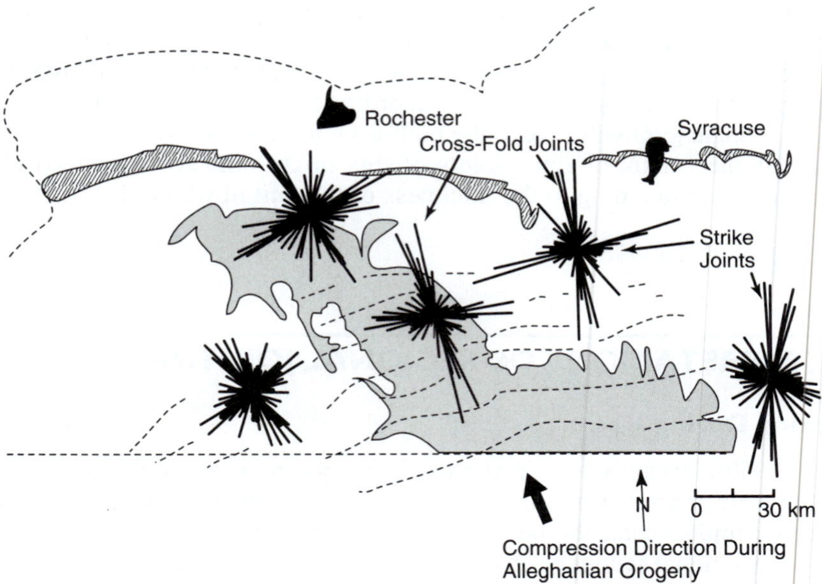

Figure 5.66 Regional pattern of cross-fold joints and strike joints within the Catskill Delta of central New York. The cross-fold joints are tectonic and hydraulic. The strike joints are due to release and unloading. (Published with permission from *Geology,* vol. 13, Engelder, T., and Oertel, G., The correlation between undercompaction and tectonic jointing within the Devonian Catskill Delta, 1985, The Geological Society of America.)

joints, but they display orientations that are not systematically related to the geometry of the fold belt. Instead, they have orientations that are related to the present-day stress field within the Appalachian Plateau region.

The care with which the regional analysis of jointing on the Appalachian Plateau has been carried out is reflected in illustrations from Engelder and Geiser (1980), which reveal the map orientations of cross-fold joints (tectonic and hydraulic) and strike joints (release joints) compared with the trends of folds and with the tectonic compression direction as revealed by deformed fossils and cleavage (Figure 5.67).

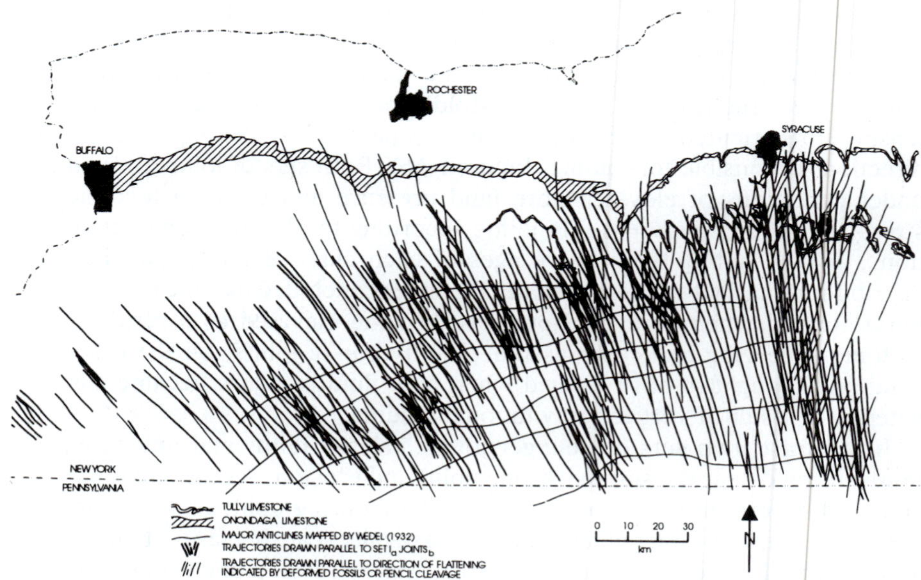

Figure 5.67 Just one example of the maps made by Engelder and Geiser (1979) to capture regional fracture patterns in relation to folding and strain in the Appalachian Plateau. (From T. Engelder and P. Geiser, *Journal of Geophysical Research,* v. 85, fig. 4b, p. 6325, copyright © 1980 by American Geophysical Union.)

Jointing Related to Local Stress Fields

The regional stress patterns that create joints of tectonic origin can be locally perturbed by the mechanical characteristics of faults that are also forming as a response to the regional stress. Rawnsley and others (1992) have described a wonderful example of this at Robin Hood's Bay on the coast of North Yorkshire in south England. There, parallel, evenly spaced, northwest-trending joints, oriented parallel to greatest horizontal stress (σ_1), abruptly converge toward an intersection of two major faults (Figure 5.68*A*,*B*). In all probability, high friction at the intersection of faults distorted the stress field in the surrounding rock. Mode I tension joints, ever sensitive to changing stress fields, simply curve into "proper" orientations.

Rawnsley and others (1992) reproduced the dynamic state that results in joint convergence by creating a stress field in a Plexiglas model prepared in a special way (Figure 5.68*C*). An open fracture was cut through the Plexiglas, and a small plug of metal was inserted into one location along the fracture. When the Plexiglas was end-loaded and viewed in polarized light (Figure 5.68*C*), it revealed a continuously curved stress field, which could produce in nature a continuously curved joint system.

Jointing and Shear Fracturing Accompanying Intrusion

Primary joints and shear fractures evolve in plutons as a natural response to the thermodynamics of intrusion. The primary nature of joints and shear fractures formed before final crystallization is disclosed in a number of ways: the geometrical coordination of the fracture system with primary flow foliation and lineation, the filling of fractures by aplite and pegmatite

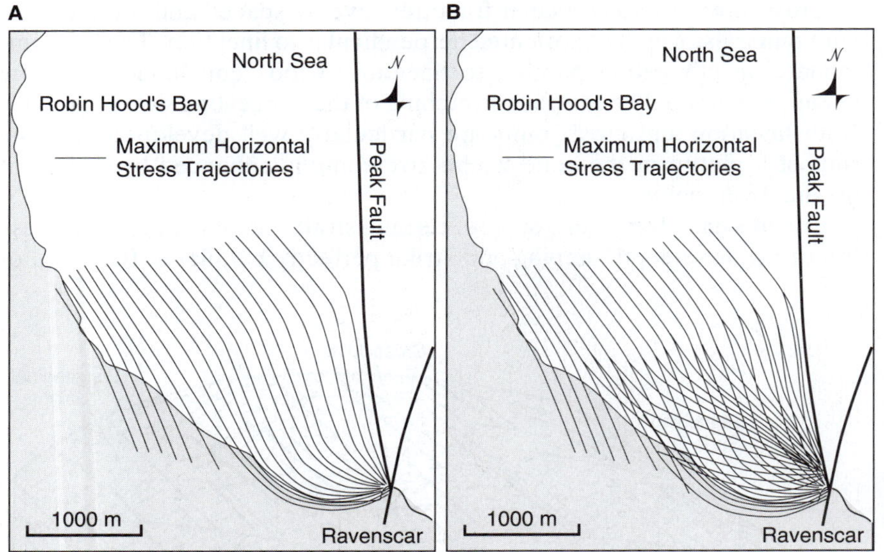

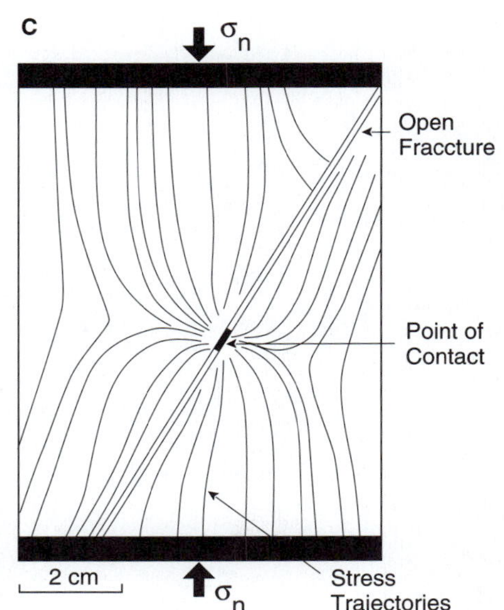

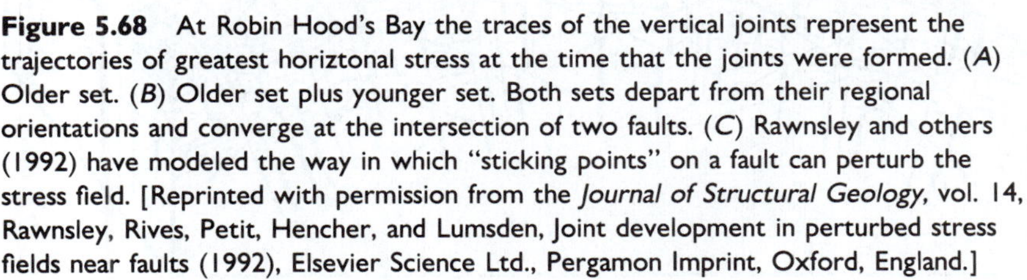
Figure 5.68 At Robin Hood's Bay the traces of the vertical joints represent the trajectories of greatest horiztonal stress at the time that the joints were formed. (*A*) Older set. (*B*) Older set plus younger set. Both sets depart from their regional orientations and converge at the intersection of two faults. (*C*) Rawnsley and others (1992) have modeled the way in which "sticking points" on a fault can perturb the stress field. [Reprinted with permission from the *Journal of Structural Geology*, vol. 14, Rawnsley, Rives, Petit, Hencher, and Lumsden, Joint development in perturbed stress fields near faults (1992), Elsevier Science Ltd., Pergamon Imprint, Oxford, England.]

Figure 5.69 "Crack" in surface of a mountain pool. [From *Structural Behavior of Igneous Rocks* by R. Balk (1937), Geological Society of America.]

dikes genetically related to the magma, and the presence of veins and/or alteration assemblages that form by late-stage hydrothermal solutions circulating within the still-hot body.

The image that best illustrates the compatibility of flow and fracture is the frontispiece of Robert Balk's (1937) memoir on the *Structural Geology of Igneous Rocks* (Figure 5.69). The foam layers form a broad arc in the plane of the surface of the water. At the crest of the arc are discrete fractures and fissures that cut the flow layers. The fractures developed through stretching of the foam layers on the convex part of the arc. If water and foam layers are capable of such fracturing, so too must be viscous granitic magma in its late stages of flow.

It is desirable to try to distinguish primary fractures related to the thermodynamics and kinematics of intrusion from fractures created by regional stresses. Making such a distinction is not always possible. Regional stresses that operate long after the crystallization of a particular pluton may fracture the pluton, just as they can fracture any other rock body (Norton and Knipe, 1977). Moreover, when a pluton is intruded into a regional setting marked by active differential stresses, primary fractures may form that are systematically oriented in relation to the regional stresses, not to the geometric and kinematic order of intrusion (Rehrig and Heidrick, 1976).

Plutons that contain a primary flow structure lend themselves to fracture analysis, for the lineation and/or foliation can be "played" geometrically against the fracture pattern to see if a "kinematic coordination" exists. On this basis, Hans Cloos recognized fundamental types of primary fractures, including **cross fractures**, **longitudinal fractures**, and **stretching surfaces**. The configuration of these primary fractures is represented in the idealized block diagram of the Strehlen, a German granite body analyzed by Cloos (1922) (Figure 5.70). The fractures are shown in relationship to primary foliation and lineation within the body.

Cross joints are long, planar fractures, evenly spaced and often coated with minerals, which are oriented perpendicular to lineation. They accommodate an elongation parallel to lineation. Cloos concluded that flow lineation commonly expresses stretching of the congealing skin of granite. Both lineation and cross joints are particularly well developed near the roof of a pluton in the zone where stretching is enhanced by surges of magma from below.

Longitudinal fractures, of less certain strain significance than cross fractures, are steeply dipping and strike perfectly parallel to the trend of

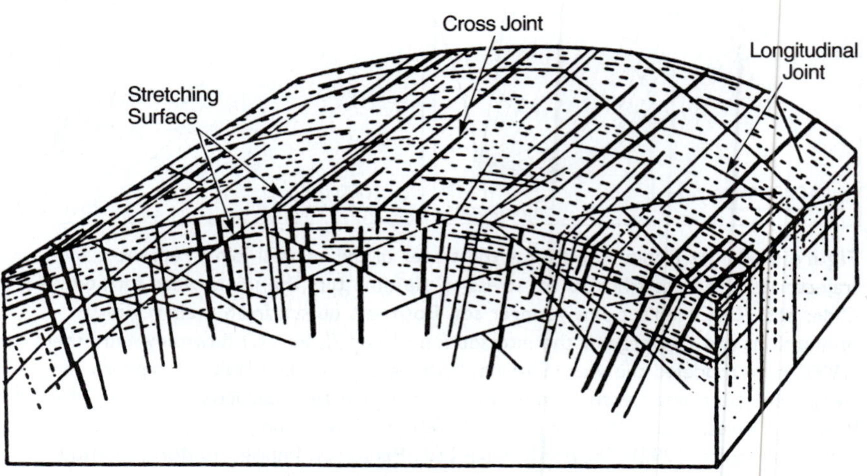

Figure 5.70 Block diagram showing ideal primary fracture pattern in a granitic pluton. [From Cloos (1922).]

lineation. They are not as planar as cross joints, and in the example studied by Cloos they tend to have thinner mineral coatings.

Stretching surfaces occur in the upper reaches of a pluton. They are low-dipping shear surfaces (both faults and shear fractures) marked by striations that trend in the direction of lineation. Where offset can be determined through visible displacement of dikes, schlieren, clots, xenoliths, or veins, the shear fractures and faults are seen to be extensional, accommodating lengthening of the top of the intruding mass. Formation of stretching surfaces may be thought of as a pure shear. Although the stretching surfaces may initially dip moderately steeply, the rotation accompanying flattening during stretching results in a progressive decrease in the angle of the dip.

Joint Patterns as Regional Stress Gauges

Rehrig and Heidrick (1972, 1976) discovered that jointing in granitic intrusions can be a very sensitive guide to the directions of regional tectonic stress operative in the crust into which the plutons are emplaced. For example, they noted that Late Cretaceous to Eocene plutons contain systematic planar, parallel mode I tension fractures and parallel dikes and veins (Figure 5.71). At all scales it is clear that the joints occur in regular patterns over a large region. (Figure 5.72).

After measuring literally thousands of joints in Late Cretaceous to Eocene plutons in southern Arizona, Rehrig and Heidrick determined that the dominant joint set trends N70°E, plus or minus 20° (Figure 5.71). They concluded that these plutons were intruded into the upper levels of a crust and, as they cooled and congealed, cracked in directions systematically related to the regional tectonic stress field that existed at the time. Based on the patterns, they reconstructed the paleostress regime, which was marked by a horizontal north–northwest trending least principal compressive stress (σ_3). Differential stress was relatively weak. Fluid pressure buildup in the cooling pluton, manifest now in dikes and veins, had the effect of decreasing the magnitudes of the principal stresses. This produced tensional fracturing by jointing, with the dominant orientation perpendicular to the direction of least principal stress. Fracturing served to dissipate the elevation of fluid pressure for a time, and then pressure built again, culminating in yet another phase of jointing. Their work has practical implications, because the mode I joints are the dominant mineralized fractures in the large porphyry copper deposits of the region.

Mineralized Joints, Veins, Dikes & Faults

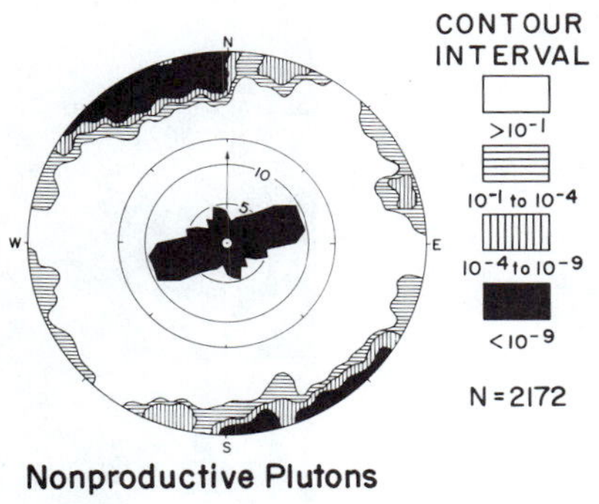

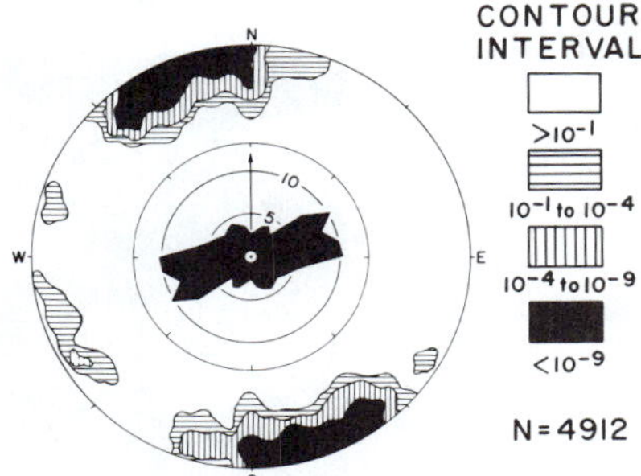

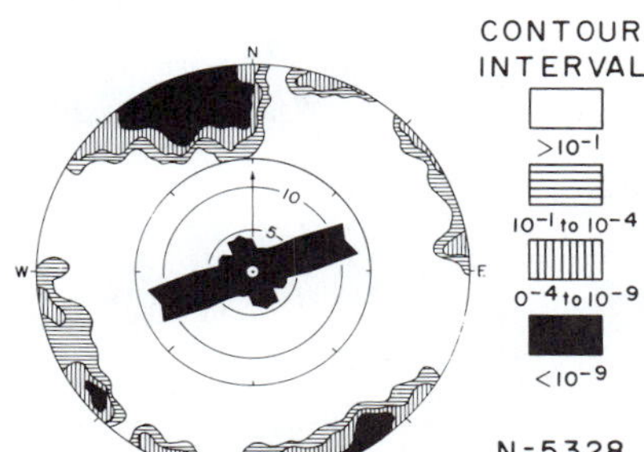

Figure 5.71 Pole-density diagrams and rose diagrams (in interior of each circle) show the orientations of mineralized joints, veins, dikes, and faults in Late Cretaceous to Eocene plutons in the Southwest. Look at the number (N) of structures measured and analyzed!! (By permission, T. L. Heidrick and S. R. Titley, "Fracture and Dike Patterns in Laramide Plutons and Their Structural Tectonic Implications," Fig. 4.6 in *Advances in Geology of the Porphyry Copper Deposits: Southwestern North America,* S. R. Titley, editor. University of Arizona Press, Tucson, copyright © 1982.)

Unusual Fractures Associated with Impacts

Shatter cones are unusual, linearly etched fractures with distinctive conical shapes (Figures 5.73, 5.74). Most are associated with craters interpreted to have formed by high-velocity impacts of meteorites, asteroids, and other planetary debris (Dietz, 1968; Grieve and Pesonen, 1992). They seem to develop best in fine-grained rocks, such as limestone and quartzite, but they can occur in a variety of other rock types as well (Dietz, 1972). The host rock may contain other impact-related features, such as impact breccias, high-pressure mineral phases (e.g., coesite), and glass formed as a result of impact-induced shock melting.

Shatter cones are interesting and informative from a kinematic point of view. The apices of the cones point *toward* the impact site (Figure 5.74). Thus the orientations of shatter cones can be used to locate buried craters.

Figure 5.72 (*A*) Distant view of joint traces in the Eocene Texas Canyon pluton, southeastern Arizona. (*B,C*) Close-up views of mineralized joints in the pluton. Long and continuous, these joints strike east–northeast, and thus are parallel to the preferred regional orientation. (Photographs by G. H. Davis.)

Figure 5.73 Photograph of shatter cones from the Wells Creek impact site in Tennessee. Rock is Knox Dolomite. It shows the telltale pattern of the diverging linear pattern so typical of shatter cones. (Photograph by John F. McHone.)

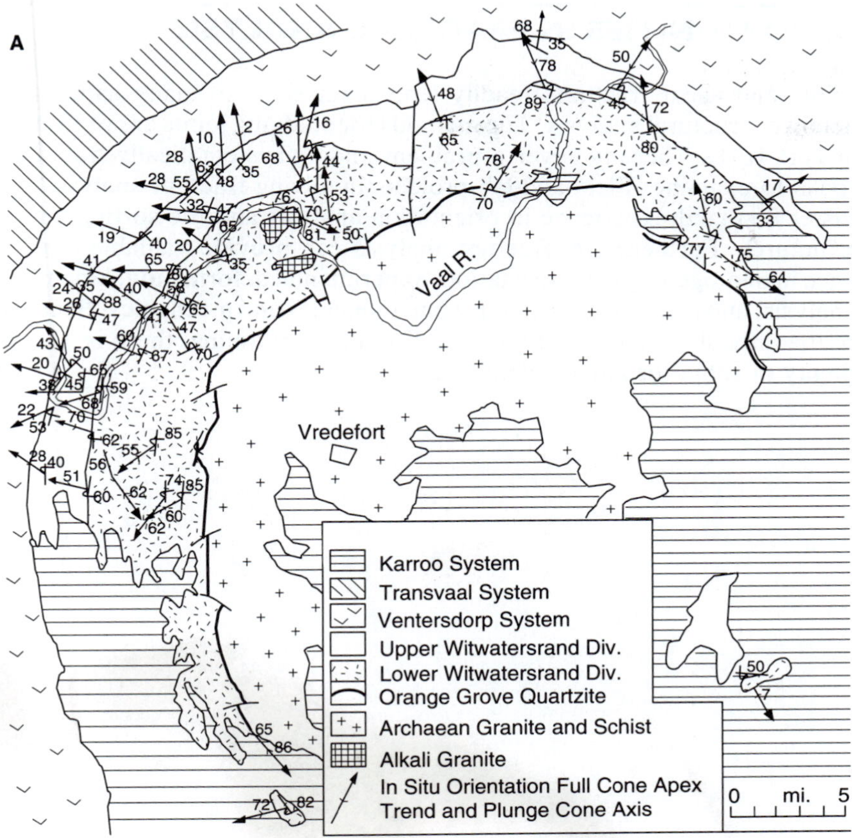

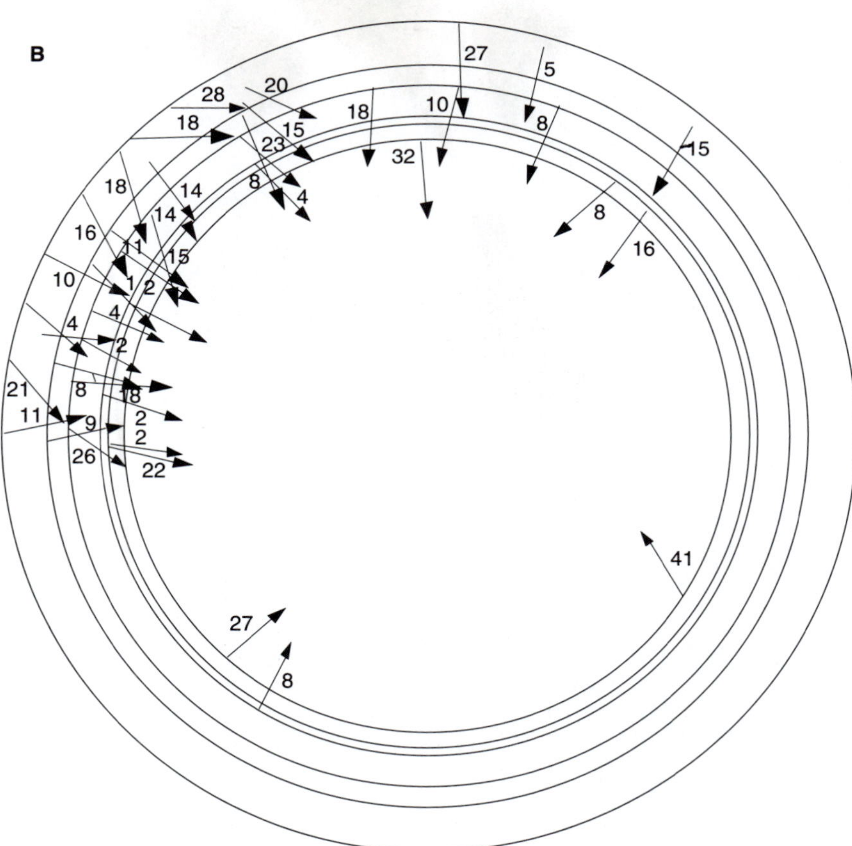

Figure 5.74 (*A*) Geologic map of the Vredefort dome, South Africa, showing the mapped orientation of full shatter cones. Note that these shatter cones occur in bedding that has been overturned. (*B*) When bedding is restored to its original orientation, the apices point to a common locus of impact. [From Manton (1965). Permission by the Annals of the New York Academy of Sciences.]

OPPORTUNITIES IN FRACTURE ANALYSIS

Joints and shear fractures readily lend themselves to many aspects of detailed structural analysis. The methods for studying joints are presented in Part III-L. There is never a problem finding systematically fractured rocks to examine. Field-and-lab projects involving fracture analysis can foster firsthand experience in orientation analysis and the sorting out of structures. Furthermore, fracture analysis can provide valuable experience in photogeologic interpretation, experimental deformation, kinematic analysis, and mathematical inquiry into dynamics and mechanics. Because of the difficulty encountered in establishing *exact* timing, there is *always* plenty of room for independent, creative inquiry.

CHAPTER 6

FAULTS

SOME DEFINITIONS AND DISTINCTIONS

Faults are fractures along which there is visible offset by shear displacement parallel to the fracture surface (Figure 6.1). Faults can occur as single discrete breaks, but where the rock has been repeatedly faulted, or where the rock is especially weak, no discrete break may be evident. What forms instead is a **fault zone** composed of countless subparallel and interconnecting closely spaced fault surfaces (Figure 6.2*A*,*B*). Faulting is fundamentally a brittle mechanism for achieving shear displacement. At deep crustal levels where rocks tend to deform plastically under conditions of elevated temperature and confining pressure, shear displacement is achieved by the development of **shear zones** (Figure 6.3). We will discuss shear zones in Chapter 9.

Faults range in length and displacement from small breaks with offsets wholly contained within individual hand specimens or outcrops (Figure 6.4*A*), to regional crustal breaks extending hundreds to more than 1000 km and accommodating offsets of tens to hundreds of kilometers (Figure 6.4*B*). On the other hand, some shear fractures are actually **microfaults**, but they generally go unrecognized as such because the tiny offset caused by shear cannot be easily resolved without aid of a microscope, or without the presence of fine-scale markers that record the displacement.

Faults should not be confused with **fissures** (Figure 6.5), which are large fractures that have accommodated conspicuous dilational opening. Fissures form at the surface of the Earth where active tectonic processes, notably earthquakes and volcanism, create local stretching and pull-apart.

Figure 6.1 Well-exposed fault in Lebung Pass, central Himalayas. The bedded units above the fault trace are Pennsylvanian quartzites that have been thrust over Permian shales. A human touch for scale. [Reproduced from photograph in Griesbach, C. L. (1891) and Wadi (1953).]

Figure 6.2 (*A*) Fault zone in Permian Chochal Limestone in northwestern Guatemala. The limestone is broken by the movement along innumerable fault surfaces. George Davis, young graduate student, for scale. (Photograph by T. H. Anderson.) (*B*) The fault zone stands in sharp contrast to this discrete fault surface, also in Chochal Limestone. Alston Boyd III (once known as "Boyd 3") for scale. (Photograph by G. H. Davis.)

Figure 6.3 India-ink sketch of close-up photograph of shear zone in metamorphic rocks in the Tortolita Mountains, southern Arizona. The zone dips from upper right to lower left. Layered units above the shear zone thin dramatically as they "move" into the zone. Some layers accommodate the movement entirely by thinning, avoiding truncation and offset. Note pencil for scale in lower left corner. [From Davis (1980). Published with permission of Geological Society of America and author.]

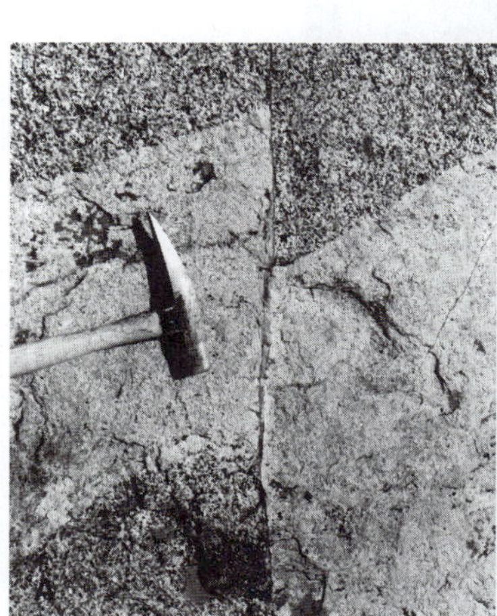

Figure 6.4 (*A*) Fault offset of a fine-grained dike. (Photograph by J. P. Lockwood. Courtesy of United States Geological Survey.) (*B*) Physiographic expression of a great regional fault, the San Andreas fault, as viewed northerly along the Elkhorn scarp, San Luis Obispo County, California. (Photograph by R. E. Wallace. Courtesy of United States Geological Survey.)

Figure 6.5 Fissure cutting the Hiliana Pali Road at Kalanaokuaiki Pali, Kilauea Volcano, Hawaii. The crack formed on December 25, 1965. Initially only 1 m wide, the crack soon enlarged to 2.4 m as a result of slumping along the edges. Note columnar jointing on far side of road. (Photograph by R. S. Fiske. Courtesy of United States Geological Survey.)

RECOGNIZING THE PHYSICAL CHARACTER OF FAULTS

Fault Scarps

Fault contacts are easy to identify in areas where faulting is presently active. Faulting that reaches the surface produces **offset** of both natural and man-made objects (Figure 6.6*A*). The offset of such features as roads, streams, and beaches simplifies the search for fault contacts. In some cases faulting may not reach the surface, but the effects of seismic energy are nonetheless evident (Figure 6.6*B*).

Figure 6.6 (*A*) "And the sea became dry land." Faulting during the Alaskan earthquake of 1964 resulted in significant vertical displacement. The northwest block (*left*) of the Hanning Bay fault was raised 4 to 5 m, relative to the southeast block. The white coating on the exposed seafloor bottom consists of the bleached remains of calcareous algae and bryozoans. The height of the fault scarp near the ponded water is 4 m. [Photograph by G. Plafker. From Plafker (1965). Courtesy of United States Geological Survey.] (*B*) Some of the 13 billion dollars of damage incurred by the magnitude 6.8 Northridge earthquake, which rocked California on January 17, 1994. Fifty-seven people were killed and 9300 injured. (Photograph by Ken Fowler.)

Fault scarps are steps in the land surface that coincide with locations of faults (Figure 6.7*A*). They are expressions of contemporary movement(s), and the height of a fault scarp will approximate the sum of the most recent displacements, although the effects of erosion of the raised block and depositional infilling atop the lowered block must be taken into account. Usually, however, scarps in the landscape that may be seen along the surface trace of a fault are not *true* fault scarps; that is, they are not a direct expression of either the dip of the fault surface or the magnitude of offset. The passing of time permits weathering and erosion to erase the original expression of the fault offset. Fault scarps thus are gradually replaced by **fault-line scarps** (Figure 6.7*B*). Fault-line scarps are located along or near the trace of a fault and are marked by a topographic relief that simply reflects the differential resistance to erosion of the rocks brought into contact by faulting. Thus fault-line scarps almost always give a false impression of the actual vertical component of displacement along a fault.

Former planar fault scarps may be dissected by erosion to yield **triangular facets**, physiographic signatures that call attention to the location(s) of fault contacts. An outstanding display of triangular facets may be seen along the Wasatch Front in northern Utah (Figure 6.8).

Figure 6.7 (*A*) Red Canyon fault scarp in the Montana earthquake area, as it looked in August 1959. Scarp height reflects the actual fault displacement. (Photograph by J. R. Stacy. Courtesy of United States Geological Survey.) (*B*) Fault-line scarp near Lake Mead. The topographic relief along the scarp does not reflect slip. Rather, it reflects differential erosion along the fault interface between resistant volcanic rock and nonresistant volcanic rock. The rock on which Ernie Anderson stands actually moved *upward* relative to the resistant volcanic rock on the other side of the fault. (Photograph by G. H. Davis.)

Fault Surfaces

Where well exposed, faults are commonly expressed by the presence of a discrete fracture break or discontinuity, a **fault surface** (Figure 6.9), across which the rocks on either side do not "match up." We use the term fault *surface* instead of fault *plane* because faults are rarely perfectly planar. Some are planar, some are made up of planar segments of different orientations, some are systematically curved, and some are highly irregular.

Figure 6.8 Triangular facets along the front of the Wasatch mountain block near Provo, Utah. (Photograph by C. Glass, G. Brogan, and L. Cluff. Courtesy of Woodward-Clyde.)

Figure 6.9 Chris Menges (mustache and glasses) proudly shows off the exceptionally well exposed, curved fault surface that he mapped near Patagonia in southern Arizona (see Menges, 1981). Chris' hands are placed on the crest of a convex groove. Richard Gillette's foot (*far left*) rests in trough of concave groove. Nancy Riggs, with feet squarely planted, measures the strike azimuth of a part of the fault surface. (Photograph by G. H. Davis.)

Individual fault surfaces do not go on forever. If we could pull discrete fault surfaces out of the ground as if they were sheets of Plexiglas, and look at them, we would find that most are roughly elliptical. Some of the evidence for this is seen in the work of Barnett and others (1987), who measured the attributes of discrete faults in the British coalfields and on seismic sections in the North Sea. They found that isolated faults have an elliptical shape with a typical aspect ratio between 2 : 1 and 3 : 1 (Figure 6.10*A*). Displacements were seen to decrease to zero from the central part of each elliptical fault surface outward to the **tip line loop**, the imaginary line formed by connecting points (i.e., **tip points**) where the fault surface comes to an end (Figure 6.10*B*).

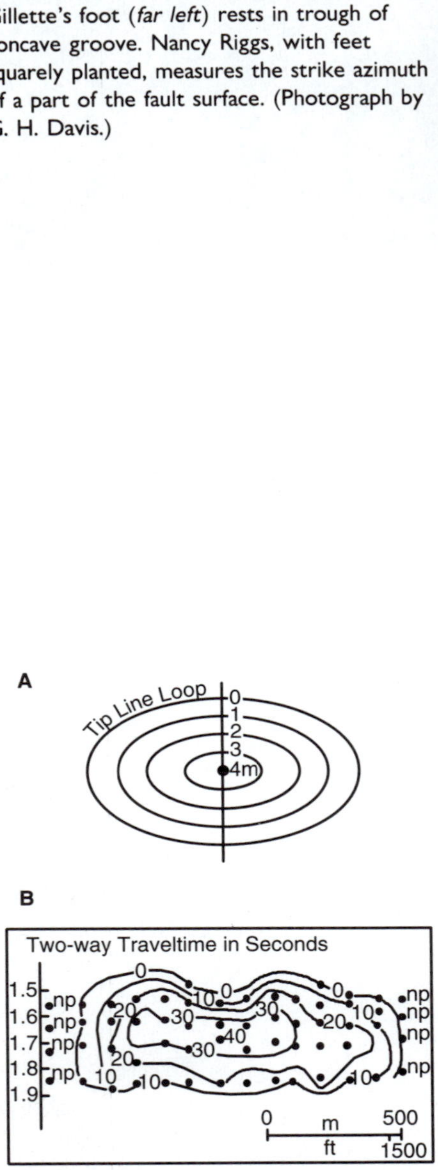

Figure 6.10 (*A*) Theoretical picture of the shape of a discrete fault surface and the gradient of displacement along it. (*B*) Displacement contour diagram of a normal fault in the North Sea. Displacement values are measured as two-way seismic wave travel times, in milliseconds; "np" denotes fault not present. [From Davison (1994), *continental deformation*, linked fault systems; extensional, strike-slip and contractional, Figure 6.5A,B, p. 126. Reprinted with permission of Pergamon Press, Oxford, England. After Barnett and others (1987).]

Slickensides and Slickenlines

Fault surfaces are commonly finely polished **slickensided surfaces**, apparently the result of the fine abrasive action of differential movement when the rock types and conditions of deformation are just right (Figure 6.11*A*,*B*). **Slickensides** are the smooth or shiny fault surfaces themselves (Means, 1987). Some surfaces are almost mirrorlike, as if "finished" by Nature's equivalent of 0000 sandpaper or steel wool. Yet many fault surfaces are not slickensided at all, either because of the host rock's inability to "take a polish" or because of removal of an original luster by weathering and erosion. We find that some slickensided surfaces owe their gloss not to frictional polishing, but to thin **neomineral coatings** (Figure 6.12) that grew and developed on the fault surface during movement.

When we examine fault surfaces in the field, it is standard procedure to inspect the surfaces closely for the presence of **slickenlines**. Slickenlines are generally straight, fine-scale, delicate skin-deep lines that occupy the fault surface itself and record the direction of slip (Figure 6.13). Slickenlines are most noticeable on slickensided surfaces, but they are by no means restricted to glossy surfaces.

Many slickenlines we examine, such as those in Figure 6.13, prove to be **striations** produced by frictional abrasion along the fault surface during differential displacement of the wall rocks (Figure 6.14*A*,*B*). Scratches and furrows are produced when hard **asperities**, projecting like metal cleats from the sole of a shoe, score lines on the opposite surface against which

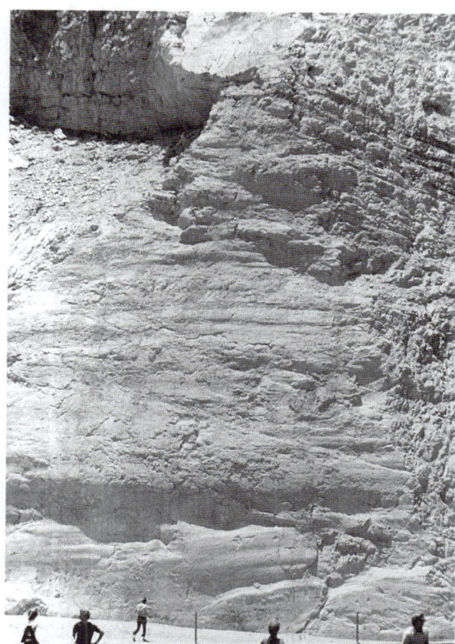

Figure 6.11 (*A*) Slickensided fault surface in interstate highway cut through Paleozoic carbonates in the Virgin Mountains, northwesternmost Arizona. Polished fault grooves disclose horizontal motion. Eric Frost goes for a closer look of the slickensided surface. (*B*) Slickensided exposure of the Hurricane fault near La Verkin, Utah. Slickenlines disclose vertical motion. (Photographs by G. H. Davis.)

it moves. When asperities break off, they can break up into pieces of hard debris that "float" along the fault surface, not unlike sand and grit underfoot. The line pattern created by scratching and furrowing is usually further enhanced by **streaks** of the fine crushed debris, which will pile up "fore and aft" of hard asperities, creating **tails** and **spikes** and other

Figure 6.12 The slickensiding (i.e., polish) on this fault surface in Jurassic sandstone (Tucson Mountains, southern Arizona) owes its expression to a neomineral coating of quartz. (Photograph by G. H. Davis.)

Figure 6.13 In bright morning light, and to the sound of clicking cameras, Edna Patricia Rodriguez points out beautifully developed slickenlines on the underside of the Rubys Inn fault surface near Bryce Canyon, Utah. The rock is Eocene Claron Formation. (Photograph by G. H. Davis.)

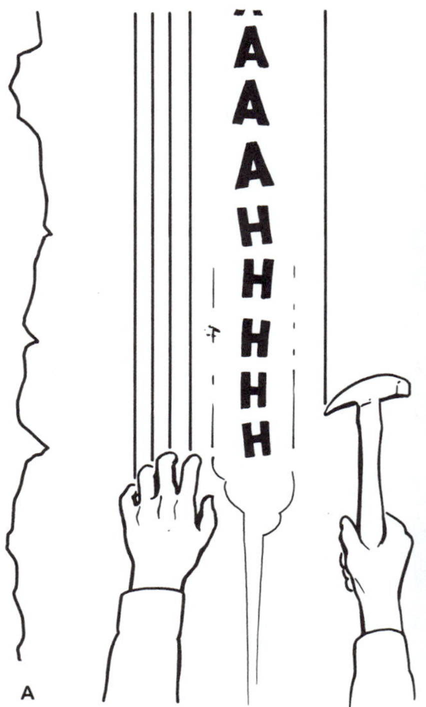

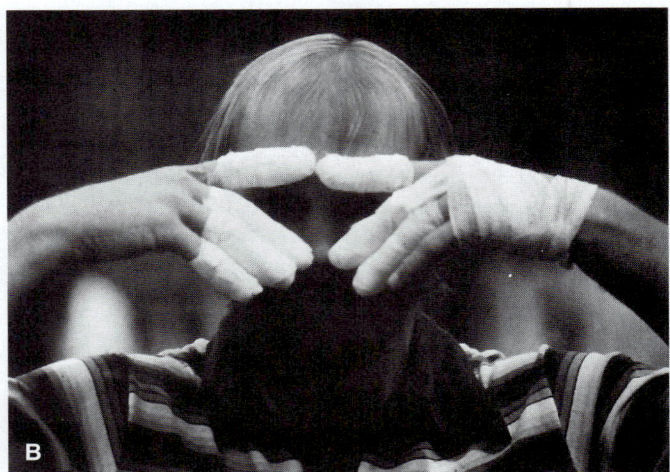

Figure 6.14 (*A*) Striae produced by differential rigid body movement. (*B*) Geologist Steve Lingrey left such striae near the Paunsaugunt fault in southwestern Utah. (Photograph by G. H. Davis.)

microridges of material (Means, 1987). A good example of the result of this process is shown in Figure 6.15. The slickenline resembles a comet trail behind a tiny chunk of resistant rock that got caught along the fault surface.

The apparent frictional development of striations can operate at a very large, coarse scale as well, producing **fault grooves**, sometimes called **fault mullions**. Fault grooves are deeply furrowed slickensided features that look exactly like bedrock surfaces that have been polished and grooved by glacial flow (Figure 6.16).

Figure 6.15 Close-up photo showing the relation of tail of microcrushed material to the tiny chunk of rock that served as a tool during the faulting that created the slickenlines. This again is the Rubys Inn thrust near Bryce Canyon, Utah. (Photograph by G. H. Davis.)

Figure 6.16 Slickensided and slickenlined fault surface in western Turkey. Note the deeply developed grooves in the left-hand part of the fault surface. The expression of the grooves is enhanced by the shadows. [Reprinted with permission from *Journal of Structural Geology,* v. 9, P. L. Hancock and A. A. Barka, Kinematic indicators on active normal faults in western Turkey (1987), Elsevier Science Ltd., Pergamon Imprint, Oxford, England.]

Figure 6.17 Win Means faulted a small block of white wax. Movement along the fault was merely 2 mm. When he opened up the block along the fault surface, he discovered the white wax had turned dark, and that each "wall" of the fault surface had well-developed slickenlines. The field width of the photos is 8.5 mm. Why on earth are the surfaces so heavily slickenlined, given the small amount of movement??? [Reprinted with permission from *Journal of Structural Geology*, v. 9, W. D. Means, A newly recognized type of slickenside striation (1987), Elsevier Science Ltd., Pergamon Imprint, Oxford, England.]

Means (1987) cautions that there are things about slickenlines that we simply do not understand. Through the creation of fault surfaces in paraffin wax, he demonstrated that well-developed slickenlines on slickensided surfaces can form even when displacement is negligible. By shearing "snow-white" wax by just a few millimeters, Means produced shiny dark surfaces with heavily pronounced slickenlines in the form of ridges and grooves that nested perfectly (Figure 6.17). Although the slickenlines resemble scored tool marks, they did not form this way, leaving open the question of how they really formed.

Some slickenlines on or along fault surfaces are actually **crystal fiber lineations**, or simply **slip-fiber lineations** (Steve Marshak, 1994, personal communication) produced by the preferred directional growth of minerals *during* the faulting and in the direction of movement. The slickenlines in quartz shown in Figure 6.18 represent such crystal fiber lineation on a

Figure 6.18 Crystal fiber lineations on a thrust fault surface in the Hudson Valley fold-thrust belt in New York. [(From Marshak and Tabor (1989). Published with permission of the Geological Society of America and the authors.]

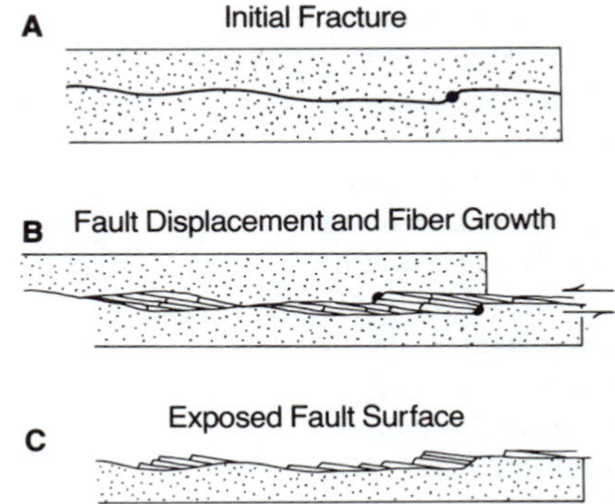

Figure 6.19 Steps in the kinematic evolution of crystal fiber lineation on a slickenlined surface. (*A*) Formation of fault surface. (*B*) Fault displacement and simultaneous growth of crystal fibers in the direction of least stress. (*C*) Striated surface as exposed by weathering and erosion. [From Durney and Ramsay (1973). Published with permission of John Wiley & Sons, Inc., New York, copyright © 1973.]

fault surface. Durney and Ramsay (1973) have explained that crystal fibers on fracture surfaces are precipitated on the leeward side of tiny ridges or bumps (Figure 6.19*A*,*B*,*C*). The microtopography in combination with shear tends to produce open space in which the crystals can grow. Narrow veins of crystal fibers grow on fracture surfaces in such a way that individual crystal fibers trend parallel to the direction of differential displacement (Figure 6.19); the sense of shear is disclosed by the plunge of the crystal fibers relative to the fracture surface. The length of individual fibers records the magnitude of slip accommodated by the fracture surface during the time of crystal fiber growth.

Slickolites, the antithesis of crystal fiber slickenlines, form on fault surfaces where fault movement in combination with stylolitic pressure solution results in removal of material through dissolution creep (Figure 6.20). The cones and teeth of the stylolites create a lineation (i.e., the slickolites) on the fault surface and thereby record the direction of slip. Whereas slip fibers are precipitated on the leeward side of tiny ridges or bumps on the fault surface, slickolites penetrate *into* the irregular fault-surface topography, reducing volume (Figure 6.20). The formation of slickolites requires lithologies that, under the conditions of deformation, can

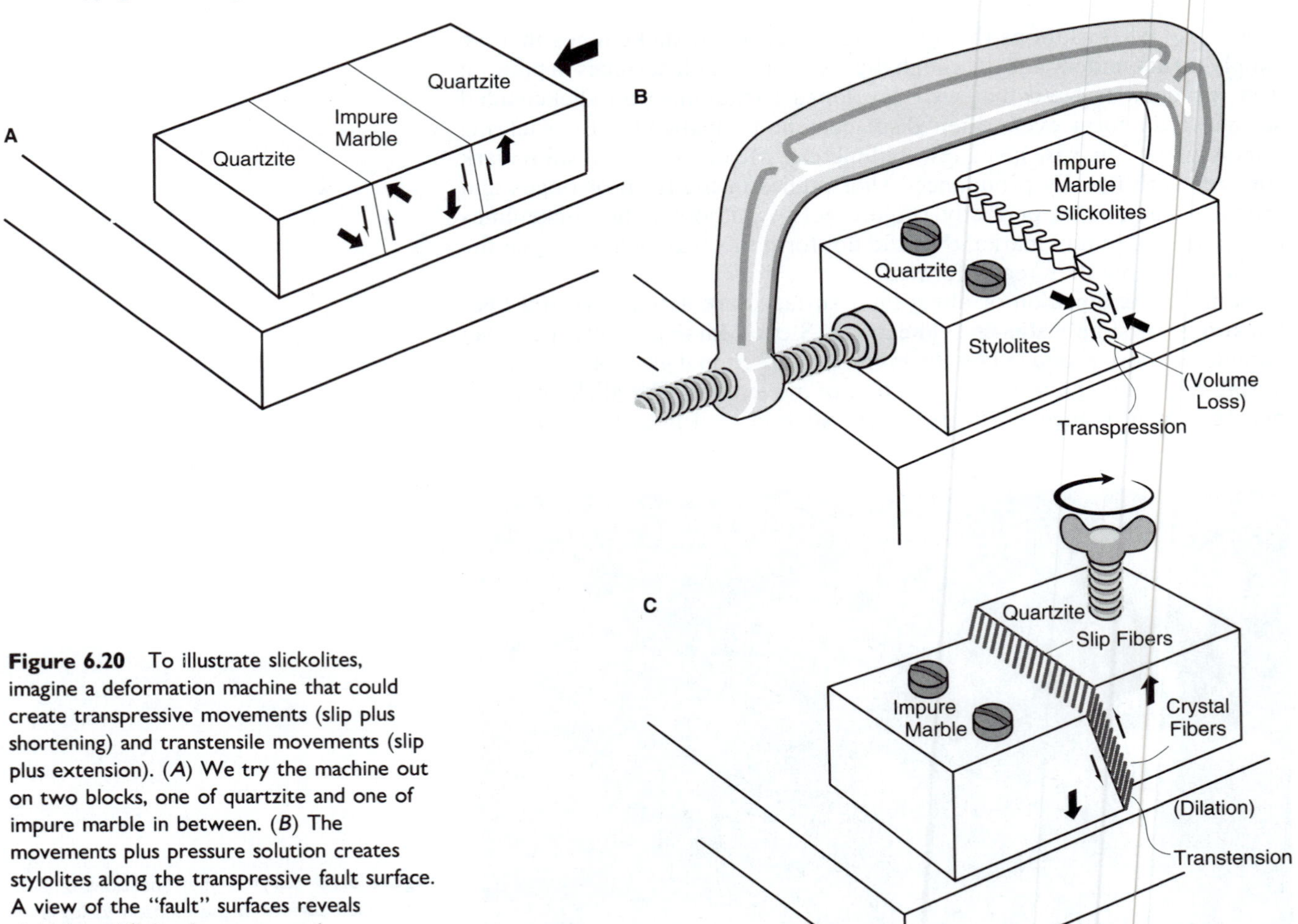

Figure 6.20 To illustrate slickolites, imagine a deformation machine that could create transpressive movements (slip plus shortening) and transtensile movements (slip plus extension). (*A*) We try the machine out on two blocks, one of quartzite and one of impure marble in between. (*B*) The movements plus pressure solution creates stylolites along the transpressive fault surface. A view of the "fault" surfaces reveals slickolites. (*C*) It takes transtensile movement (slip plus extension) to create crystal fibers oblique to a failure surface.

Figure 6.21 Photo of outcrop expression of slickolites on fault surface in the Claron Formation (Eocene) near Cedar Breaks National Monument, Utah. (Photograph by G. H. Davis.)

"switch on" pressure dissolution as a deformation mechanism. Figure 6.21 is a photograph of slickolites in outcrop.

Chatter Marks

Chatter marks commonly are formed on the surfaces of faults (Figure 6.22). These small, asymmetrical, steplike features are typically oriented perpendicular to striations. "Topographic relief" on chatter marks is usually less than 5 mm. Sense of movement can be interpreted (very cautiously) on the basis of the geometry of chatter marks. The ideal relationship of chatter-mark geometry to differential slip is shown in Figure 6.22: One face of the step is protected from mechanical grinding by the opposite block.

Although the asymmetry of chatter marks does not appear to be a dependable indicator of displacement sense, the presence of chatter marks is helpful in confirming that a particular surface is actually a fault. Moreover, in the absence of any other sense-of-movement criteria, chatter-mark orientations and asymmetries may yield the statistically most probable direction and sense of slip.

Fault Zones

Fault zones consist of numerous closely spaced fault surfaces, commonly separating masses of broken rock. Zones of intensely fractured and crushed rocks associated with faults vary in thickness from less than several centimeters to a kilometer or more. In general, the thicker the fault zone, the larger the amount of displacement on the zone (Schulz, 1990). A single fault zone, however, may thicken or thin along strike as individual fault strands merge with or bifurcate away from the zone and as the fault encounters rocks of varying mechanical properties. Changes in the width and character of fault zones are especially common near bends or jogs. The wall rocks outside of a brittle shear zone may be largely unaffected by the faulting or may show a zone of **drag folding** flanking the zone (Figure 6.23). Whether a discrete fault surface or a fault zone is produced by faulting depends on many factors, none of which can be precisely known for a given fault. Rock strength, strain rate, the physical environment of deformation, and duration of faulting are all important factors.

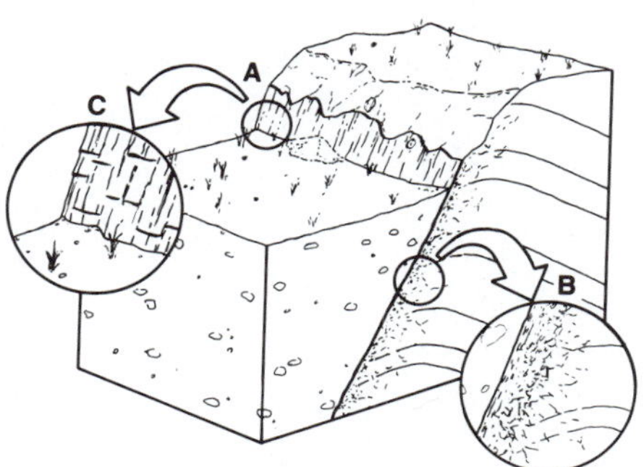

Figure 6.22 Schematic diagram showing (*A*) fault surface, (*B*) chatter marks and slickenlines, and (*C*) fault rocks like gouge and breccia.

Figure 6.23 The Lincoln Ranch fault zone in western Arizona carries strongly deformed Precambrian rocks over Miocene sandstone. Note how the sedimentary layers are bent and broken as they approach the zone of deformation. Ivo Lucchitta of the U.S. Geological Survey is admiring the result of this faulting. (Photograph by S. J. Reynolds.)

FAULT ROCKS

During frictional movement along a fault, or within a fault zone, rocks may be transformed into a sizable number of different **fault rocks**, depending upon starting materials, conditions, and deformation mechanisms (Table 6.1). The presence of fault rocks along a contact of uncertain origin

TABLE 6.1
Brittle fault rocks

Breccia Series	
These brittle fault rocks are marked by angular clasts in a finer matrix. Generally there is no preferred orientation of clasts, and thus the fabric is "random." Breccia series rocks are noncohesive to compacted, except where silicified or mineralized. The four main varieties are based on size of clasts:	
Megabreccia	clast size > 0.5 m
Breccia	clast size > 1 mm < 0.5 m
Microbreccia	clast size > 0.1 mm < 1 mm
Gouge	clast size < 0.1 mm
Cataclasite Series	
These brittle fault rocks are also marked by angular clasts in a finer matrix, and generally there is no preferred orientation of clasts. Cataclasite series rocks are cohesive and strongly indurated. The two main varieties are based on size of clasts:	
Cataclasite	clast size > 0.1 mm < 10 mm
Ultracataclasite	clast size < 0.1 mm
Pseudotachylite	
Unlike breccias and cataclasites, this brittle fault rock is created by frictional melting, not by grinding and fracturing. It is glassy to cryptocrystalline and generally brown, gray, or black in color. There may be tiny crystals (less than 1 μm) in an isotropic glassy groundmass. Pseudotachylite occurs in veinlike arrays, some of which may be spiderlike in form.	

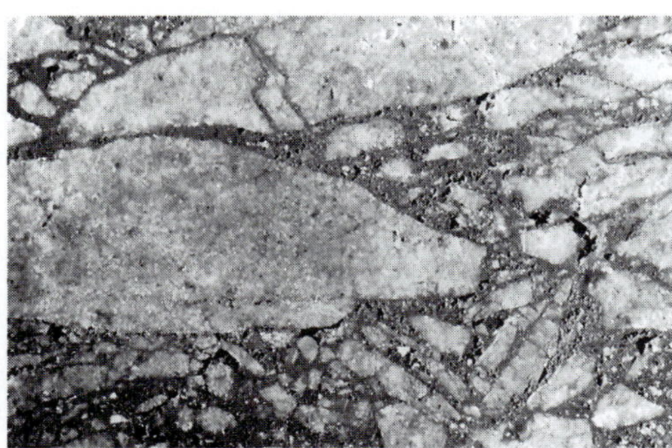

Figure 6.24 Slab of cataclastic rock, in the form of angular clasts of light-colored rock in a dark, finer grained matrix. The rock is granite, which was punished in the Pirate Fault zone, Tucson, Arizona. (Photograph by G. H. Davis.)

may indicate that the contact is indeed a fault contact, that is, not an unconformity or intrusive contact.

Fault rocks are formed by processes involving pervasive crushing and comminution of rock and minerals, brought about by repeated fracturing, grinding, and frictional sliding along networks of microcracks during recurrent fault movements.

As a result of intense fracturing, grinding, pull-apart, and comminution, cataclastic rocks are marked by unsorted, angular clasts in a finer-grained matrix (Figure 6.24). Fault rocks generally lack foliation, lineation, or any other preferred orientation of minerals, unless as relict vestiges of the original rock that was transformed into fault rock and not completely overprinted. Fault rocks have a wide range in clast size, and appear surprisingly similar when viewed at widely different scales of observation.

Gouge

Gouge is a light-colored *very* fine-grained clayey fault rock that is commonly found along fault surfaces and within fault zones (Figure 6.25). Grain size in general is less than 0.1 mm. In its dry state gouge feels like a loose to moderately compacted talcum powder, although admittedly a poor grade of talcum because it retains bits and pieces of grains that did not completely succumb to the crushing processes. When wet gouge crops out as a sticky clay, and globs of it will stick on your rock hammer. Zones of gouge may be wispy and thin, or they may be meters wide. The weak, friable character and high clay content of gouge indicates that it forms under relatively low-temperature and low-pressure conditions.

Figure 6.25 Fault gouge (white material beneath fracture surface) formed along low-dipping, nearly horizontal fault in the Rincon Mountains, Arizona. (Photograph by G. H. Davis.)

Breccia

Breccias are fault rocks composed of angular fragments (clasts) of wall rock set in a finer-grained matrix of crushed wall-rock material (Figure 6.26). Matrix is subordinate to clasts, comprising less than 30% of the breccia. Ordinarily the clasts in breccias appear to have translated and rotated with respect to one another during the faulting, but in some breccias it appears that the clasts were "frozen" during the midst of a simple three-dimensional pull-apart.

There are clast-size cutoffs for the classification of breccias. Ordinary **breccias** contain clasts that are larger than 1 mm and smaller than 0.5 m. **Megabreccias** contain clasts larger than 0.5 m (Figure 6.27). **Microbreccias** are composed of clasts that are smaller than 1 mm but greater than about 0.1 mm (Table 6.1).

Dilation and volume increase are characteristic of brecciation, which implies that the formation of breccias is favored in environments of low confining pressure and/or high fluid pressure. Voids representing pull-

Figure 6.26 (*A*) Brecciated granite. Note the slickensided slickenlined fault surface adjacent to the breccia. (Photograph by S. J. Reynolds.) (*B*) Photo of hand specimen of brecciated quartzite. (*C*) Polished slab of breccia. (Photographs by G. Kew.)

apart and dilation may not be completely filled by the finer-grained crushed matrix at the time of faulting. However, the void space may later be partially or completely filled by precipitation of minerals through circulation of groundwater and/or hydrothermal solutions. The crustification of minerals around the lining of a void is a typical product of partial open-space filling in breccias. Formation of open space during faulting is especially fortunate when mineralizing solutions have filled voids with precious or base metals.

When pore fluid pressures are elevated during the faulting, brecciated material may be partly flushed into tensional openings, forming **breccia dikes** (Figure 6.28). Such dikes do not show the telltale faulting signatures, such as slickensides or slickenlines, nor are they igneous intrusions.

Figure 6.28 Matt Davis checks out the breccias and microbreccias on Pontatoc Ridge in the foothills of the Santa Catalina Mountains, Tucson, Arizona. Dark zones in foreground are pockets and zones of breccia dikes, within which fine-grained crushed materials have been squirted along fractures and faults. (Photograph by G. H. Davis.)

Figure 6.27 Stan Ballard stands next to a spectacular megabreccia mapped by Mitch Reynolds in Titus Canyon, Death Valley, California. (Photograph by G. H. Davis.)

Figure 6.29 Photomicrograph of cataclasite derived through pervasive crushing and grinding. Width of field of view is approximately 4 mm. This sample was collected along the Catalina detachment fault in the Rincon Mountains (Arizona). (Photograph by G. H. Davis.)

Cataclasite

In some environments faulting results in the comminution of rock to a very fine-grained strongly indurated fault rock known as **cataclasite**. Grain size of cataclasite is like that of gouge, typically less than 0.1 mm, although it can contain up to 50% visible though still fine-grained clasts (Figure 6.29). More typically, cataclasites are ultra-fine-grained, nearly glassy in appearance. We use the term **ultracataclasite** in reference to the very finest, hardest, most ''glassy'' of the cataclasites.

The very fine-grain size of cataclasite reflects intense comminution and continual splitting of rock fragments and individual crystals. Cataclasites derived from granite, for example, microscopically display tiny angular chips of feldspar and quartz. The indurated cohesive nature of cataclasite reflects comminution under somewhat elevated temperature and pressure conditions. The tiny particles of crushed materials are cemented by hydrothermal or metamorphic minerals. Because of their resistance to erosion, cataclasites are often easy to spot, for they form protruding ledges and layers in the topography (Figure 6.30).

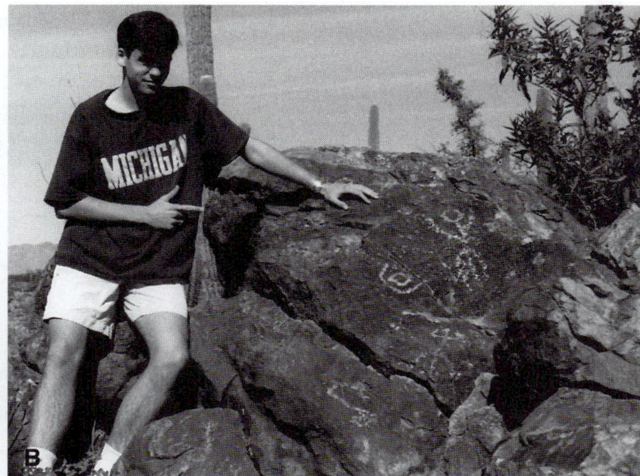

Figure 6.30 (*A*) Ledge of cataclasite in the Whipple Mountains, southeastern California. The top surface of the ledge marks the position of the Whipple Mountains detachment fault. (Photograph by S. J. Reynolds.) (*B*) Drew Davis points out extremely fine-grained ultracataclasite along a part of the Catalina detachment fault, Tucson, Arizona. He was not the first to check out the cataclasite. (Photograph by Drew's Dad.)

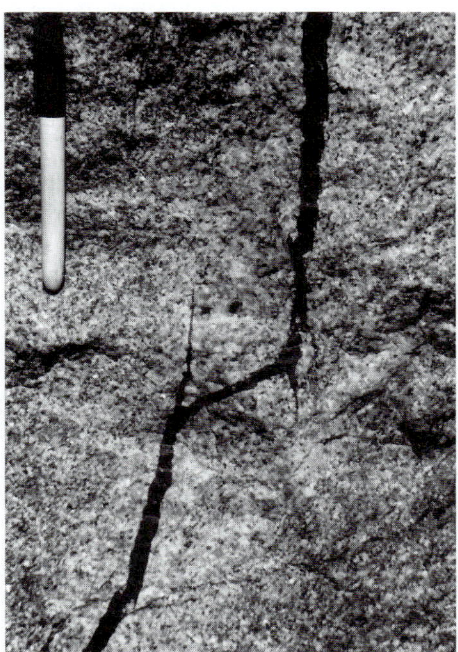

Figure 6.31 Black glassy pseudotachylite occupies a fracture in granitic rock in the Vredefort dome, South Africa. (Photograph by S. J. Reynolds.)

Pseudotachylite

The term **pseudotachylite** is reserved for a somewhat rare, but intriguing component of certain major fault zones, particularly those that have faulted the deep reaches of the Earth's crust. Tachylite is an old name for basaltic volcanic glass, but pseudotachylite is *not* a volcanic product. It is fault melt that "froze" to glass (Figure 6.31). Pseudotachylite occurrences are in essence fossil records of deep earthquakes (Sibson, 1980). To form pseudotachylite there must be a powerful shock under very high-pressure conditions and an instantaneously imposed strain.

Pseudotachylite is a dark, very fine-grained, originally glassy rock, with or without scattered clasts derived from the wall rocks. Pseudotachylite is sometimes seen microscopically to be completely glassy, but in most cases the original glass has partially to totally **devitrified** to a fine-grained crystalline matte. The matte may contain small radial and concentric clusters of crystals, analogous to **spherolites** in devitrified volcanic rocks. Pseudotachylite generally forms thin veins or selvages along discrete fault surfaces, which are the **generation surfaces** (Figure 6.32). Pseudotachylite may be restricted to the generation surface, although small injection veins typically branch out from the main generation surface into the wall rocks.

Pseudotachylite that forms along fault zones is thought to represent a small volume of melt produced by frictional heating during a seismic event—that is, an earthquake! In general, the thicker the pseudotachylite zone, the larger the amount of displacement on the generation surface, which in turn implies a larger magnitude paleoseismic event (Sibson, 1975, 1980).

Melting to form pseudotachylite probably requires high strain rates and reasonably high confining pressure. The formation of pseudotachylite may also require relatively dry rocks, because any water present would absorb the heat and carry it away from the generation surface. If the rocks are suddenly forced to deform at a much higher strain rate, such as during a seismic event or asteroid impact (Figure 6.33), ductile deformation mechanisms are not fast enough to accommodate the strain, stress builds up within the grains, and the rocks fail brittlely, by faulting. Frictional heating along the fault causes the already hot rocks to melt, resulting in the pseudotachylite. After the pseudotachylite is formed, continued low-strain rate ductile deformation may be localized in the fine-grained, and therefore relatively weak, glass. In such cases, the former presence of pseudotachylite is revealed by preserved injection veins that penetrate into the wall rocks adjacent to the generation surface.

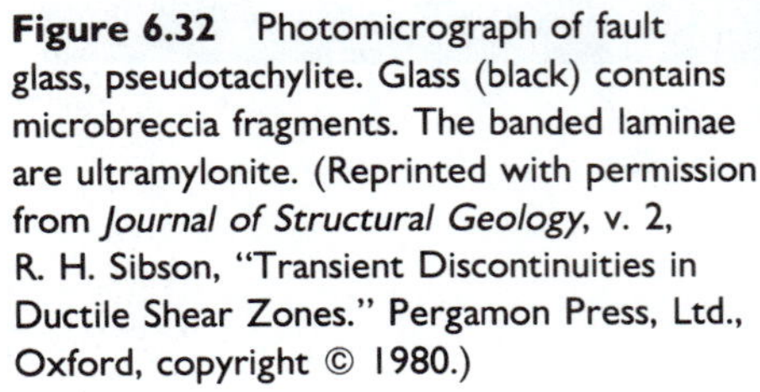

Figure 6.32 Photomicrograph of fault glass, pseudotachylite. Glass (black) contains microbreccia fragments. The banded laminae are ultramylonite. (Reprinted with permission from *Journal of Structural Geology*, v. 2, R. H. Sibson, "Transient Discontinuities in Ductile Shear Zones." Pergamon Press, Ltd., Oxford, copyright © 1980.)

Figure 6.33 Pseudotachylite (black) created during asteroid impact at Vredefort, South Africa. Precambrian rocks became thoroughly brecciated by the impact event. (Photograph by S. J. Reynolds.)

Experimental Work on the Origin of Fault Rocks

The origin of fault rocks of cataclastic origin has been explored through deformation experiments. In most such experiments, a fault is simulated by cutting a solid rock specimen into two parts along a cut inclined to the long axis of the specimen. The two halves are rejoined, inserted together in a deformation apparatus, and loaded until the ''fault'' slips. To simulate larger amounts of fault slip, two cylindrical specimens are placed end to end, and one specimen is rotated against the other. The temperatures and confining pressures can be varied to simulate different crustal levels. These experiments show that at the temperatures, pressures, and strain rates expected in the upper crust (approximately 10 km or less), deformation occurs by frictional sliding, presumably accompanied by cataclastic flow once displacement has been sufficient to produce an adequate thickness of crushed, cataclastic material.

Experiments and field studies reveal that cataclastic fault rocks have a **self-similar character** (Sammis and others, 1986). Like many *fractal* phenomena, cataclastic rocks look similar at a wide range of magnifications (Figure 6.34), from outcrop scale to microscopic. At each scale of observa-

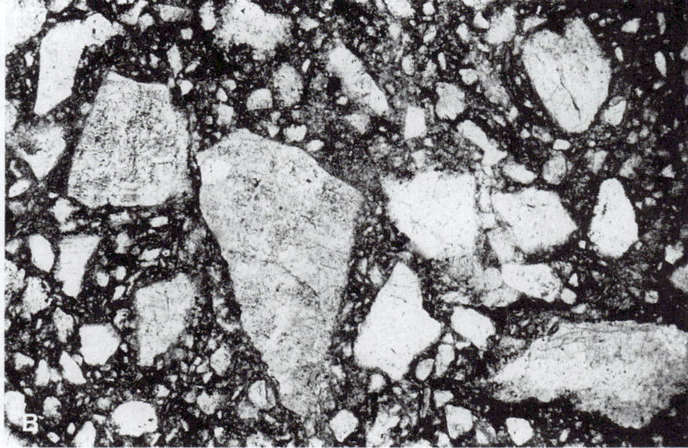

Figure 6.34 (*A*) A typical cataclastic texture, seen at outcrop scale. The texture would look the same viewed microscopically, at almost any scale. The rock happens to be from the foothills of the Santa Catalina Mountains, Tucson, Arizona. Find the penny for scale. (Photograph by S. J. Reynolds.) (*B*) Photomicrograph of cataclastic texture, revealing clasts sprinkled in a very fine-grained matrix. (Photograph by G. H. Davis.)

Figure 6.35 Photomicrograph showing results of progressive development of self-similar character of cataclastic textures.

tion, a specimen of cataclastic fault rocks has about the same overall appearance, including the same proportion of clasts to matrix. Under progressively higher magnification, what appears at lower magnification to be the fine-grained matrix is found to be clasts in a yet finer-grained matrix. This pattern repeats itself down to extremely small, microscopic scales.

The explanation for the self-similar character of cataclastic fault rocks is that fracturing preferentially "targets" any clast that abuts against a clast of comparable size. Two similarly sized, nearest neighbors concentrate the imposed stress along the points of contact, and the clast that is weakest will fail. The broken clast will now consist of two or more smaller clasts, which will then interact with each other and with other similar-sized, neighboring grains. In contrast, the larger surviving grain becomes totally surrounded by smaller grains and will not statistically experience the requisite stress concentration to cause failure. Cataclasis therefore leads to a rock with a few, relatively large clasts scattered about in a fine-grained matrix (Figure 6.35).

MAP AND SUBSURFACE EXPRESSIONS OF FAULTS

Geologic Map Expressions

Systematic geologic mapping has proved to be an extremely effective method for locating faults (see Part III-A), particularly where the faults cut a geologic column of sedimentary or volcanic rocks whose stratigraphy is well known. Geologic maps reveal the plan-view expression(s) of fault patterns, and books have been written to explore the relationships (e.g., Bolten, 1989; Spencer, 1993). Faults are identified and tracked on the basis of mapped patterns that disclose **truncation** and **offset** of one or more bedrock units (Figure 6.36).

Truncation and offset can take many forms, depending on the orientations of the faults, the movements on them, and the orientations of the rock layers that are cut and displaced. Truncation and offset usually result in an apparent horizontal (i.e., map-view) shifting of mapped bedrock units.

Where faults trend parallel to the strike of bedding, translation does not produce the simple but conspicuous horizontal shifting of layers. Instead, the presence of such faults must be recognized on the basis of

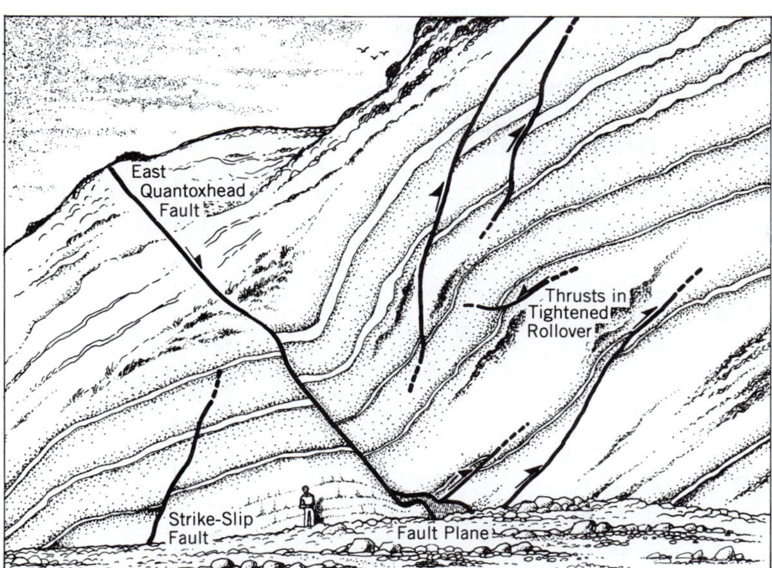

Figure 6.36 Faults can be discovered and mapped on the basis of truncation and offset of bedrock units. Here the East Quantoxhead fault and associated faults reveal themselves in the disturbed stratigraphy. [From Peacock and Sanderson (1992), Figure 7A. Reprinted with permission from The Geological Society Publishing House, Bath, England.]

inconsistent stratigraphic patterns. The rock formation shown in Figure 6.37 is composed of 10 distinctive units (numbered in the figure) and is cut by two faults that strike parallel to bedding. The two faults dip such that they cut across the bedding at steep angles. The presence of the two faults might not be immediately apparent in the geometry of the map pattern, but knowledge of the stratigraphy of the formation permits the faults to be recognized by **repetition** and **omission of strata**. If, during a traverse across the map area (Figure 6.37), we move up-section from unit 1 to unit 3, only to find ourselves crossing back unexpectedly into a *repeated* section of the same units, we would recognize that we had crossed a fault. And if during our continued traverse we were to cross directly from unit 3 into unit 8, the *omission* of units 4 through 7 would

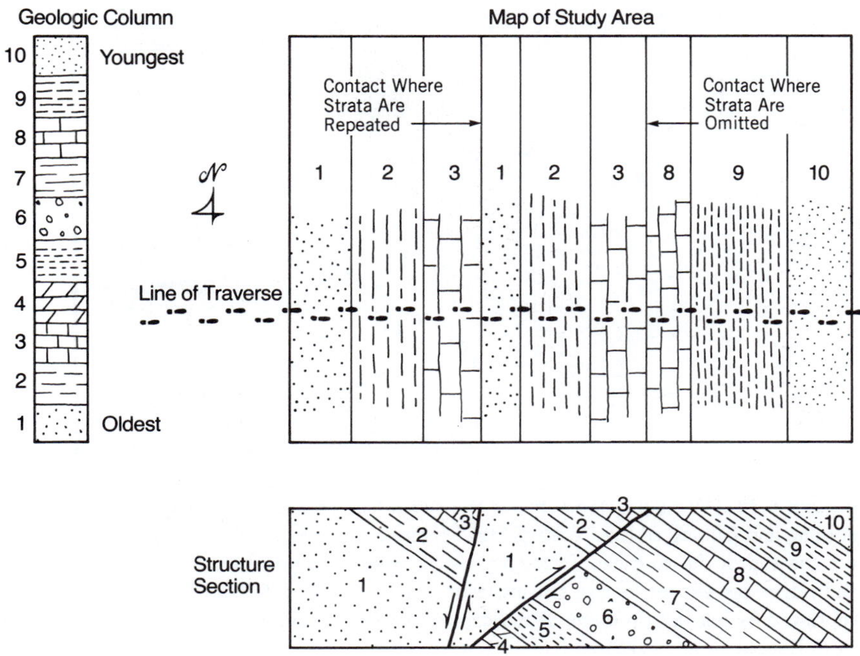

Figure 6.37 Identification of locations of faults on the basis of inconsistent stratigraphic patterns. One fault is marked by repetition of strata, the other by omission of strata.

alert us to the fact that we had crossed a second fault. Unconformities can also account for omission of strata, so be careful! Cross sections through faulted ground reveal such relationships (see Part III-E).

Expressions in Drilling Data

Petroleum explorationists have learned to spot faults in the subsurface based upon drilling data. For example, in the Gulf Coast basin, salt intru-

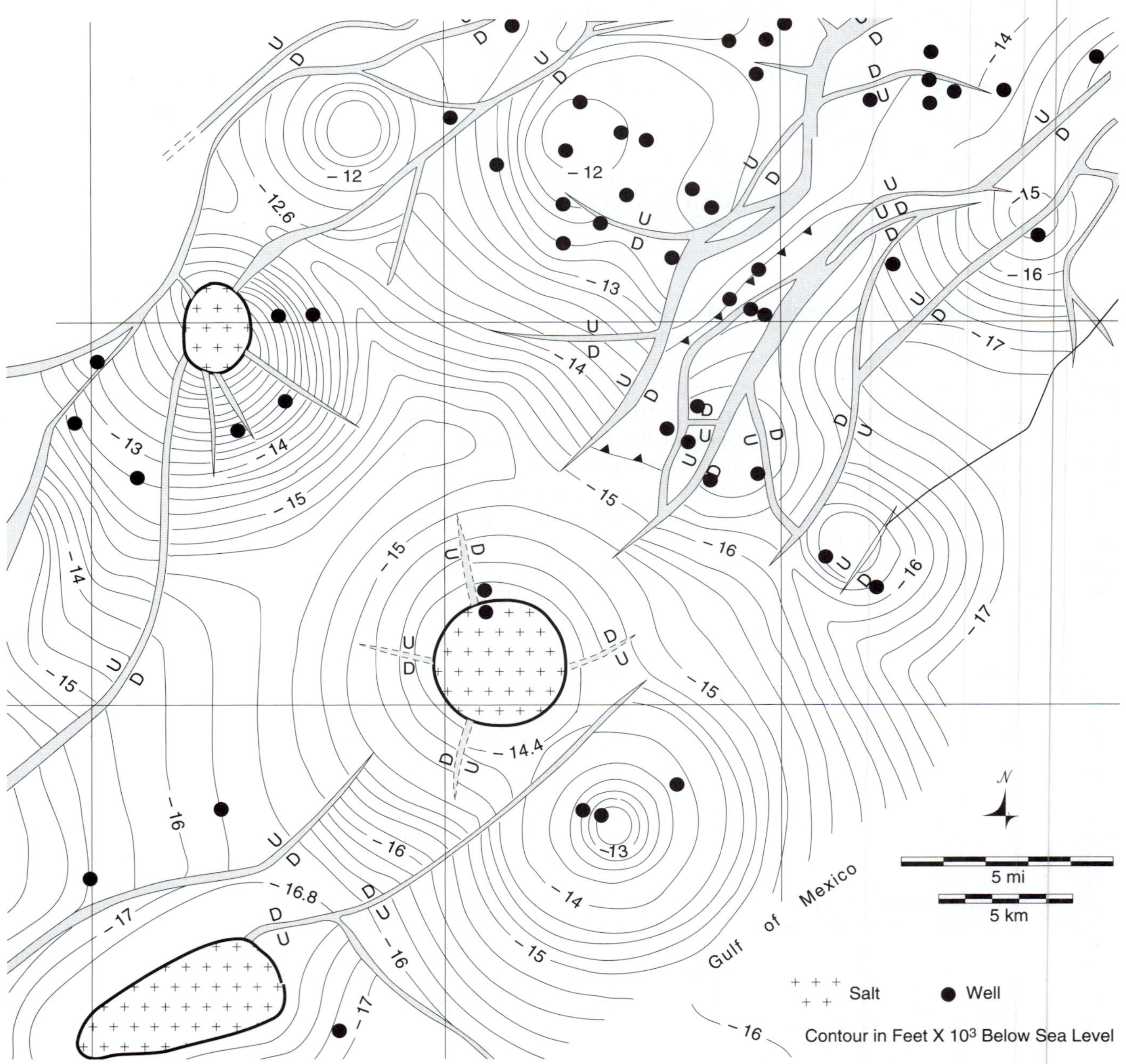

Figure 6.38 Wells provide some of the subsurface control in defining the pattern of doming and faulting of the Tertiary Frio Formation in the Gulf Coast region. Contours are the top of the so-called T5 horizon within the Frio Formation. The mounds, domes, and ridges are expressions of salt intrusion. [From Capuano (1993), Figure 1, p. 1304, reprinted by permission of the American Association of Petroleum Geologists. Modified from Bebout et al. (1978).]

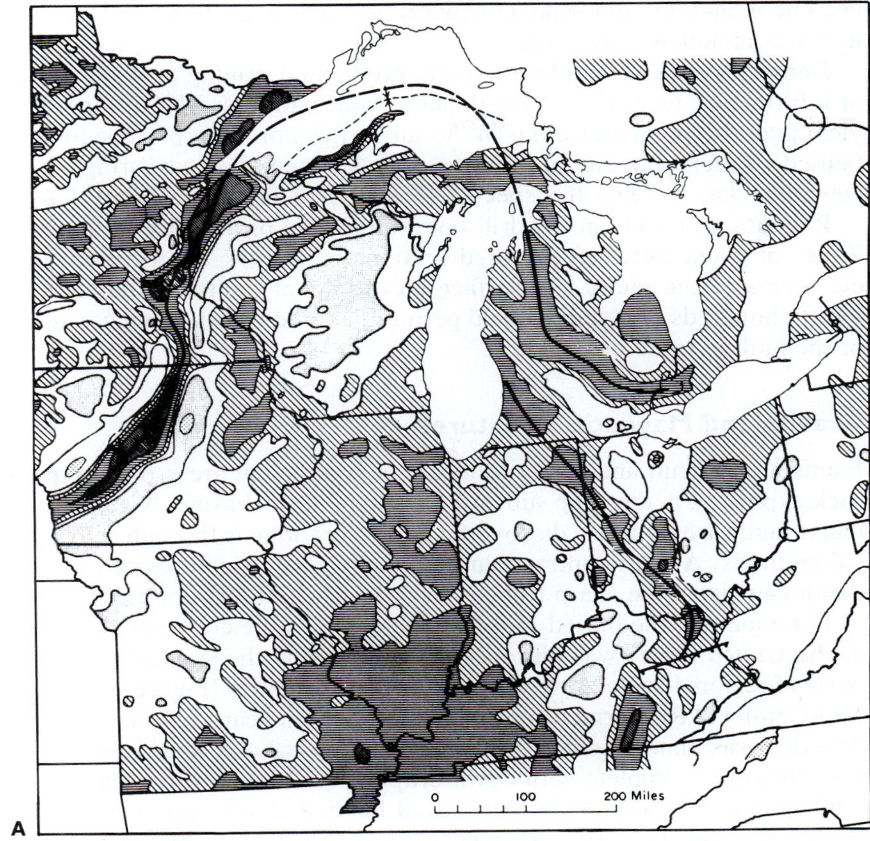

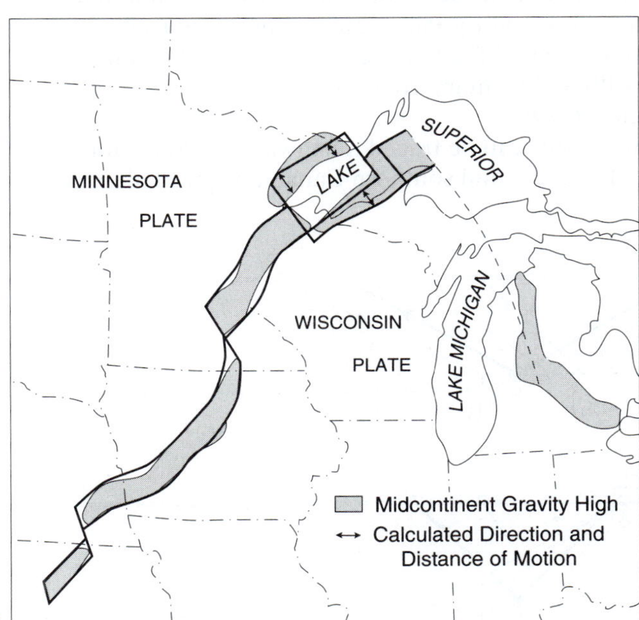

Figure 6.39 Geophysical expression of the Mid-Continent Rift, which formed in the Precambrian. (*A*) Bouguer gravity anomaly map of the midwestern United States. The map reveals a very pronounced positive anomaly (dark) with marginal negative anomalies. The positive anomaly coincides with a belt of Precambrian mafic igneous rocks estimated in places to be close to 10,000 m thick. (*B*) Interpretation of the so-called mid-continent gravity high as a Precambrian intracratonic rift filled by mafic materials. Arrows indicate direction of movement involved in the opening of the rift. [From Spencer (1969), Figures 18-15 and 18-16, p. 350–351.]

sions contribute to the trapping of oil and gas in country rock sediments that they have domed and, in some cases, pierced (Halbouty, 1969). Dome-like, pipelike, and pluglike intrusions of salt ascend hundreds to thousands of meters from their source bed(s) into and/or through the overlying sedimentary cover (Figure 6.38). Stiff strata directly above the salt intrusions become folded into domelike configurations, and become faulted in ways that accommodate extension.

Dense networks of exploration and production drilling provide a base of information to work out the detailed fault patterns. The drilling data yield point-by-point elevations in the subsurface for the top of the folded, faulted stratigraphic unit(s) of interest, and systematic contouring of the elevation data reveals the specific locations of faults (see Part III-F).

Repetition of units in the drill hole signals the probability of faulting. Zones of gouge between repeated sections may mark the locations and elevations of the faults. Subsurface elevations of faults encountered in tens to hundreds of holes in a field permit the three-dimensional geometry of the fault(s) to be modeled.

Gravity and Magnetic Signatures

Faults are difficult and costly to locate in continental areas of very poor rock exposure, in the deep subsurface, and in ocean basins. Nonetheless, major faults are routinely discovered by explorationists through geophysical methods. Abrupt contrasts in the geophysical signatures of rocks at depth can signal the sharp truncation of bedrock at locations of faults.

For example, the buried Mid-Continent Rift in the continental interior of the United States was discovered and traced in the subsurface on the basis of regional-scale gravity and magnetic anomalies (Figure 6.39). The Mid-Continent Rift formed approximately a billion years ago, when North America was almost split in half by tectonic forces. A long, continuous fracture zone, complete with down-dropped rift basins, was created, extending from Canada to Oklahoma and beyond. Mafic igneous rock intruded into the rift, and extruded onto the surface. Although now in the subsurface and blanketed by Paleozoic sedimentary rocks, the mafic igneous rocks and associated faults have not remained hidden. The high-density contrast with sedimentary rocks stands out sharply on regional gravity maps (Figure 6.39).

The Mid-Continent Rift can be traced northward to Precambrian exposures of northern Minnesota and southern Canada, where Paleozoic rocks

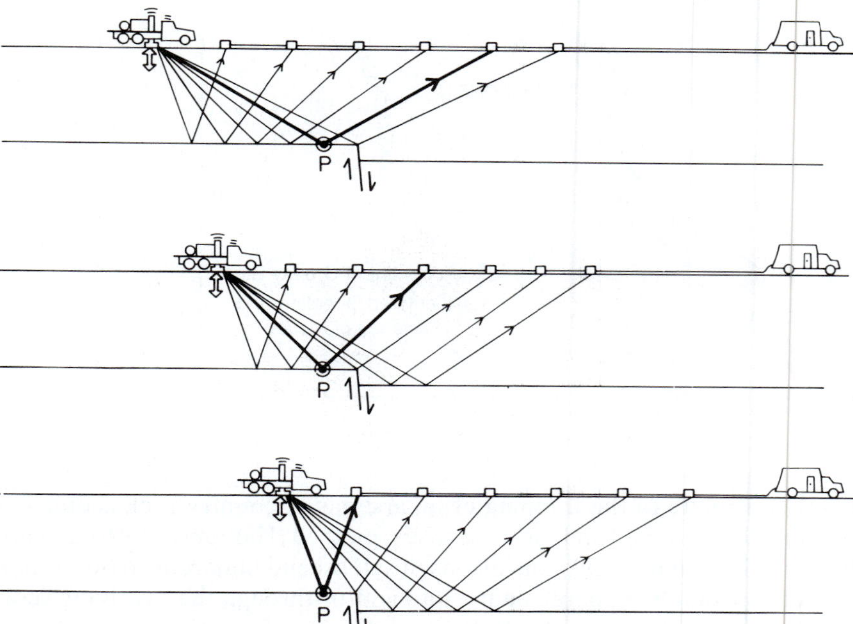

Figure 6.40 Schematic illustration of the process of seismic-reflection profiling. (From "The Southern Appalachians and the Growth of Continents" by F. A. Cook, L. D. Brown, and J. E. Oliver. Copyright © 1980 by Scientific American, Inc. All rights reserved.)

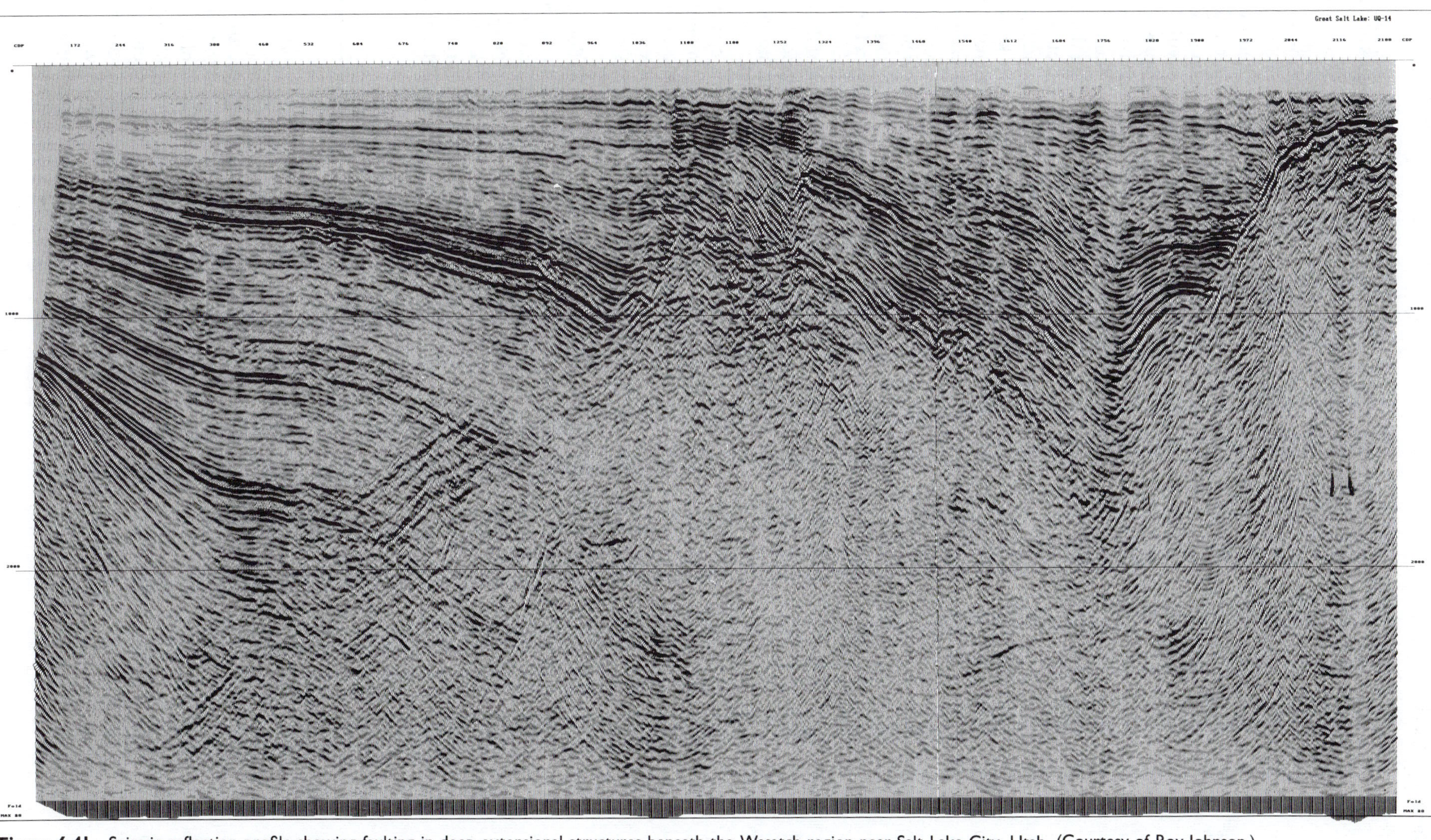

Figure 6.41 Seismic reflection profile showing faulting in deep extensional structures beneath the Wasatch region near Salt Lake City, Utah. (Courtesy of Roy Johnson.)

are stripped away and the Precambrian mafic igneous rocks and associated rift-basin sediments crop out right at the surface.

Seismic Expression

The ''bread-and-butter'' of geophysical methodology used by petroleum exploration companies is **seismics**. In carrying out seismic-reflection profiling, sound waves are propagated through the subsurface by setting off charges of dynamite or by shaking the ground surface with large truck-mounted vibrators. The waves speed to depth, where they eventually are reflected by discontinuities marking sharp contrasts in density and rigidity between rocks above and below (Figure 6.40). Since faults often bring rocks of different character into contact, they are among the discontinuities that are capable of causing sound waves to be reflected. The sound energy radiates back to the surface, where its character and time of arrival are collected and measured through vibration sensors known as geophones. The challenge is to decipher which incoming signals ''bounced'' off the same reflector, and to determine the depth of each reflector at every point along the seismic line.

Interpreting sound waves reflected from multiple levels in the crust and mantle has been compared to interpreting the ''sonar'' signals of a bat flying about in a small room. To build an accurate geological cross section of the relationships, the velocity distortions must be removed, which requires both a theoretical and a practical understanding of the fundamentals and idiosyncrasies of seismic reflection. Once seismic distortion has been removed, structural geological distortion remains, and this, after all, is the object of the probe.

Countless examples of seismic methods applications can be found in structural analysis. For example, seismic reflection techniques are absolutely essential to deciphering the subsurface fault relations in the Wasatch region near Salt Lake City, Utah. The interpreted profiles reveal extensional structure in the form of faults separating tilt-blocks of sedimentary strata (Figure 6.41).

Geophysical exploration of fault structures in the ocean floor completely changed the way in which Earth dynamics is viewed. Mapping the distribution of magnetic patterns in seafloor sediments and volcanics revealed the presence of **transform faults**, which we will study in Chapter 10. Transform faults constitute by far the most important fault network on the face of the globe. The discovery of these faults and the interpretation of their significance were integral to the revolution called plate tectonics.

THE NAMING OF FAULTS

Slip and Separation

Most faults that we see in the field, or in the subsurface, are named and described on the basis of the dip of the fault surface and the direction and sense of offset. The ideal goal is to establish the direction of displacement, the sense of displacement, and the magnitude of displacement (Figure 6.42). These three components constitute the **slip**, which is the *actual* relative displacement. Where the basis for a slip determination is unavailable, we settle for **separation**, which is the *apparent* relative displacement. Distinguishing slip from separation is one of the most important steps in fault analysis (Crowell, 1959; Hill, 1959).

Figure 6.42 The "key" to describing *slip* along a fault lies in measuring (1) the direction of displacement, (2) the sense of displacement, and (3) the magnitude of displacement. (Photograph by G. H. Davis.)

Slip Classification

If we can establish the three components of slip for a given fault, we can use special terms to "name" the fault. **Strike–slip faults** accommodate horizontal slip between adjacent blocks. They are described as **left-handed** or **right-handed**, depending on the sense of actual relative movement (Figure 6.43*A*). **Dip–slip faults** are marked by translation directly up or down the dip of the fault surface (Figure 6.43*B*). Movement on a dip–slip fault is described with reference to the relative movement of **hanging wall** and **footwall**. The hanging wall is the fault block toward which the fault dips. The footwall is the fault block on the underside of the fault. The terms "hanging wall" and "footwall" are derived from old mining jargon: a prospector working in a drift that he has tunneled along an inclined fault finds his head close to the hanging wall, his feet on the footwall (Figure 6.43*B*).

Several kinds of dip–slip faults exist (Figure 6.43*B*). A **normal-slip fault** is one in which the hanging wall moves down with respect to the footwall. Many normal-slip faults dip at moderate to steep angles, averaging 60° or so. Normal-slip faults that dip less than 45° are referred to as **low-angle normal-slip faults**. **Thrust-slip** and **reverse-slip faults** are marked by movement of the hanging wall *upward* relative to the footwall. Thrust-slip faults dip less than 45°, usually around 30° or so. Reverse-slip faults dip more steeply than 45°.

Translation on **oblique-slip faults** is inclined between strike–slip and dip–slip (Figure 6.43*C*). The rake of the displacement vector lies somewhere between 0° to 90°. To name oblique-slip faults in an informative manner, the terms "normal," "reverse," "thrust," "right-handed," and "left-handed" are combined in ways that conform with the interpreted direction and sense of translation. If the main component is strike–slip, then "right-handed" or "left-handed" is used as a modifier, preceded by "normal," "reverse," or "thrust," depending on the dip–slip movement. If the chief component is dip–slip, then "normal-slip," "reverse-slip," or "thrust-slip," is used as the modifier, preceded by "right-handed" or "left-handed," depending on the strike–slip component of movement (see Figure 6.43*C*).

Some faults are rotational or scissorslike. A rotational fault changes both in its magnitude of slip and in its sense of slip along strike, which presents quite a challenge. Thus, a rotational fault may be normal slip along part of its length and reverse slip along another part (Figure 6.43*D*).

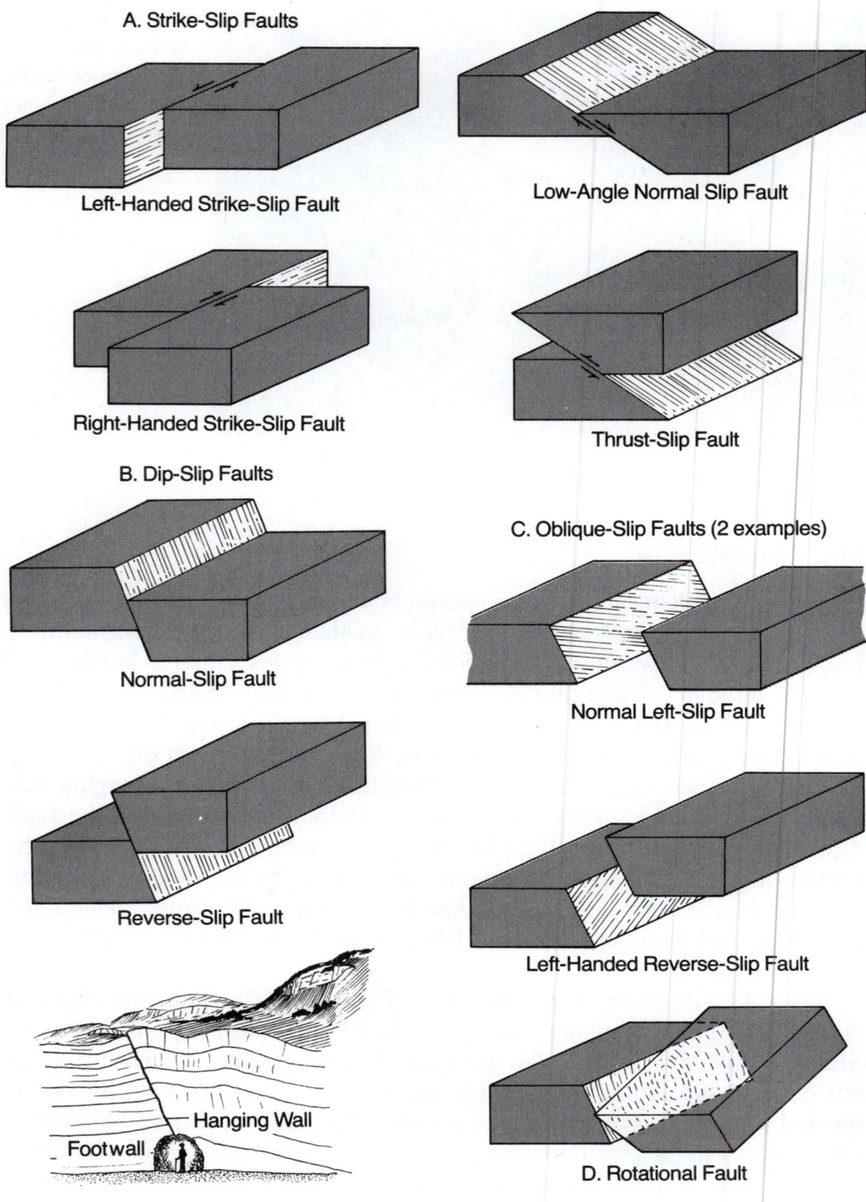

Figure 6.43 Slip classification of faulting. (*A*) Block diagrams showing left-handed and right-handed strike-slip faulting. (*B*) Block diagrams showing normal slip, reverse-slip, low-angle normal slip, and thrust-slip faulting. Prospector in tunnel pauses to think about the difference between the footwall and the hanging wall of a fault. (*C*) Examples of oblique-slip faults, including a normal left-slip fault and a left-handed reverse-slip fault. (*D*) Schematic block diagram of a rotational fault.

The Separation Problem

Interpreting the direction and sense of slip on faults is often complicated by deceptive patterns created by the interaction of structure and topography and by the absence of minor structures that, if present, would have helped to define the slip path. The offset observed along faults in outcrops, in map patterns, and in structure profiles is **separation**. Separation refers only to the apparent sense and magnitude of offset along faults; it is the *apparent* relative movement, and it is described irrespective of the *actual* relative displacement. The faulted basalt layer shown in map view in Figure 6.44*A* displays 200 m of right-lateral separation. The faulted sandstone layer seen in cross section in Figure 6.44*B* is marked by 6 m of normal separation.

Stratigraphic throw is a special measurement of separation, one that is commonly used in exploration geology as a convenient measure of the magnitude of faulting. Stratigraphic throw is the thickness of the stratigraphic interval between two beds that are brought into contact by faulting (Figure 6.45). At point *A* in Figure 6.45, faulting has brought the base of unit 3 and the top of unit 4 into contact. Throw equals the combined thickness of units 3 and 4, namely 174 ft (53 m). At point *B*, the base of unit 5 is faulted into contact with the base of unit 6. Throw equals the thickness of unit 5, namely 172 ft (52 m) (Figure 6.45).

Apparent relative movement seldom corresponds with actual relative movement. The separation that we view on geologic maps and in canyon walls is a product of many influences: orientation of layering; strike and dip of the fault surface; slip on the fault, including direction, sense, and magnitude of displacement; and the orientation of the exposure in which separation is viewed. Consider some simple examples. The tilted sedimentary rocks shown in Figure 6.46*A* are cut by a normal-slip fault whose magnitude of net slip is 600 ft (182 m). The fault strikes at right angles to the strike of bedding. Following erosional beveling of the faulted terrain to a common level (Figure 6.46*B*), the outcrop relationships convey the false impression that left-handed strike–slip faulting had taken place. Conversely, left-handed strike–slip faulting of a tilted sequence of rocks (Figure 6.47*A*) can produce structural relationships that in cross-sectional view invite interpretation by normal-slip faulting (Figure 6.47*B*). The possibilities are limitless!

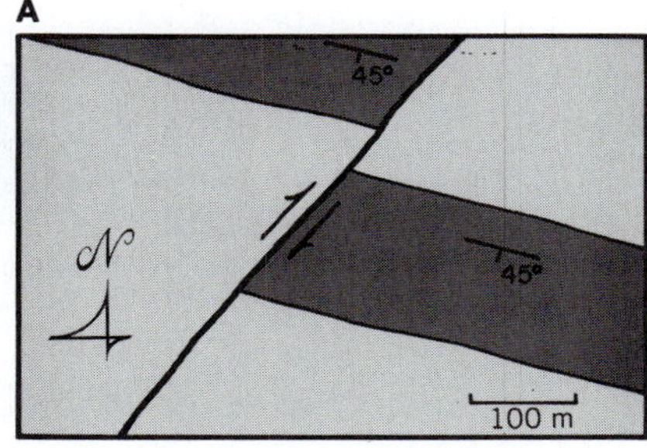

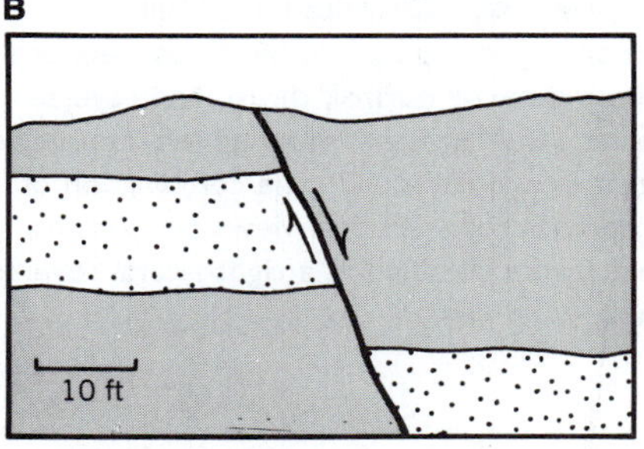

Figure 6.44 (*A*) Map view of right-lateral separation of a basalt layer. (*B*) Cross-sectional view of normal separation of a sandstone layer.

Separation Classification

In confronting the separation problem we learn that slip cannot be determined on the basis of apparent relative displacement. At best, knowledge of separation places modest constraints on what the translation might have been. Such limitations notwithstanding, evaluation of separation constitutes the main descriptive record for faulted rocks or faulted regions.

Separations viewed in cross-sectional exposures are described simply as "normal," "thrust," and "reverse" (Figure 6.48*A*). Normal separation is marked by apparent offset of the hanging wall downward relative to the footwall. Thrust and reverse faults, in the separation sense, are characterized by offset of the hanging-wall rocks upward relative to the footwall. Thrust faults dip less than 45° and reverse faults dip more steeply than 45°. Separations viewed in plan view are described as *left lateral* or *right*

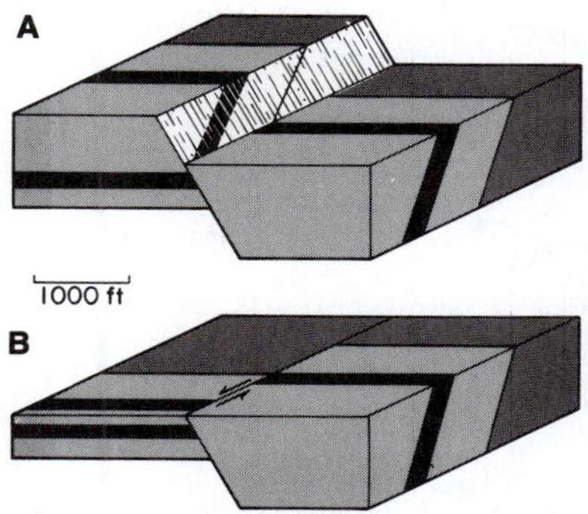

Figure 6.46 Block diagrams that underscore the difference between slip and separation. (*A*) Inclined layers are displaced along a normal-slip fault. (*B*) Erosion of the upper reaches of the footwall block creates the illusion of left-handed strike–slip faulting.

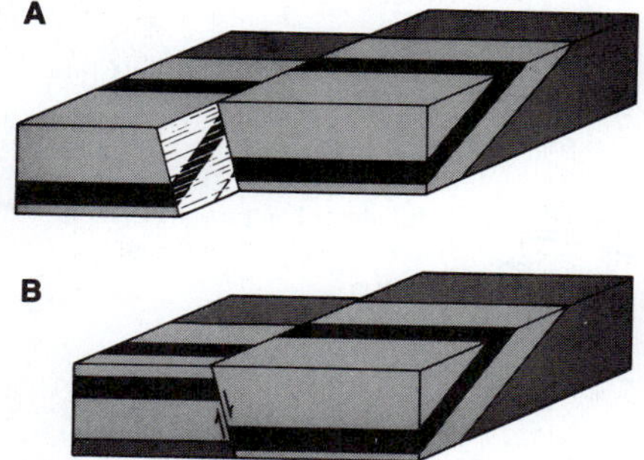

Figure 6.47 (*A*) Left-handed strike–slip faulting of inclined layers. (*B*) Erosion of the front end of the footwall block gives the impression that the faulting was normal slip.

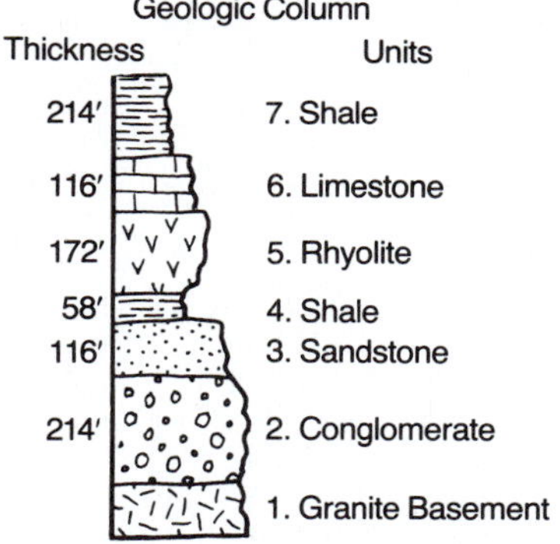

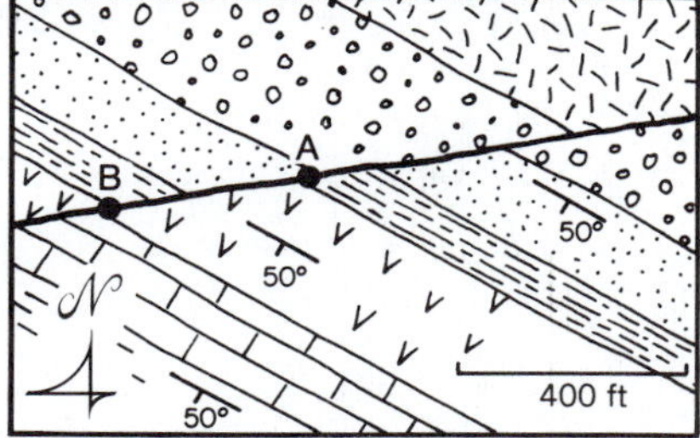

Figure 6.45 Stratigraphic throw at point *A* is equal to the combined thickness of units 3 and 4, or 174 ft (53 m). Stratigraphic throw at point *B* is simply equal to the thickness of unit 5, which is 172 ft (52 m).

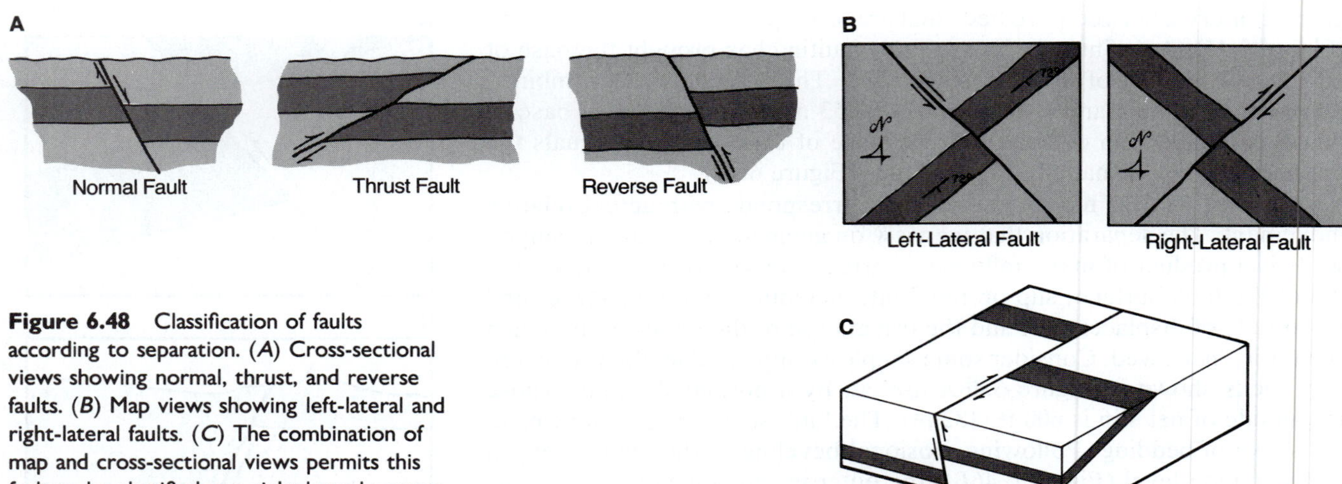

Figure 6.48 Classification of faults according to separation. (*A*) Cross-sectional views showing normal, thrust, and reverse faults. (*B*) Map views showing left-lateral and right-lateral faults. (*C*) The combination of map and cross-sectional views permits this fault to be classified as a right-lateral reverse fault.

lateral (Figure 6.48*B*). The suffix "-handed" is reserved for proclaiming slip.

Complete descriptions of separation are based both on plan-view and cross-sectional observations of offset (Figure 6.48*C*). If the main separation is lateral, *right-lateral* or *left-lateral* is used as the modifier, preceded by *normal*, *reverse*, or *thrust*, depending on the separation seen in cross-sectional view. If the chief separation is normal, reverse, or thrust, one of these terms is used as the modifier, preceded by *right-lateral* or *left-lateral*, depending on the separation observed in plan (Figure 6.48*C*). Figure 6.49 presents the separation classification and suggestions regarding map symbology for distinguishing separation and slip.

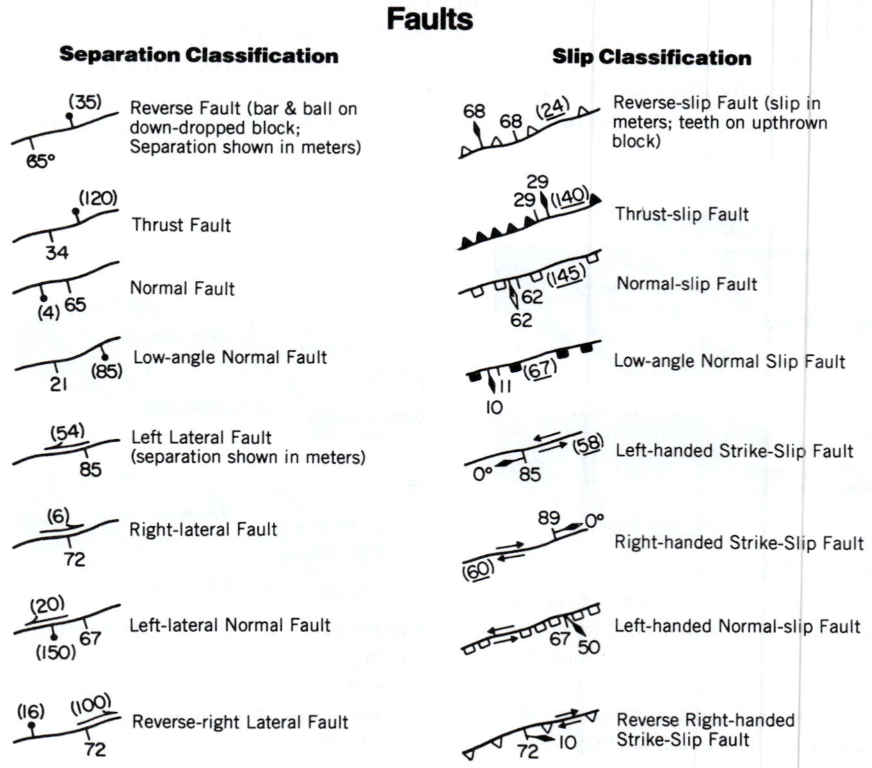

Figure 6.49 Suggested map symbols for distinguishing between separation and slip.

DETERMINATION OF SLIP ON FAULTS

Using Slickensides and Grooves as Guides to Slip

Slip is most commonly deduced on the basis of outcrop features. The rake of slickenlines and grooves on a fault surface is an inscription of the direction of net slip. Faults and fault systems that accommodate single movements and/or sustained but simple movement plans commonly reveal slickenlines and grooves of uniform orientation.

For example, consider the mapped fault relationship shown in Figure 6.50*A*. The fault strikes N30°E and dips 70°SE. It separates volcanic rocks of Jurassic age on the footwall from Permian sedimentary rocks on the hanging wall. Permian strata were faulted upward with respect to the Jurassic volcanics, but along what line? Slickenlines on the fault surface reveal the direction of slip. They rake 80°SW. This indicates oblique-slip faulting, featuring a major dip–slip component of translation. Given the orientation of the slickenlines and the age relationships of the rocks on hanging wall and footwall, the fault is classified as a *left-handed reverse-slip* fault. The actual magnitude of slip is determined by constructing a vertical cross section parallel to the trend of the slip direction, and measuring the distance between the **footwall cutoff** and the **hanging-wall cutoff** of a common geologic horizon, like a marker bed (Figure 6.50*B*).

There may be difficulties in using striations and other slickenlines as a guide to slip. Many faults undergo multiple, complex histories of movement. Although this is not common, slickensided surfaces can be marked by slickenlines of different orientations. "Scatter" in slickenline orientations may prevent simple "naming" of faults according to slip criteria. As if this were not enough, the generally held opinion among many geologists is that slickenlines on fault surfaces reflect only the latest fault movement. If indeed faults are good "erasers," heavy reliance on slickenlines alone as a guide to the direction of fault movement may lead to oversimplified or incorrect interpretations. Nonetheless, slickenlines and grooves remain an integral part of the descriptive record. They are useful kinematic guides to slip, provided they are treated with the care that the foregoing words of caution suggest.

Using Drag Folds as Guides to Slip

Drag folds also can be used to determine the direction and sense of slip during faulting. Drag folding is a distortion of bedding, or other layering, resulting from shearing of rock bodies past one another. Consider how a sequence of sedimentary rock is affected by fault movements (Figure 6.51*A*). Strata close to the fault surface are deformed by frictional drag into folds that are convex in the direction of relative slip. The truncated ends of dragged layers point away from the sense of actual relative movement. Ideally, hanging-wall strata of a thrust-slip or reverse-slip fault are dragged into an anticline, whereas the footwall strata are dragged into a syncline (Figure 6.51*A*). Similarly, under ideal circumstances, hanging-wall strata of a normal-slip fault are dragged into a syncline, while the footwall strata are dragged into an anticline (Figure 6.51*B*).

Drag folds resulting from strike–slip faulting can be spectacular, especially where steeply dipping layering is radically folded. Right-handed and left-handed patterns of drag are distinctly different (Figure 6.51*C*). Within broad fault zones marked by spaced strike–slip faults, individual layers can become completely fault bounded. And if each end of a fault-bounded layer is curled by drag, **sigmoidal drag folds** result (Figure 6.51*D*). The

A

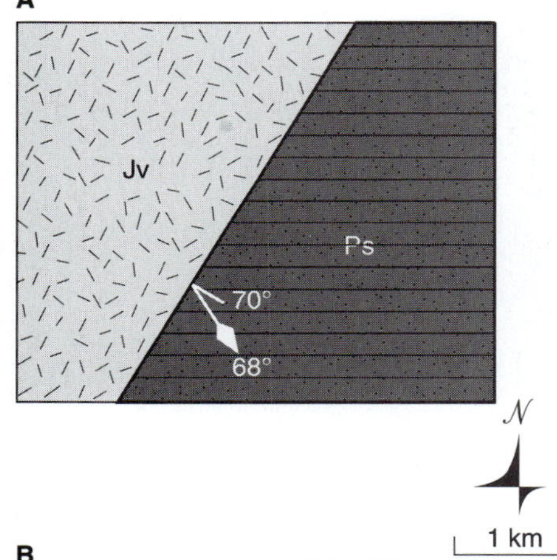

B

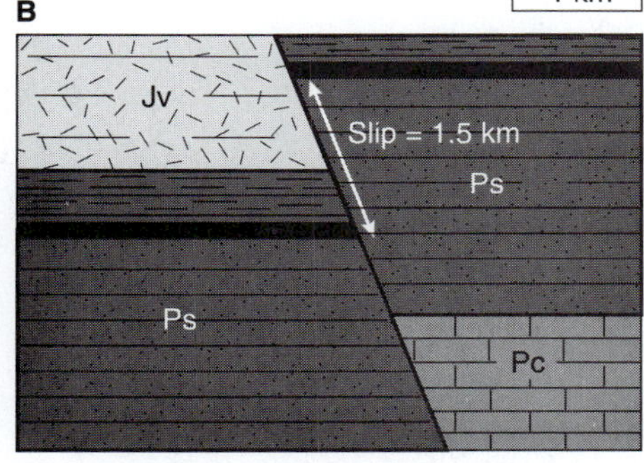

Figure 6.50 (*A*) Map of fault contact between Jurassic and Permian strata. Note that through symbols a distinction is made between the dip of the fault and the trend and plunge of slickenlines on the fault surface. The fault is a left-handed normal-slip fault. (*B*) The actual slip along the fault can be measured along a vertical cross section drawn parallel to the trend of the slickenlines, and then matching up a faulted layer.

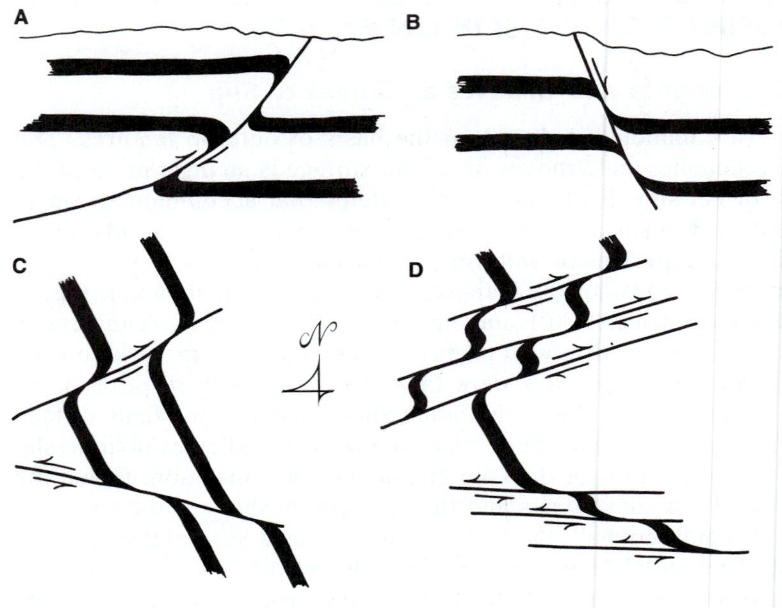

Figure 6.51 Examples of drag folds. (*A*) Strata on the hanging wall of thrust-slip and reverse-slip faults are dragged into an anticline, whereas strata on the footwall are dragged into a syncline. (*B*) Drag fold relationships along a normal-slip fault. (*C*) Patterns of drag folding along right-handed and left-handed strike–slip faults. (*D*) Sigmoidal drag folding along closed-spaced, distributed right-handed and left-handed strike–slip faults.

layers are doubly curved. Sigmoidal drag folds formed by right-handed simple shear are gently S shaped. Those formed by left-handed simple shear resemble backward S's, the kind that kids paint on signs. Folds shaped like backward S's are referred to as Z shaped. A wonderful example of sigmoidal folding within a strike–slip fault zone is shown in Figure 6.52, a detailed structure map of the Stockton Pass fault, southern Arizona, prepared by Swan (1976).

Note that where dip-slip faults (i.e., normal-slip, reverse-slip, thrust-slip) cut and drag horizontal strata, the axes of drag folds will be oriented approximately horizontal, perpendicular to the direction of slip (Figure 6.51*A,B*). Where pure strike–slip faulting operates along vertical surfaces cutting steeply dipping layers, the axes of resulting drag folds will be steep, and perpendicular to the direction of movement (Figure 6.51*C,D*).

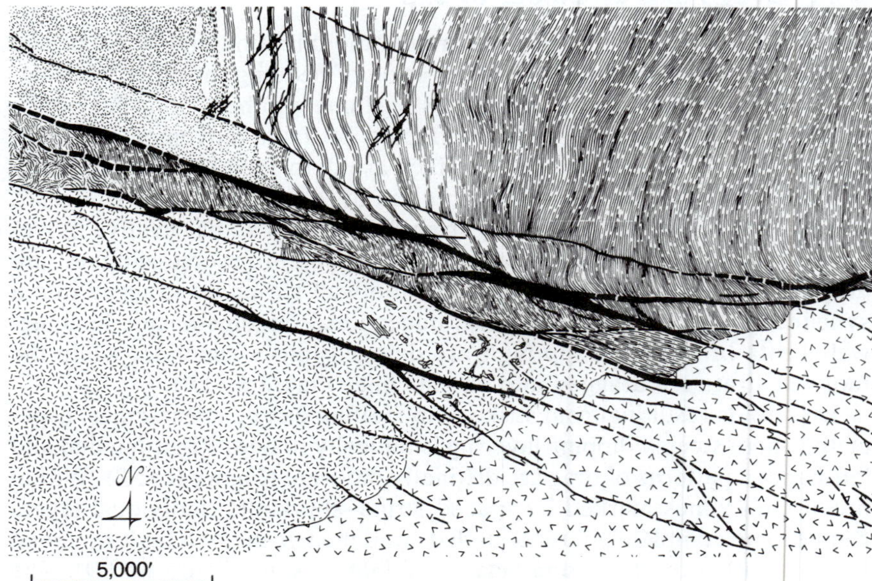

Figure 6.52 Drag folding of foliation in Precambrian gneiss, Stockton Pass, Arizona. Drag folding was caused by left-handed strike–slip faulting along the Stockton Pass fault zone. [From Swan (1976).]

The orientations of slickenlines and drag folds should be mutually perpendicular.

Tight asymmetrical drag folds commonly occur in the interior of fault zones; these are very useful in evaluating the sense of simple shear movement during faulting. The asymmetrical folds, when viewed downplunge, can be described as S shaped or Z shaped (Figure 6.53). The clockwise rotation that typifies asymmetric, Z-shaped drag folds reflects right-handed shear. The counterclockwise rotation that characterizes asymmetric S-shaped drag folds reflects left-handed shear. Learning to interpret asymmetric drag folds is an important aid in naming faults according to slip.

Bold generalizations are dangerous when it comes to drag folding, because rocks do not always cooperate in the ideal ways just described. The most notorious exception to the rule is **reverse drag** on normal-slip faults. Originally horizontal bedding in the hanging wall of a normal-slip fault can become folded in such a way that it actually dips toward the fault surface, thus forming a **rollover anticline** (Figure 6.54). Hamblin (1965) first recognized this phenomenon in faulted Colorado Plateau strata. He reasoned that reverse drag may be unique to listric fault geometries. **Listric faults** are curved faults that flatten with depth (Figure 6.54). Sustained movement on a listric normal-slip fault favors a pulling away of the hanging-wall block from the footwall. Such a gap never actually develops because of such folding (Figure 6.54*A*), or **antithetic faulting**, into the zone of potential separation (Figure 6.54*B*). We will say more about reverse drag folding when we talk about normal faulting.

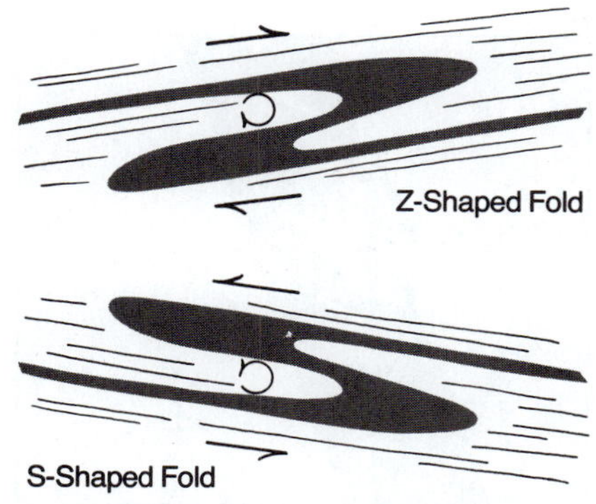

Figure 6.53 The forms of Z-shaped and S-shaped drag folds: Z-shaped fold forms result from clockwise internal rotation; S-shaped folds result from counterclockwise internal rotation.

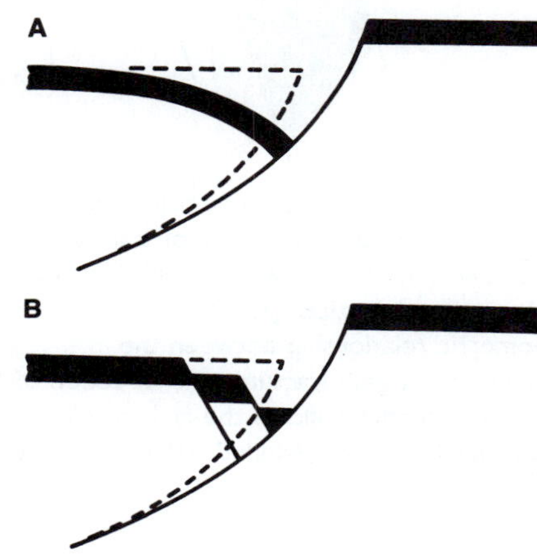

Figure 6.54 Formation of reverse drag through (*A*) downbending of strata into the zone of potential separation and/or (*B*) stepped faulting. [From Hamblin (1965). Published with permission of Geological Society of America and the author.]

Using Gash Fractures and Tight Drag Folds as Guides to Slip

There is yet another way in which the direction and sense of slip due to faulting can be evaluated on the basis of minor structures. We will touch on it here, saving the full details for Chapter 9 (Shear Zones and Progressive Deformation). The method is a direction application of strain theory, in particular the interplay of *finite* strain and *instantaneous* strain. Here's how it works.

Think of fault zones as brittle zones of noncoaxial shear that distort rocks. During shear, at each instant of time, distortion stretches the rock at a 45° angle to the fault zone, and shortens it at right angles to the direction of **maximum instantaneous stretch ($\dot{S}_1$)**. If you could sit and watch the shearing for a time, you would see that the directions of instantaneous stretching and instantaneous shortening do not shift relative to the fault zone. You might also see that **en echelon tension fractures** and **en echelon veins**, which are mode I features created in noncoaxial shear, form **gash fractures** at right angles to the direction of maximum instantaneous stretching ($\dot{S}_1$) (Figure 6.55*A*). In addition, you might see **asymmetric folds** develop, leaning in a direction consistent with the sense of shear. The folds initially oriented themselves at right angles to the direction of **minimum instantaneous stretch ($\dot{S}_3$)**.

Interestingly, as shearing continues, the earlier formed gash fractures rotate in a way that is sympathetic with the sense of shear within the fault zone (Figure 6.55*B*), although brand new gash fractures form perpendicular, as always, to the direction of maximum instantaneous stretch ($\dot{S}_1$). Meanwhile, the asymmetric folds rotate and become tighter and tighter, maintaining an orientation perpendicular to the direction of minimum *finite* stretch (S_3). Given the geometry of this progressive deformation, the combination of gash fractures and tight folds virtually establishes the sense of slip that took place during faulting.

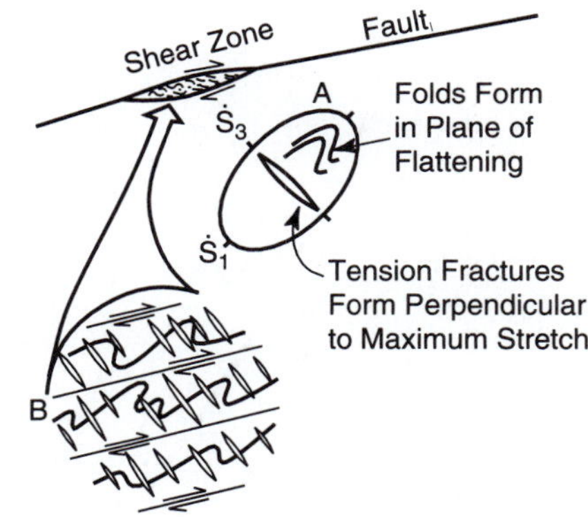

Figure 6.55 Simple shear origin of tension fractures and tight folds within a shear zone. (*A*) Instantaneous strain. (*B*) Finite strain. See text for details.

Analysis of minor structures, such as gash fractures, in fault zones has

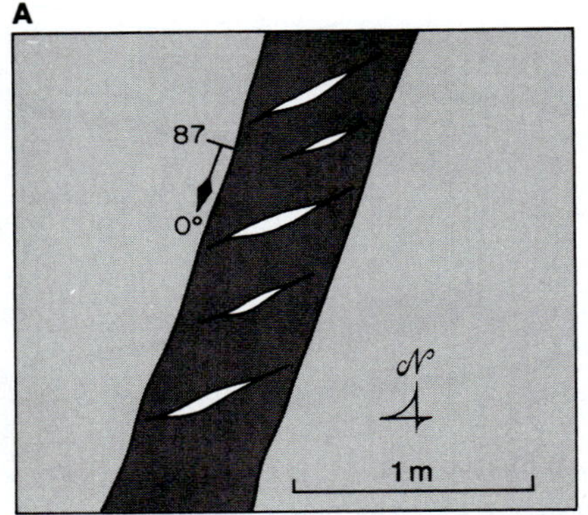

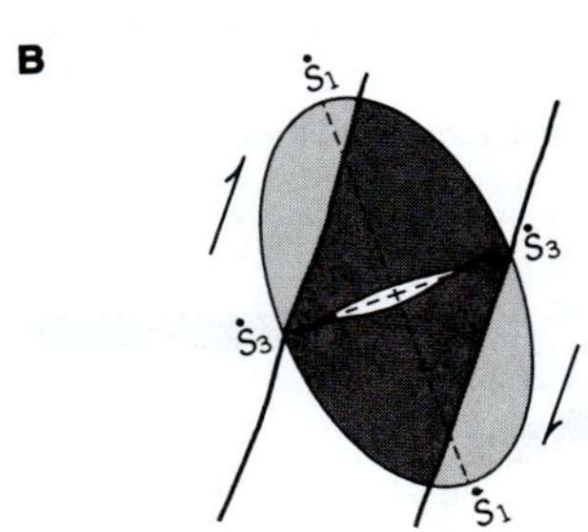

Figure 6.56 Use of gash fractures to determine direction and sense of translation on a fault. (*A*) Map showing steeply dipping fault zone and vertical gash fractures. (*B*) Geometric relationship between the orientation of gash fractures and direction of greatest stretching allows the fault to be recognized as a right-handed strike–slip fault.

important practical application in mining geology. Suppose a rich ore vein is abruptly intercepted by a fault and translated safely out of sight. How can its offset position be found using gash fractures as a guide?

Consider the steeply dipping N15°E-striking fault zone shown in the map view in Figure 6.56*A*. Within the zone are vertical gash fractures filled with quartz. They strike approximately N70°E. Horizontal slickenlines on the fault surfaces indicate that the fault zone accommodated strike–slip movement. However it is not known whether translation was right-handed or left-handed. To solve for sense of slip, assume that the gash fractures formed at right angles to the direction of maximum instantaneous stretching ($\dot{S}_1$) of the rock during faulting (Figure 6.56*B*). The attitude of the gash fractures in this example is compatible with right-handed strike–slip faulting.

Part III-J provides some additional material regarding determining the slip on faults.

DETERMINING RECURRENCE INTERVALS ON FAULTS

Today it is common practice to more fully expose and map the structural relations along active faults, or along potentially active faults, by digging trenches along and across them. The power of the approach was demonstrated by Kerry Sieh (1978) in a classic paper, and his continued work has permitted quantitative understanding of specific dates of major faulting events (i.e., earthquake events) within parts of the San Andreas fault system during a history of more than several thousands of years. The trenching and mapping permit the construction of detailed geologic cross sections, which picture not only the fault geometry and fault rocks, but a stratigraphy that helps constrain the timing, rate, and recurrence interval(s) of the faulting. The stratigraphy is not a conventional stratigraphy, because it includes the delineation of soil horizons (past and present) and unconsolidated alluvial materials. Intensive efforts are typically made to date each horizon by whatever means possible, especially through carbon dating.

An example is shown in Figure 6.57, which features the results of trenching at Pallett Creek, California, across the trace of the San Andreas fault (Figure 6.57*A*). Quite a number of trenches were excavated by Kerry Sieh (1978), and the structural and stratigraphic relationships permitted him to sort out the details of the interrelationships of faulting and sedimentation. The descriptive detail in Kerry's sections is extraordinary (see Sieh, 1978). Results here are presented in simplified cross sections. For example, the geologic relations in trench 11 (Figure 6.57*B*) disclose a paleoseismic event (event V) that resulted in 400 mm of separation. The fault scarp was then buried by sediments. Subsequently (Figure 6.57*B*), a younger seismic event (event X) formed a scarp about 200 mm high as well as a fissure, which became filled in by sediments as they slumped and collapsed into the fissure.

Such an approach is providing an avenue for interpreting **paleoseismic** activity, even including estimates of magnitude and slip on earthquakes that took place thousands of years ago. By understanding past activity (including recurrence intervals), the projection of future activity, and the preparation for future activity, can be carried out with greater understanding.

STRAIN SIGNIFICANCE OF FAULTS

Faults exist because the rocks they occupy were distorted by deforming stresses. *Normal-slip faults, thrust- and reverse-slip faults, and strike–slip faults have a common function: to stretch the crust in one direction and shorten it in another.* The directions of stretching and shortening are at right angles to one another. In examining how this is accomplished, we generally assume plane strain. In other words, we assume that volume is conserved during faulting and that there is neither stretching nor shortening in the direction of intermediate finite stretch (S_2).

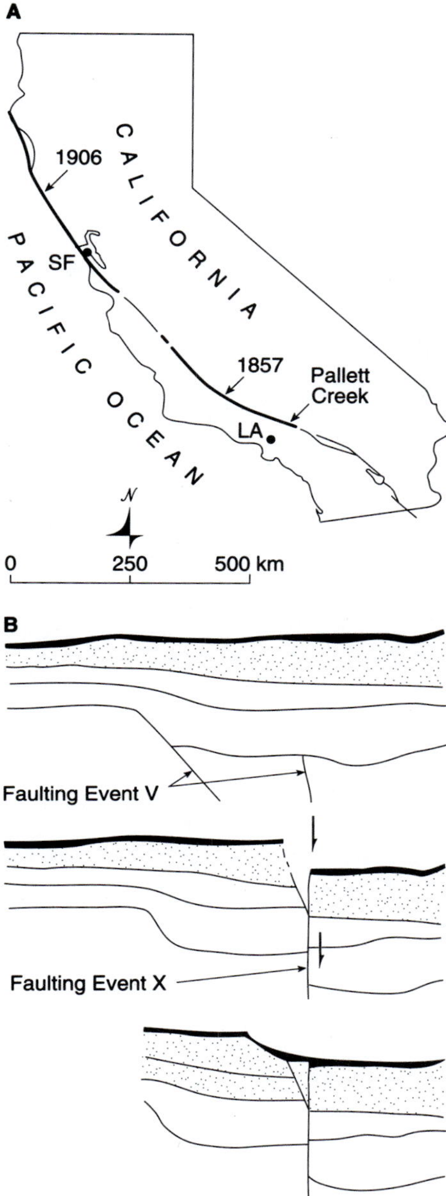

Figure 6.57 (*A*) Location of Pallett Creek relative to fault rupture along the San Andreas during the great earthquakes of 1857 and 1906. (*B*) Relationships in exposure 11, revealing the influence of faulting event V on the structural and stratigraphic relations. Faulting event X produced displacement and opened a fissure. The fissure became filled from above through slumping. [From Sieh, K. E., *Journal of Geophysical Research,* v. 83, fig. 1, p. 3908; and fig. 21, p. 3925; copyright © 1978 by American Geophysical Union.)

Normal-Slip Faults

Normal-slip faults accommodate horizontal extension and vertical shortening. Consider the flat-lying sedimentary layer 120 m long and 10 m thick shown in Figure 6.58. When this layer is cut and displaced by twenty 60°-dipping normal-slip faults, the end points of the layer are shifted to positions 132 m apart (Figure 6.58). The layer is said to be stretched, even though we all know that it was not stretched like a rubber band or a rubber sheet. The 12 m of stretching is the sum of the **gaps** between offset layers. Stretch (S) measured parallel to the layer is easily computed:

$$S = \frac{l_f}{l_0} = \frac{132\text{ m}}{120\text{ m}} = 1.1$$

$$\%\text{ stretching} = (S - 1) \times 100\% = (1.1 - 1.0) \times 100\% = 10\%$$

If the layer had been uniformly and penetratively stretched to its 132-m final length, its thickness would have been reduced from 10 m to 9 m. The final thickness, $t_f = 9$ m, can be calculated by assuming that the cross-sectional area (A) of the layer remained the same before and after deformation.

$$A_0 = A_f$$

$$(l_0)(t_0) = (l_f)(t_f)$$

$$(120\text{ m})(10\text{ m}) = (132\text{ m})(t_f)$$

$$t_f = \frac{1200\text{ m}^2}{132\text{ m}} \cong 9\text{ m}$$

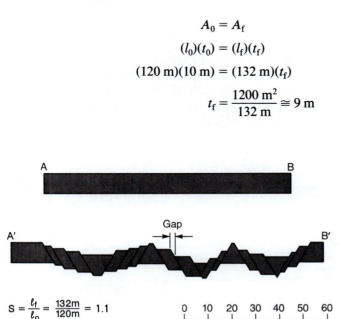

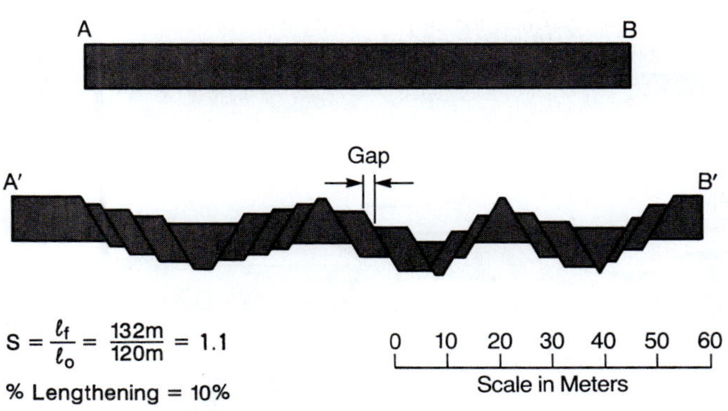

Figure 6.58 "Stretching" of a layer by normal-slip faulting. Stretch (S) measured parallel to the layer is 1.1. Total lengthening of the layer is 10%.

Final thickness (t_f) calculated in this way can be called the **virtual thickness** of the layer. It is the average thickness that a faulted layer would possess if the lengthening (or shortening) of the layer had been achieved by a homogeneous penetrative flow equivalent in magnitude to the sum of the offsets along spaced, discrete faults.

Thrust- and Reverse-Slip Faults

Thrust-slip and reverse-slip faults perform the opposite function of normal-slip faults. Thrust-slip and reverse-slip faults allow a layer of rock to be shortened horizontally and extended vertically. Figure 6.59 shows how this is achieved. A layer 188 m long and 10 m thick is cut and displaced by a combination of thrust-slip and reverse-slip faults (Figure 6.59). Translation on each of the faults creates an **overlap** between offset layers that effectively shortens the initial layer. The sum of the overlaps in this example is 35 m. Final length (l_f) of the layer is 153 m. Stretch (S) measured parallel to the layer can be calculated as follows:

$$S = \frac{l_f}{l_0} = \frac{153 \text{ m}}{188 \text{ m}} = 0.81$$

$$\% \text{ shortening} = (0.81 - 1.00) \times 100\% = 19\%$$

Virtual thickness is computed by equating the cross-sectional areas of the layer, before and after deformation.

$$A_0 = A_f$$

$$(l_0)(t_0) = (l_f)(t_f)$$

$$(188 \text{ m})(10 \text{ m}) = (153 \text{ m})(t_f)$$

$$t_f = \frac{1880 \text{ m}}{153 \text{ m}} \cong 12.3 \text{ m}$$

Strike–Slip Faults

If we view the cross sections shown in Figures 6.58 and 6.59 as if they were plan-view maps of fault patterns, we see the effects of two-dimensional strain produced by strike–slip faulting. Strike–slip faulting results in stretching and shortening in the plane of horizontal layering. The faulting has no effect on thickness of the deformed layer. Although the directions

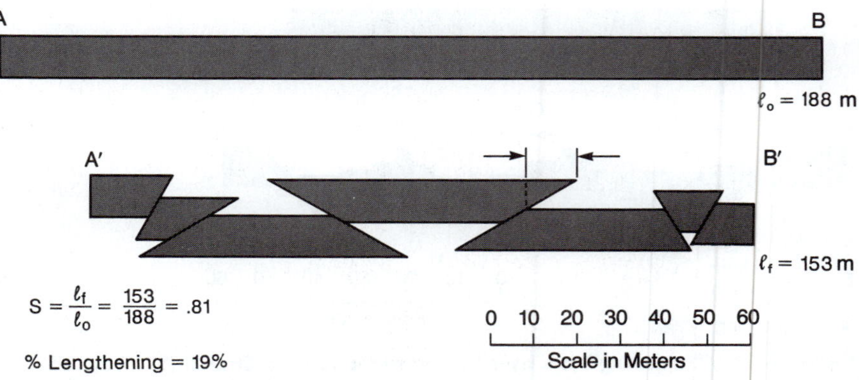

Figure 6.59 "Shortening" of a layer by thrust-slip and reverse-slip faulting. Stretch (S) measured parallel to the layer is 0.81. Shortening of the layer is 19%.

of extension and shortening are mutually perpendicular, the absolute orientations of these directions depend on the relative degree of development of right-handed versus left-handed strike–slip faults. If translations along right-handed and left-handed strike–slip faults are balanced, strain is coaxial. On the other hand, if one sense of strike–slip faulting predominates, the strain is non-coaxial.

Distinctions Among Classes of Faults

The main classes of faults are distinctive because the absolute directions of stretching and shortening are different for each. Normal-slip faulting accommodates horizontal stretching and vertical shortening, whereas thrust-slip and reverse-slip faulting accommodate horizontal shortening and vertical stretching. With regard to strike–slip faulting, stretching and shortening *both* take place in the horizontal plane. Taken together, the main classes of faults are a versatile array of structures: they can distort the crust in a variety of ways, as the requirements of plate tectonic configurations and/or local stresses dictate.

Because the main fault classes all have the same strain function, to achieve simultaneous stretching and shortening, we might guess that each brand of faulting would produce fault systems of *identical* physical properties, irrespective of actual orientations. Such is not the case, however. Each class of faults has its own special and distinctive properties. The differences result largely from the way each of the major classes of faults interacts with preexisting structures, especially bedding and preexisting faults.

Relation of Faults to Principal Finite Strain Directions

It is not uncommon for faults to occur in **conjugate** sets (Figure 6.60*A*) that intersect in an acute angle, commonly approximately 60°, except where modified by flattening brought about by internal rotations. Whether they are thrust-slip, normal-slip, or strike–slip, the conjugate faults generally intersect in a line that is parallel to the direction of intermediate finite stretch (S_2) (Figure 6.60*A*). The directions of maximum and minimum finite stretch (S_1 and S_3) occupy a plane that is perpendicular to the direction of intermediate finite stretch (S_2). The direction of minimum finite stretch (S_3) bisects the conjugate angle between faults. Slickenlines are parallel to the line of intersection of each fault with the S_1–S_3 plane (Figure 6.60*A*). Figure 6.60*B* presents these relationships stereographically.

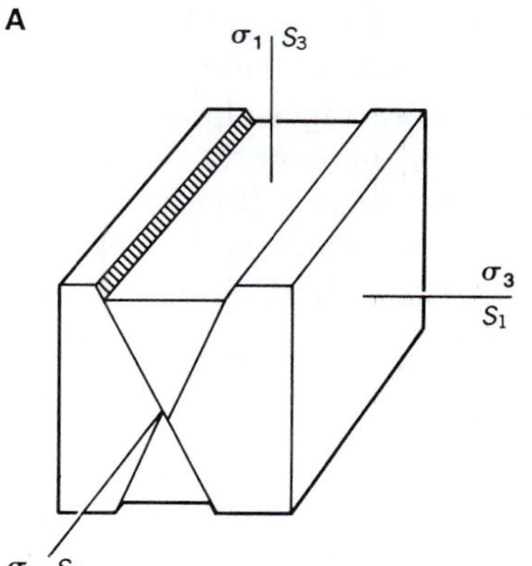

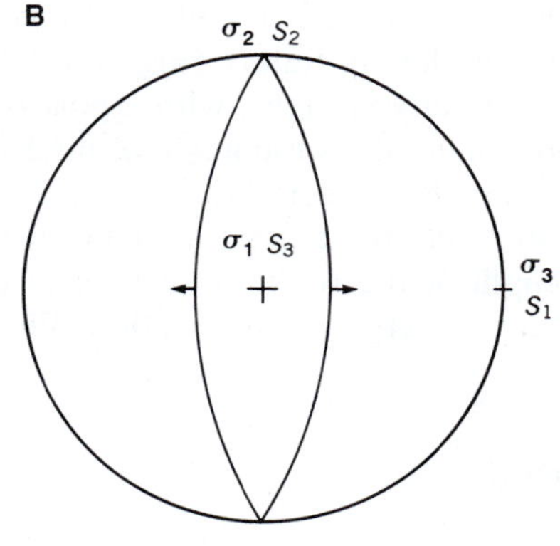

Figure 6.60 (*A*) Relation of conjugate faults to the principal finite stretch directions. (*B*) Stereographic representation of the principal finite stretch directions to the orientations of conjugate faults.

DYNAMIC ANALYSIS OF FAULTING

Dynamic analysis of faulting allows us to understand why faults can be so conveniently separated into normal-slip, thrust-slip, and strike–slip categories. We learn that there is a dynamic basis for the fact that, on average, strike–slip faults dip vertically, normal-slip faults dip at 60°, and thrust-slip faults dip at 30°. Dynamic analysis of faulting is concerned both with stress conditions under which rocks break and with the orientations of faults relative to stress patterns.

Anderson's Theory of Faulting

The Premise

Anderson (1951) recognized that the properties of principal stress directions, in combination with the Coulomb law of failure (see Chapter 5), implies that only strike–slip, thrust-slip, and normal-slip faults form at or near the surface of the Earth. Considering the Earth as a perfect sphere, Anderson reasoned that the discontinuity between air and ground at any point on the Earth's surface is a plane along which shear stress is zero. (During hurricanes, roofs are blown off houses and pedestrians may be swept off their feet, but as strong as such wind-shears might feel, they are not of sufficient intensity to cause faulting of bedrock.)

Since the principal stress directions are directions of zero shear stress, *the surface of the Earth must be a principal plane containing two of the three principal stress directions*. The third principal stress direction is oriented perpendicular to this principal plane and thus, at any point, is vertical, i.e., perpendicular to the surface of an ideally spherical Earth.

Anderson explained the average dip and sense of slip on faults at or near the surface of the Earth on the basis of the **Coulomb law of failure**. He used the Coulomb criterion to predict the orientation of faulting with respect to principal stress directions, using as a basis the angle of internal friction.

Coulomb's Law of Failure

$$\tan\phi = \text{coefficient of internal friction}$$

$$\sigma_N = \text{normal stress}$$

As described in Chapter 5, countless strength tests have been carried out through the faulting of small cylinders of rocks under controlled laboratory conditions. As a result of testing it has been determined that the stress conditions for which faulting in a given rock type will occur abide by the Coulomb law of failure (Figure 6.61*A,B*). The **Coulomb failure envelope** is essentially straight, with a positive slope that averages about 30°. The slope angle (ϕ) is the **angle of internal friction** for the material. When the envelope is projected to the σ_S axis of the Mohr diagram (Figure 6.61*C*), it intersects the axis at a small positive value, σ_0, which is the **cohesive strength** of the rock. As we recall from Chapter 5, the Coulomb law of failure is stated as an equation (Figure 6.61*C*):

$$\sigma_c = \sigma_0 + \tan\phi(\sigma_N)$$

where,

$$\sigma_c = \text{critical shear stress required for faulting}$$

$$\sigma_0 = \text{cohesive strength}$$

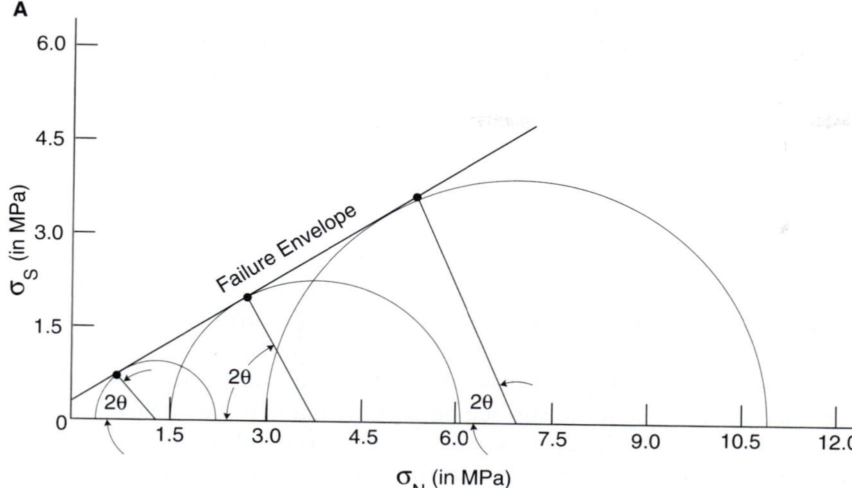

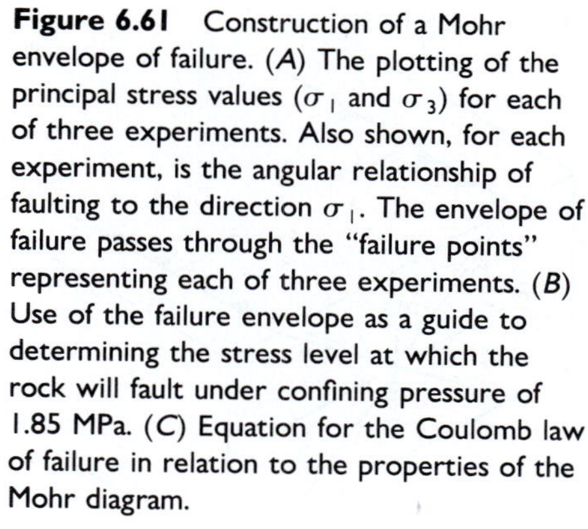
Figure 6.61 Construction of a Mohr envelope of failure. (*A*) The plotting of the principal stress values (σ_1 and σ_3) for each of three experiments. Also shown, for each experiment, is the angular relationship of faulting to the direction σ_1. The envelope of failure passes through the "failure points" representing each of three experiments. (*B*) Use of the failure envelope as a guide to determining the stress level at which the rock will fault under confining pressure of 1.85 MPa. (*C*) Equation for the Coulomb law of failure in relation to the properties of the Mohr diagram.

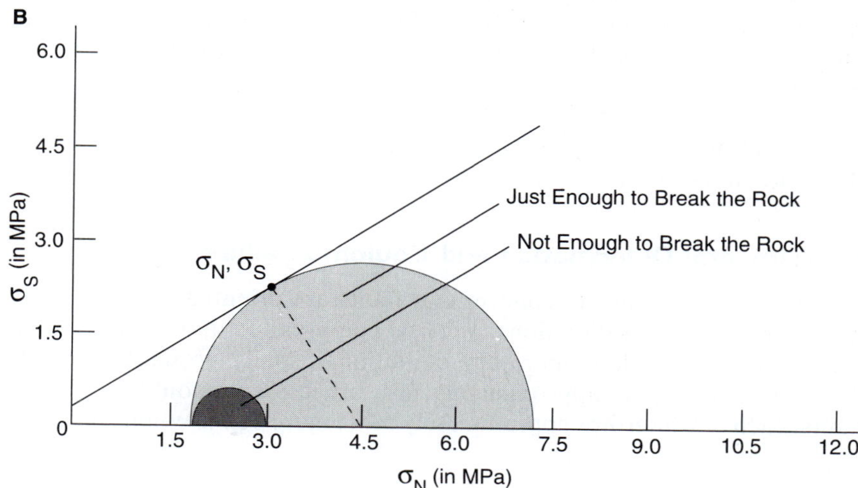

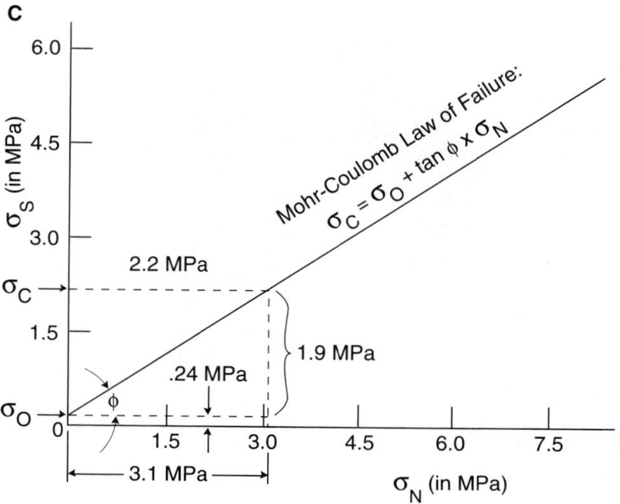

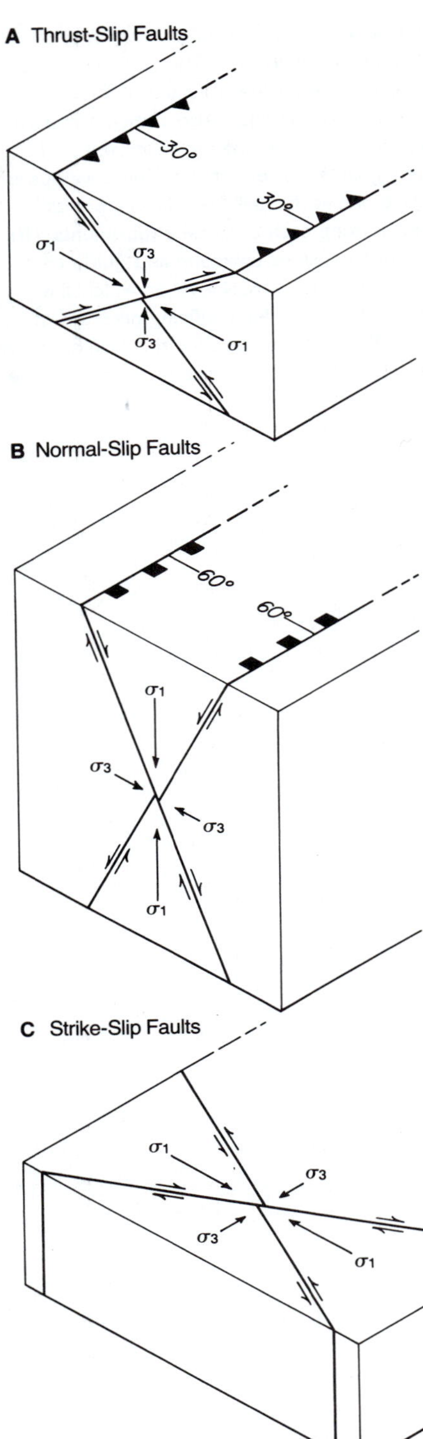

Figure 6.62 Schematic representation of (*A*) thrust faults, (*B*) normal faults, and (*C*) strike–slip faults at or near the surface of the Earth. These are the likely orientations since each of the three principal stress directions at or near the surface of the Earth is either horizontal or vertical, and since the angle of internal friction for rocks is almost always approximately 30°.

Anderson emphasized that the angle of internal friction (ϕ) determines the angle (θ) between the fault surface and the direction of greatest principal stress (σ_1). Based on the geometry of the failure envelope, we have

$$\phi = 90° - 2\theta$$

Thus,

$$2\theta = 90° - \phi$$

Since most rocks in nature possess an angle of internal friction of about 30°, the value of θ for most fault relationships is also 30°.

Anderson thus concluded that if principal stress directions are vertical or horizontal at or near the surface of the Earth, and if the angle of internal friction for most rocks is about 30°, only normal-slip, strike–slip, and thrust-slip faults should be able to form at or near the Earth's surface (Figure 6.62). Thrust-slip faults form when σ_3 is vertical; normal-slip faults form when σ_1 is vertical; and strike–slip faults form when σ_2 is vertical. Stereographic relationships of faults to principal stress directions are presented in Part III-K.

The Coulomb law of failure predicts the stress conditions under which faults form, in the same way that it describes the conditions under which shear fractures form.

Experimental Deformation and Coulomb Failure

Deformation experiments confirm that faults are oriented systematically with respect to stress directions. Vertical compression of hard rock specimens under controlled laboratory conditions usually produces a single discrete fault dipping approximately 60°, or, less commonly, conjugate faults, each of which dips approximately 60° (Figure 6.63). When conjugate

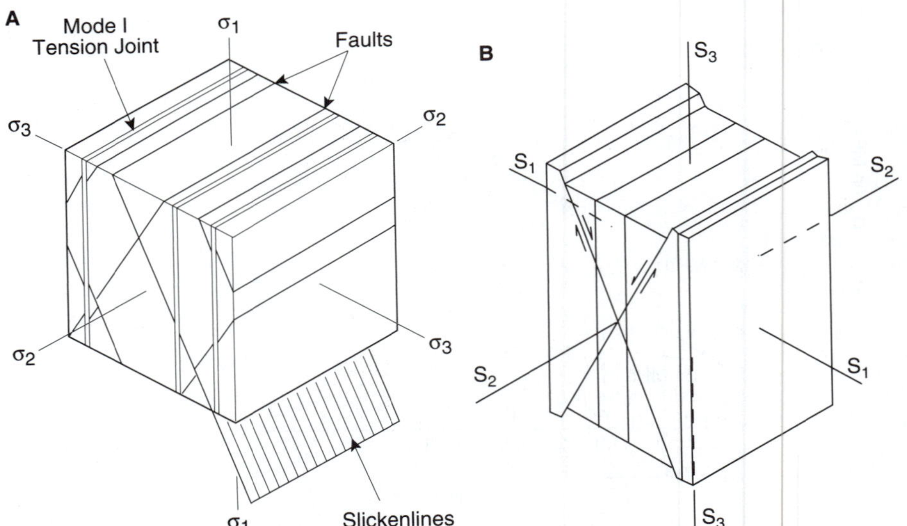

Figure 6.63 Drawings of conjugate faults as well as mode I tension fractures in a block of rock that had been subjected to length-parallel shortening. Note the orientations of the principal stress directions with respect to the orientations of faults, slickenlines, and mode I joints. Referenced with respect to (*A*) principal stress directions, (*B*) principal strain directions.

faults form, the **conjugate angle** between the faults is bisected by the direction of greatest principal stress (σ_1), which is the direction that the piston bears down on the rock cylinder. Slickenlines on the fault surfaces are visible when the specimen is pulled apart. Slickenline orientations on a given fault are defined by the intersection of the fault surface with the σ_1/σ_3 plane. If Mode I tension fractures open up, they form parallel to the direction of greatest principal stress (σ_1) and perpendicular to the direction of least principal stress (σ_3) (Figure 6.63).

The systematic relationships that exist among conjugate faults, tension fractures, and principal stress directions provide a basis for interpreting paleostress directions in rocks deformed millions of years ago. Descriptive Analysis (Part III-K) focuses on how to interpret principal stress directions on the basis of fault attributes, and how to predict fault attributes on the basis of principal stress directions. This work involves stereographic projection as an aid to dealing with the three-dimensional geometries.

M. King Hubbert's Sandbox Illustrations of Coulomb Theory

Appreciation of Anderson's theory of faulting and the Coulomb law of failure can be gained in a simple but elegant set of experiments described by Hubbert (1951). The experiments allow us to develop a working knowledge of constructing failure envelopes and applying the results to fault analysis. Sand is substituted for rock, but otherwise all conditions hold.

Constructing an Envelope for Sand

The first step is to construct an envelope of failure for sand. Two small, lightweight wooden or metal frames are stacked one on top of the other (Figure 6.64). The surfaces of the frames should be very smooth so that the frames can effortlessly slide past each other. The interior of the double-frame box is filled with fine white sand. A piece of Masonite is cut to fit snugly into the interior of the upper frame, to distribute the load. When positioned, it rests entirely on sand.

To construct an envelope of failure for sand, we determine the level of shear force that is required to cause differential movement of the upper and lower frames under a variety of conditions of normal force. For the frames to move with respect to each other, the sand must actually fault.

Normal force (F_N) is applied to the Masonite by placing rock samples of known weight on top of the Masonite (Figure 6.64). The rocks are

Figure 6.64 Construction of the failure envelope for sand. Experimental setup consists of two wooden frames, sand, a piece of Masonite, rocks of known weight, a good spring scale, and a baseball cap. The experiment itself involves measuring the amount of shear force required to move the upper frame for a number of given conditions of normal force. Andrew Arnold demonstrates the technique. (Photograph by R. W. Krantz.)

weighed as the experiment proceeds, and the values (in ounces or grams, pounds or kilograms) are written directly on them using a marking pen. Once a given amount of normal force is loaded onto the Masonite, the **shear force (F_S)** required to induce faulting of the sand is determined. This is achieved by pulling on a "spring" scale (we call it the "meathook" scale) attached to an eyescrew on the upper frame (Figure 6.64). At the instant the upper frame begins to move, the load registered on the spring scale is read and recorded. This load constitutes the critical shear force (F_S) required for faulting.

Each combination of normal force (F_N) and shear force (F_S) is plotted on the *x*- and *y*-axes of a Mohr diagram, respectively (Figure 6.65). With careful attention to weighing and plotting, an amazingly straight-lined envelope of failure can be fit to the half-dozen or so failure points. Because sand is cohesionless, the failure envelope ought to pass through the origin of the *x*-*y* coordinate system. In point of fact, the intercept typically lies just above the origin because of the effect of sliding friction between the frames. Hubbert found that the failure envelope for loose sand displays an angle of internal friction (ϕ) of 30°. For compacted sand the angle of internal friction (ϕ) is 35°.

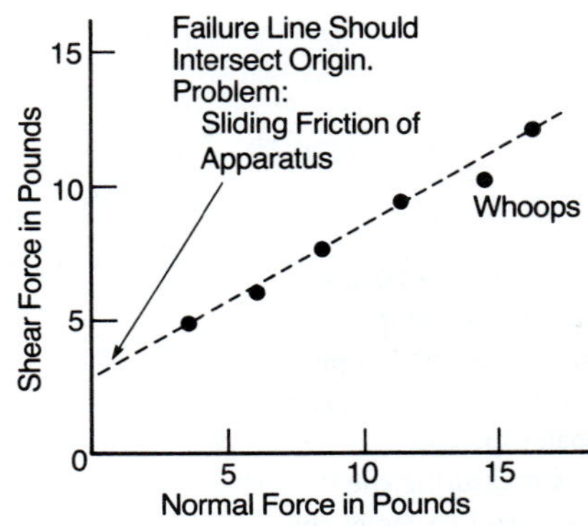

Figure 6.65 Mohr diagram showing the envelope of failure for sand based on paired measurements of shear force (F_s) and normal force (F_n).

The Sandbox Experiment

The second part of Hubbert's experiment involves generating normal-slip and thrust-slip faults in sand. A sandbox of the type shown in Figure 6.66 is filled with layers of fine white sand separated by marker horizons of white or colored dry powdered clay. A vertical wooden or metal partition serves to separate two compartments, a smaller one on the left in which normal-slip faults form, and a larger one on the right in which thrust-slip faults form. Deformation of the sand simply requires moving the partition to the right by means of a manually driven worm-screw arrangement. The instant that the partition moves, a normal-slip fault develops in the left-hand compartment (Figure 6.66). Its dip is typically 60°, that is, the complement of the angle of internal friction for sand. As the partition is forced to move farther and farther to the right, the first-formed normal-slip fault increases in displacement. Other normal-slip faults form as well. As a result of the normal faulting, the upper surface of the sand in the left-hand compartment develops a fault-scarp topography.

Compressional shortening of sand in the right-hand compartment eventually forces the development of a thrust-slip fault. The sand first arches slightly and then is cut by a thrust-slip fault, which slices up-section at an angle of approximately 30°. After a certain amount of translation has been accommodated by the thrusting, a second thrust-slip fault develops; this fault typically forms beneath the first, cutting up to the surface beyond it in the direction of tectonic transport. Translation on the first thrust ceases the instant that translation is initiated on the second. Steady translation on the second thrust results in folding of the first. Sometimes **back-limb**, **antithetic** thrusts develop that dip oppositely to the main thrust-slip faults.

The structural relationships produced in the sandbox experiment conform to what we have learned about dynamic analysis of faulting. Before movement of the partition in the sandbox, both compartments of sand are marked by a state of lithostatic stress in which stress is equal and all sided (Figure 6.66). Movement of the partition to the right relieves horizontal stress in the left compartment. At the same time, horizontal stress intensifies in the right-hand compartment. Thus, very early in the experiment, different states of stress evolve in the two compartments. In the

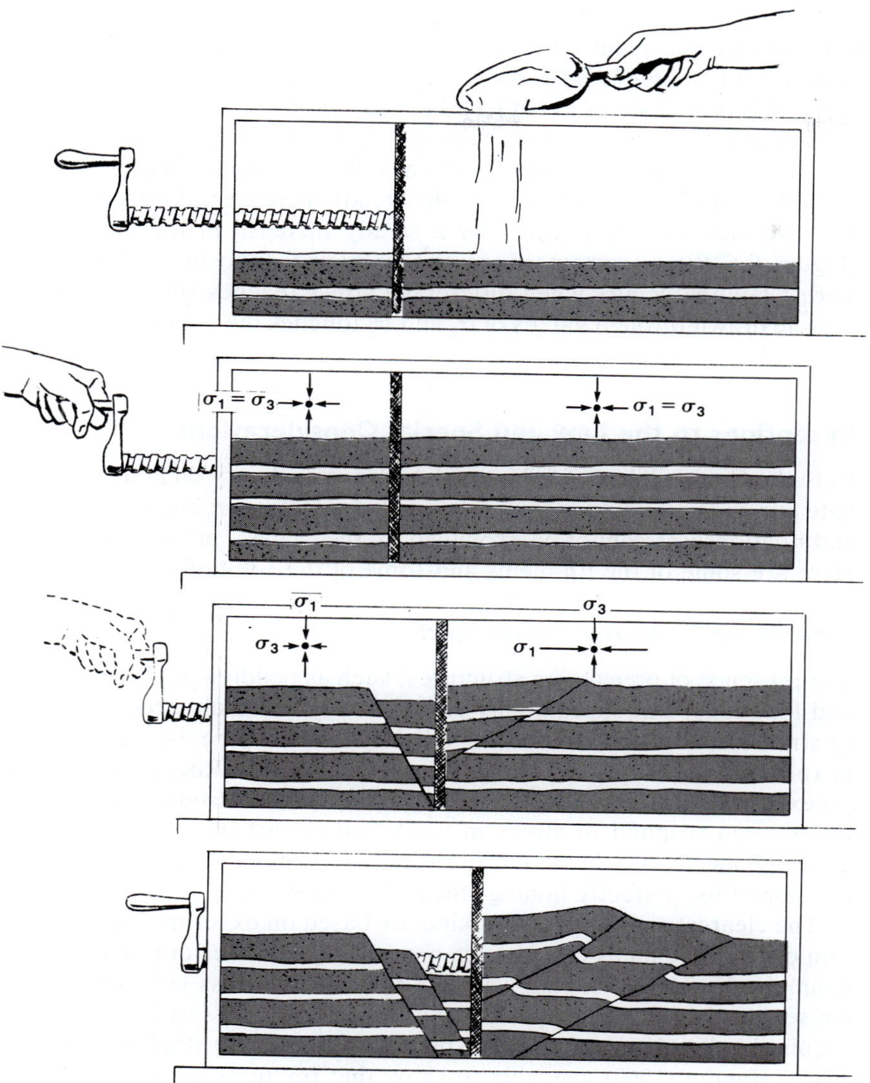

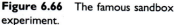

Figure 6.66 The famous sandbox experiment.

left compartment, σ_1 is vertical and σ_3 is horizontal. In the right compartment, σ_1 is horizontal and σ_3 is vertical. Given these principal stress directions and our knowledge of the angle of internal friction (ϕ) for sand, it becomes possible for us to predict the orientations of the faults that must develop in the sandbox experiment. Faults form in loose sand at angles of $\theta = 30°$ to the direction of greatest principal stress. Since σ_1 is vertical in the left compartment during ongoing deformation of the sand, the faults that form there *must* dip 60°. Conversely, since σ_1 is horizontal in the right-hand compartment during deformation, the faults that form there *must* dip 30°.

Mohr Diagram Portrayal of the Sandbox Experiment

The buildup of differential stress during the sandbox experiment can be pictured by means of a Mohr diagram. We start with a figure that shows only the Coulomb envelope of failure and the lithostatic state of stress that exists in the sand, before deformation (Figure 6.67*A*). The lithostatic stress state is one in which $\sigma_1 = \sigma_3$; it is represented on the Mohr diagram by a single point on the σ_N axis. As soon as the partition begins to move,

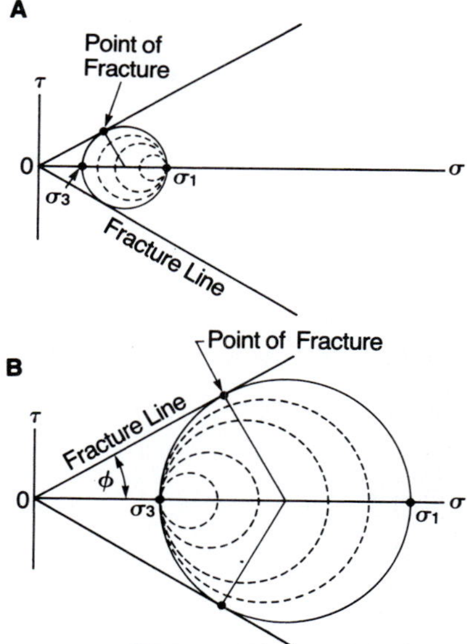

Figure 6.67 Mohr diagram portrayal of the dynamic conditions of the sandbox experiment. (*A*) Differential stress conditions leading to normal faulting in the left-hand compartment. (*B*) Differential stress conditions leading to thrust faulting in the right-hand compartment.

the lithostatic state of stress is altered to one of differential stress. In the left-hand compartment σ_3 becomes increasingly weaker, but σ_1 remains constant (Figure 6.67*A*). Because of the progressive decrease in the value of σ_3, differential stress eventually becomes large enough to "break" the sand. The differential stress at failure is represented by values of σ_1 and σ_3, which define a circle that just touches the envelope of failure for sand.

In the right-hand compartment, the steady increase in horizontal compressive stress is represented by a steady increase in the value of σ_1 (Figure 6.67*B*); σ_3 remains fixed at the original lithostatic stress level. Thrust faulting occurs when differential stress reaches such a level that a circle drawn through values of σ_1 and σ_3 touches the envelope of failure for sand.

Exceptions to the Law and Special Considerations

Both the Coulomb law of failure and Anderson's theory on faulting illuminate our understanding of fault relationships. However, a number of facts and relationships cannot be explained by the "laws" presented thus far. Here are some of the problems and some of the solutions.

What if the Rocks Are Anisotropic?

The presence of preexisting structures, such as bedding, faults, fractures, and foliations, can substantially alter the way in which a rock responds to stress. Like grain in wood, or cracks in metal, preexisting structures in rocks can be exploited as agents of deformation. Rocks that possess preexisting mechanical weaknesses are called **anisotropic rocks**. These rocks often respond to stress in ways that depart from the "normal," expected mechanical behavior, for "normal" behavior is based upon laws developed for perfectly homogeneous, isotropic rocks.

The clearest examples of this kind are based on experimental deformation of test specimens that contain preexisting fractures or foliations. For example, Handin (1969) showed that the critical stress level required to cause reactivation along preexisting fracture surfaces is *less* than that required to break an unfractured specimen of the same lithology. Furthermore, Handin (1969) was able to show that fractures oriented at angles as high as $\theta = 65°$ to the direction of greatest principal stress (σ_1) can be reactivated as faults. $\theta = 65°$ is a long way from $\theta = 30°$, which is the expected angle of failure for rocks in a perfect world. Thus if we were to disregard the possible influence of anisotropy and assume that all faults form close to $\theta = 30°$, we would get into deep trouble in our attempts to reconstruct the paleo-orientations of greatest principal stress directions.

As yet another example, Donath (1961) demonstrated the degree to which the orientation of foliation (e.g., schistosity or slaty cleavage) influences the orientation of faults that develop in test specimens (Figure 6.68). When foliation is oriented at a very high angle to the greatest principal stress direction (σ_1), the specimen will fault, as usual, at an angle such

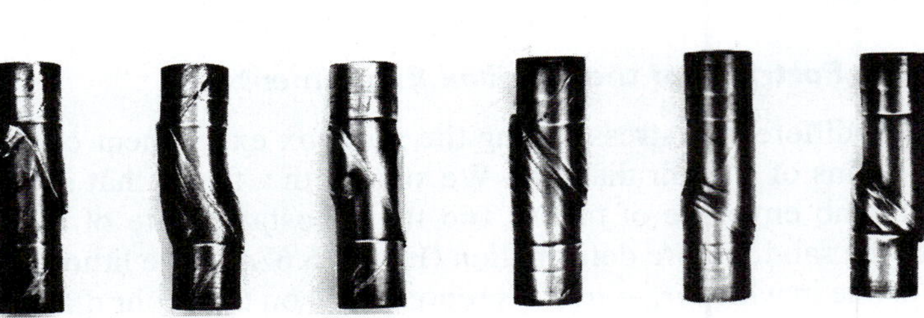

Figure 6.68 Specimens of anisotropic rock compressed at various angles to foliation. The orientation of faulting, in each case, is influenced by the angular relationship between the orientation of foliation and the direction of greatest principal stress (σ_1). [From Donath (1961), Geological Society of America.]

that θ, the angle between σ_1 and the fault, is approximately equal to the angle of internal friction (ϕ) for the faulted rock. But if foliation is at a closer angle to σ_1, the orientation of the fault will differ from what we would expect for a homogeneous specimen. For example, if foliation is parallel to σ_1, the angle θ will be very small, perhaps 10° or 20°. If foliation is inclined between 25° and 45° to σ_1, the fault surface will commonly develop right along the foliation, thereby overriding the influence of internal friction. If foliation is inclined more than 45° to σ_1, the fault surface will form at an angle (θ) that obeys Coulomb's law of failure, almost as if the rock were homogeneous.

What if There Are Preexisting Fractures?

Determining whether or not preexisting structures such as joints, faults, and foliations will reactivate by sliding to form faults requires revisiting frictional behavior and **Byerlee's law** (see Chapter 5). **Friction** is the measure of resistance to sliding along a surface. Friction increases as normal force on the surface increases, but not as a function of the area of the surface (Suppe, 1985). Normal force, as we know from Chapter 3, varies with the orientation of the surface relative to principal stress directions. As a consequence of these factors, the level of stress required to activate a preexisting fracture surface depends upon the combination of two factors: the **friction** along the surface and the **orientation** of the surface.

The frictional character of a given rock type is expressed by the **coefficient of sliding friction** (μ_f), which is the shear stress (σ_S) required to activate sliding divided by the normal stress (σ_N) acting on the surface along which sliding takes place (Engelder and Marshak, 1988).

$$\mu_f = \frac{\sigma_S}{\sigma_N}$$

Recall that the coefficient of sliding friction is the same for almost all rock types. Except for conditions of very low confining pressure, friction is essentially independent of rock type!! Byerlee (1967, 1978) reached this rather amazing conclusion on the basis of countless frictional-sliding experiments and creative insight regarding what was happening at the microscopic scale. His plots of shear stress versus normal stress are the basis for **Byerlee's law of rock friction** (Figure 6.69) (Suppe, 1985). There are two expressions of the law. Where the effective normal stress (σ_N^*) ranges from 0 MPa to 200 MPa (Figure 6.69),

$$\sigma_S = 0.85\sigma_N^*,$$

and this corresponds to an **angle of sliding friction** (ϕ_f) of 40°. Where the effective normal stress (σ_N^*) is greater than 200 MPa (Figure 6.69),

$$\sigma_S = 0.5 + 0.6\sigma_N$$

and this corresponds to an angle of sliding friction (ϕ_f) of 31°.

Like the other laws of behavior that we discovered in Chapter 5, Byerlee's law can be added to the grand envelope of failure to create the "ultimate" description of conditions of brittle failure (Figure 6.70). The combination of this diagram and the known orientations of preexisting structures, as well as the magnitudes of the principal stresses, permit informed predictions of whether stresses will fault a rock anew or activate preexisting surfaces (see Chapter 5).

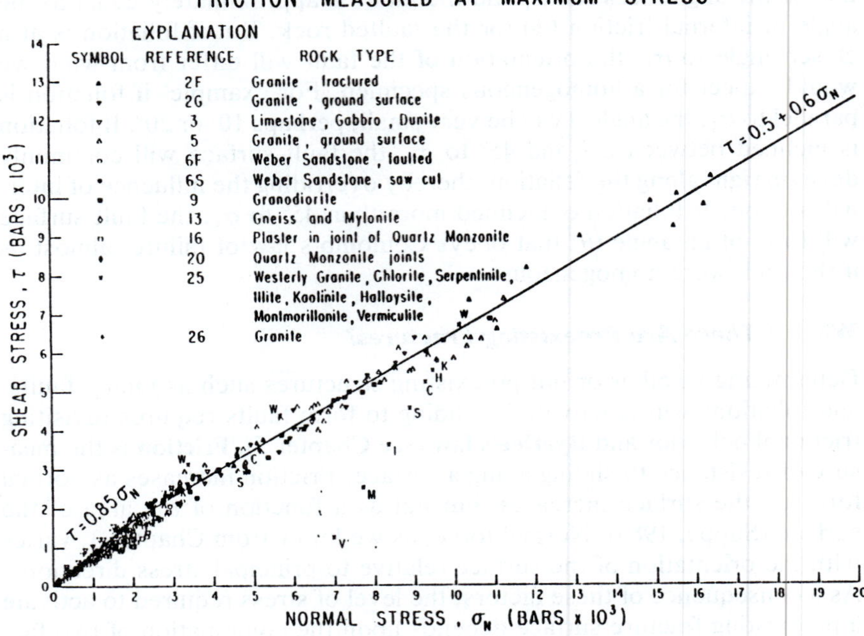

Figure 6.69 Plot of Byerlee's law of sliding friction, which is based on hundreds of sliding friction experiments on a wide variety of rock types. (From Byerlee, J. D., Friction of rocks, 1978, Pure and Applied Geophysics, v. 116, Birkhauser Verlag Ag, Basel, Switzerland.)

We might think that rock friction will change dramatically when rock is crushed and broken, and that this would require us to modify the grand envelopes diagram still further. This in fact is the case under conditions of very low confining pressure; the coefficient of sliding friction for surfaces lined by gouge tends to be lower than predicted by Byerlee's law. But

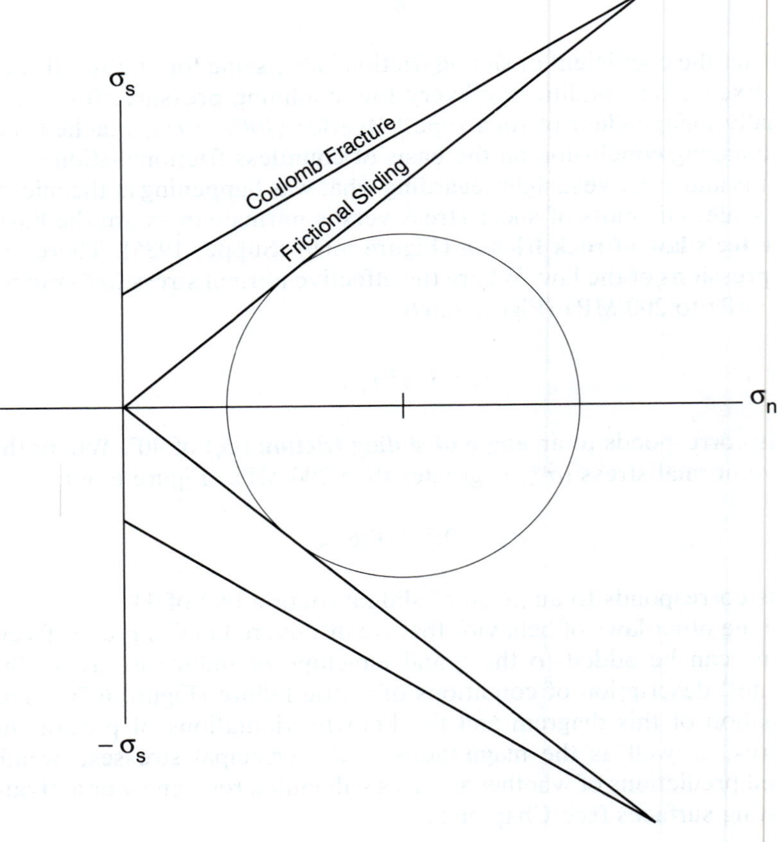

Figure 6.70 The merging of the standard envelope of failure with the frictional law of failure on a Mohr diagram. In this example, greatest principal stress (σ_1) has been raised to the point where the differential stress circle touches the frictional sliding envelope. No need to form brand-new fractures. Faulting will take place along suitably oriented preexisting fracture surfaces, and differential stress may never grow to reach the Coulomb failure envelope. (From Suppe, J., *Principles of Structural Geology,* © 1985, p. 163. Reprinted by permission of Prentice-Hall, Upper Saddle River, New Jersey.)

where confining pressure is higher, the gouge found along fault surfaces generally has a coefficient of sliding friction that conforms to Byerlee's law. The only notable exceptions occur when certain water-rich clay minerals such as montmorillonite or vermiculite are present (Byerlee, 1978; Suppe, 1985).

Predicting the Direction of Slip on Preexisting Fractures

Angelier (1979; 1994) has emphasized that the actual rake of slip along a reactivated fracture or fault surfaces will be dependent upon the magnitude of the intermediate principal stress (σ_2) relative to the magnitudes of the greatest (σ_1) and least (σ_3) principal stresses. Imagine that a given region of rock contains a preexisting fracture surface that strikes north–northwest and dips moderately steeply to the northeast (Figure 6.71*A*). Let us subject the rock to a stress environment marked by a greatest principal stress direction (σ_1) that is vertical, an intermediate principal stress direction (σ_2) that is horizontal and trending east–west, and a least principal stress (σ_3) direction that is horizontal and trending north–south (Figure 6.71*B*). If the stresses cause slip along the preexisting fracture surface, the actual slip direction will depend upon the magnitude of the intermediate stress (σ_2) relative to the magnitudes of σ_1 or σ_3. Angelier (1979) demonstrated that the slip direction will be parallel to the direction of greatest shear stress resolved along the preexisting fracture surface.

How do we predict the direction of slip? Following Angelier, if the magnitude of the intermediate principal stress (σ_2) is the same as the magnitude of the greatest principal stress (σ_1), then the slip direction is determined stereographically in two steps: (1) by finding the great circle that contains σ_3 and the pole (*n*) to the fracture surface, and (2) by finding the intersection of this great circle with the great circle representing the orientation of the fracture surface (Figure 6.71*C*). This intersection will coincide with the direction of slip on the reactivated fracture surface. If,

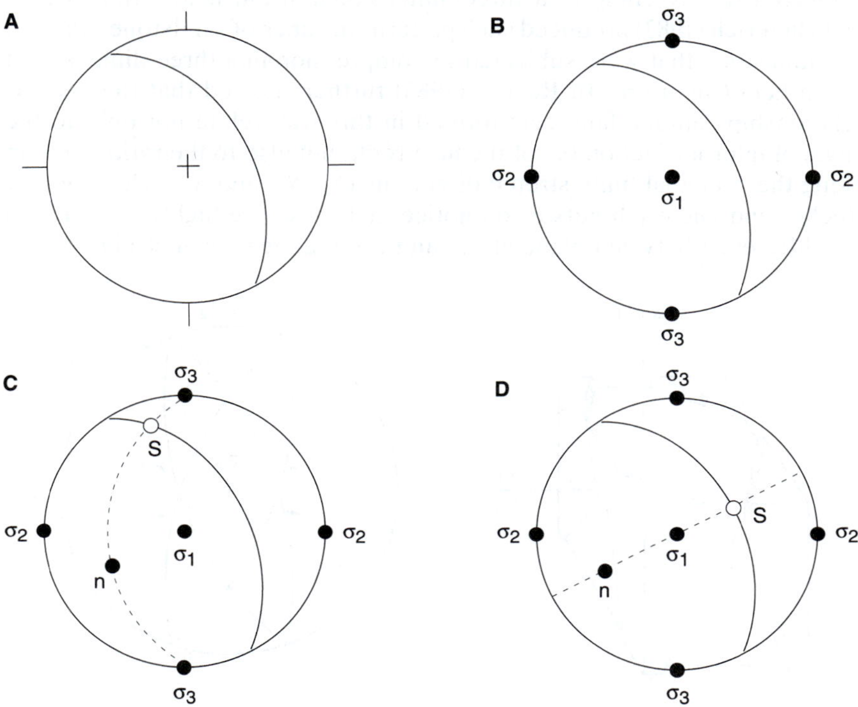

Figure 6.71 Lower-hemisphere stereographic projections that illustrate the heart of the Angelier method. (*A*) A fracture surface strikes north–northwest and dips moderately steeply to the northeast. (*B*) The rock body within which the preexisting fracture surface is found is subjected to stress such that greatest principal stress (σ_1) is vertical, intermediate principal stress (σ_2) is horizontal, east–west, and least principal stress (σ_3) is horizontal, north–south. (*C*) Slip direction (S) on preexisting fracture surface, assuming that $\sigma_2 = \sigma_1$ (n = normal to fracture surface). (*D*) Slip direction (S) on preexisting fracture surface, assuming that $\sigma_2 = \sigma_3$. [From Angelier (1994), Fault slip analysis and palaeostress reconstruction, Figure 4.29, p. 67. Reprinted with permission of Pergamon Press, Oxford, England.]

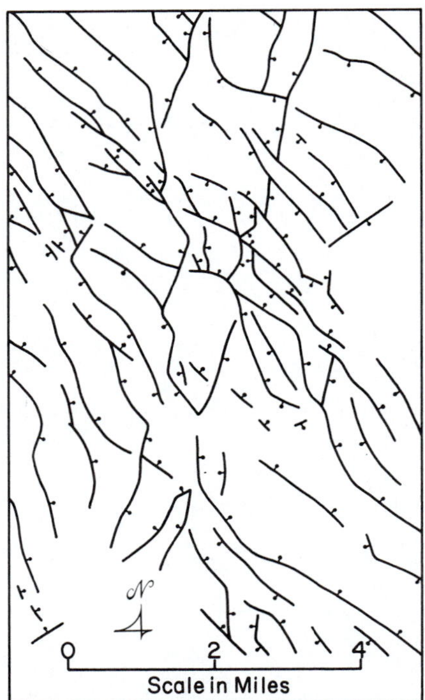

Figure 6.72 Geologic map pattern of faults mapped in south-central Oregon. [From Donath (1962). Published with permission of Geological Society of America and the author.]

on the other hand, the magnitude of the intermediate principal stress (σ_2) is the same as the magnitude of the least principal stress (σ_3), then the slip direction is determined stereographically in a slightly different way: (1) by finding the great circle that contains σ_1 and the pole (n) to the fracture surface, and (2) by finding the intersection of this great circle with that representing the orientation of the fracture surface (Figure 6.71*D*).

Jacques Angelier "inverted" this theory to produce what is now known as the Angelier method. This is a method of reconstructing stress directions and stress ratios on the basis of slip directions disclosed by slickenlines on preexisting fracture surfaces in deformed rocks. Angelier has worked in regions throughout the world testing the validity of his method and interpreting paleostresses (e.g., Angelier, Colletta, and Anderson, 1985; Angelier, 1990, 1994).

Role of Three-Dimensional Strain

The Coulomb theory predicts that faults should form in conjugate sets. This theory seems to be supported by the results of triaxial deformation experiments. Yet faulted rocks in nature commonly display *two* pairs of conjugate fault sets. Donath (1962) mapped such a pattern in south-central Oregon (Figure 6.72). One way to explain the presence of four fault sets is to call on two episodes of faulting, each of which yields conjugate pairs of faults. However, it is now clear that four fault sets can indeed be generated in a single event. How is this possible?

Reches (1978a) emphasized that the presence of three or four fault sets in a given region of study is the natural result of faulting within a three-dimensional strain field. The patterns seem peculiar only because rock deformation experiments traditionally have been carried out under conditions of two-dimensional coaxial stress and strain. Where specimens are allowed to shorten (or stretch) by different amounts in *three* mutually perpendicular directions, the characteristic fault pattern that emerges is one of three or more sets arranged in orthorhombic symmetry (Figure 6.73). Such a fault pattern was produced by Oertel (1965) in a clay cake subjected to stretching in a three-dimensional strain field. And Reches and Dieterich (1983) produced such patterns in cubes of sandstone, granite, and limestone that were subjected to compression in a three-dimensional strain field (Figure 6.74). Reches (1983) further showed that the angular relationships among fault sets formed in this way relate not only to the angle of internal friction (ϕ) of the host rock, but also to the ratio of strain along the principal finite stretch directions (S_1, S_2, and S_3). The work by Reches and Dieterich puts us on notice that rocks are highly sensitive to the difference between plane strain and three-dimensional strain.

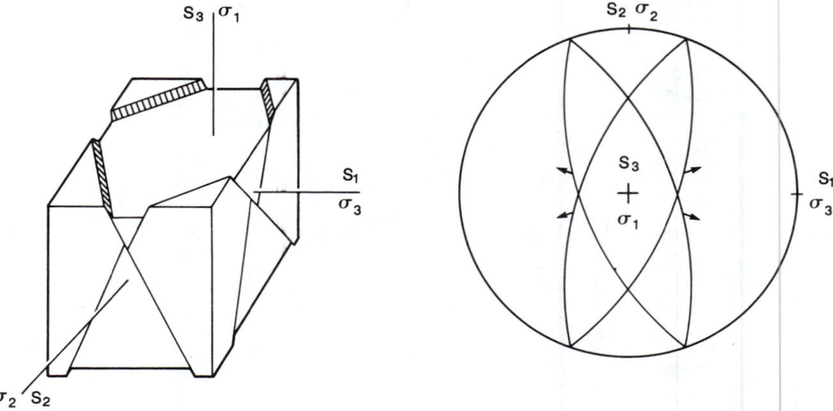

Figure 6.73 Fault sets produced in three-dimensional strain field. [From Reches (1983), *Tectonophysics*, v. 95. Published with permission of Elsevier Scientific Publishing Company, Amsterdam.]

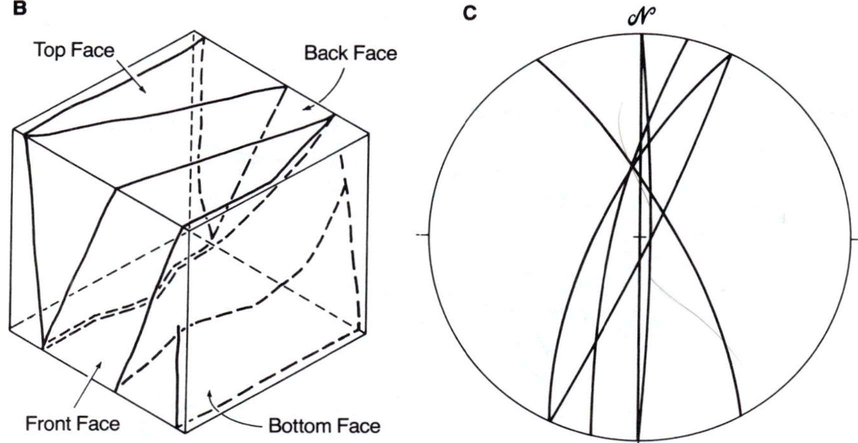

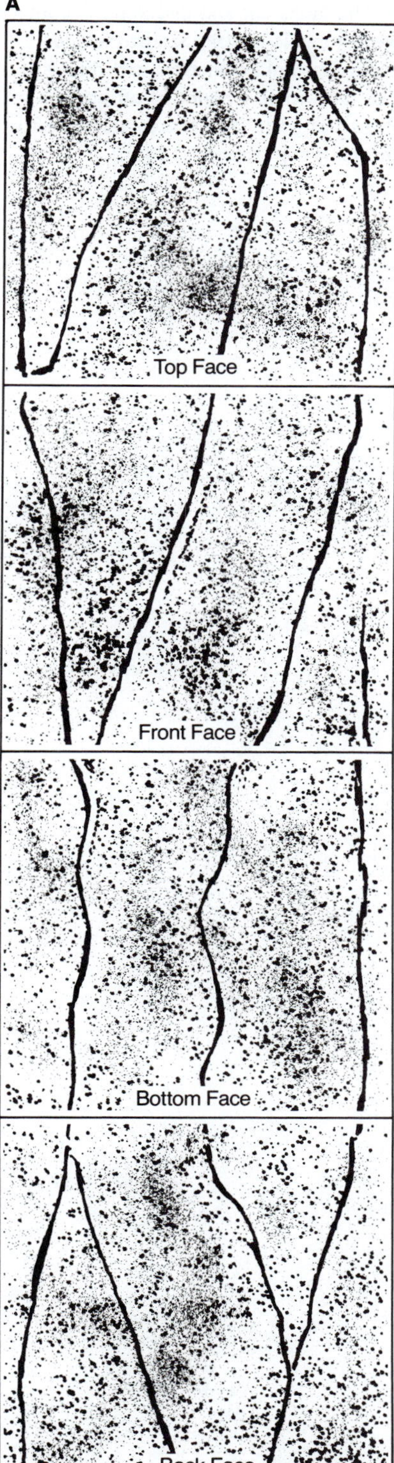

Figure 6.74 Fault pattern in cube of rock subjected to three-dimensional strain. (*A*) "Mapped" traces of faults on faces of deformed cube. (*B*) Three-dimensional portrayal of the faulted specimen. (*C*) Stereographic projection of the faults. [From Reches and Dieterich (1983), *Tectonophysics*, v. 95. Published with permission of Elsevier Scientific Publishing Company, Amsterdam.]

Bob Krantz (1988) carried out an elegant analysis of faulting in a three-dimensional strain field. His field area was within the San Rafael Swell, a major Colorado Plateau uplift. Exposures there are exceptional, and he was able to map out the full lengths and displacement gradients along the faults within the system (Figure 6.75). He determined the slip directions on the faults through slickenlines on slickensided surfaces. On the basis of the fundamental theory developed by Reches, as well as a clever stereographic methodology he developed himself, Krantz determined both the directions and magnitudes of the principal strains responsible for the faulting. Testing these results with those calculated from line-length measurements in detailed cross sections, he found there was a very close correspondence.

The Problem of Reverse Faults

Where does reverse faulting fit into the Anderson model of faulting and Coulomb theory? Reverse-slip faults typically dip 60° or more and accommodate crustal shortening. They are found in orogenic belts around the world, yet they are not featured in Anderson's fundamental classes of faults.

One obvious explanation for the origin of some reverse-slip faults is that they occupy former sites of normal faulting. Where reverse faulting is a result of **fault reactivation**, the dip of the reverse-slip fault is largely inherited from a previous event (Figure 6.76).

Another explanation of reverse faulting is that principal stress directions are not necessarily vertical and horizontal at depth. Rather, the orientations of the principal stress directions may be inclined. **Stress trajectories** become inclined and/or curved as the result of changes in the state of stress both laterally and vertically. Hafner (1951) demonstrated this theoretically. He showed, for example, that the dissipation of horizontal compressive stresses can result in a strain field marked by curved stress trajectories (Figure 6.77*A*). Curved compressive stress trajectories give rise to continuously curved faults, parts of which are thrust-slip faults, and parts of which are reverse-slip faults (Figure 6.77*B*).

Figure 6.75 Bob Krantz prepared an extraordinary map of normal faults in the Chimney Rock area of the San Rafael Swell, Utah. The faults cut and displace the contact between Jurassic Navajo Sandstone and overlying Carmel Formation, and thus it is easy to determine amount of offset. The numbers along the fault traces denote offset, in meters. [From Krantz (1986).]

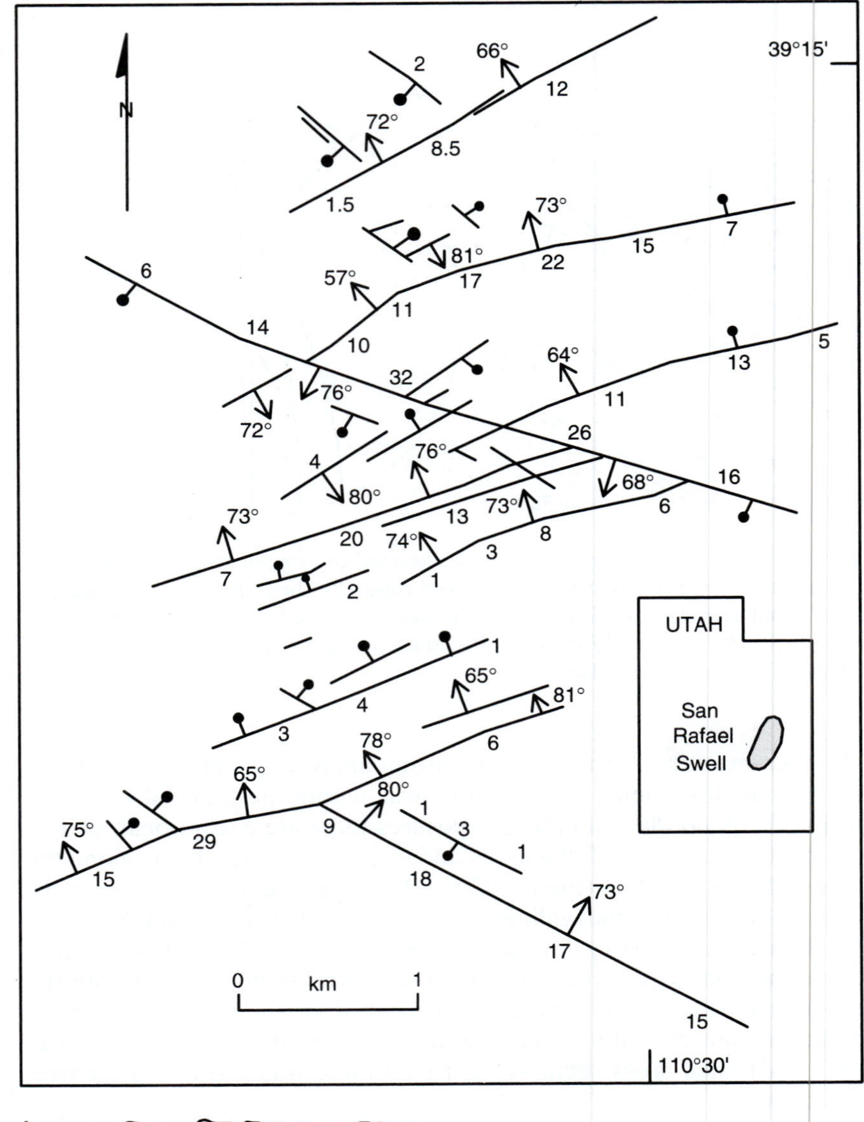

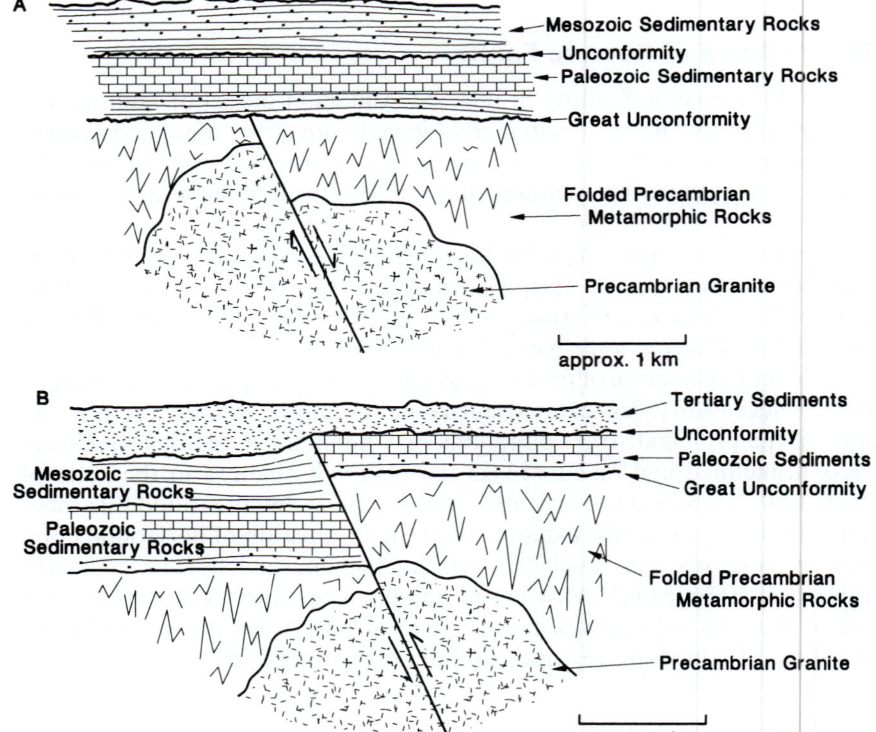

Figure 6.76 Stratigraphic relations in these schematic cross sections reveal (*A*) Precambrian normal faulting followed by (*B*) post-Mesozoic reactivation of the basement faults as reverse faults.

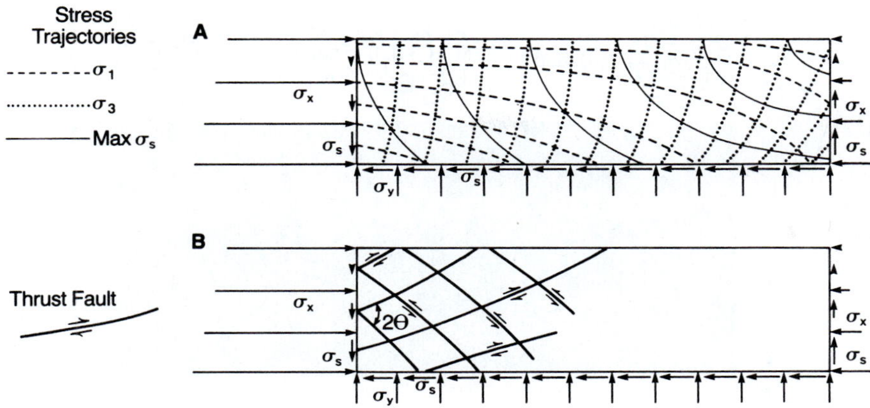

Figure 6.77 (*A*) Pattern of stress trajectories that would be produced in a block subjected to strong horizontal compressive stresses that die out laterally. (*B*) The pattern of curved faults that would emerge within the field of curved stress trajectories. [From Hafner (1951), Geological Society of America.]

Hafner's work (1951) showed that compression-induced thrusts at depth may *steepen upward* into reverse-slip faults. In contrast, Sanford (1959), Stearns (1978), and Friedman and others (1976) have carried out experiments indicating that upper-level thrusts can *steepen downward* into reverse-slip faults (Figure 6.78). Interpreting the dynamics of formation of reverse faulting thrust requires full three-dimensional views of faults and fault systems (Figure 6.79).

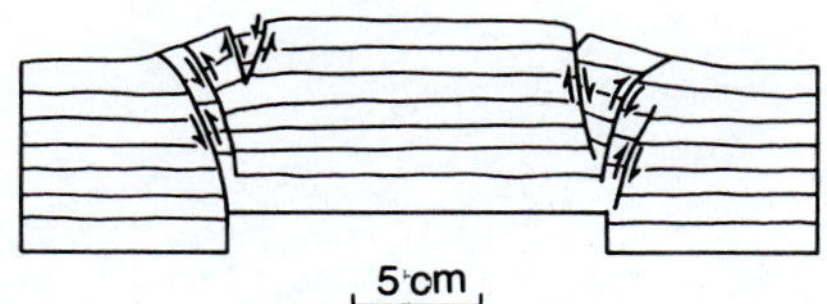

Figure 6.78 Curved reverse-slip faults produced by differential vertical uplift in deformation experiment. [From Sanford (1959), Geological Society of America.]

Faulting Accompanied by Changes in Rock Volume

Coulomb theory assumes that faulting proceeds without any significant change in volume of rock at the site that is deformed. Yet there are certain rocks, notably porous pure-quartz sandstones, that fault in a special way. The physical and geometric characteristics of such faulting are different from what we are accustomed to seeing.

Credit Atilla Aydin and Arvid Johnson (Aydin, 1978; Aydin and Johnson, 1978) for bringing this kind of faulting to the attention of the structural geology community. **Deformation bands**, also known as **band faults**, occur preferentially in highly porous sandstones (e.g., poorly cemented aeolian sandstones) that have sheared under conditions that would produce in more highly cemented, much less porous sandstones, slickenlined shear surfaces, and faults with very minor offset. They are ultra-thin brittle shear zones that, by virtue of loss of porosity and volume during shearing (Figure 6.80*A*), are strongly resistant to erosion and thus stand out like veins, ribs, or fins in outcrop (Figure 6.80*B–D*). Commonly they occur in concentrated zones of deformation bands. I have been mapping deformation bands along the East Kaibab monocline in southern Utah that project like walls from the Navajo Sandstone (Figure 6.81*A*). Differential resistance to erosion displayed by bands versus the normal porous sandstone in between the bands creates magnificent "inside views" of the deformation band architecture (Figure 6.81*B*).

The formation of individual deformation bands begins with a stress-induced collapsing of pore spaces. Grains move closer and closer together, and eventually come into contact, creating stress concentrations (Figure 6.82*A*). Internal friction will have increased to the point that further deformation must involve grain-scale microfracturing and cataclasis within the inner zone of the deformation band. Grains become demolished, and ultimately the deformation band strain hardens to the point that if continued porosity-reduction and grain-fracturing are to continue, they will take place by initiating a new band alongside (Figure 6.82*B*). A *zone* of deforma-

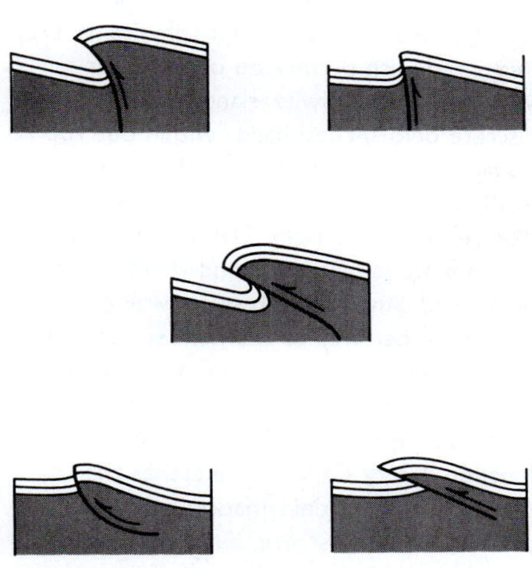

Figure 6.79 Possible configurations of the Wind River fault (Wyoming) at depth. Seismic reflection studies revealed that the actual fault geometry is one of low-angle (~30°) thrust faulting. [From Smithson and others (1978). Published with permission of Geological Society of America and the authors.]

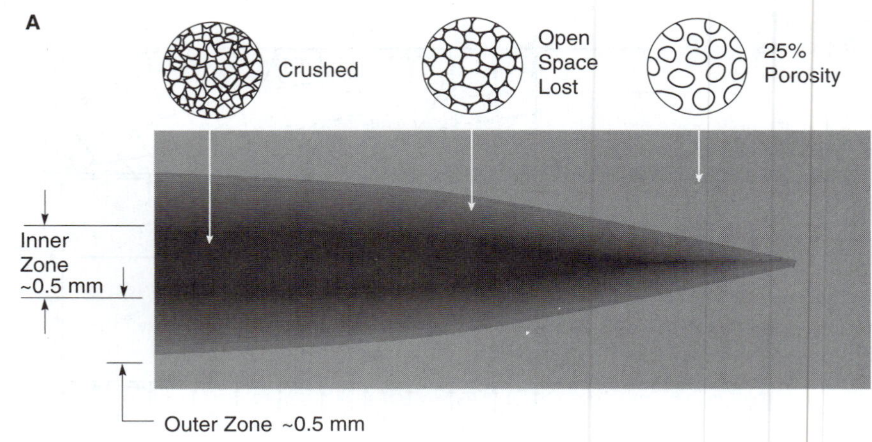

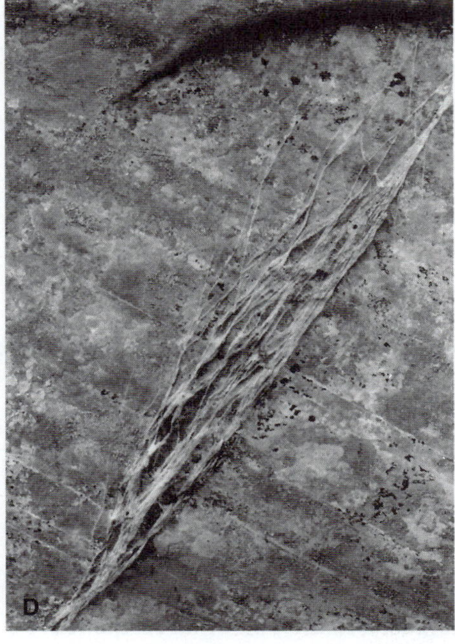

Figure 6.80 (*A*) Idealized anatomy of a deformation band, showing inner zone of crushed material, outer zone of compressed material, and porous host rock. [From Aydin (1978), Pure and Applied Geophysics, v. 116. Reprinted with permission of Birhauser Verlag Ag, Basel, Switzerland.] (*B*) Single, discrete deformation bands within outcrop of Navajo Sandstone in southern Utah just east of Zion Canyon. Cross-bedding in Navajo dips gently to the right. Deformation bands cut bedding at high angle and stand out in bold relief. Angela Smith scales wall of Navajo on her way to the next set of strike and dip measurements. (*C*) Conjugate family of deformation bands in a wall of Navajo Sandstone in southern Utah near Mt. Carmel Junction. Kerry Caruthers sketches the pattern. (*D*) This deformation band is, in places, thin and discrete, but more commonly branches and braids, as here, to form a zone of deformation banding. [Photographs by G. H. Davis.]

Figure 6.81 (*A*) Outcrop expression of resistant wall of a zone of deformation bands. (*B*) Lesley Perg stands next to the nearly vertical resistant wall of a zone of deformation bands known as "radiator rock." Navajo Sandstone has been intensely and systematically deformed along closely spaced bands. (Photographs by G. H. Davis.)

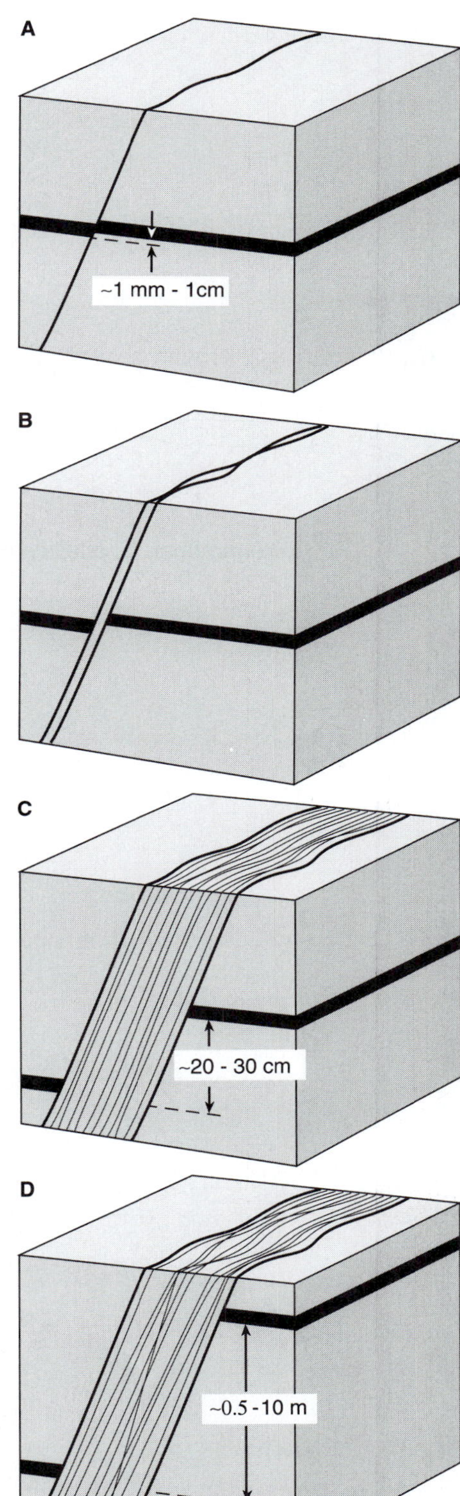

Figure 6.82 Progressive development of a zone of deformation bands, starting with a single band (*A*) and ending with full zone (*D*). [From Aydin and Johnson (1978), *Pure and Applied Geophysics,* v. 116. Reprinted with permission of Birhauser Verlag Ag, Basel, Switzerland.]

tion bands will evolve in this way (Figure 6.82*C,D*) (Aydin 1978; Aydin and Johnson 1978, 1983). These are no ordinary faults; they are faults that form through a combination of shear and volume reduction.

A Transitional Thought

Given what we now know about the physical and geometric characteristics of faults in general, and given what we now know about kinematics and dynamics, let us take a look at the major classes of faults and how they operate.

THRUST FAULTING

Regional Characteristics

The structural characteristics of thrust belts are beautifully portrayed in regional structural profiles, like those constructed for the Canadian Rockies of Alberta and British Columbia (Figure 6.83). Great expanses of strata

Cc: Chatter Creek Fault
Mo: Mons Fault
Sp: Simpson Pass Thrust
Pp: Pipestone Pass Thrust
Bo: Bourgeau Thrust
Sm: Sulphur Mountain Thrust
Mc: McConnell Thrust
Bi: Bighorn Thrust
Bz: Brazeau Thrust

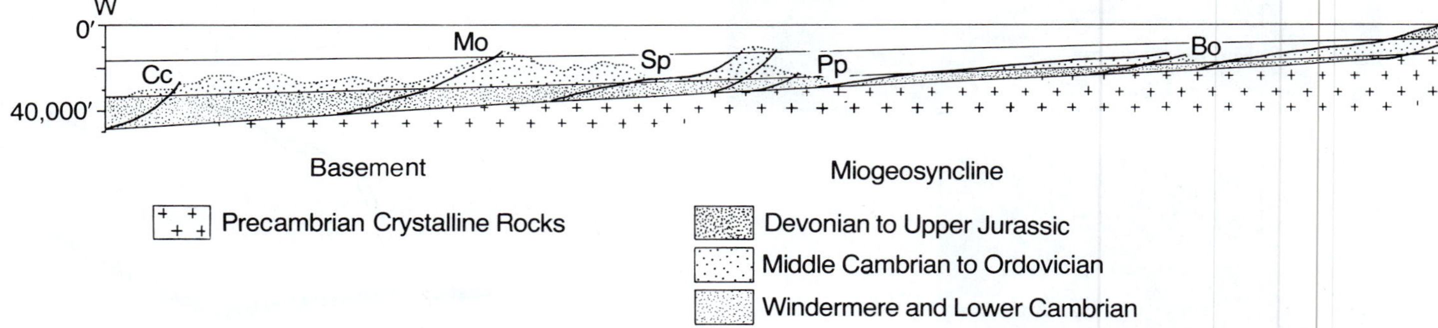

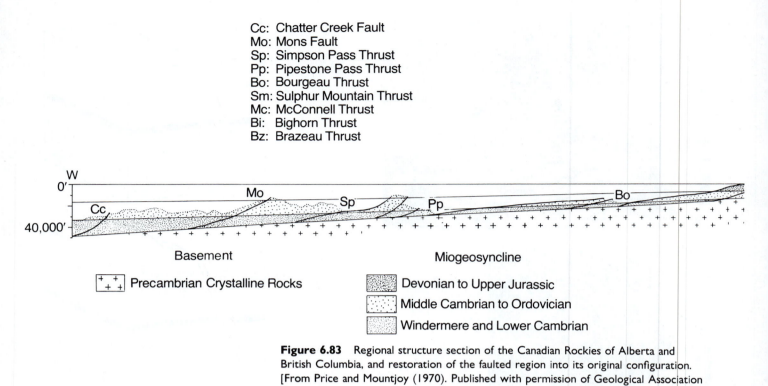

Figure 6.83 Regional structure section of the Canadian Rockies of Alberta and British Columbia, and restoration of the faulted region into its original configuration. [From Price and Mountjoy (1970). Published with permission of Geological Association of Canada.]

have been deformed by thrusting in this region, as a result of tectonic processes related to plate convergence. Price and Mountjoy (1970) estimated that the original **miogeoclinal prism** in the Canadian belt was about 500 km wide, tapering from a thickness of 12 km on the west to 2 km on the east. The westernmost strata were moved at least 125 mi (200 km) to the northeast. The nonconformity between sedimentary cover and crystalline basement is preserved underneath the thrusted and folded mass (Figure 6.83); it dips gently westward. The thrust-slip faults do not cut into basement, but feed into ductile sedimentary layers right above the basement/cover interface.

The eastern Idaho/western Wyoming thrust belt is similar to the Canadian belt. Royse, Warner, and Reese (1975) estimate that the miogeoclinal prism was originally about 180 km wide, tapering from 21 km on the west end to 3 km on the east. These thicknesses include **syntectonic sedimentary basin deposits** derived from fault-bounded uplifts and dumped during the course of thrusting. Strata in the westernmost margin of the belt were moved about 80 km eastward to their present site. As in the Canadian belt, the nonconformity separating basement and sedimentary cover dips gently and is not broken by the thrusts.

The regional packages of rocks above thrust faults are called **thrust sheets**. Regional thrust sheets that have moved great distances are considered to be **allochthonous**. Allochthonous rocks rest in thrust contact on **autochthonous** rocks, which retain their original location because they have not been thrusted. Regional thrust faults serve to separate rocks of the **allochthon** from rocks of the **autochthon** (Figure 6.84). **Windows** through allochthonous cover can provide a deep look into autochthonous rocks, which are otherwise concealed. Isolated **klippen** of allochthonous rocks can disclose the former extensiveness of overthrust strata (Figure 6.84). The Rombak window in northern Norway provides a glimpse into autoch-

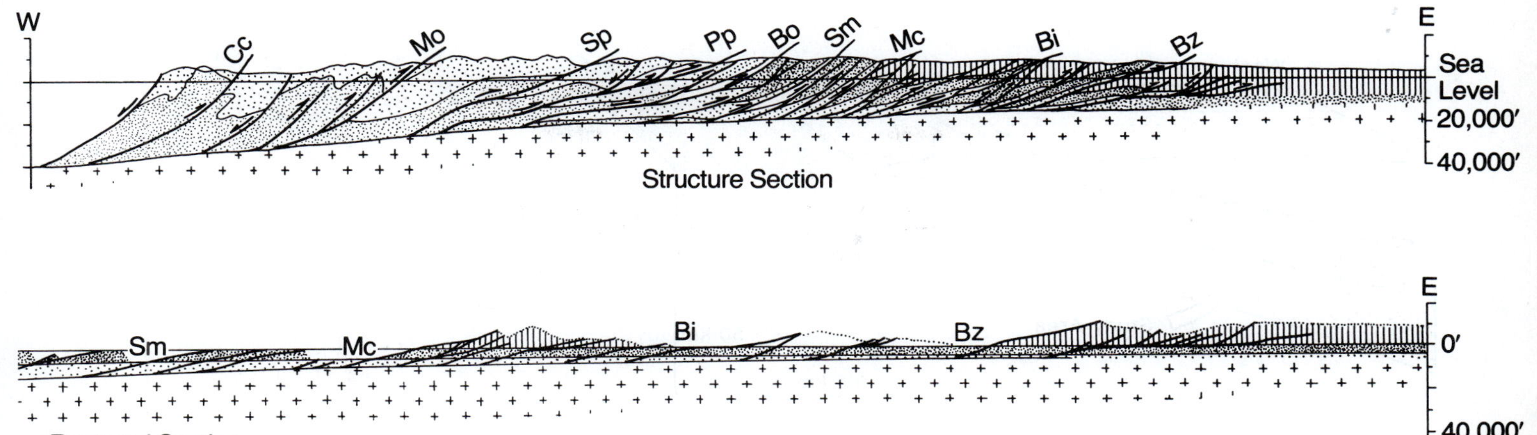

thonous Scandinavian basement rocks, which are otherwise covered by extensive allochthonous thrust sheets (Figure 6.85).

Thin-Skinned Deformation and Decollement

The contradiction of overthrusting is revealed in the structure profile of the Canadian Rockies (see Figure 6.83): strongly folded, faulted sequences of miogeoclinal strata rest atop basement that clearly has not been involved in the distortion. Dahlstrom, in reference to the Canadian Rockies belt, wrote:

> . . . there is never a balance between the length of Mesozoic and Paleozoic beds and the length of basement. The "cover" beds are always too long, which is explained in cross section by a sole fault along which the upper beds have been moved (shoved? glided?) into the cross section from the west. (From "Balanced Cross Sections" by D. C. A. Dahlstrom, 1969, p. 746. Reproduced by permission of National Research Council of Canada from *Canadian Journal of Earth Sciences*.)

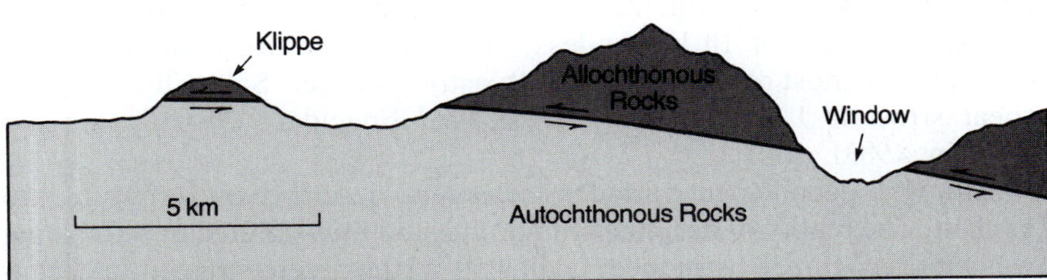

Figure 6.84 Cross-sectional view of allochthonous and autochthonous rocks, as well as klippe and window.

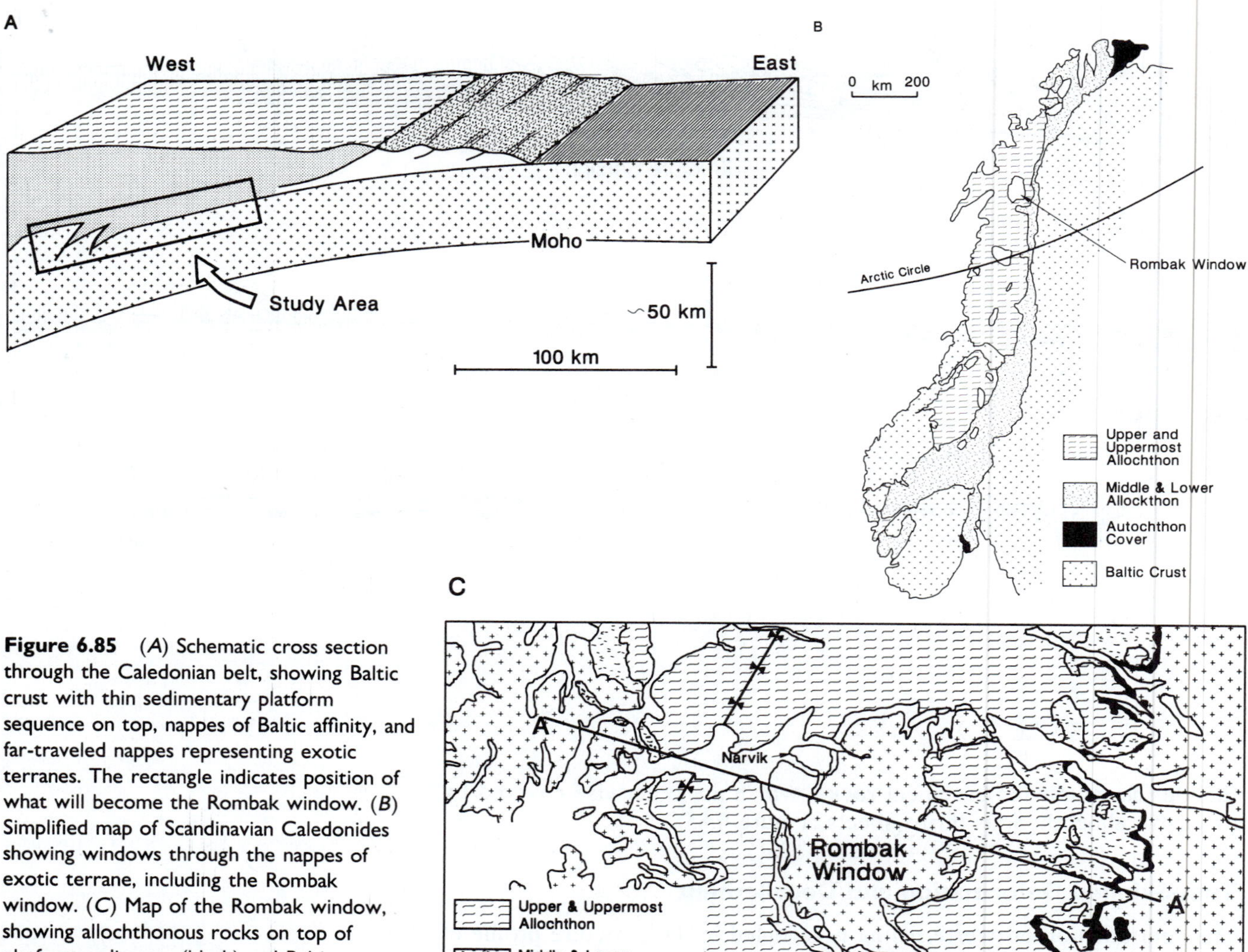

Figure 6.85 (*A*) Schematic cross section through the Caledonian belt, showing Baltic crust with thin sedimentary platform sequence on top, nappes of Baltic affinity, and far-traveled nappes representing exotic terranes. The rectangle indicates position of what will become the Rombak window. (*B*) Simplified map of Scandinavian Caledonides showing windows through the nappes of exotic terrane, including the Rombak window. (*C*) Map of the Rombak window, showing allochthonous rocks on top of platform sediments (black) and Baltic basement. [Reprinted with permission from *Tectonophysics*, v. 231, Rykkelid and Andresen, Late Caledonian extension in the Ofoten area, northern Norway (1994), Elsevier Science, Amsterdam.]

''Sole fault'' as used by Dahlstrom has other names as well. Some call it a **basal shearing plane** (DeSitter, 1964). The commonly used term is **decollement**, although as DeSitter points out the term refers not to a structure but to a process: ''the detachment of the upper cover from its substratum.'' The position of decollement is normally at a discontinuity marking major ductility contrast, permitting cover to become distorted independently of what is underneath. The actual basal decollement itself is commonly a weak horizon, like clay, or shale, or salt (Davis and Engelder, 1985). At depths of 10 km or less, clay and shale are relatively weak compared to most rocks, and salt is ''vastly'' weaker. Salt will flow when shear stress is 1 MPa or less! (Davis and Engelder, 1985; Carter and Hansen, 1983).

Although decollement surfaces are easy to describe, they are tough to explain. The reality of **decollement faulting** and **thin-skinned overthrusting** was first recognized by pioneer geologists of the nineteenth century mapping in the Jura Mountains and the Alps. They mapped enormous, far-traveled **nappes** of folded and faulted strata resting on crystalline rocks of an entirely different structural character. The image of thin-skinned overthrusting haunted them with questions regarding the dynamics of overthrusting: If the thrusts formed by compressional shortening, why was the basement not affected? Or was it? If the deformed cover moved as a whole, what magnitude of force was required to translate it by ~150 km?

Rich's Model of Bedding-Step Thrusting

It was Rich (1934) who recognized the key to modern understanding of the kinematics of thrusting in regional terranes. While studying the geology of the Pine Mountain region in the Central Appalachian Mountains, he recognized that the Pine Mountain thrust, a major thrust-slip fault, did not simply slash up-section as a planar fault of uniform orientation. Instead, he demonstrated that it *stepped* up-section from one incompetent layer to another.

The geometry of the Pine Mountain thrust is pictured in Figure 6.86. The fault steps up-section westward, in the direction of **tectonic transport**, in the direction of translation of the hanging wall relative to footwall. The fault consists of two layer-parallel segments occupying ductile shale layers. These are connected by a **step** or **ramp** where the fault cuts obliquely across competent beds. Rich estimated the net slip on this fault to be 6 mi (10 km).

Recognition of the stepped or ramped nature of thrust-slip faults, like the Pine Mountain, opened the door to understanding the structural geology of deformed strata in many thrust belts. Harris (1979) and Suppe (1980a,b) have described the geometric requirements of this model beautifully. They have emphasized that the segmented nature of step thrusts forces a distinctive geometry upon layered rocks of the hanging wall. When faulted hanging-wall strata are forced to move up a thrust ramp, they are flexed into an angular syncline (Figure 6.87*A*). The form of this syncline is propagated upward into the hanging-wall strata. Its position remains fixed as long as the ramp thrust remains active. When faulted hanging-wall strata are translated to the top of the step or ramp, to the position where the fault once again assumes a layer-parallel position, they are flexed into an anticline (Figure 6.87*B*). The anticline is angular, almost kinklike. Called **snakehead folds** (L. D. Harris, personal communication, 1980) or **fault-bend folds** (Suppe, 1980a,b), these structures grow in amplitude as faulting continues until the lowest unit of the hanging-wall sheet has reached the summit of the ramp. From that time on, the fold grows laterally, but not in amplitude, as long as translation continues (Figure 6.87*C*).

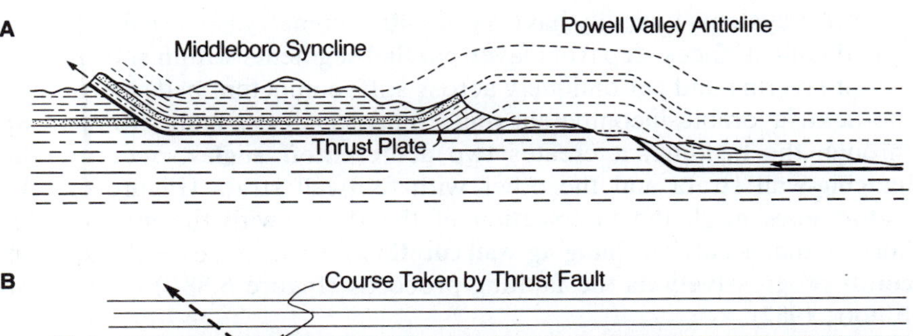

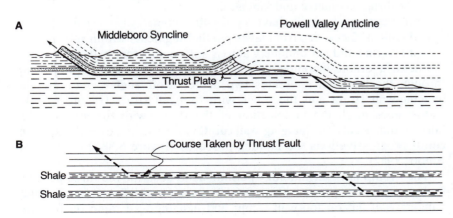

Figure 6.86 (*A*) Structure section of the Pine Mountain thrust in the Central Appalachian Mountains. (*B*) Path of fault through the stratigraphic column as seen before thrusting. [From Rich (1934). Published with permission of American Association of Petroleum Geologists.]

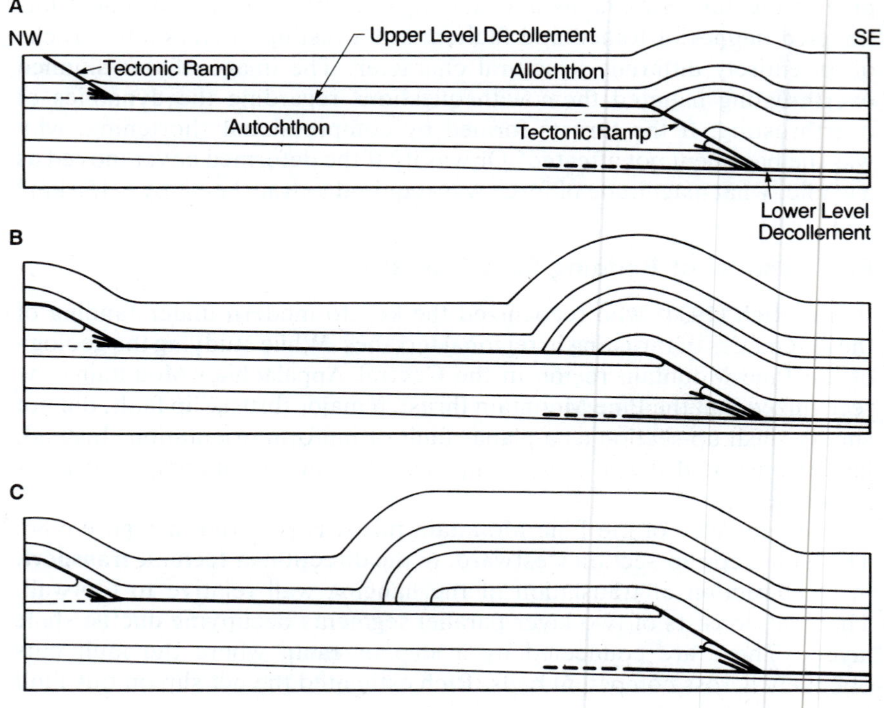

Figure 6.87 Kinematic evolution of regional bedding plane step thrusts. Diagrams show progressive westward movement of upper plate strata. [From Harris (1979). Published with permission of Geological Society of America and the author.]

It is a tribute to Rich's geometric perception that he was able to recognize that the flat-topped anticlines and synclines in the hanging wall of the Pine Mountain thrust could be explained by a fault that steps in and out of layer-parallel orientations.

Ramp-Flat Geometry and Kinematics

Much of our present understanding of the geometry and kinematics of thrusting has been derived from structural geologists, like Rich, who have applied their skills in the arena of petroleum exploration. Explorationists like Bally, Gordy, and Stewart (1966), Dahlstrom (1969), and Royse, Warner, and Reese (1975) developed an unusual appreciation for the full three-dimensional nature of thrust systems. They achieved this understanding through insights derived from geologic mapping, subsurface drilling, and seismic reflection profiling carried out in the search for petroleum. Furthermore, a very important paper, by Boyer and Elliott (1982), got to the very heart of thrust geometry and kinematics.

Thrust systems typically have a **ramp-flat** geometry (Figure 6.88). Individual fault surfaces step from layer-parallel segments within soft, incompetent layering and cut obliquely across stiff, competent beds en route to the next favorable incompetent unit. Where a thrust fault ''ramps'' up through the bedding, it creates two acute **cutoff angles**, one with the hanging-wall strata and the other with footwall strata (Figure 6.88*A*). **Cutoff lines** mark the intersection of the thrust with the stratigraphic horizon that is cut. The **hanging-wall cutoff** moves up and over the **footwall cutoff** progressively as the faulting proceeds (Figure 6.88*B*) (Boyer and Elliott, 1982).

The ''ramp-flat'' terminology permits us to describe the three possible fault-contact arrangements (Figure 6.89): hanging-wall flat on footwall flat; hanging-wall flat on footwall ramp; and hanging-wall ramp on footwall

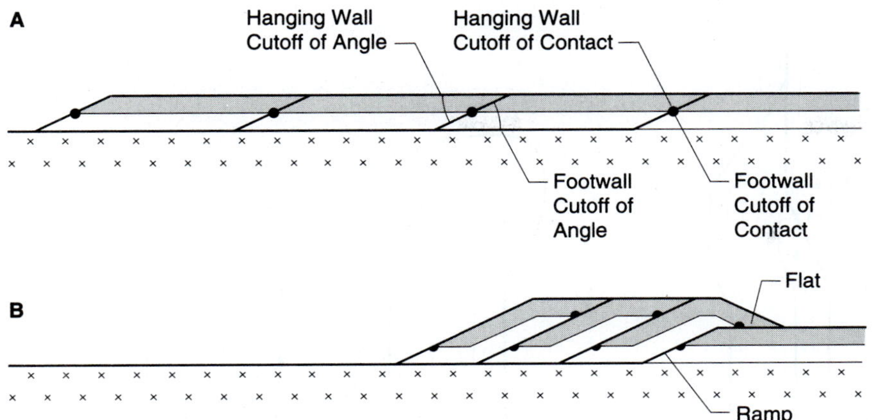

Figure 6.88 The ramp-flat geometry of a typical regional thrust. (Boyer and Elliot, 1982). (*A*) Before it starts, showing cutoff angles and the hanging wall and footwall cutoffs. (*B*) After it has progressed. [From Boyer and Elliot (1982), Figure 18, p. 1207, reprinted by permission from the American Association of Petroleum Geologists.]

flat, Woodward, Boyer, and Suppe (1985) have emphasized that movement along a **flat** (i.e., a flat-fault segment parallel to bedding) is not reflected in stratigraphic separation until the **hanging-wall flat** moves onto another unit at the **footwall ramp**.

Some ramps are **lateral ramps** (Figure 6.90). These represent places where a thrust "flat" abruptly cuts up-section laterally as a ramp until it reaches a higher glide horizon, where it once again becomes a flat. Fault movement along a lateral ramp tends to be strike–slip, for the trend of the ramp is usually subparallel to the overall direction of thrusting.

Typically ramp-flat thrusts evolve in such a way that the major faults cut up-section, one at a time, in the direction of tectonic transport (Figure 6.91*A*). Overall, the major thrusts split off from one another at **branch points**, and the resulting pattern is marked typically by **in-sequence** faulting in which successively younger faults form in the direction of tectonic transport (Dahlstrom, 1969; Royse, Warner, and Reese, 1975). We actually can see this "in-sequence" progression in the sandbox experiment (see Figure 6.66). Not uncommonly, however, thrust faults will form **out of sequence** (Figure 6.91*B*). Boyer and Elliott (1982) have emphasized that the overall characteristics of a thrust system will be influenced by the order in which the various faults and fault slices form. If a thrust does not make it to the surface, but instead is replaced by a fold, the tip line of the thrust will be underground, and the thrust will be called a **blind thrust**. Blind thrusts beneath Quaternary folds in the Los Angeles basin are like hidden time bombs (see Concluding Thoughts).

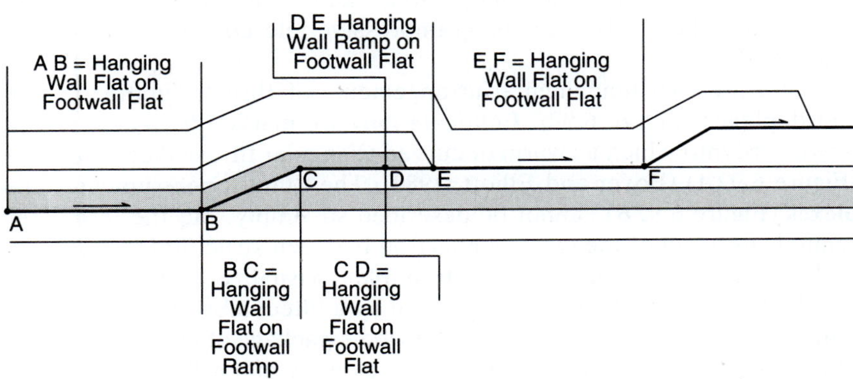

Figure 6.89 Representation of the three types of fault-contact relationships associated with ramp-flat thrust systems. (From Marshak/Mitra, BASIC METHODS OF STRUCTURAL GEOLOGY, © 1988, p. 305. Reprinted by permission of Prentice Hall, Upper Saddle River, New Jersey.)

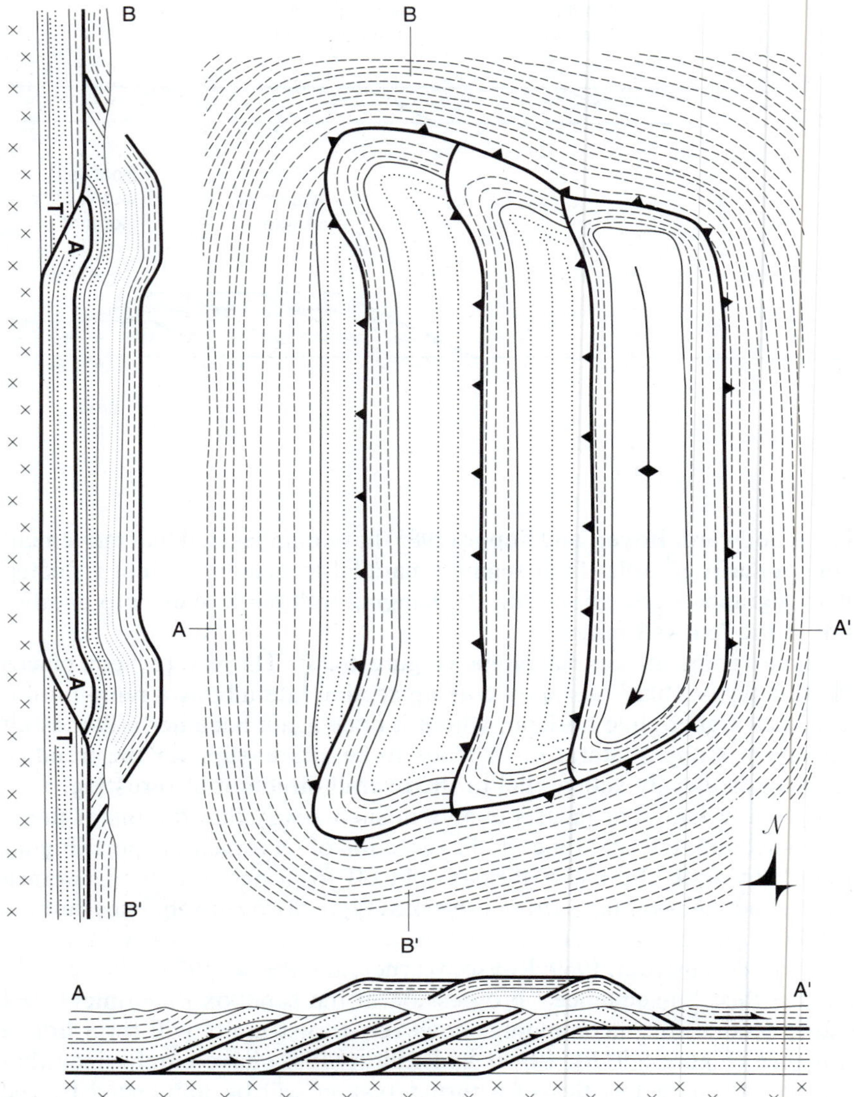

Figure 6.90 Geologic map and cross sections of structural window into a thrust system. Thrust movement is to the east, as portrayed in cross section *A–A′*. Cross section *B–B′* features hanging wall moving away ("*A*") from the viewer (i.e., to the east). Relative movement of footwall is toward ("*T*") the viewer. Note that hanging-wall material has not only moved laterally in strike–slip fashion, but it has also "ramped" up and over the footwall on the flanks. [From Boyer and Elliot (1982), Figure 25, p. 1212, reprinted by permission from the American Association of Petroleum Geologists.]

Horses, Imbricate Fans, and Duplexes

Thrust faults seldom are solitary. Instead, the thrusts, and the **horses** of rock enveloped by thrusts that branch and merge around them, overlap "like roof tiles, all dipping in the same general direction" (Bailey, 1938, p. 607). The amount of movement along each of the thrusts may not be much, but collectively there is an impressive telescoping of the section (Bailey, 1938).

There are two fundamental arrangements of thrusts: imbricate fans and duplexes (Figure 6.92). **Imbricate fans** are marked by a sole thrust, downward into which a swarm of curved triangular thrust slices converge (Figure 6.92*A*) (Boyer and Elliott, 1982). The "fan" opens upward. **Duplexes** (Figure 6.92*B*) cannot be described so simply, but the term itself reminds us again of the close connection between structural geology and architecture. To picture a complete duplex, visualize a tilted imbricate set of horses, each of which is separated by a tilted imbricate set of thrust faults (Figure 6.92*B*). Now picture the whole package of imbricate "tiles" of horses and thrusts bound above and below by major flat-lying thrusts,

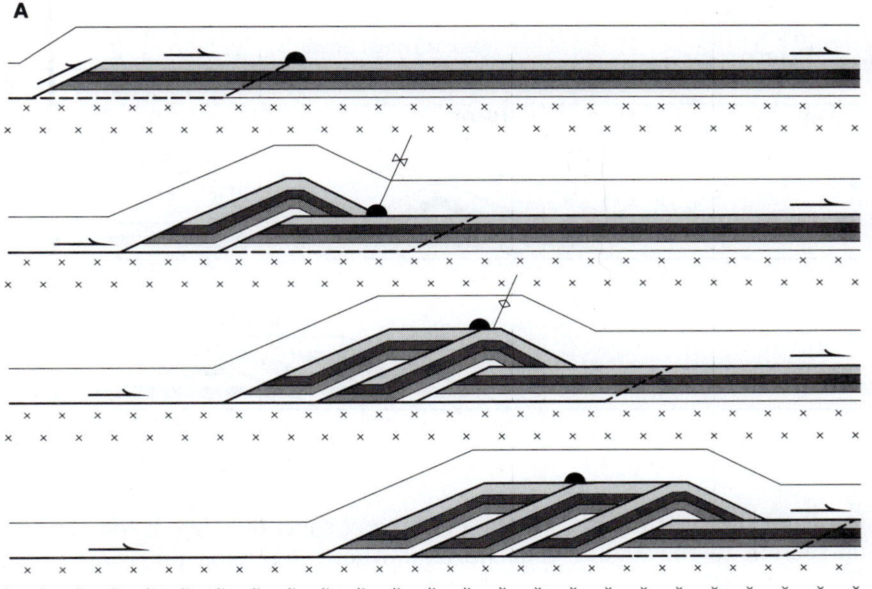

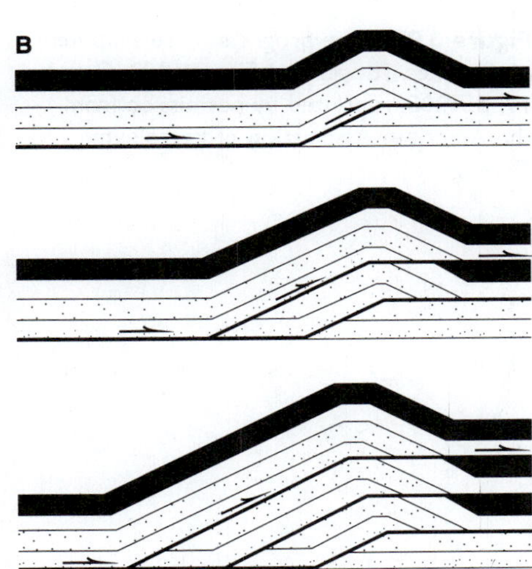

Figure 6.91 (*A*) In-sequence thrusting. (*B*) Out-of-sequence thrusting. [Boyer and Elliot (1982), Figure 19, p. 1208, and Figure 21, p. 1210, reprinted by permission from the American Association of Petroleum Geologists.]

a **floor thrust** below, and a **roof thrust** above. Above the roof thrust are mostly flat-lying, seemingly undisturbed strata. Below the floor thrust are mostly flat-lying, seemingly undisturbed strata (Figure 6.92*B*). Within each horse is an anticline and a syncline (i.e., a "fold pair"); the fold pairs are approximately the same size and shape (Figure 6.92*B*) (Boyer and Elliott, 1992).

Duplex geometries form in the following way (Boyer and Elliott, 1982): A thrust climbs upward from a lower to an upper "glide" horizon, that is, from the floor thrust to the roof thrust (Figure 6.93*A*). The thrust in effect transfers slip from one glide horizon to the next one above, thereby causing structural shortening. As a thrust climbs upward, it cuts steeply across the more competent sequence that lies between the glide horizons, creating a footwall ramp (Figure 6.93*A*). Hanging-wall strata move up the ramp, bending first into a synclinal fold, then (at the top of the ramp) into an anticlinal fold. After a certain amount of slip is accomplished, a new thrust propagates from the base of the ramp in a way that extends the

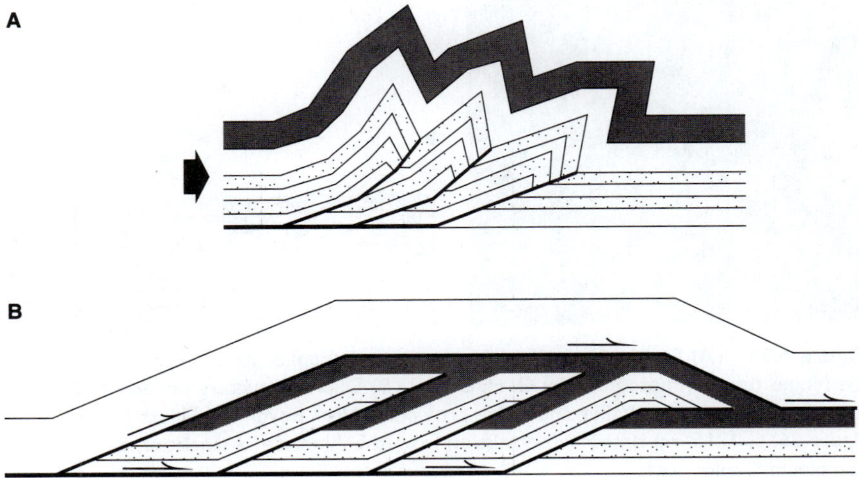

Figure 6.92 Two types of thrust systems. (*A*) Imbricate fan. (*B*) Duplex. [After McClay (1992a), Figure 3, p. 419, and Figure 30, p. 426, reprinted by permission of Chapman and Hall.]

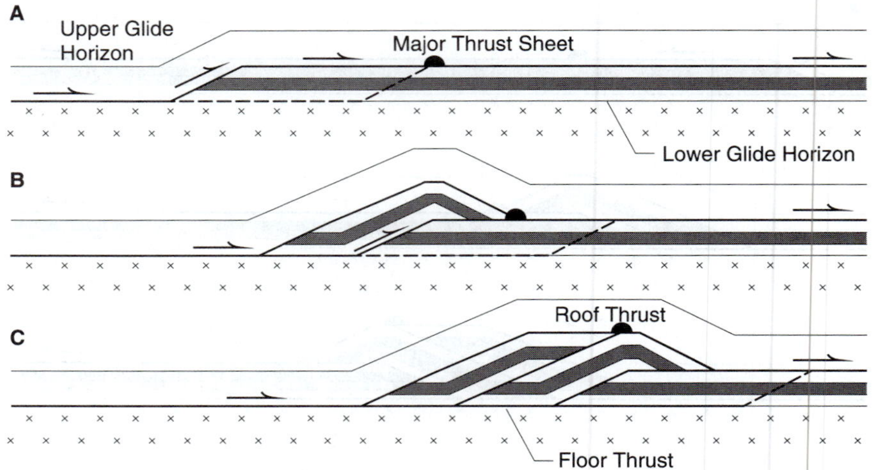

Figure 6.93 The progressive development of duplexes. [Boyer and Elliot (1982, Figure 19, p. 1208, reprinted by permission from the American Association of Petroleum Geologists.]

length of the floor thrust. Within a relatively short distance it too slashes up-section to the higher glide horizon where it rejoins the roof thrust (Figure 6.93*B*). Just as soon as the new thrust begins to propagate, the former ramp thrust dies, and from this point on "just goes along for the ride," deforming passively by folding and tilting as soon as the "new" hanging-wall strata begin moving up the "new" footwall ramp, and so it continues (Figure 6.93*C*). The net effect of duplex development is not unlike the piggyback arrangement of truck cabs on flatcars (Figure 6.94), an efficient arrangement that, compared to an end-to-end lineup, shortens the length of the line of truck cabs. Outcrop examples of duplexes can be stunning (Figure 6.95*A*), as can examples of regional duplexes (Figure 6.95*B*)!

Figure 6.94 Duplex image of pickup trucks arranged like imbricate tiles on flatcars.

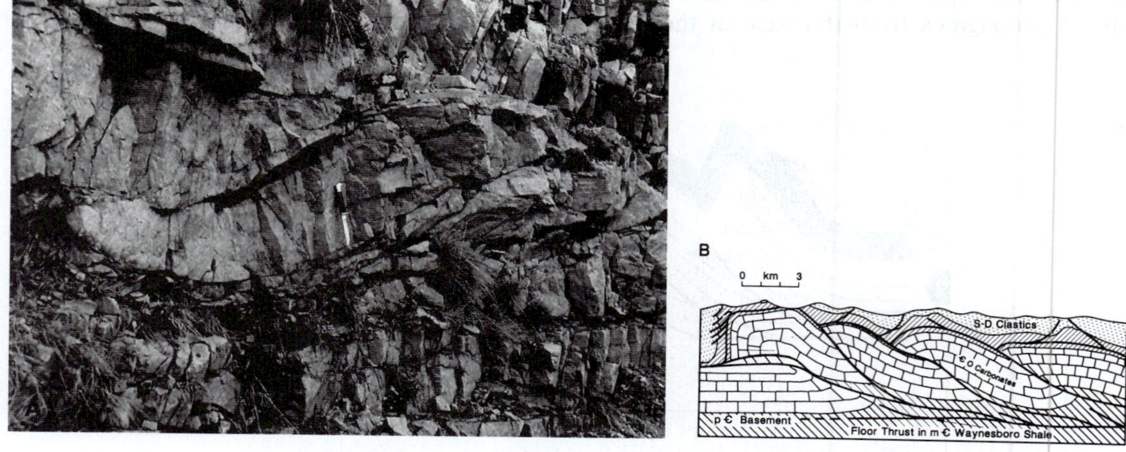

Figure 6.95 (*A*) Several horses within outcrop-scale duplex structure associated with the Moine thrust. [Photograph by G. H. Davis.] (*B*) Duplex structure underlying folds in the central Appalachian Valley and Ridge province. [After Boyer and Elliott (1982), Figure 26, p. 1215, reprinted by permission from the American Association of Petroleum Geologists.]

The magnitude of **slip** on a thrust fault in comparison with the **length** of the next lower thrust slice will determine the actual shape and geometry of a duplex (Figure 6.96). Where slice length is greater than fault slip, a **normal duplex** forms in which the horses dip gently backward, that is, opposite the direction of overall thrust transport (Figure 6.96*A*). Where slice length and fault slip are about the same, an **antiformal duplex** develops in which the horses are actually arched (Figure 6.96*B*). Where fault slip exceeds slice length, a **forward-dipping duplex** develops (Figure 6.96*C*) (Boyer and Elliott, 1982).

The Moine Thrust, Northern Scotland

Boyer and Elliott (1982) credited Dahlstrom (1970) as the person who first used the term "duplex" structure. However, Dahlstrom made it clear that the discovery of what he called duplex thrust geometry was made much earlier by Peach and Horne of the Geological Survey of Great Britain, who between 1883 and 1896 mapped what is now *revered* as the Moine thrust system in northern Scotland.

The Moine thrust is quite simply a classic structural locality. It was here that Peach and Horne (e.g., Peach et al., 1907) carried out phenomenal structural mapping of duplexes and imbricates (Figure 6.97); where mylonites were described for the first time, by Lapworth (1885); and where outstanding British workers have, over the decades, picked apart the fine detail of the systems of structures. We have drawn mainly from the work of McClay and Coward (1981) in presenting the description that follows.

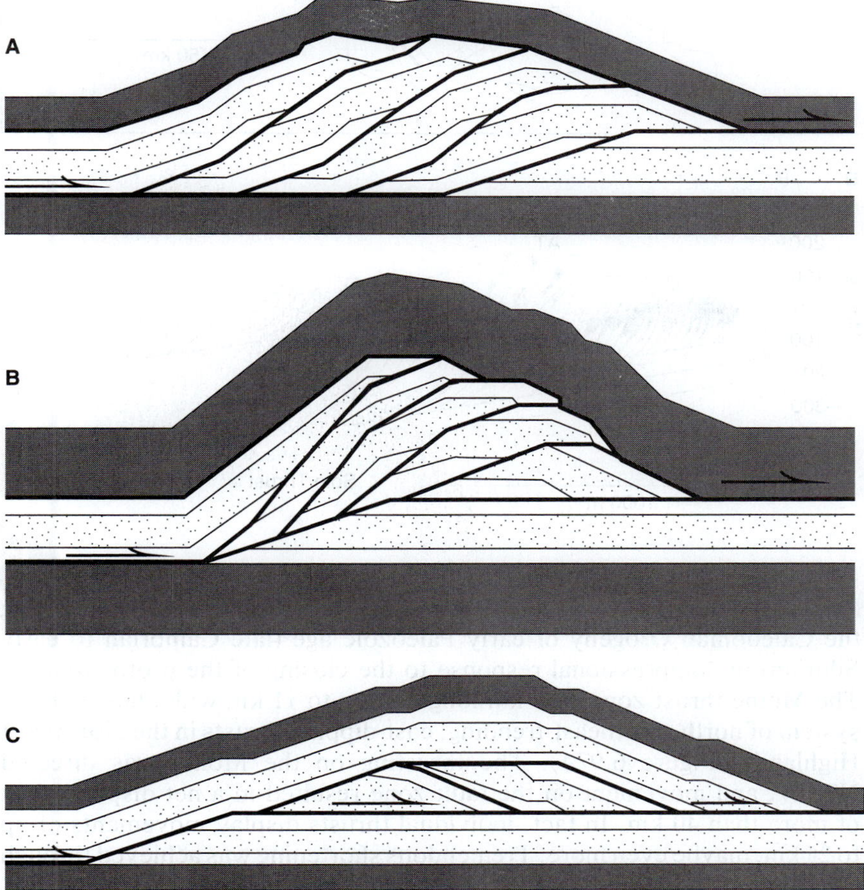

Figure 6.96 Form of duplexes depends on spacing of ramps and the amount of slip. (*A*) A normal duplex develops where slice length exceeds fault slip. (*B*) An antiformal duplex develops where slice length and fault slip are essentially equal. (*C*) A forward-dipping duplex develops where fault slip is greater than slice length. Terminology from Boyer and Elliot (1982). [After McClay (1992a), Figures 24–26, p. 426, reprinted with permission from Chapman and Hall.]

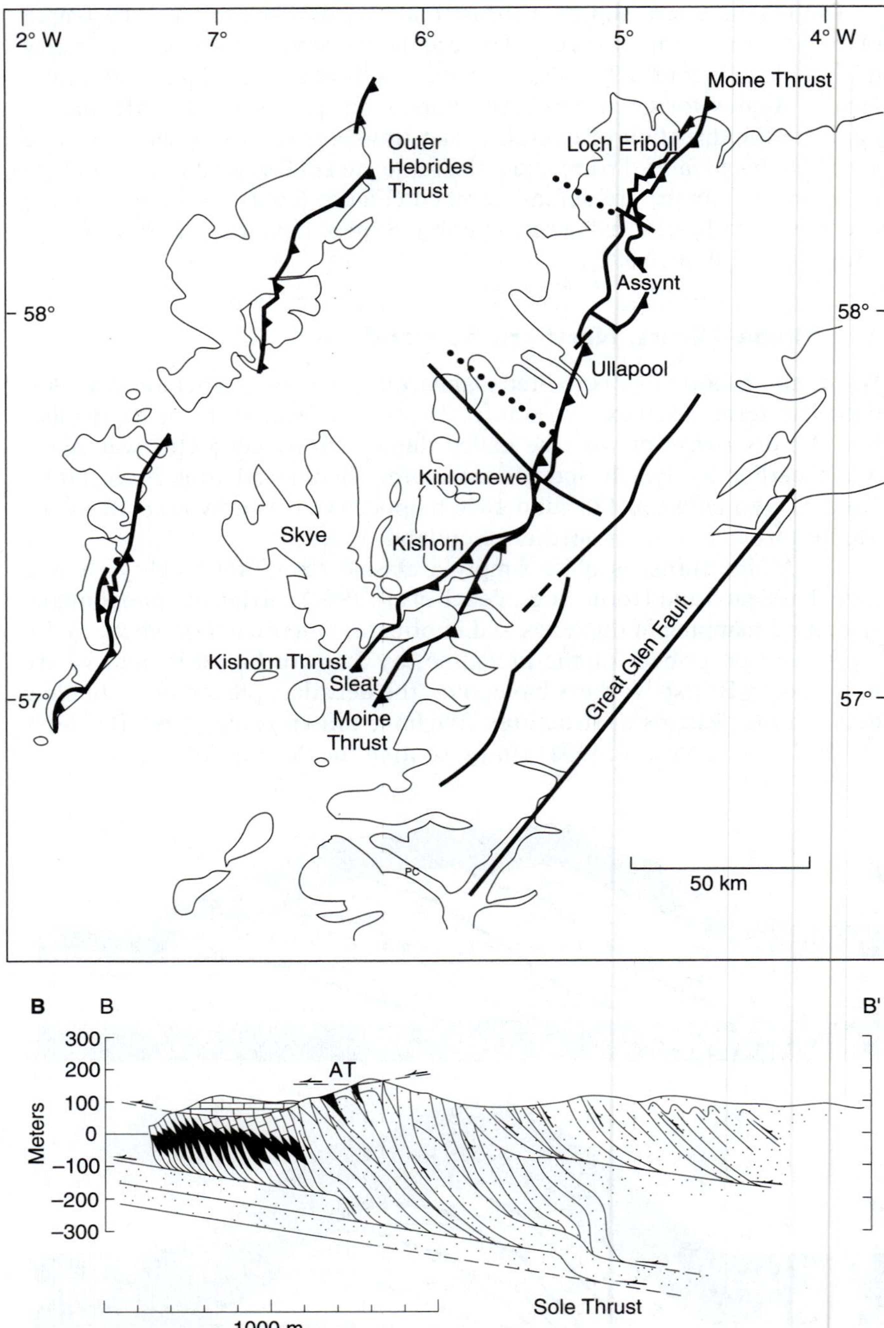

Figure 6.97 (*A*) Location of the Moine thrust zone in the NW Highlands of Scotland. Also shown are two other crustbusters, the Outer Hebrides thrust and the Great Glen fault. (*B*) Cross section across the Moine thrust zone in the vicinity of Loch Eriboll. [From McClay and Coward (1981), Figure 1, p. 242, and Figure 3b, p. 244, reprinted with permission from Chapman and Hall.]

Located in northern Scotland, the Moine thrust zone developed during the Caledonian Orogeny of early Paleozoic age (late Cambrian to early Silurian) in compressional response to the closing of the proto-Atlantic. The Moine thrust zone, 190 km long and up to 11 km wide, lies within a system of north–northeast-trending, east-dipping thrusts in the Northwest Highlands (Figure 6.97*A*). The thrusting on the Moine was directed N70°W, and movement on the fault zone resulted in a net displacement of more than 40 km. In fact, individual thrusts display movements of up to 25 km, maybe even more. Tremendous shortening was achieved through the generation of duplexes (Figure 6.97*B*).

The mapped trace of the Moine thrust zone is by no means perfectly straight. There are bends and curves that result in a nesting of fault surfaces with **megamullion** antiforms and synforms (Figure 6.98) that separate lens-shaped nappes above and below. The megamullions are oriented parallel to tectonic transport, which is beautifully disclosed by slickensided, slickenlined surfaces. The flanks of the foldlike surfaces are lateral ramps, and are associated with strike–slip movements.

The basic arrangement consists of an autochthonous footwall of Precambrian rocks unconformably overlain by Cambro-Ordovician sedimentary rocks, and an overlying, imbricated allocthonous stack of thrusted nappes composed of Precambrian and Cambro-Ordovician rocks (Figure 6.98). The relationships are incredible, including, for example, clean outcrop exposures of flat segments of the Moine thrust separating Precambrian rock *above* from Cambrian rock *below* (Figure 6.99). Imagine hiking up the long ridge shown in Figure 6.100. You start out in Precambrian basement and after a while cross the great unconformity into Cambrian. Then as you hike up the ridge still farther, you find yourself back in Precambrian, having crossed the very gently dipping Moine thrust.

Classic Moine thrust localities, such as in the Northwest crags of the Stack of Glencoul (wonderful names!), feature mylonitic Precambrian rocks in low-angle fault contact on Cambrian sedimentary rocks that are transformed to quartzite mylonites from place to place. The mylonites in the upper-plate Precambrian rocks were fashioned at deep structural levels (e.g., 10 or 12 km) during part of their upward transport to their present structural location. The Cambro-Ordovician sediments were never as deep as the Precambrian rocks now above them, and thus the degree of mylonitization in them is much less.

Because the Moines is such a geological sanctuary, when you go there be prepared to read official signs that in official language say to the public: "Leave dogs and rock hammers in the car."

Dahlstrom's Guidelines on Thrust Kinematics

Although the general kinematics of the thrusting process are now reasonably well known, it is the interpretation of specific fold/thrust configurations in the subsurface that is fundamental to regional tectonic interpretation and petroleum exploration. In spite of all that is known about thrusting, it remains very difficult on the basis of geologic mapping alone to confidently interpret subsurface structural details. In theory, an infinite variety of subsurface geometric configurations of beds, folds, and thrusts can be drawn to fit the surface control (geologic map patterns) and subsurface control (well data and seismic models).

Dahlstrom (1969) developed some kinematic rules that provide a means to test and improve interpretations of subsurface geology in thrust belts. The rules are summarized in the construction of **balanced cross sections**, structural profiles drawn to avoid violating the original cross-sectional area of the rocks that were subjected to deformation (See Part III-E). If it can be shown that deformation did not alter bed thickness, balanced structural profiles simply require that the thickness of each bed be held constant and that the beds shown in profile view be drawn of equal length (Figure 6.101). This is called **bed-length balancing**. The very first balanced cross sections to appear in print were cross sections of the Canadian Rockies published by Bally, Gordy, and Stewart (1966). These geologists as well as Dahlstrom (1969) and Royse, Warner, and Reese (1975) emphasized that thrust faulting and folding in the Canadian Rockies and the eastern Idaho/western Wyoming thrust belt indeed took place without

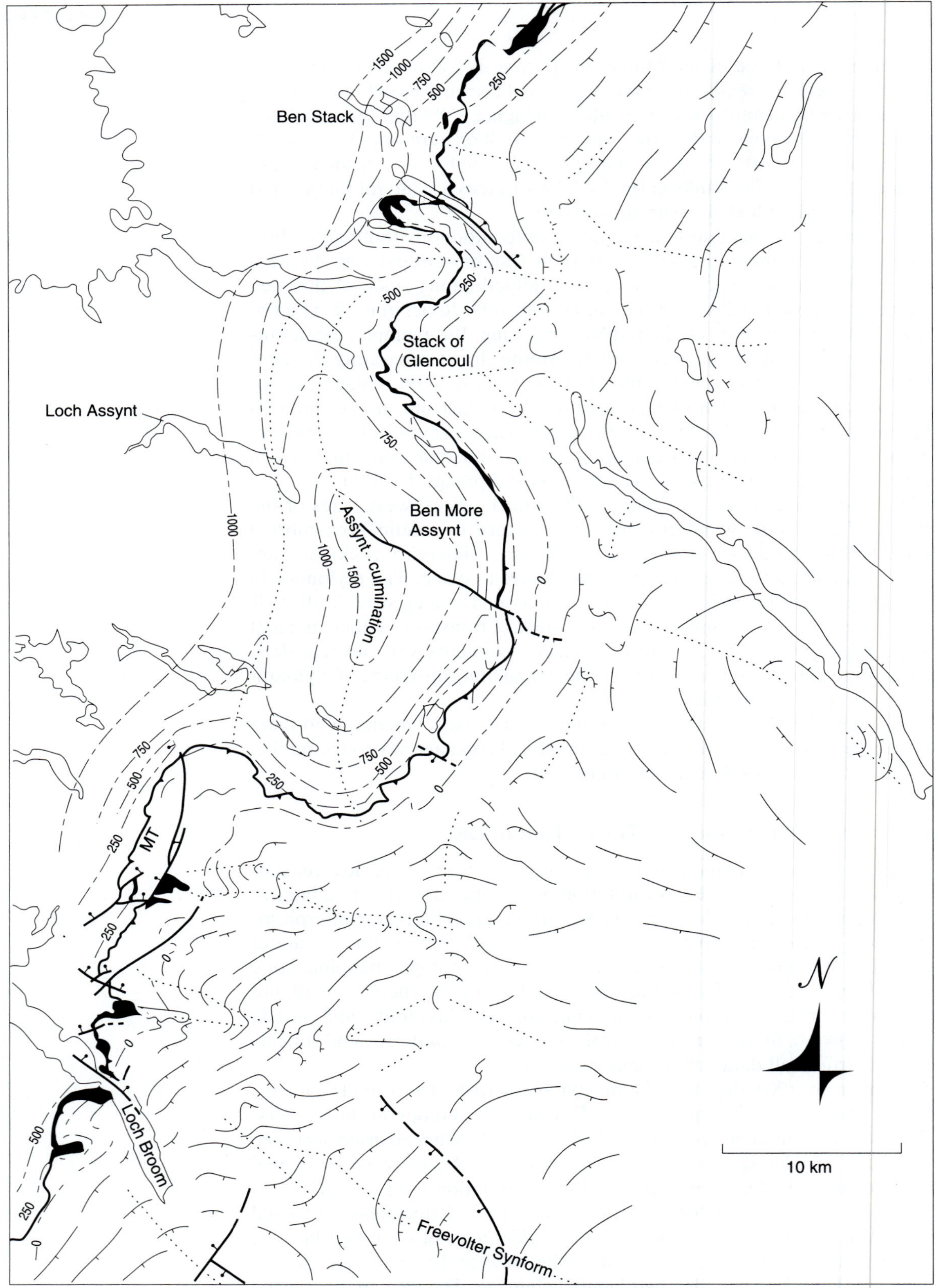

Figure 6.98 Map showing the trace of the Moine thrust zone, as well as contours on the thrust itself. Long curved lines with dip symbols represent the attitude of foliation in Moine metasediments. Crests and troughs of major folds are shown by the dotted lines. Note the structural relief on the Moine thrust at the Assynt culmination. [Reproduced by permission of the Royal Society of Edinburgh and M. R. W. Johnson from *Transactions of the Royal Society of Edinburgh: Earth Sciences,* Volume 71, part 2 (1980), pp. 69–96.]

Figure 6.99 Ken McClay checks out a beautiful exposure of the Moine thrust. Here it separates Precambrian rock above from Cambrian rock below. (Photograph by G. H. Davis.)

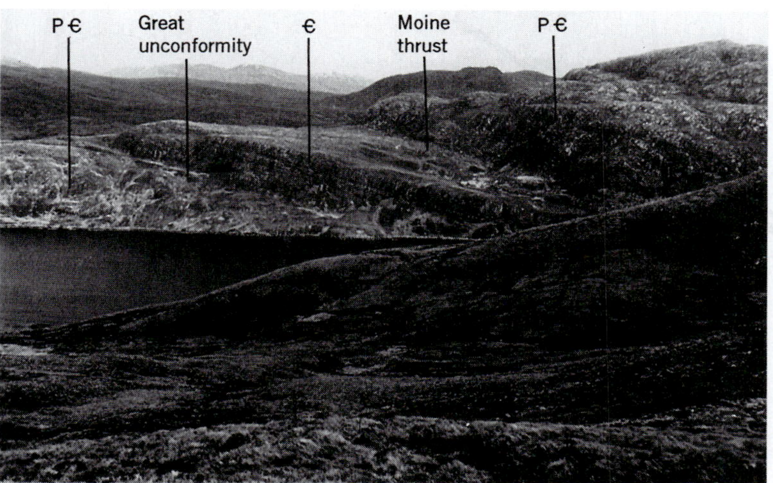

Figure 6.100 Distant view of the Moine thrust, showing thrust repetition of the Great Unconformity. (Photograph by G. H. Davis.)

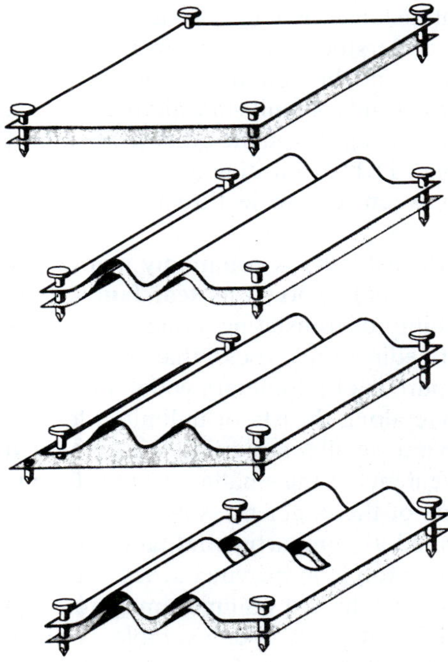

Figure 6.101 Balanced cross sections are constructed in a way to preserve the original cross-sectional areas of beds that are involved in the deformation. [From Dahlstrom (1969). Reproduced with permission of National Research Council of Canada from *Canadian Journal of Earth Sciences.*]

appreciable volume change. Furthermore, they demonstrated that bedding was neither thickened nor thinned significantly during the deformation.

Balanced cross sections of regional thrust relationships can provide a strong foundation for estimating regional strain. In the eastern Idaho thrust belt, the sedimentary prism was shortened by about 50% because of thrusting and associated folding that took place between latest Jurassic and early Eocene. On the basis of carefully balanced geologic cross sections, Royse, Warner, and Reese (1975) computed a total shortening of 83 km for the region between the Wasatch Mountains of Utah and the Green River Basin of Wyoming. Their estimate of the original width of the rocks in the belt, based on their sections, is 178 km. These data permit stretch (S) to be calculated along the direction of shortening:

$$S = \frac{l_f}{l_0} = \frac{178\text{ km} - 83\text{ km}}{178\text{ km}} = \frac{95}{178} = 0.53$$

$$\%\text{ shortening} = (0.53 - 1.00) \times 100 = 47\%$$

Price and Mountjoy (1970) reported a similar magnitude of crustal shortening by thrust faulting between late Jurassic and Eocene in the Canadian Rockies, again based on the balancing of sections. Reconstruction shows an original miogeoclinal sediment package 475 km wide that was reduced by folding and thrusting to 240 km.

$$S = \frac{240\text{ km}}{475\text{ km}} = 0.51$$

$$\%\text{ shortening} = (0.51 - 1.00) \times 100 = 49\%$$

The main kinematic principle emphasized by Dahlstrom (1969) in the preparation of balanced cross sections is **consistency of displacement**. Slip measured along any given fault should be uniform in sense and magnitude, assuming that the rocks are indeed rigid (Figure 6.102). If data show that the slip varies from place to place along a fault (Figure 6.102*A*), this circumstance must be explained by interpretations that involve the **inter-**

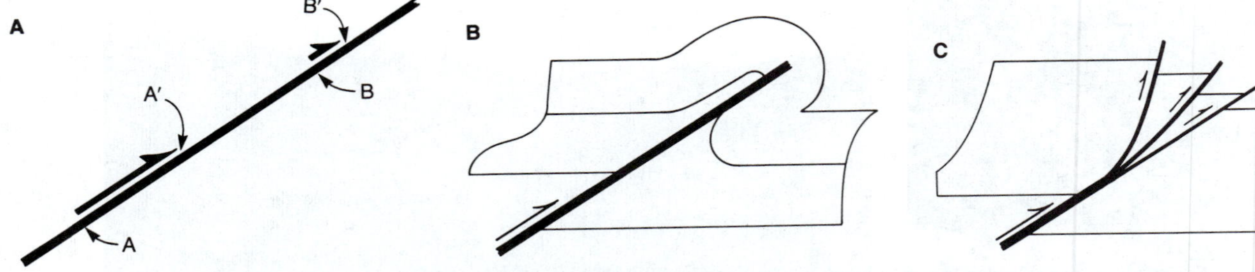

Figure 6.102 (*A*) Fault showing an inconsistency of displacement. The slip along the lower reaches of the fault is much greater than that toward the surface. The inconsistency can be accommodated by (*B*) the interchange of faulting and folding or (*C*) the replacement of the single fault by several splay faults. [From Dahlstrom (1969). Reproduced with permission of National Research Council of Canada from *Canadian Journal of Earth Sciences.*]

change of different degrees of folding and faulting to accommodate a common shortening (Figure 6.102*B*), or the **replacement** of one fault by a series of **imbricate splay faults** (Figure 6.102*C*). Balanced cross sections must stay within the bounds of all subsurface, seismic, and geologic data on hand, and they should be constructed with an understanding and appreciation of kinematic principles. The balancing of sections provides a way to place realistic constraints on the structures that might exist at depth, and helps to explain the map-view expressions of thrust belts as well.

Displacement on a single fault may decrease along strike to zero, even though regional shortening along that distance remains strong and robust. This apparent contradiction can be understood by recognizing that displacement (slip) can be transferred from one thrust fault to another (Figure 6.103). **Transfer zones** are the overlapped ends of faults in which decreasing slip on one fault is compensated by increasing slip on the other (Dahlstrom, 1969). Figure 6.104 presents a classic Dahlstrom diagram, this time in cross-sectional view, of transfer between low-angle thrusts above and below a salt sequence.

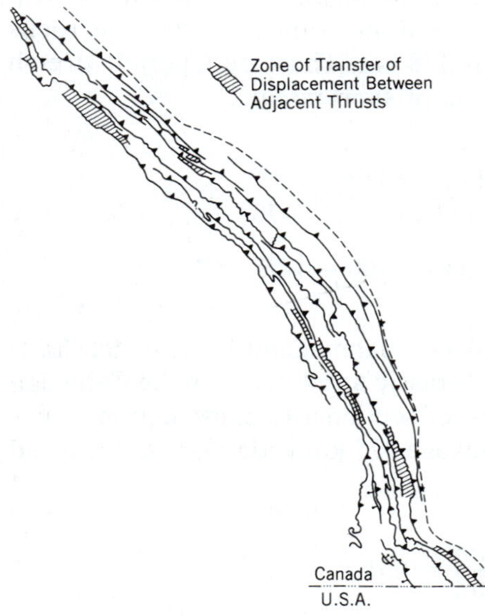

Figure 6.103 Map of regional thrust faults in the Canadian Rockies. Cross-hatched symbols identify transfer zones where decreasing translation on one "dying" fault is compensated by increasing translation on an "emerging" fault. [From Dahlstrom (1969). Reproduced with permission of National Research Council of Canada from *Canadian Journal of Earth Sciences.*]

The plan-view expression of overthrust belts is commonly marked by transverse strike–slip faults (Figure 6.105). Known as **tear faults**, they form mainly because of the impossibility of translating a huge rock mass as a single unit. A regional mass of sedimentary rocks that is hundreds or thousands of kilometers long, hundreds of kilometers wide, and many kilometers thick, simply cannot move along thrusts of unlimited length. Instead, larger masses are broken up into smaller structural units bounded by thrust faults and tear faults. Movements of the smaller structural units are orchestrated with the movements of the larger mass as a whole.

Tear faults have also been described as **compartmental faults**. Recognized by Brown (1975) in the Wyoming foreland province, these serve as partitions between domains of rocks in which a common magnitude of shortening has been achieved in different ways. For example, a given compartmental fault might separate a fold-dominated compartment from another that is fault dominated (Figure 6.106). Compartmental faults are unlike classic strike–slip faults in that the sense of slip need not be uniform along the length of the fault. Moreover, compartmental faults are surprisingly short when compared to the magnitude of slip they accommodate.

The art of balancing sections has become very sophisticated, in part because of the success in their use in petroleum exploration. Some of the tediousness of detailed balanced section restoration has been reduced

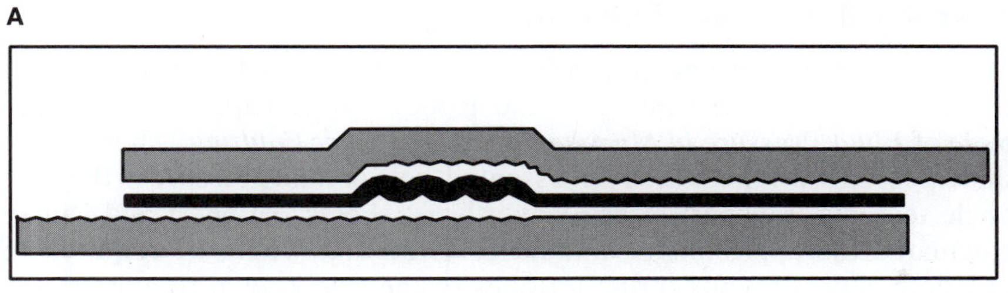

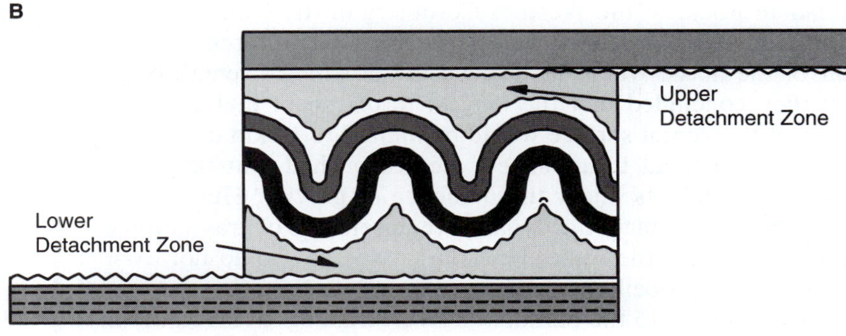

Figure 6.104 Transfer zone above and below a salt layer. (*A*) This regional schematic features a lower and upper detachment plane whose ends overlap above and below a folded salt sequence. The slip along the lower detachment plane is transferred to the upper detachment plane by folding (shortening) of the salt layers. (*B*) Close-up view of the transfer. [After Dahlstrom (1960), reproduced with permission of the Canadian Society of Petroleum Geology.]

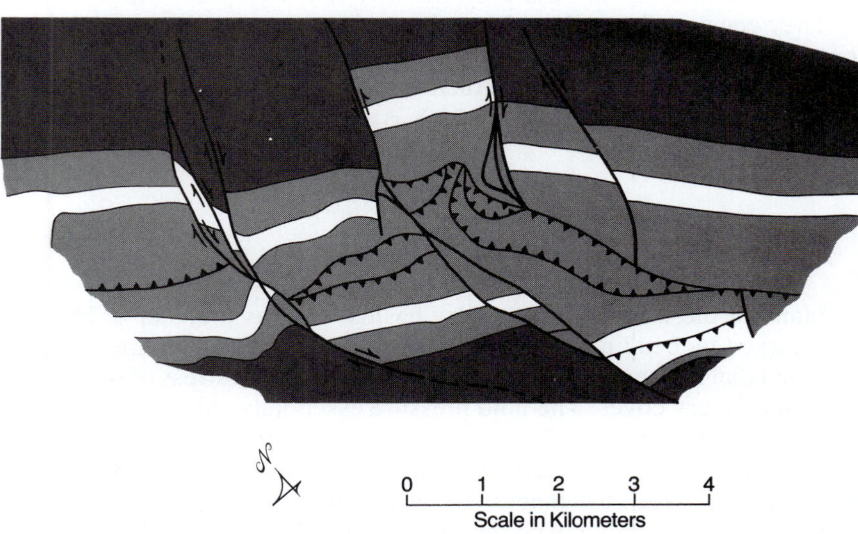

Figure 6.105 Map showing tear faults within a regional system of folds and thrust faults. [Adapted from Price, Mountjoy, and Cook (1978).]

through harnessing of computer technologies (Geiser, Kligfield, and Geiser, 1986). For example, once the basic geologic and geometric data are entered, it is possible to use an iterative process of trial-and-error time and again to optimize the interpretation. Again, in Part III-E we provide some of the nuts and bolts of balancing, and provide a starter kit with which you can try your hand.

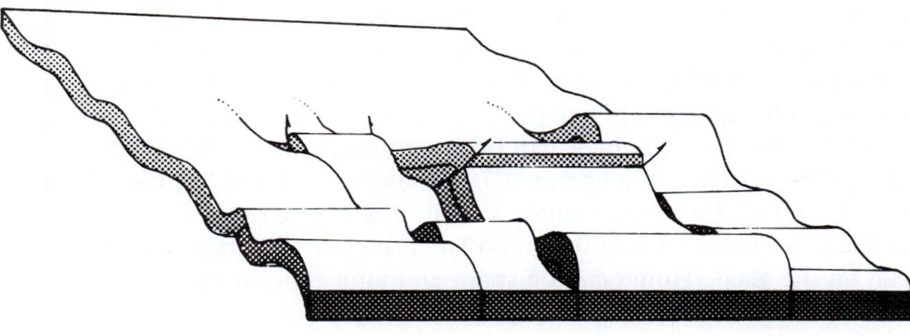

Figure 6.106 Schematic block diagram illustrating compartmental faulting. (Artwork by R. W. Krantz.)

Mechanical Paradox of Overthrusting

One of the most provocative papers ever written on the mechanics of thrust faulting was the Hubbert and Rubey (1959) paper entitled, *The Role* of *Fluid Pressure in Mechanics* of *Overthrust Faulting*. The authors underscored the dominant, if not essential, role of fluid pressure in the low-angle tectonic transport of great overthrust sheets. Hubbert and Rubey calculated the approximate amount of force that would be required to translate allochthonous thrust terranes of the size known to exist in the Canadian Rockies, the Western Cordillera of the United States, and the Appalachians. They concluded that the calculated force, if applied to the rear of the mass, would far exceed the crushing strength of granite. In fact they concluded that the longest thrust sheet that could be pushed across a horizontal surface would be limited to 10 km or so. Considering the possibility that thrust sheets are not pushed, but rather slide down structural gradients under the influence of gravity, Hubbert and Rubey proceeded to calculate the dynamic requirements of gravitational models. The results were the same: large thrust sheets should not exist, even if gravity is the propelling mechanism.

Their solution to the paradox of overthrusting involved modifying the Mohr–Coulomb law of failure, in the way that we examined in Chapter 5. They concluded that the critical shear stress (σ_c) required to cause slip on a thrust fault, which has no cohesive strength (σ_0), must be equal simply to the product of the coefficient of internal friction ($\tan \phi$) and the effective normal stress ($\sigma_N{}^*$), which is normal stress (σ_N) minus pore fluid pressure (P_f):

$$\sigma_C = \sigma_0 + \tan \phi \, (\sigma_N^*)$$

$$\sigma_C = \sigma_O + \tan \phi \, (\sigma_N - P_f)$$

Many of the world's sedimentary basins, such as the Gulf of Mexico, are marked by bedding-parallel zones of fluid pressure that are so highly elevated that they approach the value of the lithostatic stress of the overlying sedimentary cover. The fluid pressure essentially supports the weight of all the rock above. When effective normal stress approaches zero, resistance to thrusting becomes almost inconsequential. All that is required to move a thrust sheet is to overcome sliding friction.

Hubbert and Rubey (1959) demonstrated the fundamentals of the fluid pressure model in their now-famous beer can experiment. Sample preparation consists of drinking two beers, preferably out of nonaluminum cans (Figure 6.107*A*). Place one of the empties in the freezer (Figure 6.107*B*), and remove a window from your house or apartment or lab (Figure 6.107*C*). Clean the glass with detergent, rinse, and leave it wet with a thin film of water. Place the can that is not in the freezer, top down, on the pane of glass. Now lift one end of the glass to form an inclined plane, and, with protractor in hand, measure the angle at which the beer can commences movement down the plane (Figure 6.107*D*). Hubbert and Rubey report typical angles of about 17° corresponding to a coefficient of sliding friction of metal on wet glass of 0.3. After the can in the freezer has been chilled, quickly pull it out and perform the same exercise (Figure 6.107*E*). This time the beer can begins to move down the inclined plane at negligible angles of slope (~ 1°). It moves easily not because the glass is wet; rather it moves because a fluid pressure derived from expansion of the warming air inside the can offsets the normal stress exerted by the can on the glass. Hours can be spent enjoying experiments on the role of fluid pressure in overthrusting (Figure 6.107*F*).

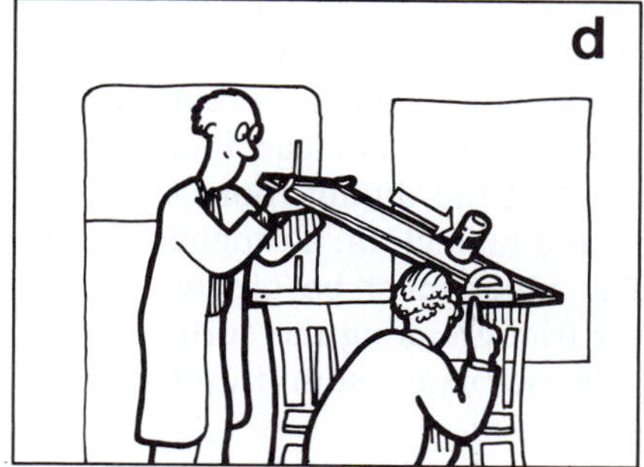

Figure 6.107 The famous beer can experiment. (Artwork by D. A. Fischer.)

Figure 6.108 Wedge-shape nature of thrust belts, as illustrated by the Canadian Rockies. [From D. M. Davis, J. Suppe, and F. A. Dahlen, *Journal of Geophysical Research,* v. 88, figure 1a, p. 1154, copyright © 1983 by American Geophysical Union.]

The Wedge Model of Overthrusting

In 1978 Chappel pointed out to the structural geological community that thin-skinned fold and thrust belts are **wedge shaped** (Figure 6.108), thicker at the back end from which the thrusts come, resting on a weak layer along the base of the wedge, and strongly internally deformed (Chappel, 1978, p. 1189). Deep down inside, the structural geology community knew all of this, but Chappel actually incorporated this fundamental characteristic into his modeling of the mechanics of thin-skinned thrusting. No more blocks sliding on tabletops, or beer cans down glass.

Gradually it became apparent that the mechanics of thin-skinned folding and thrusting is analogous to the movement of a mass of soil that builds up into a wedge shape as it is pushed by the blade of a bulldozer (Figure 6.109) (Dahlen, Suppe, and Davis, 1984). Next time you are standing around with three or four others watching a bulldozer in action, you will see that the blanket of material (soil or snow) will not slide as a mass until a wedge shape is first attained (Figure 6.109). Then the whole mass moves. And if you look even more closely, squinting through a plastic protractor, you will see that not just *any* wedge shape will do. Instead, before moving as a mass, the wedge must have a certain **critical taper** whose dip relates to the strength of the material being shoved (snow, soil, sand, gravel), the friction of the surface over which the material is about to slide (concrete aggregate, blacktop, cobblestone), and the dip of the surface (flat, gently

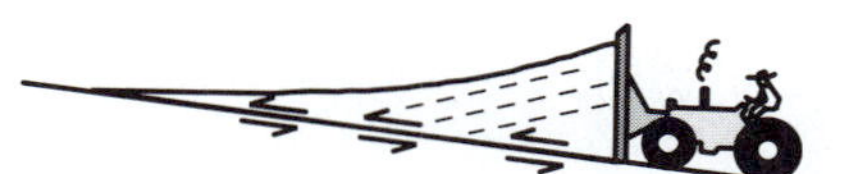

Figure 6.109 The mechanics of development of a thrust belt is analogous to the movement of a mass of snow, sand, or dirt as it piles up and eventually slides *en masse* in front of the blade of a tractor. This is a far different image than pushing a large rectangular block of rock in front of the blade. [From F. A. Dahlen, J. Suppe, and D. M. Davis, *Journal of Geophysical Research,* v. 89, figure 3, p. 10,088, copyright © 1984 by American Geophysical Union.]

inclined, more steeply inclined). The critical taper builds through internal deformation of the material. (With the cooperation of the physical resources group on your college campus, you might want to compare and contrast the critical tapers for topsoil, wet leaves, snow, sand, sawdust—different materials in different seasons!)

I run the wedge-model experiment at breakfast, pouring Cheerios onto our highly polished butcher-block table in the kitchen, patting the tiny letter ''O's'' into a sheet, and then slowly plowing the cereal box into the rear of the layer. What happens? The rear end of the sheet first thickens as the tiny letter ''O's'' rearrange themselves, but while this happens the front of the sheet does not move. Only when the sheet builds to a critical taper does the whole sheet move *en masse*, and when this happens internal deformation ceases. I can disturb the system by spooning ''O's'' out of the sheet, which stops movement at the front and requires reattainment of the critical taper; or by pouring on more cereal. I can increase the required critical taper by eating breakfast at the dining room table and pouring out the cereal onto our embroidered tablecloth, a rougher surface with some microtopographic relief. The taper must be greater for the internal stresses to overcome the higher friction along the surface.

Davis, Suppe, and Dahlen, in their classic paper, sum it up this way:

> Material deforms until a critical taper is attained, after which it slides stably, continuing to grow at constant taper as additional material is encountered at the toe. The critical taper is the shape for which the wedge is on the verge of failure under horizontal compression everywhere, including the basal decollement.'' (D. Davis, Suppe, and Dahlen, 1983, p. 1153.)

Wedge Apparatus Experiments

Davis, Suppe, and Dahlen (1983) built a deformation apparatus to create wedge models of thrusting in the laboratory. The results of their experiments were so compelling that other workers followed their lead right away. For example, Huiqui, McClay, and Powell (1992) built the apparatus shown in Figure 6.110*A*. Sand is stratified into color-coded layers atop a Mylar sheet, which serves as a decollement. The Mylar is wound around a spool in such a way that the Mylar can be ''reeled in,'' thereby transporting the overlying sand against the back wall of the wooden frame. A little graphite on the side wall reduces friction. The sand thickens initially at the wooden buttress, and then the locus of thickening moves forward until a uniform smooth taper is achieved (Figure 6.110*B*). Thickening is achieved through thrusting and folding.

Ultimately, when just the right taper is achieved, internal deformation of the sand ceases and the Mylar slips freely beneath the load of sand. Swapping Mylar for sandpaper increases the taper. Sprinkling graphite on the Mylar is like adding fluid pressure, or replacing a shale decollement zone with one of salt, which reduces the taper. Among the experiments carried out by Huiqui, McClay, and Powell (1992) were ones to illustrate how the taper changes under conditions of high versus intermediate versus low basal friction (Figure 6.110*C*). The frictional failure law controls the system: the critical shear stress required for sliding to occur is equal to the product of the coefficient of sliding friction and the effective normal stress!

Changing the dip of the decollement changes the critical taper as well, for with each different inclination of the decollement there is a different magnitude of normal force on the decollement. The original Davis-Dahlen-Suppe apparatus design permits changes in the dip of the decollement (Figure 6.110*D*).

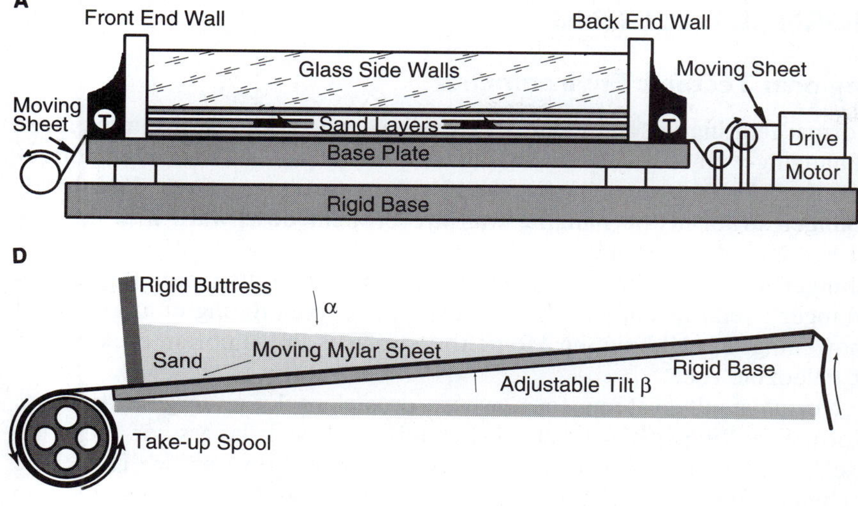

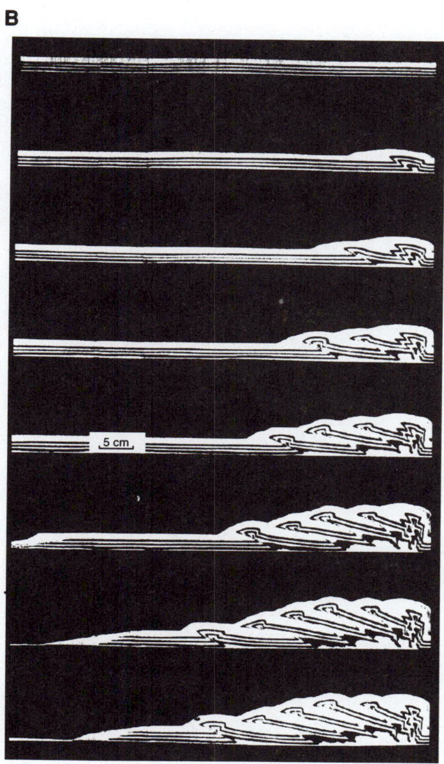

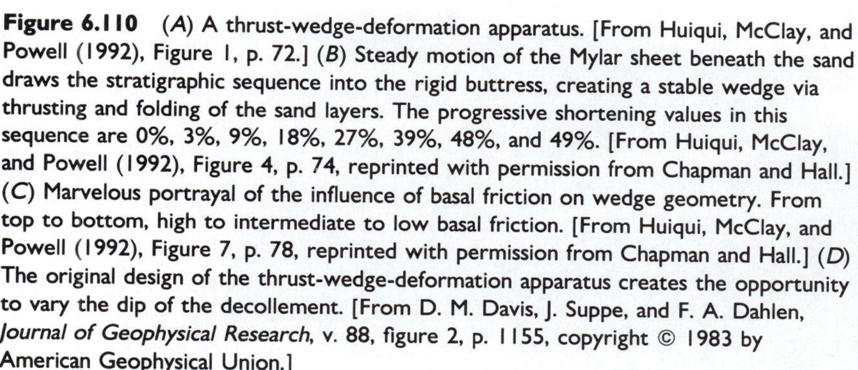

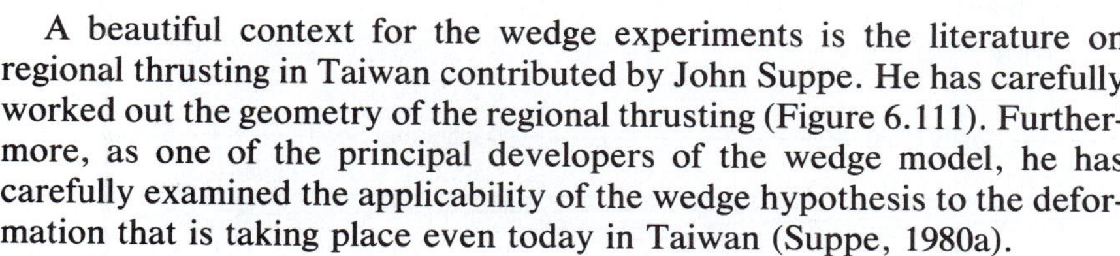

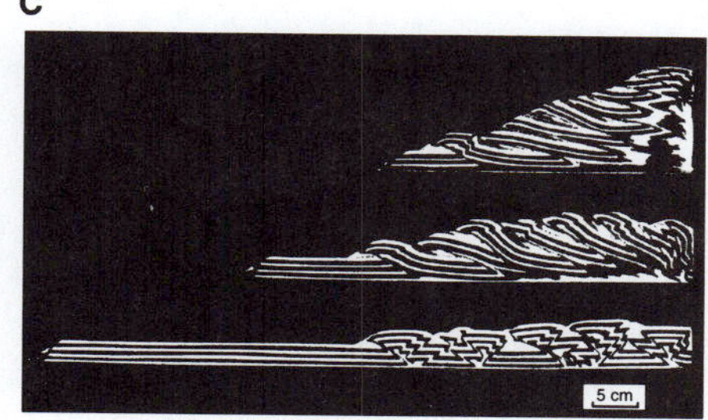

Figure 6.110 (*A*) A thrust-wedge-deformation apparatus. [From Huiqui, McClay, and Powell (1992), Figure 1, p. 72.] (*B*) Steady motion of the Mylar sheet beneath the sand draws the stratigraphic sequence into the rigid buttress, creating a stable wedge via thrusting and folding of the sand layers. The progressive shortening values in this sequence are 0%, 3%, 9%, 18%, 27%, 39%, 48%, and 49%. [From Huiqui, McClay, and Powell (1992), Figure 4, p. 74, reprinted with permission from Chapman and Hall.] (*C*) Marvelous portrayal of the influence of basal friction on wedge geometry. From top to bottom, high to intermediate to low basal friction. [From Huiqui, McClay, and Powell (1992), Figure 7, p. 78, reprinted with permission from Chapman and Hall.] (*D*) The original design of the thrust-wedge-deformation apparatus creates the opportunity to vary the dip of the decollement. [From D. M. Davis, J. Suppe, and F. A. Dahlen, *Journal of Geophysical Research*, v. 88, figure 2, p. 1155, copyright © 1983 by American Geophysical Union.]

A beautiful context for the wedge experiments is the literature on regional thrusting in Taiwan contributed by John Suppe. He has carefully worked out the geometry of the regional thrusting (Figure 6.111). Furthermore, as one of the principal developers of the wedge model, he has carefully examined the applicability of the wedge hypothesis to the deformation that is taking place even today in Taiwan (Suppe, 1980a).

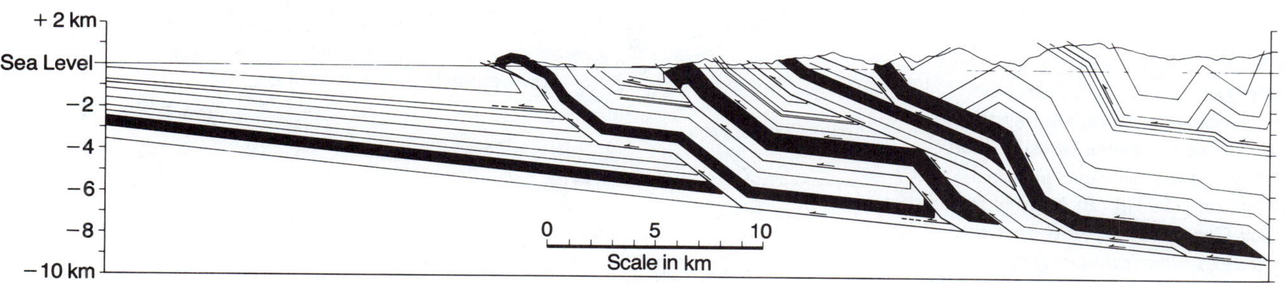

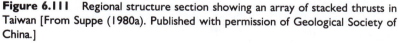

Figure 6.111 Regional structure section showing an array of stacked thrusts in Taiwan [From Suppe (1980a). Published with permission of Geological Society of China.]

NORMAL FAULTING

Regional Tectonic Environments

Normal faulting accommodates extension, and it does so by movement of hanging-wall rock downward with respect to footwall, and in the direction that the fault dips (Figure 6.112). As a result, relatively high-level, younger strata in the hanging wall are brought downward into contact with relatively deep-level, older strata in the footwall. We call this a **younger-on-older** relationship. In contrast, thrust faults create ''older-on-younger'' relationships in the process of accommodating compressional shortening. Recall how the Moine thrust places Precambrian rock on top of Paleozoic rock.

Normal fault systems form where crustal rocks undergo extension. Normal faulting is a way in which individual rock layers, or the crust itself, can lengthen and stretch in brittle or semibrittle fashion. The Rhine graben of Germany, the East African rift valleys, and the Basin and Range province of western North America emerged as among the foremost early classical examples of regions extended by normal faulting. With the discovery of seafloor spreading and plate tectonics in the 1960s, normal faulting

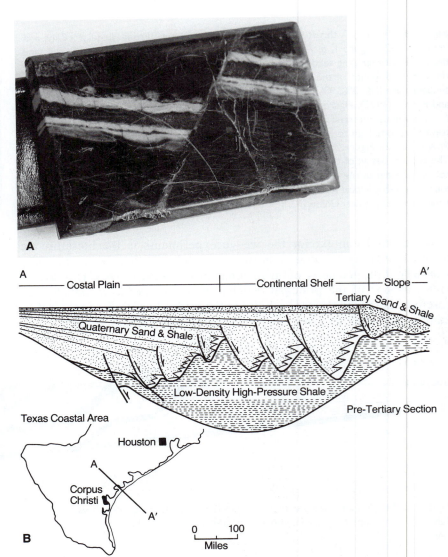

Figure 6.112 (*A*) Close-up of a George Davis belt buckle, a present from John Guilbert. Among structural geologists, the buckle gives rise to severe cases of "belt-buckle envy." [John Guilbert, personal communication (1994).] (*B*) Normal faulting in the Gulf Coast region. Note that the faulting creates older-on-younger relationships, e.g., Quaternary sand and shale is in fault contact with older shale. [From Bruce (1972). Published with permission of Gulf Coast Association of Geological Societies.]

became recognized as the dominant fault class along midoceanic spreading centers (see Chapter 10). The character of normal faulting of oceanic crust along spreading centers is now known in detail based on insights gained from detailed bathymetric studies, underwater geologic mapping in submersibles, on-land geologic mapping of portions of ridge systems exposed in volcanic islands, and seismic investigations.

Of course, supercontinents have been rifted apart by seafloor spreading, and the **rifted continental margins** reveal the normal faulting by which rifting is achieved. The interesting parts of rifted continental margins are typically offshore, underwater, and under post-rifting sediments. Thus it has required offshore drilling and seismic work to disclose, often with exceptional clarity, the structural details of the rifting, extension, and normal faulting.

Salisbury and Keen (1993) have published exceptionally detailed seismic reflection images of normal faulting off the coast of Nova Scotia (Figure 6.113). The faults cut and displace oceanic basalts and sediments of Jurassic age, and were formed during seafloor spreading and the opening of the Atlantic. The seismic sections distinguish flat-lying, post-faulting mid-Cretaceous strata; post-faulting lower Cretaceous strata cut by nontectonic growth faults produced by settling, compaction, and consolidation; basalt flows and sedimentary strata of Jurassic age that were deposited during the faulting and subsequently cut and displaced by continued faulting; and interlayered basalt and sedimentary strata that were deposited prior to the faulting. At a depth of 4 km, which is interpreted to have been the brittle-ductile transition at the time of faulting, the faults "sole" into the base of crust.

The Red Sea region is a place where an oceanic ridge has "stepped" into a continental region. Details of normal faulting in the Red Sea region have been described beautifully by Lowell and Genik (1972) and by Lowell and others (1975). The tectonic environment of the Red Sea region is essentially that of the Carlsberg Midoceanic Ridge, except that the ridge underlies continental crust (Figure 6.114*A*). Like normal midoceanic spreading centers, the Red Sea region is marked by high heat flow, active seismicity, symmetrical magnetic anomaly patterns, a gravity high, and even some midocean ridge basalts. Stretching of the Red Sea region was achieved by penetrative, distributed normal-slip faulting (Figure 6.114*B*). Thinning by normal faulting resulted in subsidence and the formation of a complex graben system (Figure 6.115). Subsidence was so significant that the land surface dropped below sea level. Transgression(s) of ocean waters resulted in deposition of oceanic sediments, including 5 to 7 km of evaporites. Ultimately, the lithosphere was stretched and thinned to the point of rupture, permitting entrance and ascent of oceanic basalts (Figure 6.115). Oceanic crust continues to be emplaced in the innermost part of the Red Sea at a rate consistent with the spreading rate of Saudi Arabia with respect to Africa.

Accurate calculation of the displacement vector for faulting requires summing up the displacements on all the normal-slip faults in the system. Based on structure profiles, like that shown in Figure 6.114, Lowell and Genik estimated that the width of the region was initially 315 km; normal faulting stretched it to 450 km. Moreover, there may be another 100 km of stretching bound up in plastic deformation and translations on minor faults for which data are not available. The interpretation of a 135-km increase in length is thus a conservative estimate. Even so,

$$S = \frac{450\text{ km}}{315\text{ km}} = 1.43$$

$$\%\text{ stretching} = (1.43 - 1.00) \times 100 = 43\%$$

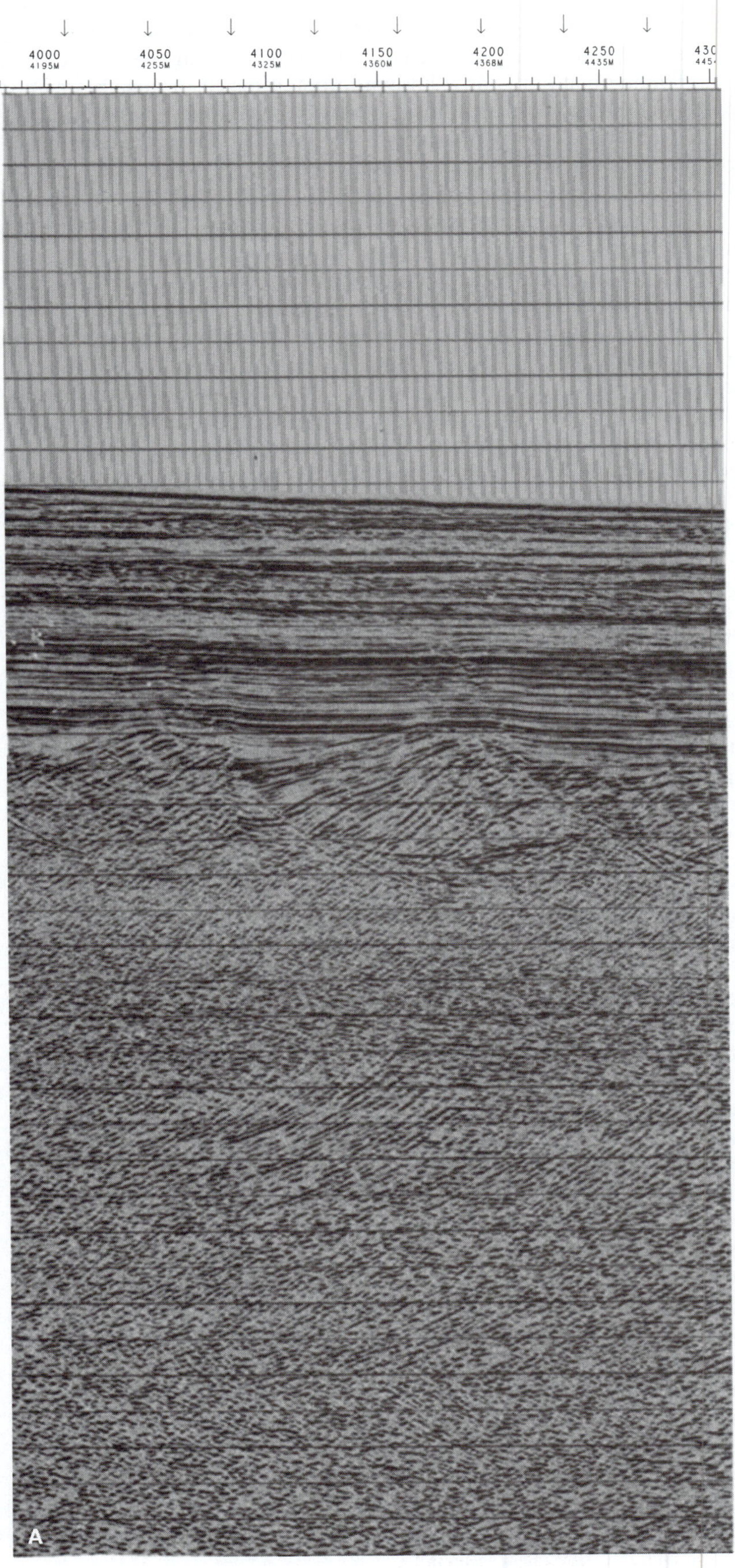

Figure 6.113 Seismic reflection profiles for region off the coast of Nova Scotia. Normal faulting occurs in oceanic basement of Jurassic age. (*A*) Seismic reflection profile showing internal structure. (*B*) Interpretation of the internal structure including (1) lower crust of magmatic origin, (2) massive plutonic rocks at base of rotated crustal block, (3) interlayered basalt and sedimentary strata deposited prior to tectonic rotation, (4) syntectonically deposited basalt flows and sedimentary strata, and (5) post-tectonic sedimentary strata cut by consolidation-induced growth faults. The growth faults cut all the way up into middle Cretaceous strata. At the time of their formation, the normal faults entered "mush" (technical term used by Salisbury and Keen) at a depth of 4 km. [From Salisbury and Keen (1993), Published with permission of the Geological Society of America.]

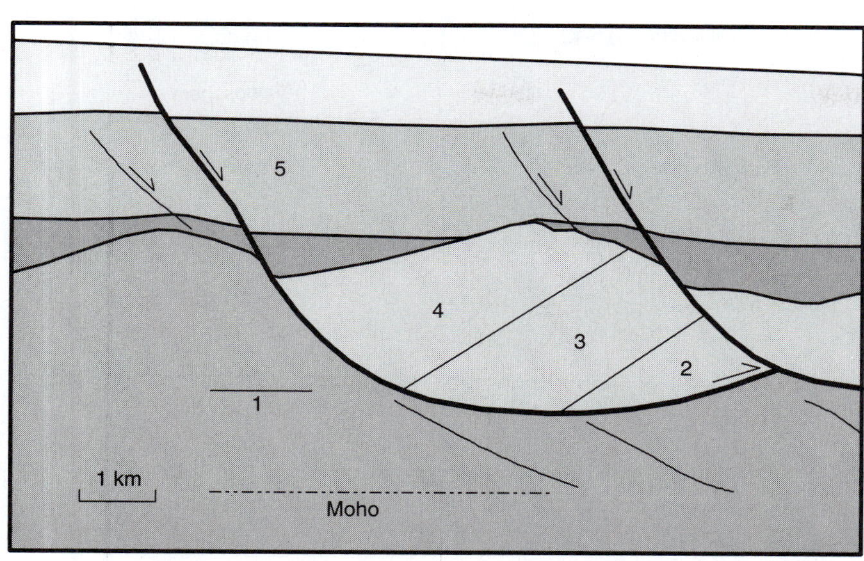

Figure 6.113 (*continued*).

Kinematic reconstruction of the Red Sea region is shown in Figure 6.116. Pull-apart is oblique, not perpendicular, to the trend of the Red Sea itself.

Some systems of normal faulting in continental regions are not related to **spreading centers.** The Basin and Range province of the western United States and northern Mexico is a case in point. It occupies a huge region

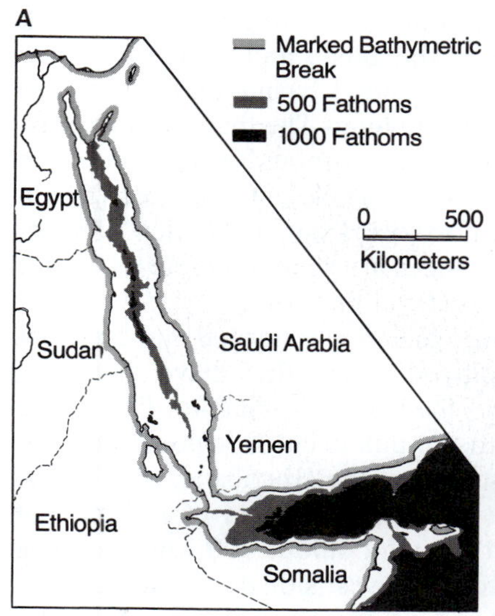

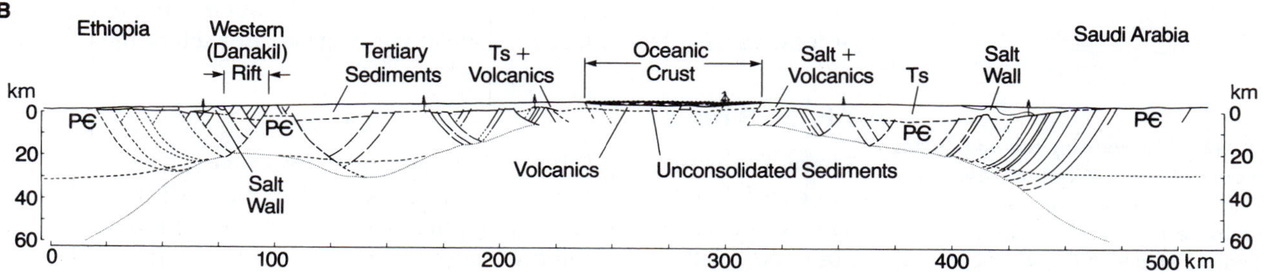

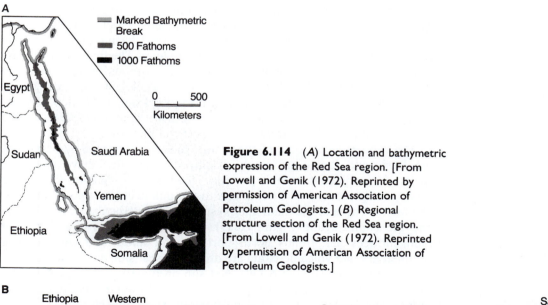

Figure 6.114 (*A*) Location and bathymetric expression of the Red Sea region. [From Lowell and Genik (1972). Reprinted by permission of American Association of Petroleum Geologists.] (*B*) Regional structure section of the Red Sea region. [From Lowell and Genik (1972). Reprinted by permission of American Association of Petroleum Geologists.]

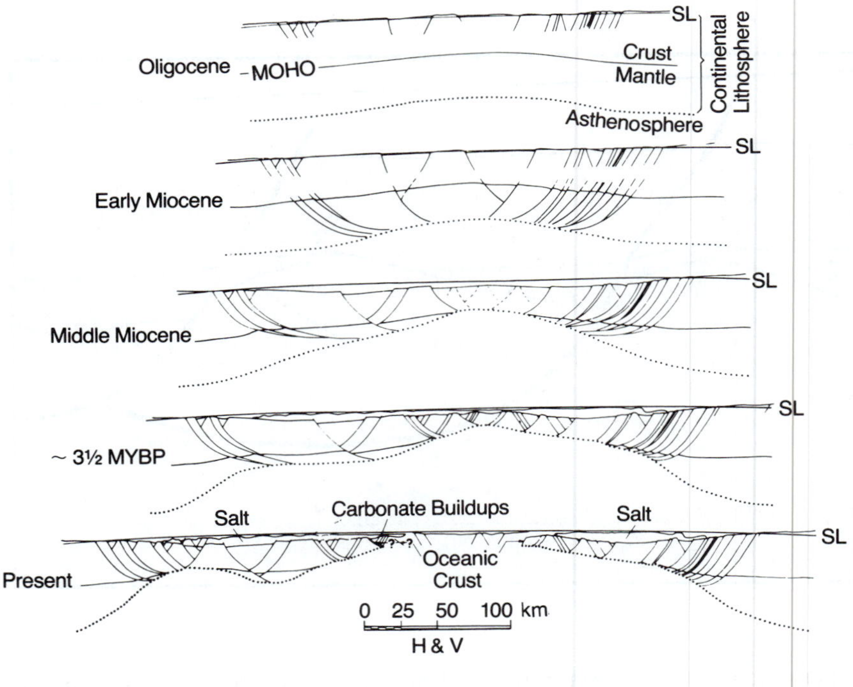

Figure 6.115 Kinematic evolution of faulting of the Red Sea region by distributed normal-slip faulting. [From Lowell and Genik (1972). Reprinted by permission of American Association of Petroleum Geologists.]

where continental crust, which formerly had been shortened through compression, has been stretched by distributed normal faulting. Extension exceeds 100%. The latest major phase of the faulting took place since about 12 m.y. ago. As portrayed in Eaton's (1980) tectonic map of the Basin and Range province, the faults trend north–northeast, north–south, or north–northwest (Figure 6.117). The normal faulting serves to block out the mountains and valleys that give the Basin and Range province its distinctiveness (Figure 6.118). Fault-bounded basins and ranges are generally 25 to 70 km broad and 50 to 300 km long. The map-view expression of the ranges in the southern part of the Basin and Range was once described by United States Geological Survey geologist Clarence Dutton as "an army of caterpillars marching northward out of Mexico" (King, 1959, p. 152). In detail, the basins and ranges are seen to divide and split, forming complex physiographic and structural patterns.

The properties of some of the young, high-angle normal-slip faults of the Basin and Range are nicely displayed where they have disturbed rocks of the adjacent tectonic province, the Colorado Plateau. Part of the breathtaking scenery of Utah is due to the influence of regional normal-slip faults. Three of the main fault systems are the Hurricane, the Sevier, and the Paunsaugunt (Figure 6.119). Each is composed of a family of fault surfaces that trends north–northeast for about 300 km. Colorful sedimentary and volcanic strata are dropped, west side down, along these normal-slip faults (Figure 6.120). Net slip on the Hurricane fault locally exceeds 3000 m! Slip along the Sevier and Paunsaugunt faults is 600 m or less, varying from place to place along the trace of each of these faults.

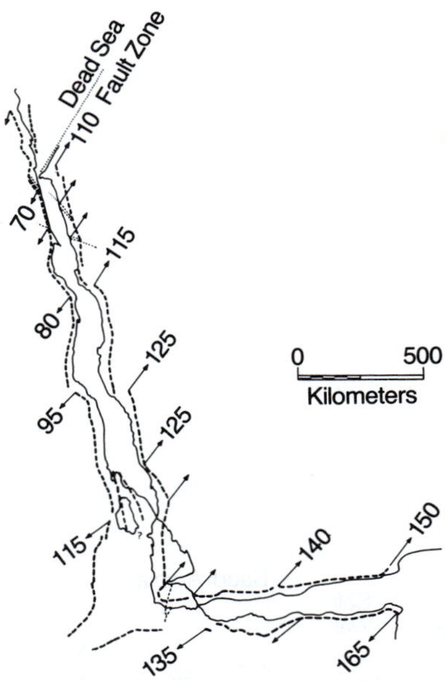

Figure 6.116 Map showing the magnitudes and directions of slip required to close the Red Sea rift. [From Lowell and Genik (1972). Reprinted by permission of American Association of Petroleum Geologists.]

Ah, So Simple

There was a time, and not so long ago, when normal faulting was easy to describe and understand. Structural geologists pictured normal faults as discrete surfaces or zones dipping steeper than 45°, *almost always at about 60°,* with the hanging wall down with respect to the footwall. Such

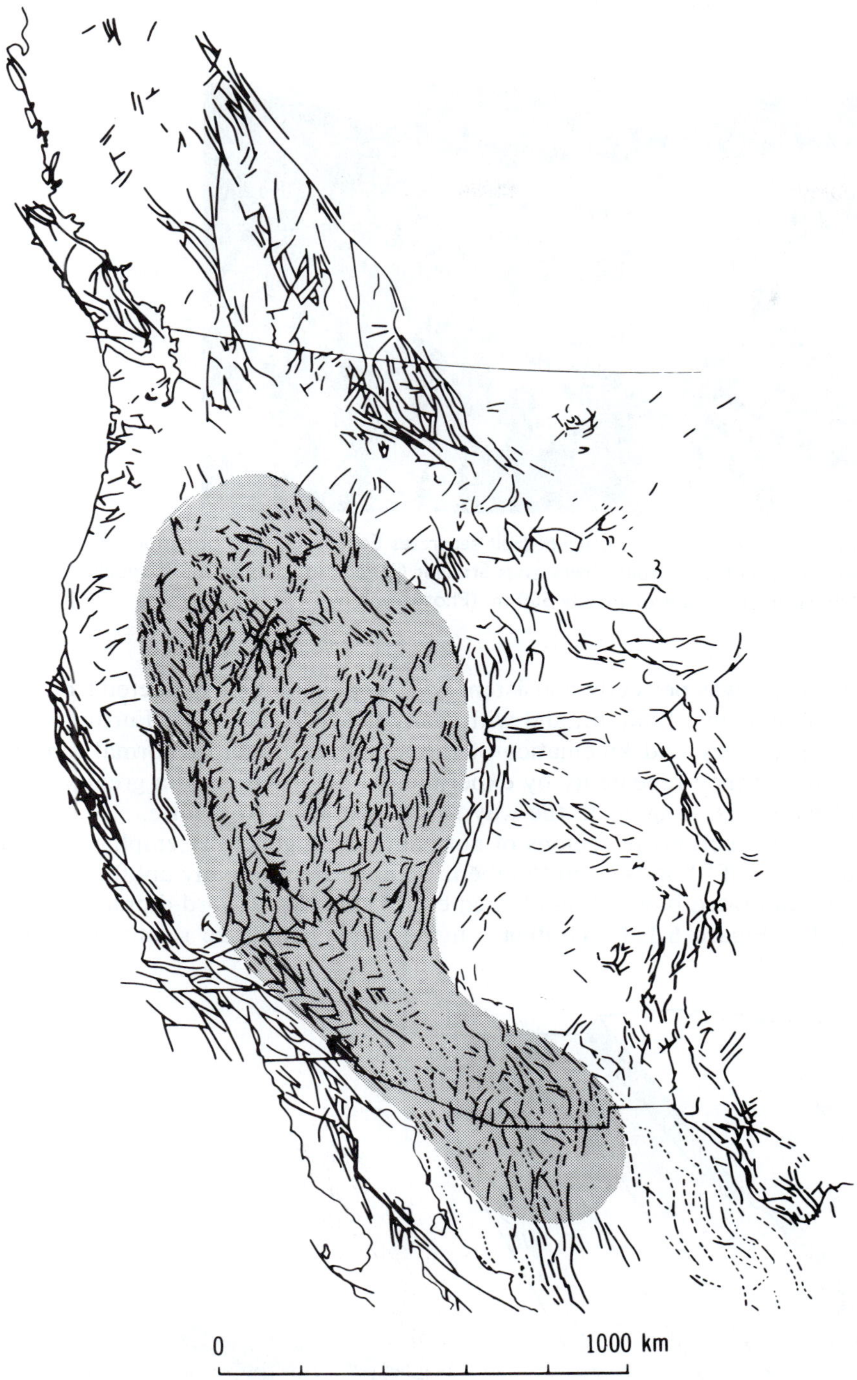

0 1000 km

Figure 6.117 Regional structure map showing the distribution and orientations of faults in western North America, including normal faults in the Basin and Range province of the American Southwest. [Reproduced from Eaton (1980) in *Continental Tectonics*, National Academy Press, Washington, D.C.].

Figure 6.118 Typical Basin and Range physiography, Tinajas Altas Mountains, southwestern Arizona. (Photograph by W. B. Bull.)

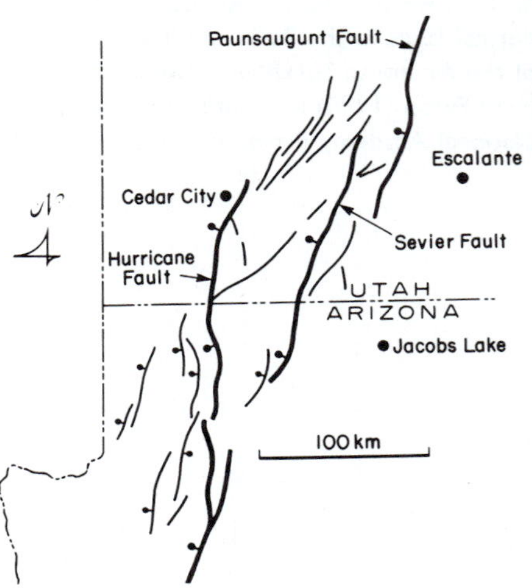

Figure 6.119 Map view of the fault pattern in the High Plateaus of western Utah.

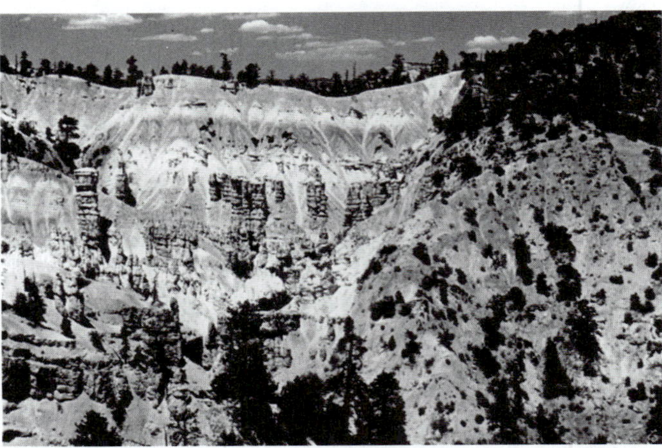

Figure 6.120 The Paunsaugunt fault separates Eocene Claron Formation on the hanging wall from Upper Cretaceous Straight Cliffs Formation on the footwall. Displacement is approximately 500 m. (Photograph by G. H. Davis).

portrayal was perfectly consistent with what should be expected from the Coulomb law of failure, and with many field observations. Furthermore, the geometry and kinematics of the standard model of normal faulting were affirmed repeatedly by experimental deformation of a great variety of materials, ranging all the way from loose sand to granite.

Cross sections of regions of normal faulting generally emphasized the presence of "horsts" and "graben." **Horsts** are relatively uplifted, generally unrotated blocks bounded on either side by outward-dipping normal faults (Figure 6.121). **Graben**, on the other hand, are relatively down-

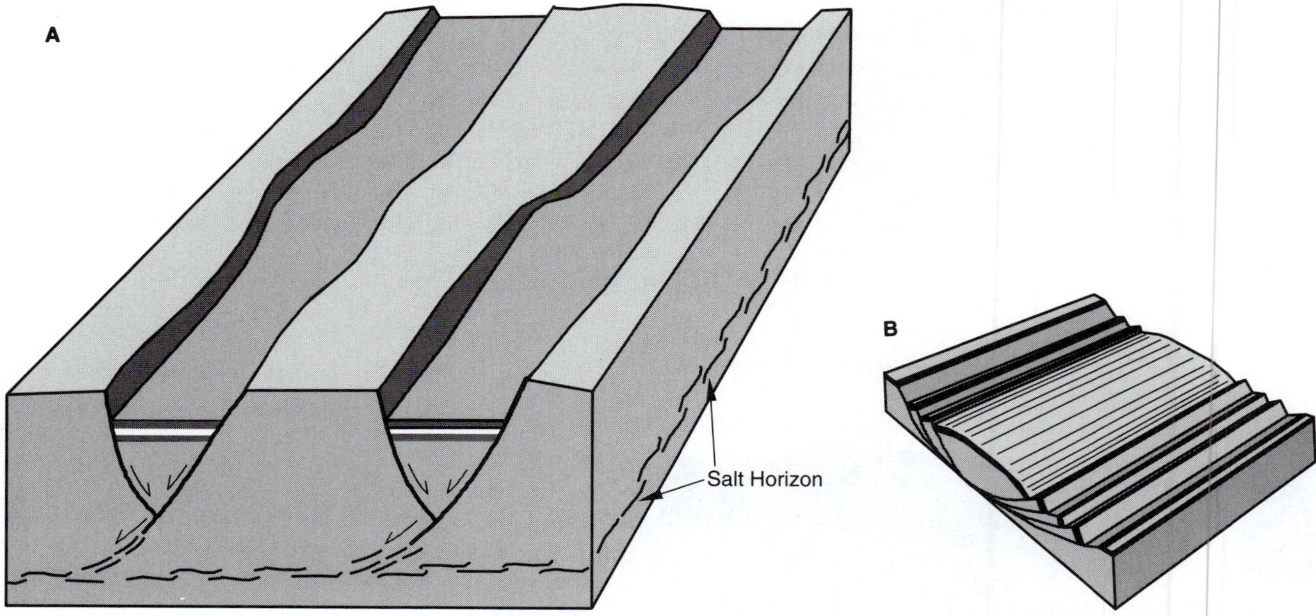

Figure 6.121 (*A*) Horsts and graben are classical terms describing fault-bounded uplifted and down-dropped blocks, respectively, in extended regions. These are located in the Canyonlands graben system, Utah. [After Trudgill and Cartwright (1994), Published with permission of the Geological Society of America.] (*B*) Normal faults that curve with depth are known as listric normal faults. Movement along these causes rotation of the hanging-wall blocks.

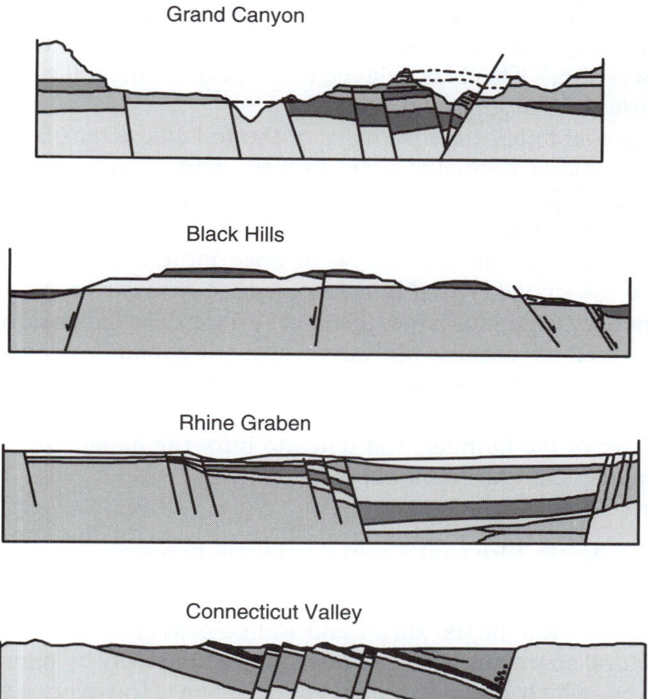

Figure 6.122 These cross sections convey the classical view of extended regions, i.e., marked by relatively high-angle normal faults. The concept of and evidence for low-angle normal faulting did not begin to emerge until the late 1960s. (From Billings, M. P., *STRUCTURAL GEOLOGY*, © 1972, p. 247, 249, 251, 523. Reprinted by permission of Prentice-Hall, Upper Saddle River, New Jersey.)

dropped, relatively unrotated blocks bounded on either side by inward-dipping normal faults (Figure 6.121*A*). Significant rotation of hanging-wall or footwall strata was seldom part of the picture, except in the case of **listric normal faulting**, a type of faulting which often requires that the hanging wall rotate from horizontal to steeper dips (Figure 6.121*B*).

With some significant exceptions, cross sections of normal faulting prepared before the 1970s showed none of the elegant fault detail and complexity that Peach and Horn recognized a full century ago in thrust-fault systems. Normal faults in textbooks of the day were simply of one type: planar surfaces dipping approximately 60° (Figure 6.122). Cross sections that were exceptions to the standard model were typically constructed by explorationists in petroleum and mining who were able to peer into the subsurface by means of seismics and drilling. For example, petroleum geologists working in the Gulf Coast region of North America have discovered nested families of listric normal faults with rotated hanging-wall strata and **growth fault sedimentation** in the hanging wall (Figure 6.123). Yet, portraits of Gulf Coast normal faulting were perceived for the longest time as exceptional, certainly not what occurs in surface exposures on dry land.

The possibility that some shallow-dipping faults could be normal faults (as opposed to thrust faults) was not entertained: at face value it violated

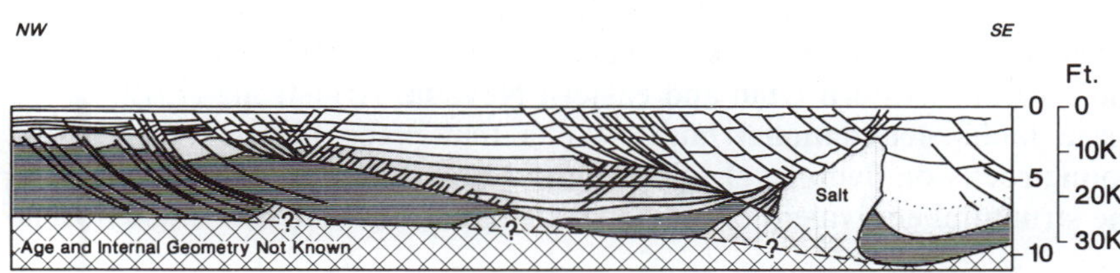

Figure 6.123 Geologic cross section through a part of the Gulf of Mexico. [From Worral and Snelson (1989), Figure 44, p. 130, Published with permission of the Geological Society of America.]

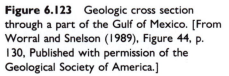

Coulomb's law. This in itself created a subliminal bias in field observation and interpretation: "Any fault dipping 30° or less must be a thrust fault produced by compression." This bias delayed recognition of the full complement of geometric details of normal faulting.

Some other factors inhibited the discovery of the full character of the geometry and kinematics of normal faulting as well. Thrust faulting and associated crustal thickening produce elevation in the form of uplift. Erosion into regional uplifts produces mountains, the haven for geologists, and exposes the character of thrusts that were once more deep-seated. Normal faulting and associated crustal thinning produce subsidence in the form of basins. Where extension is large, the land surface descends below sea level and is covered by oceans. Moreover, thrusting tends to cover and "protect" the fundamental geometry of faulting, by moving hanging wall over footwall. In contrast, normal faulting tends to expose to erosion the footwall character of the faulting, and tends to bury the hanging-wall character of faulting beneath basin-fill deposits.

Discovery of Low-Angle Normal Faulting/Detachment Faulting

The early conventional view of the Basin and Range province was that its distinctive structural character had been fashioned exclusively by high-angle normal faulting, which produced horsts and graben. However, detailed geologic mapping and analysis revealed that the Basin and Range province contains both high-angle normal faults *and* low-angle normal faults as well, the combination of which has produced profound crustal stretching. Many of the low-angle faults, the largest of which are called **detachment faults**, are Oligocene–Miocene in age. They are cut and offset by bona fide high-angle normal faults.

In their simplest form the low-angle fault relationships consist of moderately to steeply tilted hanging-wall strata resting on expansive flat to gently dipping fault surfaces (Figure 6.124). Systems of such faults create **tilted fault blocks**. Some low-angle normal faults started out at a higher angle but were rotated progressively to shallower dips as the crust extended. On the other hand, it is clear that some low-angle normal faults actually originated as shallow-dipping faults. In the Basin and Range province, some of the most impressive detachment faults contain Precambrian crystalline rocks in the footwalls and steeply tilted Oligocene and Miocene strata in the hanging walls (Figure 6.125). Intervening rocks (e.g., Paleozoic and Mesozoic) may be missing entirely, having been eroded or tectonically removed from the footwall, and having been transported to depth in the hanging wall. Thus, relationships are strikingly "younger on older."

Directly beneath such faults, on the very top of the footwall, the rocks are typically transformed over several meters or several tens of meters into microbrecciated and highly fractured cataclastic fault rocks. Formation of the microbreccias is the natural result of the frictional effects of kilometers of slip as the footwall is "raised" from depth to the surface (Figure 6.126).

Fault relationships of this type were "co-discovered" in the late 1960s by Ernie Anderson (1971), who mapped them in the Lake Mead region of southeastern Nevada; John Proffett (1977), who mapped them in the Yerington district of Nevada; and Dick Armstrong (1972), who recognized them within western Utah and eastern Nevada. Armstrong (1972) called these faults **denudational faults**: faults that place high-level relatively young rocks on deep-level relatively old rocks. Denudational faulting is the structural equivalent of erosional stripping. Indeed, low-angle normal

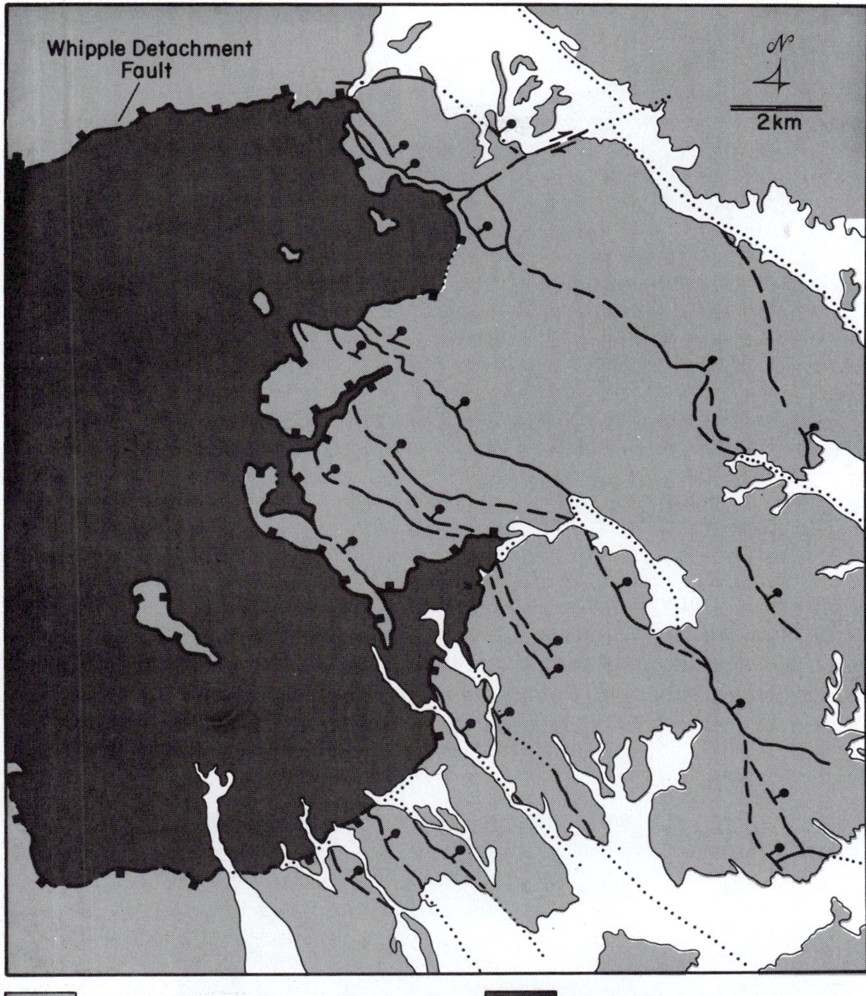

Figure 6.124 (*A*) Geologic map of detachment faulting in the Whipple Mountains, southeastern California. [From Davis and others (1980). Published with permission of Geological Society of America and the authors.] (*B*) Photograph of the fault relationship between volcanics (dark rock) above and mylonite (light rock) below. (Photograph by G. A. Davis.)

Figure 6.125 View of southwest-tilted Tertiary strata separated from underlying mylonitic gneisses by a detachment fault. [From Davis and others (1980). Published with permission of Geological Society of America and the authors.]

faulting can be thought of as a tectonic stripping of cover to expose progressively deeper levels of crustal rocks.

The magnitude of individual low-angle normal faults in the Basin and Range is captured in the extraordinary work of Wright and Troxel (1973) in Death Valley. They suggested that the great **turtleback structures** on the eastern margin of Death Valley (Figure 6.127*A*) are enormous fault grooves or **fault mullions** preserved along the upper surface of Precambrian

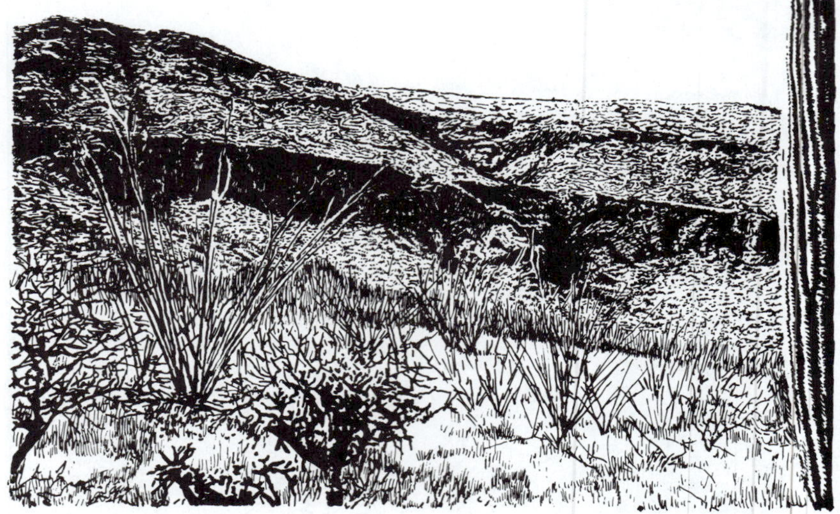

Figure 6.126 The Catalina fault in the Rincon Mountains separates microbreccias (dark cliff) from overlying Pennsylvanian–Permian limestones. [From Davis (1980). Published with permission of Geological Society of America and the author.]

crystalline footwall rock. As smooth, polished mountain ridges, they plunge northwestward in the direction that hanging-wall strata were translated during faulting. Remnants of the faulted, rotated hanging-wall strata make up the Panamint Range, which forms the western flank of Death Valley. The thick pile of strata there is dipping moderately to steeply

Figure 6.127 (*A*) One of the great turtleback structures in Death Valley. The fault surface accommodated major normal-slip translation. (Photograph by G. H. Davis.) (*B*) Bennie Troxel on the outcrop.

eastward. The original position of this section of rock was above and east of the present location of the turtlebacks. Net slip was probably several tens of kilometers. Precambrian rock along the upper surface of the turtlebacks is strongly cataclastically deformed (Figure 6.127*B*).

Proffett's Discovery of Low-Angle Normal Faults at Yerington

John Proffett's (1977) study of low-angle normal faulting in the Yerington district of Nevada was very revealing. Geometric and kinematic details of the fault system emerged from exploration and development of a major porphyry copper deposit. The site of Proffett's study was southeast of Reno in west-central Nevada. Radiometric potassium-argon age dating indicates that the faulting there began in the Miocene, 17 to 18 m.y. ago. East-dipping concave-upward fault surfaces, with large troughlike grooves, accommodated normal-slip translation and accompanying rotation of hanging-wall strata (Figure 6.128). During the faulting, ignimbrite volcanics of Oligocene age were back-tilted from horizontal to steep westward dips. Some rocks rotated by more than 90°! Relatively old normal-slip faults were rotated to shallow dips and became inactive, later to be cut and rotated to even shallower dips by relatively young faults. Indeed, the youngest faults in the Yerington district are the steepest, whereas the oldest faults typically dip shallowly (Figure 6.128).

Because the bias against low-angle faulting in the Basin and Range was so strong in the late 1960s and earlier, John Proffett had great difficulty in convincing his supervisors that the low-angle structures he discovered were indeed faults, and not unconformities or bedding contacts. The moment of truth came when John asked to be held by his ankles, upside down, and lowered into the top of a deep exploration shaft, where he reached his hand into the trace of the key low-angle horizon and extracted a handful of sticky gouge!

Proffett constructed detailed structural profiles of the fault relationships on the basis of geological mapping and subsurface drilling. From these he computed slip for the major faults. The largest slip on any fault was found to be 4000 m. Proffett determined this value of net slip on the basis

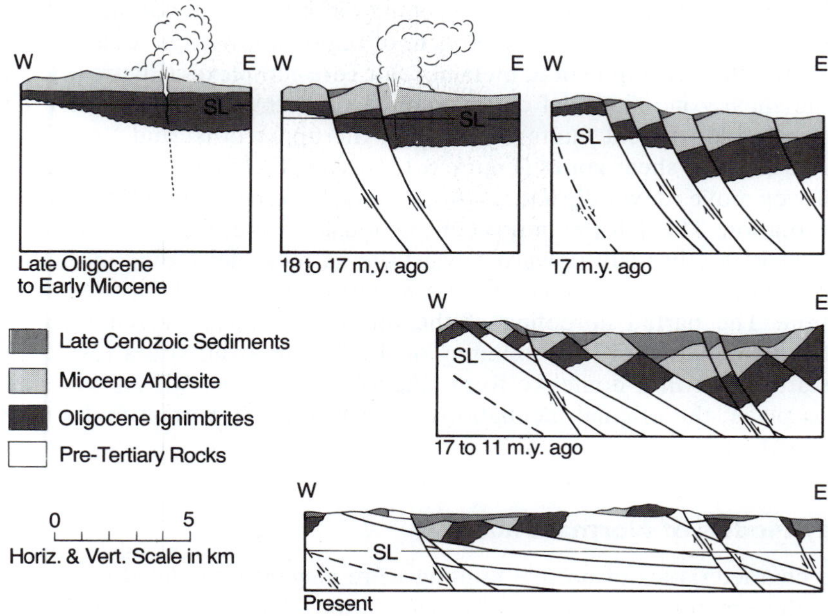

Figure 6.128 Kinematic evolution of normal faulting in the Yerington district, west-central Nevada. [From Proffett (1977). Published with permission of Geological Society of America and the author.]

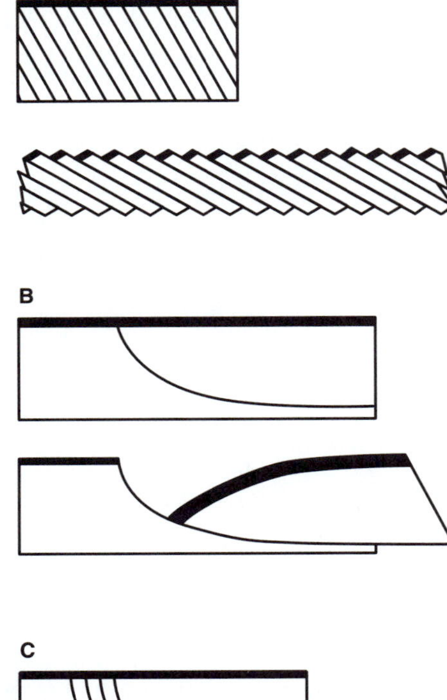

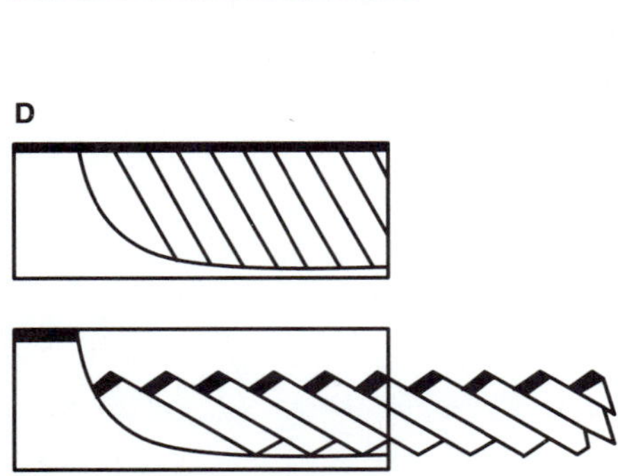

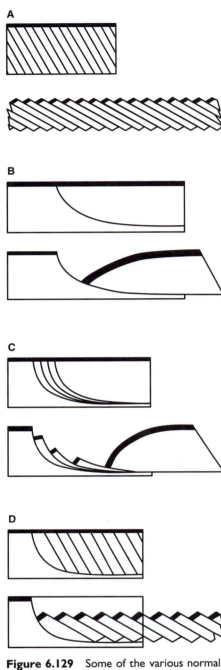

Figure 6.129 Some of the various normal fault geometries employed by nature to accomplish extension and stretching of the crust. (*A*) Domino-style normal faulting; (*B*) listric normal faulting with reverse drag; (*C*) imbricate listric normal faulting; and (*D*) listric normal faulting bounding a family of planar normal faults. [Reprinted with permission from *Journal of Structural Geology*, v. 4, B. Wernicke and B. C. Burchfiel, Modes of extension tectonics (1982), Elsevier Science, Ltd., Pergamon Imprint, Oxford, England.] Burchfiel (1982), Figs. 1, 4, 5, 7.]

of a truncated and offset geometric line, a line formed by the pinch-out of a bed against an unconformity.

The strain implications of the fault system in the Yerington district are astounding: more than 100% stretching. Based on structure profiles, Proffett estimated the original east–west width of the study site to have been 7.3 km. The extended width, after normal faulting, is 17.3 km. On this basis, stretch and percent stretching could be calculated:

$$S = \frac{17.3 \text{ km}}{7.3 \text{ km}} = 2.37$$

$$\% \text{ lengthening} = (2.37 - 1.00) \times 100 = 137\%$$

Proffett's estimate of total stretching across the entire Basin and Range province is very conservative, much less than the 137% cited previously. Based on regional cross sections, he estimated that the Basin and Range has been stretched by about 170 km in Nevada. This amounts to a 35% stretching. Other workers have suggested values closer to 100% (Hamilton, 1978). Reconstructions for stretching in the Lake Meade region, carried out by Wernicke, Axen, and Snow (1988), suggest crustal lengthening of 250%.

Current Cross-Sectional Pictures of Normal Faulting

Based on detailed geologic mapping and subsurface exploration, it is now known that "younger-on-older" normal faulting accommodating extensional strain can be accomplished in any number of ways (Wernicke and Burchfiel, 1982). These include: **high-angle normal faulting** along planar faults without any rotation of the faults or the strata, as in Figure 6.122; **domino-style normal faulting** in which planar normal faults rotate to progressively shallower dips and the originally flat-lying strata, now fault-bounded, rotate to progressively steeper dips (Figure 6.129*A*); **listric normal faulting** accompanied by reverse drag accomplished by collapse with or without antithetic faulting (Figure 6.129*B*); **imbricate listric normal faulting,** again with rollover (Figure 6.129*C*); and listric normal faulting bounding a family of planar normal faults (Figure 6.129*D*).

The most extreme cases of extensional deformation in continental regions have resulted in the development of regional low-angle **detachment faults** and the development of **metamorphic core complexes** (Figure 6.130). In the shear zone model of core complexes (Wernicke, 1981 model) the low-angle detachment faults cut through the upper crust and "root" in middle crust as shear zones (Figure 6.131*A* and *B*). Upper-level hanging-wall rock moves down-dip and, in the process, becomes thinned by faulting and rotation. The deformation is conspicuously brittle. Deeper level footwall rocks are brought upward toward the surface, deforming in ductile fashion at depth but becoming progressively overprinted by brittle deformation. The partial unroofing of the footwall block achieved through "tectonic denudation" and thinning of the hanging-wall strata results in isostatic uplift in a domelike form (Figure 6.131*C* and *D*). As rebound takes place, the original inclinations of early formed normal faults may be modified.

Clay Models of Normal Faulting

The characteristics of normal faulting were elegantly modeled in simple clay deformation experiments by Hans Cloos (1936), and later by his

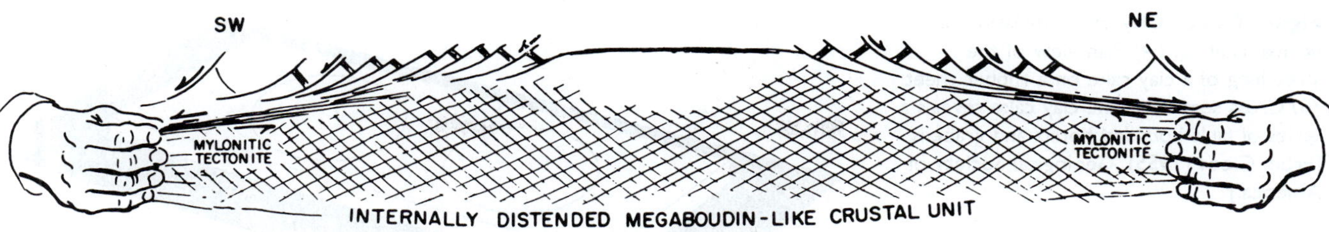

Figure 6.130 Schematic portrayal of the development of metamorphic core complexes and detachment faults through regional extension. Normal ductile shear at deep levels creates mylonite. [From Davis and Hardy (1981). Published with permission of Geological Society of America and authors.]

brother Ernst Cloos (1955). The delicate, intricate fault patterns that can emerge in deformed clay that is properly prepared and deformed are astonishingly similar to natural patterns.

The clay cake to be deformed is prepared by mixing dry kaolin with water until a soft buttery consistency is achieved. The clay is spread as a layer ±4-cm thick, approximately 15 × 15 cm in size, and smoothed with a putty knife. A reference circle may be gently impressed onto the surface of the clay cake to serve as a strain marker.

Hans and Ernst Cloos recommended two different methods for deforming the clay. In the first experiment the clay cake is built on top of an elastic rubber sheet (Figure 6.132*A*). Then the rubber sheet is stretched slowly and uniformly. Since the base of the clay cake sticks to the rubber sheet, the entire clay layer experiences the effects of the stretching. As stretching commences, normal-slip faults, along with tear faults, emerge to accommodate the extension (Figure 6.132*B*). Gradually the faults become uniformly distributed throughout the clay cake.

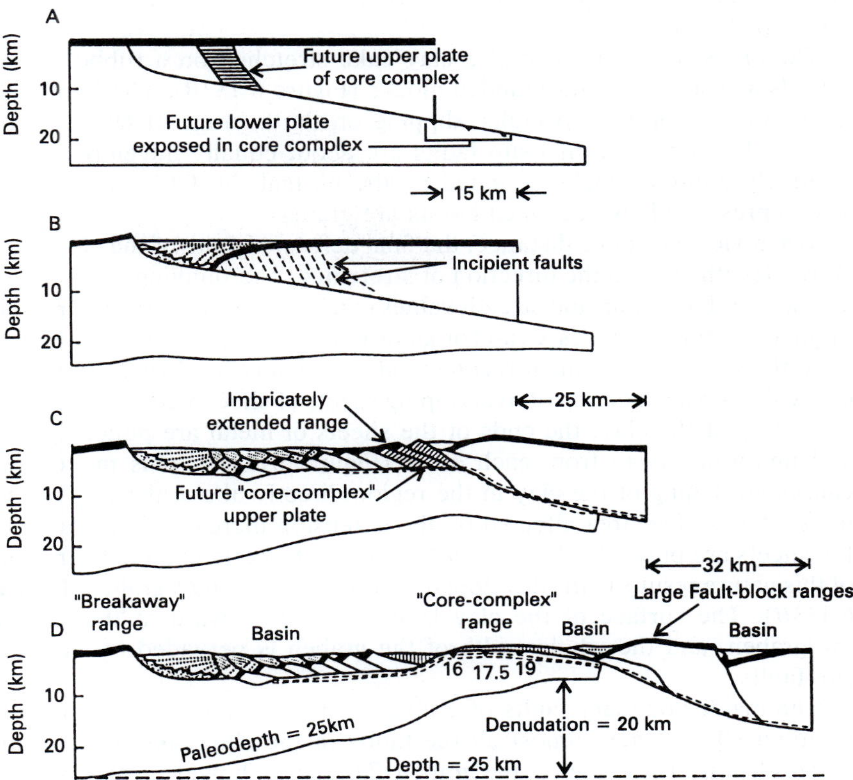

Figure 6.131 Low-angle detachment faulting and the formation of a metamorphic core complex. (*A*) Starting configuration, including original locations of domains of rock that will eventually become upper plate and lower plate. (*B*) Low-angle normal faulting is marked by breakaway zone at surface. Deeper reaches of the fault penetrate the middle crust. At that depth level, displacement is achieved by shearing and the formation of mylonites. Growth fault basins at surface steadily form and steadily fill with sediment. (*C*) Shear zone rocks are "drawn" closer to surface via continued normal displacement. Unroofing of the upper level rocks triggers isostatic adjustment and doming. (*D*) The final configuration of core complex and detachment faulting. [After Wernicke (1985), v. 291, reprinted courtesy of the Canadian Journal of Earth Sciences.]

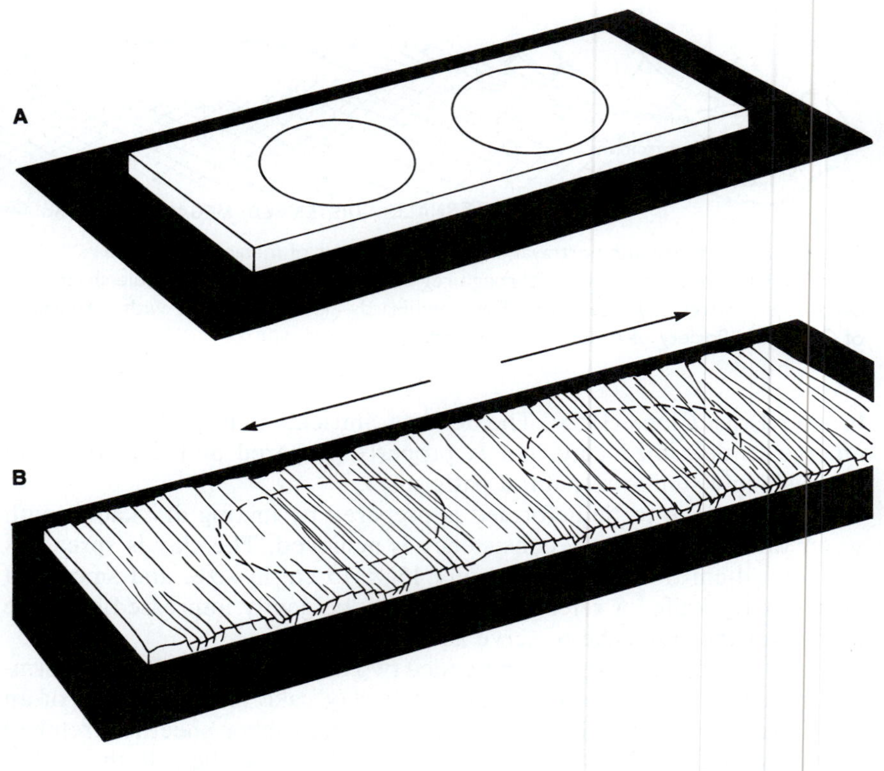

Figure 6.132 Clay cake simulation of normal faulting. (*A*) Plan view of the stretching of a clay cake on a rubber sheet. (*B*) Cross section of the clay cake reveals a series of fault-bounded horsts and grabens. [From Cloos (1955), Geological Society of America.]

The top of a clay cake stretched in this way reveals numerous closely spaced faults and fault scarps, which typically are oriented at a high angle to the direction of stretching. Some fault traces are straight, whereas others are curved. Fault surfaces exposed along the fault scarps are marked by dip–slip striations.

The cross-sectional view of a clay cake stretched on a rubber sheet reveals a system of fault-bounded blocks (Figure 6.132*B*). The bounding faults are all normal-slip faults, dipping on the average of 60°. Sets of oppositely dipping normal-slip faults are about equally developed. The relatively uplifted blocks bounded by the normal-slip faults are horsts. The depressed blocks between horsts are graben.

When viewed from a distance, the final deformed state of the clay cake shows lengthening in the direction of stretching, and thinning of the layer as a whole. Extension and stretch values can be determined by comparing original and final thicknesses and original and final lengths of the clay cake.

In the second experiment recommended by Hans and Ernst Cloos, the clay cake is placed on top of overlapping sheets of sheet metal. To achieve stretching of the clay, the ends of the sheets of metal are pulled slowly and uniformly away from each other (Figure 6.133*A*). This movement causes stretching of the clay in the region directly above the join of the metal sheets. The area affected by the extension increases in breadth as the sheets are pulled farther and farther apart. Experimental deformation of this nature results in the development of a single complex graben (Figure 6.133*B*). The surface of the clay layer bows downward at the site of the graben, and the whole width of the graben is pervaded by normal-slip faults.

The major boundary faults of graben formed in this way dip inward, at about 60°. In fact, almost all the faults in the clay cake tend to dip inward toward the center of the graben. There are, however, exceptions.

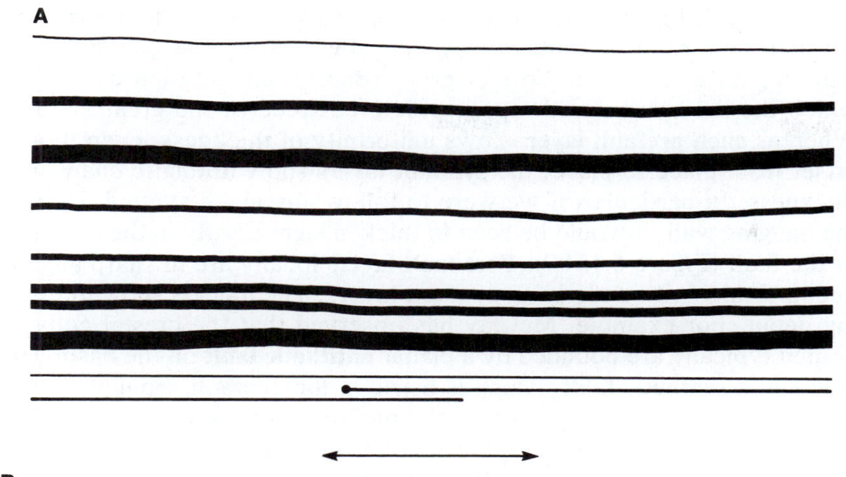

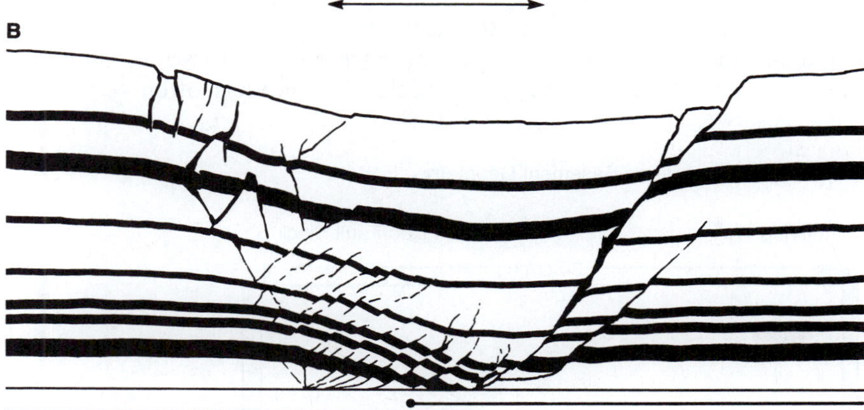

Figure 6.133 A second clay cake simulation of normal faulting. (*A*) Deformation of a clay cake on top of overlapping panels of sheet metal. Stretching and normal faulting are achieved by pulling apart the sheet metal panels. (*B*) Cross-sectional view of the deformed clay cake shows a series of tilted fault blocks. [From Cloos (1968). Published with permission of American Association of Petroleum Geologists.]

Antithetic faults form close to the major boundary faults but dip in the opposite direction (Figure 6.133*B*). The role of antithetic faults is to eliminate the gap(s) that would otherwise be produced by displacements on curved fault surfaces. Antithetic faults and reverse drag perform the same function; they fill in potential voids.

Ultimately, the surface of a clay cake stretched on sheet metal is distended into a series of **tilted fault blocks** (Figure 6.133*B*). The step blocks dip toward the surface on which they are rotated. The strike of each tilted step block is about perpendicular to the direction of stretching.

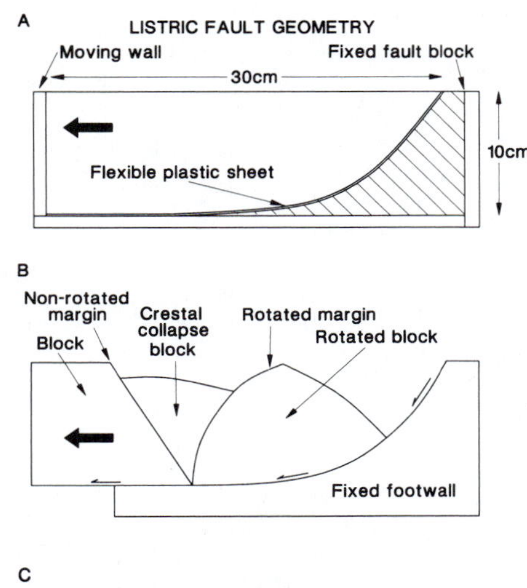

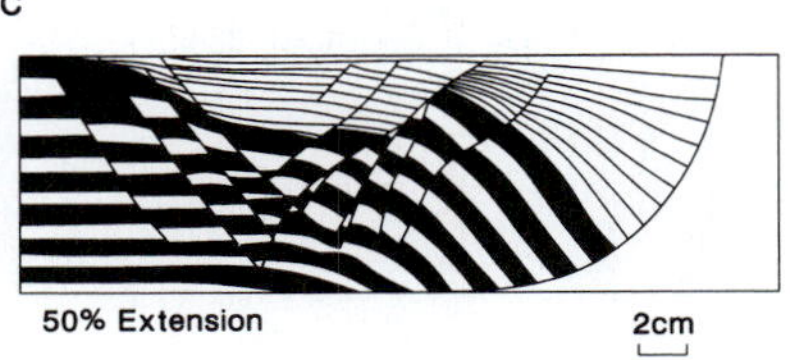

Figure 6.134 (*A*) Experimental setup for producing simple listric fault geometries. [From Ellis and McClay (1988), Figure 2c, p. 57.] (*B*) Structural elements produced during simple listric normal faulting. Two of the major features are the rollover anticline and the crestal collapse graben. [From Ellis and McClay (1988), *Basin Research* Figure 3, p. 59, reprinted with permission of Blackwell Science, Ltd.] (*C*) Tracing of photograph of model produced through simple listric faulting, showing prefaulting sediments in black and white bands, with synfaulting sediments stippled. [From Ellis and McClay (1988), *Basin Research,* v. 1 Figure 4b, p. 59, reprinted with permission of Blackwell Science Ltd.]

McClay Models of Normal Faulting

Building on the experimental approach originally developed by Peter Cobbold and Jean Pierre Brun, Ken McClay and his colleagues have performed world-class **scaled analogue modeling** of normal faulting (McClay and others, 1991a,b). His VCR recordings of normal faulting are outstanding. McClay uses very fine-grained sand to simulate brittle upper crust, and after dyeing the sand he deposits a distinctive stratigraphy of alternating colored and white sands. In order to create listric faulting, which is one of his specialties, he builds a rigid footwall block with a concave curved upper surface (Figure 6.134*A*). He then drapes a movable plastic sheet over the footwall block, and attaches the sheet to a moving wall. Movement and strain rate are computer automated. McClay's models of simple listric faults show the development of rollover in the hanging-wall strata, as well as the crestal collapse of the rollover to produce graben structures (Figure 6.134*A–C*).

The models that McClay creates are especially realistic and informative because new sand is continuously deposited during the course of the faulting. As a result, patterns of **growth fault sedimentation** develop in which the most depressed topographic areas receive the greatest sand. Whereas each **prefault layer** shows uniformity of thickness, even though offset from place to place, the **synfault layers** show dramatic changes in thickness. In particular, if we were to follow any given synfault layer in the hanging wall, it would be seen to thicken significantly in the direction of the fault (Figure 6.134*C*). **Post-fault layers** locally are in sharp angular unconformity with underlying synfault layers. The models show interesting detail. For example, McClay has observed that the crestal collapse graben typically are bounded by a planar **antithetic fault** on the basinward side and a **synthetic fault**, which is listric in the synfault sequence but is **antilistric** (i.e., convex upward) in the prefault sequence.

The structures and stratigraphic relationships produced in McClay models bear uncanny similarity to the structures seen in seismic sections of rifted continental margins. As a result, petroleum exploration companies have a keen interest in McClay models. The models often reveal fine

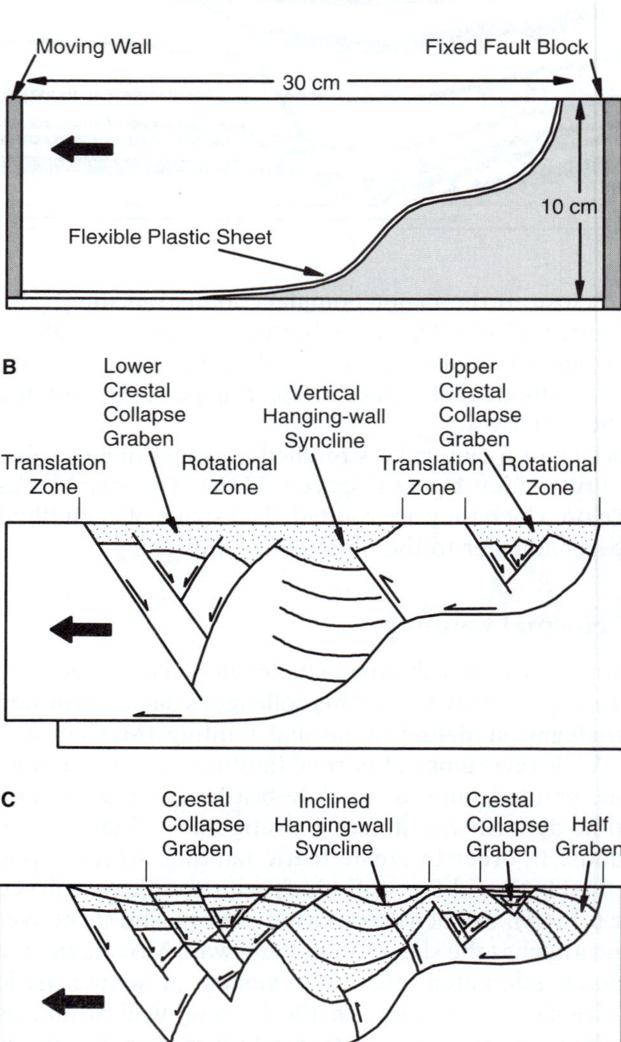

Figure 6.135 (*A*) Experimental setup for normal faulting accommodated by a ramp-flat basement geometry. (*B*) Ramp-flat listric fault geometry showing hanging-wall deformation features in the prefaulting sediments. Low extension. (*C*) Ramp-flat listric fault geometry showing deformational features in the *syn-rift* sediments (stippled). High extension. [From Ellis and McClay (1988), *Basin Research,* v. 1, Figure 2b, p. 57, Figure 22a, p. 69, and Figure 22b, p. 69, reprinted with permission of Blackwell Science Ltd.]

structure that is not necessarily visible in structure sections, and some of this fine structure will influence whether a given structure will produce. As a clever step, McClay and his colleagues prepare **synthetic seismograms** of the models, which are computer simulations of the signals that such a stratigraphy and structure would likely produce in nature. These can be compared with the real thing.

McClay's work does not stop with single listric faults. For example, he has created and observed the kinematic development of normal fault systems with **ramp-flat geometries** and **extensional duplexes** (Figure 6.135). He has found that the sprinkling of mica along the top of a bed, which is later buried by more sand, can set the stage for development of a "flat" by reducing internal friction. He has also modeled the reactivation of normal faults.

INVERSION TECTONICS

In the last decade a whole new field of structural analysis called **inversion tectonics** suddenly emerged. During "inversion," normal faults become thrust and reverse faults. During "inversion," a regional geologic system formed in extension is forced to shorten; it does so in part by "reversing" the displacement on the preexisting faults. The recognition of the unique and peculiar structural geometries that result from turning an extensional tectonic system inside out already has had profound tectonic and practical value. A rifted continental margin can later be the site of plate convergence. When this occurs, the original normal faults (Figure 6.136*A*) are reutilized as surfaces of reverse faulting or thrust faulting (Figure 6.136*B*). Shortening by such fault reactivation can cancel out all of the original extensional deformation in the "basement" and can ultimately convert the faults to reverse or thrust faults. Differences in amount of shortening in the basement versus cover reveal that the faults were reactivated.

An example of inversion tectonics may be seen in the Broad Fourteens Basin, located offshore of Holland. This is a Mesozoic extensional basin that formed in Triassic to late Cretaceous (Figure 6.137); then it inverted in late Cretaceous (Hayward and Graham, 1989). Pre-normal-faulting units at depth might, in some cases, show a net extension, whereas post-normal-faulting, pre-reverse-faulting units above may show a net shortening (Williams, Powell, and Cooper, 1989). McClay (1989), once again, has produced elegant models that capture the kinematics of inversion tectonics (Figure 6.138).

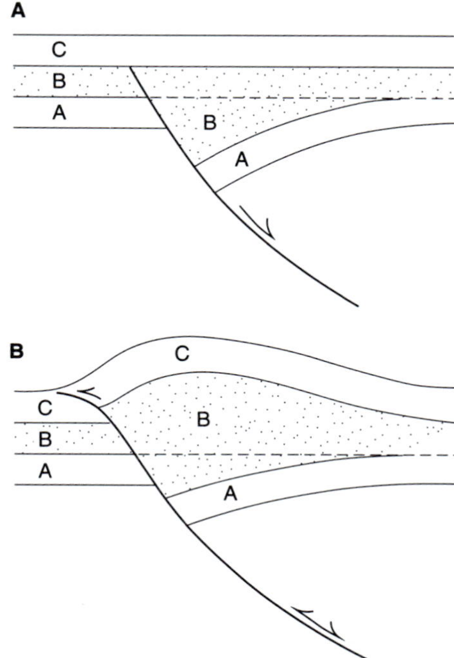

Figure 6.136 Inversion tectonics, in action. *A*, *B*, and *C* are stratigraphic sequences, where *A* is prerift; *B* is synrift; and *C* is postrift. (*A*) Original normal faulting during extensional deformation creates offset of older sequence (unit *A* and the basement on which it rests) and growth-fault sedimentation of syntectonic units (unit *B*, which is thicker on the downthrown block). (*B*) Compressional deformation "inverts" the extensional basin by causing reactivation of the normal fault(s). A point may be reached where the original normal displacement is completely "canceled out." [After Williams, Powell, and Cooper (1989), *Inversion tectonics* Figure 1, p. 3, reprinted with permission from Blackwell Science Ltd.]

STRIKE–SLIP FAULTING

Regional Tectonic Settings

Strike–slip faults are generally steeply dipping faults along which horizontal slip has occurred. They comprise a fundamental class of faults whose geometric and kinematic properties are quite distinctive. Major strike–slip faults in continental settings are "landlubber" counterparts to the mid-oceanic **transform faults** (see Chapter 10). Like oceanic transform faults, most of the truly large strike–slip faults in continental terranes are fundamental plate boundaries. The San Andreas fault in California, the Motagua fault in Guatemala, and the Alpine fault in New Zealand are examples.

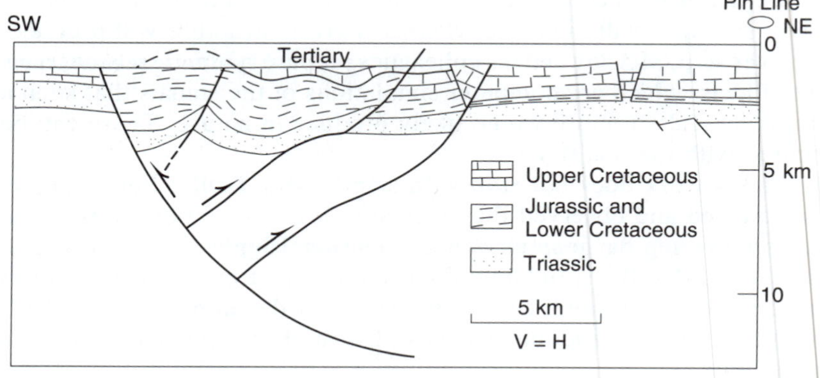

Figure 6.137 Inversion tectonics as revealed in the Broad Fourteens Basin located offshore of Holland. The Broad Fourteens Basin was produced as an extensional basin during Triassic to late Cretaceous. During Late Cretaceous it underwent inversion, and thus the present-day cross section shows the combination of extension and compression reflected in structures. [After Hayward and Graham (1989), *Inversion tectonics* Figure 8, p. 23, reprinted with permission from Blackwell Science Ltd.]

These faults, and some others like them, are thousands of kilometers long and have accommodated hundreds of kilometers of strike–slip displacement. Although strike–slip faults occur as **compartmental faults** and **tear faults** within systems of thrust, reverse, and normal faulting, these cannot compare in size, slip, and overall importance with those that currently are, or formerly were, major plate boundaries.

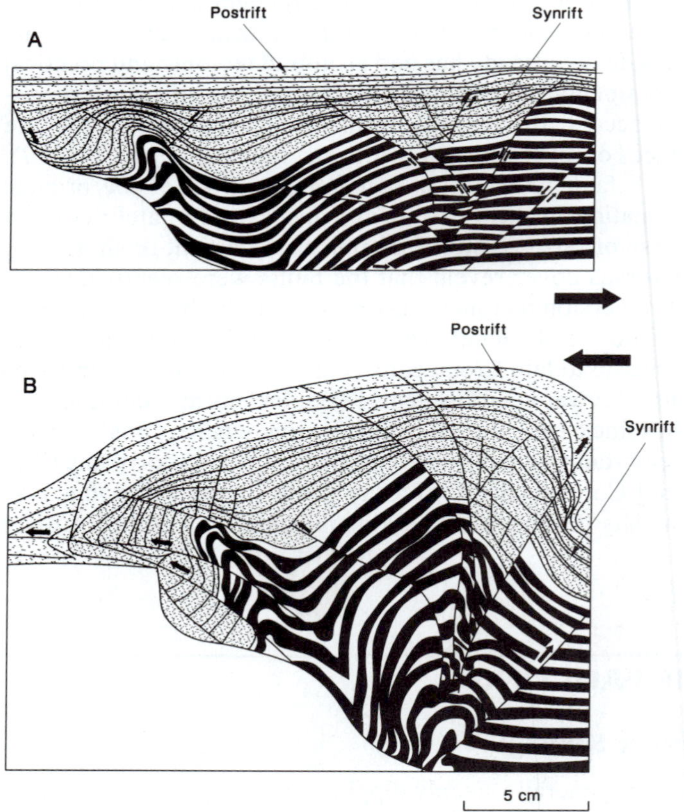

Figure 6.138 McClay models that capture the kinematics of inversion tectonics. (*A*) 50% extension above a ramp-flat listric detachment. The top four layers (undeformed) are postfaulting sand layers. In the deformed model the upper 10 layers (light gray) are sand/mica and mica synfaulting sediments deposited during extension. (*B*) 40% contraction. The model is now shorter than the original preextension geometry. [From McClay (1989), *Inversion tectonics* Figures 9B and 9D, p. 57, reprinted with permission from Blackwell Science Ltd.]

Strike–Slip Faulting in Guatemala

Guatemala has been ripped by strike–slip faulting. Signatures of tectonic unrest in the country are abundant: long, straight to gently curved lineaments; stream valleys offset by recent strike–slip faulting; marine Pliocene strata perched at elevations as high as 14,500 ft (4400 m); earthquake shocks; smoking volcanoes; crater lakes fed by hot springs; and imposing topographic relief. Oversteepened slopes in northwestern Guatemala are locally planted in corn. Some fields are so steep that they must be climbed on hands and knees.

Physiographic and geologic indicators of tectonic unrest are academic compared with the events of February 4, 1976, when sudden strike–slip movement along the Motagua fault resulted in more than 74,000 injuries and the loss of 23,000 lives. The Motagua fault is a transform fault boundary between the Caribbean and the North American plates (Figure 6.139). Fully one fifth of the country's population were victims of the tragic earthquake. George Plafker of the United States Geological Survey reported that the Guatemalan earthquake produced more surface faulting than any earthquake event in the Western Hemisphere since the San Francisco earthquake of 1906 (Plafker, 1976).

According to Plafker (1976), 230 km of the Motagua fault zone was activated during the earthquake (Figure 6.140). Movement was left-handed strike–slip, with an average slip of 108 cm (Figure 6.141). The maximum net slip observed, calculated on the basis of an offset road, was 340 cm. From place to place along the fault zone, dip–slip movement was observed, but nowhere was sense of displacement seen to be especially systematic. The minor structures that formed along the fault include synthetic left-slip faults oriented subparallel to the fault trace, en echelon tension fractures trending at a high angle to the fault trace, and left-handed en echelon fold structures.

The Alpine Fault in New Zealand

As part of the boundary between the Pacific plate and the Indian–Australian plate, the Alpine fault cuts across the South Island of New Zealand for a trace length of 600 km (see Figure 10.44). Within the past 40 m.y., the Alpine fault has accommodated approximately 460 km of right-handed strike–slip movement (Allis, 1981). But during the past 10 to 15 m.y., slip along the Alpine fault has not been purely strike–slip. Instead, relative plate motions have created a strong east–west convergence across it.

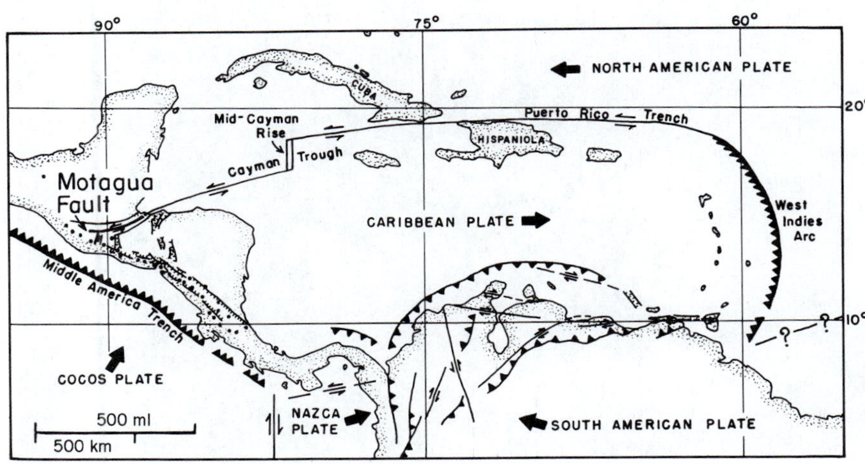

Figure 6.139 Tectonic setting of the Motagua fault zone in Guatemala. [From Plafker (1976), *Science*, v. 193, fig. 6. Copyright © 1976 by American Association for the Advancement of Science.]

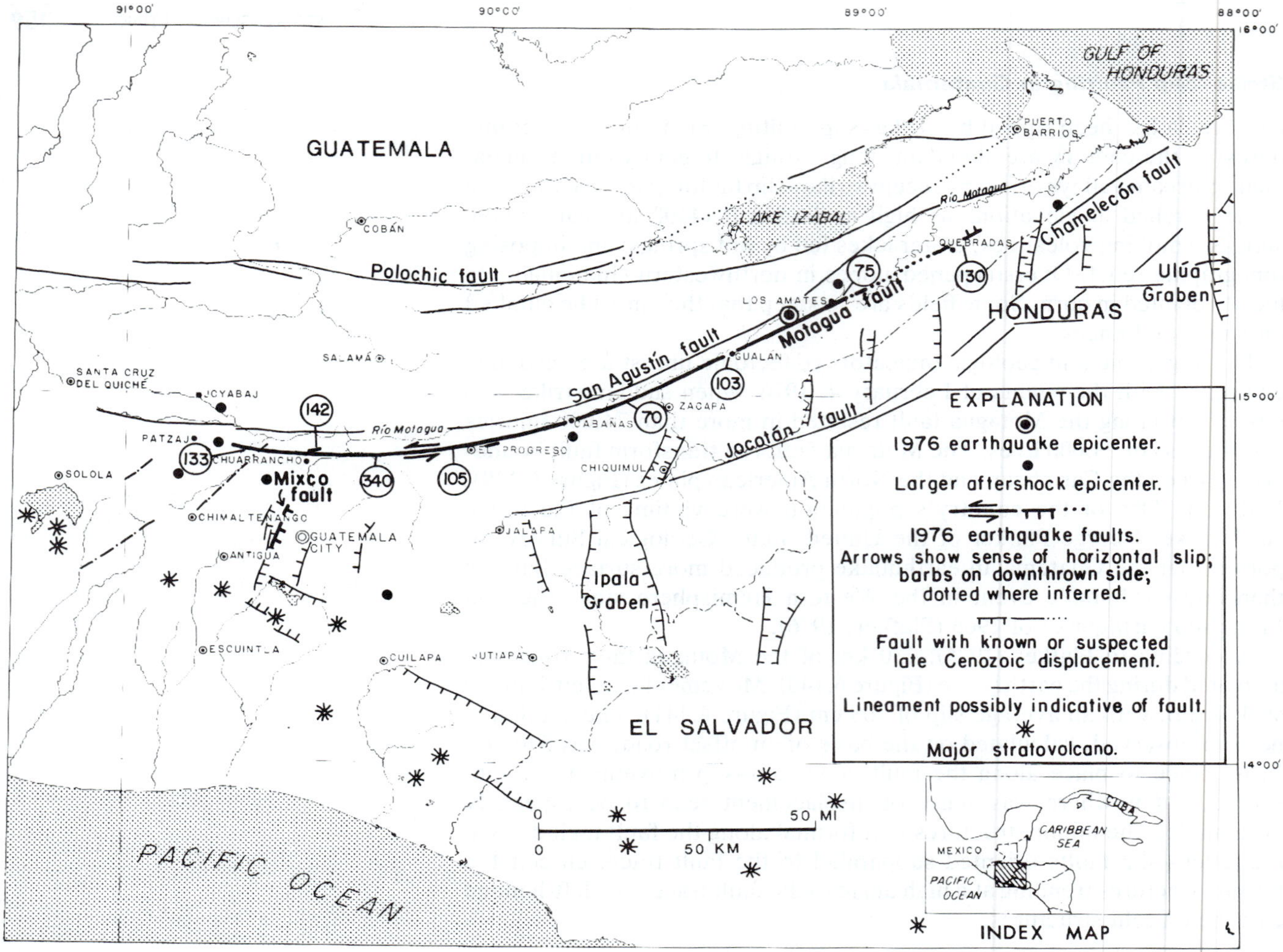

Figure 6.140 Geologic map showing the trace of the Motagua fault, as well as measured horizontal slip (circled numbers in centimeters) along the fault during the 1976 earthquake. [From Plafker (1976), *Science*, v. 193. fig. 1. Copyright © 1976 by American Association for the Advancement of Science.]

Figure 6.141 Left-lateral offset of a line of trees by left-handed slip along the Motagua fault. [From Plafker (1976), *Science*, v. 193, fig. 4. Copyright © 1976 by American Association for the Advancement of Science.]

Figure 6.142 Marine terraces rise like a flight of stairs at Kaikoura Peninsula, an active anticline on the northeast coast of the South Island of New Zealand. The peninsula has been rising at a rate of 0.6 m every thousand years. Not too far away the rates are as high as 6 m per thousand years. [Photograph kindly provided by W. B. Bull.]

Differential uplift (10 mm/yr!) by compression-induced reverse and thrust faulting is changing the face of South Island. Flights of old beach terraces are notched into the western flank of South Island, testifying to the uplift (Figure 6.142).

The San Andreas Fault

The San Andreas fault has been studied as carefully as any fault in the world, in part because of its huge threat to the major metropolitan centers in California. The San Andreas fault was not always recognized as a strike–slip fault. The first serious proposals that the San Andreas fault had accommodated very significant strike–slip displacement was made a surprisingly short time ago, during the early 1950s.

J. Tuzo Wilson (1965) was the first to perceive the San Andreas system as a transform fault linking the East Pacific Rise to the Juan de Fuca Ridge. Tanya Atwater's exquisite interpretation of the evolution of the San Andreas fault system made the fault truly understandable (Atwater, 1970). Her interpretation "washed" plate tectonics to shore. Until the time of her work, most geologists regarded seafloor spreading and plate tectonics as an ocean-bound concept, only vaguely applicable to understanding structural deformation in rocks within continental settings. By interpreting the structural history of western California in the direct context of plate interactions, Atwater demonstrated just how encompassing the revolution called plate tectonics would be.

Atwater showed that the role of the San Andreas fault as a plate boundary between the North American and Pacific plates can best be appreciated when the fault is shown in a special map projection, drawn about the pole of relative motion between the Pacific and North American plates (Figure 6.143). As Morgan (1968) and Atwater (1970) have explained, the trace of the San Andreas fault in this projection is horizontal, occupying the trace of a small circle about the pole of relative motion of the two plates (see Chapter 10). Relative motion between the Pacific and North American plates is about 6 cm/yr, right-handed, and oriented parallel to the trace of the San Andreas fault system. The San Andreas fault accommodates part of the motion (approximately 4 cm/yr). The rest of the motion is distributed on other faults and in other ways (Atwater, 1970).

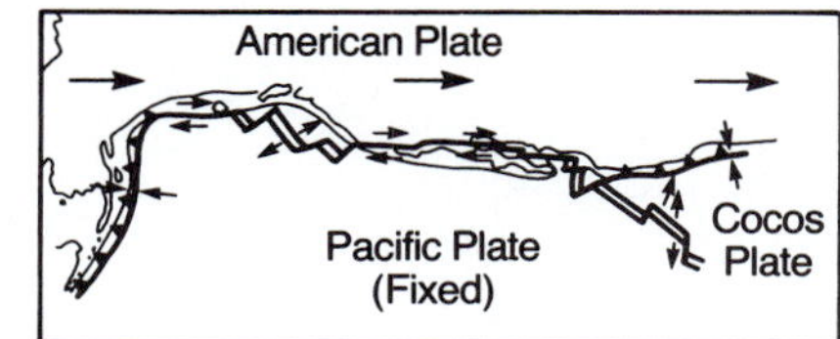

Figure 6.143 Map of the trace of the San Andreas fault as portrayed on a Mercator projection. [From Atwater (1970). Published with permission of Geological Society of America and the author.]

Direct contact between the Pacific and North American plates was achieved through a steady, systematic evolution of the plate configuration through time. Forty million years ago and earlier, the Pacific and North American plates were everywhere separated by the Farallon plate (McKenzie and Morgan, 1969) (Figure 6.144). In the early Tertiary, a trench

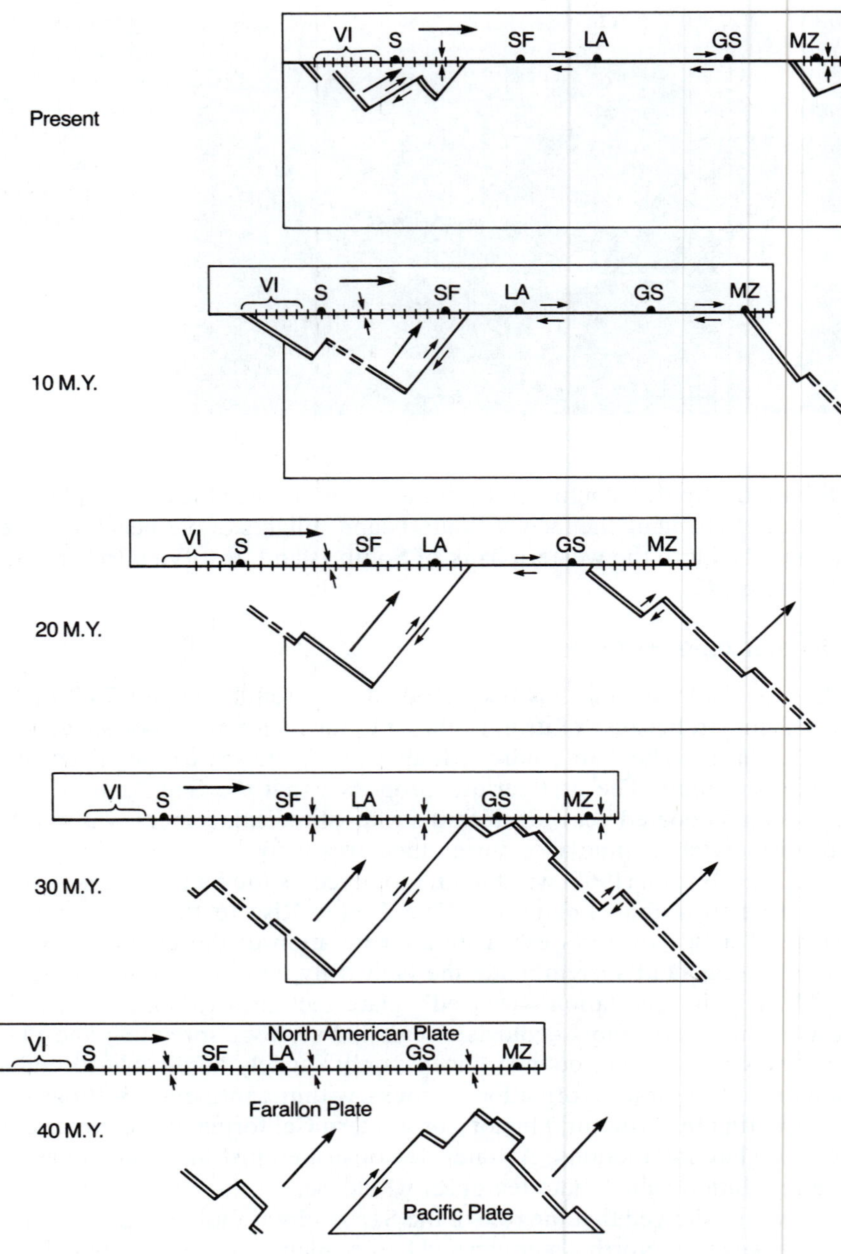

Figure 6.144 Plate tectonic evolution of the San Andreas fault. S = Seattle; SF = San Francisco; LA = Los Angeles; GS = Guaymas; and MZ = Mazatlan. [From Atwater (1970). Published with permission of Geological Society of America and the author.]

between the Farallon and North American plates continuously consumed Farallon plate lithosphere. Subduction took place faster than new Farallon lithosphere was being created. Consequently, most of the Farallon plate was subducted along the trench. In effect, the birth of the San Andreas fault system required elimination of the Farallon plate and the oceanic ridge from which it was spawned. Once eliminated, the Pacific and North American plates were permitted to come into contact (Figure 6.144).

"Ernst Cloos" Clay-Cake Modeling of Strike–Slip Faulting

Clay model experiments illustrate some of the geometric characteristics of strike–slip fault zones quite nicely, and they provide a framework for introducing terminology. Ernst Cloos (1955) described some simple experiments in which clay cakes become penetratively deformed by closely spaced, distributed strike–slip faulting. The experimental setup involves fastening a wire-mesh cloth to a hinged wooden frame that is capable of being deformed from a square to a rhomb. A 2-cm-thick clay cake is constructed on top of the wire-mesh screen, and a 5-cm-diameter reference circle is impressed on the surface of the clay, so that strain can be monitored during the course of ensuing deformation (Figure 6.145*A*). Deformation of the clay cake can be achieved in one of two ways: (1) pulling two opposing corners of the frame away from each other, or (2) shifting two opposite sides of the wooden frame in simple shear fashion. Both motions distort the wire mesh and the clay cake such that the reference circle is transformed to an ellipse (Figure 6.145*B*). In the first case, the clay is distorted by pure shear, and the axes of the strain ellipse remain fixed in orientation throughout the course of the experiment. In the second case, the clay is distorted by simple shear non-coaxial strain, and the axes of the strain ellipse rotate systematically and continuously during the deformation.

Distortion of the clay is accomplished largely by **conjugate strike–slip faulting**. The term "conjugate" means that the faults occur in two intersecting sets that are coordinated kinematically. Each set is distinctive both in orientation and in sense of shear. One set is right-handed; the other is left-handed. The expression "conjugate faulting" is not used exclusively in reference to strike–slip faults. Normal faults and thrust faults generally occur in conjugate sets as well. Conjugate faulting brings about simultaneous stretching and shortening. Triangular, wedge-shaped

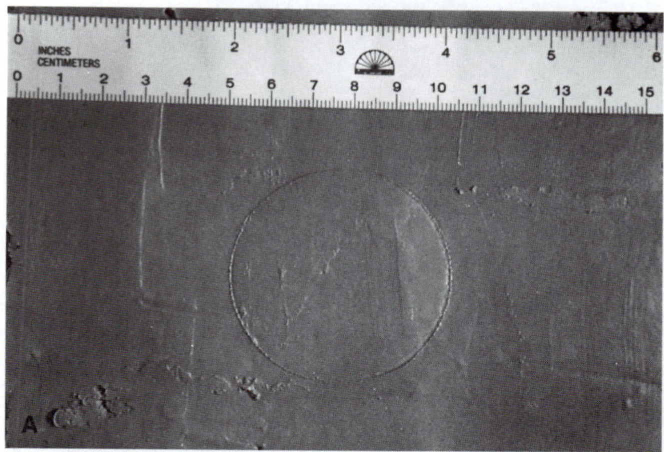

Figure 6.145 Clay cake simulation of strike–slip faulting. (*A*) Experimental setup consists of clay cake on a hinged wire-mesh screen. (*B*) Distortion of the clay cake.

fault-bounded blocks move toward one another in the direction of minimum extension. Complementary wedge-shaped fault-bounded blocks move outward to bring about an extension of the clay. The initial acute angle of intersection between the conjugate sets of strike–slip faults that emerge in the clay cake is about 60°. The line that bisects this conjugate angle is parallel to the direction of minimum finite stretch (S_3) of the strain ellipse.

Conjugate strike–slip faults that form in the clay cake are not fixed in orientation through time. Instead, they rotate as deformation proceeds. If the clay is distorted to an extreme degree, the original 60° conjugate angle can expand by flattening to 120° or more.

Tension fractures can also develop in clay cakes that are deformed in the manner described by Cloos (1955). The tension fractures are approximately at right angles to the direction of maximum finite stretch (S_1). In all of the clay cake experiments, such as the stretching of a clay cake atop a rubber sheet (see Figure 6.132*A*), tension fracture development can be enhanced by moistening the surface of the clay by a thin film of water, thus breaking the surface tension of the clay. Open tensional gash fractures will emerge as the dominant structure, very similar in appearance to gash fractures seen in naturally deformed rocks (Figure 6.146).

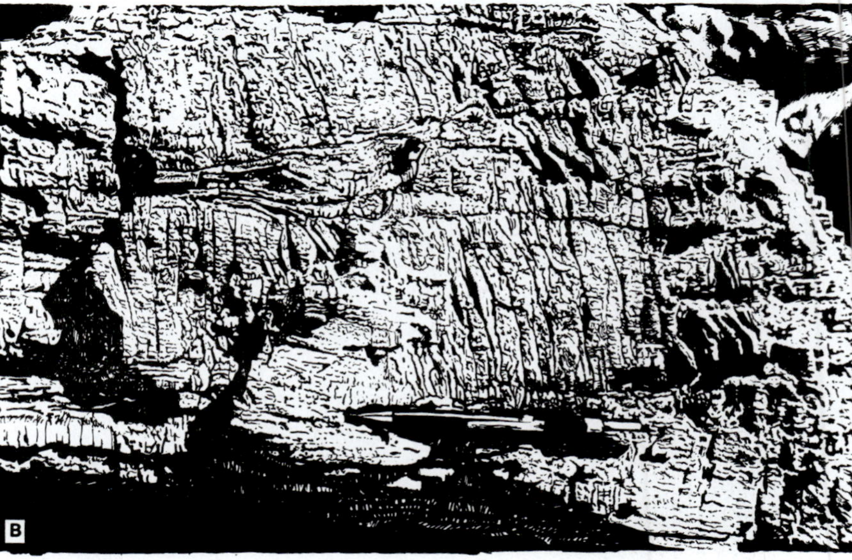

Figure 6.146 Very closely spaced gash fractures in outcrop. (*A*) Stretched upper surface of basalt in the Pinacate volcanic field, Sonora, Mexico. The tension fractures formed as primary structures during late stage movements of highly viscous basalt. (Photograph by G. H. Davis.) (*B*) Penetrative, parallel tension gashes in calc-silicate layer interbedded in marble, Happy Valley, Arizona. [Tracing of photography by G. H. Davis, rendered by D. O'Day. From Davis (1980). Published with permission of Geological Society of America and the author.]

Modeling of Strike–Slip Faulting by Wilcox, Harding, and Seely

Wilcox, Harding, and Seely (1973) designed some clay cake experiments to evaluate the structural patterns that develop in sedimentary strata above deep-seated strike–slip faults (Figure 6.147). Clay cakes are prepared on panels of sheet metal that can be moved oppositely and uniformly past one another during the course of an experiment. The line of contact between the metal panels predetermines the location and orientation of the linear zone of strike–slip faulting that develops in the overlying clay. Reference circles are impressed on the surface of the clay cakes before deformation so that strain, translation, and rotation can be evaluated (Figure 6.147*A*). Initial strike–slip movement of the metal panels results in distortion of the clay in such a way that the reference circles are transformed to ellipses (Figure 6.147*B*). Eventually the clay begins to fault within a zone parallel to the underlying metal panels (Figure 6.147*C*). As this happens the distorted reference circles become progressively faulted (see Figures 6.147*D–F*).

Both right-handed and left-handed strike–slip faults emerge in clay deformed in this way (Wilcox, Harding, and Seely, 1973). The faults combine to form conjugate sets marked by an initial conjugate angle of intersection of about 60° (Figure 6.147*C*). Of the two conjugate fault sets,

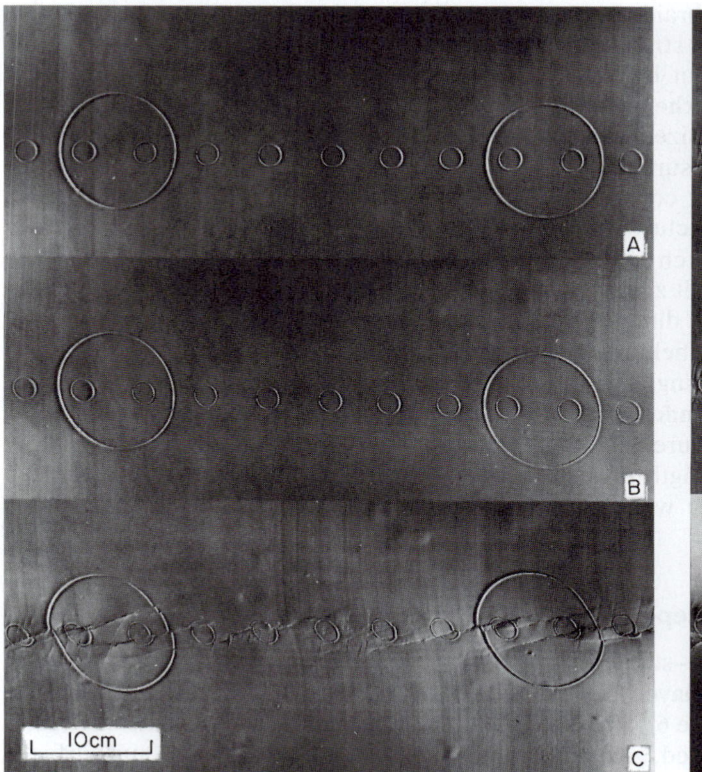

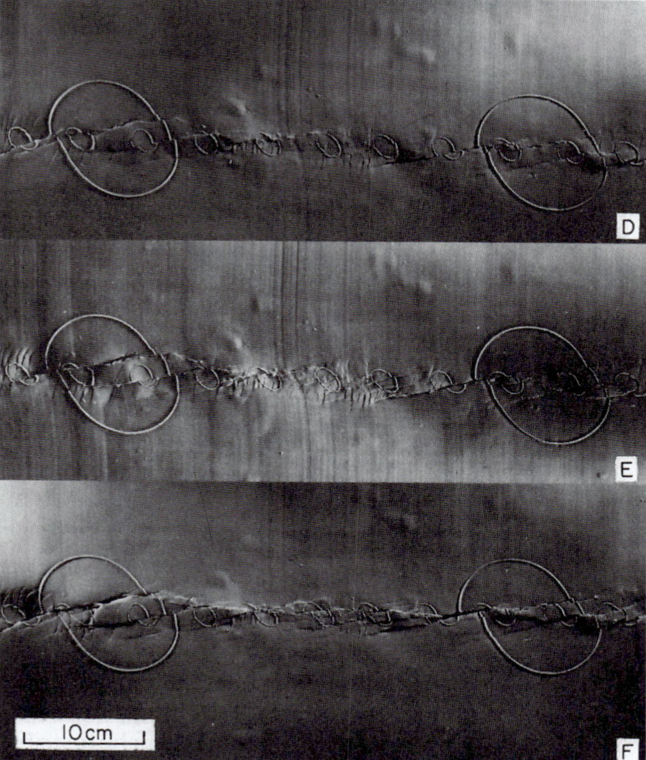

Figure 6.147 Clay cake deformation experiments simulating strike–slip faulting. Clay cake is placed on adjoining panels of sheet metal. Strike–slip faulting is achieved by shifting the panels horizontally past one another. (*A*) Starting configuration. (*B*) Initial distortion of clay. (*C*) Onset of faulting and the formation of synthetic and antithetic faults. (*D*) to (*F*) Continued faulting. Folds that develop become oriented parallel to the direction of maximum finite stretch (S_1). [From Wilcox, Harding, and Seely (1973). Published with permission of American Association of Petroleum Geologists.]

the one whose sense of slip is identical to that of the main zone of faulting is called synthetic. The one whose sense of slip is opposite that of the main zone is called antithetic. The synthetic faults are typically oriented at a small acute angle to the trace of the main fault zone; the antithetic faults are oriented at a very high angle to the main zone (Figure 6.147*E*).

The faulting produced in the experiments carried out by Wilcox, Harding, and Seely beautifully exemplifies the geometry and kinematics of **Riedel shearing**, which is characteristic of, but not limited to, strike–slip faulting. The synthetic strike–slip faults are **Riedel shears (*R*-shears)**, and they form at an acute angle of about 15° to the main line of faulting (Figure 6.148*A*). Their arrangement is **en echelon**, which means that they are parallel to one another and arranged along a common line of bearing. The antithetic strike–slip faults are **conjugate Riedel shears (*R'*-shears)**, and they form at a high angle of about 75° to the main line of faulting (Figure 6.148*A*). The direction of greatest principal stress (σ_1) bisects the angle between R and R'. As strike–slip faulting along the main zone proceeds, R-shears achieve a closer angle with the main line of faulting, and R'-shears may be rotated to a higher angle. Furthermore, a new set of synthetic shears known as ***P*-shears** develop, and these form at a small acute angle of about 10° to the main line of faulting (Figure 6.148*A*). The beauty of the arrangement of R-shears, R'-shears, and P-shears is that as "minor" faults they can together be used to independently interpret the sense of movement for the main line of faulting as a whole. Additionally, the whole concept of Riedel shearing gives us a better sense of why strike–slip fault zones have a braided appearance (Figure 6.148*B*).

Not only do strike–slip faults form in the clay cake experiments of the type carried out by Wilcox, Harding, and Seely (1973), but **en echelon folds** and **en echelon tension fractures** can develop as well. The folding can be emphasized by adhering a thin paper film (e.g., onionskin typing paper) to the surface of the clay cake. The tension fracturing can be emphasized by coating the surface of the clay with a thin film of water. The tension fractures form perpendicular to the direction of instantaneous maximum stretch ($\dot{S}_1$), and become arranged en echelon along the trace of the main fault zone. The folds, on the other hand, are aligned with axes *parallel* to the direction of maximum finite stretch (S_1). They are also arranged en echelon to the main zone of faulting.

Folds resulting from strike–slip faulting are described as **right-handed folds** or **left-handed folds**, depending on the sense of shear along the main fault zone (Figure 6.149). "Right-handed folding" means that if we were to walk the length of one of the en echelon folds, to the point where it disappears, we would turn to the right to search and discover the next fold in line.

Bends and Stepovers Along Strike-Slip Faults

Faults in strike–slip fault zones may be arranged in any number of ways: **en echelon**, **relayed**, **anastomosing** or **braided**, **left-stepping**, and **right-stepping** (Figure 6.150). The patterns express different ways in which slip can be distributed, given the frictional character of individual fault surfaces and the overall strain budget that needs to be satisfied. Some of the most interesting structural consequences of strike–slip faulting are seen in the places where individual fault surfaces either **bend** or **step**.

Crowell (1974) emphasized that bends in strike–slip faults invite high concentrations of strain. The strain is distributed within distinctive suites of structures. Movement along perfectly planar strike–slip faults results in coherent structural patterns. Wall rocks slide past each other without

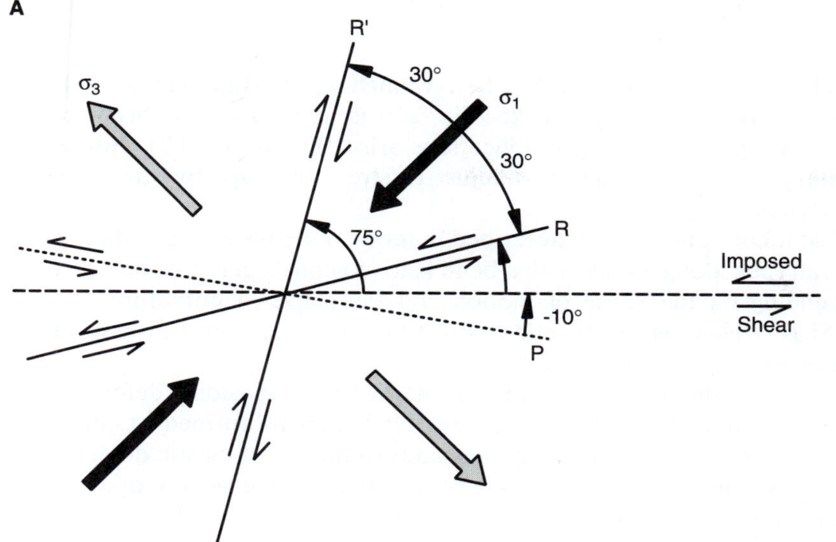

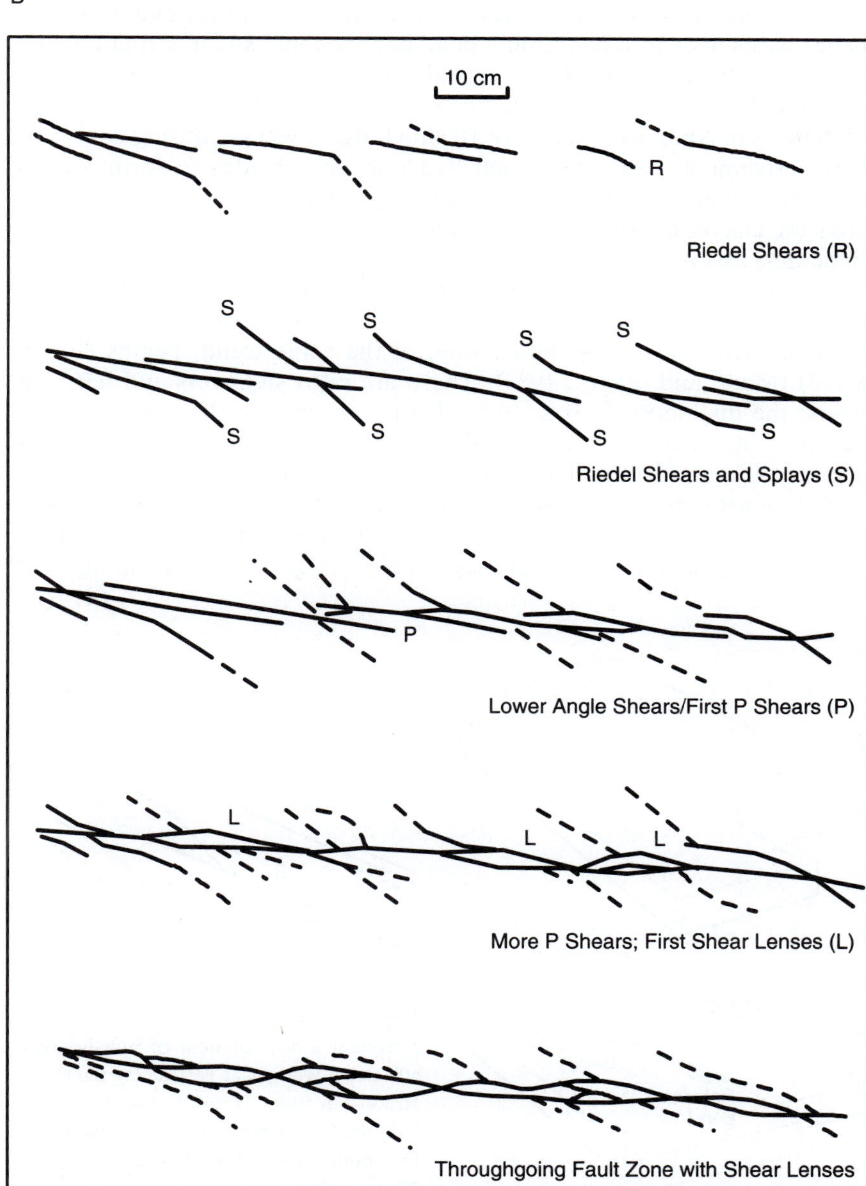

Figure 6.148 (*A*) In this *left-handed* strike-slip zone, Riedel shear arrays for both left-handed and right-handed faulting. *R* and *R′* are conjugate Riedel shears; *R* is synthetic, whereas *R′* is antithetic to the main movement. *R* and *R′* form first. *P* shears "come in" a little later. (*B*) In this *right-handed* strike-slip zone the interrelationships of Riedel shears, splays, and *P*-shears combine to produce a braided shear pattern [After Woodcock and Schubert (1994), *Continental deformation* Figure 12.6, p. 256, reprinted with permission from Elsevier Science, Ltd., Pergamon Imprint.]

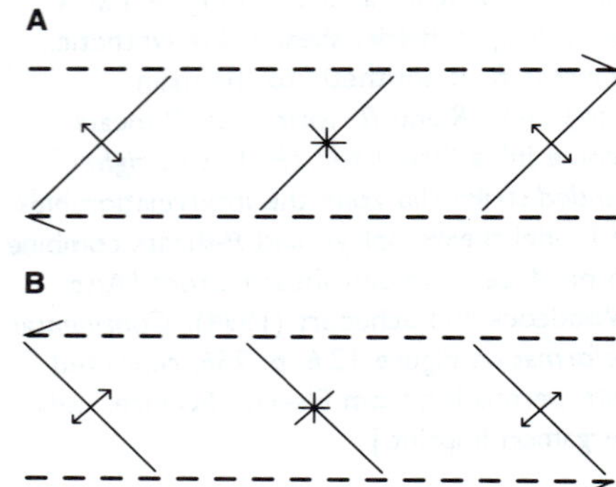

Figure 6.149 (*A*) Right-handed en echelon folds created within right-handed strike–slip fault zone. (*B*) Left-handed en echelon folds created within left-handed strike–slip fault zone.

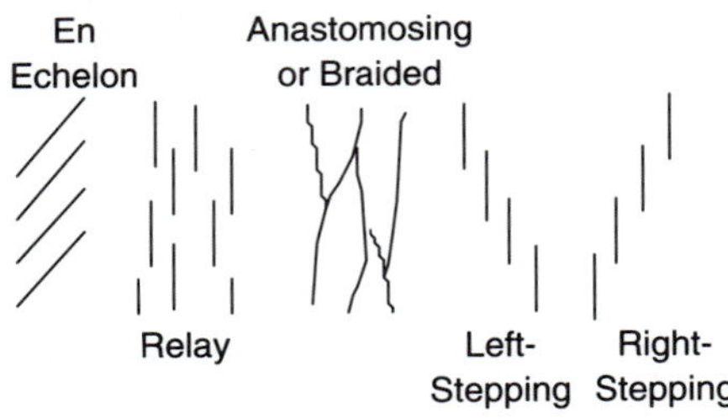

Figure 6.150 Map-view geometric arrangements of strike–slip faults: en echelon, relayed, anastomosing or braided, left-stepping, and right-stepping. [After Woodcock and Schubert (1994), *Continental deformation,* Figure 12.8, p. 257, reprinted with permission from Elsevier Science, Ltd., Pergamon Imprint.]

much interference (Figure 6.151*A*). **Branching** and **braiding** of faults is minimal. In contrast, if a strike–slip fault is marked by an abrupt bend, or even a gradual bend, complications arise (Figure 6.151*B*). Bordering country rocks are required to adjust to stress buildups by stretching or shortening.

Fault curvature can be described in terms of **bends or jogs**. Movement of wall rock along a fault with a bend leads to convergence or divergence, depending on the sense of motion and the sense of curvature (Figure 6.151*B*). **Releasing bends** tend to create open space, whereas **restraining bends** are sites of crowding.

Deformation at releasing bends is marked by extensional deformation, especially normal faulting. Fault-bounded grabens formed at releasing bends are capable of receiving thousands of meters of clastic deposits. In mineral deposit settings, releasing bends are probable sites of open-space filling and ore deposition. Restraining bends are marked by shortening, which is achieved by thrusting and folding. A common response to such crustal shortening is vertical uplift of the thickened block.

Fault-bounded **wedges** of rock form during the natural evolution of a bend. Resistance at a restraining bend may become so large that a newly formed fault cuts around the restraint (Figure 6.152), thus isolating a wedge of country rock. Wedges can be large or small; some associated with the San Andreas system are 100 km long. A wedge, during continued fault movement, can rotate such that one tip subsides to form a basin while the other tip rises to produce an uplift. The uplift becomes a source area for clastic debris, which inevitably is deposited in the nearest tectonic depression.

In addition to bends, or jogs as they are sometimes called, there may be **stepovers** within strike–slip fault zones. At a stepover, one strike–slip fault segment ends and another one, of the same trend, begins (Figure 6.153) (Mann and others, 1983). There are **right stepovers** and **left stepovers**; the distinction is based on whether we turn to the right or left to find the "stepped fault" as we walk to the end of a given strike–slip fault segment (Figure 6.153) (Twiss and Moores, 1992).

Pull-apart basins and **pop-ups** form at stepovers, at releasing locations and restraining locations, respectively. Left stepovers along left-handed strike–slip faults are sites of pull-apart, as are right stepovers along right-

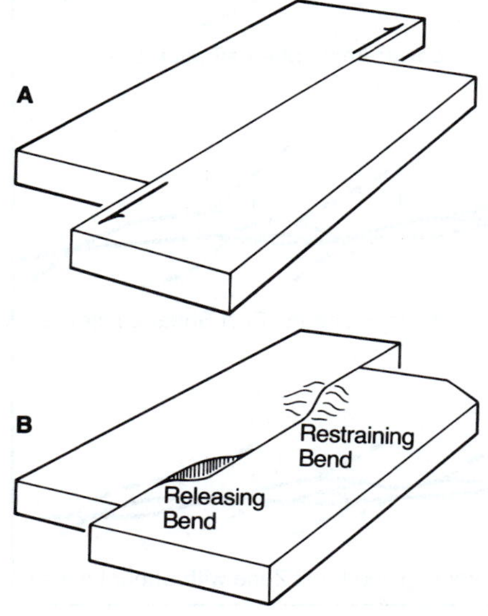

Figure 6.151 (*A*) Strike–slip movement along perfectly planar faults produces neither gaps nor overlaps. (*B*) Strike–slip movement along irregularly curved faults produces gaps at releasing bends and crowding at restraining bends. [From Crowell (1974). Published with permission of Society of Economic Paleontologists and Mineralogists.]

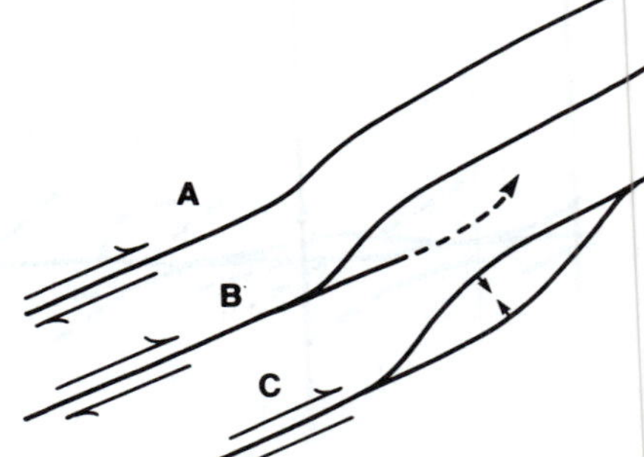

Figure 6.152 (*A*) to (*C*) The progressive development of fault-bounded structural wedges at restraining bends of strike–slip faults. [From Crowell (1974). Published with permission of Society of Economic Paleontologists and Mineralogists.]

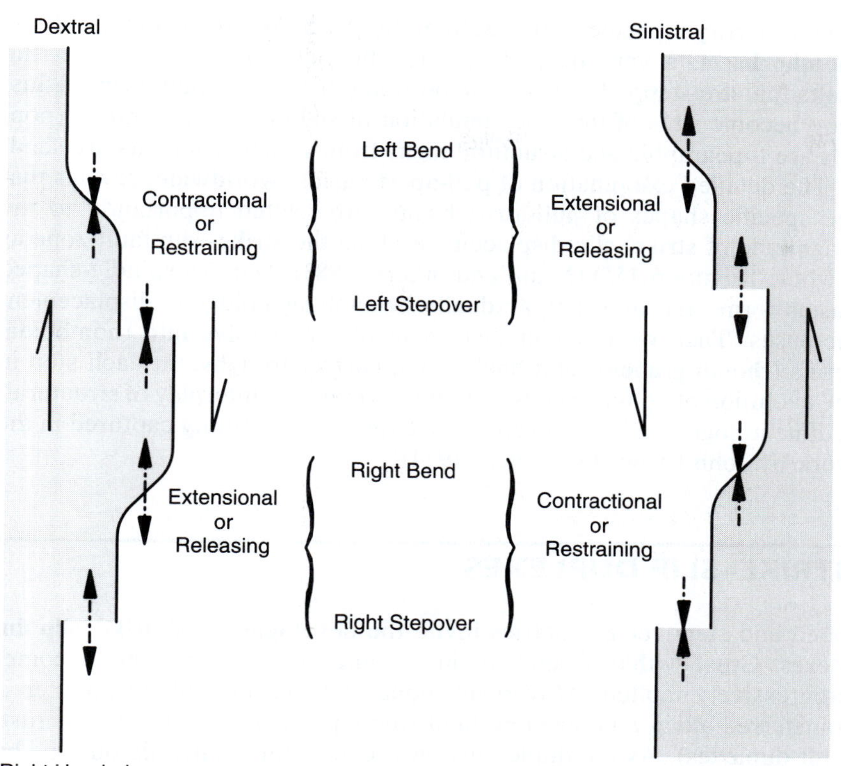

Figure 6.153 Portrayal of the difference between bends and stepovers, both in right-handed and left-handed strike–slip systems. (From *Structural Geology* by Twiss and Moores, Figure 7.5, p. 116. Copyright © 1992 by W. H. Freeman and Company.)

handed strike–slip faults (Figure 6.154*A*). In contrast, left stepovers along right-handed strike–slip faults are sites of pop-ups, as are right stepovers along left-handed strike–slip faults (Figure 6.154*A*) (Mann and others, 1983).

The term **pull-apart basin** was introduced by Burchfiel and Stewart (1966) in interpreting the tectonic origin of Death Valley (Figure 6.154*B*).

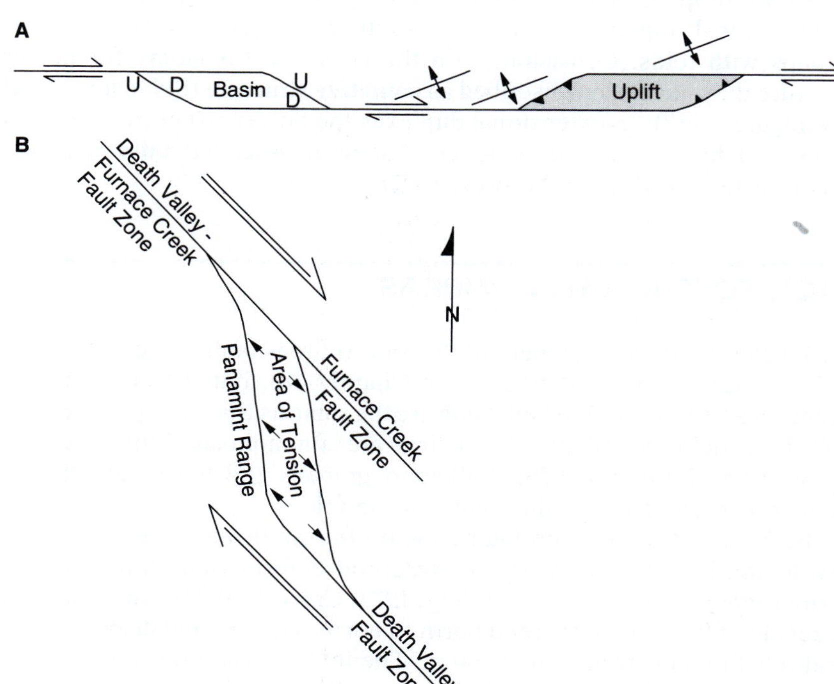

Figure 6.154 (*A*) Map views of pull-apart basins and pop-ups and their relations to bends in right-handed vs. left-handed strike–slip systems. (From Suppe, J., *Principles of Structural Geology,* © 1985, p. 280. Reprinted by permission of Prentice-Hall, Upper Saddle River, New Jersey.) (*B*) Giant pull-apart in Death Valley, California. [After Burchfiel and Stewart (1966), Published with permission of the Geological Society of America]

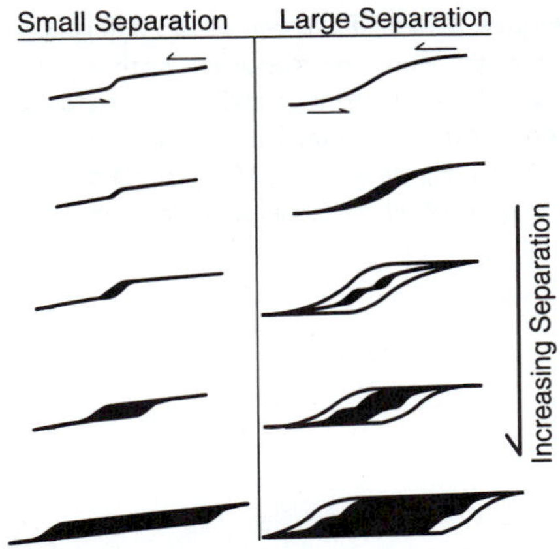

Figure 6.155 There appears to be a relationship between the shapes of pull-apart basins to the amount of strike–slip along master fault. [From Mann and others, 1983, Development of pull-apart basins, *The Journal of Geology* v. 7, p. 97–116, by permission of the University of Chicago Press.)

A more terrifying name, introduced earlier by S. Warren Carey (1958), is **rhombochasm**, a very deep depression bounded by master strike–slip faults that are stepped. Since "the bottom falls out" of pull-apart basins, they become sites of thick accumulation of sediments. In contrast, pop-ups are topographic and structural highs from which sediments are shed.

The detailed examination of pull-apart basins, worldwide, reveals that the specific shapes of pull-apart basins are related importantly to the magnitude of strike–slip displacement along the strike–slip fault zone as a whole (Figure 6.155) (Mann and others, 1983). Narrow spindle-shaped basins form first, and they gradually become sigmoidal as displacement increases. Then with further displacement they evolve into rhomboidal forms (**rhomb grabens**) and finally long narrow troughs. At each step in the evolution of pull-apart basins there is a complex interplay of structural, sedimentologic, and geomorphic processes, as beautifully captured in the work by John Crowell (Crowell, 1974).

STRIKE–SLIP DUPLEXES

Bend and stepover geometries invite the development of **strike–slip duplexes**. Crust within a bend, or in the interior of a stepover, becomes progressively faulted. Movement along one master fault to another is transferred along a connecting fault (the equivalent of a ramp in thrust-fault duplexes). As the duplex evolves, connecting faults die out one by one, and new ones form (Figure 6.156). Each fault "connects" with the master fault at depth, thus assuring kinematic organization. The geometric evolution of strike–slip duplexes is akin to what we would see if we turned an evolving thrust–fault duplex or normal–fault duplex on end, so that it could be seen in a horizontal view. Depending upon whether there is shortening or stretching across the zone, the strike–slip duplex will be compressional or extensional. The fault patterns evident in cross sections of strike–slip duplexes have been described as **flower structures**: bouquets of steeply dipping fault splays that diverge upward (Figure 6.157). In compressional duplexes the flower structures consist of reverse faults, perhaps with folds. Consistent with the botanical taxonomy for flower structure they have been described as **palm tree structure** (Sylvester, 1984) (see Figure 6.157). In extensional duplexes the flower structure is formed by normal faults, and the bouquet is sometimes called **tulip structure** (Figure 6.157) (Twiss and Moores, 1992).

A

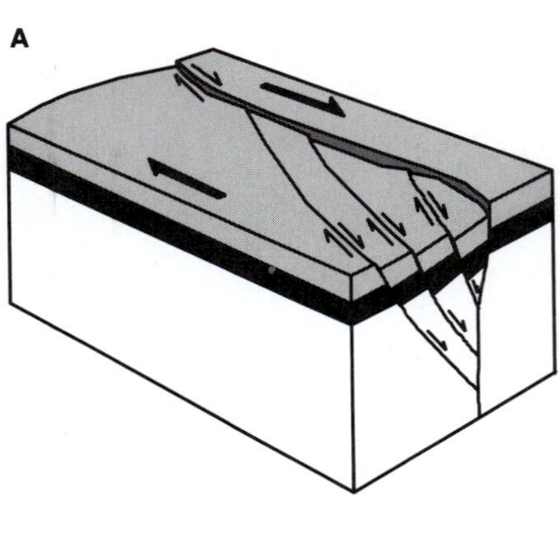

B

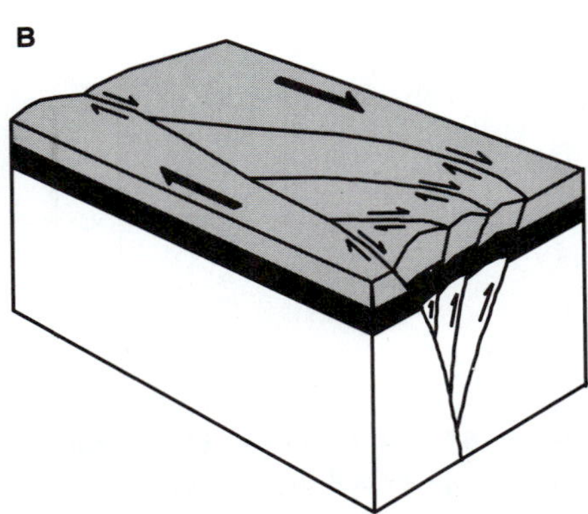

Figure 6.156 Strike–slip duplexes may form at bends along the major strike–slip fault. (*A*) Extensional duplexes form at releasing bends. (*B*) Compressional duplexes form at restraining bends. (From *Structural Geology* by Twiss and Moores, Figures 7.6 and 7.7, p. 118 and 119. Copyright © 1992 by W. H. Freeman and Company.)

BACK TO THE SAN ANDREAS

The kinematics of development of the San Andreas fault make sense only in the context of plate tectonics (see Chapter 10), but the physical and geometrical features of the fault zone are the same as that of any strike–slip fault. En echelon right-handed anticlines are commonplace along the fault (Crowell and Ramirez, 1979). Pull-apart grabens and tensional rifts are oriented perpendicular to the trends of the folds.

The Newport–Inglewood fault zone on the southwest side of the Los Angeles basin is a microcosm of structural deformation along the San Andreas fault (Figure 6.158) (Harding, 1973; Crowell, 1974). Right-handed en echelon fold structures trend north to northeastward and display structural relief ranging from 180 to 760 m. The folded structures are offset by right-handed synthetic en echelon strike–slip faults. The folds, although

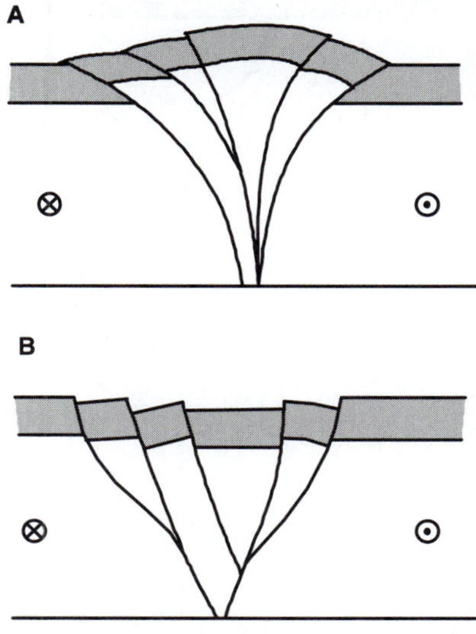

Figure 6.157 The two main varieties of "flower" structures associated with strike–slip duplexes: (A) Palm tree structure, (B) tulip structure. [After Woodcock and Schubert (1994).]

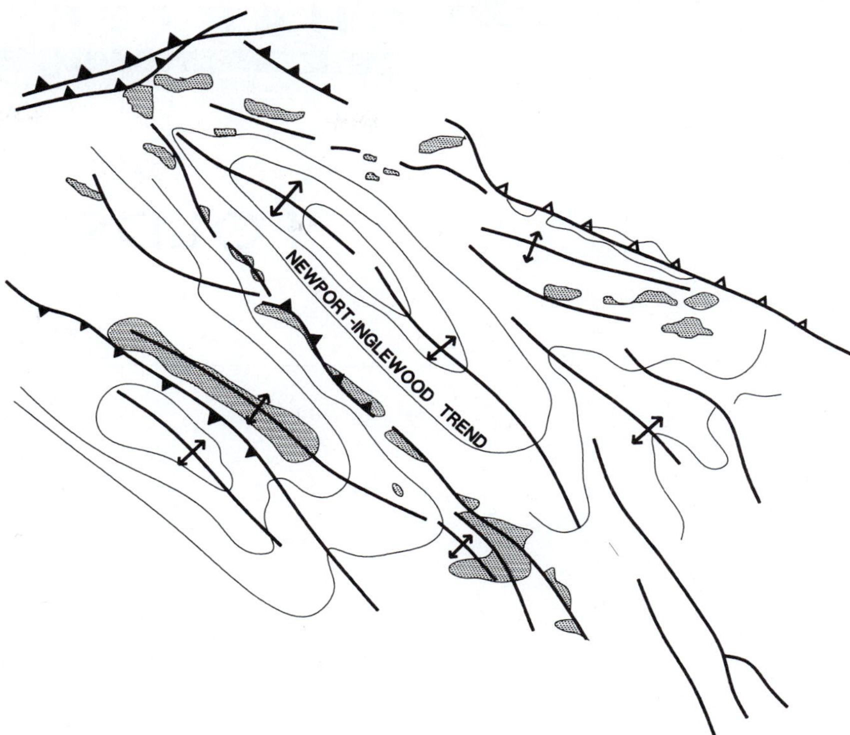

Figure 6.158 Some of the major structures and oil fields in the Los Angeles basin region lie along the Newport–Inglewood trend. Shaded regions are oil fields. Structure contours are generalized. [After Harding (1973), Figure 1, p. 98, reprinted by permission of the American Association of Petroleum Geologists.]

faulted, are great oil traps. Oil is obtained from sediments ranging in age from mid-Miocene to Holocene.

Basins along the Newport–Inglewood fault zone (Figure 6.158) trend northwest-southeast and range in length from 50 to 100 km. They too are arranged en echelon. Average depth of the basins is about 3 km. The basins are rhombic to lens shaped in plan view, defined by straight margins on the northeast and southwest sides. The straight margins are traces of en echelon synthetic right-handed strike–slip fault zones; the irregular margins are defined by normal-slip faults, oriented perpendicular to major extension.

CONCLUDING REMARKS

The study of faulting encompasses geometric, kinematic, and dynamic analysis in the broadest possible way. If we are to understand faulting, we need to integrate field, experimental, and theoretical research. The rewards for such activities are significant. Faulting is one of the most important mechanisms of rock deformation; it produces regional distortion; it traps petroleum and controls ore deposition; it provides clues to rearrangements of the Earth's architecture through time. Furthermore, faulting and the associated earthquakes present one of the most significant natural hazards to society. Experimental deformation and theoretical studies help us immensely in understanding the properties of faults that we find in nature.

CHAPTER 7

FOLDS

INCENTIVES FOR STUDY

Visual Impact

Folds are visually the most spectacular of Earth's structures. They are extraordinary displays of strain, conspicuous natural images of how the original shapes of rock bodies can be changed during deformation. The physical forms and orientations of folds seem limitless. Some are upright (Figure 7.1*A*); some lie on their sides (Figure 7.1*B*); some are inclined (Figure 7.1*C*). Some show neatly arranged, uniformly thick layers; others are sloppy. Fold size varies too, from anticlines that fit into the palm of a hand to regional folds best seen through the eyes of a satellite (Figure 7.2). Mapping the forms of folds is pure pleasure, unless of course the folds turn into a geometric nightmare. Constructing cross sections of folded terrains becomes a fundamental tool for structural geologic insight (Part III-E).

Mechanical Contradiction of Folding

It is almost impossible to view waves of solid rock without wondering how materials we regard as strong can be folded in a manner that makes them seem so weak. Bailey Willis (1894), in trying to model the structures of the Appalachian fold belt, found it necessary to represent sedimentary strata with layers of very soft materials like clay, putty, cheese, and wax (see Figure 1.39). Regional rock bodies yield effortlessly to folding by penetrative movements along preexisting weaknesses, like bedding, and/or by the generation of secondary penetrative weaknesses, like cleavage.

Figure 7.1 Some of the many geometric renditions of folds. (*A*) Upright anticline in Silurian sandstones and shales, 3 mi west of Hancock, Maryland. (Photograph by I. C. Russell. Courtesy of United States Geological Survey.) (*B*) Folds in Oligocene-Miocene turbidites in Kii Peninsula, southwest Japan. (Photograph by W. R. Dickinson.) (*C*) Kinklike fold in limestone at Durdle Door in the Purbeck Country, England. (Photograph by G. H. Davis.)

Where strata are very well bedded and relatively stiff, folding is achieved by layer-parallel slippage (Donath and Parker, 1964). Like the pages of a slick magazine, beds can easily slip past one another along bedding surfaces during folding (see Figure 2.10). When the mechanical influence of bedding or other layering is not as dominant, and the layers tend to be relatively soft, folding is achieved by flow or slip or even the pressure solution loss of material along cleavage surfaces created during the deformation (see Chapter 8, "Cleavage, Foliation, and Lineation"). The cleavages themselves testify to the weakness of rock in the face of tectonic stresses: cleavages cut across the lithological layering of folded beds as if the beds did not exist (Figure 7.3), as if the mechanical influence of layering could hardly be felt, as if the rocks had no strength to resist!

Geometric Pleasures

Analysis of the geometry of folds invites us to probe more deeply into the stereographic analysis of structural data (see Part III-H,I). This, in turn, leads to a more expansive awareness of geometrical concepts and methods. Simple stereographic procedures permit us to calculate and describe the orientations of folds, to measure fold tightness, to unfold folds and reconstruct original orientations of primary structures, and to describe the total integrated geometry of fold systems. The full use of

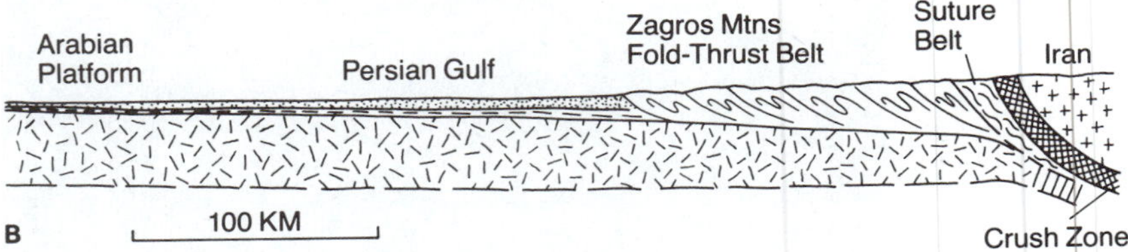

Figure 7.2 (*A*) Apollo 7 view of anticlines in the Zagros Mountains, Iran. The dark circular forms are the surface expressions of salt domes. Persian Gulf in foreground. (Courtesy of National Aeronautics and Space Administration). (*B*) Structure section shows the plate tectonic configuration responsible for the formation of the Zagros Mountains fold-thrust belt. The belt lies at the convergent boundary of the Arabian and Iran plates. (From W. R. Dickinson unpublished.)

stereographic analysis in modern geometric analysis of folds and fold-related structures was inspired by Turner and Weiss (1963) in their classic text, *Structural Analysis of Metamorphic Tectonites*.

Informative Minor Structures

Folding results in the formation of a delightful and curious array of cognate structures, typically referred to as **minor structures**. The adjective "minor" belies the usefulness of this class of structures in evaluating the kinematics

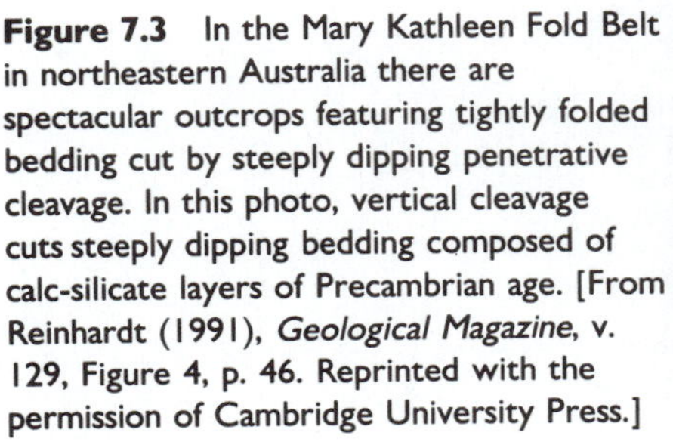

Figure 7.3 In the Mary Kathleen Fold Belt in northeastern Australia there are spectacular outcrops featuring tightly folded bedding cut by steeply dipping penetrative cleavage. In this photo, vertical cleavage cuts steeply dipping bedding composed of calc-silicate layers of Precambrian age. [From Reinhardt (1991), *Geological Magazine*, v. 129, Figure 4, p. 46. Reprinted with the permission of Cambridge University Press.]

of folding. Within the family of minor structures are small folds and faults; slickenlines and crystal fiber lineations; a variety of cleavages and penetrative lineations; joints, shear fractures, veins, and stylolites; and some other structures as well. Different kinds of minor structures form in different parts of folds because different parts of folds are marked by different local strain environments. Some parts of folded beds are stretched; others are shortened; some lose volume; some gain volume; still other parts suffer no distortion or dilation whatsoever.

Most students of geology enjoy the challenge of trying to interpret the orientations and structural locations of minor structures in light of the overall geometry and kinematics of folding. Learning to do so is fundamental to mapping and unraveling regional assemblages of folded rocks. For example, a change from *Z*-shaped to *S*-shaped minor folds within uniformly dipping strata (Figure 7.4*A*) may signal the presence of an otherwise hidden fold structure (Figure 7.4*B*).

Interpreting folds, including the formation of minor structures, is a fruitful exercise in applying principles and methods in kinematic and dynamic analysis. Kinematic analysis of folding requires integration of *all* kinematic movements: translation, rotation, dilation, and distortion. Dynamic analysis requires identifying the physical and mechanical variables that influence the nature of folding. Dynamic analysis is pursued both theoretically and experimentally. Fold analysis provides an unusual opportunity for integrating and synthesizing principles of structural geology. Friends of folds have John Ramsay (1967) to thank for *Folding and Fracturing of Rocks,* which presents the foundation for modern quantitative fold analysis.

Tectonic Considerations

To understand the origin of folding, we must move beyond geometric and kinematic analysis, beyond experimental and theoretical studies, to search at the tectonic scale for dynamic circumstances that are likely to generate folding. In this regard, it is especially useful to view the locations of the world's fold systems in the context of reconstructions of plate configura-

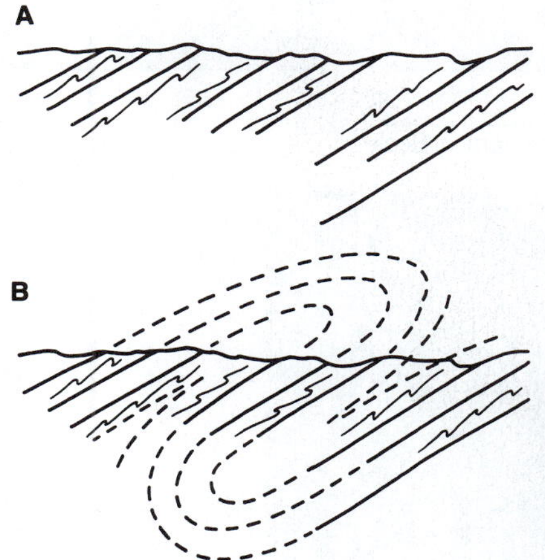

Figure 7.4 The asymmetry of folds can be used to disclose hidden structures. (*A*) *Z*-shaped and *S*-shaped minor folds in homoclinally dipping strata denote the limbs of (*B*) a hidden anticline. (*C*) Minor fold on the flank of a major fold in northern Norway near Bjornfeld. The layer is composed of Cambrian quartzite. The geologist is Lisa Rindstad. (Photograph by G. H. Davis.)

tions through time. In doing so, we find that folds and fold systems tend to appear in settings that are or were convergent plate margins.

Many different tectonic circumstances can give rise to folds. Consequently, it is generally difficult to interpret the ultimate cause of folding within any given tectonic assemblage in any given region. Folds found in subduction complexes can form as products of shearing that result from underthrusting and underplating accompanying subduction (see Figure 7.5); but they also can form as a response to gravity-induced flow of oversteepened, not-yet-consolidated sediments. Strata in forearc, interarc, and backarc basins can fold during overall crustal shortening and/or strike-slip shearing created by plate convergence. However, the same sediments and volcanics, deposited in and around magmatic arcs, may contain folds whose origins are difficult to pin down (Figure 7.6). Do they represent forceful shouldering aside of strata during emplacement of granitic plutons, gravity-induced gliding of strata off the tops of rising diapirs, or compressional thrusting not even related to intrusion? The origin of folds and fold systems in foreland settings (Figure 7.7), *inboard* from the edges of active plate margins, has been interpreted by some as the direct expression of significant crustal shortening, a shortening due to stresses generated at plate margins and transmitted through continental lithosphere for great distances horizontally. But such folds and fold systems have been interpreted by others as products of the draping of strata over the edge(s) of vertically uplifted fault blocks, in a manner not related to crustal shortening.

Thus, where strain is intense, deformations superimposed, and exposures limited, it may not always be simple to interpret the dynamic condi-

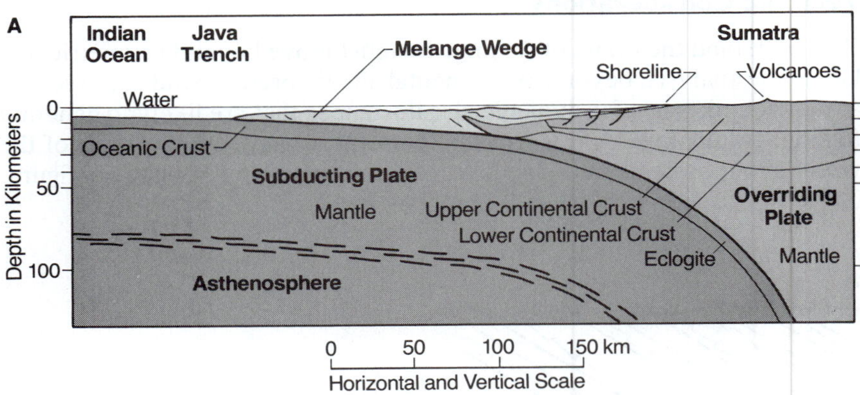

Figure 7.5 (*A*) True-scale profile view showing the relations of an accretionary prism (melange wedge) to the fundamental tectonic components of an active continental margin. [From Hamilton (1979). Courtesy of United States Geological Survey.] (*B*) Folds in foliated melange exposed in the San Juan Islands, Washington. (Photograph by G. H. Davis.)

Figure 7.6 Folded metamorphic rocks in Upper Convict Canyon in the Sierra Nevada of California. (Photograph by E. B. Mayo.)

tions responsible for folding. The geologic literature, bulging with contradictory interpretations regarding the origin of folds and fold systems in specific regions, bears this out. Increasingly, however, seismic reflection profiles across deformed belts are clarifying the nature and origin of folding.

Guides to Exploration and Mining

There are practical incentives to come to know folds and their properties. Foremost, there is a legacy of discovery of oil and gas in **structural traps** created by folding (Figure 7.8). Folds commonly serve as collection sites for oil and gas that migrate up the dip of strata from hydrocarbon source beds below. The migration of oil and gas takes place within a permeable reservoir rock, like a porous sandstone or a vuggy limestone. If the reservoir bed is overlain depositionally by a tight, impermeable layer, like shale, the top of the reservoir is effectively sealed to prevent escape of the oil and gas during up-dip migration. Where a sealed reservoir is folded into a dome or anticline, the **closure** of the fold prevents further upward migration of the oil and gas. The fluids may collect there to form a rich "pool."

Figure 7.7 Spectacular folding along Clark's Fork Corner of the Beartooth uplift, Wyoming. (Photograph by K. Constenius.)

Figure 7.8 Folds commonly trap petroleum accumulations. This fold occurs in thin-bedded Monterey formation exposed along Soto Street, north of Alhambra Avenue, Los Angeles. (Photograph by M. N. Bramlette. Courtesy of United States Geological Survey.)

Where ore deposits are known to be localized within folded rocks, prospects for exploration and mining offer clear incentives to understand the three-dimensional geometry of folds. Of special importance is learning methods for projecting the forms and trends of folds to depth as an aid in estimating reserves and in planning underground mining. **Saddle reef deposits** represent perhaps the most intimate of the associations between folding and mineralization. Saddle reefs are **lodes** of quartz and precious metals that occupy the cores of folds, in openings where bedding and/or foliation has been separated by fold-forming movements (Figure 7.9). Although saddle shaped in cross-sectional view, they are pipelike in three dimensions, trending parallel to the axis of folding. The most renowned saddle reef deposits are in the Bendigo goldfields of Victoria, Australia (McKinstry, 1961; Guilbert and Park, 1986; Park and MacDiarmid, 1964), and in the Salmon River gold district of Nova Scotia (Malcolm, 1912).

Strata-bound ore deposits are commonly associated with folded, metamorphic rocks. The lead–zinc–silver deposits of Broken Hill, Australia, constitute a fine example (Gustafson, Burrell, and Garretty, 1950) (Figure 7.10). Shaped like ordinary beds of volcanic or sedimentary rocks, these mineralogically exotic deposits are abundantly rich in sulfides and/or precious metals. Although traditionally viewed as replacements of chemically favorable host rocks, most strata-bound sulfide deposits are now considered to have been deposited originally as submarine chemical sediments and/or volcanic-exhalative precipitates. In the course of the normal geologic history of orogens, these unusually valuable rock layers become folded, faulted, and metamorphosed, just like their ordinary sedimentary and volcanic counterparts. For my dissertation research, I mapped a strata-bound sulfide deposit in eastern Canada, and discovered through the mapping and detailed structural analysis that the body had been folded three times, creating wonderful geometries (Figure 7.11) (Davis, 1972).

Folded ore bodies, and ore bodies in folds, are challenging to exploration and mining geologists. The step-by-step underground geologic mapping of ore bodies in folded rocks, carried out during the normal course of mining, has yielded illuminating documentaries on the three-dimensional anatomy of folds.

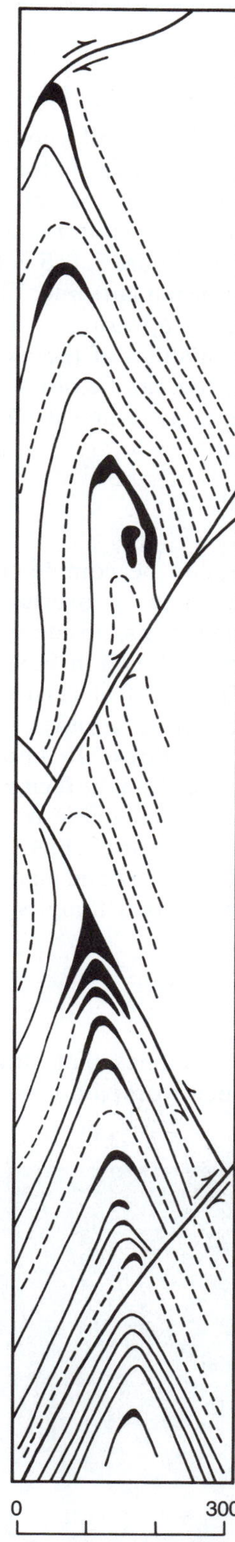

Figure 7.9 Saddle reef deposit in the Bendigo goldfield, Bendigo, Victoria, Australia. Scale in feet. (From *Mining Geology* by H. E. McKinstry. Published with permission of Prentice-Hall, Inc., Englewood Cliffs, New Jersey, copyright © 1961.)

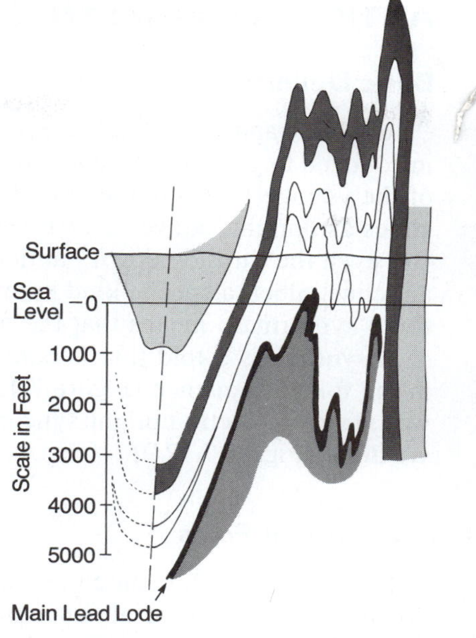

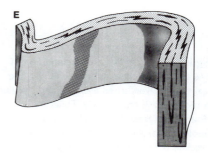

Figure 7.10 Structure section through part of the folded, strata-bound Broken Hill ore deposit. [From Gustafson, Burrell, and Garretty (1950), Geological Society of America.]

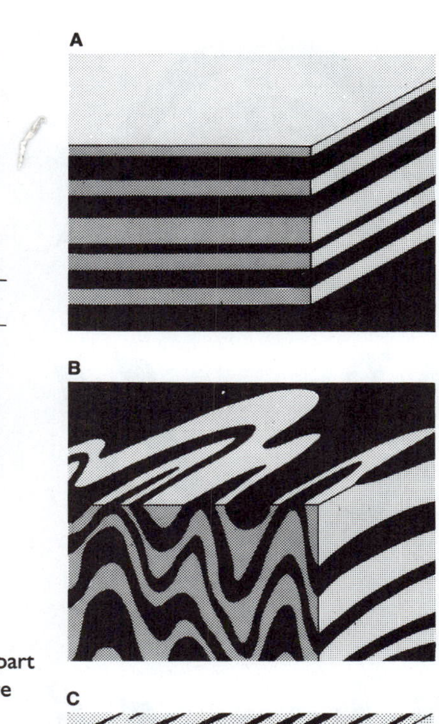

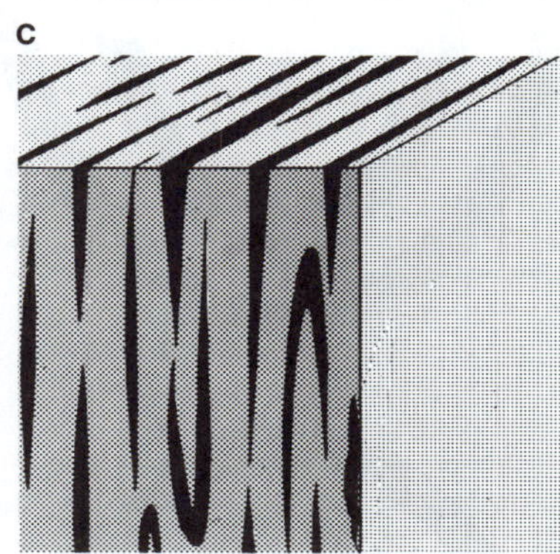

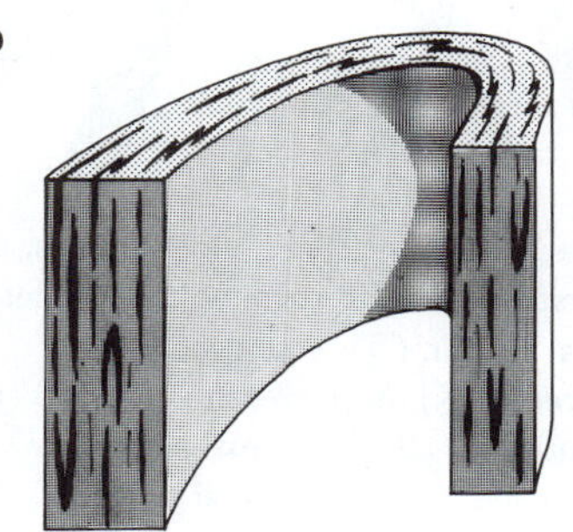

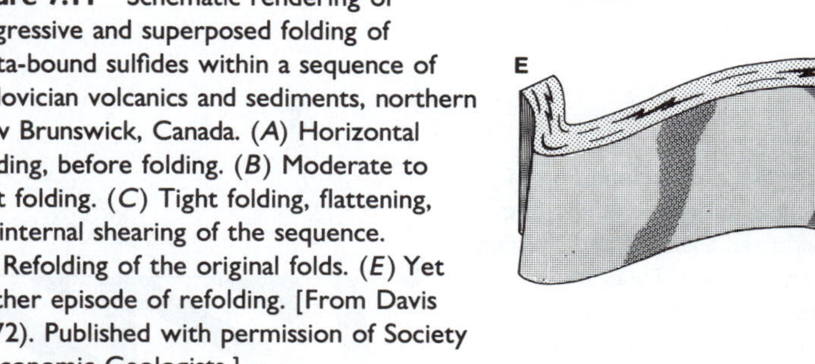

Figure 7.11 Schematic rendering of progressive and superposed folding of strata-bound sulfides within a sequence of Ordovician volcanics and sediments, northern New Brunswick, Canada. (*A*) Horizontal bedding, before folding. (*B*) Moderate to tight folding. (*C*) Tight folding, flattening, and internal shearing of the sequence. (*D*) Refolding of the original folds. (*E*) Yet another episode of refolding. [From Davis (1972). Published with permission of Society of Economic Geologists.]

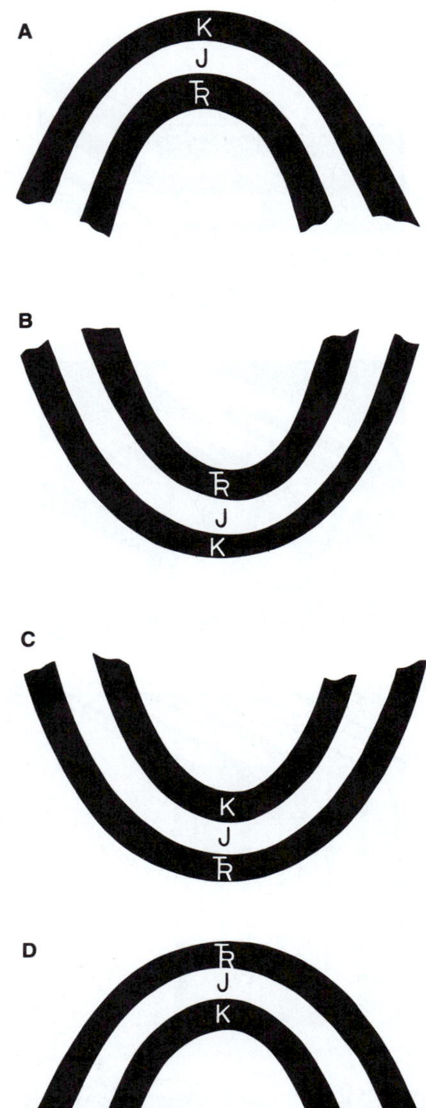

Figure 7.12 Anticlines and synclines, the cornerstones of fold terminology. Oldest layer is Triassic (Ŧ̶R); youngest layer is Cretaceous (K); in-between is Jurassic (J). (*A*) Anticline. (*B*) Synformal anticline. (*C*) Syncline. (*D*) Antiformal syncline.

ANTICLINES AND SYNCLINES

Basic Definitions

''Anticline'' and ''syncline'' are part of the A,B,C's for all who have been introduced to geology. An **anticline** is a fold that is convex in the direction of the youngest beds in the folded sequence. Ordinarily we think of anticlines as upright, convex-upward folds, like that shown in Figure 7.12*A*. But even the ''upside-down'' anticline shown in Figure 7.12*B* is called an anticline, albeit a special kind known as a **synformal anticline**. The adjective ''synformal'' means that the fold is concave upward.

A **syncline** is a fold that is convex in the direction of the oldest beds in the folded sequence (Figure 7.12*C*). These too can be oriented in any way, from perfectly upright synclinal folds to convex-upward **antiformal synclines** (Figure 7.12*D*).

Overturned Folds

It is very rare for anticlines and synclines to be completely upside down, but it is not at all uncommon for them to be **overturned**. The distinction between upside down and overturned is this: upside-down folds are totally inverted, like the antiformal syncline shown in Figure 7.12*B*. In contrast, a fold is considered to be *overturned* if at least one of its **limbs** (i.e., flanks) is overturned. Saying that the limb of a fold is overturned does not mean that the fold is completely upside down. It simply means that one limb has been rotated beyond vertical such that the **facing direction** of the limb points downward at some angle. A schematic rendering of overturned anticlines and synclines is shown in Figure 7.13*A*. Each fold is marked by a right-side-up limb dipping at about 30°W and an overturned limb dipping at about 80°W. An example of the real thing is shown in Figure 7.13*B*.

Antiforms and Synforms

Use of the term ''anticline'' (or ''syncline'') implies that stratigraphic succession within the folded sequence has been worked out on the basis

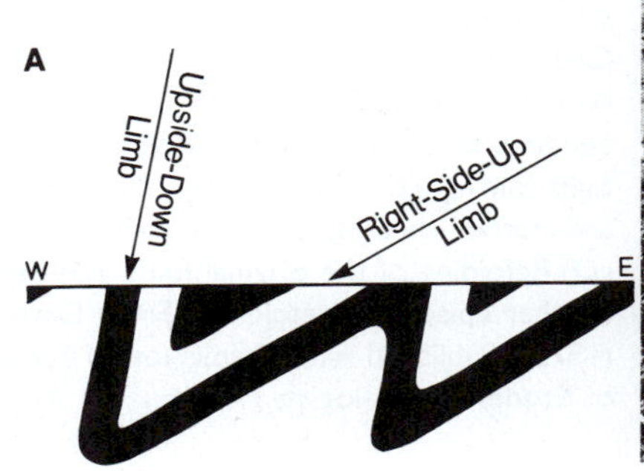

Figure 7.13 (*A*) Schematic rendition of overturned anticline and syncline. (*B*) The real thing! Overturned anticline and syncline in sedimentary rocks in the Funeral Mountains, Death Valley, California. Geologist barely visible in core of main anticline is Stan Ballard. (Photograph by S. J. Reynolds.)

of the established geological column or the determination of facing. If facing and stratigraphic order cannot be established, these terms must be scrapped, at least temporarily, in favor of **antiform** and **synform**. An antiform is, very simply, a fold that is convex upward. A synform is a fold that is concave upward.

"Antiform" and "synform" are normally used in reference to folds in sedimentary and/or volcanic sequences within which facing and/or stratigraphic order is either unknown or uncertain. But these terms are also appropriately used in describing folds in metamorphic rocks such as gneisses and schists (Figure 7.14*A*), and in igneous rocks where, for example, folds may have formed in stiff magma along the margin of a dike as the magma, during emplacement, moved past wall rock. If layering and/or foliation in a rock body is not related to normal depositional process, nor to conventional stratigraphic succession, it is meaningless to attempt to describe facing and stratigraphic order (Figure 7.14*B*).

The terms "anticline" and "syncline" may have only limited application in describing **superposed folds** in a sedimentary and/or volcanic sequence. By way of example, the early-formed fold shown in structure profile view in Figure 7.15*A* may be described as a **recumbent, isoclinal** anticline. "Recumbent" means that the fold lies on its side. "Isoclinal" means that the limbs of the fold are equally inclined. Figure 7.15*B* shows

B

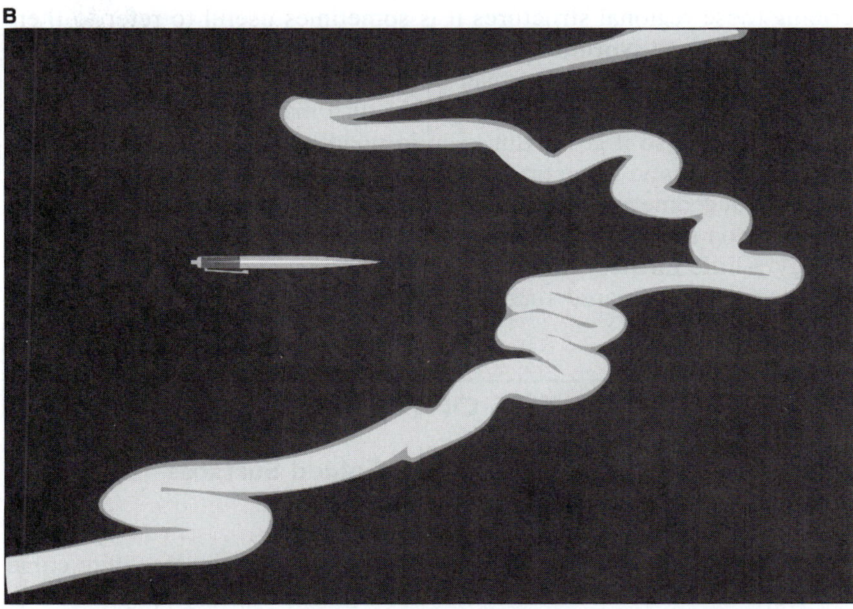

Figure 7.14 (*A*) This beautiful, tight overturned antiform (white layer at left) occurs within a sequence of high-grade metamorphic rocks in the Napier Complex, Antarctica. The cliff face is approximately 250 m high. [From Sandiford (1989). Published with permission of Geological Society of America and the author.] (*B*) The aplite dike shown here has been folded back and forth on itself like an intestine. We would not call the individual folds anticlines and synclines. Instead, we would call them overturned antiforms and synforms.

Figure 7.15 An example of regional-scale superposed folding. (*A*) Recumbent, isoclinal anticline, of pre-Jurassic age. (*B*) Antiformal and synformal folding of the recumbent, isoclinal anticline. (*C*) Superposed folding at the outrop scale. In the lower central portion of this photograph there is a well-exposed overturned antiform that *refolds* an isoclinal antiform. This outcrop is in the Ruby Mountains, Nevada. (Photograph by G. H. Davis.)

the early-formed isoclinal fold after it has been modified by a second folding. The anticline is no longer perfectly recumbent. Rather, it is deflected into convex-upward and convex-downward superposed folds. These late folds cannot be described as anticlines and synclines because there is no overall internal coherence of stratigraphic order and facing within the layers that define the folds. The second-stage folds are appropriately described as antiforms and synforms. An outcrop example that presents the same dilemma is shown in Figure 7.15*C*.

Anticlinoria and Synclinoria

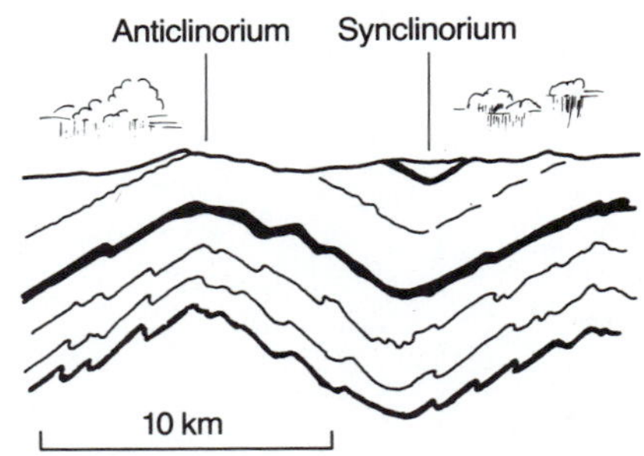

Figure 7.16 Schematic rendering of an anticlinorium and a synclinorium.

Regional fold belts contain very large anticlines and synclines, kilometers across, that are themselves marked by the presence of reasonably systematically spaced smaller anticlines and synclines (Figure 7.16). When describing these regional structures it is sometimes useful to refer to them as **anticlinoria** and **synclinoria**. The flank of an **anticlinorium** (or **synclinorium**) is typically marked by a set of approximately equal-sized **second-order** anticlines and synclines. These in turn may contain sets of **third-order** folds, and so it goes. But keep in mind that folds in nature do not look like systematic wave trains, like the ones we see in physics class or in analytical geometry. Instead, they can change abruptly in size and form depending on rock type and strain environment. To push the wave train analogy a little further—folds are commonly *decoupled* by faulting, and thus cannot be compared simplistically to sine waves.

GEOMETRIC ANALYSIS OF FOLDS

Geometric Properties of Individual Folded Surfaces

Limbs, Hinges, and Inflections

Folded surfaces vary greatly in size and form. Furthermore, they come in every conceivable orientation and configuration. They may occur as single isolated structures, or as part(s) of a system of repeated waveforms.

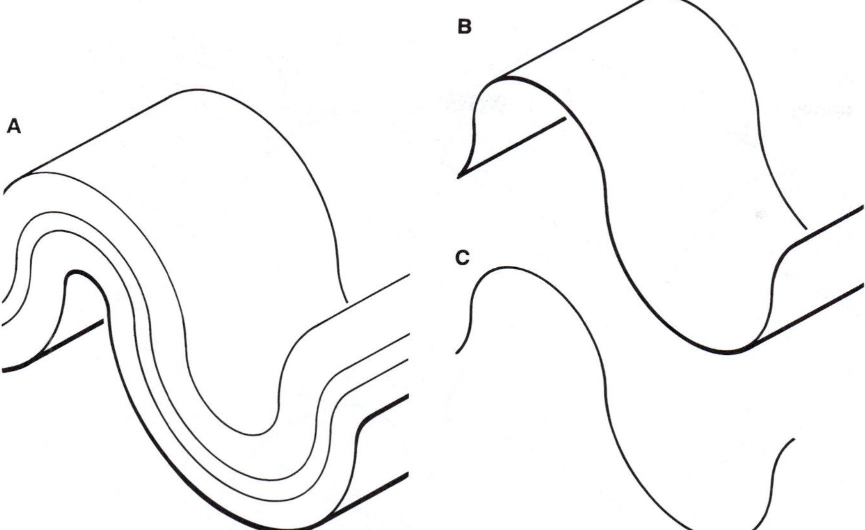

Figure 7.17 Form of an individual folded surface. (*A*) Sequence of folded layers. (*B*) Single folded surface within the folded sequence. (*C*) Normal profile view of the single folded surface.

The infinite variety of folds, and their three-dimensional complexity, compel us to learn a special vocabulary for descriptive analysis and geologic mapping.

We can begin to dissect the anatomy of a fold by extracting a single **folded surface** from the deformed sequence of which it is a part (Figure 7.17*A,B*). By stripping away the rock layers that serve as boundaries to the folded surface, top and bottom, we can focus on the geometric properties of the folded surface itself. We start the process two-dimensionally, by describing the folded surface as seen in **normal profile view**, at a right angle to the axis of folding (Figure 7.17*C*).

Considered two-dimensionally in normal profile view, folded surfaces can be subdivided into **limbs** and **hinges** (Figure 7.18). Limbs are the flanks of folds, and these are joined at the hinge. The hinge of a folded surface is sometimes a single point, called the **hinge point**. More commonly, the hinge is a zone, **the hinge zone**, distinguished by the maximum curvature achieved along the folded surface (Ramsay, 1967). Figure 7.18*A* shows an angular fold marked by planar limbs and an easily identifiable hinge point that separates the limbs. Figure 7.18*B*, on the other hand, pictures a fold marked by planar limbs connected by a hinge zone within which the folded surface displays uniformly high curvature. Strictly speaking, the hinge zone of a folded surface does not possess a unique hinge point. But for descriptive purposes a hinge point can be arbitrarily posted at the midpoint of the hinge zone.

Limbs of folds are commonly curved. Curved limb segments of opposing convexity join at locations known as **inflection points** (Figure 7.18*C*) (Ramsay, 1967). A special case of continuously curved folded surfaces is shown in Figure 7.18*D*, a surface that has been folded into a series of perfectly circular arcs. Folded surfaces of this type lack fold limbs, *per se*. Instead, each discrete circular arc may be thought of as a hinge zone marked by uniform curvature. The hinge point of each fold is taken as the midpoint of each circular arc. The inflection points of the folded surface separate circular arcs of opposing convexity.

The vocabulary used in describing folded surfaces in normal profile view is summarized in Figure 7.18*E*, which shows a fold marked by slightly curved to planar limbs connected by a narrow hinge zone within which the surface displays a pronounced and high degree of curvature. The hinge zone is the zone of maximum curvature, the midpoint of which is the

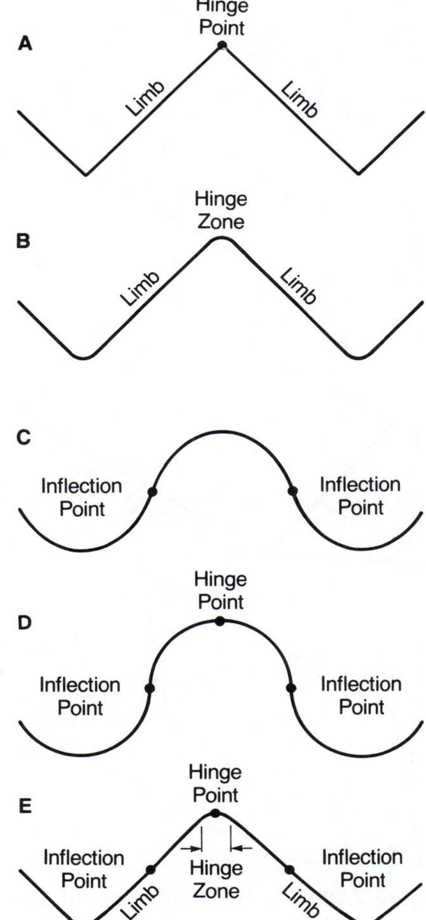

Figure 7.18 Geometric and physical elements of single folded surfaces. See text for discussion.

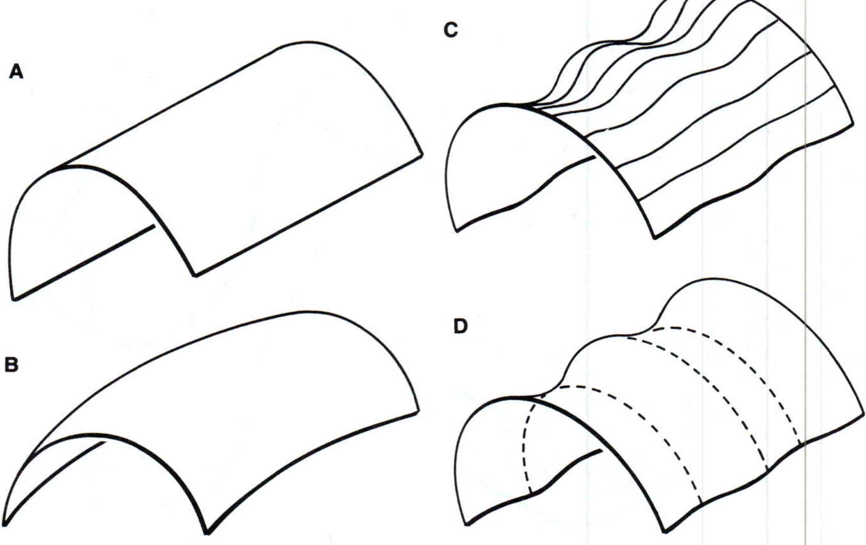

Figure 7.19 (*A*) Fold with straight hinge line. (*B*) Fold with systematically curved hinge line. (*C*) Fold with irregularly curved hinge line. (*D*) Subdivision of fold with irregularly curved hinge line into domains marked by nearly straight hinge lines.

hinge point. Inflection points are taken to be the midpoints of the planar limb segments.

Hinge Lines, Axial Surfaces, and Axial Traces

The three-dimensional geometric characteristics of folded surfaces invite a yet fuller nomenclature (Figure 7.19). The hinge points along a single folded surface, taken together, define a **hinge line**. In some cases the hinge line of a folded surface is perfectly straight (Figure 7.19*A*); more commonly hinge lines are systematically curved (Figure 7.19*B*) or, more rarely, terribly irregular (Figure 7.19*C*) (Turner and Weiss, 1963).

The orientation of a folded surface can in part be specified by the orientation of its hinge line. The orientation of a hinge line is described conventionally in terms of trend and plunge. A single measurement of trend and plunge is adequate for hinge lines that are perfectly straight. But where folded surfaces are marked by hinge lines that are not straight, it is necessary to document the variations in hinge line orientation through a number of measurements of representative segments of the hinge. In practice this is achieved by subdividing the folded surface into domains within which the hinge line approaches a straight line (Figure 7.19*D*).

Knowing the orientation of the hinge line of a fold does not uniquely establish the orientation, or **attitude**, of the fold. Folds having the same hinge line orientation can have strikingly different configurations (Figure 7.20). To describe unambiguously the attitude of a fold, it is necessary to measure yet another structural element, a geometric element known as an **axial surface**. The axial surface of a fold passes through successive hinge lines in a stacking of folded surfaces (Figure 7.21) (Ramsay, 1967; Dennis, 1972). The axial surface of a fold may be planar (Figure 7.21*A*), in which case it is called an **axial plane**. More commonly the axial surface of a fold is either systematically curved (Figure 7.21*B*) or nonsystematically irregular (Figure 7.21*C*), in which case **axial surface** is the appropriate term (Turner and Weiss, 1963).

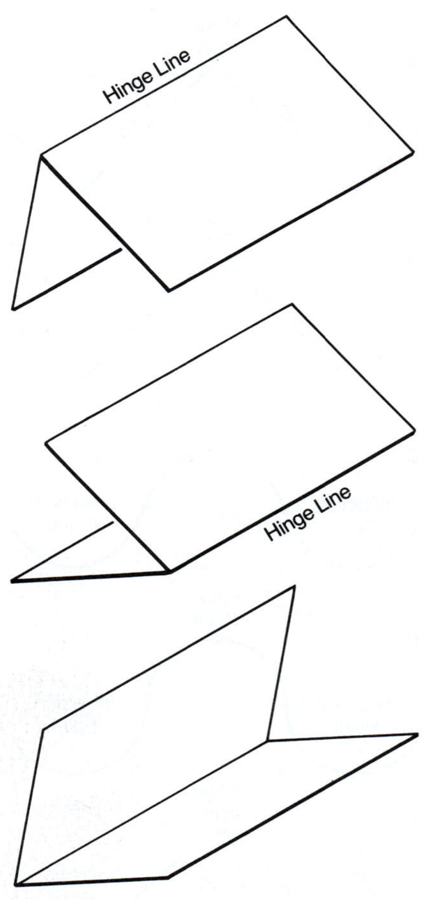

Figure 7.20 The trend and plunge of the hinge line of a fold does not uniquely define the overall orientation of the fold. See for yourself.

In normal profile view, the trace of the axial surface of a fold can be seen to pass through successive hinge points in the stacking of folded surfaces (Figure 7.21*D*). This line is called the **axial trace** of the fold in profile view. In a more general sense, ''axial trace'' refers to the line of

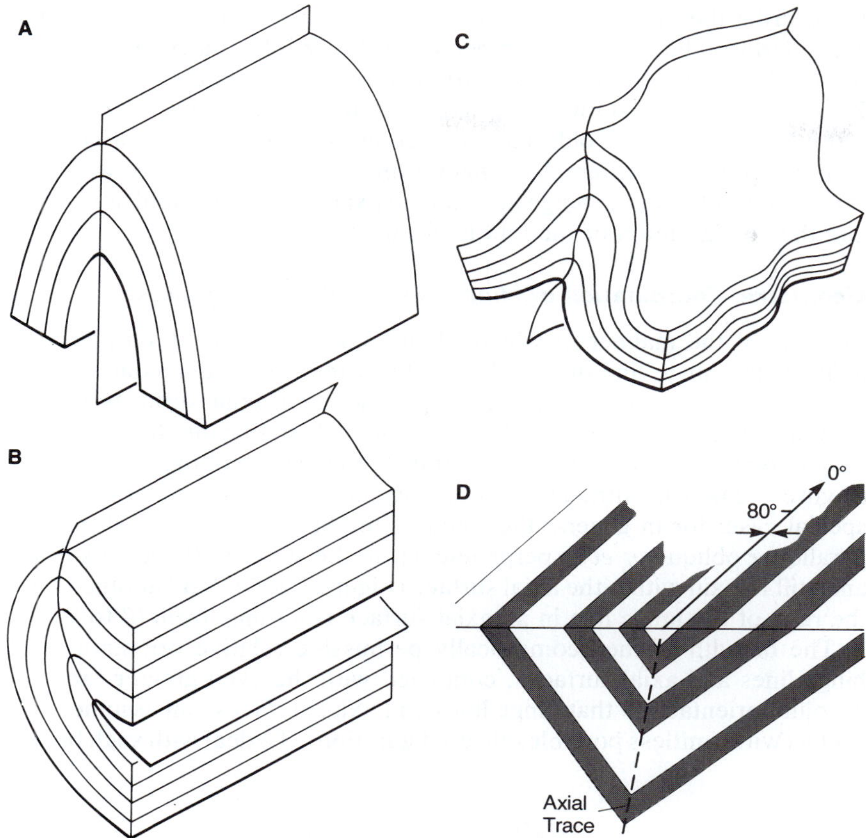

Figure 7.21 (*A*) Fold with a planar axial surface. (*B*) Fold with a systematically curviplanar axial surface. (*C*) Fold with an irregularly curviplanar axial surface. (*D*) Axial trace of a fold, as seen in cross section and in map view.

intersection of the axial surface with *any* other surface, whether it be the ground surface, the cut of an open-pit mine, the steep flank of a mountain, or the faces of a block diagram of folded layers (Figure 7.22).

The orientation of the axial surface of a fold is described in terms of strike and dip. A single strike-and-dip measurement is all that is necessary

Figure 7.22 Two examples of delineating axial traces (shown in white lines) of folds in the real world. (*A*) Axial trace of syncline in Cambrian quartzite in northern Norway near Bjornfeld. Steve Naruk stands like "Rocky" for scale. (*B*) Axial traces of anticlines and synclines in folded Pliocene sediments alongside the San Andreas fault in the Mecca Hills, southeastern California. Note graduate students mapping the folds using plane table and alidade. (Photographs by G. H. Davis.)

Figure 7.23 The strike and dip of the axial surface of a fold does not uniquely define the overall orientation of the fold.

to describe the orientation of the axial plane of a fold. For a nonplanar axial surface, however, a number of strike-and-dip measurements are required to document the full spectrum of orientations of the axial surface. Knowing the orientation of the axial surface does not fix uniquely the attitude of the fold. Folds having a common axial surface orientation can have radically different configurations (Figure 7.23). Only when the orientations of both the hinge line and the axial surface of a fold are known can the configuration of the fold be firmly established.

Geometric Coordination of Hinge Lines and Axial Surfaces

Because of the manner in which "hinge line" and "axial surface" are defined, the hinge line of a fold must lie within the fold's axial surface. This constraint notwithstanding, hinge lines and axial surfaces can be combined in many more ways than we might at first imagine. It is relatively easy to picture a hinge line that is parallel to the strike of the axial surface (Figure 7.24*A*,*B*). Although this arrangement is common, it is a very special case: for in general the trend of the hinge line of a fold may be parallel *or* oblique *or even* perpendicular to the strike of the axial surface and still remain within the axial surface (Figure 7.24*C*). Said another way, the rake of the hinge line in an axial surface can range from 0° to 90°.

The breadth of the geometrically permissive relative orientations of hinge lines and axial surfaces, combined with the even greater range of absolute orientations that hinge lines and axial surfaces can assume, present us with limitless possible fold configurations. To deal with such broad-

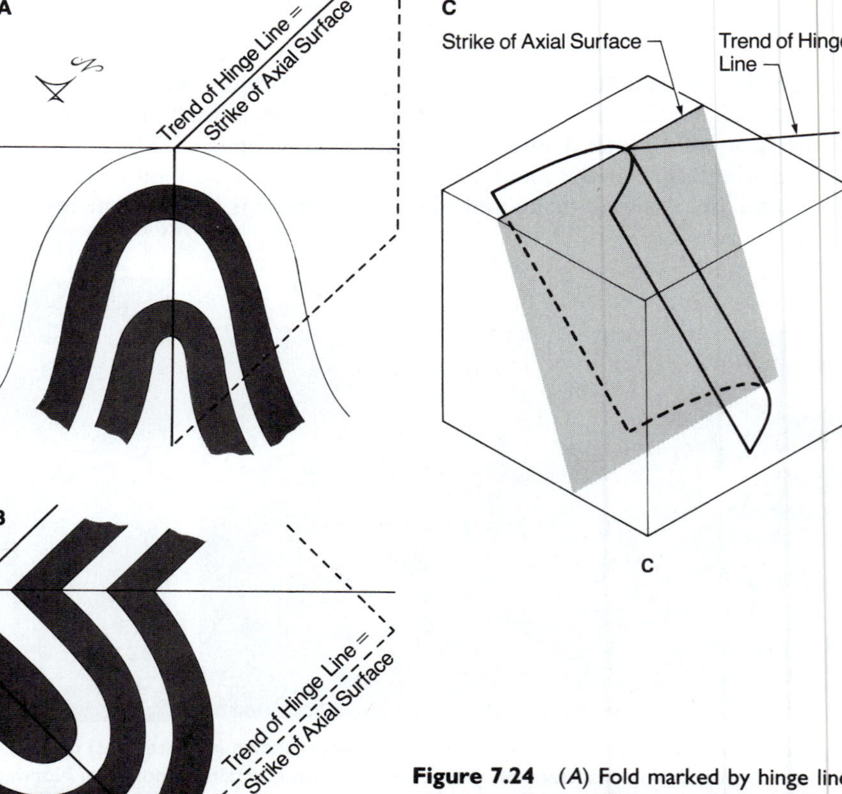

Figure 7.24 (*A*) Fold marked by hinge line whose trend is parallel to the strike of the axial surface. (*B*) Another example. (*C*) Fold marked by hinge line whose trend is discordant to the strike of the axial surface.

ranging geometries, Fleuty (1964) created a useful classification scheme, one that is based on both the relative orientations and the absolute inclinations of hinge lines and axial surfaces. The classification scheme, which he presented in the form of a diagram (Figure 7.25), permits folds to be named according to fold configuration. Along the x-axis of Fleuty's diagram is plotted the **dip** of the axial surface; along the y-axis is plotted the **plunge** of the hinge line. On the basis of cutoffs of 0, 10, 30, 60, 80, and 90°, sixteen categories of folds are identified. A fold whose hinge line plunges 5° N38°E and whose axial surface orientation is N40°E, 82°NW can be described as a **subhorizontal upright fold** (see Figure 7.25). A fold whose hinge line plunges 50° S72°W and whose axial surface orientation is N80°W, 70°SW is classified as a **moderately plunging, steeply inclined fold**. A fold with hinge line and axial surface orientations of 20° N65°E and N30°W, 20°NE, respectively, is called a **gently plunging, gently inclined, reclined fold**. A **reclined** fold is one whose hinge line plunges directly down the dip of the fold's axial surface.

The Difference Between a Hinge Line and a Fold Axis

It is easy to fall into the trap of using the terms **hinge line** and **fold axis** interchangeably. But beware. Strictly speaking, a fold axis is a geometric (thus imaginary) linear structural element that does not possess a fixed location. It is the closest approximation to a straight line that when moved parallel to itself, generates the form of the fold (Figure 7.26) (Donath and Parker, 1964; Ramsay, 1967).

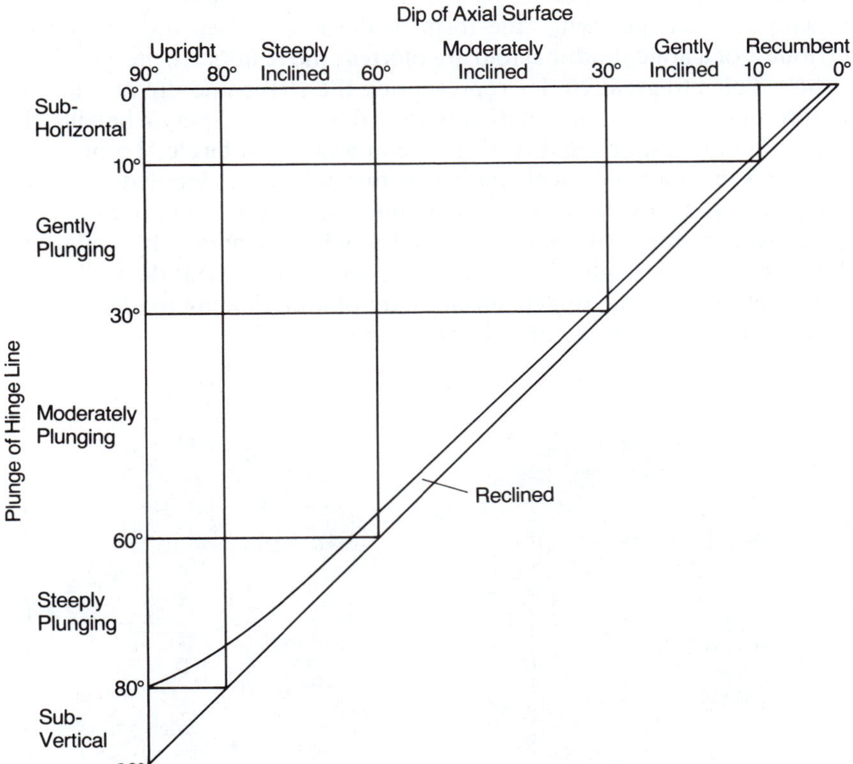

Figure 7.25 Fleuty diagram used for describing folds on the basis of the geometric interrelationship of axial surface and hinge line. [After Fleuty (1964). From *Folding and Fracturing of Rocks* by J. G. Ramsay. Published with permission of both the Geologists Association and McGraw-Hill Book Company, New York, copyright © 1967.]

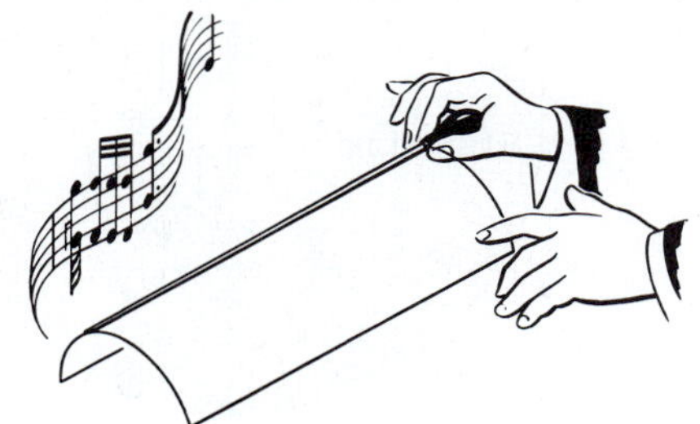

Figure 7.26 Fold axis, an imaginary straight line, which when moved parallel to itself, can define the form of the fold.

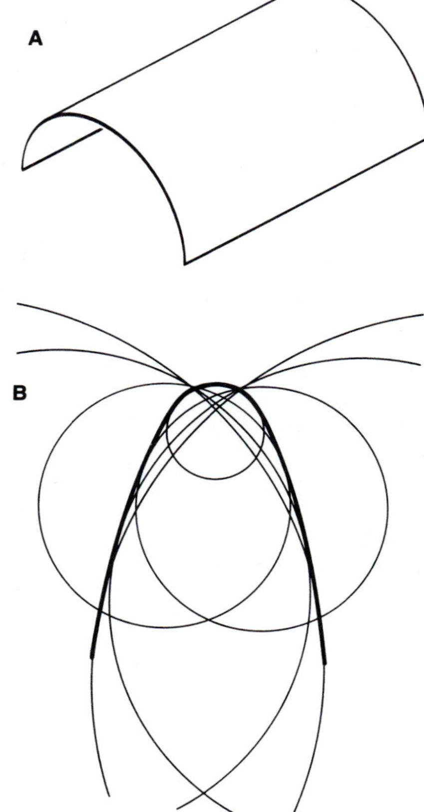

Figure 7.27 Two perfectly acceptable cylindrically folded surfaces. (*A*) One conforms to the outline of a single, perfect cylinder. (*B*) The other (bold dark folded surface) is composed of a colinear arrangement of parts of cylindrical surfaces (parts of circles in this view) of different curvatures.

Folds that possess axes are **cylindrical folds**. The word "cylindrical" folds should bring to mind parts of cans and pipes. A cylindrical fold can have a form best described as *part* of a single, perfect cylinder (Figure 7.27*A*). But more commonly a cylindrical fold has the form of a colinear arrangement of parts of cylinders of different diameters (Figure 7.27*B*). The distinctive geometric characteristic of cylindrical folds is that every part of the folded surface is oriented such that it contains a line whose orientation is identical to that of the hinge line. The orientation of this line is the orientation of the fold axis.

A close geometric coordination exists between the form of a perfectly cylindrical fold and the orientation of its fold-generating axis. The coordination is best pictured stereographically. Poles to great circles representing the strike-and-dip orientations of cylindrically folded bedding lie exactly on a common great circle, perpendicular to the trend and plunge of the hinge line (Figure 7.28*A*). This means that "every part of the folded surface is oriented such that it contains a line whose orientation is identical to that of the hinge line" of the fold.

Truly cylindrical folds are rare in nature, but many folds so closely approximate purely cylindrical forms that they are considered to be cylindrical, and thus they are considered to possess axes. The term "cylindrical," then, can be broadened to include **near-cylindrical folds**. Poles to great circles representing the strike-and-dip orientations of nearly cylindrically folded bedding do not lie exactly on a common great circle (Figure 7.28*B*).

Noncylindrical folds do not possess fold axes. The limbs and hinge zones are so irregular in orientation, and the hinge lines so crooked and/ or curved, that there does not exist a single straight line that when moved parallel to itself, generates the form of the fold. When the geometrical attributes of a noncylindrical fold are plotted, the results are messy (Figure 7.28*C*). Poles to great circles representing the strike-and-dip orientations of different parts of noncylindrically folded bedding display a bewildering array of points that cannot be fit to a common great circle. To penetrate the chaos of noncylindrical folding in the course of detailed structural analysis, it is necessary to subdivide noncylindrical folds into domains of cylindrical or near-cylindrical folds, each of which is marked by a relatively short, relatively straight hinge line. When this is done, and the strike-and-dip orientations of the folded surfaces are plotted domain by domain, the stereographic patterns clean up immeasurably.

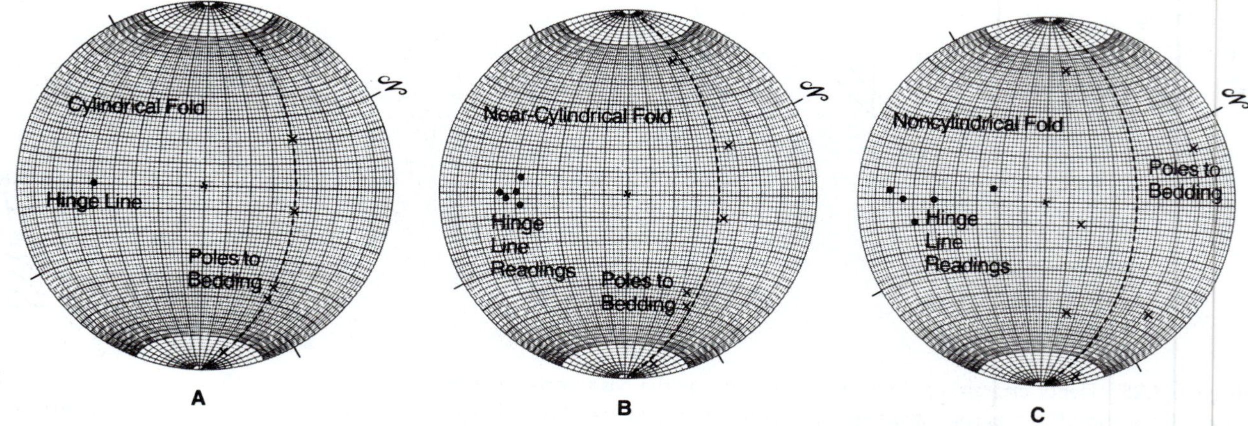

Figure 7.28 Stereographic geometry of (*A*) a perfectly cylindrical fold, (*B*) a nearly cylindrical fold, and (*C*) a noncylindrical fold.

Stereographic Determination of Fold Orientations

It is usually impossible to measure directly in an accurate way the axis and axial surface orientations of a large fold that spills beyond the expanse of a single outcrop. Instead, the trend and plunge of the fold axis and the strike and dip of the axial surface must be calculated stereographically on the basis of representative strike-and-dip measurements of the folded surface. A stereographic plot of the axis and axial surface orientations measured for a single fold at a single outcrop displays the geometric coordination of these elements. An axial surface, by definition, passes through the hinge lines of successive folded surfaces within a given fold. Expressed stereographically, the point representing the trend and plunge of the hinge line (or fold axis) lies on the great circle that describes the orientation of the axial surface (Figure 7.29).

In the simplest case, the trend and plunge of the axis of a fold can be determined stereographically on the basis of merely two strike-and-dip measurements, one for each limb of the folded surface. The orientations of the two limbs are plotted stereographically as great circles (Figure 7.30). The intersection of the great circles, labeled **β**, represents a close approximation to the trend and plunge of the hinge line of the folded surface. The hinge line orientation, in turn, is taken to be a close approximation to the fold axis orientation. This specific stereographic construction is called a **beta (β) diagram**, where β refers to the trend and plunge of a fold axis deduced stereographically in the manner described.

Another way to calculate stereographically the orientation of a fold is through the construction of a **pi (π) diagram**. This requires plotting the limbs of the folded surface as poles, not as great circles (Figure 7.31*A*). Once plotted, the poles to each limb are fitted to a common great circle, known as a **π circle** (Figure 7.31*B*). The special geometric property of a π circle is that it represents the strike and dip of a plane that is perfectly perpendicular to the hinge line of the fold. The pole to this great circle, known as the **π-axis** (Figure 7.31*B,C*), expresses stereographically the orientation of the fold axis; "π-axis" refers to the trend and plunge of a fold axis as deduced stereographically in this manner.

The key to comfort in understanding π-diagram construction is visualizing that poles to a cylindrically folded surface indeed lie geometrically in a plane oriented at a right angle to the hinge line of the fold. Figure 7.32, an extraordinary view along the mined-out hinge of an anticline, can help

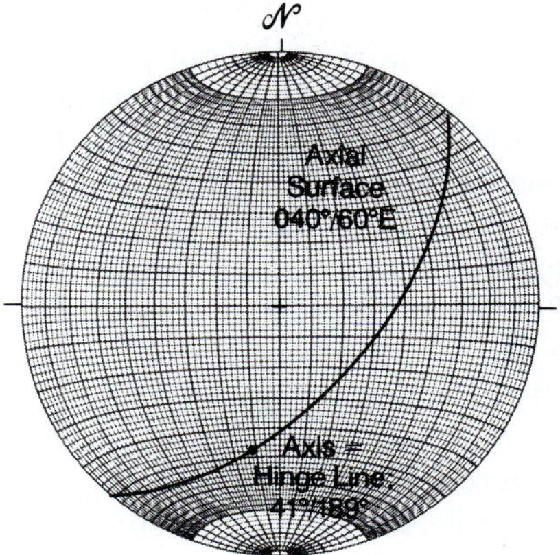

Figure 7.29 Geometric coordination of the axis and axial surface of a fold, as portrayed stereographically.

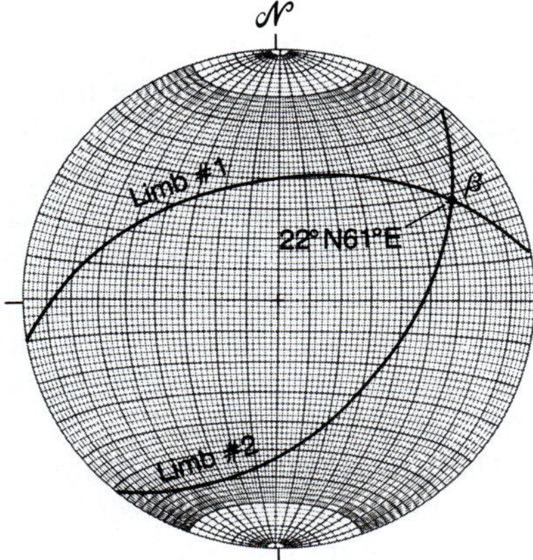

Figure 7.30 Simple β diagram. The orientation of the axis of the fold (β) is the line of intersection of the fold limbs.

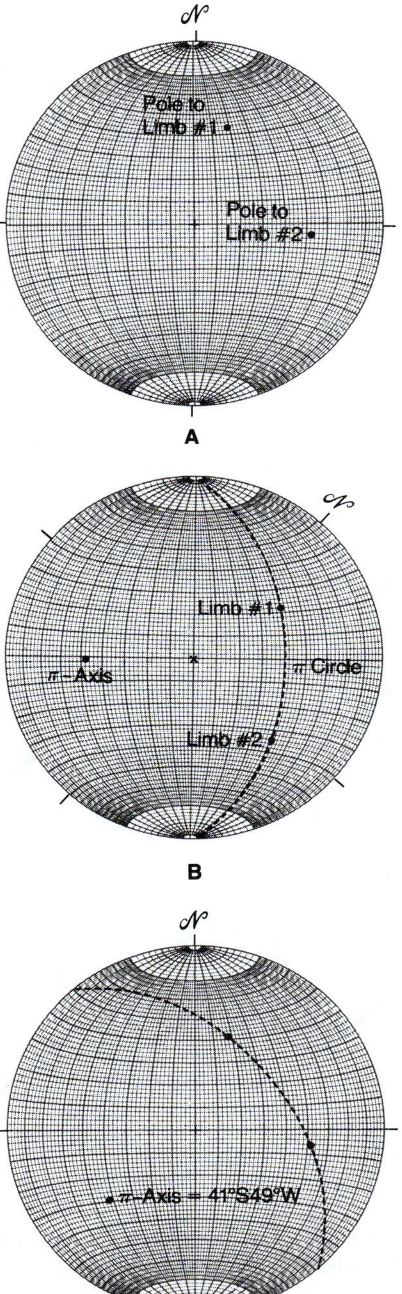

Figure 7.31 Steps in the construction of a π diagram. (*A*) Stereographic portrayal of poles to limbs. (*B*) The fitting of the poles to a common great circle (the π circle). Identification of the pole to the π circle (the π axis). (*C*) Trend and plunge of the π axis.

Figure 7.32 Visual image of the geometry of a π diagram. (Photograph by C. D. Walcott. Courtesy of United States Geological Survey.)

us picture this relationship. Each of the timbers that supports the mine roof is oriented as a pole to the bedding it supports. The orientations of the timbers, taken together, define a plane at right angles to the hinge line. The miner with hands on hips in the deep recesses of the tunnel is smiling because he recognizes how closely his timber support system captures the inherent stereographic geometry of π-diagrams.

Although the axial surfaces of small folds can normally be measured directly in outcrop, stereographic procedures are required to evaluate the axial surface orientations of large folds. The simplifying premise in the stereographic calculation is that the **bisecting surface** of a fold is a close approximation to the axial surface. For a given folded surface, the bisecting surface passes through the hinge line and splits the angle (the **interlimb angle**) between the limbs (Figure 7.33*A*). When the bisecting surface of a single folded surface is compared to the axial surface of the fold as a whole (Figure 7.33*B*), minor differences in orientation may sometimes be evident. But generally the differences are so slight that the strike and dip of the bisecting surface can be taken as the strike and dip of the axial surface.

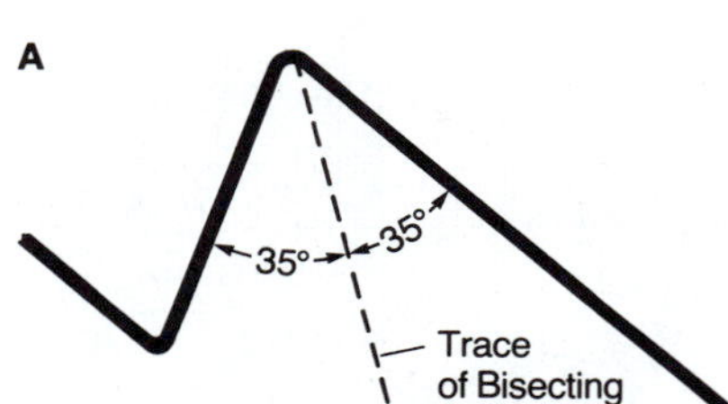

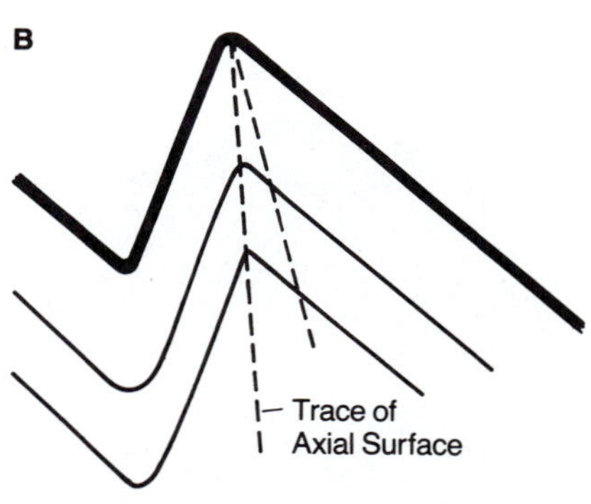

Figure 7.33 (*A*) The bisecting surface of a fold splits the angle between the limbs. (*B*) The axial surface of a fold passes through the hinge points of successive fold surfaces. The bisecting surface of a given fold is not necessarily the same as the axial surface. (From *Structural Analysis of Metamorphic Tectonites* by F. J. Turner and L. E. Weiss. Published with permission of McGraw-Hill Book Company, New York, copyright © 1963.)

The stereographic procedure in computing the orientation of the bisecting surface of a folded surface is reasonably straightforward. Although shortcuts can be taken, the full flavor of the method emerges by combining a β-diagram and a π-diagram on a common projection. First a simple β-diagram is constructed by plotting the attitudes of the fold limbs as great circles, then identifying the intersection of the great circles as β (Figure 7.34*A*). Next a π-diagram is added, by plotting the two poles to the two fold limbs and fitting these poles to a common great circle (the π circle) (Figure 7.34*B*). The π-axis of this great circle is coincident with β. The bisecting surface of the fold is the great circle that passes through the hinge line (through β) and perfectly bisects the angle between the two poles as measured along the π circle (Figure 7.34*C*). In the example we are considering, the strike and dip of the bisecting surface proves to be N60°W, 66°SW (Figure 7.34*D*).

There is always some ambiguity in stereographically computing the orientation of a bisecting surface. There are actually two different points on the π circle that serve as bisectors to the limbs (Figure 7.34*C*): one bisects the acute angle between the bedding traces; the other bisects the obtuse angle. One of these bisectors, but not both, must be used along with β as a reference point for constructing the great circle representing

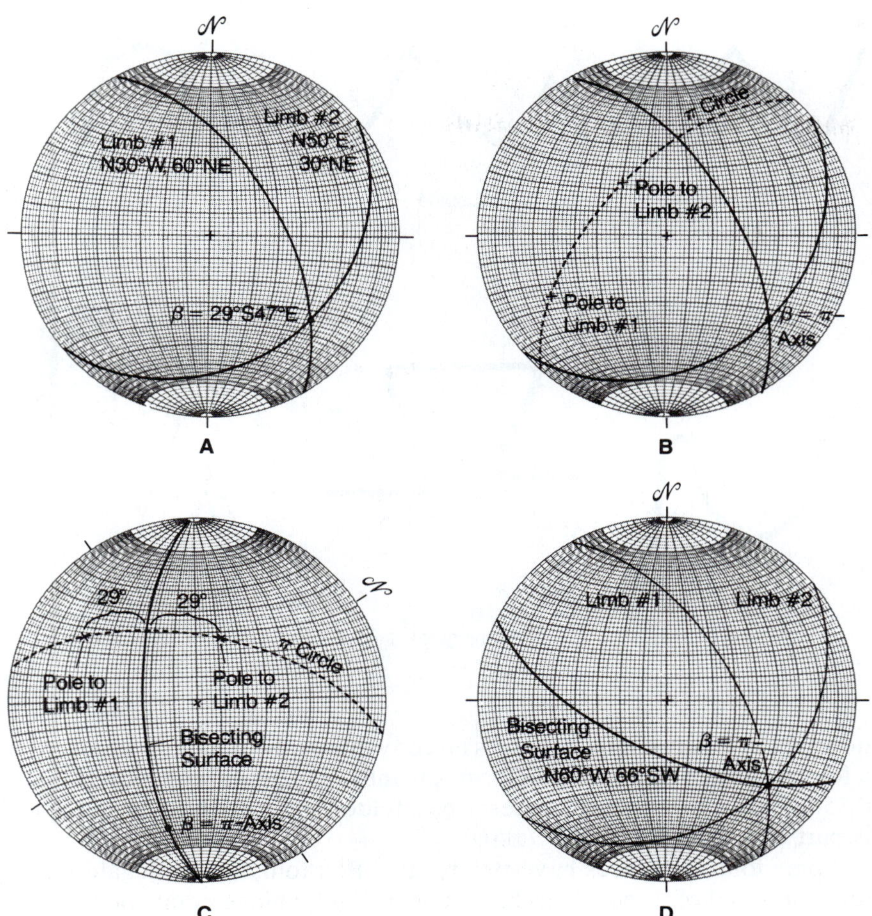

Figure 7.34 Steps in stereographically determining the orientation of the bisecting surface of a fold. (*A*) Plot fold limbs as great circles and identify β. (*B*) Plot poles to limbs. (*C*) Measure the angle between the poles to the limbs. Fit a great circle to the bisector of this angle and to β. (*D*) Stereographic configuration of the bisecting surface, in proper orientation.

the orientation of the bisecting surface. To select the appropriate bisector, it is necessary to keep in mind the fold form whose orientation is being sought. If the fold is upright, the proper bisector is one that yields a relatively steeply inclined bisecting surface. If the fold is overturned or recumbent, the proper bisector is one that yields a relatively low-dipping bisecting surface.

Describing the Shape and Size of a Folded Surface

Common Fold Shape

As part of the overall description of a folded surface, it is useful to convey a sense of the shape of the fold, including its tightness. Fold shape is described in normal profile view. Normal profile views of folded surfaces are afforded by appropriately oriented outcrop exposures, photographs, geologic cross sections, and rock slabs or thin sections.

All the conventional terms for describing the profile shape of a folded surface attempt to convey a picture of the form and the configuration of limbs and hinge (Figure 7.35). A **chevron fold**, for example, is marked by planar limbs that meet at a discrete hinge point or at a very restricted subangular hinge zone (Figure 7.35*A*). A **cuspate fold** exhibits curved limbs that are opposite in sense of curvature to those of most ordinary folds (Figure 7.35*B*). An upright, cuspate anticline displays limbs that are concave upward; an upright, cuspate syncline has limbs that are concave

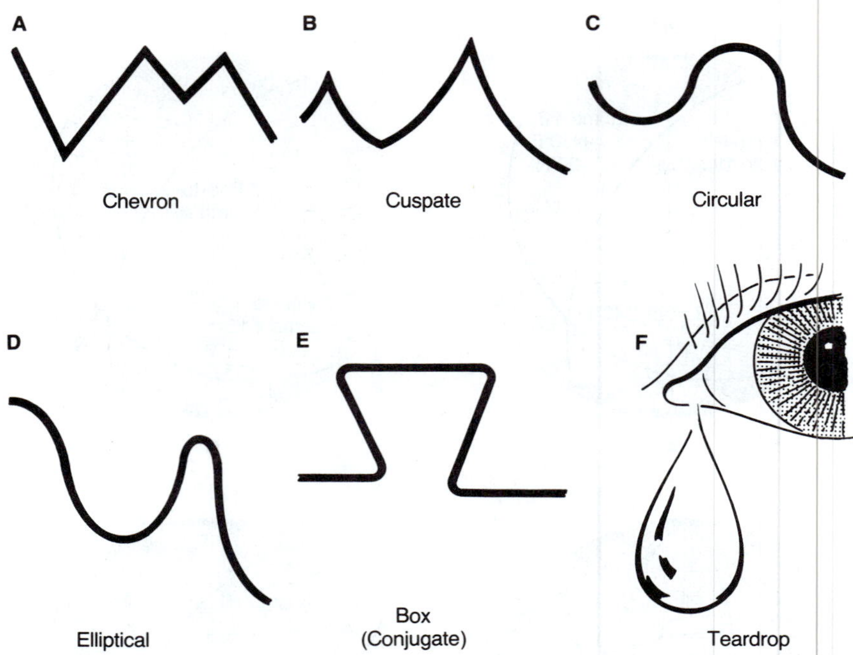

Figure 7.35 Some common fold shapes.

downward. Oddly enough, there is no conventionally used term to describe a folded surface whose profile form is wholly part of a circular arc (Figure 7.35*C*), nor is there a term to describe a folded surface whose profile form is part of an ellipse (Figure 7.35*D*).

Some folded surfaces have two hinges. Box folds (or conjugate folds) are composed of three planar limbs connected by hinge points or narrow, restricted subangular hinge zones (Figure 7.35*E*). **Teardrop folds** are continuously curved folded surfaces shaped, of course, like teardrops or mushrooms. They are involuted and curve back on themselves (Figure 7.35*F*).

Fold shapes in nature are intriguing, no matter what the scale of view (Figure 7.36).

Figure 7.36 (*A*) Elliptical fold shapes exposed at a grand scale in the Canadian Rockies. (Courtesy of the Canadian Geological Survey.) (*B*) Circular, elliptical, and boxy folds exposed in a small outcrop in the Ruby Mountains, Nevada. (Photograph by G. H. Davis.)

Fold Tightness

Fold tightness is described in terms of **interlimb angle** (Ramsay, 1967), the internal angle between the limbs of the folded surface. Although the interlimb angle of a folded surface can be measured with a protractor on the surface of a profile exposure of a small fold, or from a profile-view photograph of a large fold, profile views of folds are the exception, not the rule. Consequently it is usually necessary to calculate interlimb angles stereographically. This is achieved by taking the strike and dip of the folded surface at each inflection point, plotting the orientations stereographically as poles, fitting the poles to a common great circle, and measuring the angle between the poles along the common great circle (Figure 7.37). To know whether the acute or obtuse angle between the poles is the appropriate interlimb angle, it is necessary to keep clearly in mind the general form of the fold. The interlimb angle of a very tight fold is acute. The interlimb angle of a very open fold is obtuse.

The measured value of interlimb angle provides a basis for choosing an adjective that describes fold tightness. Figure 7.38 shows a classification scheme adapted but slightly modified from the nomenclature proposed by Fleuty (1964). **Gentle folds** are marked by interlimb angles ranging from

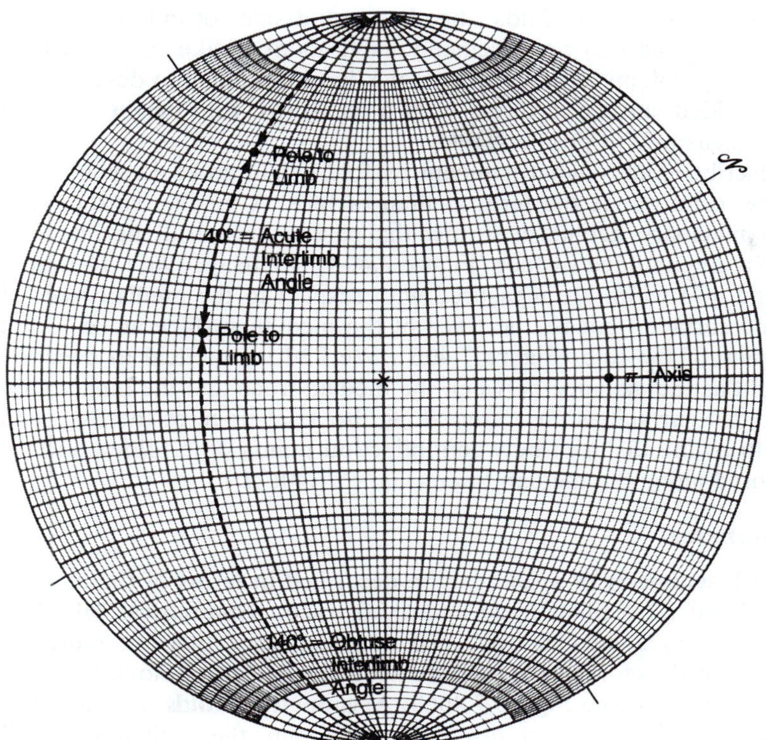

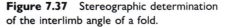

Figure 7.37 Stereographic determination of the interlimb angle of a fold.

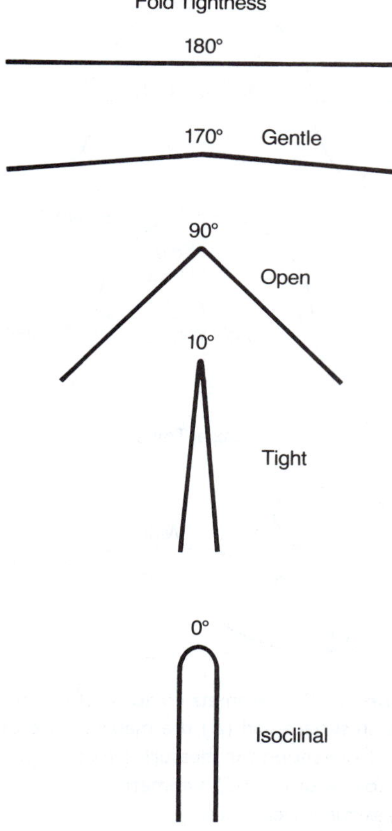

Figure 7.38 Classification of folds according to tightness, based on the size of the interlimb angle. [Modified from Fleuty (1964). Published with permission of the Geologists Association.]

170 to 180° (Figure 7.39). **Open folds** have interlimb angles ranging from 90 to 170°. Folds are considered to be **tight** if they display interlimb angles in the range of 10 to 90°. And **isoclinal folds** are marked by interlimb angles in the range of 0 to 10°. The cutoffs for isoclinal, tight, open, and gentle are easy to remember: 10, 90, and 170°.

Fold Size

Fold size is surprisingly difficult to describe. The standard measures of **wavelength** and **amplitude** can seldom be employed because so many folds occur as solitary, isolated, "decoupled" structures, and not as obvious parts of continuous, repeated, **sinusoidal** waveforms. Many folds encountered in the field are not linked structurally to other folds: they are **rootless,** cut off on either side by faults and/or shear zones. Some folds that appear to be rootless may in fact be continuous with other folds, but the connection cannot be demonstrated because of the fortunes of erosion and/or the quality of exposure. Even if a fold can be shown to occur within an

Figure 7.39 Gently folded Pennsylvanian sedimentary rocks exposed in a highway cut just east of Morgantown, West Virginia. Full-blown anticlines and synclines of the Valley and Ridge province of the Appalachians begin just a few kilometers to the east. (Photograph by G. H. Davis.)

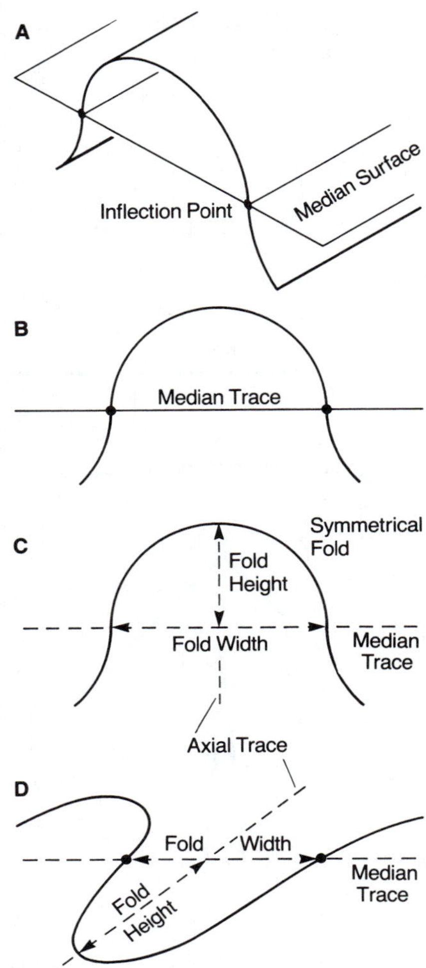

Figure 7.40 Geometric nature of (*A*) the median surface and (*B*) the median trace of a fold. Convention for measuring fold height and fold width of (*C*) symmetrical and (*D*) asymmetrical folds.

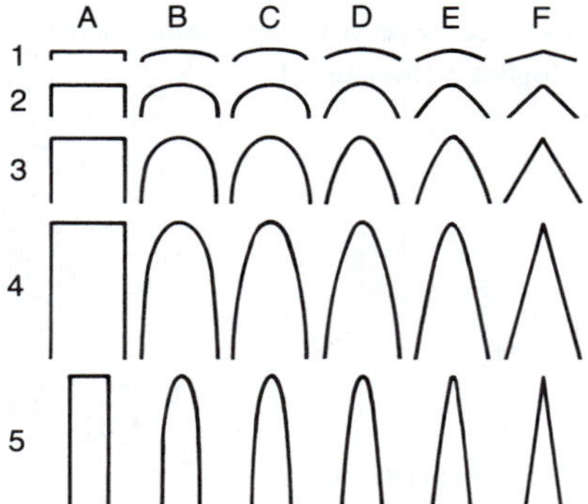

Figure 7.41 Visual classification of the shape(s) of individual folded surfaces. [From Hudleston (1973), *Tectonophysics*, v. 16. Published with permission of Elsevier Scientific Publishing Company, Amsterdam.]

interconnected system of folds, the shapes and sizes of individual folds within a wave train may vary tremendously, not at all like an ideal wave. Given these problems and limitations, we find it practical to describe the size of a folded surface in terms of two measures: **fold height** and **fold width**, as measured in profile view.

To describe exactly what is meant by fold height and fold width, it is necessary to introduce the term **median surface**. A median surface of a fold is an imaginary, geometric surface that passes through all the inflection points of a given folded surface (Figure 7.40*A*) (Ramsay, 1967). Both fold height and fold width are measured with respect to the **median trace** of the folded surface, that is, the trace of the median surface as seen in a profile view of the fold (Figure 7.40*B*). We find it practical to describe **fold height** as the distance between the median trace and the hinge point of the folded surface as measured along the axial trace of the fold, and **fold width** as the distance between inflection points on a folded surface as measured along the median trace (Figure 7.40*C,D*).

Fold Symmetry

The overall **symmetry** of a fold can be described in terms of the angular relationship of its median trace and axial trace. **Symmetrical folds** are characterized by a median trace and an axial trace that are mutually perpendicular; thus fold height and fold width are measured along mutually perpendicular lines (see Figure 7.40*C*). **Asymmetrical folds**, on the other hand, are marked by limbs of different lengths; thus the median trace and the axial trace of an asymmetric fold intersect at some oblique angle (see Figure 7.40*D*).

Overall Form

The overall form of a folded surface owes its character to a combination of factors, including shape, tightness, symmetry, and ratio of height to width. Thus, strings of adjectives are normally required to describe adequately the profile form of a given folded surface: for example, "the fold is best described as a tight symmetrical cuspate anticline." If we add adjectives describing the geometric configuration of the folded surface, the string of adjectives becomes even longer: "the fold is best described as a gently plunging, moderately inclined, tight attenuated, symmetrical cuspate anticline."

With the goal of trying to convey more detail about fold form in fewer words, Hudleston (1973) devised a **visual classification scheme** that aids in categorizing the forms of folded surfaces. Using the Hudleston classification, the shape of a folded surface, from hinge point to inflection point, is compared to 30 idealized fold forms arranged systematically by number (1 to 5) and letter (*A* to *F*) (Figure 7.41). In using Hudleston's scheme it is simplest to reproduce the 30 basic fold forms on a plastic or Mylar template, then compare the forms to the folded surface in question by peering through the template toward the outcrop, photograph, or geologic cross section portraying the fold. The payoff in using this technique comes in discovering that certain rock types and/or structural domains are characterized by certain specific fold shapes.

Classifying Folds on the Basis of Changes in Layer Thickness

Thickness Changes, Reflections of Distortion

Some **folded layers** maintain uniform thickness across the full profile view of the fold. Other layers show striking, systematic variations in layer

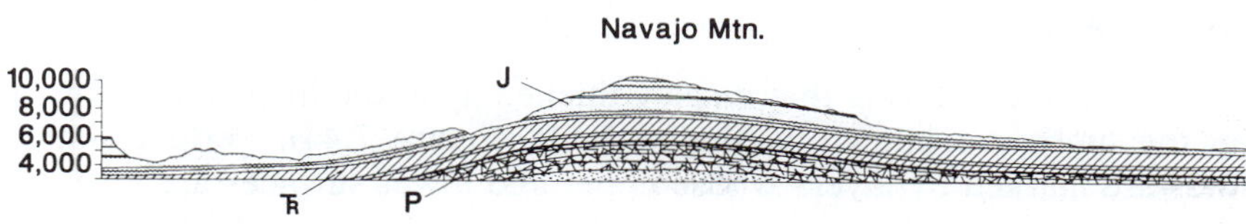

Figure 7.42 At Navajo Mountain in southern Utah, Paleozoic and Mesozoic sedimentary rocks have been domed by a middle Tertiary intrusion. The fold is represented as perfectly concentric in this geologic cross section by Baker (1936). (Courtesy of United States Geological Survey.)

thickness. Whether the thickness of a rock layer is modified during folding depends on the internal stresses it is forced to bear and the rock's strength to resist. Remarkably, the degree of distortion from one layer to the next is nearly always somehow perfectly regulated to assure a perfectly compatible fit among layers within the folded sequence. Sometimes noticeable gaps and overlaps testify to the difficulty in achieving a perfect fit. The achieving of **strain compatibility** from folded layer to folded layer is yet more of the magic of strain.

Concentric Folds

Individual folded layers that are marked by uniform thickness are known as **concentric folds** (Figure 7.42) (Van Hise, 1896). The profile forms commonly are circular or elliptical. Surfaces that separate individual folded layers in an ideal concentric fold are perfectly parallel, like the rails of a curved train track at a bend in the line. Because of this distinctive geometric characteristic, concentric folds are also known as **parallel folds**.

An unexpected geometric peculiarity arises from parallel folding: the profile form of folded layers must continuously change upward and downward within the folded sequence, until the folds gradually disappear altogether. For example, an upright anticline becomes progressively tighter downward within a concentrically folded sequence, ultimately transforming into a narrow, pinched, cuspate anticline before completely dying out (Figure 7.43). Upward, the concentric anticline progressively flattens into a very gentle arc before vanishing. Synclines behave in the opposite manner. Upright synclines pinch out upward in very tight cuspate folds. Downward they gradually become gentle dishlike folds before subtly merging with deeper, unfolded layers.

The geometric idiosyncrasies of concentric folding can be more fully appreciated by graphically constructing a structure profile view of an upright, circular concentric anticline. Concentric circular arcs representing folded surfaces are drawn with a drafting compass (as in Figure 7.43). The circular arcs serve to define the boundaries of individual folded layers, which maintain uniform thickness as *measured perpendicular to layering*. As the arcs are drawn, one by one, it becomes more and more difficult to propagate the form of the anticline to depth. A space problem develops, and it becomes impossible to fit a decent circular arc into the available space. The space problem is satisfied by replacing the folded layers above with unfolded flat-lying layers below. In essence, the folded layers are **detached** from their underlying foundation. Nature achieves detachment through formation of a surface of ''unsticking,'' a **decollement zone** of layer-parallel slippage and rock flowage. The last remaining vestige of the concentric anticline that can be constructed is a tiny cuspate fold (Figure 7.43). Between the cuspate anticline and the flat-lying strata below, a small amount of open space is created. In natural systems this open space is immediately filled by soft incompetent rock, capable of distortional flow during folding.

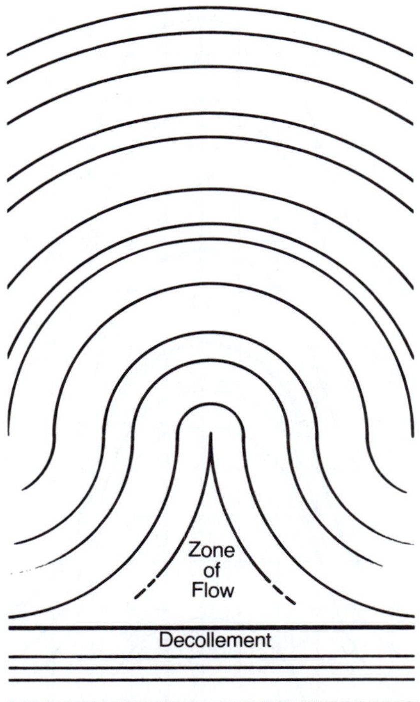

Figure 7.43 Geometric properties of an ideally concentric anticline.

Figure 7.44 Similar folding in Precambrian banded gneisses exposed in Gjeroy Island, Nordland, Norway. [Reprinted with permission from the *Journal of Structural Geology*, v. 14, Lisle, R. J., Strain estimation from flattened buckle folds (1992). Elsevier Science Ltd., Pergamon Imprints, Oxford, England.]

Similar Folds

Individual folded layers that display thickening in the hinge and thinning on the limbs are known as **similar folds** (Figure 7.44). Perfect similar folds are marked by layers whose upper and lower surfaces are virtually identical in shape (Van Hise, 1896). Because this is so, the form of an ideally similar fold can be propagated upward and downward for any distance without change. The secret to the geometry of an ideal similar fold is that layer thickness measured parallel to the axial trace of the fold remains constant.

The geometric intrigue of similar folding can be appreciated through another graphical construction (Figure 7.45). The first step is to draw the profile form of a single folded surface in the middle of a long sheet of paper. Any fold form will do. The next step is to construct folded layers above and below this folded surface, carefully building each folded layer by measuring and maintaining a constant thickness *parallel to the axial trace*. When this construction is carried out with care and precision, there is *never* a departure of individual folded surfaces from the starting profile form.

Full Range of Shape of Folded Layers

Concentric folds and similar folds are simply two special cases within a broad range of possible shapes of folded layers. Ramsay (1967, pp. 359–372) was able to demonstrate that fundamental classes of folded layers can be distinguished on the basis of **relative thickness** of the folded layer in the hinge versus the limbs. He showed that the **relative curvature** of the upper and lower bounding surfaces of an individual folded layer is also a sensitive index to systematic variations in layer thickness.

Three main classes of folds were distinguished by Ramsay on the basis of the relative curvature (i.e., the upper and lower "arcs," of a folded layer) of the upper and lower bounding surfaces (Figure 7.46). **Class 1 folds** are marked by a curvature of the inner arc that is greater than that of the outer arc. **Class 2 folds** are ideal similar folds, distinguished by identical curvatures of the inner and outer arcs. **Class 3 folds** are marked by curvature of the outer arc that is greater than that of the inner arc.

Ramsay further subdivided class 1 folds into three types on the basis of thickness variations (Figure 7.46). **Class 1A folds** are marked by a layer thickness in the hinge that is less than layer thickness on the limbs. **Class 1B folds** are ideal concentric folds, distinguished by uniform layer thickness across the whole fold profile. **Class 1C** are intermediate between ideal concentric folds (class 1B) and ideal similar folds (class 2). They show a modest thickening in the hinge, and a modest thinning on the limbs.

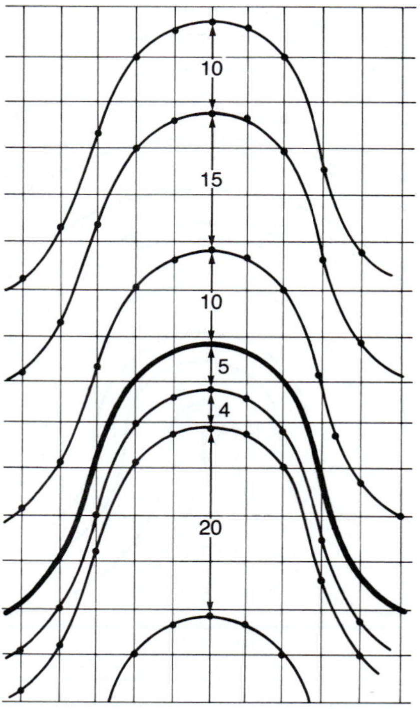

Figure 7.45 Geometric construction of an ideally similar anticline.

Distinguishing Fundamental Fold Classes

The assignment of a given folded layer to one of the five fundamental fold classes can be carried out in a qualitative way on the basis of an eyeball estimate of relative curvature and relative thickness. But the power of Ramsay's approach is best appreciated through the actual measurement of relative curvature and relative thickness. Normal profile views of the folded layers are used as the database for carrying out the necessary constructions and measurements.

By convention, the fold profile under study is first rotated into the orientation of a perfectly upright antiform (Figure 7.46). Next, **dip isogons** connecting points of equal inclination on the outer and inner bounding surfaces of the folded layer are constructed graphically. Once constructed, the dip isogon pattern sensitively reveals differences in outer arc and inner arc curvature, thus providing a basis for assigning the folded layer to class 1, 2, or 3 (Figure 7.46).

Class 1 folds are distinguished by dip isogons that converge downward, signifying that the curvature of the outer arc is less than that of the inner arc. Dip isogons drawn for class 2 folds are strictly parallel, revealing that curvature of the outer arc matches exactly the curvature of the inner arc of the fold. Class 3 folds are marked by dip isogons that diverge downward, because outer arc curvature exceeds inner arc curvature (Figure 7.43).

Dip isogon patterns are especially revealing when they are drawn for a series of folded layers of different shapes (Figure 7.47). The divergence, convergence, and parallelism of dip isogons, as they cut through a folded sequence of layers, draws attention to the variety of classes of folded layers that can be represented in a single structure. Dip isogon "maps" of folds call attention to layer shape distortion as a function of rock type. A clever approach to using fold layer shape as a guide to strain is presented in Part III-M.

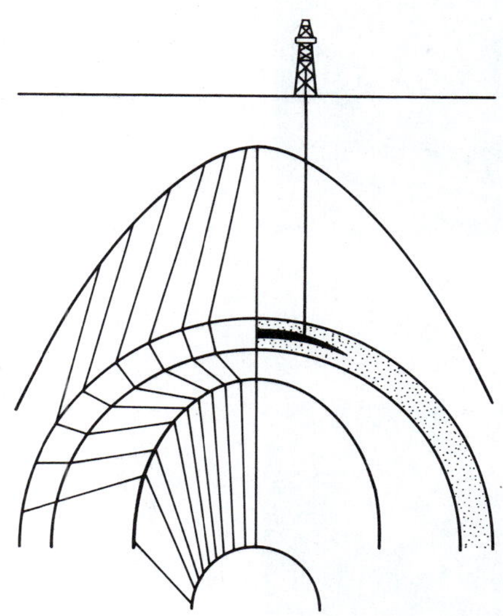

Figure 7.47 Schematic diagram showing how changes in inclination of dip isogons reflect changes in shape(s) of folded layers.

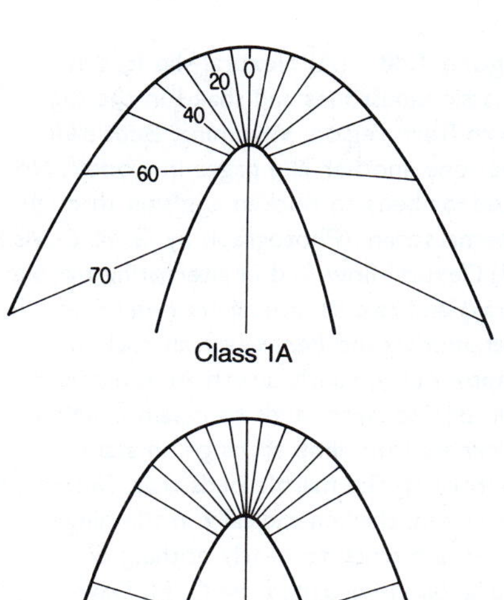

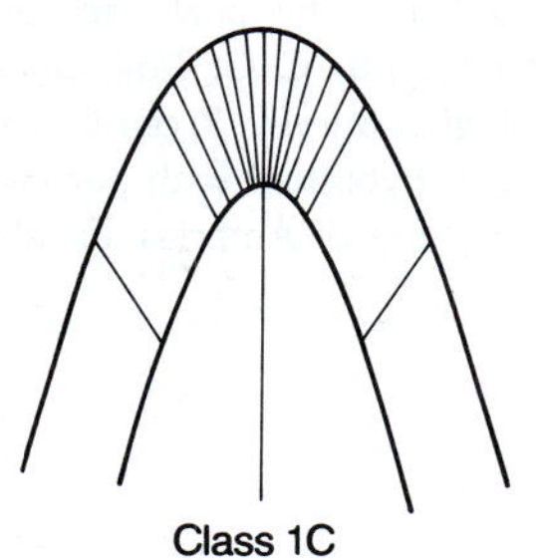

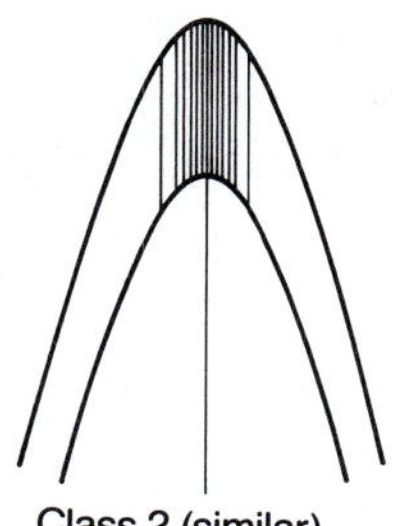

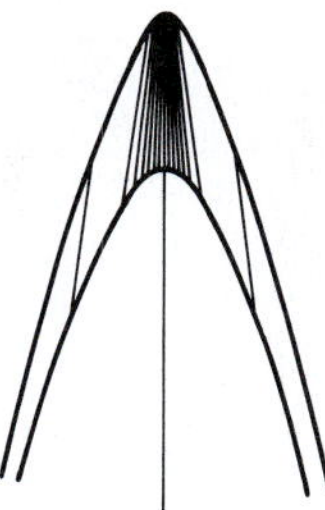

Figure 7.46 The fundamental classes of shapes of folded layers. See text for explanation. (From *Folding and Fracturing of Rocks* by J. G. Ramsay. Published with permission of McGraw-Hill Book Company, New York, copyright © 1967.)

KINEMATIC ANALYSIS OF FOLDING

Flexural Folding Versus Passive Folding

Donath and Parker (1964) recognized two fundamental mechanisms of folding: **flexural folding and passive folding**. Flexural folding takes place when the mechanical influence of layering in a rock is very strong. The layers *actively* participate in the folding by bending and flexing. Flexural folding can take place by flexural slip, by flexural flow, or by a combination of these. Depending on the mechanical properties of the layered sequence, one or both of these mechanisms are initiated when layer-parallel resistance to shortening is overcome and the layers of rock begin to actively buckle. **Flexural-slip folding** accommodates the buckling by **layer-parallel slip** along contacts between layers (Figure 7.48*A*). The layers slip like pages in a slick magazine when the magazine is rolled up. **Flexural-flow folding** accommodates the buckling by **layer-parallel flow** or shear within

Figure 7.48 (*A*) Flexural-slip fold in Triassic sandstones and shales in the Big Horn Basin region, Wyoming. Beds slide over one another like pages in a book. No need for beds to thicken and thin through internal strain. (Photograph by G. H. Davis.) (*B*) Flexural-flow fold in alternating marble (gray) and calc-silicate layers (white) in metamorphosed Pennsylvanian rock in Happy Valley, southeastern Arizona. Each calc-silicate layer tends to retain uniform thickness from limb to hinge, in stark contrast to the marble beds that, almost like ice cream, thicken radically in the hinge zones and pinch to nearly nothing on the limbs. (Photograph by G. H. Davis.) (*C*) Passive fold in Barton River Slate, north-central Vermont. Original beds show up as alternating light and dark bands and are thickened in the hinge of the fold. If you look closely, you will see the trace of cleavage oriented parallel to the axial trace of the fold. [Photograph by C. G. Doll, United States Geological Survey. From Donath and Parker (1964). Published with permission of Geological Society of America and the authors.]

mechanically soft units sandwiched between stiff units (Figure 7.48*B*). Think of the folding of an ice-cream sandwich.

In contrast to flexural folding, **passive folding** is the favored mechanism when the mechanical influence of layering in a sequence of rocks is very weak. Passive folding can be thought of as a ''fake'' folding: layers take on a folded form without really having been bent. The layering is passive. It is not active, but is acted upon. It submissively endures distortion, apparently without much resistance. Passive folds are commonly marked by the presence of penetrative **cleavage** (Figure 7.48*C*), that is, the presence of an array of closely spaced aligned secondary discontinuities that cut the folded layers in a direction parallel or subparallel to the axial surfaces of folds. Full discussion of passive folding is deferred to Chapter 8, where it is presented in the context of the physical properties and origins of cleavages and foliations.

Flexural-Slip Kinematics

Whenever I use a telephone book, I cannot resist flexing it and thinking about flexural-slip kinematics. If I flex the book by flexural-slip folding into the form of an upright antiform, each page moves up-dip with respect to the page(s) beneath (Figure 7.49*A*). *The direction of relative slip of the pages is perfectly perpendicular to the hinge of the fold* (Figure 7.49*B*).

Flexural-slip displacements between layers (or pages) are tiny when viewed individually, but the sum of the displacements is always enough to accommodate a true bending of a rock body (or book). The actual amount of slippage along the top of any layer is easy to calculate (Ramsay, 1967, pp. 392–393). As in the analysis of layer shape, the fold form to be analyzed is rotated into the orientation of a perfectly upright antiform. Then the locations where slip is to be calculated are specified by the inclination values (α) of the top of the folded layer at the chosen sites (Figure 7.50). Slip is then determined using the following formula:

$$s = t\alpha \tag{7.1}$$

where s = slip
t = thickness of the folded layer
α = inclination in radians (1° = 0.0175 radian)

For the fold shown in Figure 7.50, we can use Equation 7.1 to calculate slip at 10 sites on the top of layer *A* (thickness = 9 cm) and 10 more sites along the top of layer *B* (thickness = 3 cm). The sites are located at 10° dip interval values, from α = 0° to α = 90°. The calculations demonstrate that the amount of layer-parallel slip increases both with layer thickness and with distance from the hinge. In fact, the calculations show that no interlayer slip whatsoever takes place at the actual hinge point of a folded layer.

The shear strain (γ) due to interbed slip can be calculated too (Ramsay, 1967, p. 393),

$$\gamma = \alpha \tag{7.2}$$

where again α is the inclination in radians.

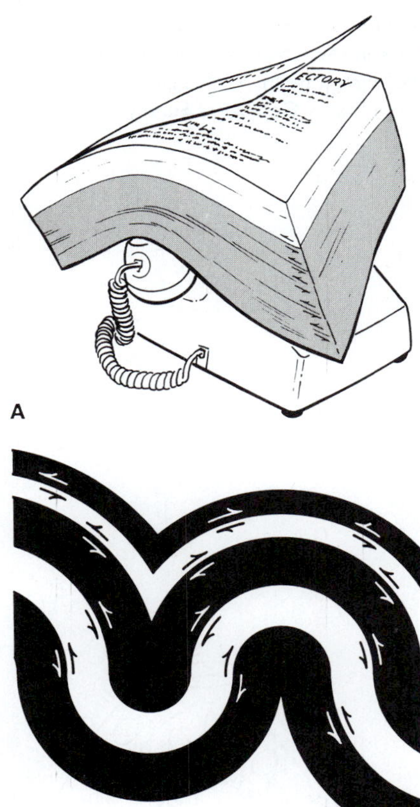

Figure 7.49 (*A*) Flexural slip of pages in a book. The direction of interbed slip is perpendicular to the axis of folding. (*B*) The kinematic character of flexural-slip folding.

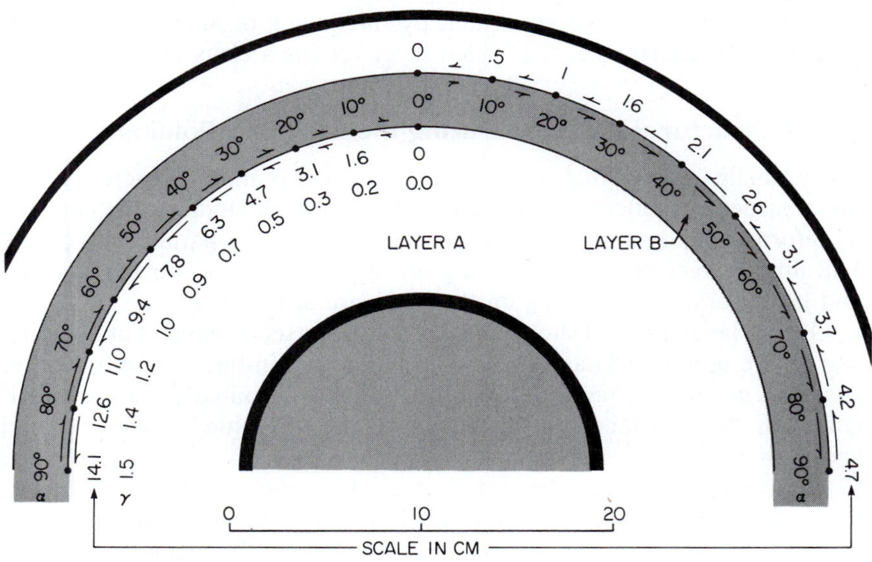

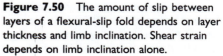

Figure 7.50 The amount of slip between layers of a flexural-slip fold depends on layer thickness and limb inclination. Shear strain depends on limb inclination alone.

Figure 7.51 Flexural-slip folding of ribbon cherts in the Cook Inlet region, Alaska. (Photograph by M. W. Higgins. Courtesy of United States Geological Survey.)

Figure 7.50 shows calculations of shear strain for layer A. The distribution of values of shear strain reveal that shear strain due to flexural-slip folding is greatest at the inflection of a fold but is negligible at the hinge.

Donath and Parker (1964) have emphasized that layered sequences that readily fold by flexural slip are marked by strong, stiff layers, the contacts of which are marked by **low cohesive strength**. Thin- to medium-bedded sandstone, siltstone, and limestone sequences are especially susceptible to flexural slip (Figure 7.51). Individual layers that are folded by the flexural-slip mechanism tend to retain their primary, original thicknesses, in the same way that the pages of a telephone directory neither thicken nor thin out when the directory is flexed. Thus layer shape of flexural-slip folds tends to be class 1B, that is, concentric.

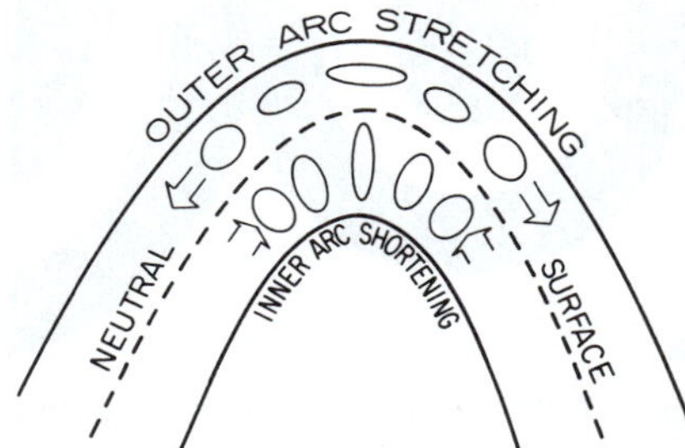

Figure 7.52 Layer-parallel stretching and layer-parallel shortening associated with folding.

Even though individual layers tend to retain their original thickness during flexural-slip folding, they nonetheless generally endure some internal distortion. The distortion takes place mainly in the hinge zone of the folded layer, where curvature is greatest. When an individual layer is actively buckled, rock on the outer arc of the hinge undergoes **layer-parallel stretching**, and rock on the inner arc of the hinge experiences **layer-parallel shortening** (Kuenen and DeSitter, 1938) (Figure 7.52). Layer-parallel strain decreases toward the middle of each folded layer, toward the **neutral surface** of no strain. The neutral surface separates an outer arc domain of layer-parallel stretching from an inner arc domain of layer-parallel shortening. When thinning of outer arc rocks by layer-parallel stretching is perfectly compensated by thickening of inner arc rocks by layer-parallel shortening, the folded layer retains a class 1B form.

Minor Structures Created During Flexural-Slip Folding

Flexural-slip folding creates an informative array of minor structures. The minor structures reflect a combination of four complementary mechanisms of deformation: overall layer-parallel shortening, layer-parallel slip on the fold limbs, layer-parallel stretching on the outer arc of the hinge zone, and layer-parallel shortening on the inner arc of the hinge.

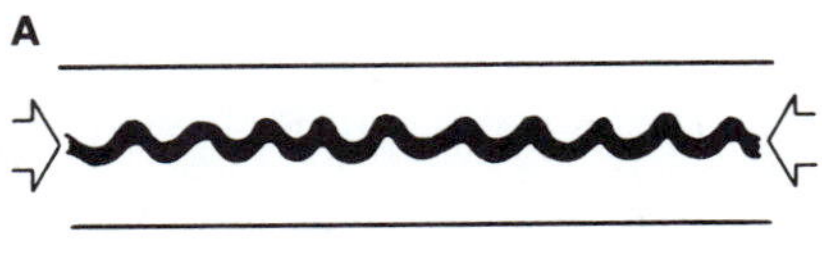

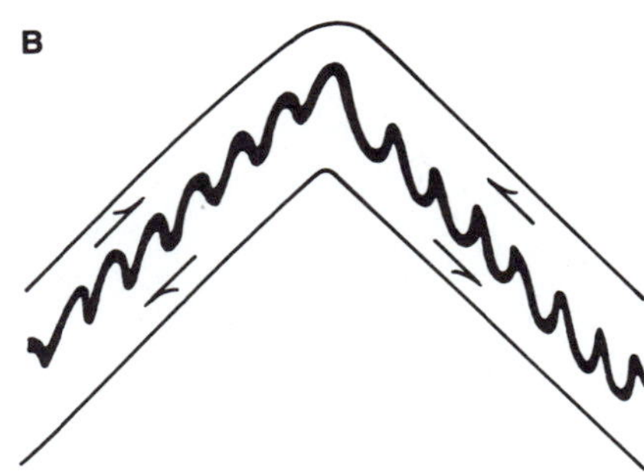

Figure 7.53 The generation of drag folds. (A) Layer-parallel shortening before buckling creates an array of upright, symmetrical anticlines and synclines. (B) Buckling and the onset of flexural-slip folding transforms the symmetrical folds into asymmetrical folds.

Overall layer-parallel shortening before the onset of significant buckling can create minor folds and thrust-slip faults in thin-but-stiff units within a layered sequence (Figure 7.53A). The minor folds initially are symmetrical, with axial surfaces perpendicular to the direction of layer-parallel shortening.

When buckling ensues, and with it layer-parallel slip between layers, minor folds that formed during overall layer-parallel shortening may be transformed into asymmetrical folds on the limbs of the major structure(s)

(Figure 7.53*B*). The opposite limbs of a given fold are marked by minor folds of contrasting asymmetry because the sense of layer-parallel slip is different on opposing limbs.

Called **drag folds** (or **parasitic folds**), asymmetric minor folds formed in this way are valuable for (at least) three reasons. First, the axis of a drag fold is subparallel to the axis of the larger fold with which it is associated. Thus, even when the hinge of the major fold is not exposed, an approximation of its axis orientation can be deduced in the field, before going to the trouble of constructing a π- or β-diagram. Second, in terranes characterized by isoclinal and poorly exposed folds, the location of the axial trace of a major fold can be identified on the basis of the shift in minor-fold asymmetry from *Z* to *S*, or *S* to *Z*. Third, upon discovery of a "hidden" isoclinal fold, its antiformal versus synformal nature can be interpreted on the basis of the **sense** of layer-parallel slip reflected by the drag folds on each limb.

The use of drag folds to interpret fold patterns can be pictured more easily than described in words. Figure 7.54*A* shows the predicament. The axial trace of a major isoclinal fold is discovered and mapped on the basis of asymmetry of drag folds. What remains uncertain is whether the fold is an antiform or a synform. When an antiformal fold form is fitted to the configuration of the axial trace and the fold limbs (Figure 7.54*B*), the expected sense of layer-parallel slip on each limb of the antiform is contradicted by the sense of asymmetry of the drag folds. On the other hand, when a synformal fold form is fitted to the axial trace–fold limb configuration, the observed drag fold pattern is wholly consistent with the expected sense of layer-parallel slip (Figure 7.54*C*).

Minor structures that form in response to flexural-slip folding conform in orientation, location, and strength of development with the state of strain from inflection to hinge (Figure 7.55*A*). En echelon tension fractures commonly develop as a response to lengthening parallel to the local stretching direction (Figure 7.55*B*). Early-formed tension fractures may become distorted into sigmoidal gash fractures and veins as the fold becomes tighter and tighter. Such distortion is a response to **progressive noncoaxial deformation** (e.g., simple shear). The axial surfaces of drag folds remain perpendicular to the direction of minimum finite stretch (S_3) during progressive deformation (Figure 7.55*B*). Cleavage may develop in weak layers such that its orientation is subparallel to the axial surfaces of the minor folds (Figure 7.55*B*). The preferred cleavage orientation is perpendicular to the direction of shortening; the formation of cleavage accommodates shortening (Chapter 8).

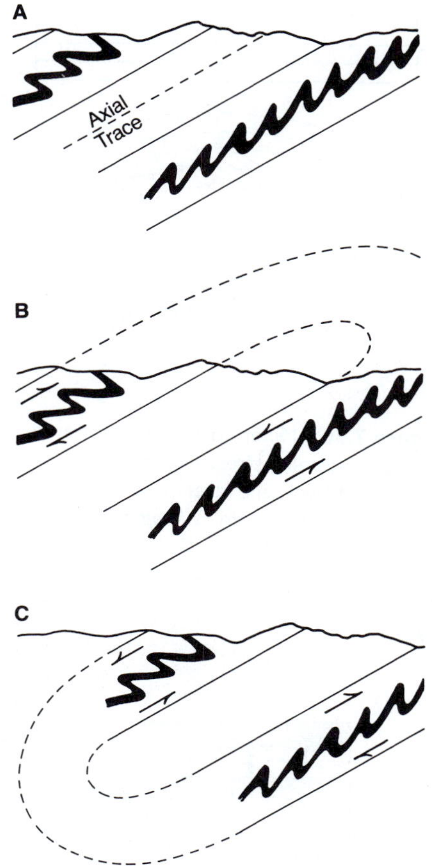

Figure 7.54 (*A*) Is the fold an overturned synform or an overturned antiform? (*B*) The fold is *not* an antiform, for the asymmetry of drag folds contradicts the sense of bedding-plane movements that would characterize the limbs of an antiform. (*C*) The drag fold pattern conforms perfectly to flexural slip on the limbs of an overturned synform.

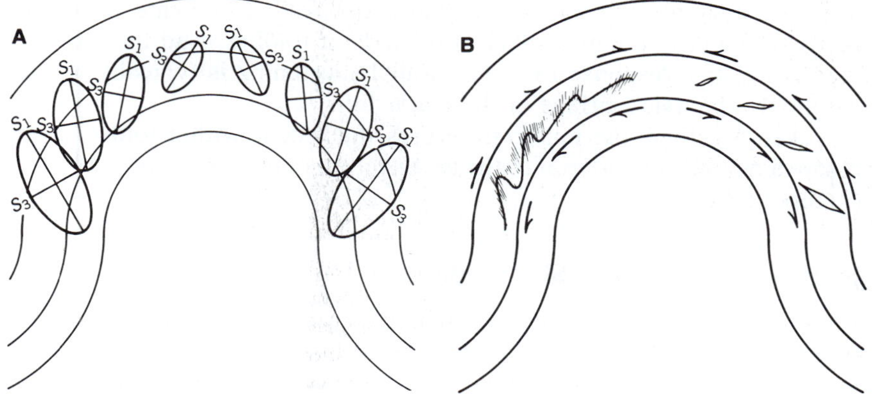

Figure 7.55 (*A*) Schematic portrayal of the state of strain that is generated in layers subjected to flexural-slip folding. (*B*) Minor structures naturally emerge as expression of the state of strain produced by flexural-slip folding. Schematically shown on the left limb are asymmetric folds cut by axial plane cleavage. Sigmoidal gash veins and tension fractures are pictured on the right limb.

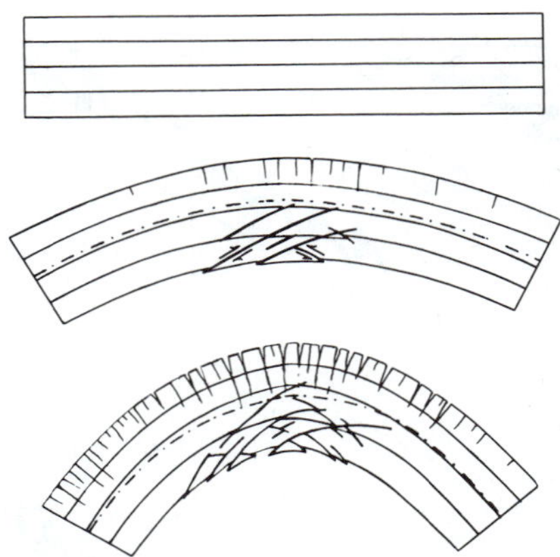

Figure 7.56 Minor structures associated with layer-parallel stretching in the outer arc of a folded layer, and layer-parallel shortening in the inner arc. (From *Folding and Fracturing of Rocks* by J. G. Ramsay. Published with permission of McGraw-Hill Book Company, New York, copyright © 1967.)

Figure 7.57 On the outer arc of this tight fold there are tiny keystone grabens and incipient boudins in the stiff layer. One chickletlike chunk of the extended outer arc has detached and is engulfed in marble. Happy Valley area, southeastern Arizona. (Photograph by G. H. Davis.)

Layer-parallel stretching on the outer arc of a folded layer can be accommodated in a number of ways, depending on the strength of the layer (Figure 7.56). Stiff layers respond to the stretching by the formation of tension fractures and normal-slip faults. Tension fractures, including veins, form perpendicular to the direction of layer-parallel stretching. Conjugate normal-slip faults form in such a way that their line of intersection is parallel to the axis of folding. **Keystone grabens** are classic expressions of stretching on the outer arc of a folded layer.

If the layering in the outer arc of a fold is a composite of soft and stiff layers, stretching is commonly achieved by boudinage and pinch-and-swell. **Boudins** form in sequences of alternating soft and stiff layers that have been subjected to flattening and extension. Stiffer layers tend to break or neck, and the softer layers tend to flow and fill in, wherever required (Figure 7.57). The forms of boudins are endlessly variable, depending on the ductility contrast between the layers that are flattened and stretched. Some boudins are symmetrical **pinch-and-swell structures** (Figure 7.58). Some, however, are like bricks that have been pulled apart, the soft layers filling in like mortar (Figure 7.59).

Layer-parallel shortening on the inner arc of a folded layer gives rise to symmetrical folds, thrust-slip faults, and/or cleavage (see Figure 7.56). These structures work together to accommodate the room problem created when the inner arc of a layer closes in on itself. The minor symmetrical folds are coaxial with the axis of the major fold. Conjugate thrust-slip faults intersect in a line parallel to the axis of folding. And the cleavage that forms as a response to layer-parallel shortening is typically aligned parallel to the axial surface of the major fold.

Taken together, the minor structures that occur within folded layers create a marvelous addition to the architecture of deformed layered rocks (Figure 7.60).

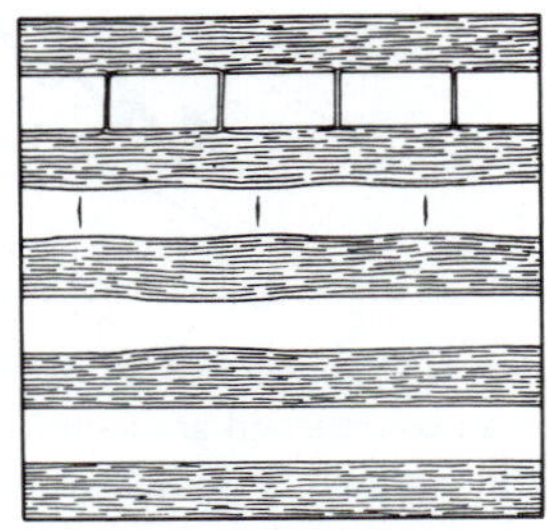

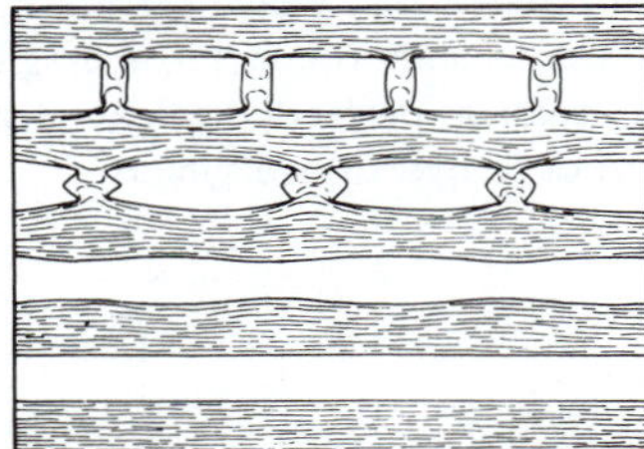

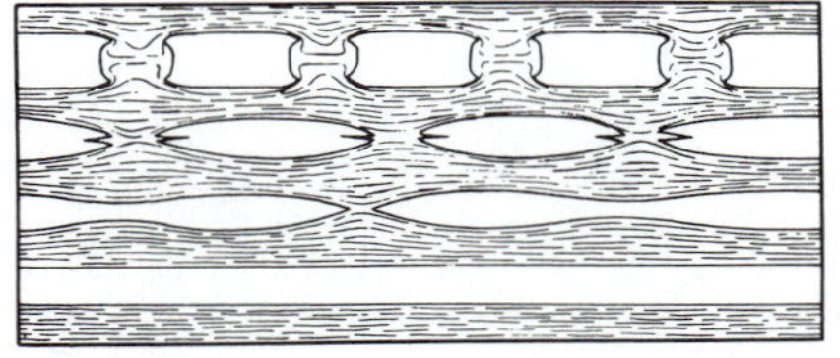

Figure 7.58 Formation of pinch-and-swell structure and boudins by layer-parallel stretching. Ductility contrast between layers determines the extent to which the stiffer layers pinch, neck, and/or break. (From *Folding and Fracturing of Rocks* by J. G. Ramsay. Published with permission of McGraw-Hill Book Company, New York, copyright © 1967.)

Jointing and Flexural-Slip Folding

Jointing commonly is closely coordinated, geometrically, with the orientation properties of the folded layers. Three classes of joints can be distinguished: cross joints, longitudinal joints, and oblique joints (Figure 7.61).

Figure 7.59 Boudins in a stretched calc-silicate layer (white) covered top and bottom by marble (gray). The boudins are bounded by tiny faults. Movement on the faults have permitted the layer to stretch its length. Happy Valley area, southeastern Arizona. (Photograph by E. G. Frost.)

Figure 7.60 The architects who fashioned this elegant entry to this cathedral would be interested in learning about minor structures in folded layers. (Photograph by G. H. Davis.)

Cross joints are mode I joints that ideally are aligned perpendicular to the axis of folding. They reflect extensional stretching of brittle rock during hinge-parallel elongation of folded layers. Hinge-parallel elongation partly compensates for the room problems that can develop in the inner arc of a folded layer as it becomes more and more tightly appressed. Cross joints are unusually planar and are *unusually* regularly spaced (Figure 7.62). Many are vein filled, clearly expressing stretching.

Longitudinal joints are mode I joints oriented subparallel to the axial surfaces of folds (Figure 7.61). They tend to be through-going, planar, continuous structures. The kinematic reason for the development of longitudinal joints is not always clear. Billings (1972) interpreted longitudinal joints as **release joints** that open up in folded layers when fold-forming shortening stresses are relieved.

Oblique joints ideally comprise two conjugate sets of shear joints (mode II or III) that are symmetrically disposed to the hinge and axial surface of a given fold. The oblique joint sets are arranged such that the axial surface of the fold bisects the obtuse angle of intersection of the joint sets (Figure 7.61). They form in folded layers as a response to shortening perpendicular to the axial surface of a fold. The acute angle of intersection of the joints is thus bisected by a line that describes the direction of shortening.

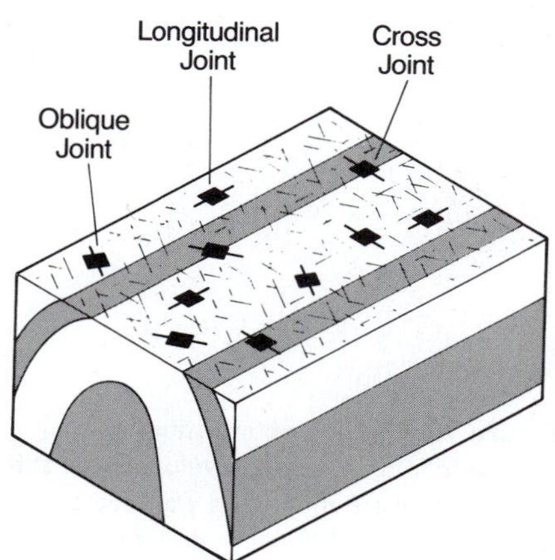

Figure 7.61 Idealized picture of the relation of joints to folds. Cross joints, longitudinal, and oblique joints are distinguished.

Figure 7.62 (*A*) Cross joints in folded metamorphic rocks along the Bay of Fundy, New Brunswick, Canada. (*B*) Close-up view of one of the joint faces shows the nature of the folded layering. After taking this shot, I was overly eager to remove one of the beautiful folds. The resounding blow of the 3-pound hammer was cushioned by my finger, splitting my fingernail along a mode I fracture. The fold remains to this very day. (Photographs by G. H. Davis.)

MECHANICS OF BUCKLING

Instability and Dominant Wavelength

Mechanical analysis of folding traditionally has focused on analysis of **buckling.** The very best analyses of buckling have combined theory and experiment (Biot, 1957; Biot, Ode, and Roever, 1961; Ramberg, 1967; Johnson, 1977). It can be shown both theoretically and experimentally that **instability** develops when layers of different mechanical properties are subjected to layer-parallel stresses (Biot, 1957). The instability gives rise to a buckling of the stiffest layer(s) in the sequence of rocks, like a stiff pegmatite dike within a plastically deforming granite (Figure 7.63).

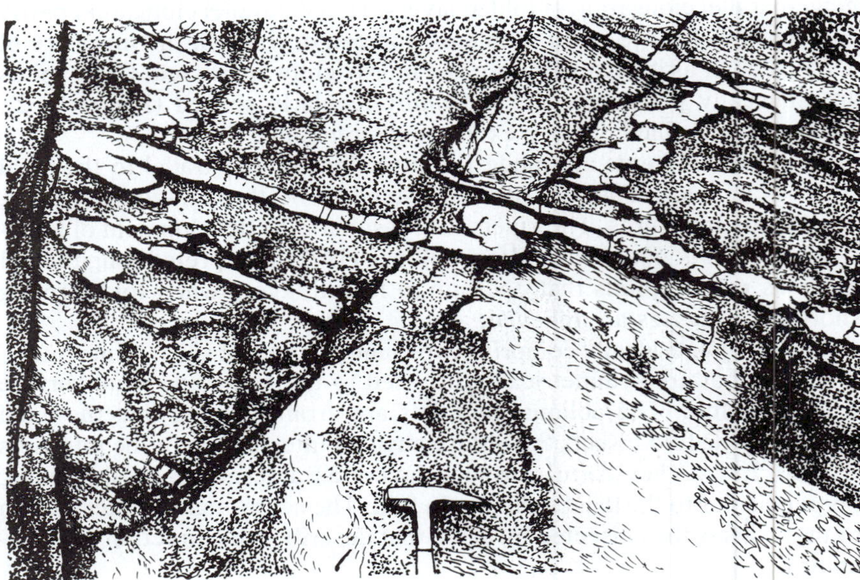

Figure 7.63 Buckled, intestinelike ptygmatic fold. The fold developed in a stiff pegmatite dike (white) that was free to shorten within a deforming granite body. Santa Catalina Mountains near Tucson, Arizona. [Tracing by D. O'Day of photograph by G. H. Davis. From Davis (1980). Courtesy of Geological Society of America and the author.]

The fold that emerges through buckling of a stiff layer is of some particular **dominant wavelength**, the fold wave that can be created with the least amount of layer-parallel stress. Buckling instability is not confined to rocks. An interesting buckle emerged in the trolley tracks of the San Francisco streets during the great earthquake of 1906 (Figure 7.64).

Knowledge gained from mechanical analysis makes it possible to predict the dominant wavelength that will emerge when a single folded layer, or a multilayer sequence, is shortened. Predictions are based on hard-earned mathematical descriptions that relate dominant wavelength to the strength and thickness properties of the layers to be deformed.

A word of caution: the sites of specific folds may not relate so much to predictable dominant wavelengths as to unpredictable sites of flaws in the multilayer sequence. Willis (1894) recognized through experimental modeling that 1–2° changes in the initial dip of sedimentary layers can predetermine the sites at which fold hinges will emerge (see Johnson, 1970). Such observations underscore one of the great contradictions that emerges from the mechanical analysis of folds: *buckling cannot occur in perfectly planar multilayers that are shortened by stresses that are perfectly layer parallel* (Biot, 1959). Fortunately for fold enthusiasts, the smallest imperfections in the primary geometry of layering can trigger the fold-forming process(es).

Figure 7.64 Buckling of rails by compression on Howard Street (South Van Ness Avenue) near 17th Street, San Francisco. The buckling was caused by movements related to the earthquake of 1906. (Photograph by T. L. Youd. Courtesy of United States Geological Survey.)

Simple Buckling of a Single Layer, in Theory

The dependence of wavelength on layer thickness and strength is most simply expressed in equations that describe the buckling of a stiff layer embedded in a softer medium. Dominant wavelength depends not only on the thickness and strength of the stiff layer, but also on the strength of the weak, confining medium. Thickness is easy to deal with both mathematically and experimentally. But how is "**stiffness**" of a rock modeled quantitatively? How stiff are those rocks shown in Figure 7.65?

As it turns out, the mathematical description of layer strength depends on whether the mechanical properties of the single stiff layer and its confining medium are viewed as elastic or viscous. If an elastic model of deformation is applied, the strengths of layers are described in terms of

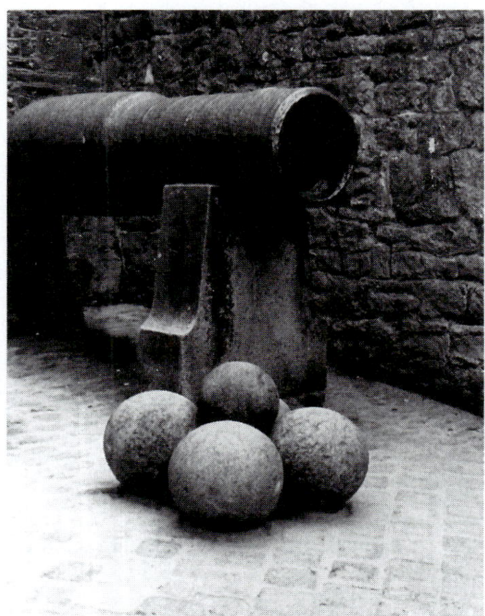

Figure 7.65 Personally, if I had a choice, I would rather get hit by a Nerf cannonball with a much lower Young's modulus. These granite cannonballs are stacked next to a cannon that once guarded the entrance of Le Mont St. Michelle. (Photograph by G. H. Davis.)

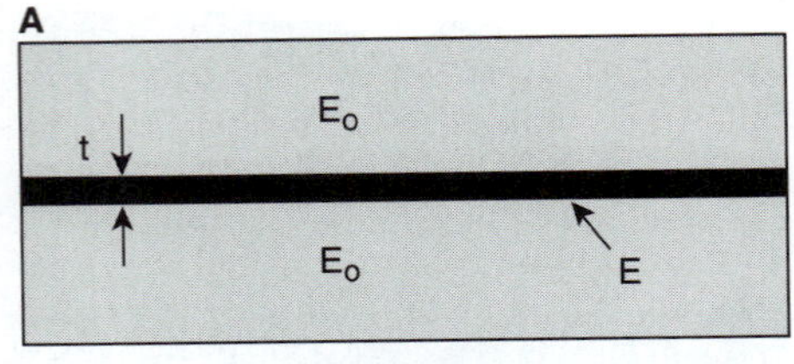

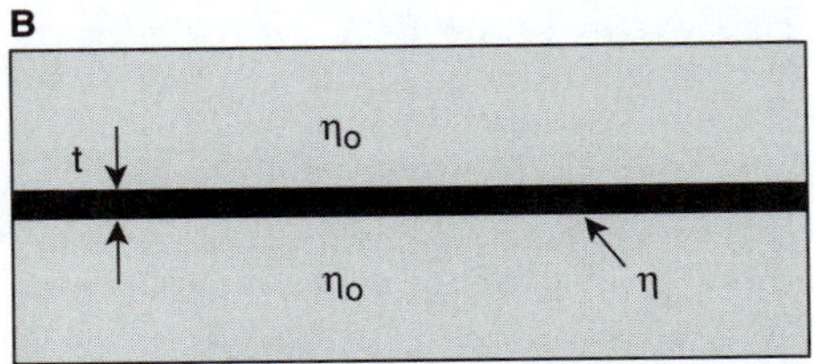

Figure 7.66 The modeling of the mechanical properties of layers about to be deformed by layer-parallel shortening and buckling. (*A*) Elastic model. (*B*) Viscous model.

the fundamental elastic moduli: **Young's modulus (*E*)** and **Poisson's ratio (*ν*)** (Figure 7.66*A*). However, if a viscous model of deformation is used, the strengths of the layer and its confining medium are expressed in terms of **viscosity coefficients** (Figure 7.66*B*). In hybrid models where a stiff elastic layer is considered to be embedded in a soft, nonelastic confining medium, the strength of the stiff layer is described in terms of elastic moduli, and the strength of the confining medium is specified by a coefficient of viscosity.

Bijlaard (1946) modeled the mechanics of folding of a single layer in terms of a stiff elastic plate in a soft elastic medium. What he discovered was a surprisingly straightforward relationship:

$$L = 2\pi t \left(\frac{B}{6B_0}\right)^{1/3} \tag{7.3}$$

where L = dominant wavelength
t = thickness of stiff layer
B = elastic modulus of stiff layer
B_0 = elastic modulus of confining medium

Elastic moduli B and B_0 are not mystery variables pulled from the sky. Rather they express strength in terms of Young's modulus and Poisson's ratio:

$$B = \frac{E}{1 - \nu^2} \tag{7.4}$$

where, as usual, E is Young's modulus and ν is Poisson's ratio.

Currie, Patnode, and Trump (1962) reexamined the mechanics of folding of a stiff elastic layer in a soft, elastic confining medium. In doing so they chose to eliminate Poisson's ratio as a variable, arguing that the influence of this modulus on folding is small, especially considering the uncertainty in trying to describe precisely the value of ν for real rock layers at the time of folding. The resulting equation has the form of Equation 7.3

$$L = 2\pi t \left(\frac{E}{6E_0}\right)^{1/3} \tag{7.5}$$

where Young's moduli of the stiff layer and the confining medium are E and E_0, respectively.

Biot (1959) and Ramberg (1959) treated the folding of a single folded layer in the perspective of viscous deformation. Their independently derived mathematical analyses uncovered the same kind of relationship reported for the case of elastic deformation:

$$L = 2\pi t \left(\frac{\eta}{6\eta_0}\right)^{1/3} \tag{7.6}$$

where L = dominant wavelength
t = thickness of stiff layer
η = coefficient of viscosity of stiff layer
η_0 = coefficient of viscosity of confining medium

Coefficients of viscosity are expressed in **poises**, the standard measure of resistance to flow of a viscous material.

Adding Layer-Parallel Internal Strain, in Theory

Just because the various buckling equations have the same form does not mean that they are the last word on the folding of single layers. Folds and folding continue to keep us humble. Sherwin and Chapple (1968), for example, have shown that the dominant fold wavelength that arises during layer-parallel shortening is responsive to the amount of layer-parallel strain absorbed by the layer *before* buckling. Thus in addition to strength and thickness, **layer-parallel strain** emerges as an important variable that must be taken into account.

Layer-parallel strain is specified in terms of a parameter that is familiar to us: stretch (S). Sherwin and Chapple found it necessary to describe stretch in two directions within the plane of layering, both parallel and perpendicular to the direction of layer-parallel shortening. Hudleston (1973) rewrote the Sherwin–Chapple equation in a form that can be directly compared with the Biot (1959) and Ramberg (1962) equations,

$$L = 2\pi t \left[\frac{\eta(s-1)}{6\eta_0(2s^2)} \right]^{1/3} \tag{7.7}$$

where t = thickness of stiff layer
η = coefficient of viscosity of stiff layer
η_0 = coefficient of viscosity of confining medium
$s^2 = \dfrac{S_1^2}{S_2^2}$

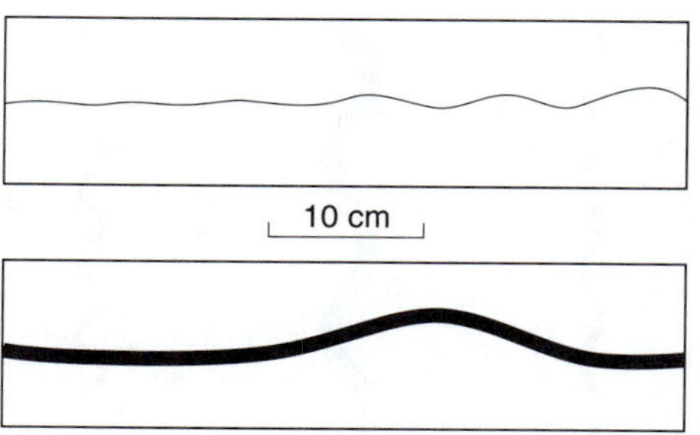

Figure 7.67 Layer-parallel shortening of pitch layers of different thicknesses in a medium of syrup. [From Biot, Ode, and Roever (1961), published with permission of Geological Society of America.]

As in all matters of science, closer and closer scrutiny of the Earth at work always seems to lead to a greater appreciation of the delicacy of dynamic process. We learn that Earth processes are influenced by a much broader range of variables than originally perceived.

Buckle Folding of a Single Layer, in Practice

Part of the fun of mechanical analysis is testing equations to see if they really work. Biot did not wait for others to test the equation he derived for the folding of a single layer, viewed viscously (Equation 7.6). Instead, he teamed up with two colleagues to check it himself (Biot, Ode, and Roever, 1961). Together the investigators set up a series of experiments that included the layer-parallel shortening of single layers of stiff pitch (i.e., tar), which they deformed in a confining medium of corn syrup (Figure 7.67). Layers of pitch of different thicknesses were fabricated in molds of different depths. Viscosities of both the pitch and the syrup were carefully measured before the start of the experiments.

On the basis of strength and thickness data (Table 7.1), Biot, Ode, and Roever calculated the dominant wavelengths predicted by Biot's equation.

TABLE 7.1
The testing of Biot's equation, $L = 2\pi t \sqrt[3]{\mu/6\mu_0}$

Thickness (t) of Pitch Layer	*Viscosity (μ) of Pitch Layer*	*Viscosity (μ_0) of Corn Syrup*	*Predicted Fold Wavelength (L_p)*	*Observed Fold Wavelength (L_0)*
0.35 cm	3×10^7 poise	1.35×10^4 poise	15.78 cm	12.4–18.0 cm
0.87 cm	3×10^7 poise	1.35×10^4 poise	39.24 cm	34.0–41.0 cm
1.08 cm	3×10^7 poise	1.35×10^4 poise	48.71 cm	38.0–52.0 cm

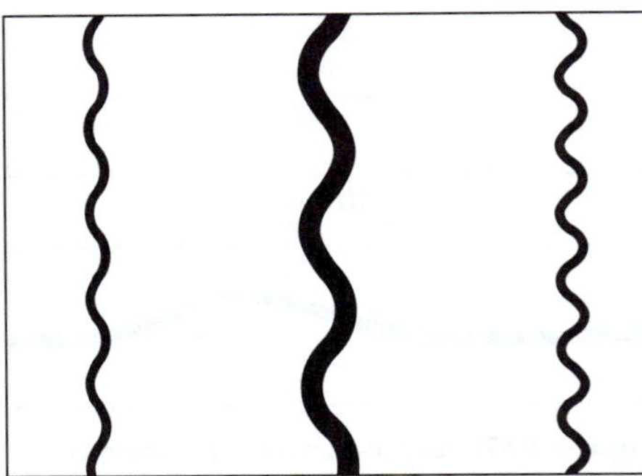

Figure 7.68 Layer-parallel shortening of gum rubber strips in a medium of gelatin. The outside strips are 4 mm thick. The middle strip is 8 mm thick. [From Currie, Patnode, and Trump (1962). Published with permission of Geological Society of America and the authors.]

Then they subjected each of the three pitch layers to layer-parallel shortening and measured the range of wavelengths of folds that emerged in each buckled layer. Experimental results were found to be quite consistent with the predictions of theory!

As part of their research, Currie, Patnode, and Trump (1962) experimentally tested their equation for folding of a single elastic plate in an elastic medium (Equation 7.5). They found it practical to deform thin gum rubber strips of known thickness within a medium of gelatin (Figure 7.68). The gum rubber used for their experiments yielded, on testing, a Young's modulus E of -69 kPa. The gelatin in which the gum rubber layers were embedded was mixed from scratch, in such a way that the Young's modulus E_0 for each gelatin specimen could be predetermined, within a range of -6.9 kPa to -69 kPa.

Geology majors at Carleton College, led by Dave Bice, perform buckling experiments with everyday materials, such as clay, Silly Putty, and even cheese and sandwich meats. They videotape the action so that they can replay the experiments and talk about the effects of different variables.

The computer may lend itself to the most effective and efficient "experimentation." Using **finite element modeling** techniques, structural geologists apply the fundamental buckling equations to materials whose strengths can be prescribed through the appropriate elastic or viscous parameters. Computer graphics then transform the incremental movements into realistic fold forms and in elegant fashion (Figure 7.69).

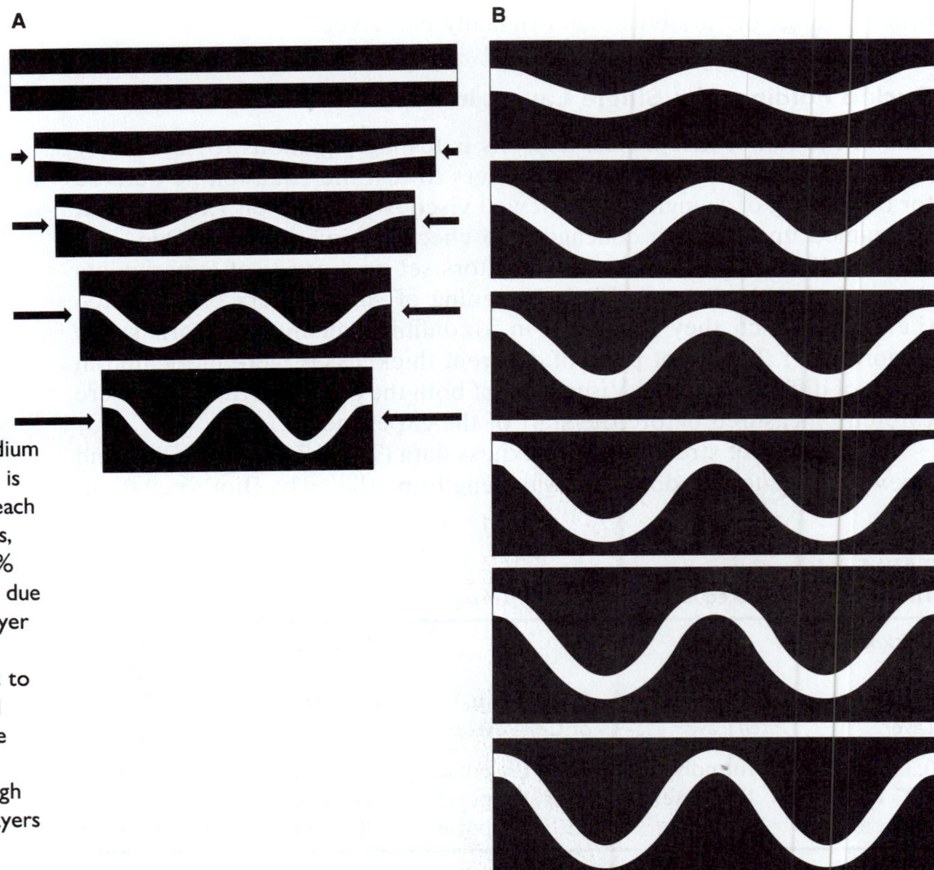

Figure 7.69 Computer simulations of buckling carried out by Cruikshank and Johnson (1993). (*A*) A stiff layer (white), 50 times stiffer than the surrounding medium (black), is progressively shortened. There is an equal amount of shortening between each step. (*B*) In each of these six experiments, the stiff layer (white) is shortened by 40% ($S = 0.6$). Differences in profile form are due to the viscosity ratio between the stiff layer (white) and its surrounding medium. The viscosity ratios, from the top experiment to the bottom, were 10, 20, 30, 40, 50, and 100. [Reprinted with permission from the *Journal of Structural Geology*, v. 15, Cruikshank, K. M. and Johnson, A. M., High amplitude folding of linear-viscous multilayers (1993), Elsevier Science Ltd., Pergamon Imprint, Oxford, England.]

Influence of Competency Contrast on Fold Form

Ramsay and Huber (1987) have done a beautiful job summarizing the influence of **competency contrast** (between a competent layer and the incompetent material above and below) on fold form (Figure 7.70). Where the competent layer is much, much stiffer than the surrounding materials, the **amplification rate** of buckling is very high and the competent layer deflects robustly into the material above and below. Large wavelength, rounded forms are produced, the best example of which is the **ptygmatic fold** (Figure 7.70). On the other hand, where competency contrast is low, the amplification rate of buckling is very small. As a result, the folds that are created are of short wavelength, and the typical forms are **cuspate–lobate folds** (Figure 7.70).

The deflection of less competent (i.e, softer) rock into more competent (i.e., stiffer) rock produces pointed, cuspate fold forms. The points "point" into the stiffer rock. Thus in outcrops dominated by cuspate–lobate fold forms, it is possible to know at a glance whether *at the time of folding,* a given layer was relatively stiff or relative soft, compared to beds on either side (Figure 7.71).

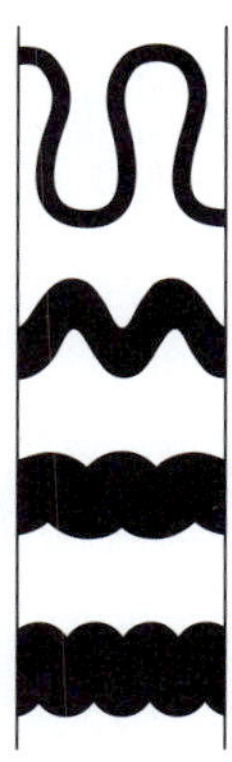

Figure 7.70 The shapes of buckle folds reflect the competency (stiffness) contrast between the stiffer layer and the less competent medium that it occupies. High contrast leads to ptygmatic folds. Low contrast leads to cuspate–lobate folds. Medium contrast creates fold forms that are intermediate between ptygmatic and cuspate–lobate. [Reprinted from *The techniques of modern structural geology, v. 2, folds and fractures,* J. G. Ramsay and M. I. Huber (1987), Figure 19.14. Copyright © by Harcourt Brace and Company Limited, with permission.]

Buckle Folding of Multilayers

Complicated mathematical expressions are required to describe the behavior of **multilayer sequences** containing layers of widely different strength and thickness. The mathematical expressions must include variables above and beyond those already mentioned, notably the spacing of stiff layers within the sequence and the degree of cohesive strength between layers within the sequence.

The ratio of dominant wavelength to thickness of a folded stiff layer is greatly reduced when the layer belongs to and is analyzed as part of a multilayer sequence (Bijlaard, 1946; Johnson, 1977). Gum rubber and gelatin experiments carried out by Currie, Patnode, and Trump (1962) reveal this quite clearly (Figure 7.72). Widely separated gum rubber strips display short-wavelength fold waves, but the dominant wavelength

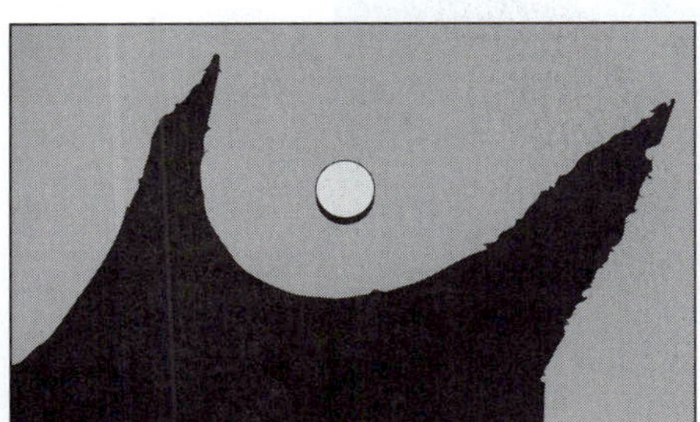

Figure 7.71 Rendering of cuspate–lobate folds at the interface between mica schist (black) and sandstone (light gray), Nufenpass, central Switzerland. The cuspate–lobate pattern looks like the shadow of Batman. [Tracing of photograph from Ramsay and Huber (1987), Figure 19.16, p. 395. Reprinted from *The techniques of modern structural geology, v. 2, folds and fractures* (1987), copyright © by Harcourt Brace and Company Limited, with permission.]

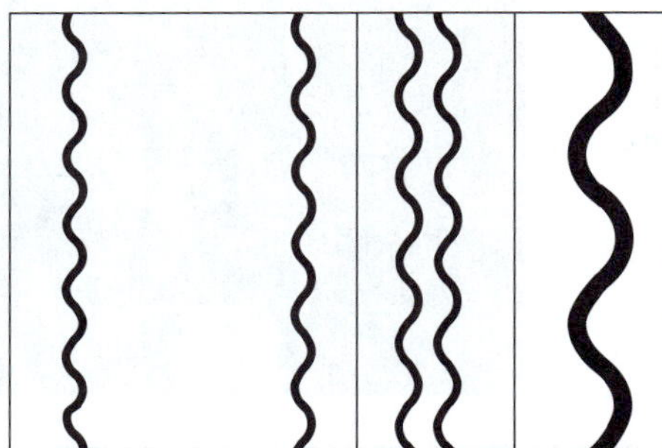

Figure 7.72 Experiments by Currie, Patnode, and Trump demonstrated that the spacing of stiff layers within a multilayer sequence has a significant influence on dominant wavelength. As the separation between stiff multilayers becomes smaller and smaller, the dominant wavelength gets bigger and bigger. [From Currie, Patnode, and Trump (1962). Published with permission of Geological Society of America and the authors.]

steadily increases as the gum rubber strips are brought into closer and closer contact.

One of the mechanical idiosyncrasies of layer-parallel shortening of multilayers is that thinner layers in the sequence may buckle into short-wavelength folds before the folding of the entire sequence (Ramberg, 1963). Such "minor" folding is an expression of the layer-parallel strain that constitutes the preliminary step in the formation of most drag folds. As discussed earlier, when buckling of the entire multilayer sequence occurs, and flexural-slip folding is initiated, these originally symmetric, short-wavelength minor folds are transformed into asymmetric drag folds by layer-parallel simple shear.

KINK FOLDING

Importance of Preexisting Foliation and Loading Direction

Strongly foliated rocks like schists and phyllites commonly display **kink folds** (Figure 7.73) that deform preexisting foliation. Kink folds are typically quite small, with fold heights and fold widths on the order of centimeters or fractions of centimeters. They are marked by sharp hinges, straight limbs, and an asymmetry expressed by a short limb connecting two longer limbs. Superficially, kink folds resemble buckle folds or flexural-slip folds, but they are really a distinct class.

Figure 7.73 (*A*) Kink bands in Devonian phyllite exposed near Morthoe, Devonshire, England. (From *The Minor Structures of Deformed Rocks: A Photographic Atlas* by L. E. Weiss. Published with permission of Springer-Verlag, New York, copyright © 1972.) (*B*) Photomicrograph of kink folds in schist. (Photograph by A. L. Albee. Courtesy of United States Geological Survey.) (*C*) Close-up view of kink fold collected from Ordovician quartz sericite schist in northern New Brunswick, eastern Canada. (Photograph by G. H. Davis.)

Z-shaped kink folds are called **dextral**, whereas *S*-shaped kink folds are called **sinistral** (Figure 7.74). Axial surfaces of kink folds are referred to as **kink planes** (Figure 7.74). The narrow zones where foliation is kinked are called **kink bands** (Figure 7.74), and they show up distinctively in outcrop views (see Figure 7.73*A* and Figure 7.75).

Paterson and Weiss (1966) eliminated most of the mystery of kink folding by successfully producing the phenomenon in highly foliated, real rock specimens, which they subjected to layer-parallel shortening under confining pressure. They demonstrated that there is a close relationship among the geometry of kink fold systems, the orientation of the strongly developed **planar anisotropy** (i.e., strong foliation), and the direction of loading.

An excellent example of this is presented by Ramsay and Huber (1987). They picture how the finite strain ellipse is oriented with respect to the overall orientation of foliation for three different situations: equally developed dextral and sinistral kink folds (Figure 7.76*A*); sinistral kink folds only (Figure 7.76*B*); and dextral kink folds only (Figure 7.76*C*). Figures 7.76*B*,*C* reveal that when shortening is layer-inclined, either dextral or sinistral kink folds will predominate, depending on the sense of shear.

Ironically, a right-handed component of shear creates sinistral kink folds (Figure 7.76*B*) and a left-handed component of shear generates dextral kink folds (Figure 7.76*C*). This is opposite to what we learned for asymmetric "drag" folds and reflects the different mechanism of folding for kink vs drag folds.

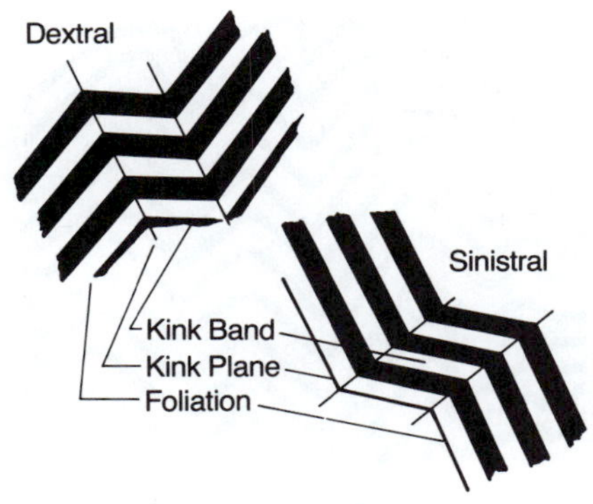

Figure 7.74 Schematic representation of the elements of dextral and sinistral kink folds.

Importance of Cohesive Bonding Between Layers

Ghosh (1968) discovered through experiments the degree to which the formation of kink folds depends on the "right" cohesive bonding between layers. Again there is irony. The conditions for kink folding include strong cohesion between layers. By simply spreading different amounts of grease between layers of modeling clay as he built his multilayer models for end-on loading, Ghosh was able to create a striking array of fold forms, without even changing the strength or the thickness or the spacing of layers within the sequence. When layers are liberally greased, layer-parallel shortening creates smooth, rounded, sinusoidal folds (Figure 7.77*A*). But when layers

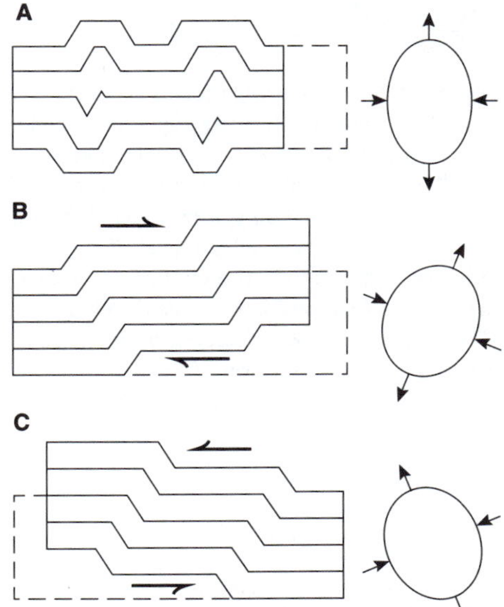

Figure 7.76 Relationship of total finite strain to the formation of kink folds. (*A*) Symmetrical sets of equally developed dextral and sinistral kink folds form when the direction of least stretch is oriented parallel to the direction of foliation. (*B*) Sinistral kink folds or (*C*) dextral kink folds develop when shortening is inclined to the layering. Note that sinistral kink folds form when there is a right-handed shear component parallel to foliation, and that dextral kink folds form when there is a left-handed shear component parallel to foliation. [Adapted from Ramsay and Huber (1987), Figure 20.30, p. 428.]

Figure 7.75 Large-scale kink banding and related folding, glistening in the sunlight in the southern Andes, Patagonia, Argentina. Scale is deceptive. The block in the upper right-hand corner is the size of a VW bug. (Photograph by S. J. Reynolds.)

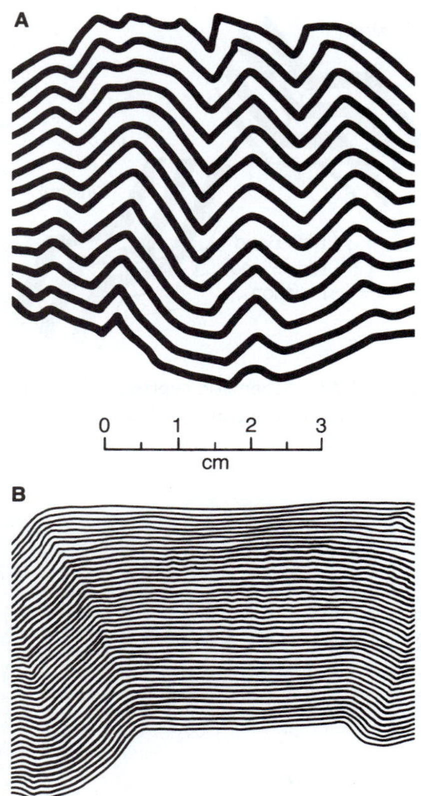

Figure 7.77 The profile forms of folds in multilayer sequences are strongly influenced by the degree of cohesion between layers. (*A*) Where little cohesion between layers exists, folds tend to be smooth, rounded, and sinusoidal. (*B*) Where a high level of cohesion characterizes the contacts between layers, kink folds and kink bands develop. [From Ghosh (1968), *Tectonophysics*, v. 6. Published with permission of Elsevier Scientific Publishing Company, Amsterdam.]

are placed in frictional contact with each other, without grease in between, layer-parallel shortening creates kink folds (Figure 7.77*B*), identical to those so abundantly found in nature.

Modes of Kink Folding

Physical modeling of kink folding, including time-lapse photography, has been very revealing. Kink folds initiate as asymmetric folds *within* **kink bands** (Figure 7.78*A*,*B*) that develop at an angle to the direction of shortening and obliquely to the penetrative planar aniosotropy (e.g., cards in a deck; foliation in a schist). As shortening proceeds, the kink bands and kink folds interfere in ways to bring about a pervasive, symmetrical, chevron folding (Figure 7.78*C*–*E*).

Twiss and Moores (1992) have nicely summarized different ways in which kink folds evolve. In some cases the kink bands migrate through the material (Figure 7.79*A*,*B*). For example, a kink band can nucleate perpendicular to foliation and then progressively rotate and expand in width to accommodate layer-parallel shortening (Figure 7.79*A*). Or the kink band may form at an oblique angle to foliation and simply expand in width (without changing orientation) to accomplish layer-parallel shortening (Figure 7.79*B*). In other cases, the kink bands do not migrate through the material. Boundaries remain fixed in orientation and spacing throughout the shortening event. Layer-parallel shortening is accomplished by simple shear parallel to the fixed boundaries (Figure 7.79*C*) or by rigid

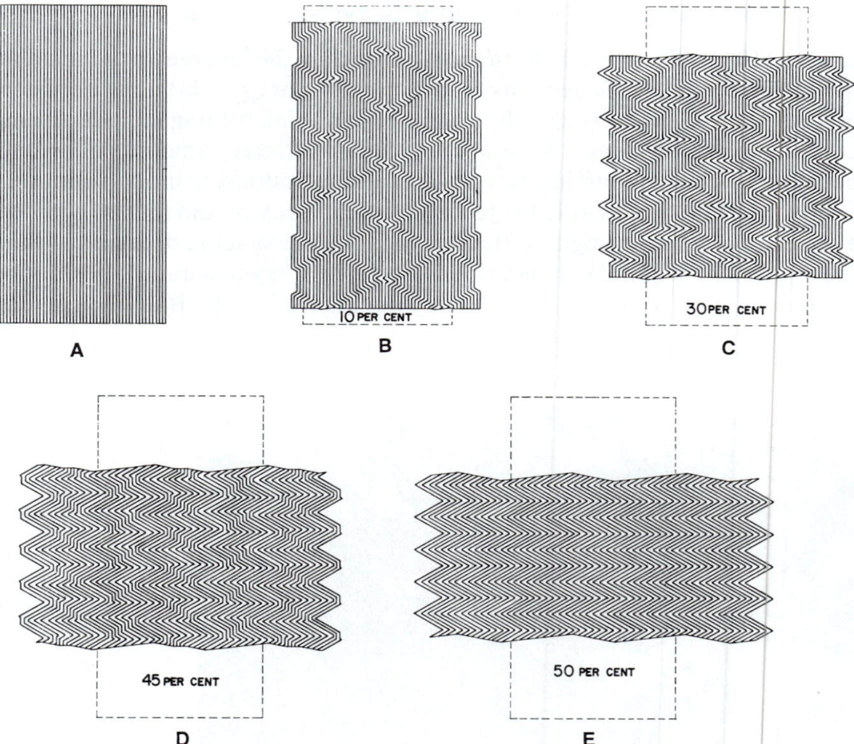

Figure 7.78 (*A*) Think of this block either as a card deck or a rock pervaded by foliation. (*B*) Shortening parallel to the direction of foliation first results in the formation of conjugate kink bands. Conjugate folds form at the intersection of the kink bands. (*C*–*E*) With further shortening the widths of the kink bands increase to the point that kink folds and kink bands are "replaced" by chevron folds. [Adapted from Paterson and Weiss (1966), with permission from Geological Society of America.]

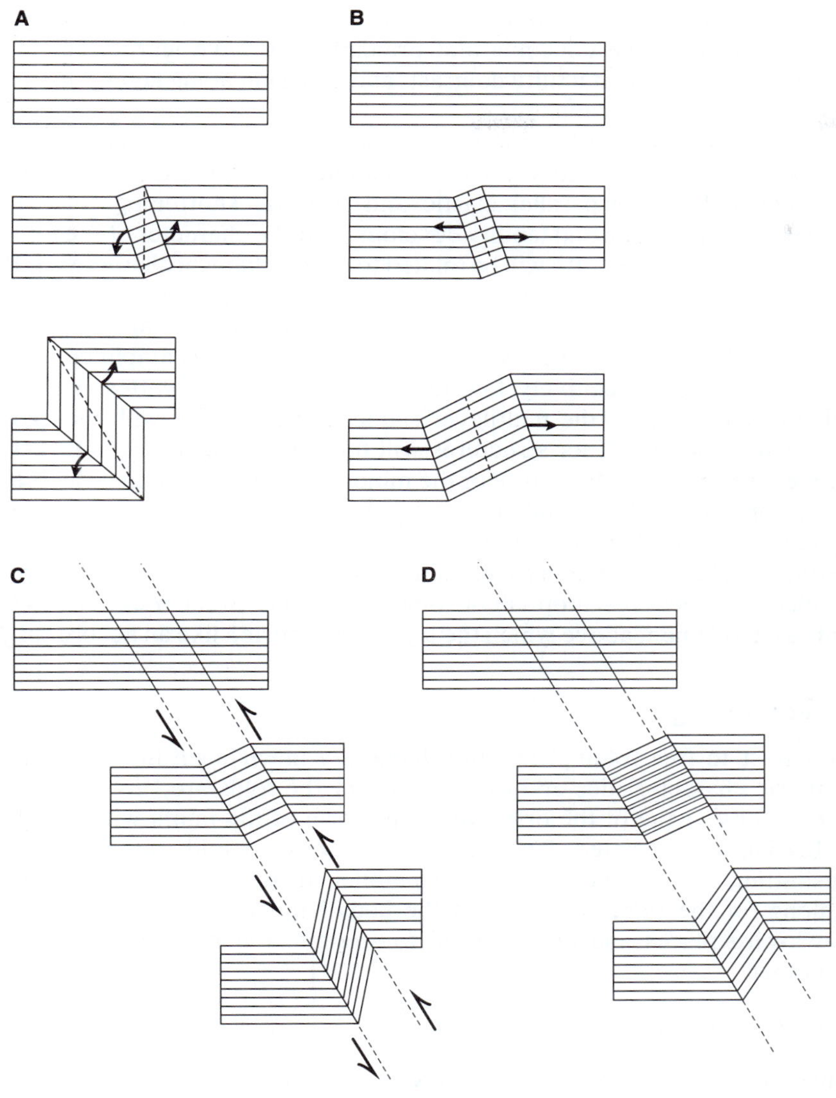

Figure 7.79 Four different ways in which kink folds can form. (*A,B*) Two ways in which kink folds form by migrating through the rock. (*C,D*) Two ways in which kink folds form within kink bands that are fixed in orientation and in width. [From *Structural geology* by Twiss and Moores, Figures 12.20 and 12.21. Copyright © 1992 by W. H. Freeman and Company, with permission.]

Figure 7.80 (*A*) The Sheep Mountain anticline in the Big Horn basin, Wyoming. (*B*) Folding at Sheep Mountain, Wyoming. (Photographs by K. Constenius.)

rotation of the laminations within the kink band (Figure 7.79*D*). Telltale geometries can be identified in outcrops to distinguish among the possible origins.

REGIONAL TECTONIC FOLD MECHANISMS

Tectonic loading at the regional scale has produced spectacular folds in mountain belts the world over. Asthetically, they are the favorite structure of most geologists (Figure 7.80). They can form in a variety of ways. Interpretations must be based on a solid descriptive foundation, including effective cross sections.

Free Folding

Buckle folding is a **free folding** in which the folds display profile forms that depend entirely on the physical–mechanical properties of the layers

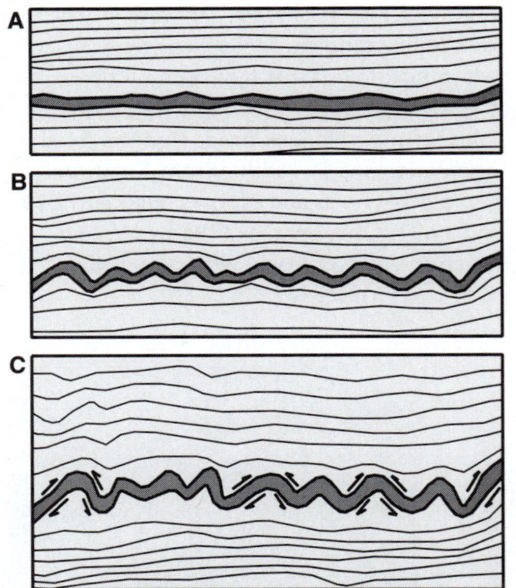

Figure 7.81 Simulation of free folding at the subregional to regional scale. (*A*) Layer-parallel tectonic loading just begins to compress and shorten a thick sequence of horizontal sedimentary strata. (*B*) A buckling instability develops, with a dominant wavelength that relates to the stiffness of the thickest, strongest unit. (*C*) Initial buckling and continued folding are accompanied by flexural-slip and flexural-flow folding. "Minor" structures form as a result of flexural-slip folding, outer arc extension, and inner arc compression.

that are compressed. Free folding occurs on all scales, from outcrop to regional, and thus the concepts and principles we have learned about buckling, flexural-slip, ductility contrast, and cohesion along layer boundaries are broadly applicable.

The evolution of free folds that affect sequences hundreds of meters in thickness proceeds as follows. Layer-parallel stresses are initially accommodated by layer-parallel elastic and inelastic shortening (Figure 7.81*A*). Next, a buckling instability develops, and the character of the fold (e.g., dominant wavelength) will be related to the mechanical character and locations of the thickest, strongest layers within the sequence (Figure 7.81*B*). Depending on ductility contrast between the most competent layers and the less competent rocks on either side, the fold is amplified at a relatively high or a relatively low rate. As the thickest, stiffest layers buckle as single units, the multilayer sequence as a whole will undergo flexural-slip folding. Minor structures will form in the relatively stiff and relatively soft layers. The nature of the minor structures, and their degree of development, will depend on the local strain environments created during the folding (Figure 7.81*C*).

Some of the best regional free-fold environments are located where stratified sequences rest atop salt or some other utterly weak rock. The salt forms a decollement above which the layers can shorten like an accordion.

Forced Folding

In contrast to free folding, **forced folding** is a mechanism in which the geometric characteristics and overall form of the folds are "forced on" the layers by virtue of the orientation and form of the faults with which the folding is associated. The beds are not *free* to fold, nor are they transmitting much in the way of layer-parallel stresses. Instead, they just go along for the ride, and some of the beds happen to find themselves in awkward places and are required to stretch or bend from one step to another.

Fault–Bend Folding

Fault–bend folding represents the predominant regional tectonic folding mechanism in thin-skinned thrust belts in miogeoclinal and foreland settings. Fault–bend folds are formed when beds are displaced along thrust faults with ramp–flat geometries (Figure 7.82). The geometric character of the folds is directly relatable to the stairstep geometry of the thrust faults and the magnitudes of displacement along them. We touched on this kind of folding in Chapter 6 ("Faults").

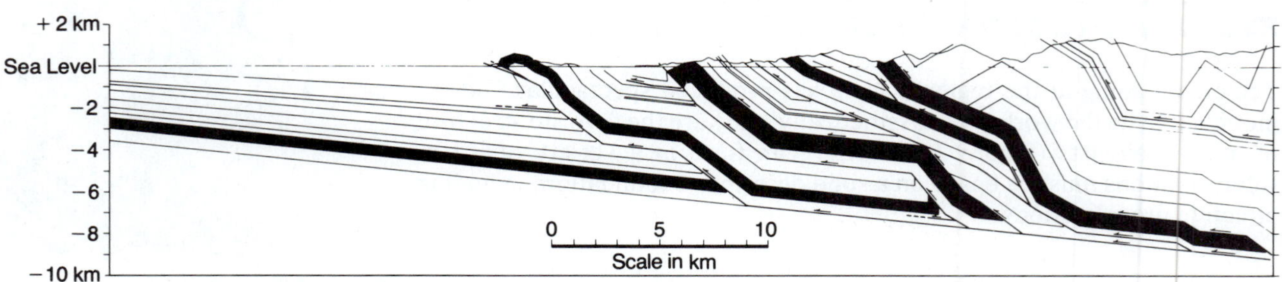

Figure 7.82 Fault–bend fold geometry "forced" upon beds as they are moved along ramp–flat fault geometries. Taiwan. [Adapted from Suppe (1980a). Published with permission of Geological Society of China.]

Based on exhaustive geometrical modeling and the study of fault–bend fold relationships in the Taiwan thrust belt, Suppe (1985) thoroughly unveiled the geometry and kinematics of fault–bend folding. The stage is set for fault–bend folding when a thrust cuts up through the stratigraphic section from one lithologically controlled flat to another (Figure 7.83*A*,*B*). Hanging-wall units and footwall units are "cut off" by the **ramp** portion of the thrust. Following Suppe, it is useful to call special attention to the **footwall cutoffs** (*X* and *Y*, Figure 7.83*B*) at the top and bottom of the ramp; and the **hanging-wall cutoffs** (X' and Y', Figure 7.83*B*) as well. When movement along the thrust begins, hanging-wall strata begin moving up the ramp. As this happens, the orientation of the ramp is "forced upon" the orientation of the beds that move up the ramp (Figure 7.83*B*). As beds in the hanging wall move from a "flat" orientation to a "ramp" orientation, they become creased into the form of a chevron syncline that has the form of a giant kink fold. The whole hanging-wall section is affected by the creasing at the footwall cutoff (*Y*). Thus, even at the earliest stages of movement up the ramp, a narrow kink band forms and extends from the footwall cutoff (*Y*) to the top of the section. At the top of the ramp there is another kink band, and it also begins to form (as a narrow band) just as soon as hanging-wall strata begin to move up the ramp (Figure 7.83*B*). As thrusting progresses, and as the hanging-wall cutoff (Y') moves further and further up the ramp (Figure 7.83*C*), the kink band above the footwall cutoff at the base of the ramp continuously grows in breadth. In the same fashion, as the hanging-wall cutoff (X') at the top of the ramp progressively moves outward along the upper flat, the kink band above the footwall cutoff (*X*) at the top of the ramp grows in breadth (Figure 7.83*C*). The net effect is the growth of a large anticline, a so-called **snakehead fold**. It reaches a maximum amplitude when Y', the hanging-wall cutoff formed at the base of the ramp, reaches *X*, the footwall cutoff formed at the top of the ramp (Figure 7.83*D*). From that point on, the fold grows in breadth but not in height as thrusting continues.

Suppe (1983) has worked out equations that describe all the angular interrelationships in fault–bend folding, including equations that permit construction of folds in duplexes. As a consequence, it is possible to generate compelling computer models showing the evolution of the fold forms. More importantly, the equations permit interpreting the subsurface ramp–flat geometries on the basis of limb dips in the various sectors of the snakehead folds, as revealed in surface outcrops or drilling information.

Fault Propagation Folds

Another mechanism of regional tectonic folding is **fault propagation folding** (Suppe, 1985). The fold shape for fault propagation folds is determined by fault shape, and thus once again we are dealing with forced folding. A fault propagation fold is like a "process zone" at the advancing tip of a fault. At each stage of development, hanging-wall strata move along a lower flat and up a ramp, but the ramp does not tie into an upper flat (Figure 7.84). Instead, the ramp thrust is replaced upward by an asymmetric fold, which is overturned in the direction of transport. As long as the thrust proceeds, a given unbroken fold just beyond the tip will eventually be overtaken by the fault as it advances. Thus the lowest reaches of a fault propagation fold system will be marked only by thrusting, the middle reaches by a faulted fold, and the upper reaches by an unfaulted fold (Figure 7.84). Suppe (1985) has worked out the precise geometric relationships that relate the fold geometry to the fault geometry.

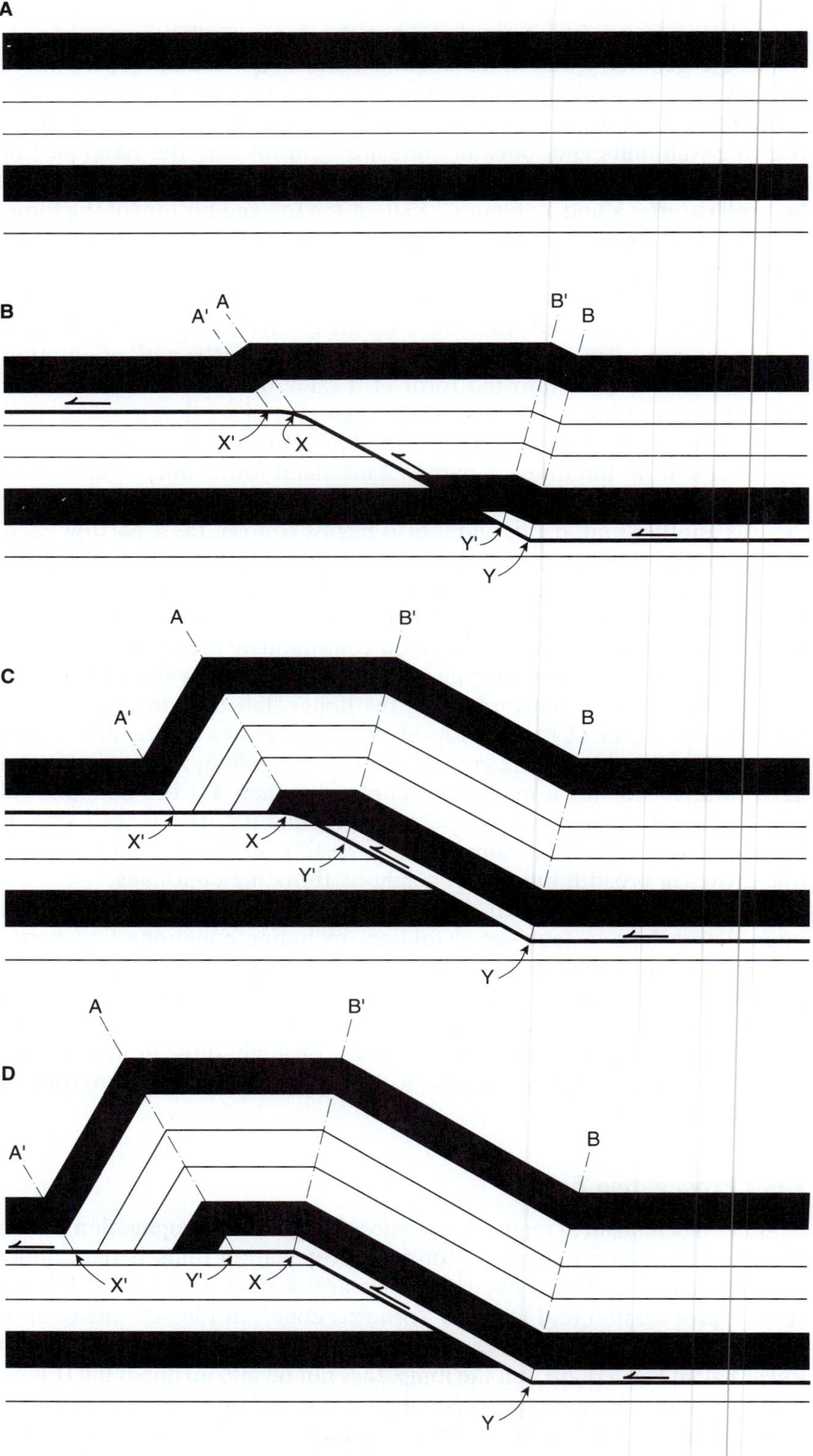

Figure 7.83 Progressive evolution of fault–bend fold geometries. (*A*) Flat-lying sequence of sedimentary rocks. (*B*) Thrusting begins. *X* and *Y* are footwall cutoffs; *X′* and *Y′* are the respective hanging-wall cutoffs. Beds flex into kinklike folds as they move from flat to ramp positions, and from ramp to flat positions. (*C*) Anticline grows in height and width. Width of kinklike fold gets larger and larger. (*D*) After hanging-wall cutoff *Y′* makes it to the top of the ramp, the fold ceases to grow in height; instead it grows in width. [From Suppe (1983), *American Journal of Science*, v. 283. Reprinted by permission of American Journal of Science.]

Monoclinal Folding

A very distinctive class of folds occupies the Colorado Plateau of the American Southwest. These folds are huge regional structures that commonly trend for more than 150 km. They are broad anticlinal, steplike folds that cause otherwise horizontal or very shallowly dipping strata to

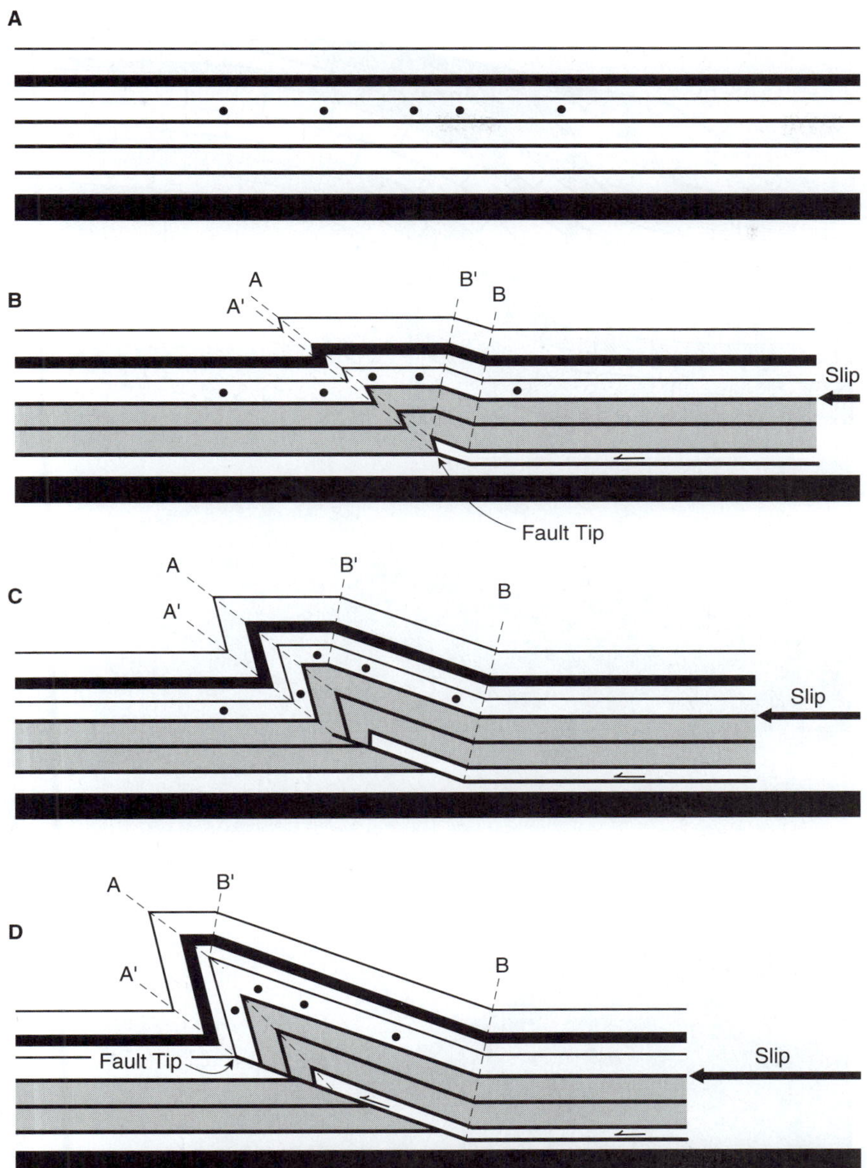

Figure 7.84 Progressive evolution of fault propagation fold geometries. (*A*) Flat-lying sequence of sedimentary rocks. (*B,C,D*) As thrusting takes place, the fault tip migrates steadily upward. Beyond the tip the fault is "replaced" by an overturned syncline, i.e., a fault propagation fold. Above the fault tip in the hanging wall there is a complementary fault propagation fold: an overturned anticline whose axial plane dips more steeply than the thrust. (From Suppe, J., *Principles of Structural Geology*, copyright © 1985, Figure 9.47, p. 351. Reprinted by permission of Prentice-Hall, Upper Saddle River, New Jersey.)

bend abruptly to steeper inclinations within very narrow zones (Figure 7.85). Asymmetric in profile form, monoclines are marked by two hinges (one anticlinal, one synclinal) connected by a middle limb. The middle-limb strata are generally smoothly curved and continuous, but sometimes they are broken by faults. The most spectacular monoclines show off middle limb dips of 90°! Structural relief on the Colorado Plateau monoclines commonly exceeds 1 km. The very largest monoclines reflect displacements that approach 3 km.

The pattern of monoclines in the Colorado Plateau is marked by sinuous multidirectional, branching folds (Figure 7.86) (Kelley, 1955; Davis, 1978). The major monoclines serve to mark the boundaries between the great uplifts and basins of the Colorado Plateau (Figure 7.86). For example, the Waterpocket monocline in Utah marks the eastern edge of the Circle Cliffs uplift and the western edge of the Henry Basin (Figures 7.86 and 7.87*A*). The Nutria monocline marks the western edge of the Zuni uplift (Figures

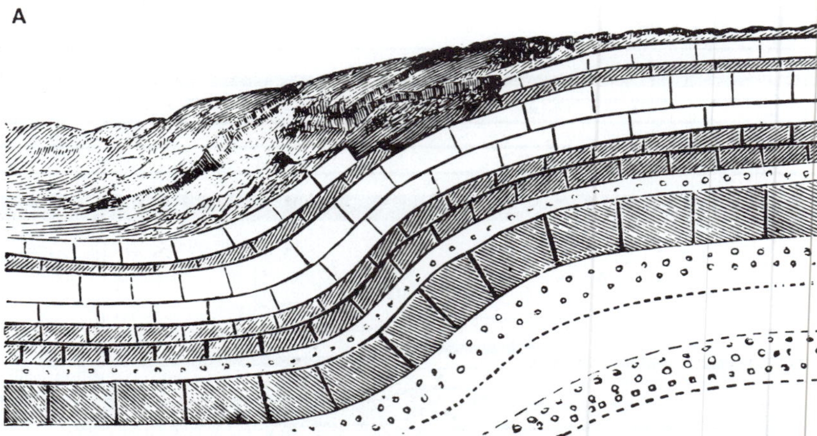

Figure 7.85 (*A*) First illustration of a monocline, by John Wesley Powell (1873). [From Kelley (1955), Geological Society of America.] (*B*) Distant view of the Hunters Point monocline along the eastern margin of the Defiance uplift in northeastern Arizona. (*C*) Close-up view of the Hunters Point monocline. Pennsylvanian sandstone abruptly bends from a horizontal to a vertical attitude. (Photographs by G. H. Davis.)

7.86 and 7.87*B*). Locally monoclines interfere with one another and constructively compound their respective displacements (Barnes, 1974; Barnes and Marshall, 1974).

Monoclines appear to be associated with ancient, reactivated, steeply dipping fault zones. This association is especially clearly revealed in the Grand Canyon, where deep erosion has exposed the "roots" of the East and West Kaibab monoclines. Separation relationships disclose the presence of ancient, reactivated Precambrian faults. Offsets among markers of different ages are inconsistent in magnitude of separation and sometimes in sense of separation as well.

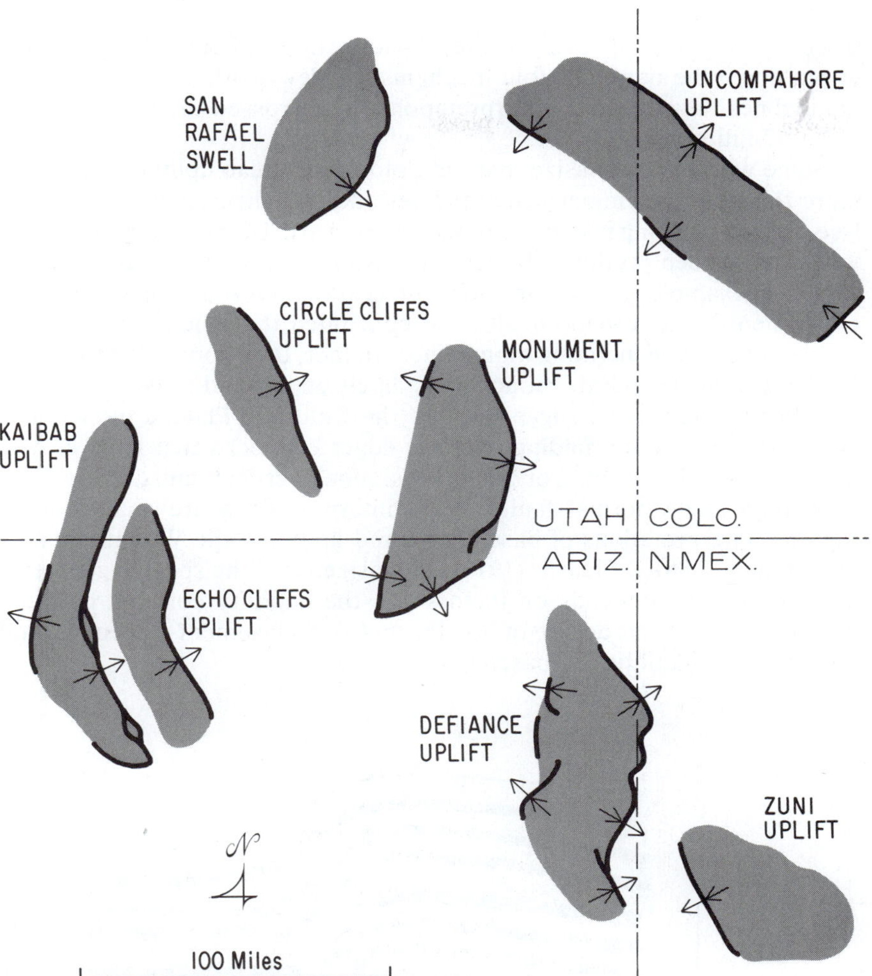

Figure 7.86 Pattern of monoclinal folding in the Colorado Plateau. Monoclines mark the boundary between the great uplifts and basins of the plateau.

Figure 7.87 (*A*) A ground-level view of the Waterpocket fold along the eastern margin of the Circle Cliffs uplift, Utah. (Photograph by G. H. Davis.) (*B*) An aerial view of the Nutria monocline along the western margin of the Zuni uplift, New Mexico. (Photograph courtesy of Vince Kelley.)

Monoclines are perhaps the ideal structure with which to close down this chapter on folding, because they illustrate how difficult it is sometimes to pin down the nature of fold mechanisms. Few workers agree on how monoclines have formed. Interpretations have crossed the full spectrum of possibilities.

Some workers emphasize that the Colorado Plateau uplifts are nothing more than giant asymmetrical anticlines and that their spacing reflects a free-folding buckling instability of the upper crust. The buckling instability would have been produced by tectonic loading during regional compression. For example, Reches and Johnson (1978) analyzed monoclines both experimentally and theoretically and concluded that buckling may be a viable option for generating monoclines. In fact, they concluded that kink folding can also generate monoclines, albeit on a grand scale.

Other workers have suggested that the Colorado Plateau monoclines are products of **drape folding** over the edges of blocks that have moved up and down like pistons, or piano keys, along vertical faults, or perhaps even high angle normal faults. Such interpretations are analogous to Stearns' interpretations of basement-cored uplifts in the Wyoming province (Figure 7.88*A*). Stearns (1978) emphasized that the specific geometry of drape folding depends on factors like the ductility contrast between basement and cover, degree of bonding between basement and cover, and the absolute ductilities of basement and cover.

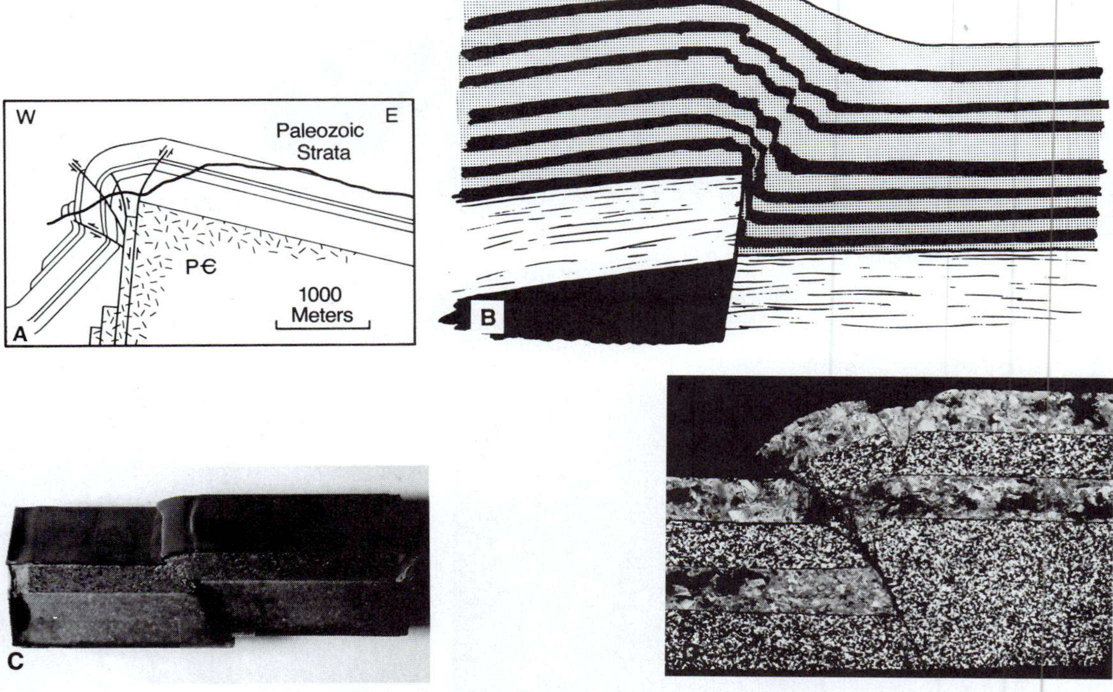

Figure 7.88 (*A*) Basement-cored uplift in Wyoming. [From Stearns (1978). Published with permission of Geological Society of America and the author.] (*B*) Experimental model picturing the relation between monoclinally folded layers and an underlying fault in basement. [From Davis (1978). Published with permission of Geological Society of America and the author.] (*C*) Experiment in forced folding. Precut basement block is forced to accommodate differential vertical uplift. Overlying layered materials passively fold. (*D*) Details of the deformation. [From Friedman and others (1976). Published with permission of Geological Society of America and the authors.]

Variations on the drape-folding theme include compression-induced uplift of basement blocks along high angle reverse faults. For example, I once made a monocline in the lab by deforming alternating layers of modeling clay and dry, powdered kaolinite that I had "deposited" on a basement of pine board. Before depositing the layers, I cut an ancient fault zone through the pine board with my saw. A little Vaseline on the sawcut combined with compressive layer-parallel loading of the pine board produced the fold shown in Figure 7.88*B*. Friedman and others produced drape folding through triaxial deformation of real rock samples (Figure 7.88*C*). Samples consisted of a "basement" of stiff lithologies (like sandstone), precut in a reverse-fault orientation. Cover consisted of softer lithologies. The deformation is such that the overall size, shape, and trend of the fold in the layered sequence reflects the size, shape, and trend of the basement block (Figure 7.88*D*).

More recent work on interpreting monoclines has focused on themes analogous to fault–bend folding and fault propagation folding. The most provocative images of monoclinal folding and basement-cored uplifts are embodied in the detailed drawings of specimens deformed by Chester, Logan, and Spang (1991). They prepared real-rock specimens consisting of sandstone basement cut by a 20° sawcut, simulating a low-angle thrust climbing up through Precambrian basement (Figure 7.89). Overlying the sandstone they placed thinly layered limestone, in turn interlayered with lead or mica. The specimens, which measured 11 × 3.2 × 2.9 cm, were subjected to layer-parallel loading under confining pressure conditions of 50 MPa. The results are fascinating (Figure 7.90): deformation maps of structural configurations that closely resemble the monoclines and basement-cored uplifts of the Colorado Plateau.

Multiple Mechanisms

In the long run, it is probably unrealistic to assume that folds, in a given structural system, were formed by a single mechanism. Take, for example, folds in the Alps. Structural geologists worldwide recognize the Alps as the type locality for **fold nappes**, that is, giant overturned to recumbent folds occupying thrust sheets that have moved more than approximately 10 km relative to footwall rocks (Ramsay and Huber, 1987). Classic fold-nappe geometries are presented, for example, in cross sections by Heim (1921) (Figure 7.91). Fold nappes in fundamentally unmetamorphosed rocks in the Alps are commonly "rootless," that is, the thrust sheets, and the folds within the sheets, do not connect anywhere with rocks and structures in the authochthonous bedrock.

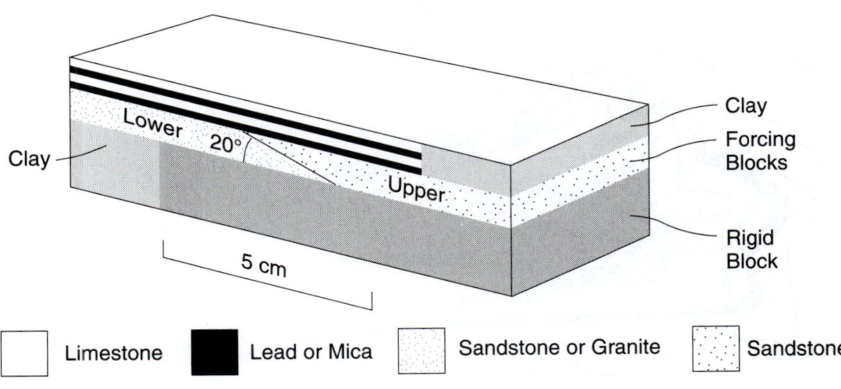

Figure 7.89 Real-rock sample, with precut thrust. [From Chester, Logan, and Spang (1991), with permission from Geological Society of America.]

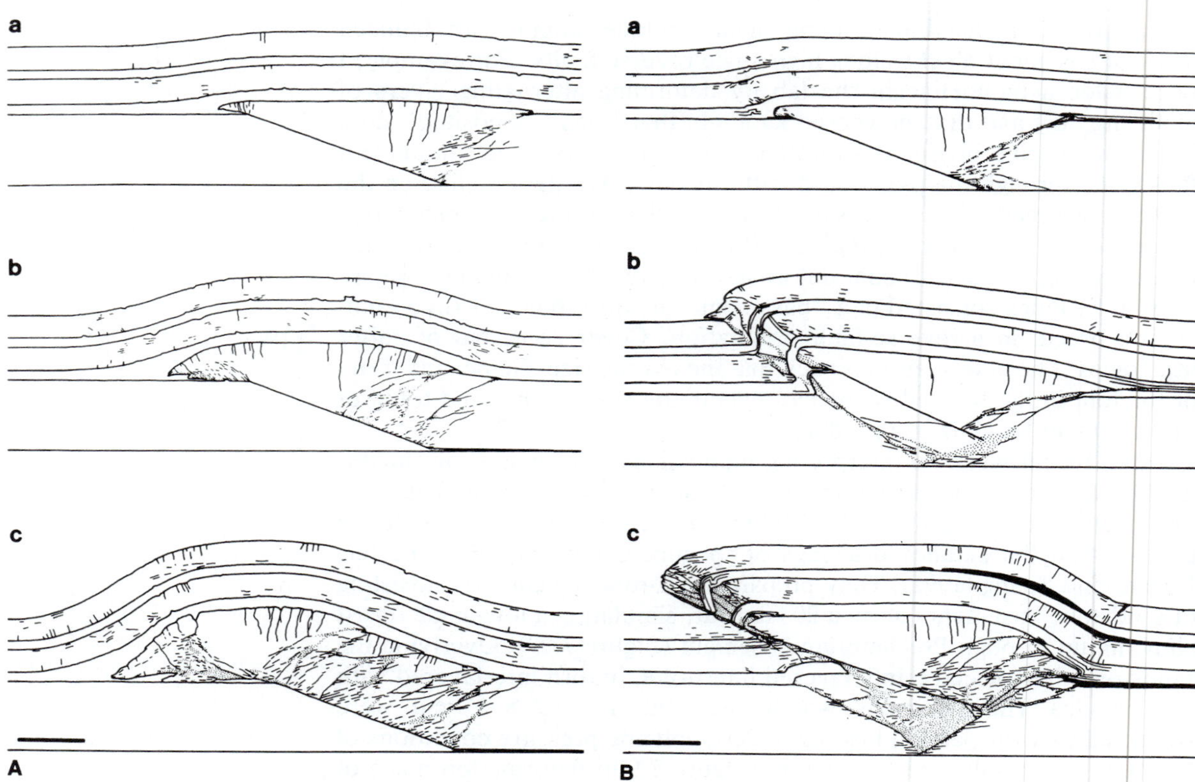

Figure 7.90 Results of the triaxial deformation by Chester, Logan, and Spang (1991). (*A*) Geometry controlled primarily by fault–bend folding. (*B*) Geometry controlled primarily by fault propagation folding. [From Chester, Logan, and Spang (1991), with permission from Geological Society of America.]

Butler (1992) points out that throughout the 1980s, there was a raging debate among Alpine geologists on the mechanisms by which the folds in the Alpine nappes were generated. Fault-bend folding? Buckling? Fault-propagation folding? The debate is still alive. For example, Rowan and Kligfield (1992) proposed that the only mechanism capable of explaining the wedge-shaped nappe geometries, as well as distribution of internal strain, is simple shear between the thrust faults bounding the nappes. And Ramsay and Huber (1987) have emphasized that a broad range of fold geometries can evolve in nappes as a consequence of the action of several mechanisms at work through time: for instance, pure fault-bend folding; a little buckling followed by pure fault-bend folding; fault-bend folding accompanied by penetrative shearing of the footwall beneath ramps;

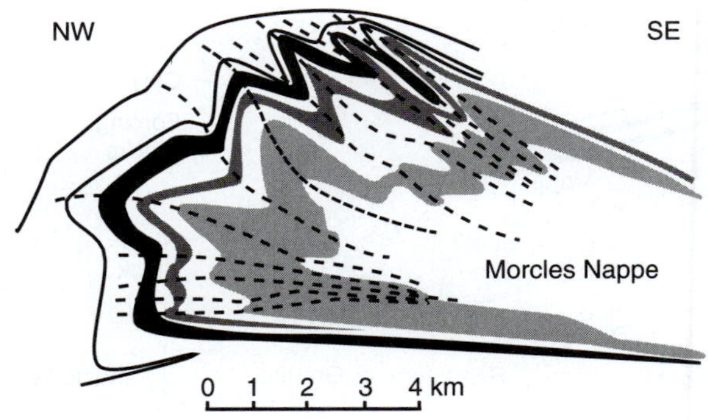

Figure 7.91 Cross section of Helvetic fold nappes near Lake Lucerne, Switzerland. (After Heim, 1921).

wholesale simple shear of the entire nappe; fault-propagation folding that takes place beyond the tip of brittle-ductile shear zones.

Thus we learn once again that we do not have to decide upon just one fold mechanism as we try to interpret folding within a given region. Instead, we acknowledge that a given regional geologic column, when subjected to stresses, has at its disposal a broad variety of mechanisms to use to accommodate shortening and transport. It is up to us to try to figure out how the fold mechanisms are *partitioned* to create a product marked by three-dimensional structural compatibility.

CONCLUSIONS

Folds are beautiful geological structures. The amazing variety of forms and styles reflects conditions of deformation, including the mechanical character of the layers that are folded. The ductility contrast between layers is especially sensitively recorded in the details of fold profiles. Folds have special practical value. Fold geometries and orientations can be used to interpret stress directions and tectonic transport directions. Fold closures can trap oil and gas, or mineralizing fluids. But there is more to come. Metamorphic rocks and shear zones are "loaded" with folds, with forms that reflect more plastic, even viscous, conditions of deformation.

CHAPTER 8

CLEAVAGE, FOLIATION, AND LINEATION

NATURE OF CLEAVAGE

General Outcrop Appearance

Folded sedimentary and metamorphic rocks often display a fundamental internal grain known as cleavage. The presence of cleavage in a rock permits the rock to be split into thin plates and slabs. The term "cleavage" is difficult to define: it broadly refers to closely spaced, aligned, planar to curviplanar surfaces that tend to be associated with folds and oriented parallel to subparallel to the axial surfaces of folds (Figure 8.1). As will become apparent, the penetrative parallel surfaces can take many physical forms. Cleavage is commonly penetrative at both the outcrop and microscopic scales (Figure 8.2). It typically cuts bedding discordantly, without much regard to the orientation of bedding.

When a rock possessing cleavage is smacked with a hammer, the rock will typically break along the cleavage. Similarly, when rocks possessing cleavage are subjected to scores of centuries of persistent weathering, the worn-down rock that survives in outcrop is commonly marked by sharp-edged, finlike projections that express the presence and general orientation of its internal grain (Figure 8.3). The slabby, platy nature of cleaved outcrops sometimes misleads us into thinking that cleavage is akin to fracturing. In truth, cleavage forms *without apparent loss of cohesion,* and in this respect alone cleavage surfaces are much different from fracture surfaces.

Figure 8.1 Well-developed cleavage exposed in folded rocks from the South Stack formation in North Wales. (Reprinted with permission from *Journal of Structural Geology,* vol. 2, "The Tectonic Implications of Some Small Scale Structures in the Mona Complex of Holy Isle, North Wales," J. W. Cosgrove. Pergamon Press, Ltd., Oxford, copyright © 1980.)

Figure 8.2 Penetrative cleavage as seen in photomicrograph of quartz-sericite schist from the Caribou mine area, New Brunswick, Canada. Folded black layer is composed of fine-grained pyrite. (Photograph by G. H. Davis.)

Geometric Relationship of Cleavage to Folding

Cleaved rocks are generally folded rocks, but folded beds are not always cleaved. Yet, almost always, when we do see cleavage in an outcrop, we can be certain that the beds in which the cleavage occurs have been strongly folded. A close geometric coordination exists between the orientation(s) of cleavage surfaces and the configuration of folded bedding. Ordinarily, cleavage sufaces either are perfectly parallel to the axial surface of folding, or they are disposed symmetrically about the axial surface in a **fan** of orientations (Figure 8.4). In either case, the cleavage surfaces comprise an **axial plane cleavage**. Folded bedding in an upright fold is cut by cleavage surfaces that everywhere are steeper than the inclination of

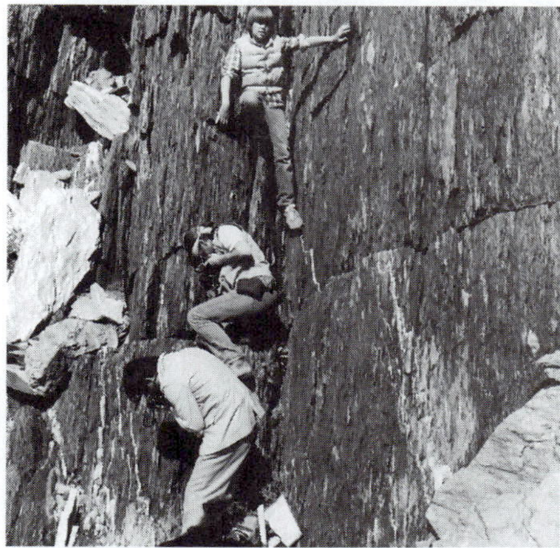

Figure 8.3 Weathered expression of cleaved metaconglomerates in Precambrian basement of northwestern Arizona. Stretched and flattened pebbles in the cleaved rocks are unusually interesting. From top to bottom of the photo are Mike Davis, Sue Beard, Ji Xiong, and "Mom," the dog. (Photograph by G. H. Davis.)

Figure 8.4 Syncline with axial planar cleavage. The cleavage "fans" symmetrically about the axial trace of the fold. These folded Silurian sandstones and shales are exposed approximately 3 mi (5 km) west of Hancock, Maryland. Geologist is C. W. Hayes. (Photograph by C. D. Walcott. Courtesy of United States Geological Survey.)

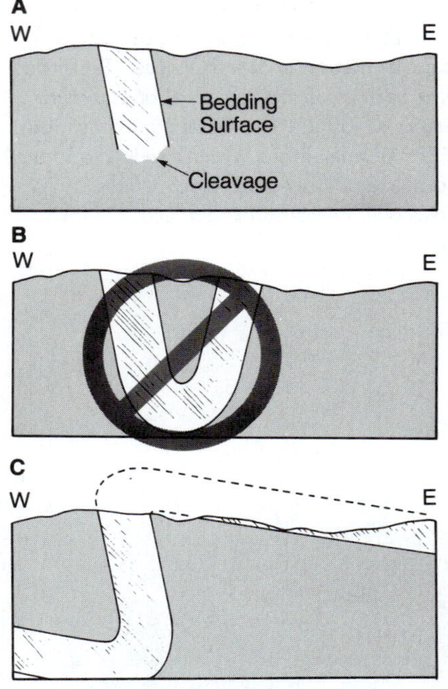

Figure 8.5 Use of the orientations of bedding and cleavage to construct the form of the fold with which the bedding and cleavage are associated. (*A*) The outcrop relationships. (*B*) Misfit between the cleavage orientation and the interpreted fold form. (*C*) A good fit!

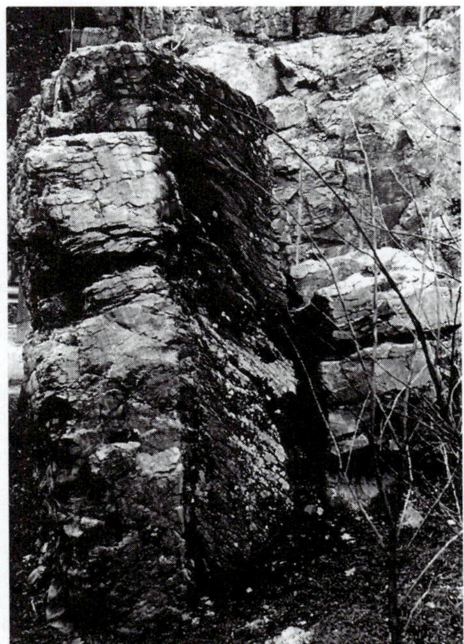

Figure 8.6 Photograph of part of Van Hise Rock, an official historic landmark in Wisconsin. Bedding is nearly vertical. Cleavage is confined to the shale layer (dark) and dips at approximately 40° from upper left to lower right. (Photograph by G. H. Davis.)

bedding. In overturned folds, cleavage can dip less steeply than bedding. *An axial plane cleavage that dips in the same direction as bedding, but less steeply than bedding, is a warning signal that bedding may be overturned.*

The relationship of axial plane cleavage to bedding in folded rocks can often be used to evaluate the likely **facing** of a bed, and to construct the configuration of folds in profile view. A useful application of the fundamental geometric relationship between bedding and axial plane cleavage is illustrated in Figure 8.5. In the outcrop shown in Figure 8.5*A*, bedding is cut by cleavage surfaces that are known to be "axial planar." Bedding dips 80°E, and the cleavage surfaces dip 45°E. Knowing that the cleavage is an axial plane cleavage, it is possible to determine the fold configuration of which the bed is a part. This is achieved by drawing a folded surface that maintains an axial planar relationship to the cleavage surfaces. If a fold profile is drawn in such a way that the east-dipping bedding represents the west limb of an upright syncline (Figure 8.5*B*), the form of the syncline cannot be fit in an axial planar manner to the cleavage surfaces. If, on the other hand, the east-dipping bedding is considered to be part of the overturned west limb of an overturned anticline (Figure 8.5*C*), the form of the fold is perfectly compatible with a 45°E-dipping axial plane cleavage.

A reference location for recognition of this relationship is Van Hise Rock in Wisconsin (Figure 8.6), where the difference in orientation between bedding and cleavage is striking. The outcrop is like a vertical sandwich, with cross-bedded sandstone layers on the outsides and dark shale in the middle. Cleavage in the shale dips from upper left to lower right, in the same manner pictured in Figure 8.5.

Bedding and cleavage surfaces are carefully distinguished on geologic maps. The common map symbol for cleavage is shown in Figure 8.7, a simplified geologic map of a plunging anticline/syncline pair. Cleavage symbols in combination with bedding symbols serve to highlight the interrelationships of bedding and cleavage across folds. Where cleavage surfaces cut through the hinge of a fold, there is a maximum discordance between bedding and cleavage. At the hinge point proper, the discordance is fully 90°. At each point on the limb of a fold, cleavage surfaces generally cut bedding at some small, acute angle. The angle of intersection steadily decreases from the hinge to the inflections of a fold. Isoclinal folds present the special case in which cleavage surfaces and bedding on the fold limbs are perfectly parallel to each other.

The orientation of cleavage surfaces in the hinge of a fold is a close approximation to the axial surface of the fold. However, the orientation of cleavage surfaces at any one point on the limb of a fold usually does not reflect the orientation of the axial surface of folding, simply because

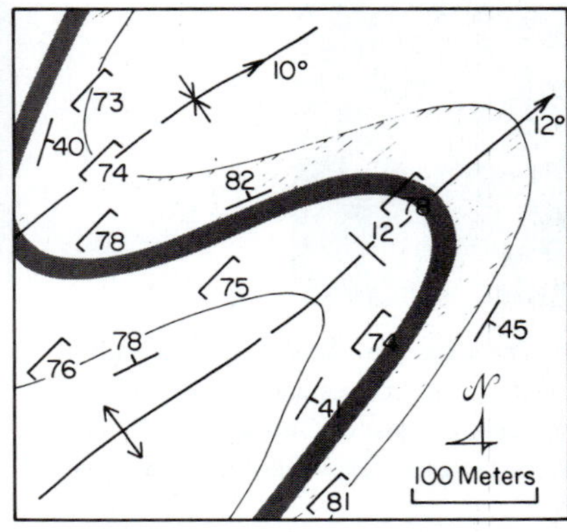

Figure 8.7 Geologic map expression of the relationship(s) between cleavage and folded bedding.

cleavage surfaces generally display a fanning of orientations across the folded surfaces. Also, cleavage can bend, or **refract**, slightly as it passes between two rocks with differing mechanical properties, such from shale to sandstone and back to shale.

The clear geometric harmony between cleavage and folding leads to the conclusion that cleavage forms as a response to shortening and flattening. The surfaces of cleavage lie in the S_1S_2 plane of the strain ellipsoid perpendicular to the direction of minimum finite stretch (S_3). More on this later.

Geometric Relationship of Cleavage to Shearing

Fault zones and shear zones may contain sheared rocks, even when no folds are present. We sometimes see subtle, delicate, penetrative cleavage surfaces in gouge zones along faults (Figure 8.8*A*). When cleavage is found in this kind of structural setting, its orientation is typically aligned at a small acute angle to the fault zone itself. Brittle–ductile shear zones can show the same relationship (see Chapter 9).

As we shall see, cleavage in fault zones and shear zones occupies an orientation of flattening, corresponding to the S_1S_2 plane of the strain ellipsoid (Figure 8.8*B*). This plane generally "leans over" in the sense of shear, thereby providing us with a way to interpret the sense of movement of fault zones and brittle–ductile shear zones, even in outcrops that lack offset marker units (see Chapter 9).

Domainal Character of Cleaved Rocks

When any cleaved rock is examined closely, the property called cleavage is found to be an expression of systematic variations in mineralogy and fabric. The term **fabric** refers to the total sum of grain shape, grain size, and grain configuration in a rock (Sander, 1930, 1970). The systematic variations in mineralogy and fabric that give rise to cleavage are not primary features related to the formation of the rock. Rather, they are expressions of the changes in mineralogy and fabric that were required to accommodate distortion of the rock body within which the cleavage is found.

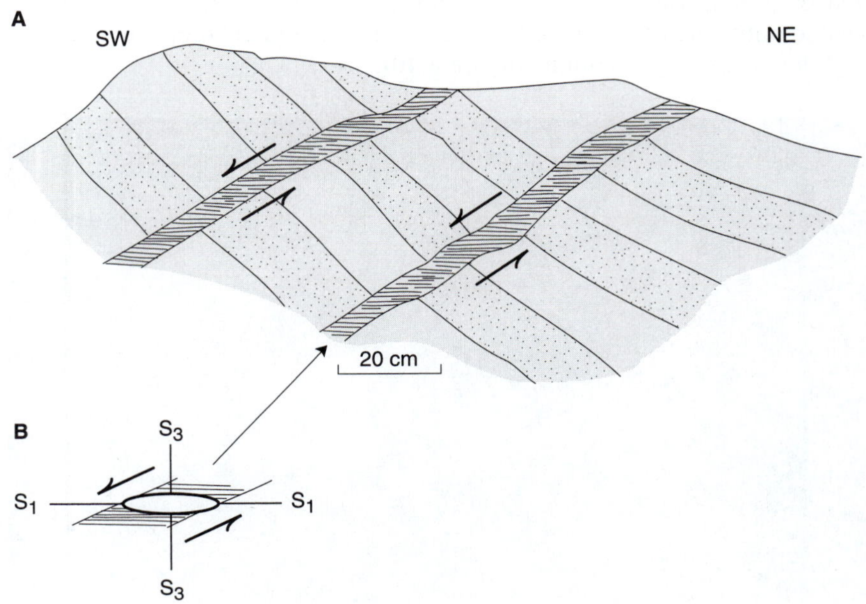

Figure 8.8 (*A*) Sketch based on outcrop relations in the San Manuel Formation (Miocene) near Tucson, Arizona. Sandstone and siltstone of the San Manuel Formation is cut by normal faults. The faults contain clayey gouge, and some of the gouge zones display a delicate penetrative cleavage. (*B*) The orientation of the cleavage in the gouge discloses the sense of movement, aligning itself with the direction of maximum finite stretch (S_1) and perpendicular to the direction of minimum finite stretch (S_3).

Figure 8.9 Excellent example of domainal structure in a quartz-mica schist exposed near Loch Leven, Inverness-shire Scotland. The cleavage domains are the dark, fine-grained micaceous zones. The microlithon domains are the light-colored, coarser grained zones of crenulated laminae of quartz and mica. (From *The Minor Structures of Deformed Rocks: A Photographic Atlas* by L. E. Weiss. Published with permission of Springer-Verlag, New York, copyright © 1972.)

The systematic variation in mineralogy and fabric in cleaved rock gives expression to the presence of what may be called **domainal structure**, that is, a kind of structural lamination composed of alternating **cleavage domains** and **microlithon domains** (Figure 8.9). Cleavage domains are thin, anastomosing to subparallel, mica-rich laminae within which the fabric of the original host rock has been strongly rearranged and/or partially removed. Minerals and mineral aggregates within cleavage domains show a strongly preferred dimensional and/or crystallographic orientation. Microlithon domains, or simply **microlithons**, are narrow lensoidal to trapezoidal slices of rock within which the mineralogy and fabric of the original host rock remain essentially preserved. Unless the microlithons are composed of rock that contains a preexisting cleavage, minerals and mineral aggregates in microlithons tend to be equigranular, lacking a conspicuous preferred orientation. Microlithons are sharply or gradationally bounded on either side by cleavage domains.

The domainal structure of some cleaved rocks is apparent in outcrop and/or thin section. However, in many cases the domainal structure is visible only when the cleaved rock is scrutinized microscopically at very high levels of magnification (Figure 8.10).

Figure 8.10 Photomicrograph showing domainal structure in mica schist. Oriented micas comprise cleavage domains. The cleavage domains separate microlithon domains of quartz, feldspar, and mica. (Photograph by D. M. Sheridan. Courtesy of United States Geological Survey.)

TYPES OF CLEAVAGE

There are many ways to name and classify cleavage. Most classifications are now anchored in the insightful work of Dennis (1972) and Powell (1979). Both Dennis and Powell recognized that it is practical to subdivide cleavage into two classes on the basis of the **scale** at which the domainal character of cleavage can be recognized. Where domainal structure (i.e., the distinction between cleavage domains and microlithons) can be seen with the unaided eye, the cleavage can be described as **disjunctive cleavage**, meaning interrupted. Where the domainal character of a cleaved rock is too fine to be resolved without the aid of a petrographic or an electron microscope, the cleavage is described as **continuous cleavage**.

Continuous Cleavage

The main types of continuous cleavage are **slaty cleavage**, **phyllitic structure**, and **schistosity**. All three are generally associated with strongly folded and distorted metasedimentary and metavolcanic rocks. Slaty cleavage, phyllitic structure, and schistosity are reasonably similar in outcrop expression, but they differ in grain size and in the scale of development of domainal structure. There is a complete gradation from slaty cleavage to phyllite and then schist, largely reflecting an increase in grain size.

Slaty cleavage is typically associated with very fine-grained (<0.5 mm) pelitic (shaley) rocks metamorphosed to low grade. Where slaty cleavage is well developed, it imparts to rocks an exquisite splitting property. Indeed, the presence of slaty cleavage allows a rock to be cleaved into perfectly tabular, thin plates or sheets (Figure 8.11). Roofing slates and old-fashioned slate blackboards owe their existence and usefulness to slaty cleavage.

Figure 8.11 (*A*) The phenomenal splitting capacity of slates, as displayed in the "middle quarry" of the Penrhyn Slate Company, Washington County, New York. (Photograph by C. D. Walcott. Courtesy of United States Geological Survey). (*B*) Specimen of Martinsburg Slate from the Delaware Water Gap region, New Jersey. The planar surface on the left is a natural break controlled by the cleavage. The texture of the cross section resembles wood fiber; it is the trace of cleavage. Note that a band can be traced across the flank of the specimen. This is relict bedding, oriented discordantly to the slaty cleavage. (Photograph and © by Peter Kresan.)

In Wales, there are large underground slate mines, and the mining is carried out in a way that exploits the differences between bedding and cleavage. In Machynlleth, Wales, I photographed a diagram of the mining operation (Figure 8.12); the slate beds are considered to be "veins" whose walls are parallel to bedding. The "cuts" are made parallel to cleavage, taking advantage of the direction of weakness.

The splitting capacity of schist is not nearly as elegant as that of slates, but it is nonetheless very pronounced (Figure 8.13). Rocks with **schistosity** are typically medium-grained (1–10 mm), containing flakes of mica that are visible in hand specimen. The grain size, which is larger than that of slates, mostly reflects greater recrystallization accompanying metamorphism. The most obvious outcrop characteristic of schistosity is the parallel, planar alignment of micas, including muscovite, biotite, chlorite, and sericite. Schists seldom split cleanly and evenly when struck with a hammer. Instead, they break off in the form of discoidal to crudely tabular hand specimens or slabs. Many outcrops of schist have a lovely sheen (Figure 8.14).

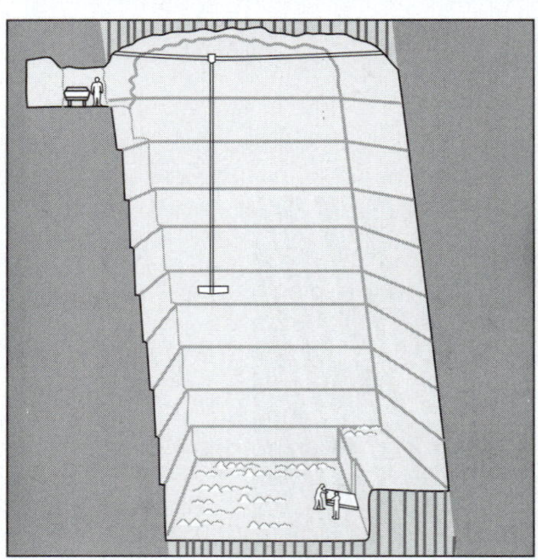

Figure 8.12 "Diagramau yn dangos y whthien o'r lechen a'r dull o'i chloddio." ("Diagrams showing the vein of slate and how it is extracted.") This illustration, based on a display I photographed in the Visitor's Center in Machynlleth, Wales, shows beautifully the difference between bedding and cleavage. The slate occurs in what the miners call "veins." Veins are the beds themselves. Cleavage is discordant to the bedding. Thus when the miners split out the slate, they do so along a direction that is oblique to bedding.

Figure 8.13 (*A*) Outcrop expression of schistose rock in northern Norway near Bjornfeld. Schistosity dips steeply from upper left to lower right. Steve Naruk is the proud owner. (*B*) More distant view of the expression of the schistosity in the landscape. (Photographs by G. H. Davis.)

Figure 8.14 Mica schist with porphyroblasts of andalusite, Pioneer Mountains, Idaho. (Photograph by S. J. Reynolds.)

Schistosity is best developed in pelitic metasedimentary rocks and certain volcanic rocks metamorphosed to medium or high grade. It is locally present in some granitic rocks, in which shearing under fluid-rich conditions caused feldspar to be converted into abundant white mica.

Phyllitic structure is intermediate in grain size and overall character between slaty cleavage and schistosity. In outcrop, phyllites display a soft, pearly, satiny luster. They glisten in the sun, but lack the distinct individual mica grains seen in schists. Phyllites exhibit the capacity to split neatly but not perfectly.

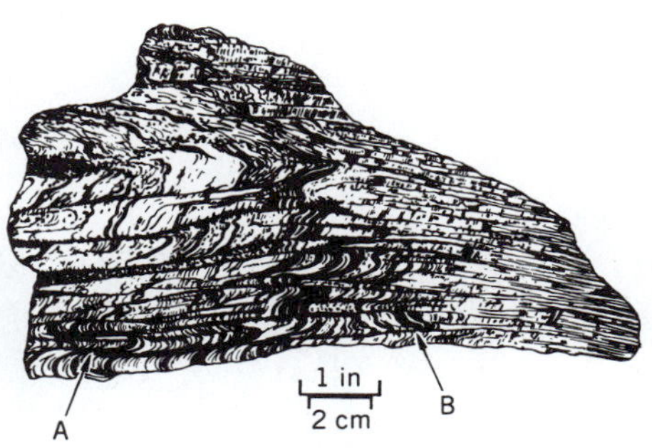

Figure 8.15 Specimen displaying crenulation cleavage. [From *Structural Geology of Folded Rocks* by E. T. H. Whitten, after Balk (1936). Originally published by Rand-McNally and Company, Skokie, Illinois, copyright © 1966. Published with permission of John Wiley & Sons, Inc., New York.]

Disjunctive Cleavage

There are two main types of disjunctive cleavage: crenulation cleavage and spaced cleavage. **Crenulation cleavage** is very distinctive in that it cuts a host rock that possesses a preexisting continuous cleavage, especially phyllitic structure or schistosity (Figure 8.15). In rocks that contain crenulation cleavage, a preexisting continuous cleavage is typically "*crenulated*" into microfolds. Two kinds of crenulation cleavage are recognized. One is discrete; the other is zonal (Gray, 1977a). **Discrete crenulation cleavage** is a disjunctive cleavage in which very narrow cleavage domains sharply truncate the continuous cleavage of the microlithons, almost like tiny faults (Figure 8.16*A*). **Zonal crenulation cleavage**, on the other hand,

Figure 8.16 (*A*) Discrete crenulation cleavage in metamorphosed tuffaceous rocks (Mesozoic), Granite Wash Mountains, western Arizona. Thin white bands are calcite veins, formed during a previous deformation event. The cleavage developed during thrusting. (*B*) Zonal crenulation cleavage in Mesozoic metasedimentary rocks, Granite Wash Mountains, western Arizona. (Photographs by S. J. Reynolds.)

is marked by wider cleavage domains that coincide with tight, appressed limbs of microfolds in the preexisting continuous cleavage preserved within microlithons (Figure 8.16*B*). Whether discrete or zonal, cleavage domains in rocks possessing crenulation cleavage are closely spaced, generally between 0.1 mm and 1 cm. Discrete crenulation cleavage tends to form in slate. Zonal crenulation cleavage tends to form in schist and phyllite.

Spaced cleavage is a second type of disjunctive cleavage. It consists of an array of parallel to anastomosing, stylolitic to smooth, fracturelike partings that are often occupied by clayey and carbonaceous matter (Nickelsen, 1972). Spaced cleavage is typically found in folded but unmetamorphosed sedimentary rocks, especially impure limestone and marl (Figure 8.17*A*,*B*), and in some impure sandstones as well. Spacing of the partings (i.e., the cleavage domains) typically ranges from 1 to 10 cm, and thus the microlithons are quite thick compared with all other cleavages. Thickness of the partings often is on the order of 0.02–1 mm, although they may be as thick as 1 cm or more.

Figure 8.17 (*A*) Outcrop expression of spaced cleavage in folded Permian limestone in the Agua Verde area near Vail, Arizona. Outcrop is being assaulted by a structure class from the early '70s. (*B*) Close-up view of the spaced cleavage, revealing clayey partings along cleavages. (*C*) Even closer view showing truncation of bedding laminations along cleavage surfaces. (Photographs by G. H. Davis.)

A fundamental characteristic of spaced cleavage is the **offset** (i.e., separation) of bedding markers along the cleavage. Offset of bedding along spaced cleavage is commonly seen in outcrop (Figure 8.17*C*). Although the offsets associated with spaced cleavages are faultlike, the cleavage surfaces are certainly not faults. Cleavage domains associated with bedding offsets are never marked by slickenlines or polish. And truncation of fossils at the boundaries of microlithons cannot be completely restored by fault-slip motions, for material has been lost. (Groshong, 1975a).

MICROSCOPIC PROPERTIES OF CLEAVAGE

Slaty Cleavage

Microscopic, high magnification examination of rocks possessing slaty cleavage reveals a fabric marked by discoidal to lenticular aggregates of quartz, feldspar, and minor mica enveloped by anastomosing, discontinuous mica-rich laminae (Figure 8.18). The micaceous laminae known as **M-domains**—that is, **mica-rich domains**—constitute cleavage domains. The discoidal, lenticular quartz–feldspar aggregates known as **QF-domains**—that is, **quartz–feldspar domains**)—comprise microlithons. The scale of development of domainal structure in slaty cleavage is mighty small. Thickness of the QF-domains typically ranges from 1 mm to less than 10 μm. The M-domains are typically only 5 μm thick (Roy, 1978).

The QF-domains in rocks possessing slaty cleavage provide a glimpse of the nature of the original host rock. Except for micas, individual grains and mineral aggregates tend to be equigranular, lacking a conspicuous preferred orientation. In sharp contrast to the QF-domains, the M-domains are zones within which the original fabric of the rock is almost completely reconstituted, transformed into strongly oriented intergrowths of aligned mica, quartz, and feldspar. The micas show the most conspicuous alignment, but hidden among the micas are flat to lensoidal quartz and feldspar grains, aligned parallel to the overall orientation of micas and the M-domains.

The lenslike, "flattened" nature of individual grains in slaty cleavage is further accentuated by **overgrowths** of chlorite and quartz. The overgrowths are like **beards**, growing from the "chins" of relatively large

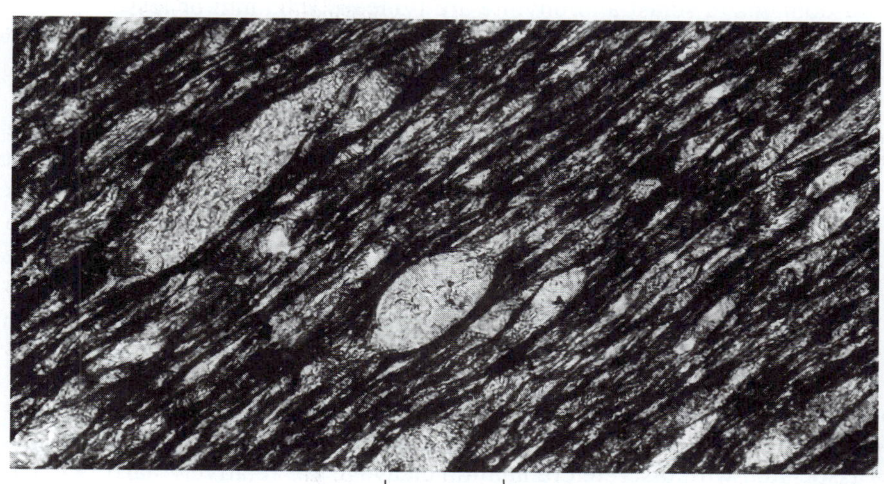

Figure 8.18 Domainal microfabric in slaty cleavage from the Ribagorzana Valley area, Spanish Pyrenees. Mica-rich domains (M-domains) are the black laminae that 'anastomose' around large quartz grains and aggregates of the QF-domains. (Photograph by W. C. Laurijssen. From *An Outline of Structural Geology* by B. E. Hobbs, W. D. Means, and P. F. Williams. Published with permission of John Wiley & Sons, Inc., New York, copyright © 1976).

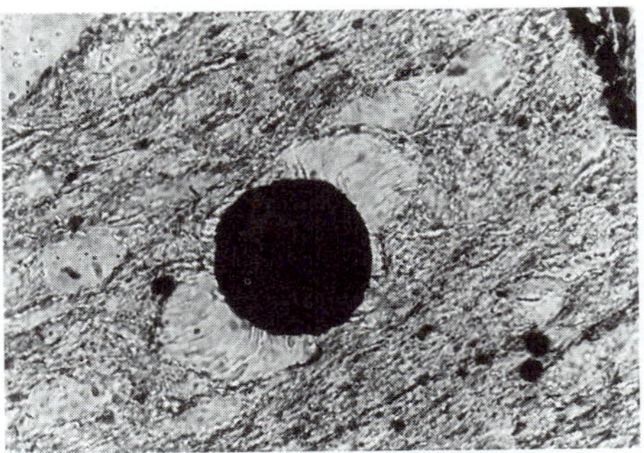

Figure 8.19 Photomicrograph of overgrowths of chlorite and quartz on pyrite (black). The "beards" are oriented subparallel to slaty cleavage. They grew under the protection of the strong pyrite grain that refused to flatten parallel to cleavage. Trace of slaty cleavage is from lower left to upper right. Martinsburg Slate, Delaware Water Gap, New Jersey. (Photograph by E. C. Beutner.)

grains of quartz, feldspar, and pyrite (Figure 8.19) (Roy, 1978). Crystal fiber beards grow in the ''plane'' of cleavage, as a response to the influence of directed stress.

One of the surprising revelations of high magnification examination of slaty cleavage is that the M-domains are somewhat curved and anastomosing. This irregularity at the microscopic scale seems to be inconsistent with the capacity of slates to split along ''perfectly'' planar and parallel surfaces. The inconsistency is only ''apparent'' and reminds us once again of the influence of scale on our geologic observations.

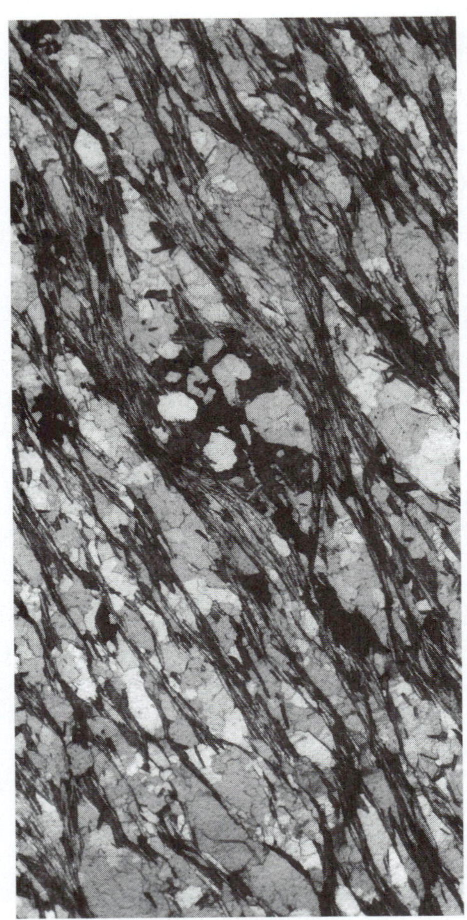

Figure 8.20 Domainal microfabric in schist from Ducktown, Tennessee. Micas form films that envelope aggregates composed principally of quartz. (Photograph by W. C. Laurijssen. From *An Outline of Structural Geology* by B. E. Hobbs, W. D. Means, and P. F. Williams. Published with permission of John Wiley & Sons, Inc., New York, copyright © 1976.)

Schistosity and Phyllitic Structure

Like the fabric of slaty cleavage, the microscopic fabric that gives expression to schistosity and phyllitic structure is composed of anastomosing M-domains and lenticular QF-domains, the cleavage domains and microlithon domains, respectively. The parallelism of micas within the M-domains imparts to schists and phyllites their fundamental splitting capacities (Figure 8.20). These mica-rich cleavage domains contain lenslike and disklike quartz and feldspar grains, which are commonly overgrown by chlorite and quartz at their tips. Quartz and feldspar grains within the QF-domains contain a reasonably preserved record of the fabric of the original host rock, albeit one that may be slightly reconstituted by the effects of recrystallization.

The fundamental distinction between phyllitic cleavage and schistosity is simply one of grain size. Phyllites tend to be fine-grained, with average grain diameter less than 1 mm. Schists tend to be medium-grained, with average diameter ranging from 1 to 10 mm. Wispy anastomosing M-domains in schist and phyllite are typically 0.05 mm or less.

Crenulation Cleavage

The microscopic fabric of crenulation cleavage is quite distinctive. The cleavage domains are M-domains packed with aligned, interlocking micas surrounding lensoidal quartz and feldspar grains as well as opaque minerals and clots of carbonaceous material (Gray, 1979) (see Figure 8.16). The microlithons are QF-domains composed of a preexisting continuous cleavage, like slaty cleavage, phyllitic cleavage, or schistosity. The continuous cleavage that makes up the microlithons of crenulation cleavage is typically ''crenulated'' into unbroken waveforms of tiny folds. Axial surfaces of the folds are subparallel to crenulation cleavage.

The physical and geometric relation of M-domains to QF-domains depends on whether the crenulation cleavage is discrete or zonal. M-Domains associated with discrete crenulation cleavage are relatively narrow micaceous laminae along which the continuous cleavage of adjacent microlithons is abruptly and sharply truncated. They are faultlike (Figure 8.21*A*).

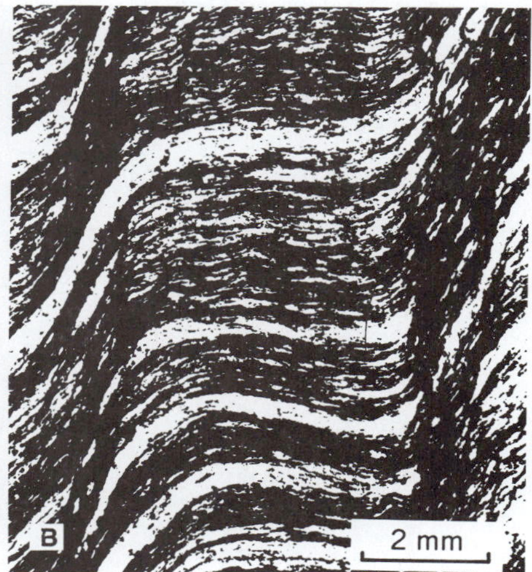

Figure 8.21 (*A*) Example of discrete crenulation cleavage. Folded schistosity is abruptly truncated along crenulation cleavage. Vishnu Schist in the Grand Canyon. (Photograph by S. J. Reynolds). (*B*) Photomicrograph of zonal crenulation cleavage (vertical) coincident with the steep limbs of asymmetric folds in schistosity. The zonal cleavage domains are carbonaceous and micaceous. They have a distinctively lower proportion of quartz than that of the initial fabric. [From Gray (1979), *American Journal of Science*, v. 279.]

In contrast, zonal crenulation cleavage is marked by M-domains that are relatively wide and serve as unbroken fold limbs connecting microfolds in the continuous cleavage of adjacent microlithons (Marlow and Etheridge, 1977; Gray, 1979) (Figure 8.21*B*). Whether associated with discrete or zonal crenulation cleavage, micas within M-domains are oriented within 5° of the orientation of the cleavage domains as a whole.

Microlithons in rocks marked by crenulation cleavage are very rich in quartz and feldspar but very poor in micas (Marlow and Etheridge, 1977; Gray, 1979). Cleavage domains in rocks marked by crenulation cleavage are very rich in micas but very poor in quartz and feldspar. To be sure, the segregation in crenulation cleavage of mica-rich domains and domains rich in quartz–feldspar is strikingly conspicuous, reflecting a strain-induced differentiation that we consider shortly.

Spaced Cleavage

Microscopic examination of cleavage domains in rocks cut by spaced cleavage reveals that they are fracturelike discontinuities lined with seams or films of clayey and/or carbonaceous material (Figure 8.22). The clayey

Figure 8.22 (*A*) Photomicrograph of anastomosing, dark, undulating spaced cleavage seams. (*B*) Scanning electron micrograph of the spaced cleavage seams. Composed of densely packed clays, these seams are markedly straight in their trace expression (vertical). Clays in intervening microlithons are more loosely packed, not as preferentially oriented. [From Gray (1981), *Tectonophysics*, v. 78. Published with permission of Elsevier Scientific Publishing Company, Amsterdam.]

material in the seams and films tends to be aligned parallel to the cleavage itself. The spaced cleavage may be stylolitic, but more commonly it is anastomosing, or smooth.

Microlithons between cleavage domains in rocks cut by spaced cleavage typically lack a preexisting continuous cleavage. Rather, the host rock for most spaced cleavage is simply unmetamorphosed impure sedimentary rock, especially limestone, marl, and clayey sandstones.

STRAIN SIGNIFICANCE OF CLEAVAGE

The Issues

The exact role of cleavage has been debated vigorously for more than a century. There are many things that require explanation! These include the geometric coordination of folds and cleavage, the mechanical role of cleavage in the folding process, the kinematic development of preferentially oriented mineral grains and cleavage domains, the "division" of micas and quartz–feldspar aggregates into domainal structure, the presence of clay-filled partings in rocks possessing spaced cleavages, the development of oriented beardlike overgrowths of chlorite and quartz, the whereabouts of the missing parts of truncated fossils, and the overall strain significance of cleavage.

The debate on the origin of cleavage once focused on the relative contributions of rigid body rotation of individual platy minerals within the original host rock; preferred directional recrystallization of individual minerals within the original host rock; and simple-shear faultlike translation along extremely closely spaced fracturelike surfaces. Hardly a thought was given to whether the deformation might have been accompanied by volume loss or volume gain. Thus constant volume deformation was an underlying but unspoken assumption in the debates. Although each of these three mechanisms contributes to cleavage development, it now seems clear that the dominant mechanism for cleavage formation is pressure dissolution removal of original host rock. Wholesale removal of rock by pressure solution is perhaps the supreme strain accommodation to directed stress. In other words, when all else fails, get out of the way!

Strain Significance of Slaty Cleavage

Expression of Shortening

There is unmistakable evidence that slaty cleavage forms as a distortional response to extreme crustal shortening. Slaty cleavage is generally associated with folded rock (Figure 8.23). The intimate coordination of the geometry of folding to the orientation(s) of cleavage leaves little doubt that folding and slaty cleavage development are, at least in part, synchronous. Thus, if folds are considered to be products of shortening, slaty cleavage must be considered a product of shortening as well.

Fossils, reduction spots, and other primary structures that are found preserved in slates typically are *flattened* in the "plane" of cleavage. Flattened fossils and reduction spots thus provide a dramatic statement that slaty cleavage is indeed an expression of severe shortening.

The deformation of fossils in slates was recognized and appreciated more than a century ago in the slate quarries of Wales. Phillips (1844) was the first to point out the close relation between fossil distortion and slaty cleavage. Sharpe (1847) went a step further than Phillips, pointing

Figure 8.23 Syncline cut by strongly penetrative cleavage. These slates are located at "old quarry 2" at Slatington, Lehigh County, Pennsylvania. (Photograph by E. B. Hardin. Courtesy of United States Geological Survey.)

out that the most highly distorted fossils are associated with the most highly cleaved slates. Noting that the fossils are flattened in the cleavage surfaces and that the cleavage surfaces tend to be parallel and/or symmetrically disposed about the axial surfaces of associated folds, Sharpe concluded that *slaty cleavage forms perpendicular to the direction of greatest shortening*.

To estimate the amount of shortening accommodated by the formation of slaty cleavage, Sorby (1853, 1856) cleverly used reduction spots as a guide to distortion. He concluded that the presence of slaty cleavage can signal levels of distortion as great as 75%. Always ahead of his time, Sorby was able to demonstrate through strain analysis that the plane of cleavage in slates is statistically perpendicular to the direction of greatest shortening.

Since the time of the classic work by Phillips, Sharpe, and Sorby, many other geologists have addressed the strain significance of cleavage. The results are the same, time and time again. Oertel (1970), for example, analyzed slaty cleavage in a volcanic tuff unit in the Lake district of England, using ellipsoidal objects within the tuff as guides to strain. Oertel assumed that the objects were initially spherical, but were transformed into ellipsoids during folding and the development of slaty cleavage. He proceeded to show that cleavage surfaces in the tuff developed perpendicular to the direction of greatest shortening (Figure 8.24). Tullis and Wood (1975), like Sorby long before, used reduction spots to evaluate the state of strain in Cambrian slates in northern Wales. They concluded that the direction of greatest shortening responsible for the formation of the slaty cleavage was oriented precisely perpendicular to the cleavage. Shortening averaged about 65% in the rocks they examined.

Early Attempts to Explain the Alignment of Mica

Measuring the nature and degree of distortion in rocks possessing slaty cleavage is one thing. Determining exactly how the development of slaty cleavage allows rocks to shorten is yet another. Flattened fossils in slate apparently reflect some form of non–rigid body deformation. But how is it achieved?

Early studies on the origin of slaty cleavage focused on explaining the alignment of platy minerals, notably mica and clay minerals, because the preferred orientation of platy minerals was considered to be the chief

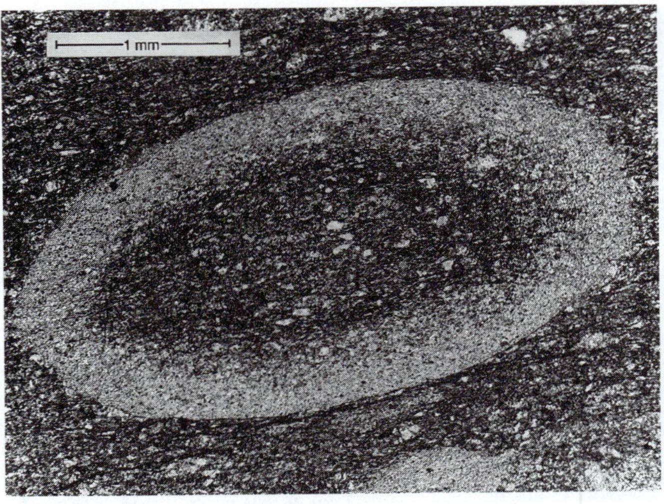

Figure 8.24 Distorted lapillus in slate derived from tuff. Lake district, England. [From Oertel (1970). Published with permission of Geological Society of America and the author.]

contributing factor to the physical expression of slaty cleavage. Explaining how platy minerals in slate are brought into preferred alignment was perceived as the key to understanding how the development of slaty cleavage achieves distortion.

Some of Sorby's contemporaries believed that mica and clay minerals in slate are brought into preferred, parallel alignment by recrystallization. Advocates of the role of recrystallization in the formation of slaty cleavage emphasized that micas and clay minerals grow preferentially in a common direction as a response to directed stress during metamorphism. The micas and clay minerals were pictured as growing perpendicular to the direction of greatest stress.

Although Sorby believed that recrystallization contributed to the development of slaty cleavage, he emphasized that the alignment of platy minerals was mainly due to rigid body rotation of these minerals during distortion of the host rock. To demonstrate the mechanism that he envisioned, Sorby (1856) experimentally deformed wax blocks containing evenly distributed and randomly oriented flakes of metals. Sorby verified that the rigid metal flakes progressively rotated into subparallel alignment as the shortening was accomplished. The greater the shortening of the wax block, the better the alignment. By measuring the initial and final orientations of the metal flakes with respect to the direction of shortening, and by comparing the initial and final orientations of the flakes in light of the magnitudes of shortening, Sorby anticipated one of the fundamental strain equations. Sorby clearly understood that each rigid metal flake, and perhaps each mica or clay in an evolving slate, rotates by an amount that is related to the initial orientation of the metal flake, the direction of shortening, and the percentage of shortening.

Although rigid body rotation of platy minerals into an alignment perpendicular to shortening provides a way to explain alignment of micas in M-domains, mechanical rotation *alone* fails to explain why micas are so densely concentrated in the M-domains. Mechanical rotation alone also fails to explain the presence of flattened, discoidal to lensoidal quartz in the M-domains.

Role of Recrystallization

In seeking additional explanations of the characteristics of slaty cleavage, we return to recrystallization as a mechanism for consideration. The role of **directional crystallization** in the formation of slaty cleavage consists of altering grain shape and enhancing the flattened, elongated appearance of minerals and mineral aggregates. In essence, new mineral growth takes place in the plane of cleavage—for example, within **pressure shadows** next to relatively large rigid mineral grains that can provide shelter from the harsh directed stresses that would otherwise inhibit the growth of new minerals. Pressure shadows of chlorite and fiber quartz are very common in M-domains in slates (Figure 8.25*A*). The shadows grow as microscopic beards from the tips of pyrite, feldspar, and quartz grains. The direction of crystal growth is the direction of incremental extension, regardless of the attitude of bedding cut by the cleavage (Figure 8.25B).

Like Sorby's model of grain rotation, the mechanism of recrystallization is by itself inadequate to explain many of the distinctive characteristics of slaty cleavage. It fails to explain the domainal character of slaty cleavage, and it fails to explain the extreme thinness of quartz and feldspar grains within the M-domains of slaty cleavage.

Role of Slip

The geometric resemblance of cleavage surfaces to tiny faultlike surfaces in certain outcrops of cleaved, folded beds suggested to some geologists that the fundamental deformation mechanism in forming cleavage is *slip*. "Folded layers" in a deck of cards would be the conceptual image (see Figure 1.37). But this proposed deformational mechanism could never be taken seriously because it called for fortuitous unbelievable slip patterns. Nor could this mechanism account for the microscopic properties of cleavage.

Pressure Solution Origin of Slaty Cleavage

Significant insights regarding the formation of slaty cleavage have been derived from fuller appreciation of the role of pressure solution in crustal

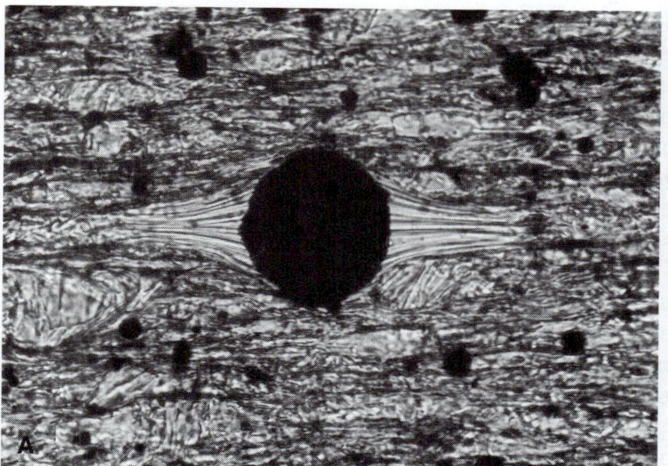

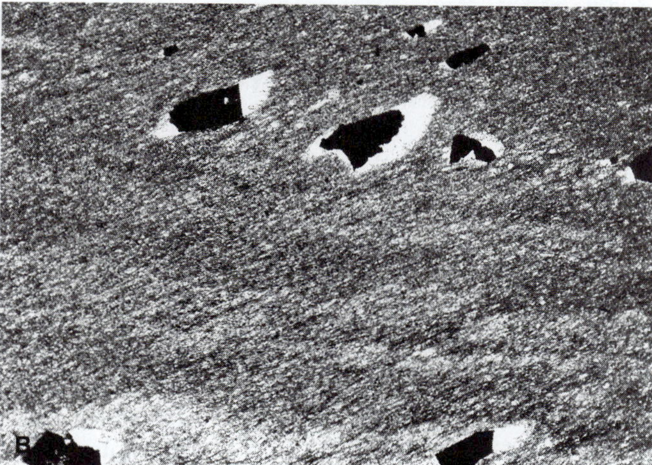

Figure 8.25 (*A*) Photomicrograph of pressure shadows containing fibrous quartz and chlorite. The pressure shadows are "attached" to a spherical pyrite aggregate. Diameter of pyrite is 36 μm. From fold in Martinsburg Slate, Delaware Water Gap, New Jersey. (Photograph by E. C. Beutner.) (*B*) Photomicrograph of feathery pressure shadows (crystal fiber beards) of quartz at the ends of pyrite crystals and calcareous slate. Note the faint horizontal trace of bedding in the matrix of this rock. The pyrite occurs mostly along the bedding, but the pressure shadows have formed parallel to cleavage. (Photograph by L. Pavlides. Courtesy of United States Geological Survey.)

deformation (recall Chapter 4). Based on an explosion of research since the early 1970s, it is now recognized that ''pressure solution'' can accommodate significant distortion in sedimentary and metamorphic rocks. Many of the attributes of slaty cleavage seem to be mineralogical and textural by-products of a shortening achieved through pressure solution.

We should not be disheartened to learn that Sorby's emphasis on the dominant role of rigid body rotation in the formation of slaty cleavage is now dwarfed by models emphasizing the pressure solution removal of material. In truth, Sorby was the first to point out the role of pressure solution in geological process, and he was the first to suggest that pressure solution contributes to the formation of slaty cleavage. Sorby's insights on the role of pressure solution were not derived simply from examination of hand specimens or outcrops of slate. Rather, they were based on microscopic observation, the primary source for fundamental appreciation of the structural significance of slaty cleavage. The saga of scientific achievements by Sorby continues to be imposing: Sorby prepared the very first thin section of a rock, and thus was the first to study the petrography of rocks microscopically using transmitted light. He attacked the microscopic characteristics of slate not by accident, but by design!

Awareness that minerals or parts of minerals can be removed from rock in solution undermines the commonly held premise of **constant volume deformation**. Volume losses of 50% and more are not unusual in highly cleaved rocks! One example of clear insight into the role of pressure solution is afforded by the deformational characteristics of quartzite–pebble conglomerates. It has been known for some time that when neighboring quartzite pebbles in a conglomerate are forced into contact during strong, penetrative deformation in a low grade metamorphic environment, the quartz in one or both pebbles is capable of dissolving (or diffusing) at the site of contact. The quartzite pebbles thus interpenetrate one another as a means of accommodating the requisite shortening (Figure 8.26). Pebbles that have experienced pressure dissolution creep, when extracted from the bedrock, display concave indentations not unlike chin-dimples. The concavities testify to removal of material. Rims of the dimples may be marked by **stylolitic halos**, where dark insoluble residue accumulated. It is the presence of a stress-induced chemical potential gradient that drives quartz away from sites of high stress concentration. The quartz reprecipitates into ''sheltered'' areas of low stress concentration, where the quartz can recrystallize as pressure shadows, veins, and/or beardlike crystal fiber overgrowths.

Figure 8.26 This specimen contains pebbles of Barnes Conglomerate (Precambrian) that experienced pressure dissolution creep during deformation. Pebbles were forced into contact with one another. The upper pebble seems to have had greatest resistance to pressure dissolution and it penetrated into the pebble below it. Stylolitic surfaces and insoluble residue can be seen at the contact. (Photograph and © by Peter Kresan.)

Microscopic characteristics of slaty cleavage show all the signs of pressure solution. Lenslike and trapezoidal grains of quartz and feldspar are corroded relics of what were once larger, more equant grains. The cleavage-parallel flanks of the lensoidal grains are facets marking the extent of advancement of pressure solution. So are the cleavage-parallel flanks of the lensoidal mineral aggregates that comprise QF-domains. The formation of the ultrathin quartz and feldspar grains in the M-domains reminds me of the Life-saver game my brothers and I would play as kids while riding for hours in the backseat of the car on family vacation. The object was to see who could hold a Life-saver on his tongue the longest before the Life-saver completely vanished. At the critical end stages of the game, we would stick our tongues out so that the competition could check to see if there was any Life-saver left. It is amazing how thin and transparent a Life-saver could become before disappearing altogether!

Some of the material missing from individual grains and grain aggregates may be accounted for in the presence of overgrowths and pressure shadows. But much of the dissolved rock must have passed completely out of the system, perhaps along the M-domains and along fractures. The densely packed concentrations of micas and carbonaceous matter in **cleavage seams** represent the accumulations of less soluble to insoluble residue of the original host rock. The strong preferred orientation of the micas in the M-domains, in some cases, may be the natural result of strain-induced progressive rotation of micas in response to the stress-induced removal of the surrounding rock matrix.

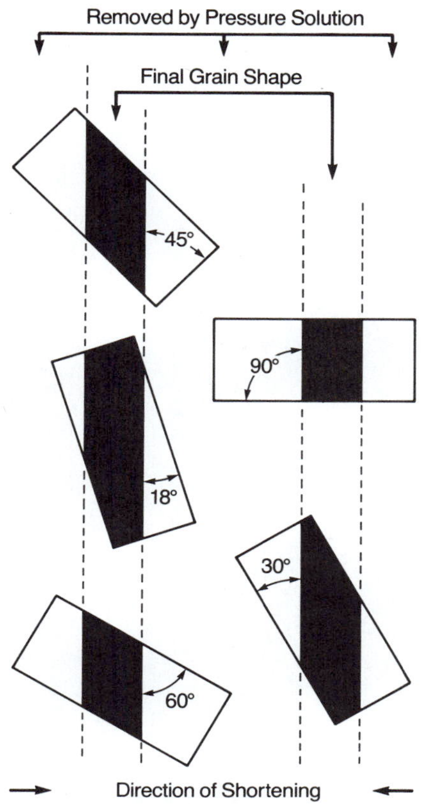

Figure 8.27 The formation of a preferred dimensional alignment of chlorite through progressive pressure solution. [From Beutner (1978), *American Journal of Science*, v. 278.]

Alignment of Mica Without Grain Rotation

One of my childhood next-door neighbors, Ed Beutner (1978), demonstrated that the preferred orientation of micas in slaty cleavage can take place without grain rotation. Beutner analyzed the structural geology and structural petrology of the Martinsburg Slate, a favorite target of structural geologists working in the eastern part of the central Appalachians.

Beutner's work focused on the chlorite grains, which he interpreted to be part of the mineral assemblage of the original pelite from which the slate was derived. Orientation analysis of the chlorite grains led him to conclude that the grain alignment of the chlorite was a by-product of the progressive systematic, selective corrosion of each original chlorite grain. The general mechanism is pictured in Figure 8.27. Beutner originally suggested that randomly oriented chlorite grains can become dimensionally aligned by virtue of preferential grain size reduction at right angles to the direction of greatest shortening. Depending on the original orientation of an individual chlorite grain with respect to the direction of greatest shortening, final grain shape due to corrosion can be a parallelogram, a rectangle, or a diamond. The resulting fabric is marked by preferred dimensional orientation of grains, but not by a preferred crystallographic orientation of grains.

Estimates of Volume Loss

The amount of shortening and volume loss accommodated by the pressure solution removal of host rock can be evaluated quantitatively in rocks containing abundant primary objects of known original shape and size. Wright and Platt (1982) analyzed shales in the Martinsburg Formation, using the geometric and dimensional properties of fossil graptolites in the slate as a guide to distortion. The deformed graptolites proved to be magnificent strain indicators. When Wright and Platt compared the sizes and shapes of the deformed graptolites with those of the undeformed counterparts, they found that graptolites oriented parallel to the trace of cleavage in the plane of bedding were narrower than normal (Figure 8.28). Graptolites oriented perpendicular to cleavage in the plane of bedding were found to be shorter than normal. (We would be too if acted on by the stresses that created the slates of the eastern Appalachians.)

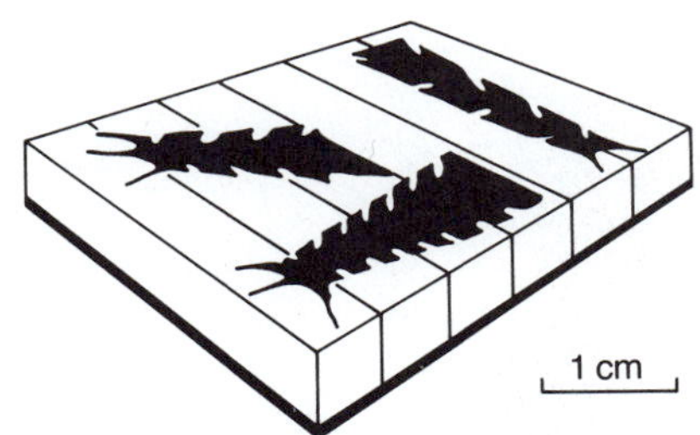

Figure 8.28 Telltale signs of pressure solution of graptolite-bearing slate. The observed size and shape of each distorted graptolite are systematically related to the orientation of each graptolite with respect to the direction of shortening. Graptolites in the plane of bedding that are oriented parallel to the trace of cleavage are long and narrow. Graptolites in the plane of bedding that are perpendicular to the trace of cleavage are short and fat. [From Wright and Platt (1982), *American Journal of Science*, v. 282.]

Examining deformed graptolites oriented at *all* angles to the direction of cleavage, Wright and Platt were able to show that the Martinsburg Formation shales were shortened by an average of 50%, perpendicular to the orientation of the slaty cleavage. Because shortening was accommodated by volume loss, not by constant volume deformation, the rock was not required to stretch out in any direction as a compensation for the shortening. Consequently, the shortening by an average of 50% reflects an average volume loss of 50%!

Continued studies of slaty cleavage will undoubtedly take structural geologists and structural petrologists into the twilight zone of submicroscopic investigations and thermodynamic considerations. Surely there are some surprises yet to surface. In the meantime we can rest assured that the presence of slaty cleavage, as well as schistosity and phyllitic structure, testifies to significant levels of crustal shortening and, in some cases, significant amounts of volume loss. Upon coming to know slaty cleavage, we can no longer limit our vision to processes of constant volume deformation. Slaty cleavage teaches us to recognize the interplay of **dilational** and **nondilational strain**.

Strain Significance of Crenulation Cleavage

The evaluation of the strain significance of crenulation cleavage serves to further underscore the importance of pressure solution in the formation of cleavage. Pressure solution nicely explains the conspicuous domainal fabric of crenulation cleavage, both discrete and zonal. Furthermore, it provides a means of understanding why micas are concentrated in thin bands and laminae (the M-domains), whereas the quartz–feldspar mineralogy is for the most part physically separated from the micas in separate domains (the QF-domains).

One of the most enlightening papers about the strain significance and mechanisms of formation of crenulation cleavage is that by Gray and Durney (1979). These authors emphasized that the development of crenulation cleavage involves a physical/chemical redistribution of minerals as a function of relative solubilities and chemical mobilities. They picture the cleavage domains as sites of the removal, by pressure solution, of substantial amounts of host rock, leaving behind insoluble residues of clayey and carbonaceous material. According to Gray and Durney, the pressure solution takes place on grain and/or layer boundary discontinuities oriented perpendicular to the direction of minimum finite stretch (the S_3-direction of the strain ellipsoid). Movement of dissolved material follows paths controlled by chemical potential gradients that relate in magnitude and direction to the local stress environment. Cleavage domains emerge along the limbs, or along the *former* positions of limbs, of microfolds in the continuous cleavage of the host rock (Figure 8.29). Spacing of the cleavage is related to the dominant wavelength of the microfolds and to the amount of solution-induced shortening across the limbs of the microfolds (Gray, 1977b, 1979).

Shortening accommodated by the formation of crenulation cleavage is achieved through a kind of progressive deformation through time, as illustrated in Figure 8.29. Continuous microfold waveforms are *buckled* into existence by layer-parallel and/or layer-inclined shortening of the

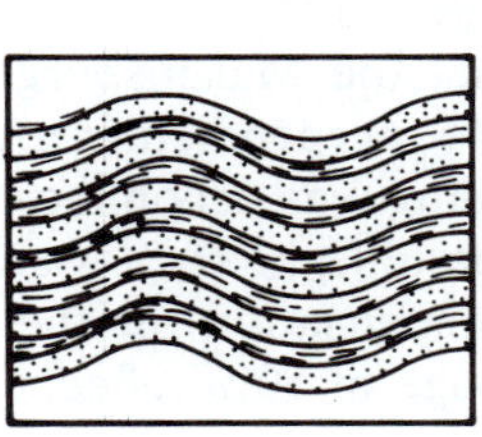

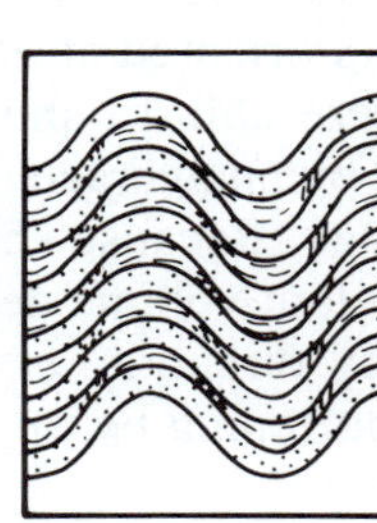

Figure 8.29 Accommodation of shortening through the development of crenulation cleavage. [From Gray (1979), *American Journal of Science*, v. 279.]

preexisting continuous cleavage (Marlow and Etheridge, 1977; Gray, 1979). The fold forms that emerge reflect the influence of thickness and mechanical character (including ductility contrast) of continuous cleavage laminae, degree of cohesion between the multilayers, and magnitude of shortening. If the shortening required by the strain environment surpasses what can be achieved through folding alone, the rock will begin to shorten by pressure solution loss of material. Dissolution takes place along the loci of fold limbs, and soon cleavage domains emerge within which quartz and feldspar become relatively depleted compared to that found in adjacent microlithon (QF) domains. When soluble mineral phases are dissolved along the fold limbs, they are transported in solution along chemical potential paths to fold hinges, where new minerals are deposited in the form of overgrowths and/or thin laminae. Some of the dissolved material exits the system altogether.

Cleavage domains (M-domains) may initially form at a variety of angles with respect to the direction of overall shortening, but progressive strain eventually brings the cleavage domains into subparallel alignment, perpendicular to the direction of greatest shortening. Substantial pressure solution can lead to the complete removal of fold limbs, leaving faultlike truncations of continuous cleavage within the microlithon (QF) domains.

Strain Significance of Spaced Cleavage

Stylolitic Surfaces

Spaced cleavage appears to be yet another product of pressure solution. Indeed, a revolution of thought has emerged from the discovery of the significant role of pressure solution in the formation of spaced cleavage in unmetamorphosed strata, especially impure limestones and marls. It is not surprising to find spaced cleavages associated with tightly folded strata. It *is* surprising, however, to discover that spaced cleavage can develop in essentially flat-lying strata, lacking conspicuous folds.

One of the clearest indications of pressure solution origin of spaced cleavage is the presence of geometrically and kinematically compatible **stylolitic surfaces** (Figure 8.30), telltale signs of dissolution and volume loss. Stylolitic cleavage surfaces tend to be occupied by clayey and carbonaceous matter (Nickelsen, 1972), the insoluble residue of pressure solution. Teeth and cones of stylolites are aligned parallel to the direction of greatest principal stress (σ_1).

Stylolites can be parallel or transverse to bedding. Both varieties are shown in Figure 8.30. Bedding parallel stylolites with teeth oriented perpendicular to bedding are typically primary stylolites formed during burial and compaction. Vertical gravitational loading on horizontal layering creates horizontal stylolitic surfaces and vertical teeth. In contrast, transverse stylolites are of tectonic origin.

An excellent description of tectonic stylolites is provided by Dean, Kulander, and Skinner (1988). They documented a perfect geometric and kinematic relation between tectonic stylolites and Appalachian folds and thrusts in southeastern West Virginia. They found abundant tectonic stylolites in the Greenbrier Limestone (Mississippian) (Figure 8.31*A*) and measured the strike and dip of the stylolitic surfaces and the trend of plunge of the teeth. When plotted on a structural geologic map (Figure 8.31*B*), these data were seen to be systematically arranged with respect to folds (e.g., the overturned Glen Lyn syncline) and thrust faults (e.g., the St. Clair fault). Outcrop expressions of the stylolites are quite impressive (Figure 8.31*C*).

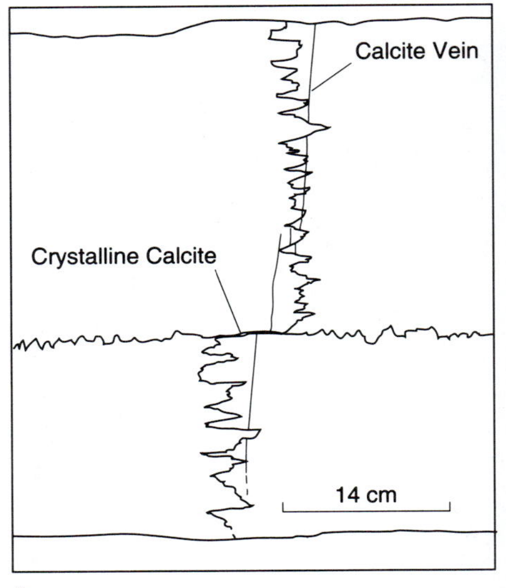

Figure 8.30 (*A*) Detailed sketch of bedding stylolite (horizontal) and transverse stylolite (vertical) in the Permian Pinery Limestone, McKittrick Canyon, in the Guadalupe Mountains, Texas. [Redrawn from Rigby (1953), Figure 6, p. 269.] (*B*) Stylolites in a large slab of Tennessee marble. The digitations are variously known as "teeth," "cones," or "columns." The black linings of the stylolites are composed of carbonaceous and/or clayey residue. (Photograph by T. N. Dale. Courtesy of United States Geological Survey.)

Stylolitic surfaces produced by tectonic stress (as opposed to burial and compaction) typically are axial planar with respect to associated folds, a geometric relationship that is quite compatible with the notion that stylolitic surfaces accommodate shortening (Nickelsen, 1972). **Teeth** and **cones** of stylolitic surfaces tend to be oriented perpendicular to axial surfaces of folding (Alvarez, Engelder, and Lowrie, 1976), an observation that supports the premise that dissolution proceeds in the direction of greatest principal stress (σ_1).

Not all spaced cleavage surfaces are stylolitic. In fact it would seem that spaced cleavage surfaces become smoother and smoother as dissolution proceeds.

Dissolution Removal of Fossils

Even in the absence of stylolitic surfaces, the role of pressure solution can be recognized along spaced cleavage surfaces. For example, a clear signature of pressure solution is the abrupt truncation of fossils along cleavage surfaces (Figure 8.32). The incompleteness manifest in *truncated fossils* is due to a stealing away of material by pressure solution, not to faulting or extensional fracturing. Missing parts of the fossils are never to be found; they go into solution and are reprecipitated as veins or overgrowths.

Strain Response of Insoluble Beds

Another clear signature of the pressure solution origin of spaced cleavage is the strain response of insoluble layers, like chert, in sequences of rock containing spaced cleavage. Whereas pressure-soluble beds like marl and

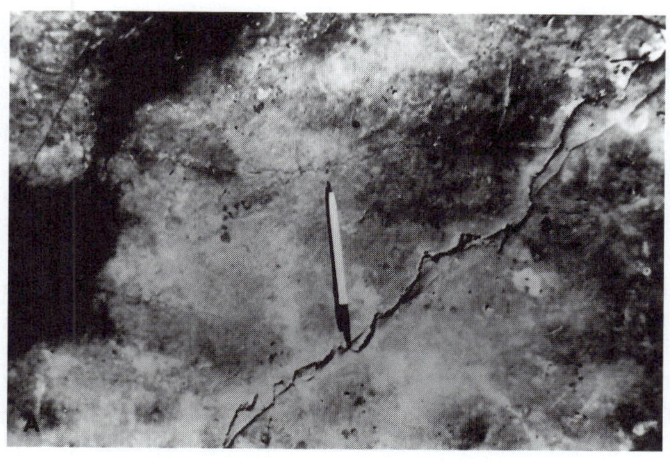

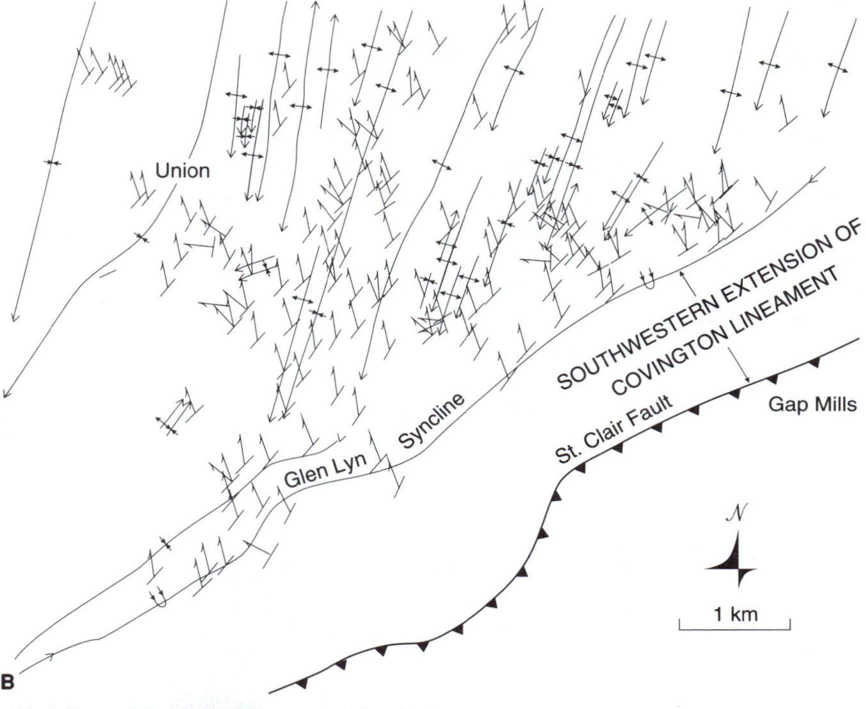

Figure 8.31 (*A*) Bedding plane view of stylolites in Greenbrier limestone. The major tectonic stylolite that runs from upper right to lower left trends N30°E. Teeth trend approximately N10°W. (*B*) Structure map showing the orientations of stylolitic surfaces and teeth in relation to the Glen Lyn syncline and St. Clair thrust, both of which were formed during Appalachian deformation. Note that the teeth are essentially perpendicular to the trace of the major fold and thrust. (*C*) Outcrop expression of Greenbrier limestone in the overturned limb of a syncline. Overturned bedding dips from upper left to lower right. Stylolites, at right angles to bedding, dip from upper right to lower left. Byron Kulander, dressed for the weather. (*D*) Close-up view of vertical face of the Greenbrier limestone. Note penny for scale. Bedding (overturned) dips from upper left to lower right. Early formed extension joints, that were later modified into stylolites by lateral compressive stresses before folding, dip from upper right to lower left [From Dean, Kulander, and Skinner (1988). Published with permission of Geological Society of America and the authors.]

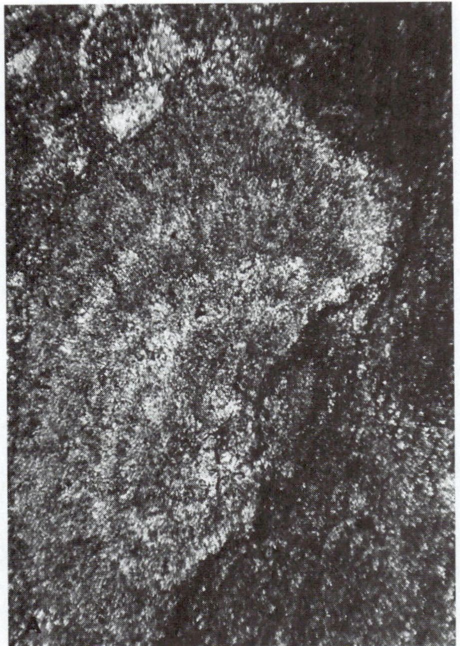

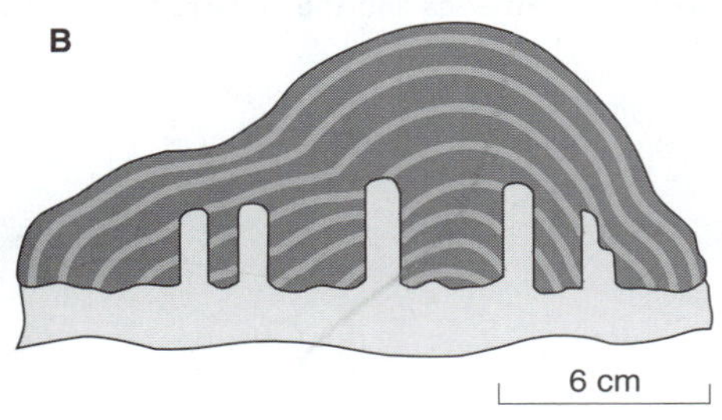

Figure 8.32 (*A*) Photomicrograph of incomplete fusulinid fossil truncated along dark stylolitic seam of insoluble residue. (Photograph by A. Bykerk-Kauffman.) (*B*) Diagram of stromatoporoid fossil into which several stylolitic columns have penetrated. [Redrawn from Stockdale (1943), Figure 2, p. 10.]

impure limestone shorten through loss of rock volume and the synchronous development of spaced cleavage, insoluble chert layers are obliged to shorten by conventional constant volume deformational mechanisms, namely folding and thrusting (Figure 8.33*A*). The amount of thrust imbrication and/or folding of chert layers is proportional to the amount of volume loss (Figure 8.33*B*).

Where faulted chert layers are driven into adjacent pressure-soluble rock, the pressure-soluble rock responds by dissolving away, leaving a pod of preferentially oriented insoluble residue as a record of its former existence. The pressure-soluble layer responds to the stress at the leading edge of an advancing chert layer in the same way that a glacier, feeling the "stress" of rising temperatures, retreats by melting and dumping its "unmeltable" residue of rocks.

Kinematic Significance of Offset Bedding

Yet another signature of pressure solution in rocks with spaced cleavage is the offset of bedding or other laminae along the spaced cleavage surfaces (Figure 8.34). The offset is due to removal of material along a spaced cleavage surface. The magnitude and sense of offset mostly depend on a number of factors, including the orientation of the spaced cleavage surface relative to the orientation of the marker bed, the orientation of the spaced cleavage surface relative to the direction of minimum finite stretch (S_3), and the amount of dissolution. Only in two conditions will pressure solution along a cleavage surface fail to cause offset of the marker bed that is cut by the cleavage: when cleavage and bedding are (1) mutually perpendicular or (2) strictly parallel, assuming that the direction of shortening

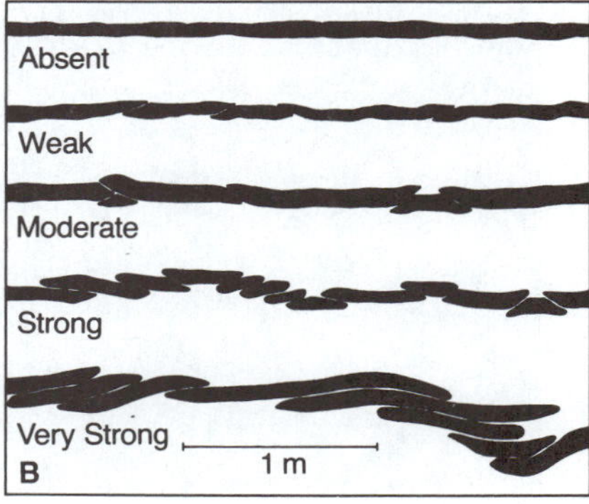

Figure 8.33 (*A*) Spaced cleavage in a strongly cleaved impure limestone. Arrows point out thrust-fault imbrication of insoluble black chert layer that was incapable of shortening by pressure solution. [From Alvarez, Engelder, and Lowrie (1976). Published with permission of Geological Society of America and the authors.] (*B*) Degree of fault imbrication of insoluble chert layers (black) corresponds to the intensity of development of cleavage. [From Alvarez, Engelder, and Geiser (1978). Published with permission of Geological Society of America and the authors.]

in each case is perpendicular to the cleavage surface. Otherwise, anything goes!

An example of bedding offset due to pressure solution is portrayed in Figure 8.35. Figure 8.35*A* is a geologic map of a nonplunging, overturned anticline cut by an axial planar spaced cleavage. At station 36 on the right-side-up western limb of the fold, there is a beautiful exposure of the interrelationship of bedding and cleavage (Figure 8.35*B*). Both bedding and cleavage strike 350°, but they dip westerly by different degrees. The bedding dips 20°W and the cleavage dips 50°W. Bedding is repeatedly offset along the cleavage by very small amounts. Separation is in all cases normal, averaging 0.5 cm.

Assuming that the direction of shortening was perpendicular to the orientation of cleavage, it is possible to graphically determine the amount of dissolution required to account for the offset along any given cleavage surface. Figure 8.35*C* shows the relationships that need to be explained. Lines *AB* and *A′B′* are bedding traces that formerly were in alignment. Distance *AA′ is* the magnitude of normal separation, namely 0.5 cm.

To determine the magnitude of dilational closing, simply 'back off' bedding trace *A′B′* along the direction of dilational closing (path *DC*, Figure 8.35*D*), maintaining the 20° dip of *A′B′*. When aligned with the projection of bedding trace *AB*, line *A′B′* is situated in its restored position, *relative to AB*, before deformation. The magnitude of dilational closing is measured along the direction of dilational closing (*DC*) between the restored location of *A′* and the trace of cleavage. It measures 0.35 cm.

Calculations of this type, when averaged over an array of cleavage surfaces in a fold, provide the basis for estimating total volume loss due to dissolution (Crespe, 1982).

Figure 8.34 Photomicrograph showing offset of once-continuous calcite lamina (white) by pressure solution along stylolitic cleavage seam (black). (Photograph by A. Bykerk-Kauffman.)

A

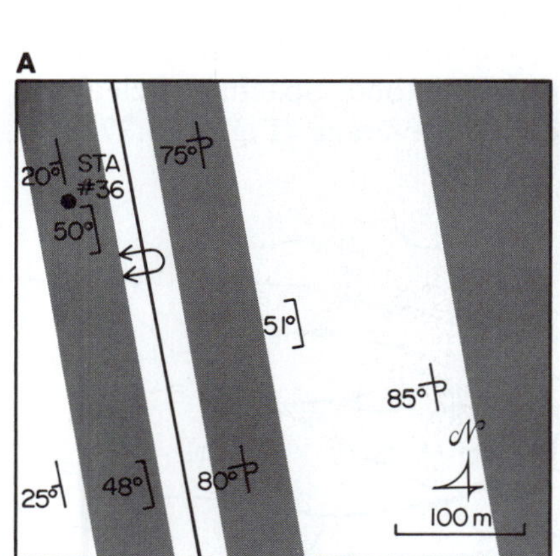

B

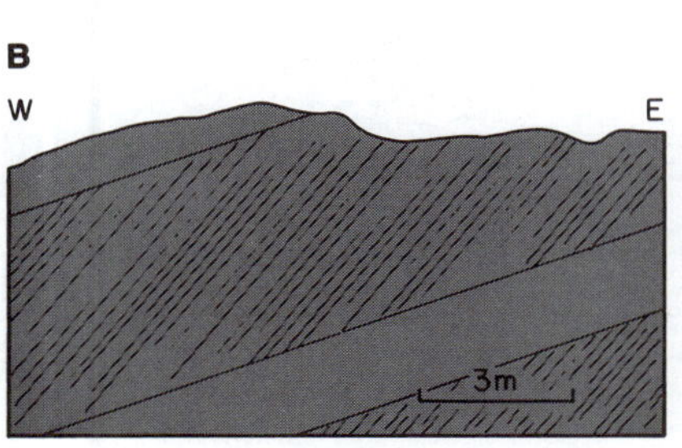

C

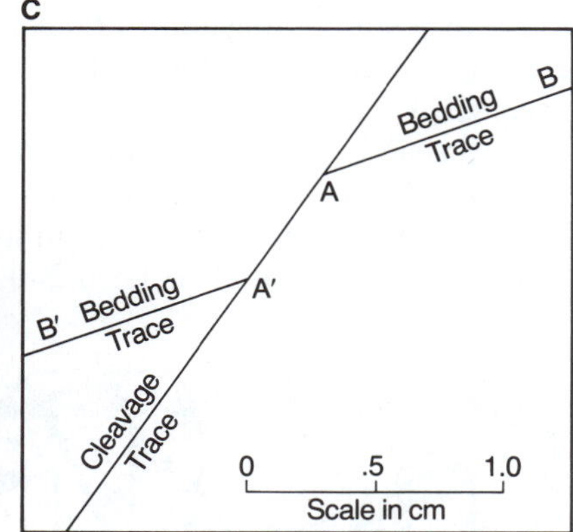

D

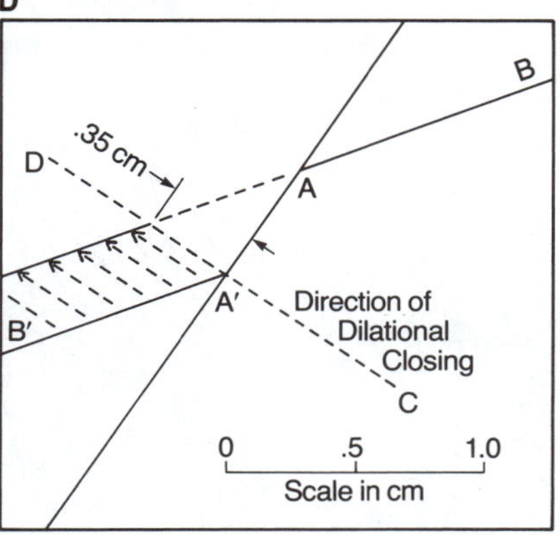

Figure 8.35 Kinematic analysis of bedding offset due to shortening by pressure solution. (*A*) Map showing overturned fold cut by axial plane cleavage. (*B*) Bedding/cleavage relationship exposed at station 36. (*C, D*) Graphical determination of the magnitude of "dilational closing" required to explain the offset of bedding along the cleavage surface.

Classification of Spaced Cleavage

Alvarez, Engelder, and Geiser (1978) designed a classification of spaced cleavage, one that provides the means to estimate shortening within a rock layer on the basis of the nature and spacing of cleavage surfaces. Their approach was to systematically correlate the properties of spaced cleavage with quantitative estimates of shortening deduced from bedding offsets, truncated fossils, and the degree of folding and thrusting of insoluble chert layers.

The categories of spaced cleavage that Alvarez, Engelder, and Geiser recognized are described as **weak**, **moderate**, **strong**, and **very strong**. These correspond to shortening percentages of 0–4, 4–25, 25–35%, and greater than 35%, respectively. Cross-sectional and bedding-plane views of two of the intensities of cleavage (moderate and strong) are shown in Figure 8.36.

The strongest cleavages are the most closely spaced. The weakest cleavages are the most stylolitic. Moderate and strong spaced cleavages tend to display clear-cut intersecting sets that are symmetrically disposed about the direction of greatest shortening. Very strongly developed cleavage is marked by sigmoidal cleavage and abundant calcite veining. Veining is yet another expression of pressure-induced mobilization of rock constituents, in this case, calcite.

Palinspastic Reconstruction

Palinspastic reconstruction of folds and fold belts marked by spaced cleavage requires not only the unfolding of folded layers, it also requires an accounting of losses due to pressure solution. The folds must be stretched out like an accordion, to account for pressure solution losses, before daring to rotate bedding up to the horizontal, thereby restoring deformation due to flexural-slip folding and/or buckling. (where the development of spaced cleavage preceded folding, the palinspastic reconstruction steps are reversed.)

A classic example of palinspastic reconstruction of folded layers cut by spaced cleavage was presented by Groshong (1975a). His analysis focused on a single buckle fold in impure limestone. Spacing of cleavage in the fold was found to be 0.2–0.5 cm, on average (Figure 8.37*A*,*B*).

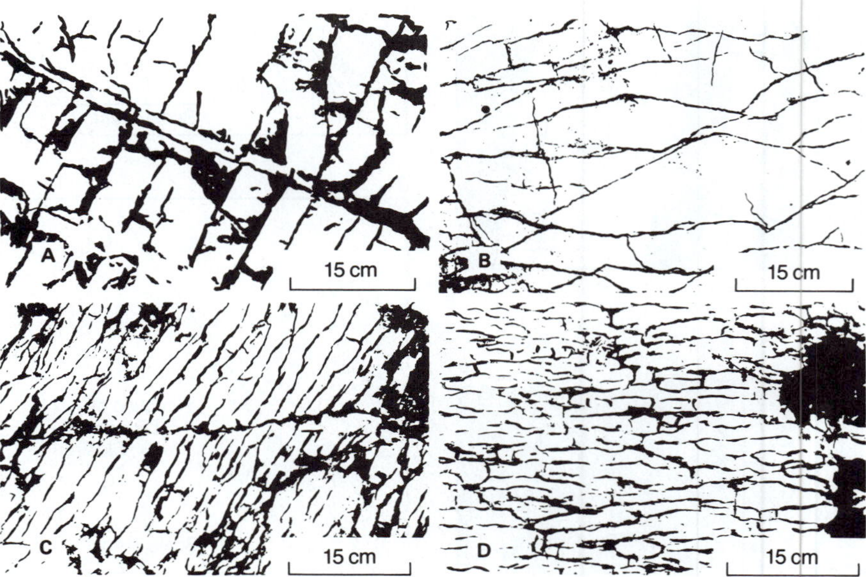

Figure 8.36 Geometry and spacing of moderate and strongly developed spaced cleavage. (*A*) Cross-sectional view of "moderate" cleavage; (*B*) expression of cleavage on the bedding surface. (*C*) Cross-sectional view of "strong" cleavage; (*D*) expression of cleavage on the bedding surface. [From Alvarez, Engelder, and Geiser (1978). Published with permission of Geological Society of America and the authors.]

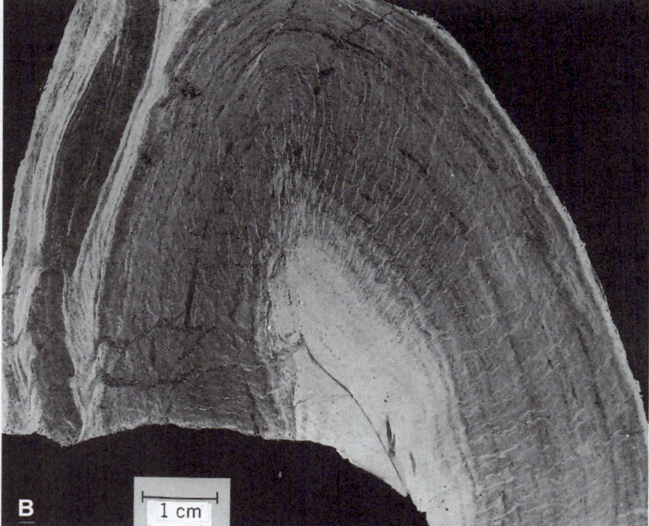

Figure 8.37 Small fold in impure limestone. (*A*) Photograph taken in normal lighting emphasizes nature of bedding. (*B*) Photograph taken in polarized light emphasizes nature of cleavage. [From Groshong (1975a). Published with permission of Geological Society of America and the author.]

Bedding in the fold is repeatedly offset along the cleavage surfaces, especially in the inner arc, core region of the fold. Small fossils in the limestone were seen to be truncated, corroded along the cleavage surfaces. After evaluating the average displacements and pressure solution losses along the cleavage surfaces, Groshong palinspastically restored the fold to a more open configuration by separating the microlithons. Minimum volume loss was estimated to be 18%.

Volume losses well above 18% are recorded in highly deformed folded terranes. Magnitudes of 40–50% dissolution are not at all uncommon. Significant volume losses are also recorded in certain foreland terranes, like the Appalachian foreland. Bedding in the Appalachian foreland region of Pennsylvania and New York is essentially flat lying, folded about very gentle folds with limb dips less than 3–4°. The innocent-looking, flat-lying nature of these foreland strata is sharply contradicted by the measured strain state of these rocks. Engelder and Engelder (1977) have demonstrated a layer-parallel shortening of 10–15%, most of which has been accommodated by the formation of spaced cleavage.

Any attempts to prepare balanced cross sections of folded and thrusted terranes must take into account the realities of pressure solution loss of material. Line length balancing alone will not suffice. It does no good to simply unfold folds if analysis of spaced cleavage in the cores of some folds reveals a 40% loss of volume. The shortening component of the loss of volume *must* be taken into consideration.

PASSIVE FOLDING

Passive folding is commonly intimately associated with cleaved rocks. In fact, the understanding of passive folding requires an understanding of the development of cleavage. Thus, the time is ripe to think about passive folds.

Characteristics of Passive Folds

Passive folds characteristically display profile forms that are class 1C, 2, or 3 (see Chapter 7), typified by some degree of hinge thickening and limb attenuation (Figure 8.38). The axial plane cleavages associated with passive folding run the full spectrum of slaty cleavage, phyllitic structure, schistosity, and spaced cleavage. The axis of a passive fold is simply the trend and plunge of the line of intersection of cleavage and bedding.

Conditions Favoring Passive Folding

Sequences that are especially susceptible to passive folding are distinguished by uniformly soft, weak layers. Donath and Parker (1964) describe this mechanical condition as one of **high mean ductility** and **low ductility contrast**. Rocks in this state are capable of profound internal distortion without the loss of cohesion.

There are several common geologic circumstances that can lead to the development of layered rock bodies marked by high mean ductility and low ductility contrast. Unconsolidated water-rich **hydroplastic** sediments are uniformly weak, devoid of the mechanical influence of layering. Given half a chance, such weak sediments deform readily by passive-slip and/or passive-flow folding (Figure 8.39*A*). Certain lithified rocks, like salt or gypsum, are so inherently weak that they behave viscously and deform by passive folding (Figure 8.39*B*). Some rocks that are perfectly stiff and strong under "normal" temperature and pressure conditions may lose most of their strength under metamorphic conditions of elevated tempera-

Figure 8.38 (*A*) Passive fold in metasedimentary rocks in the Salt River Canyon region, Arizona. Height of fold is approximately 1.5 m. (Photograph by F. W. Cropp.) [From Davis and others (1981), fig. 32, p. 83. Published with permission of Arizona Geological Society.] (*B*) Passive fold in polished slab of pyritic ore from the Caribou strata-bound sulfide deposit in the Bathurst mining district of New Brunswick, Canada. Cleaved black layers represent original bedding. Cleavage is axial planar to the folded layering. (Photograph by G. Kew.) (*C*) Recumbent passive folds in marble derived from Pennsylvanian-Permian limestone in Happy Valley, southeastern Arizona. (Photograph by G. H. Davis.)

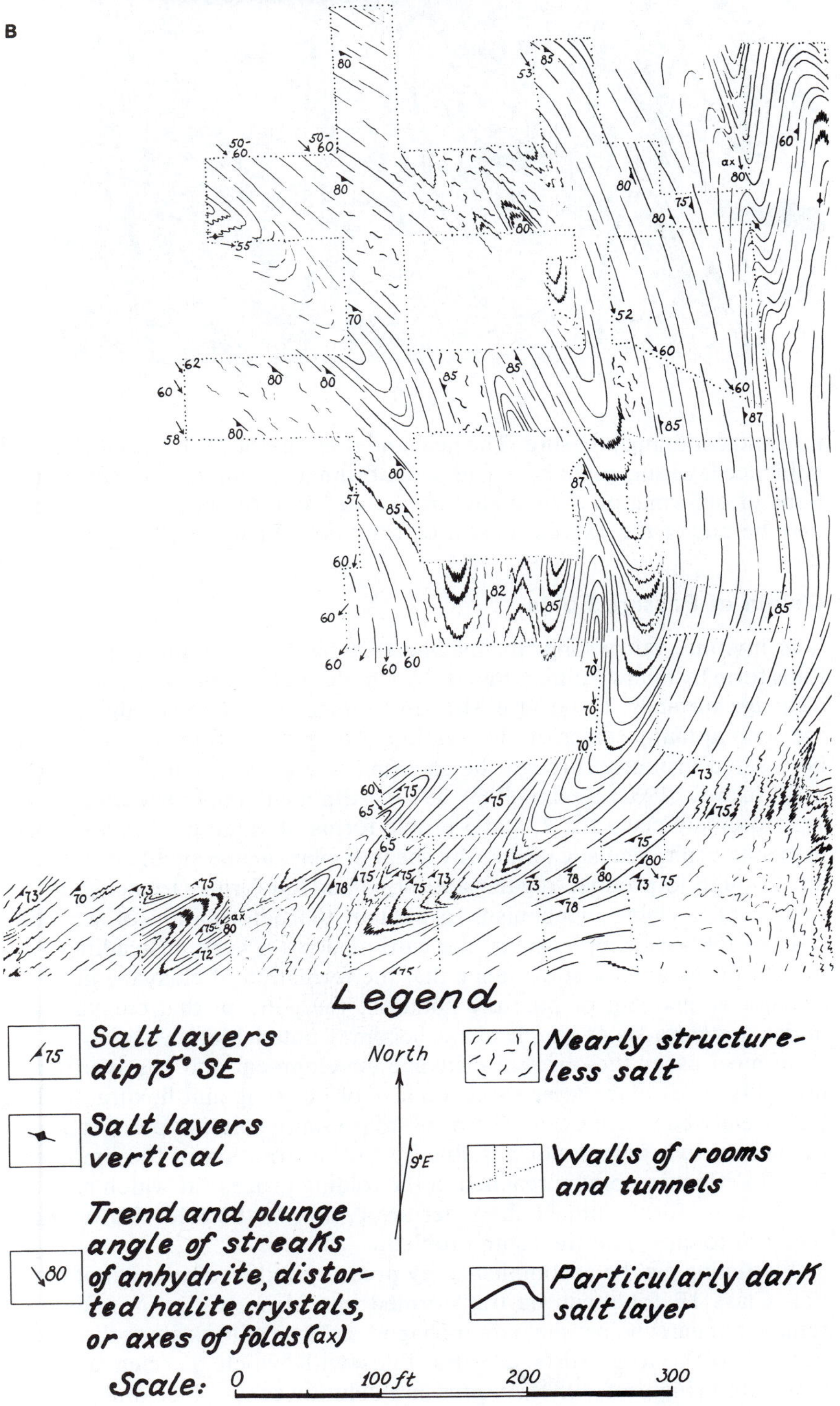

Figure 8.39 (*A*) Primary fold in the Green River Shale (Eocene) in western Colorado. The fold is intraformational, overlain and underlain by undistorted laminae of rock. (Photograph by G. H. Davis.) (*B*) Geologic map of passive folds in salt in part of the subsurface of the Grand Saline salt dome, Texas. [From Balk (1949). Published with permission of American Association of Petroleum Geologists.]

B

Legend

Salt layers dip 75° SE

Salt layers vertical

Trend and plunge angle of streaks of anhydrite, distorted halite crystals, or axes of folds (ax)

North

9°E

Nearly structureless salt

Walls of rooms and tunnels

Particularly dark salt layer

Scale: 0 100 ft 200 300

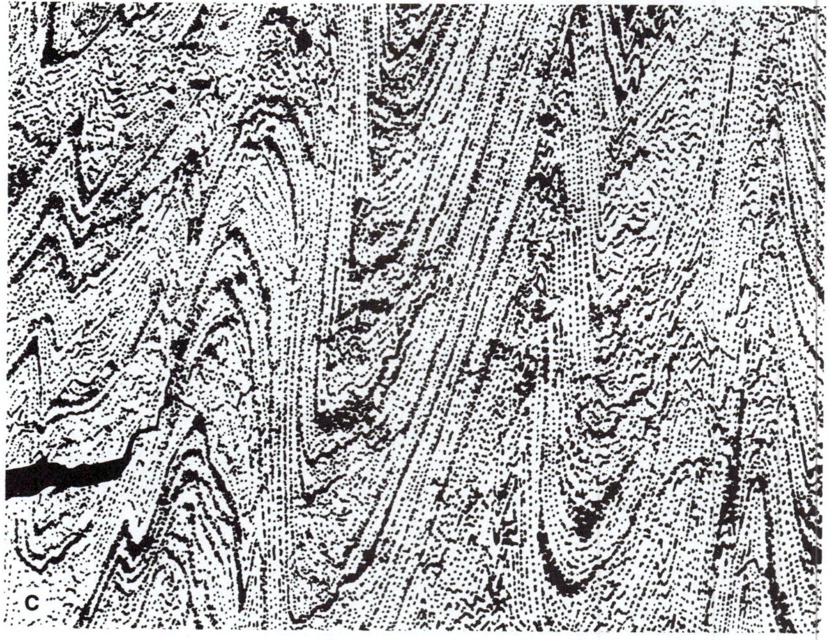

Figure 8.39 (*continued*) (*C*) Outcrop-scale passive-flow folds in hornblende-plagioclase gneiss, Medicine Bow Mountains. [From Donath and Parker (1964). Published with permission of Geological Society of America and the authors.]

ture and confining pressure. The heat and pressure blot out the mechanical influence layering might have had under ordinary, nonmetamorphic conditions of deformation. When such rocks are compressed, passive folding may be one of the expressions of deformation (Figure 8.39*C*).

Origin of Passive Folds

It is inviting, but wrong, to imagine that passive-slip and passive-flow folds form by microfaulting, that is, by simple shear along the close-spaced cleavage surfaces. To be sure, shearlike offsets of bedding along cleavage are seen at many scales of observation. And passive fold geometries can be created instantly through the shearing of a deck of cards (see Figure 1.37) (Carey, 1962; Ragan, 1969, 1973). Although simple shearing is one way to create class 1C, 2, and 3 profile forms, it is impossible to explain in mechanically realistic terms why shear displacements should be systematic enough to produce regular antiforms and synforms (Johnson, 1977). In lieu of a viable mechanism to explain the deformation of wavelike passive folds by simple shear, the card analogy is very misleading. And in light of the abundant evidence that the formation of cleavage involves a major component of pressure solution, the concept that cleavage domains are surfaces of simple shear becomes untenable.

In most cases the onset of cleavage development and passive folding probably takes place after some amount of buckling and flexural folding has already been achieved. The onset of pressure solution permits a rock to shorten to a degree not possible by further rotation of the limbs of a fold. In effect, a point is reached in the folding process at which material needs to be forced out of the inner arc region of the fold. **Flattening** is required to eliminate the room problem.

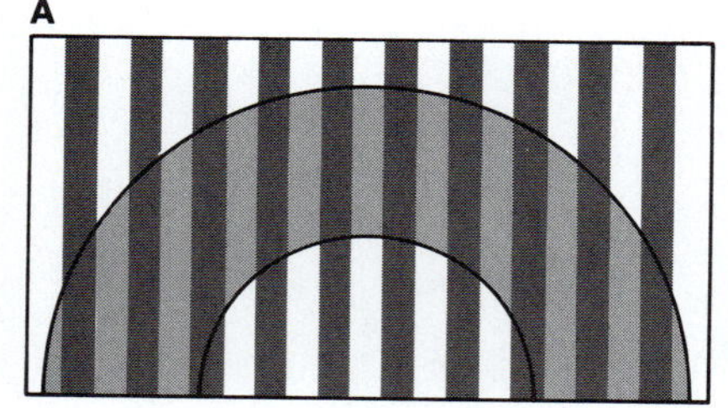

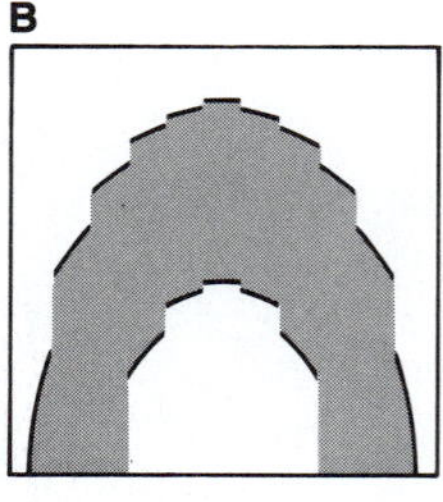

Figure 8.40 Schematic rendering of the transformation of (*A*) a class 1B fold to (*B*) a class 1C fold by pressure solution. Such a transformation can be simulated easily with a deck of cards, not by displacing the cards in simple shear fashion, but by removing domains of material at spaced intervals within the deck.

Flattening can be accomplished by pressure-induced removal of material. Class 1B folds can be transformed into class 1C, 2, or 3 folds by removal of narrow, parallelogram-shaped sections of rock along directions parallel to the axial surface of the folds with which the cleavages are associated (Figure 8.40*A*). As pressure solution takes place and material is removed, shortening keeps pace to prevent the creation of open space. Adjacent microlithons bordering on a common cleavage domain move toward one another and interpenetrate one another along a direction that is parallel to S_3, the direction of minimum finite stretch. Fold limbs naturally steepen to greater inclinations. Fold profiles change from concentric to similar. The effect of pressure solution imposed on an early fold form is the creation of shearlike separations of bedding along cleavage surfaces (Figure 8.41).

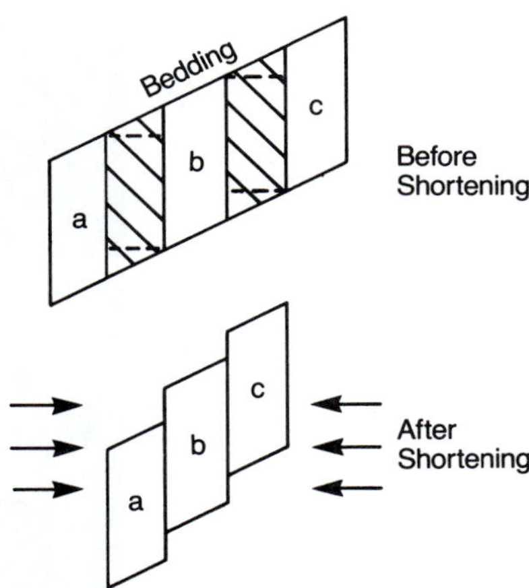

Figure 8.41 The steepening of the inclination of overall bedding attitude by pressure-solution loss of material. [From Alvarez, Engelder, and Lowrie (1976). Published with permission of Geological Society of America and the authors.]

Transposition

Strongly cleaved sequences of metasedimentary and metavolcanic rocks in fold belts characteristically display a parallelism of foliation and lithologic layering. The lithologic layers, distinguished on the basis of color, texture, mineralogy, and general outcrop appearance, tend to pinch out along strike, giving way to other combinations of lithologic layers (Figure 8.42). The mapped relationships give the impression that foliation developed parallel to bedding in the original host rock, a host rock that was marked by abundant abrupt facies changes.

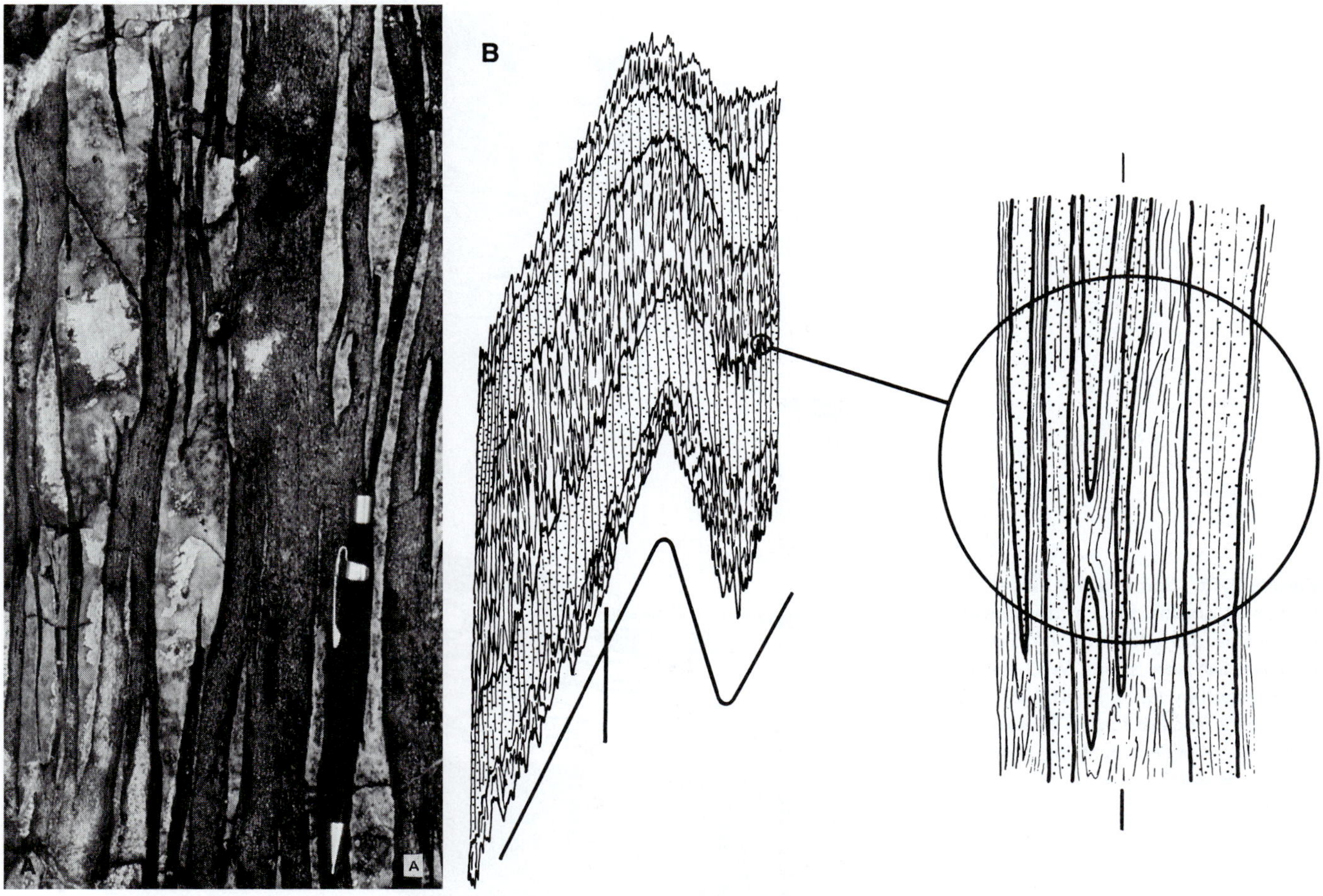

Figure 8.42 (*A*) Pseudostratigraphy in metasedimentary rocks in the Happy Valley region of the Rincon Mountains, near Tucson, Arizona. (Photograph by S. H. Lingrey.) (*B*) Pseudostratigraphy within folded rock. (From *Structural Analysis of Metamorphic Tectonites* by F. J. Turner and L. E. Weiss. Published with permission of McGraw-Hill Book Company, New York, copyright © 1963.)

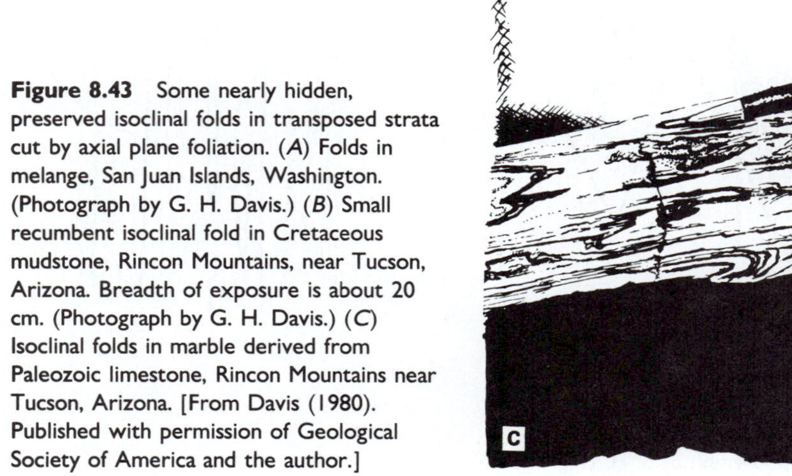

Figure 8.43 Some nearly hidden, preserved isoclinal folds in transposed strata cut by axial plane foliation. (*A*) Folds in melange, San Juan Islands, Washington. (Photograph by G. H. Davis.) (*B*) Small recumbent isoclinal fold in Cretaceous mudstone, Rincon Mountains, near Tucson, Arizona. Breadth of exposure is about 20 cm. (Photograph by G. H. Davis.) (*C*) Isoclinal folds in marble derived from Paleozoic limestone, Rincon Mountains near Tucson, Arizona. [From Davis (1980). Published with permission of Geological Society of America and the author.]

Structural relationships are not always what they seem. Careful inspection of terranes containing "bedding-plane foliation" usually uncovers rare exposures of **tight to isoclinal intrafolial folds** that are cut by the dominant foliation in axial planar fashion (Figure 8.43). The folds are passive, class 1C to 3 in form, and they are sandwiched by through-going foliation and lithologic layering. According to Turner and Weiss (1963) and Whitten (1966), intrafolial folds in metamorphic tectonites commonly reflect **bedding transposition** in which tight folding of the original beds is accompanied by slip parallel to the axial planes of developing flexures (Figure 8.44*A*,*B*). Individual fold limbs are attenuated by progressive slip and pressure solution. Ultimately they are separated from their hinge zones (Figure 8.44*C*,*D*). The planes of slip and dissolution are in fact the foliation, occupying the S_1S_2 plane of the finite strain ellipsoid.

Transposition creates a **pseudostratigraphy** containing disrupted and rotated segments of once-continuous beds. The entire sequence is pervaded by structural discontinuities, one expression of which is apparent "bedding-plane cleavage." Transposed sequences show no internal consistency of facing: right-side-up and upside-down beds are stacked together (Whitten, 1966). The main clue that all this has happened is the presence of intrafolial folds. The concept of bedding transposition teaches us to be very cautious in trying to identify original bedding. Unless it can be firmly demonstrated that the layering represents an internal, undisrupted, coherent, original stratigraphy, lithologic layering in metamorphic tectonites is simply described as "layering" rather than bedding.

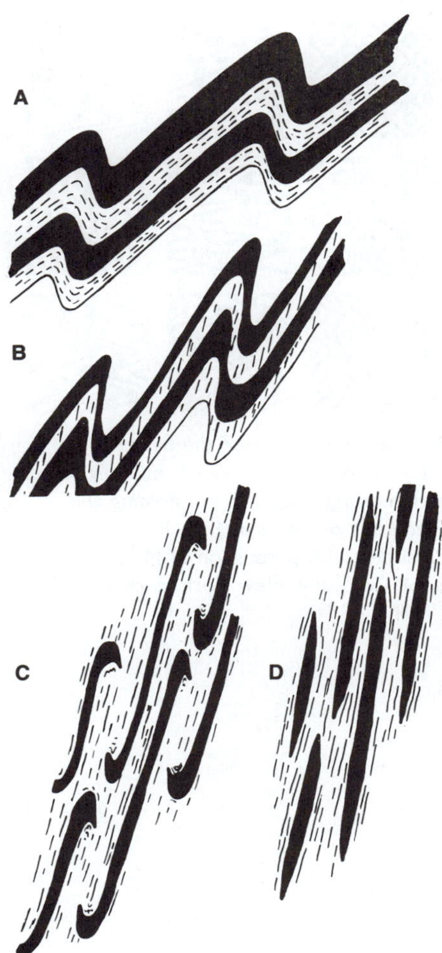

Figure 8.44 Transposition of bedding. (*A*) Flexural folding of bedded sequence of stiff (black) and soft (white) layers. (*B*) Tight folding and onset of cleavage development. (*C*) Attenuation and rupture of fold limbs. *(D)* Flattening of sequence and creation of pseudostratigraphy. (Modified from *Structural Analysis of Metamorphic Tectonites* by F. J. Turner and L. E. Weiss. Published with permission of McGraw-Hill Book Company, New York, copyright © 1963.)

Figure 8.45 The classic work in analyzing refolded folds was carried out by John Ramsay (1958) in the Loch Monar region in the Northern Highlands of Scotland. Here are two examples of outcrop displays at Loch Monar: (*A*) Dome and basin pattern produced by interference of two sets of horizontal, upright, tight to isoclinal folds trending at right angles to one another. The folds interfere constructively and destructively, just like wavetrains in physics. (*B*) Upright folding of recumbent isoclinal folds. (Photographs by G. H. Davis.)

Figure 8.46 Some of the patterns of refolding that can emerge in multiply folded metamorphic rocks. (From *Folding and Fracturing of Rocks* by J. G. Ramsay. Published with permission of McGraw-Hill Book Company, New York, copyright © 1967.)

Superposed Folding

One of the challenges in mapping regional terranes of passively folded rock is trying to unravel the geometrical artwork of superimposed folding. The superposition of passive folding can be the product of repeated tectonic deformations separated in time, or the product of a single progressive deformation over time. Systems of refolded folds present us with imposing geometric challenges (Figure 8.45). The interference patterns are almost unbelievable: egg carton patterns, boomerang folds, isoclinally folded isoclines . . . each a special product of the geometric relationship of the marker layers to the overall kinematic picture. The different patterns that develop reflect the different ways fold generations of different orientations interfere (Figure 8.46). Ramsay (1967) and Ramsay and Huber (1987) spell out the details for all who are interested.

FOLIATION

Definition of Foliation

Cleavage is just one brand of foliation. **Foliation** is any *mesoscopically penetrative parallel alignment of planar fabric elements in a rock,* usually metamorphic rock (Figure 8.47). The definition of foliation is full of code words.

"Planar fabric element" can refer to any number of features, such as domainal structure in slaty cleavage, flattened pebbles in a metaconglomerate, or compositional banding in a granite gneiss. "Parallel alignment" means *roughly* parallel, with a lot of latitude. It can refer to perfectly parallel compositional bands in granitic gneiss, approximately parallel pebbles in a flattened pebble metaconglomerate, or converging/diverging microscopic domainal structure in slaty cleavage. "Planar" can mean perfectly planar, like a slaty cleavage surface in outcrop; curviplanar, such as the face of a flattened, discoidal pebble in metaconglomerate; or irregular, like the margin of a compositional band in granite gneiss when viewed microscopically.

The code words in the definition of foliation attempt to set limits and to be encompassing, both at the same time. The definition of "foliation" must be broad enough to include the extraordinarily wide variety of foliation fabrics that exist.

Primary Versus Secondary Foliation

Some foliations are **primary foliations**, and these form in lava or magma or wet sediment before consolidation into rock. **Flow banding** in rhyolite or granite (Figure 8.48), and crude **axial plane cleavage** in slump folds in sediments are examples of primary foliation. Volcanic ash flow tuffs are distinguished by **eutaxitic structure**, a foliation created through compaction flattening of pumice fragments and gas bubbles as the volcanic unit flattens under its own weight during cooling and congealing (Figure 8.49). Igneous rocks along the margins of dikes, sills, and plugs acquire a **flow foliation** as a result of viscous flow of magma against wall rock (Figure 8.50). Granitic rocks may retain a record of internal flow movements in the form of subtle or not-so-subtle foliations defined by aligned crystals, phenocrysts, inclusions, xenoliths, and **schlieren** (Figure 8.51), which are wispy dark bands representing the segregation of minerals within the magma. In general, primary foliations reflect a viscous flow and internal deforma-

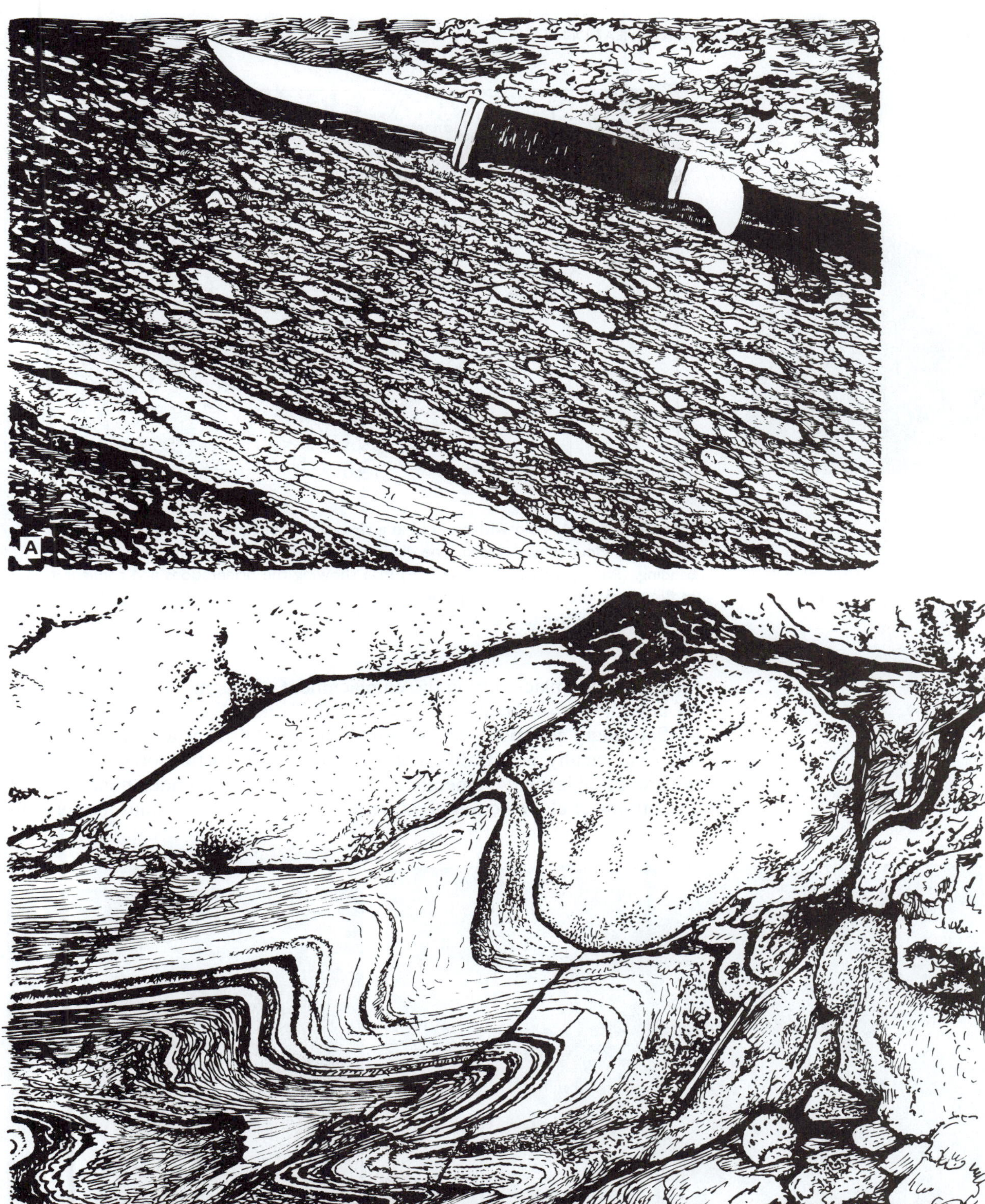

Figure 8.47 Two examples of foliated rock. (*A*) Augen gneiss in the Tanque Verde Mountain, Tucson, Arizona. The aligned augen ("eyes") are composed of feldspar (white). The matrix is composed of quartz, feldspar, and biotite mica. The white layer that contributes to the gneissic structure is composed of quartz. (*B*) Banded gneiss composed of layers and laminae of quartz-feldspar (white) and biotite mica (black), Little Rincon Mountains, Arizona. [Tracings by D. O'Day of photographs by G. H. Davis. From Davis (1980). Published with permission of Geological Society of America and the author.]

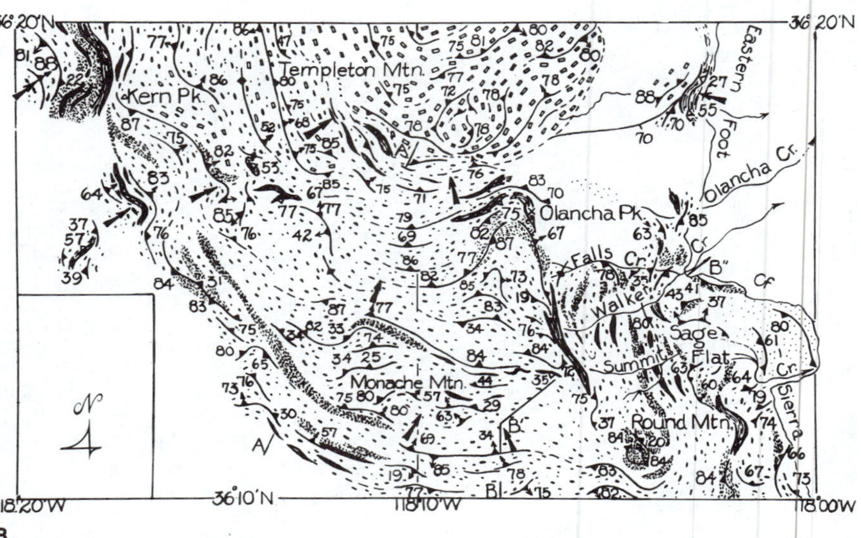

Figure 8.48 (*A*) Fold within a fold in a rhyodacite specimen collected in the Chiricahua Mountains, Arizona. The distinctive lamination is an expression of flow banding. (Photograph by G. Kew.) (*B*) Map showing the orientations and configuration of flow foliation defined by alignment of feldspars, mica, and inclusions in granitic rocks in the Temple Mountain area of the Sierra Nevada. (Mapping by E. B. Mayo.)

tion that took place during the emplacement of magma or during the flow of lava or wet sediments.

Secondary foliations will be our focus. They form as a result of the microscopically penetrative deformation and distortion of sedimentary, volcanic, or intrusive igneous rocks, almost always under metamorphic conditions of elevated temperature and pressure. Throughout the remain-

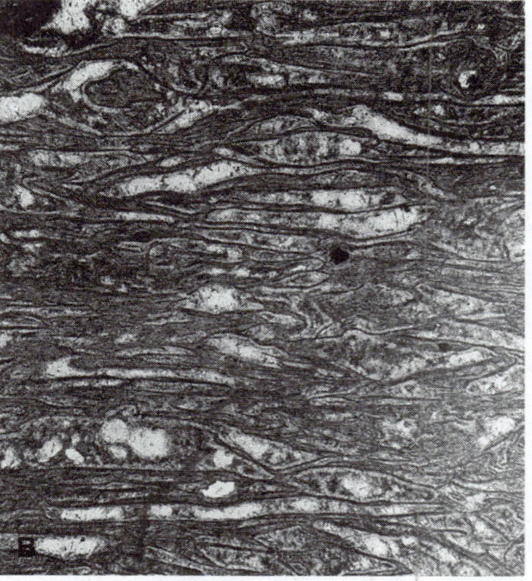

Figure 8.49 (*A*) Eutaxitic structure within a Tertiary ash flow sheet in Nevada. Lower right half of outcrop is composed of densely welded tuff. Partings in the tuff follow concentrations of pumice lapilli, which have been preferentially weathered out. (Photograph by P. W. Lipman. Courtesy of United States Geological Survey.) (*B*) Photomicrograph of primary igneous foliation in rhyolite from the Creede caldera, San Juan Mountains, Colorado. (Photograph by J. C. Ratte. Courtesy of United States Geological Survey.)

der of this chapter, we will continue to concentrate on foliations that typify classic regionally deformed and metamorphosed rocks, such as slates, phyllites, schists, and gneisses. We will also examine **mylonitic foliations**, which are mainly associated with *shear zones* formed by highly concentrated strain.

Foliations in Typical Metamorphic Rocks

Foliation is a fundamental characteristic of regionally metamorphosed rocks. We have already discussed in some detail the foliation known as cleavage: slaty cleavage, phyllitic cleavage, and schistosity. **Slaty cleavage** develops as fine-grained sedimentary and volcanic rocks are transformed into slate during low grade metamorphism. **Phyllitic structure** and **schistosity**, on the other hand, form when fine- to medium-grained sedimentary and volcanic rocks are deformed during low to high grade metamorphism. (My brother the dentist shouts "*SCHISTOSITY*!" indiscriminately whenever we drive by any bold outcrop of any rock anywhere.)

Gneissic structure, another major category of foliation, typically develops as a result of recrystallization of igneous or sedimentary rocks during middle to high grade metamorphism. It is generally composed of medium- to coarse-grained minerals and can be defined by compositional banding, by a preferred planar orientation of platy, tabular, or prismatic minerals, and by subparallel lenticular mineral grains and grain aggregates (Figure 8.52*A*). The compositional banding occurs at all scales, from thick, continuous bands or layers that can be mapped across an entire field area, to discontinuous lenses and laminae that pinch out within individual outcrops, hand specimens, or thin sections (Figure 8.52*B*). Gneissic structure occurs in metamorphic rocks of almost every composition. In high grade rocks, it is commonly associated with thin layers and pods of granitic material, in some cases forming a swirled, "mixed up" appearing rock called **migmatite** (Figure 8.52*C*).

Figure 8.50 (*A*) Basalt dike cutting Tertiary conglomerate in northwestern Arizona. Peter Coney is seen checking out a hand specimen of the dike rock. (*B*) Close-up of margin of the dike. Note the flow foliation expressed by banding and lamination in the dike. The foliation is oriented parallel to the contact with wall rock. (Photographs by G. H. Davis.)

Figure 8.51 Schlieren in granite in the Sierra Nevada. Geologist is Dan Lynch. (Photograph by D. Trent.)

Figure 8.52 (*A*) Mylonitic granitic gneiss, with gneissic foliation defined by feldspar porphyroclasts (white) and "ribbons" of quartz with mica. (Photograph by Carol Simpson.) (*B*) Banded granitic gneiss. (*C*) Banded gneiss and migmatite. The original rock was Precambrian, but it acquired a strongly foliated fabric during Cretaceous regional metamorphism and intrusion. Harquavar Mountains, western Arizona. (Photographs by S. J. Reynolds.)

Foliation in typical metamorphic rocks is also expressed through preferentially oriented "odds and ends." For example, strongly flattened and metamorphosed conglomerates may contain pancake-shaped pebbles all in parallel, planar alignment, thus defining a conspicuous **flattened-pebble foliation** (Figure 8.53). Strongly flattened and metamorphosed volcanic rocks may contain flattened lapilli inclusions, all in parallel, planar alignment, thus expressing foliation (see Figure 8.24). The flattening that accompanies the formation of most foliation can cause stiff compositional layers surrounded by softer layers to neck and pull apart into **boudins**,

Figure 8.53 Flattened pebbles define a foliation in highly deformed Barnes Conglomerate (Precambrian), Tortolita Mountains, Arizona. (Photograph by G. H. Davis.)

Figure 8.54 Boudins in layered gneiss. (Photograph by S. J. Reynolds.)

sausage-shaped structures whose interesting forms can accentuate the appearance of the foliation (Figure 8.54).

Foliations in Mylonitic Rocks

A structurally important family of foliated rocks are the strongly deformed rocks of the **mylonite series** (Table 8.1). Mylonitic rocks are a type of "fault rock" in which the grain size has been reduced as a result of intense shearing (Figure 8.55). The grain size reduction is a result of ductile or mixed brittle–ductile deformation mechanisms, especially dislocation creep, dynamic recrystallization, and fracturing of brittle grains. The formation of mylonitic rocks, called **mylonitization**, commonly occurs within the high strain environment of ductile shear zones (Chapter 9).

Most mylonitic rocks have a distinctive, somewhat platy appearance because of a well-developed foliation. The foliation, which has been likened to the flattening fabric seen in welded ash flow tuffs, is defined by a planar-parallel arrangement of flattened grains, mineral aggregates, broken mineral grains, and small shear surfaces. Mylonitic foliation commonly contains very lenticular, individual crystals, termed **ribbons** (Figure 8.55). Foliation can be fairly planar in monomineralic rocks, like mylonitic quartzite, but more commonly has a distinctly lensoidal or anastomosing aspect. In part, the lensoidal foliation reflects how different minerals responded to mylonitization. If a rock contains more than one mineral, ductile strain may be accommodated preferentially by the mineral that is weakest. This can result in lenses or ribbons of the weaker mineral wrapped across less deformed crystals of the stronger one. Such an appearance is common in mylonitized granitic rocks, where strongly flattened quartz ribbons drape less deformed feldspars. In such rocks, quartz typically looks as if it has flowed like butter, whereas the feldspars have behaved brittlely, breaking into chips, which may become oriented within the foliation (Figure 8.55). Through the process of mylonitization, a coarse-

TABLE 8.1
Mylonite series rocks

Rock Name	*Matrix Grain Size*	*Percentage Matrix*
Protomylonite	$>50\ \mu m$	$<50\%$
Mylonite	$<50\ \mu m$	50%–90%
Ultramylonite	$<10\ \mu m$	$>90\%$

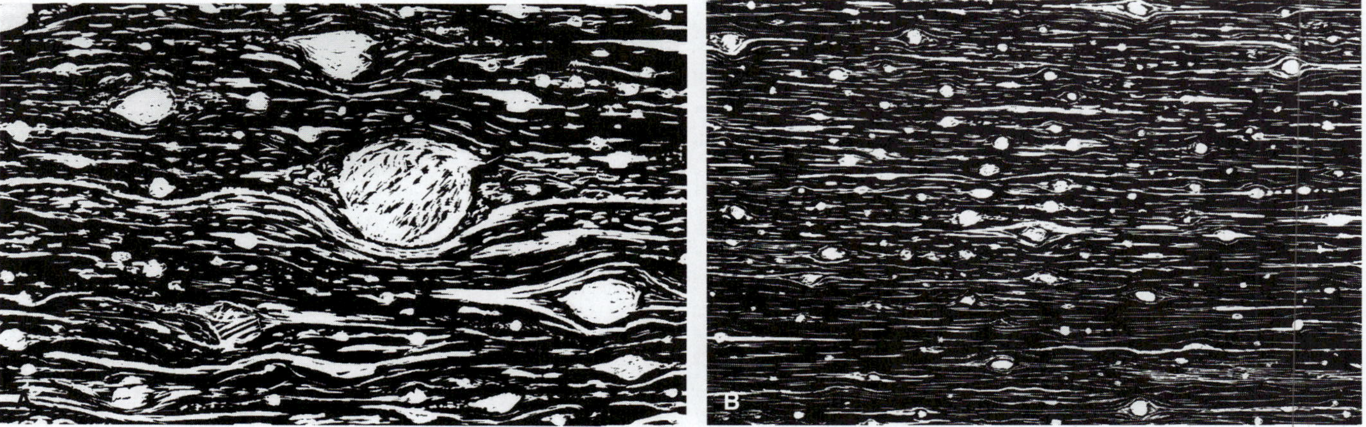

Figure 8.55 (*A*) The microscopic texture of mylonite. The white represents fragmented grains (porphyroclasts) and breccia streaks. The black represents plastically deformed quartz. (*B*) The microscopic texture of ultramylonite. [From Higgins (1971). Courtesy of United States Geological Survey.]

grained protolith, such as a granite, can be metamorphosed into an ultra-fine-grained, laminated rock.

The nomenclature of mylonitic rocks is based mainly on the grain size and the proportion of fine-grained matrix to larger relict grains (Table 8.1). In essence, the nomenclature scheme establishes arbitrary boundaries within the progression from undeformed protolith to an extremely deformed and uniformly fine-grained mylonitic rock. The initial stages of mylonitization produce a weakly to moderately mylonitic rock, called a **protomylonite**. Protomylonite contains less than 50 percent fine-grained matrix (i.e., is at least 50 percent relict grains), and in thin section commonly displays **mortar texture**, where small recrystallized grains surround larger relict crystals. Foliation ranges from very subtle to well defined and is defined by flattened, inequant grains and locally by very thin, nonpenetrative zones of high shear strain. With further deformation and grain-size reduction, a protomylonite is converted into a **mylonite**, which contains 50 to 90 percent matrix. Most mylonites are very strongly foliated and lineated, and contain porphyroclasts scattered throughout the fine-grained, laminated matrix. Mylonitization culminates in the formation of an **ultramylonite**, a more thoroughly deformed and fine-grained rock containing more than 90 percent matrix and less than 10 percent relict grains. **Ultramylonites** are mylonites taken to the edge of recognition.

Under certain conditions, shearing results in rocks that are intermediate in character between mylonitic rocks and more-common metamorphic rocks, such as gneiss and schist. A rock intermediate in character between a mylonite and a gneiss is a **mylonitic gneiss**. A rock that is transitional between a mylonite and a schist is a **mylonitic schist** or, if finer grained, a **phyllonite**.

Foliation in mylonitic rocks is also commonly defined by lenticular grains or clasts called **augen** or **porphyroclasts** (Figure 8.56). Augen are commonly relics from the protolith, being relatively strong minerals that resisted mylonitization while surrounding grains were progressively converted into the fine-grained matrix. Many mylonitized granites, for example, contain lenticular feldspar augen floating in a fine-grained matrix of ductilely deformed or recrystallized quartz and mica. With continued mylonitization, even the resistant augen may be destroyed, leaving a fine-grained rock with few clues about its heritage.

Figure 8.56 View of outcrop of augen mylonite (and white pegmatitic sill) in Tanque Verde Mountain, southeastern Arizona. White feldspars float as "augen" in quartz-mica ribbons. (Photograph by G. H. Davis.)

Some foliation in mylonitic rocks is actually the expression of thin shear zones. Within these zones, the "normal" foliation of the mylonitic rock is smeared out into ultrathin laminations. These zones may be at an acute angle to the main foliation, imparting a lensoidal aspect to the overall rock (see Chapter 9).

The wide range of appearances of foliations in typical metamorphic rocks and in mylonitic rocks reflects differences in the composition, mineralogy, and texture of the original **protolith** from which the metamorphic rock was derived; differences in the temperature–pressure environments of metamorphism and deformation; and differences in strain. Each of these factors influences the appearance of deformed metamorphic rocks, and such knowledge helps us to better interpret the protolith and history of metamorphic rocks we encounter in the field.

LINEATION

Definition and Expression

The definition of lineation, like the definition of foliation, is loaded with code words. **Lineation** is the *subparallel to parallel alignment of elongate, linear fabric elements in a rock body, commonly penetrative at the outcrop and/or hand specimen scales of observation* (Figure 8.57), and commonly at the microscopic scale as well. Some lineations are so penetrative that the lineated rock looks like driftwood with a pronounced etched grain. Other lineations are expressed in the form of such large aligned parallel elements that it is best to refer to the lineation as **linear structure**. Just like foliation, lineation has innumerable physical expressions and occurs in metamorphic, igneous, and sedimentary rocks (Cloos, 1946; Turner and Weiss, 1963; Weiss, 1972).

Some lineation is **primary**, produced in sediments, lava, or magma *prior to lithification*. For example, **parting lineation** is a subtle primary structure that commonly occurs in siltstone and sandstone. Expressed as a faint linear grain on bedding surfaces, it records the current direction at the time

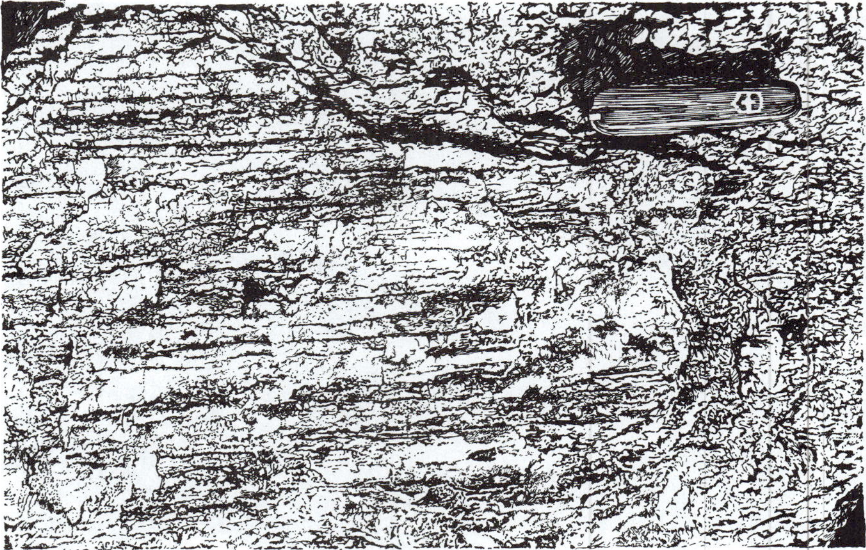

Figure 8.57 India-ink rendering of lineation on foliation surface in mylonitic gneiss in the Santa Catalina Mountains, Tucson, Arizona. [Tracing by D. O'Day of photographs by G. H. Davis. From Davis (1980). Published with permission of Geological Society of America and the author.]

of deposition of the sand and silt (Figure 8.58). The physical expression of parting lineation is due to the subparallel alignment of the longest dimensions of silt or sand grains within bedding laminae. The grains are aligned in the direction of paleocurrent (Conybeare and Crook, 1968).

Ropy lava (pahoehoe) is an example of primary linear structure (Figure 8.59*A*). It forms during the flow of relatively low viscosity basalt when local lava currents drag the plastic skin of the flow, contorting it into a series of nested arcs, which tend to be convex in the sense of current flow.

There are primary linear features in intrusive igneous rocks as well (Figure 8.59*B*). In the Big Bend country of West Texas, there is an outcrop that looks just like a petrified tree! It represents a small part of the exposed "neck" of a volcano. The grain of the "tree" reflects penetrative lineation that formed when magma flowed upward through a tight constricted neck toward the surface.

Primary lineation and linear structure can provide a sense of paleocurrent in an ancient depositional basin, as well as flow directions in ancient volcanic fields and magma ascent patterns in ancient igneous intrusions. Secondary lineations give us different information: the nature of strain within rocks forced to change size or shape during deformation. We will

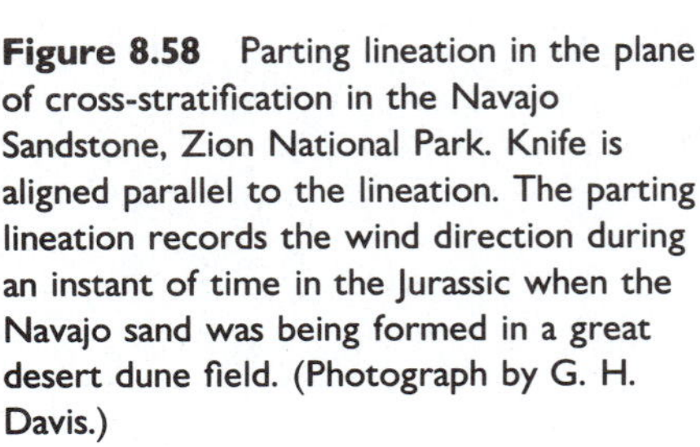

Figure 8.58 Parting lineation in the plane of cross-stratification in the Navajo Sandstone, Zion National Park. Knife is aligned parallel to the lineation. The parting lineation records the wind direction during an instant of time in the Jurassic when the Navajo sand was being formed in a great desert dune field. (Photograph by G. H. Davis.)

Figure 8.59 Two examples of *primary* lineation in igneous rocks. (*A*) Ropy lava (pahoehoe) in a basalt flow on the island of Hawaii. (Photograph by S. J. Reynolds.) (*B*) This thing that looks like a tree trunk is actually igneous rock containing penetrative mineral lineation. The lineation formed in stiff magma ascending as if through a pipe toward the surface. Big Bend country of West Texas. (Photograph by G. H. Davis.)

focus on the **secondary** lineation and linear structure that forms when sedimentary, igneous, or metamorphic rocks are deformed during metamorphism and/or shearing.

Telling the Difference Between Lineation and Foliation

Once when I was doing field work in northern Norway, Arild Andresen showed me an unusually interesting outcrop of white medium-grained granite containing distinct feldspar crystals. In one part of the outcrop the feldspar was distributed and arranged in a way that had resulted in foliation. In another part of the same outcrop the feldspar was arranged as a lineation.

I collected two samples, one of foliated granite, the other of lineated granite (Figure 8.60*A*). At first glance the samples looked identical. They both "appeared" to be foliated. However, when I turned each specimen through my fingers, I saw an important difference (Figure 8.60*B*). In the first sample, the feldspars are arranged in crude planes, not lines, thus defining foliation. (I was able to measure the *strike and dip* of this foliation in outcrop.) In the second sample, the feldspars are arranged in lines, not planes, thus defining lineation, even though the flanks of this specimen look *just like* the trace of foliation. Thus, it was only when I looked at the ends of the specimens that I recognized that the feldspar crystals are aligned like lines, intersecting the top and bottom surfaces of the sample as black dots. (I was able to measure the *trend and plunge* of this lineation in outcrop.)

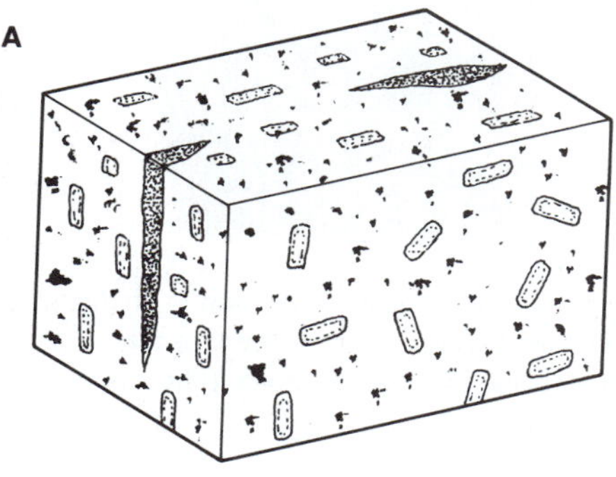

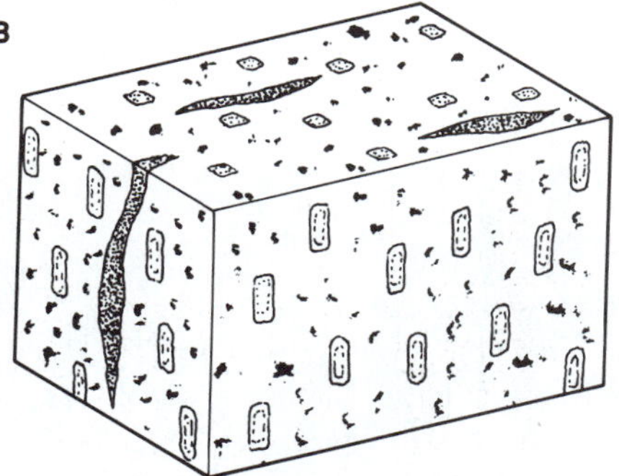

Figure 8.60 These feldspar drawings by Evans B. Mayo illustrate the fabrics I saw in a granite outcrop in Norway. Both rocks appear to be foliated, yet one is and one is not. (*A*) This drawing shows a granite that contains foliation defined by feldspars, but no lineation. (*B*) This drawing shows a granite that contains lineation defined by feldspars, but no foliation.

Types of Lineation

The dominant classes of lineation are intersection lineation, crenulation lineation, and mineral lineation.

Intersection Lineation

Perhaps the most common lineation in metamorphic rocks is intersection lineation, which consists of geometric lines created by the intersection of

Figure 8.61 Crenulation lineation, defined by crests and troughs of minor folds in quartz-sericite schist. (Photograph by G. Kew.)

two (or more) foliations, the intersection of foliation and any compositional layering, or the intersection of foliation or compositional layering with the outcrop surface. As we will soon learn, multiple foliations are a characteristic of strongly deformed metamorphic rocks. The intersection of two closely spaced, penetrative, parallel foliations can result in well-developed lineation, a linear grain if you will, marked in outcrop by a myriad of closely spaced subparallel to parallel lines. The orientation of intersection lineation quite naturally is the trend and plunge of the intersection of the "planes" of foliation; this can be confirmed stereographically.

In some ways, intersection lineation is not true lineation. But when intersection lineation is identified as such, it leads quite naturally to the recognition of two or more intersecting planar fabrics.

Crenulation Lineation

Crenulation lineation is a lineation expressed in the form of bundles of tiny, closely spaced fold hinges (Figure 8.61). Crenulation lineation is especially well developed in phyllites and schists that have been repeatedly deformed by folding. Viewed in the plane of cleavage, crenulation lineation is an array of straight to slightly curved, discontinuous **crests** and **troughs** of folds, more folds than anyone could ever hope to measure, even for "fold junkies" like us.

Crenulation lineation is especially pronounced in mica schists, where the conspicuousness of the structure is enhanced by contrasts in light reflected from the variously oriented micaceous surfaces. It is not surprising that phyllites and schists so characteristically display crenulation lineation. Phyllitic structure and schistosity impart to rocks an **anisotropy** that makes the rock highly vulnerable to kink folding in the presence of stresses that are approximately layer parallel. Mica-rich phyllites and schists readily crinkle, almost like paper.

Figure 8.62 Mineral lineation in strongly deformed quartzite from the Coyote Mountains, southern Arizona. (Photograph by G. Kew.)

Mineral Lineation

Mineral lineation forms in the plane of foliation of many metamorphic rocks, especially slates, phyllites, schists, and gneisses. Mineral lineation is typically marked by a *streaky,* fiberlike lineation (Figure 8.62), the actual physical expression of which is hard to discern in outcrop or hand specimen. When viewed microscopically, however, mineral lineation is typically found to be of varied composition: aligned aggregates of fine-grained minerals, especially quartz and mica; aligned inequant mineral grains, like hornblende; beardlike pressure shadows with associated crystal fibers growing in a preferred orientation from the tips of a sheltering mineral grain; or a subtle lineation produced by the wrapping of thin mineral laminae over relatively large, aligned, inequant grains or mineral aggregates.

Much of the expression of mineral lineation is due to preferred directional crystallization of minerals. However, the linear alignment of minerals and mineral aggregates can in part be derived by mechanical breakdown, that is, **comminution**, of once-larger elements. The relative roles of recrystallization and comminution are a function of the mineralogy of the rock and the conditions under which deformation was achieved.

Figure 8.63 Streaky mineral lineation on the plane of foliation in augen gneiss in Tanque Verde Mountain, Tucson, Arizona. (Photograph by G. H. Davis.)

Mineral lineation in mylonitic rocks reflects this battle between recrystallization and comminution, commonly resembling slickenlines. But unlike slickenlines, the lineation is not restricted to a single surface or a thin zone of faulting or shearing. Instead it pervades a substantial part of the body of the mylonitic rock (Figure 8.63). Mylonitic lineations are defined by an alignment of elongated grains, mineral aggregates, trains of broken grains, and streaks of smeared-out minerals. Where initially equant objects, such as pebbles or inclusions, were present in the rocks prior to mylonitization, they are elongated parallel to lineation and flattened into parallelism with foliation. In other words, lineation in most mylonitic rocks is a stretching lineation parallel to S_1.

Figure 8.64 Elongate flattened pebbles in outcrop of strongly deformed Barnes Conglomerate (Precambrian) in the Tortolita Mountains, southern Arizona. (Photograph by G. H. Davis.)

Types of Linear Structure

The most common examples of linear structure are stretched-pebble conglomerate, rodding, mullion, pencil structure, and boudins.

Stretched-Pebble Conglomerate

Stretched-pebble conglomerate is composed of closely packed elongate clasts (Figure 8.64), a linear structure fashioned through the distortion of cobbles and pebbles, and in some cases boulders. Many different stretched-pebble shapes are possible: cigar-shaped pebbles (Figure 8.65) and pebbles shaped like tongue depressors (ahhhh, what pebbles!) are especially fun to collect. The expression of stretched-pebble linear structure is often enhanced by the presence of mineral lineation in the rock matrix and by the development of crystal fiber beards emanating from the tips of the stretched clasts. The longest stretched pebble I ever extracted from an outcrop measured 20 in. (50 cm). Longer ones are common, but they are difficult to remove from outcrop in one piece.

Rodding

Rodding is a linear outcrop-scale structure that is defined by a penetrative array of parallel, highly elongate bodies of rock (Figure 8.66) (Wilson, 1961). Some are made simply of milky or icy quartz. Rods typically vary in size from flashlights to the length of an arm, although some resemble long tree trunks. Viewed end-on, they are circular, oblate, or lensoidal.

Quartz rods and stretched quartz-pebble conglomerate clasts can resemble each other to a remarkable degree. Quartz rods can be formed in a number of ways. Some may be the boudined, necked expressions of once-continuous layers of quartz. Some may be thought of as the linear equivalent of veins, products of open-space filling and/or replacement, not along fractures but rather along the hinge zones of penetrative folds. Quartz in such rods may in part represent the reprecipitation of quartz

Figure 8.66 This rod occurs in a strongly deformed outcrop of Devonian Excellsior Phyllite in the central Andes of Peru. The landscape within this general outcrop area was strewn with rods, many of them streamlined like bombs. (Photograph by G. H. Davis.)

Figure 8.65 Cigar-shaped stretched pebble, enjoyed by Stan Keith in the Tortolita Mountains, southern Arizona. (Photograph by G. H. Davis.)

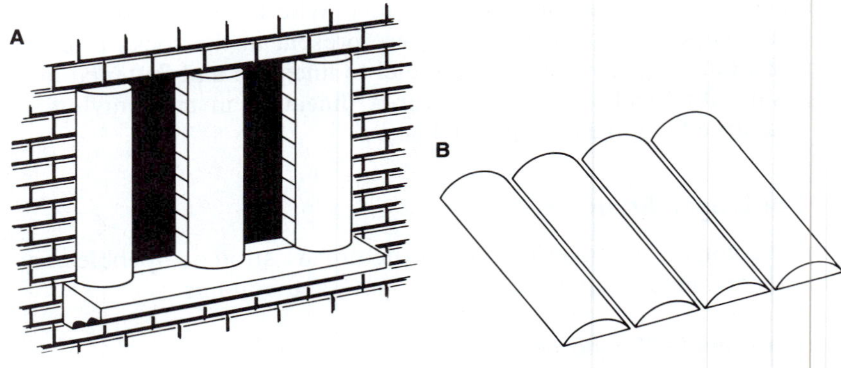

Figure 8.67 Architectural mullions (*A*) adorning a Gothic church and (*B*) lined up on the ground in a way resembles geologic mullions.

made available by local pressure solution of the same host rock in which the rods were found.

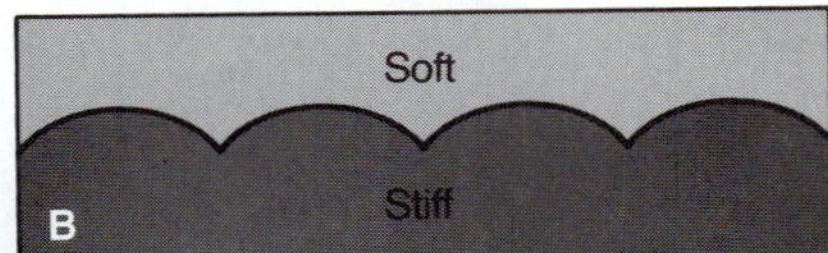

Figure 8.68 (*A*) Outcrop expression of mullion structure, formed by cuspate–lobate folds along the interface between sandstone and slate. The locality is North Eifel, Germany. [Reprinted from *The techniques of modern structural geology,* V. 2: folds and fractures. J. G. Ramsay and M. I. Huber (1987), © by Harcourt Brace and Company Limited, with permission.] (*B*) Mullions form preferentially at the interface between mechanically soft vs. mechanically stiff rocks. Buckling instability due to layer parallel compression produces the cuspate–lobate pattern.

Mullion

The term **mullion** reminds us again of the kinship of structural geology and architecture. Architecturally, mullions are the long, vertical stone members that separate adjacent window openings of Gothic churches (Figure 8.67*A*) (Holmes, 1928; Hobbs, Means, and Williams, 1976). The mullion face that projects into open air is convex outward. If the mullions were laid out and aligned on the ground in the manner shown in Figure 8.67*B*, the array of stone would look exactly like mullion structure in rock.

Viewed in outcrop (Figure 8.68*A*), mullion structure displays regular, repeated, foldlike forms, ranging in wavelength from centimeters to meters. The cuspate–lobate forms are very distinctive, consisting of linked circular or elliptical arcs. They bear a likeness to oscillation ripple marks but are much more linear and systematic, and they are usually larger. Mullions are not composed of newly introduced minerals like quartz, but rather are always fashioned from the host rock itself.

Perhaps the most important descriptive relationship bearing on mullion structure is its occurrence along the **interface** between a mechanically soft and a mechanically stiff layer—argillite and quartzite, respectively, for example. The foldlike mullion forms are convex in the direction of the mechanically soft layer, with the pinched, cuspate foldlike forms pointing toward the mechanically stiff layer (Figure 8.68*B*). Mullion structures arise from **buckling instability** produced by layer-parallel shortening of a contact separating two rock layers of contrasting mechanical strength. Verification of this can be achieved in the laboratory by subjecting two-layer models to strong layer-parallel shortening (Ramsay, 1967). Experiments of this type demonstrate that **dominant wavelength** of mullion structure is determined by the viscosity ratio of the stiff and soft layers.

Pencil Structure

Pencil structure is a very distinctive linear structure associated with folded and cleaved mudstones and siltstones. Outcrops pervaded by pencil structure are strewn with unnatural-looking "pencils" of rock (Figure 8.69*A*). Tiny ones are more like short toothpicks. Larger ones are like magic wands (Figure 8.69*B*). After publication of the first edition of this book, Larry Rogers sent me a picture of himself striking an heroic pose while standing on cleaved rock in Spitsbergen, Norway, supporting part of his weight on a "walking stick" pencil structure 1.5 m long (Figure 8.69*C*).

Figure 8.69 Pencil structure. (*A*) Outcrop of pencil structure in fine-grained calcareous siltstone at Agua Verde near Tucson, Arizona. (*B*) A former record-breaking pencil structure, proudly displayed by Ralph Rogers, the former record holder. The pencil was extracted from a wonderful display in Devonian phyllite in central Peru. (Photographs by G. H. Davis.) (*C*) The new record-breaking pencil structure is helping to support its discoverer, Larry Rogers (no relation to Ralph), on the slopes of Spitsbergen. This outcrop of pencil structure occurs in a graded mud/sandstone of Proterozoic age from the Hecla Hock succession. (Photograph by friend of Larry Rogers.)

He claims the new record. Pencils of rock are irresistible to collectors of nature's oddities.

Pencil structure is almost always found in folded strata. In fact, the orientation of pencil structure is dependably subparallel with the axes of associated folds. The actual physical expression of pencil structure is formed by the intersection of **bedding fissility** and cleavage (Reks and Gray, 1982). Bedding-parallel fissility is a common characteristic of incompetent layers within folded strata. The fissility is an expression of layer-parallel shear. Where layers containing bedding fissility are cut by penetrative axial plane cleavage, the intersection of the two foliations serves to isolate millions of parcels of pencils. The shapes and sizes of the pencils in cross-sectional view reflect the spacing and geometric characteristics of the fissility and cleavage surfaces. The lengths of the pencils are determined by spacing of cross fractures. Individual pencils are not hexagonal like "USA Readibond Wallace CONQUEST pencils"; rather, they are irregularly faceted along smoothly curved interfering surfaces.

Reks and Gray (1982) have pointed out that pencil structure is a potentially useful strain marker. Their work indicates that pencils form parallel to the direction of intermediate finite stretch (S_2) in rocks that have been shortened by 9–26%. It is within this strain range that bedding fissility and axial plane cleavage are equally well developed.

Boudin [Fr., "Sausage"]

A final linear structure of significant importance is one we have already encountered—the **boudin** (Figure 8.70). Boudins form as a response to layer-parallel extension (and/or layer-perpendicular flattening) of stiff layers enveloped top and bottom by mechanically soft layers. The way in which the stiff layer stretches depends mainly on the ductility contrast of the participating layers and on the magnitude of the stretches (S_1, S_2, S_3) describing the level of strain parallel and perpendicular to the layering (Ramsay, 1967). Boudinage is most common in highly deformed sequences with interlayered rock types of different strengths (Figure 8.71).

Boudinage of the strong layer may occur via tension fracturing, shear fracturing, ductile necking (pinch-and-swell structures), or some combination, depending on the ductility contrast between the boudinaged layer and its matrix. Once formed, individual boudins may behave rigidly or may deform somewhat along with the matrix, and this behavior influences the final shape of the boudin (Ghosh, 1993).

Rectangular and **rhombic** boudins (Figure 8.72) form where there is a large ductility contrast between a rigid boudinaged layer and the adjacent rocks. Rectangular boudins separate via tension fractures, whereas rhombic boudins form by shear fracturing or by tension fracturing followed by rotation and deformation of individual boudins. Openings between boudins of both types are filled by plastic flow of the enveloping soft layers or by infillings of "vein" material, such as calcite and quartz. Rectangular and rhombic boudins, because they develop from layers that are very strong compared to the flowing matrix, generally remain relatively rigid and unstrained after separation.

Where the strength contrast is moderate, the stiff layer will tend to deform into boudins whose cross-sectional forms are stubby lenses or barrels. **Barrel-shaped** boudins reflect some amount of ductile necking of the stiff layer prior to tension fracture, and they tend to deform somewhat after separation (Figure 8.72). **Fish-head** boudins begin as rectangular or barrel-shaped boudins, but are strongly deformed after separation, causing the lateral walls of the boudin to collapse inward. Some material that flows into the opening between boudins is trapped in the mouth of the fish (Figure 8.72). Where strength contrast is very small, boudinage occurs via complete ductile necking of the relatively stiff layer, without the involvement of tension fractures. This forms simple **pinch-and-swell** structure and **lenticular** boudins (Figure 8.72). Once formed, lenticular boudins may deform similarly to the matrix material.

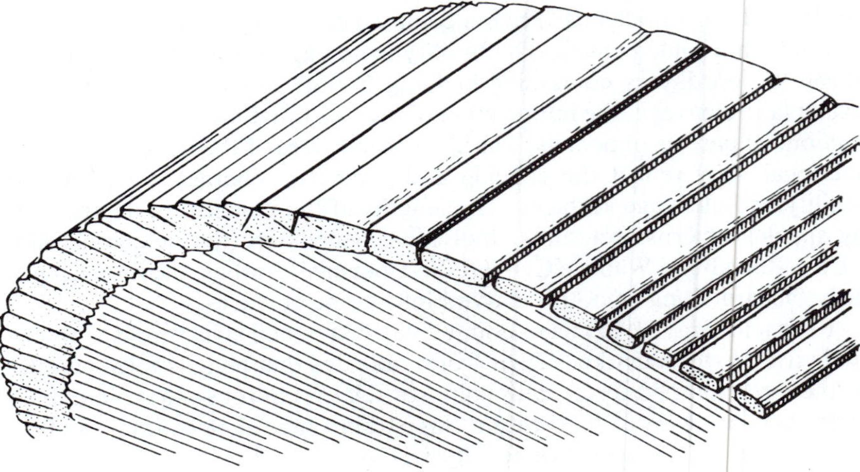

Figure 8.70 Boudins on the flank of a fold. [From *Introduction to Small-Scale Geological Structures* by G. Wilson. Published with permission of George Allen & Unwin (Publishers) Ltd., London, copyright © 1982.]

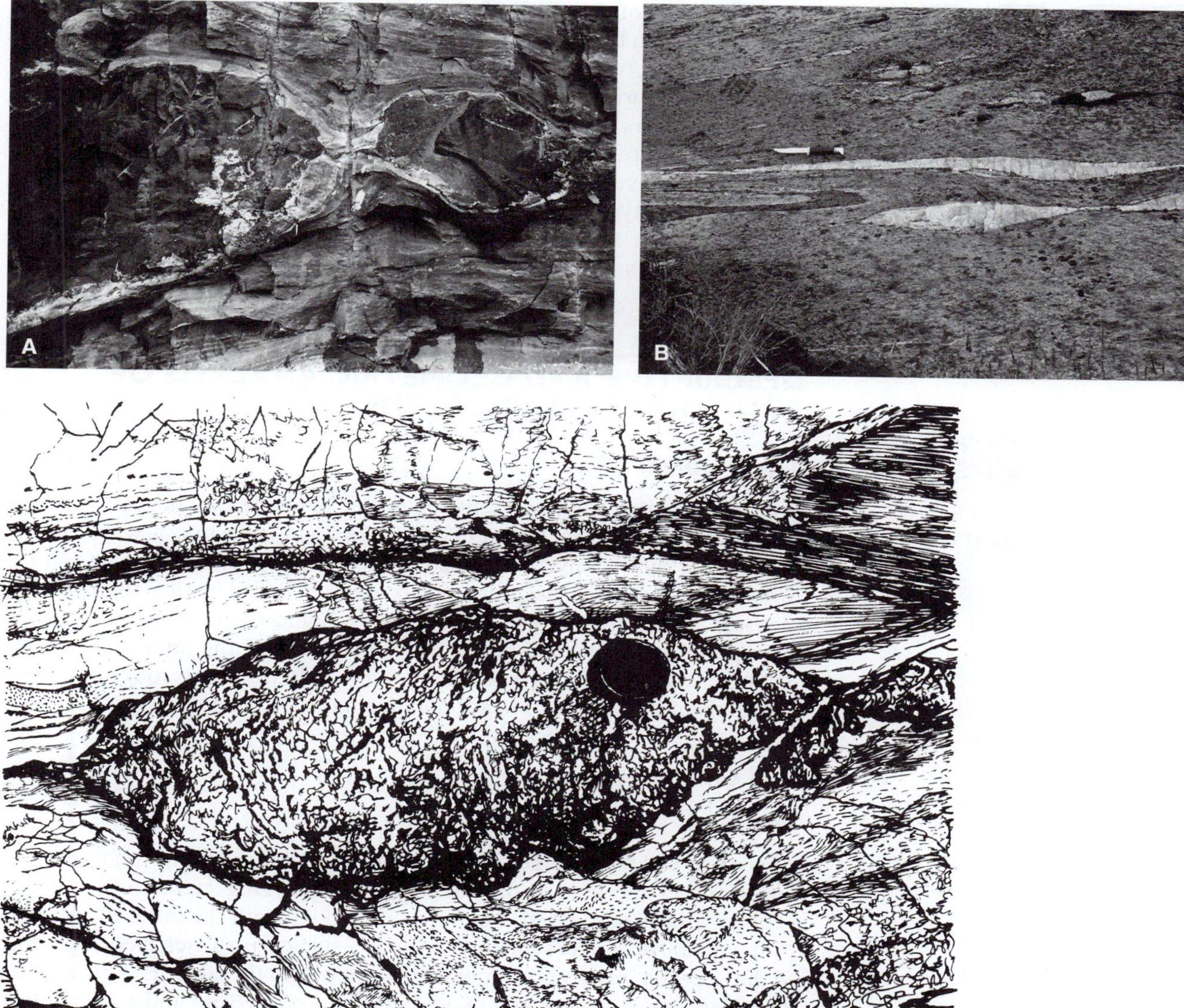

Figure 8.71 (*A*) The relatively stiff, dark mafic igneous rock in northern Norway was flattened and stretched to the point that it "necked," creating boudins. Surrounding rock is a mechanically softer, highly foliated metasedimentary/metavolcanic sequence. (*B*) Another northern Norway example of boudins, but this time affecting a concordant quartz vein in schist. Note the isoclinal fold on the left, reflecting the high degree of deformation of the sequence. (Photographs by G. H. Davis.) (*C*) Quartzite boudin surrounded by mylonitic marble. Note that the lower margin of the boudin has in part deformed by faulting. [India-ink tracing by D. O'Day of photograph by G. H. Davis. From Davis (1980). Published with permission of Geological Society of America and the author.]

Although their shape is quite variable, boudins have a more consistent aspect ratio, being commonly one to four times longer than they are wide when viewed on surfaces cut parallel to lineation and perpendicular to foliation. Boudins longer than this tend to be broken again by the shear stresses generated as the surrounding matrix flows past the top and bottom of the boudin. Once the boudin has become shorter than about twice its thickness, the shear stresses are insufficient to break it again (Ghosh, 1993).

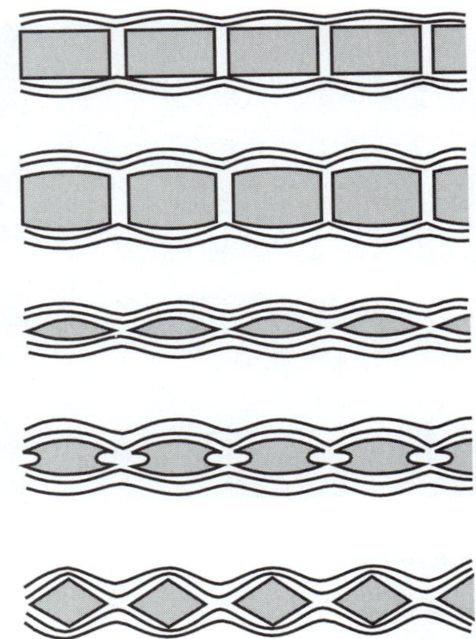

Figure 8.72 Some shapes of boudins. [After Ghosh (1993), Figure 17.6, p. 387.]

Irrespective of the profile forms of boudins, the process of boudinage creates in rock a linear structure in the plane of layering (see Figure 8.70). The linear structure produced by pinch and swell is an array of parallel to subparallel furrows. The linear structure produced by true boudins, in which stiff layers are repeatedly disrupted, consists of a subparallel array of stripes that reflect the presence and structural configurations of the stiff and soft layers. Where the stiff layer is extended in two directions parallel to the layering, the boudins are more equant or blocky, as viewed on the foliation surface. This is called **chocolate tablet** boudinage and commonly looks like rows of lumps, like a colony of sleeping fur seals.

DESCRIPTIVE AND GEOMETRIC ANALYSIS OF FOLIATION AND LINEATION

The Problem

Foliated, lineated rocks present some special challenges in geologic mapping and detailed structural analysis. We have seen that a wide variety of physical and geometric forms can give rise to foliation(s) and lineation(s), and we must be prepared to identify and describe them and to measure their orientations. Furthermore, some rocks possess multiple, **superposed** foliations and lineations. Unscrambling the interrelationships of these is sometimes geometrically very difficult. Some rocks display a half-dozen foliation and lineation elements in a single exposure!

The Coding System

Foliations

There are established methods for sorting and classifying foliations and lineations in the course of mapping and analyzing metamorphic rocks (Turner and Weiss, 1963). Foliations are coded with the letter **S**, meaning "planar" *surface,* like schistosity. Each S-surface is subscripted according to the apparent relative order of formation within the rock. For example, a given outcrop might contain two foliations: a schistosity and a crenulation cleavage. Suppose that close examination of the outcrop reveals that bedding is cut by cleavage (Figure 8.73). Bedding is entered into the field notebook as S_0. Cleavage is entered as S_1, not to be confused with the

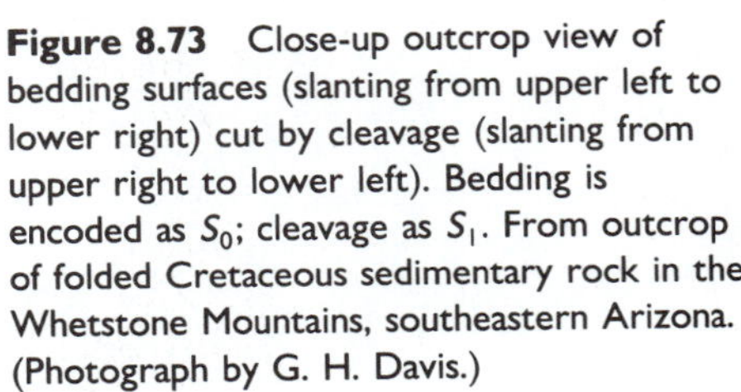

Figure 8.73 Close-up outcrop view of bedding surfaces (slanting from upper left to lower right) cut by cleavage (slanting from upper right to lower left). Bedding is encoded as S_0; cleavage as S_1. From outcrop of folded Cretaceous sedimentary rock in the Whetstone Mountains, southeastern Arizona. (Photograph by G. H. Davis.)

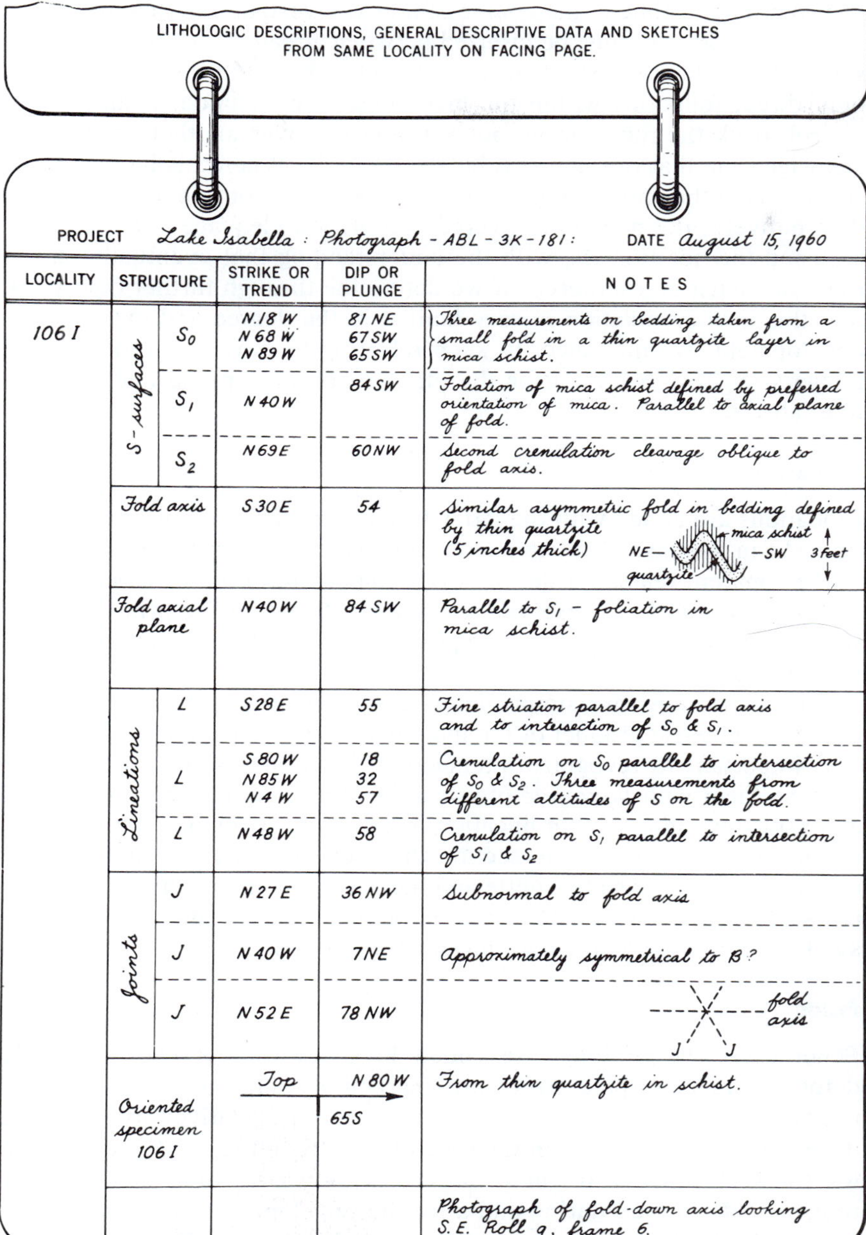

LITHOLOGIC DESCRIPTIONS, GENERAL DESCRIPTIVE DATA AND SKETCHES FROM SAME LOCALITY ON FACING PAGE.

PROJECT *Lake Isabella : Photograph - ABL - 3K - 181 :* DATE *August 15, 1960*

LOCALITY	STRUCTURE		STRIKE OR TREND	DIP OR PLUNGE	NOTES
106 I	S-surfaces	S_0	N.18 W N 68 W N 89 W	81 NE 67 SW 65 SW	Three measurements on bedding taken from a small fold in a thin quartzite layer in mica schist.
		S_1	N 40 W	84 SW	Foliation of mica schist defined by preferred orientation of mica. Parallel to axial plane of fold.
		S_2	N 69 E	60 NW	Second crenulation cleavage oblique to fold axis.
	Fold axis		S 30 E	54	Similar asymmetric fold in bedding defined by thin quartzite (5 inches thick)
	Fold axial plane		N 40 W	84 SW	Parallel to S_1 - foliation in mica schist.
	Lineations	L	S 28 E	55	Fine striation parallel to fold axis and to intersection of S_0 & S_1.
		L	S 80 W N 85 W N 4 W	18 32 57	Crenulation on S_0 parallel to intersection of S_0 & S_2. Three measurements from different altitudes of S on the fold.
		L	N 48 W	58	Crenulation on S_1 parallel to intersection of S_1 & S_2
	Joints	J	N 27 E	36 NW	Subnormal to fold axis
		J	N 40 W	7 NE	Approximately symmetrical to B?
		J	N 52 E	78 NW	
	Oriented specimen 106 I		Top	N 80 W 65 S	From thin quartzite in schist.
					Photograph of fold-down axis looking S.E. Roll 9, frame 6.

Figure 8.74 Field notebook entries showing the record keeping of foliation, lineation, and folding. (Modified from *Structural Analysis of Metamorphic Tectonites* by F. J. Turner and L. E. Weiss. Published with permission of McGraw-Hill Book Company, New York, copyright © 1963.)

direction of greatest finite stretch (S_1) (Figure 8.74). (The distinction is always made obvious in context.) Or an outcrop might reveal that schistosity is cut by crenulation cleavage (Figure 8.75). Schistosity is entered in the notebook as S_1; crenulation cleavage is entered as S_2 (see Figure 8.74). The physical properties and orientations of each foliation are described and posted before moving on to the next outcrop. If, at the next outcrop, the crenulation cleavage (S_2) is seen to be cut by yet another foliation, perhaps a spaced cleavage, the crenulation cleavage retains the status S_2 and the spaced cleavage is awarded the symbol S_3. If, at yet another outcrop, schistosity (S_1) is seen to cut across the original bedding of the rock from which the schist was derived, the bedding is symbolized as S_0. Again, the subscript "0" is reserved for original bedding.

Figure 8.75 Schistosity (S_1) is essentially horizontal, and is cut by a nearly vertical crenulation cleavage (S_2). Devonian phyllites in the central Andes, Peru. (Photograph by G. H. Davis.)

From time to time in the course of a field investigation, new foliations are discovered that must be inserted between already "established" subscripted S-surfaces. Just when we feel confident that there are three, and only three, foliations within the metamorphic rock under examination, we roll back the moss on an outcrop and discover another crenulation cleavage, one that postdates S_2 but predates S_3. When this happens, the subscripts in the entire notebook must be edited according to the change. Alternatively, the newly discovered crenulation cleavage is temporarily assigned the notation S_{2A} with the understanding that eventually all the foliations must be reordered. If we determine through further field work that this new crenulation is very limited in the degree and extent of its development, we may elect to keep referring to it as S_{2A}, retaining the integers (S_1, S_2, S_3, . . .) for the other more widespread and penetrative fabrics.

Lineations

Lineations are coded with the letter **L**, meaning lineation or linear structure. As in the case of foliations, each lineation is subscripted according to the apparent relative order of development within the rock being studied. The symbol L_0 is reserved for primary lineations or primary linear structure within the rock (i.e., the protolith) from which the metamorphic rock was derived.

Some lineations are simply intersection lineations, produced as the passive geometric product of the intersection of two (or more) foliations. It is important to recognize intersection lineations as such, and not to confuse them with *bona fide* physical lineations. Again, the presence of a conspicuous intersection lineation forces us to recognize the presence of the foliations whose intersection they reflect, foliations that in some cases might otherwise be subtle and hard to spot. Intersection lineations are seldom entered as part of the hierarchy of subscripted L's. The L-symbols are reserved for real, physical lineations.

Folds

Penetrative sets of folds are coded **F**. Since most bodies of highly deformed metamorphic rocks are marked by superposed folds, it is necessary to assign subscripts to identify the relative order of the development of folds. As in the case of the "S and L's," F_0 is reserved for folds of primary origin preserved within low grade metamorphic rocks. Folds formed during and after the development of the metamorphic rock in which they are found are coded F_1, F_2, F_3, . . . , in the order in which they formed.

Within fold belts in regionally metamorphosed terranes, it is the rule, not the exception, to find at least two clearly defined fold sets in phyllite and schist. The earliest fold set (F_1) is typically composed of tight to isoclinal passive folds that are axial planar to the earliest formed foliation (S_1). These F_1 folds, along with S_1, are typically refolded by a set of younger folds (F_2) that is composed of tight to open flexural-slip folds, especially kink folds. Key outcrops show the interrelationship and relative timing between two (or more) fold sets in multiply deformed areas.

Fold structures entered into the field notebook as F_1, F_2, and F_3 are described carefully according to physical and geometric properties. Furthermore, the orientations of representative folds are posted. Axial surface orientations are presented in terms of strike and dip; axis orientations are recorded in terms of trend and plunge.

The Grouping of Structural Elements

Foliations, lineations, and folds that are interpreted to have formed at the same time are given the same subscript rank. If a field notebook bears entries that discuss and describe S_1, L_1, and F_1, it is understood that all these structural elements formed contemporaneously in the deformed rock. A typical scenario is this: S_1, a schistosity, formed as an axial plane cleavage to F_1, a passive fold; and L_1, a mineral lineation, formed parallel to the axis of F_1 folding. Structural elements of a given discrete event, or phase of a progressive *deformation,* are gathered into headings D_1, D_2, D_3 ... or D_1, D_{2A}, D_{2B}, D_3, and so on.

Lineations (or foliations or folds) that are quite dissimilar in physical expression may be assigned the same subscript rank on the basis of compatibility of physical expression as well as compatibility of geometric orientations. A given lineation may have a physical expression that varies from lithology to lithology because of the texture and mineralogy of the host rocks from which the metamorphic rock was derived. The clue that the physically dissimilar lineations formed as a set is expressed in the compatibility of geometric orientations. The geometric compatibility may be disclosed as a constancy of preferred orientation, or as a systematically changing array of orientations. The final check on whether the lineations should be grouped together is based on interpreting whether all the lineations could have formed in the same temperature–pressure environment by a common structural process. Even though slickenlines and mineral lineation in high grade gneiss may have a common orientation, they are not compatible physical elements. Slickenlines are products of brittle faulting; mineral lineation in high grade gneiss is a product of deformation and/or recrystallization under metamorphic conditions of elevated temperature and pressure (Figure 8.76).

Yet another sorting process is required to decide which foliations, lineations, and folds may have formed together as a system in response to a common event. Decisions for this brand of sorting are based on the geometric interrelationships among the elements and, once again, the compatibility of physical forms. Correctly interpreting the structural compatibility of foliations, lineations, and folds is a critical step in beginning to evaluate the strain significance of particular fabrics.

Correlation of Fabric and Fold Elements in Time and Space

The subscripted rankings of foliation, lineation, and folds give the impression that structures of different rankings formed during different structural

Figure 8.76 This penetrative mineral lineation that formed under upper greenschist conditions in pegmatite superficially resembles brittle slickenlines that form on faults under "cold" conditions. Coyote Mountains, southern Arizona. (Photograph by G. H. Davis.)

events, perhaps even different tectonic events. This is not necessarily the case. A given set of penetrative folds (F_1) might be systematically refolded by a second set of folds (F_2) during a continuum of **progressive deformation**. In similar fashion, a given cleavage (S_1) that develops in the early stages of progressive simple shear may be systematically rotated to near-parallelism with the plane of shear and be cut by a newly formed cleavage (S_2) that is genetically related to the same simple shear continuum of deformation. Determining whether structural elements are products of a continuum of deformation or are products of discrete events widely separated in time is based, once again, on physical and geometric compatibility. In some cases radiometric dating can help resolve the problem.

Subscripted rankings may be misleading in yet another way. It is tempting to assume that an S_2 structure defined by a geologist working in one part of a deformed belt may correlate with an S_2 structure mapped and defined by some other geologist working in the same deformed belt, but in a different area. Keep in mind that one person's S_2 may be another person's S_3! Correlation of structural elements from place to place within a region requires careful appraisal of descriptive and geometric properties of the elements.

A final interpretive pitfall is linked to the assumption that a foliation–lineation–fold suite in one area formed synchronously with a physically identical foliation–lineation–fold suite in another area within the same orogen. This is not necessarily the case. The development of fabric elements can act in a **diachronous** fashion. A cold front is a good example of a diachronous event. A cold front delivers the same orderly sequence of events at every point along its path across a region—sudden drop in temperature, snow flurries, 3 feet of snow, . . . first in Chicago, then in Cleveland, then in Pittsburgh, then in Philadelphia. The compaction foliation fabric at the base of the 3 feet of snow in Chicago is not the same age as that at the base of the new snow in Cleveland, even though it looks identical physically and geometrically.

TECTONITES

The Concept

Rocks that are *pervaded* by cleavage, foliation and/or lineation are known as **tectonites** (Figure 8.77). Tectonites are rocks that have flowed in the solid state in such a way that no part of the rock body escaped the

Figure 8.77 Rocks that are penetratively foliated and/or lineated are known as tectonite. This tectonite, replete with isoclinal recumbent folds, occurs near Bjornfeld in north Norway. (Photograph by G. H. Davis.)

distortional influence of flow, at least when observed at the scale of a single hand specimen and/or outcrop. Tectonites are like the penny that has been flattened by a train (see Figure 1.23).

Foliation and lineation are structures of the kinds that can accommodate the distortion a tectonite is forced to endure. The extraordinary alignment of foliation and/or lineation in a tectonite is an expression of the geometric requirements of the **state of strain**. Although tectonites, by definition, are rocks that have been able to flow in the solid state (Turner and Weiss, 1963), we now are fully aware that what we perceive as flow is scale dependent. The flow of tectonites is seen microscopically to be a combination of slip, crystallization, or dissolution along exceedingly closely spaced discontinuities.

Most tectonites, and thus most foliations and lineations, form in environments of elevated temperature and confining pressure. Metamorphic and igneous environments are ideal. However, tectonites also form through distortion of soft sediments, excessively weak lithologies like salt or gypsum, or rock types that are especially vulnerable to stress-induced dissolution.

Types of Tectonite

Several classes of tectonites can be distinguished on the basis of whether the tectonite contains foliation, lineation, or both. These three types are known as **S-tectonites**, **L-tectonites**, and **LS-tectonites** (Figure 8.78), where "L" and "S" refer to lineation and foliation, respectively. S-tectonites are tectonites marked by foliation but not lineation. L-tectonites are tectonites marked by lineation but not foliation. LS-tectonites contain both. Of all the tectonites, LS-tectonites are the most common. These are the tectonites marked both by foliation and lineation. Lineation in LS-tectonites lies in the plane of foliation.

The use of the letter **S** is based on the long-established convention of employing **S-surface** in reference to the penetrative, planar, parallel elements that constitute foliation (Turner and Weiss, 1963). The expression "S-surface" is handy because it places no limits on the physical and/or geometric element that may contribute to the expression of a foliation in a rock. S-surfaces in a schist are the aligned micas and cleavage domains. S-surfaces in a gneiss are the planar, parallel, compositional bands. S-surfaces in a flattened-pebble conglomerate are the aligned discoidal pebbles. And so it goes.

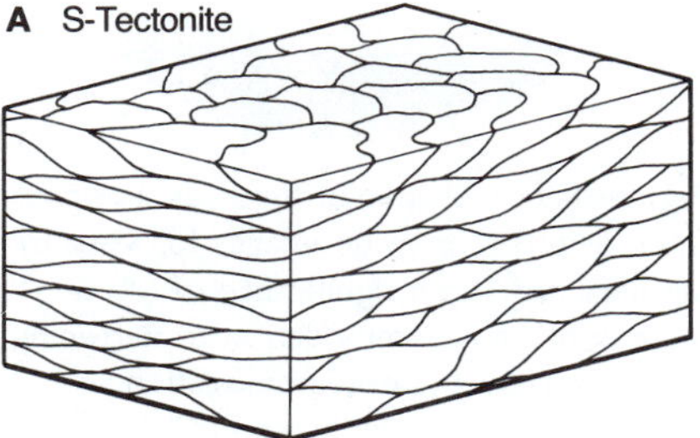

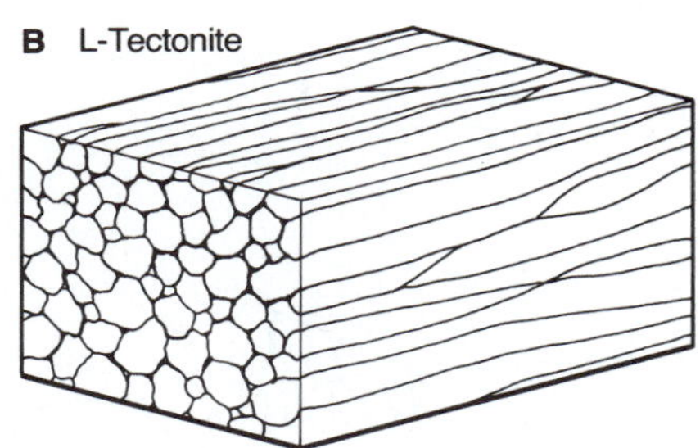

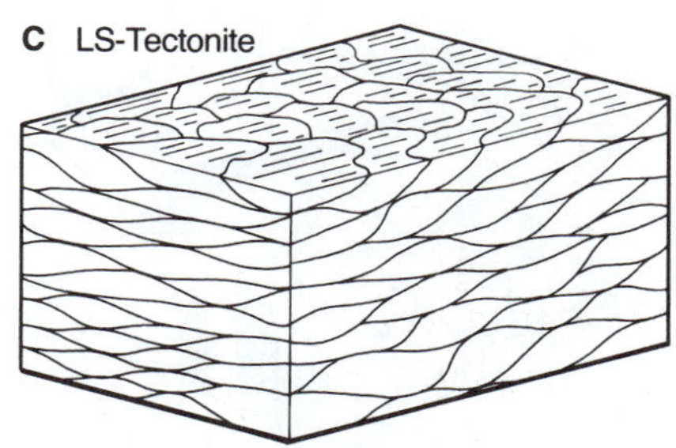

Figure 8.78 Schematic portrayal of S-, L-, and LS-tectonites. (*A*) S-tectonites are marked by a single, penetrative foliation. (*B*) L-tectonites are marked by pervasive lineation, but no foliation. (*C*) LS-tectonites are marked both by foliation and lineation. The lineation in LS-tectonite lies in the plane of foliation.

Strain Significance of Tectonites

Overall Objective

The presence of foliation or lineation in a rock is a signal that the rock in which these structural elements are contained has undergone significant distortion, with or without dilation. One of the goals of detailed structural analysis is to interpret the strain significance of tectonites, that is, to try to interpret the magnitudes and directions of distortion and/or dilation that were accommodated by the development of foliation and lineation. This, of course, is not a simple task. Success in interpreting the strain significance of tectonites depends to a large extent on discovering and analyzing distorted primary objects, of known original shape and/or size, as guides to the extent of dilation and/or distortion. Finding such treasures is more the exception than the rule.

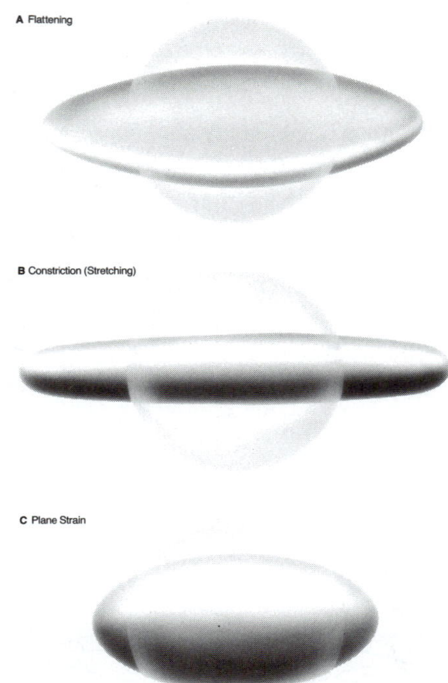

Figure 8.79 Shapes of three-dimensional strain ellipsoids provide images for visualizing the idealized strain significance of (*A*) S-tectonites (an accommodation to flattening), (*B*) L-tectonites (an accommodation to unidirectional stretching), and (*C*) LS-tectonites (commonly an expression of plane strain).

Figure 8.80 A thick viscous pancake batter, plus the right equipment, can result in early-morning experimental production of stretching fabrics (L-tectonites), flattening fabrics (S-tectonites), and combo fabrics (LS-tectonites). I prefer the S variety.

Flattening, Constriction, and Plane Strain

The evaluation of distorted primary objects in tectonites has revealed that **S-tectonites** tend to be products of **flattening**, a state of finite strain in which the magnitudes of finite stretch are such that $S_1 = S_2 > S_3$. Flattening is the kind of distortion that transforms an original sphere into an **oblate strain ellipsoid** (Figure 8.79*A*). This type of strain, where shortening in the S_3 direction is accommodated by extension equally along S_1 and S_2, is most easily explained by volume loss due to pressure solution. Beware, however, because the lack of lineation in some rocks may be due to a predominantly micaceous character or other lithologic influence that either obscures the lineation or prevents it from forming even though the strain conditions are appropriate.

L-tectonites tend to be products of unidirectional stretching or **constriction**, a state of strain in which $S_1 > S_2 = S_3$. Constriction transforms an original sphere into a **prolate strain ellipsoid** (Figure 8.79*B*). L-tectonites, although not exceedingly common, are present in some shear zones, probably because extension parallel to S_1 was accommodated by semiequal shortening along S_2 and S_3.

LS-tectonites, probably the most common type of tectonite, ideally form through **noncoaxial strain**, such as plane strain, where stretching in one direction is compensated by flattening at right angles to the direction of stretching, with neither stretching nor shortening in the intermediate direction. Simple shear can transform an original sphere into a **triaxial ellipsoid** in which the state of strain is marked by $S_2 = 1$, and $S_1 > S_2 > S_3$ (Figure 8.79*C*). Flattening, constriction, and plane strain are distortional strains that may or may not be accompanied by gains or losses in volume of the rock that is converted to tectonite.

We can create a kitchen image that distinguishes between L-tectonites, S-tectonites, and LS-tectonites. Imagine mixing up a thick pancake batter and pouring it from a pitcher onto the griddle (Figure 8.80). As the batter hangs in midair between pitcher and griddle, it stretches itself out into a narrow column of L-tectonite (Figure 8.80). Air bubbles in the batter are prolate ellipsoids: one long axis parallel to the direction of stretching, and two smaller axes equal in length. When the batter hits the griddle, flattens, and spreads out equally in all directions, a thin, planar, S-tectonite is created (Figure 8.80). Air bubbles in the flattened batter are oblate spheroids: one short axis perpendicular to the direction of flattening, and two long axes equal in length. Now imagine pouring the batter from a pitcher onto an inclined griddle, dipping 10° (Figure 8.80). The batter will flow downslope via simple shear. Stretching lineation will be evident in the plane of the batter; the lineation will be oriented down-dip. The air bubbles will evolve into true triaxial ellipsoids: three axes each of different length; the long axis becomes progressively longer, the short axis becomes progressively shorter, and the intermediate axis stays the same length throughout the down-griddle journey. Now if only we had some fresh blueberries!

It is not difficult to imagine structural environments in nature in which tectonites can be formed through flattening, constriction, and plane strain. Some slaty cleavage is undoubtedly a product of distortion by flattening. An axial planar slaty cleavage could form through significant stress-induced shortening perpendicular to the ''plane'' of cleavage, accommodated by pressure solution rather than appreciable stretching or shortening *within* the plane of cleavage. Examples of the development of an L-tectonite through constriction could include the emplacement of a salt diapir, and the magmatic intrusion of a rhyolitic plug. Plane strain is a state of strain that typifies many shear zones (see Chapter 9). Progressive simple shear can create LS-tectonites whose physical and geometric characteristics express the nature of the simple shear process itself. Foliation would occupy the direction of flattening within the shear zone. Lineation would form in the plane of flattening (S_1S_2), oriented parallel to S_1.

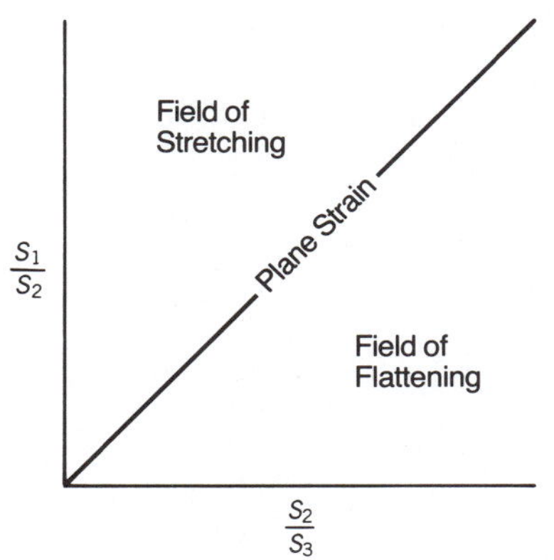

Figure 8.81 The Flinn diagram, a device for portraying the state of strain of deformed rock.

Flinn Diagrams

The full range of three-dimensional strains that may be reflected in the physical and geometric properties of tectonite goes well beyond the special cases of pure flattening, pure constriction, and plane strain. Each of these strain states can be achieved with or without volume changes. Furthermore, stretching and flattening may team up in a spectrum of combinations that are limitless.

A convenient way to visualize the possibilities of three-dimensional strain, with or without volume change, is through a device known as a **Flinn diagram**. Introduced by Zingg (1935) and expanded by Flinn (1962), the Flinn diagram is a simple *x*-*y* graph that pictures ellipsoids that result from distortion and/or dilation of an original reference sphere. Along the *y*-axis of a Flinn diagram is plotted the strain ratio S_1/S_2 (Figure 8.81). Along the *x*-axis of a Flinn diagram is plotted the strain ratio S_2/S_3. Values of S_1, S_2, and S_3 are derived from strain analysis of distorted primary objects in the tectonite under study, objects like pebbles, fossils, and reduction spots.

If a tectonite forms through pure flattening, the coordinates (S_2/S_3, S_1/S_2) that describe the state of strain of primary objects in the tectonite will plot along the *x*-axis of a Flinn diagram. Conversely, if a tectonite forms through pure constriction, the coordinates (S_2/S_3, S_1/S_2) will plot along the *y*-axis. If a tectonite forms by plane strain, the coordinates (S_2/S_3, S_1/S_2) will plot along a 45°-sloping line that intersects the origin of the plot. The lines of pure flattening, pure constriction, and constant volume plane strain are simply three in number. There is plenty of room in a Flinn diagram to plot the limitless combinations of flattening and constriction that one might encounter in nature.

Logarithmic Flinn Diagram

An even more useful way to present the state of strain of tectonite is through the use of a logarithmic Flinn diagram. This modification of the Flinn diagram was introduced by Ramsay (1967), and it can be used to keep track of changes in volume that might accompany distortion.

Logarithmic strain (ε), otherwise known as **natural strain** or **true strain**, is equal to the logarithm of the finite stretch S (Ramsay, 1967):

$$\varepsilon = \log_e(S) \tag{8.1}$$

where ε is logarithmic strain and S is stretch.

Logarithmic strain (ε) can be instantly calculated if values of stretch are known. To represent on a logarithmic Flinn diagram the state of strain of a given tectonite, values of S_1, S_2, and S_3 must first be transformed into values of ε_1, ε_2, and ε_3. Once these conversions have been made, ε_1–ε_2 is plotted along the *y*-axis of the logarithmic strain Flinn diagram, and ε_2–ε_3 is plotted along the *x*-axis (Figure 8.82). As in the ordinary Flinn diagram, pure flattening is described by values of ($\varepsilon_2 - \varepsilon_3$, $\varepsilon_1 - \varepsilon_2$) that plot along the *x*-axis, and pure constriction is represented by values of ($\varepsilon_2 - \varepsilon_3$, $\varepsilon_1 - \varepsilon_2$) that plot along the *y*-axis of the diagram. Points

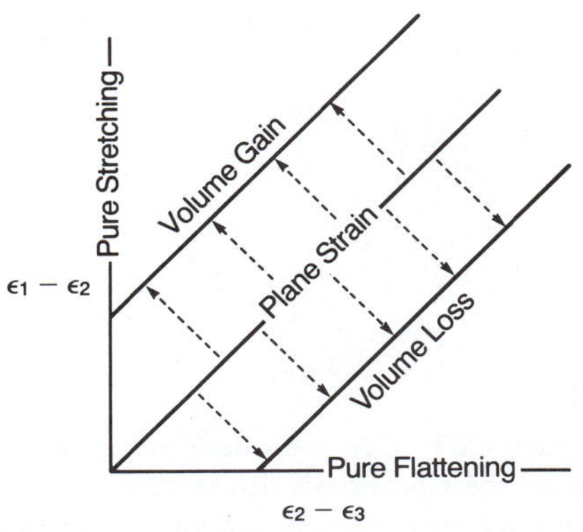

Figure 8.82 The logarithmic Flinn diagram is even more useful than the Flinn diagram because it can be used to evaluate changes in volume.

representing the condition of constant volume plane strain fall along a 45°-sloping line that intersects the origin of the plot (Figure 8.82). If a tectonite forms through plane strain accompanied by dilation, the coordinates $\varepsilon_2 - \varepsilon_3$ and $\varepsilon_1 - \varepsilon_2$ describing the state of strain of primary objects in the tectonite will plot along 45°-sloping lines that do not intersect the origin of the plot. Instead they intersect the *x*- or *y*-axis, depending on whether the dilation was accompanied by a volume decrease or a volume increase.

Deformation Paths

Flinn diagrams and logarithmic Flinn diagrams take on value and meaning when we put them to use. Let us consider a simple example of plane strain that we can model with a card deck. Figure 8.83 shows progressive simple shear of a deck embossed with a reference circle. Five stages of incremental, progressive deformation are shown—*A* through *E*. At each stage of the deformation, the value of S_2, measured perpendicular to our view of the simple shear, is 1.0. In other words, there is neither shortening nor stretching in the third dimension. If we know the size of the original

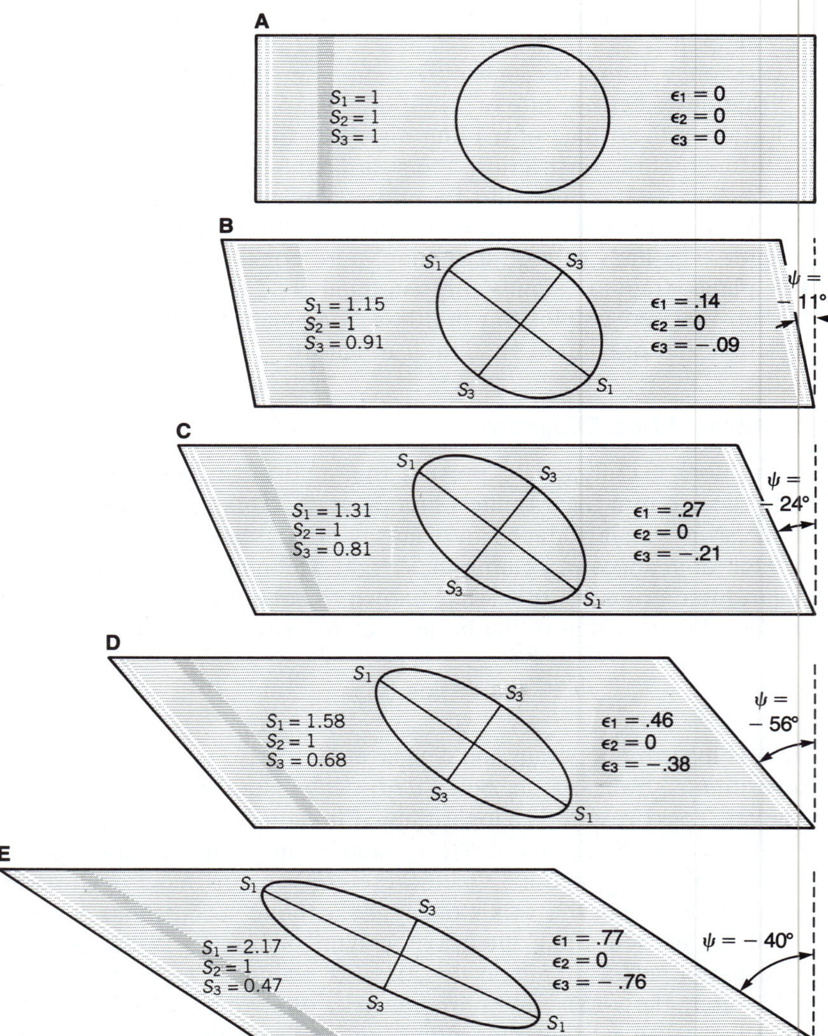

Figure 8.83 Computer deck modeling of simple shear provides a database for preparing Flinn diagrams and logarithmic Flinn diagrams, to see how they work. Stretch (S) values and logarithmic strain (ε) are computed for five stages of progressive simple shear—*A–E*.

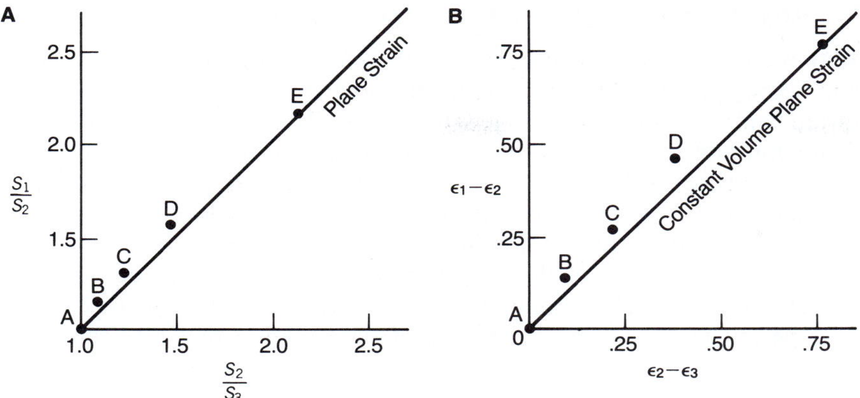

Figure 8.84 (*A*) Flinn and (*B*) logarithmic Flinn diagrams showing the deformation path of the progressive simple shear illustrated in Figure 8.83. Departures from the straight line are simply due to measurement errors.

reference circle, we can calculate the values of S_1 and S_3 at each stage of the deformation. These values in turn can be converted to logarithmic strain E (Figure 8.83).

With these strain data in hand, it becomes a simple matter to plot on Flinn and logarithmic Flinn diagrams the state of strain representing deformational stages A through E (Figure 8.84). Portrayal of these states of strain on both diagrams illlustrates what we know to be true: that the deformation was achieved by constant volume plane strain. Points A through E approximate a 45°-sloping line that intersects the origin of the graph. These points, taken together, represent the **deformation path** that was followed during the progressive deformation.

Progressive simple shear accompanied by volume changes would follow a deformational path different from that represented by path A–E in Figure 8.84. By way of example, the logarithmic Flinn diagram pictured in Figure 8.85 shows two deformation paths, each representing a different combination of distortion and dilation. Path 1 describes a progressively increasing distortion accompanied by steady volume loss. Path 2 portrays a progressively increasing distortion accompanied by steady volume gain.

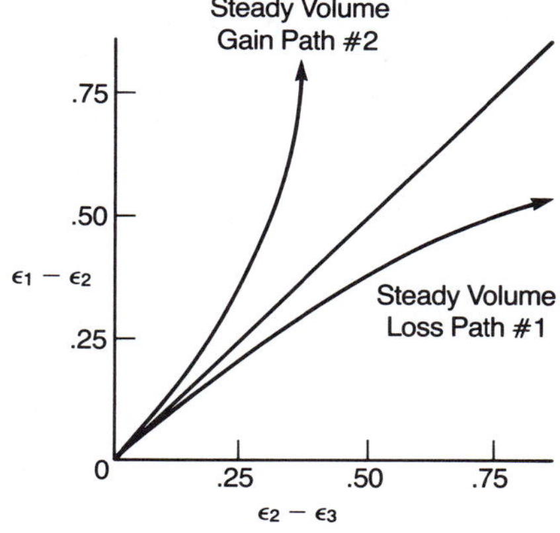

Figure 8.85 Logarithmic Flinn diagram picturing progressive simple shear accompanied by progressive volume loss (path 1), and progressive simple shear accompanied by progressive volume gain (path 2).

By using sleight of hand, it is possible to simulate deformation paths of the type represented in Figure 8.85 with card decks. For example, to simulate a 20% volume loss followed by simple shear, first remove every fifth card from the card deck. This automatically transforms the original reference circle on the flank of the deck to an ellipse. Then deform the deck by progressive simple shear. To simulate an increase in volume, add cards at an appropriate even spacing through the deck. To portray simple shear accompanied by steady volume loss, prepare a card deck by drawing a reference circle on the flank of the deck when the deck is "leaning" in the configuration shown in Figure 8.86. Then begin to deform the deck by progressive simple shear, while at the same time steadily rotating the cards to an upright position. The effect of rotating the deck to an upright position is to decrease the surface area of the deck, and thus to decrease the area of the original circular object.

With the help of a friend, you can measure the lengths of the axes of the strain ellipse at each stage of the progressive simple shear. These data can be converted to values of logarithmic strain (ε) describing each stage of the progressive deformation. When plotted in the proper manner on a logarithmic diagram, the array of points defines a deformation path reflecting a deformation marked by steadily increasing distortion and steadily decreasing volume.

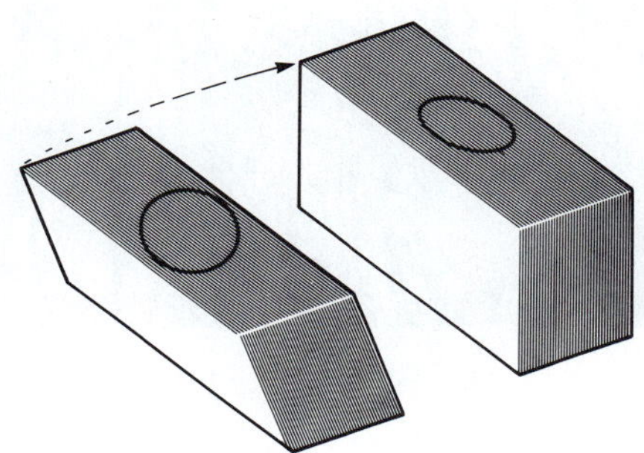

Figure 8.86 Card trick to simulate volume loss.

Identifying Deformation Paths of Natural Tectonites

Within any tectonite found in nature, the state of strain is never uniform and homogeneous. Rather, different parts of a body of tectonite are distorted and/or dilated by different amounts. The specific magnitude of distortion and/or dilation at each point in the body will never be fully known, but representative values can be gleaned from preserved but distorted primary objects. When the states of strain thus derived are plotted on a common logarithmic Flinn diagram, it becomes possible to evaluate the strain significance of the tectonite, whether it be an S-tectonite, an L-tectonite, or an LS-tectonite. Furthermore, the array of plotted points may permit us to interpret the path of deformation that was following during the distortion and/or dilation.

ESTIMATING STRAIN IN TECTONITES AND OTHER METAMORPHIC ROCKS

When we walk up to an outcrop of a strongly deformed tectonite, such as that shown in Figure 8.87, we are generally impressed with the amount of strain the rocks must have experienced. The question naturally arises—how much strain really occurred? Specifically, we are interested in the shape and orientation of the finite strain ellipse. There are a number of methods available for determining strain in a tectonite or any other deformed rock, depending on the preserved suite of **strain markers**—objects whose shape and/or distribution record the strain. In this section, we provide a "taste" of several commonly used strain determination methods.

Deformed Shapes of Initially Spherical Objects

One possible way we can gauge the amount of strain a rock has undergone is to examine how objects within the rock have changed shape during deformation. For example, reduction spots start out spherical (Figure

Figure 8.87 (*A*) Outcrop of highly cleaved sandstone (light) and argillite (dark) of the Meguma Group (Ordovician) of Newfoundland. (*B*) Close-up view shows the vertical cleavage along which pressure-solution removal of material has taken place. Moreover, the close-up view reveals distorted cross-bedding and ripples in the sandstone. (Photograph by S. J. Reynolds.)

Figure 8.88 (*A*) Undeformed reduction spots (white) in red sandstone of Proterozoic age in the Grand Canyon. Note that some of the spots have been eroded in a way to reveal their exact centers, i.e., black specks of reducing agents, such as pyrite or organic fragments. (Photograph by S. J. Reynolds.) (*B*) Deformed reduction spots. The distortion reflects a secondary strain. (Photograph by O. T. Tobisch.)

8.88*A*) but may become ellipsoidal if the rock within which they occur becomes distorted (Figure 8.88*B*). When this happens we can measure the **aspect ratios** (e.g., length to height) of such objects on three different surfaces, ideally those cut parallel to the principal axes of the finite strain ellipsoid (Figure 8.89). The ideal surfaces include one parallel to foliation, one parallel to lineation and perpendicular to foliation, and one perpendicular to both lineation and foliation. For each surface, we determine the ratio of two principal stretches, and then combine results for the three surfaces to define the shape, but not the absolute dimensions, of the three-dimensional finite strain ellipsoid (Figure 8.89). We can discover the true dimensions (actual values of the principal finite stretches) only if we know the average initial diameter of the objects, which is unlikely. If we assume constant volume deformation, however, we can use the volume of the ellipsoid to determine the diameter of a sphere with the same volume. Each principal stretch is then determined by dividing its length by the diameter of the sphere.

R_f/ϕ Method

We need to use a slightly more involved approach if the objects were originally elliptical, rather than spherical. The so-called R_f/ϕ method (Ramsay, 1967) is commonly used to measure the strain in deformed conglomerates, most of which start with nonspherical clasts. The basis of the method is recognizing that final shapes and orientations of ellipsoidal pebbles in the deformed rock (Figure 8.90*B*) are the product of the original shapes and orientations in the undeformed rock (Figure 8.90*A*) and the shape and orientation of the finite strain ellipsoid to which the undeformed rock was subjected (Figure 8.90*B*).

For a number of deformed objects exposed on a single planar surface, we measure present-day ellipticity and long-axis orientation, which are a function of (1) the original ellipticity and orientation of the objects and (2) the orientation and ellipticity of the finite strain ellipse. **Ellipticity** is defined as the ratio of the long axis of the ellipse to the short axis, as

Figure 8.89 This slab of "rock" containing deformed reduction spots has been cut in a way to reveal the "aspect ratios" of ellipses on each of the three principal planes. By combining the ratios derived from each plane, we can construct the shape of the strain ellipsoid for the outcrop from which the specimen was collected.

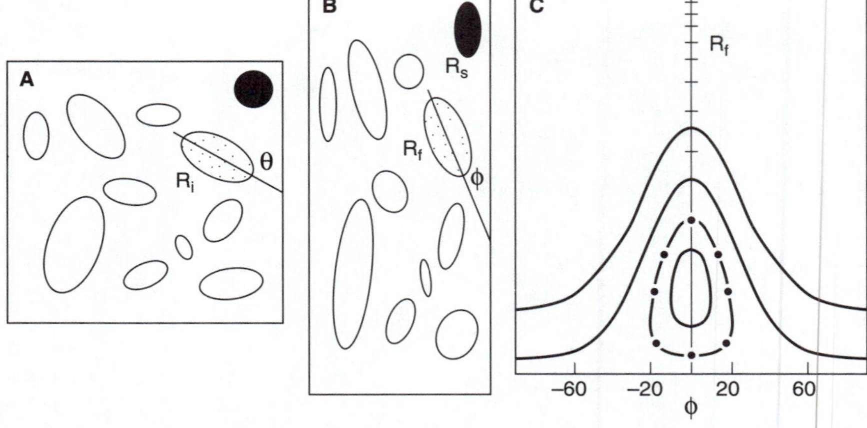

Figure 8.90 The final shapes and orientations of ellipsoidal objects in deformed rock is a product of (*A*) the original shapes and orientations of the ellipsoidal objects in the undeformed rock and (*B*) the shape and orientation of the finite strain to which the undeformed rock was subjected. Thus an original elliptical object of initial axial ratio R_i and initial orientation θ becomes transformed by a finite strain of R_s (oriented at $\phi = 0°$) to an elliptical object of final axial ratio R_f and final orientation ϕ. (*C*) The shapes of these onion curves of the R_f/ϕ plot reveal the interrelationships between R_f, R_i, and R_s. [From Lisle (1994), Paleostrain analysis, Figure 2.4, p. 32. Reprinted with permission of Pergamon Press, Oxford, England.]

measured on the surface we are observing. Using these measurements we can derive, either graphically or mathematically, the orientation and shape of the finite strain ellipse on the exposed surface (Figure 8.90*C*). To more fully characterize the strain, we repeat the technique on several surfaces, especially those three special directions: parallel to foliation, parallel to lineation and perpendicular to foliation, and perpendicular to both foliation and lineation.

Thinning of Layers of Known Thickness

In other cases we can get an idea of the amount of strain by seeing how deformation has thinned or thickened stratigraphic units or other layers whose initial thickness we know. A classic example of this is found in the Big Maria Mountains of southeastern California (Figure 8.91*A*), where Warren Hamilton (1964, 1987) recognized that an upright Paleozoic section

Figure 8.91 In the Big Maria Mountains, southeastern California, the familiar Paleozoic sequence of the Grand Canyon (from the Cambrian Tapeats to Permian Kaibab) has been reduced in places to a tiny percentage of its original thickness. (*A*) View looking east showing overturned fold. Thicknesses are close to normal (1000 m) in the lower upright limb (right-hand side). The overturned upper limb (thin band on left in photo) is as thin as 10 m, i.e., 1% of its original thickness! (Photograph by S. J. Reynolds.) (*B*) In the nearby Granite Wash Mountains, the entire thickness of the middle part of the Paleozoic section can be measured with an on-the-outcrop photo scale. All that is left of the mighty Redwall Limestone is the thin white band to the right of the scale. We have Warren Hamilton to thank for recognizing this profound thinning, and then convincing the world. You haven't lived until you have taken three or four giant steps across the entire Grand Canyon Paleozoic. (Photograph by S. J. Reynolds.)

1 km thick could be followed around the hinges of huge synmetamorphic folds and into the overturned limbs, where it was attenuated to 30 m or less! The units are thinned to 3% of their normal thickness, thus representing a stretch of 0.03 (S = 30 m/1000 m). Incredibly, each of the distinctive formations is still present and can be recognized, even in the most extremely attenuated sections. These formations are equivalent, in both name and thickness, to the well-known stratigraphic section of the Grand Canyon. Imagine the entire Grand Canyon section flipped upside down and thinned to less than 30 meters! And it gets better in nearby mountain ranges, where the entire Paleozoic section can be measured with a pocket photo scale (Figure 8.91). The only way to do stratigraphy!

Center-to-Center Technique

The **center-to-center technique**, also called the **Fry method** after its inventor, Norman Fry (1979), is used to calculate the strain in rocks that before deformation contained randomly scattered objects or particles, such as phenocrysts in a deformed granitic rock or oolites in a limestone. During deformation, the centers of adjacent objects become closer in directions in which the rock is being shortened and farther apart in directions in which the rock is being lengthened. The average distance between adjacent particles, called **nearest neighbors**, should be least in directions parallel to S_3 and greatest in those parallel to S_1 (Figure 8.92*A*). To perform the analysis, we commonly use as a data base a photograph of an exposed surface or thin section. We overlay the photo with a piece of tracing paper, the center of which is marked by a tiny reference cross. The process is simple. Place the reference cross on a given object and mark, on the overlay, points representing the locations of the neighboring objects. Then slide the tracing paper such that the reference cross lies on a second object, and again mark the locations of nearest neighbors. Repeat and repeat and repeat, being careful *never* to rotate the tracing paper (Figure 8.92*B*). The shape of the finite strain ellipse magically appears! The method works as long as the objects we are studying were initially randomly distributed and were not clustered in any way.

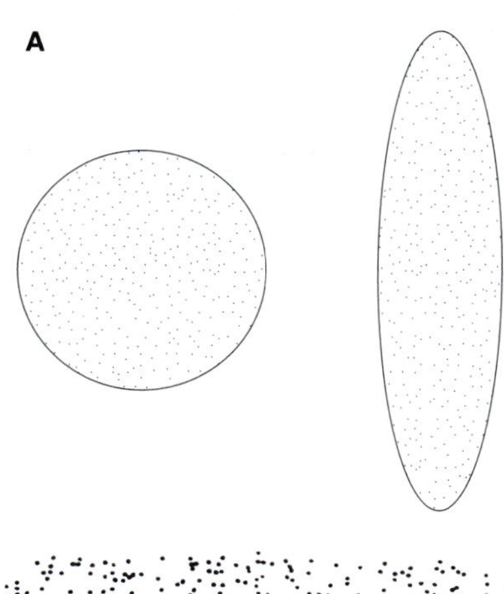

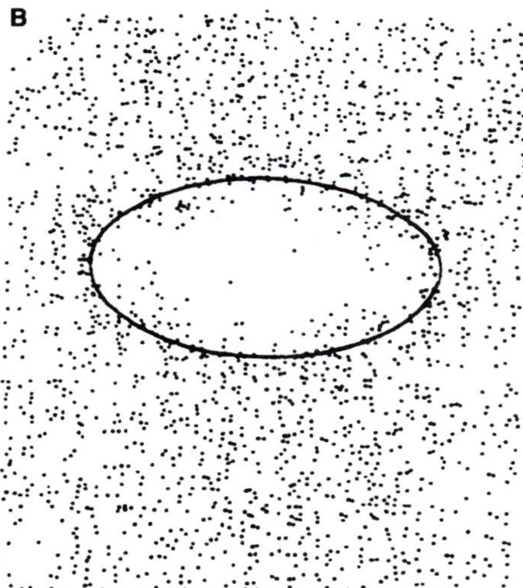

Figure 8.92 (*A*) The success of the Fry method depends on the initial presence of an "anticlustered" distribution of points within the undeformed body. When the body is deformed (as represented by the change in the circle to an ellipse), the spacing between points systematically changes. [From Simpson (1988), in Marshak (Mitra, BASIC METHODS OF STRUCTURAL GEOLOGY, © 1988, p. 352. Reprinted by permission of Prentice Hall, Upper Saddle River, New Jersey.] (*B*) Example of plot produced by the Fry method as applied to an oolitic limestone. [From Lisle (1994), Paleostrain analysis, Figure 2.6, p. 34. Reprinted with permission of Pergamon Press, Oxford, England.] Based on Dunne and others (1990).]

Other Methods

There are many other methods for determining strain in special circumstances (see Ramsay, 1967; Ramsay and Huber, 1983.) One involves evaluating changes in the angles between initially perpendicular lines, such as for deformed brachiopods (see Chapter 2). Other methods are specific to strain within shear zones and are introduced in Chapter 9. The entire collection of methods is quite diverse and useful, providing that nature has been kind enough to preserve the appropriate strain makers.

RELATION BETWEEN DEFORMATION AND METAMORPHISM

Deformation that forms tectonites occurs within the context of metamorphism, and so we are clearly interested in the relationship between structural processes and metamorphic ones. As in any deforming rock mass, several processes are all occurring at once, often in competition with one another. The processes involved in the formation of tectonites are the same processes that form more "normal" metamorphic rocks, such as gneiss and schist. In tectonites, though, the processes operate in different relative amounts or are carried out to an extreme degree. A process that is very important in most tectonites, especially mylonitic rocks, is the **reduction of grain size** during deformation via ductile dislocation creep, dynamic recrystallization, and fracturing of more brittle minerals.

The process of grain size reduction is counteracted by **grain growth via recrystallization**. We are witnessing the results of grain growth when we walk a shale unit across a regional metamorphic gradient and see it successively turn into slate, phyllite, and then schist.

Grain size reduction dominates at low temperatures or high strain rates, where the introduction of strain into crystals outpaces recrystallization, producing fine-grained rocks. As a result, tectonites deformed under greenschist and lower amphibolite facies conditions are commonly platy mylonitic rocks.

At higher temperatures or lower strain rates, recrystallization of existing minerals, the growth of new minerals, and other forms of annealing become dominant, leading to typical high grade metamorphic rocks, such as gneiss and schist. Such rocks represent middle amphibolite and granulite–facies conditions, where metamorphism may also be accompanied by the formation of granitic layers and pods via metamorphic differentiation, in situ melting, and the influx and **lit-par-lit injection** of granitic magma parallel to foliation.

Between these extremes are conditions that are just right to form rocks intermediate in character between true mylonitic rocks and "normal" metamorphic rocks. These rocks are usually medium grained and include mylonitic gneiss, mylonitic schist, and phyllonite.

The overall fabric of a rock strongly depends on its *subsequent* metamorphic and structural history. During **prograde metamorphism**, for example, rocks become progressively deeper and hotter as a result of tectonic burial, perhaps related to the emplacement of an overlying thrust sheet. In this case, tectonite fabrics that formed early in the deformation history may be overprinted by higher grade metamorphic conditions; or, they may become annealed during the higher temperatures and overgrown by new, higher grade metamorphic minerals.

Some tectonites are instead formed during **retrograde metamorphism**, such as when rocks are uplifted in the footwall of major normal faults (Davis, 1983; Wernicke, 1985; Reynolds and Lister, 1987). In this case early ductile fabrics may be overprinted by lower grade mineral assemblages, but they should not be extensively annealed. Many retrograde mineral reactions are hydration reactions and cannot proceed in the absence of water. Since high grade metamorphic rocks have generally lost most of their water during dehydration reactions that accompanied the *prograde* part of their pressure–temperature path, many retrograde reactions cannot occur unless water is somehow reintroduced into the metamorphic rock. Often this does not happen, which is why we find relatively pristine high grade metamorphic rocks and structures exposed at the surface. If retrograde reactions were more successful, metamorphic petrologists and structural geologists would have a much harder time studying processes that occur at depth.

The predictable differences in the character of the total fabric of tectonites for different **metamorphic paths** provide the means to distinguish fabrics formed in different tectonic settings. This approach has been used to distinguish tectonites formed along retrograde paths during crustal extension from those formed along prograde paths during thrusting (Reynolds et al., 1988).

The differences in the availability of fluid between prograde and retrograde metamorphic paths influence the formation of tectonites in yet one other way. Dehydration reactions during prograde metamorphism typically liberate large amounts of fluid during deformation. These fluids may promote deformation and/or may deposit veins of quartz, calcite, and other minerals. Multiple generations of veins in various states of deformation are common in tectonites, attesting to the important interplay between deformation and fluids. The passing fluids, most of which are composed of variable proportions of water, CO_2, methane, and other volatiles, may leach some elements and deposit others in a process called **metasomatism**. Metasomatism may obscure or obliterate previously formed fabrics and change the way the rock deforms as old minerals are dissolved and new minerals are formed.

RELATION BETWEEN DEFORMATION AND PLUTONISM

When studying tectonites and other metamorphic rocks in the field, we inevitably encounter plutons whose relation to deformation turns out to be an important, if not critical, part of the story (Figure 8.93). Plutons are incredibly important because they generally provide us with the clearest, most unambiguous way to determine the age of deformation. Also, because they are emplaced in a relatively short time period compared to deformation, plutons may help us determine whether any time intervened between two phases of deformation (D_1 versus D_2). Plutons may also represent an important potential source of heat responsible for the metamorphism.

We are primarily concerned with determining whether a given intrusion (e.g., pluton, dike, sill) was emplaced before, during, or after deformation (Figure 8.94), i.e., prekinematic, synkinematic, or postkinematic. An intrusion emplaced before deformation and metamorphism is called **prekinematic** and should reveal evidence of having experienced deformation *after* it had become totally solidified. Such an intrusion might contain foliation and lineation parallel to that of the country rock it intrudes, indicating that the intrusive rock and the wall rock experienced a common deformational history. Furthermore, a prekinematic intrusion might be transformed into an elliptical shape much different than its original shape immediately following intrusion. Some prekinematic intrusive rocks are deformed only

Figure 8.93 Dark roof pendant of metamorphosed sedimentary rock (highly foliated tectonite) surrounded by granite of the Sierra Nevada batholith. (Photograph by E. B. Mayo.)

Figure 8.94 Aplite sills in foliated granite. Note that the sills are parallel to the elongated dark inclusions in the granite. (Photograph by G. H. Davis.)

along their margins, remaining undeformed in their centers. The degree to which the center of a prekinematic intrusive rock is deformed will depend on such factors as ductility, contrast, metamorphic conditions, and intensity of the local or regional strain.

An intrusion emplaced *during* deformation is **synkinematic** and should display evidence that it was partially molten during deformation. Criteria for recognizing synkinematic intrusion include (1) evidence for injection of melt into synchromously developing structures in the country rock; (2) recognition that early phases of the pluton are deformed whereas later phases crosscut the fabrics; and (3) geometric and kinematic coordination between primary magmatic fabric and the deformation of the country rock (Passchier, Myers, and Kröner, 1990; Paterson and Tobish, 1992; Karlstrom, 1989; Ghosh, 1993).

An intrusion emplaced *after* deformation is **postkinematic** and should lack the fabric observed in the surrounding deformed country rock. Postkinematic intrusions commonly crosscut the fabric of the wall rock and may contain inclusions of previously foliated wall rocks. Contact metamorphic aureoles associated with a postkinematic intrusion should be superimposed on deformation-related fabric and metamorphic minerals in the wall rocks.

As always, Nature has some playful little tricks Postkinematics intrusions may inject *along* foliation, giving the impression that the igneous rock and the metamorphic country rock became foliated at the same time. Some prekinematic intrusions of the ''right'' petrology, e.g., coarse-grained pegmatites, do not easily ''pick up'' fabric, even though they were around for the entire deformation history. Also, as we will see in the next chapter, a prekinematic dike in one orientation relative to the imposed strain may appear essentially undeformed, whereas a nearby prekinematic dike of the same age but a different orientation will be strongly folded.

Plate Tectonic Environments of Tectonite Formation

The three fundamental classes of plate tectonic movements (see Chapter 10) are all quite capable of producing regional terranes of tectonites. Ridge spreading is achieved to some extent by stretching of upwelling mafic intrusions that issue from asthenosphere and move into the dynamically active lower part of oceanic lithosphere at spreading centers (Figure 8.95).

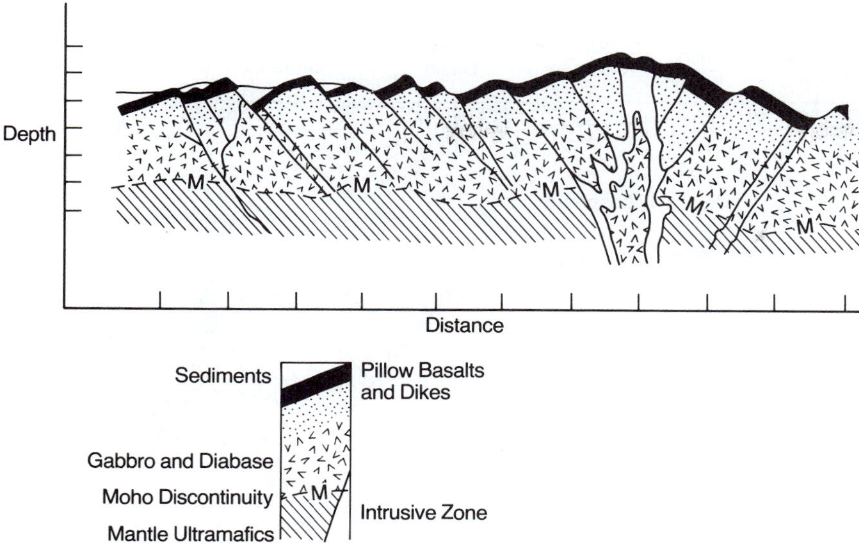

Figure 8.95 Vertically exaggerated schematic rendering of extensional processes at work in an ocean ridge system. [Modified from Anderson and Noltimier (1973), v. 34, fig. 5, p. 144. Published with permission of *Geophysical Journal of the Royal Astronomical Society.*]

Undoubtedly S- and LS-tectonites with low-dipping foliation are fashioned in these zones. Where spreading centers unabashedly cut across continents, the deep reaches of continental crust may be rendered into tectonite as a response to penetrative layer-parallel stretching. If we could get down there to look, we might discover that the deep continental crust beneath the Red Sea rift is distinguished by LS-tectonites.

Strike-slip movements along transform boundaries between plates can create regional terranes of tectonite through simple shear. Tectonites thus fashioned tend to be LS-tectonites characterized by steep foliation and subhorizontal lineation. The formation of tectonites along transform faults and strike-slip faults is especially favored at depth, where rocks reside in an environment of elevated temperature and pressure (Figure 8.96). Shear zone kinematics apply beautifully to transform boundaries. If the deformational conditions are right, the simultaneous flattening and stretching so characteristic of simple shear will be recorded in the formation of thick zones of mylonitic LS-tectonites.

The most abundant tectonites on the face of the Earth seem to have been created along convergent plate margins. The tectonic movements that operate at convergent margins all favor the development of tectonite (Figure 8.97). These include the formation of magmatic arcs through voluminous intrusion of granitic batholiths, the simple shear subduction of oceanic slabs beneath continental margins, the accretion of plates or parts of plates against continental edges, and the accommodation of major crustal shortening in response to continental collision and/or subduction-related processes. How a given part of a continental margin responds—by simple shear or pure shear deformation, by constant volume or variable volume deformation, or by mild or substantial distortion—is ultimately a function of many, many factors, including structural position within the tectonite-forming system and the scale at which the products of strain are being observed and described.

Some parts of fold belts that are created in convergent settings are marked by S-tectonites, which may include axial planar cleavages in slate and schist, and axial planar spaced cleavages in sedimentary rocks. Such S-tectonites might be regarded as products of simple flattening, a response to horizontal crustal shortening generated by plate collision. In contrast to flattening fabrics in fold belts, some parts of fold belts are distinguished

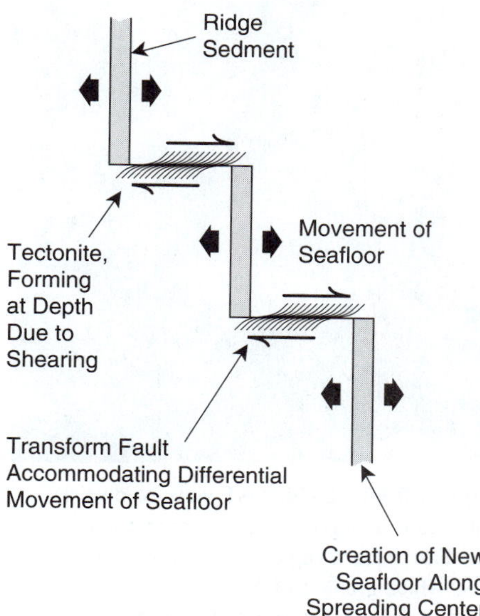

Figure 8.96 Transform fault linkage of one spreading center to another, and the formation of tectonite at depth where differential movement of seafloor is accommodated plastically.

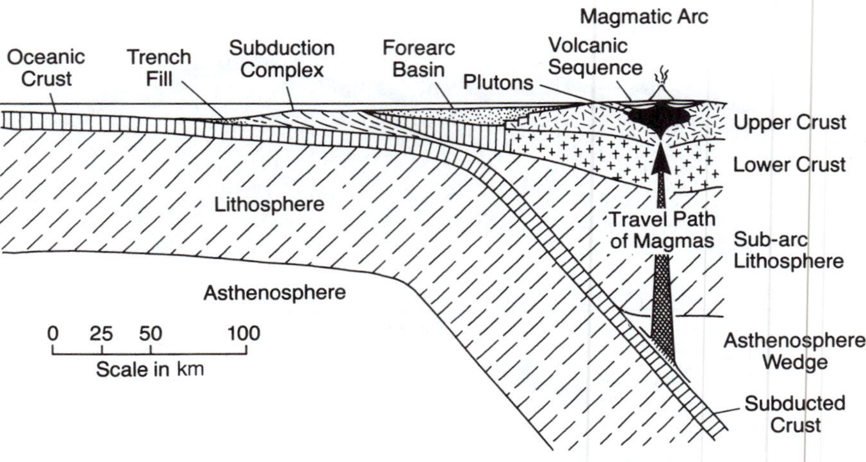

Figure 8.97 Diagram of a convergent margin, where there is plenty of opportunity to form tectonites, e.g., within the subduction complex; along the interface between the overriding plate and the subducting plate; and within the magmatic arc. [From Dickinson and Payne (1981), cover illustration. Published with permission of Arizona Geological Society.]

by LS-tectonites in the form of expansive terranes of low-dipping schistosity and low-plunging, unidirectional lineation. These tectonites are intimately associated with imbricate thrust nappes and fold nappes. Structural systems like these reflect thrust-related simple shear on a crustal scale (Mattauer, 1975), a structural accommodation to profound crustal shortening. LS-tectonite terranes of this type mark young and old orogens alike, such as the tectonically active Himalayan orogen (Mattauer, 1975) and the ancient Moine thrust complex in Scotland (Elliot and Johnson, 1980; McClay and Coward, 1981).

The formation of tectonite terranes in some convergent settings is influenced significantly by the role of igneous intrusion, including diapirism. The Precambrian shield provinces that lie within the world's continents contain, among other things, **greenstone belts** distinguished by strongly foliated and deeply infolded regionally metamorphosed metavolcanic and metasedimentary rocks flanked by deep exposures of enormous batholithic complexes of granitic gneiss (Figure 8.98). Batholithic rocks in **magmatic arc terranes** of continental margins contain voluminous granitic rocks, some of which are tectonites by virtue of penetrative internal flowage, either accompanying emplacement or due to subsequent solid-state deformation. And within the batholiths are extensive roof pendants of folded and foliated metasedimentary and metavolcanic rocks, tectonites fashioned by the combination of prebatholith deformation and the rigors of intrusion. **Mantled gneiss domes**, classically described by Eskola (1949), are tectonite complexes that occur from place to place along the lengths of many orogenic belts, like the eastern Appalachians. They are characterized by domes and arches of high grade gneissic tectonites that are concordantly overlain by strongly deformed, high grade layered gneisses and schists (Figure 8.99). In some ways mantle gneiss domes appear to be the tops of enormous diapiric intrusions of magma and remobilized basement. However, workers who have addressed the geology of gneiss domes uniformly conclude that the histories of some are long and complex, and may involve crustal shortening, not just diapirism.

Figure 8.98 LANDSAT view of diapiric intrusions in a granite-greenstone terrane in the East Pilbara region of western Australia. The white granitic batholiths are separated by dark infolded greenstones of the Warrawoona Series. The size of the area shown is approximately 180 × 100 km. [From Lowman (1976). Copyright © 1976 by the University of Chicago. All rights reserved.]

To sum up, regional terranes of tectonites force us to apply principles of strain on a vast scale. They require us to think of flattening, constriction, and plane–strain simple shear at scales that are quite dramatic, almost unimaginable. Furthermore, they require us to think of the progressive, step-by-step development of the structures in foliated rocks (Figures 8.100 and 8.101).

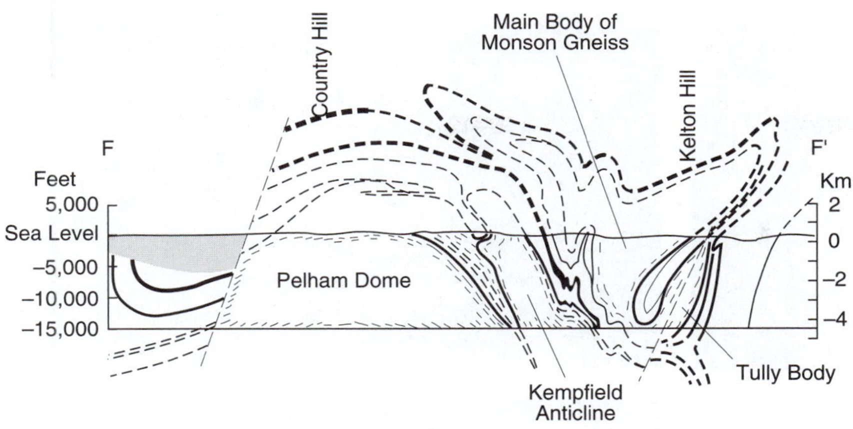

Figure 8.99 Cross section of gneiss dome complex in Massachusetts. [Modified from Ramberg (1973), Model studies in gravity-controlled tectonics by the centrifuge technique, Figure 18, p. 63. Reprinted by permission of John Wiley and Sons, © 1973. Based on Zen and others (1969).]

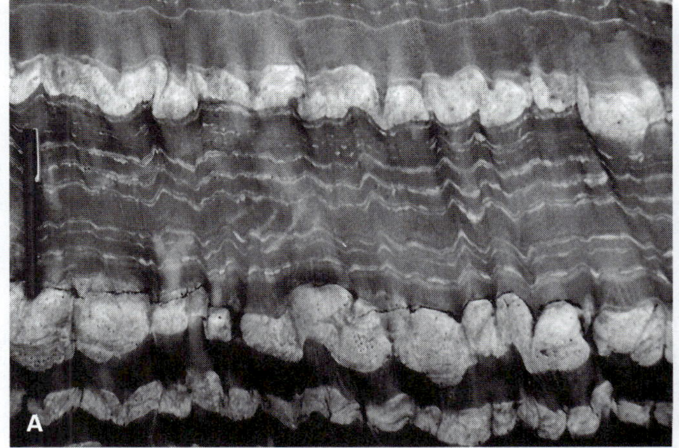

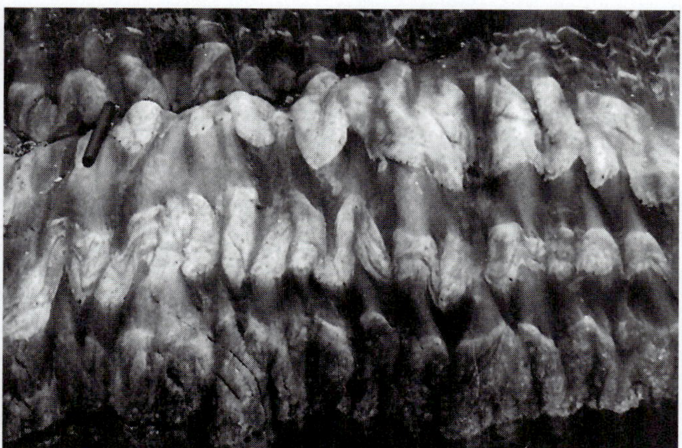

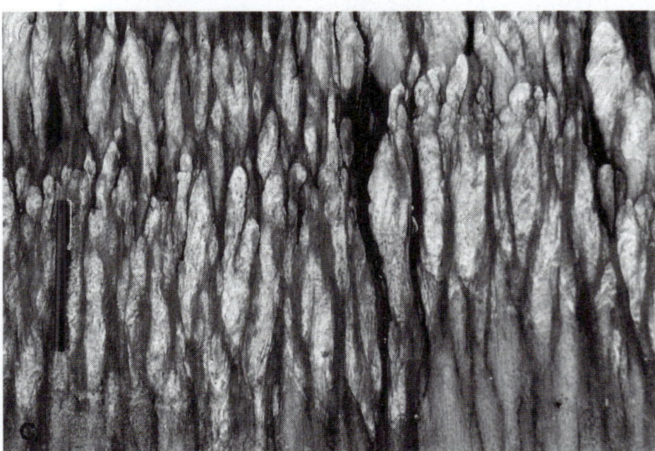

Figure 8.100 Progressive development of axial plane cleavage, and transposition, within sequence of interbedded argillites and sandstones of Meguma Group (Ordovician), east coast of Newfoundland. (*A*) Open folding of sandstone layers (light) and slaty cleavage development in argillite (dark). (*B*) Tight folding of sandstone layers, strong slaty cleavage development in argillite, and incipient spaced cleavage development in sandstone. (*C*) Pervasive cleavage development in both argillite and sandstone. Few fold hinges recognizable. (Photographs by S. J. Reynolds.)

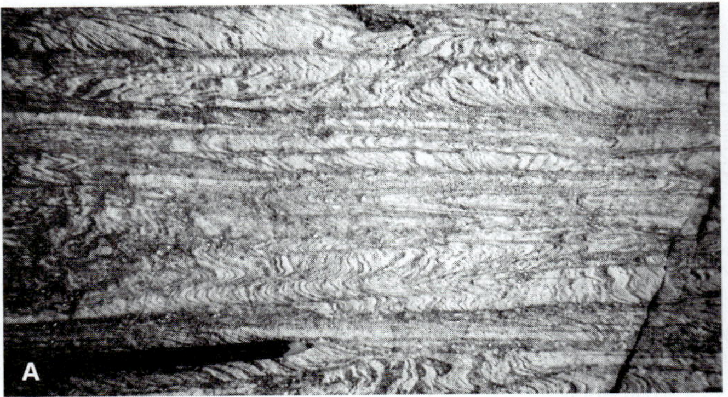

Figure 8.101 These photos of Vishnu Schist in Vishnu Canyon, Grand Canyon, are marvelous displays of progressive transposition. (*A*) The original schistosity of the Vishnu is preserved in the microlithon domains (white). Note that the schistocity has become crenulated with axial surfaces parallel to the newly developed cleavage. (*B*) Progressive deformation builds cleavage domains (dark bands) at the expense of microlithons. (*C*) Where transportation is nearly complete, you need to look long and hard for vestiges of the original schistocity, for most of the rock is composed of dark cleavage domains. (Photographs by S. J. Reynolds.)

ON TO SHEAR ZONES

Some of the most beautiful foliated and lineated rocks in the world lie concentrated along ductile shear zones. Ductile shear zones are "hot faults" along which displacement is achieved through distributed shear and distortion across a zone of deformation. Within shear zones, the components of the array of structures and fault rocks lie oriented sympathetically to the progressive strain history the rocks have experienced. Thus, they provide valuable information that can be used to decipher structural and tectonic history.

Equipped with some knowledge about deformation mechanisms, strain, cleavage, foliation, and lineation, we will now enter the realm of shear zones and put this knowledge to further use. We will discover in Chapter 9 that not all shear zones are "hot" and ductile; some are "cold" and brittle. And we will be back to looking even deeper into the nature of fault zones. The study of shear zones provides a wonderful opportunity for synthesis.

CHAPTER 9

SHEAR ZONES AND PROGRESSIVE DEFORMATION

THE NATURE OF SHEAR ZONES

A **shear zone** is a tabular to sheetlike, planar or curviplanar zone composed of rocks that are more highly strained than rocks adjacent to the zone (Figure 9.1). The intensity with which rocks can be deformed in shear zones is astonishing. Granites can be so strongly and thoroughly sheared that they resemble, and have been mistakenly mapped as, metarhyolite or metasedimentary schist (Figure 9.2*A*). Conglomerates can be smeared out to such a degree that individual clasts resemble thin sedimentary layers (Figure 9.2*B*).

Shear zones have certain characteristics that permit us to recognize them in the field, in thin sections, and on geologic maps and cross sections. The distinguishing characteristics vary, depending on whether the shear zone formed under brittle, ductile, or intermediate conditions. A fault zone is a shear zone formed under brittle conditions. Displacement is taken up on a network of closely spaced faults (Figure 9.1*A*). When shear zones form under ductile conditions, deformation is accompanied by metamorphism and produces rocks with foliation, lineation, folds, and related features (Figure 9.1*B*). Some shear zones develop under conditions that are intermediate between strictly brittle and strictly ductile deformation. These may consist of zones that are partly faults and partly ductile shear zones (Figure 9.3*A*), and may have formed in interlayered rocks with contrasting strengths (Figure 9.3*B*).

For many geologists, including the two of us, shear zones are some of the most fun and fascinating structures to work with. Shear zones contain the most intensely deformed rocks known on Earth! The geometry and interplay among different structural elements within shear zones can be

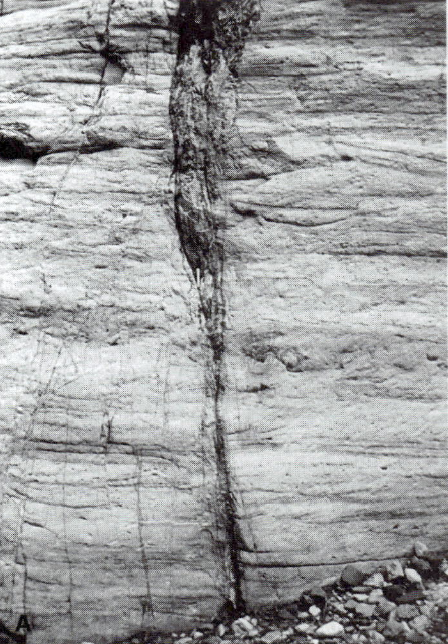

Figure 9.1 Shear zones in the field: (*A*) Brittle shear zone cutting Proterozoic quartzite, 75 Mile Canyon, Grand Canyon, Arizona; note marking pen in center for scale. (Photograph by S. J. Reynolds.) (*B*) Steeply inclined shear zone cutting Scourie pyroxenite, Ballcall Bay, north Scotland. (Photograph by Carol Simpson.)

truly elegant, providing a startlingly detailed record of the history of deformation. Most shear zones contain features that permit us to determine the *sense of displacement* along the zone. In favorable situations, we may be able to reconstruct the *amount of displacement*. Deformed objects within shear zones can be used to quantify the *amount of strain* imposed on the rocks during shearing. Finally, shear zones present us with a wonderful "mystery story" that challenges us to reconstruct from the available clues.

General Characteristics

Most shear zones that we encounter are much longer and wider than they are thick, commonly having relative dimensions somewhere between a

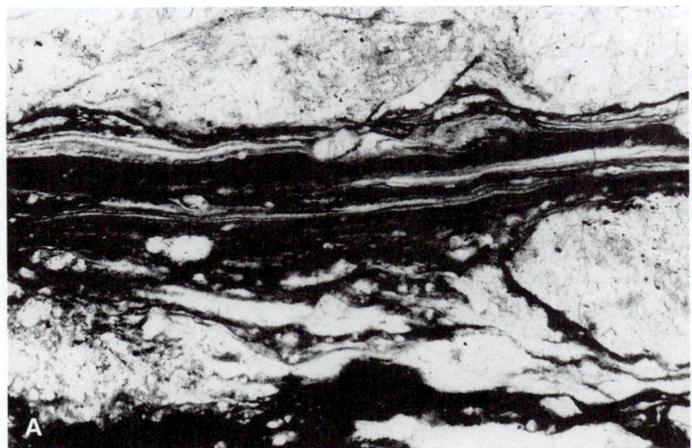

Figure 9.2 Rocks deformed in shear zones. (*A*) Thin ductile (mylonitic) shear zone cutting Tertiary granodiorite, South Mountains, central Arizona. Width of dark shear zone is 1 mm. (*B*) Strongly deformed Mesozoic conglomerate within a thrust zone in the Granite Wash Mountains, western Arizona. (Photographs by S. J. Reynolds.)

Figure 9.3 Brittle–ductile shear zones: (*A*) Brittle–ductile shear zone cutting metasedimentary rocks, Mosaic Canyon, Death Valley, California. Width of area shown is approximately 1 m. (*B*) Brittle–ductile shear zone produced by interlayered ductilely deformed marble and brittlely fractured quartzofeldspathic gneiss, Plomosa Mountains, western Arizona. (Photographs by S. J. Reynolds.)

sheet of paper and an audio compact disc (of *soft*-rock music, of course). They exist at all scales. The largest are hundreds of kilometers long and tens of kilometers thick, with displacements of tens to hundreds of kilometers (Figure 9.4*A*). The smallest shear zones observable in outcrop are typically several centimeters long and one millimeter thick, and may have a centimeter or so displacement. Even smaller shear zones, with appropriately scaled relative dimensions, are visible in thin section (Figure 9.4*B*).

All shear zones reflect a localization or concentration of deformation into a narrow zone. The presence of a shear zone indicates that within a given deforming rock mass, the distribution of strain was heterogeneous rather than homogeneous. As a result, shear zones are characterized by spatial gradients in the amount of strain. The amount of strain is generally highest within the center of a shear zone, decreasing outward into the wall rocks adjacent to the zone. If the decrease in strain away from the zone is gradual without any distinct physical break, the shear zone is considered to be **continuous** (Figure 9.5*A*). Continuous shear zones most commonly form under ductile conditions, where the rocks flow in the solid state without loss of cohesion (Figure 9.6*A*). If the decrease is more abrupt, the zone is considered to be **discontinuous** (Figure 9.5*B*). In most discontinuous shear zones, strongly deformed rocks within the zone are juxtaposed against much less deformed rocks along a sharp physical break or a very thin band along one or both margins of the shear zone (Figure 9.6*B*). This occurs, for example, when a discrete fault surface forms that accommodates all further shearing. Whether a strain gradient appears abrupt or gradual, and whether deformation *appears* homogeneous or localized, depends on the scale at which we observe the structures (Figure 9.7). A shear zone that appears discontinuous in outcrop may show, in thin section, a continuous gradient from weakly to strongly deformed rock over several millimeters.

Geometries

Shear zones are typically planar to gently curved, but some can have complex geometries. Most shear zones have subparallel margins and retain

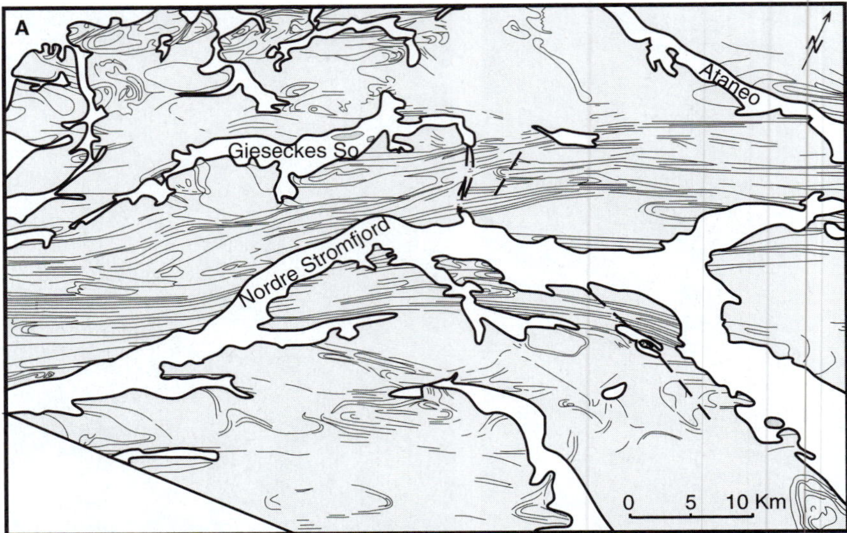

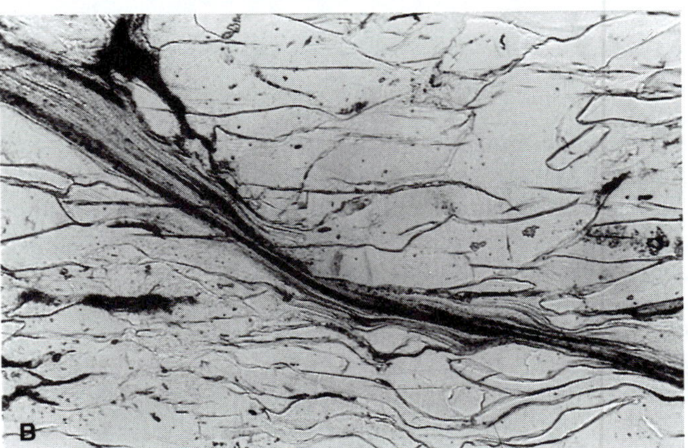

Figure 9.4 The big and little of shear zones: (*A*) Map of lithologic boundaries within and near the Nordre Strømfjord shear zone, Greenland. [After Sørensen (1983). Published with permission of American Geophysical Union]. (*B*) Shear zone in a thin section of experimentally deformed aplite. Shear zone is approximately 0.1 mm wide. (Photomicrograph courtesy of Jan Tullis.)

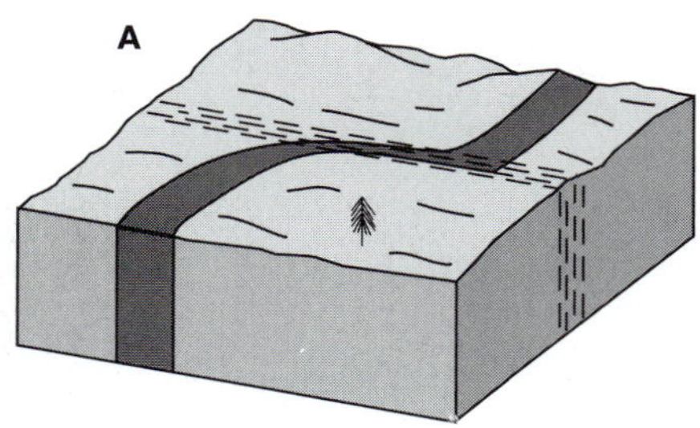

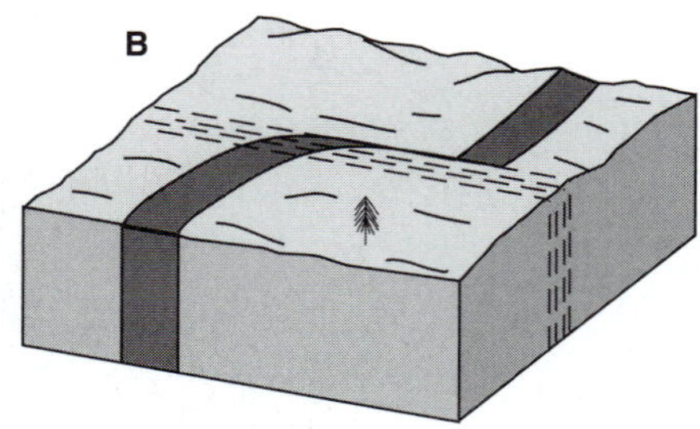

Figure 9.5 Continuous and discontinuous shear zones: (*A*) Continuous shear zone deflecting a marker that passes uninterrupted through the shear zone. (*B*) Discontinuous shear zone that truncates a marker.

a fairly consistent thickness over much of their length (Figure 9.8*A*). Where the margins diverge, the shear zone becomes wider (Figure 9.8*B*). Widening is most common near the ends of a shear zone, where strongly deformed rocks within the zone grade into a wider zone of less deformed rocks. A shear zone may also thin or taper as the margins converge, such as where a shear zone passes near or between rigid objects (Figure 9.8*C*). This thinning or tapering is generally accompanied by an increase in the degree of deformation as strain becomes concentrated in a progressively thinner and thinner zone.

Shear zones are commonly arranged in networks or sets composed of a number of individual shear zones. They may occur in subparallel sets, may deflect toward one another and link up in an anastomosing pattern, or may crosscut and displace one another (Figure 9.9).

Some shear zones have a curviplanar or folded geometry. Such a geometry may indicate that an originally planar shear zone (Figure 9.10*A*) was folded or warped by subsequent deformation (Figure 9.10*B*). Alternatively, many shear zones form with an original curviplanar geometry, encompassing and wrapping around more rigid, less deformed objects (Figure 9.10*C*). On a regional scale, such objects may include a relatively rigid pluton or volcanic pile surrounded by less competent shale, schist, and marble (Figure 9.11). Examples of smaller rigid objects include inclusions or

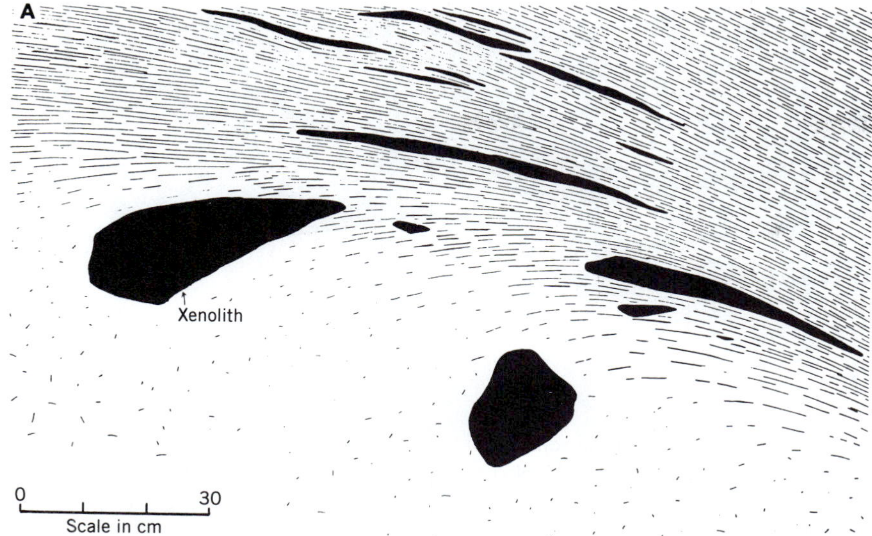

Figure 9.6 Continuous and discontinuous shear zones in the field: (*A*) Continuous shear zone cutting granite and inclusions, Lepontine Alps, Switzerland. [Reprinted from *The techniques of modern structural geology, v. 1: Strain Analysis,* J. G. Ramsay and M. I. Huber (1983), © by Harcourt Brace and Company Limited, and the authors.] (*B*) Thin discontinuous, ductile shear zone cutting foliation in mylonitic granite, Harcuvar Mountains, western Arizona. (Photograph by S. J. Reynolds.)

pebbles of a strong rock in a weaker matrix, and strong crystals of one mineral in a matrix of other, weaker minerals.

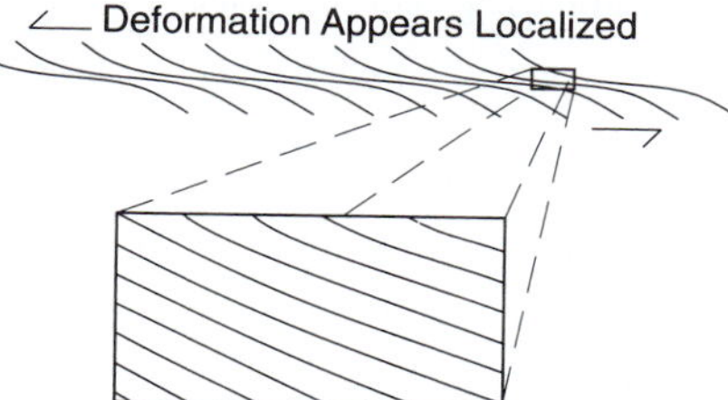

Figure 9.7 Whether deformation appears homogeneous or localized depends on the scale of observation.

Offset and Deflection of Markers

When a shear zone cuts across a preexisting feature, such as compositional layering or a dike, it generally offsets or deflects the feature (Figure 9.12).

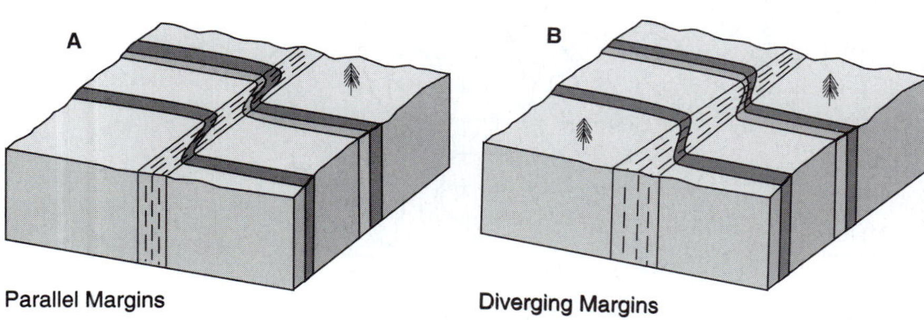

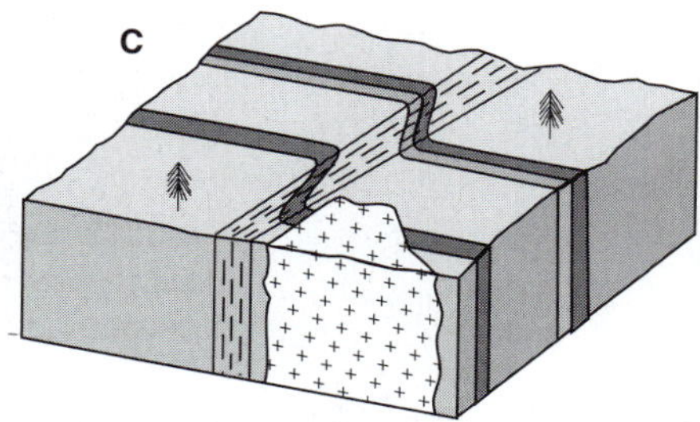

Figure 9.8 Shear zone margins: (*A*) parallel, (*B*) diverging, and (*C*) converging near a rigid pluton.

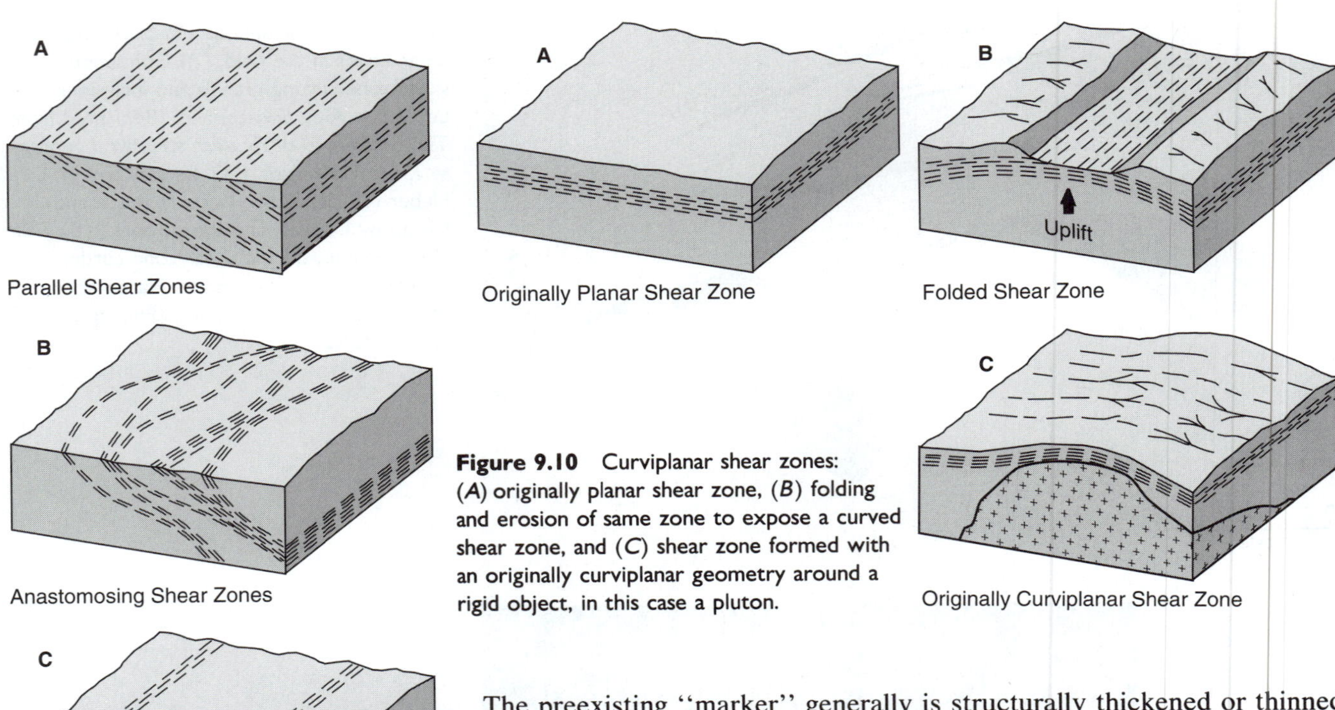

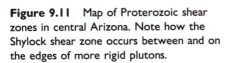

Figure 9.9 Shear zone sets: (*A*) parallel, (*B*) anastomosing, and (*C*) conjugate.

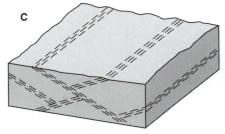

Figure 9.10 Curviplanar shear zones: (*A*) originally planar shear zone, (*B*) folding and erosion of same zone to expose a curved shear zone, and (*C*) shear zone formed with an originally curviplanar geometry around a rigid object, in this case a pluton.

The preexisting "marker" generally is structurally thickened or thinned and changed in orientation within the shear zone. A marker cut by a *continuous* shear zone will generally show a gradual deflection in its orientation and a corresponding gradual change in thickness across the zone (Figure 9.5*A*). Where cut by a *discontinuous* shear zone, the marker will generally show a discrete offset across the abrupt boundary between the shear zone and the wall rocks (Figure 9.5*B*). The amount of deflection or offset depends on the magnitude and type of strain within the zone.

We are especially interested in determining relative displacement of rocks on *opposite* sides of a shear zone, which reveals the **sense of shear** within the zone. In keeping with fault terminology (Chapter 6), there are

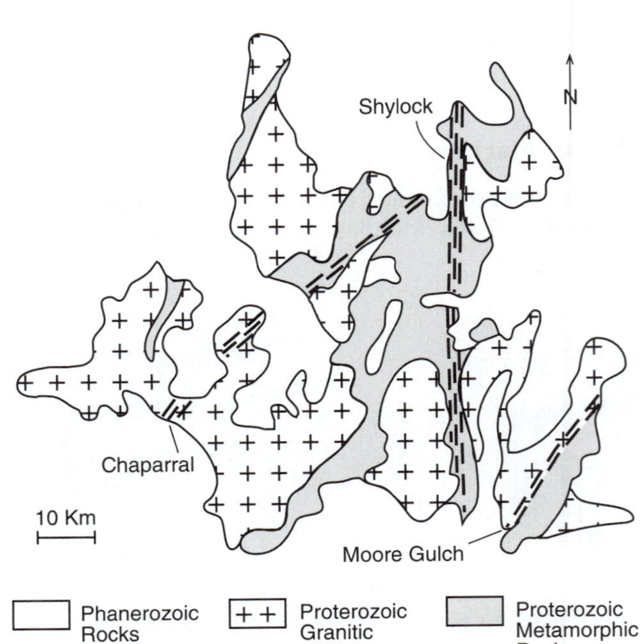

Figure 9.11 Map of Proterozoic shear zones in central Arizona. Note how the Shylock shear zone occurs between and on the edges of more rigid plutons.

Figure 9.12 Brittle shear zone cutting Tertiary sandstone, Waterman Hills, Mojave Desert, California. (Photograph by S. J. Reynolds.)

strike-slip, normal-, low-angle normal-, reverse-, thrust-, and oblique-slip shear zones. *Strike-slip* shear zones may be right-handed (*dextral*) or left-handed (*sinistral*) (Figure 9.13*A*,*B*). *Normal-slip* shear zones are marked by hanging wall displacement *downward* relative to the footwall (Figure 9.13*C*). *Reverse-* and *thrust-slip* shear zones are marked by hanging wall displacement *upward* relative to the footwall (Figure 9.13*D*). **Oblique** shear zones have components of both strike-slip and dip-slip.

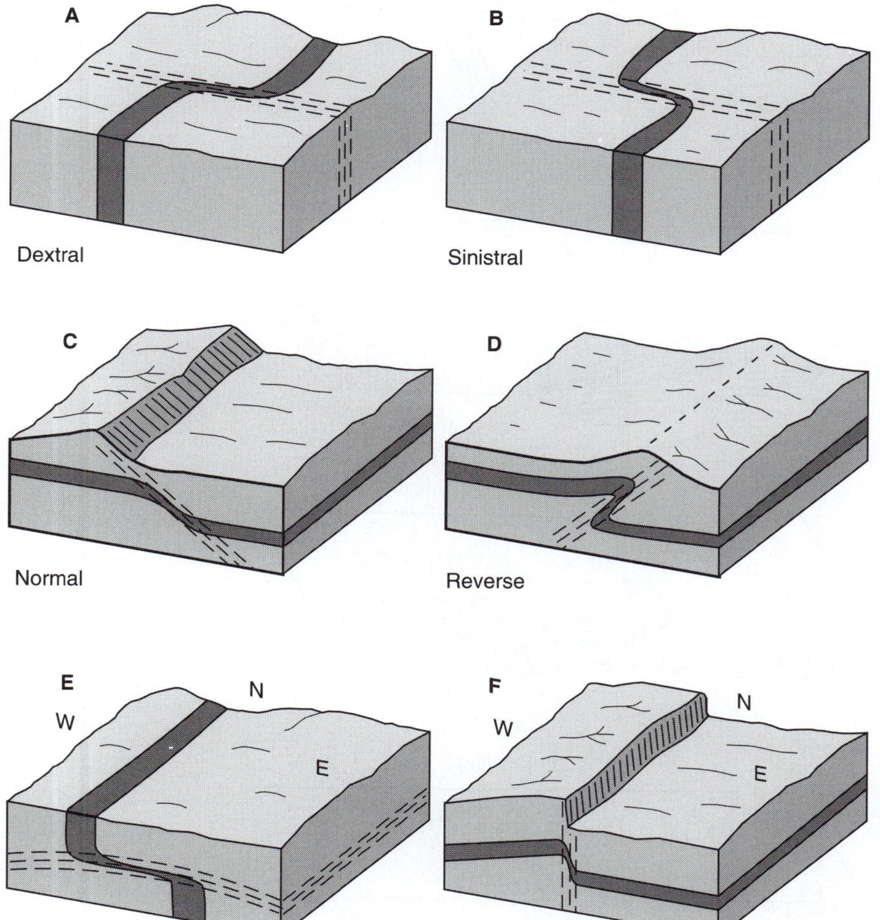

Figure 9.13 Deflection and offset across shear zones: (*A*) right-handed or dextral, (*B*) left-handed or sinistral, (*C*) normal, (*D*) reverse, (*E*) top to the west, and (*F*) west side up.

Another way of describing the sense of shear on subhorizontal or variably dipping shear zones is by specifying which way the hanging wall moved, such as "top to the west" (Figure 9.13*E*). For vertical shear zones with a dip-slip component of motion, we use phrases such as "west-side up" (Figure 9.13*F*) or "northeast-side down" to convey the sense of shear.

Tectonic Settings

Shear zones form in a wide variety of tectonic settings (Figure 9.14), including plate boundaries of all types. They are undoubtedly forming at depth today in any region with abundant earthquakes or other manifestations of active deformation.

Shear zones are present along seismically active strike-slip zones, such as the San Andreas fault of California, the Alpine fault of New Zealand, and the numerous strike-slip faults that dissect China and Tibet north of the India–Asia continental collision. Shear zones also mark the sites of

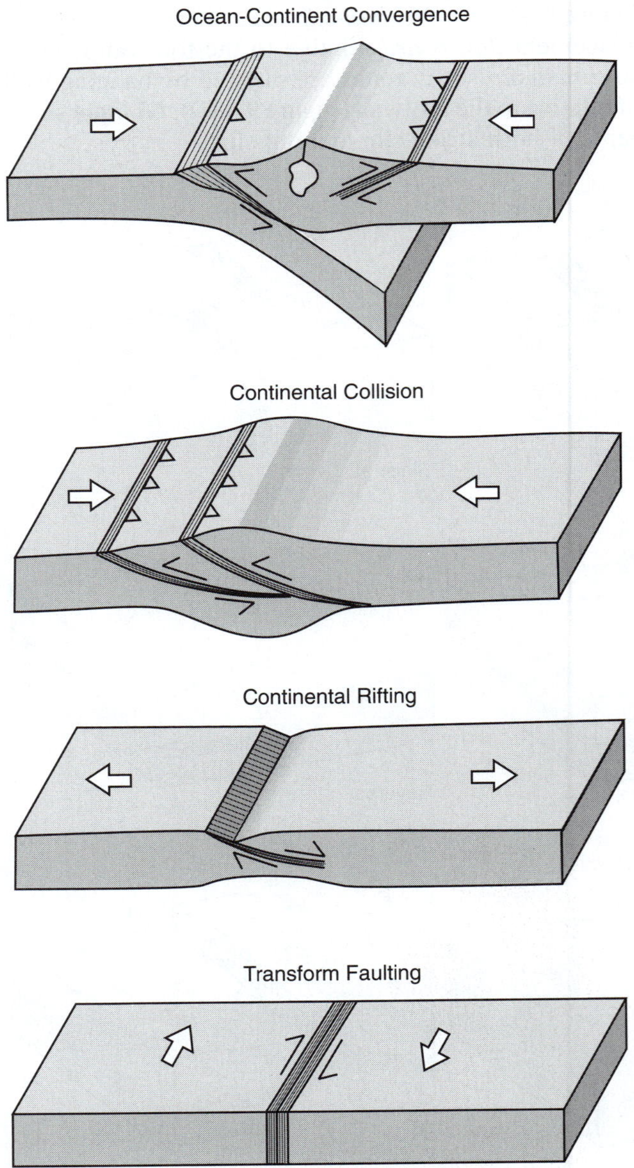

Figure 9.14 Plate tectonic settings of some shear zones.

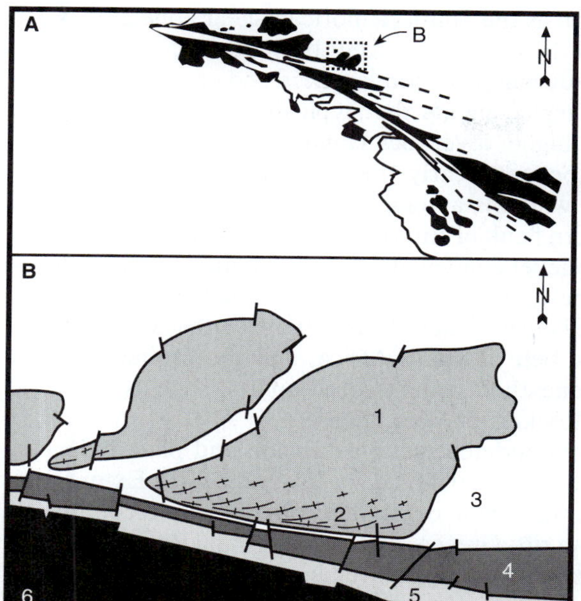

Figure 9.15 South Armorican shear zone in Brittany, western France. Rock types include: (1) undeformed Hercynian (Paleozoic) granite and (2) its deformed equivalent, showing the trend of foliation, (3) Precambrian sedimentary and metasedimentary rocks, (4 and 5) Paleozoic sedimentary and metasedimentary rocks, and (6) Precambrian granite. (Reprinted with permission from *Journal of Structural Geology*, v. 2, Berthe, D., and Brun, J. P., Evolution of folds during progressive shear in the South Amorican shear zone, France, 1980, Elsevier Science Ltd., Pergamon Imprint, Oxford, England.)

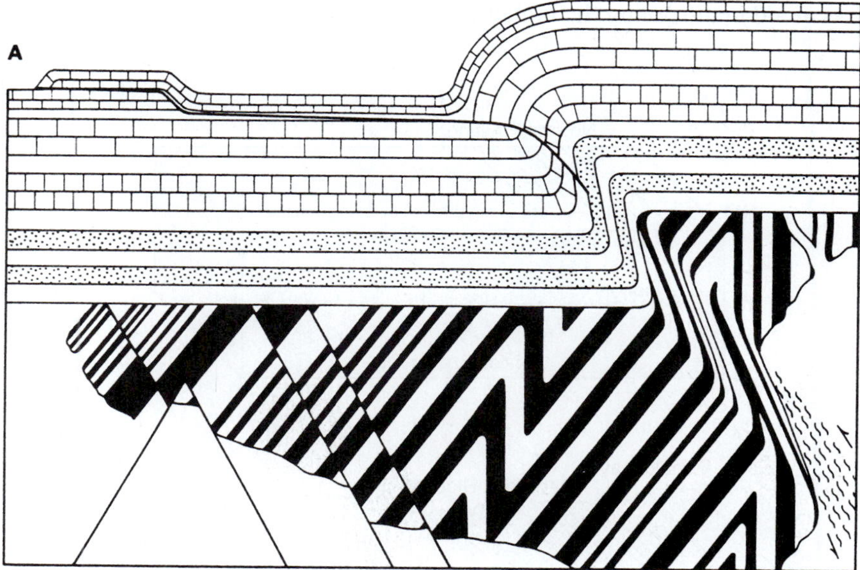

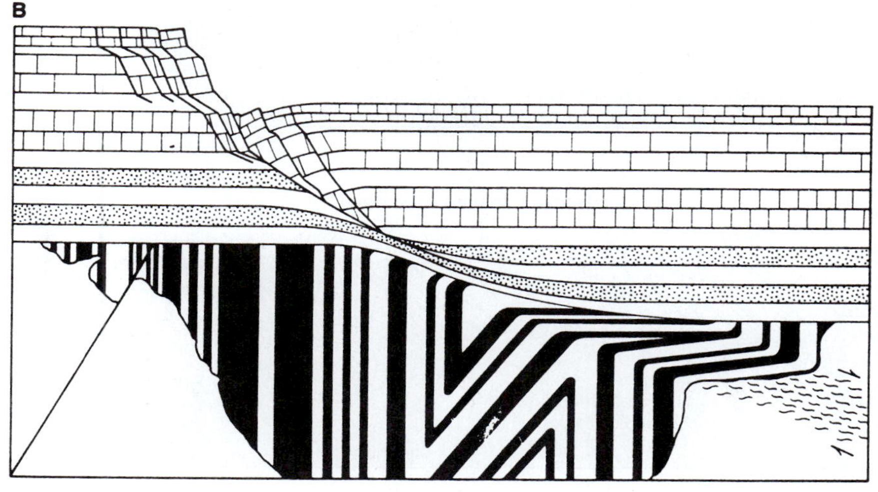

Figure 9.16 Shear zones that continue up dip into faults: (*A*) Reverse displacement during crustal shortening, (*B*) Normal displacement during crustal extension. (Reprinted with permission from Journal of Structural Geology, v. 2, Ramsay, J. G., Shear zone geometry: a review, 1980, Elsevier Science Ltd., Pergamon Imprint, Oxford, England.)

past strike-slip zones, such as the South Armorican shear zone of western France (Figure 9.15) and the Great Glen fault of Scotland.

Shear zones that form during plate convergence and crustal shortening commonly have thrust displacements that typically bring older, deeper rock up against younger, higher level rock (Figure 9.16*A*). Huge, impressive thrust shear zones occupy nearly the entire length of the Alpine–Zagros–Himalayan belt, which is associated with the collision of Africa and India with the southern flank of Europe and Asia. Some ductile thrust zones represent the deep-level equivalents of thin-skinned fold and thrust belts, such as those in the Canadian Rockies, the Appalachian and Ouachita Mountains of the eastern United States, the ''Outback'' of central Australia, the Cape Fold belt of South Africa, and the Moine thrust of Scotland. Others form beneath basement-cored uplifts, such as the Wind River Mountains in the Rockies of Wyoming.

Shear zones that accommodate crustal extension and place younger, high level rock down onto deeper, older rocks (Figure 9.16*B*). Extensional shear zones are no doubt forming at depth in regions of active continental rifting, such as the African rift, Greece, and the Basin and Range province of western North America. Sites of past extreme extension are represented by shear zones in metamorphic core complexes of western North America and the Aegean (Figure 9.17), and by the South Tibetan detachment, which has aided in unroofing the high grade metamorphic rocks of the Himalaya.

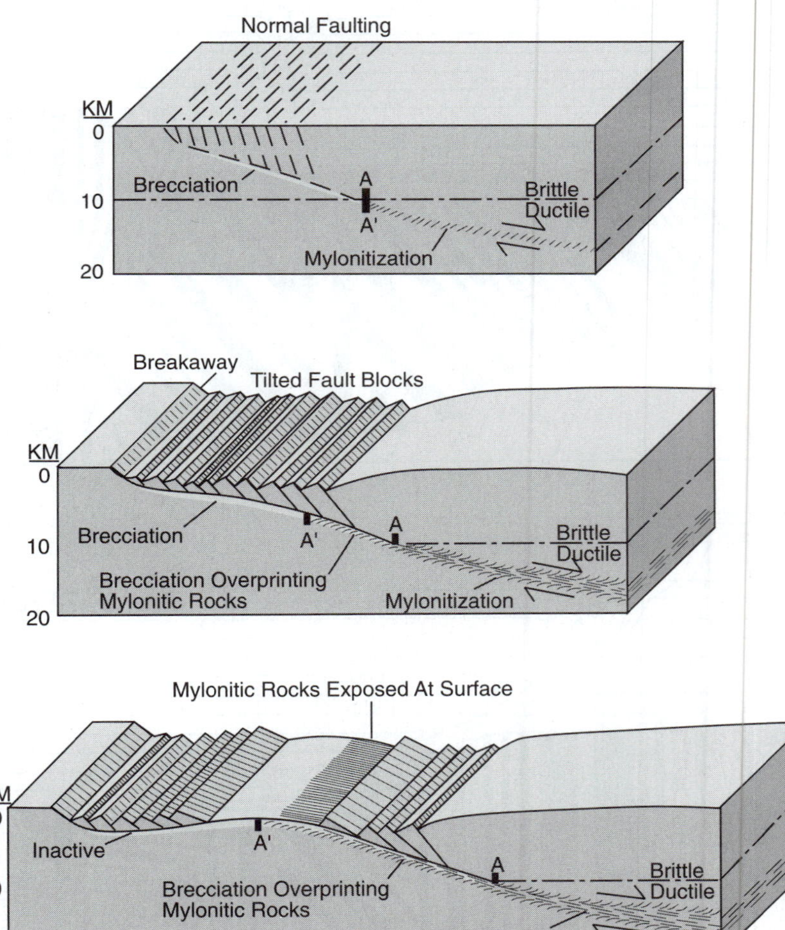

Figure 9.17 Schematic evolution of a low-angle shear zone with both ductile and brittle segments. Normal displacement along the shear zone progressively unroofs footwall rocks, causing early ductile fabrics to be overprinted by brittle ones as the rocks are isostatically uplifted and cooled. This is one model for the origin of metamorphic core complexes. [Modified from Reynolds and others (1988), in Ernst, W. G. (ed.), METAMORPHISM AND CRUSTAL EVOLUTION OF THE WESTERN UNITED STATES, © 1988, p. 493. Reprinted by permission of Prentice Hall, Upper Saddle River, New Jersey.]

TYPES OF SHEAR ZONES

When we walk up to a shear zone in the field, we generally are impressed with the variety of structures and the many ways that rocks deform. Such variety exists because shear zones form under many different conditions and from a variety of preexisting rock types. The deformation mechanisms that operate within a shear zone as it forms depend on the mineralogy and grain size of the affected rock and on the physical conditions that prevailed during deformation. Brittle deformation mechanisms tend to dominate when temperatures are cooler, pressures are lower, strain rate is faster, and fluid pressure is higher. Ductile deformation mechanisms become dominant when temperatures and pressures are higher, strain rate is slower, and fluid pressure is lower. Of course, the threshold from brittle to ductile depends on rock type: rock units composed of halite and gypsum will deform ductilely under conditions in which quartz and feldspar are decidedly brittle.

We subdivide shear zones into four general types, based on the characteristic type of deformation. A **brittle shear zone** contains fractures and other features formed by brittle deformation mechanisms. A **ductile shear zone** displays structures, such as foliation and lineation, that have a metamorphic aspect and record shearing by ductile flow. **Semibrittle shear zones** include en echelon veins and stylolites, and involve mechanisms such as pressure solution and cataclastic flow. **Brittle–ductile shear zones**, which show evidence for both brittle and ductile deformation, form where conditions during shearing either were intermediate between brittle and ductile or changed from ductile to brittle or from brittle to ductile.

Brittle Shear Zones

Brittle shear zones form in the shallow parts of the crust, generally within 5–10 km of the Earth's surface, where deformation is dominated by brittle mechanisms, such as fracturing and faulting. Brittle deformation is also favored by the relatively rapid strain rates that occur during seismic events (most earthquakes occur within the upper 10–15 km of the crust). Accordingly, shear zones formed in this environment are characterized by closely spaced faults, numerous joints and shear fractures, and brecciation (Figure 9.18). Brittle shear zones are in effect **fault zones**, and they are marked by fault gouge and other rocks of the **breccia series** (see Chapter 6).

The dominance of faulting and fracturing in brittle shear zones results in abrupt, discontinuous margins that truncate and offset markers. Closely

Figure 9.18 Brittle shear zone cutting Miocene volcanic rocks, River Mountains, Nevada. Slickenlined fault surface forms an overhang. (Photograph by S. J. Reynolds.)

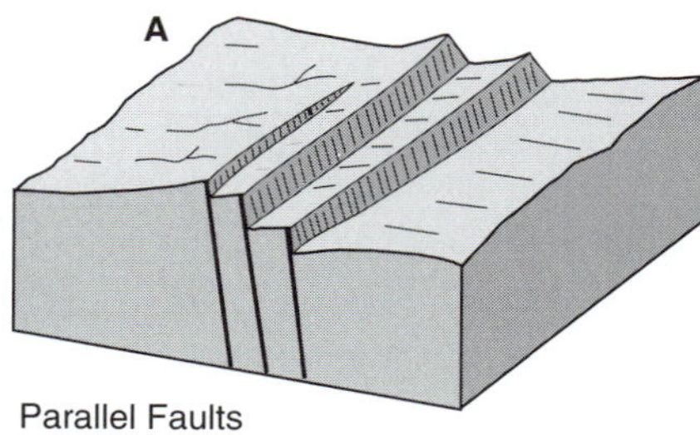

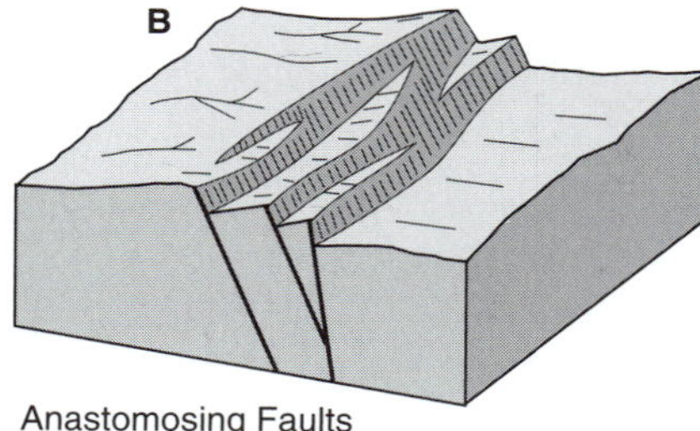

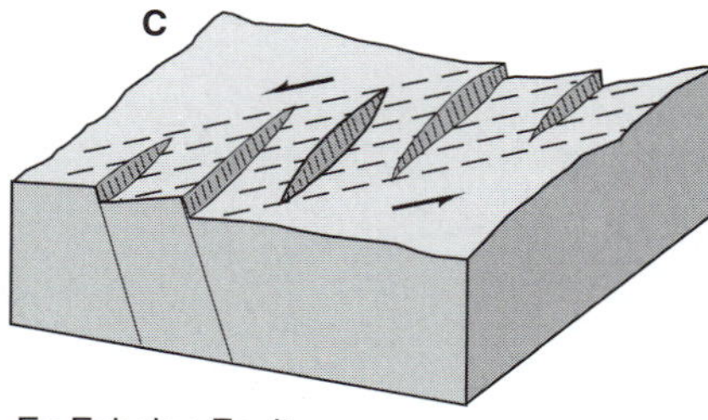

Figure 9.19 Sets of brittle shear zones (faults and fault zones): (*A*) parallel, (*B*) anastomosing, and (*C*) en echelon.

spaced faults define brittle shear zones composed of numerous discrete fault surfaces (Figure 9.19) and a chaotic assemblage of strongly fractured and brittlely disrupted rocks. Zones of intensely fractured and crushed rocks associated with faults vary in thickness from less than a millimeter to a kilometer or more. In general, the thickness of a brittle shear zone increases with the amount of displacement accommodated by the zone. A single brittle shear zone, however, may thicken or thin along strike as individual fault strands merge with or bifurcate away from the zone, such as when the zone encounters rocks of varying mechanical properties. Changes in the width and character of fault zones are especially common near bends or jogs. The wall rocks outside a brittle shear zone may be largely unaffected by the faulting or, alternatively, may show a zone of drag folding flanking the zone.

Ductile Shear Zones

Ductile shear zones are formed by shearing under ductile conditions, generally in the middle to lower crust and in the asthenosphere. For the most common crustal rocks (e.g., granite), brittle deformation at shallow crustal levels gives way downward into ductile deformation at the **brittle–ductile transition** (see Chapter 4). A similar brittle-to-ductile transition is present within the mantle, probably near the **lithosphere–asthenosphere boundary**. In both cases, rocks below the brittle–ductile transition are at temperatures and pressures so high that they respond to imposed stresses by ductile flow rather than by faulting and brittle fracture. Accordingly, we generally see ductile shear zones developed at the expense of rocks we would expect to find in the middle crust and deeper—gneiss, schist, marble, amphibolite, granulite, migmatite, large intrusions, pegmatites, and deep level mafic and ultramafic rocks. A few exceptionally weak rock types, such as salt, gypsum, and clay, are able to deform ductilely at relatively low temperatures and at very shallow crustal levels.

Most ductile shear zones form under *metamorphic conditions*, and the resulting sheared rocks are *metamorphic* in character, typically possessing foliation and metamorphic minerals (Figure 9.20). Rocks within a ductile

Figure 9.20 Ductile shear zones in the field: (*A*) Thin shear zones cutting Proterozoic gneiss, South Mountains, central Arizona. (Photograph by S. J. Reynolds.) (*B*) Shear zone cutting folded metasedimentary rocks, Cap de Creus Peninsula, axial zone of the Pyrenees. (Photograph by Carol Simpson.)

shear zone may be so changed by the intense shear, by metamorphism, and by fluids passing through the shear zone that it becomes very difficult, if not impossible, to decipher the original rock—the protolith. Some rock types that form in ductile shear zones are sufficiently distinctive in appearance to merit a terminology separate from that for "normal" metamorphic rocks. Such rocks are sometimes simply called **tectonites**, with appropriate adjectives to further indicate what type of rock is being described (e.g., marble tectonite). More commonly, the deformed rocks are assigned to the important family of metamorphic tectonites called **mylonitic rocks** (see Chapter 8).

Ductile shear zones, unlike typical fault surfaces or brittle shear zones, commonly do not display any discrete physical break (Figure 9.21). Instead, differential translation of rock bodies, separated by the shear zone, is achieved entirely by ductile flow. Markers pass through ductile shear zones without necessarily losing their continuity, but the effects of the shearing are recorded by distortion of the markers and by the development of foliation, lineation, and other shear-related fabrics. A continuous ductile "shear zone" is formed when we step on a piece of gum on a hot sidewalk (Figure 9.21*C*). When we step too far for the gum to stretch, the gum breaks and the "shear zone" becomes discontinuous.

Between Brittle and Ductile: The Middle Ground

Many shear zones have characteristics intermediate between those of brittle and ductile shear zones (Ramsay, 1980a). These shear zones display some distinctly brittle aspects, like fractures, in combination with more ductile aspects. Brittle versus ductile character may change along a shear zone as it encounters rocks of contrasting mechanical properties. Some rock types affected by the shear zone may respond brittlely, whereas others respond by ductile flow. In other cases, a shear zone may operate under progressively changing physical conditions, from ductile to brittle (e.g., progressive uplift accompanied by a drop in temperature and pressure). Alternatively, a shear zone may be reactivated under physical conditions totally different from those under which it first formed.

All these situations can produce a shear zone that is neither strictly brittle nor strictly ductile. We subdivide such shear zones into two general types: **semibrittle** and **brittle–ductile**.

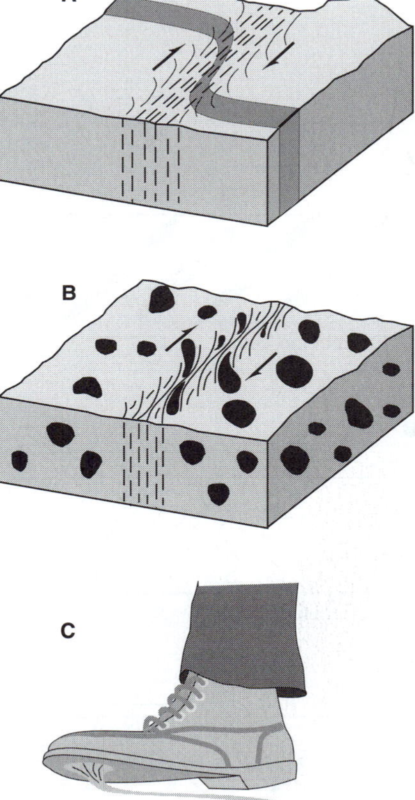

Figure 9.21 Ductile shear zones: (*A*) marker offset by continuous, dextral shear zone, (*B*) shear zone cutting plutonic rocks with inclusions and isotropic initial fabric (compare with Figure 9.6*A*), and (*C*) sinistral shear of chewing gum.

Semibrittle Shear Zones

Although dominated by brittle deformation mechanisms like fracturing and cataclastic flow, **semibrittle shear zones** contain some ductile aspects as well (Figures 9.22, 9.23). A common example of a semibrittle shear zone is a zone of **en echelon veins** or **en echelon joints** (Figures 9.22, 9.23*A*). Deformation along the zone is accommodated by brittle mode I fractures, now filled by veins, and by distributed deformation between the veins. Another common example is a zone of **en echelon stylolites**, formed by pressure solution (Figure 9.23*B*). Some shear zones contain both veins and stylolites, so arranged that the shortening direction for the stylolites is approximately perpendicular to the extension direction indicated by the veins.

Shear zones defined by **en echelon folds** (Figure 9.23*C*) can be either semibrittle or ductile, depending on the conditions under which they form and on the character of associated structures. Many zones of en echelon folds are associated with faults and are probably best classified as semibrittle shear zones. The faults are brittle features, but the folding may occur

Figure 9.22 Semibrittle shear zones in the field defined by en echelon quartz veins. (*A*) A single set of veins defining a shear zone crossing diagonally across the photograph. (Photograph by G. H. Davis.) (*B*) Conjugate sets of en echelon calcite veins crossing in left center of photograph, as well as subvertical tension veins bisecting the acute angle between the shear zones. Jasper Park, Canada. (Photograph by Shelby J. Boardman.)

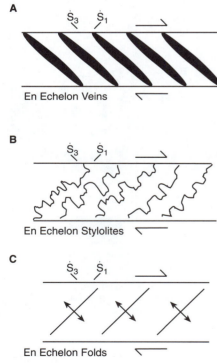

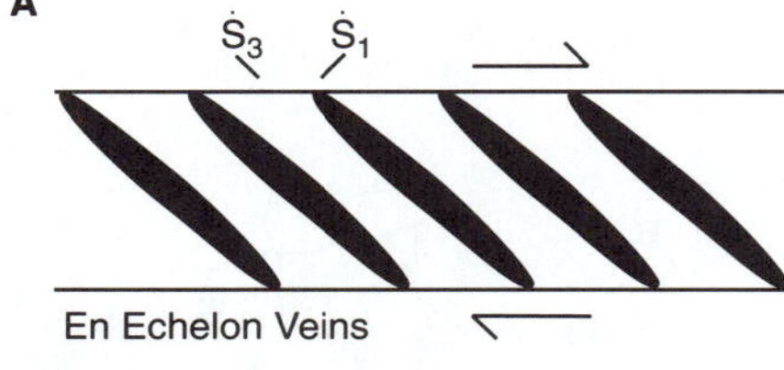

B
$\dot{S}_3$ $\dot{S}_1$
En Echelon Stylolites

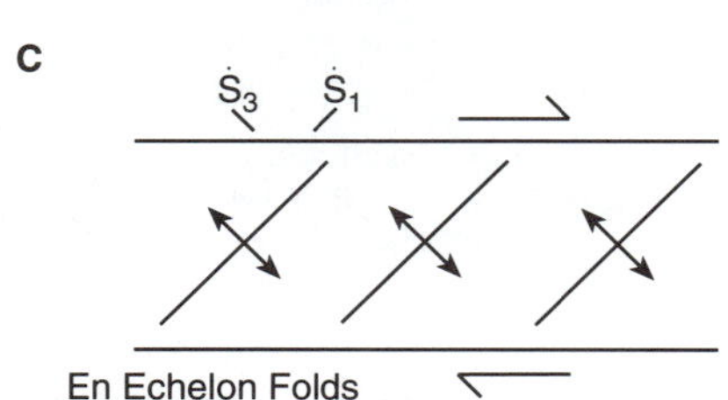

Figure 9.23 Semibrittle shear zones: (*A*) en echelon extension veins, (*B*) en echelon stylolites, and (*C*) en echelon folds. $\dot{S}_1$ is the axis of maximum instantaneous stretching, and $\dot{S}_3$ is the axis of maximum instantaneous shortening.

by ductile mechanisms, such as pressure solution, without loss of cohesion of the rocks. Alternatively, folding may be accommodated by brittle or semibrittle mechanisms, including layer-parallel slip and pervasive small-scale faulting, jointing, and cataclasis. The folds appear ductile from a distance, but up close the rocks look as if they have deformed in brittle fashion. Some zones of en echelon folds are formed at greater depths, where truly ductile mechanisms prevail, but such environments more commonly favor the formation of classic ductile shear zones composed of mylonitic rocks.

Brittle–Ductile Shear Zones

Brittle–ductile shear zones contain evidence of deformation by both brittle and ductile mechanisms and come in many flavors. Nearly all contain some type of tectonite fabric (Figure 9.24), such as mylonitic foliation and lineation, but this fabric may be well developed only in some mineralogic phases of the rock or in the more easily deformed rock types in a lithologically diverse sequence. Many brittle–ductile shear zones contain boudins, rock fragments, and porphyroclasts of the more brittle minerals and rock types, all floating in a tectonite matrix of more easily deformed minerals and rocks. In some brittle–ductile shear zones, tectonite fabric may resemble a mylonitic fabric in outcrop, but have many brittle aspects in thin section, including microfaults, grain-scale fractures, and zones of microbreccia and cataclasite. The margins of many brittle–ductile shear zones are sharply defined and faultlike, or are zones of concentrated fracturing and brecciation.

Brittle-ductile shear zones form when (1) the physical conditions permit brittle and ductile deformation to occur at the same time, (2) different

Figure 9.24 Brittle–ductile shear zones cutting Archean metavolcanic rocks along the Cadillac Break, Malartic Hygrade Mine, Quebec, Canada. (Photograph by S. J. Reynolds.)

parts of a rock have different mechanical properties, (3) a shear zone "strain hardens", (4) a short-term change in physical conditions, such as strain rate, causes the rock to switch from ductile to brittle mechanisms or vice versa, (5) physical conditions change systematically during deformation, or (6) a shear zone is reactivated under physical conditions different from those in which the shear zone originally formed. We discuss each of these conditions below.

Many brittle–ductile shear zones form under physical conditions that permit brittle and ductile deformation mechanisms to be active at the same time (Figure 9.25*A*). This might appear to require an unlikely combination of just the right temperature, pressure, strain rate, and so on, but actually the situation is very common. In large part this is because the different deformation mechanisms overlap appreciably in the physical conditions under which they operate. Even in a monomineralic rock, like quartzite or marble, it is possible to have adjacent grains deforming by different mechanisms. Some calcite grains in a marble may deform by twin gliding while other nearby grains deform by dissolution creep or by microcracking. Such different responses could be controlled by variations in *crystallographic orientation* of the grains; the slip planes may be favorably oriented for slip in one grain but in a "hard" orientation in an adjacent grain. *Different-sized* grains of the same mineral may also accommodate deformation by different mechanisms. Even within a single grain, deformation may proceed by one mechanism until the grain experiences strain hardening and begins deforming by fracturing or some other mechanism.

A mixed brittle–ductile style of deformation is also expected if different parts of a rock have different mechanical properties (Figure 9.25*B*). Most rocks contain more than one mineral, and each mineral may deform by a different mechanism, even under the same physical conditions. In a marble containing both calcite and dolomite, the calcite begins behaving ductilely at temperatures at which dolomite is still brittle. Deformation at this temperature therefore yields a rock with brittlely fractured dolomite grains floating in a matrix of ductilely flowed calcite. A similar discrepancy in style of deformation may occur where a shear zone cuts through a lithologically, and therefore mechanically, heterogeneous sequence of rocks. Shear zone fabric may be well developed where the zone cuts a relatively weak rock type, such as a calcite marble, but poorly developed or of a different structural style where it cuts a stronger or more brittle rock unit, like quartzite or pegmatite. A shear zone cutting a lithologically heterogeneous sequence of rocks may contain boudins of relatively stiff rocks floating in a matrix of more easily deformed ones (Figure 9.25*B*).

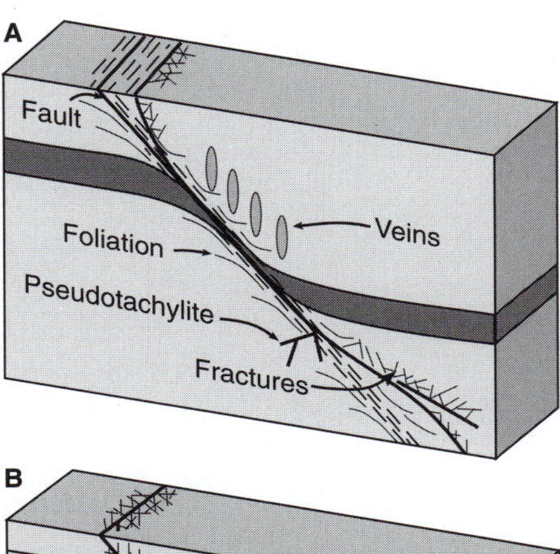

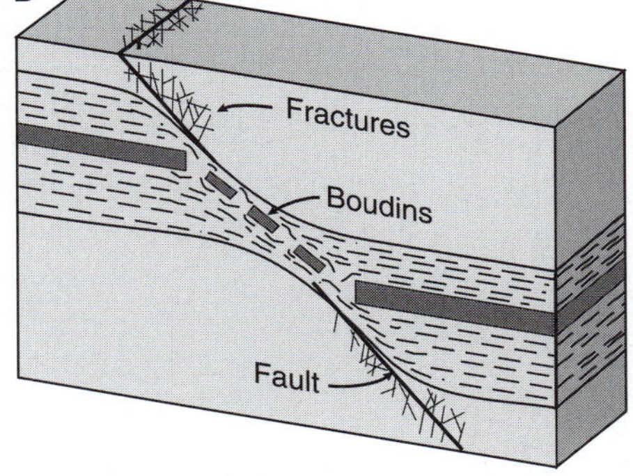

Figure 9.25 Brittle–ductile shear zones. (*A*) Formed by intermediate (brittle–ductile) conditions, where brittle, ductile, and semibrittle deformation occurred during the same event, in part due to variations in strain rate and fluid pressure. (*B*) Formed in interlayered rocks with differing rheologies and responses to deformation. The thin, middle gray layer was relatively rigid and formed boudins because it is enveloped in a less competent and more foliated unit that deformed ductilely. The top and bottom layers were more competent and deformed by faulting and fracturing.

Some shear zones may switch from ductile to brittle simply as a result of strain hardening. Some shear zones are fundamentally ductile in character but have abrupt margins and faultlike attributes. Such shear zones may begin as continuous ductile shear zones but develop into sharp, faultlike features when the shear strain is too great or too fast to be accommodated by strictly ductile means. Just like our "gum shear zone."

A mixed brittle–ductile character to a shear zone may also indicate that physical conditions fluctuated during deformation. The most likely short-term changes are in strain rate or fluid pressure. A short-term increase in strain rate can cause a rock to switch from ductile to brittle behavior. Likewise, an increase in fluid pressure acts to reduce the effective confining stress and can cause a ductilely deforming rock to fracture. Ductile deformation may resume after this short-term event.

In many cases, the brittle–ductile character of a shear zone indicates either that the physical conditions systematically changed during deformation or that the shear zone formed under one set of conditions and was later reactivated under much different conditions. When conditions change from *ductile to brittle,* brittle structures, such as fractures, will overprint an earlier ductile fabric in the shear zone (Figure 9.26*A*). We may have a more difficult time recognizing a shear zone formed during a change from *brittle to ductile* conditions, because early, brittle structures may be totally overprinted and "healed" by later ductile fabric and metamorphic minerals (Figure 9.26*B*).

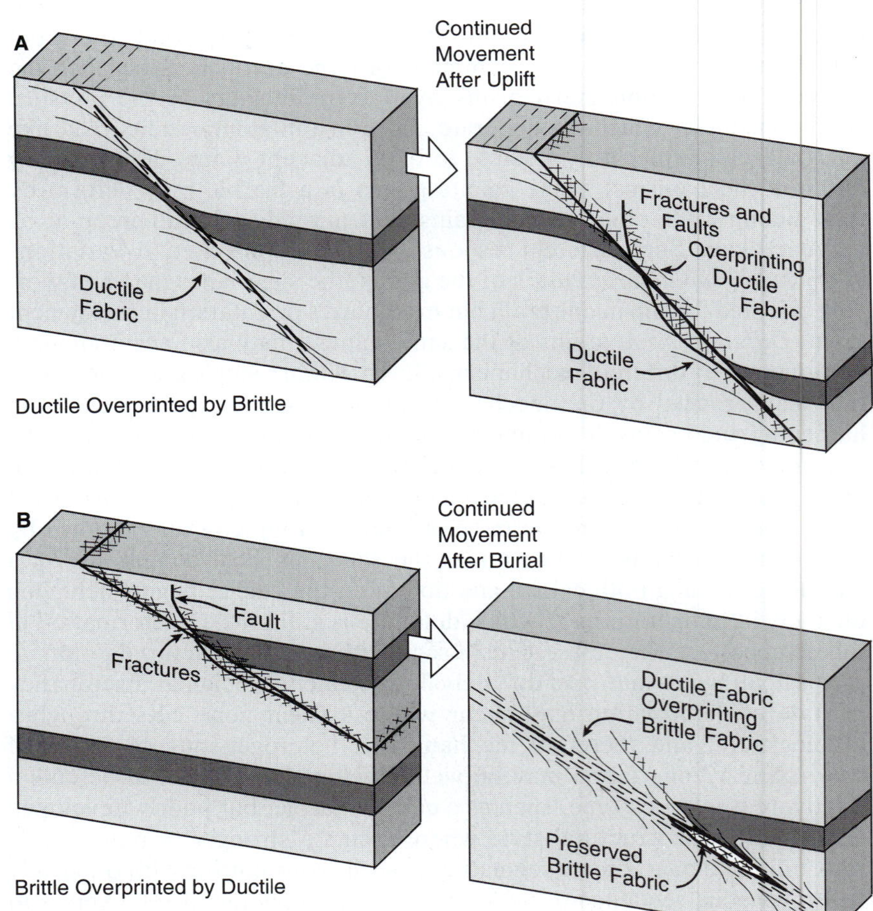

Figure 9.26 Brittle–ductile shear zones formed by a change in physical conditions during shearing or by reactivation of the shear zone under conditions different from those in which it first formed. (*A*) Ductile structures overprinted by brittle ones. A similar shear zone is shown in Figure 9.17. (*B*) Brittle structures overprinted and largely obliterated by a ductile overprint as rocks are buried.

WHY SHEAR ZONES FORM, THIN, AND THICKEN

The formation of a shear zone requires the concentration of deformation into a relatively thin zone, rather than a dispersion throughout the entire rock mass. As a shear zone develops and accommodates displacement over a long period of time, the *active* part of the zone, where structures are forming and deformation mechanisms are operating, may become restricted to a very thin part of the zone or may become broadly diffuse by migrating outward into unsheared protolith. What accounts for these behaviors? The explanation resides in how the different deformation mechanisms proceed and interact.

Softening Processes in Ductile Shear Zones

The localization of deformation into shear zones indicates that it is easier to continue deforming rocks *within the zone* than it is to broaden the zone by deforming the wall rocks. This is somewhat surprising, since the continued deformation of grains commonly can lead to strain hardening. When strain softening occurs, there must be some processes that cause rocks within the zone to be weakened or "softened" relative to rocks outside the zone (White et al., 1980). Four main types of **strain softening** operate in ductile shear zones: softening due to grain size reduction, geometric softening, reaction softening, and fluid-related softening. Shear zones also may become softer if they become hotter than surrounding rocks because of shear heating (White et al., 1980; Pavlis, 1986; Molnar and England, 1990).

Softening due to Grain Size Reduction

In most ductile shear zones, there is a reduction of grain size of the affected rocks during deformation. Since many deformation mechanisms are most efficient in fine-grained rocks, a reduction in grain size can lead to strain softening and continued localization of shear in the most deformed, finest grained parts of the shear zone (Figure 9.27). A reduction in grain size may also permit a change in the dominant deformation mechanism. For example, a mylonitic rock may switch from dislocation creep to superplastic creep once the grain size has reached an acceptably small size. Superplastic creep can operate at high strain rates because it (1) involves grain boundary sliding and (2) leaves the grains relatively defect-free, so there is no strain hardening. Once a ductile shear zone has switched to superplastic creep, it can continue deforming and can accommodate very high shear strains without strain hardening.

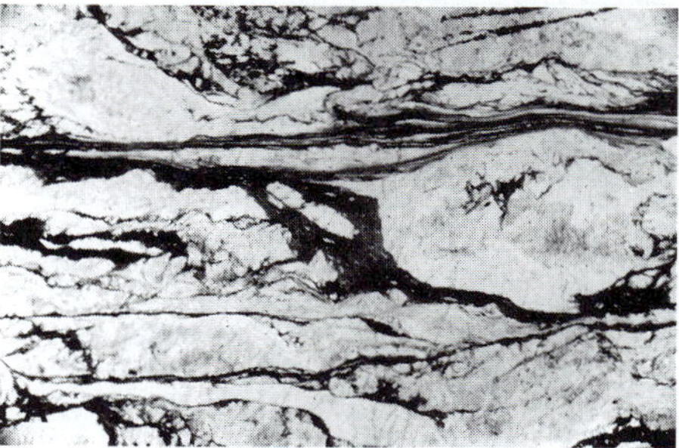

Figure 9.27 Localization of strain by fine-grained zones within Tertiary granodiorite, South Mountains, central Arizona. Dark shear zones in both (*A*) and (*B*) represent glass (pseudotachylite) formed by frictional melting during movement along small faults. The broad spectrum of deformational style, from elongated, ductilely deformed grains to angular, fractured, brittlely deformed ones, reflects the differing mechanical responses of feldspar (brittle) versus quartz and glass (both ductile), as well as likely variations in strain rate. Both photographs display an area 1.1 cm wide. (Photographs by S. J. Reynolds.)

Geometric Softening

In **geometric softening**, grains rotate during deformation until their slip systems are more favorably oriented for slip. Favorably oriented grains can easily continue slipping, whereas grains in "hard" orientations may rotate by rigid body rotation or may shatter, dissolve, or accommodate deformation in some other way. Eventually, most grains end up with their slip systems aligned favorably for slip, and this leads to a **crystallographic preferred orientation** and an overall weakening of the rock.

Reaction Softening

Reaction softening occurs when deformation is accompanied by the formation of new minerals that deform more easily than the minerals from which they were derived (White and Knipe, 1978; Beach, 1980; White, Bretan, and Rutter, 1986). Serpentine, an exceptionally weak mineral, commonly forms in shear zones cutting otherwise strong, ultramafic rocks. In rocks of felsic to intermediate composition, mylonitization under greenschist–facies conditions can cause feldspar to break down into white mica and cause amphibole to break down into biotite or chlorite, provided there is sufficient water for these *hydration* reactions to proceed. The newly formed micas, which slip easily parallel to their well-developed basal cleavage, are generally easier to deform than either feldspar or amphibole. Therefore, as a rock becomes more micaceous during mylonitization, it becomes weaker and will tend to concentrate additional strain. The additional strain may cause even more feldspar and amphibole to break down into mica, further weakening the rock. This process may continue until all the feldspar and amphibole have been consumed or until there is insufficient water for the reactions to take place.

Fluid-Related Softening

A ductilely deforming rock may also become weakened by other types of interaction with fluids. For example, fluids may dissolve and remove grains that otherwise resist ductile deformation. The dissolved material may not form new minerals, as in reaction softening, but may instead be reprecipitated as the same mineral in pressure shadows or veins, or else be entirely lost from the rock as the fluid flows out of the shear zone. Alternatively, fluids entering a shear zone may deposit new minerals, such as calcite, that are weaker than the preexisting minerals. Such fluids are responsible for many synkinematic veins of calcite and quartz that form parallel to foliation in ductile shear zones. The veins may concentrate further strain because they are (1) composed of weaker minerals than the wall rocks, (2) finer grained than the wall rocks, and (3) initially less deformed and therefore easier to deform than strain-hardened grains in the host rocks.

The presence of fluids may also cause a change in the dominant deformation mechanism. It is well known that fluids cause **hydrolytic weakening** of quartz, thereby permitting quartz to deform by dislocation creep. An increase in the *abundance of fluids* may cause dissolution creep or grain boundary diffusion creep to become relatively more important deformation mechanisms. An increase in *fluid pressure* can cause strong grains (i.e., those that are relatively resistant to ductile flow) to rupture by hydrofracture. This in turn may cause the entire rock to become weaker as the strong grains accommodate strain by fracture and are dispersed as fragments into the ductilely flowing matrix.

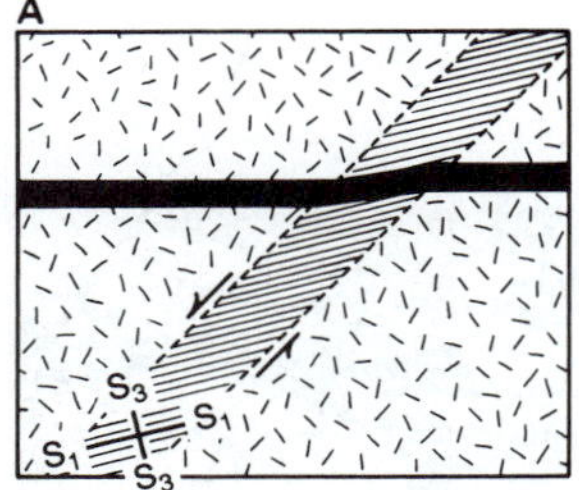

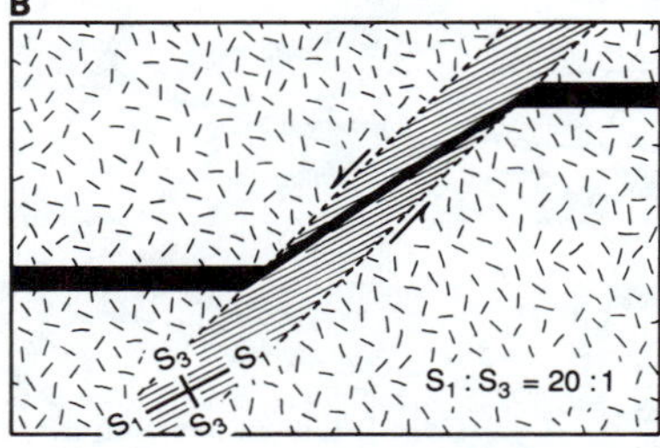

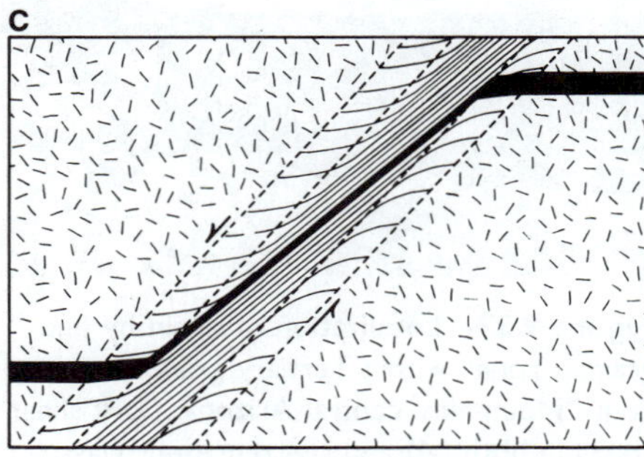

Figure 9.28 Kinematic development of a zone of simple shear. (*A*) Shearing results in stretching and thinning of the marker and initial development of foliation at 45° to the shear zone. (*B*) With progressive deformation, foliation leans over in the direction of shear and toward parallelism with the shear zone. (*C*) The shear zone widens, probably due to strain hardening, and foliation attains a sigmoidal form.

Strain Hardening in Ductile Shear Zones

If left unabated, the various types of strain softening should cause deformation to be continually concentrated into the same, very narrow zones, and we should see only thin shear zones. However, we actually observe some shear zones that are tens of kilometers wide. Also, we observe that shear zones with large displacements are generally wider than those with less displacement. This implies that ductile shear zones, like fault zones, widen as they accommodate increasing amounts of displacement (Figure 9.28). For a shear zone to widen, it must be *more difficult* to continue deforming rocks in the shear zone than it is to deform the adjacent wall rocks. In other words, the shear zone eventually undergoes **strain hardening** relative to the wall rocks. Strain hardening occurs when highly deformed grains attain a high density of dislocations that become pinned or tangled and thereby impede further dislocation glide (see Chapter 4). It also can result from mismatches along grain boundaries as adjacent grains attempt to deform against one another. The balance between strain softening and strain hardening is controlled by factors such as strain rate and temperature, and it determines whether a shear zone develops as a thin zone of extremely strained rocks or as a wider zone of less strained rocks.

THE STRAIN IN SHEAR ZONES

Deformation within shear zones can be accommodated by *displacement on discrete breaks,* by *straining of rocks* within or adjacent to the shear zone, and by *rotation* of relatively rigid objects entrained in the shear zone. Displacement on discrete breaks is easily visualized and can be evaluated via the offset of a marker, such as a dike, across the break (Figure 9.28). Strain within shear zones may have components of **distortion** (change in shape) and **dilation** (volume change), and is typically heterogeneous. It can involve **coaxial deformation**, like pure shear, or **noncoaxial deformation**, like simple shear, or components of both. Objects entrained within a shear zone can exhibit complex, and perhaps unexpected, rotations.

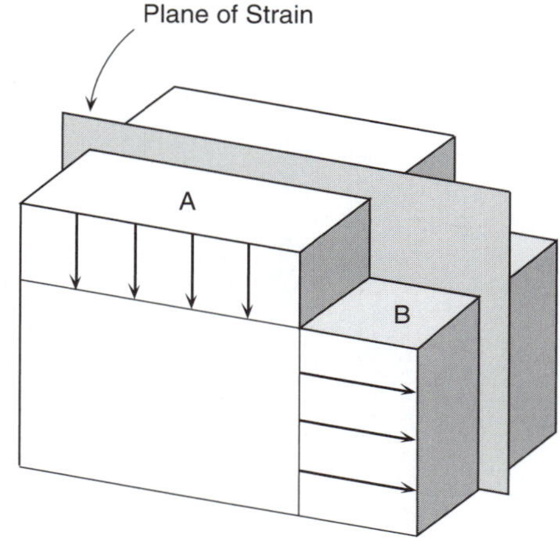

Figure 9.29 Deformation of a cube by plane strain, in this case, pure shear. Vertical shortening of initial cube "*A*" is accommodated by horizontal lengthening, resulting in the boxlike shape of "*B*." The motion of all material particles during deformation occurs along a family of parallel planes, represented by the shaded plane of strain cutting through the material.

In trying to understand strain in shear zones, we seek to compare the size, shape, and orientation of a deformed rock mass with its initial condition. Since size, shape, and orientation are three dimensional aspects, we ideally are interested in a three-dimensional representation of strain—the strain ellipsoid. For many shear zones, however, it is convenient to consider the special two-dimensional type of strain called **plane strain** (see Chapter 2). In plane strain, the motion of particles is restricted to a family of parallel planes (Figure 9.29). All strain occurs within these planes and no strain occurs perpendicular to the planes. Maximum finite extension (S_1) in one direction is accommodated by maximum finite shortening (S_3) in a perpendicular direction within the *plane of strain;* no strain occurs in the third (S_2) dimension, normal to the plane. Because no strain occurs in the third dimension, plane strain can be fully characterized with a two-dimensional strain ellipse. Equations and concepts for analyzing three-dimensional strain and volume change are presented in more detail elsewhere (Ramsay, 1977; Ramsay and Huber, 1983, 1987; Twiss and Moores, 1992).

Coaxial or Noncoaxial

When geologists converge on an outcrop that has experienced shear zone deformation, everyone starts talking about whether the deformation was coaxial or noncoaxial. So that you too can become part of the festive mob, we will explore the differences between coaxial and noncoaxial strain by examining plane strain examples of each: pure shear, which is coaxial, and simple shear, which is noncoaxial.

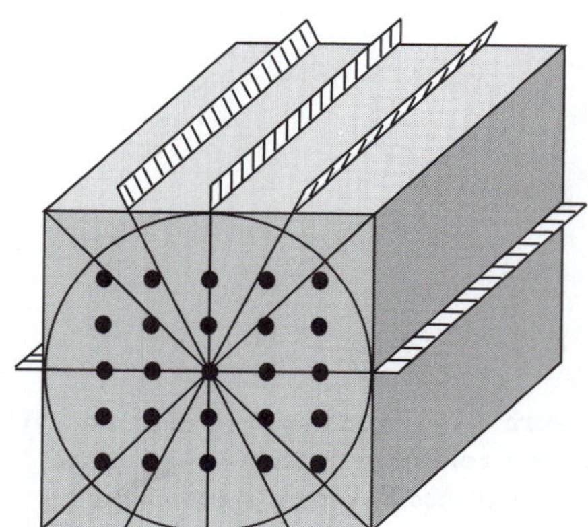

Figure 9.30 Cube before deformation by pure and simple shear. Points represent material particles and can be used to compare the displacement paths for pure and simple shear. Note that each line on the front, square face of the cube represents a family of lines that define a plane striking into the cube.

We start with a cube of rock inscribed with a circle and deform it by both types of strain (Figures 9.30–9.32). A grid of points, representing material points in the deforming rock mass, are shown on the front of the cube so that we can track the motion of individual particles during deformation. For plane strain with no volume change, there is no motion of particles in the third dimension (into the page). We will therefore concentrate on how the front face of the cube, which is a two-dimensional square, is affected by deformation. We need to remember, however, that we are dealing with a three-dimensional rock mass, and when we refer to a line on the square, we are really referring to a *family of parallel lines* that define a plane that projects along strike *into* the cube (Figure 9.30).

To represent **pure shear**, our cube of rock is shortened vertically and extended horizontally on the page (Figure 9.31*A* to *D*). The front square is progressively converted into a rectangle, and the circle is converted into an ellipse. Individual particles within the square are squeezed closer in a vertical direction and stretched farther apart in a horizontal one. Displacement paths for particles on the edge of the original cube (Figure 9.31*E*) show that particles flow inward and curve toward a common plane, termed the **shear plane** or **flow plane**. Flow is symmetric relative to the shear plane and its normal. By **shear plane**, we do not mean a single plane or a discrete structural break. Rather, we use the term for any one of a family of planes that is oriented parallel to the shear zone and to the flow plane of the deformation.

In **simple shear**, a *noncoaxial* plane strain, the cube of rock is sheared like a deck of cards, whereby the square is converted into a parallelogram (Figure 9.32*A* to *C*). The top and bottom of the parallelogram have not

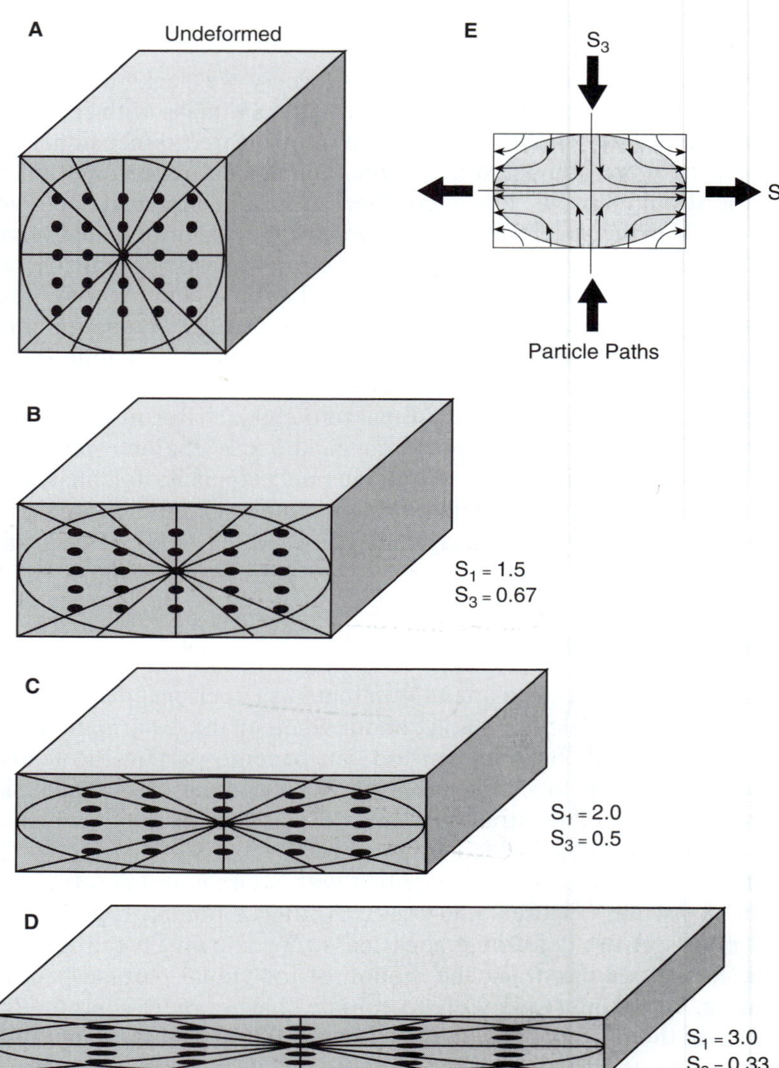

Figure 9.31 Pure shear of a cube. (*A–D*) Cube is converted to progressively flatter, boxlike shapes by vertical shortening and horizontal extension. The successive diagrams, from top to bottom, represent progressive increments of deformation and increasing amounts of strain. S_1 and S_3, the principal finite stretches, are shown for each stage. Note that they are reciprocals of one another. (*E*) Overall displacement paths for particles during deformation [Part E modified from Twiss and Moores (1992)].

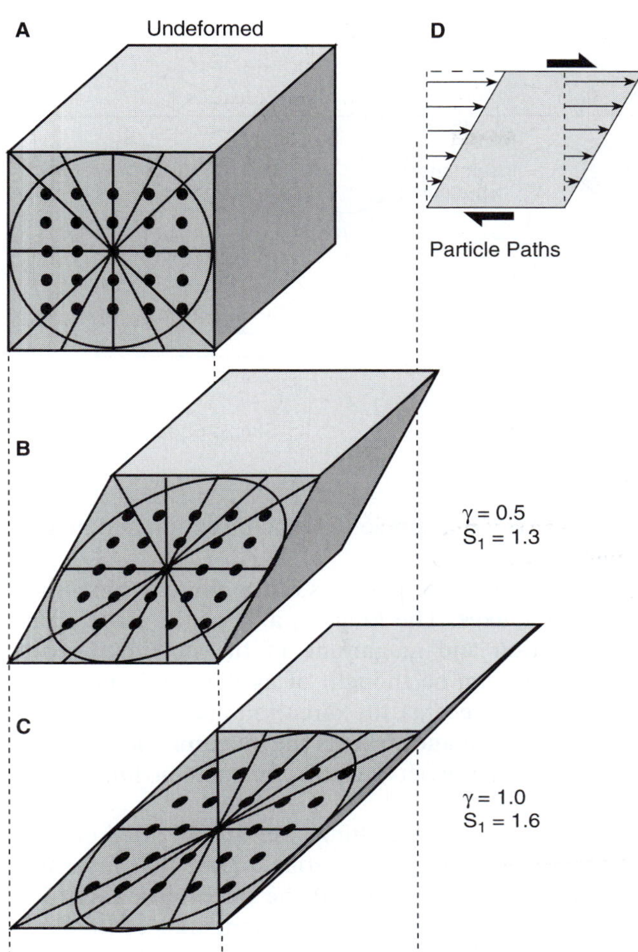

Figure 9.32 Simple shear of a cube. (*A–C*) Cube is converted to progressively more angular, rhomblike shapes by dextral simple shearing. The successive diagrams represent progressive increments of simple shear and increasing amounts of strain. Shear strain (γ) and S_1 are shown for each stage. (*D*) Overall displacement paths for particles during deformation [Part D modified from Twiss and Moores (1992)].

changed length or orientation, but both sides have been lengthened and sheared over in the direction of shearing (dextral, or top to the right, in this case). The height does not change with deformation. All particles move parallel to the shear plane (parallel to the cards), but the flow is asymmetric with respect to the shear zone and its normal. The relative velocity of a particle varies with its position normal to the shear plane. If the base of the square is considered to be fixed, the fastest relative velocities are for particles at the top (Figure 9.32*D*).

The Instantaneous and Finite Strain Ellipses

In addition to particle paths, we can explore the differences between coaxial pure shear and noncoaxial simple shear by examining the strain produced by very small (infinitesimal) increments of deformation. An **incremental** or **instantaneous strain ellipse** is used to portray how a circle is affected by a small increment of deformation. It should be nearly circular because it represents infinitesimally small amounts of strain, but we will exaggerate its elliptical aspect for clarity. Because this ellipse represents the strain accrued in one small period of time, it also contains information about the relative *rates* of strain in different orientations (Figure 9.33). The principal axes are the **instantaneous stretching axes**, with the long axis ($\dot{S}_1$) representing the direction and magnitude of most rapid extension

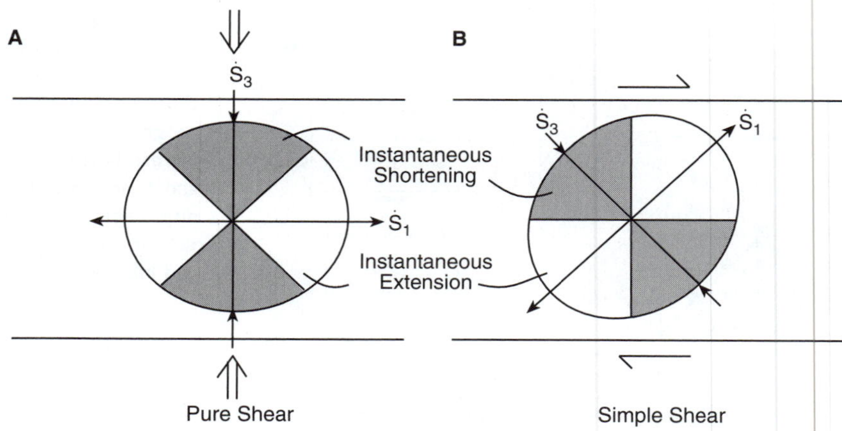

Figure 9.33 Instantaneous strain ellipses: (*A*) pure shear, and (*B*) simple shear. The ellipses display the instantaneous stretching rate along all radial directions and can be subdivided into quadrants of positive instantaneous stretching rates (i.e., extension or lengthening) and negative instantaneous stretching rates (i.e., shortening). The instantaneous stretching axes are directions of maximum and minimum stretching rate, where the maximum extension rate is along $\dot{S}_1$ and the maximum shortening rate is along $\dot{S}_3$.

(lengthening) and the short axis ($\dot{S}_3$) representing the direction and magnitude of most rapid shortening.

Recall from Chapter 2 that the principal axes of the **finite strain ellipse** are called the **finite stretching axes**. The long, S_1 axis of the finite strain ellipse represents the direction and magnitude of the maximum finite stretch, which for plane strain can be thought of as the **maximum finite extension**. The short, S_3 axis represents the direction and amount of the minimum finite stretch, which for plane strain is the **maximum finite shortening**. As we shall see, the distinction between the *finite* and *instantaneous* strain ellipses is very important.

For pure shear, the *instantaneous* stretching axes are oriented parallel and perpendicular to the shear plane (shear zone), with the maximum shortening rate normal to the shear plane and the maximum extension rate parallel to the shear plane (Figure 9.33*A*). The long axis of the *finite* strain ellipse (S_1) starts out parallel to the shear plane from the onset of deformation and remains so during successive increments (Figure 9.31). The deformation is **coaxial** because the finite stretching axes do not rotate; they instead stay aligned with the instantaneous stretching axes throughout the entire history of deformation.

For simple shear, the *instantaneous* shortening axes are inclined 45° to the shear plane (Figure 9.33*B*). For this reason, the S_1 axis of the *finite* strain ellipse *first* appears at 45° to the shear zone during the first increment of deformation (Figure 9.32). With additional increments of deformation, S_1 is successively rotated over, out of parallelism with the instantaneous stretching axis and toward parallelism with the shear zone. In other words, it "leans over" in the sense of shear. The axis of maximum finite shortening (S_3) progressively rotates toward the normal to the shear plane. For simple shear, the finite stretching axes rotate away from parallelism with the instantaneous stretching axes as deformation proceeds, and the strain is therefore **noncoaxial**. In noncoaxial deformation, the directions of maximum *finite* extension and maximum *finite* shortening do not coincide with their *instantaneous* counterparts.

The Coaxial–Noncoaxial Distinction

There are clearly important differences between coaxial and noncoaxial deformation. The motion of particles is asymmetrical about the shear zone for noncoaxial deformation, but symmetrical for coaxial deformation. The instantaneous and finite stretching axes are inclined to the shear zone for noncoaxial deformation, but not for coaxial deformation. Finally, for noncoaxial deformation, the finite stretching axes rotate with increasing strain.

A measure of whether and how the finite stretching axes rotate is the **vorticity** (Means et al., 1980; Lister and Williams, 1983). If the axes do not rotate, as in coaxial deformation, the vorticity is zero. If the axes do rotate, as in any noncoaxial deformation, the vorticity is not zero. Vorticity is especially useful in discussing **general shear**, a type of deformation that has both coaxial and noncoaxial components (Hanmer and Passchier, 1991; Simpson and De Paor, 1993; Weijermars, 1993). General shear is analogous to shearing a deck of cards, except the entire deck is composed of modeling clay and so can be stretched (thinned or thickened) during shearing. The noncoaxial component causes shearing of the cards, and the coaxial component causes the thinning or thickening. Vorticity measures the relative contributions of the coaxial and noncoaxial components. The closer a general shear is to a purely coaxial deformation, the closer to zero is the vorticity.

At this point, we suspect that we should be able to distinguish coaxial and noncoaxial deformation (i.e., whether vorticity is zero) based on whether the associated small-scale structures are inclined to the shear zone. We also suspect that we should be able to tell the sense of shear on a noncoaxial shear zone by determining which way S_1 is inclined with respect to the zone. Specifically, S_1 will lean over to the right for dextral sense of shear and to the left for sinistral shear. As we shall see, we can indeed often do this by carefully observing the small-scale structures.

DETERMINING SENSE OF SHEAR

One of our main goals in studying shear zones is to determine the overall **sense of shear**—the direction one side of a shear zone is displaced laterally relative to the other side. The recognition and application of shear-sense indicators in ductile shear zones represents one of the major advances in structural geology in the last decade. **Shear-sense indicators** are those features that reveal the sense of shear for a deformation. They come in a variety of expressions and are most commonly observed at the scale of an outcrop or thin section; however, they range in scale from geologic map patterns of large regions to X-ray analyses of tiny parts of a single sample. The new gospel of **shear-sense indicators** has revolutionized how we view shear zones and has answered many long-standing questions about the relative roles of coaxial and noncoaxial deformation. Determining the sense of shear has proven to be the key to unraveling the tectonic history of many regions, in some cases demonstrating that major shear zones formerly interpreted as thrusts actually had normal displacement instead.

A Frame of Reference

To determine the sense of shear for a shear zone, we need to have a convenient frame of reference and we need to know in which direction we should look (Figure 9.34). This is true whether we are examining a natural outcrop or a thin section. For most ductile shear zones, the shear zone itself is the most convenient reference frame because generally it is a tangible feature that can be seen in the field and is parallel to the shear plane of the deformation. If we cannot see enough of the shear zone to determine its orientation—for example, when neither margin is ex-

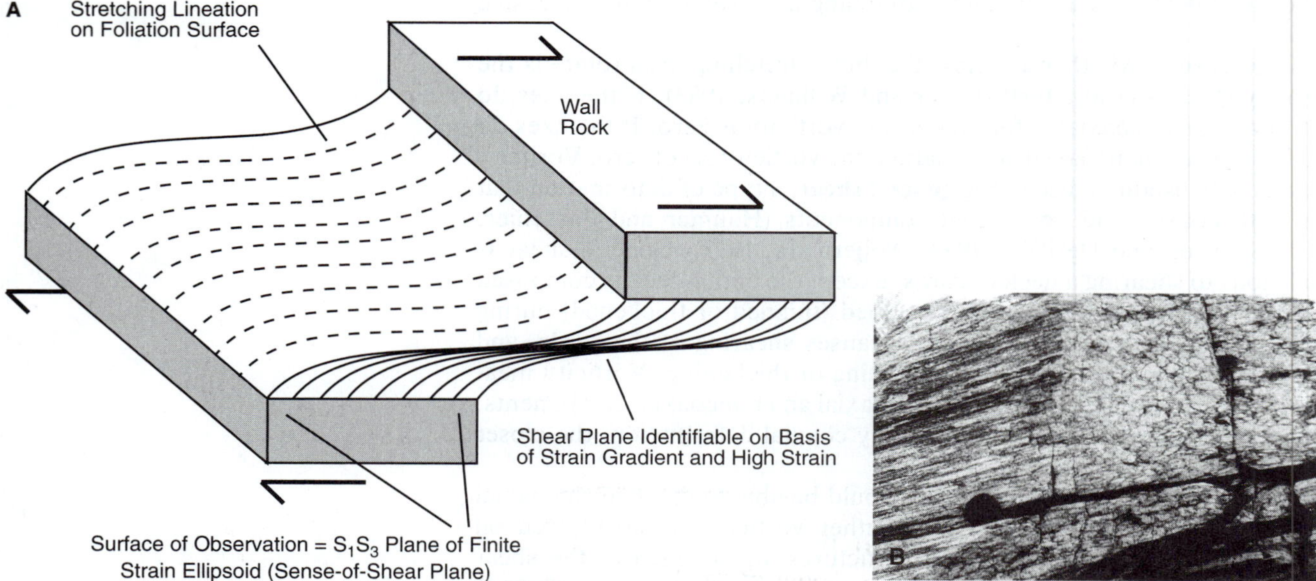

Figure 9.34 Relation of the sense-of-shear plane to structures in the rock. (*A*) Choose a rock face cut parallel to lineation and perpendicular to foliation to determine the sense of shear. This face will be parallel to the S_1S_3 plane of the finite strain ellipsoid. [Modified from Hanmer and Passchier (1991). Courtesy of the Geological Survey of Canada.] (*B*) Mylonitic quartz-pebble conglomerate, Tortolita Mountains, southern Arizona. Sense-of-shear plane is face on left, which is cut parallel to lineation and perpendicular to foliation. Face on right, cut perpendicular to lineation, reveals the conglomeratic nature of the protolith. (Photograph by S. J. Reynolds.)

posed—we can use the foliation as a reference frame, keeping in mind that foliation will generally be inclined to the shear zone.

Once we have determined the orientation of the shear zone or foliation, we need to find the **line of transport**—the direction within the shear zone along which relative displacement occurred. If deformation resulted in a stretching lineation parallel to S_1, as is generally the case, then a plane cut *parallel* to the stretching lineation and *perpendicular* to the shear zone or foliation will contain the line of transport (Figure 9.34). This is the plane in which all displacements occur in plane strain (S_1S_3), and is the plane we should inspect for shear-sense indicators. For convenience, we will call this the **sense-of-shear plane**, or simply the ''SOS'' plane—shear-sense indicators save our souls.

Below, we briefly describe the most common shear-sense indicators. We restrict the discussion to shear-sense indicators found in ductile, brittle–ductile, and semibrittle shear zones, since criteria for sense of slip on faults (i.e., brittle shear zones) were discussed in Chapter 6. We will revisit some key shear-sense indicators later in this chapter to explore how they form and how they can be used to estimate the *amount of strain* in a shear zone. Even more detail on shear-sense indicators can be found in Ramsay and Huber (1983, 1987), Simpson and Schmid (1983), Lister and Snoke (1984), Simpson (1986b), Cobbold et al. (1987), and Hanmer and Passchier (1991).

Offset Markers

Perhaps the most unambiguous shear-sense indicator is an offset marker, such as a distinctive dike, lithologic layer, or other rock unit (Figure

9.35*A*). In this case, we can determine both the amount and sense of displacement, as long as we are certain that the similar-appearing features on opposite sides of the shear zone are indeed equivalent and were originally continuous across the zone. To illustrate the potential pitfalls of this method, I (SJR) commonly take my students to a site in western Arizona where displacement on a major normal fault has coincidentally aligned two lithologically similar, but not identical, dikes. This outcrop is one of nature's more cruel tricks, but we have great fun debating whether the fault cuts the dike or vice versa.

Deflection of Markers

Although sometimes we cannot find the actual offset of a marker, it is possible to observe how a marker is deflected as it goes from the wall rocks into the shear zone (Figure 9.35*A*). The sense of the deflection reflects the sense of shear of the zone, as long as we are looking at the SOS plane. It is very easy to be fooled by an apparent sense of deflection on the wrong outcrop face (Figure 9.35*B*). We can even use the geometry of a deflected marker to determine the amount of displacement along a shear zone.

Deflections of markers are not restricted to outcrop-scale features but may be visible on a larger scale, such as on a local or regional geologic map (Figure 9.36). A dike swarm, or any other linear feature, that trends into the shear zone at a high angle is especially diagnostic, provided the shear zone actually cuts the dikes! Post-shearing dikes may bend into a shear zone, even though they postdate all displacement, because they were preferentially intruded along the platy shear zone fabric, rather than breaking a new path across the zone. Another of nature's cruel tricks.

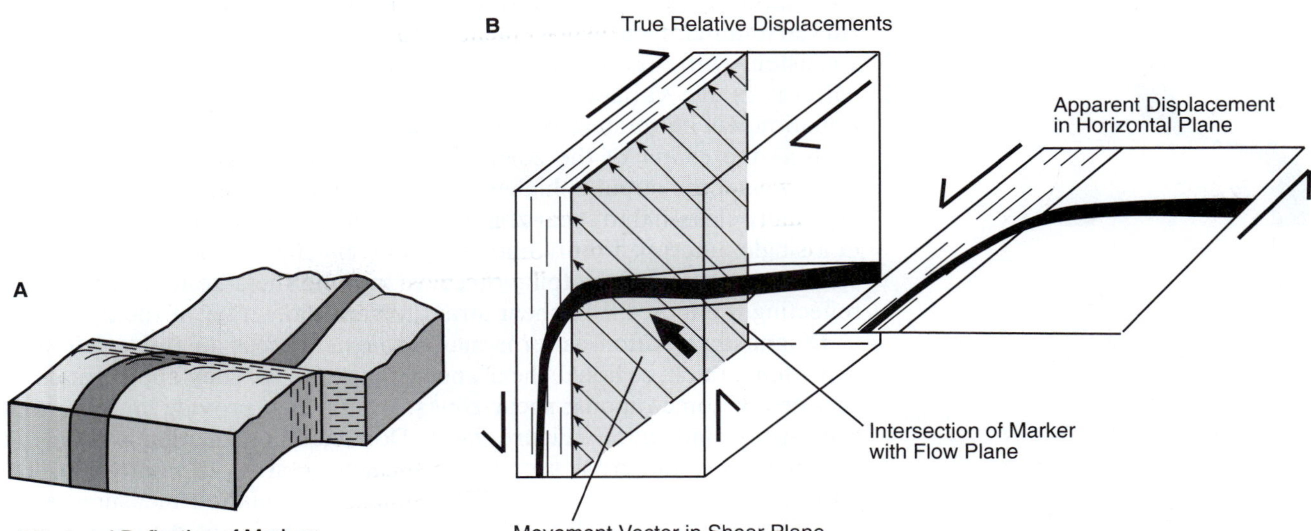

Figure 9.35 Offset and deflection of markers in ductile shear zones. (*A*) Discontinuous shear zone with dextral offset. Note deflection of marker into shear zone prior to its truncation. (*B*) Apparent relative displacement on a single improperly oriented surface (horizontal plane on right) may not reflect true displacement (arrows within plane of shear zone on left). [Modified from Hanmer and Passchier (1991). Courtesy of the Geological Survey of Canada.]

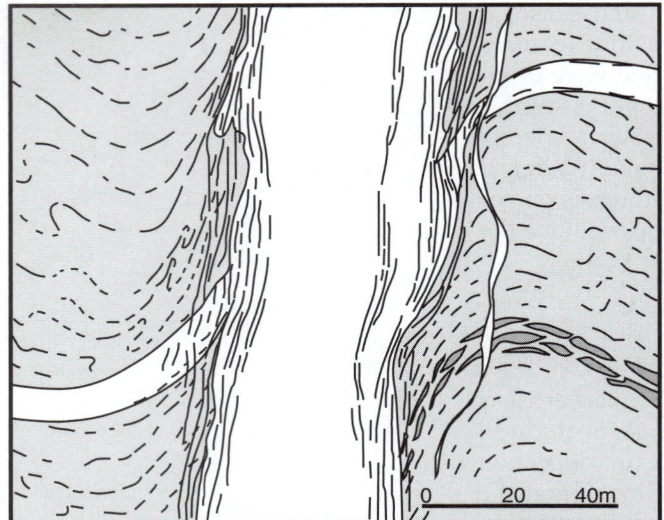

Figure 9.36 Deflection of dikes: (*A*) Map view of deflection of mafic dikes and other features into shear zones along the margins of a dike, Kangâmiut dike swarm, West Greenland. [From Escher, Escher, and Watterson (1975), Canadian Journal of Earth Sciences Published with permission of the National Research Council of Canada.] (*B*) Deflection and dextral offset of granitic dikes along a thin, discontinuous shear zone cutting Proterozoic granite, Harquahala Mountains, western Arizona. Note the shear zone fabric developed in rocks directly above the zone. (Photograph by S. J. Reynolds.)

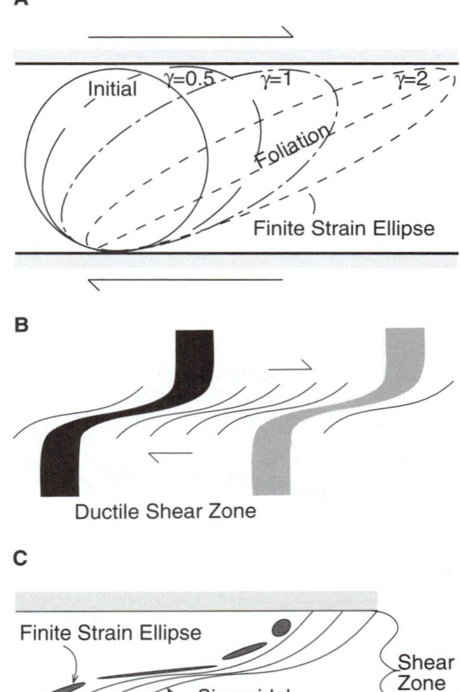

Figure 9.37 Foliation patterns in shear zones. (*A*) Finite strain ellipse and associated foliation lean over in the sense of shear and rotate toward the shear zone during progressive deformation. (*B*) Sigmoidal foliation patterns in a shear zone. (*C*) Variations in finite strain ellipse based on foliation patterns.

Foliation Patterns

Systematic variations in the orientation of foliation are common in ductile shear zones and provide one of the most useful shear-sense indicators. Foliation, because it generally reflects the S_1S_2 plane of the finite strain ellipsoid, is inclined to the shear zone for any noncoaxial deformation and leans over in the sense of shear (Figure 9.37*A*). It also commonly has a curved or a sigmoidal shape *across* a shear zone, with a sense of deflection consistent with the overall sense of shear (Figure 9.37*B*). Such foliation patterns mimic variations in the orientation of the finite strain ellipse. Foliation will be rotated toward parallelism with the shear zone in areas, such as the center of the zone, where the rocks are more strongly deformed, and the strain is higher (Fig. 9.37*C*). Such foliation patterns are especially clear in thin shear zones, where the entire shear zone is exposed in a single outcrop. If only one margin of the shear zone is exposed, we should expect to see the foliation curve into the shear zone (Figure 9.38), reflecting the increase in shear strain toward the center of the zone.

Diagnostic foliation patterns may also exist across an entire field area or within a thick, regional shear zone (Figure 9.39). They appear in cross sections of some dipping shear zones, where they provide an important shear-sense indication (Figure 9.40). Documenting foliation patterns in regional maps and cross sections, especially in areas of poor exposure, may require many traverses, and *very careful* foliation measurements. But when a pattern emerges that nails the kinematics of the entire zone, the close encounters with cactus and brush in the pursuit of foliation measurements become suddenly worth it.

So what do we do if we find that foliation is subparallel to the shear zone? We should entertain at least three possible explanations. First, foliation may parallel the shear zone because it reflects coaxial deformation, such as pure shear perpendicular to the zone (Figure 9.41*A*). Alternatively, the zone may be noncoaxial, but shear strains are so high that foliation was rotated into subparallelism with the shear zone (Figure 9.41*B*). A third alternative is that the observed foliation itself represents thin shear zones, not the S_1S_2 plane of the finite strain ellipsoid (Figure 9.41*C*). In our experience, the last two alternatives are most likely, especially if the shear zone rocks are strongly deformed. The issue of coaxial versus noncoaxial deformation can generally be resolved by observing other shear-sense indicators, such as S-C fabrics.

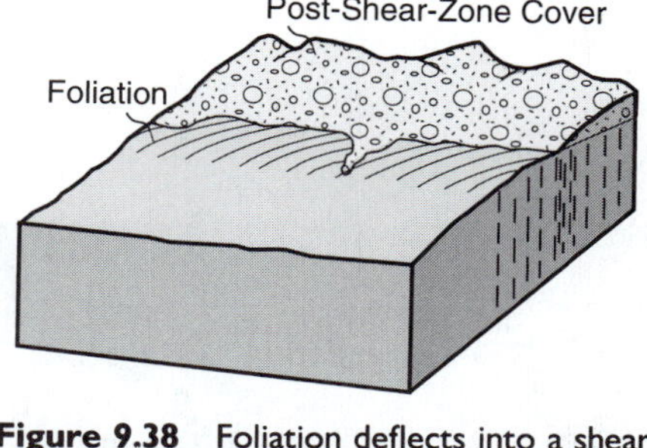

Figure 9.38 Foliation deflects into a shear zone, the far margin of which is covered by a rock unit deposited after movement on the shear zone.

Shear Bands, S-C Fabrics, and Oblique Microscopic Foliation

Shear bands are thin zones of very high shear strain within the main shear zone (Figure 9.42). They are shear zones within a shear zone. Most are less than 2–3 mm thick and less than 10–20 cm long, and many are microscopic. Shear bands can be either parallel or oblique (Figure 9.43*A*) to the main shear zone. A shear band is **synthetic** if it is inclined in the *same* direction as the overall sense of shear, and it is **antithetic** if inclined in the *opposite* direction (Figure 9.43*B*).

Shear bands commonly crosscut foliation within the shear zone (Figure 9.43*B*), displacing it in a normal, or less commonly a thrust, sense (Platt and Vissers, 1980; Behrmann, 1987; Dennis and Secor, 1987). Shear bands with normal displacement are commonly called **extensional shear bands**.

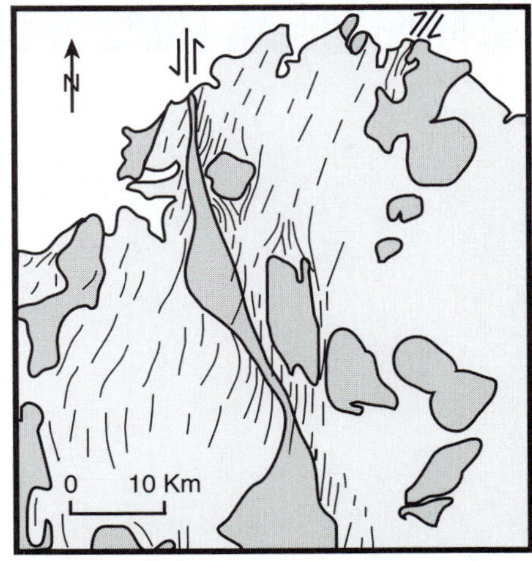

Figure 9.39 Foliation patterns in regional shear zones in the Iberian Arc, Galicia, Spain. (Reprinted with permission from *Journal of Structural Geology*, v. 2, Ponce de Leon, M. I. and Choukroune, P., Shear zones in the Iberian Arc, 1980, Elsevier Science Ltd., Pergamon Imprint, Oxford, England.)

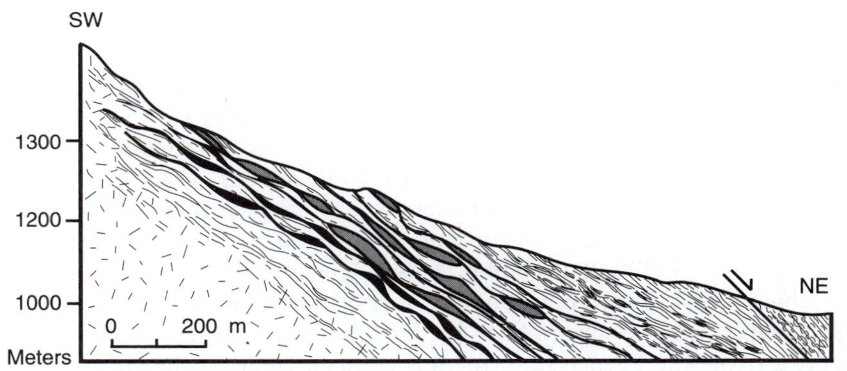

Figure 9.40 Foliation patterns in cross section, Coyote Mountains, southern Arizona. (Reprinted with permission from *Journal of Structural Geology*, v. 9, Davis, G. H., Gardulski, A. F., and Lister, G. S., Shear zone origin of quartzite mylonite and mylonitic pegmatite in the Coyote Mountains, Arizona, U.S.A., 1987, Elsevier Science Ltd., Pergamon Imprint, Oxford, England.)

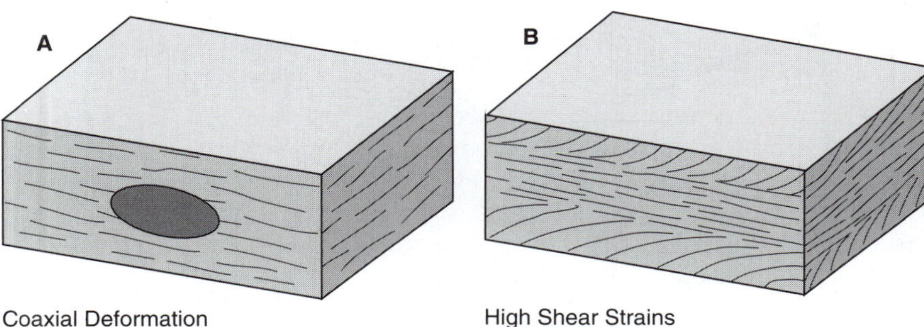

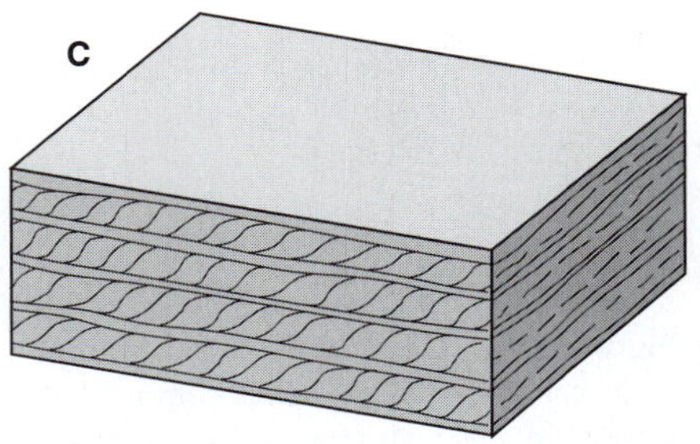

Figure 9.41 Foliation oriented parallel to a shear zone may be the result of (*A*) coaxial deformation, (*B*) high shear strain, or (*C*) the presence of thin discrete shear zones.

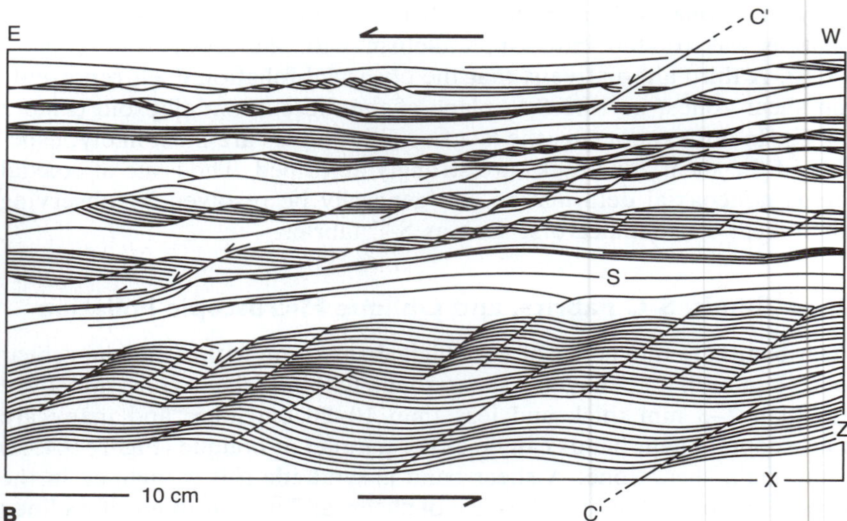

Figure 9.42 Shear bands: (*A*) Cutting Proterozoic granite, Maricopa Mountains, central Arizona. (Photograph by S. J. Reynolds.) (*B*) Cutting mica schist and quartzite, Raft River Mountains, Utah. Shear bands are labeled C'. (Reprinted with permissions from *Journal of Structural Geology*, v. 9, Malavieille, J., Kinematics of compressional and extensional ductile shearing deformation in a metamorphic core complex of the northeastern Basin and Range, 1987, Elsevier Science Ltd., Pergamon Imprint, Oxford, England.)

As the foliation approaches the shear bands, it is deflected toward parallelism with the bands because of the associated increase in shear strain. The foliation patterns associated with shear bands therefore mimic those observed in a typical shear zone, and the same sense-of-shear principles apply. The warping of foliation is similar to that seen in **crenulations**, and some geologists use this term for shear bands as well.

S-C fabrics are among the most useful sense-of-shear indicators in ductile shear zones (Berthé, Choukroune, and Jegouzo, 1979; Simpson and Schmid, 1983; Lister and Snoke, 1984; Simpson, 1986b; Hanmer and Passchier, 1991). They consist of two sets of planes or surfaces: foliation and shear bands (Figure 9.44*A–C*). The foliation planes are called **S-surfaces** (from the French term for schistosity), whereas the shear bands are denoted **C-surfaces** because they are zones of high shear strain (*cisaillement* is French for shear). C-surfaces typically are discrete zones of high shear strain 2–20 cm long and less than 1 mm thick. Most are aligned

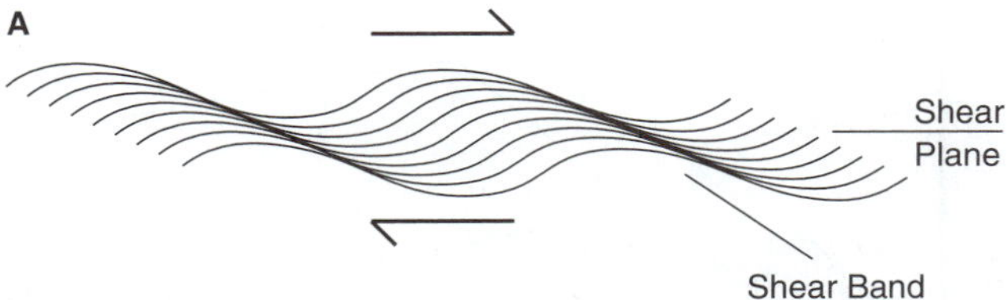

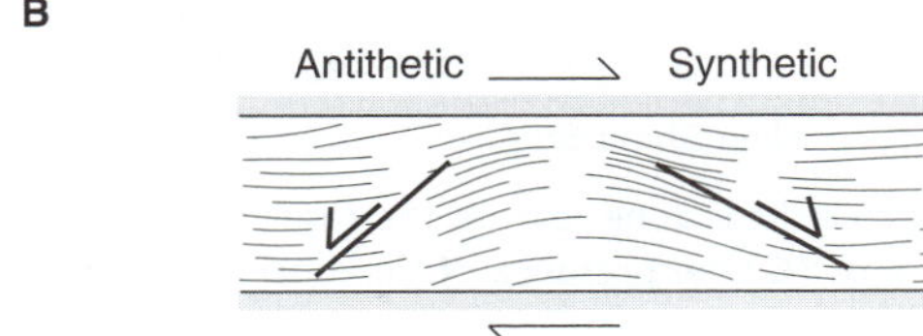

Figure 9.43 Geometries of extensional shear bands: (*A*) Shear bands commonly cut across the shear plane at approximately 30°. [Modified from Hanmer and Passchier (1991). Courtesy of the Geological Survey of Canada.] (*B*) Synthetic versus antithetic extensional shear bands.

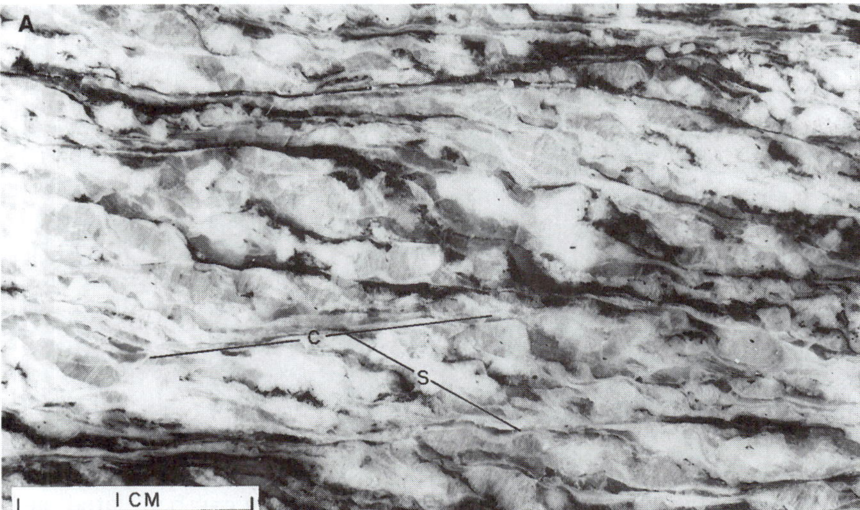

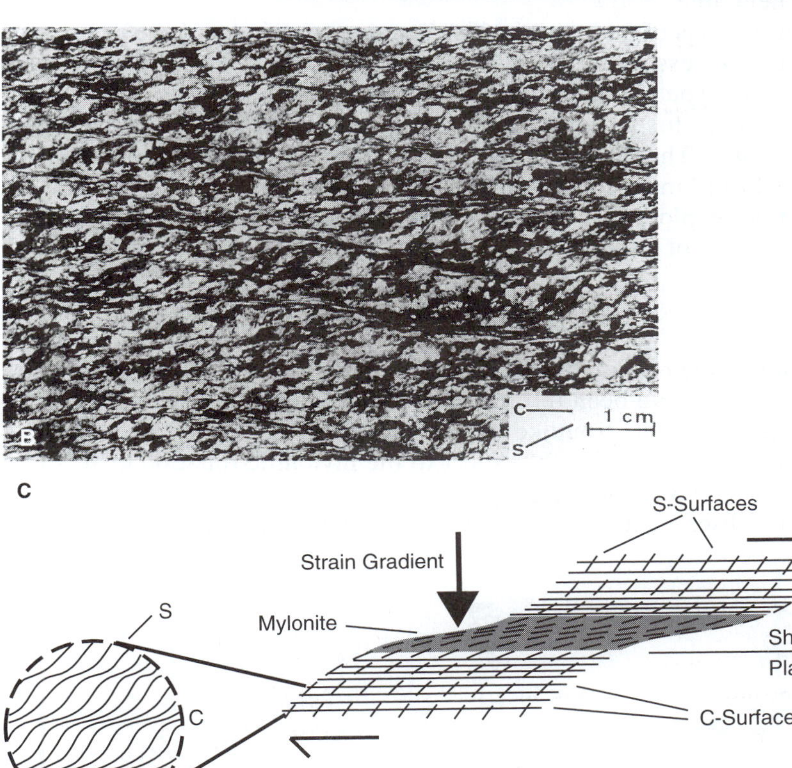

Figure 9.44 S-C fabrics. (*A*) S-C fabric in polished slab of Tertiary granodiorite, South Mountains, Arizona. Sense of shear is sinistral. (Photograph by S. J. Reynolds.) (*B*) S-C fabric in late Cretaceous pluton within the Santa Rosa mylonite zone, southern California. Sense of shear is dextral. [From Simpson and Schmid (1983). Published with permission of the Geological Society of America and the authors.] (*C*) S-C fabrics mimic the sigmoidal foliation patterns in the host shear zone. [From Hanmer and Passchier (1991). Courtesy of the Geological Survey of Canada.]

parallel to the shear zone and crosscut foliation. S-surfaces (i.e., foliation) deflect toward parallelism with a C-surface as they approach it and have a distinctive sigmoidal shape between adjacent C-surfaces. S-surfaces lean over in the direction of shear, relative to the C-surfaces. S-C fabrics are commonly visible in outcrop, once we have trained our eyes to recognize them. The clearest examples are in mylonitic granitic and gneissic rocks that contain coarse porphyroclasts or augen of feldspar. In some rocks, S-C fabrics are not obvious in outcrop but are well developed in thin section. Such thin sections should be cut parallel to the SOS plane (parallel to lineation, perpendicular to foliation), and from oriented samples, so

Figure 9.45 Photomicrograph of oblique foliation in quartz ribbons in mylonite. Sense of shear is sinistral. (Photograph by Carol Simpson.)

that the sense of shear interpreted from thin section can be related back to the field area.

Thin sections of mylonitic rocks may also contain another shear-sense indicator, expressed as a microscopic foliation oblique to the main mylonitic foliation. The **oblique microscopic foliation** is defined by aligned subgrains oblique to the long axis of larger individual grains and ribbons (Figure 9.45). The oblique foliation leans over in the direction of shear relative to the main foliation defined by the larger grains. Later in the chapter, we explore how these fabrics form and how their geometry relates to the amount of strain.

Mica Fish

Many mylonitic rocks contain lenticular porphyroclasts of muscovite and biotite, which have been termed **mica fish** (Lister and Snoke, 1984). The name refers to their troutlike shape (Figure 9.46). Mica fish, which are commonly asymmetric with respect to the mylonitic foliation or to shear bands, make excellent shear-sense indicators. Asymmetric mica fish are generally observed in thin section but can also be visible in hand specimen.

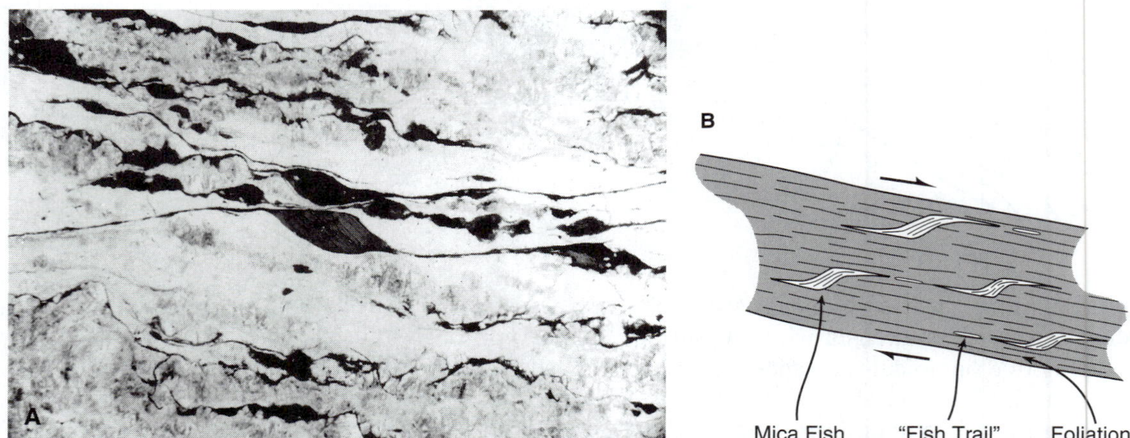

Figure 9.46 Mica fish. (*A*) Mylonitic Tertiary granodiorite, South Mountains, Arizona. Thin, dark tails streaming off the dark, 1 mm long, fish-shaped biotite crystal in the center of the photograph reflect sinistral shear. (Photograph by S. J. Reynolds.) (*B*) Mica fish lean over in the sense of shear, and their tails are commonly parallel to C-surfaces and to the shear zone.

They are commonly aligned with S-surfaces in S-C fabrics, and thus lean over in the sense of shear.

While working on some spectacular mylonitic dikes in the South Mountains metamorphic core complex of Arizona, my enthusiastic Australian friend, Gordon Lister, and I (SJR) discovered that we could determine the asymmetry of mica fish by observing their reflections in the sunlight (Reynolds and Lister, 1987; Simpson, 1986b). We call it the **fish-flash** method. Simply examine the outcrop for a potential hand specimen with a visible stretching lineation on foliation or a shear band. Mark a north arrow on the top, so that later you can relate your shear-sense results back to the outcrop. With your back to the sun and the sample held in front of you, look down the foliation surface and along the lineation, tilting the sample forward and backward, and note whether the sample looks dull or ''flashy'' (Figures 9.47 and 9.48). From the position of the sun and the rock, it is easy to figure out the dominant inclination of the mica fish and thus the sense of shear. When you see the flash, you are looking in the same direction as the sense of shear (Figure 9.48*A*).

The fish-flash technique works best on deformed rocks with several percent mica, including mica-bearing quartzites, dikes, and many fine- to medium-grained granitic rocks. It is less useful in rocks with abundant mica, such as mica schists and phyllonites, where the flash is commonly scattered or is nearly perpendicular to the foliation. In some cases, the fish-flash technique is the only one available. ''Fishing'' permits are not required!

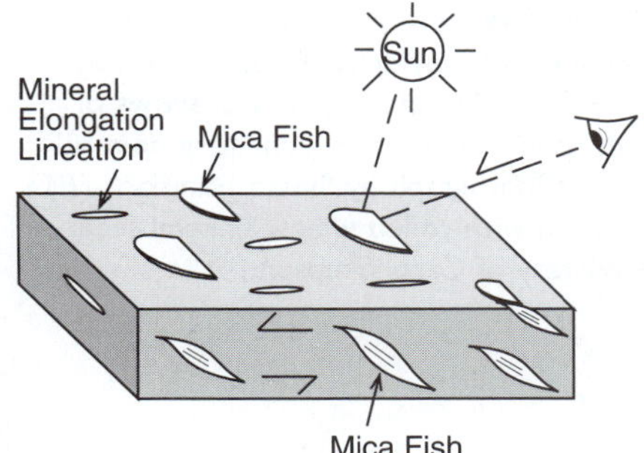

Figure 9.47 Fish-flash. Most mica fish in a rock will lean over in the same direction, resulting in a maximum reflectivity when looking down the lineation, in the same direction as the sense of shear. [From Simpson, C. (1986). Reprinted with permission of the Journal of Geological Education and the National Association of Geology Teachers.]

Inclusions

Within outcrops of shear zones we may see various **inclusions**—discrete objects that are lithologically or mechanically distinct in some way from the main mass, or **matrix**, of the shear zone (Figure 9.49). Inclusions range from nearly equant to strongly elliptical, lenticular, or sigmoidal. They may be single crystals that are fragments of older, large crystals or they may be crystals that grew during metamorphism. Many inclusions are fragments of one rock type in another, such as original *igneous* inclusions, or boudins derived from fracturing of a relatively rigid, competent layer. Inclusions in shear zones are mostly centimeters to meters in diameter, but range from single grains less than a centimeter in diameter to fragments of rock tens to hundreds of meters long.

Many inclusions are *more rigid* than the surrounding, ductilely deformed matrix and resist distortion, whereas others deform similarly to the matrix. The term *inclusion,* as used in the context of shear zones, generally refers to an object that responded to deformation in a *more rigid* manner than the surrounding, ductilely deformed matrix (Hanmer and Passchier, 1991). There is a complete spectrum, however, from rigid inclusions to those that deform as if they were matrix. The shape of an inclusion both reflects its original shape and indicates whether it was rigid or deformable during shearing.

Figure 9.48 Fish-flash viewed parallel to the lineation in mylonitic schist, Rincon Mountains, southern Arizona. Keeping the lighting conditions and the orientation of foliation constant, the sample shows (*A*) strong flash when viewed in the same direction as the sense of shear but (*B*) much less flash when viewed in the opposite (180°) direction. (Photographs by S. J. Reynolds.)

Rotation of Inclusions

Inclusions provide sense-of-shear information by the way they rotate, deform, recrystallize, and interact with their matrix. The specifics of how an inclusion rotates relative to the matrix are more complex than we might suspect. Whether an inclusion rotates depends on its shape, rigidity, and orientation relative to the imposed strain, and on the type of strain (e.g., coaxial versus noncoaxial).

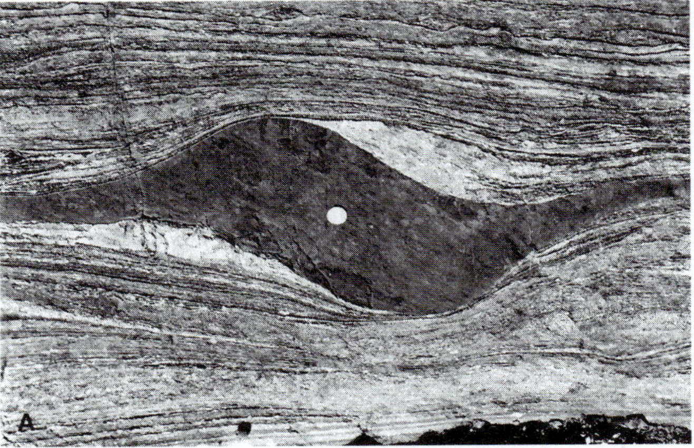

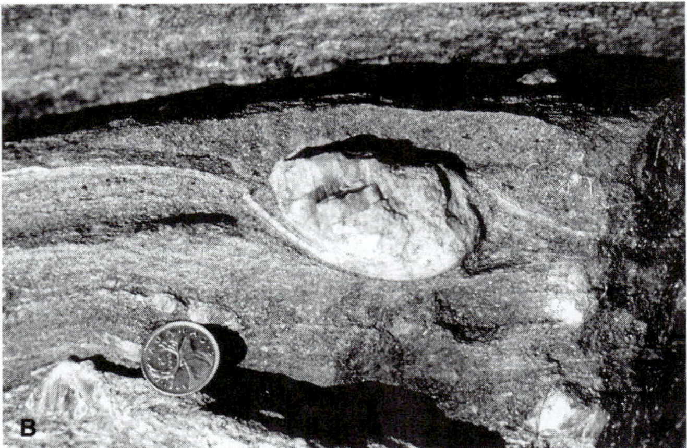

Figure 9.49 (*A*) Inclusion of mafic dike within granite mylonite, Great Slave Lake shear zone, Canada. Pressure shadows of melt flank the inclusion, reflecting dextral shear. (Photograph by Simon Hanmer.) (*B*) Inclusion with coiled wings, Sinistral shear. (Courtesy of Carol Simpson.)

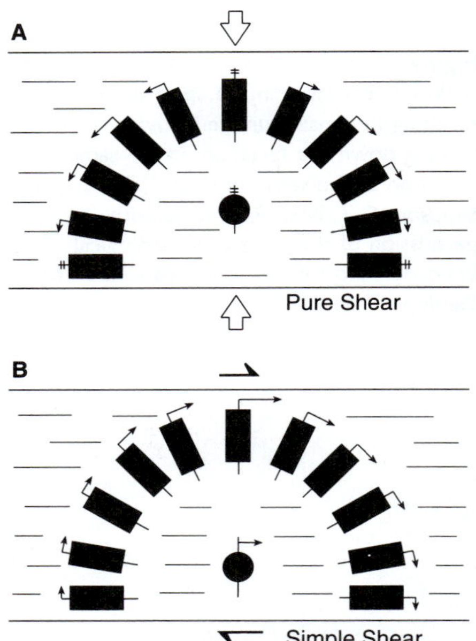

Figure 9.50 Rotation rates of inclusions as a function of orientation. The sense of rotation (clockwise versus counterclockwise) for each inclusion is shown by the direction of the attached arrow, with the relative rotation rates being proportional to the lengths of the arrows. Lines with double ticks do not rotate. (*A*) In a zone of coaxial strain, in this case pure shear, inclusions oriented parallel to either instantaneous stretching axis (parallel to and perpendicular to the shear zone) do not rotate. An equant inclusion (center of figure) does not rotate. Rotation rates are fastest for inclusions oriented at 45° to the shear zone. Note that some inclusions rotate clockwise, whereas others rotate counterclockwise. Rotation rates shown are for rigid inclusions with a 2 : 1 aspect ratio; longer inclusions may rotate faster. (*B*) In a zone of noncoaxial simple shear, all rigid inclusions, including equant ones, rotate in the same direction as the sense of shear. The slowest rate is for inclusions oriented nearly parallel to the shear zone, and the fastest rate is for those oriented perpendicular to the zone. Nonrigid, passive markers (not shown) do not rotate if they are strictly parallel to the shear zone. Equations are presented in Ghosh (1993).

For coaxial deformation, a *rigid, equant* inclusion does not rotate, but an *inequant* or *deformable* inclusion *may* rotate, depending on its orientation (Figure 9.50*A*). For coaxial deformation and an initially random distribution of orientations of inequant or deformable inclusions, equal numbers should rotate with clockwise and counterclockwise senses. An individual inclusion, therefore, may have an asymmetry reflecting its sense of rotation, but the sense of rotation for a number of inclusions should be inconsistent and statistically symmetric with respect to the shear zone.

The story is different for a noncoaxial shear zone. During simple shear, inclusions will rotate with a uniform sense but at different rates, depending on their shape, orientation, and rigidity compared to the matrix (Figure 9.50*B*). Shear-sense indicators recording the rotation of a single inclusion and of the entire population of inclusions should be asymmetric and should reflect the overall shear sense of the zone.

Pressure Shadows

When inclusions are strong compared to the matrix, they help shield the matrix on the flanks of the inclusion from strain. These shielded areas, or **pressure shadows**, are wedge-shaped areas composed of less deformed matrix or of minerals that grew or recrystallized during deformation (Figure 9.51). Most pressure shadows are microscopic, and those visible in outcrop are typically less than 1 cm long. They are thickest adjacent to the associated inclusion and taper outward, commonly becoming aligned parallel to or at a low angle to foliation and lineation in the surrounding matrix.

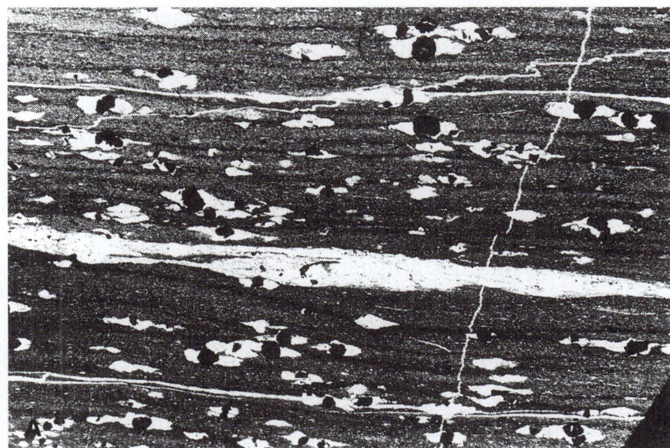

Figure 9.51 Pressure shadows. (*A*) Pressure shadows of quartz around pyrite grains in chloritic shist, Caribou sulfide deposit, Bathurst, New Brunswick, eastern Canada. (Photograph by G. H. Davis.) (*B*) Fibers adjacent to rigid pyrite grains in limestone, Helvetic nappes, Engelberg, Switzerland. (Photograph by Carol Simpson.)

Pressure shadows tend to be the locus of crystallization of quartz, calcite, chlorite, and other materials that are relatively mobile during deformation. These materials are deposited either as void fillings or as metamorphic–metasomatic replacements. Many pressure shadows, especially those of quartz, contain aligned fibrous or platy minerals connecting the inclusion and matrix (Ramsay and Huber, 1987).

Growth of minerals in pressure shadows largely occurs within zones of extension and decoupling between the rigid inclusion and the flowing matrix. Mineral growth occurs gradually, accompanying each small increment of extension. For kinetic reasons, it may be easier to continue growing an existing crystalline grain than it is to form a new grain. As deformation proceeds, therefore, existing grains grow into long fibers or plates. Such fibers can grow from grains in the inclusion, from grains in the matrix, or as new grains unrelated to those in either the inclusion or the matrix.

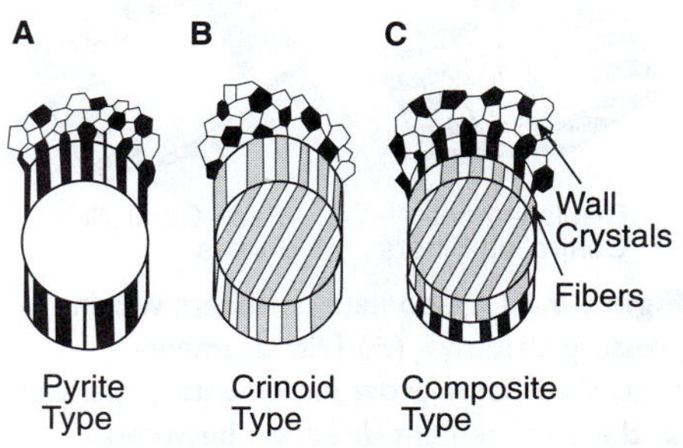

Figure 9.52 Types of pressure shadows and their fibers: (*A*) pyrite-type shadow; fibers grow in continuity with grains of the matrix, inward toward the inclusion; (*B*) crinoid-type shadow, in which fibers grow outward from the inclusion; (*C*) composite type. [Reprinted from *The techniques of modern structural geology, v. 1: strain analysis*, J. G. Ramsay and M. I. Huber (1983), © by Harcourt Brace and Company Limited, with permission.]

Types of Pressure Shadows

The style of fiber growth has been used to define three different types of pressure shadows (Figure 9.52): pyrite, crinoid, and composite (Durney and Ramsay, 1973).

A **pyrite type** of pressure shadow (Figure 9.52*A*) is composed of material that is mineralogically similar to the matrix and different from the inclusion. Fibers grow in crystallographic continuity with the *matrix* grains. Grain growth takes place at the *inner* ends of the fibers, along the *fiber–inclusion contact*. These fibers appear to grow from the *matrix inward* toward the inclusion, a process called **antitaxial** growth. To remember the *antitaxial* growth direction, just think of growth *inward* as being opposite to (''anti'') the way most things grow, including trees and us.

A **crinoid type** of pressure shadow consists of material that is similar to the inclusion rather than the matrix. Fibers grow in crystallographic continuity with grains in the *inclusion* (Figure 9.52*B*). They lengthen *outward* via grain growth at the *outer* edges of the fibers, at the *fiber–matrix contact*. Such **syntaxial** growth is common where calcite fibers grow from a crinoid or other fossil fragment. Syntaxial, or outward, growth is just the opposite of antitaxial growth.

A **composite type** of pressure shadow incorporates aspects of both the pyrite and crinoid types (Figure 9.52*C*). Such shadows contain an *outer* fringe of *antitaxial* fibers growing *inward* from grains in the matrix, and an *inner* zone where *syntaxial* fibers are growing *outward* from the inclusion (Ramsay and Huber, 1983).

Other pressure shadows represent newly formed mineral grains that are unrelated to grains in the inclusion or the matrix. This generally requires an external derivation of the constituents of the fibers—for example, redistribution of material from one rock type to another as a result of fluid flow during deformation.

Geometry of Fibers

The geometry of fiber growth in pressure shadows can be controlled by the incremental displacement direction between the inclusion and its matrix, or by the orientation of the faces of the inclusion.

Fibers within pressure shadows commonly grow parallel to the local displacement direction between the inclusion and its matrix (Figure 9.53*A*). Such fibers are called **displacement-controlled fibers** and, because they record the displacement history between the inclusion and its matrix, they directly provide information about the progressive strain history of the rock. The shape of the inclusion controls the distribution, but not the orientation, of the fibers.

In other cases, fiber growth is controlled by the orientation of the faces of the inclusion, rather than local displacement directions (Figure 9.53*B*). Such **face-controlled fibers** generally grow perpendicular to the faces of the inclusion. Fibers growing from two adjacent faces meet at a **suture line** that extends outward from the corners of the inclusion (Figure 9.53*B*). The geometry of individual fibers does not provide a direct record of progressive displacement, but the suture line tracks the displacement path of the associated corner of the inclusion as it was progressively pulled away from the matrix. Short fiber segments truncated by the suture line represent fibers that stopped growing when they were no longer in contact with the inclusion.

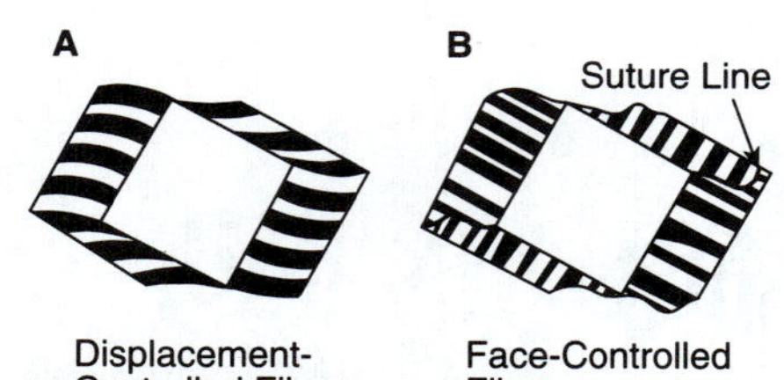

Figure 9.53 Geometry of fibers within pressure shadows. (*A*) Displacement-controlled fibers grow incrementally, parallel to the displacement direction between inclusion and matrix. (*B*) The geometry of face-controlled fibers is governed by the faces of the inclusion. The suture line records the progressive displacement of the corner of the crystal away from the matrix. [Reprinted from *The techniques of modern structural geology, v. 1: strain analysis*, J. G. Ramsay and M. I. Huber (1983), © by Harcourt Brace and Company Limited, with permission.]

Determining Sense of Shear from Pressure Shadows and Fibers

Pressure shadows form incrementally, generally approximately parallel to the axis of maximum *instantaneous* extension. They should be parallel to the shear zone for *coaxial* deformation, but they will grow obliquely to the shear zone for *noncoaxial* simple shear (Figure 9.54), leaning over in the sense of shear.

The geometry of fibers within the shadows is also a shear-sense indicator (Figure 9.54). Displacement-controlled fibers grow incrementally, parallel to the direction of maximum instantaneous extension, and so lean over in the sense of shear. Face-controlled fibers can provide a sense of which way an inclusion is rotating during deformation, and therefore also reflect the sense of shear.

Beautiful, but geometrically complex pressure shadows and fibers can arise when the inclusion rotates during deformation (Figure 9.55). Etchecopar and Malavieille (1987) have successfully used computer models to explain how such ornate, batlike creatures grow. Other geometries can develop around inclusions that rotate, and these require some care when interpreting sense of shear (Hanmer and Passchier, 1991).

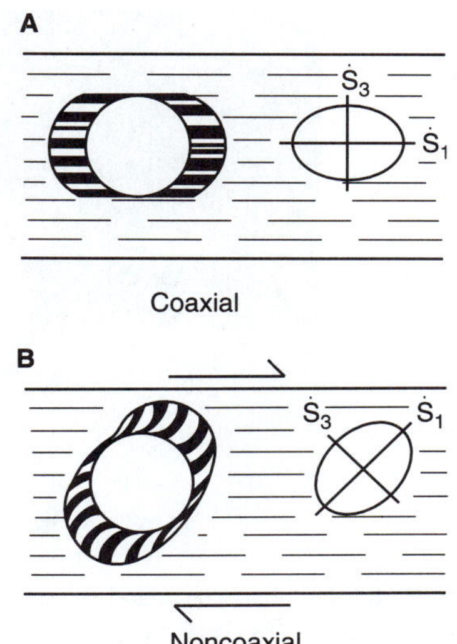

Figure 9.54 Pressure shadows, fibers, and instantaneous strain ellipses for coaxial and noncoaxial deformation. (*A*) Pressure shadows and fibers grow parallel to the shear zone during coaxial deformation, but (*B*) are inclined over in the direction of shear for noncoaxial deformation. In both cases, the shadows and fibers grow parallel to $\dot{S}_1$, the long axis of the instantaneous strain ellipse. [In part from *The techniques of modern structural geology, v. 1: strain analysis*, J. G. Ramsay and M. I. Huber (1983), © by Harcourt Brace and Company Limited, with permission.]

Porphyroclasts and Porphyroblasts

Ductile shear zones commonly contain relatively rigid grains of one mineral within a more strongly deformed, fine-grained matrix having a different mineralogy. Such rigid grains may represent relics from the original protolith, or they may be new metamorphic grains that grew during or after deformation. Grains interpreted to be *relics from the protolith* are called **porphyroclasts** because they commonly represent fragments or clasts of original phenocrysts or detrital grains (Figure 9.56*A*). Many ductilely deformed granites, for example, contain large feldspar porphyroclasts, which represent original feldspar phenocrysts, scattered in a more deformed, fine-grained matrix of quartz, feldspar, and mica. Likewise, metavolcanic and metasedimentary rocks may contain deformed original phenocrysts or detrital clasts of feldspar, mafic minerals, or quartz, which "float" in a strongly deformed, fine-grained, micaceous matrix. Under the proper conditions, porphyroclasts may flow ductilely or recrystallize during deformation.

If the large grains instead are new metamorphic grains that *grew during or after deformation,* they are called **porphyroblasts** (Figure 9.56*B*). Porphyroblasts that grow during deformation are **synkinematic** and those that grow after deformation are **postkinematic**. Porphyroblasts, once formed, may be very rigid compared to the matrix, as is the case for garnet. Other porphyroblasts, including andalusite and muscovite, are more deformable, and their response to deformation may be similar to that of the matrix. The distinction between a deformable and a rigid inclusion is important, because the two types of inclusion may be associated with different suites of small-scale structures. Deformable porphyroblasts are less likely to be associated with pressure shadows and may be less instructive as shear-sense indicators.

Some rigid porphyroblasts contain sigmoidal or spiral **inclusion trails** of another mineral (Figure 9.56*B*, and *C*). These trails represent a succession of small crystals that were entrapped as the porphyroblast grew and rotated during deformation. The shape of the inclusion trail reveals which way the inclusion rotated relative to the matrix, and documents the sense of shear. Alternatively, some porphyroblasts may grow around and entrap fabric formed during an earlier deformation event or successive increments of one event, possibly providing the only record of fabrics that have been largely or totally obliterated by continued deformation in the matrix (Bell, 1985).

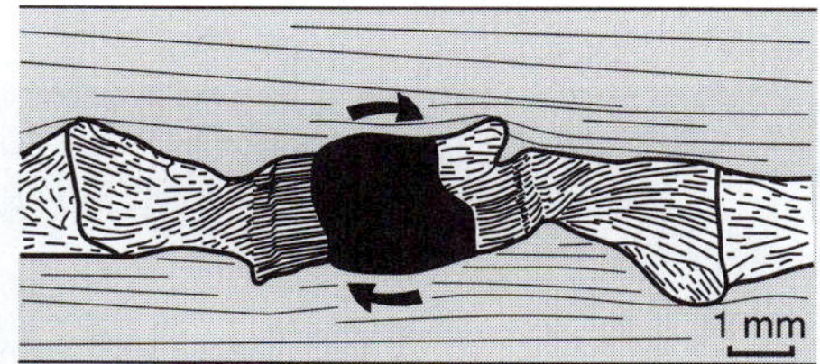

Figure 9.55 Bat-winged pressure shadow around pyrite. (Reprinted with permission from *Journal of Structural Geology*, v. 9, Etchecopar, A. and Malavieille, J., Computer models of pressure shadows; a method for strain measurement and shear sense determination, 1987, Elsevier Science Ltd., Pergamon Imprint, Oxford, England.]

As porphyroclasts, porphyroblasts, and other inclusions interact with the shear zone matrix, their outer edges may preferentially accumulate crystalline strain and recrystallize (Simpson and Passchier, 1986; Passchier, 1987, 1994). The recrystallized **mantle** is composed of numerous

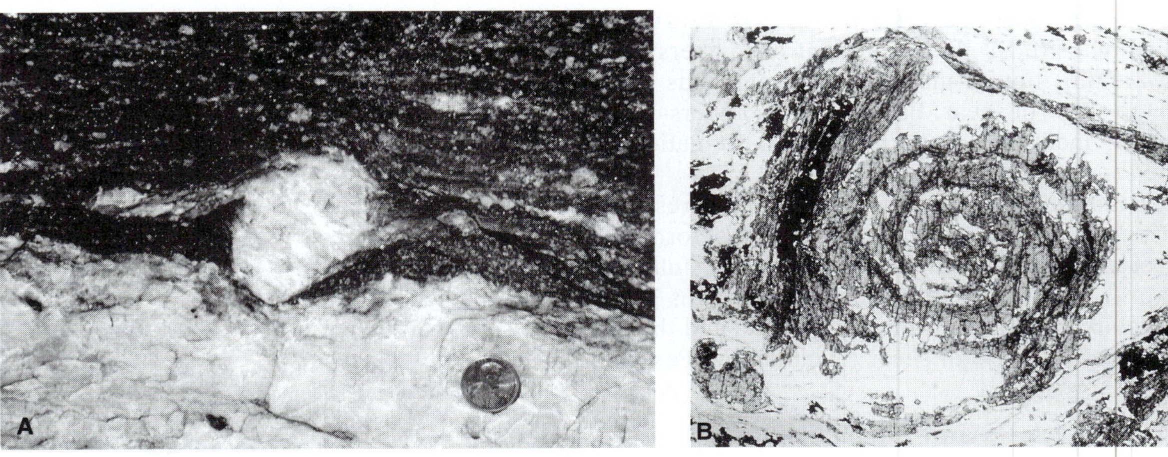

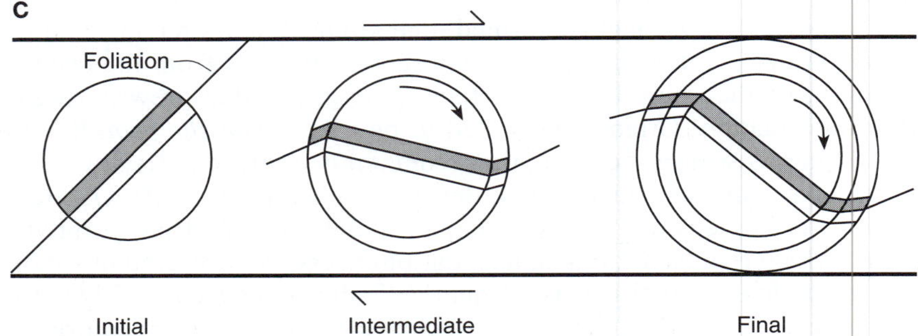

Figure 9.56 Porphyroclasts and porphyroblasts. (*A*) Feldspar porphyroclast derived from phenocryst in mylonitic granite, Santa Catalina Mountains, Arizona. (Photograph by S. J. Reynolds.) (*B*) Garnet porphyroblast with spiral inclusion trails. [From Hanmer and Passchier (1992). Courtesy of the Geological Survey of Canada and the authors]. (*C*) Sequential rotation and growth of a porphyroblast, from left to right. [After Ghosh (1993). Published with permission of Butterworth-Heinemann Ltd.]

small crystals and is weaker than the monocrystalline core. As a result, the mantle tends to be sheared out away from the porphyroclast, forming "wings" or "tails" that extend out into the matrix (Figure 9.57). The centerlines of wings on opposite sides of the porphyroclast may be straight and aligned, or they may be curved and markedly asymmetric with respect to the porphyroclast.

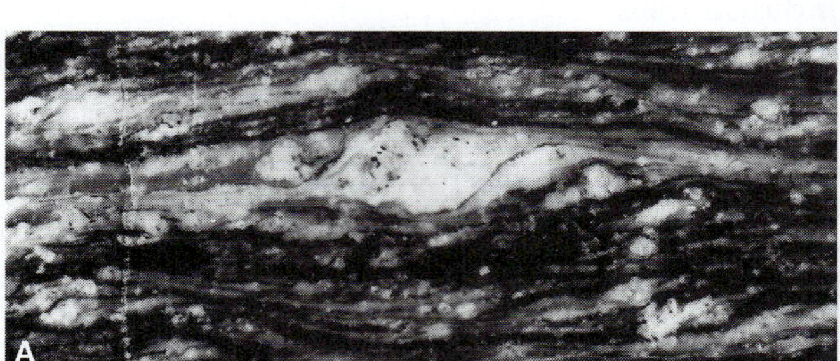

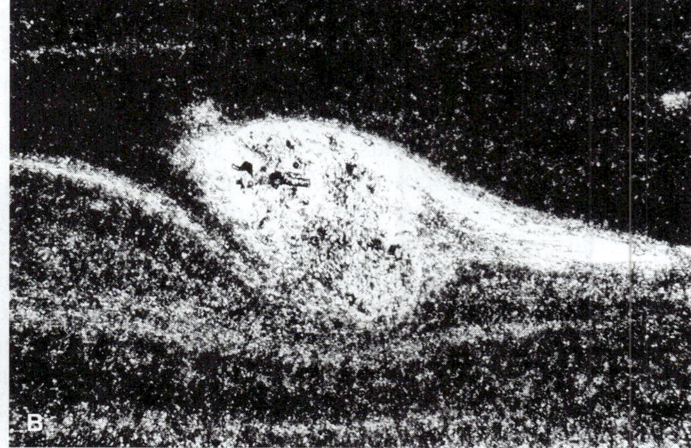

Figure 9.57 Sigma- and delta-type porphyroclasts. (*A*) σ-type porphyroclast in polished slab of mylonitic granite, Santa Catalina Mountains, Arizona. Dextral shear. (Photograph by S. J. Reynolds.) (*B*) δ-type porphyroclast in mylonitic granite. Sinistral shear. (Photograph by Carol Simpson.)

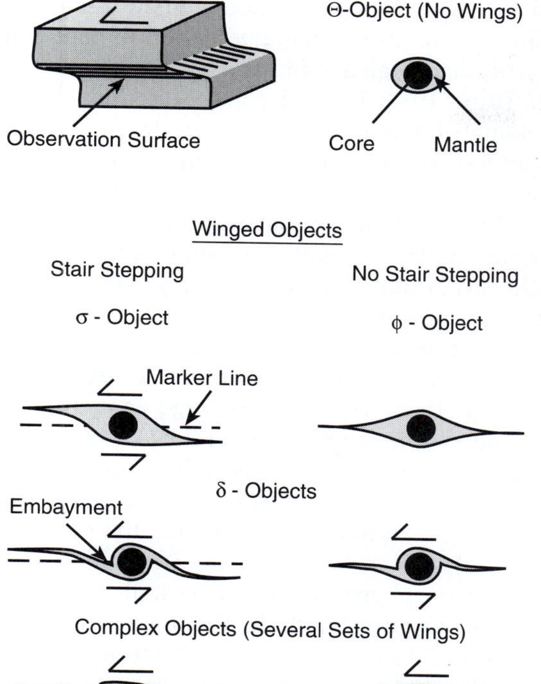

Figure 9.58 Greek squadron of winged porphyroclasts. The darkly shaded center is the rigid core and the more deformable recrystallized mantle is lightly shaded. Views are cross sections as observed on the sense of shear plane. (Reprinted with permission from *Journal of Structural Geology*, v. 16, Passchier, C. W., Mixing in flow perturbations: a model for development of mantled porphyroclasts in mylonites, 1994, Elsevier Science Ltd., Pergamon Imprint, Oxford, England.].

The Greek Squadron of Winged Porphyroclasts

Five main geometries of winged porphyroclasts have been identified based on whether the centerlines of the wings are straight or curved, and symmetrical or asymmetrical with respect to an imaginary reference line through the center of the clast (Figure 9.58). Four of these have been named after Greek letters whose shape they resemble (Simpson and Schmidt, 1986; Passchier, 1994). **Theta (*θ*)** objects have round to elliptical mantles, but no real wings. Wings on **phi (*ϕ*)** type porphyroclasts have fairly straight centerlines and are largely symmetrical with respect to the porphyroclast; they resemble the Greek letter *phi* turned onto its side. **Sigma (*σ*)** type have wings with gently curved centerlines and are asymmetric; the wing extends off the top of one side and the bottom of the opposite side, a pattern referred to as "stair stepping" (Figure 9.57*A*, 9.58). **Delta (*δ*)** type porphyroclasts have strongly curved wings that are asymmetric with respect to the porphyroclast (Figure 9.57*B*). The final type consists of **complex porphyroclasts** with several sets of wings. The general shear-sense implication of each type is illustrated in Figure 9.58. In three dimensions, wings are lenticular to rodlike (Figure 9.59).

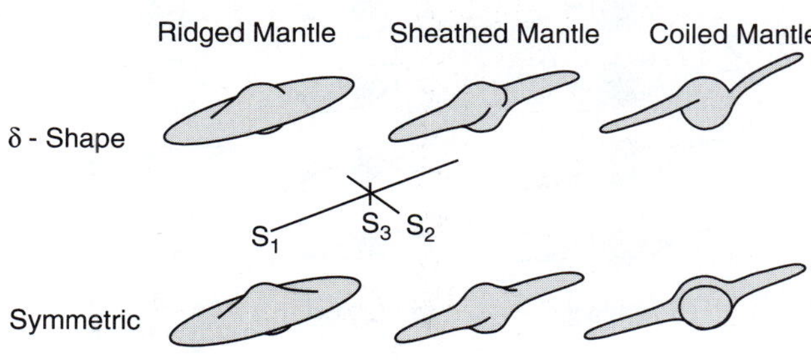

Figure 9.59 Three-dimensional view of winged porphyroclasts. (Reprinted with permission from *Journal of Structural Geology*, v. 15, Passchier, C. W., and Sokoutis, D., Experimental modelling of mantle porphyroclasts, 1993, Elsevier Science Ltd., Pergamon Imprint, Oxford, England.)

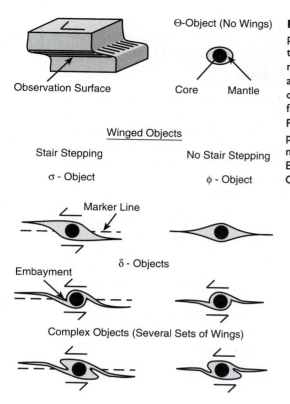

The formation of the different wing types has been studied in natural rocks and in experimental simulations using analogue materials that can be deformed at low temperature and high strain rates (Passchier, 1987, 1994; van den Driessche and Brun, 1987; Passchier and Sokoutis, 1993). From the ingenious experiments of Passchier (1994), we learn that all wings initially develop parallel to the maximum *instantaneous* extension direction and subsequently become drawn out in the direction of maximum *finite* extension (Figure 9.60). Several distinct geometries of wings evolve with additional strain, depending on the amount of strain and the rate of recrystallization relative to the strain rate. The rate of recrystallization is key because it controls the relative proportions of mantle material to rigid core. A rigid core, like a huge jagged outcrop in the middle of a stream, causes flow to diverge around it, forming a little protected "eddy" of slowly moving material around it. If the rigid core is large compared to its mantle, the eddy protects the mantle from the main flow of the shear zone, whereas a relatively smaller core does not cause a big enough eddy to protect its mantle. Unlike the fixed, jagged outcrop, however, both the rigid core and the material inside the eddy rotate, reflecting the sense of shear of the zone.

According to the experiments, a porphyroclast that recrystallizes slowly or not at all becomes a θ object (Figure 9.60). In this case, the eddy protects the relatively small mantle. At rates of recrystallization that are higher, but still low relative to the strain rate, the mantle is smeared out into wings, which become rolled along with the porphyroclast to form δ-type wings. Where recrystallization is fast compared to the strain rate, the supply of recrystallized mantle material is fast enough to permit each wing to develop into a continuous wedge-shaped mass that is smeared out from the porphyroclast, either symmetrically (ϕ-type) or asymmetrically (σ-type).

The shape of the inclusion also influences the type of wings that form, in part because it affects the rate at which the inclusion rotates compared to the matrix. For example, δ-type wings are commonly associated with equant inclusions, whereas complex wings are more likely to be found around elliptical ones. The experimental results of Passchier and others are exciting because they help us appreciate why so many different kinds of winged porphyroclasts may "fly" within a single shear zone.

Foliation Fish

A different kind of inclusion forms when part of the shear zone becomes structurally isolated between discrete shear bands, fractures, or some other type of structural discontinuity (Figure 9.61). Once formed, such discontinuity-bounded material may behave mechanically differently from the surrounding matrix, being deformed into lenticular or sigmoidal masses

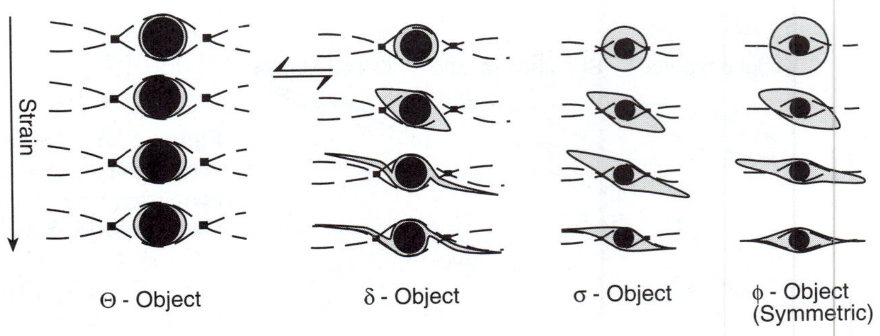

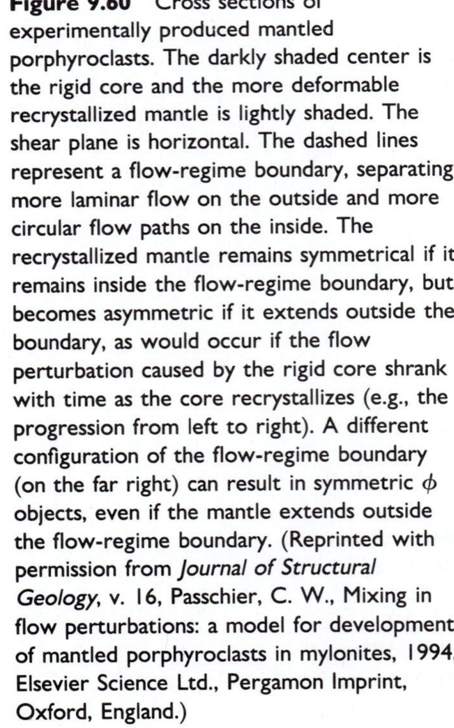
Figure 9.60 Cross sections of experimentally produced mantled porphyroclasts. The darkly shaded center is the rigid core and the more deformable recrystallized mantle is lightly shaded. The shear plane is horizontal. The dashed lines represent a flow-regime boundary, separating more laminar flow on the outside and more circular flow paths on the inside. The recrystallized mantle remains symmetrical if it remains inside the flow-regime boundary, but becomes asymmetric if it extends outside the boundary, as would occur if the flow perturbation caused by the rigid core shrank with time as the core recrystallizes (e.g., the progression from left to right). A different configuration of the flow-regime boundary (on the far right) can result in symmetric ϕ objects, even if the mantle extends outside the flow-regime boundary. (Reprinted with permission from *Journal of Structural Geology*, v. 16, Passchier, C. W., Mixing in flow perturbations: a model for development of mantled porphyroclasts in mylonites, 1994, Elsevier Science Ltd., Pergamon Imprint, Oxford, England.)

Figure 9.61 Foliation fish: (*A*) With back-rotated foliation, within mylonitic gneiss, Chemihuevi Mountains, southeastern California. Sinistral shear. (Photograph by S. J. Reynolds.) (*B*) Back-rotated foliation within fish. (Reprinted with permission from *Journal of Structural Geology*, v. 14, Stock, P., A strain model for antithetic fabric rotation in shear band structure, 1992, Elsevier Science Ltd., Pergamon Imprint, Oxford, England.)

called **tectonic fish** or **foliation fish**. As you might suspect, the sigmoidal shape of foliation within tectonic fish reflects the sense of shear (Figure 9.61). In many cases, the foliation in tectonic fish undergoes differential rotation, becoming "back-rotated" with respect to foliation in the surrounding matrix (Dennis and Secor, 1987; Hanmer and Passchier, 1991).

Fractured and Offset Grains

Porphyroclasts and other rigid inclusions may accommodate deformation by becoming sliced up by small-scale or grain-scale faults (Figure 9.62). The sense of displacement on such faults can be a shear-sense indicator, but not a totally reliable one. When such faults are at low angle to the shear zone, they generally have a sense of shear that is **synthetic** (i.e., the same) as the shear zone (Figure 9.62*B*). Like shear bands, synthetic faults can displace the inclusion with either a normal or a thrust sense when viewed in the SOS plane. When the faults are at a high angle to the shear zone, they commonly have a sense of shear that is **antithetic**, or opposite, to the shear zone (Figure 9.62*B*). This antithetic slip helps accommodate synthetic rotation of the inclusion, much as the slip surfaces between books on a shelf permit the books to topple over when the bookend is removed, or knocked over by the cat (Figure 9.62*C*).

The kinematic significance of such features can be ambiguous, hence should not be the sole basis for interpreting the sense of shear (Hanmer and Passchier, 1991). For example, a structure that can be mistaken for an offset inclusion arises where adjacent inclusions rotate into one another, piling up in a process called **tiling** (Figure 9.63). We might get the wrong sense of shear if we misinterpret tiled grains as a single offset grain.

Veins

As we map and analyze shear zones, we are generally impressed with the abundance of veins (Figure 9.64). Most shear zone–related veins contain quartz and calcite, but feldspar, mica, iron oxide, and gypsum are also common in certain settings and rock types. These minerals are deposited

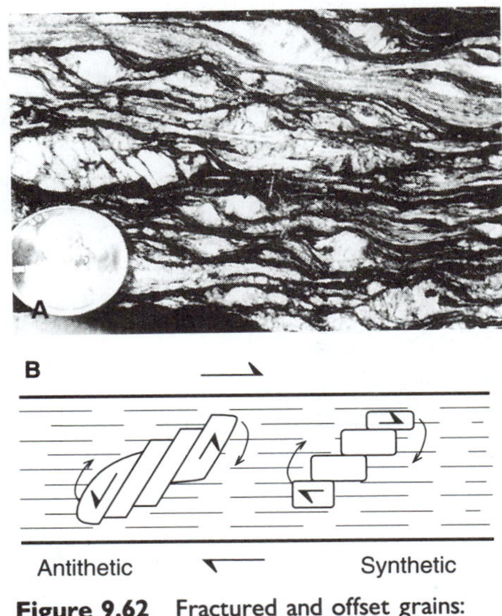

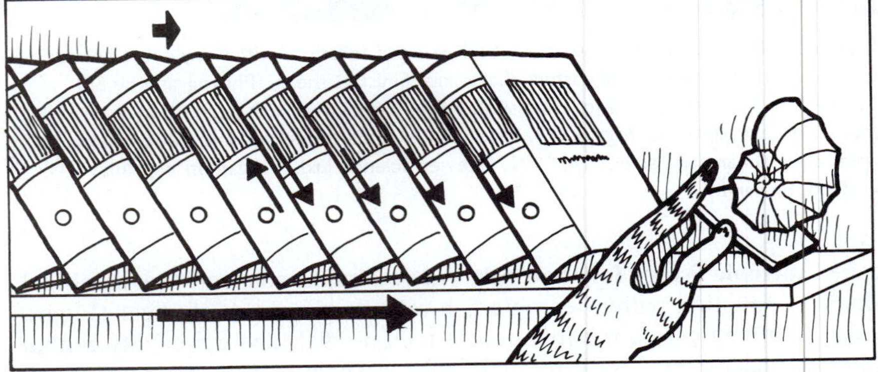

Antithetic Synthetic

Figure 9.62 Fractured and offset grains: (*A*) Offset feldspar grain (directly above coin in lower left corner) in S-C mylonite. (Photograph by Carol Simpson.) S-C fabrics document sinistral shear. (*B*) Synthetic and antithetic offset of grains. Curved arrows adjacent to grains indicate direction of rotation of long axis of grain. (*C*) Antithetic rotation of books due to "cat tectonics." (Artwork by D. A. Fisher.)

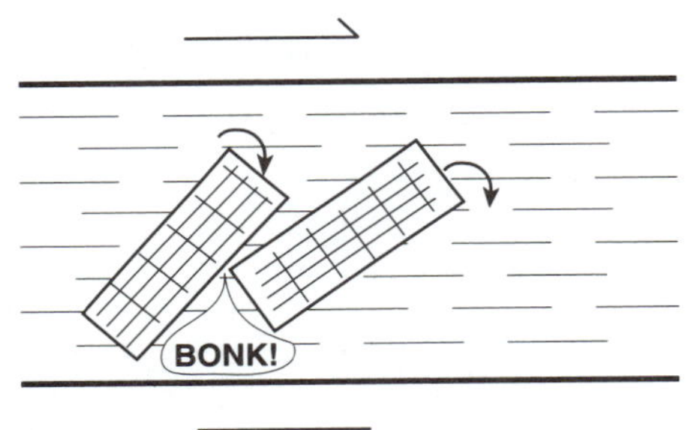

Figure 9.63 "Tiling," ... when one grain rotates into another one.

from the fluids that helped "prop" open the fracture filled by the vein material.

Veins can be excellent and reliable shear-sense indicators because their orientations are commonly controlled by the *instantaneous* stretching axes. Most veins form *perpendicular* to the axis of maximum *instantaneous* extension, because this is the direction in which tension fractures form. Veins should be oriented perpendicular to foliation and lineation for coaxial deformation, but not for noncoaxial deformation (Figure 9.65). For simple shear, veins will form at 45° to the shear zone—that is opposed to the direction of foliation and to the inclination of most other shear-sense indicators we have examined so far. Once formed, the veins may be shortened and partially rotated over, toward the direction of shear (Figure 9.66). New veins, however, will continue to form in the same

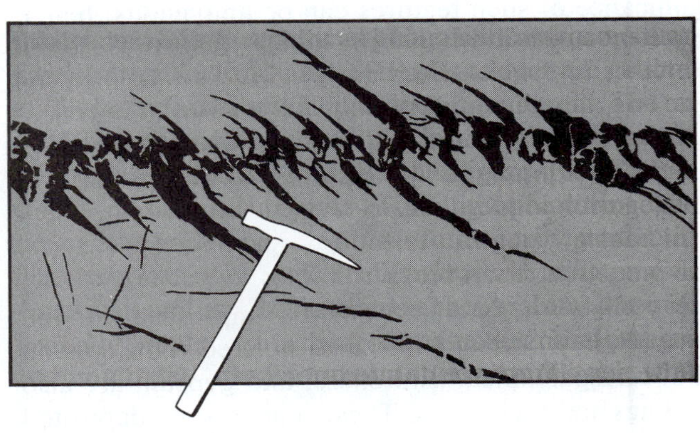

Figure 9.64 En echelon, sigmoidal quartz veins defining a dextral shear zone. (Reprinted with permission from *Journal of Structural Geology*, v. 2, Ramsay, J. G., Shear zone geometry: a review, 1980, Elsevier Science Ltd., Pergamon Imprint, Oxford, England.)

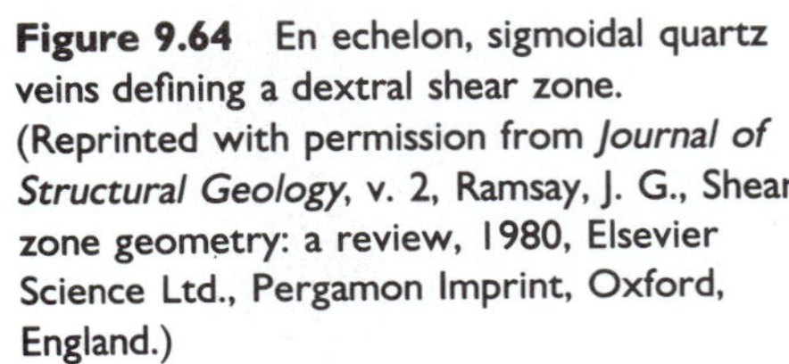

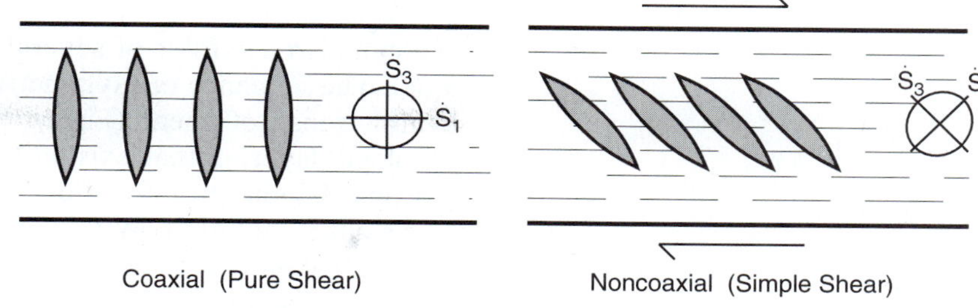

Figure 9.65 Orientation of veins compared to the instantaneous stretching axes for coaxial and noncoaxial deformation. Veins form parallel to the maximum shortening rate ($\dot{S}_3$) and perpendicular to the maximum stretching rate ($\dot{S}_1$).

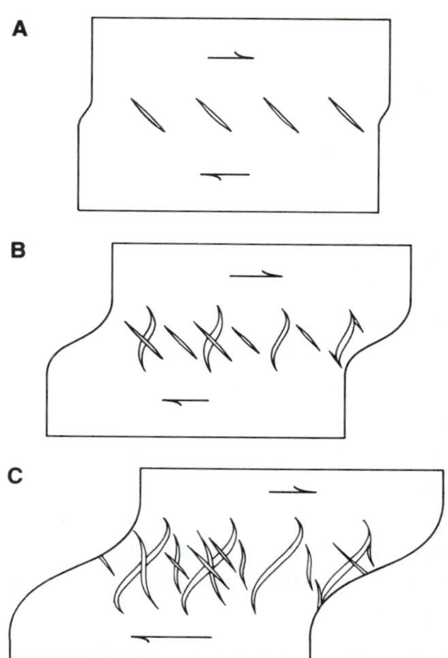

Figure 9.66 Progressive formation and folding of en echelon veins within a zone of simple shear. Tension fractures are initially oriented at 45° to the shear zone and progressively rotate in the direction of the sense of shear during subsequent deformation. Later veins form in the original orientation. (From Durney, D. W. and Ramsay, J. G., Incremental strains measured by syntectonic crystal growths, *in* DeJong, K. A. and Scholten, R. (eds.), *Gravity and tectonics*, 1973. Reprinted by permission of John Wiley & Sons, Inc.)

original orientation, with their tips aligned in accordance with the instantaneous stretching axes, *irrespective of changes in the orientation of the finite strain ellipse with time*. Multiple generations of earlier, deformed veins and younger, less deformed ones are commonly present in a single outcrop, and the deflection of the older veins and how they are buckled are kinematic indicators (Figures 9.64, 9.66). A curved vein geometry can also locally arise because of complex stress fields created by interactions between the tips of adjacent, closely spaced veins.

Veins, like pressure shadows, may contain crystal fibers that grow one increment at a time during opening of the vein (Figure 9.67; see Chapter 5). The fibers grow via a **crack–seal mechanism**: a small increment of extension fracturing and opening of the vein is followed by growth of fibers from existing crystals along the walls of the fracture (Ramsay, 1980b). The new fiber increment will be in optical continuity with the existing crystal, but their boundary may be marked by a thin train of fluid inclusions that were trapped as the fiber grew into the fluid that filled the fracture. The fibers grow parallel to the direction of maximum instantaneous (incremental) extension and therefore lean over in the direction of shear. The fibers may be straight or may exhibit curved shapes that record shearing parallel to the walls or changes in the orientation of the vein with respect to the instantaneous stretching axes (Figure 9.67*B*).

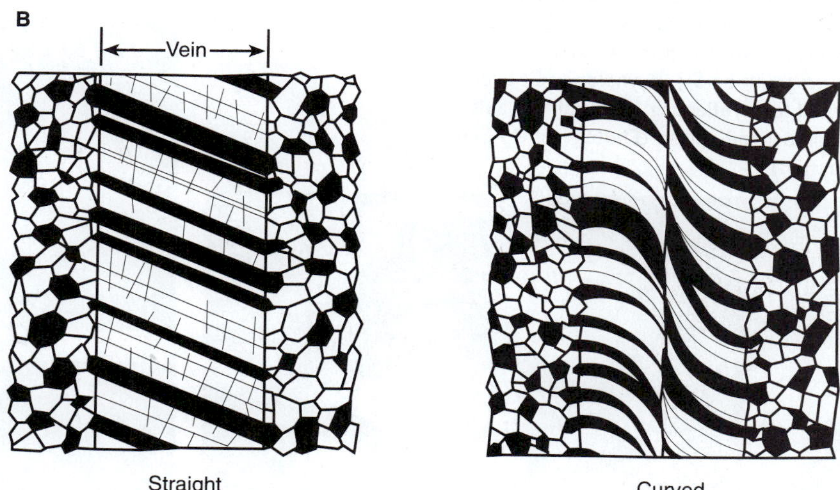

Figure 9.67 Fibers in veins: (*A*) Photomicrograph of antitaxial calcite fibers in veins within Martinsburg Formation, western Maryland. (Photograph by Carol Simpson.) (*B*) Straight and curved syntaxial fibers, which grow inward from grains in the wall rocks. [Reprinted from *The techniques of modern structural geology, v. 1, strain analysis*, J. G. Ramsay and M. I. Huber (1983), © by Harcourt Brace and Company Limited, with permission.]

Folds

We can also use folds of several types as shear-sense indicators (Figure 9.68). The vergence of asymmetric **intrafolial folds** commonly reflects the overall sense of shear (Figure 9.68*A*). Intrafolial folds may be folded lithologic layering that existed before deformation (e.g., bedding, metamorphic banding) or foliation that formed during the earlier phases of shear zone evolution. With progressive deformation, either type of fold may be rotated into subparallelism with lineation, at which point it becomes undiagnostic in terms of sense of shear. Crenulations and other folds that *cut across* the shear zone fabric should not be used to determine shear sense of the shear zone because they may be caused by a younger and unrelated episode of deformation. The geometry and asymmetry of folds in a single outcrop may not accurately indicate the sense of shear if they are (1) parasitic to a larger scale fold structure, or (2) controlled by the *pre–shear zone orientation* of the folded layer relative to the shear zone. In our experience, interpreting folds in shear zones can be a little tricky unless the folds show a consistent asymmetry over a large region.

Sheath folds are unusual noncylindrical folds formed in some strongly deformed rocks (Carreras et al., 1977; Quinquis and others, 1978; Cobbold and Quinquis, 1980; Ramsay, 1980a). They have strongly curved hinge lines and a rounded, conical shape that looks like a wind sock (Figure 9.68*B*). In planar outcrop faces, they may appear as elliptical "eyes" defined by rings of lithologic layers (Figure 9.68*C*). Their long axis is generally parallel to lineation, and their shape reflects the sense of shear (Figure 9.69). They are formed by lateral variations in particle velocities within the flow regime, much like those that exist between the center of a flowing river and its banks. To envision how a sheath fold forms, imagine the shape of a curtainlike fishing net that is strung across a river, anchored at both banks and on the bottom. The faster flow velocities in the center of the river would make that part of the net bulge farther downstream, and the bulge would point in the direction of flow. It seems that rivers and fish are marvelous analogies to shear zones.

Figure 9.68 Folds in shear zones. (*A*) Asymmetric interfolial fold in mylonitic granite, Santa Catalina Mountains, southern Arizona. Dextral sense of shear indicated by fold is supported by the asymmetrical wings of the overlying, round porphyroclast of feldspar. (Photograph by G. H. Davis.) (*B*) Looking down the axis of a tubular shealth fold, Boyer Gap area, Dome Rock Mountains, west-central Arizona. (Photograph by S. J. Reynolds.) (*C*) Eye-shaped fold, reflecting a curved fold axis, within banded gneiss. (Photograph by Simon Hanmer.)

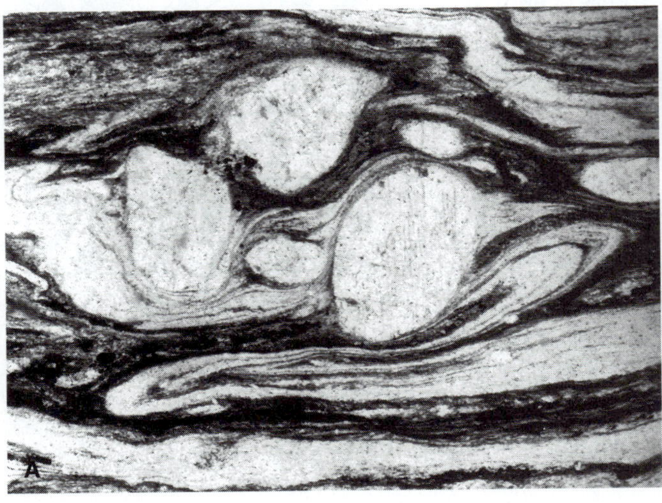

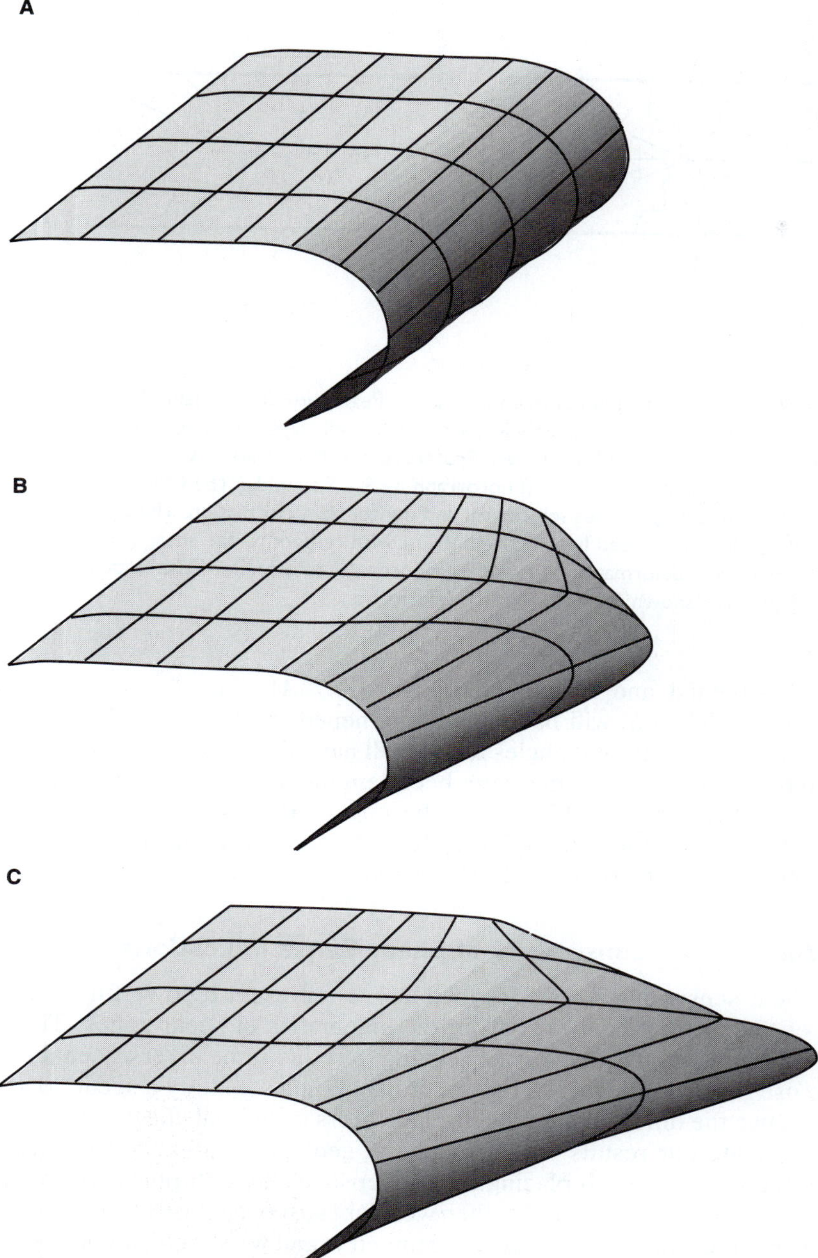

Figure 9.69 Formation of sheath folds. Differential flow causes the middle part of this fold to progressively bow out, forming a sheath fold.

Folds with strongly curved axes can also reflect superposed folding due to multiple episodes of deformation or to progressive deformation during a single event (Ramsay, 1967; Ramsay and Huber, 1987). Overprinting fabrics, if present, will usually be evident after careful work in a number of outcrops. As with any aspect of structural geology, it is always best to see as many outcrops, representing the broadest geographic spread possible, before jumping to conclusions.

Orientation of Folded and Boudinaged Layers

Finally, we can often infer the sense of shear by comparing the orientations of layers or other features that have been shortened versus those that

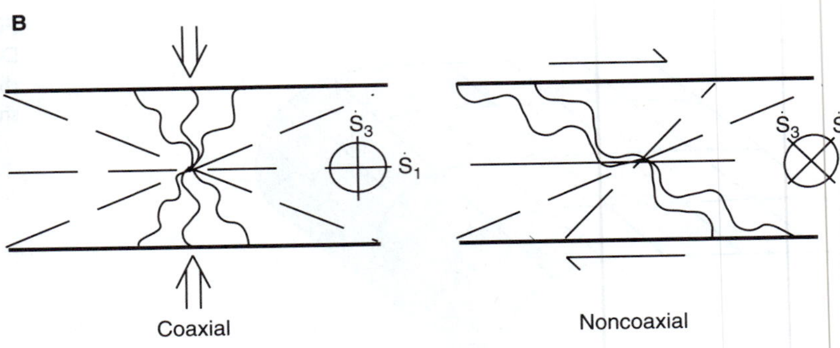

Figure 9.70 Folded and boudinaged layers: (*A*) Pegmatite dikes oriented at a high angle to the gneissic foliation are folded, whereas a mafic layer and pegmatite dikes parallel to the foliation are boudinaged. Proterozoic Vishnu Schist in Clear Creek, Grand Canyon, northern Arizona. (Photograph by S. J. Reynolds.) (*B*) Orientations of folded versus boudinaged lines for coaxial and noncoaxial deformation. The distribution of boudinaged versus folded lines is symmetrical with respect to the shear zone and its normal for coaxial deformation, but not for noncoaxial deformation. The instantaneous strain ellipses are shown for each type of deformation.

were lengthened and boudinaged (Figure 9.70*A*). Any feature that is aligned parallel to S_1 will have been lengthened, perhaps by boudinage, whereas a feature at right angles to this will have been folded or otherwise shortened. The features that have been lengthened will lean over in the sense of shear (Figure 9.70*B*). At the end of this chapter, we see an example of using the orientation of folded versus boudinaged layers to reconstruct the type of strain in shear zones.

A Word About Consistency of Shear-Sense Indications

The shear-sense indicators described above represent a powerful collection of tools with which to analyze the kinematics of shear zones. Their diversity may seem a little overwhelming, but this same diversity enables us to use them confidently, albeit carefully. When we study a shear zone, we can use the different shear-sense indicators as mutual checks. We can also evaluate our results with the general geologic context of the shear zone. If our shear zone is placing older, deep level metamorphic rocks over younger, shallow-crustal rocks, then we might suspect a thrust history. But there are other ways to produce this geometry, and we should not discount the *observed* sense of shear if it appears inconsistent with the geometry. The observed sense of shear may be telling us that the shear zone started out as a thrust but has been reactivated with a normal sense of displacement. The recognition of such a series of geologic events may have major implications for our understanding of the tectonic evolution of the region or for modeling the distribution of petroleum or other natural resources.

Thankfully, most shear zones that have been studied in the last decade display a fairly consistent sense of shear, reflecting a clear *noncoaxial* component of deformation. Other shear zones have contradictory shear-sense indicators or lack obvious indicators, possibly because they record coaxial deformation, overprinting episodes of deformation with different senses of shear, or an usual strain regime. But even these should be decipherable with careful work and a trained eye. The complexities only add to the "shear joy" we feel in analyzing a shear zone and finally figuring it out.

THE GEOMETRY AND DISTRIBUTION OF STRAIN IN SHEAR ZONES

We now turn our attention to the *distribution* of strain in shear zones. The mere presence of a shear zone indicates that the distribution of strain was heterogeneous rather than homogeneous—strain is clearly localized preferentially in the shear zone relative to the wall rocks. But how about strain *within* the shear zone? Is it also heterogeneous? And if so, are the variations in strain at all systematic? How exactly does strain decrease from the shear zone into the wall rocks?

Thanks to the pioneering work of John Ramsay and others, we can address these questions and many more. In examining the geometry and distribution of strain in shear zones and their wall rocks, we will learn that strain, although heterogeneous, can vary in some remarkably systematic and useful ways. These systematic variations contain important clues about the history of a shear zone, including its sense of shear and amount of displacement, and the kinds of strain (simple shear, dilation, etc.) that were involved. We begin by exploring the geometry and implications of heterogeneous strain.

Heterogeneous Strain

For deformation to be **homogeneous**, strain must be uniform in orientation and magnitude throughout the entire volume of material being considered (Figure 9.71). Lines that were straight before deformation are straight after deformation, and any two lines that were parallel before deformation remain so during the entire deformation history. Circles are transformed into ellipses, and spheres become ellipsoids.

In **heterogeneous** deformation, strain is not uniform (Figure 9.71). Some parts of the rock mass are more strongly deformed than others. Lateral variation in strain causes some lines that were originally straight to become curved, and may rotate two lines out of parallelism. A circle may be deformed into the outline of a teardrop or a boomerang, and a sphere may be converted into a shape that would be right at home in the produce section of your neighborhood grocery store.

The Geometry of Heterogeneous Simple Shear

The distribution of strain within a zone of heterogeneous simple shear after two large increments of strain is shown in Figure 9.72*A*,*B*. Because strain in the zone is *heterogeneous*, we have subdivided the rock mass into a number of small **domains** that are small enough to allow us to consider the deformation within each domain to be more or less *homogeneous*. We represent the state of strain within each domain with a finite strain ellipse, the size, shape, or orientation of which will vary between different domains.

The shear zone is defined by a zone of concentrated deformation cutting horizontally through the center of the material. The wall rocks are undeformed away from the shear zone, but strain increases systematically inward toward the center of the zone. The finite strain ellipses become progressively more elliptical toward the center of the zone, without major discontinuities, recording a continuous strain gradient *perpendicular* to the shear zone. There is, however, no strain gradient *parallel* to the shear zone; strain ellipses are the same for all points that are the same distance from the center of the shear zone. The concept of a **strain gradient** is illustrated in Figure 9.73.

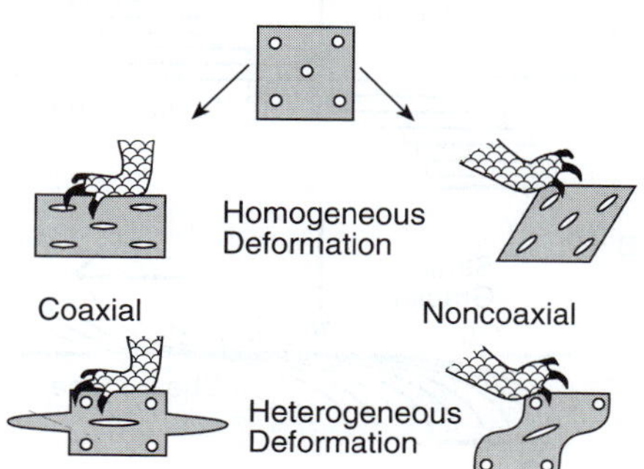

Figure 9.71 Homogeneous versus heterogeneous, and coaxial versus noncoaxial strain. [Modified from Passchier, Myers, and Kröner (1990). Reprinted with permission of Springer-Verlag.]

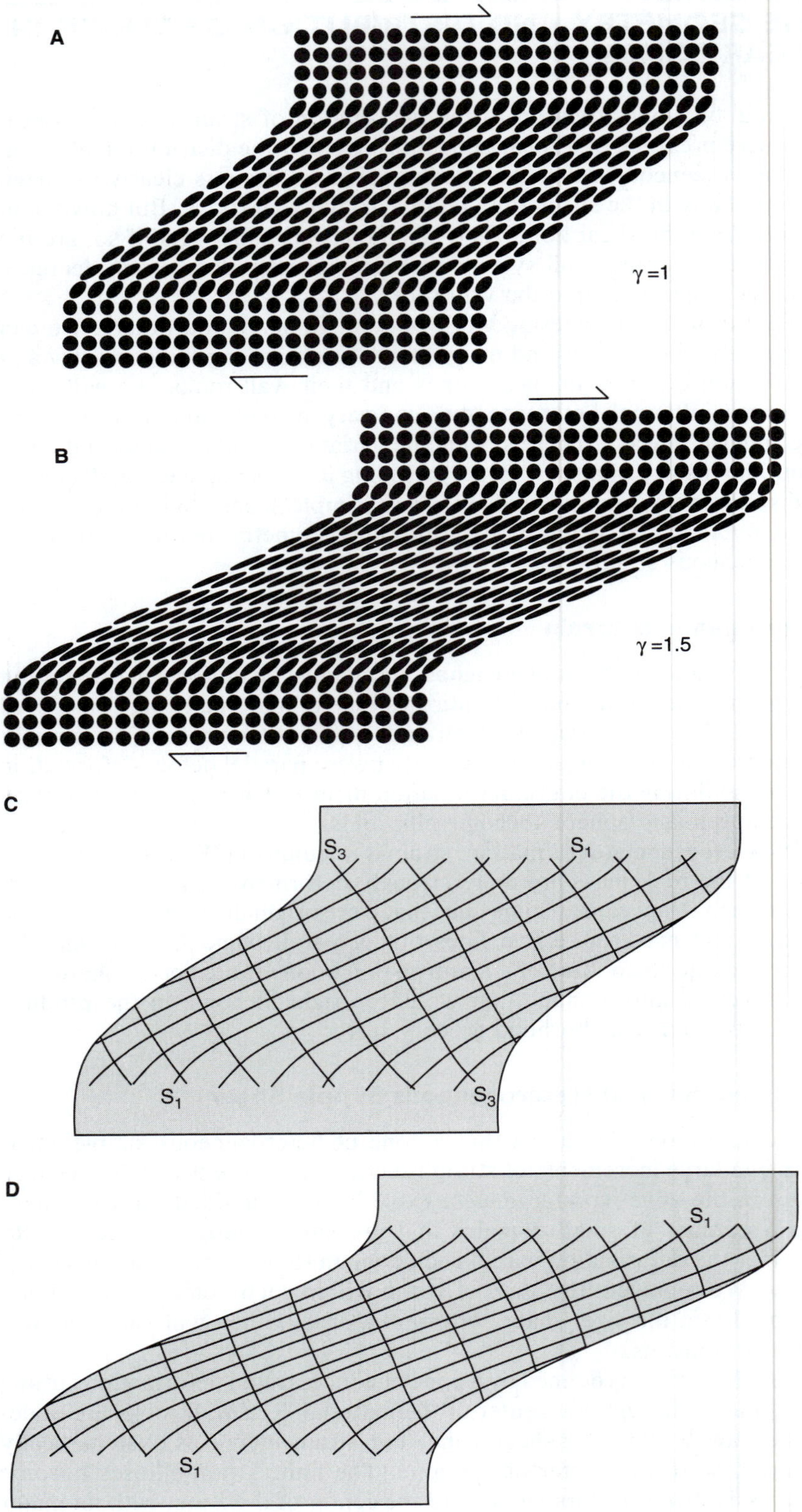

Figure 9.72 Heterogeneous strain in a dextral zone of simple shear. (*A*) Initially square grid of circles after a shear strain of approximately 1. (*B*) Same shear zone after a greater amount of shear strain. (*C*) Finite strain trajectory for (*A*); (*D*) Finite strain trajectory for (*B*). Note that the trajectories do not simply connect the centers of adjacent ellipses. [Modified from *The techniques of modern structural geology, v. 1, strain analysis,* J. G. Ramsay and M. I. Huber (1983), © by Harcourt Brace and Company Limited, with permission.]

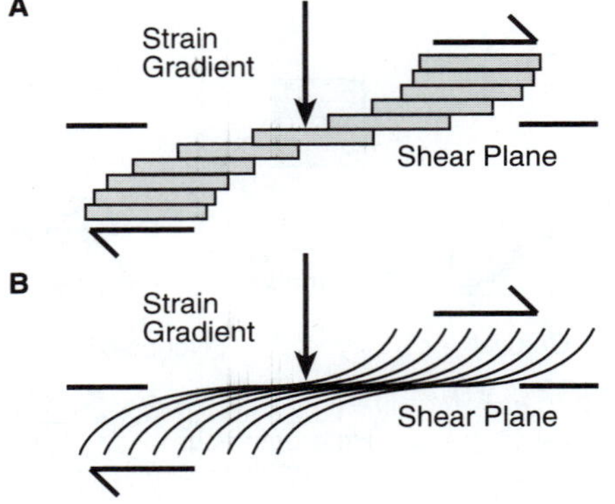

Figure 9.73 Strain gradient normal to a (*A*) discontinuous or (*B*) continuous shear zone [Modified from Hanmer and Passchier (1991). Courtesy of the Geological Survey of Canada.]

If we were to overlay the shear zones in Figure 9.72*A* and *B* with tracing paper, we could mark the orientation of the long S_1 axis of the finite strain ellipse for each domain (Ramsay and Huber, 1983). We could then interpolate between the domains and draw smooth curves, called **finite strain trajectories**, showing how the orientation of S_1, the maximum finite stretch, varies across the shear zone (Figure 9.72*C*). Repeating this process for the short axis of the ellipse yields a second set of curves representing the trajectories of S_3. The trajectories representing S_1 are at 45° to the shear zone at the margins of the zone, where the strain is lowest, and curve toward lower angles with the shear plane as they approach the more highly strained center of the zone. It is this increase in strain toward the center of the shear zone that causes the trajectories to be **sigmoidal**, especially after a greater amount of shear strain on the entire zone (Figure 9.72*D*).

The S_3 trajectories will be perpendicular to the S_1 trajectories at any point because the two strain axes are by definition always perpendicular. Accordingly, the S_3 trajectories first form at 45° to the shear zone, but with the opposite inclination as S_1. They rotate toward the normal to the shear zone with increasing strain. Because the S_1 and S_3 directions of the finite strain ellipse have the special properties of being originally perpendicular before deformation, the S_1 and S_3 trajectories were two sets of straight, perpendicular lines prior to deformation (Ramsay and Huber, 1987). When we look at finite strain trajectories, we are therefore seeing what a grid of originally perpendicular lines would look like after deformation.

In addition to being sigmoidal, the S_1 trajectories converge and diverge, being most closely spaced in the center of the zone. The convergence of S_1 trajectories reflects the increase in strain within the zone and becomes more pronounced at higher overall shear strains (Figure 9.72*D*). S_1 trajectories, and therefore foliation patterns, will always converge as they pass from regions of low strain to those of higher strain, except for some special geometries of dilation (Ramsay and Huber, 1983, 1987).

Finite strain trajectories are an incredibly useful way to study shear zones. For example, any feature in the rock that develops parallel to S_1 should display the same trajectory as S_1 across the shear zone. The trajectories reveal the underlying explanation for the sigmoidal geometry of foliation across a shear zone and of S-surfaces in S-C fabrics. We can also use the trajectories to determine the *amount* of shear strain across the entire zone (Ramsay and Huber, 1983).

Strain Compatibility in Shear Zones

The lateral variations in strain are commonly quite systematic, even for deformation that is very heterogeneous. The reason for this is **strain compatibility**, which specifies how strain can vary within a heterogeneously deformed material without causing structural discontinuities, holes between domains, or abrupt changes in the type of strain (Ramsay and Huber, 1983). In our example, strain in two adjacent domains must be compatible; otherwise, room problems, such as gaps and overlaps, will develop along the interface. If material in one domain stretches or rotates more than the adjacent material, an intervening structural discontinuity must form to alleviate the resulting room problems and mismatches.

The requirement of strain compatibility places important constraints on the types of strain that can occur in shear zones. Using strain compatibility and other arguments, Ramsay and Graham (1970) demonstrated that

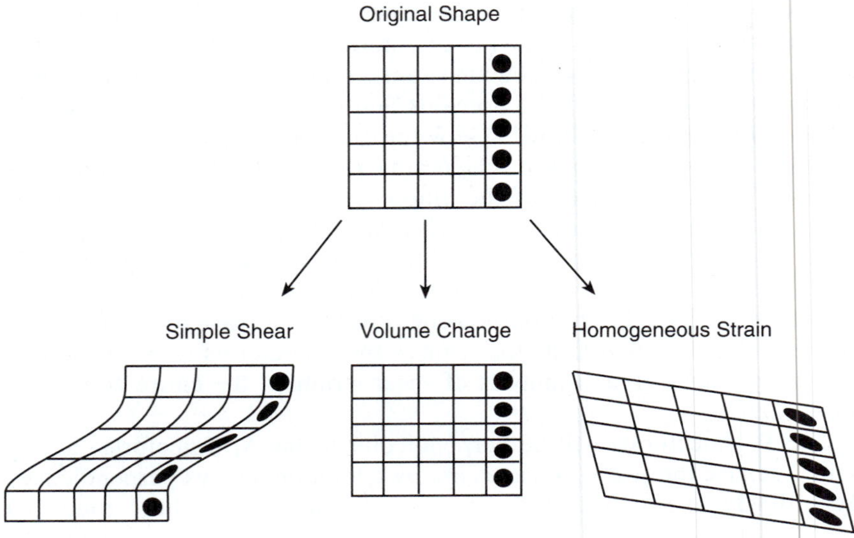

Figure 9.74 Three components of strain possible in a planar, parallel-sided shear zone with no discontinuities along its margins. Components are simple shear, volume change perpendicular to the shear zone, and homogeneous strain affecting the zone and its wall rocks uniformly. [Reprinted from *The techniques of modern structural geology, v. 2: folds and fractures,* J. G. Ramsay and M. I. Huber (1983), © by Harcourt Brace and Company Limited, with permission.]

only three main components of strain are possible in planar, parallel-sided shear zones. They are illustrated in Figure 9.74:

- Heterogeneous simple shear, with the shear plane oriented parallel to the shear zone walls
- Heterogeneous dilation, with the volume change occurring via displacement perpendicular to the shear zone walls
- Homogeneous strain of any type that affects the shear zone and its wall rocks

These three types of strain can be combined in any proportions without violating the requirement of strain compatibility (Figure 9.75). A combination of simple shear plus dilation normal to the shear zone walls is a type of general shear. In this case, the dilational component of shortening or extension normal to the shear zone is accommodated by volume loss or volume gain, respectively.

We can also think of these three types of strain as the three *components* of strain possible within a parallel-sided shear zone. By treating each component separately, a concept known as **strain factorization** (Ramsay and Huber, 1987), we can mathematically isolate and predict the strain effects for each component. Equations describing the displacement caused by each component can then be mathematically combined, most easily through the use of matrix multiplication, to define the total state of strain. Ramsay and Huber (1987) describe this methodology and provide some applications of strain factorization.

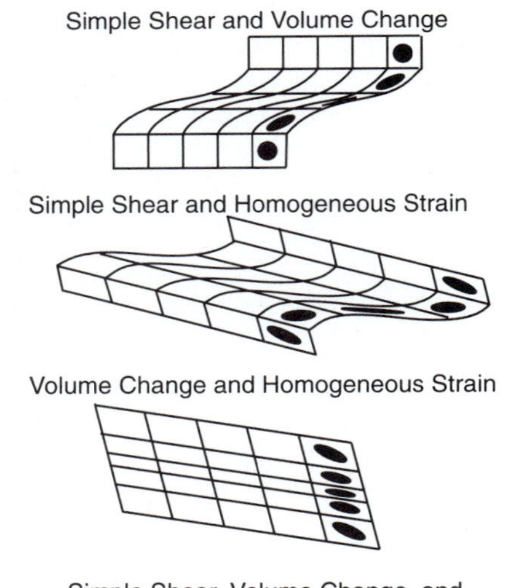

Figure 9.75 Various combinations of the three components of strain in Figure 9.74. [Reprinted from *The techniques of modern structural geology, v. 1, strain analysis,* J. G. Ramsay and M. I. Huber (1983), © by Harcourt Brace and Company Limited, with permission.]

Although all three components of strain are possible, only the first two are *heterogeneous,* hence actually able to *form* a shear zone. Because the second type of strain involves a volume change, simple shear is left as the only *constant volume* deformation that can form a parallel-sided continuous shear zone.

To explore why strain compatibility makes certain types of strain geometrically unlikely in shear zones, we consider a shear zone formed by heterogeneous pure shear (Figure 9.76). The shear zone forms parallel to S_1 and perpendicular to S_3. Extension along S_1 is greatest in the center of the shear zone and decreases away from the center until the wall rocks are undeformed. This heterogeneous distribution of extension causes the

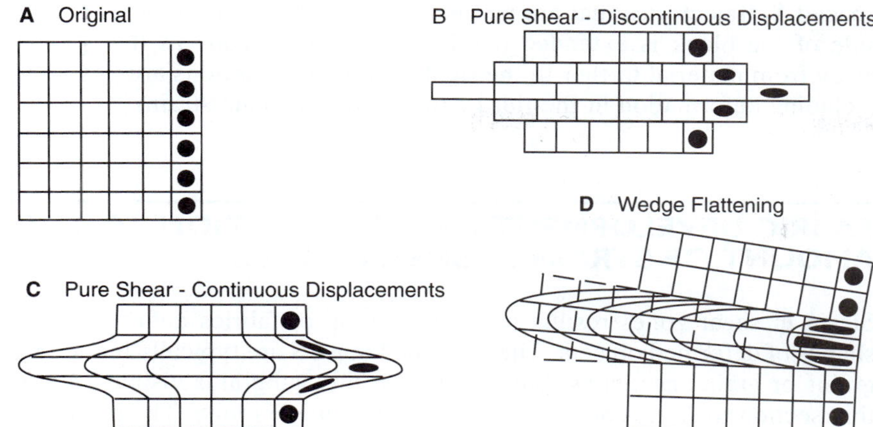

Figure 9.76 Strain incompatibility of a shear zone formed by heterogeneous pure shear. (*A*) Original square showing domains. (*B*) Differing amounts of coaxial strain between adjacent domains requires the formation of structural discontinuities between the domains. (*C*) Strain must be very heterogeneous and noncoaxial if discontinuities do not form. (*D*) Heterogeneous deformation required in a wedge-shaped shear zone. [Reprinted from *The techniques of modern structural geology, v. 1: strain analysis,* J. G. Ramsay and M. I. Huber (1983), © by Harcourt Brace and Company Limited, with permission.]

edges of the original square of material to bulge out, in what Ramsay and Huber (1987, p. 610) have called the "cream-cake effect—the extrusion of a central layer of cream when the top and the bottom of the cake are pressed together."

If strain within each domain is to remain coaxial, then discontinuities must develop between domains (Figure 9.76*B*). Otherwise, fairly complex *noncoaxial* strain patterns result, where the bulge amplifies outward toward the sides of the original square (Figure 9.76*C*). The outward flow of material displaces only air in the cream-cake example, but in the Earth it causes room problems because the adjacent material that must be displaced is rock. Similar strain incompatibilities exist for such shear zones with nonparallel walls (Figure 9.76*D*).

A planar zone of simple shear (Figure 9.77) does not have such strain compatibility problems because (1) the noncoaxial strain in each domain is compatible with that in the adjacent domains, and (2) strain within the shear zone can be accommodated by the relative translation of blocks on opposite sides of the shear zone. This does not mean that discontinuities will not form within a zone of simple shear, only that they are not required.

The ends of shear zones pose some special problems in strain compatibility. A shear zone either is terminated by another shear zone or somehow dies out into undeformed or homogeneously deformed material. In Figure 9.78, we can see how a zone of simple shear cutting undeformed wall rocks might change laterally into a complex strain pattern (Ramsay and Allison, 1979). The thin lines represent the deformed shapes of originally square domains, and the thicker lines represent the S_1 finite strain trajectory. As the block above the shear zone is translated to the left, its left side is shortened parallel to the shear zone, becoming a zone of closely

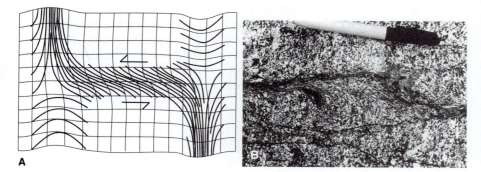

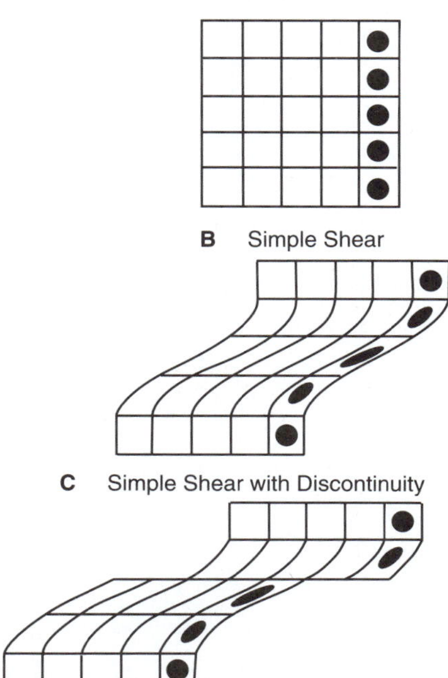

Figure 9.77 Strain compatibility of a shear zone formed by heterogeneous simple shear. (*A*) Original square showing domains. (*B*) Simple shear in each domain is compatible with that in adjacent domains, even though a strong strain gradient may exist. (*C*) Strain in adjacent domains is still compatible, even if structural discontinuities develop between domains. [Parts *A* and *B* from *The techniques of modern structural geology, v. 1: strain analysis,* J. G. Ramsay and M. I. Huber (1983), © by Harcourt Brace and Company Limited, with permission.]

Figure 9.78 End of shear zones. (*A*) Strain pattern that can accommodate the termination of a planar zone of sinistral simple shear cutting undeformed wall rocks. (After Ramsay and Allison, 1979.) (*B*) A natural example in a granitic rock, Sudbury, Ontario, Canada. (Photograph by S. J. Reynolds.)

spaced S_1 trajectories that are perpendicular to the shear zone. The right side of the block is extended parallel to the shear zone as it is pulled away from material farther to the right. Many other options are possible, including deformation in the third dimension (nonplane strain).

FABRIC DEVELOPMENT AND ITS RELATION TO THE AMOUNT OF STRAIN IN SHEAR ZONES

Strain in shear zones results in the formation of fabrics defined by the shapes of deformed objects. The deformed objects are typically individual grains or grain aggregates, but range from tiny subgrains visible only in thin section to slivers or boudins hundreds of meters long. The fabrics in shear zones can be subdivided into three general types based on how well the fabrics reflect the total strain in the shear zone. Some fabrics, referred to as **strain-sensitive fabrics**, fully record the finite strain and so can be used to determine the *amount* of strain in a shear zone. In contrast, **strain-insensitive fabrics** do not record the total finite strain because some competing process prohibits a fully strain-sensitive fabric from developing. **Composite shape fabrics** may or may not record the finite strain; typically, they develop when discrete, subsidiary shear zones form parallel to or at an angle to the main shear plane.

In this section, we explore how each type of fabric forms, and how we can use such fabrics to estimate the *amount* of strain within a shear zone. We present only methods of determining strain that are applicable to ductile and brittle–ductile shear zones. Additional methods suitable for determining strain in metamorphic rocks in general are covered in Chapter 8.

Strain-Sensitive Shape Fabrics

Strain-sensitive shape fabrics are parallel to the finite strain ellipse and therefore record the total finite strain of the material in which they are found. They include the typical foliation and lineation we have already considered. An example of this type of fabric is a deformed conglomerate that contained initially spherical clasts (Figure 9.79). As long as the clasts and their matrix responded identically to deformation, the shape and orientation of each clast would reflect the finite strain ellipse. For any plane strain, the clasts would be flattened in the S_3 direction and correspondingly lengthened parallel to S_1. The clasts would define a foliation parallel to the S_1S_2 plane and perpendicular to S_3. A stretching lineation, if present, is parallel to S_1. The amount of strain in such a rock can be evaluated by the R_f/ϕ, center to center, and other conventional strain methods discussed separately (see Chapter 8) (see also Ramsay and Huber, 1983).

Within shear zones, we can also use the orientations of foliation relative to the shear zone to determine the amount of strain. Recall that in noncoaxial deformation, foliation starts forming at an angle to the shear zone and rotates into low angles with the shear zone at high shear strains. It turns out that for simple shear, the angle between foliation and the shear zone, as measured parallel to the shear direction on the SOS plane, is *directly related to the shear strain* via the extremely useful little equation (Figure 9.80) (Ramsay and Graham, 1970):

$$\gamma = \frac{2}{\tan 2\theta} \tag{9.1}$$

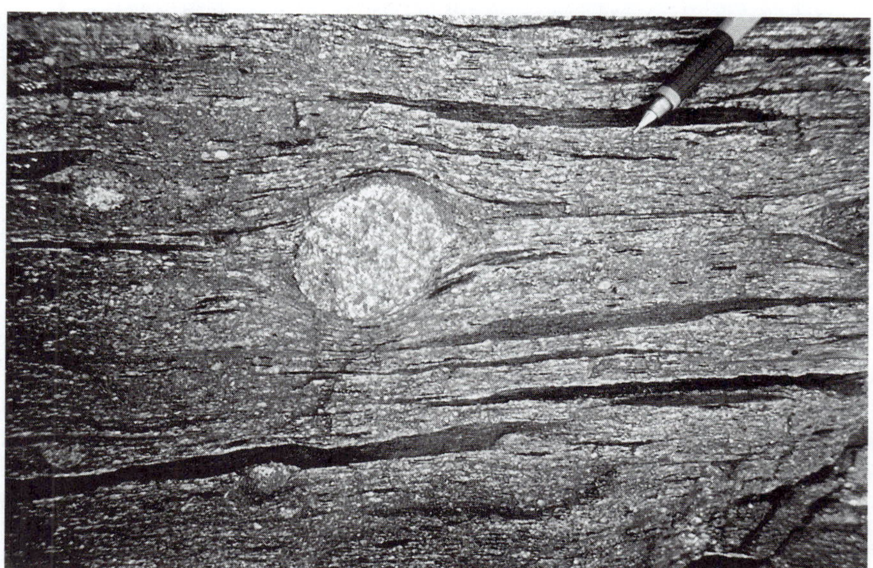

Figure 9.79 Shape fabric defined by the preferred orientation of clasts in a strongly deformed Proterozoic metaconglomerate, Phoenix Mountains, central Arizona. Note the less deformed, rigid granitic cobble, which does not fully reflect the finite strain of the entire rock. (Photograph by S. J. Reynolds.)

where γ is the shear strain and θ is the angle between the foliation and shear zone. If we know simply the angle between the foliation and shear zone, we can use Equation 9.1 to calculate the shear strain at any one site within a shear zone. In our own studies of shear zones, we both use this equation a lot, even right in the field.

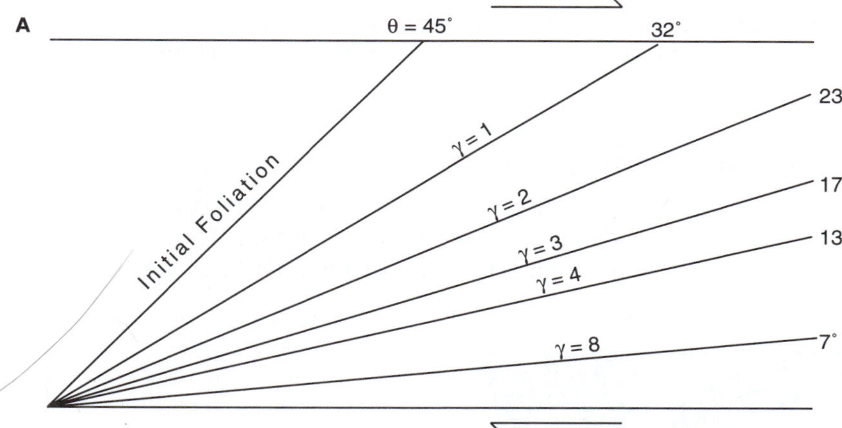

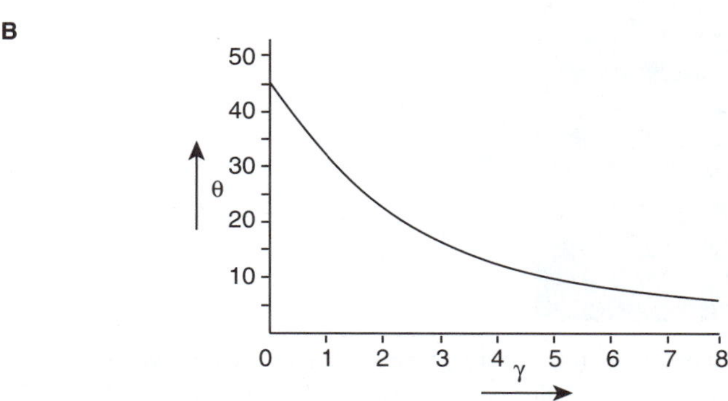

Figure 9.80 The angle between foliation and shear zone (θ) as a function of shear strain (γ): (*A*) a zone of simple shear, (*B*) plot of θ versus γ for simple shear. [Part *B* from Ghosh (1993), with permission from Butterworth-Heinemann Ltd.]

We can also calculate the principal stretches directly from θ (Ramsay and Graham, 1970):

$$S_1 = \cot\theta$$

$$S_3 = \tan\theta$$

Alternatively, if we can determine the principal stretches by some other method, we can use them to calculate θ, which then gives us the shear strain. Note that these equations are valid only for simple shear. The angle between foliation and shear zone will change if there is any non–simple shear component, such as a dilational or pure shear component of shortening or extension *perpendicular* to the shear zone. If these other components cause the shear zone to be *thinned* during deformation, θ will decrease compared to its expected value for simple shear and the equations will overestimate the shear strain. If extension normal to the shear zone causes the shear zone to *thicken*, θ will be greater than its expected value for simple shear, and the equations will underestimate the shear strain.

An ideal ductile shear zone shows a gradual decrease in strain away from its center, as is reflected by the sigmoidal shape of foliation, lineation, and other strain-sensitive fabrics across the zone (Figure 9.81*A*). The pattern of foliation we observe in the field can be portrayed as **foliation trajectories** (Figure 9.81*B*), which are generally equivalent to the S_1 finite strain trajectories we explored earlier. Foliation trajectories provide us with an opportunity to estimate the total amount of strain across the entire zone and variations in shear strain across the zone. The angle between foliation and the shear zone (θ) is 45° near the walls, decreasing progressively as strain increases toward the center of the zone. If we determine θ values for a transect normal to the shear zone, we can estimate the shear strain, and therefore the amount of displacement, for the entire zone. Ramsay and Huber (1983, 1987) describe several methods by which to do this.

Strain-Insensitive Shape Fabrics

Strain-insensitive shape fabrics do not reflect the finite strain ellipse and are not therefore an accurate record of the total strain. They record some of the finite strain, but not all. Their orientation relative to the shear zone

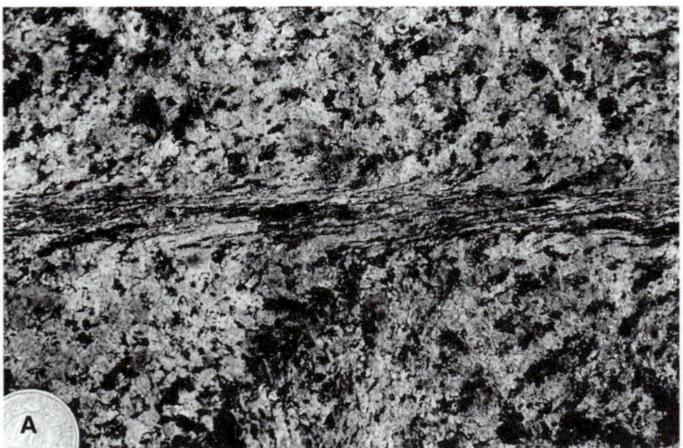

B

Figure 9.81 Dextral shear zone: (*A*) Cutting granite, Laghetti, Ticino, Switzerland. (Photograph by Carol Simpson.) (*B*) Foliation trajectories for shear zone in (*A*).

or foliation, which reflects whether strain was coaxial or noncoaxial, is an important sense-of-shear indicator.

The most common *strain-insensitive* fabric is the **microscopic oblique foliation** we used as a shear-sense indicator (Simpson and Schmid, 1983; Lister and Snoke, 1984). In most shear zone rocks, including mylonites, the main foliation is largely a *strain-sensitive* fabric defined by a preferential orientation of lenticular grains and aggregates of grains, ribbons, small subsidiary shear zones, and the shapes of deformed inclusions. In thin section, however, a second, oblique foliation, defined by *aligned subgrains,* is present within the individual lenticular grains, aggregates, and ribbons (Figure 9.82*A*). Relative to the main foliation, the oblique foliation leans over in the sense of shear. Curiously, though, the angle between the two types of foliation is a fairly consistent 20–30°, even for rocks with very different finite strains. Also, the microscopic oblique foliation is at a higher angle to the shear zone than is the main strain-sensitive foliation (Figure 9.82*B*); it cannot record the same finite strain as the main foliation. Therefore, the angle between the oblique foliation and either the shear zone or the main foliation cannot be used to estimate the shear strain. Instead, the angle remains approximately constant, irrespective of the finite strain.

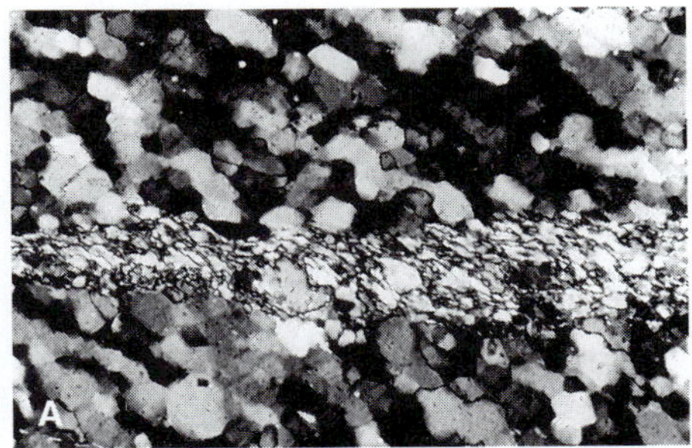

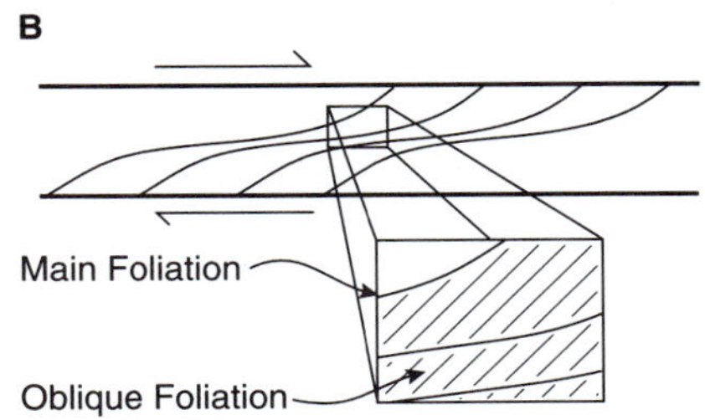

Figure 9.82 Strain-insensitive oblique fabric in mylonitic rocks. (*A*) Photomicrograph of deformed pegmatite, Borrego Springs mylonite zone, southern California. Main foliation is subhorizontal, parallel to the strongly deformed feldspar grain that defines a band across the center of photograph. The oblique foliation is expressed by the alignment of quartz and feldspar subgrains from upper left to lower right (sinistral sense of shear). The quartz and feldspar subgrains have a slightly different orientation because the quartz subgrains recrystallize more easily than do the feldspar grains and so are more nearly parallel to the direction of maximum instantaneous extension. (*B*) Geometric relations between strain-sensitive main foliation, strain-insensitive oblique foliation, and the shear zone.

Strain-insensitive fabrics require some process that can form a foliation that is oblique to the finite strain ellipse and does not rotate with increased finite strain. Win Means (1980) proposed that for a shape fabric to avoid rotation during noncoaxial deformation, processes that form the fabric must be balanced by those that try to obliterate it. As grains begin to deform and rotate in accordance with the finite strains, dynamic recrystallization modifies the resulting fabric (Figure 9.83). As the grains accrue stored strain energy (in the form of dislocations), they may recrystallize into "new" grains or subgrains with less stored strain energy, with more equant or more irregular shapes, and with less of a preferred orientation. Such recrystallization effectively "resets" the finite strain "clock" for each recrystallized grain, or at least causes the grain to have an imperfect "memory" of the total finite strain. Recrystallization therefore decreases the overall preferred orientation of the grains and weakens the appearance of the fabric. This process is aided by the tendency of recrystallization to affect preferentially the most deformed, strain-hardened grains, the very ones that will have the most preferred orientation and best define the fabric.

It does not stop there, however. The newly recrystallized grain, once formed, will again experience strain as deformation continues. In fact, such grains, because they are relatively strain-free, will deform preferentially relative to adjacent grains that are still strain hardened. As the newly recrystallized grain begins to deform, it changes shape incrementally, in accordance with the *instantaneous* stretching axes, *not* the *finite* strain ellipse. In simple shear, the first increment of strain will form a foliation that will be at 45° to the shear plane and will be oblique to the finite-strain-sensitive foliation in the surrounding rock.

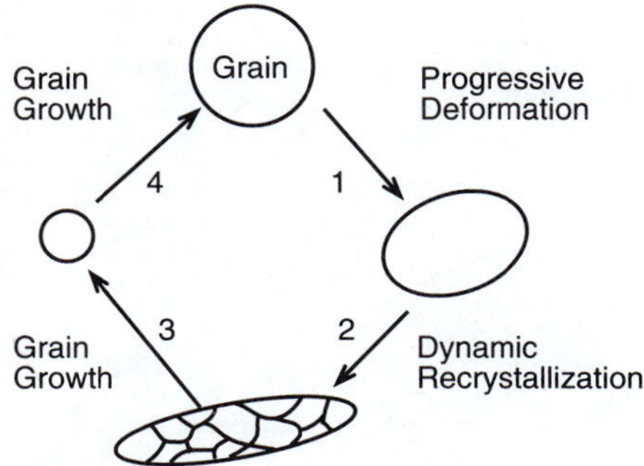

Figure 9.83 Cycle of progressive deformation leading to dynamic recrystallization, leading to grain growth and grain boundary migration, resulting in a "new" less-strained grain. [From Hanmer and Passchier (1991). Courtesy of the Geological Survey of Canada.]

The entire process represents a cycle in which grains are continually being strained, recrystallized, and strained again. At any given time, different grains will be in different phases of the cycle and the overall fabric will represent a composite or hybrid state of strain. The oblique foliation will be intermediate in position between the finite and instantaneous strain ellipses at any given time, and its angle to the shear plane may partially reflect the relative success of the fabric-forming and fabric-weakening processes. The process has been modeled using computer simulations (Jessel, 1988a,b).

The temperature during deformation influences the degree of development of such strain-insensitive fabrics, since it controls the rate of recrystallization. At higher temperatures, recrystallization is relatively fast and should be most efficient at "resetting the finite strain clock." It will be less able to keep up with the imposed finite strain at lower temperatures or higher strain rates, where the resulting fabrics should more accurately reflect the finite strain.

Composite Shape Fabrics

Composite shape fabrics contain shear bands in addition to either a strain-sensitive or strain-insensitive fabric (usually foliation and lineation). The shear bands are subsidiary to the main ductile shear zone and can be either parallel or oblique to it. The term *shear band* is used by some geologists only for thin shear zones that are oblique to the main shear zone; we believe, however, that it is more useful as a purely nongenetic, descriptive term. In part, this is because we do not always know the orientation of the main shear zone relative to an outcrop we are examining. The two main types of composite fabric that contain shear bands are **S-C fabrics** and **extensional shear bands**.

S-C Fabrics

S-C fabrics consist of foliation (S-surfaces) and shear bands (C-surfaces). The S and C surfaces are at an angle to one another (Figure 9.84), indicating the sense of shear (Berthé et al., 1979). In most S-C fabrics, the S-surfaces are a typical foliation, being defined by flattened lenticular grains and aggregates of grains. They are interpreted to be a *strain-sensitive* fabric parallel to the S_1S_2 plane of the finite strain ellipsoid. S-surfaces in some other S-C fabrics are the oblique *strain-insensitive* foliation described in the preceding section. These fundamental differences in the character of

Figure 9.84 S-C mylonites: (*A*) Type I S-C mylonite, Santa Catalina Mountains, Arizona. C-surfaces are defined by the thin zones of very high shear strain, whereas S-surfaces are defined by the obliquely inclined feldspar porphyroclasts. Dextral shear sense. (Photograph by S. J. Reynolds.) (*B*) Photomicrograph of type II S-C mylonite, Coyote Mountains, southern Arizona. S-surfaces are defined by mica fish and oblique subgrains in quartz, whereas C-surfaces are defined by the long, subhorizontal tails (bright in photograph) extending off the tips of mica fish. Dextral shear sense. (Reprinted with permission from *Journal of Structural Geology*, v. 9, Davis, G. H., Gardulski, A. F., and Lister, G. S., Shear zone origin of quartzite mylonite and mylonitic pegmatite in the Coyote Mountains, Arizona, U.S.A., Elsevier Science Ltd., Pergamon Imprint, Oxford, England.)

the S-surfaces can be used to subdivide S-C fabrics into two general types (Lister and Snoke, 1984). **Type I** are the dominant S-C fabrics in most mylonitic quartzofeldspathic rocks. S-surfaces in type I fabrics are normal *strain-sensitive* foliation and C-surfaces are crosscutting shear bands (Figure 9.84*A*). **Type II** S-C fabrics are most common in micaceous quartzite and some mylonites. S-surfaces in type II fabrics are defined by *strain-insensitive* microscopic oblique foliation and by mica fish, whereas C-surfaces are thin shear zones marked by trails of deformed mica sheared (scaled?) off the fish (Figure 9.84*B*).

In both types of S-C fabric, the C-surfaces are commonly aligned parallel to the shear zone and are associated with the same strain gradients as any continuous shear zone (Figure 9.84). These gradients cause the sigmoidal shape of the accompanying S-surfaces as they approach and rotate toward parallelism with the C-surfaces. Some C-surfaces display a more discontinuous strain gradient, whereby they offset individual grains along a very abrupt contact. Such sharp C-surfaces are most common adjacent to a grain that behaved as a rigid object during deformation.

We can use the angle between S- and C-surfaces in type I S-C fabrics to estimate the amount of strain, using the same equation we used for the angle between foliation and the main shear zone. For simple shear, S-surfaces, like any strain-sensitive foliation, start to form at 45° to the shear zone and the future orientation of C-surfaces, but at this stage they are too weakly developed to be seen. As the strain increases, they become better developed and rotate toward the shear zone along with S_1. They first become visible in the field when they are approximately 35° to the shear zone, equivalent to a shear strain (γ) of 0.7. With further strain of the entire rock or as they approach a C-surface, the S-surfaces rotate toward parallelism with and become asymptotic to the C-surfaces.

The presence of both S- and C-surfaces accentuates the already-lenticular appearance of most mylonitic rocks. Both fabrics can contribute to the overall foliated appearance of a rock, and the "main" foliation that we tend to see and measure in the field can be either one, or an "average" of both, depending on how each fabric is expressed and how closely we look at the rocks. In our experience, we tend to see and measure S-surfaces in weakly deformed rocks and C-surfaces in strongly deformed ones. In moderately deformed rocks, we may see and measure either one, or, if we are not careful, some average of the two which has with no real kinematic significance.

Ideally, we would like to measure the strike and dip of both S- and C-surfaces in the field. We would then know something about the orientation of the main shear zone, even if its margins are not exposed, because C-surfaces generally should be subparallel to the overall shear zone. We could also use a stereonet to compute the angle between S and C in the line of transport, and thereby estimate the associated shear strain. In many cases, however, we cannot accurately measure the strike and dip of both types of surfaces because we see them exposed on only a single rock face and we cannot tell how the planes continue into the rock. In such a case, we can still assess the angular relation between C and S along the line of transport. The method that Gordon Lister and I (SJR) have developed consists of measuring the apparent dips of S and C on a rock face cut parallel to the SOS plane (Figures 9.85 and 9.86). In this way, we directly measure the angle between S and C in the field and immediately calculate the shear strain, assuming simple shear (Equation 9.1). Because we are able to evaluate the variations in shear strains as we hike along, questions about the geometry and kinematics of the shear zone arise while we are still in the field, where the key observations and answers surely lie.

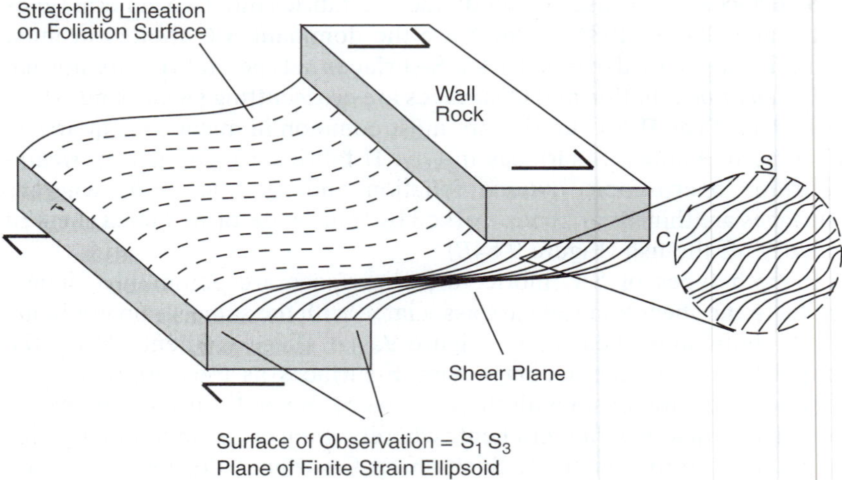

Figure 9.85 Measuring the angle between S and C on the sense of shear plane. [Adapted from Hanmer and Passchier (1991). Courtesy of the Geological Survey of Canada.]

The Origin of S-C Fabrics

The origin of S-C fabrics has been primarily evaluated by tracing undeformed protoliths into shear zones and documenting how the S- and C-surfaces develop (Berthé et al., 1979; Lister and Snoke, 1984). These studies reveal that the S- and C-surfaces are clearly related, rather than being two totally unrelated, superimposed fabrics. As a rock is progressively deformed, the S-surfaces generally form first and in weakly deformed rocks may be the only type of surface present. With increasing shear strains, C-surfaces start to form, and then S and C continue to develop together (Figure 9.87*A*). The S-surfaces progressively rotate toward C, and the C-surfaces may become more closely spaced and better developed (Figure 9.87*B*). At very high shear strains, S becomes subparallel to and indistinguishable from C.

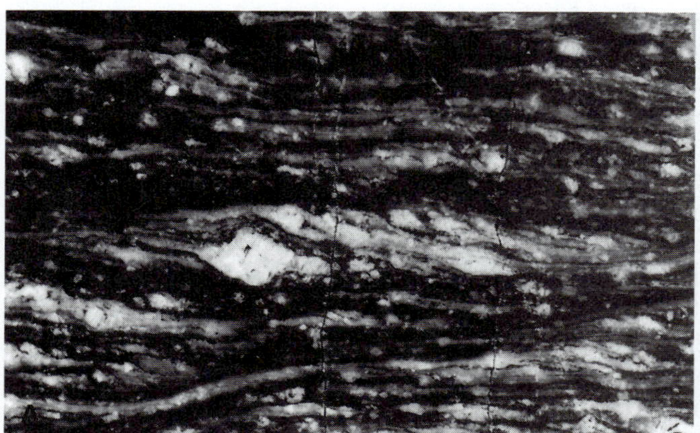

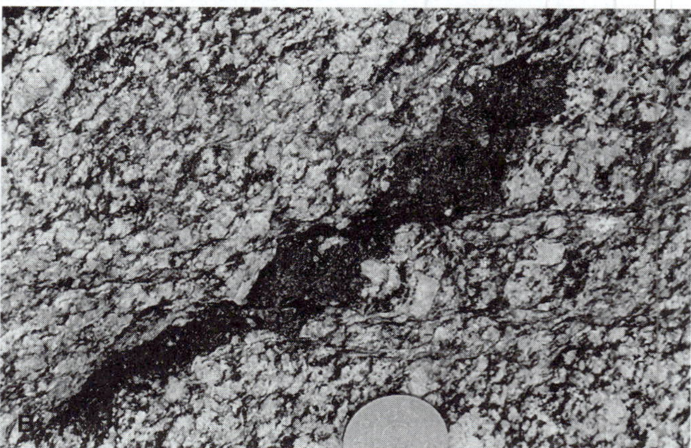

Figure 9.86 Try measuring the angle between S and C on the sense of shear plane in actual rocks. (*A*) Polished slab of mylonitic granodiorite, South Mountains, Arizona. C-surfaces are expressed as the dark, continuous, strongly deformed bands, whereas S-surfaces are defined by the individual elongated, light-colored, feldspar grains. The relatively low angle (15–20°) between S and C is consistent with the highly strained appearance of the rock. (Photograph by S. J. Reynolds.) (*B*) Outcrop of mylonitic late Cretaceous tonalite containing a mafic inclusion, which is parallel to S. The high angle (45°) between S and C is inconsistent with magnitude of shear strain. This may indicate that the C surfaces formed after much foliation was already developed. (Photograph by Carol Simpson.)

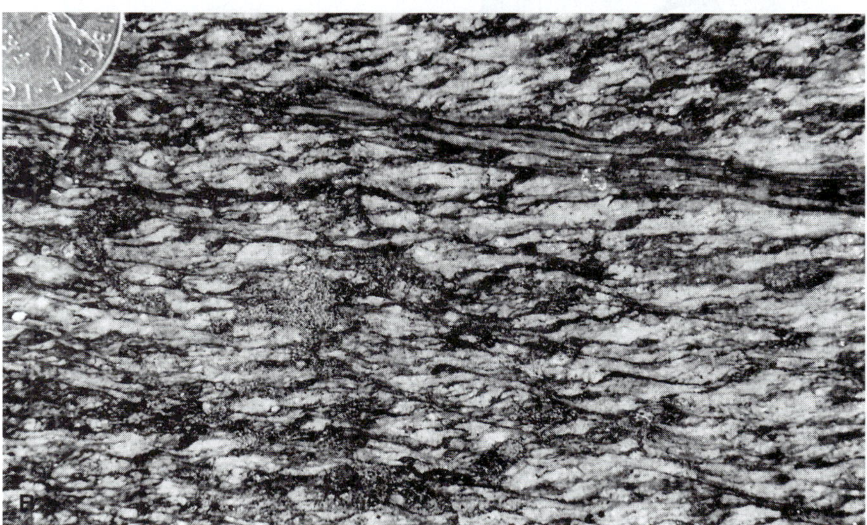

Figure 9.87 Formation of S-C fabrics in granite, South Armorican shear zone near Lescastel, Brittany, France. (*A*) Initial formation of S-C fabric, with weakly developed S-foliation, thin C-surfaces, and a large angle between S and C. Sinistral sense of shear. (*B*) Strongly developed S-C fabric in which S and C are nearly parallel. The thin, well-developed shear zone is a late-stage shear band (*C*′). Dextral shear sense. (Photographs by Carol Simpson.)

The formation of S-C fabrics may be triggered by rigid grains that cause the strain distribution to become heterogeneous on a small scale (Hanmer and Passchier, 1991). These grains act as stress raisers to concentrate stress, perhaps causing a C-surface to nucleate where the boundaries of adjacent rigid grains are aligned parallel to the shear plane. When strain starts to become concentrated into a zone adjacent to the rigid grains, strain softening takes over, causing additional localization of strain, and a C-surface is born. This sequence of events may explain why S-C fabrics are generally found in rocks that contain some rigid grains interspersed with ductile ones; such rocks include quartzofeldspathic rocks (e.g., granite, gneiss, metarhyolite) deformed at moderate temperatures, where quartz flows ductilely but feldspar is more rigid. S-C fabrics are less common in monomineralic rocks and in quartzofeldspathic rocks (granites and gneisses) deformed at higher temperatures, where both quartz and feldspar deform by ductile mechanisms and dynamic recrystallization.

Extensional Shear Bands

Extensional shear bands are oblique to the main shear zone (White et al., 1980; Hanmer and Passchier, 1991). Shear bands of this type crosscut the main foliation, displacing it in a normal sense when viewed on the SOS plane (Figure 9.88). Although sporadically developed in some outcrops, they may be the most obvious shear bands in others, imparting a second foliation to the rocks. As the main foliation approaches the shear bands, it is rotated into parallelism with the bands. This results in an apparent folding or crenulation of the main foliation, and for this reason extensional shear bands are also called **extensional crenulation cleavage** (Platt and Vissers, 1980). They are designated by a **C′** (''C prime'') to differentiate them from normal C-surfaces and to emphasize that they generally form after most of the main foliation has been established.

Extensional shear bands locally form conjugate sets, but more typically they occur in one dominant set that is *synthetic* to the overall shear sense of the main shear zone. The sense of shear on this set of shear bands is

Figure 9.88 Extensional shear bands: (*A*) Cutting pegmatitic layer in banded gneiss. [David O'Day tracing of photograph by G. H. Davis. From Davis (1980). Published with permission of Geological Society of America and the author.] (*B*) Cutting mylonitic granite, Harquahala Mountains, western Arizona. (Photograph by S. J. Reynolds.)

generally the same as the overall sense of shear for the zone. Thus, they are asymmetrical and may be a sense-of-shear indicator for the main zone. In certain cases, however, extensional shear bands or other late-stage shear bands have the sense of shear opposite to that of the main shear zone (Behrmann, 1980; Reynolds and Lister, 1992). We therefore suggest that extensional shear bands be used very cautiously, especially if they appear from crosscutting relations to be very late-stage features that formed after most of the main shear zone fabric. Such late-stage shear bands may have metamorphic mineral assemblages different from those of the main shear zone fabric, or such other distinctive characteristics as associated veins or a brittle–ductile style.

The origin and significance of extensional shear bands are somewhat less well understood than for S-C fabrics. Extensional shear bands are so named because motion along them results in apparent extension of the shear zone parallel to the line of transport. Most appear to form in the waning phases of displacement on a shear zone, after the main foliation is well developed. They may represent strain hardening of the shear zone to the point where homogeneous deformation was no longer possible (White et al., 1980; Passchier, 1986; Dennis and Secor, 1987). Small perturbations of the main foliation may form and be amplified into crosscutting shear bands that attempt to continue accommodating displacement (Cobbold, 1976).

INSIDE THE ELLIPSE: PROGRESSIVE DEFORMATION

So far, we have considered strain using the finite and instantaneous strain ellipses, not really accounting for what is happening to lines of different orientations *within* each ellipse. In this section, we go inside the ellipse to further explore the differences between coaxial and noncoaxial deformation. We also consider progressive deformation and its rather amazing consequences, such as layers that shorten and then stretch during a single deformation event. Progressive deformation in shear zones has rightly received much attention by geologists and is one of the most exciting and evolving aspects of structural geology (Ramberg, 1975; Means et al., 1980; Ramsay and Graham, 1980; Lister and Williams, 1983; Ramsay and Huber, 1983, 1987; Bobyarchick, 1986; Hanmer and Passchier, 1991; Simpson and De Paor, 1993; Weijermars, 1993).

The Instantaneous Strain Ellipse

We have used the **instantaneous strain ellipse** to represent the *instantaneous stretching rates* along every radial line within the plane of the ellipse (Figure 9.89*A*). If we superimpose a circle representing the unstrained state on top of this ellipse (Figure 9.89*B*), positive stretching rates (extension) occur along radii where the ellipse is *outside* the initial circle and negative stretching rates (shortening) occur where the ellipse is *inside* the circle. Two radial lines passing through the intersections of the ellipse and circle represent orientations with *zero* rates of instantaneous stretching and are called **lines of zero stretching rate**.

We will designate the principal axes of the instantaneous strain ellipse, or the **instantaneous stretching axes**, by $\dot{S}_1$ and $\dot{S}_3$, with the dot over the letter indicating that we are referring to stretching rates and the *instantaneous* strain ellipse, rather than the *finite* strain ellipse. The long, $\dot{S}_1$ axis of the ellipse represents the direction and magnitude of most rapid

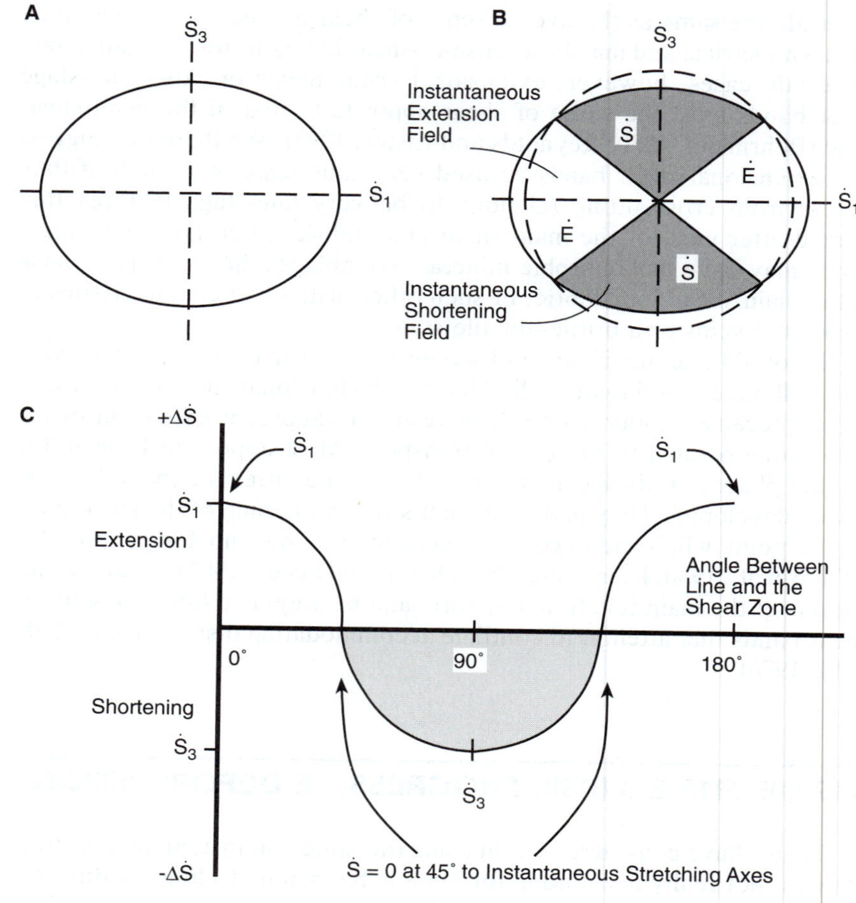

Figure 9.89 Instantaneous strain ellipse: (*A*) ellipse showing instantaneous stretching axes, (*B*) superimposed circle (dashed line) on ellipse, showing instantaneous stretching fields, and (*C*) stretching rates of lines as a function of orientation relative to the stretching axes. The maximum extension rate is along $\dot{S}_1$ and the maximum shortening rate is along $\dot{S}_3$. [Modified from Hanmer and Passchier (1991). Courtesy of the Geological Survey of Canada.]

extension. The short, $\dot{S}_3$ axis of the ellipse is, for plane strain, the direction with the maximum instantaneous shortening rate. The stretching rate progressively decreases for lines at a higher angle to the instantaneous stretching axes, reaching zero along the lines of zero stretching rate at 45° to the instantaneous stretching axes (Figure 9.89*C*).

The **lines of zero stretching rate** subdivide the instantaneous strain ellipse into four quadrants (Figure 9.89*B*), which we will refer to as the **instantaneous stretching fields**. The shaded quadrants in Figure 9.89*B* encompass all radial lines that are undergoing instantaneous shortening and comprise the **instantaneous shortening field**. The unshaded quadrants include all lines that are undergoing instantaneous extension and comprise the **instantaneous extension field**. The instantaneous stretching fields are symmetrical to and bisected by the instantaneous stretching axes.

Instantaneous Strain Ellipses for Pure and Simple Shear

An important distinction between pure and simple shear, or between any coaxial and noncoaxial deformation, is the inclination of the instantaneous stretching axes to the shear plane. The instantaneous strain ellipses for pure and simple shear are shown in Figure 9.90.

For pure shear, the maximum extension rate ($\dot{S}_1$) is parallel to the shear plane and the maximum shortening rate ($\dot{S}_3$) is normal to the shear plane (Figure 9.90*A*). The lines of zero stretching rate are at 45° to the shear plane, and both the instantaneous shortening and extension fields are symmetrical with respect to the shear plane. The instantaneous extension

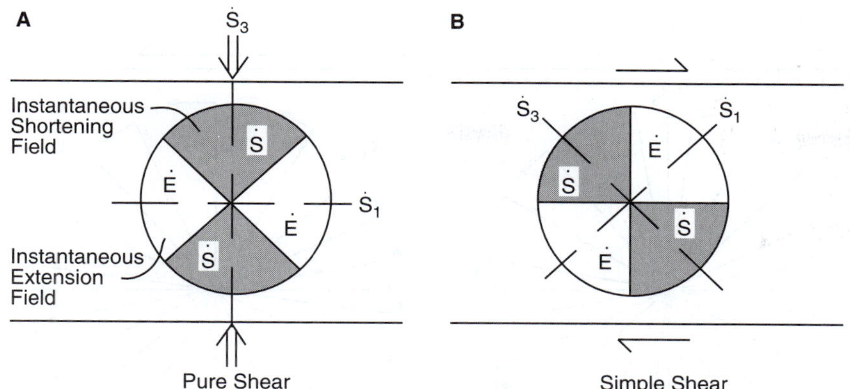

Figure 9.90 Instantaneous strain ellipses, showing instantaneous stretching axes and stretching fields, for (*A*) pure shear and (*B*) simple shear.

field straddles the shear plane, whereas the instantaneous shortening field straddles the normal to the shear plane. This is why structures formed within a zone of pure shear are symmetrical with respect to the zone.

For simple shear, the instantaneous shortening axes are inclined 45° to the shear plane (Figure 9.90*B*) and the instantaneous stretching fields are asymmetrical with respect to the shear plane. One line of zero stretching rate is parallel to the shear plane and the other is along the normal to the shear plane. We can compare these observations with a card deck analogy of simple shear. The maximum instantaneous shortening at 45° to the cards represents the direction we would have to push down on the deck to shear its top to the right. A zero stretching rate parallel to the shear plane is consistent with individual cards in our sheared deck not changing size or shape. A zero stretching rate perpendicular to the shear plane is consistent with the deck not changing in thickness, except at the ends where we run out of cards (which would be unlikely in real rocks).

Rotation Rates of Lines

We can also evaluate the *instantaneous rotation rates* (angular velocities) of lines, rather than their stretching rates. With rotations, we need to choose a convenient frame of reference that we can consider to be fixed in position. Possible reference frames include the shear zone, the instantaneous stretching axes, the axes of the finite strain ellipse, or some "absolute" geographic reference frame. In the field, the shear zone commonly is the most practical reference frame.

Like the stretching rates, instantaneous rotation rates of lines vary systematically with respect to orientation for both pure and simple shear. In pure shear (Figure 9.91A), lines parallel to the instantaneous stretching axes do not rotate. The rotation rate is very slow for lines at a low angle to the instantaneous stretching axes; such lines are nearly parallel to or perpendicular to the shear plane. The rotation rate progressively increases for lines at higher angles to the instantaneous stretching axes, reaching a maximum for lines at 45° to the axes. These lines of maximum rotation rate are also the lines of zero stretching rate. In other words, the lines that rotate the fastest are those that stretch the slowest, and vice versa.

In simple shear, all lines rotate except those parallel to the shear plane (Figure 9.91*B*). The rotation rates are very slow for lines at low angles to the shear plane, but they progressively increase for lines at higher angles to the shear plane, reaching a maximum for lines normal to the shear plane. In simple shear, all lines that rotate do so in the same direction.

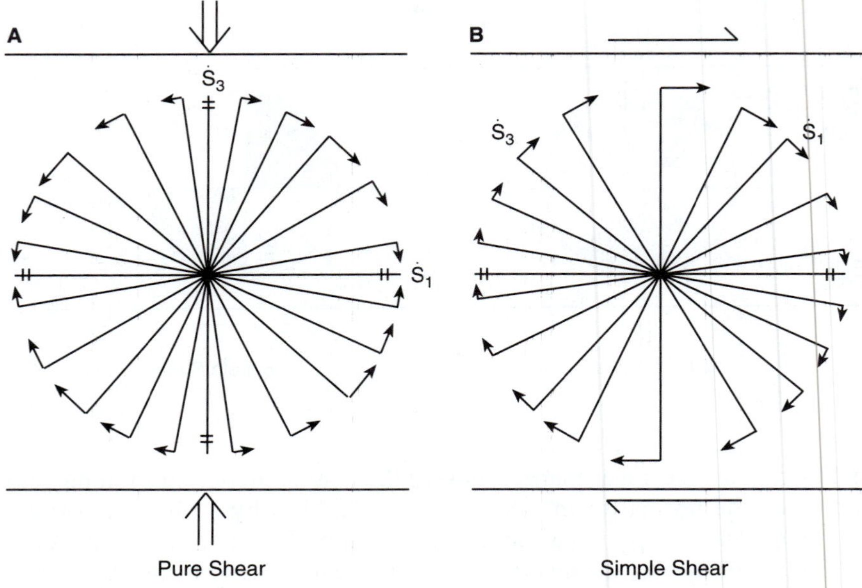

Figure 9.91 Rotation rates of different orientations of lines (passive markers). The relative rate of rotation of each line is represented by the length of the arrow on the end of that line. Lines that do not rotate are shown with double ticks. (*A*) In coaxial strain, in this case pure shear, lines parallel to either instantaneous stretching axis (parallel to and perpendicular to the shear zone) do not rotate. Rotation rates are fastest for lines oriented at 45° to the shear zone, with some lines rotating clockwise and others rotating counterclockwise. (*B*) In noncoaxial simple shear, all lines that rotate do so in the same direction as the sense of shear. The fastest rate is for those perpendicular to the zone. Lines parallel to the shear zone do not rotate.

The Finite Strain Ellipse

As small increments of stretching and rotation add up, they result in significant strain that can be represented with the **finite strain ellipse**, whose principal axes S_1 and S_3 are the **finite stretching axes** (Figure 9.92). If we superimpose the original circle on the finite strain ellipse, there are two radial lines that pass through points where circle and ellipse intersect (Figure 9.92*A* and *B*). These lines are the same length as they were at the start and are called **lines of no finite stretch**. They separate the finite strain ellipse into four sectors, called the **finite stretching fields**. The shaded sectors represent families of radial lines that have been shortened, whereas the unshaded sectors represent lines that have been extended. The sectors of shortening on the finite strain ellipse are called the **finite shortening field** and the sectors of extension are called the **finite extension field**.

We can use the positions of the finite stretching fields to predict the structures we would see from deformation of a rock containing dikes of various initial orientations (Figure 9.92*C* and *D*). Dikes with orientations in the finite shortening field are represented as being folded, whereas dikes oriented within the finite extension field are shown as boudinage. The geometric disposition of folded versus boudinaged dikes can be used in some circumstances to determine, in the field, the approximate orientations of the finite shortening and extension fields, and the sense of shear. What a concept!

We can explore the finite strain differences between pure and simple shear by progressively deforming identical circles by each mechanism and comparing the end results (Figures 9.93 and 9.94). The undeformed circle contains four radial lines spaced 45° apart, which we can watch to see how they rotate. In each case, we will deform the circle in such a way that the resulting shear plane (shear zone) will be horizontal, parallel to line *P* on each figure. Line *N* begins normal to the shear plane, and the other two lines, labeled *L* and *R*, begin at 45° to the shear plane. The shaded sectors between lines *L* and *R* on the initial circle represent the finite shortening field that will exist after the first infinitesimal increment of deformation, which is too small to show with an ellipse.

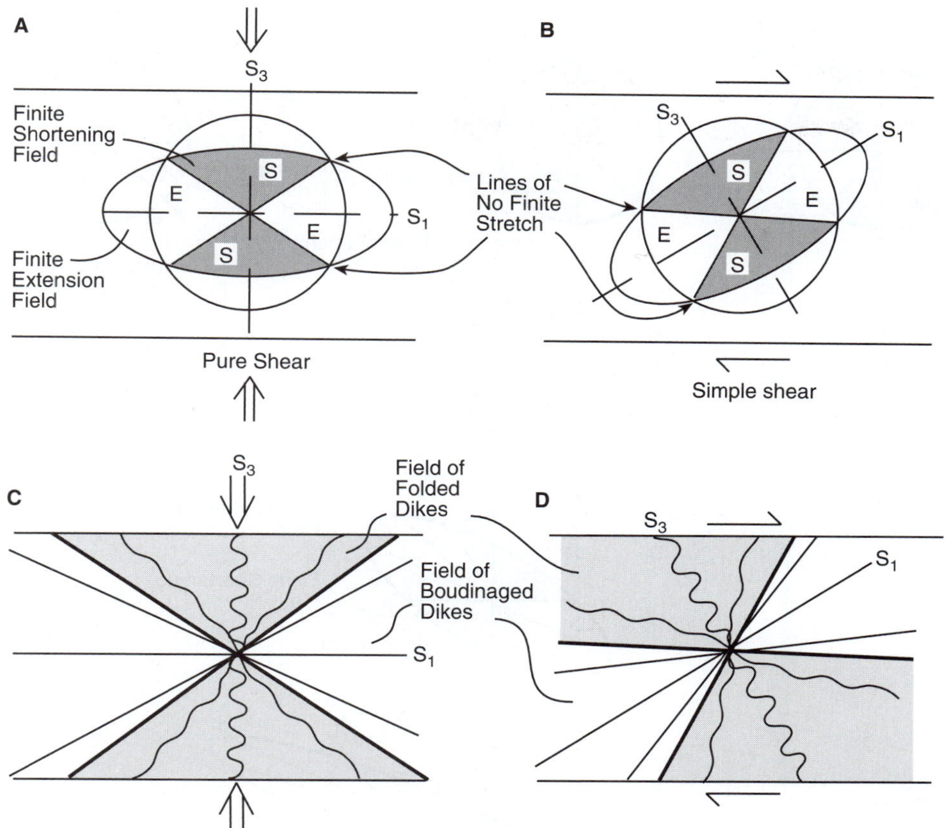

Figure 9.92 Finite strain ellipses and expected orientations of structures for pure and simple shear. Finite strain ellipses are shown for (*A*) pure shear and (*B*) simple shear, along with finite stretching axes (S_1 and S_3), lines of no finite stretch, and finite stretching fields. Expected orientations of folded and boudinaged dikes for a zone of (*C*) pure shear differ from those for (*D*) simple shear.

The Finite Strain Ellipse During Pure Shear

For pure shear, deformation consists of shortening perpendicular to the shear plane (parallel to line *N*) and associated extension parallel to the shear plane (parallel to line *P*). At the start of deformation (Figure 9.93*A* and *B*), the instantaneous stretching axes are parallel to lines *P* and *N*, and the lines of zero stretching rate are parallel to *L* and *R*. After some strain, lines *L* and *R* have lengthened only slightly and rotated toward the shear plane (Figure 9.93*C*). In contrast, lines *N* and *P* have substantially changed length but have not rotated at all. With continued deformation (Figure 9.93*D* and *E*), lines *L* and *R* continue rotating toward the shear plane and lengthening with time. At very high strains, lines with diverse original orientations end up at a low angle to the shear plane.

The lines of no finite stretch, marked by the intersections of the finite strain ellipse and the initial circle, also change orientation as the finite strain ellipse becomes more elliptical. As they rotate, the finite shortening field widens in angle and the finite extension field narrows. The lines of no finite stretch are not fixed to any material lines during deformation. They start out parallel to material lines *L* and *R*, but rotate more slowly toward the shear plane. In contrast, the finite stretching axes (the axes of the ellipse) do not rotate, but instead stay aligned with the instantaneous stretching axes throughout the entire history of deformation. This is char-

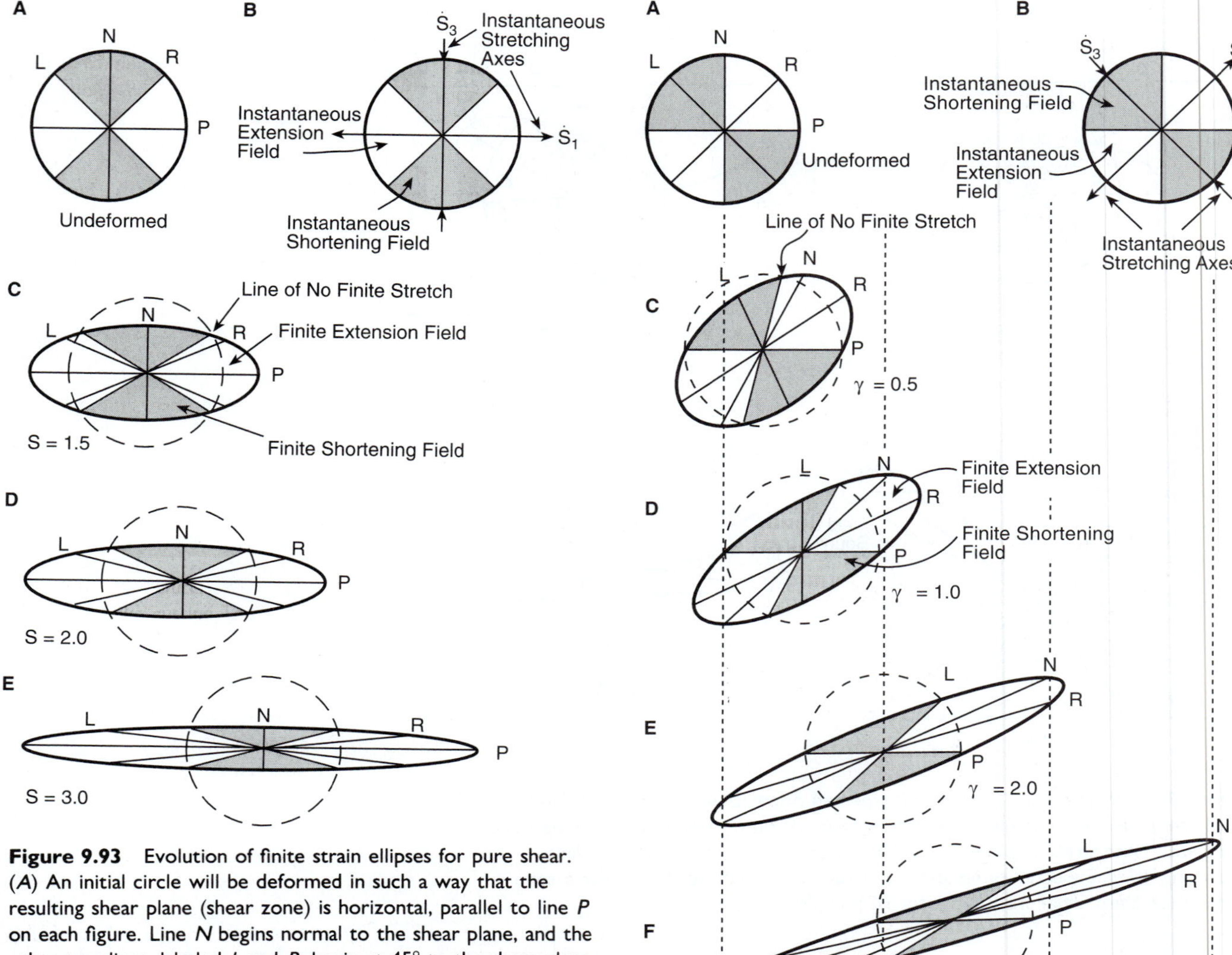

Figure 9.93 Evolution of finite strain ellipses for pure shear. (*A*) An initial circle will be deformed in such a way that the resulting shear plane (shear zone) is horizontal, parallel to line *P* on each figure. Line *N* begins normal to the shear plane, and the other two lines, labeled *L* and *R*, begin at 45° to the shear plane. (*B*) The shaded sectors between lines *L* and *R* on the initial circle represent the finite shortening field that will exist after the first infinitesimal increment of deformation, which is too small to show with an ellipse. The instantaneous stretching axes are along $\dot{S}_1$ and $\dot{S}_3$. (*C–E*) Progressive deformation causes lines *L* and *R* to rotate into the finite extension fields, whereby they are lengthened. The intersection of the finite strain ellipse and the initial circle (dashed line) defines lines of no finite stretch, which have the same length with which they started. The finite shortening field is shaded and grows with time as the lines of no finite stretch rotate.

Figure 9.94 Evolution of finite strain ellipses for simple shear. Symbols and lines are same as described in Figure 9.93. Note that line *N*, normal to the shear zone, rotates, but line *P*, parallel to the shear zone, does not. All other lines rotate in the same direction as the sense of shear. Also, the long axis of the finite strain ellipse progressively leans over in the direction of the sense of shear.

acteristic of coaxial deformation, where the material lines that at first were stretched the fastest ended up being stretched the most.

Let's examine what happens to lines *L* and *R* as they rotate (Figure 9.93). At the start of deformation, both lines were parallel to the lines of zero instantaneous stretching rate, which are also lines of maximum rotation rate. Accordingly, lines *L* and *R* rotate relatively fast, but change length slowly. As deformation proceeds, their new orientations progressively cause them to rotate more slowly and to lengthen faster. Their rate of rotation steadily decreases as they approach the shear plane. At very high strains, such lines will be greatly lengthened and will be at a very low angle to the shear plane.

The Finite Strain Ellipse During Simple Shear

In simple shear, S_1 first appears at 45° to the shear plane, parallel to $\dot{S}_1$ (Figure 9.94*A* and *B*). With additional increments of deformation, S_1 is successively rotated over, out of parallelism with $\dot{S}_1$ and toward parallelism with the shear plane (Figure 9.94*C, D,* and *E*). S_3 becomes increasingly shortened and progressively rotates toward the normal to the shear plane and away from $\dot{S}_3$. Both finite stretching axes therefore rotate away from parallelism with the instantaneous stretching axes as deformation proceeds, a characteristic of a *noncoaxial* deformation. An implication of noncoaxial deformation is that different material lines pass through both the *instantaneous* stretching axes and the *finite* stretching axes during deformation. S_1 starts out subparallel to line *R* at the onset of deformation but ends up nearly aligned with line *N* after a shear strain of 3 (Figure 9.94*F*).

There are two lines of no finite stretch that pass through the intersection of the finite strain ellipse and initial circle. One is fixed parallel to the shear plane and does not rotate or stretch during deformation (line P in Figure 9.94). The other line of no finite stretch starts out perpendicular to the shear plane at the onset of deformation but then rotates in the same direction as the finite strain ellipse. It rotates more slowly than material lines; it started out parallel to line *N* but becomes increasingly discordant to *N* with progressive deformation. As this line of no finite stretch rotates, the finite shortening field expands in angular width at the expense of the finite stretching field. The other boundary of the finite shortening field, parallel to line *P*, remains parallel to the shear plane and does not rotate.

Note that lines *L, N,* and *R* all rotate in the same direction as the sense of shear. Line *N* starts out normal to the shear plane and rotates the fastest during the initial stages of deformation. Line *R* rotates progressively more slowly as it rotates into a lower angle with the shear plane. Line *L* starts out rotating at the same rate as *R,* but accelerates until it has reached the normal to the shear plane, after which it rotates more slowly. As with pure shear, lines rotate very slowly as they approach the shear plane.

Progressive Deformation

The behavior of lines during the two types of shear illustrates a fundamental concept in structural geology—**progressive deformation**. Strain is built into rocks one small increment at a time, and the finite strain of the rock is the sum of all these small increments. A material line, because it may rotate during deformation, may not be affected in the same way by each increment of deformation. A line may start out in the instantaneous shortening field but progressively rotate into the instantaneous extension field; in other words, it may shorten and then lengthen during the course of a single kinematically consistent deformation.

To explore the implications of progressive deformation, we superimpose the instantaneous stretching fields onto the finite strain ellipse (Figure 9.95). After any amount of shear, one or both lines of no finite stretch will have rotated out of parallelism with its originally corresponding line of zero stretching rate. The finite stretching fields, therefore, will not coincide with the instantaneous stretching fields even after the first increment of strain.

The lack of correspondence of the finite and instantaneous stretching fields results in three different sectors of the finite strain ellipse that contain

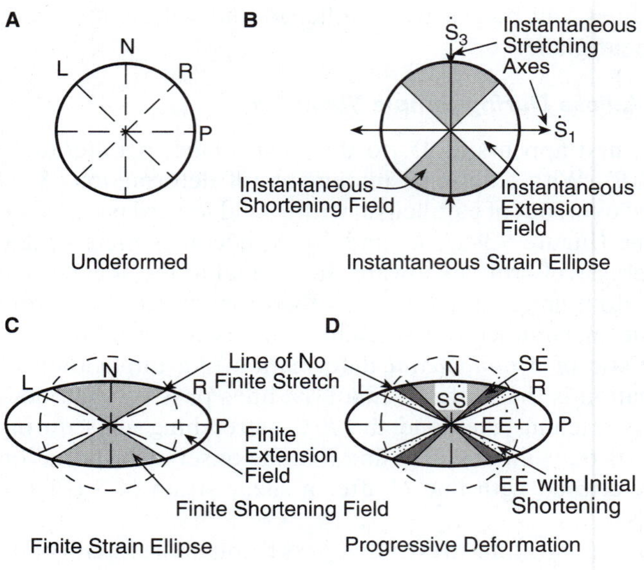

Figure 9.95 Strain ellipses for progressive deformation. (*A*) Undeformed circle containing lines *P* (parallel to shear zone), *N* (normal to shear zone), *L*, and *R*. (*B*) Instantaneous strain ellipse, showing the instantaneous stretching axes and stretching fields. (*C*) Finite strain ellipse with finite stretching fields bounded by the lines of no finite stretch. (*D*) Ellipse for progressive deformation, derived by superimposing the instantaneous and finite strain ellipses. The lightly shaded $S\dot{S}$ sectors indicate where the finite and instantaneous extension fields overlap. The darkly shaded $S\dot{E}$ sectors indicate where lines are shortened overall (inside the lines of no finite stretch) but are now being extended. The $E\dot{E}$ sectors indicate the orientation of lines that are extended overall and are still being extended. Some such lines may have had an initial history of shortening ($E\dot{E}$ sectors with initial shortening).

lines with similar stretching histories (Figure 9.95*D*). The sectors are labeled with two letters: the first indicates whether radial lines that fall within that sector have been shortened or lengthened overall. Lines with orientations that fall within the *finite shortening field* have been shortened overall and are designated *S*. Lines that fall into the *finite extension field* have been lengthened (extended) overall and are designated *E*.

The second letter denotes whether a line is being *instantaneously* shortened or lengthened. Lines that fall in the *instantaneous shortening field* are in the process of being shortened and are designated with an $\dot{S}$ (the dot over the *S* indicates that we are referring to a rate). Lines that fall in the *instantaneous extension field* are in the process of being lengthened and are designated with an $\dot{E}$.

The $S\dot{S}$ sector is inside both the *instantaneous* and *finite shortening fields*, and it represents radial lines that have been shortened overall and are still being shortened (Figure 9.95*D*). The sector retains the same 90° angular width during progressive deformation because the instantaneous shortening field does not widen with time. In the field, dikes, layers, or other long features with orientations within this sector would be folded or would display some other evidence of continual shortening.

The unshaded and lightly stippled $E\dot{E}$ sector is within the *instantaneous* and *finite extension fields* and contains lines that have been lengthened overall and are still being lengthened. It decreases in angular width during deformation as the lines of no finite stretch rotate.

The intervening $S\dot{E}$ sector is within the *finite shortening field* but *instantaneous extension field*. It encompasses lines that have been shortened overall but are being instantaneously lengthened. It widens with progressive deformation as the lines of no finite stretch rotate toward S_1. Material lines that fall in this sector have moved from the instantaneous shortening field into the instantaneous extension field. They were originally shortened but later rotated into orientations where they could be lengthened. They are still shorter than they were at the start because they are in the finite shortening field. This sector might be expressed in the field as folds that have been partially unfolded or dissected into incipient boudinage.

A more surprising stretching history is represented by the lightly stippled area of the $E\dot{E}$ sector. This area is within both the *instantaneous* and *finite extension fields*, and so lines within it have been lengthened overall and are still being lengthened. They began, however, in the instantaneous shortening field, but were rotated out of this field and into the $S\dot{E}$ sector, where they began to become lengthened. With enough strain, they were rotated into the $E\dot{E}$ sector, where they were lengthened to more than their original length!

Note that no lines rotate from the instantaneous extension field into the instantaneous shortening field. As a result, we do not expect to see boudins that have been folded. If such features occur, they may indicate that the rocks experienced more than one deformation event.

Progressive Pure Shear

To examine how progressive deformation works for pure shear, we superimpose the instantaneous stretching fields on the finite strain ellipses (Figure 9.96). As expected, the finite stretching axes (S_1 and S_3) do not rotate out of parallelism with the instantaneous stretching axes. In contrast, both lines of no finite stretch rotate toward the shear plane, resulting in a widening of the finite shortening field compared to the instantaneous shortening field. This produces the three types of sectors containing lines with similar stretching histories ($S\dot{S}$, $S\dot{E}$, and $E\dot{E}$ in Figure 9.78). The sectors

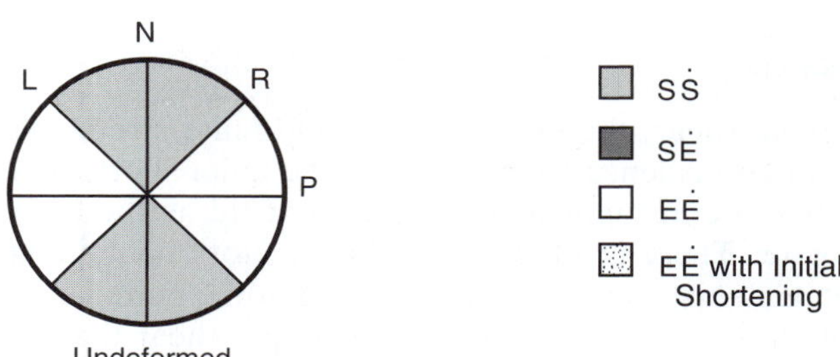

Figure 9.96 Progressive deformation during pure shear. Same initial conditions and sequence of deformation as in Figure 9.93. The different sectors ($S\dot{S}$, $S\dot{E}$, and $E\dot{E}$) are described in Figure 9.95.

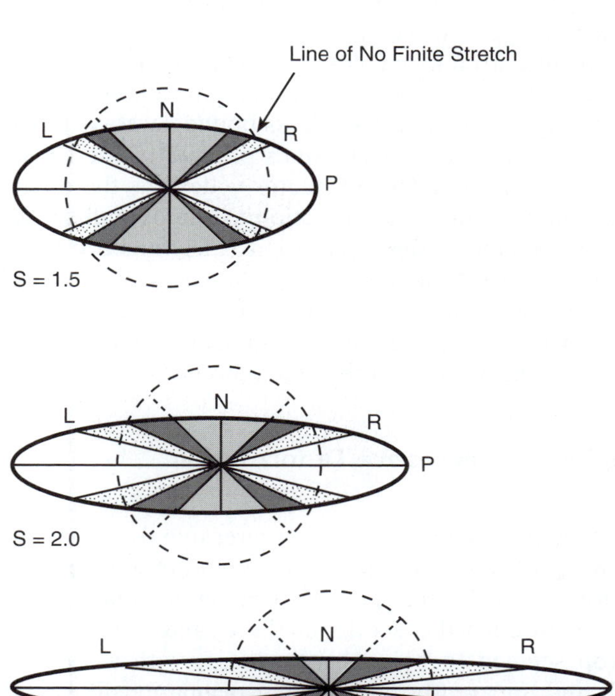

remain symmetrical with respect to the shear plane and to the instantaneous stretching axes because the two lines of no finite stretch rotate in opposite directions at the same rate. Note that there is an $S\dot{E}$ sector on both sides of the normal to the shear zone.

All material lines, except those parallel to the stretching axes, rotate toward the shear plane and $\dot{S}_1$. Lines rotate out of the instantaneous and finite shortening fields and into the instantaneous and finite extension fields. Lines in the stippled area within the $E\dot{E}$ sector were initially shortened but have become lengthened overall.

During any stage of deformation, the lines of no finite stretch always fall within the instantaneous extension field. In other words, although these lines are the same length as they started, they are now being extended. They have experienced the history described above—that is, they were originally shortened but are now being extended. The implication is this: in pure shear, a line can have its original length only if it has been shortened and then extended. The members of a whole succession of previously shortened lines will regain their original lengths as they rotate out of the finite shortening field and into coincidence with the lines of no finite stretch. With continued deformation, however, they are rotated away from the lines of no finite stretch and into the instantaneous extension field, where they will be lengthened further.

Progressive Simple Shear

During progressive simple shear, the instantaneous and finite stretching fields again rotate out of coincidence (Figure 9.97). The finite shortening field becomes wider than the instantaneous shortening field as the lines of no finite stretch rotate. There are the same three sectors containing lines with similar stretching histories, but the evolution of the boundaries of the sectors is much different during progressive simple shear than in progressive pure shear. This is because in simple shear, the finite stretching fields are asymmetric with respect to the shear plane.

The instantaneous and finite stretching fields share a common boundary parallel to the shear plane, where one line of no finite stretch coincides with a line of zero stretching rate (Figure 9.97). This boundary stays fixed during deformation because, in simple shear, lines parallel to the shear plane never stretch and never rotate. The $S\dot{E}$ sector widens as the other line of no finite stretch rotates in the same direction as the overall sense of shear. From the first increment of deformation, it is asymmetric with respect to the instantaneous stretching axes and to the shear plane. The adjacent $E\dot{E}$ sector narrows with time and is similarly asymmetric. The asymmetry reflects the sense of shear and the rotation of all material lines, except those parallel to the shear zone, in the same direction.

Applying the Principles of Progressive Deformation to the Field

As an example of how to apply the principles of progressive deformation in the field, Figure 9.98*A* portrays a composite of several exquisite outcrops of mylonitically foliated gneiss that I (SJR) discovered while rafting down the Colorado River through the depths of the Grand Canyon. At first glance, the outcrop seemed to be a chaotic mess, with a variety of folded, boudinaged, and apparently undeformed pegmatite dikes. In addition to the dikes, a lone thin shear zone and a few S-C fabrics cut across the foliation, also at a low angle. Luckily, lineation within the thin

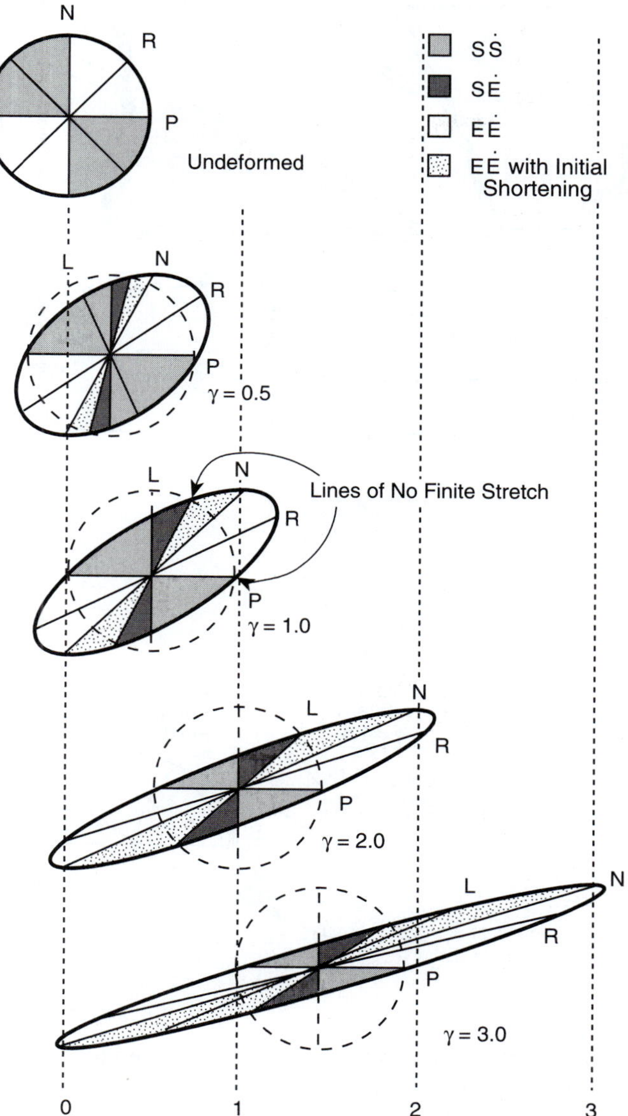

Figure 9.97 Progressive deformation during simple shear. Same initial conditions and sequence of deformation as in Figure 9.94. Symbols are the same as described for Figure 9.95. Note the difference in symmetry between this figure and the one for pure shear (Figure 9.96).

shear zone and in the mylonitic host rock was roughly parallel to the outcrop surface (i.e., I was observing the SOS plane).

So what can we tell about the strain that affected this outcrop? Where do we start? After describing and sketching the outcrop in detail, we might decide to make a plot of how each dike was strained versus its apparent trend on the outcrop (Figure 9.98*B*). This plot shows that the folded dikes all trended from upper left to lower right across the outcrop, whereas dikes that had been boudinaged had a more restricted range of trends in the opposite orientation. A few dikes, with a narrow range of orientations, were boudinaged folds, and some dikes that were largely undeformed were parallel to the lone thin shear zone. We can see that the plot has different sectors, each confining orientations of dikes with a specific structural style (Figure 9.98*C*). The folded dikes are assigned to the finite shortening field, whereas the boudinaged dikes must be in the finite extension field. Dikes containing boudinaged folds fall in an intervening sector where they have been shortened and then lengthened.

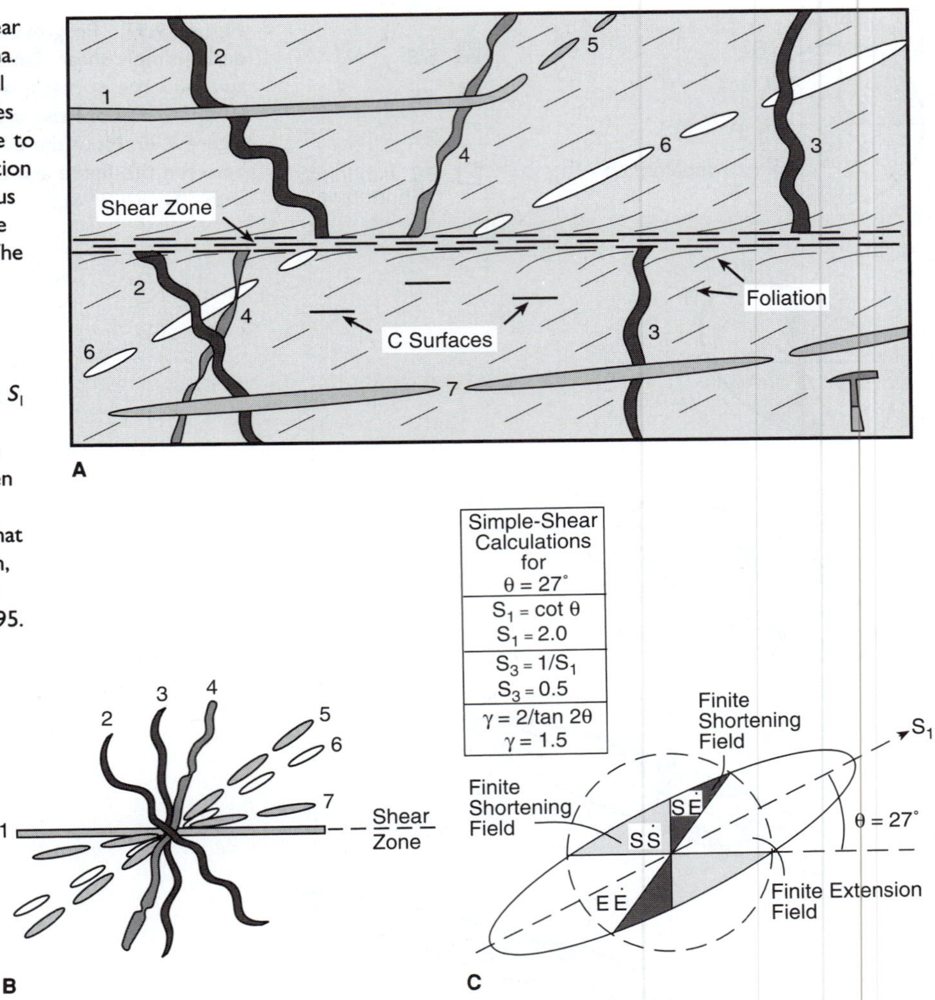

Figure 9.98 Progressive deformation near a shear zone in the Grand Canyon, Arizona. (*A*) Sketch of relations observed in several adjacent outcrops, showing numbered dikes of differing orientations and their response to deformation. (*B*) Plot of style of deformation (folded, boudinaged, or undeformed) versus orientation of dike. (*C*) Orientation of the finite stretching fields derived from (*B*). The asymmetric distribution of the stretching fields with respect to the shear zone indicates noncoaxial deformation and a dextral sense of shear. If we assume that deformation was by simple shear and that S_1 bisects the finite extension field, we can calculate the shape and orientation of the finite strain ellipse from the angle between S_1 (foliation) and the shear zone. The undeformed character of dike 1 implies that it coincides with a line of no finite stretch, which for simple shear lies parallel to the shear plane. Symbols same as in Figure 9.95.

This figure looks strikingly similar to ones we just examined for progressive deformation. But does the distribution of our sectors look more like coaxial deformation (pure shear) or noncoaxial deformation (simple shear)? And what is the sense of shear, if any? By inspection, our sectors appear to resemble simple shear because they have a clear asymmetry, displaying only one sector of boudinaged folds, rather than the two we expect for pure shear (Figure 9.96). The pattern suggests a dextral sense of shear, along a shear plane at a low angle to foliation. This calculated shear plane is approximately parallel to the thin shear zone and the C-surfaces, which is reassuring. It is also approximately parallel to the least deformed dikes, which must have escaped significant deformation because they were parallel to the shear plane during simple shear. The inclination of foliation to the thin shear zone and the way it is deflected across the zone are consistent with a dextral shear sense (Figure 9.98A). By the way, did we forget to mention that the S-C fabrics showed a dextral sense of shear?

This example, although nearly ideal, illustrates how an understanding of progressive deformation may help us make sense of outcrops, which otherwise might be perplexing. It also emphasizes how we use every shear-sense indicator available to compare the individual results for consistency. Finally, it conveys just a taste of the incredible geometric elegance of shear zones and progressive deformation.

ON TOWARD PLATES

Displacements on shear zones and their associated faults make headlines every time an earthquake strikes, in China, Iran, Turkey, Greece, Mexico, or California. Such manifestations of *active deformation* are not distributed uniformly across the face of the planet. Instead, earthquakes, just like shear zones, are extremely localized in their distribution, being concentrated in relatively narrow belts that coincide with our modern plate boundaries. Between these belts of active deformation are broad regions where, from a structural standpoint, not much is happening. The present-day tectonic situation further illustrates that deformation commonly becomes concentrated in narrow zones rather than being homogeneously distributed throughout the entire rock mass. This principle applies at all scales of observation, whether the rock mass we are considering is a thin section or a continent.

We have now completed our tour of the main types of structures we will find in outcrop: joints and shear fractures, faults, folds, cleavage, foliation, and lineation, and shear zones. We gather up this new and important knowledge and step back, pulling our focus away from the outcrop-scale and regional features we have grown familiar to recognize, and concentrate instead on the entire blue planet we call home. We now enter a new realm, one in which structures are thousands of kilometers long and along which deformation occurs that overnight can disrupt major cities and impact millions of lives. We now enter the realm of plate tectonics, the driving force for nearly all large- and small-scale deformation on Earth.

CHAPTER 10

PLATE TECTONICS

PLATES ON THE MOVE

Plate tectonics provides a backdrop for understanding the origin and significance of geologic structures, especially regional structures. Plate tectonic analysis is the essential basis for interpreting the dynamic circumstances that give rise to deformational movements. According to plate tectonic theory, the Earth can be subdivided into discrete fundamental "rigid" plates that move in relation to one another (Figure 10.1). A lot happens at the boundaries between plates, and within the margins of plates. The boundaries and margins of plates typically are sites of tectonic deformation. Notably, mountains 'build' at plate boundaries and plate margins.

The study of contemporary **active tectonics** permits us to observe plates in action. We can "see" firsthand some of the cause-and-effect relationships between plate motions and plate-induced deformations. Thus actively deforming regions within the Earth constitute unrivaled natural laboratories for studying the basic science of structural processes.

There is no doubt that plate motions are real. The emerging technologies that can be used to describe contemporary plate tectonic activity and deformation are extraordinary. For example, Global Positioning System (GPS) studies permit precise satellite monitoring of the four-dimensional (x,y,z,t) space–time locations of "targets" placed in networks on plates. The reference system is the geoid itself, not some small part of it. Steel spikes constitute the reference targets, and they are embedded in the ground, preferably in bedrock. Geoscientists then set up a tripod-supported GPS instrument above the spike (Figure 10.2). The instrument is wired to a portable computer and calls in radio signals from three or more

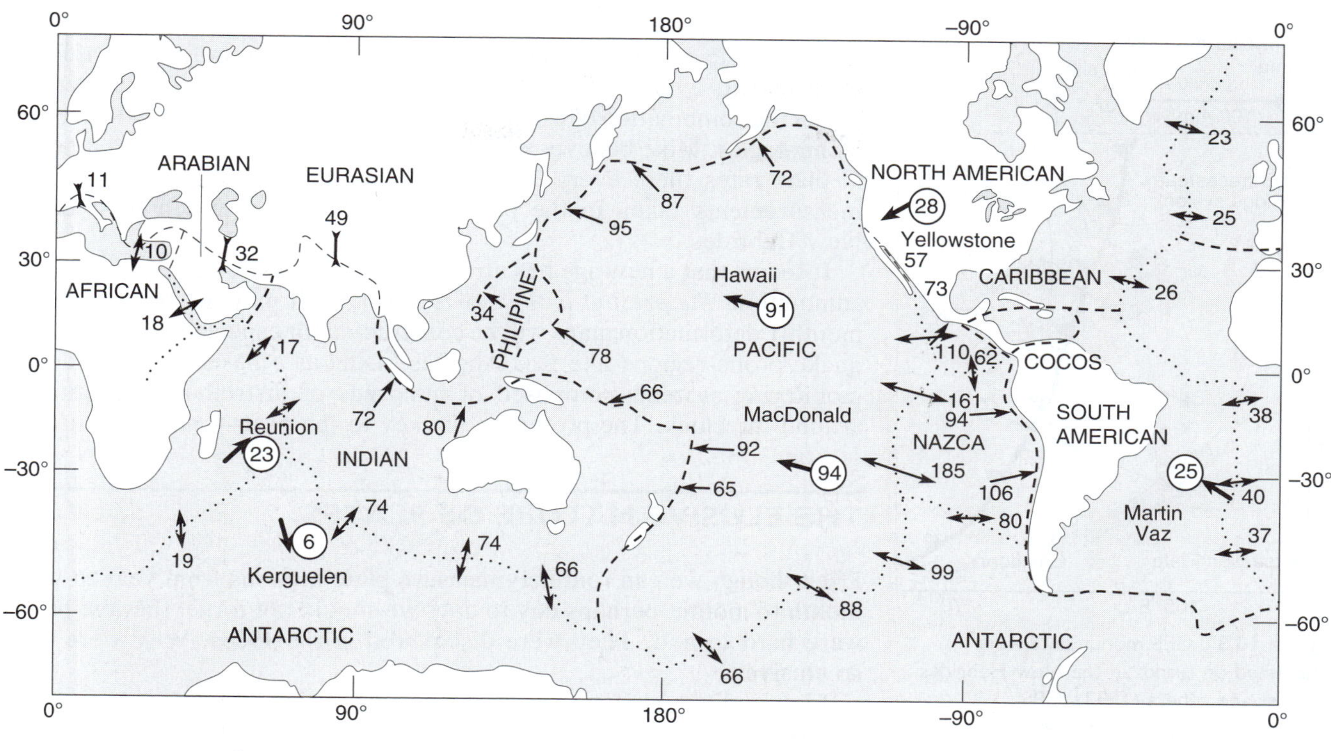

Figure 10.1 The directions and rates of relative motions of the plates of the Earth. Velocities are given in mm/yr. Circled numbers (also in mm/yr) are the absolute motions of plates relative to hotspots. [From Chase (1978), Earth and Planetary Sciences Letters, v. 37. Reprinted with permission of Elsevier Science.]

satellites to triangulate the position of the spike in space and time. Upon revisiting the reference targets periodically over a period of months or years, changes in horizontal and vertical positions can be detected at a confidence level of millimeters.

Mike Bevis and his colleagues (1993) placed such a network of spikes on islands in the Tonga and New Hebrides island arcs in the southwest

Figure 10.2 Bob Smith, Rick Hutchinson, and Chuck Meertens (right to left) stand in the mist at one of their GPS stations in the Yellowstone region, Wyoming. (Wouldn't you love to be there!) The tripod-mounted high precision receiver is positioned directly above the monument target, consisting of a spike-nail secured into the ground. The head of the nail contains at its exact center a tiny indentation, which is the target. The main objective of the Yellowstone GPS project is to understand the dynamics and space–time signature of active crustal deformation associated with the Yellowstone hot spot. (Photograph courtesy of Bob Smith.)

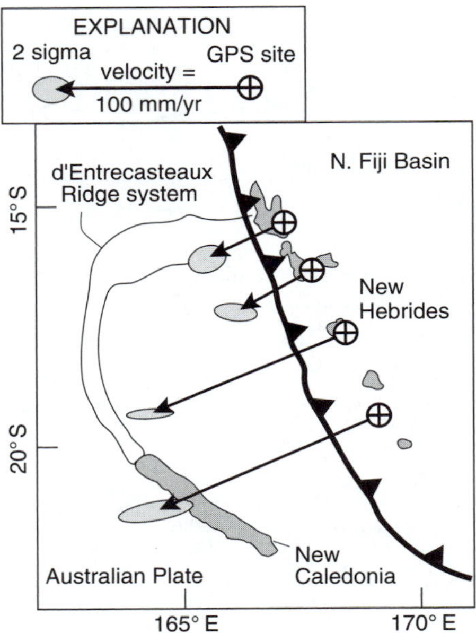

Figure 10.3 GPS monuments were established on islands in the New Hebrides by Bevis and others (1993). GPS measurements taken in the period 1990–1992 have revealed velocities that are locally in excess of 100 mm per year, and even these are not the very fastest rates. (Diagram courtesy of Mike Bevis.)

Pacific to measure the motions across a plate boundary between the Australian and Pacific plates. They in fact measured the most rapid relative plate velocity ever observed on Earth: 22 cm/yr across the Tonga Trench!! Given the combination of extraordinarily high rates and steadily improving technologies, Mike believes that he will soon be able to solve for changes in plate rates there every 3 weeks. Figure 10.3 presents the results of measurements taken in the period 1990–1992 in a trench west of the New Hebrides.

It seems that a new age has arrived in exploiting the power of Hutton's admonition: *the present is the key to the past*. It may not be long before monthly deformation maps will be constructed for especially active, earthquake-prone regions like Los Angeles, using as a basis permanent GPS monitoring systems composed of hundreds of instruments functioning around the clock. The present is the key to the past—and the future.

THE ELUSIVE NATURE OF PLATES

Even though we can routinely measure plate motions from year to year, month to month, perhaps day to day, we should not forget that the plates were hard to find. They were discovered in the 1960s. Why were they so elusive?

First, individual plates are not composed simply of continents or ocean basins. If this were true, the existence of plates probably would have been discovered long ago. We now know that individual plates most commonly encompass *parts* of continents and *parts* of ocean basins. Depending on the locations of the plate boundaries, an individual continent may occur on one or more plates; the same is true of ocean basins. Since continents generally are just parts of plates, detailed studies of continental geology did not lead, and perhaps never would have led, to the discovery of the plate concept. It was exploration of the ocean basins that allowed the sum of the parts to emerge into full view.

A second factor—this one a peculiarity of the Earth's mechanical properties with depth—further prevented early recognition of the existence of plates. Alfred Wegener (1915) was extraordinarily insightful in recognizing the geometric congruence of the outlines of continents (Figure 10.4*A*), which he interpreted as resulting from "continental drift." In fact, long after the work of Wegener (1915), Bullard and others (1965) demonstrated that the fit of continents around the Atlantic Ocean is nearly perfect (except for the absence of Mexico) when the match is made along the edge of the continental shelves at 500 fathoms (927 m) depth (Figure 10.4*B*). In trying to explain **continental drift**, geologists emphasized the mechanical importance of the seismic boundary between the crust and mantle, the so-called **Mohorovicic discontinuity**, better known to us as the **Moho**. Continents of crust were envisioned by Alfred Wegener (1915; 1936) as plowing through continuously deforming mantle, like ships through the sea. This interpretation was formidably resisted, for Wegener's mechanical model simply was not compatible with interpretations of the strength of the Earth's crust. Now it is understood that the base of the plates does *not* coincide with the Moho. Rather, the lower boundary of the Earth's lithospheric plates lies *in* the mantle of the Earth, well below the Moho.

In short, the boundaries of individual plates do not exist where we might have expected them to be. Boundaries between continents and ocean basins generally are not plate boundaries, nor is the interface between the Earth's mantle and the Earth's crust. And since plates move so slowly by human standards, they never attracted much attention.

Lithosphere and Asthenosphere, Crust and Mantle

Plates are composed of **lithosphere**—crust and upper mantle material that is rigid enough to withstand very low levels of differential stress indefinitely without flowing (Figure 10.5). The upper reaches of individual lithospheric plates are crust, both continental and oceanic. **Oceanic crust** is relatively thin, ranging from about 4 to 9 km in thickness. It is composed predominantly of rocks of basaltic composition that are relatively high in density (average $\rho = 2.9$). **Continental crust** is relatively thick, ranging from approximately 25 to 70 km, and composed of rocks of granitic composition having relatively low density (average $\rho = 2.7$).

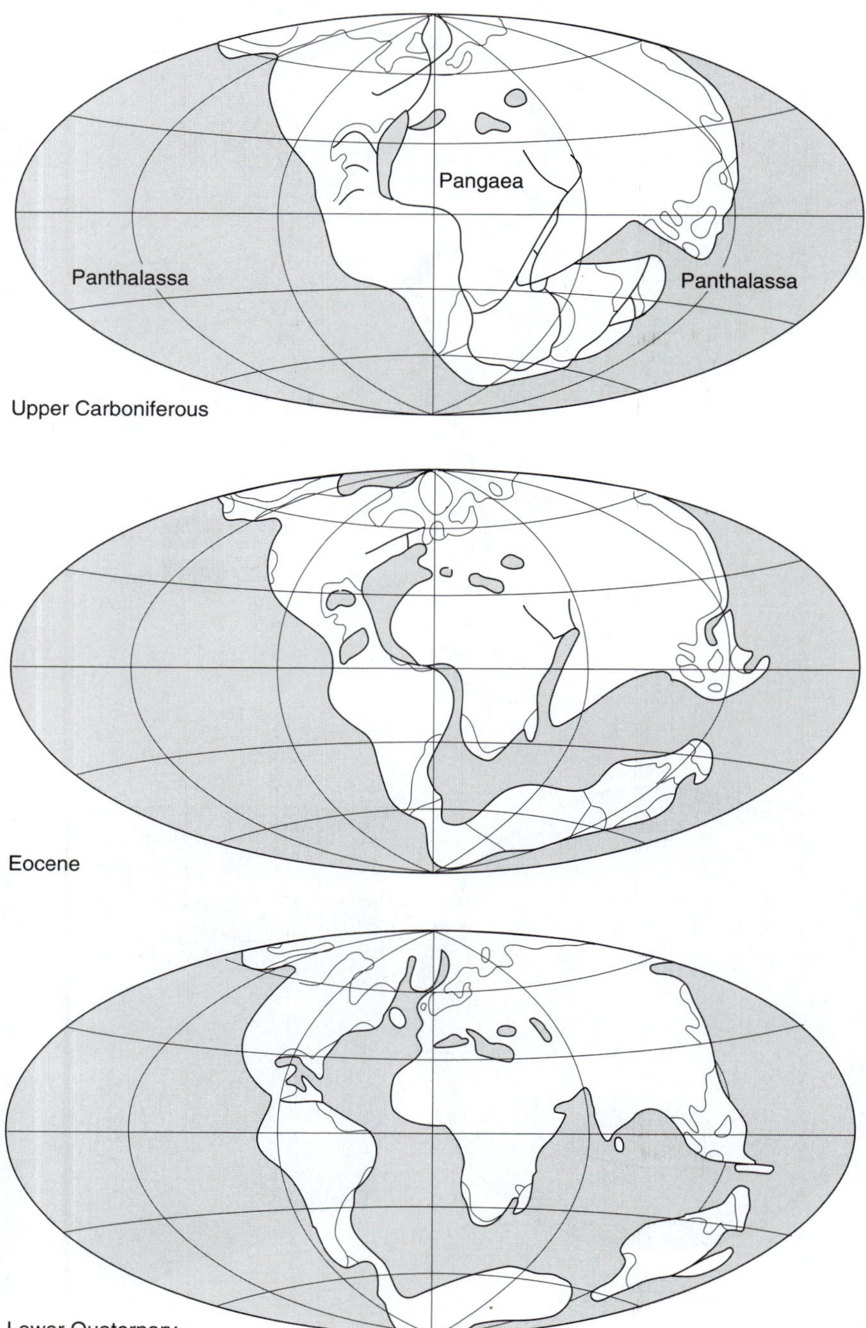

Figure 10.4 (*A*) Wegener's representation of continental drift. His mechanical model of continental drift featured continental crust plowing through oceanic crust like a ship through sea. [From Kearey and Vine (1990), global tectonics, Figure 1.3, p. 3, based on Wegener (1915). Reprinted with permission from Blackwell Science Ltd.]

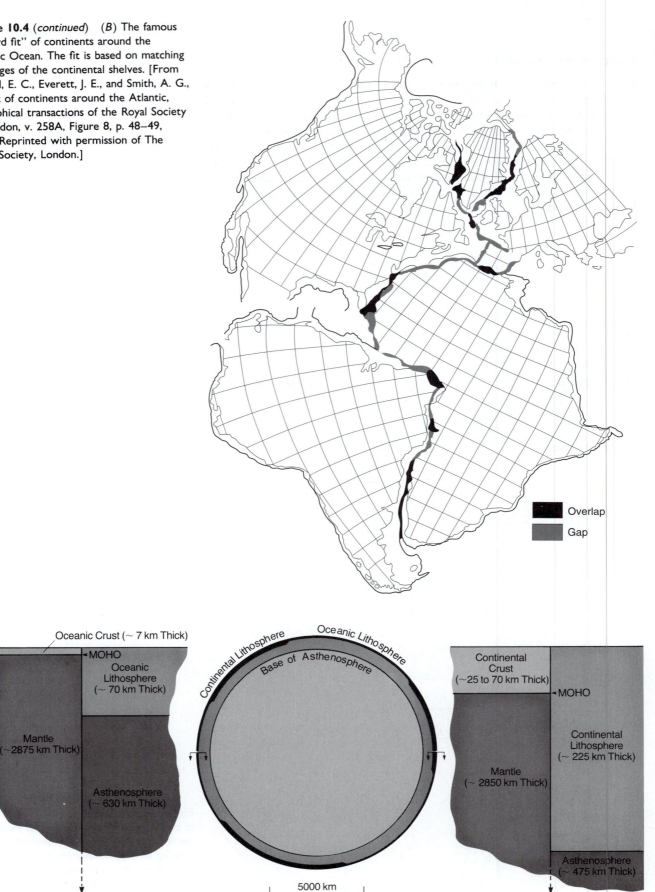

Figure 10.4 (*continued*) (*B*) The famous "Bullard fit" of continents around the Atlantic Ocean. The fit is based on matching the edges of the continental shelves. [From Bullard, E. C., Everett, J. E., and Smith, A. G., The fit of continents around the Atlantic, Philosphical transactions of the Royal Society of London, v. 258A, Figure 8, p. 48–49, 1965. Reprinted with permission of The Royal Society, London.]

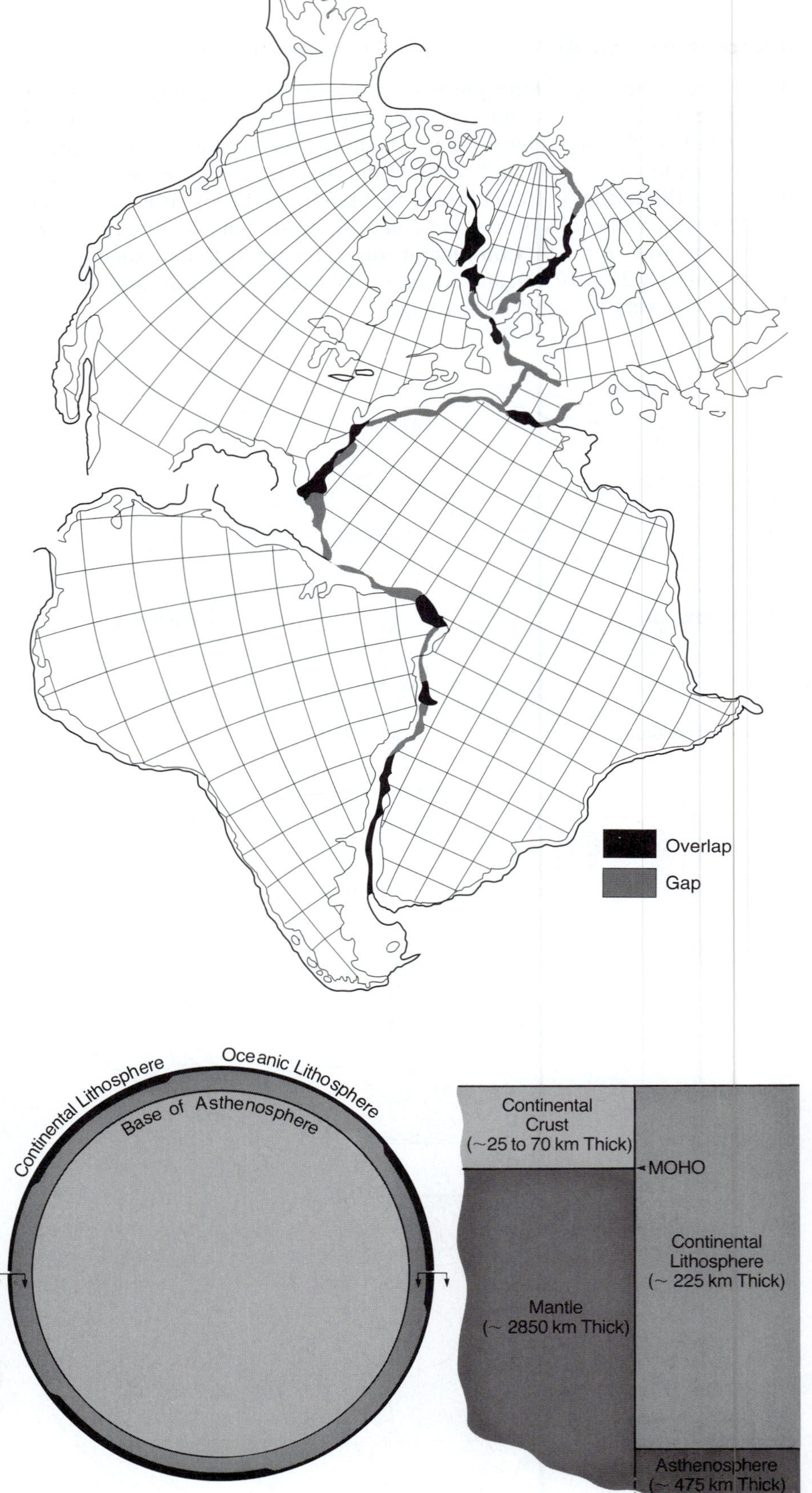

Figure 10.5 Physical/mechanical components of the crust and mantle.

The **Moho**, which marks the base of the crust and the top of the mantle, lies *within* lithospheric plates (Figure 10.5). In fact, it normally lies at a high structural level within lithosphere. The position of the Moho can be identified by a seismic velocity discontinuity. The Moho is thought to mark a lithological transition to underlying ultramafic rocks. It is a *compositional* boundary.

Lithosphere probably is thickest under continents (Figure 10.5), but this is still somewhat controversial. The relatively thick, convex-downward bulges of lithosphere under continental terranes may serve as "viscous anchors" that may halt, or slow down, plate motions (Chapman and Pollack, 1977; Jordan, 1988). Even so, the thickest lithosphere, when viewed at the scale of the Earth as a whole, is mighty thin. Lithospheric plates are like contact lenses on the surface of an eye: exceedingly thin and gently curved. Like contact lenses, the plates move and slip readily.

The lithosphere rides on **asthenosphere** (Figure 10.5), upper mantle material that is capable of flowing continuously, even under the lowest levels of differential stress (Karato, 1993). The ability of the asthenosphere to flow continuously permits movement of the overlying plates of lithosphere. The boundary between lithosphere and asthenosphere is one of major and significant ductility contrast. It is a *mechanical* boundary, and not a compositional one. The asthenosphere may contain a small fraction of partial melt, thus enhancing its capacity to flow in the solid state. The viscosity of the lower lithosphere is approximately 10^{23} poise, whereas the viscosity of the asthenosphere is approximately 10^{21} poise, two orders of magnitude lower!

The depth of the structural transition between lithosphere and asthenosphere is vague and for now is impossible to define precisely. The top of the asthenosphere is generally placed at the top of the **low velocity zone** in the upper mantle, where seismic waves undergo a significant drop in velocity (Figure 10.5). Under the ocean basins, the depth to this discontinuity is approximately 75 km. Under continental crust, it lies at an average depth of approximately 225 km, but its position is truly variable (Jordan, 1988).

The base of the asthenosphere may lie at a depth of approximately 700 km. The low velocity zone resides in the upper quarter of the asthenosphere. The 700 km depth to the base of the asthenosphere coincides with the depth level of the very deepest earthquakes.

Plate Motions and Plate Boundaries

The lithosphere is not a flawless outer shell. Rather, it is fragmented into many discrete plates that move relative to one another (Figure 10.6). **Plate boundaries** are the edges of plates, the contacts between adjacent plates. Plate boundaries in modern settings are marked by abundant earthquake activity. In fact, the historic record of earthquake activity around the globe is nearly a tracing of present-day plate boundaries (Figure 10.7). Depth levels of earthquakes at or near plate boundaries vary from very shallow to very deep, depending on the nature of the plate boundary and the mechanical conditions of the plates brought into contact.

The motions of plates can be described ideally as rigid body motions involving combinations of translation and rotation. Plate motions can be pictured in terms of convergence, divergence, and transform (Figure 10.8). Relative velocities between plates range from 1 to 22 cm/yr. Plate movements may combine in many ways, depending on the overall plate interactions.

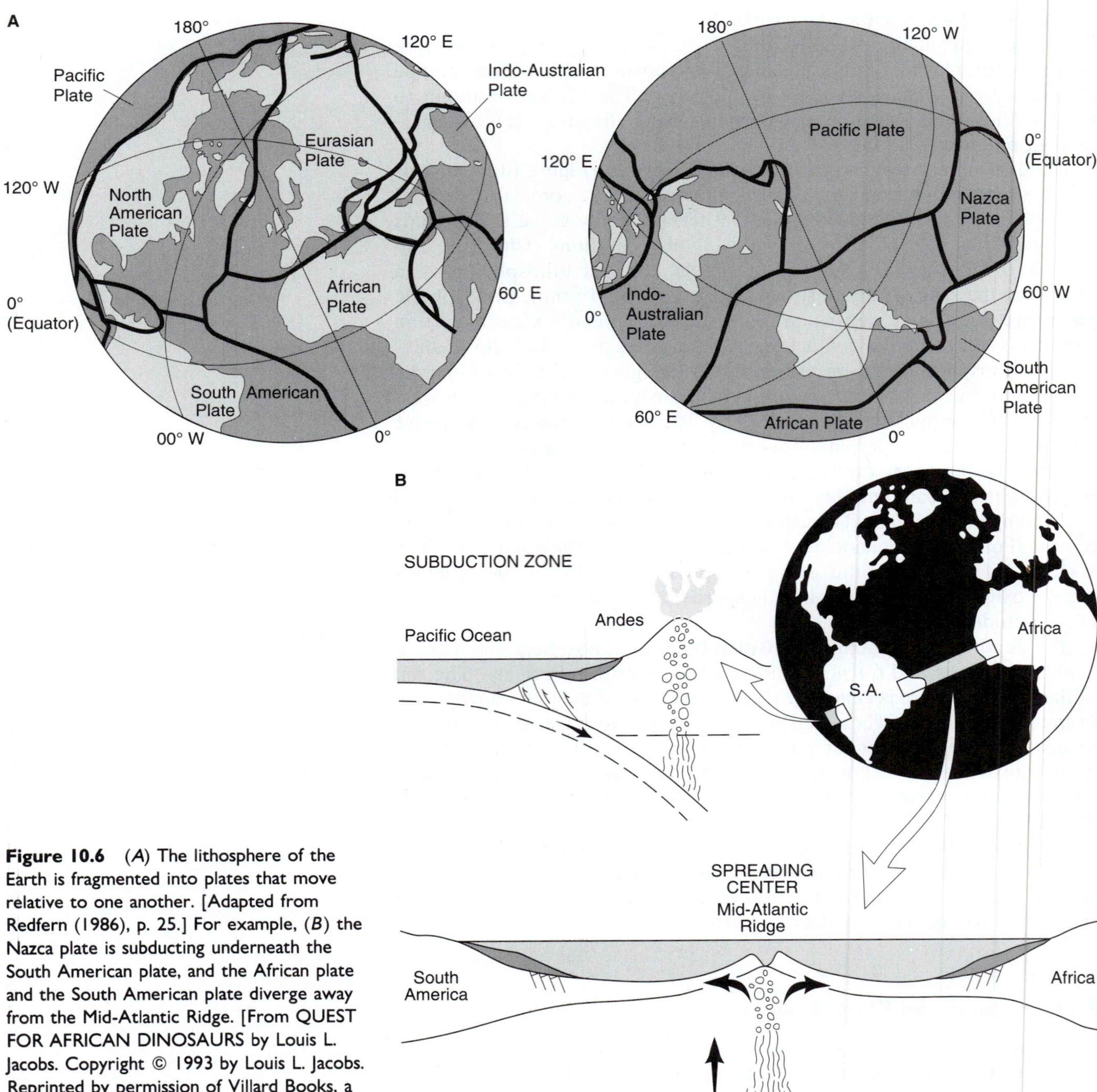

Figure 10.6 (*A*) The lithosphere of the Earth is fragmented into plates that move relative to one another. [Adapted from Redfern (1986), p. 25.] For example, (*B*) the Nazca plate is subducting underneath the South American plate, and the African plate and the South American plate diverge away from the Mid-Atlantic Ridge. [From QUEST FOR AFRICAN DINOSAURS by Louis L. Jacobs. Copyright © 1993 by Louis L. Jacobs. Reprinted by permission of Villard Books, a division of Random House, Inc.]

Convergence is marked by a relative movement that brings adjacent plates toward one another (Figure 10.8). Plates in convergence are in constant competition for space. A common response to the space problem is the structural descent of one plate beneath the other. In effect, rock is ''swallowed'' to greater depth (Bally and Snelson, 1980), through a tectonic process known as **subduction**. An alternative response of plates to the space problem of convergence is **collision**. Plate collision is like a slow-motion, head-on collision on a slippery highway between two cars whose brakes have locked. Converging plates in collision can be thought

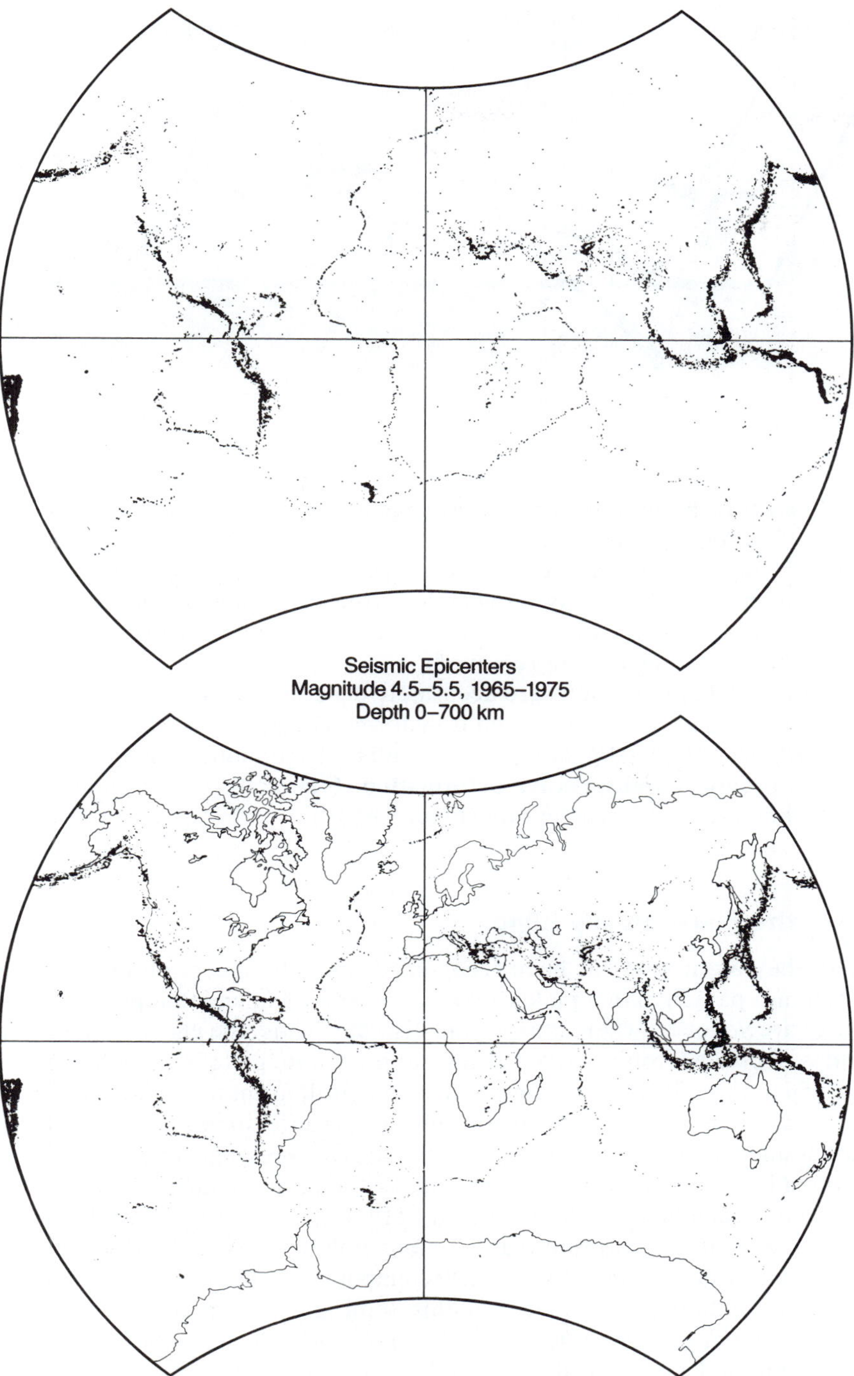

Figure 10.7 Earthquake epicenters are a tracing of plate boundaries. [From Lowman (1981). Courtesy of Goddard Space Flight Center, National Aeronautics and Space Administration.]

of as equally buoyant. Still, two plates cannot occupy the same space, and so regional-scale non–rigid body shortening must occur.

Divergence is marked by the movement of plates away from one another (Figure 10.8). The actual relative movement may be perpendicular or oblique to the boundary between adjacent plates. Without compensation, some void or opening would surely develop between diverging plates. But "would-be" voids are simultaneously and continuously filled in by upwelling igneous intrusions. When solidified, the intrusions and freshly made volcanic and sedimentary accumulations constitute new additions to the lithosphere.

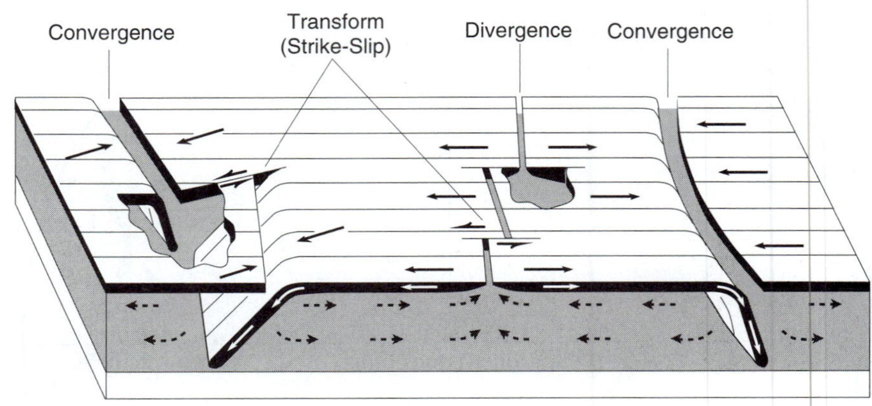

Figure 10.8 Convergence, divergence, and transform (strike-slip) of lithospheric plates. (From Isacks, Oliver, and Sykes, *Journal of Geophysical Research,* v. 73, fig. 1, p. 5857, copyright © 1968 by American Geophysical Union.)

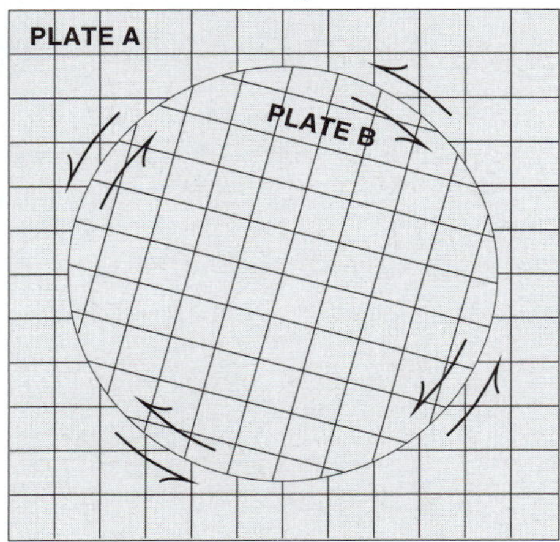

Figure 10.9 Ideal transform faulting between two plates is pure strike-slip. [From Cox, A. and Hart, R. B., Plate tectonics—how it works, 1986, Box 1-2, p. 13. Reprinted by permission of Blackwell Science, Inc.]

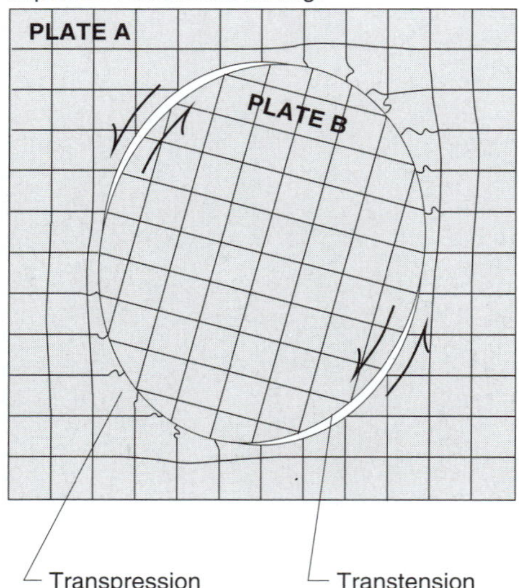

Figure 10.10 Where transform faulting has a component of compression, the environment is transpressive. Where transform faulting has a component of extension, the environment is transtensive.

Transform boundaries are where one plate scrapes past another horizontally (Figures 10.8 and 10.9). Steeply dipping fault zones and shear zones absorb the mechanical effects of stresses generated during the frictional movements. Material may be **accreted** from one plate to another as the plate interaction takes place. Alternatively, material may be sliced off a plate boundary region and removed.

Some plate interactions are accurately described as some combination of convergence, divergence, and transform. The notion is quite similar to describing oblique-slip motions on faults. An oblique *convergence* of plates produces **transpressive deformation** (Figure 10.10). An oblique *divergence* produces **transtension** (Figure 10.10).

Are the Plates Really Rigid?

No, they are not. It has been recognized since the inception of the plate tectonics paradigm that plates are *not* perfectly rigid and that plate boundaries are *not* everywhere narrow and discrete. Thus, the elegant kinematic rules of plate tectonics do not strictly apply within plate boundary regions in which there is a broad zone of distributed deformation. The challenge is to combine the ideal kinematic rules of plate tectonics with a realistic assessment of how stresses, deformational processes, and strains are *partitioned* laterally and vertically through the semirigid plates.

Gene Humphries at the University of Oregon has a wonderful way of viewing the implications of plate motions involving plates whose strength variations range from stiff to soft. He emphasizes that the ''partitioning'' and distribution of deformation within semirigid regions is an expression of the *interplay of stress fields and strength fields*. Because the lithosphere is strong and mechanically decoupled from the Earth's interior by the weak asthenosphere, tectonic forces focus stresses within lithosphere that can be transmitted great distances. Permanent strain or deformation occurs whenever and wherever the stress exceeds the strength of the material. The strength is determined by the rheology, which is in turn a function of a host of internal properties, such as composition and mineralogy, and external conditions, such as temperatures and stress.

Within complex systems there is, *at all scales,* a ceaseless competition between stress and strength. Where stresses overcome strengths, deformation processes bring about change in countless ways. We will see this as we explore the main plate tectonic boundary regions.

DIVERGENT BOUNDARIES

Seafloor Spreading

The path to discovery of seafloor spreading and plate tectonics was illuminated during World War II, when Harry Hess, while a member of the submarine service, engaged in mapping the bathymetry of the oceans. Out of this work came the discovery of oceanic ridges, notably the Mid-Atlantic Ridge. Hess and others also engaged in measuring the magnetic properties of the seafloor. The mapping and measuring were strategic efforts to support tracking and locating enemy submarines. Following World War II, Bruce Heezen and his colleagues at Lamont–Doherty Geological Observatory carried out exhaustive mapping of the seafloor bottom and discovered the world-encircling **ocean ridge system** (Figure 10.11), a network of ridges that branch and fork, marked by long linear segments broken by faults oriented transversely to the ridges. The ocean ridge system, exposed to partial view in places like Iceland, is about 60,000 km long and ranges from 1000 km to 4000 km in width. It is the longest mountain range on Earth! It displays a topographic relief of approximately 2 km. Rocks within the system are highly faulted, composed

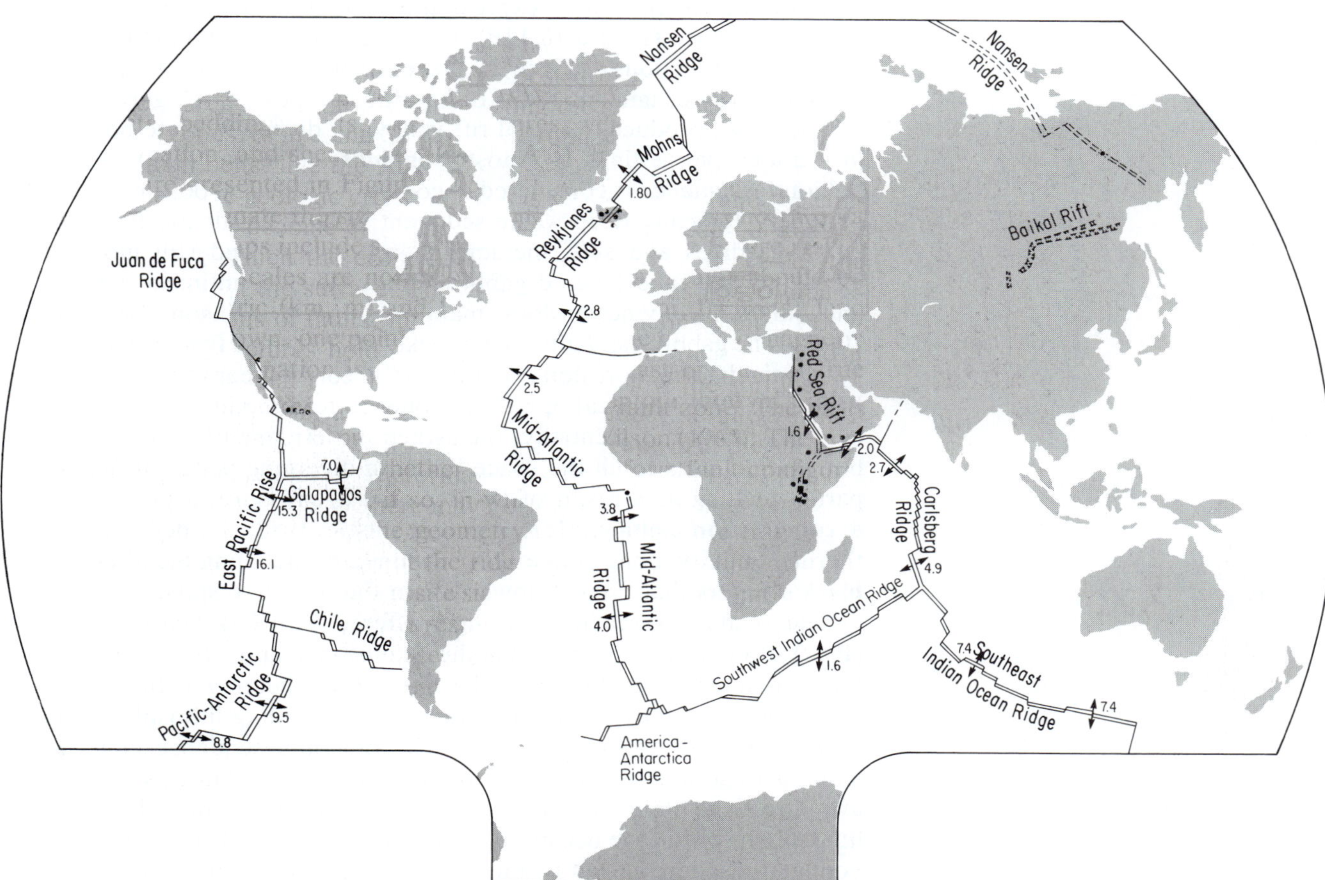

Figure 10.11 Ocean ridge systems of the world. Directions and magnitudes of spreading are shown as well. Magnitudes of spreading are given in total spreading rate in centimeters per year. [From Lowman (1981). Courtesy of Goddard Space Flight Center, National Aeronautics and Space Administration.]

mainly of basaltic volcanic rocks with interbedded oceanic sediments and sills, dikes, and thick-layered intrusions of mafic igneous rocks.

Insight regarding interpretation of the significance of the ocean ridge system stemmed from the revolutionary thinking of Bob Dietz and Harry Hess. Dietz (1961) and Hess (1962) both reached the conclusion that the oceanic ridges are sites where new lithospheric ocean floor is generated. Their model of **seafloor spreading**—the idea that newly created seafloor "spreads" in opposite directions outward from the ridges—radically transformed geological thinking about Earth dynamics.

The ocean ridge system is very broad and is marked by ridge-symmetrical topographic relief. The dynamically active part of the ridge is very restricted, lying along the axis of the system. Within a zone no wider than 30 km, new oceanic lithosphere is created by the combination of intrusion, extrusion, and extensional faulting. Ongoing dynamic activity is expressed partly by shallow-focus earthquakes of small to moderate magnitude. The shallowness of the earthquakes reflects thin lithosphere and shallow depth to asthenosphere. The relatively small magnitudes of the earthquakes reflect the overall weakness of such thin lithosphere.

Along the axis of the ocean ridge system is a **rift valley**, ranging commonly 2–10 km wide and displaying hundreds of meters of topographic relief. The center of the rift valley generally features a central high point, an edifice of volcanic accumulations (Figure 10.12*A*). Rocks and structures within the axial rift valley express the structural processes involved in plate divergence (Figure 10.12*B*). Cross sections of the axial rift vary in detail, but all sections disclose the presence of extensionally faulted volcanic and intrusive igneous rocks of basaltic composition (Figure 10.12*C*).

The rocks produced in axial rift valleys at divergent plate boundaries in intraoceanic settings are mostly **ophiolites** (Figure 10.13). A typical ophiolitic sequence is considered to consist, from top to bottom, of cherty or limey, typically deep-water sediment; pillow basalt locally intruded by mafic dikes and sills; metamorphosed gabbro and still more mafic dikes; massive and layered gabbros and some serpentinized peridotite; and underlying olivine peridotite mantle material (Dickinson, 1980) (Figure 10.13). The gabbros and ultramafic rocks are deformed from place to place by plastic deformation along subhorizontal zones shear (Nicholas and Le Pichon, 1980). A typical ophiolite sequence forming in axial rift settings is 5–10 km thick. Essentially, the asthenospheric mantle is mostly solid, but when it rises to fill the space between diverging plates, it undergoes partial melting as a result of the **decrease** in pressure, a process called **decompression melting** (McKenzie and Bickle, 1988). Evidence of hydrothermal activity along active rifts is abundant, including metal-encrusted **black smokers** that project upward from the floor like stalagmites.

The seafloor spreading paradigm offered by Dietz (1961) and Hess (1962) allows us to recognize that the extensional faults, the vertical dikes, the intrusions of gabbro, and the horizontal shearing within ophiolite sequences all contribute to the process of spreading lithospheric plates away from a divergent boundary. The ophiolite sequence constructed along a ridge axis is welded to the edges of both plates as they move away from the plate boundary. There it congeals to become the youngest lithosphere within the ocean ridge. Simultaneously and continuously, this young lithosphere is split by faulting and intrusion and is translated away laterally from the ridge axis, thus making room for yet younger lithosphere.

Vine and Matthew (1963) showed that geomagnetic field reversals are symmetrically imprinted on the seafloor lithosphere that spreads from ocean ridge plate boundaries (Figure 10.14*A*). Age data from many sources demonstrate that **magnetic stripes** at successively greater distances from

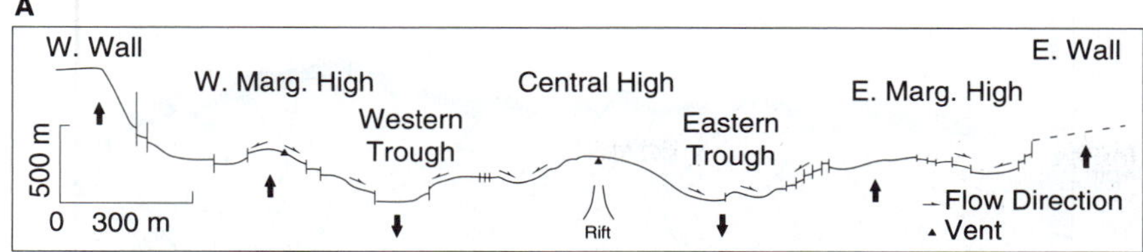

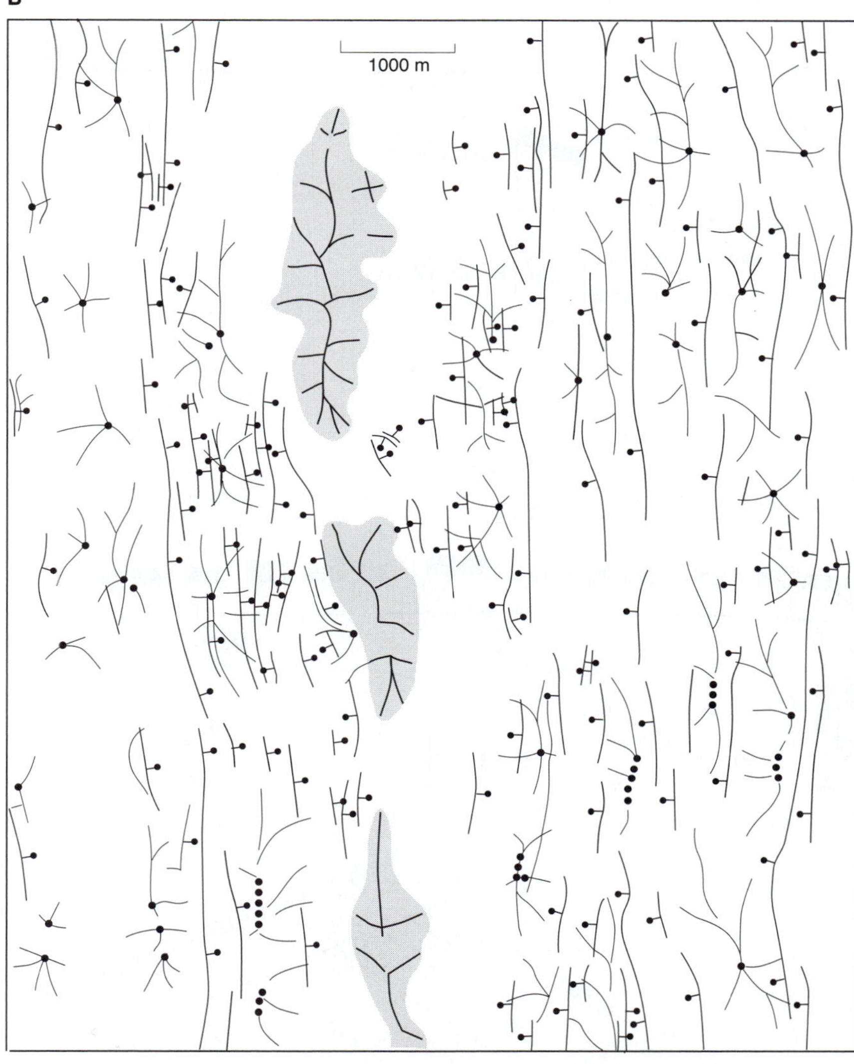

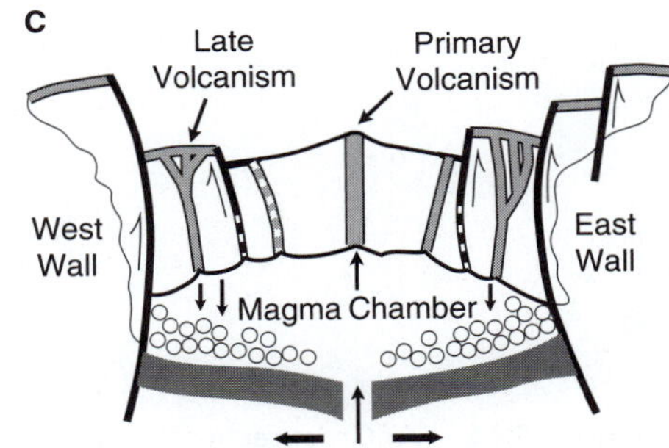

Figure 10.12 Details of the inner rift zone of the Mid-Atlantic Ridge at 36°N latitude. (*A*) Topographic profile showing relation of central high to the east and west walls. (*B*) Map showing faults (bar and ball symbols), volcanic bodies (shaded), and vents and crest lines of volcanoes (rows of black dots). (*C*) Schematic illustration of magmatic and volcanic processes at work in the inner rift zone. [From Ballard and van Andel (1977). Published with permission of Geological Society of America.]

ridge crests are increasingly older. Rates of seafloor spreading can be calculated on the basis of the age of ocean floor in combination with the distance of the dated ocean floor from the ridge crest (Figure 10.14*B*) (Isacks, Oliver, and Sykes, 1968).

Rates of divergent movement along the axis of the ridge system may seem slow, ranging from 1 to 20 cm/yr. Yet in geological terms, the rates of spreading are astonishingly fast. A simple way to make the connection is to express the velocity in mm/yr. A rate of 1 mm/yr corresponds to 1 km/m.y. Thus, if lithosphere moves at a rate of 100 mm/yr away from a ridge axis, it will cover 100 km in 1 m.y., or 1000 km in 10 m.y.

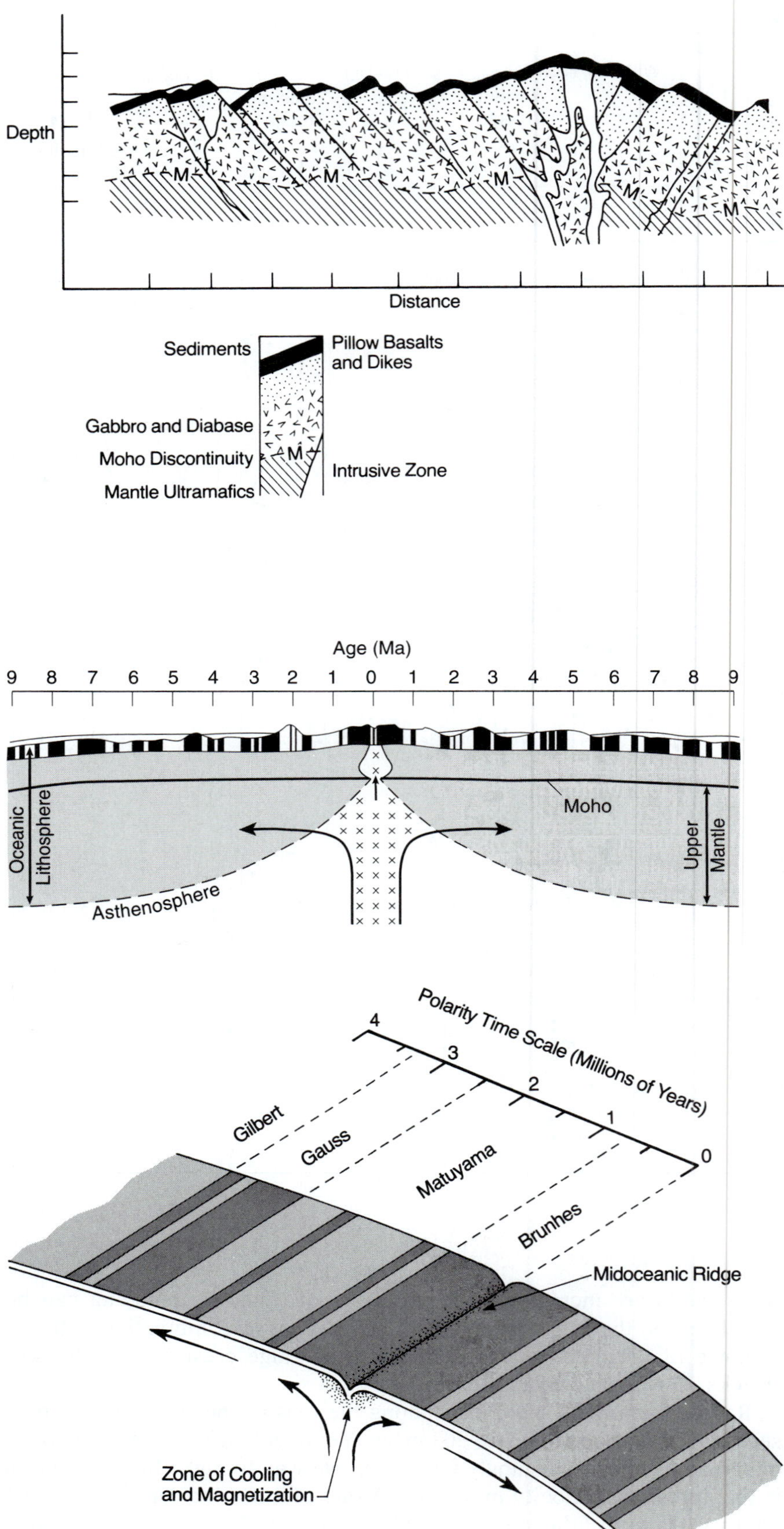

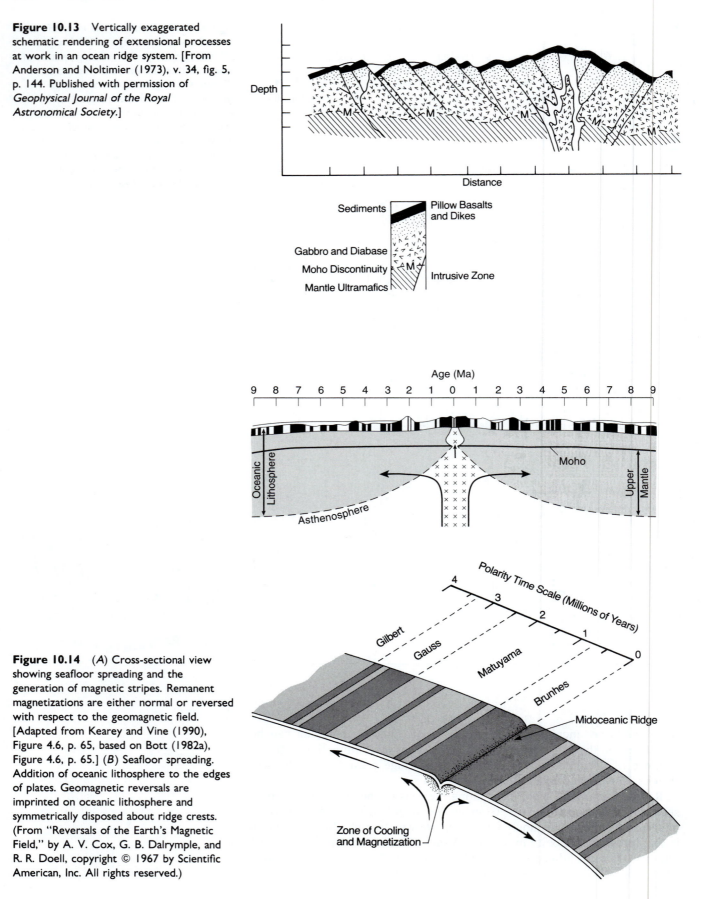

Figure 10.13 Vertically exaggerated schematic rendering of extensional processes at work in an ocean ridge system. [From Anderson and Noltimier (1973), v. 34, fig. 5, p. 144. Published with permission of *Geophysical Journal of the Royal Astronomical Society.*]

Figure 10.14 (*A*) Cross-sectional view showing seafloor spreading and the generation of magnetic stripes. Remanent magnetizations are either normal or reversed with respect to the geomagnetic field. [Adapted from Kearey and Vine (1990), Figure 4.6, p. 65, based on Bott (1982a), Figure 4.6, p. 65.] (*B*) Seafloor spreading. Addition of oceanic lithosphere to the edges of plates. Geomagnetic reversals are imprinted on oceanic lithosphere and symmetrically disposed about ridge crests. (From "Reversals of the Earth's Magnetic Field," by A. V. Cox, G. B. Dalrymple, and R. R. Doell, copyright © 1967 by Scientific American, Inc. All rights reserved.)

Geologic Development of Passive Continental Margins

The natural evolution of a divergent plate boundary is recorded not only in oceanic crustal rocks, but also in rock assemblages deposited along the margin of an original supercontinent, split and spread apart by the seafloor spreading process (Ingersol, 1988). Initial rifting and normal faulting of an original supercontinent is attended by stretching and subsidence of the lithosphere (Figure 10.15*A*) (Dickinson, 1980; Le Pichon and Sibuet, 1981; Bond and Kominitz, 1988). Extensional faults accommodate continental crustal pull-apart. Down-dropped blocks become sites of continental basins that are filled by clastic debris derived from adjacent high-standing blocks. The sedimentary deposits that form in such environments are called **rift basin deposits** (Figure 10.15*A*).

As extension proceeds, the lithosphere is thinned until eventually basaltic magma intrudes into and extrudes through the deformed lithosphere. Oceanic lithosphere is thus emplaced in the developing zone of separation between adjacent diverging plates (Figure 10.15*B*). In this way, oceanic crust is welded directly onto continental crustal rocks. The mixture of original continental crust and the added oceanic component at the continental edge produces a hybrid **transitional crust** that becomes nestled between bona fide continental crust and young oceanic crust (Figure 10.15*B*). The thermal input is so great during this initial rifting stage that the margin of the rifted continent is buoyantly uplifted.

With sustained seafloor spreading, the continental edge moves farther and farther from the **spreading center** represented by the midoceanic ridge. The edge of the continent is called a **passive margin** because it is no longer a plate boundary (Figure 10.15*C*). Structures and rock associations of the rifting phase are frozen into rocks of the continental margin, and they are passively conveyed laterally along with the rest of the plate.

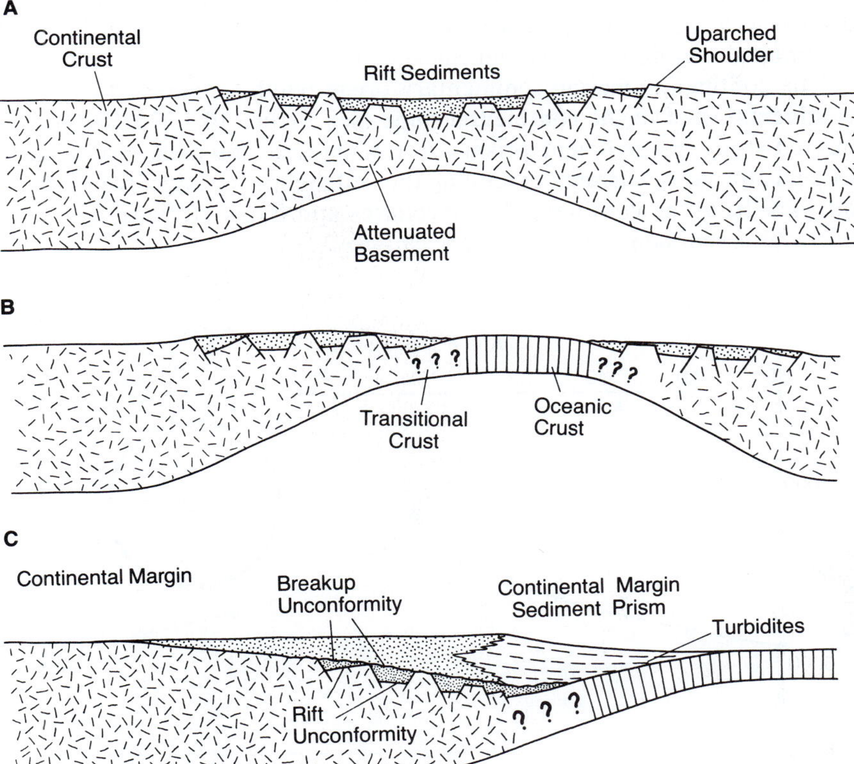

Figure 10.15 Structural evolution of a passive margin. (*A*) Stretching and subsidence. (*B*) Emplacement of new oceanic crust; buoyant uplift. (*C*) Subsidence of the passive margin and formation of continental margin prism. [From Dickinson (1980). Published with permission of Geological Association of Canada.]

Systematic massive subsidence of the passive continental margin takes place as the attached ophiolitic rocks cool off (Dietz, 1963; Dietz and Holden, 1966; Dewey and Bird, 1971; Bond and Kominitz, 1988). Oceanic lithosphere moves away from the ridge, cools, contracts, and increases in density. As this takes place, there is a progressive increase in water depth to the top of the lithosphere away from the ridge. The relationship is extraordinarily systematic for oceanic lithosphere younger than 80 Ma (Parsons and Sclater, 1977) (Figure 10.16):

$$D = 2500 + 350\sqrt{T}$$

where

$$D = \text{depth (m)}$$
$$T = \text{age (Ma)}$$

Imagine a column of oceanic lithosphere as it moves laterally from the spreading center. It will increase in density and will subside as it cools. In addition, the lithosphere will thicken as the boundary between lithosphere and asthenosphere moves to deeper and deeper levels with cooling.

Subsidence and down-flexing of the passive continental margin due to subsidence make room for deposition of a **continental margin sedimentary prism** (also known as a **miogeoclinal prism** or **miogeocline**) that builds outward from the continental interior (see Figure 10.15*C*). The basal part of the miogeoclinal prism contains conglomerates, coarse sandstones, and even some volcanics, all reflecting the waning rifting stage. These are overlain by great thicknesses of quartz-rich clastic rocks derived from the ancient continental margin. With time, a **carbonate platform** may build seaward from the continental margin, prograding over the basal sands. Such carbonate sequences of limestone and dolomite, along with interbedded shales and siltstones, record transgression of the sea onto the subsided continental margin (Stewart and Suczek, 1977).

The continental margin sedimentary prism builds oceanward from the **continental shelf** to the **continental slope** (see Figure 10.15*C*). Sediments pirated from the sediment prism are transported down the continental slope as turbidity currents extending the sedimentary wedge even farther outward from the continent. The **turbidites** interfinger seaward with thin deepwater sediments that rest on ocean crust.

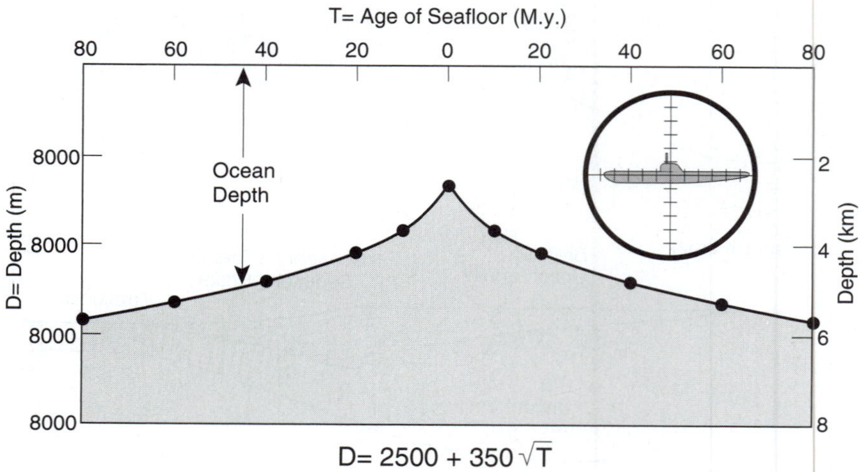

Figure 10.16 Graph showing the age–depth relationships for oceanic lithosphere younger than 80 Ma. A few simple calculations produced the control points on which the graph is based.

TRANSFORM BOUNDARIES

Discovery of Transform Faults

Transform faults, discordant to the trend of ocean ridge segments, are an integral part of the ocean ridge system. The topographic expression of these faults is boldly revealed in the bathymetry of the oceans (Figure 10.17). The faults appear to offset ridge segments within the otherwise continuous ocean ridge system. Shallow-focus earthquakes emanate from parts of these faults. The seismic activity expresses faulting and frictional sliding of thin lithosphere. Details of topographic and structural expression partly depend on spreading rates (Figure 10.18).

It was J. Tuzo Wilson who unlocked the geometric and kinematic significance of transcurrent intraoceanic faults. The stepping-stone for Wilson's interpretation was the model of seafloor spreading as set forth by Dietz (1961) and Hess (1962). Wilson (1965) reasoned that the function of transcurrent intraoceanic faults was to connect the end of one ridge segment (spreading center) to another, enabling the entire movement system along the spreading axis to be integrated (Figure 10.19). Because the faults *transform* the motion from one ridge to another, Wilson named these faults **transform faults**. Transform faults are bona fide plate boundaries that serve as zones of strike-slip accommodation between opposite-traveling plates.

Rather than considering transform faults to be strike-slip faults that systematically *displaced* a once-continuous ocean ridge, Wilson recognized that the repeatedly offset nature of ridges was inherited from initial breakup of lithosphere. He argued that the original configuration of ridge crests and transform faults probably was not characterized by the mutually perpendicular ridge–transform patterns displayed by most mature ocean ridge systems today. Instead, formerly irregular, nonsystematic configurations of ridge crests and transform faults gradually adjusted to the kinematics of spreading.

Confirmation of Transform Kinematics

Wilson's model has stood the test of time. Isacks, Oliver, and Sykes (1968) were able to show that **first motions** on earthquakes along transform faults display the exact sense of movement predicted by Wilson. Their investigations showed that movements are indeed strike-slip, and that the relative displacements conform to the relative motions of seafloor spreading along ridges joined by the transforms. Like Wilson (1965), Isacks, Oliver, and

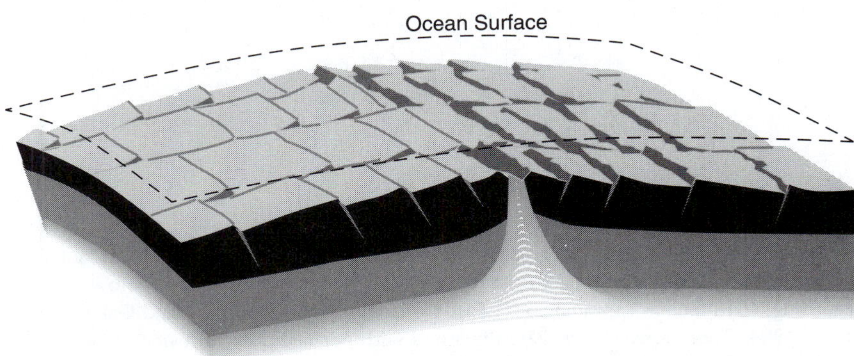

Figure 10.17 Spreading at a ridge axis. [Adapted from Redfern (1986), p. 9.]

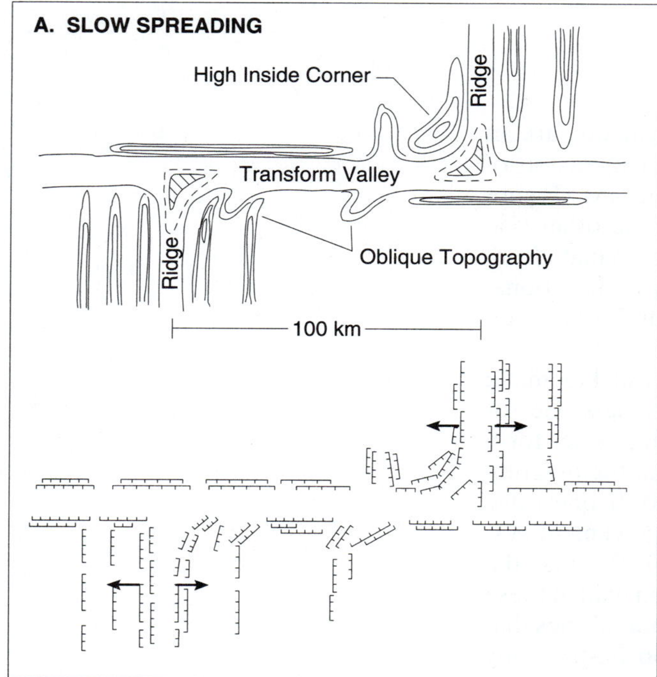

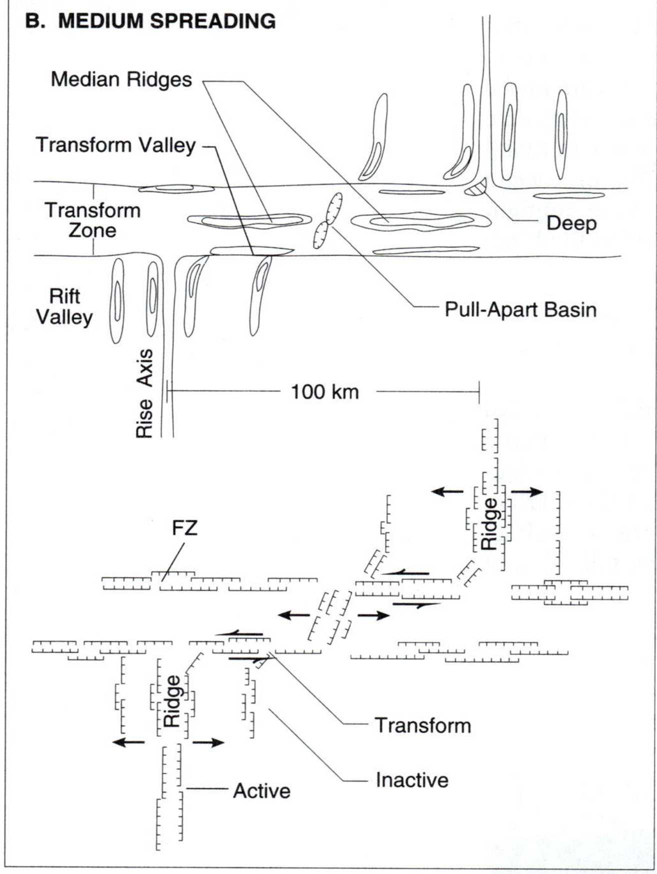

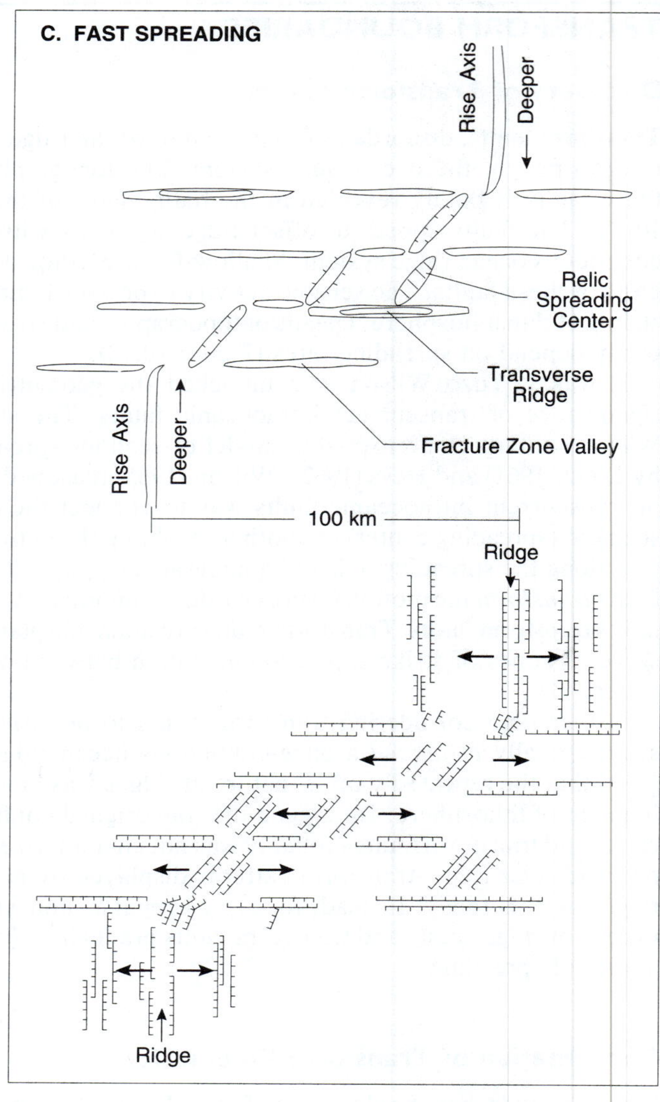

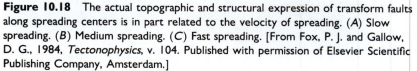

Figure 10.18 The actual topographic and structural expression of transform faults along spreading centers is in part related to the velocity of spreading. (*A*) Slow spreading. (*B*) Medium spreading. (*C*) Fast spreading. [From Fox, P. J. and Gallow, D. G., 1984, *Tectonophysics*, v. 104. Published with permission of Elsevier Scientific Publishing Company, Amsterdam.]

Sykes (1968) emphasized that the displacement vector for a given transform fault is opposite to the sense of separation that would be required to explain the "offset" of the ridge crest (Figure 10.19).

Additional insight regarding the inner workings of transform faults was provided by Sykes (1967). He demonstrated that the only seismically active part of a **ridge-to-ridge transform fault** is the interior segment between the ridge crests connected by the transform. This is the only segment of a transform that separates *oppositely moving* domains of seafloor (Figure 10.19), even though **fracture zones**, ("fossil" records of once-active transform fault segments) are evident beyond the tips of transforms. No frictional interference arises, and thus these "fossil" transform faults are inactive. If one were to figuratively "stand" in the active part of a transform fault zone, it would be like standing on the narrow medial strip of a Los Angeles highway at 5:25 P.M., watching traffic slowly go by on each side in opposite directions (Figure 10.20). The brushing of the flanks of oppositely moving cars, as well as shouts of irate drivers, would create the "seismic" noise.

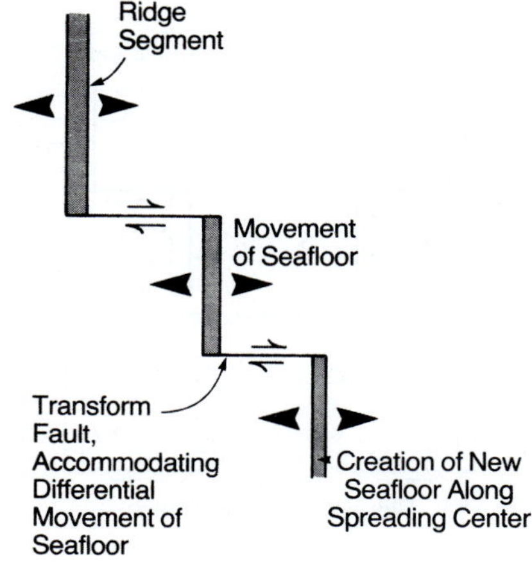

Figure 10.19 Transform fault linkage of one spreading center to another, of one part of the movement system to another.

The Different Kinds of Transform Faults

Transform faults that connect offset ridge segments are the most abundant and perhaps the easiest to understand kinematically. However, two other fundamental types of transform fault exist as well, and these too were recognized by Wilson (1965). One links the end of a ridge with the end of a **trench**, the surface expression of a subduction zone; the second connects the ends of two subduction zones. Like **ridge-to-ridge** transform faults, **ridge-to-trench** and **trench-to-trench** transform faults exist because they serve to accommodate differential strike-slip movement of adjacent plates. Figure 10.21 presents the full spectrum of transform faults and the movements they accommodate.

Not all transform faults are intraoceanic. Tanya Atwater in 1970 stunned the geologic community with her interpretation that the San Andreas fault is a transform fault that slashes through continental crust. It has accommodated major strike-slip faulting and continues to threaten the population of coastal California.

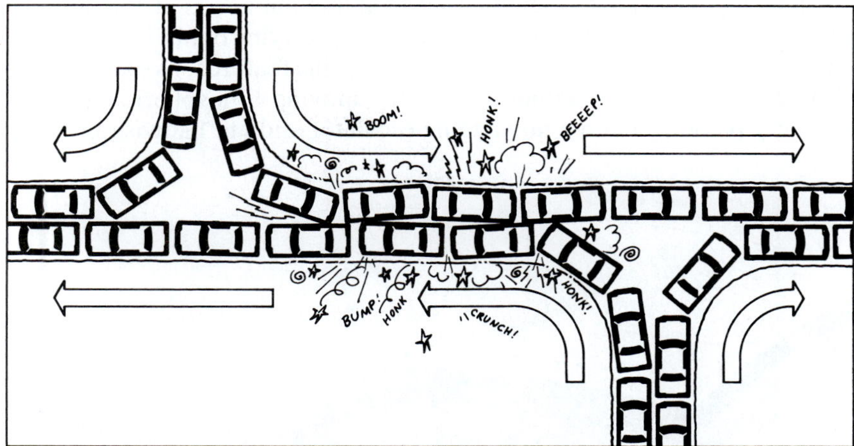

Figure 10.20 The noisy part of a transform fault is restricted to the interior segment, which lies between the ends of spreading centers connected by the fault. (Artwork by D. A. Fischer.)

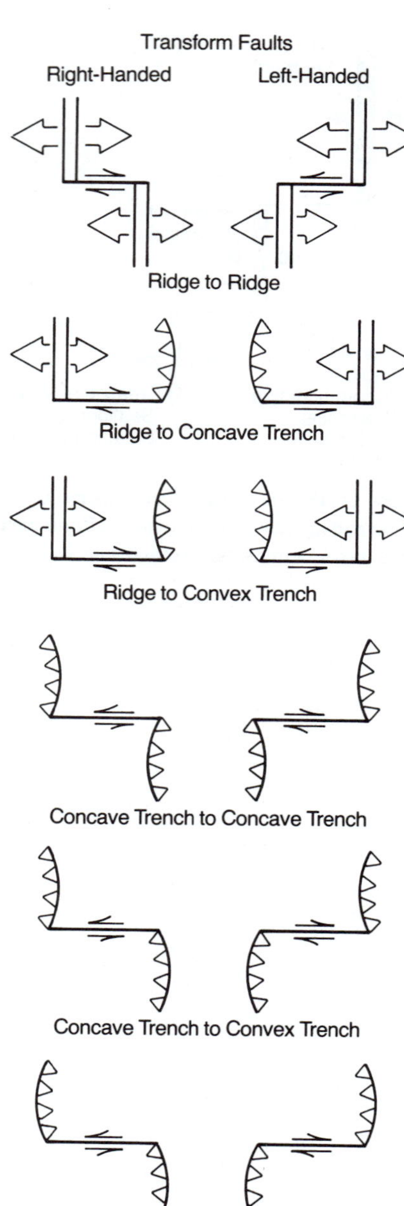

Figure 10.21 The classes of transform faults. Teeth are on overriding plate. [From Wilson (1965). Reprinted by permission from *Nature*, v. 207, pp. 343–347, copyright © 1965 by Macmillan Journals Ltd., London.]

CONVERGENT BOUNDARIES

Subduction Zones

Convergent plate boundaries are commonly marked by the slow descent of one plate beneath the other, an underthrusting or **subduction** of the denser, heavier plate. In effect, one of the plates is swallowed down a subduction zone into the Earth's mantle (Figure 10.22). The descent of one plate beneath another during subduction creates frictional interference and promotes phase changes disclosed seismically by severe earthquake activity. Earthquake foci describe an inclined zone, the inferred **subduction zone**, which projects as deep as 700 km (Figure 10.23). These **Benioff zones** or **Benioff–Wadati zones**, named after the scientists who first recognized them (e.g., Benioff, 1954), are believed to mark the sites of convergent plate boundaries at depth.

An oceanic **trench** is created along the surface trace of a subduction zone (Figure 10.24). Trenches mark the positions where the **slab**, i.e., the part of the plate to be underthrust, begins to flex and go under. Trenches are deep, averaging 8 km with water depths as great as 12 km. The average breadth of a trench is about 100 km. Cross-sectional topographic profiles of trenches reveal slightly asymmetrical forms (Figure 10.25*A*), with the steepest trench wall, the **inner wall**, facing the plate that is being subducted. The **outer wall** of a trench may rise gently away from the trench for distances of 100–200 km. Profiles of trench topography, like that shown in Figure 10.25*A*, are normally constructed with significant vertical exaggeration to identify features that would be subtly expressed, at best, on ordinary profiles. Profiles drawn without vertical exaggeration and viewed in a regional perspective reveal that trenches are actually very slight depressions (Figure 10.25*B*). Such profiles understate the extraordinary mechanical processes taking place beneath the trenches.

Trenches in the world marking sites of present-day subduction are shown in Figure 10.26. One manifestation of the dynamics of these sites of subduction is the presence of active volcanoes—an even clearer signal of the dynamic state of subduction zones is the earthquake activity along them.

Action in the Trenches

It is amazing what junk is jammed into subduction zones at the sites of trenches. Deepwater sediments and slabs of underlying ophiolitic crust are conveyed into the subduction zones. Topographic features, like submarine **volcanic plateaus** and **seamounts**, ride the moving lithospheric slab like stationery passengers on a downgoing escalator and are likewise brought

Figure 10.22 Schematic illustration of subduction in action. Here the Aleutian Islands are forming as a consequence of the subduction of the Pacific plate beneath the North American plate. [Adapted from Redfern (1986), p. 89.]

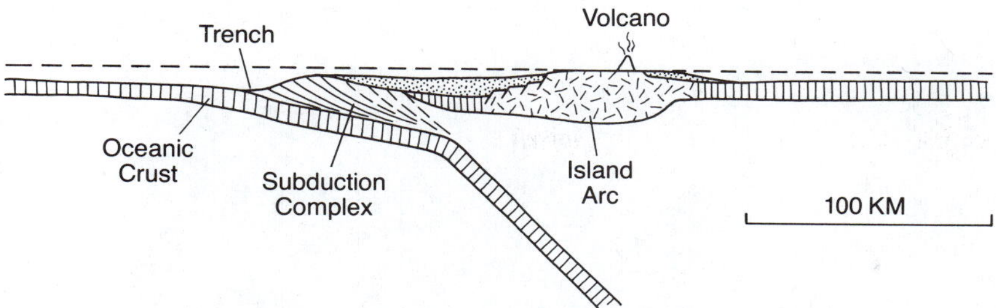

Figure 10.24 Intraoceanic subduction. (From Dickinson, *American Geophysical Union Maurice Ewing Series 1*, fig. 1.A, p. 34, copyright © 1977 by American Geophysical Union.)

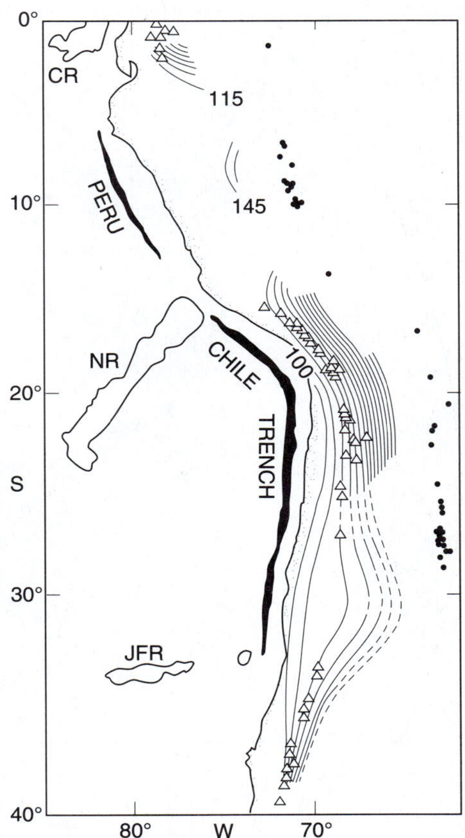

Figure 10.23 Structure-contour map of the geometry of the Benioff–Wadati zone separating Pacific oceanic lithosphere of the Nazca plate from the overriding South American plate. The contour interval is 15 km. Triangles represent Quaternary volcanoes in South America. Black circles are deep focus earthquake events at depths ranging between 525 and 625 km. (From Bevis and Isacks, *Journal of Geophysical Research*, v. 89, fig. 11, p. 6164, copyright © 1984 by American Geophysical Union.)

into trenches. **Turbidity currents**, triggered by earthquake shocks and composed of high-density slurries of mud and sediment, flow into trenches under the influence of gravity. These too contribute to the mass that must be swallowed. Most extravagant of all, the steady movement of plates may eventually feed the leading edge of a continent into a trench. The buoyancy of continental masses precludes significant subduction. When a continent is fed into a subduction zone, the subduction zone may become inoperative, **sutured** by collisional tectonics. Alternatively, the dip direction (**polarity**) of the subduction zone may reverse (Figure 10.27), and the former overriding plate may become the subducted plate. In any event, the leading edge of the continent becomes an **active margin** of deformation.

Normally during subduction, uppermost sediments and seafloor brought into the trench are partly scraped off and plastered or accreted to the inner wall of the trench (Figure 10.28). The result is an **accretionary prism** of low density ($\rho = 2.3$), water-saturated materials distorted by the chaotic structure of fault imbrication and plastic folding (Figure 10.29). The topographic expression of the accretionary prism forms a lip, which may dam

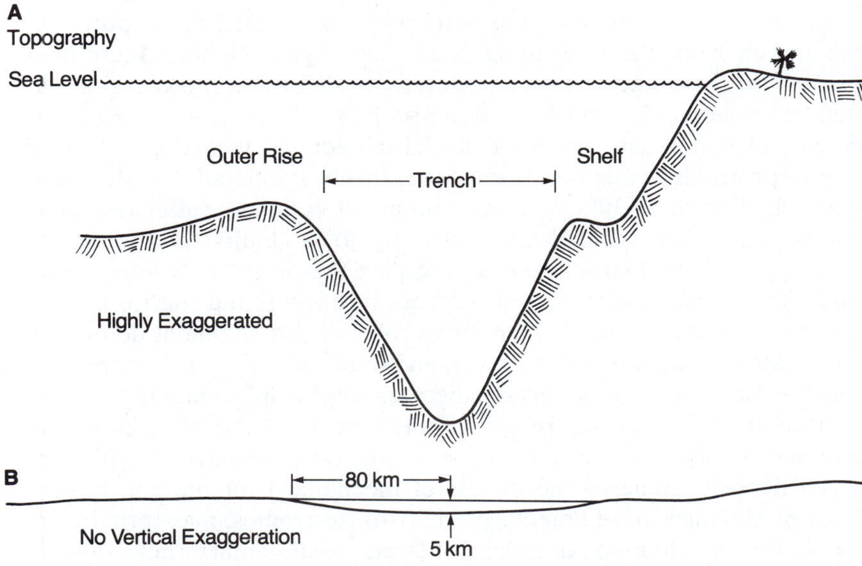

Figure 10.25 (*A*) Profile view of a typical trench, with vertical exaggeration. (*B*) Profile view of a typical trench, without vertical exaggeration. [From Melosh and Raefsky (1980), v. 60, fig. 1, p. 344. Published with permission of *Geophysical Journal of the Royal Astronomical Society*.]

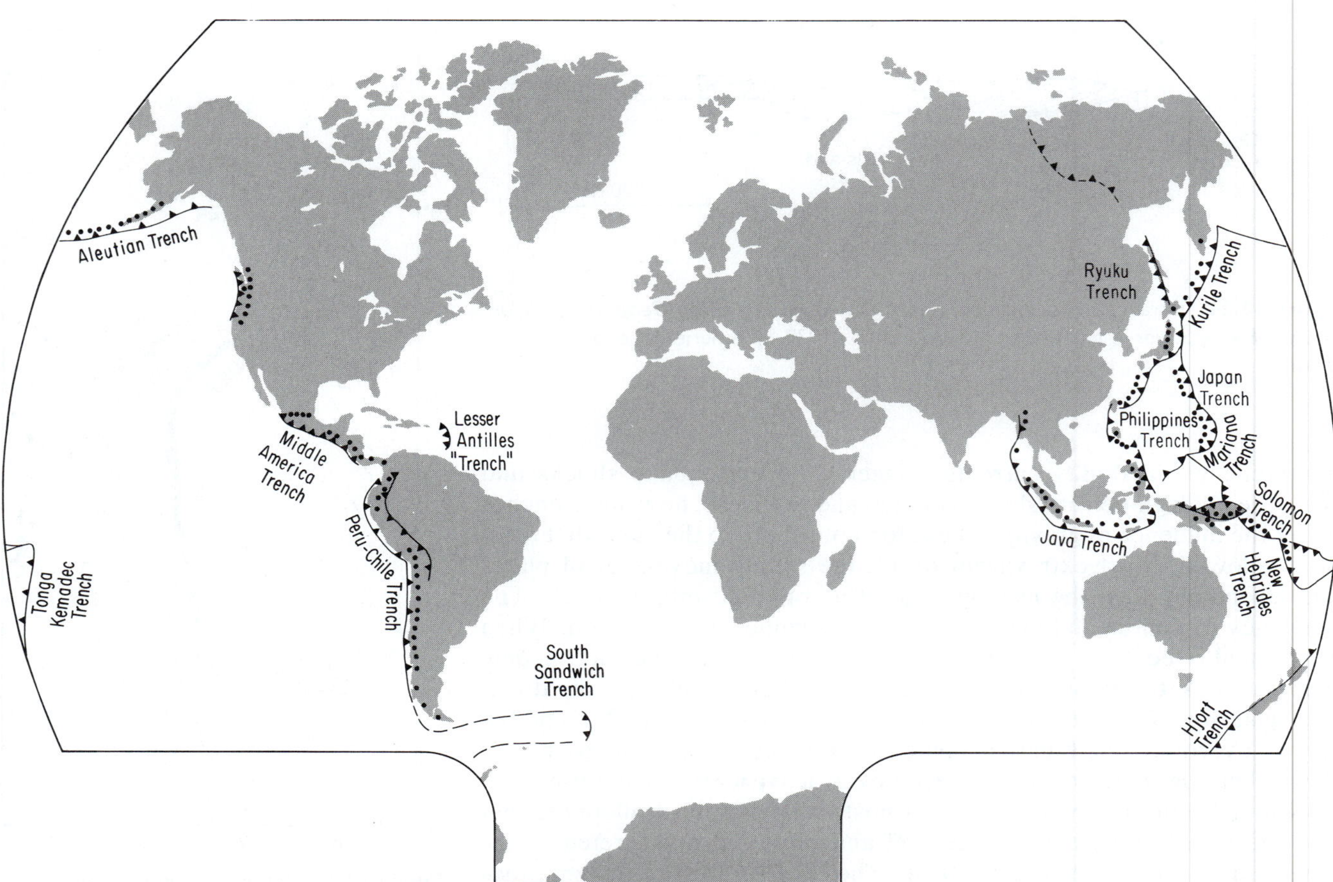

Figure 10.26 Trenches of the world, marking sites of subduction. Volcanoes active within the last 1 m.y. are also shown. [From Lowman (1981). Courtesy of Goddard Space Flight Center, National Aeronautics and Space Administration.]

basins of accumulated materials that otherwise would be transported into the trench from the overriding plate (see Figure 10.28). Accretionary prisms are the home of **melange**, intensely deformed packages of rocks that lack coherent internal layering and coherent internal order. Exposures consist of competent, resistant blocks of tectonic inclusions of diverse origins protruding from poorly exposed, sheared, foliated, "shaly" matrix material (Figure 10.30). Melange commonly contains **radiolarian ribbon cherts**, which are typically disrupted by folds, faults, and penetrative cleavages. Internal structure is so complex that it is nearly impossible to map. The constant dilemma of workers in interpreting the formation of melange lies in evaluating the relative roles of soft sediment deformation and postconsolidation deformation, and in ascertaining the extent of tectonic versus gravitational dismemberment and emplacement.

Mixing of the various rocks within a melange can be achieved in a number of ways, both sedimentary and tectonic (Goodwin, 1976). For a given deposit, some of the details of mechanical mixing will never be known. Melanges of sedimentary origin (**olistostromes**) may form by gravity sliding and slumping of blocks of largely sedimentary rocks down the inner slopes of a trench, where they are incorporated into muddy sediments. Melanges of a tectonic origin may form by the movement of a subducting plate past the trench inner wall, in simple shear fashion. The simple shear would tend to flatten the rocks in one direction and stretch

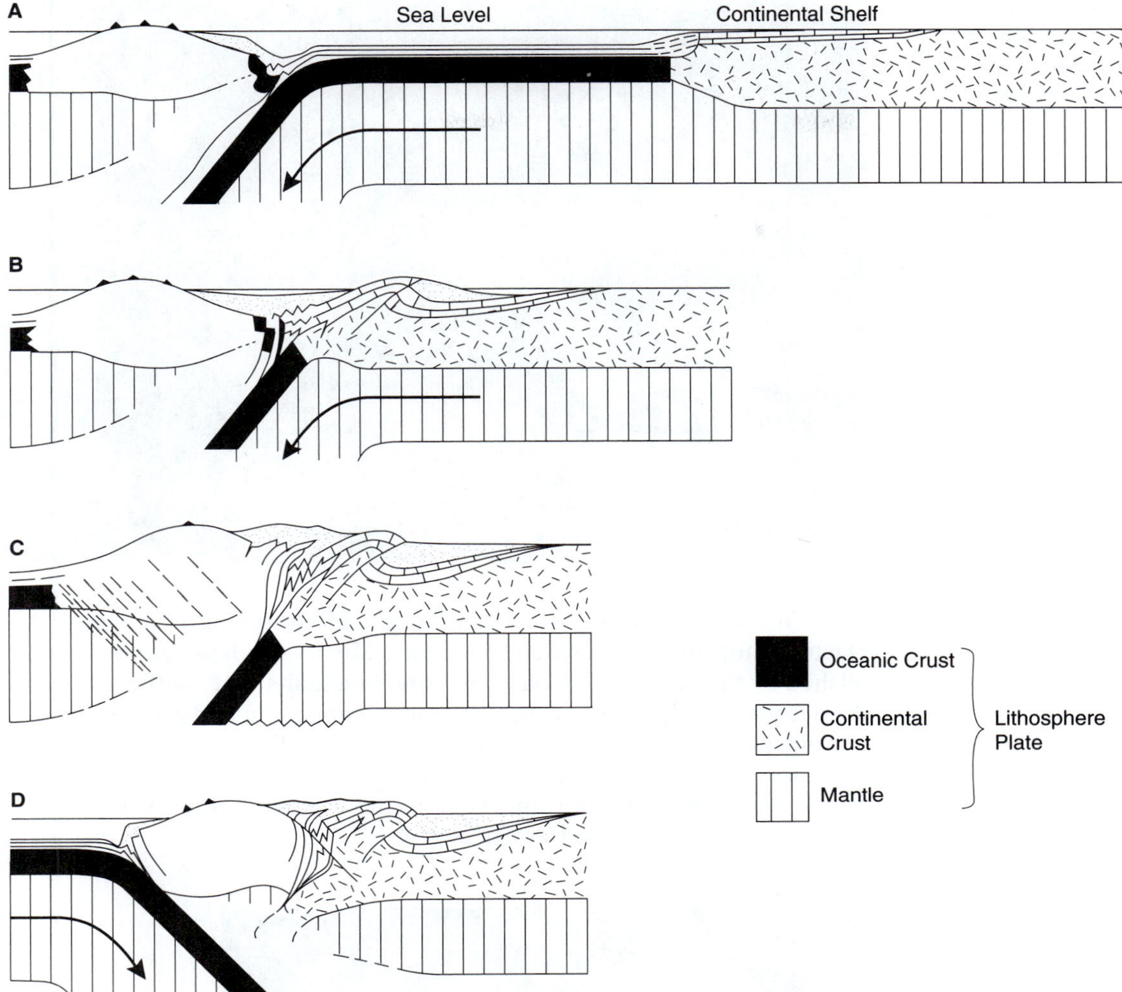

Figure 10.27 Schematic rendition of a change in polarity of subduction. (*A*) Oceanic lithosphere is being subducted from right to left beneath an island arc of basalt and andesite. (*B*) Continental margin makes contact with island arc. (*C*) Profound deformation takes place at location of collision, and continent begins to be overridden by island arc. (*D*) Shift in polarity of subduction results in the formation of new trench to the left of the island arc, and the descent of oceanic lithosphere beneath the island arc and continental margin, from left to right. [Adapted from Dewey and Bird (1970), Figure 46-12, p. 627.]

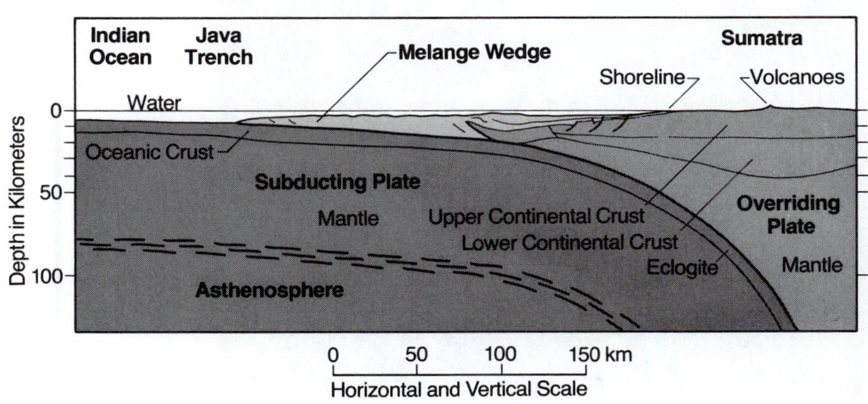

Figure 10.28 True-scale profile view showing the relations of an accretionary melange wedge to the fundamental tectonic components of an active continental margin. [From Hamilton (1979). Courtesy of United States Geological Survey.]

Figure 10.29 Faulting and shearing in highly deformed melange in the San Juan Islands, Washington. Margi Rusmore is the geologist. (Photograph by G. H. Davis.)

them in another. Some of the flattening and stretching is achieved by plastic distortion of water-rich sediments. Much of it, however, is accomplished by the action of subparallel, **imbricate** faults that shuffle and stack the deforming sequence. Tectonic burial, compaction, and prograde metamorphism of the sediments causes dewatering. The resulting high fluid pressures decrease the effective normal stress on the numerous imbricate faults, permitting the entire package to deform according to the critical taper theory we discussed in Chapter 6.

Figure 10.30 Tectonic inclusions in melange, Kodiak Islands, Alaska. Largest inclusion is 2 m long. [From Moore (1978). Published with permission of Geological Society of America and the author.]

Sediments on the outer slope of a trench are pristine and undeformed, consisting of **abyssal plain sedimentation**, commonly in the form of thin, undeformed, sheetlike muds and siliceous oozes. These sediments are drawn ever slowly into the trench, where eventually they are transformed into melange.

Slab Descent

Beneath a trench, the down-going slab descends at an angle determined primarily by several factors: convergence rate, absolute motion of upper plate, and age of oceanic lithosphere (Cross and Pilger, 1982) (Figure 10.31). The deepest trench in the world's oceans, the Marianas Trench (11,000 m below sea level), is linked to a Benioff zone that dips fully 90°. It is receiving, through subduction, some of the oldest ocean floor of the Pacific (150 Ma). In contrast, the Mexico Trench, which is swallowing oceanic lithosphere only 20 Ma in age, is the mouth of a subduction zone that is inclined merely 15–20° (England and Wortel, 1980). In general, relatively old oceanic lithosphere may be subducted with relative ease because it is cold, dense, and negatively buoyant. Old, cold, relatively heavy lithosphere sinks ''like a rock.'' Conversely, young, warm, relatively light oceanic lithosphere is less negatively buoyant and resists subduction, thus creating strong horizontal compressive stresses along the

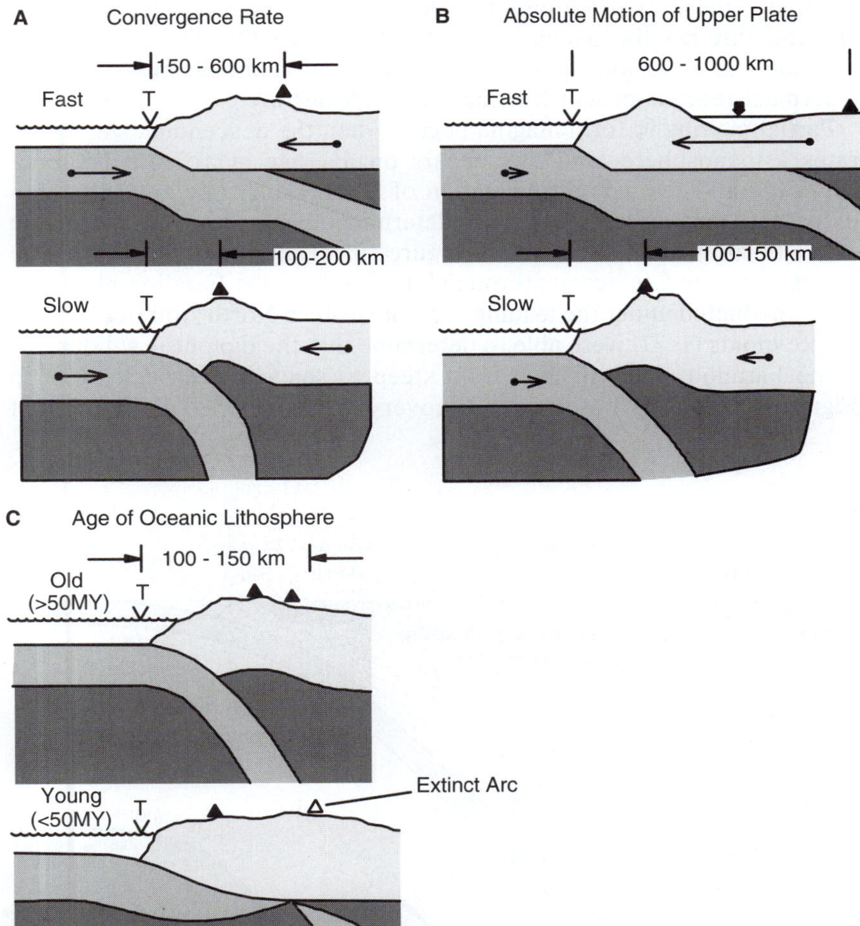

Figure 10.31 The dip of the subducting slab depends primarily on three factors: (*A*) convergence rate, (*B*) absolute motion of upper plate, and (*C*) age of oceanic lithosphere. [Adapted from Cross and Pilger (1982), Figure 1, p. 547. Published with permission of Geological Society of America.]

strongly coupled zone of contact (England and Wortel, 1980). For these reasons and others, the dip of Benioff zones beneath trenches is highly variable, and descent velocities of subducted slabs are likewise variable, commonly ranging from 2 to 10 cm/yr. The dip and descent velocity of present-day subduction zones can be determined through the combination of seismology and GPS geodesy (Figure 10.31).

Downward movement of a subducted slab not only generates deep-focus earthquakes but it also leads to metamorphism at high pressure and relatively low temperature (Figure 10.32). Characteristic **blueschist metamorphic assemblages** are formed at the expense of partially subducted melange sediments and underlying, ophiolitic oceanic crust (Miyashiro, 1973). The low temperature character of blueschist metamorphic assemblages is attributed to the down-sinking of relatively cool rocks into relatively hot regions (Ernst, 1975; Peacock, 1993). In essence, cool rocks are tectonically shoved to depth faster than they can be conductively warmed up by the surrounding hot mantle. The high pressure metamorphism takes place at depths of 20–35 km or greater. Tremendous vertical faulting, including extensional faulting, is required to lift these rocks to the level of surface exposures (Platt, 1986). The presence of blueschist facies metamorphic rocks at the surface is yet another testimony to the great structural relief bound up in orogenic belts.

Ultimately a subducted slab descends to the point at which metamorphic dehydration reactions take place in the down-going slab, and the released volatiles cause partial melting of the overlying mantle. Under the influence of gravity, these high temperature, low density magmas buoyantly rise into and through the overriding plate (Figure 10.33). The products of intrusion and extrusion contribute to the formation of a **magmatic arc**, the uppermost reaches of which comprise a **volcanic arc**.

Partial melting to form magma begins when the descending slab penetrates asthenosphere, and this occurs on average at a depth of about 100–150 km. The angle of inclination of the Benioff zone and the depth to the asthenosphere combine to determine the location at which the magmatic arc will be constructed (Figure 10.34). For example, during the period from late Cretaceous through late Tertiary, the Farallon plate was being subducted under the leading edge of western North America. Coney and Reynolds (1977) were able to determine that the dip of the subduction of the Farallon plate changed from steep to shallow and back to steep (Figure 10.35). They based their discovery on the mapped distribution of

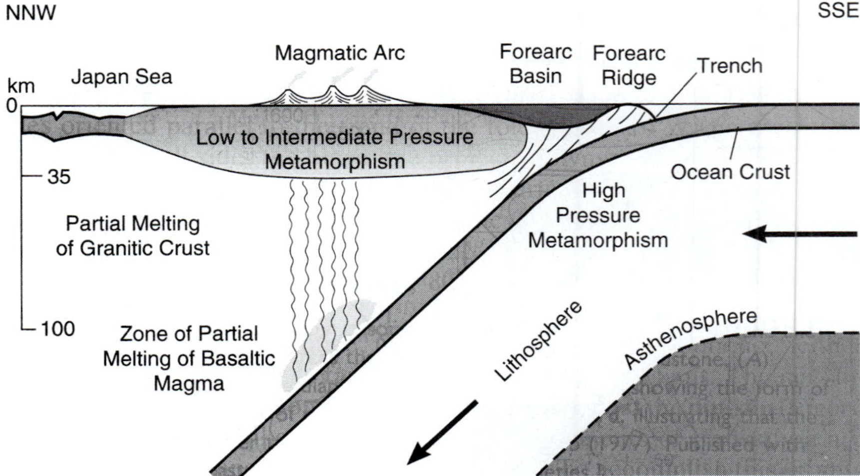

Figure 10.32 Schematic model showing the subsurface region within which high pressure metamorphism takes place. Also note the depth at which partial melting of basaltic magma occurs. Inboard are shown the regions of low pressure metamorphism, intermediate pressure metamorphism, and the partial melting of granitic crust. [Adapted from Kearey and Vine (1990), Figure 8.20, p. 158, after Barber (1982).]

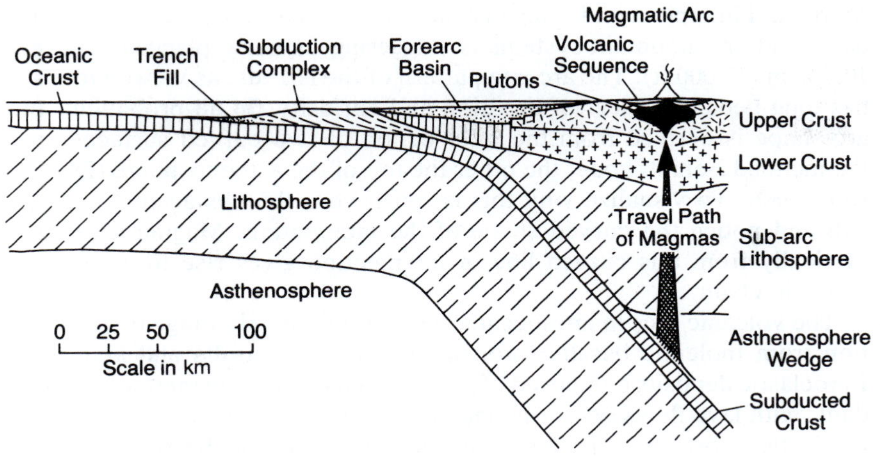

Figure 10.33 Generation of magma during subduction. [From Dickinson and Payne (1981), cover illustration. Published with permission of Arizona Geological Society.]

dated intrusive and volcanic igneous rocks. Their discovery has had a profound influence not only on models of the tectonic evolution of western North America, but on understanding of mineralization and the formation of ore deposits as well.

The magmas generated within zones of partial melting in the mantle rise through the continental crust to form large intrusive igneous bodies of dominantly granodioritic to granitic composition. The flooding of the crust by magma and the associated heat produce a belt of *low pressure, high-temperature* regional and/or contact metamorphism. Geothermal gradients in belts of low pressure metamorphism are on the order of 25°C/km or higher, in contrast to the average 10°C/km gradient that characterizes sites of high pressure metamorphism (Miyashiro, 1974).

Volcanic arcs are produced near sites of subduction. One of the best examples of island arc activity is in the western Pacific, from the Aleutian arc southward through a belt of archipelagoes that includes the Japan,

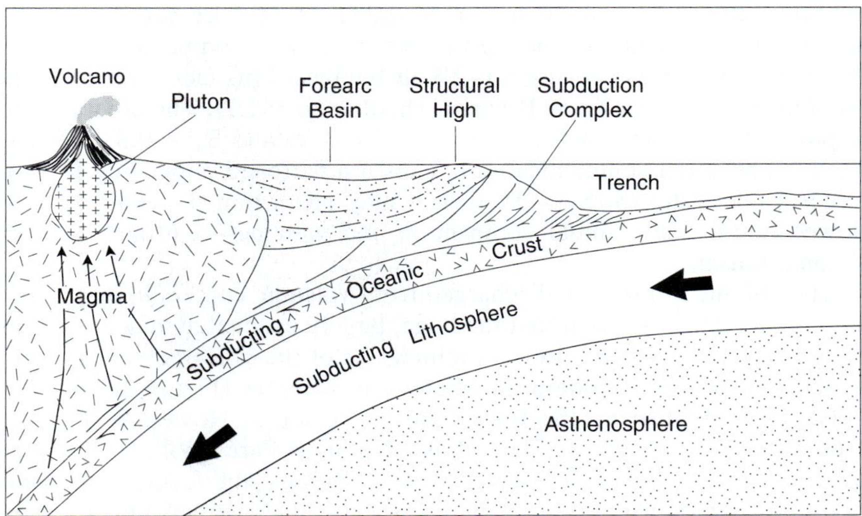

Figure 10.34 Schematic diagram showing relation between subducting slab and the location of magmatic arc in overriding plate. [From Cox, A. and Hart, R. B., Plate tectonics—how it works, 1986, Figure 1-19, p. 32. Reprinted by permission of Blackwell Science, Inc.]

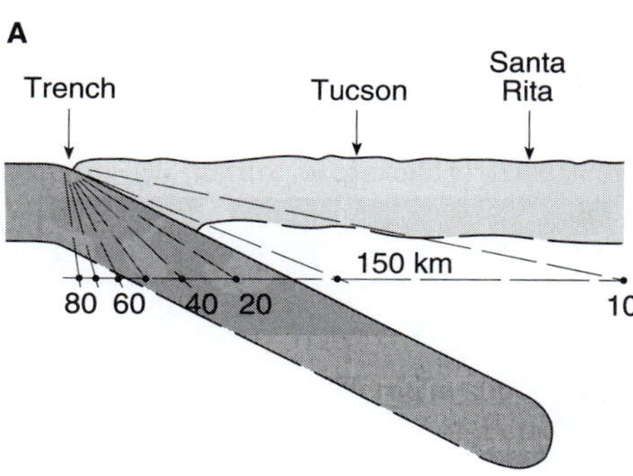

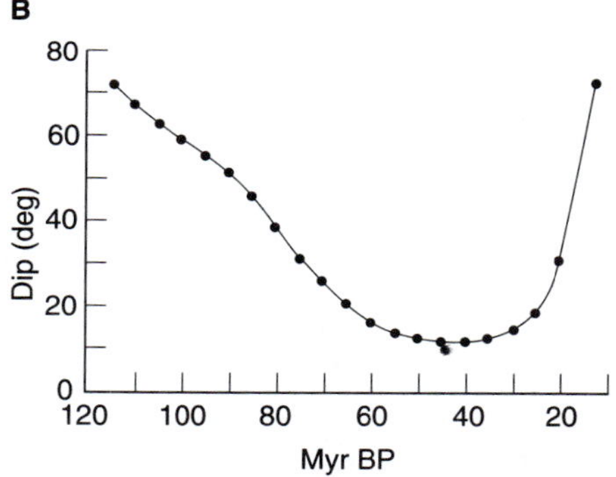

Figure 10.35 Coney and Reynolds (1977) determined that the angle of descent of the subducting slab under southwestern North America progressively decreased from Cretaceous to Eocene, and then steepened again. The signature of this "variably dipping Benioff zone" is a steady migration of the magmatic arc from west to east, followed by a sudden retreat. (*A*) Hypothetical arc–trench system scaled to southwestern North America. Various dip-angles of Benioff zones shown with points of intersection with 150 km depth line. (*B*) Benioff zone dip angle as a function of time beneath southwestern North America, based on distribution of radiometric ages of magmatic arc rocks. (Reprinted with permission from *Nature*, v. 270, Cordilleran Benioff zones by P. J. Coney and S. J. Reynolds, p. 403–406. Copyright 1977, Macmillan Magazines Limited.)

Mariana, Philippine, and Tonga islands. The topographic relief of individual island arc mountain systems is staggering, in some places exceeding 10,000 m (33,000 ft). The arcuate pattern of the islands, as viewed in plan, has long been an object of question and curiosity. In all probability the arc shape is derived from the intersection of two curved surfaces, one the inclined Benioff zone, the other the top of the asthenosphere (Dickinson, 1980). To visualize why this is, push your thumbs against a tennis ball and notice the curved shape of the indentation. Magma ascending vertically from this curved line of intersection gives rise to an arcuate volcanic chain.

The volcanic arcs in intraoceanic settings commonly range in composition from tholeiitic basalts through calc-alkaline basalts and andesites. Pyroclastic deposits composed of gas-charged particulate matter are abundant, with lava flows occurring mainly near the eruptive centers. Aprons of clastic debris run off the composite volcanoes of the islands, to be deposited in adjacent deep basins as volcaniclastic turbidites, marine tuffs, and submarine ash deposits.

Volcanic arcs are produced in great abundance at active continental margins where subduction has been long-lived. The circum-Pacific Ring of Fire is the most often cited example. Magma extruded in such settings is of intermediate composition, largely because of contamination of the primary magmas during ascent through the continental crust. Classically, the volcanic products are andesitic.

Volcanic edifices in arcs of continental margin orogens are large **composite volcanoes** featuring interlayered flows and pyroclastic deposits. In addition, there are **cinder cones**, formed almost exclusively of pyroclastic material, and local **volcanic domes** of rhyolitic materials. Some of the most impressive products are **ash flows**, or nuée ardente. Ash flows form through processes that involve the high velocity of hot, glowing, gas-charged avalanches of particulate material, notably glass shards and pumice fragments in a very fine-grained matrix (Smith, 1960a,b). Single sources have produced flow fields larger than 20,000 km^2, containing more than 1000 km^3 of material. Single flows may travel 100 km and more from their source. Ash flows produce tuffs, which, if thick enough, may compact into **welded tuff**. The strain involved in the welding process is a negative dilation, reflecting a volume loss as air-filled vesicles are squashed. The amount of compaction is about 3 : 1. Pumice can be compacted to 1/3 of its original thickness because it is 2/3 air bubbles. This fact is reflected in the density of pumice ($\rho = 1$) versus rhyolite ($\rho = 2.7$). The compaction is pure volume loss, such that $S_1 = 1$, $S_2 = 1$, and $S_3 = 0.333$ (Peter Lipman, personal communication). Thus if a flattened pumice clot has a larger aspect ratio ($S_1/S_3 > 3$), then it may mean that the ash flow has flowed ductily after initial compaction and welding—a process called **rheomorphism**.

The volume of material discharged from volcanic centers in magmatic arcs is sometimes so great that the huge, largely evacuated magma chambers collapse inward and cause a foundering of the volcanic edifice. The result is a collapsed **caldera**, kilometers in diameter (Figure 10.36). An outstanding treatise on the formation of a caldera, Howell Williams's report on the geology of Crater Lake National Park (Williams, 1942), describes the geological history of Mount Mazama, the former occupant of the site of Crater Lake (see Figure 1.4). The volcano blew out so much magma in the form of pyroclastic material and flows that its foundation gave way. As a result, a volcanic mountain, estimated to be 2000 m high and 30 km^3 in volume, vanished into the collapsed foundation. This event occurred only 5000 years ago.

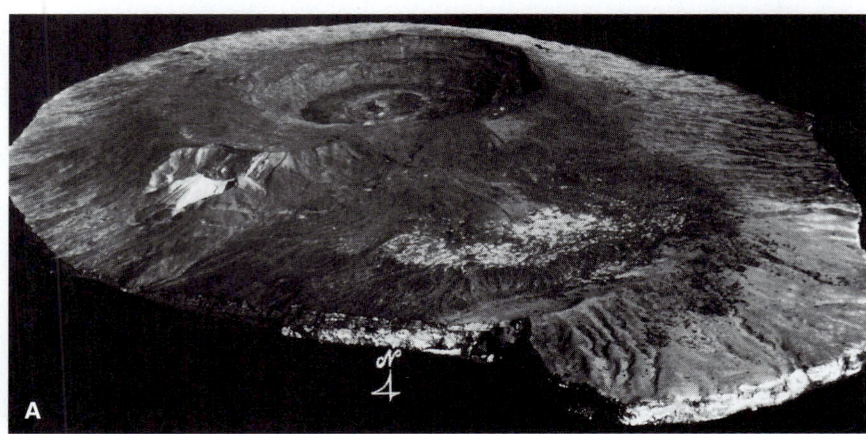

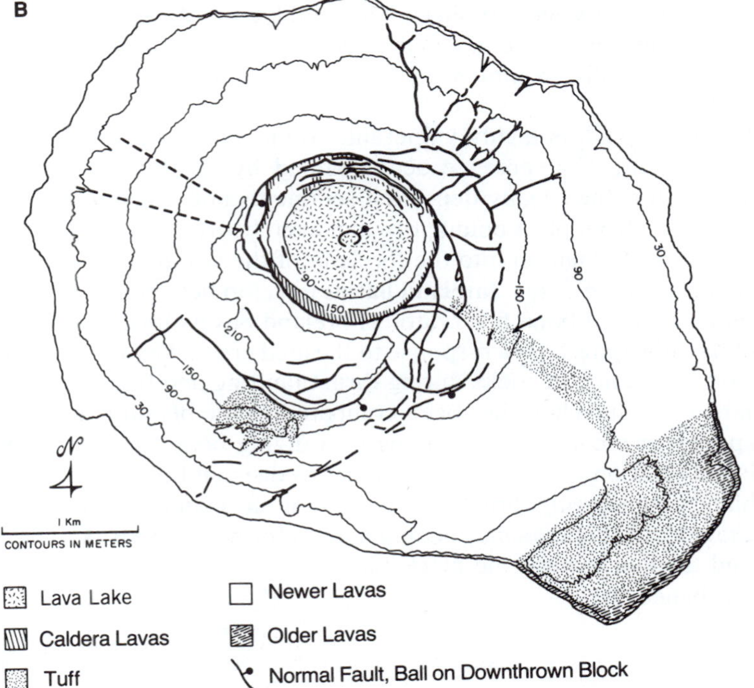

Figure 10.36 Photograph and geologic map of Isla Tortuga, a recently formed volcanic island in the Gulf of California. The volcanic island is composed mainly of basalt and vitric tuff. During its latest stage of activity, the volcano suffered caldera collapse, accompanied by extrusion and the formation of a lava lake. Collapse was accommodated by the concentric rings of inward-dipping faults. [From Batiza (1978). Published with permission of Geological Society of America and the author.]

By virtue of their topographic prominance, volcanic arcs are vulnerable to rapid erosion. Enormous quantities of eroded debris spill toward opposite sides of the arc complex and may be deposited in a variety of basins, the names of which are commonly referenced with respect to the position of the magmatic arc (Figure 10.37). **Forearc basins**, **interarc basins**, and

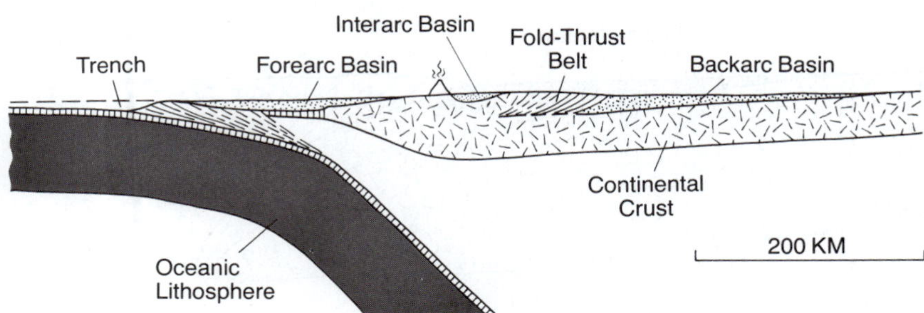

Figure 10.37 Sedimentary basins at the site of an active continental margin. (From Dickinson, *American Geophysical Union Maurice Ewing Series 1*, fig. 1 B, p. 34, copyright © 1977 by American Geophysical Union.)

backarc basins form within or immediately adjacent to the active magmatic arc, whether it be continental or intraoceanic. The sediments accumulate in thick wedge-shaped to prism-shaped deposits whose specific forms reflect the structural and depositional environment (Ingersol, 1988). Deposits of forearc basin settings include fluvatile–deltaic deposits, continental shelf depositional prisms, and deepwater turbidity current deposits and submarine fans (Dickinson and Seely, 1979). Interarc basins typically are wedge-shaped to lens-shaped deposits of very locally derived clastic and volcanic material. Deposition is largely subaerial. Some of the interarc deposits are simply alluvial fans and lahars (volcanic mud flows) interlayered with volcanic flows and ash deposits of the arc. Others are basinal accumulations at sites of down-dropped blocks.

Tectonic stresses generated by plate convergence and/or magmatic processes within the arc can lead to stretching and/or shortening of the framework behind the arc. Extensional faulting may open up rifted basins that become **backarc basins** of accumulation of poorly sorted deposits of locally derived sedimentary and volcanic detritus. Backarc spreading is especially common in island arc intraoceanic settings (Karig, 1971).

Each basinal deposit is highly vulnerable to mechanical distortion by folding and faulting. Convergence accompanied by subduction and/or collision can distort the freshly deposited sediments and volcanic rocks. Stresses that issue from plate boundaries can even be transmitted beyond the active continental margin into the foreland of the continental interior (Figure 10.38). In some cases, **foreland basins** develop between the active continental margin and bona fide craton. Foreland basins form by down-bending of the lithosphere, as a response to the load created by the vertical stacking and thickening of rocks during thrust faulting and folding. Such down-bending of lithosphere has been compared to the bending down of the leaf springs in a car, or the bending of a diving board when two or three kids play out on the end. Sedimentary wedges and prisms derived from highlands built during thrusting are shed into foreland basins. Thousands of meters of sediments may accumulate to form the deposits, which coarsen and thicken toward the plate margin, toward the source of dynamic disturbance.

Accretion

Subduction is not a perfect disposal system. Like any ecology-conscious citizen, subduction systems recycle and reuse some material. Subduction appears to have been responsible for destroying all coherent terranes of oceanic crust older than about 150 million years, but there is growing

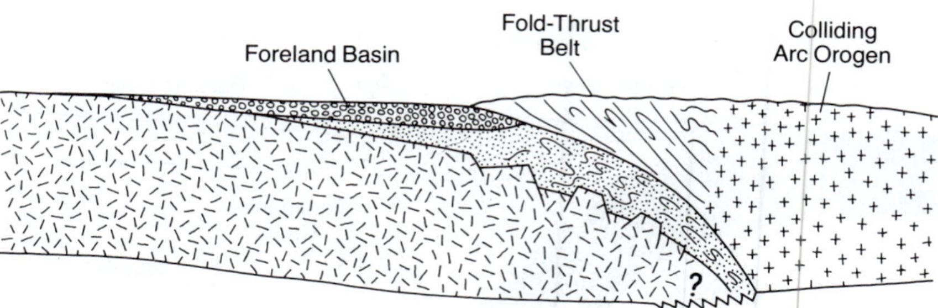

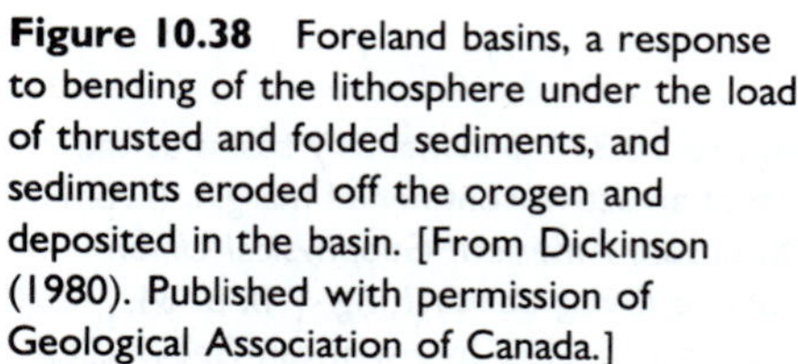
Figure 10.38 Foreland basins, a response to bending of the lithosphere under the load of thrusted and folded sediments, and sediments eroded off the orogen and deposited in the basin. [From Dickinson (1980). Published with permission of Geological Association of Canada.]

evidence that convergent plate boundaries have been unable in some cases to swallow up, or keep down, major masses of rock. For example, in much of Alaska and western North America, continental margin mountain systems are dominated by the deformed remains of Paleozoic–Mesozoic ocean ridges, volcanic islands, and deep-sea sediments. These appear to have been plastered against the western edge of North America along with other scraps and pieces of oceanic floor (Cox, Dalrymple, and Doell, 1967; Coney et al., 1981). The mapped distribution of **terranes**, some of which have been accreted to North America and Central America is enormous. They comprise one-half to two-thirds of the Western Cordillera of North America (Figure 10.39). The intensity of deformation within some of the accreted terranes is staggering, whereas others are not greatly deformed at all.

Peter Coney coined the term **suspect terrane** for regional rock assemblages that *may* be exotic to the home continent in which they now reside. Some suspect terranes are composed of rocks that have been transported thousands of kilometers from their place of origin. The accretion process amounts to international kidnapping. Rocks that started out within or atop one plate are forced to come aboard some other plate at a convergent or transpressive–convergent boundary. The mechanical transfer takes place through accretion, which is a combination of aborted subduction and strike-slip "smearing" (Figure 10.40). Suspect terranes represent an enormous challenge!

Collision

If a subduction zone lies along the leading edge of a continent, it may be only a matter of time before continuous subduction of the neighboring

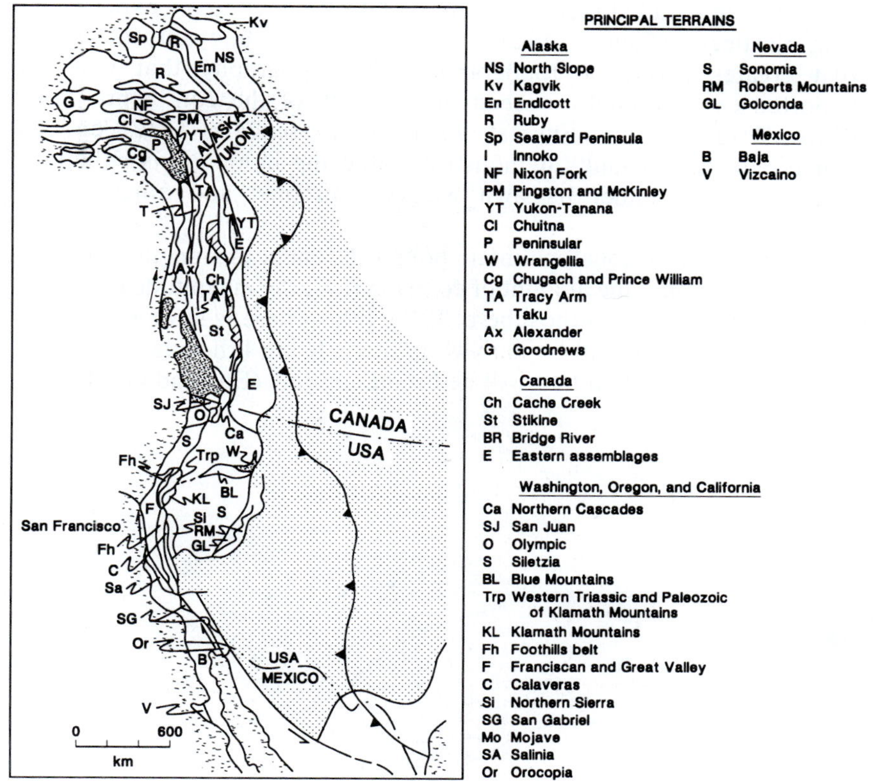

Figure 10.39 Accreted and suspect terranes of western North America. [After Coney and others (1981). Courtesy of United States Geological Survey.]

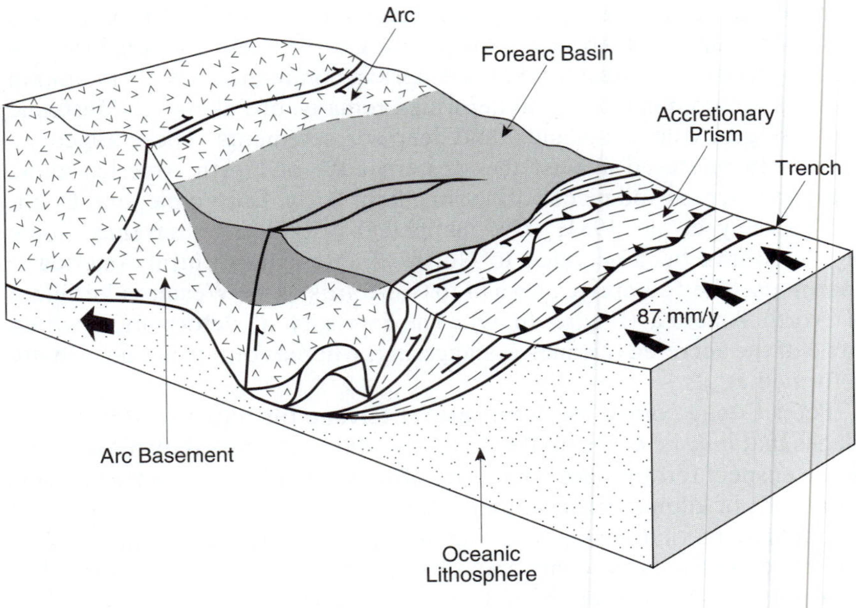

Figure 10.40 One way in which terranes can be accreted is through transpressional subduction. This schematic diagram portrays transpressional subduction along the Central Aleutian trench, where an accretionary prism–arc system is being sliced and moved by right-lateral strike slip faulting during oblique plate convergence. [Adapted from Gibbons (1994), originally based on Ryan and Scholl (1989).]

plate ''delivers'' the passive margin of a continent. When this occurs, collision results as the continents, in effect, slam into each other (Figure 10.41). Conventional subduction may cease, and lithospheric-scale continental thrusting may begin. The collision of India and Asia to form the Himalayas is an example (Figure 10.42*A*). Tapponier and others (1982) have captured the fundamental kinematic relations beautifully by combining detailed tectonic analysis with ''indentation'' experiments using plasticene (Figure 10.42*B,C*). Thickening of buoyant continental crust creates enormous topography. The topography that must be supported can eventually exceed the strength of the continental lithosphere. Witness Mount Everest, where the presence of Late Cenozoic normal detachment faults demonstrates that the mountain is collapsing (i.e., spreading) under its own weight (Burchfiel and others, 1993) (Figure 10.43).

The form of collisional mountain belts will reflect the length and shape of the contact zone, the amount of collision (i.e., the magnitude of strain), and the strength of the lithosphere. If the lithosphere is strong, the collisional mountain belt will be narrow and tall. If the lithosphere is weak, the collisional mountain belt will be broad and low (England and Houseman, 1986).

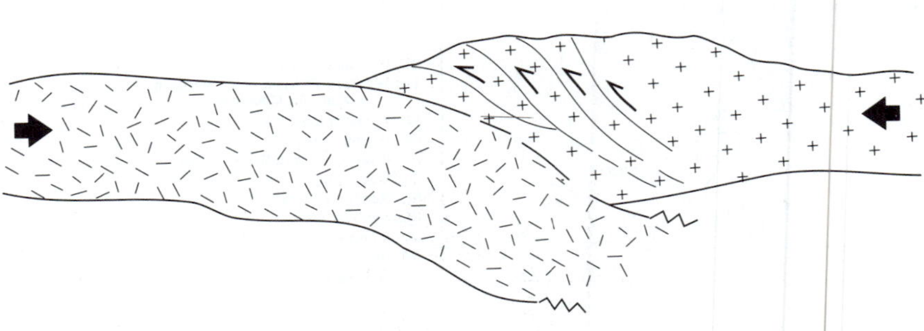

Figure 10.41 Schematic rendition of continental collisional tectonics.

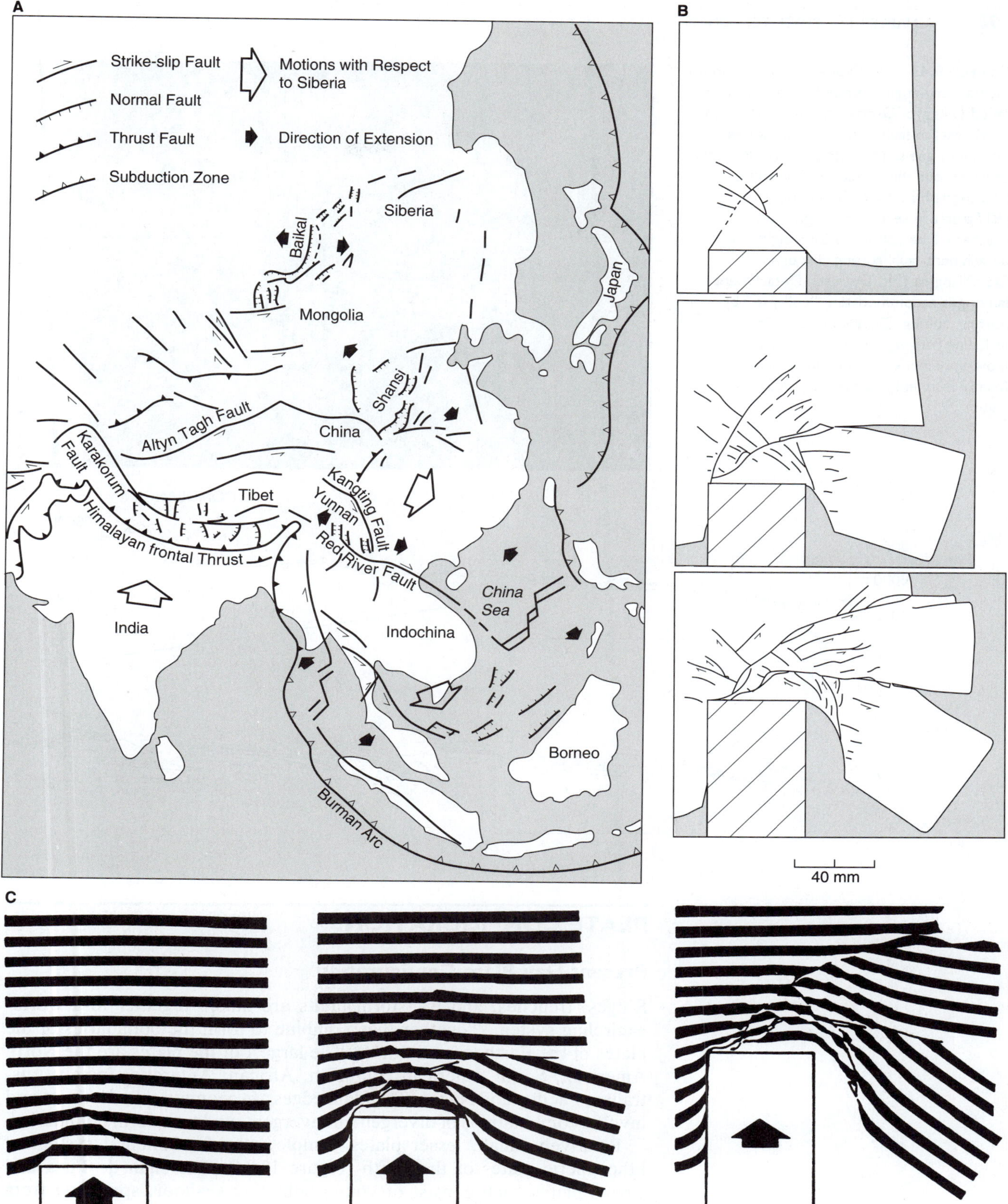

Figure 10.42 (*A*) Map showing the collisional tectonic framework of eastern Asia, where the northward indention of the Indian–Australian plate into the Eurasian plate has caused thrusting in the Himalayas, north of which and east of which strike-slip faulting and extensional faulting accommodates extrusion of the Eurasian plate eastward. (*B*) A rigid Indian-Australian plate indents a plasticene Eurasian plate, simulating the extrusion tectonic patterns in eastern Asia. (*C*) Close-up tracing of the indention experiments in plasticene. [From Tapponnier and others (1982). Published with permission of Geological Society of America.]

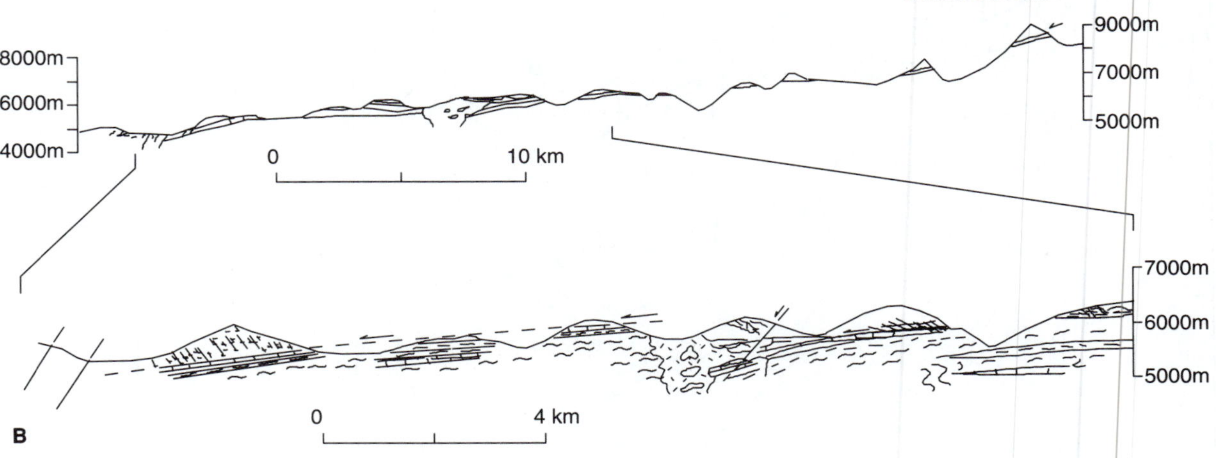

Figure 10.43 (*A*) Expression of the South Tibetan low-angle normal fault at a place called Dinggye. Ordovician limestone (light gray) rests in fault contact on a footwall of sillimanite gneiss (dark gray). Leigh Royden is checking out the contact relationships. (Photograph by Clark Burchfiel.) (*B*) Structure section through a part of southern Tibet, showing low-angle detachment fault cutting through Qomolangma (Mt. Everest). Detachment faulting is accommodating "collapse" by normal faulting. Details of the deformation, including hanging-wall rotation, shown in blow-up. Check out the elevation! [Adapted from Burchfield and others (1992), Figure 19.]

PLATE CONFIGURATIONS

Present-Day Plate Configuration

Ridges, trenches, and transform faults are linked together in a world-encircling system. Together they combine to form the boundaries of the plates of the Earth (Figure 10.44). The largest of the plates are the North American, South American, Eurasian, African, Antarctic, Indian–Australian, Pacific, and Nazca. The plate edges are geometrically intricate and involve combinations of divergent, convergent, and transform boundaries.

Relatively small, lesser plates complete the fundamental system of lithospheric plates of the Earth (Figure 10.44). The Juan de Fuca and Cocos plates each consist of young ophiolitic seafloor, spreading from ridges and guided by transforms into nearby subduction zones. The Scotia plate consists of oceanic seafloor transforming past the parts of the South American and Antarctic plates. The Arabian plate is composed of continental lithosphere diverging from the Red Sea Rift, guided both by oceanic and continental transforms into the suture zone where continental crust of the Eurasian and Arabian plates collide. The Philippine, Bismarck,

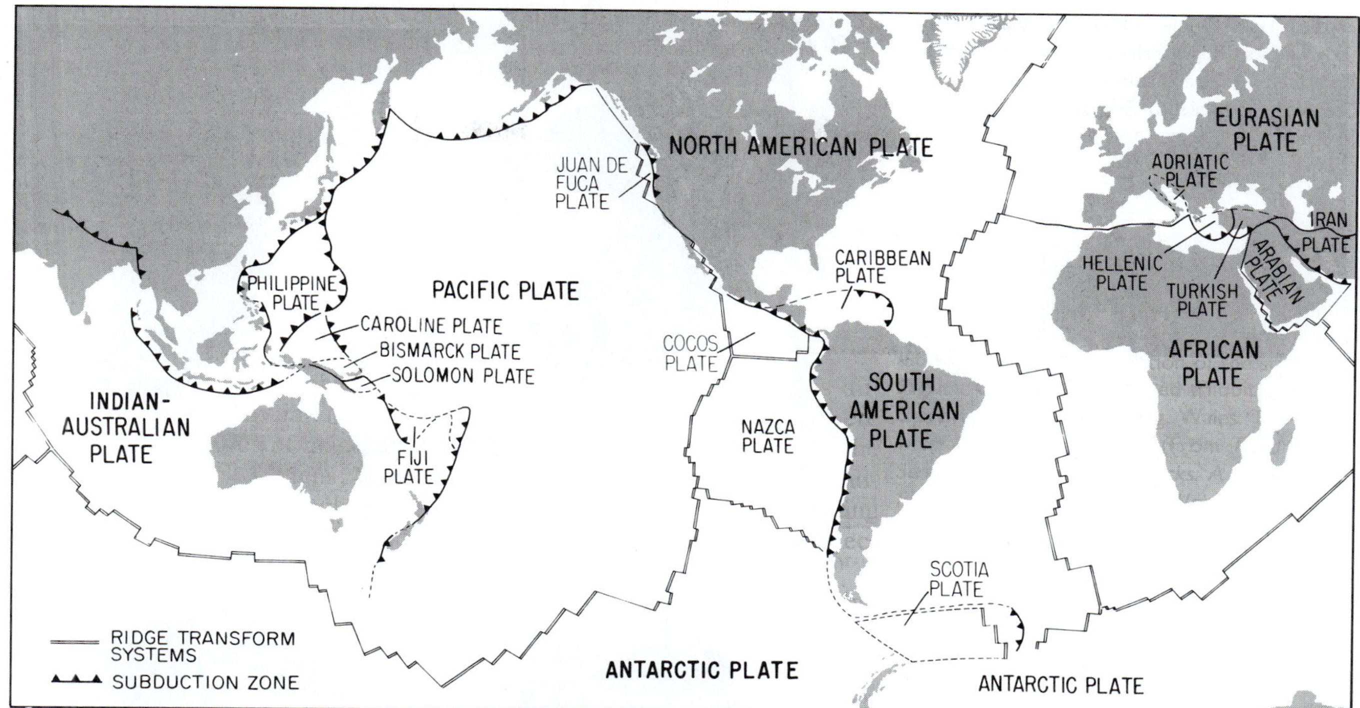

Figure 10.44 Map of the world's plates. (From "Plate Tectonics" by J. F. Dewey, copyright © 1972 by Scientific American, Inc. All rights reserved.)

Solomon, and Fiji plates in the western Pacific occupy marginal seas bounded by combinations of trenches and transform faults.

Triple junctions occur where three plates come together. There are a number of types (Figure 10.44). The Nazca, Cocos, and Pacific plates intersect in a **ridge–ridge–ridge triple junction**, as do the African, Antarctic, and South American plates and the African, Antarctic, and Indian–Australian plates. In contrast, the Pacific, Caroline, and Philippine plates intersect in a **trench–trench–trench triple junction**, as do the Eurasian, Philippine, and Pacific plates. There are other varieties as well. The Nazca, Cocos, and Caribbean plates intersect in a **ridge-trench-trench triple junction**. And the North American, Cocos, and Pacific plates meet at a **trench–ridge–transform triple junction**.

Plate Configurations in the Past

The movement of plates through time is fascinating. Reconstructions of the positions of continents and the plates they occupy impart a sense of awe regarding the degree to which our "stable foundations" are active. We see in plate reconstructions, such as those by Bally and Snelson (1980) and Engbretson and others (1985), the fundamental plate movements that took place in the Mesozoic and Cenozoic eras (Figure 10.45).

The breakup of the supercontinent of Pangaea began with rifting between North America and Africa in the mid-Jurassic period, at about 180 Ma. The rifting initiated seafloor spreading and the formation of the central Atlantic Ocean and the Gulf of Mexico. At about 138 Ma ago, in the early Cretaceous, South America and Africa began to split apart, forming the beginnings of the South Atlantic. At approximately the same time India, Antarctica, and Australia began to break away from the east flank of what

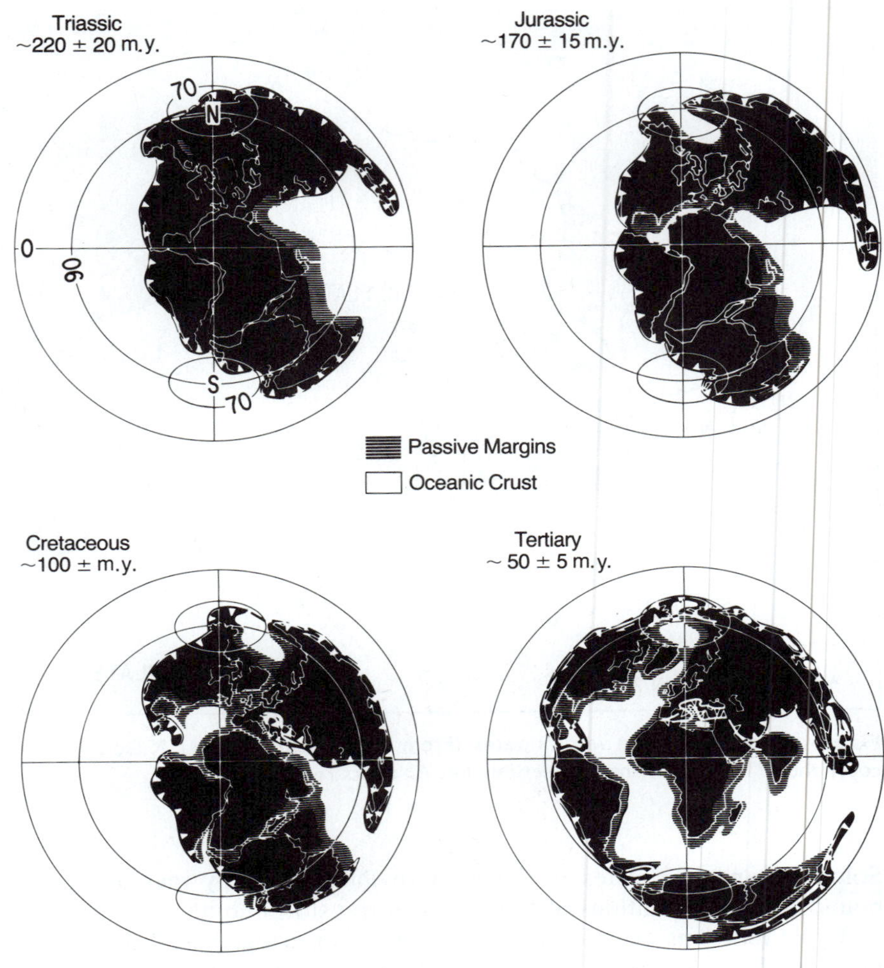

Figure 10.45 The drama of past plate movements. [From Bally and Snelson (1980). Published with permission of Canadian Society of Petroleum Geologists.]

is now Africa. In the late Cretaceous, 90 m.y. ago, Greenland and North America separated. At 50 Ma Australia and Antarctica separated from each other, while India continued on its flight toward Eurasia, with which it finally began to collide at 40 Ma. In upper Miocene, 10 m.y. ago, the Red Sea and the Gulf of Aden opened. The Gulf of California opened at 5 Ma when the Baja Peninsula pulled away from the western part of Mexico.

Before the theory of plate tectonics was accepted, all these movements and shifts of continents were described as continental drift. But more than just the continents are moving. The continents are simply passengers on *lithospheric* ships moving on mantle asthenosphere. What an unpredictable drama! What will the major scenes be like in the next 200 million years?

Cross-Sectional Views of Plate Interactions

Map-view reconstructions, like those by Bally and Snelson (1980) and Engbretson and others (1985), provide one kind of insight into plate kinematics. A complementary perspective can be derived from cross-sectional views of plate interactions. Dewey and Bird (1970), in a classic paper describing plate tectonic configurations, provided geologists with a clearer picture of the impact of seafloor spreading and plate tectonics on the fashioning of continental geology through time. In essence, they con-

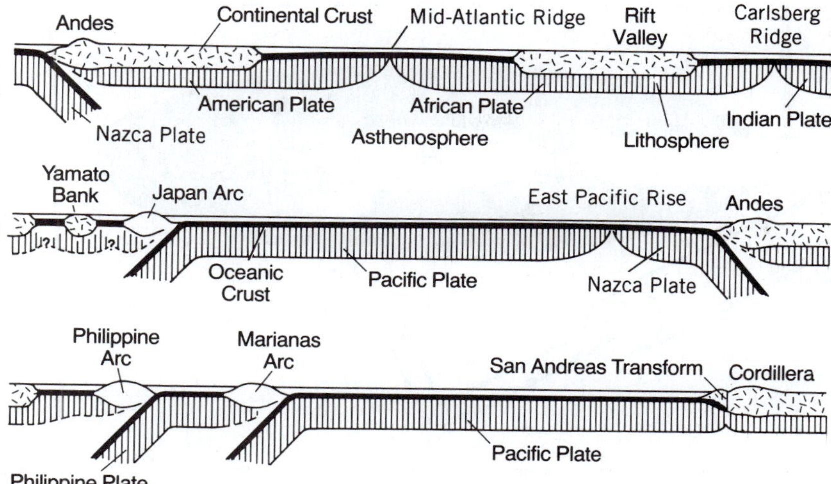

Figure 10.46 Worldwide cross sections of plate interactions. (From Dewey and Bird, *Journal of Geophysical Research*, fig. 2, p. 2627, copyright © 1970 by American Geophysical Union.)

structed schematic world-view cross sections that display the interactions of the existing plates (Figure 10.46).

The South American plate is shown diverging westward from the Mid-Atlantic Ridge (Figure 10.46). The continent of South America is positioned on the plate such that its eastern edge is a **passive margin**. In contrast, the western edge of South America is an **active margin** beneath which the Pacific Ocean floor of the Nazca plate is continually being subducted. The African continent resides in the center of the African plate. Both its eastern and western edges are passive margins. Seaward from the coasts of Africa, ophiolitic oceanic crust is welded to the continental crust. The oceanic crust has spread from the Carlsberg Ridge and the Mid-Atlantic Ridge in symmetrical fashion.

The Pacific plate spreads westward from the East Pacific Rise over a broad expanse (Figure 10.46). Near Asia it descends into the trench complexes of the western Pacific. Volcanic arcs continue to build above the subducted lithosphere. Off western South America, the Nazca plate is subducted beneath the South American plate. Off western North America, the Pacific plate comes in direct contact with the North American plate along the San Andreas transform fault. The San Andreas fault accommodates strike-slip faulting between the North American and Pacific plates.

Such examples give color to some of the many possible plate tectonic interactions that can take place.

PLATE KINEMATICS

Determining Absolute Motions of Plates

Absolute motions of plates were very difficult to determine. But Jason Morgan proposed that fixed hot spots in the mantle may constitute absolute points of reference for describing plate motions (Morgan, 1971, 1972). A hot spot is regarded as a thermal plume, probably of deep mantle origin, that sears the overriding lithosphere. An image that comes to mind in visualizing hot spots is one of slowly passing an old phonograph record over a blowtorch, thus steadily melting and blistering the record along the line of movement (Figure 10.47).

Figure 10.47 The Hawaiian Islands were generated by the translation of lithosphere of the Pacific plate over a fixed hot spot. (Artwork by D. A. Fischer.) →

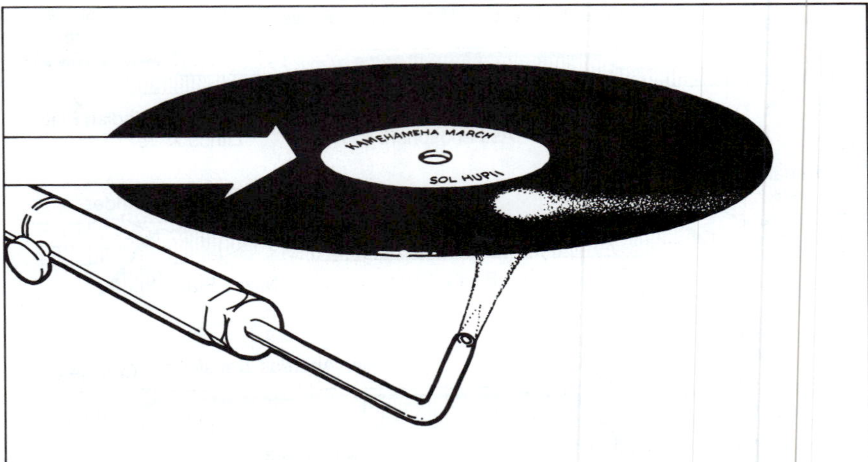

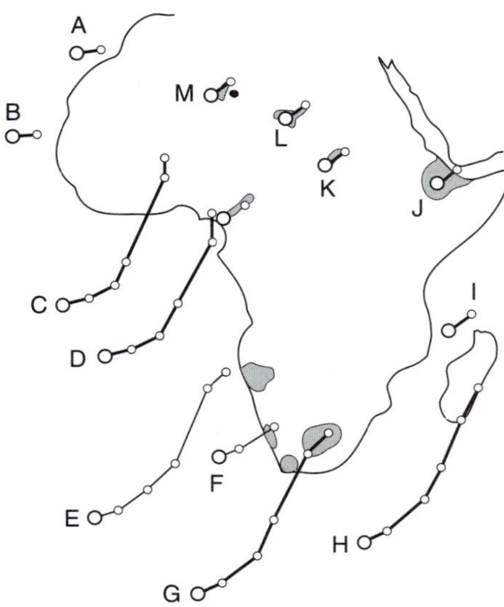

Figure 10.48 Africa hot spot motion since 100 Ma. The presently active hot spots are shown by relatively large open circles (110 km in diameter). Smaller circles mark probable paths of movement. Shaded areas are continental volcanic provinces. A = Canaries; B = Cape Verde; C = Ascension; D = St. Helena; E = Tristan da Cunha; F = Vema; G = Bouvet; H = Prince Edward; I = Comores; J = Afar; K = Jebel Mara; L = Tibesti; and M = Ahaggar. [Adapted from Duncan (1981), *Tectonophysics*, v. 74, Figure 1, p. 31. Published with permission of Elsevier Scientific Publishing Company, Amsterdam.]

The path of movement of plates over hot spots is recorded in the form of a linear scar of volcanoes in the lithosphere (Figure 10.48). The ages and mapped locations of the volcanic rocks yield the time-and-place data necessary to unravel absolute motions. Continental flood basalts and large oceanic plateaus represent the initial massive outpourings of basaltic lava produced when an ascending **mantle plume** burns a hot spot into the overlying plate. (Figure 10.49).

A fascinating geological record of the passing of a plate over a hot spot is the Hawaiian Islands–Emperor Seamounts chain in the Pacific Ocean (Figure 10.50). Oriented west–northwest, the Hawaiian Islands systematically increase in age from east to west. The island of Hawaii, at the eastern end of the chain, lies directly above or just northwest of the hot spot and thus still contains active volcanoes. In contrast, the Midway Islands, at the western end of the chain, are composed of 40 Ma volcanic rocks. The Midway Islands formed when that part of the lithosphere was located where Hawaii is located today. Knowing that the distance between Hawaii and the Midway Islands is 3000 km, the actual relative movement of the Pacific plate with respect to the fixed hot spot in asthenosphere can be determined. The average velocity is 3000 km/40 m.y. (7.5 cm/yr).

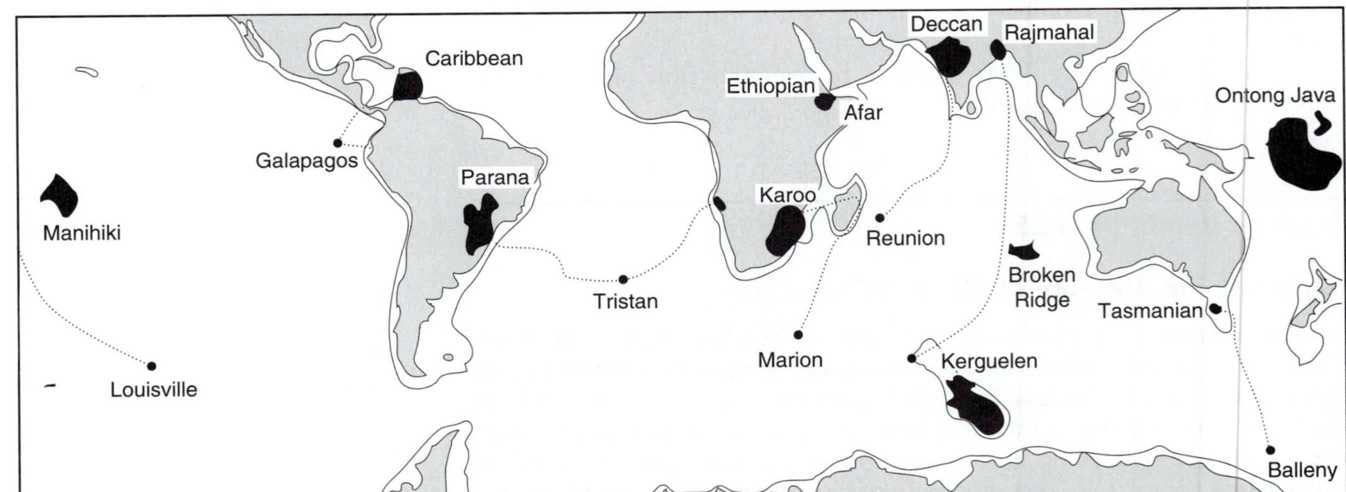

Figure 10.49 Map showing part of the world's flood basalt provinces and oceanic plateaus in relation to hot spots. [From Farnetani and Richards, *Journal of Geophysical Research*, v. 99, fig. 1, p. 13,814, copyright © 1994 by American Geophysical Union.]

Northward from the Midway Islands, the Emperor Seamounts trend along a north–northwest line (Figure 10.50) for a distance of 3000 km, where they encounter the Aleutian Trench. Volcanic rocks of the seamounts increase in age from 40 to 80 Ma from south to north along the chain. Thus we conclude that the Pacific plate moved north–northwest with respect to a fixed hot spot in the mantle during the period 80 to 40 Ma. Combining the Emperor Seamounts and Hawaiian Islands data, Morgan postulated that 40 m.y. before present was a moment of dramatic shift in the direction of relative movement of the Pacific plate lithosphere with respect to deeper mantle beneath the plate.

Hot spots also can show up under continental lithosphere. The Yellowstone hot spot is a case in point. Bob Smith from the University of Utah tells the story (personal communication, 1994):

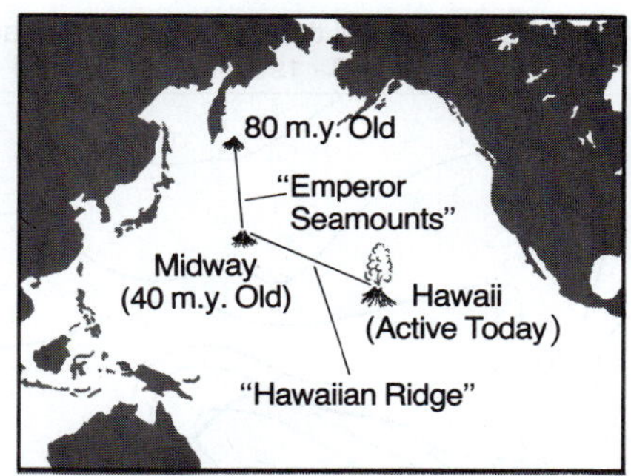

Figure 10.50 Map showing the Hawaiian Islands–Emperor Seamounts chain, an inferred hot spot track.

> At Yellowstone National Park, massive outpourings of hundreds to thousands of cubic kilometers of rhyolite and ash flows have been released in climactic eruptions unheard of in recorded historic times and which have covered the entire western U.S. with airborne ash deposits. The source of this massive volcanism is ascribed to the passage of the North American plate across a mantle hot spot which wedged apart the crust like a giant northeastward-opening zipper along the Snake River Plain of Idaho. The passage of the North American plate across the hot spot produced voluminous silicic centers as far back as 16 Ma in northern Nevada and southeastern Oregon—the original "Yellowstone National Park." Progressively older silicic volcanic deposits extend along a SW–NE track at rates of 4.5 cm/yr along the 800 km-long Snake River Plain of Idaho, and vents for future eruptions are expected to pop up across southern Montana and perhaps into the Great Plains. (Figure 10.51)

Bob Smith and his colleagues have been carrying out GPS studies of the structural effects over an area of the entire Yellowstone hot spot, 600 km in diameter. The hot spot is making its presence felt, for Bob and others have measured up to 22 cm of subsidence during the period 1987–1993 (Figure 10.52). The maximum subsidence is centered on the Yellowstone caldera. The topographic and seismicity signatures of the Yellowstone hot spot are like a "bow-wave" parabola-shaped region of active tectonics (see Figure 10.51).

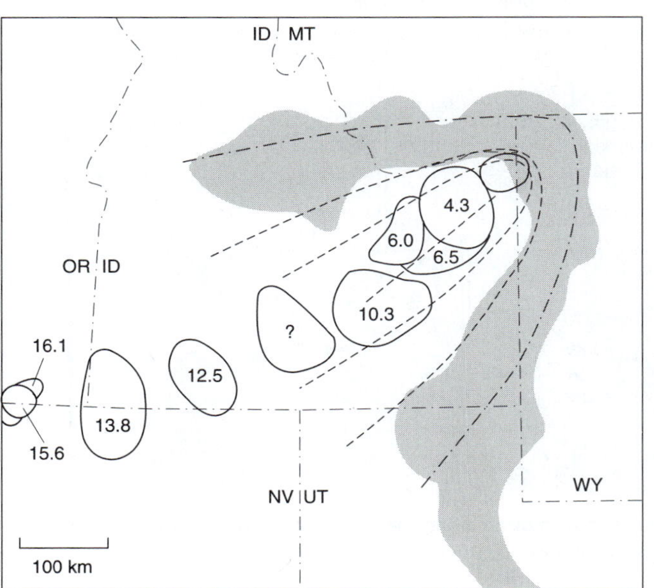

Figure 10.51 The North American plate has moved southwest at a velocity of 4.5 cm/yr relative to the Yellowstone hot spot. The velocity is determined through the time–space distribution of silicic volcanics extruded along the way. Shaded region encompasses the predominance of present-day earthquake distribution. The earthquake swath resembles a "bow wake" flowing around the Yellowstone basin. [From Smith and Braile (1994), *Journal of Volcanology and Geothermal Research*, v. 61, fig. 8, p. 141. Published with permission of Elsevier Scientific Publishing Company, Amsterdam.]

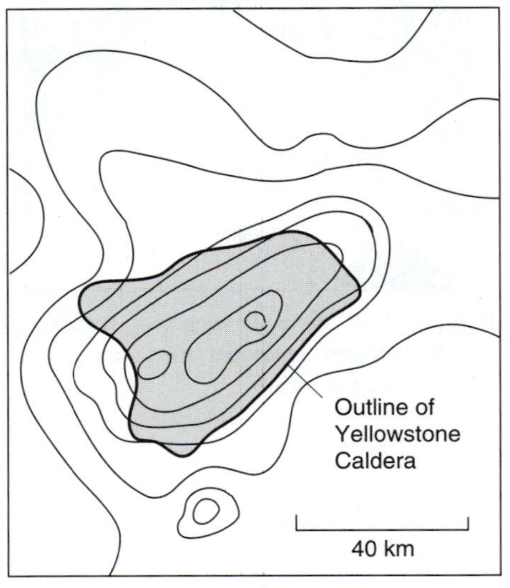

Figure 10.52 Contour map showing the ground subsidence (in millimeters) in the Yellowstone Basin from 1987 to 1993. The values are relative to a base station in West Yellowstone, on the west side of Yellowstone National Park. Maximum subsidence values correspond to a rate of approximately 2 cm/yr centered on the axis of the Yellowstone caldera. Results were determined from 150 benchmarks observed by high precision GPS surveys in 1987, 1989, 1991, and 1993. [From Smith and Braile (1994), *Journal of Volcanology and Geothermal Research*, fig. 28, p. 167. Published with permission of Elsevier Scientific Publishing Company, Amsterdam.]

Determining Relative Motions and Migration of Plate Boundaries

Relative motions are the basis for evaluating the migration of plate boundaries. Consider, for example, a segment of the Mid-Atlantic Ridge at latitude 25°S, where oceanic floor of the South American plate and the African plate diverge at a total spreading rate of 4 cm/yr (Figure 10.53*A*). The **half-spreading rate** of 2 cm/yr describes the movement of the South American plate relative to the Mid-Atlantic Ridge, imagining the ridge as fixed. Alternatively, the 2 cm/yr velocity might be viewed as the motion of the African plate relative to the Mid-Atlantic Ridge. If a volcanic island that had been constructed along the ridge were "split" by seafloor spreading along the plate boundary, each half would migrate laterally away from the ridge at 2 cm/yr. In 1 m.y., the two halves of the original volcano would be 40 km apart (Figure 10.53*B*).

Let us now change our perspective in viewing this system. Let us fix the westernmost edge of the African plate as it exists today, and then

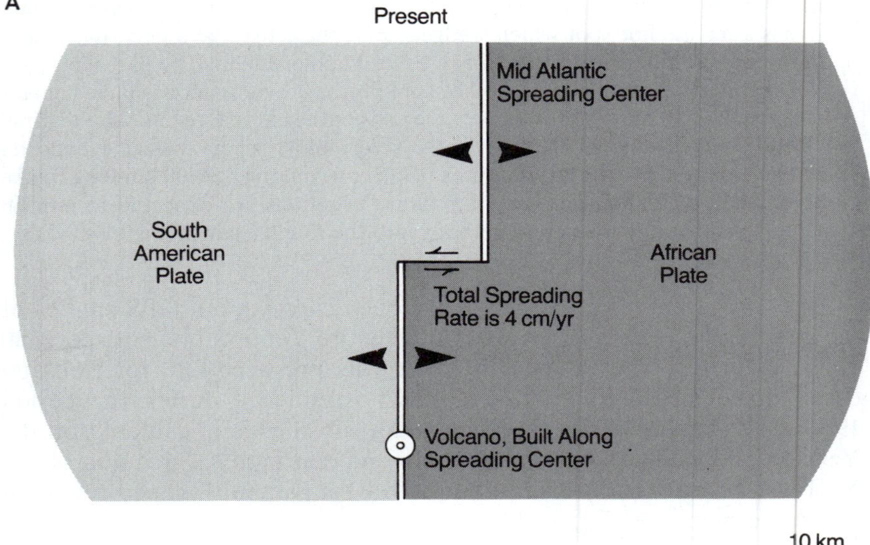

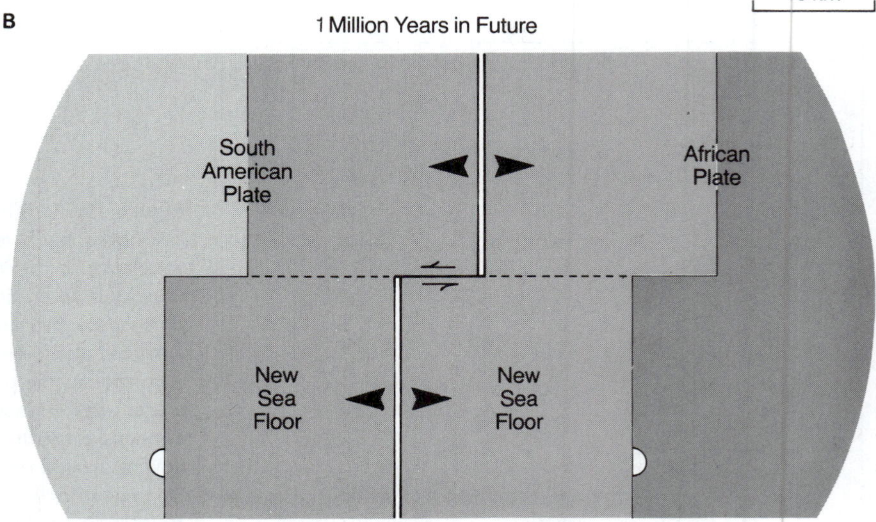

Figure 10.53 Actual relative motion along the Mid-Atlantic Ridge in the South Atlantic. (*A*) Present configuration. (*B*) Configuration 1 m.y. hence.

describe the movement of both the Mid-Atlantic Ridge and the South American plate relative to the African plate. To visualize this perspective, consider a "ridge-shaped" tape measure (not available at most hardware stores) that is capable of feeding out steel tape from both sides (Figure 10.54*A*). Consider the emerging tape as new seafloor lithosphere that is fed out at the rate of 2 cm/yr (Figure 10.54*B*). Relative to a fixed African plate, the ridge moves westward at a rate of 2 cm/yr. However, a point in the interior of the South American plate will move at twice that velocity, namely 4 cm/yr, westward from the African plate. In 1 Ma, the ridge will have moved 20 km westward from the present western edge of the African plate. A point on the eastern edge of the South American plate today will move 40 km westward during the next million years, as measured with respect to the African plate.

Relative movements are calculated for convergent plate boundaries as well, like the tectonically active western continental margin of South America (Figure 10.55*A*). Pacific seafloor of the Nazca plate is underthrusting the South American plate at an inferred rate of 5 cm/yr. This means that if the Peru–Chile Trench is considered to be fixed in space, a point on the Nazca plate moves toward the trench, eastward, at the rate of 5 cm/yr (Figure 10.55*A*). Another way to look at this relative motion is to hold the Nazca plate fixed and visualize the Peru–Chile Trench moving westward over the Nazca plate at the rate of 5 cm/yr (Figure 8.55*B*). Will the trench overtake the East Pacific Rise? This depends on how rapidly the rise moves westward from the western edge of the Nazca plate. Since

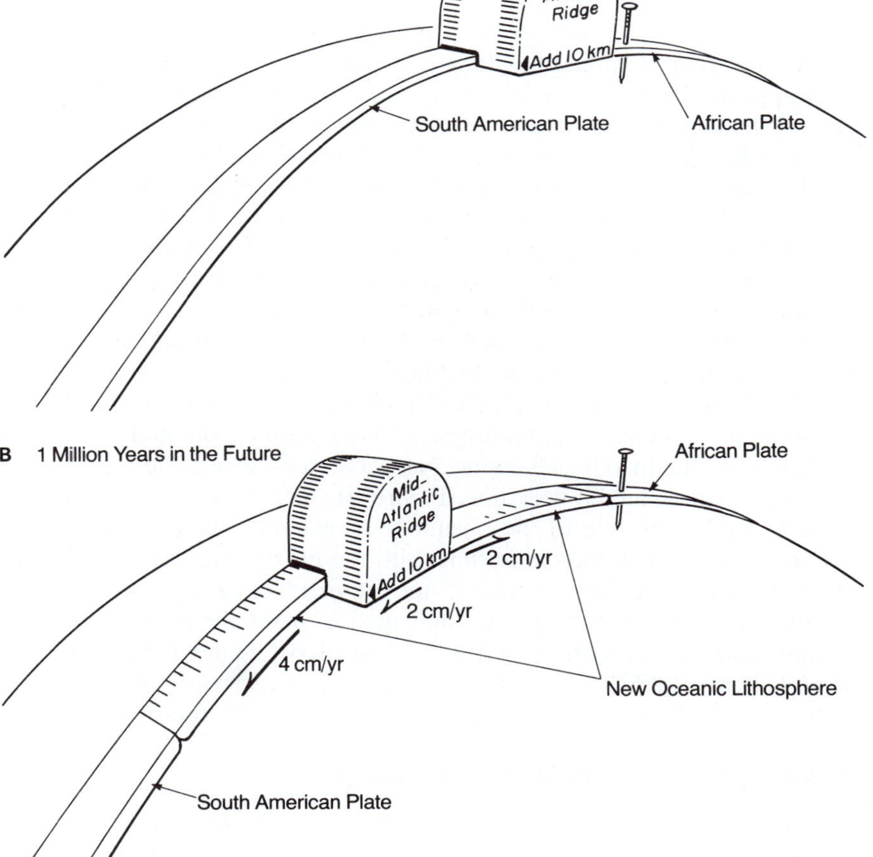

Figure 10.54 Tape measure imagery of relative motions. (*A*) Present configuration of the African and South American plates and the Mid-Atlantic Ridge. (*B*) Configuration 1 m.y. hence. New oceanic lithosphere has been added to the South American and African plates. With respect to a *fixed* African plate, the Mid-Atlantic Ridge moves away at 2 cm/yr and the South American plate moves away at 4 cm/yr.

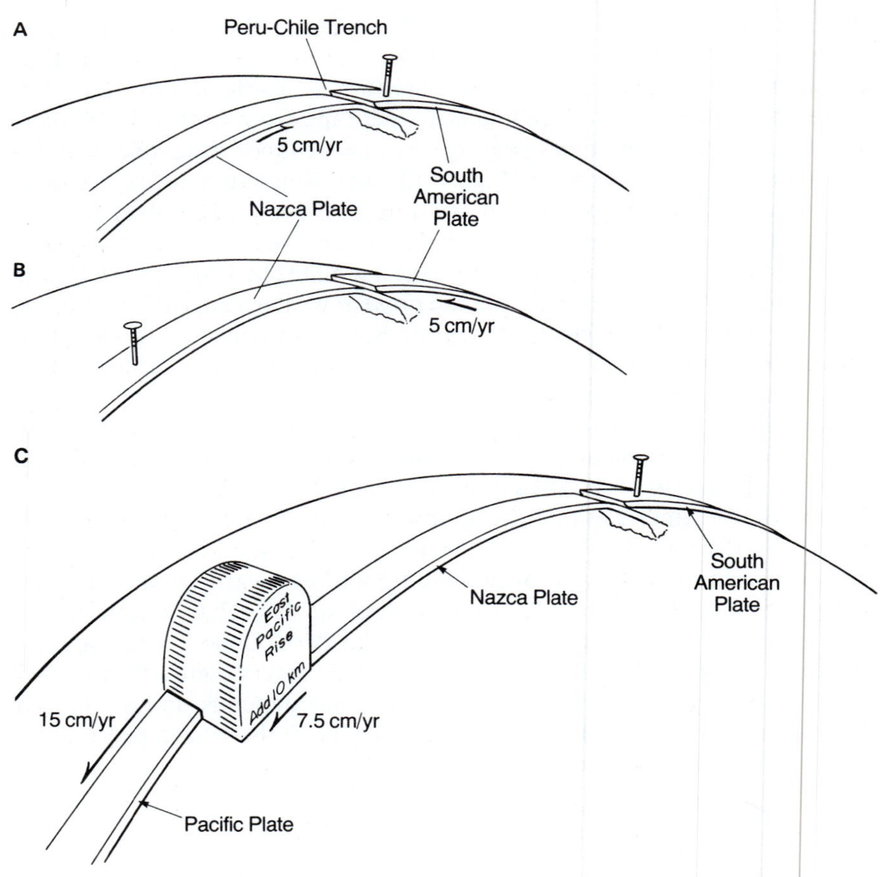

Figure 10.55 (*A*) Relative motion of the Nazca plate with respect to the South American plate (fixed). (*B*) Relative motion of the South American plate with respect to the Nazca plate (fixed). (*C*) Relative motion of the East Pacific Rise and the Pacific plate with respect to the South American plate (fixed).

the present rate of spreading along that part of the rise is about 7.5 cm/yr, it appears that the rise will outdistance the trench and that the breadth of the Nazca plate will actually expand (Figure 10.55*C*).

When I taught geology for one term at Carleton College, we had some fun dramatizing relative motions in two different ways. In the first arrangement, two "ridge people" each held in one hand a stick inserted into a roll of paper towels; and in the other hand two magic markers, one for normal magnetic polarity (red) and the other for reverse polarity (green) (Figure 10.56*A*). Two "seafloor spreaders" stood back to back at the ridge, grabbed the leading edge of paper towel from behind and, in time to a recording of Pachelbel, walked heel–toe–heel–toe (the shoe sizes of the "seafloor spreaders" must be *exactly* the same) (Figure 10.56*B*). As the seafloor "spread," the stationary "ridge people" painted "magnetic stripes" on the towels, changing marker colors when the "keeper of reversals" announced it was time (Figure 10.56*B*). In the second arrangement of Pachelbel, one of the "seafloor spreaders" stayed put, the two "ridge people" each moved in time with the music (heel–toe–heel–toe), and the second "seafloor spreader" moved in advance, *double-time* (Figure 10.56*B*). The "ridge people" managed to paint the magnetic stripes as they walked. The symmetry created in each of the movements was the same (Figure 10.56*C*), yet in the first case the ridge was stationary, and in the second case the ridge was on the move. It's all relative!

Tracking the Evolution of Transform Faults

Keeping the principles of "relative motion" in mind, we can take the next step in plate kinematics: computing the changes in distance between ridges

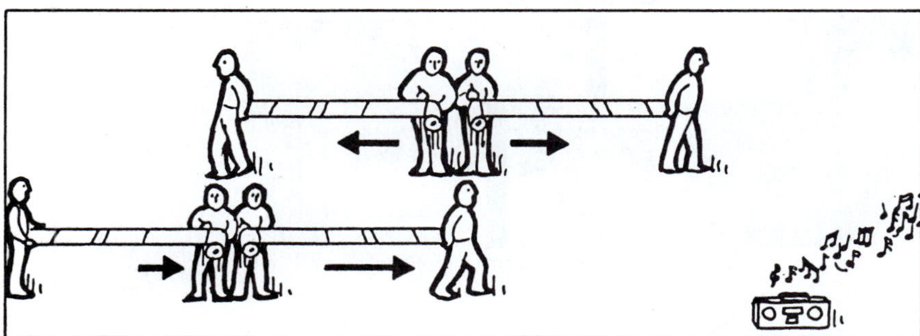

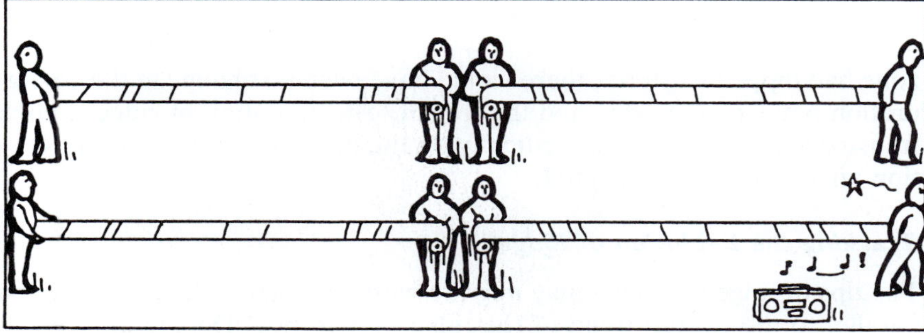

Figure 10.56 Seafloor spreading to Pachelbel. See text.

or trenches that are linked by a common transform fault zone. The basis for this kind of analysis was presented by J. Tuzo Wilson (1965). The goal of the exercise is to determine whether a given transform fault configuration changes through time and, if so, in what way.

Figure 10.57*A* is a map of the geometry of a ridge–ridge transform. New seafloor is being generated at the ridge at a half-spreading rate of 5 cm/yr. The ridge segments on opposite sides of the transform fault migrate eastward at a rate of 5 cm/yr with respect to plate *A*, which is assumed to be fixed. In 1 m.y., each ridge segment shifts eastward a distance of 50 km (Figure 10.57*B*). Even though the total translation is significant, the configuration of the ridge–ridge transform does not change.

Such a static configuration does not hold for the ridge–trench transform shown in Figure 10.58*A*. The trench in this example is the mouth of an eastward-dipping subduction zone. If plate *A* is held fixed, and assuming a spreading rate of 5 cm/yr, the ridge migrates eastward at a rate of 5 cm/yr (Figure 10.58*A*). Plate *B* moves away from the ridge at a rate of 5 cm/yr, and at the subduction zone it underthrusts a fixed plate *A* at the rate of 10 cm/yr. The net effect is a steady decrease in the distance between the ridge and the trench, and thus a steady decrease in the length of the ridge–trench transform. The rate of movement of the ridge toward the trench (assumed to be fixed) is 5 cm/yr (Figure 10.58*B*). If the subduction

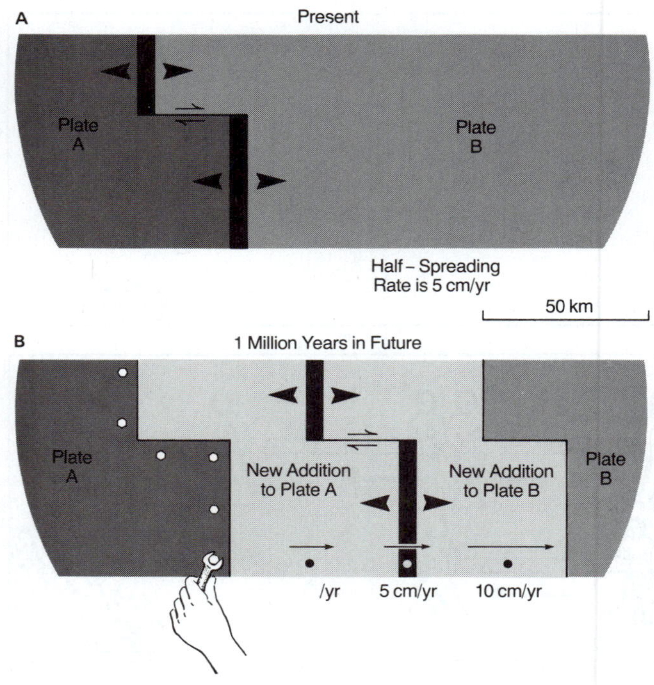

Figure 10.57 (*A*) Ridge–ridge transform system. (*B*) Configuration of system remains the same 1 m.y. hence. [From Wilson (1965). Reprinted by permission of *Nature*, v. 207, pp. 343–347, copyright © 1965 by Macmillan Journals Ltd., London.]

zone had dipped westerly, there would have been no change in the configuration between the ridge and the trench. The evolution of ridge–trench transform systems depends critically on the dip directions of the subduction zones that are connected.

Tracking the Evolution of Ridges

The tips of ridge segments may migrate with time, generally in a direction parallel to the overall trend of the ridge system (MacDonald and others 1991). In a manner not unlike the progressive evolution of mode I joints, the leading migrating ends or tips of two different ridge segments may approach each other and become so close that each "feels" the stress field of the other. When this happens, **overlapping spreading centers** develop, especially where lithosphere is too thin and weak to accommodate trans-

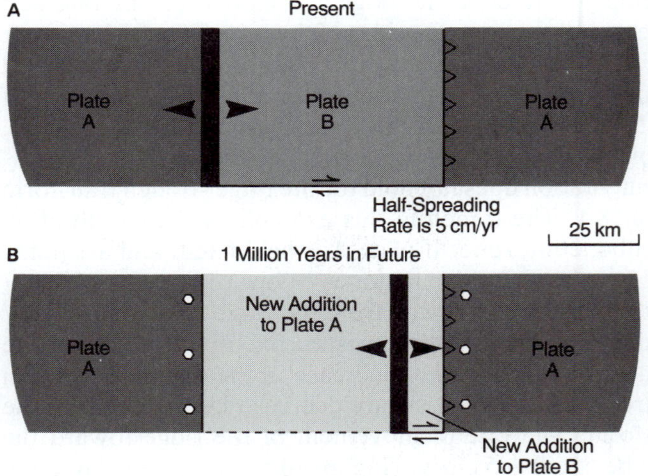

Figure 10.58 (*A*) Ridge–trench transform system. (*B*) Configuration of system dramatically altered 1 m.y. hence. [From Wilson (1965). Reprinted by permission of *Nature*, v. 207, pp. 343–347, copyright © 1965 by Macmillan Journals Ltd., London.]

form faulting (MacDonald and Fox, 1983). Simple models of the development of overlapping spreading centers can be produced on the stove (or in the lab) by first melting wax in a saucepan, and allowing a solid wax film to develop and float on the molten wax. When this condition is achieved, pull out your Swiss Army knife and cut two parallel but nonaligned slits in the wax lithosphere (Figure 10.59*A*). Pull-apart tension in the wax lithosphere will cause the tips to migrate toward each other (Figure 10.59*B*). When each feels the stress influence of the other, the overlapping ridges will curve toward each other and encircle what will become a zone of shear and rotational deformation (Figure 10.59*C*). Eventually a connection will be achieved between the two ridge segments, leaving a continuous ridge and an abandoned overlapping spreading center (Figure 10.59*D,E*). Yet additional challenges in interpreting the evolution of ridge spreading stem from rift propagation as an adjustment to changes in spreading direction. This can involve the replacement of an old spreading center by a new one perpendicular to the new spreading directions (Hey and others, 1980). What results is a V-shaped pattern pointing in the direction of propagation of the new rift (Figure 10.60).

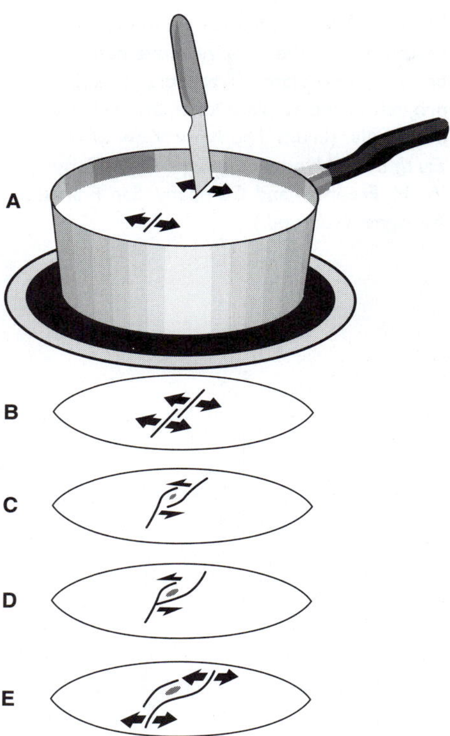

Figure 10.59 Experimental analogue of the development of overlapping spreading centers. (*A*) A thin wax film atop molten wax is cut in the manner shown. (*B*) The cuts are subjected to tension orthogonal to the slits. (*C*) The tension causes the slits to propagate and overlap. (*D*) Eventually the tips of the overlapping propagating tips bend toward one another. The area surrounded by the overlapping spreading centers is marked by shear and rotational deformation. (*E*) One of the overlapping spreading centers breaks through and links with the other. The former overlap zone "drifts" away. [From MacDonald and Fox (1983). Reprinted by permission of *Nature*, v. 302, pp. 55–58, copyright © 1983 by Macmillan Journals, Ltd., London.]

Poles and Plate Motions

The relative motion between two plates that meet along a spreading center can be described quite handily. Principles for the constructions were developed by Morgan (1968) and Le Pichon (1968). The rules are elegant in their simplicity. The divergent motions between two plates at any point along a ridge can be described with respect to a pole of rotation (Figure 10.61). A pole of rotation can be thought of as an imaginary point on the surface of the Earth representing the surface projection of an axis of rotation passing through the center of the Earth. Adjacent plates at a spreading center can be thought of as pie-shaped wedges whose tips attach to the pole.

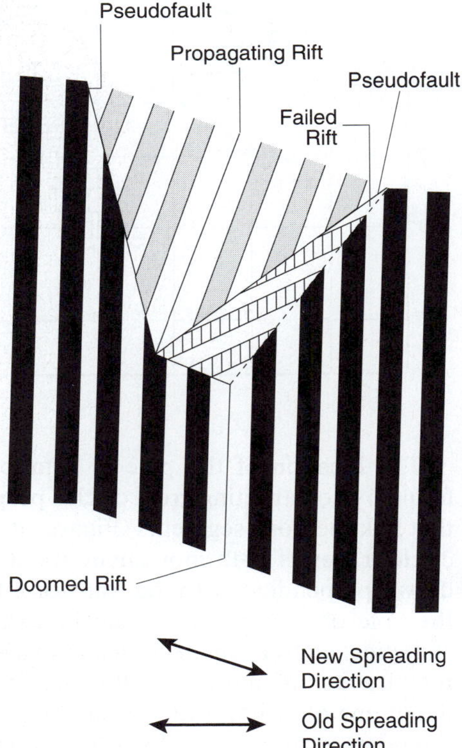

Figure 10.60 V-shaped pattern that develops on the ocean floor when a change in the spreading direction is accommodated by propagation of a new rift, which causes the old failed rift to rotate out of the way. New "stripes" of seafloor are misaligned with respect to the old stripes, but perfectly aligned with respect to the new spreading direction. (From Hey, Menard, Atwater, and Caress, *Journal of Geophysical Research*, v. 93, fig. 1B, p. 2804, copyright © 1988 by American Geophysical Union.)

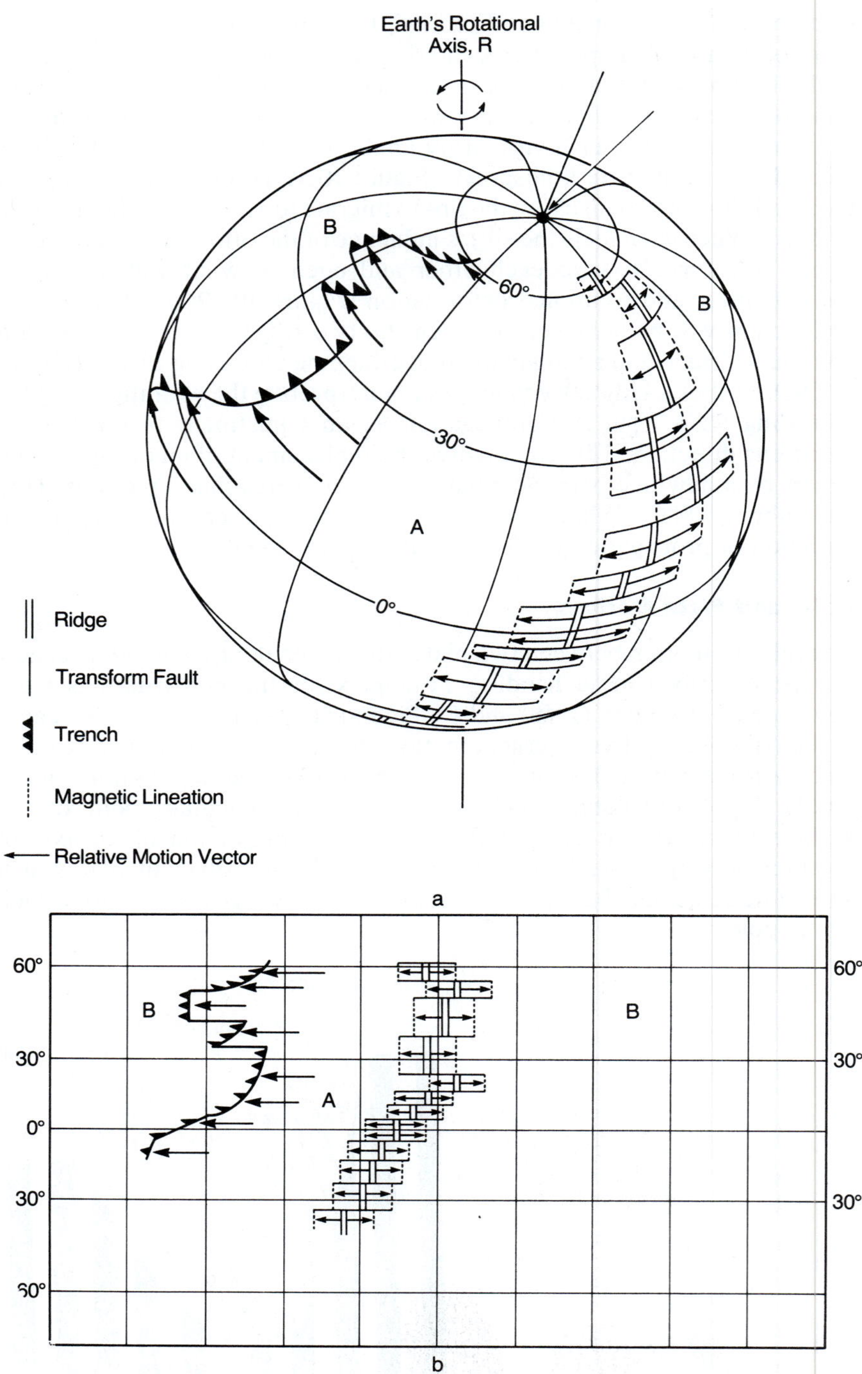

Figure 10.61 Spherical geometric description of the relative movement between two plates. Transverse Mercator projection shows plate kinematic relations beautifully. (From *The New View of the Earth* by S. Uyeda, copyright © 1978 by W. H. Freeman and Company, San Francisco. All rights reserved.)

The location of the pole of rotation for any two diverging plates is found by constructing great circles perpendicular to transform fault traces that link the ridge segments (Figure 10.61). The transform faults are small-circle traces of path movement about the pole of rotation. Great circles drawn perpendicular to the transform fault traces will intersect ideally at the pole of rotation. Defining the exact location of the pole becomes a statistical matter. Small circles drawn about the pole of rotation represent the "tracks" of movement. By combining these kinematics with age determinations of seafloor at various locations, the **average linear velocity** of the relative movement of seafloor from a ridge can be determined. Maximum rates of spreading are found farthest from the pole of rotation; minimum rates of spreading are closest to the pole of rotation. It is just the same as playing "crack the whip." Coupling linear velocity with direction of movement becomes the basis for establishing displacement vectors of movement for any given time period.

It is instructive to portray this geometry stereographically, on an equal area net. The stereographic projection used is an **upper hemisphere projection** such that constructions reflect what would be seen on the surface of a globe. Using stereographic projection, it is possible to construct a hypothetical ridge system, complete with ridge–ridge transform faults (Figure 10.62*A*) and to evaluate what the ocean ridge system would look like at some time in the future, say 15 m.y. hence. The pole of rotation that describes the relative motion between any two diverging plates, like plates *A* and *B* in Figure 10.62*A*, is positioned at the *N* point of the projection. Small circles of the projection describe appropriate orientations of transform faults that would help accommodate the spreading. Ocean ridge segments, in contrast, trace out on great circles that pass through the pole of rotation.

To evaluate how the plate configuration shown in Figure 10.62*A* would look 15 m.y. in the future, it is necessary to represent the spreading rate not as linear velocity (in km/m.y.), but rather as **angular velocity**. Angular velocity is expressed in degrees per million years. In this example, we assume that the angular velocity (ω) is 1°/m.y.

In representing the spreading stereographically, the individual ridge segments as well as plate *B* are moved with respect to a fixed plate *A* (Figure 10.62*B*). To add interesting detail to the construction, newly generated seafloor is shaded dark or light, depending on the assumed polarity of the Earth's magnetic field during the time of formation of new

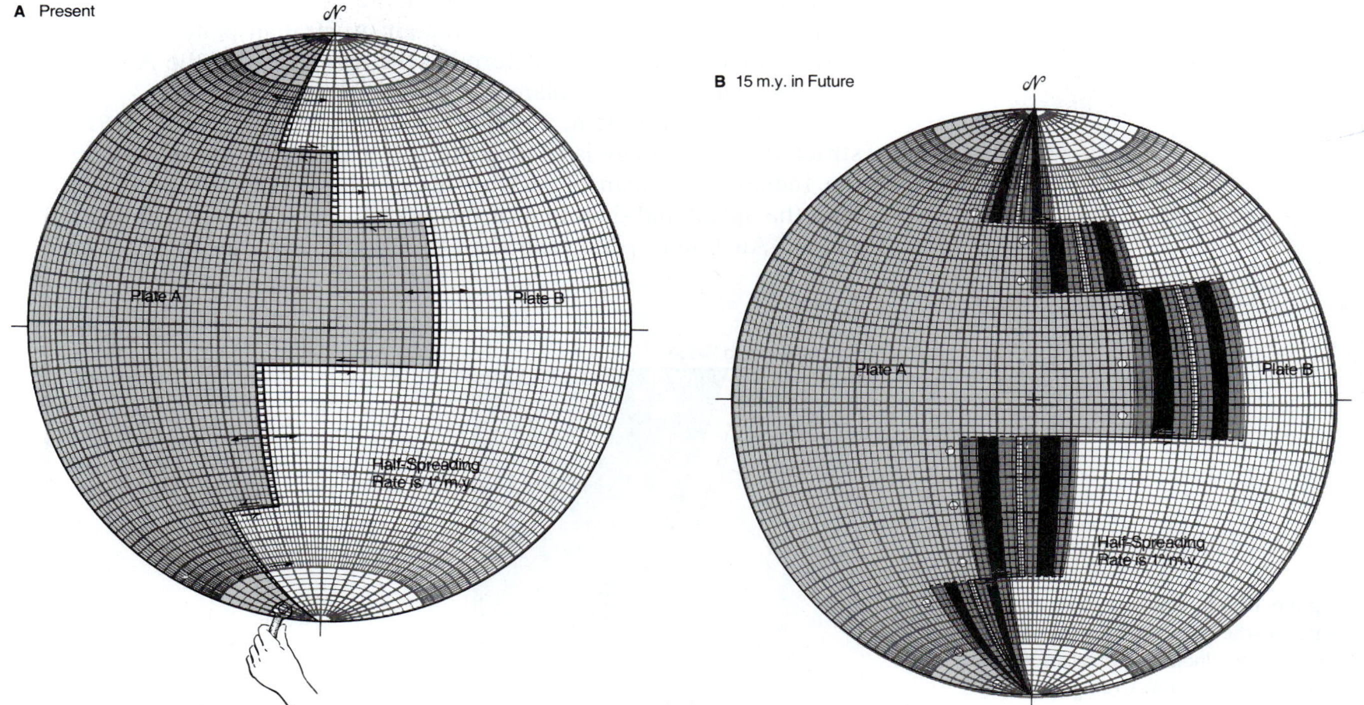

Figure 10.62 Stereographic display of the relative motion between plates. (*A*) Starting configuration of ridge segments and transform faults. (*B*) Configuration 15 m.y. later, assuming a half-spreading rate of 1°/m.y.

oceanic crust. For convenience and simplicity, Figure 10.62*B* is drawn such that the magnetic polarity reverses each 5 m.y. Points in plate *B* move with respect to *A* at an angular velocity of 2°/m.y., and the ridge moves eastward from the present-day eastern boundary of plate *A* at an angular velocity of 1°/m.y. The configuration of magnetic stripes and plate boundaries issues systematically from the movements, with the final pattern frozen in lithosphere in the manner shown in Figure 10.62*B*.

Vector Diagrams and Plate Motions

When we look at the plate configuration map of the world (see Figure 10.44), we are reminded instantly that the computing of plate interactions between just two plates is only a small part of the analysis. It becomes necessary to consider multiple plates and the relative motions between them. **Triple junctions**, the points at which three plates come together, become the focus of study. Our objective is to learn how to determine the relative motions between *any two* of the plates that meet at a triple junction. It is possible to do this, *provided* information is available regarding the relative motions along two of the three boundaries. Facts regarding linear velocity and direction of relative movement are represented as vectors, and vector circuit diagrams are constructed to determine relative motion along the third boundary.

Consider the ridge–ridge–ridge triple junction that brings the Indian–Australian, African, and Antarctic plates into contact (Figure 10.63*A*). The ridges that meet at this triple junction in the Indian Ocean are the Carlsberg Ridge, whose half-spreading rate is 2.5 cm/yr along a N60°E line; the Southeast Indian Ocean Ridge, whose half-spreading rate is 3.7 cm/yr along a N30°E line; and the Southwest Indian Ocean Ridge, whose half-spreading rate and direction we would like to determine. These starting facts describe the movement of the Antarctic plate relative to the Indian–Australian plate, and the Indian–Australian plate relative to the African plate. We want to determine the movement of the African plate relative to the Antarctic plate.

To solve this problem, a **vector circuit diagram** (Figure 10.63*B*) is constructed. The strategy is to map the speed and direction of movement of the Indian–Australian plate with respect to a fixed Antarctic plate, to map the speed and direction of the African plate with respect to a fixed Indian–Australian plate, and then to measure directly from the current

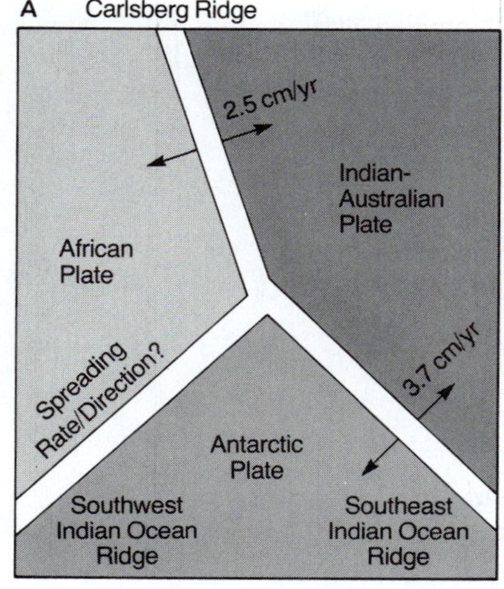

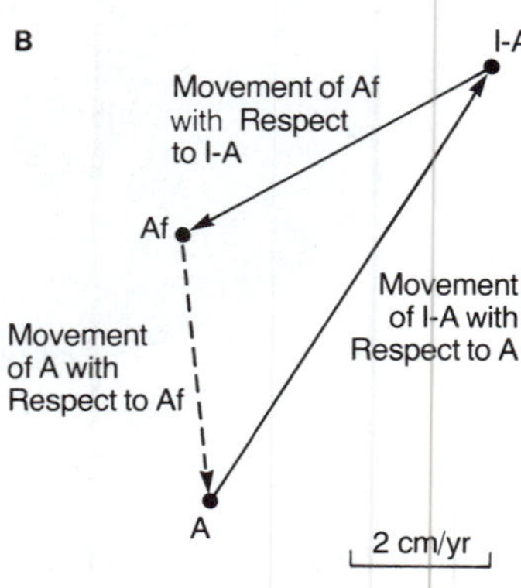

Figure 10.63 Relative motion among plates at the ridge–ridge triple junction between the Indian–Australian, African, and Antarctic plates. (*A*) Configuration and half-spreading rates. (*B*) Vector circuit diagram to determine the relative velocity of the Antarctic plate with respect to the African plate.

diagram the movement of the Antarctic plate with respect to a fixed African plate. When this approach is used, the circuit of movement is counterclockwise around the triple junction. (Using a different strategy, a clockwise circuit could be employed.)

The Antarctic plate is fixed in vector space by plotting the point labeled *A* in Figure 10.63*B*. Points in the Indian–Australian plate move according to the total spreading rate (2 × 3.7 cm/yr) from the Antarctic plate along an azimuth of N30°E. After a reasonable map scale has been chosen for the vector diagram, a vector is constructed having a scale length of 7.4 cm and a trend of N30°E. An arrowhead is placed at the tip of the vector, denoting the direction sense of the movement. At the tip of the arrowhead, the label *I-A*, for Indian–Australian plate, is marked. A second vector is constructed, which describes the movement of the African plate relative to the Indian–Australian plate. Its trend is S60°W. Its scaled length of 5 cm is the total spreading rate. Its southwestward tip is marked by an arrowhead labeled *Af* for African plate (Figure 10.63*B*). Thus with the location of the African plate fixed in the vector circuit diagram, we can measure the relative movement of the Antarctic plate relative to the African plate directly. Using the map scale, it is found to be 3.9 cm/yr, signifying a half-spreading rate of 1.9 cm/yr.

Let us work through a second example. Consider the transform–transform–trench triple junction that marks the common point between the Juan de Fuca, Pacific, and North American plates at the Mendocino triple junction (Figure 10.64*A*). The rate of right-slip transform faulting between the North American and Pacific plates along the San Andreas fault is about 5.6 cm/yr. The rate of right-slip transform faulting between the Juan de Fuca and Pacific plates is 5.9 cm/yr. We would like to determine the nature of the relative motion between the Juan de Fuca and North American plates where they converge along the subduction zone that separates them.

The proper vector circuit can be constructed by moving clockwise from the North American plate about the triple junction. Mark first on the vector diagram the point *NA*, symbolizing the North American plate fixed

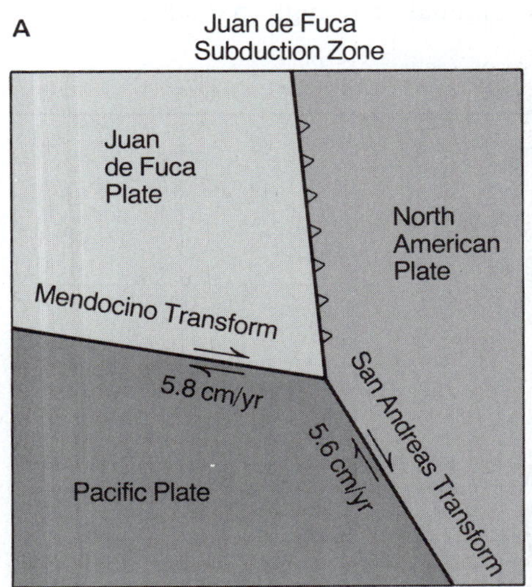

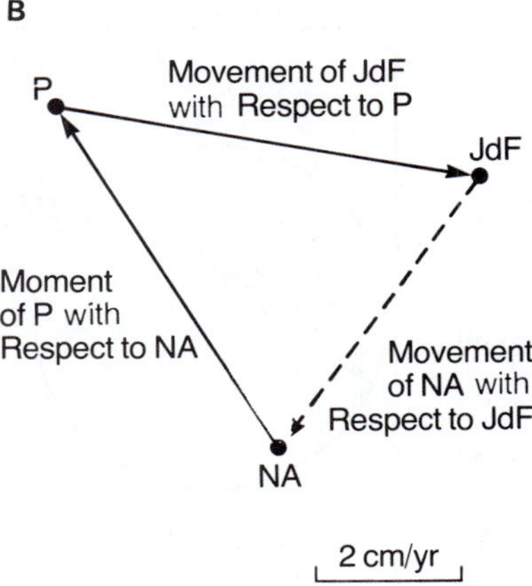

Figure 10.64 Relative motion among plates at the transform–transform–trench triple junction between the Juan de Fuca, Pacific, and North American plates. (*A*) Configuration and half-spreading rates. (*B*) Vector circuit diagram to determine the relative velocity of the North American plate relative to the Juan de Fuca plate.

in vector space (Figure 10.64*B*). The movement of the Pacific plate with respect to the North American plate is 5.6 cm/yr. This rate is plotted along a direction that is N30°W, parallel to the trace of the San Andreas fault. The tip of the vector arrow is marked *P*, for Pacific plate. Keeping the Pacific plate fixed, we next plot the movement of the Juan de Fuca plate relative to the Pacific plate along the transform fault boundary (the Mendocino fracture zone) (see Figure 10.64*B*). Its rate is 5.9 cm/yr. Its direction, parallel to the trace of the transform, is S80°E. The tip of the arrow is marked *JdF* for Juan de Fuca. With these two vectors drawn, we can measure the movement of the North American plate relative to the Juan de Fuca plate directly. Its magnitude is 5.1 cm/yr, and its direction is S35°W. Thus, the vector diagram describes seafloor of the Juan de Fuca plate being subducted under North America at a rate of 5.1 cm/yr.

The construction of vector circuit diagrams requires accurate data on rates and directions of movement. Such data are abundantly available for ridge boundaries, less available for transform boundaries, and difficult to extract from trenches. At trenches, the record that would otherwise provide information on relative motion is swallowed. Consequently, in practice, subduction rates are derived secondarily from construction of vector diagrams using data describing motions at transform and ridge boundaries. Alternatively, the relative-motion vectors are determined by comparing plate motions to a number of hot spots, which are assumed to be essentially fixed in position with respect to the lower mantle.

There is a wonderful example of the power and application of these methods. Peter Coney (1978, 1981) made a pioneering contribution in relating plate motions to regional deformations. On the basis of hot spots and magnetic anomalies, he reconstructed the position of North America with respect to "Pacific" plates to the west for specified time slots between the Mesozoic era and the present. For each time slot he calculated the relative velocity vector between western North America and the immediately adjacent "Pacific" plate, in the hope of identifying a correspondence between the relative velocity vectors and the patterns of deformation in western North America. He struck paydirt within the time frame of 80 to 40 Ma, the time of the Laramide orogeny (Figure 10.65). Coney was able to show that the relative velocity between western North America and the oceanic plates to the west was anomalously high (namely, 14 cm/yr)

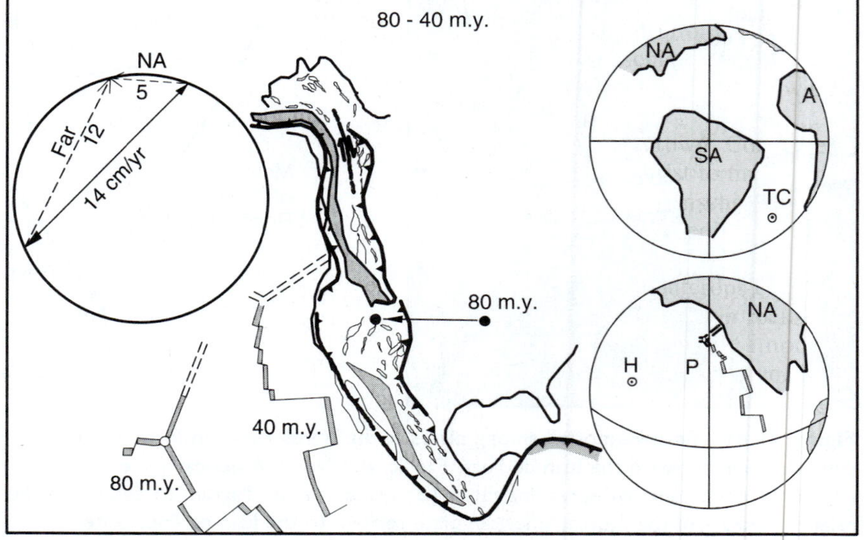

Figure 10.65 Plate configurations from 80 Ma to 40 Ma. The absolute motion of the Farallon plate was 12 cm/yr in a north–northeasterly direction. The absolute motion of the North American plate was 5 cm/yr westerly. Thus, the North American and Farallon plates converged on one another at a rate of 14 cm/yr, creating the compressional stresses responsible for the Laramide orogeny. [From Coney (1978). Published with permission of Geological Society of America.]

during the interval 80 Ma-40 Ma (see also Engebretson et al., 1985). Moreover, his calculations revealed that the 14 cm/yr relative velocity vector was oriented almost perfectly perpendicular to the boundary between the North American plate and the Farallon plate to the west (see Figure 10.65). Plate motion was purely convergent! Coney concluded that strong, head-on plate convergence generated the stresses that created the fold-and-thrust belts and Rocky Mountain uplifts that characterize the Laramide deformation pattern of western North America.

THE FORCES AND STRESSES OF PLATE TECTONICS

Moving from Kinematic to Dynamic Analysis

A revolution in understanding regional deformation has emerged from examination of the nature of plate boundaries and from consideration of the relationships between directions and velocities of plate motions to the deformation observed. Month by month, journal by journal, book by book, the march of kinematic models of cause-and-effect relations between plate movements and regional deformation becomes more and more sophisticated, more and more comprehensive. We might begin to ask: "What *more* is there left to do?"

Plenty! Each region of deformation in the Earth is unique and distinctive. This is not to say that there are not first-order similarities among systems of structures that develop within convergent boundaries, within transform boundaries, or within divergent boundaries. But in detail, no two systems are exactly alike. Why is this? Part of the answer lies in the pre-deformation history. Does the region contain a thick miogeoclinal sedimentary prison, or does the region consist entirely of crystalline basement?

The other part of the answer lies in dynamic analysis, in this case the relationship of the **force field** generated in plate tectonics to the **resistance field** of the region. The character of the resistance field, in turn, derives from strength, which in turn is related to the conditions of deformation (e.g., temperature, confining pressure, fluids, strain rate), the presence or absence of preexisting structures (notably faults), and the microscopic and submicroscopic deformational mechanisms available to the rocks. Thus, to really understand the inner workings of the **tectonic processes** at all scales that are responsible for regional deformation, we must work at the interface between kinematic and dynamic analysis. There we can interpret deformational movements in terms of force and resistance, and stress and strain. When one day we truly understand the details of the interplay between force and strength, and stress and strain, we will know *why* structures form *where* they do *when* they do. This knowledge has considerable scientific and strategic societal value.

Forces and Resistances

Advances in the understanding of the origin of plate forces have issued from the research of Forsyth and Uyeda (1975), Elasser (1968), Bott (1982a), and Kearey and Vine (1990). Much of this work is summarized effectively by Park (1988), and we have drawn from Park's work in the discussion that follows. The principal forces and resistances that control plate movements have interesting names, such as **ridge push**, **slab pull**, and **slab suction**. To picture the whole array of forces and resistances, let us imagine a spreading center that is feeding out new oceanic lithosphere (Figure 10.66). The new oceanic lithosphere at the ridge thickens outward as it ages and cools, and the whole plate moves laterally across a vast expanse before descending into a subduction zone. Beyond the subduction zone is continental lithosphere belonging to a different plate.

The origin of **ridge push** along spreading centers is the gravitational *head* created by topography, which reflects the warm buoyant character of the lithosphere along the ridge. Ridge push is applied to the separating plates, causing the plates to move (Figure 10.66). Resistance to movement is manifest in shallow-focus earthquakes that originate above asthenosphere in thin brittle lithosphere. Ridge push generates compression, with stress levels typically in the 20–30 MPa range.

A **mantle drag** resistance operates on the underside of a moving plate (Figure 10.66), like the drag on the underside of a barge moving on a river. Kearey and Vine (1990) point out that the mantle drag beneath continental lithosphere is approximately eight times greater than the mantle drag beneath oceanic lithosphere, probably because of the deep lithospheric roots of continents (Jordan, 1988).

Where the oceanic plate enters the subduction zone, it must flex to get in (Figure 10.66). The resistance to flexing is a **bending resistance**, which slightly impedes plate motion.

Other forces arise when oceanic lithosphere is in effect pulled down a subduction zone. **Slab pull** is caused by negative buoyancy, created by the density contrast between the cold slab going down and the hot mantle into which the slab descends. A distinction is made between the **negative buoyancy force**, which acts vertically, and slab pull, which acts parallel to the descending slab (Figure 10.66). According to Park (1988), slab pull is capable of creating stresses of magnitudes up to 50 MPa.

As you might expect, there is resistance to the descent of a slab of oceanic lithosphere. First of all, there is a **slab resistance** force oriented perpendicular to the leading edge of the slab as it descends (Figure 10.66). In addition, the upper and lower surfaces of the descending slab are marked by **slab drag**. Slab resistance is actually much larger (5–8 times) than slab drag.

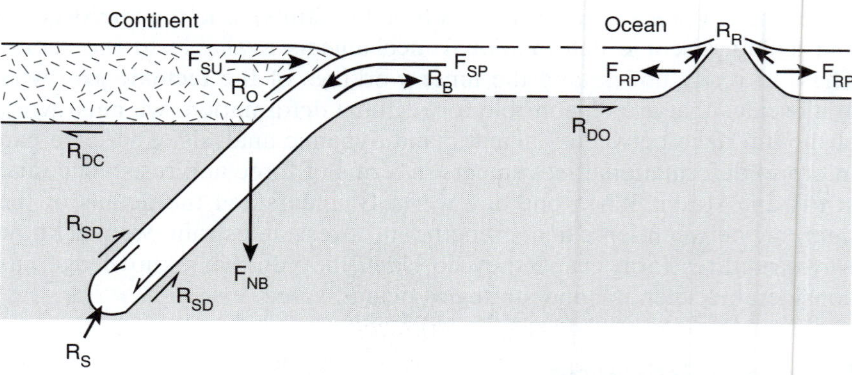

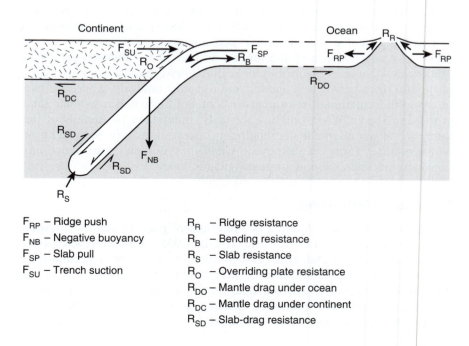

Figure 10.66 The forces acting on plates. [Adapted from Kearey and Vine (1990). Global tectonics, Figure 5.17, p. 89. Reprinted with permission from Blackwell Science Ltd.]

The overriding plate above the subduction zone imposes additional dynamic factors. For example, **trench suction** is a force that tends to drive the overriding plate into the trench at a faster rate (Figure 10.66). Tension results. There are debates over the exact origin, but the effect is expressed in a number of ways, including collapse of the overriding plate into the trench and backarc spreading. Stresses of magnitudes up to 20 MPa are created by slab suction. In contrast, the strength and velocity of movement of the overriding plate create **overriding plate resistance** (Figure 10.66). It is this resistance that gives rise to major earthquakes at the active plate margin.

The actual state of stress in the descending slab depends on a number of factors. If slab pull exceeds the combination of slab resistance and overlying plate resistance, the slab assumes a dynamic state of tension. If the opposite holds, the slab finds itself in length-parallel compression. The proof of the pudding lies in the earthquake record within descending slabs. The earthquake records can be used to monitor slab-parallel extension versus slab-parallel shortening as a function of depth (Figure 10.67).

If the resisting forces are relatively low at the location of subduction zones, both the slab-pull forces and the suction forces will produce tension in the adjacent lithosphere. This result is in some ways counterintuitive and should make us cautious about trying to equate convergence-related subduction with collision and compression. Recall that ridge-push forces at spreading centers, induced by buoyancy forces, generate compressional stresses of approximately 20–30 MPa. There is a certain irony in the generation of compressive stresses at spreading centers and tensional stresses in subduction zones (Park, 1988).

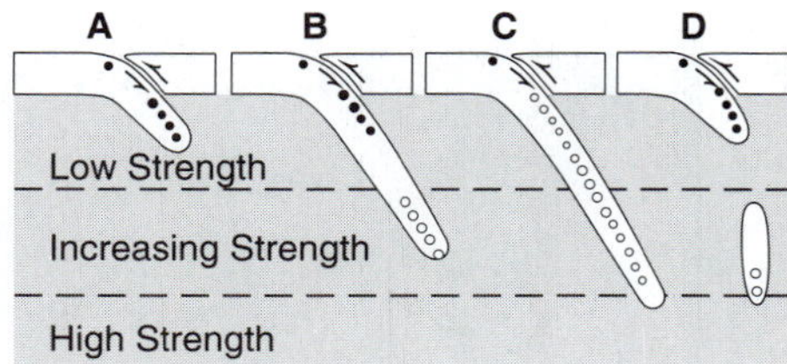

Figure 10.67 Portrayal of how stresses within a descending slab of lithosphere may change with depth of projection into mantle of steadily increasing strength. (*A*) Where the slab begins to subduct relatively rapidly into weak asthenosphere, the slab experiences tensional stress. (*B*) Where the slab penetrates through the asthenosphere and feels increasing resistance to descent, the lower reaches of the slab experience compressional stress while the upper reaches continue to experience extension stress. (*C*) Deep penetration into high strength regimes throws the whole slab into compression. (*D*) If the slab breaks in half, the upper "autochthonous" part within asthenosphere returns to an environment of extensional stress, whereas the descending "allochthonous" part endures slab-parallel compressional stress. [Adapted from Kearey and Vine (1990), Global tectonics, Figure 8.15, p. 154, based on Isacks and Molnar (1969).]

Present-Day Stresses

To help us to go from the theory to the actual, we can look at maps and interpretations of the present-day global stresses that are being measured. In 1986, under the auspices of the International Lithosphere Program, the World Stress Map project was initiated to compile and interpret the orientations and relative magnitudes of present-day stresses throughout the world. As reported by Mary Lou Zoback (1992), six types of geological and geophysical data sets were used in the compilation: earthquake focal mechanisms, well bore breakouts, in situ hydraulic fracturing, in situ overcoring, Quaternary fault slip data, and Quaternary volcanic alignments.

It was discovered that uniformly oriented horizontal stresses can be traced in continental regions over distances of up to 5000 km, and these are correlative with absolute and relative plate motions (Zoback, 1992). These compressional intraplate stresses are the result of compressional forces applied at the plate boundaries, especially ridge push and continental collision. Interestingly, where topography is high, such as in the Andes, in the Tibetan plateau, and in the western United States, intraplate stresses are extensional. This suggests that plate-driving forces cause the buoyancy forces, which are related to crustal thickening, to exceed the magnitudes of the compressional stresses. This leads to gravitationally driven collapse and extension of the overthickened crust (e.g., Coney and Harms, 1984).

The generalized map of the primary stress field for the globe is interesting (Figure 10.68). It summarizes more than 7300 data points! Different arrows distinguish stress configurations that would produce thrusting ver-

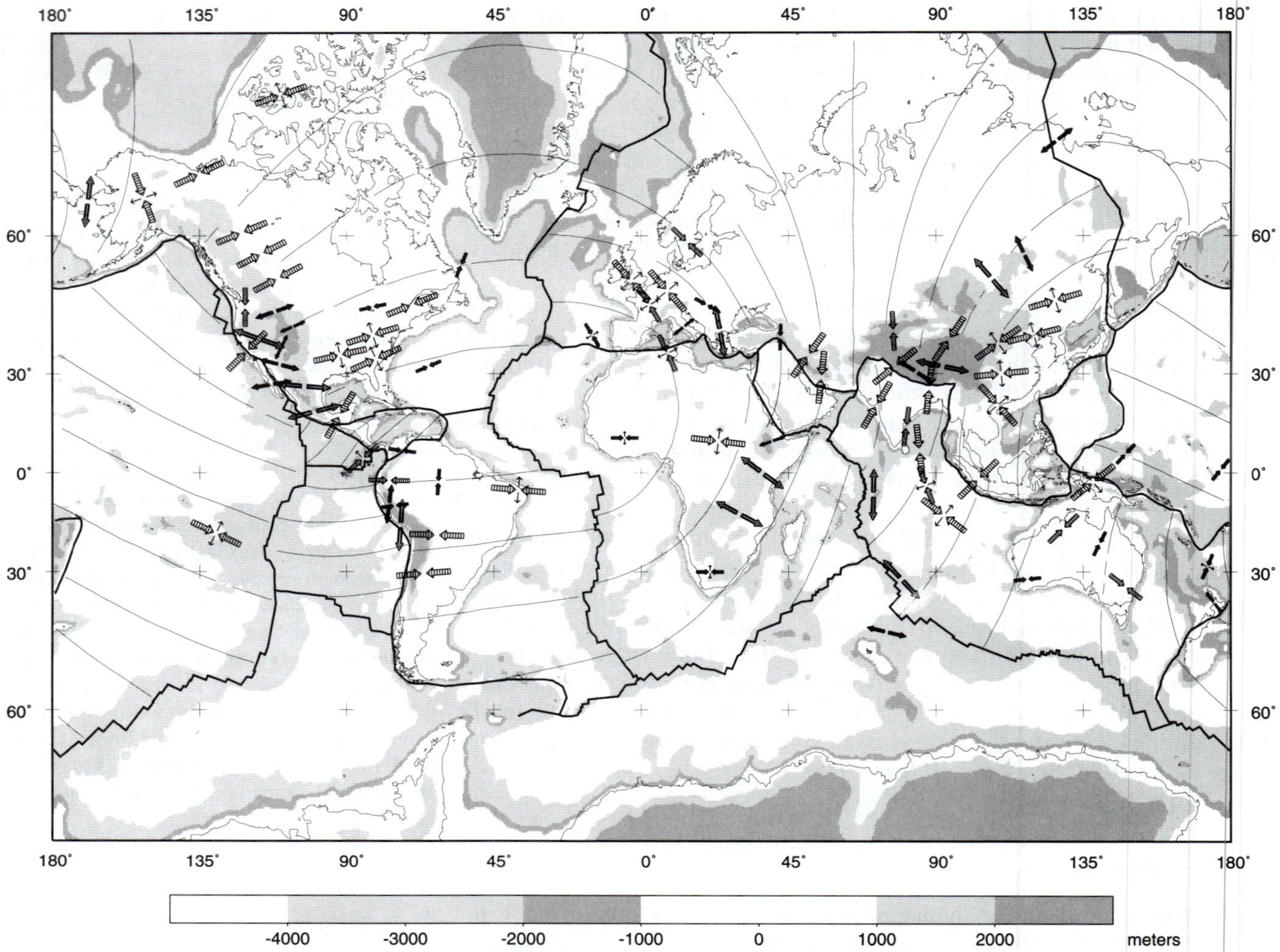

Figure 10.68 Mary Lou Zoback's Generalized World Stress Map summarizing present-day stresses, measured worldwide. The compilation is based on stress measurements determined in a variety of ways, including earthquake focal mechanisms, well bore breakouts, in situ hydraulic fracturing, in situ overcoring, Quaternary fault slip data, and Quaternary volcanic alignments. Thin lines show absolute velocity trajectories for individual plates. Single set of inward-directed arrows indicates a thrust faulting regime. Single set of outward-directed arrows indicates normal faulting regime. Sets of inward- and outward-directed arrows indicate a strike-slip regime. (From M. L. Zoback, *Journal of Geophysical Research*, v. 97, fig. 4, p. 11,714, copyright © 1992 by American Geophysical Union.)

sus normal faulting versus strike–slip faulting. Stresses in the midplate regions of North America and South America are primarily attributable to ridge push. Stresses in the high Andes are due to trench suction or buoyancy due to thick crust and/or thinned lithosphere. Stresses in western Europe are attributable to the combined effects of ridge push and continental collision with Africa. Stresses in China and eastern Asia reflect primarily the indention of the Indian plate into Asia. Along the East African Rift, buoyancy overcomes ridge-push compression. In North Africa, continental collision with Europe dominates the stress pattern. Stresses in India are reflecting continental collision (Zoback, 1992).

CONCLUSIONS

The plate tectonics paradigm took the world by storm in the 1960s and 1970s. There has been nothing like it in the Earth sciences. It is fair to say that plate tectonics is to solid-earth evolution as Darwin's models of the origin of species and natural selection are to biological evolution. Plate tectonics illuminates the origins of structural deformation. It gives us a detailed basis for understanding global kinematics and much of the deformation we see recorded in the continents and oceans, through time (Figure 10.69). As we explore the structure of mountain belts, both new and old, we discover again and again undeniable cause-and-effect relations between plate movements and crustal deformation. And as we interpret the structure of mountain belts, we find that we can use plate tectonic theory to achieve a degree of specificity that the founders of geology could never have imagined.

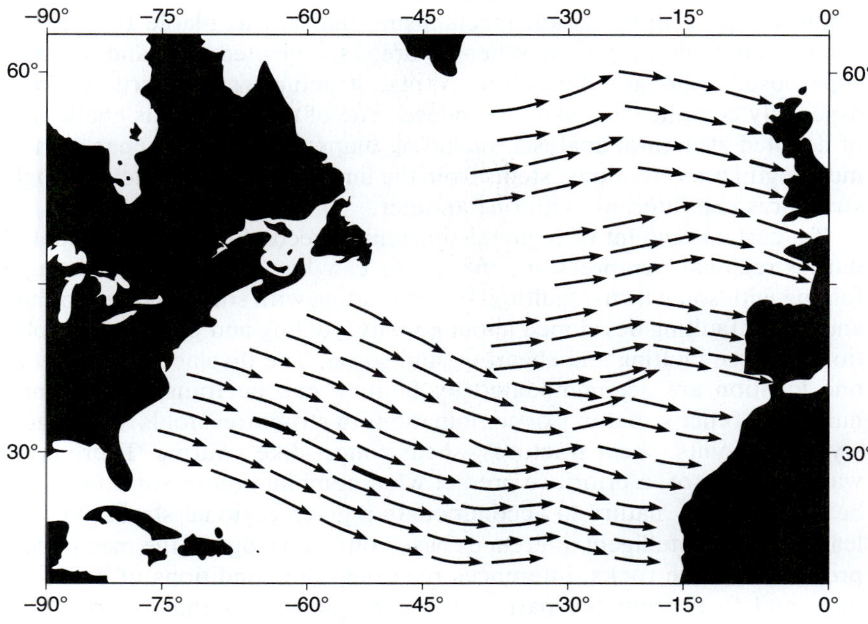

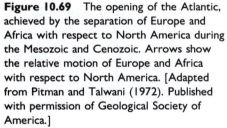

Figure 10.69 The opening of the Atlantic, achieved by the separation of Europe and Africa with respect to North America during the Mesozoic and Cenozoic. Arrows show the relative motion of Europe and Africa with respect to North America. [Adapted from Pitman and Talwani (1972). Published with permission of Geological Society of America.]

CONCLUDING THOUGHTS

In conclusion, we first present a rapid-fire series of figures that reemphasize the degree to which structures, at all scales, are arranged in systems (Figures CT.1–CT.6). Geologic structures comingle. Rocks utilize the whole range of deformation mechanisms that are available to them to accommodate deformation. When an area is subjected time and again to superposed regional deformation events, structures may accrue to form especially complicated systems. Indeed, one of the marvelous challenges of detailed structural analysis, including mapping and the preparation of meaningful cross sections, stems from the limitless ways in which geologic structures can interlink with one another.

If a certain amount of regional shortening needs to be accommodated during regional compression, the strain can be "partitioned" through folding with some thrust faulting; thrust faulting with some folding; folding and thrust faulting developed about equally; folding and pressure dissolution; reverse faulting and shearing; and so on. The displacements that at one location are accommodated by faulting can be transferred to any number of other structures or combinations of structures: folds, cleavages, stylolites, joints, shear fractures, shear zones, dike swarms. There is no way we know of, a priori, to predict what combination of structures will be employed by nature to accommodate a given regional strain. Yet we learn to make intelligent inferences based on knowledge of the mechanical properties of the rocks, inferences regarding the conditions of deformation, and facts from the part of the geologic record that is preserved and accessible.

There are practical incentives to understand how structures link up with one another. As surprising as it once may have seemed, some large magnitude earthquakes in California do not produce faulting at the surface. Displacement may "give out" along "blind faults" that never reach ground level. As a result of certain California earthquakes, such as the Northridge earthquake (January 1993), news commentators now use the term "blind fault" as comfortably as structural geologists. Displacement is transferred upward to the surface through folding (Figure CT.1). In these situations, faulting *in combination with* folding accommodates the total displacement expressed in the release of seismic energy that "rocks" the surface. Knowing exactly how a blind thrust may link with a fault propagation fold thus becomes extremely important to seismologists and structural geologists who, in Los Angeles and San Francisco and other regions in the world, are working to mitigate the loss of life and property related to earthquake hazards. Structural systems, like those near Colingua (Figure CT.1), are parts of other structural systems, like southwestern North America (Figure CT.2). You can't understand the one without understanding the other.

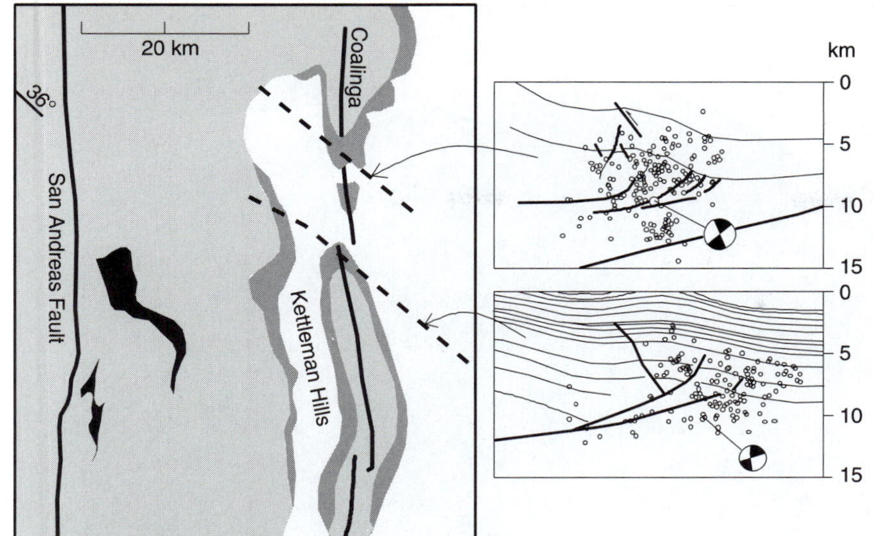

Figure CT.1 Structure sections of blind faults, and their relationship to folding, near Coalinga and the Kettleman Hills, just east of the San Andreas fault. Small open circles are aftershocks, with beachball (black-compression) showing the nature of the main shock. Arrowheads on the surface show positions of peak coseismic uplift. [From Stein and Ekstrom, *Journal of Geophysical Research*, fig. 2, p. 4868, copyright © 1992 by American Geophysical Union.]

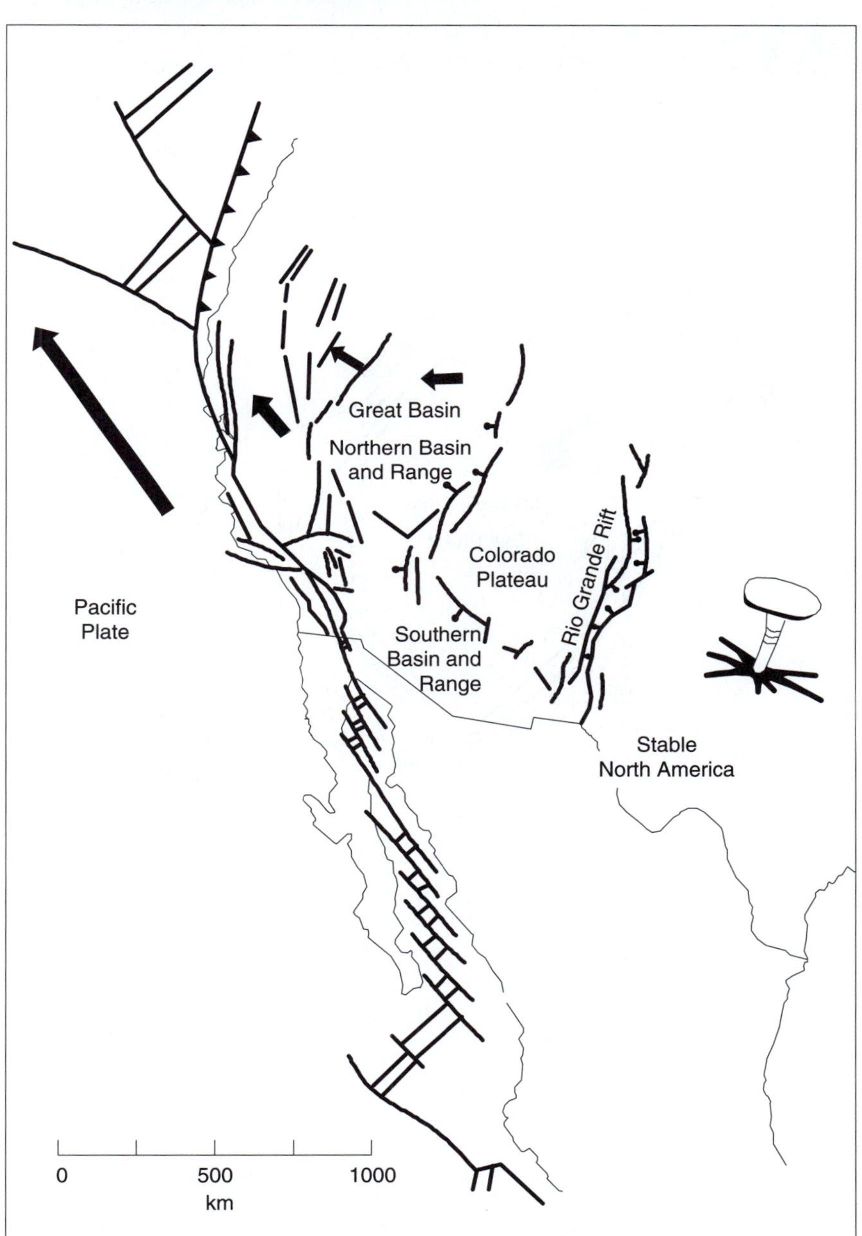

Figure CT.2 Understanding the big picture across the Pacific–North America plate boundary, all the way into the stable interior, is essential to understanding the Kettleman Hills (Figure CT.1). (From Humphreys and Weldon, *Journal of Geophysical Research*, fig. 1, p. 19,976, copyright © 1994 by American Geophysical Union.)

Figure CT.3 Landslide on the face of a sand dune in Sonora, Mexico, breakaway normal faulting at the top, strike-slip tear faults along margins, and thrust at toe. (Photograph and copyright © by Peter Kresan.)

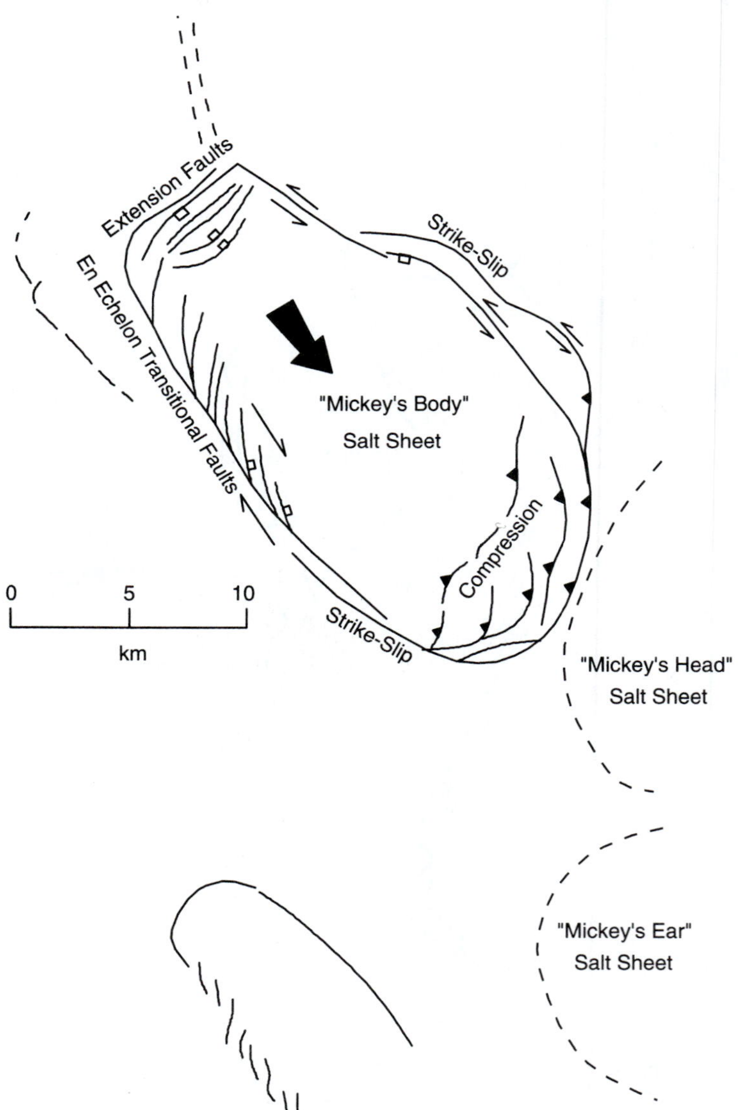

Figure CT.4 Here at Mississippi Canyon in the seafloor bottom of the Gulf of Mexico, sediments are deformed in ways that reflect gliding of underlying salt sheets. "Mickey's body" salt sheet is a beautiful example of structural compatibility: extensional faults at the breakaway, strike-slip faults and en echelon transtensional faults along the margins, and thrust faults at the toe. Just like the landslide in Figure CT.3, only much larger. (Courtesy of Kerry Inman and the British Petroleum Company.)

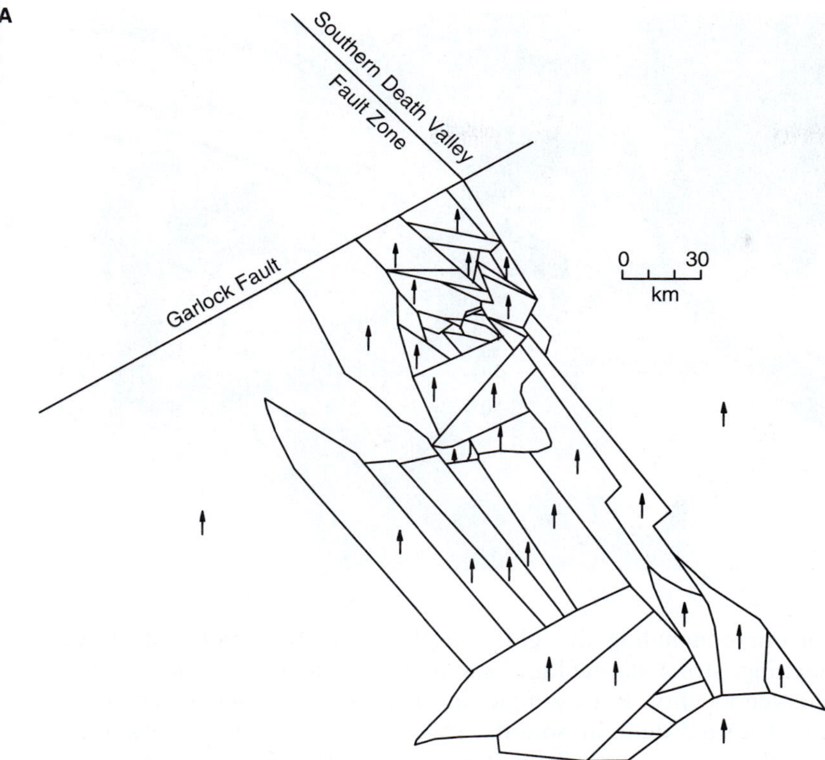

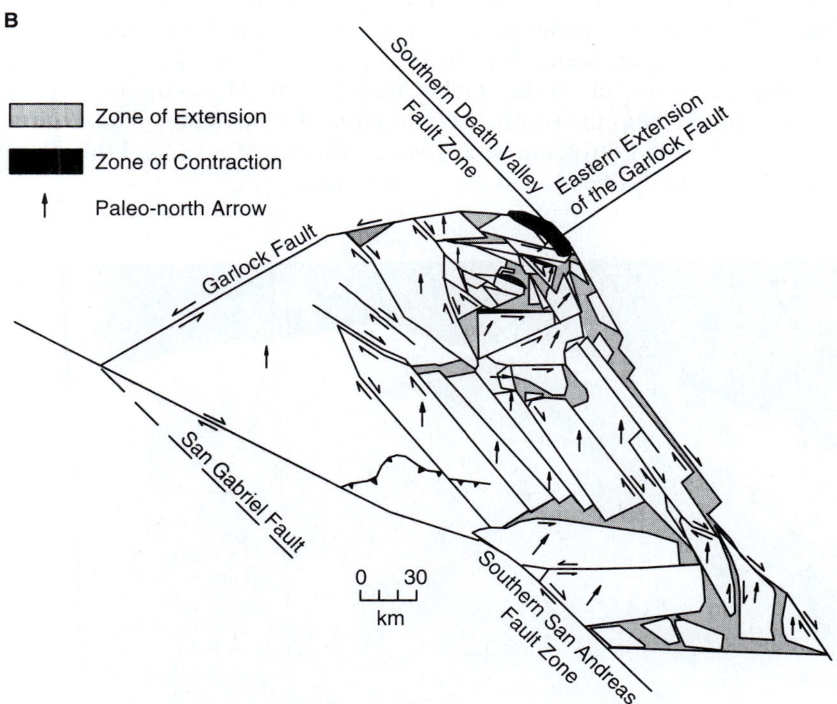

Figure CT.5 Reconstruction of the fault block pattern in a part of the Mojave Desert region, based on structural analysis, paleomagnetics, and GPS. (*A*) The way things looked 10 million years ago. (*B*) The way things look today. [From Dokka (1993), Figures 5A and 5B. Published with permission of Geological Society of Nevada.]

Figure CT.6 A. Kronenberg and R. Christoffersen mapping on the TEM scale. Kronenberg is about to sample a subgrain boundary while Christoffersen consults his selected area diffraction pattern map for location. (Special microphotography carried out under tough lighting conditions. (Courtesy of Jan Tullis.)

In our concluding thoughts we also want to recall that shortening, stretching, strike-slip faulting, and more, can all operate at the same time at a given locality. For example, in Figure CT.3 we see a landslide on the flank of a sand dune in Sonora. This ''system'' of deformation, marvelously photographed by Peter Kresan, is bounded laterally by strike-slip faults (i.e., tear faults) along which, if you look closely, you will see releasing bends and restraining bends. The toe of the slide is a thrust fault. The back of the slide is a breakaway zone featuring normal faulting. In Figure CT.4 we see a giant geologic system for which the landslide in sand is a miniature analogue. This linked system of extension, strike-slip, and compression occurs in the Gulf Coast region (Mississippi Canyon) and has resulted from the southeastward glide of a salt sheet. The system is known to British Petroleum geologists as Mickey Mouse, and this illustration reveals body, head, and one of the ears.

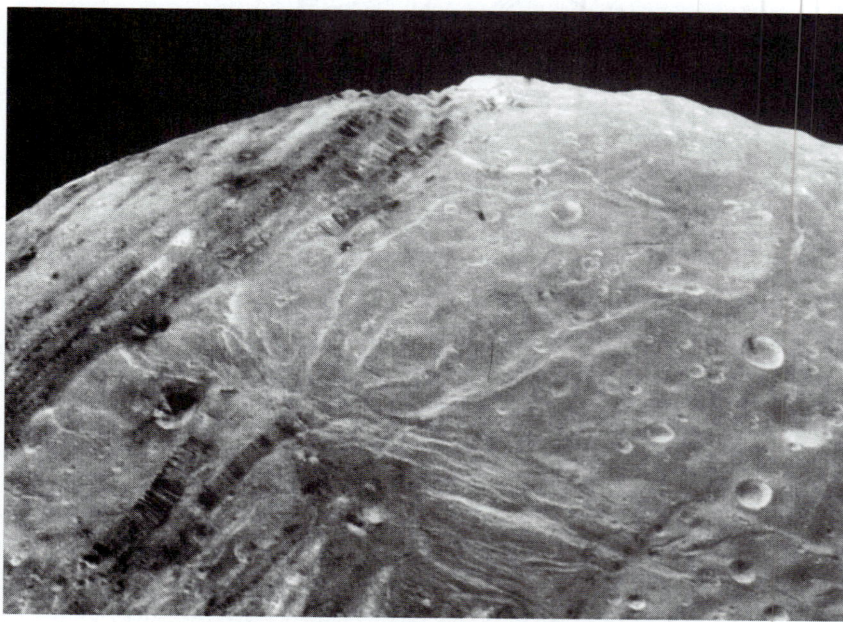

Figure CT.7 View of Miranda, revealing along the "skyline" the profiles of normal fault scarps, some 2 km high. (Photograph courtesy of Robert Pappalardo.)

The complexities of reconstructing complex systems are challenging indeed, but if you like jigsaw puzzles, this is the place to be. Take for instance Figure CT.5, an amazing reconstruction by Roy Dokka of a part of the fault-block pattern in the Mojave Desert region as it looked 10 million years ago, and how it looks today.

Where might your fieldwork take you? Perhaps you will hike in virtual reality, mapping a subgrain boundary in quartz on the TEM scale, keeping track of location using not your topo base but your selected area diffraction map pattern (Figure CT.6). Or possibly you may one day end up on Miranda (Figure CT.7), that icy satellite of Uranus. The radius of Miranda is just 236 km, and yet there you will find normal fault scarps 2 km in height that accommodated the collapse of Arden Corona, an ancient volcanic complex.

Finally, please take a look at the "photo block" illustration that frames "Concluding Thoughts", p. 618, as well as all the other chapter openings. This image is intended to illustrate how scales of observation can be played off, one against the other: structural geology of *rocks* and *regions*. Although the photo is a close-up outcrop image of metamorphic rocks in Norway, we can imagine the left-hand part to be the profile of part of the Alps, complete with giant fold nappe. We love the way small structures mimic large ones!

Time to go. We wish you well.

PART III

DESCRIPTIVE ANALYSIS

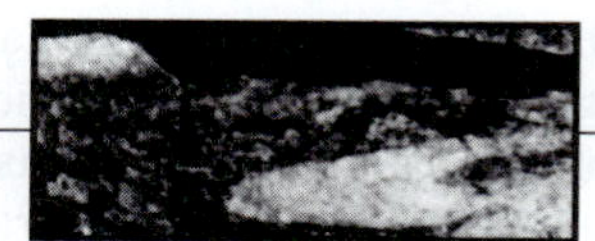

HOW TO FUNCTION IN THE FIELD, AND HOW TO REDUCE THE DATA

A. GEOLOGIC MAPPING

Philosophy and Mind Set

Geologic mapping is the heart of descriptive analysis in structural geology. The geologic map provides an image of the distribution of rock formations within an area (Figure A.1). At the same time, it discloses the form and structural geometry of rock bodies. Even the internal structures within the rocks are portrayed on geologic maps. In effect, the geologic map is a three-dimensional description of rocks, structures, and contacts in a given area. The "flattening" of the description into a two-dimensional sheet of paper is achieved through the use of special symbols.

Geologic mapping is more than an activity. It is a powerful method for systematic structural analysis and scientific discovery. A proper geologic map cannot have loose ends. If a geologist decides that one of the essential formations to be mapped is the Nugget Sandstone, then the Nugget must be tracked across the entire map area. Its upper and lower contacts with adjacent formations must be followed (figuratively speaking) step by step. Where the contacts are covered with alluvium, this must be shown. Where the contacts are offset by faults, or intruded by dikes, the offsets or intrusions also must be shown. If the formation "suddenly disappears," the geologist must cope until the formation is "found" or until a satisfactory explanation is discovered that can be portrayed through the use of standard geological symbols.

Hundreds of decisions have to be made in the course of a single day of geologic mapping. Consider the map area shown in Figure A.2. The geology consists of northeast-trending sedimentary rocks that first were faulted, then intruded by granite and pegmatite. But what if the geology

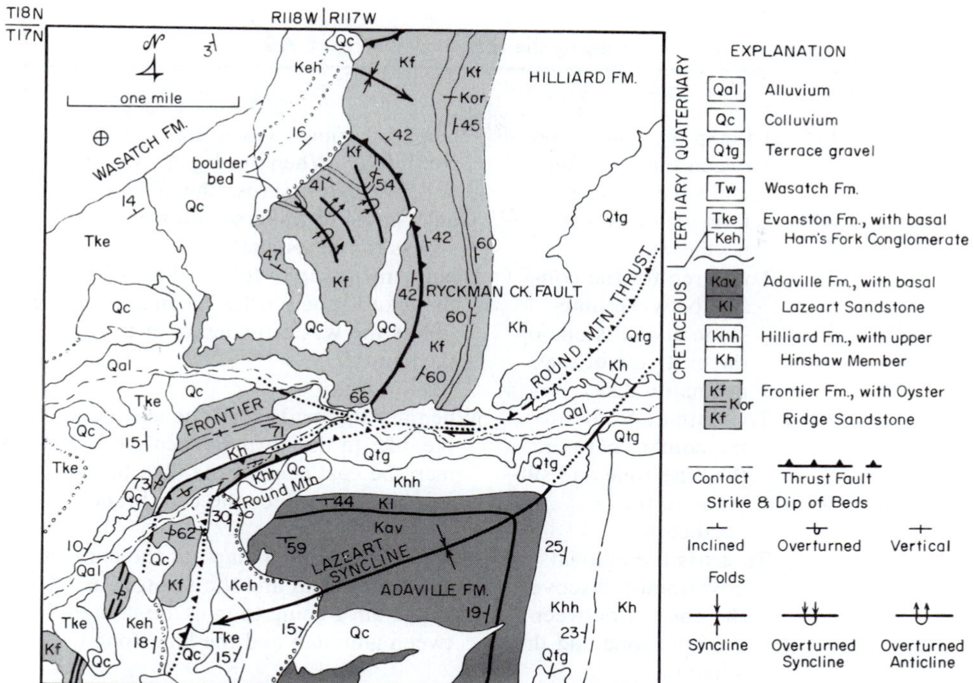

Figure A.1 Example of informative geologic map, Round Mountain area, Wyoming. [From Worrall (1977). Published with permision of Wyoming Geological Association.]

of this area had never been mapped, and we chose to map it? As we begin traversing the area, we might as well be walking through a corridor with walls so high that we are cut off from all geologic relationships except those exposed directly underfoot. Problems and nagging worries can accumulate exponentially in even the simplest geologic circumstances (see Table A.1). We need to remember to keep a sense of humor and a sense of purpose during the initial stages of mapping and not be overwhelmed by loose ends.

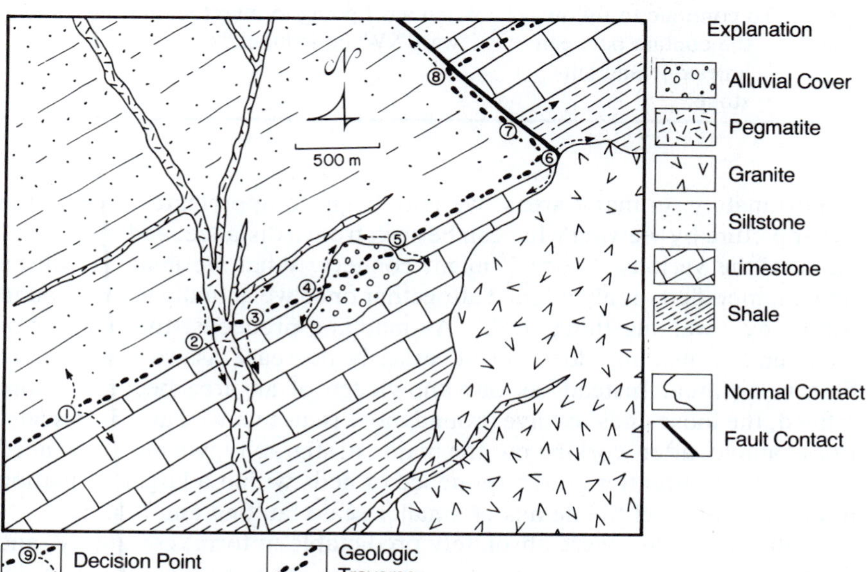

Figure A.2 Decisions, decisions, decisions in geologic mapping. Table A.1 presents the specific nagging worries that might surface at the numbered locations.

TABLE A.1
Nagging worries while mapping the area shown in Figure A.2

Point	*Decision*	*Nagging Worries*
1	To map the contact between the limestone and siltstone units, following it to the northeast.	When am I going to map the contact to the southwest? When am I going to map northwestward across the siltstone? When am I going to traverse southeastward across the limestone?
2	To search for the contact between limestone and siltstone on the east side of the pegmatite dike.	When am I going to follow the west edge of the dike to the north? When am I going to follow the western contact of the dike to the south?
3	To continue to follow the contact between the limestone and siltstone to the northeast.	When am I going to map the east edge of the dike to the north? When am I going to map the east edge of the dike to the south? Where does the dike branch?
4	To cross the alluvial cover and rediscover the contact between the limestone and the siltstone.	When am I going to map-out the contact between alluvium and the siltstone unit? When am I going to map the contact between alluvium and the limestone unit?
5	To continue to follow the contact between limestone and siltstone to the northeast.	When am I going to traverse northwestward across the siltstone? When am I going to traverse southeastward across the limestone?
6	To map the fault northwestward and to search for the offset limestone unit.	When am I going to map the contact between limestone and granite to the southwest? When am I going to map the contact between shale and granite to the southeast? Where is the offset limestone bed? What if I don't find it??
7	To continue to map the fault contact and to search for the siltstone/limestone contact.	When am I going to map the contact between limestone and shale to the east?
8	To continue to follow the contact between limestone and siltstone.	When am I going to map the fault northward? When is lunch??

Fortunately, in many areas, we may begin to appreciate some of the larger picture by viewing what can be seen from a distance, and by imagining how the various "pieces" might connect. What we imagine will be refined, modified, or abandoned altogether once we actually gain "ground truth" by "walking things out." As mapping proceeds, the number of loose ends remaining does not necessarily decrease as we spend more time in an area. Instead, as our knowledge of an area becomes more refined, the loose ends revolve around more issues viewed at a finer scale. For example, after working out the major stratigraphy, we may now need to pay closer attention to the sequence of beds within a larger formation in order to unravel the details of smaller scale faulting and folding.

If all rock bodies were absolutely predictable in form, continuity, and distribution, the geologic mapping process might not be essential to detailed structural analysis. However, given the unpredictable twists in geometry of rock formations, geologic mapping is fundamental to understanding. Through mapping, the investigator who is concerned about tying up all loose ends is led into outcrop areas of real discovery.

Base Maps

The choice of base maps, and the scale of mapping, is influenced by any number of factors. What is the intended purpose of the mapping? How large is the area? How much time do we have to complete the mapping? What level of detail is required? Will the rocks and structures show up clearly on aerial photographs?

The physical, geographic, political, and logistical accessibility is another factor that may influence choice of base maps. Let's face it, there are different degrees of inaccessibility of geology: Manhattan Island; the Amazon; Mount St. Helens; Mt. Everest; Bosnia; Olympus Mons. Spending a month mapping a single square mile as part of a million-dollar drilling program will result in a different choice of base map than a class project northwest of town.

The initial step in preparing for geologic mapping is selecting a suitable base on which mapping can be carried out. **Control** for establishing the exact locations of contacts and structures can be achieved in many ways. The measurement of compass bearings and the pacing (or taping) of distances can afford reasonable location control in some projects. Paced distances over irregular topography are converted to true horizontal map distances (Figure A.3).

Mapping by the **pace-and-compass** or **tape-and-compass system** is reserved primarily for mapping very small areas at **large scale** (e.g., 1:100 or 1:1000). However, these methods may also be used as a complement to other mapping approaches. One of my first remote mapping projects was in northern Ontario, Canada, where the only base available to me was aerial photography. Aerial photographs usually provide a sufficient basis for control, but in this case the terrain was so monotonously uniform in photogeologic expression (i.e., unending swampy expanses) and so "closed-in" by vegetation (i.e., the bush), I found it impossible to locate myself on the basis of photos alone.

The solution to the dilemma was traversing into the bush from some known location on a lake shore, with my teammate pacing along a fixed bearing for control (Figure A.4). Maintaining a straight and accurate pace-and-compass traverse for distances up to 3 km or so is practically impossible in thick bush. Consequently, traverses were planned so that large, distinctive control checks (like a clearing or an isolated hill) would be crossed along the way. Outcrops were positioned on the map according to bearing and paced distance, and modified slightly (or substantially) according to the accuracy of the traverse as determined through meeting or missing the checkpoints. When we found ourselves within areas of iron formation, which was magnetic, the compass needle became undependable and we used the position of the sun, as best we could see it through the canopy, as a rough guide to direction.

Mapping within a **grid-line** or **picket-line system** is common practice in exploration. It is the chief method used around mining camps when companies are in the advanced stages of exploring a prospect or in the early stage of developing an ore body. It is an essential mapping approach for detailed work in areas of heavy vegetation. Picket-line paths are guided by engineering surveying and cut along two perpendicular directions oriented sensibly with respect to topography and geology (Figure A.5*A*).

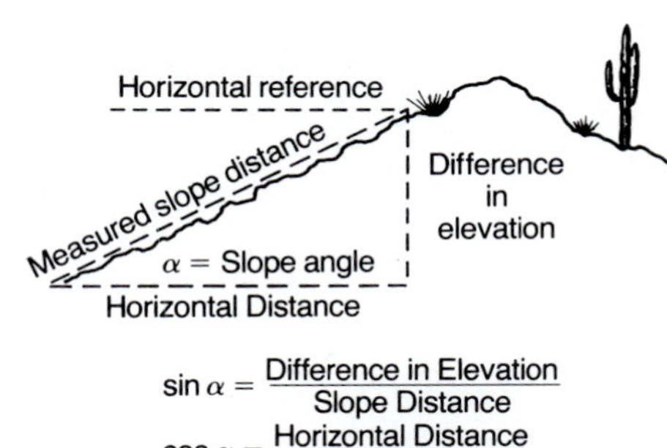

Figure A.3 Trigonometric relationships for converting slope distance to true map distance (i.e., horizontal distance).

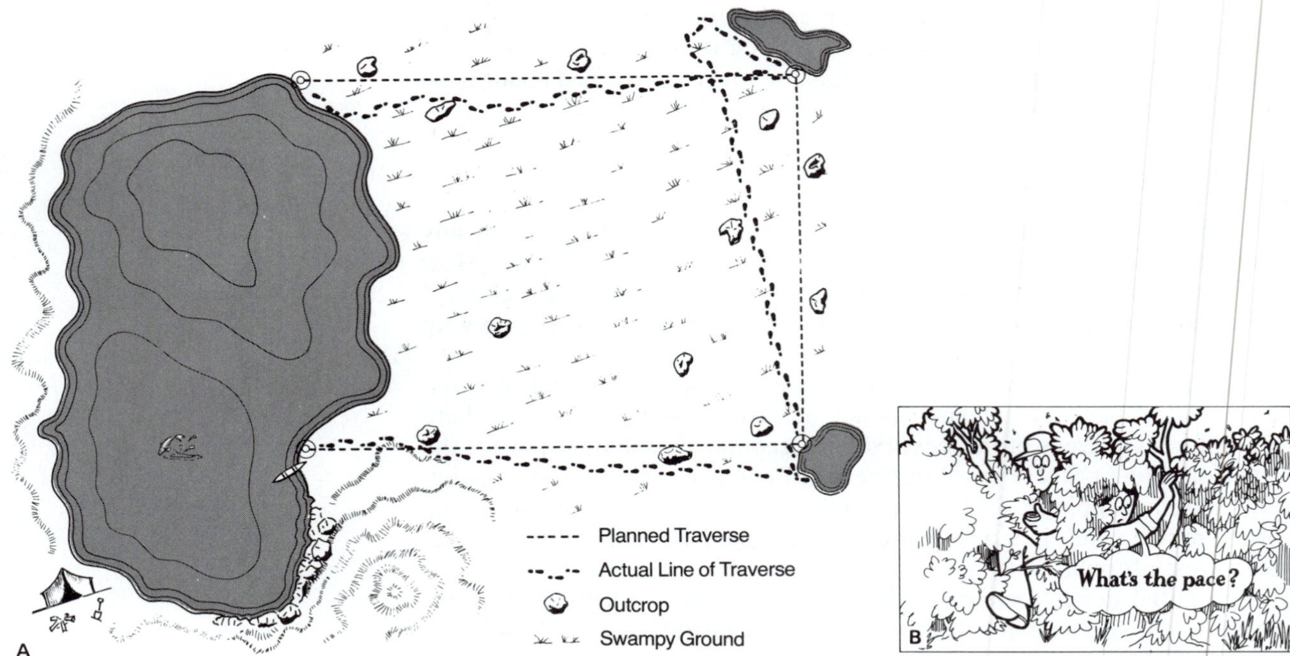

Figure A.4 Pace-and-compass mapping in northern Ontario, Canada. (Cartoon artwork by D. A. Fischer.)

Flagged picket posts are positioned and marked according to location every 20 m or so. Upon completion of the gridwork, the geologist traverses the complete course, mapping the locations and geologic characteristics of bedrock outcrops and **float** (rocks out of bedrock position but inferred to be short traveled). The map begins as an outcrop map (see Figure A.5*A*) but may be expanded into a geologic map by projecting and following contacts through the interior parts of the gridwork (Figure A.5*B*). Geophysicists may walk the same lines with instruments to monitor magnetic, electrical, and gravitational properties of the rocks in the subsurface. Additionally, soil samples for geochemical analysis are collected at locations along the grid. The combined geological, geophysical, and geochemical data, all assembled and plotted on the same grid map, yield useful integrated descriptive data regarding the ore potential of an area.

Underground mapping proceeds in a similar fashion, except that traverses are made through solid earth. Mining engineers direct the tunneling of drifts, crosscuts, raises, and shafts and furnish vertical and horizontal map control for the entire mine complex (Figure A.6). Survey markers are positioned at close intervals throughout the mine. Mapping is carried out commonly at scales of 1 in. = 100 ft (1:1200) and 1 in. = 50 ft (1:600). Exposures available to the geologist are on the walls and ceiling (**back**) of each tunnel, the floor being covered with water, mud, or dust. Underground mapping is made difficult by the dirty and/or oxidized nature of the rocks, the dim lighting, and the unconventional mode of projection that is used in the mapping. Since rock relationships on the floor are not exposed, the mapped locations of contacts and structures are projected from walls and the back to an imaginary waist-high horizontal plane. Thus an inclined fault whose trace is exposed near the top of a wall might project to the center of the drift (Figure A.7). Results of underground mapping, at all levels inside the mine and in all vertical shafts and raises, provide a three-dimensional skeleton of descriptive and geometric facts.

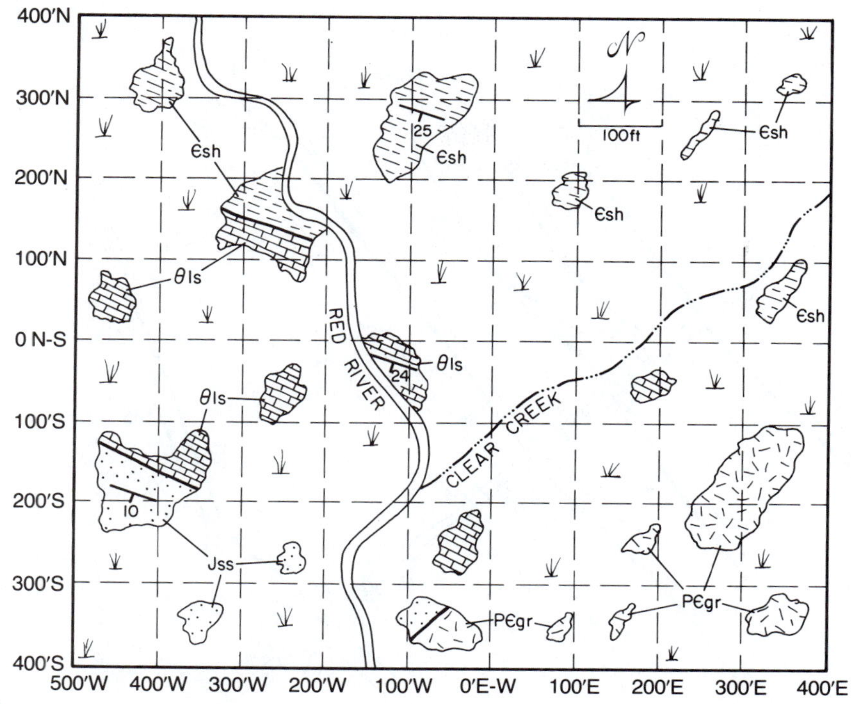

A

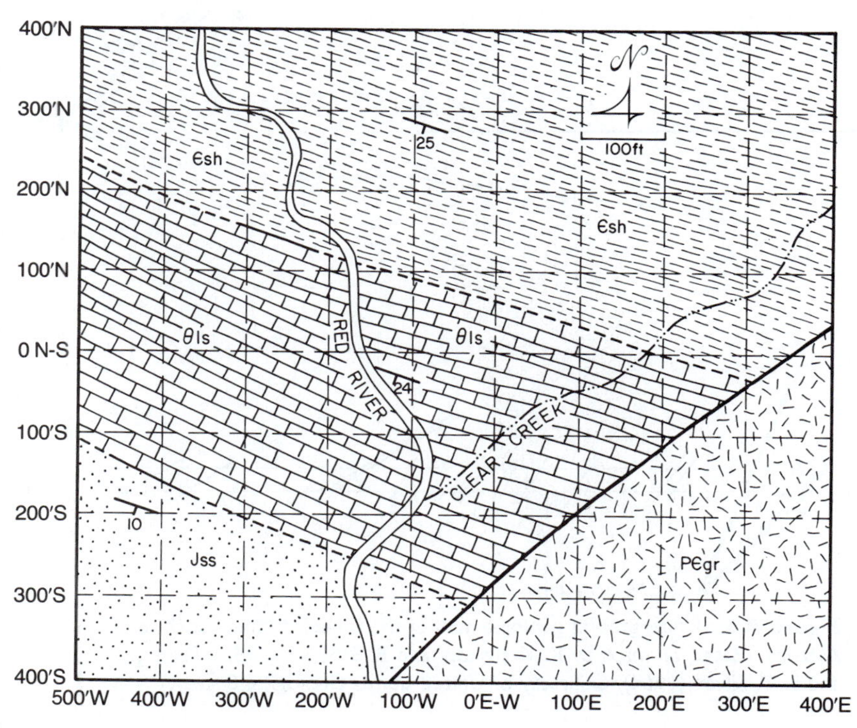

B

Figure A.5 Gridline mapping. (*A*) Outcrop map showing locations of rock exposures relative to distances within surveyed grid. (*B*) Geologic map interpreted on the basis of the outcrop relationships shown in *A*.

When supplemented by selective drilling, it furnishes the support for a geological model of the deposit.

The traditional base for geologic mapping is the **topographic map.** Topographic maps show details of physiography and culture, which permit accurate positioning of contacts and structures. The United States Geological Survey publishes both 15- and 7.5-minute maps, scaled at 1:62,500

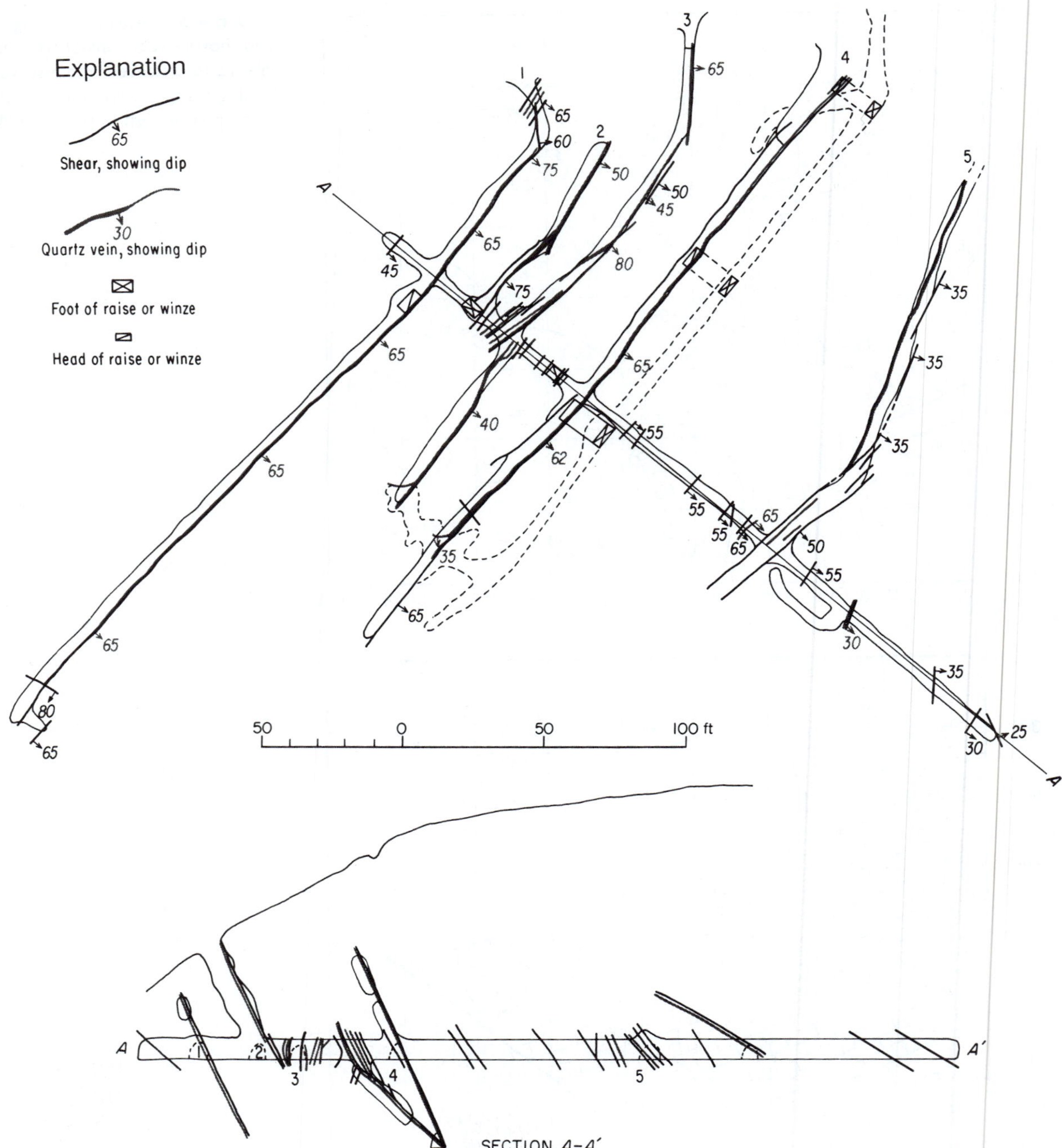

Figure A.6 Underground geologic map of the Bluebird Mine, Cochise County, Arizona. [Based on maps by K. Krauskopf and R. Stopper (1943). From J. R. Cooper and L. T. Silver (1964). Courtesy of United States Geological Survey.]

and 1:24,000, respectively. These maps cover areas corresponding to 15 or 7.5 minutes of latitude and longitude. For regional mapping and compilation, 30 minute × 60 minute (i.e., 30′ × 60′) topographic sheets scaled at 1:100,000, or 1° × 2° topographic sheets scaled at 1:250,000 are extremely valuable. A geologic map of an entire state or country may be at a scale of 1:500,000 or 1:1,000,000 or 1:5,000,000.

In some cases it is advantageous to map on a copy of the topographic map that is enlarged compared to the intended presentation of the final

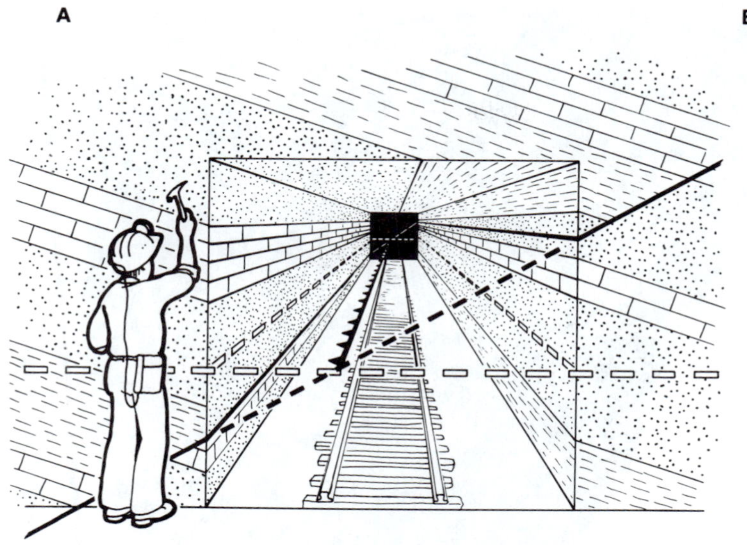

Figure A.7 (*A*) The convention in underground mapping is to project contacts to a waist-high projection plane. (*B*) Map showing the location and orientation of the fault (heavy line with teeth) relative to the walls of the tunnel. (Art design by R. W. Krantz.)

map. This permits extra room for detail that may be essential for documentation and understanding, but that is not necessary for the final map presentation. Reducing/enlarging copying machines and optical scanners connected to personal computers can provide inexpensive reproductions at the "right" scale with only modest "kinematic" distortion.

There are some circumstances in which topographic maps must be prepared as part of the geologic mapping process. This is a time-consuming process, warranted only by the need to carry out large-scale mapping of some small area containing structures for which elevation and topographic control is necessary to unravel the three-dimensional structural forms. Large-scale topographic maps can be constructed through the use of the **plane table** and **alidade** (see Compton, 1985). The basis for this approach is surveying the elevations and locations of an array of points that afford optimum geological and topographic control (Figure A.8).

In some instances the most valuable base for geologic mapping is a set of **aerial photographs,** preferably with enough overlap to afford stereoscopic coverage. Aerial photographs of some areas reveal the bedrock distribution and structure so explicitly that the mapping of contacts can be done effortlessly (Figure A.9). The total display of rock, vegetation, and physiographic and cultural features permits the locations of control points and contacts to be posted easily and accurately. For regional studies, high-quality space-satellite imagery of large regional tracts is the answer (Figure A.10), for they can reveal the major through-going regional structures. One major disadvantage of mapping on aerial photographs is the time-consuming step of transferring contacts and structures (i.e., the "line-work") from the photos onto a topographic base map. This must be done very carefully to avoid inaccuracies.

For special studies of small areas, we have found it useful to contract low-altitude flyovers. Through this approach it is possible to obtain high-quality imagery of the study areas at scales as large as 1:1200. Enlargement is achieved through preparing Mylar positives approximately 1 m × 1 m in size. Blue-line or black-line copies can be produced inexpensively from a Mylar positive.

Ideal control for geologic mapping is the combination of topographic maps and aerial photographs. The photographs reveal rock relationships

Figure A.8 Plane table mapping, in progress. Target of mapping is strata in Aikens Corner in the Mecca Hills of southeastern California. (*A*) Ken Yeats (left) and Steve Wust on instrument. (*B*) Jeff Moorehouse and Barbara Beatty on geology. (Photographs by G. H. Davis.)

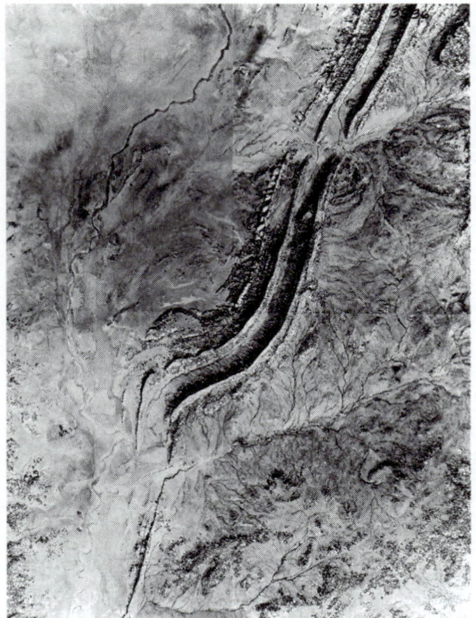

Figure A.9 Aerial photograph showing conspicuous expression of steeply inclined resistant beds on the east side of the Defiance uplift, northeastern Arizona–northwestern New Mexico.

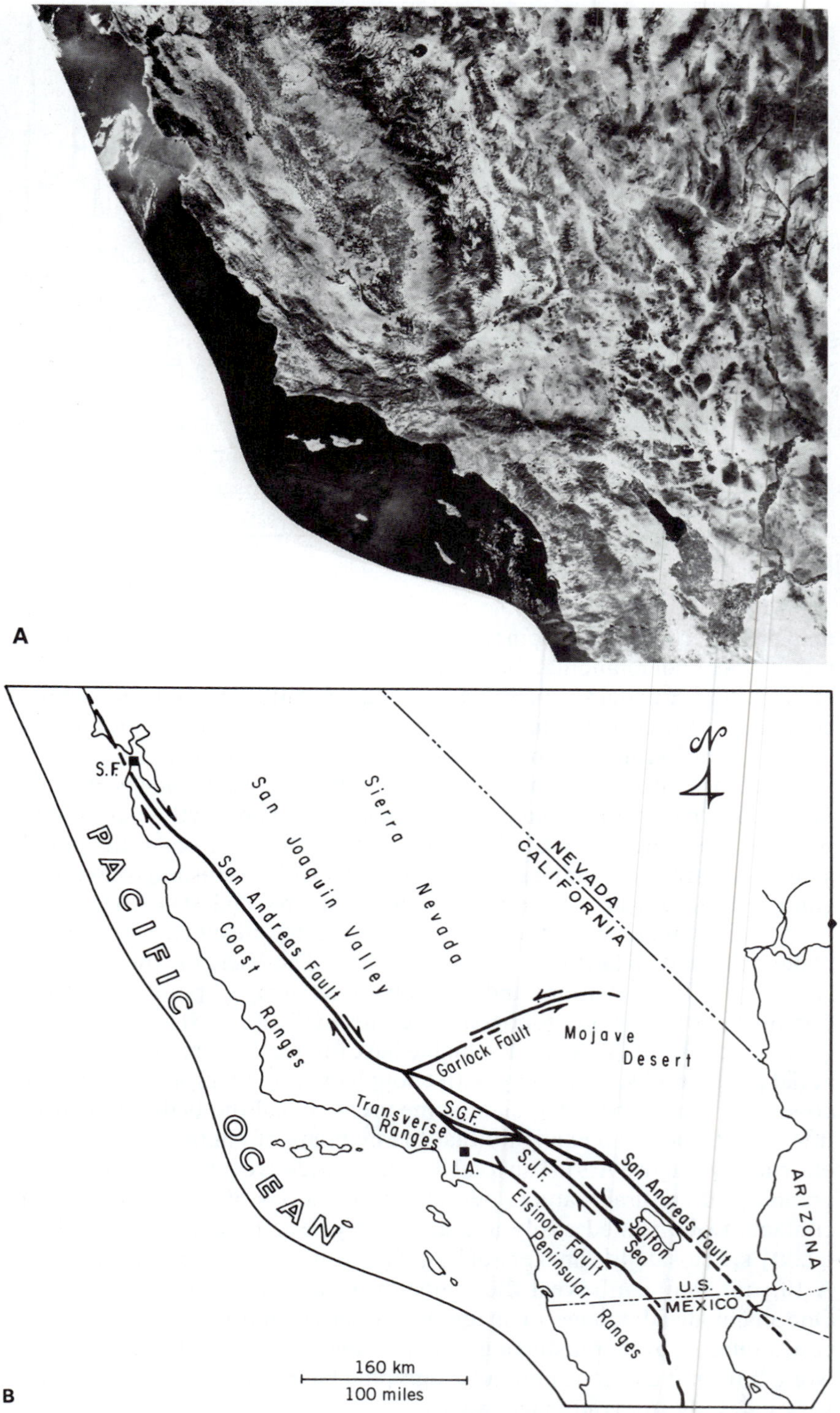

A

B

Figure A.10 (*A*) LANDSAT-1 mosaic of the southwestern United States. (*B*) Geographic index map for some of the structures revealed in the LANDSAT mosaic. S.G.F. = San Gabriel fault; S.J.F. = San Jacinto fault. [From Lowman (1981). Courtesy of Goddard Space Flight Center, National Aeronautics and Space Administration.]

in ways that aid in planning traverses and interpreting structures. The topographic map, on the other hand, provides a stable base of uniform scale and permits the critical third dimension, the vertical dimension, to be evaluated during the mapping process.

Components of a Geologic Map

Learning how to make a geologic map is aided by learning the fundamental components of such a map. Ridgeway (1920) long ago described the components (Figure A.11). The map shows the distribution of rock formations by means of color, patterns, and letter symbols. Each color, pattern, and symbol on the map should coincide exactly with the distribution of the corresponding rock formation. The meaning of the colors, patterns, and symbols is provided in the **Explanation**: the series of boxes, suitably colored, patterned, and labeled, which provide the identification of formation name and usually a brief lithologic description.

The official list of map colors used by the United States Geological Survey is presented in Table A.2. Different colors can of course be used to emphasize one aspect of the geology over another. For example, if a map area contains 20 Precambrian units and nothing else, a full range of colors should be used, rather than limiting the choices to similar-appearing shades of brown and gray. Our preference is to choose pleasant pastel

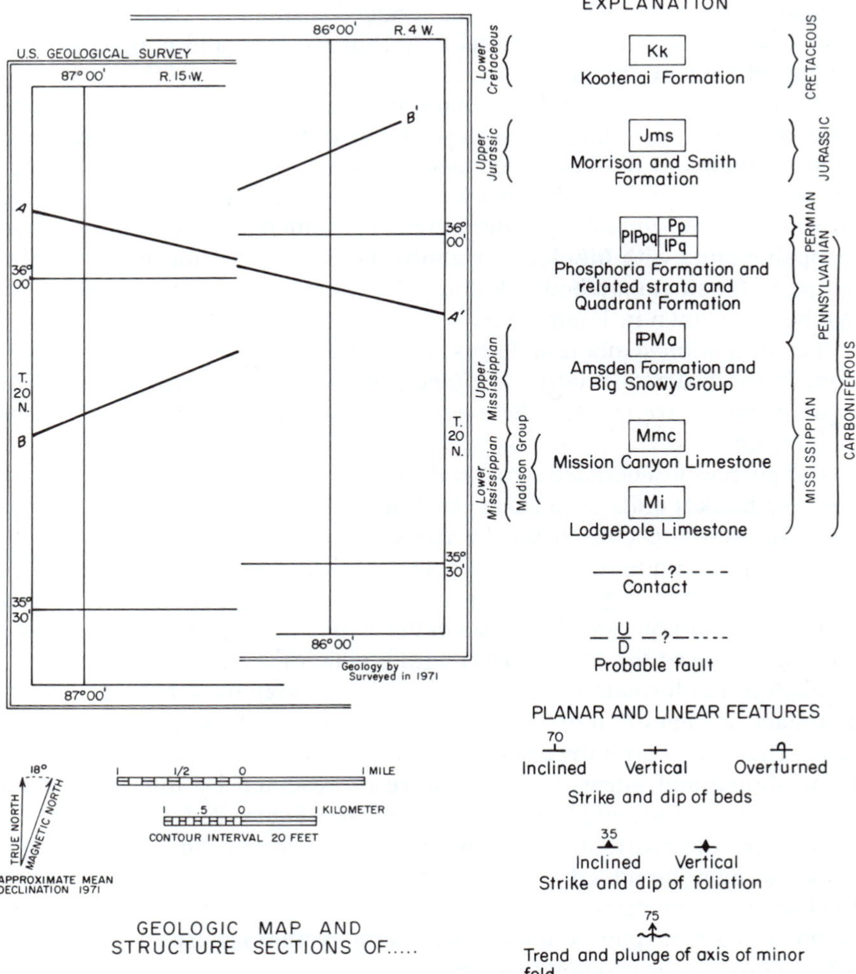

Figure A.11 Elements of the layout of a geologic map. [From Ridgeway (1920). Courtesy of United States Geological Survey.]

Table A.2
Official United States Geological Survey map colors

Age of Rock Unit	*Map Color*
Quaternary	orange
Tertiary	yellow ocher
Cretaceous	olive-green
Jurassic	blue-green
Triassic	bluish gray-green
Permian	blue
Pennsylvanian	gray
Mississippian	blue-violet
Devonian	heliotrope (helio*what??*)
Silurian	purple
Ordovician	red-violet
Cambrian	brick red
Precambrian	pink or gray-brown

colors for rock units that cover large areas on the map and to use bright or otherwise strongly contrasting colors for thinner "key" marker units. When I (SJR) prepared the Geologic Map of Arizona (1988), I gave careful attention to which rock units would be next to one another, and then tried to choose colors and shades that would complement and not clash. Also, I bypassed the standard choice of colors for Quarternary–Tertiary surficial deposits, choosing instead various subdued shades of light to medium gray. As a result, the bedrock geology "jumps out" and is not lost in a sea of yellow and orange.

If geologic maps are printed in black and white, line and symbol patterns must be used in lieu of color to show the distribution of rock formations. Line symbols are difficult to choose in structural studies because most patterns interfere with and/or mask the fundamental **line work** of structural information and symbology. Therefore, it is useful to represent the various mappable units with **lithologic symbols** that convey an impression of the nature of the dominant bedrock units. An example of the use of lithologic symbols is shown in Figure A.12.

The formation symbols and ages are arranged vertically in the Explanation, in the order of relative age (see Figure A.11). Astride the vertical array of boxes are entered the ages of the rock formations, as known or as inferred.

For purposes of detailed structural analysis, it is often helpful to identify and map **marker units** as an aid in clarifying the structural style of deformation. Markers are chosen on the basis of distinctiveness, resistance to erosion, and continuity (Figure A.13). Marker beds are a godsend in complexly folded sequences of rock. The marker units are shown in a distinctive line-style and described in the Explanation of the map, in boxes corresponding to the formation(s) within which they lie (Figure A.14).

Each of the formations shown on a geologic map must be enclosed by a contact, except of course where the formations extend outside the map area. There they are intercepted (cut off) by the border of the base map. Depositional and intrusive contacts are represented by thin lines; fault contacts are represented by thicker lines (see Figure A.1).

Showing the distribution and contacts of rock formations is not enough. To these must be added symbols that disclose the major structures and the internal geometry of the rock system. The most informative structural geologic maps display abundant **structural symbology** (see Figure A.1), which conveys the geometric and physical nature of the structures present,

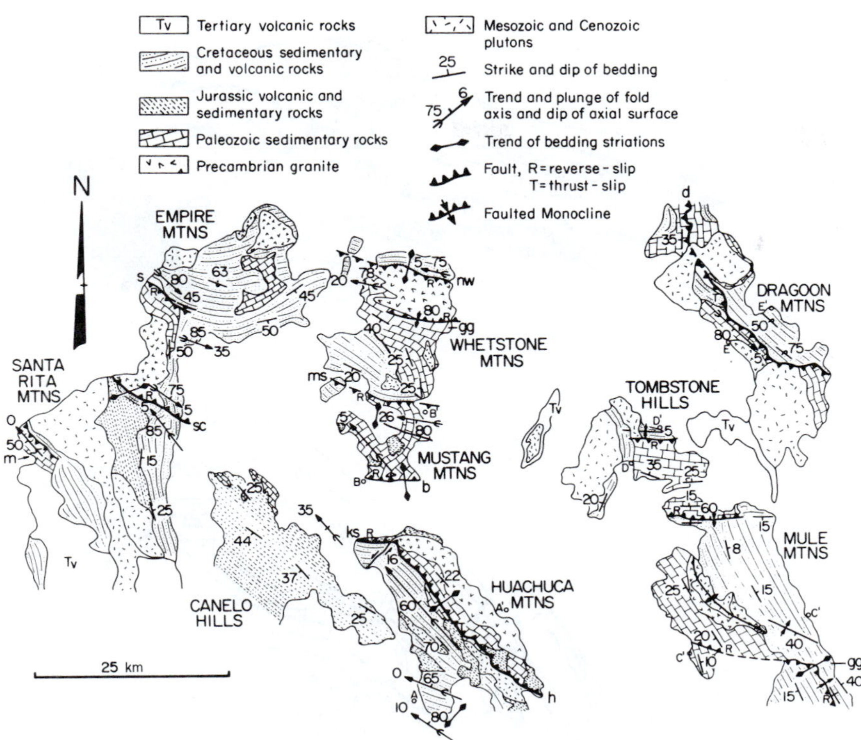

Figure A.12 Structural geologic map of southeastern Arizona showing the use of lithologic symbols to distinguish among the mapped units. [From Davis (1979), *American Journal of Science,* vol. 279.]

like contacts, bedding, joints, shear fractures, veins, faults, cleavage, foliation, lineation, and shear zones (Table A.3). Examples of the use of symbology are presented in Figure A.15.

In order to designate the size and orientations of features shown on a geologic map, all maps include **scales** and **north arrows** (see Figure A.11). Both bar and ratio scales are normally presented. Bar scales should be labeled both in metric (km, m) and U.S. customary (mi, ft) units. Two north arrows are shown, one pointing to true north, the other to magnetic north. Magnetic declination is specified in degrees east or west of true north. If the base map is a topographic map, the contour interval is also shown (e.g., 20′, 40′, . . .).

Figure A.13 Note the white limestone marker bed in the left quarter of this photograph of steeply dipping Cretaceous sedimentary rocks in the Central Andes, Peru. Peaks exceed 5500 m in elevation. [Photograph by G. H. Davis.]

Figure A.14 Complexly folded sedimentary strata in the Rincon Mountains near Tucson, Arizona. (*A*) India-ink rendering of photograph showing marker units that allow the structure to be unraveled. (*B*) Geologic map of the area and its markers. [From Davis et al. (1974). Published with permission of National Association of Geology Teachers.]

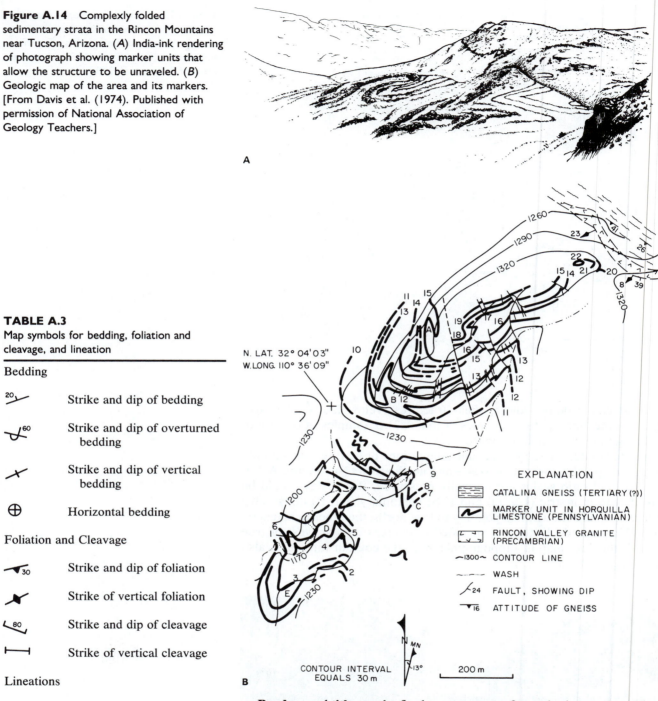

TABLE A.3
Map symbols for bedding, foliation and cleavage, and lineation

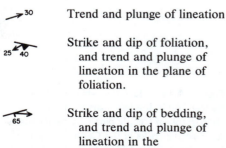

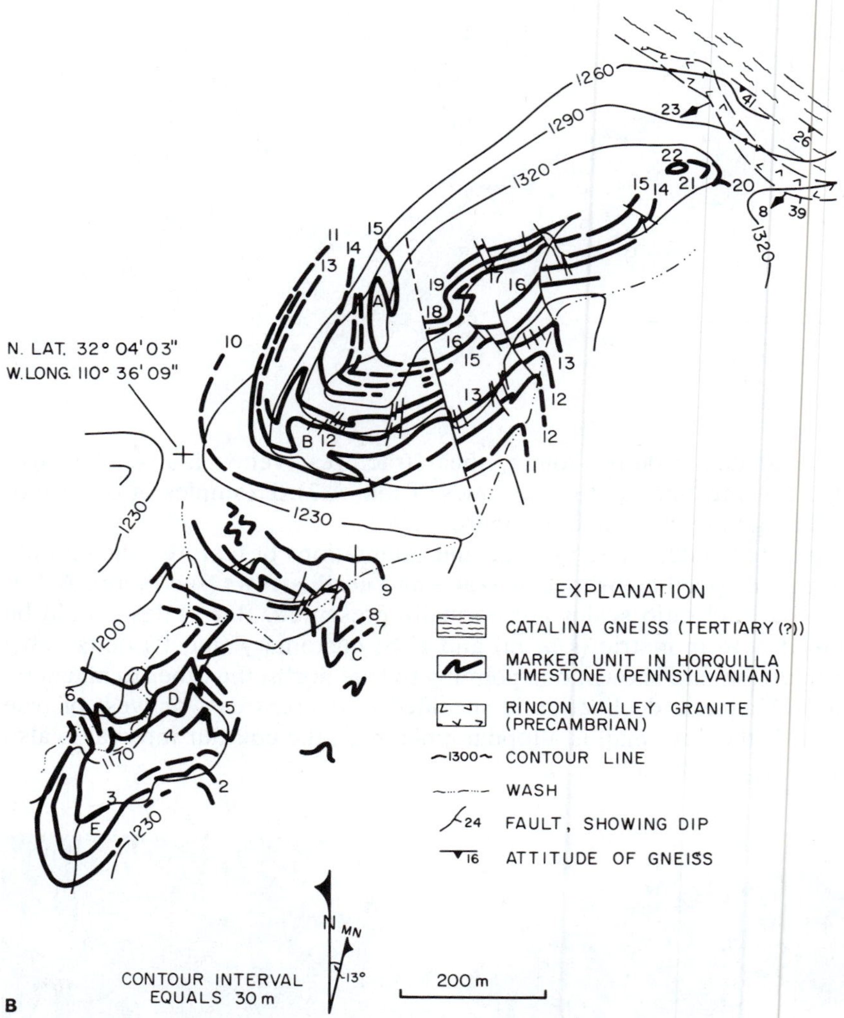

Symbol	Meaning
Bedding	
20	Strike and dip of bedding
60	Strike and dip of overturned bedding
	Strike and dip of vertical bedding
⊕	Horizontal bedding
Foliation and Cleavage	
30	Strike and dip of foliation
	Strike of vertical foliation
80	Strike and dip of cleavage
	Strike of vertical cleavage
Lineations	
30	Trend and plunge of lineation
25 40	Strike and dip of foliation, and trend and plunge of lineation in the plane of foliation.
65	Strike and dip of bedding, and trend and plunge of lineation in the plane of bedding.

Borders and **title** are the final components of a geologic map (see Figure A.11). Most geologic maps do not need an inked border; this feature is used only for maps that are very large or irregular. The title of the map is generally placed along the lower margin, and it conveys the name and general location of the map area. The name of the geologist who prepared this map appears either beneath the title or in the lower right-hand corner of the map. The source of the base map should appear in the lower left-hand corner.

Much can be learned about geologic maps and geologic mapping by studying those that have been published. Time spent in the library examining geologic maps, map explanations, and map patterns is time well spent.

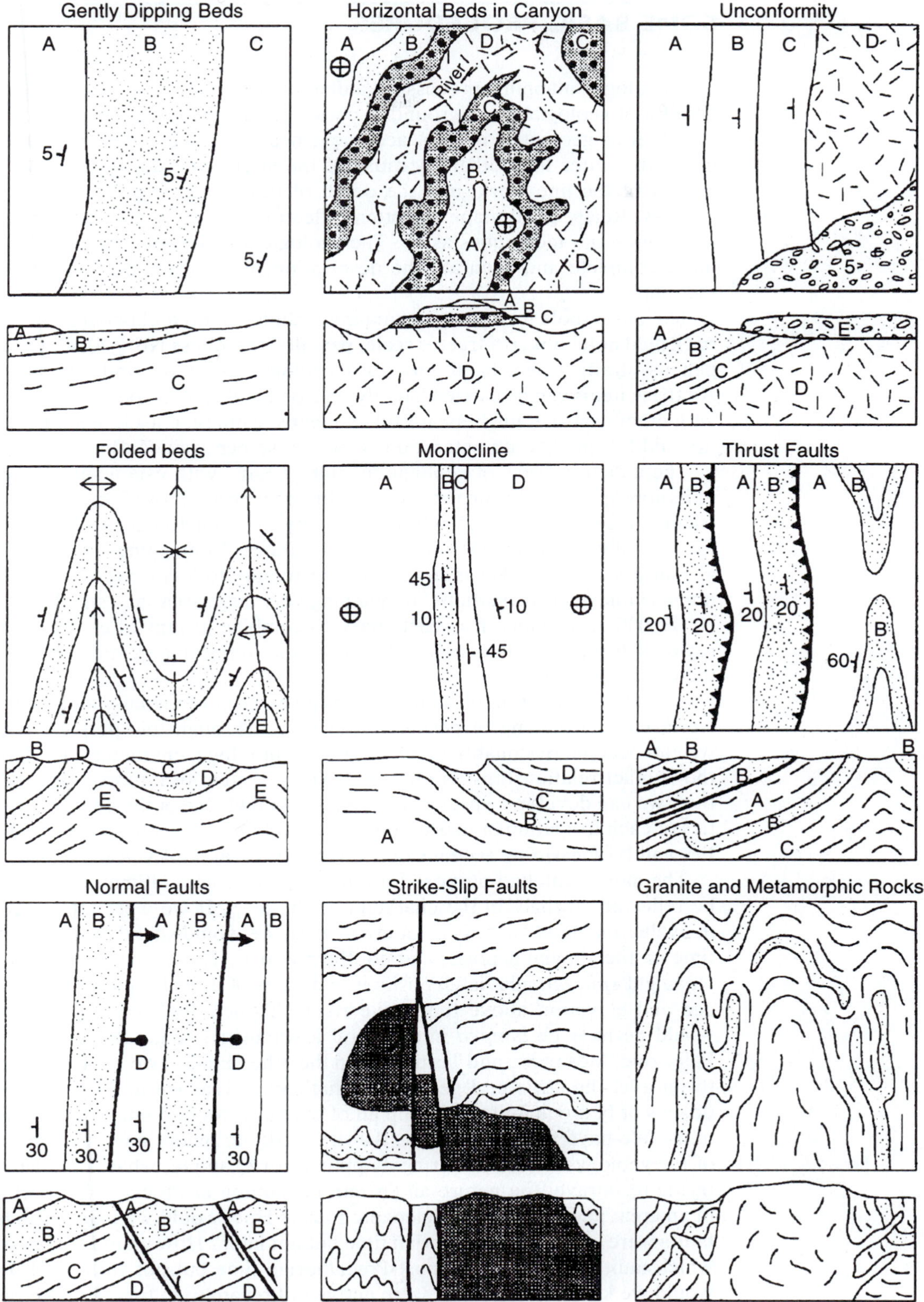

Figure A.15 Examples of symbology used in presenting structures in map view and cross section. Note how different structural associations have different and distinctive map patterns.

SOME BASIC PROCEDURES

The geologic mapping process is aided by keen observational skills, breadth of geological background, a curious and questioning disposition, attention to geometric details, accuracy, neatness, and patience. Some maps are better than others because of the attention paid to detail, the quality of the line work, and the depth of thought during the mapping process. Really good maps are often reflections of really good thinking, mechanics and techniques aside. The geologic map is a description. The more complete and accurate the description, the greater the impact of the map.

The basic tools for geologic mapping include covered clipboard, hard-back field notebook, protractor scale, pencil(s) (hardness No. 3-4), colored pencils, drafting pen, hand lens, rock hammer, 2-m tape, and compass. All these items can be carried handily on a belt and in a day pack, together with other gear that might be useful (canteens, camera, chisel, binoculars, first-aid kit, pocket altimeter, 50-ft tape, sunscreen, raincoat).

The actual process of geologic mapping begins with establishing **map units** appropriate to the project at hand. In mapping a 7.5′ or 15′ quadrangle (scales: 1:24,000, and 1:62,500, respectively), it is customary to map out the formally established stratigraphic formations. In mapping small areas at a large scale (1:12,000 or larger), it is customary to *break out* distinctive units *within* single formations. In mapping very large areas at a small scale (1:125,000 or smaller), it is customary to combine groups of established regional formations into assemblages having important stratigraphic or tectonic significance.

The actual selecting of map units involves both reconnaissance field-work and the reading of pertinent literature. Through this preparation, we can become reasonably familiar with the lithologic characteristics and the contact relationships within the geologic system of interest. Beyond that, we can develop a specific and detailed knowledge of the characteristics of each formation or marker unit or assemblage by visiting **type localities** and by **measuring sections** of the rock in the area of study.

The motivation and purpose for the mapping project ultimately should be reflected strongly in the character of the map. If we were interested in evaluating flood hazards in an urban area, we would spend most of our time on the alluvial deposits and not the bedrock. Thus, during mapping, we might *split out* alluvial deposits into dozens of units based on relative age and elevation above the active river channel. Bedrock formations would perhaps be *lumped* into several assemblages (e.g., Precambrian, Paleozoic, Mesozoic, and Tertiary). On the other hand, if we were seeking to unravel the structural geology of that same area, we might split out dozens of bedrock units and lump all of the alluvium into one or two map units. We have found, however, that it is risky to ignore any one aspect of the geology of an area during mapping. For example, what if we were trying to unravel the nature and origin of faulting in Precambrian rocks in an area, and ignored the Paleozoic rocks? How would we know which faults were active in Precambrian time but not later? How would we know which faults were active in post-Precambrian time but not earlier? How would we know which faults were active in Precambrian time, only to be reactivated in post-Precambrian time?

Geologic mapping may proceed in a number of ways, but significant time and effort will always be invested in tracing out contacts of each of the map units. It is useful early in the mapping to traverse across the grain of the rocks and structures, to become familiar with each of the rock formations, their contacts, and their internal structures. It is also important to *walk out* specific contacts, in order to follow their course, their geometry, and their continuity (or lack of) to the limits of the area.

Rock and structural data are collected along each traverse. We assign **station numbers** to localities where we collect data and where we make observations and interpretations that we believe might be included in the final map or final report. These numbers are posted on the base map at the exact locations where the measurements and observations were made. Data are entered in the field notebook under the appropriate station number. So that station numbers do not interfere with the geologic data, it is useful to poke a tiny pinhole through the base map at each station location, turn the map over, circle the pinhole, and write the station number next to the circled pinhole. In lieu of a pin, press the thin metal barrel of your mechanical pencil into the map such that a tiny ''bump'' shows up on the reverse side of the map, which can be circled and labeled.

Figure A.16 Veteran prospector-geologist mapping at close range in the field. (G. H. Davis encountered this person in the Whetstone Mountains, southeastern Arizona.)

Geologic maps are made in the field (Figure A.16), not constructed in the office at the end of the day on the basis of notebook data and memory. Contact lines are drawn on the base map as geologic mapping proceeds, and map units are colored progressively as their distributions become established. Orientations of bedding, foliation, folds, cleavage, fractures, and other structures are entered in the field notebook, and representative readings are *posted directly on the base map* as well. It is easier (but wrong) to record orientation readings in the notebook only, waiting to plot them on the map back at the office. Plotting readings immediately on the map in the field is essential in order to visually *track* the evolving patterns of orientations and geometries and thereby recognize inconsistencies and departures that will require explanation. Careful study of the map as it develops will provide direction in determining where to go next for geologic insight.

Record Keeping in Field Notebooks

Fundamental descriptive data collected in the field or laboratory are recorded in notebooks. The facts and information entered into the notebooks become an essential part of the framework for geologic mapping and descriptive analysis.. There is no one way to arrange your notebook. Structural geologists use different approaches based largely on personal preference. Turner and Weiss (1963) recommended a notebook layout that is practical when large numbers of orientation and physical measurements are to be taken (Figure A.17). One side of the notebook is reserved for the measurements. The data are recorded in an orderly way in columns (not scattered within paragraphs) to enable efficient recovery when it is time to plot data on maps or enter data into a computer. In the approach used by Turner and Weiss, rock descriptions are placed on facing pages, along with sketches or cross sections of important relationships, code numbers for specimens collected, photograph numbers, and most important, ideas that come to mind during fieldwork.

Hardback, bound notebooks, approximately 5 × 8 in., are traditionally used in geologic studies. But a practical alternative approach is to carry a small two-ring loose-leaf notebook containing punched notecards. The cards are durable. They may be coded, cataloged, and filed. When it is time to write a geologic report, the cards can be combined with cards bearing notes from library or experimental research, and arranged according to an appropriate organizational scheme. If the cards are filed regularly, all is not lost on the tragic day the notebook is left on a remote mountain peak or is swallowed by torrential canyon whitewater.

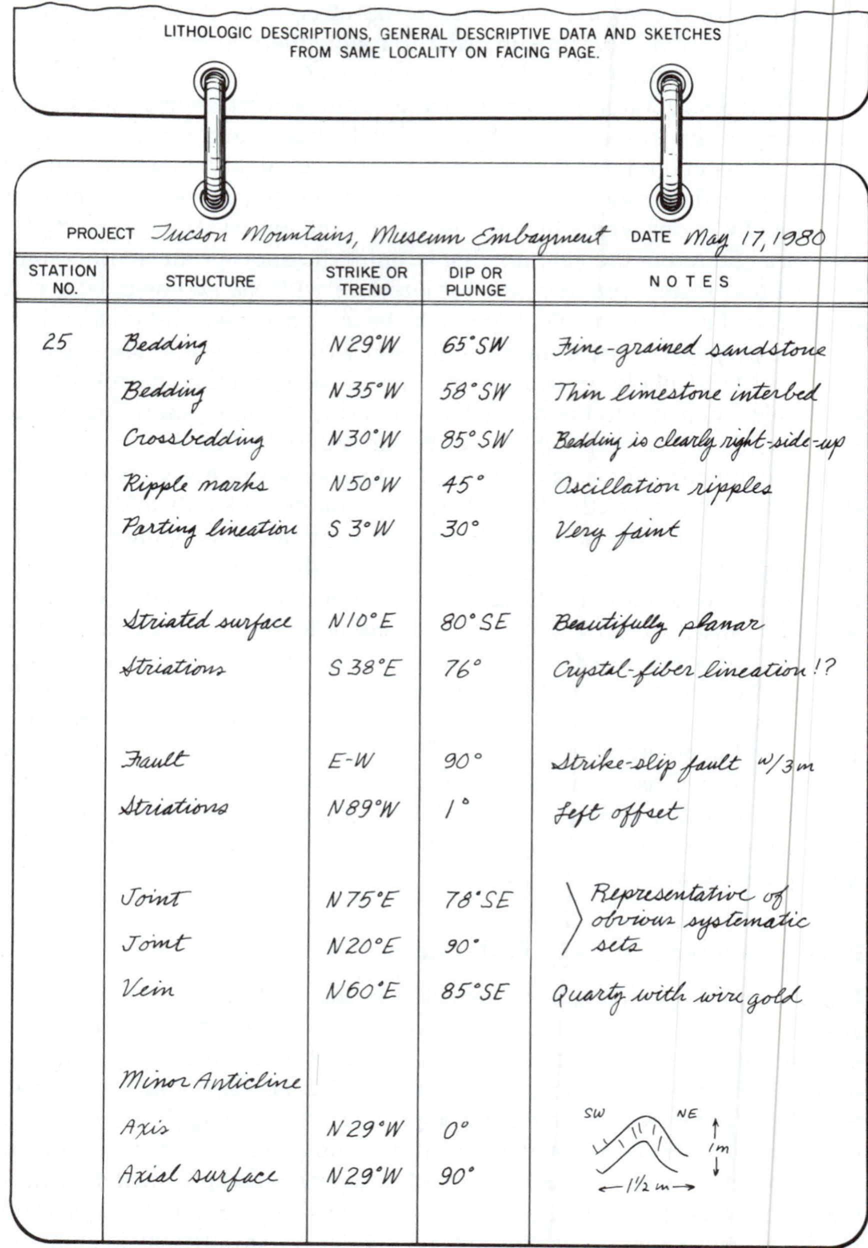

LITHOLOGIC DESCRIPTIONS, GENERAL DESCRIPTIVE DATA AND SKETCHES FROM SAME LOCALITY ON FACING PAGE.

PROJECT Tucson Mountains, Museum Embayment DATE May 17, 1980

STATION NO.	STRUCTURE	STRIKE OR TREND	DIP OR PLUNGE	NOTES
25	Bedding	N29°W	65°SW	Fine-grained sandstone
	Bedding	N35°W	58°SW	Thin limestone interbed
	Crossbedding	N30°W	85°SW	Bedding is clearly right-side-up
	Ripple marks	N50°W	45°	Oscillation ripples
	Parting lineation	S 3°W	30°	Very faint
	Striated surface	N10°E	80°SE	Beautifully planar
	Striations	S38°E	76°	Crystal-fiber lineation!?
	Fault	E-W	90°	Strike-slip fault w/3m
	Striations	N89°W	1°	Left offset
	Joint	N75°E	78°SE	Representative of obvious systematic sets
	Joint	N20°E	90°	
	Vein	N60°E	85°SE	Quartz with wire gold
	Minor Anticline			
	Axis	N29°W	0°	
	Axial surface	N29°W	90°	

Figure A.17 Notebook arrangement recommended by Turner and Weiss. (From *Structural Analysis of Metamorphic Tectonites* by F. J. Turner and L. E. Weiss. Published with permission of McGraw-Hill Book Company, New York, copyright © 1963.)

My approach (SJR) is to take notes on normal lined or gridded notebook paper carried beneath the map in a covered clipboard. I sketch, take notes, and enter measurements on the sheets in an organized way so that I can quickly extract what I need (Figure A.18) at a later time. I have found this method to be efficient, because my map and note sheets are all in one place and I never have to ''fish'' for my notebook. I like not having to carry *all* of my notes with me *all of the time* in the field.

Describing the Rocks

Preparation for the mapping should include the ''homework'' of reading about the rock units you expect to encounter in the area, and writing down in the field notebook those descriptive properties of the rock units

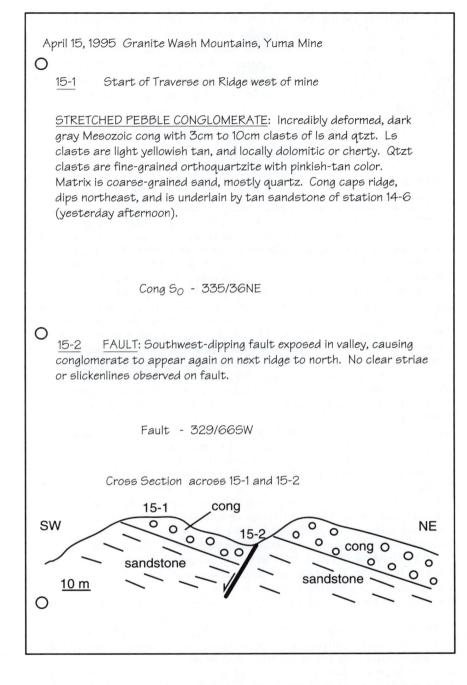

April 15, 1995 Granite Wash Mountains, Yuma Mine

15-1 Start of Traverse on Ridge west of mine

STRETCHED PEBBLE CONGLOMERATE: Incredibly deformed, dark gray Mesozoic cong with 3cm to 10cm clasts of ls and qtzt. Ls clasts are light yellowish tan, and locally dolomitic or cherty. Qtzt clasts are fine-grained orthoquartzite with pinkish-tan color. Matrix is coarse-grained sand, mostly quartz. Cong caps ridge, dips northeast, and is underlain by tan sandstone of station 14-6 (yesterday afternoon).

Cong S_0 - 335/36NE

15-2 FAULT: Southwest-dipping fault exposed in valley, causing conglomerate to appear again on next ridge to north. No clear striae or slickenlines observed on fault.

Fault - 329/66SW

Cross Section across 15-1 and 15-2

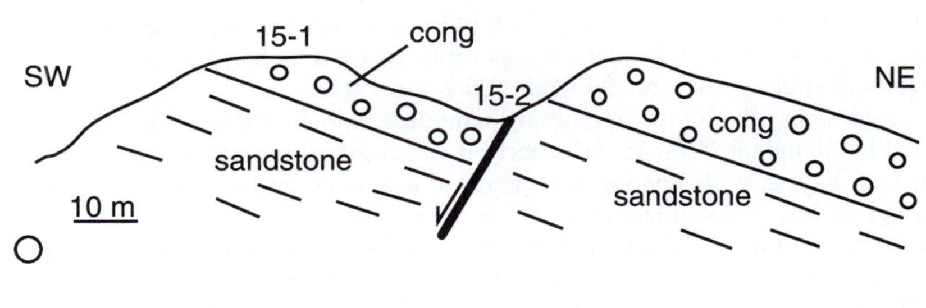

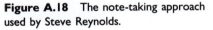

Figure A.18 The note-taking approach used by Steve Reynolds.

that will allow you to distinguish them. The "Explanation" of a published geologic map for the area is a good starting point, for it will describe, in greater or lesser detail, the geologic column. Notes regarding the ages and general lithologic characteristics of the rocks serve to anchor the study, especially in its early stages.

On-the-outcrop work brings us in touch with the full range of rock types, and for most structural studies it is essential to describe the hand specimen and formation characteristics of each mappable unit. Samples are collected for microscopic studies, as needed. Care in rock description is essential, for the lithology and texture of a given rock unit strongly influences its strength and thus its mechanical response to deforming stresses.

Descriptions should be couched in short paragraphs, organized in a standard format to guide systematic observation and to make it easy to go from the notes to clear, concise, well-written unit descriptions in the report. A useful sequence is **rock name**, **color**, **mineralogy**, **texture**, **primary features**, **gross characteristics**, **formation name**, and **contacts** (Schreiber, 1974). Examples of some well-written rock descriptions are presented in Table A-4.

TABLE A.4
Some rock descriptions

Sandstone (*quartzarenite*); white and very pale orange, weathers light brown and moderate reddish brown; very fine grained; subangular, well sorted, laminated; locally cross-bedded; bedding thickness as much as 1 ft (30 cm), mostly covered with rubble; forms steep, rounded slope. Bolsa Quartzite.

Siltstone, calcareous; grayish red, weathers pale red; coarse silt; faint horizontal laminations on weathered surface; thin bedded. Earp Formation.

Chert pebble conglomerate; grayish red purple, weathers grayish red; 70% red chert, 30% gray to white chert; ranging from 0.4 cm in diameter to 6 cm in diameter, average 1.3 cm in diameter; angular to subround, very poor sorting; cement mostly sparry calcite, minor micritic cement; rare microscopic limonite; tabular-planar and wedge-planar cross-bedding; medium bedded; forms resistant bed on slope; disconformity at base. Earp Formation.

Clay, calcareous; moderate brown, numerous limy nodules and concretions near top, crumbly and blocky; forms moderate slopes; top surface irregular. Pantano Formation.

Limestone, light gray to light yellowish gray and pale olive gray, commonly sandy and dolomitic, beds generally 1–4 ft (30–120 cm) thick, contains pelecypods, brachipods, and echinoid spines, grades laterally into sandstone. Forms steep slopes and cliffs with distinctive bench at the top. Kaibab Limestone.

Limestone (*micrite*)*;* pale red, weathers pale yellowish brown; red chert flecks; slightly silty; medium bedded; fossils replaced by red chert, crinoid stems, echinoid spines, small brachiopods. Earp Formation.

Dolomite; light olive gray, weathering same; medium crystalline; unit characterized by abundant (6 in; 15 cm) chert nodules (oolites with fossil hash), white to very light gray, becoming non-cherty at top; pock-marked, gash-like weathered surface; thin-bedded.

Porphyritic and *amygdaloidal basalt* flows and sills, dark olive green to dark gray; weather olive brown to light olive gray. Phenocrysts of olivine and augite partly to wholly altered to iddingsite and serpentine-chlorite. Amygdules consist of calcite, natrolite, analcite, or quartz. Flows are interbedded with younger gravel. Sills are in the Naco Formation.

Schist, light gray to dark gray, weathers brown or greenish-brown. Comprised of a variety of types from coarse-grained quartz-sericite schist to fine-grained quartz-sericite-chlorite schist. Low grade metamorphism greenschist facies; higher-grade metamorphism occurs locally. Relict bedding of sedimentary protolith is generally recognizable in outcrop; plunging overturned tight to isoclinal folding is pervasive. Poorly exposed, forming subdued outcrops covered with flaky chips. Overlain unconformably by younger Precambrian rocks of the Apache Group. Pinal Schist.

Granite, light gray or light pink, usually deeply weathered to light brown. Typically coarse-grained, containing large phenocrysts of pale-pink orthoclase up to 3 in (7.6 cm) long. Coarse-grained groundmass consists of pale-pink orthoclase, chalky plagioclase (albite or andesine), quartz, and books of black biotite. Probably underlies diabase and sedimentary formations in most of the region. Ruin Granite.

B. MAPPING CONTACT RELATIONSHIPS

Among the most important features to recognize, study, map, and describe in the field notebook in the course of fieldwork are **geologic contacts** between different rock formations. Rock bodies are the building blocks of the crust of the Earth. They come in all sizes, shapes, and strengths. They are modules, of sorts, that make up the whole. Through processes of deposition, intrusion, faulting, and/or shearing, they generally fit together perfectly along tight contacts. A wonderful nongeologic image of the ultimate in perfect fit is seen in the mortarless contacts between limestone blocks in walls of the Inca architectural marvel known as Saqsaywaman, Cuzco, Peru (Figure B.1). Geologic symbols for lithologic contacts (contacts of deposition or intrusion) and fault contacts are presented in Table B.1.

Contacts are simple to distinguish in theory. But where structures are complex, deformations are multiple, and exposures are poor, the job of recognizing the nature of contacts can be very difficult and challenging.

Figure B.1 Limestone blocks in Inca wall fit together perfectly along mortarless contacts. Windowlike opening is about 1 m high. Like many geologic contacts, these man-made contacts are difficult to interpret: What processes were required to bring the limestone blocks into perfect contact? (Photograph by L. A. Lepry.)

Conformable Depositional Contacts Within Formations

Within any given geologic formation, such as the Madison Limestone or the Dakota Sandstone, it is always possible to identify beds that are distinctive because of properties such as color, composition, lithology, texture, and thickness. Such beds *within* a formation of sedimentary or volcanic rocks may be nearly the same age, and are essentially parallel to one another. Such beds comprise a *conformable* sequence within the formation. The beds accumulated through an essentially continuous deposition, or intermittently in a way that the age difference(s) between adjacent beds are negligible. We refer to such contacts formed *between* layers *within* such sequences as **conformable depositional contacts** (Figure B.2). Some such contacts represent **diastems**, that is, times of nondeposition, or places within the formation where material once deposited was eroded before the next interval of deposition. Others simply represent an abrupt or gradual change in the environment of deposition. Conformable depositional contacts are usually planar to slightly irregular in form. We routinely measure the orientations of the beds and the contacts between the beds as a part of geologic mapping, but we only map such **intraformational conformable depositional contacts** if they occur on either side of a distinctive marker bed whose structure we wish to represent.

The recognition of contacts as conformable depositional contacts is sometimes achieved in a roundabout way: by noting the absence of the characteristic physical signatures of unconformities, intrusive contacts, faults, and shear zones.

TABLE B.1
Geological symbols for lithologic contacts and fault contacts

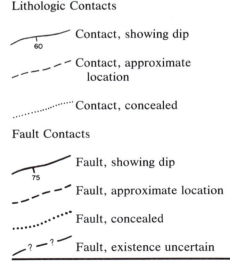

Symbol	Meaning
Lithologic Contacts	
(solid line with dip tick, 60)	Contact, showing dip
(dashed line)	Contact, approximate location
(dotted line)	Contact, concealed
Fault Contacts	
(heavy solid line with dip tick, 75)	Fault, showing dip
(heavy dashed line)	Fault, approximate location
(heavy dotted line)	Fault, concealed
(line with ? — ?)	Fault, existence uncertain

Unconformities

Rock sequences are made much more interesting because of the presence of contacts that represent gaps within the geologic column, where conformable deposition was interrupted, or where erosion during long intervals removed a part of the rock record. A critical part of geologic mapping is recognizing and mapping where those interruptions occur in the bedrock geology.

It is easy to imagine how gaps in the rock record can develop. A drop in sea level will cause a shallow sea to recede from the continental interior it once occupied, resulting in the cessation of deposition of marine lime-

Figure B.2 (*A*) Normal depositional contacts within rhyolite ignimbrite in the barranca country of Sinaloa, Mexico. Contacts are beautifully exposed in this region, the most deeply dissected part of the Sierra Madre Occidental. The volcanic rocks are part of a magmatic arc, approximately 400 × 1500 km in surface area. The volcanics of the arc erupted almost entirely during Miocene time. (Photograph by D. J. Lynch.) (*B*) Normal depositional contacts within sequence of sedimentary rocks exposed along the Goosenecks of the San Juan River, Utah. (Photograph by G. H. Davis.)

stones, sandstones, and shales, and the onset of erosion. Long afterward, when the sea advances again and renews marine deposition in the continental interior, younger sediments will be deposited *unconformably* on the eroded top of sedimentary rocks of the earlier period. The contact between the base of the younger sequence and the eroded top of the older sequence will be marked by an **unconformity**, which might later be preserved in the geologic record. The unconformity is marked by a gap in the rock record; time is missing. In this instance, the sedimentary rocks both above and below the unconformity will be parallel to the unconformity itself.

There are other ways in which unconformities can develop. Imagine the uplift of mountains in a region occupied by the sea. Both the marine sediments and the older rocks beneath will be elevated high above sea level. Tens of millions of years of erosion may eventually reduce the mountain to a coastal plain, and when the sea again advances it will deposit much younger sediments *unconformably* on the upturned ends of older sedimentary rocks, and perhaps on the deeply eroded igneous and metamorphic rocks that once made up the deep core of the mountain belt. The unconformity in places will separate young horizontal sedimentary rocks from older, tilted, deformed sedimentary rocks. In other places the young horizontal sedimentary rocks will rest directly on an erosionally carved surface on old granite or schist. Again, the unconformity marks a gap in the rock record. Time is missing.

In short, an **unconformity** is a depositional contact between two rocks of measurably different ages. Unconformities are special because they are surfaces that represent intervals of time during which either there was no deposition, or there was erosional removal of the entire rock record for the interval of time represented by the unconformity, or both.

Unconformities are divided into three major classes: **nonconformities**, **angular unconformities**, and **disconformities** (Figure B.3), and their characteristics permit their identification during geologic mapping.

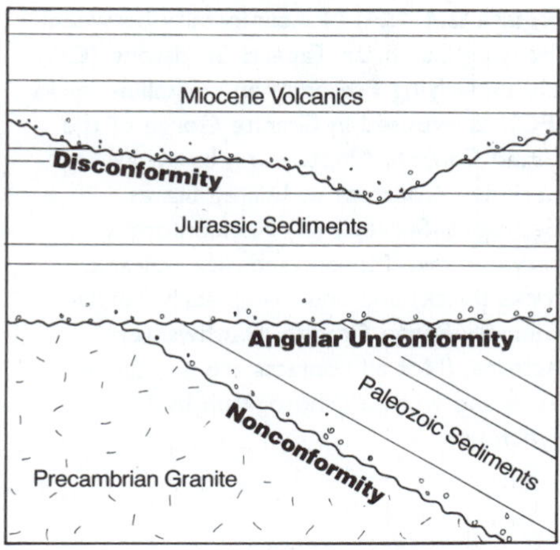

Figure B.3 Schematic portrayal of the three kinds of unconformity: nonconformity, angular unconformity, disconformity.

Nonconformities are depositional surfaces separating distinctly younger sedimentary or volcanic rocks above from distinctly older igneous or metamorphic rocks beneath (Figure B.3). Nonconformities include the **great unconformities** that separate Precambrian crystalline basement rocks from overlying Paleozoic, Mesozoic, or Cenozoic sedimentary or volcanic strata (Figure B.4*A*). But they also include the unconformities separating post-Precambrian crystalline rocks from overlying, younger, sedimentary and/or volcanic strata (Figure B.4*B*). Standing on a nonconformity is something special, particularly if the time gap between crystalline and cover rock represents a billion years.

An **angular unconformity** (Figure B.3) is an unconformity that separates layers above and below that are not parallel (Figure B.5). Classical angular unconformities are horizontal depositional surfaces separating relatively young horizontal strata above from older steeply dipping strata below (Figure B.6).

Spectacular angular unconformities are sometimes displayed in single outcrops or cliff exposures. In contrast, subtle angular unconformities marked by very slight angular discordance can be ''seen'' only through careful regional mapping. Consider a regional assemblage of sedimentary rocks tilted at an angle of just 4° (Figure B.7*A*). If this regional rock unit is beveled by a 1°- dipping erosional surface, the resulting 3° of angular discordance at the unconformity usually would not be discernible at the outcrop scale of view (Figure B.7*B*). However, the angular discordance would be conspicuous when seen in the perspective of tens of kilometers of viewing distance (Figure B.7*C*). Units below the unconformity would pinch out against it regionally. The younger strata would overlap the older strata. Such unconformable relationships are common in foreland terranes and in parts of miogeoclinal prisms.

A **disconformity** is an unconformity separating strata that are parallel to each other (Figure B.3). The physical presence of a disconformity may be hard to detect. Recognition may require complete knowledge of the ages of beds within the sequence of strata that contains the disconformity. In the Paleozoic geologic column of the Grand Canyon, Ordovician and Silurian rocks are completely missing, and Devonian sedimentary rocks are spotty. Thus there exists a disconformity separating flat-lying Mississippian Redwall Formation above from the flat-lying Cambrian Muav Limestone below.

For all three types of unconformity, the surface marking the unconformity itself is parallel to the bedding or layering of the rocks above the unconformity. The only major exception is where unconformities are marked by unusually high topographic relief, and the younger sediments or volcanics above the unconformity lap up against a steep-sided island, hill, or scarp. Such an unconformity is called a **buttress unconformity** (Figure B.8).

During the course of geologic mapping, nonconformities and intrusive igneous contacts can sometimes be confused. Angular unconformities

Figure B.4 (*A*) Nonconformity between flat-lying Cambrian Tapeats Sandstone (Ꞓt) and underlying Precambrian crystalline rocks (PꞒ), as exposed in Granite Gorge of the Grand Canyon. (Photograph by N. W. Carkhuft. Courtesy of United States Geological Survey.) (*B*) Nonconformity between mid-Tertiary andesitic volcanic rocks (black) and underlying early Tertiary Gunnery Range Granite, southwestern Arizona. (Not all contacts are as plain as black and white.) (Photograph by D. J. Lynch.)

and fault contacts also can be misinterpreted. Disconformities can be completely overlooked because they lack geometric expression. Chances for misinterpretation are enhanced by poor outcrop exposure and complex structural overprinting. In their raw unblemished state, however, unconformities possess a number of outcrop-scale physical and geometric properties that aid in their identification.

The bed directly above an unconformity commonly contains a **basal conglomerate**, normally composed of clasts of the rock directly beneath

Figure B.5 Architectural analogue to angular unconformity shows up in this view of a part of the Great Wall of China. (Photograph by Gregory A. Davis.)

Figure B.6 Angular unconformity between flat-lying Paleozoic strata and underlying tilted upper Precambrian strata, exposed in a section of wall in the Grand Canyon of the Colorado River. [From Powell (1875). Courtesy of United States Geological Survey.]

the unconformity. Basal conglomerates advertise erosional intervals. The basal conglomerate may range in coarseness from a thin, fine, granule conglomerate to a thick, coarse, boulder conglomerate. Surfaces of unconformity may locally possess **topographic relief** that can be recognized as the product of ancient erosion, perhaps including the preservation of the cross section of an old stream channel. One such channel can be seen in the sidewall of Blacktail Canyon in the Grand Canyon (Figure B.9). Under ideal conditions, fossil soil profiles, called **paleosols**, are preserved in rocks directly beneath the old erosion surface. These may be baked where overlain by lava flows.

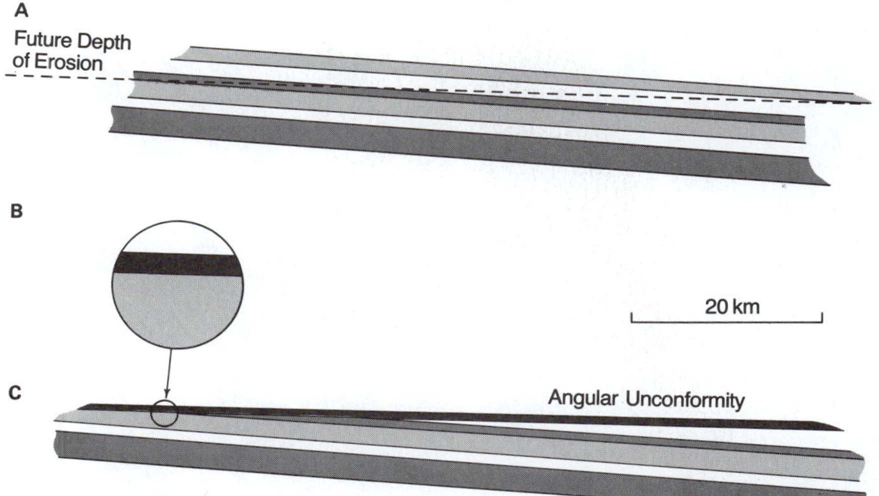

Figure B.7 Development of a subtle regional angular unconformity. (*A*) Sedimentary sequence tilted at 4°, about to be beveled by erosion to the depth of the 1°-dipping reference line. (*B*) Angular unconformity looks like normal depositional contact when only a tiny segment of the regional relationship is viewed. (*C*) Regional view shows that a number of the 4°-dipping units beneath the unconformity pinch out at the unconformity, thus drawing attention to the existence of the subtle unconformable relationship.

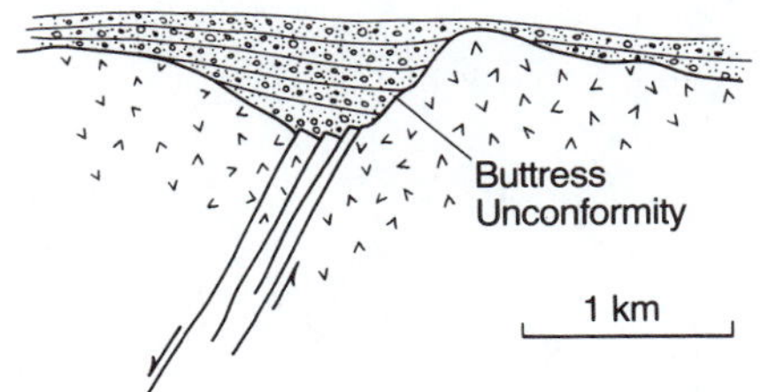

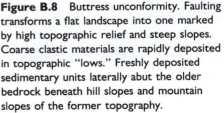

Figure B.8 Buttress unconformity. Faulting transforms a flat landscape into one marked by high topographic relief and steep slopes. Coarse clastic materials are rapidly deposited in topographic "lows." Freshly deposited sedimentary units laterally abut the older bedrock beneath hill slopes and mountain slopes of the former topography.

Figure B.9 Deep down in the Grand Canyon, a hike up Blacktail Canyon takes you to this amazing exposure of the Great Unconformity between Cambrian sandstone above, and Proterozoic schist beneath. Right at the contact is a Cambrian channel, full of cobbles derived from quartz veins in the underlying schist. This is relative dating at its best. (Fish-eye view photograph by S. J. Reynolds.)

Intrusive Contacts

The mapping of **intrusive contacts** requires identifying locations where magma has solidified against the country rocks through which it once flowed (Figure B.10). The term **country rock** refers to the rock assemblage, whatever its nature, that hosts the intruder. Country rock may be sedimentary, igneous, or metamorphic. The intrusive contact proper is the interface between country rock and the intrusive body.

Igneous intrusive contacts may be recognized on the basis of a number of features. Outcrops of the country rock close to an igneous intrusive contact may be invaded by **apophyses** of irregular tonguelike injections into country rock (Figure B.11). Pieces of fractured country rock, known as **inclusions** or **xenoliths,** may become detached from the wall of contact during intrusion to become incorporated within the main igneous intrusion (Figure B.12).

Figure B.10 Igneous dike swarm of pegmatites intruding quartz diorite near Mount Lemmon in the Santa Catalina Mountains, southeastern Arizona. Geologist is Evans B. Mayo. (Photograph by G. H. Davis.)

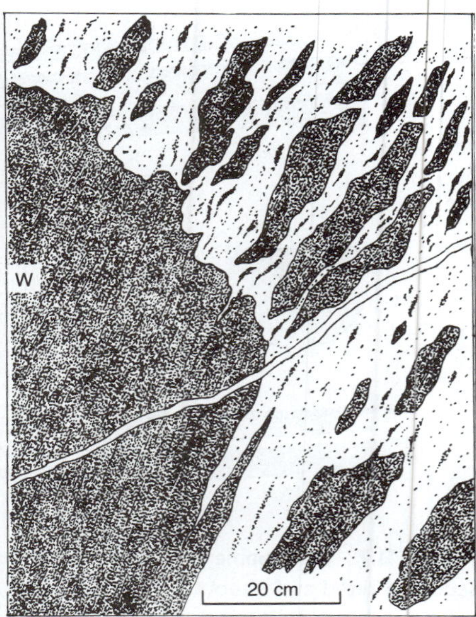

Figure B.11 Tonguelike apophyses of quartz monzonite (white) invade faintly foliated diorite (dark), Sierra Nevada, California. (Field sketch by E. B. Mayo.)

Figure B.12 Dark angular xenoliths enveloped in the younger light-colored granitic rock. (Photograph by E. B. Mayo.)

The form of igneous intrusions varies with depth, the composition of magma, and the presence or absence of structural control of the site of intrusion. As a consequence, the description of the contact relationships between igneous intrusions and country rock can be quite challenging to map and visualize. Even the vocabulary is challenging (Figure B.13): **batholith**, **laccolith**, **lopolith**, and **phacolith**, not to mention **harpolith** and **bysmalith**. Best of all is **cactolith**:

> A quasi-horizontal chonolith composed of anastomosing ductoliths, whose distal ends curl like a harpolith, thin like a sphenolith, or bulge discordantly like an akmolith or ethmolith. (From "Geology and Geography of the Henry Mountain Region, Utah" by C. B. Hunt, P. Averitt, and R. L. Miller, 1953, p. 150. Courtesy of United States Geological Survey.)

Country rocks invaded by magma respond to the heat by recrystallization and metamorphism. Sedimentary rocks, unaccustomed to hot environments, are particularly vulnerable to thermal alteration. The intrusive magma itself may quickly cool upon contact with the wall rock, forming a **chill zone** of very fine-grained igneous rock at the border of the igneous

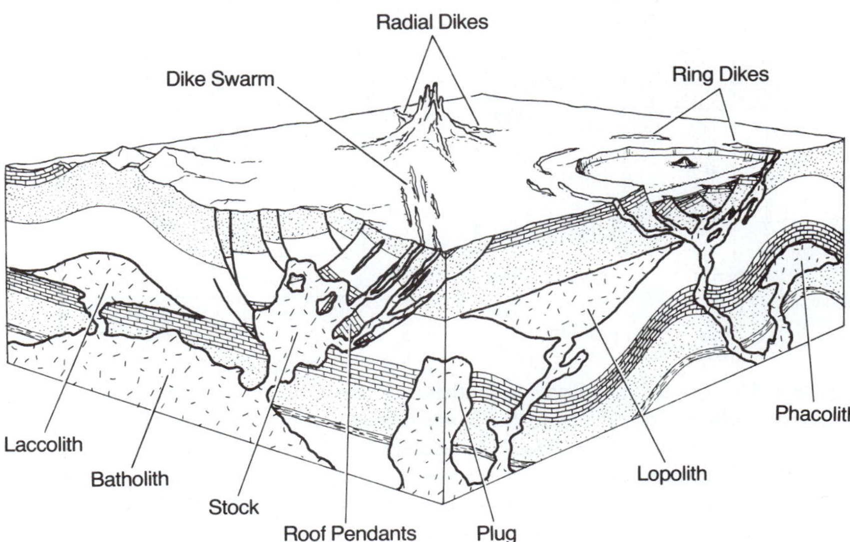

Figure B.13 Index to names of some igneous intrusive bodies and igneous intrusive structures. (Artwork by R. W. Krantz.)

Figure B.14 Outcrop example of dark sandstone dike cutting Tertiary sedimentary breccia at Camelback Mountain, Phoenix, Arizona. (Photograph by S. J. Reynolds.)

body. Igneous intrusions that penetrate country rocks at high, relatively cool levels in the crust may impart to the wall rocks a local **contact metamorphism** that produces a restricted halo or **aureole** of metamorphism in wall rocks along the contact. For example, one day you might find yourself mapping a limestone unit, only to see it progressively transform into marble, or a **skarn** (limestone altered to a calcium silicate rock). As you walk farther, you come into contact with the igneous intrusive rock that "baked" the limestone.

The term "intrusive contact" generally brings to mind the *igneous* intrusion of a hot viscous fluid (i.e., magma) into country rock. However intrusions of sedimentary rock are also found. Sedimentary intrusions are of two types: **soft-sediment intrusions** and **salt diapirs**. Soft-sediment intrusions involve the squeezing or buoyant rise of buried but yet-unconsolidated water-rich muds and sands into adjacent or overlying country rock. The mud and sand intrusions typically originate from within source beds marked by high fluid pressure. Tabular intrusions of mudstones and sandstones are referred to as **clastic dikes** (Figure B.14). These may move up or down, and thus the *material* within them may be older or younger than the rock they invade. **Mud diapirs** are pluglike or domelike masses up to 1 km in height and up to several kilometers in diameter (Figure B.15). They can form when there is a **density inversion** of relatively low-density uncompacted mud (ρ = 2.3 or less) overlain by higher-density sand and sand–mud mixtures (ρ = 2.4–2.7). The density inversion may lead to the buoyant rise of the lower-density materials. The ascent of diapirs triggered by density inversion has been replicated in the laboratory through scale-model experiments (Figure B.16). These experiments feature the disturbance of stratified soft materials (like clays and oils and asphalts) arranged in an order of density inversion (Nettleton, 1934; Parker and McDowell, 1955; Ramberg, 1967). In the same fashion, **salt diapirs**

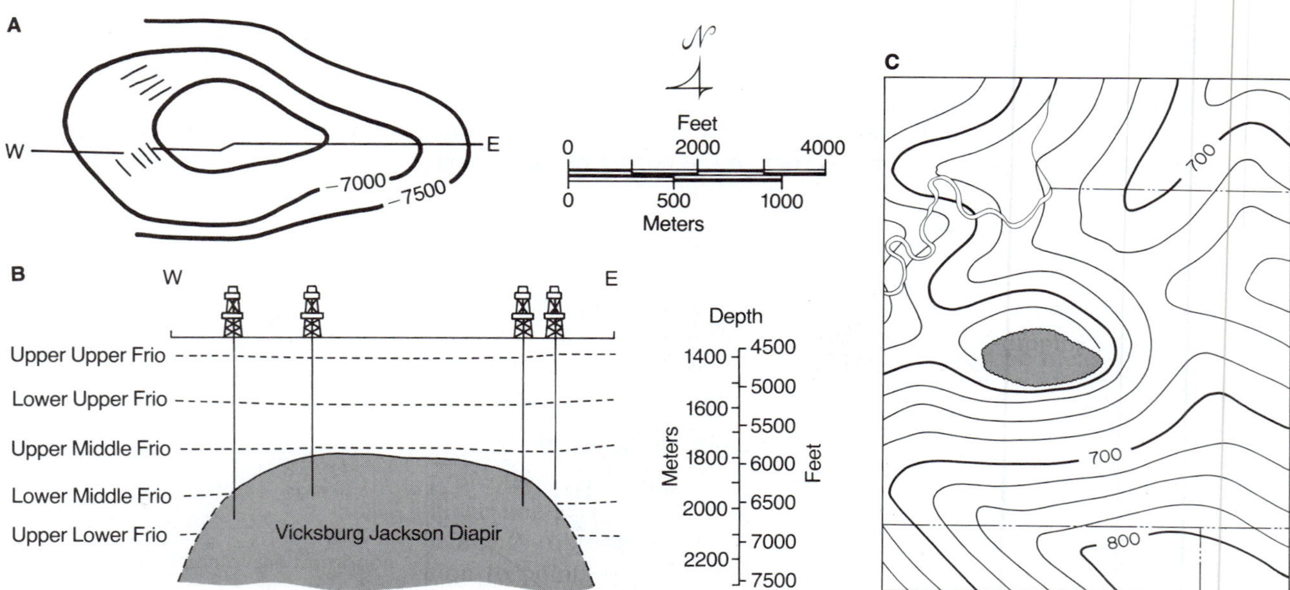

Figure B.15 Laward mud diapir in southeastern Texas. Composed of Eocene and Oligocene shale, the diapir pierces the lower part of Oligocene sandstone. (*A*) Structure contour map of the diapir. (*B*) Geologic cross section showing the form of the diapir. (*C*) Isopach map of the pierced sediments at Laward, illustrating that the sediments become thinner toward the diapir. [From Bishop (1977). Published with permission of Gulf Coast Association of Geological Societies.]

result from density inversions. Salt diapirs are domes, pillars, and walls of salt that buoyantly rise as plastic solids from thick beds of evaporites into overlying sedimentary country rock (Figure B.17). In the Gulf Coast region salt diapirs average 3 km in diameter, and some have risen 10 km(!) through the Cenozoic overburden to very shallow levels. Salt can flow rapidly, even in terms of the scale of human history:

> . . . in salt mines in the Austrian Salzburg district, old tunnels were discovered in which bodies of ancient Celtic miners were found with their mining tools, embedded in salt. (From "Structure of Grand Saline Salt Dome, Van Zandt County, Texas" by R. Balk, 1949, p. 1823. Published with permission of American Association of Petroleum Geologists.)

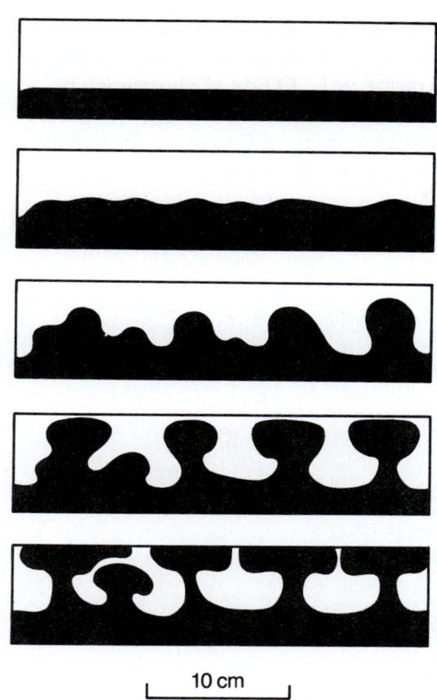

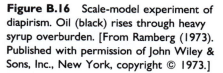

Figure B.16 Scale-model experiment of diapirism. Oil (black) rises through heavy syrup overburden. [From Ramberg (1973). Published with permission of John Wiley & Sons, Inc., New York, copyright © 1973.]

Fault Contacts

Faults rearrange conformable and nonconformable sequences of rocks in ways that can create amazing geometric puzzles. The game is to map out the fault contacts, measure the displacements, and discover the history of movements that took place.

Fault contacts can be recognized at the outcrop scale by virtue of an array of characteristic physical properties, not all of which will be assembled at any one place. Commonly we should expect to find a discrete fracture break or discontinuity, a **fault surface** (Figure B.18) whose orientation and position satisfactorily describe the contact between rocks. Alternatively, a **fault zone** may be present instead of a discrete fault surface. Fault zones consist of numerous closely spaced fault surfaces, commonly separating masses of broken rock, including **gouge** and **breccia** (see Chapter 6, "Faults"), and perhaps some minor folds. Fault surfaces may be finely polished to **slickensided surfaces** as a result of differential movement, and the slickensided surfaces are almost always marked by **striations** or **grooves** that reflect fault movement (Figure B.19).

Mapping permits faults to be found. Imagine walking parallel to the trend of a distinctive sandstone formation that is steeply tilted, only to

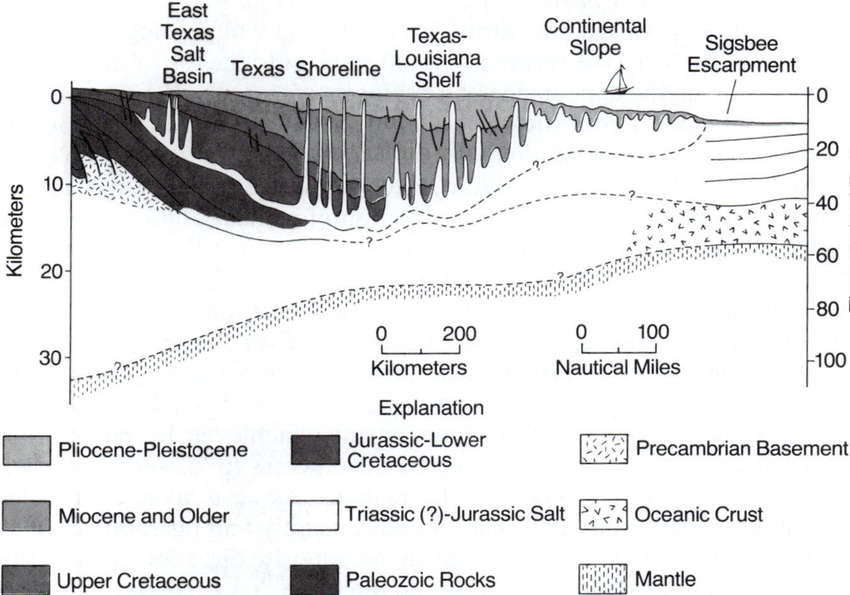

Figure B.17 Cross-sectional profile of salt domes and salt spines in the Gulf Coast region. [From Martin (1978). Published with permission of American Association of Petroleum Geologists.]

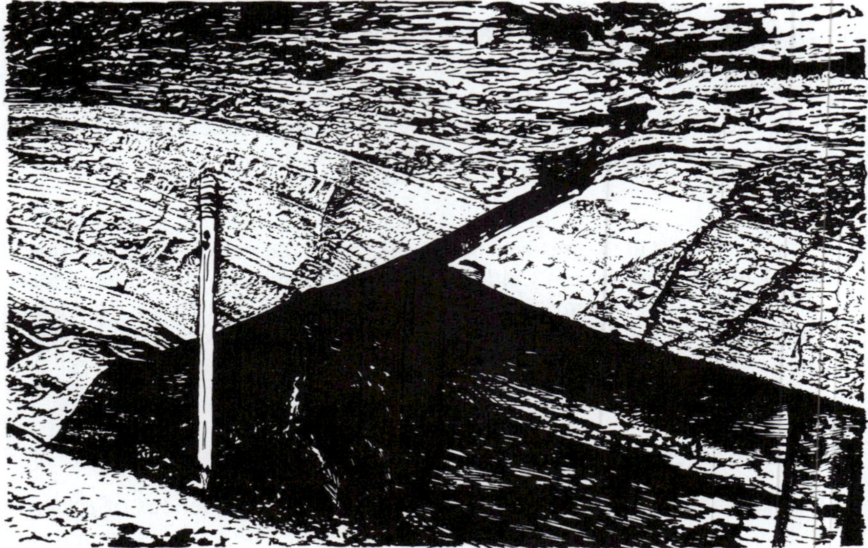

Figure B.18 Fault in metamorphic rock in the Tortolita Mountains, southern Arizona. [Tracing by D. O'Day of photograph by G. H. Davis. From Davis (1980). Published with permission of Geological Society of America and the author.]

find that it terminates against a limestone formation. As you cross the contact, you see unusually fractured and crushed bedrock as well as some slickensided surfaces. Because you know the stratigraphic sequence perfectly, you are able to recognize the limestone as the formation that normally directly overlies the sandstone. Thus, you know that the faulted location of the sandstone is to the right. Turning in that direction, you walk a short distance and find yourself back in the sandstone.

Upon discovering faults and mapping them, be sure to represent them through complete symbology, distinguishing wherever possible what you have learned by separation versus slip.

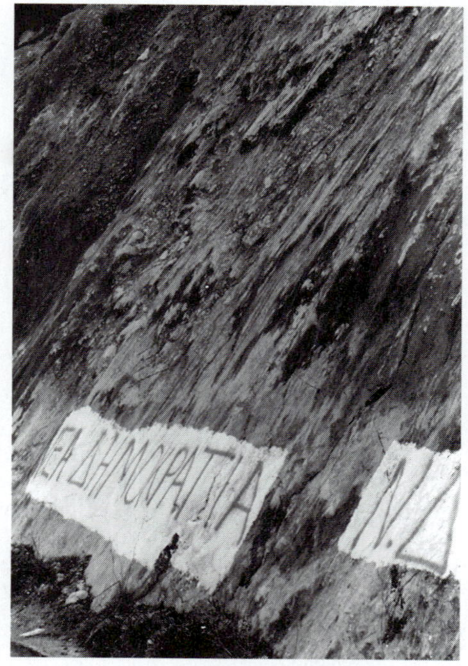

Figure B.19 Well-exposed fault surface at Delphi, Greece. It lies along the margin of Mount Parnassos, home of the gods. Slickenlines and grooves are conspicuous, as is breccia along the fault surface. Grafitti is political sloganeering. (Photograph by G. H. Davis.)

Shear Zone Contacts

In order to map shear zones, we first must be able to visualize what they look like (see Chapter 9, "Shear Zones and Progressive Deformation."). We can do this through a model. We build a very small deformation box (we call it the "butter basher") fastened to sliding panels (Figure B.20*A*). A single stick of butter (margarine if you prefer) will fit snugly into the box (Figure B.20*B*). (We sprinkle colored sugar onto the top of the stick of butter, to create the appearance of a geologic reference unit.) To simulate faulting, we place a *cold* stick of butter in the box, and shift the wooden base in opposite directions along the saw cut until the butter is completely severed by fault movement (Figure B.20*C*). The fault "cuts" the butter. Cohesion is lost. To simulate the formation of a shear zone, we place a warm stick of butter in the box, and then slowly shift the wooden base about 2 cm to create displacement (Figure B.20*D,E*). The shear zone does not "cut" the butter. This time the displacement is accommodated without loss of cohesion, through "plastic" distortion of the butter.

Shear zones are faultlike in that they accommodate displacements. Unlike ordinary fault surfaces, however, shear zones often do not display any physical break. Instead, the movement is achieved by penetrative deformation within a zone that may be centimeters to kilometers across (Figure B.21). One day you may be mapping along a layer of marble, 3 m thick, and suddenly find that it bends abruptly and tapers quickly to 6 cm, and then just as abruptly bends in the opposite direction and returns to its original thickness. Other layers do the same. You may have crossed

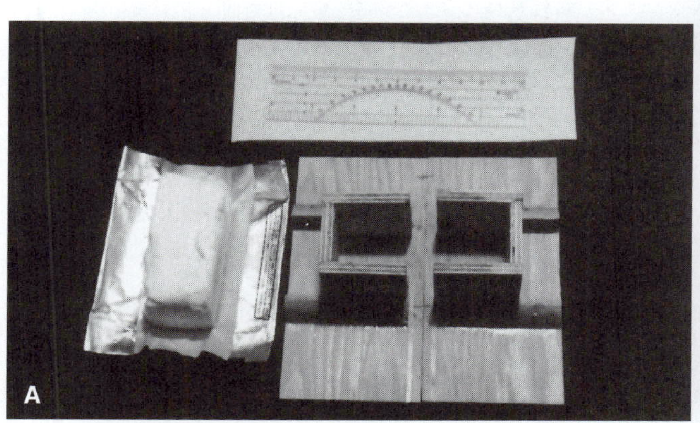

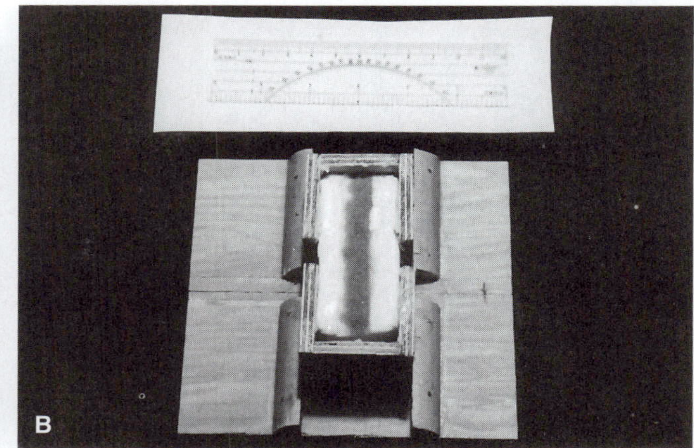

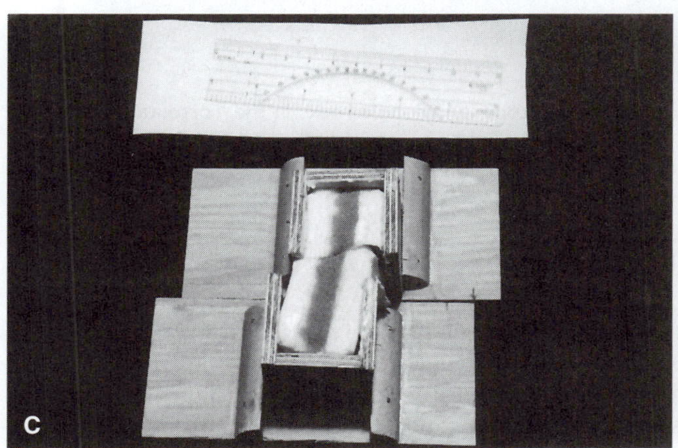

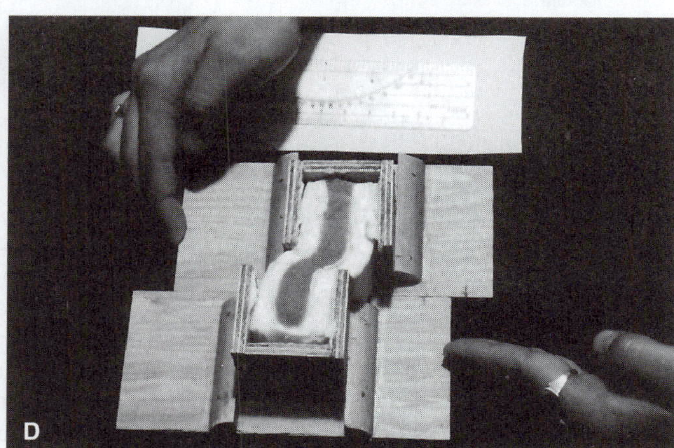

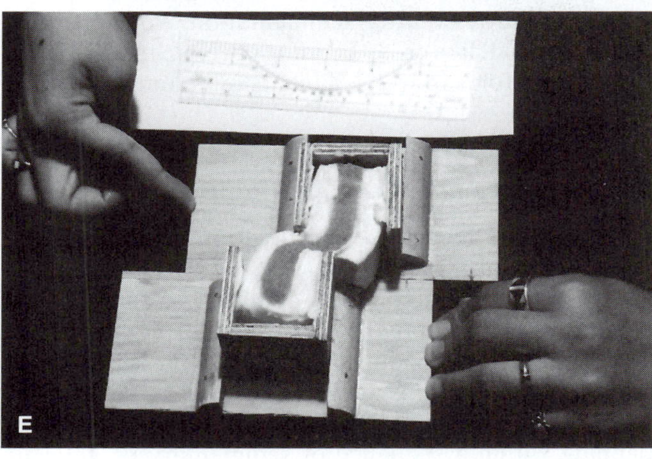

Figure B.20 Butter-basher experiments used to simulate the difference between a fault and a shear zone. (*A*) The apparatus. (*B*) The setup. (*C*) Faulting of cold stick of butter. (*D* and *E*) Shearing of warm stick of butter. Experiments carried out and photographed by Angela Smith, Brook Riley, and Brian Bennon.

into a shear zone, and you will want to map the contacts where the bending first begins.

Shear zone contacts are most commonly formed in igneous or metamorphic environments (Figure B.22), where elevated temperature and/or confining pressure renders one or both wall rocks ductile. However, shear zones can also form during soft-sediment deformation of unconsolidated sands and muds. In such environments, the plastic response is due to the water-rich nature of the sediments.

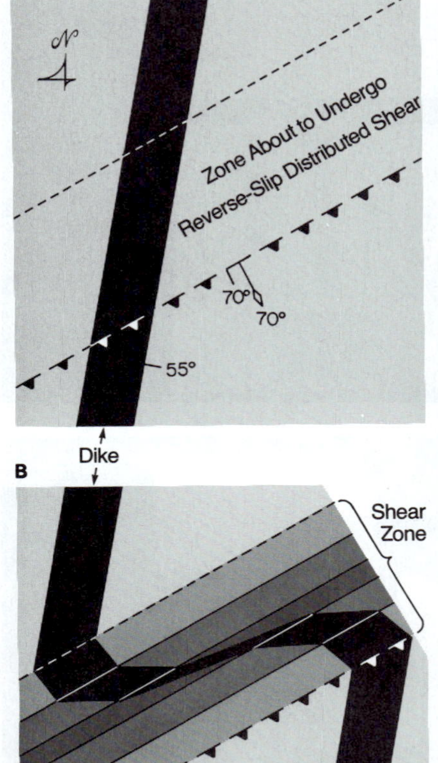

Figure B.21 (*A*) Map view of dike about to be deformed by ductile shearing. (*B*) Distortion of dike achieved by ductile shearing. Contacts of shear zones separate distorted rock from undistorted rock.

Figure B.22 Outcrop scale shear zone in mylonitic rocks in the Tortolita Mountains, southeastern Arizona. (Photograph by G. H. Davis.)

C. IDENTIFYING PRIMARY STRUCTURES

During the course of geologic mapping, we always are on the lookout for primary structures in the rocks. **Primary structures** are ones that originate during the formation of rocks: before sediments become lithified into sedimentary rocks; before lava stops moving and becomes volcanic rock; before an intruding magma solidifies and hardens into plutonic rock. Primary structures reflect conditions during sedimentation, volcanism, or intrusion. Some are depositional, like cross-bedding or ripple marks. Some are deformational, like slump folds in wet sediments or columnar jointing in basalt.

Types of Primary Structures and Their Usefulness

Primary structures can help us recognize the direction of bedding and flow layering in sedimentary and volcanic rocks, which is sometimes very difficult in poorly exposed massive units. **Bedding** is a fundamental primary sedimentary structure. Distinctive because of color, texture, composition, and resistance to erosion, bedding imparts to sedimentary rocks their fundamental architecture (Figure C.1). **Flow layering** is the counterpart in volcanic sequences (Figure C.2).

When we are mapping within a sequence of sedimentary or volcanic rocks that are "on end," we look for primary structures to determine the "way up" within the sequence (Figure C.3). The most commonly used **facing** (i.e., "way up") indicators are cross-bedding, graded bedding, oscillation ripple marks, and mud cracks. **Cross-bedding** sweeps into parallelism with the base of a bed and is sharply truncated along the top (Figure C.4). **Graded bedding**, which occurs in sandstones and conglomerates, is marked by intervals on the order of centimeters or meters within which grain or clast size decreases systematically upward toward the top of the bed (Figure C.5). **Oscillation ripple marks** have a symmetrical, concave-upward form, with cusps that point toward the top of the bed (Figure C.6). **Mud cracks**, which form when the wet mud of a river bottom or

Figure C.1 Folded bedding in sedimentary rocks in the Big Horn basin, Wyoming. (Photograph by G. H. Davis.)

Figure C.2 Close-up of flow layering in volcanic rock in the Pinacates Volcanic Field, Sonora, Mexico. (Photograph by G. H. Davis.)

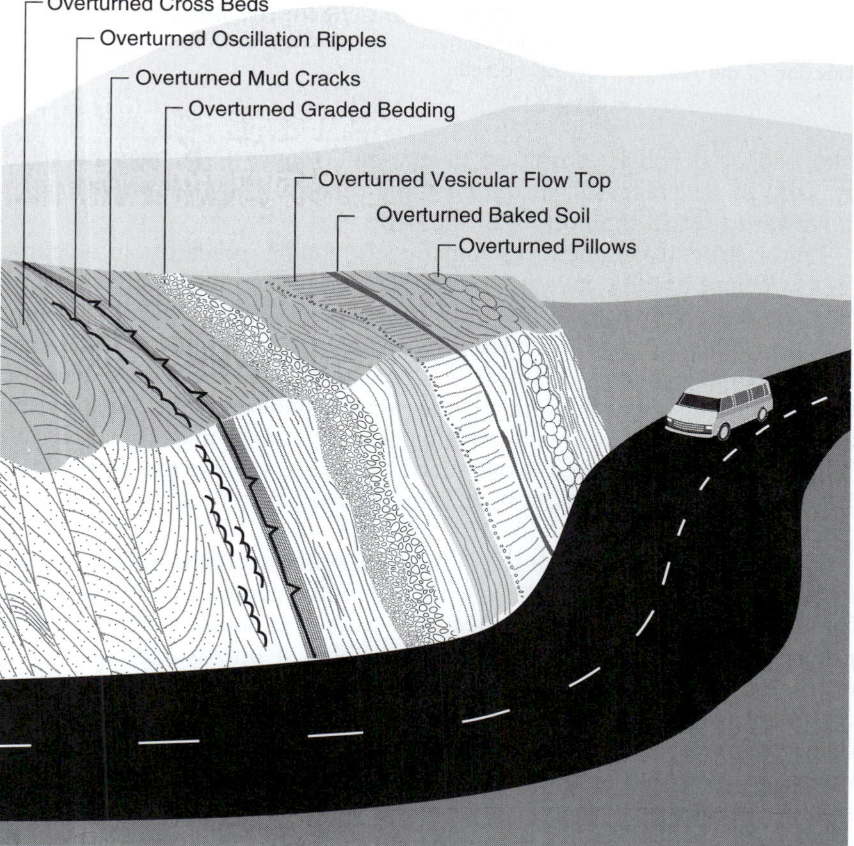

Figure C.3 Diagrammatic representation of an array of "facing indicators" in a roadcut of interlayered sedimentary and volcanic rocks. Facing indicators reveal the way up within the sequence.

Figure C.4 Cross-bedding in Navajo Sandstone near Checkerboard Mesa, Zion National Park. Geologist is Drew Davis. Truncation of cross-strata along upper bedding surface where Drew is sitting reveals that the Navajo sandstone is right-side up. (Photograph by Drew's dad.)

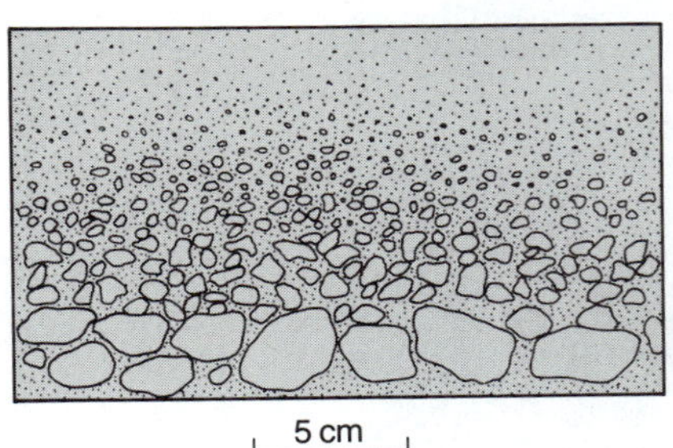

Figure C.5 Graded bedding, another indication of facing. Clast and/or grain size decreases upward within the unit, in the direction of the youngest part of the bed.

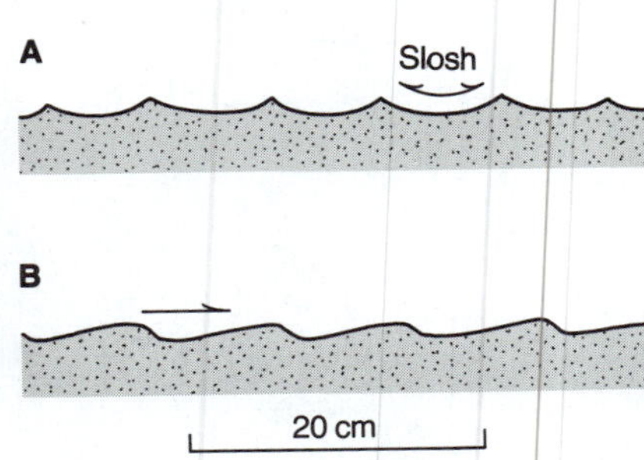

Figure C.6 (*A*) Oscillation ripple marks. (*B*) Current ripple marks.

Figure C.7 Mud cracks in the Rio Grande, west Texas. (Photograph by G. H. Davis.)

lake bed or beach is permitted to dry out (Figure C.7), may be filled by sand to be preserved in the geologic record. They commonly taper downward toward the bottom of the bed.

Pillow structure is one of the most useful facing-indicators in volcanic rocks. Where basaltic lava is extruded underwater, or where lava pours into the sea, pillow lavas instantly form (Figure C.8*A*). "Pillows" are generally flattened, with rounded tops. Where a pile has been formed, the bottoms of the pillows conform in shape to the rounded tops of those beneath them, so that in vertical cross section they show lobes projecting downward between the intersecting curved tops of the underlying pillows (Figure C.8*B*). Given their cross-sectional forms, pillows are valuable as facing indicators, even within metamorphosed basalts (i.e., greenstones) of Precambrian age.

We have to be careful not to confuse **soft sediment deformational structures** or **flow structures** in lava with "normal" geologic structures that formed much later when the rocks were fully consolidated. **Penecontemporaneous folds** can form during deformation of unconsolidated water-rich sediments. Such folds are typically **intraformational**; that is, they are restricted to a given bed or a given sequence of beds (Figure C.9). The restriction of penecontemporaneous folds to narrow stratigraphic intervals is consistent with folding on the floor of a depositional basin or at very shallow depths of burial. Most penecontemporaneous folds are simply irregularly contorted beds. The folds do not propagate upward or down-

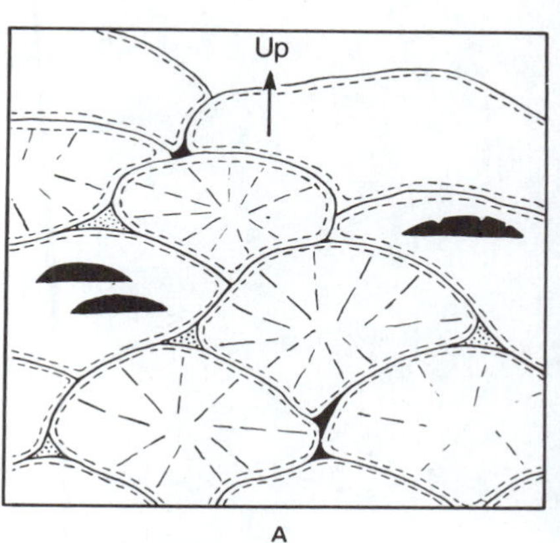

Figure C.8 (*A*) Sketch of pillow lava showing the characteristic right-side-up form of pillows. [From Macdonald (1967). Publsihed with permission of Wiley–Interscience, Inc., New York, copyright © 1967.] (*B*) Outcrop photograph of pillows in steeply dipping basalt. Top of flow is to the left; bottom is to the right. (Photograph by S. J. Reynolds.)

ward in any systematic, predictable manner, and they commonly show radical changes in layer thickness, a reflection of mobility of the mass of material thus distorted. If sedimentary material is partly consolidated, it may accommodate soft sediment deformation by **penecontemporaneous faulting and shearing**, rather than by slumping and disharmonic folding. The faults and shear zones serve to extend or shorten the not-yet-consolidated layers (Figure C.10).

Figure C.10 Normal faulting in Pleistocene glacial outwash in northern Minnesota. Faulting was produced by soft-sediment slumping of unconsolidated sands and gravels. (Photograph by G. H. Davis.)

Figure C.9 Intraformational recumbent folds in Pleistocene Flora Formation near Saskatoon, Saskatchewan. [From Hendry and Stauffer (1977). Published with permission of Geological Society of America.]

Figure C.11 Fold within a fold in rhyodacite specimen collected in the Chiricahua Mountains, Arizona. (Photograph by G. Kew.)

Figure C.12 A look at the underside of a steeply inclined bed of Pennsylvanian sandstone in the Marathon Basin, Texas. The sole of the bed displays a nice array of grooves and flute casts. Current direction responsible for deposition of the sandstone was from right to left. (Photograph by E. F. McBride.)

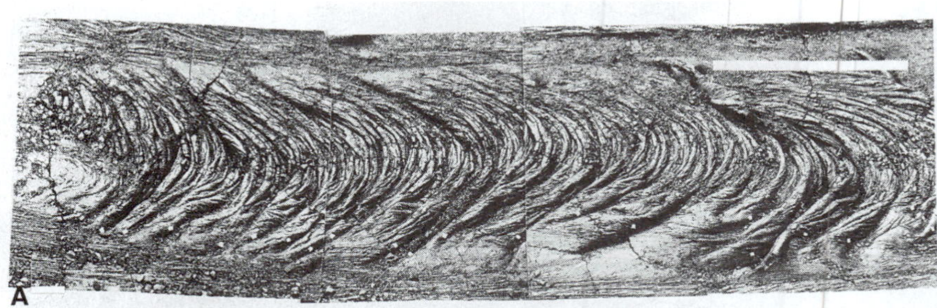

Figure C.13 (*A*) Ropy pahoehoe structure, Kilauea Volcano, Hawaii. Bar scale represents 2 m. [From Fink (1980). Published with permission of Geological Society of America and the author.] (*B*) Pressure ridges in rhyolite obsidian flow, Big Glass Mountains, Medicine Lake Highlands Volcano, California. Bar scale is 500 m. [From Fink (1980). Published with permission of Geological Society of America and the author.]

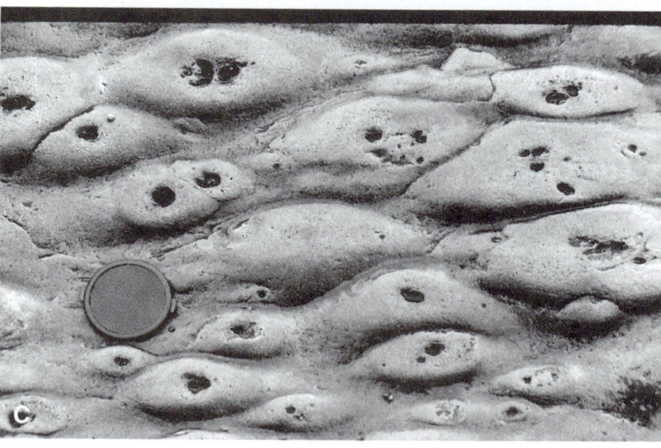

Figure C.14 (*A*) Distorted trilobite in Cambrian shale, Maentwrog, Wales. The width of the fossil is 3 cm. (From *The Minor Structures of Deformed Rocks: A Photographic Atlas* by L. E. Weiss. Published with permission of Springer-Verlag, New York, copyright © 1972.) (*B*) Slightly distorted crinoid stems in limestone, Appalachian Plateau sector of New York State. [From Engelder and Engelder (1977). Published with permission of Geological Society of America and authors.] (*C*) Distorted sand volcanoes in low-grade metasedimentary rocks, Meguma Group, Nova Scotia, Canada. (Photograph by S. J. Reynolds.)

Primary folding occurs in lavas as well. For example, rhyolitic to rhyodacitic flows are of such high viscosity that the frictional resistance to flow produces internal shear. The shear distorts the lava and produces, among other structures, **asymmetric intraformational folds**. Progressive deformation with continued flow can result in the refolding of earlier formed folds (Figure C.11).

Primary structures have other uses as well. They can be guides to the flow direction of rivers, ocean currents (Figure C.12), wind currents, lava

Figure C.15 Nearly vertical beds of Triassic Moenkopi Formation display pervasive ripple markings in the plane of bedding. These beds were "cranked up to vertical" along the confluence of the East Kaibab and Grandview monoclines, in the Grand Canyon region. Note knife for scale. (Photograph by G. H. Davis.)

(Figure C.13), or slumping. Furthermore, when primary structures *of known original shape and size* become distorted as a result of **secondary deformation** after lithification, they can provide a basis for quantitative strain analysis (Figure C.14).

Finally, the real joy of dealing with primary structures comes while mapping steeply dipping, even vertical beds, where there is complete exposure of primary structures displayed in bedding (Figure C.15). Such outcrop areas underscore the importance of being able to tell top from bottom, and in some cases they lend themselves to reconstructing paleocurrent directions. It is these kinds of exposures that really stimulate the imagination.

D. MEASURING STRUCTURAL ORIENTATIONS WITH A COMPASS

There are countless oriented structures in rocks, each with its own peculiar geometry. This is why a compass is *the* indispensable tool in geologic mapping. Measurements of the three-dimensional orientations of geological features provide the backbone of structural analysis.

The Brunton compass is the standard compass used by geologists in the United States (Figure D.1*A*), but the Silva compass (Ranger) is becoming increasingly popular (Figure D.1*B*). Each instrument is equipped with

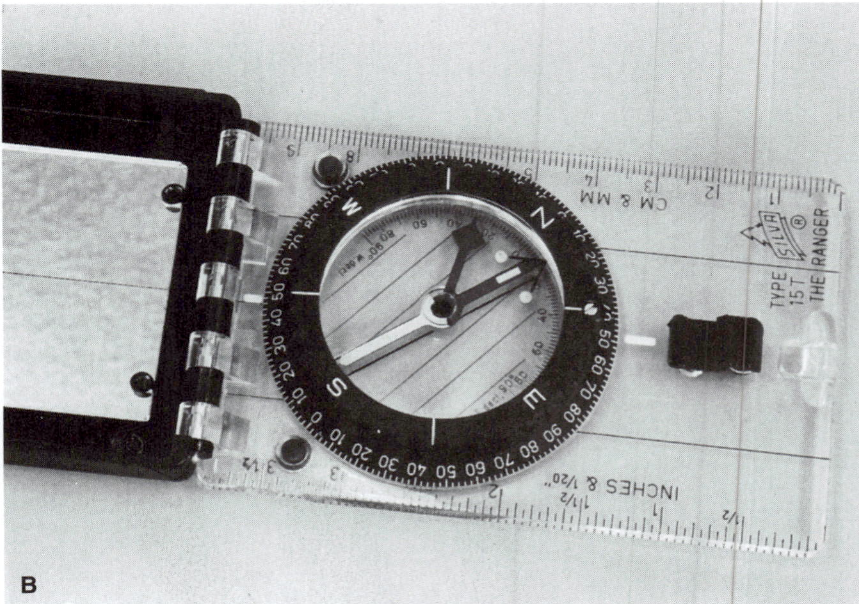

Figure D.1 (*A*) The Brunton compass. (Courtesy of the Brunton Company.) (*B*) The Silva compass (Ranger). (Photograph by G. Kew.)

the means to set **magnetic declination**, the angle between true north and magnetic north for a specific locality. When we take compass readings, we make sure that magnets, rock hammers, and metal clips are a safe distance away.

Trend and Plunge

Compasses are used in structural analysis to measure trend and inclination. **Trend** refers to the azimuth or bearing of a line. **Azimuth** is measured in degrees clockwise from north (e.g., 120°; 267°). **Bearing** is measured in degrees east or west from north or south (e.g., N60°E; S21°W). **Inclination** is the angle, measured in degrees, between an inclined line and horizontal. Its value may range from 0° to 90°.

Figure D.2 depicts the trend and inclination of a rather unusual line, the tallest human ladder ever constructed on a steeply inclined fault surface. The trend is determined by projecting both the "foot" of the ladder and the "head" of the ladder **vertically upward** into a common horizontal plane, and connecting these points of projection. The azimuth of this line is measured with a compass. The inclination of the human ladder is the angle between the line of bodies and horizontal, *as measured in a vertical plane*.

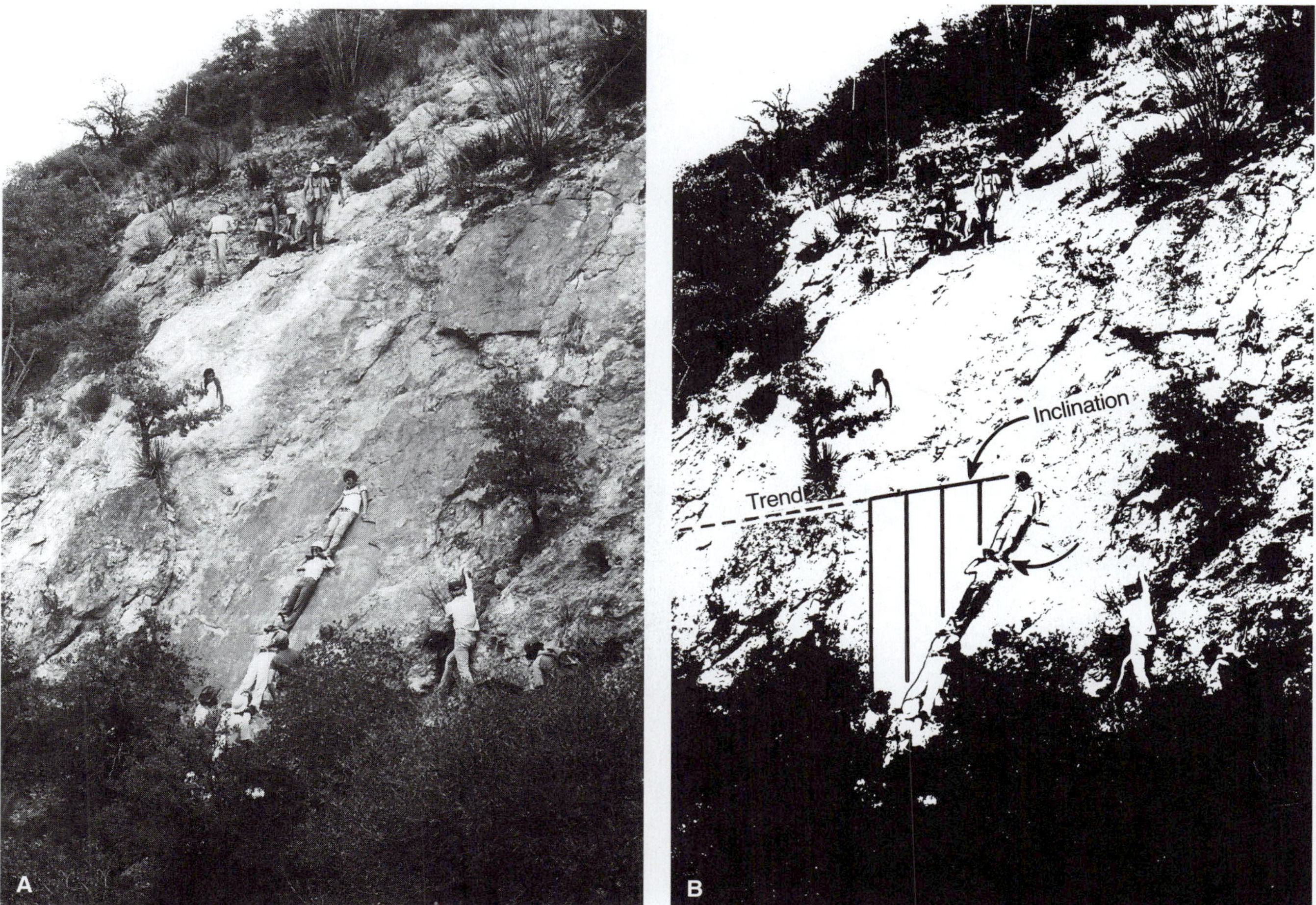

Figure D.2 (*A*) Human ladder constructed on face of a steeply inclined fault, near Patagonia, Arizona. (Photograph by G. H. Davis.) (*B*) Azimuth and inclination of human ladder.

Figure D.3 Chris Menges (mustache and glasses) proudly shows off the exceptionally well-exposed, grooved fault surface that he mapped near Patagonia in southern Arizona (see Menges, 1981). Chris' hands rest on crest of a convex groove. Richard Gillette's foot (*far left*) is squarely placed in trough of concave groove. Nancy Riggs measures the strike azimuth of a part of the fault surface. (This is the same fault surface on which the human ladder of Figure D.2 was constructed.) (Photograph by G. H. Davis.)

In practice, the orientation of a line in space is expressed in terms of **trend and plunge**, where plunge is a measure of inclination. Geologic lines, like grooves on a fault surface (Figure D.3), are called **linear elements**. The trend of a linear element is measured by holding the compass level while aligning its edge parallel to the direction of the line (Figure D.4*A*). The compass is pointed parallel to the **vertical projection** of the line onto

Figure D.4 Steps in measuring trend and plunge. See text for details.

an imaginary horizontal plane. Trend is read directly from the Brunton compass after the compass needle has come to rest (Figure D.4*B*). To measure the azimuth of trend using the Silva compass, the calibrated outer ring on the face of the compass must be rotated until the rotatable outline of the compass arrow coincides with the actual free-spinning magnetic needle (Figure D.4*C*).

Plunge is measured by turning the compass on its side and aligning its edge along the linear element, or parallel to it (Figure D.4*D*). If the Silva compass is used, a plumb-boblike inclination needle points automatically to the value of the plunge when the compass is thus oriented (Figure D.4*E*). (The outer dial of the Silva has to be set on 90° or 270° so that the inclinometer reads correctly.) The Brunton compass does not have a free-swinging inclination needle. Instead, a calibrated scale known as a **clinometer** and located inside the compass is used to measure plunge. The clinometer, attached to a small carpenter's level, can be moved back and forth by means of a lever on the outside base of the compass (Figure D.4*F*). Holding the Brunton such that its edge is positioned parallel to the line being measured, the clinometer lever is moved until the bubble in the carpenter's level becomes centered (Figure D.4*F*). The value of plunge is read directly using the inner scale embossed on the inside base of the compass.

The full description of the trend and plunge of a line in space can be recorded in two different ways. For example, 20° N60°E refers to the orientation of a line that plunges 20° along an azimuth 60° east of north; N60°E is the *sense* of direction of the down-plunge end of the line. Similarly, 45° S20°E describes the orientation of a line plunging 45° in a direction that is 20° east of south. For a compass calibrated in azimuth from 0° to 360°, these measurements would be recorded as 20°/060° and 45°/160°, respectively.

Strike and Dip

Measuring the orientation of a planar feature is handled differently. If the orientations of two lines are known, the orientation of the plane that contains these lines is also known. Two lines determine a plane. The measuring of orientations of bedding planes, fault planes, dikes (Figure D.5*A*), and other geological **planar elements** is based on this relationship. If the orientations of two lines that lie in a plane can be established, the orientation of the plane itself is established as well. Any two lines will do, as long as they are not parallel or close to being parallel. For convenience, the two lines in a plane that are chosen are a horizontal line and the line of steepest inclination (Figure D.5*B*). These two lines are at right angles to each other. The first is called the **line of strike**; the second is the **line of dip**. For the special case of a strictly horizontal plane, all lines are strike lines.

Strike and dip are the measurements required to define the orientation of a plane. **Strike** is the trend of the line of strike, that is, the trend of a horizontal line in a plane. Because its inclination is by definition 0°, the value of strike is recorded simply in terms of degrees of azimuth or bearing. For compasses with trend calibrated by quadrant, strike is always expressed in terms of north: N72°E, N68°W, N1°W. For compasses calibrated in azimuth from 0° to 360°, azimuth is best presented as degrees between 0° and 90°, or between 270° and 360° in order to keep all strike measurements in the northern half of the compass. This helps us more easily to compare successive readings in the field, instead of wondering for a moment why the last reading was 282° and this one was 102°. The

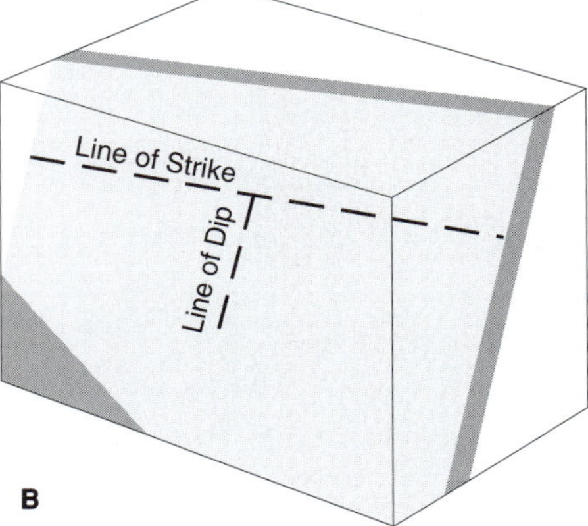

Figure D.5 (*A*) Aplite dike (white) in granite. The planar dike is exposed in both plan and cross-sectional views. (Photograph by G. H. Davis.) (*B*) Schematic view into the granite within which the dike occurs. The orientation of the dike is expressed in terms of orientations of the lines of strike and dip.

dip of a plane is the inclination of the line of dip, that is, the line of steepest inclination in a plane. It is recorded in terms of inclination angle and the dip of the plane (SW, NW, NE, SE, N, S, E, W). The specific azimuth of dip direction is not directly measured because its value can be determined from the strike of the plane. For example, a N35°W-striking plane dips either N55°E or S55°W. Only the general value of dip direction is recorded in the notebook: SW or NE. This distinction permits the two possible dip directions to be distinguished.

The procedure for measuring the strike and dip of a plane using a Brunton compass is as follows (Figure D.6). To find the line of strike, first set the clinometer to 0° so that the compass may be used as a carpenter's level (Figure D.6*A*). Place a side or edge of the compass flush against the plane (Figure D.6*B*), or against a field notebook or nonmagnetic clipboard held parallel to the plane; rotate the compass until the carpenter's level bubble is centered. When centered, the edge of the compass held against the plane is horizontal and oriented parallel to the line of strike. To determine azimuth of the strike line, rotate the compass downward (Figure D.6*C*), still keeping the lower-side edge of the compass fixed against the surface until the bull's-eye bubble is centered. The compass needle now swings freely. Dampen the compass needle to stop it from swinging and read the azimuth (Figure D.6*D*).

To measure the strike of a plane using a Silva compass, it is useful to carry an auxiliary level, or to attach a small level to the compass itself. The level can be used to quickly identify the line of strike. Once found, its orientation can be measured by aligning the edge of the Silva parallel to the line of strike and rotating the calibrated outer ring until the reference compass needle is aligned with the actual magnetic needle (Figure D.6*E*).

To measure dip with a Brunton, place the compass on a side face on the inclined plane such that the compass is aligned in the direction of the line of dip (Figure D.6*F*). Then measure the inclination of the line by rotating the clinometer until the carpenter's level bubble is centered. The Silva compass has an inclination needle for measuring dip directly, as long as the outer ring is set at E or W to properly align the inclinometer and the edge of the compass (Figure D.6*G*).

Strike-and-dip readings can be taken by the **sighting method** as well. This method is especially useful when beds or layers do not crop out as convenient resistant planes for direct measurement, and/or when attempting to measure the average strike and dip of an area of rock that is larger than outcrop size. The method is illustrated in Figure D.7. The trick is to position yourself in the proper location so that your line of sight is a strike line in the plane of the dipping layer whose orientation is being determined. When viewed in this way, the dipping layer appears in strict cross-sectional view, with no expression of the surface of the layer. The azimuth of the line of sight constitutes the strike of the dipping layer. The inclination of the layer as seen from this unique line of sight is true dip. Watch where you stand!

Recording the Measurements

Strike-and-dip and trend-and-plunge orientations measured in this way are recorded in the field notebook. Representative readings are placed on the geologic map as well. If a Brunton compass is used, a protractor or protractor scale is used to accurately plot the strike-and-dip readings. The base map is scribed lightly with penciled N–S guidelines so that strike or trend can be measured and plotted with relative efficiency. The Silva compass has the added advantage of being functional as a protractor for

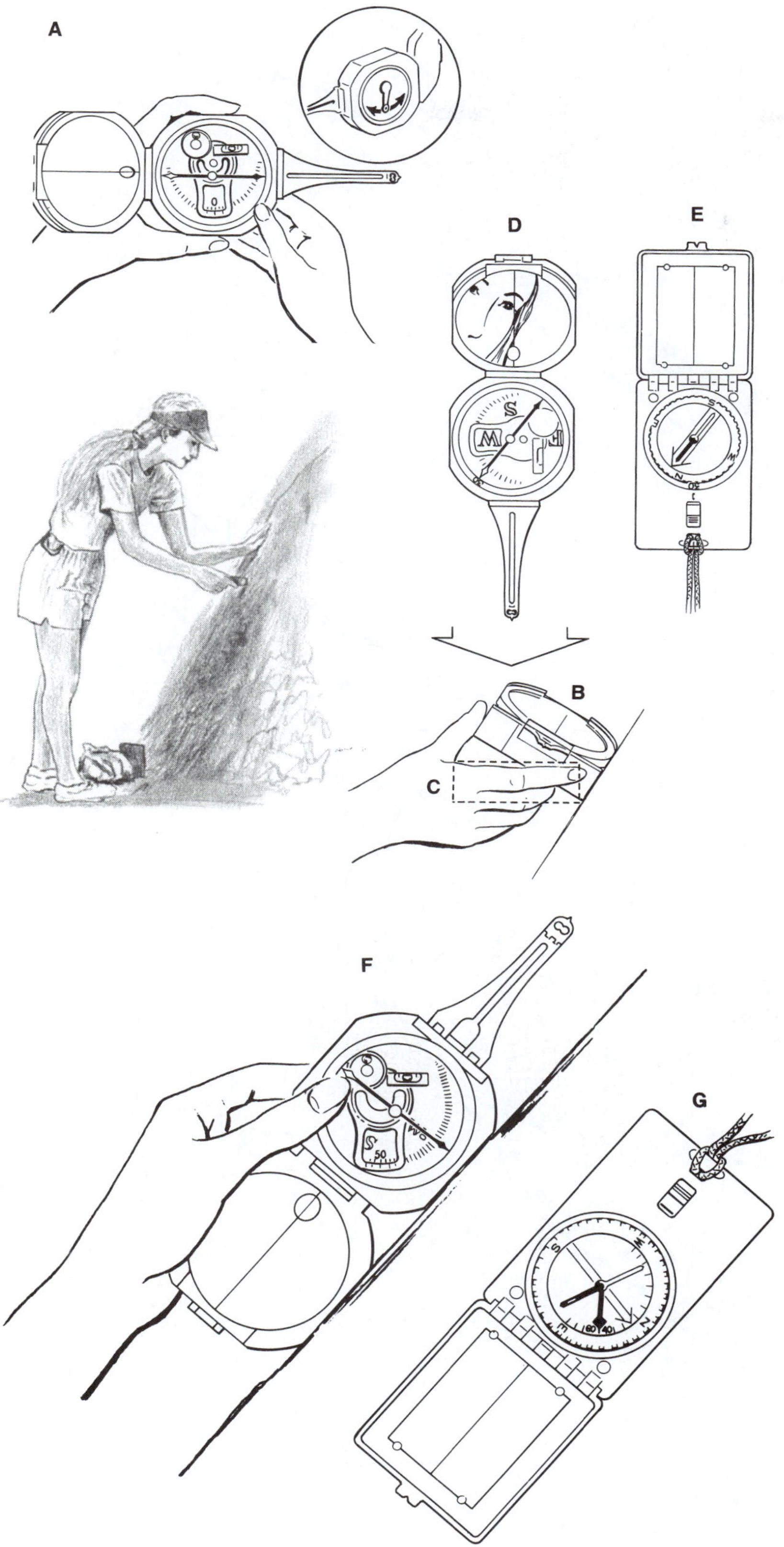

Figure D.6 Steps in measuring strike and dip. See text for details.

Figure D.7 The do's and don'ts of the sighting method for measuring strike and dip.

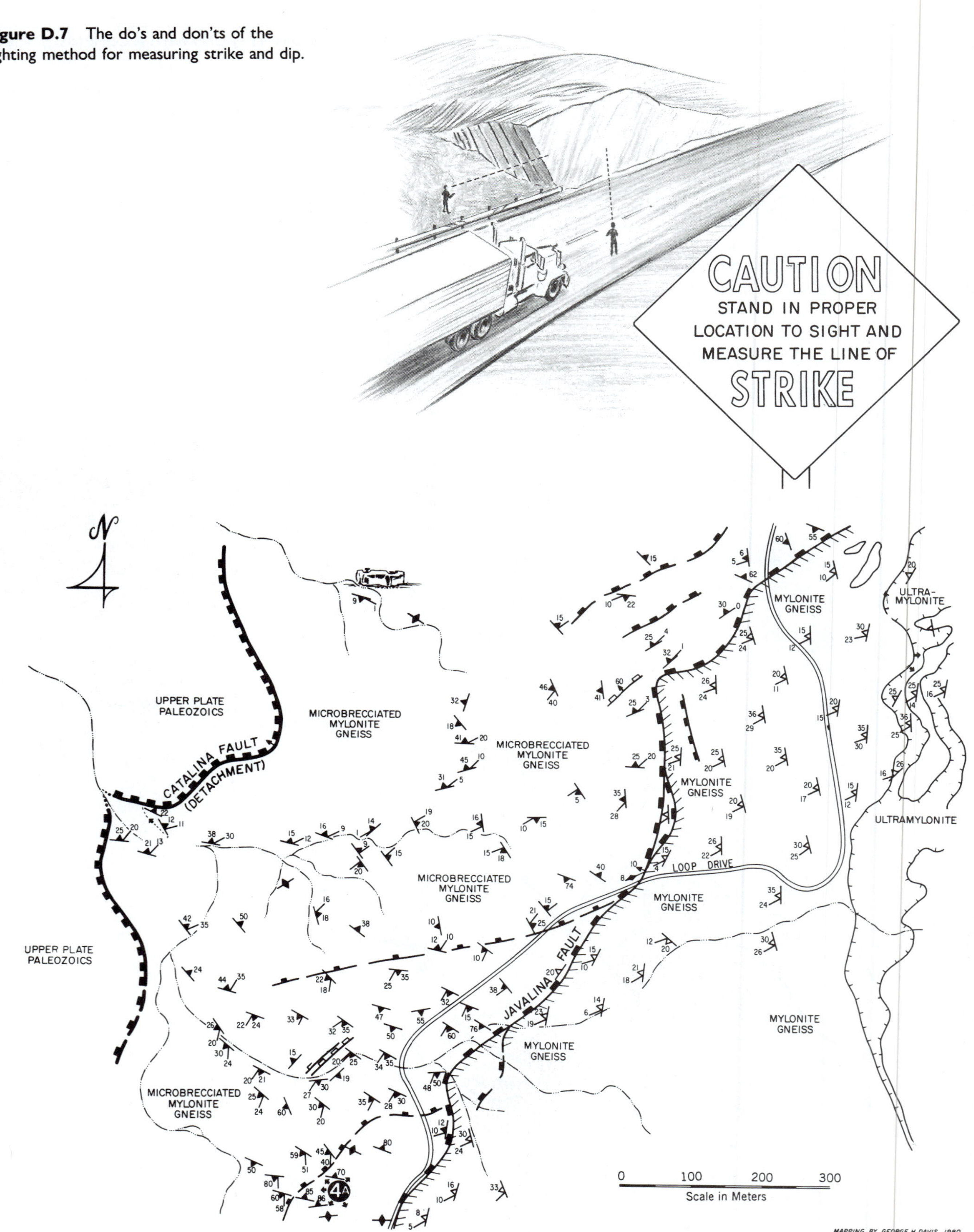

Figure D.8 A plethora of strike-and-dip readings brings geometric life to maps. This map, rendered by G. H. Davis, shows internal structure within mylonitic and cataclastic rocks in the Saguaro National Monument (East) area, Tucson, Arizona. [From Davis (1987). Published with permission of Geological Society of America.]

plotting strike or trend on the base map. Without disturbing the compass setting for trend or strike, the compass is placed on the map in such a way that the red lines on the inner base of the compass coincide with the N–S guidelines scribed on the map. When this is accomplished, the straight edge of the compass matches the azimuth of strike or trend.

Measurements collected and plotted in this way give geometric life to the geologic map (Figure D.8). There emerges from the map a physical and geometric expression of the form and internal structure of the rock formations. Furthermore, the orientation measurements stored in the field notebook become the basis for subsequent analysis and interpretation. Finally, the distribution of plotted measurements on the maps gives the map reader an idea of which areas have exposures and were "covered" by the geologist who made the map.

E. PREPARING GEOLOGIC CROSS SECTIONS

Geologic maps are used as a basis for constructing **geologic cross sections**, which can be thought of as vertical slices through a map area showing a **profile view** of the subsurface structure (Figure E.1). Geological cross sections reveal interpretations of what lies below the surface. This is where "the rubber meets the road," because geologic cross sections get most of the attention in spotting drill holes in the exploration for oil and gas or geothermal energy; in locating the best drill sites or exploration shafts in the search for metals; in interpreting the subsurface expression of active fault zones; or in communicating the fundamental structural style within mountain belts, like the Alps.

Drawing realistic geologic cross sections is very difficult, and the construction steps have become quite sophisticated. The difficulty is that the geologic map, on which the cross section is based, only describes the relationships right at the surface. Therefore, the locations and compass-measured orientations of faults, intrusive contacts, unconformities, bedding, and other structures hold only for the surface of the area, and it cannot be presumed that they hold for any depth below the surface. As a consequence, the subsurface geology somehow must be filled in in a way that is everywhere consistent with the **geologic control** at the surface, and consistent with the style of deformation seen in the map area (Figure E.2). If we are very fortunate, there may be some subsurface information

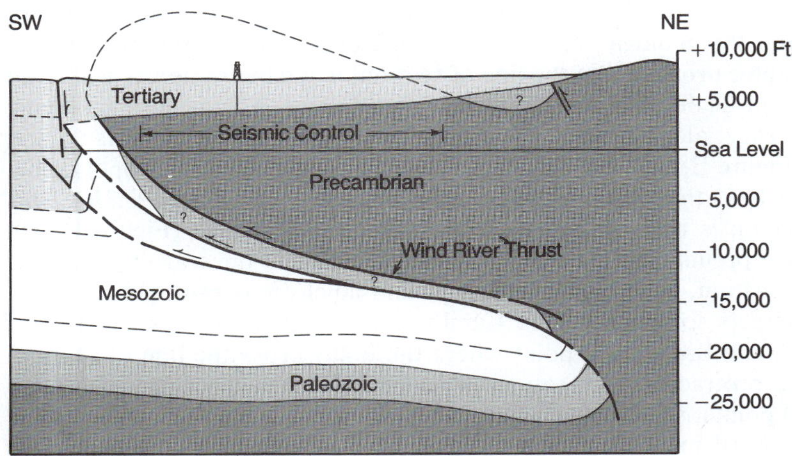

Figure E.1 Geologic cross section of the Wind River Range, Wyoming. [From Berg (1962). The "control" for the section includes surface exposure, drilling, and seismics. Published with permission of American Association of Petroleum Geologists.]

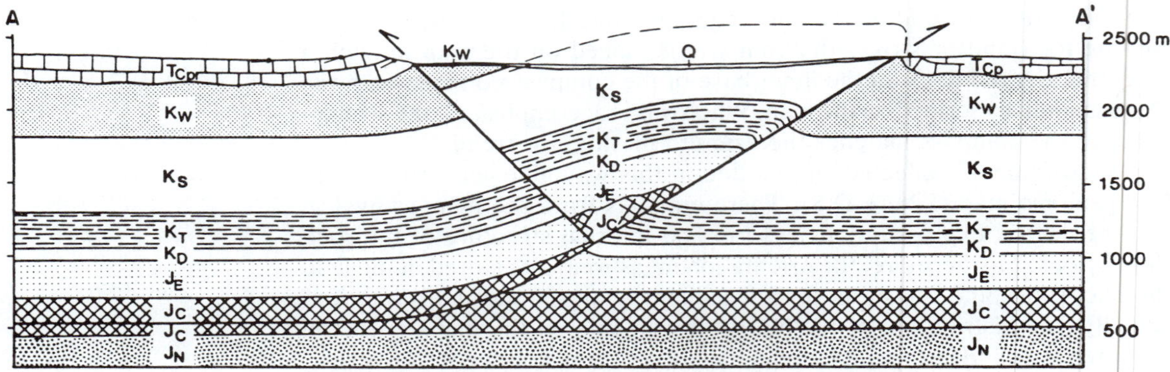

Figure E.2 Geologic cross sections constructed by Eric Lundin (1989) across thrust-faulted strata in the Bryce Canyon region, Utah. Drilling and seismic information reveal that the major thrust "soles" into weak salt horizons within the Jurassic Carmel Formation. [From Lundin (1989). Published with permission of Geological Society of America.]

available to us, such as the locations of key rock formations or structures at depth, based upon drilling information, underground mapping, or seismics. But most often it works the other way: locations of key rock formations and structures at depth are predicted exclusively on the basis of the cross section.

Steps in Constructing Cross Sections

Let us take a look at the basic steps in preparing a geologic cross section for an area marked by straightforward geologic relationships. The geologic map (Figure E.3*A*) shows that the area of interest contains three formations of sedimentary rock of Cambrian age, resting nonconformably on Precambrian granite. The Cambrian rocks and the nonconformity are inclined approximately 30° to the east. Two north–south trending faults, each inclined 60° to the west, cut and displace the rock units, west-side-down (see Figure E.3*A*). We are interested in showing the geologic relationships as we believe they might exist in the subsurface.

First, select a **line of section** (line *A-A'* in Figure E.3*A*) that is strategically placed along a line that trends east–west, perpendicular to the trend of the faults, and perpendicular to the trend of the Cambrian rock formations and the major unconformity. Let the east–west-trending line of section pass through a part of the area where the rock exposures are good and the geologic control at the surface is solid. Next, construct a **topographic profile** along the line of section, using the same vertical scale as the map scale, that is, *no vertical exaggeration*. This is done by laying a work sheet along the line of section, transferring the end points (*A* and *A'* on Figure E.3*A*), and transferring and labeling each of the topographic contour lines that crosses the line of section as well (Figure E.3*B*). This information is used to construct the topographic profile (Figure E.3*B*). The final preliminary step is to lay the work sheet once again along the line of section, and this time transfer and label the **control points** where the contacts for each of the Cambrian formations crosses the line of section, and the control point where the faults cross the line of section. Using a protractor, plot *short* **control lines** from each of the respective control points, showing the 60° westward inclination of the faults and the 30° eastward inclinations for each of the Cambrian formations and the

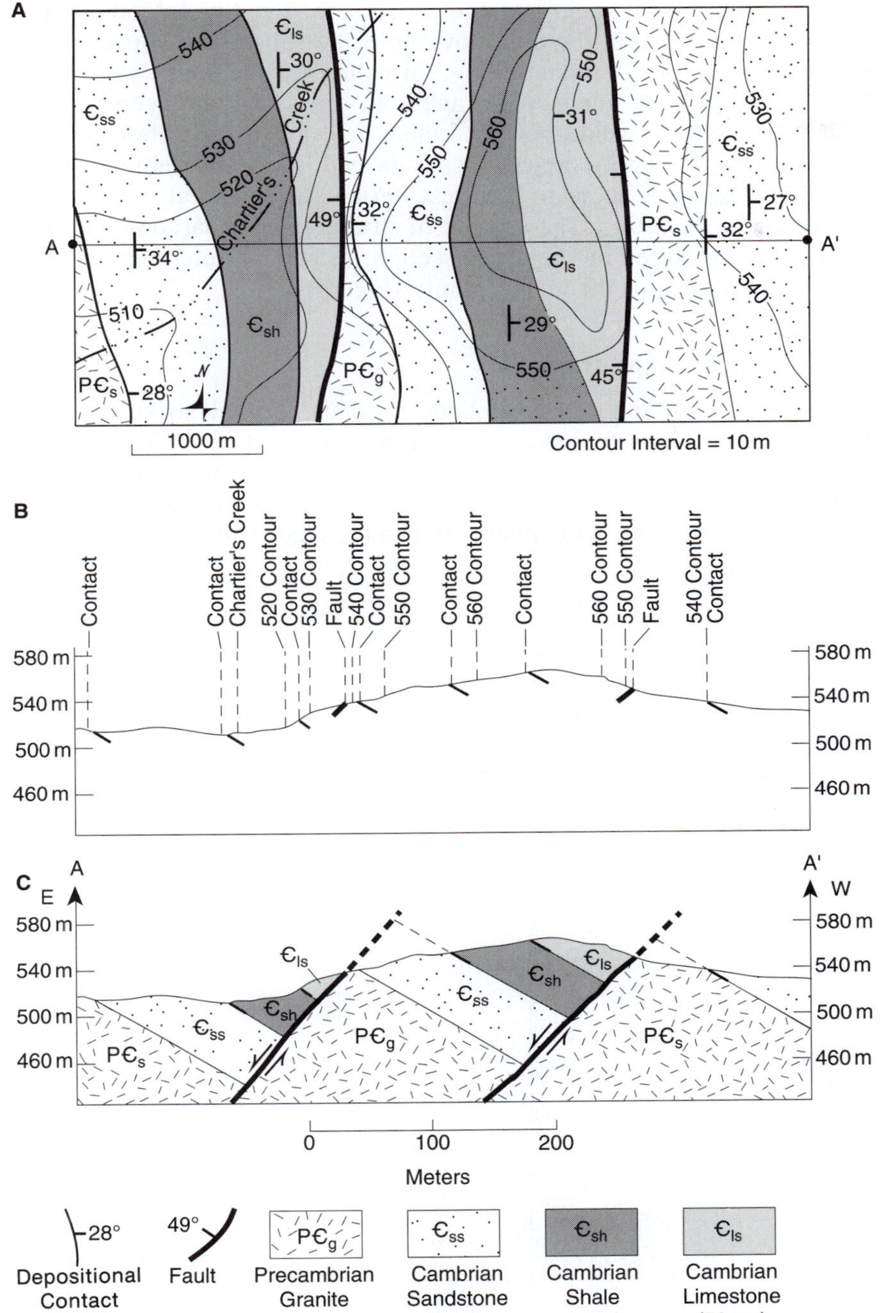

Figure E.3 Basic steps in constructing a simple geologic cross section. See text for explanation.

unconformity (Figure E.3*B*). With this step, the topographic profile starts to become a geologic cross section.

To fill in the geology at depth, extend each of the rock formations along the line of section down into the subsurface along the 30° eastward trajectory (Figure E.3*C*). Unless you have reason to do otherwise, do not change the thickness of any of the formations as you project them into the subsurface. Next you might extend the faults into the subsurface along the 60° westward trajectory, using a heavier hand, and thus a heavier line, than that for the formation contacts (see Figure E.3*C*). Again, unless you have reason to do otherwise, do not change the angle of inclination of the faults as you project them into the subsurface. Finally, extend each of the rock formations in the area down into the subsurface along the 30° eastward trajectory, terminating the formations at the faults (see Figure E.3*C*). Add color or line symbols and your cross section is complete.

If asked, "How would you determine how much separation there is on the faults?," simply project (using dashed lines) the Cambrian rock formations up into the air to the west, terminating them at the skyward projection of the faults. The amount of stratigraphic separation can then be directly measured on the cross section, using as a reference any one of the formations, or the unconformity (Figure E.3).

Correcting for Apparent Dip

Most lines of section will be oblique to some of your strike readings. Thus, it is necessary to make a correction from the true dip measured in the field to the apparent dip that should show up on your geologic cross section. There are two practical ways to do this. One is to do it stereographically (see Section H). The other is to use the chart shown in Figure E.4.

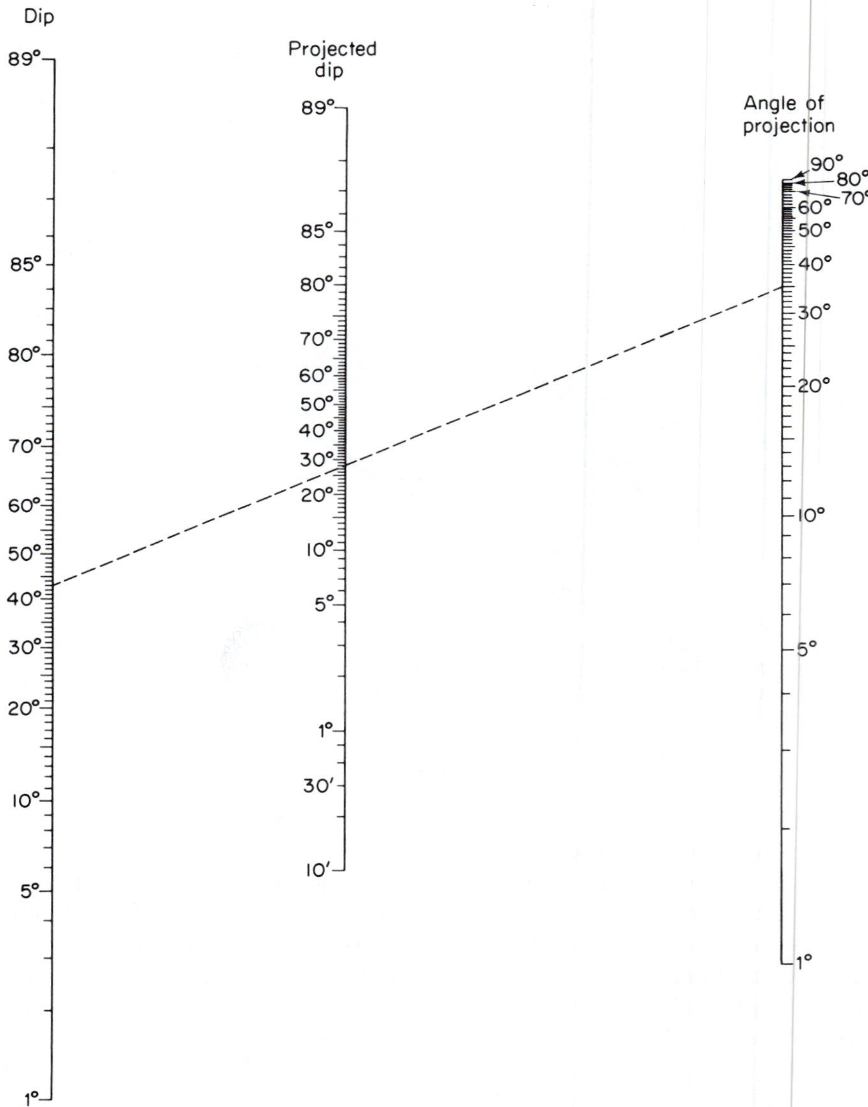

Figure E.4 Chart used to convert from true dip to apparent dip when constructing geologic cross sections. If the true dip is 43°, the apparent dip on a vertical section making a 35° angle with the strike of the bedding would be 28°. [From Billings, M. P., STRUCTURAL GEOLOGY, © 1972, p. 523. Reprinted by permission of Prentice Hall, Upper Saddle River, New Jersey.]

The point is this: Only sections that are perpendicular to strike will reveal true dip. All other sections will reveal apparent dip, which will be *less* than true dip. To use the correction chart presented in Figure E.4, measure with a protractor on the geologic map the difference in bearing between the line of section and the line of strike. Find this value in the right-hand column in Figure E.4. Lay a ruler from this value, let it pass through the magnitude of true dip in the left-hand column. The ruler passes through the middle column at the value of the apparent dip.

Preparing Vertically Exaggerated Sections

Vertically exaggerated sections are sometimes used to portray certain types of geology, such as thin, gently dipping units. These are awkward to prepare. Not only must the topography be exaggerated (i.e., stretched in a vertical direction), but the dips of all formation contacts and faults must be recalculated and increased for the new scale. An easier way to achieve an exaggerated section is to draw it at true scale, without exaggeration, and then scan it, bring it into a computer drawing program, and stretch it to the desired exaggeration (2 times exaggeration will be twice the original height). Presto! now you have a vertically exaggerated section *and* a true scale one!

Constructing Normal Profiles of Folds in Vertical Sections

Geometric analysis of the shapes of individual folded surfaces and folded layers requires views of folds in normal profile. In areas or regions where folds are nonplunging, normal profile views are obtained by constructing vertical structure sections. The base of control is a geologic map (Figure E.5). The **line of section** along which the structure section is to be fashioned is laid out perpendicular to the average trend of fold axes, an average

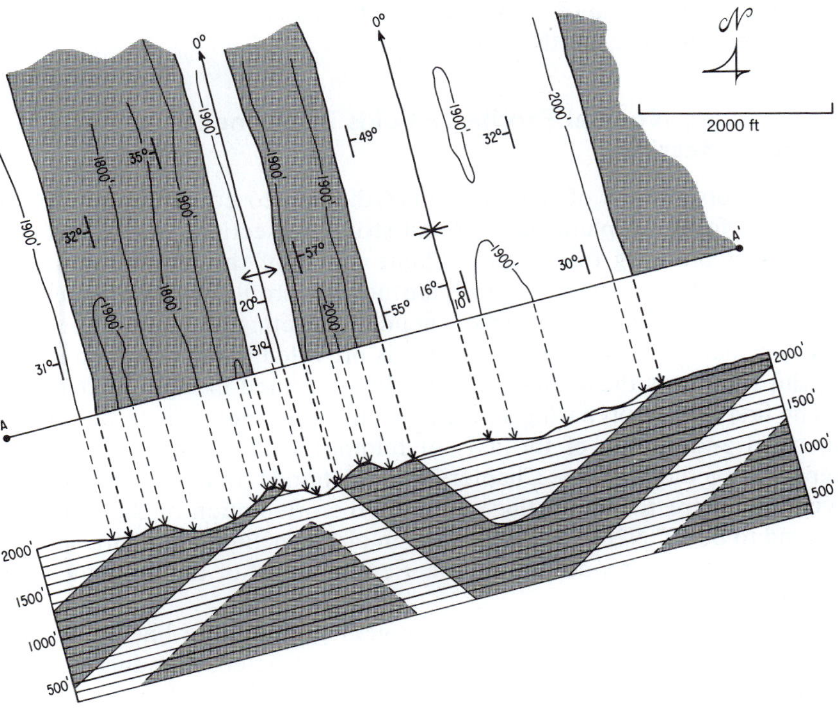

Figure E.5 Preparation of a vertical structure section on the basis of geologic map relationships. See text for details.

determined stereographically (see Section H) if necessary. End points of the line of section are marked on the map (A, A', Figure E.5). The line of section is positioned where strike-and-dip data for the folded layering are reasonably abundant and where the fold patterns are especially interesting and informative.

Once the line of section has been chosen and positioned on the geologic map, a topographic profile is constructed using elevation control afforded by topographic contour lines (Figure E.5). The topographic profile, like the structure section itself, is drawn without vertical exaggeration, to avoid distortion of the true form of the folded layers. To the topographic profile are added the exact locations where contacts between formations and marker beds cross the line of section.

Strike-and-dip data posted on the geologic map nearest the line of section provide a means to gauge the direction and angle of inclination of each geologic contact that is to be portrayed in the structure section. Where the strike of layering is perpendicular to the line of section, true dip is plotted (by protractor) in the structure section. Where the strike of layering is oblique to the line of section, apparent dip must be plotted. Apparent dip can be computed stereographically, or by using the chart presented in Figure E.4.

The inclinations of the upper and lower surfaces of each folded layer, as measured in outcrop, are represented on the structure section by short **control lines** that are plotted with a protractor at appropriate locations (Figure E.5). No matter how the pattern of folded layers is portrayed at depth, each contact of each folded layer must emerge at the surface of the section along one of the control lines.

The manner in which the folded layers are portrayed in the subsurface (e.g., amount of thickening and thinning) is guided by field observations regarding the response of each folded layer to the distortional influence of folding. Similarly the kinkiness versus the roundness of fold hinges is influenced in part by the styles of folding directly observed. As a final check on the internal consistency of the subsurface interpretation, the trace lengths of the midlines of each of the folded layers are measured to determine whether the section is **balanced**. An example of the 'balancing act' is presented later in this section.

Constructing Normal Profile or Folds in Inclined Structure Sections

Normal profile views of folds are more difficult to construct in regions where the folds are plunging. Vertical structure sections drawn through terranes of plunging folds yield oblique views of the shapes of folded layers and folded surfaces. To obtain normal profile views of plunging folds it is necessary to construct inclined structure sections. The data base for constructing inclined sections is provided by geologic map relationships, such as those diagrammatically presented in Figure E.6*A*.

The first step in preparing an inclined structure section is to determine the trend and plunge of the axis of folding. This can be achieved by preparing a π diagram (see Chapter 7 and Section H) of poles to the folded layering (Figure E.6*B*). In this example, the orientation of the fold axis is found to be 40° N26°E. The orientation of the inclined structure section that would serve as a normal profile view of a fold of this orientation is N64°W, 50° SW.

The next step is to position on the geologic map the line of section where the inclined structure section is to be constructed. Trending N64°W, the line of section is shown as line AB in Figure E.6*C*. Line AB is the

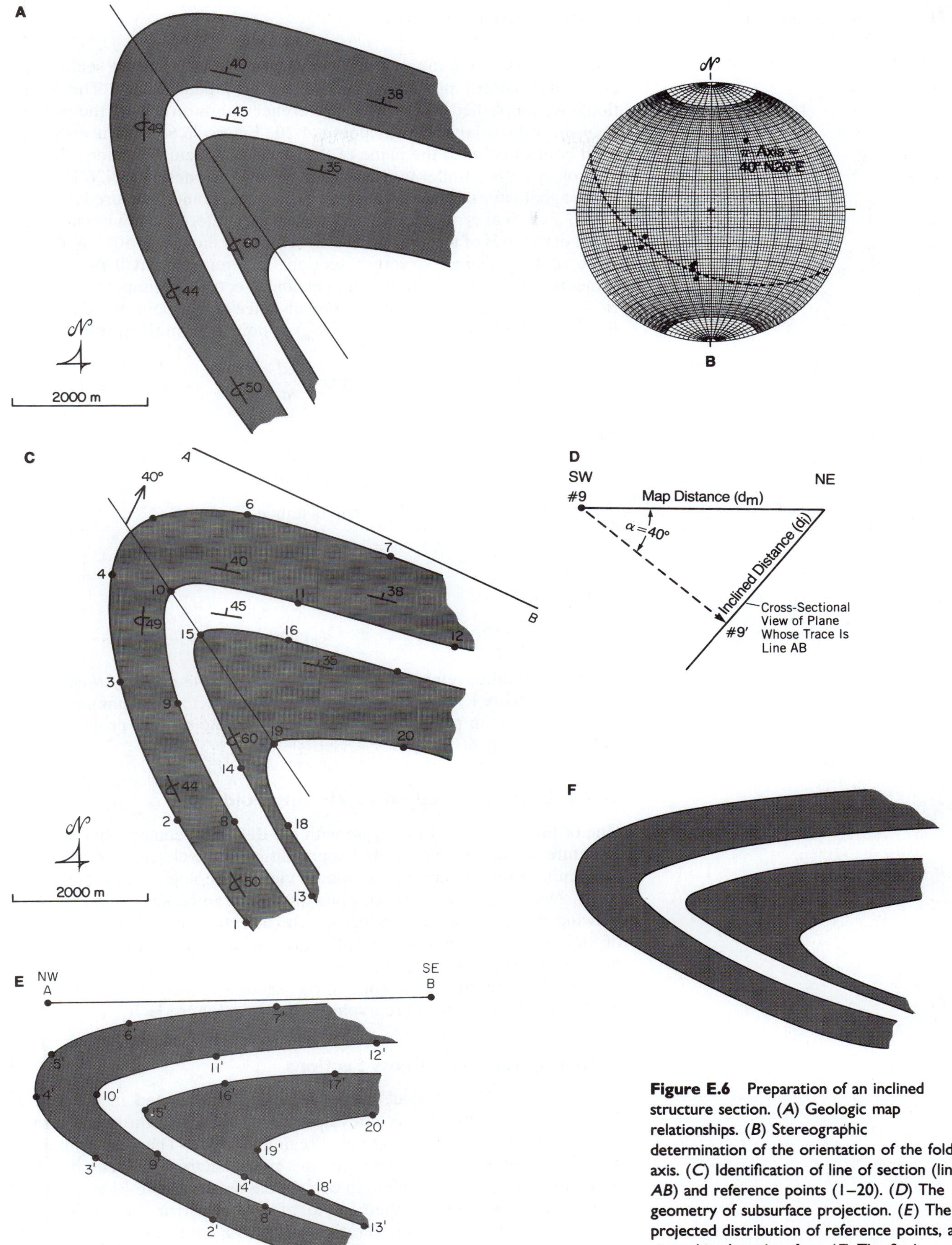

Figure E.6 Preparation of an inclined structure section. (*A*) Geologic map relationships. (*B*) Stereographic determination of the orientation of the fold axis. (*C*) Identification of line of section (line *AB*) and reference points (1–20). (*D*) The geometry of subsurface projection. (*E*) The projected distribution of reference points, as viewed in the subsurface. (*F*) The final structure section.

trace of a plane that dips 50° SW. In preparing the structure section, the entire map pattern must be projected into this inclined plane. The projection is accomplished by identifying reference points on each of the contacts between folded layers (e.g., points 1-20, Figure E.6*C*), then projecting each control point to the plane of the inclined structure section. The line of projection is parallel to the axis of the fold, namely 40° N26°E.

The geometry of projection of each reference point is pictured in Figure E.6*D*, a vertical cross section passing through reference point 9 along the trend of the axis of folding. The vertical section shows the 50° SW-dipping trace of the inclined structure section. Reference point 9 on the map projects to reference point 9′ in the inclined section. A simple trigonometric relationship connects the **inclined distance** (d_i) of point 9′ beneath the line of section to the **map distance** (d_m) of point 9 from the line of section:

$$\sin \alpha = \frac{d_i}{d_m}$$

where

$$d_i = \text{inclined distance}$$

$$d_m = \text{map distance}$$

$$\alpha = \text{plunge of fold axis}$$

Thus

$$d_i = d_m(\sin \alpha)$$

Point 9′ and other reference points are shown *in the plane of the inclined section* in Figure E.6*E*. These points can be connected in a manner consistent with the map pattern. Once this is done, the normal profile springs to life (Figure E.6*F*).

Down-Structure Method of Viewing Folds

One of the incentives to struggle with the details of constructing inclined structure sections is to gain a full appreciation of the elegance of J. Hoover Mackin's "down-structure method of viewing geologic maps" (Mackin, 1950). Normal profile views of plunging folds can be seen at a glance by viewing the geologic map patterns in the direction of plunge, at an angle of inclination of view corresponding to the amount of plunge. Try it on the map pattern shown in Figure E.7*A*, comparing what is seen to the graphically constructed geologic cross section (Figure E.7*B*). Then try out the method on the hieroglyphics shown in Figure E.7*C*.

Preparing Balanced Cross Sections

The concept of "balanced" cross sections is presented in Chapter 6, "Faults." The methods were developed for folded and thrusted regions, but they can be applied to extended regions as well. There are important guidelines in constructing even the most rudimentary balanced cross sections, not the least of which is to carry a big eraser (Dave Bice, personal communication, 1993), for there is a lot of trial-and-error work in doing the procedure by hand. It is an iterative process. Fortunately, sophisticated computer programs have been developed to make the work more manageable, but nonetheless it takes a major effort to get it right.

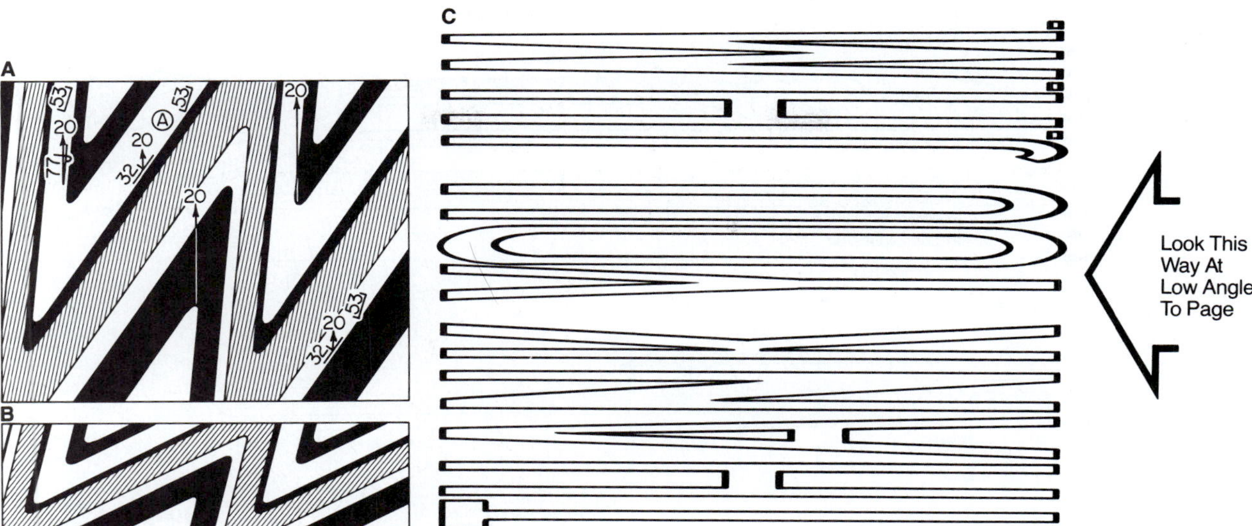

Figure E.7 (*A*) Map pattern of folds and (*B*) down-structure view of same. [From *Structural Geology of Folded Rocks* by E. T. H. Whitten, after Mackin (1950). Originally published in 1966 by Rand-McNally and Company, Skokie, Illinois. Published with permission of John Wiley & Sons, Inc., New York]. (*C*) Glance at this pattern in the down-structure view.

Perhaps the best way to visualize the preparation of balanced sections is to see examples of some good ones, such as those prepared by Shankar Mitra (1988) for the Pine Mountain thrust region in the southern Appalachians. The locations of his cross sections with respect to the major regional structures are shown in Figure E.8*A*, and three representative sections are presented in Figures E.8*B–D*. Woodward, Boyer, and Suppe (1985) and Marshak and Woodward (1988) have presented guidelines for preparing such sections. To begin, choose a line of section that lies parallel to the direction of thrusting and avoids major lateral structures, like tear faults and lateral ramps. Establish the thickness (and changes in thickness) of each stratigraphic unit. Estimate the depth to the floor thrust, sole thrust, or decollement based upon seismic information, regional geologic

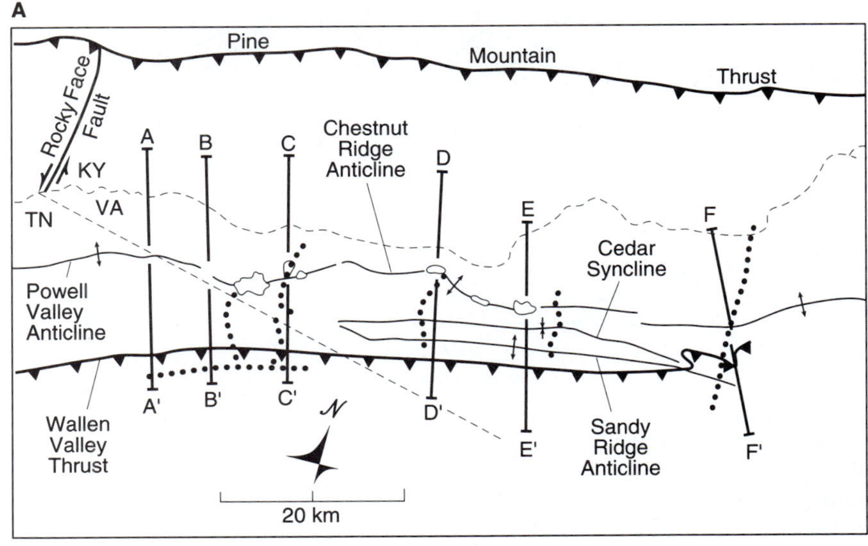

Figure E.8 (*A*) Map showing locations of the cross sections prepared by Mitra (1988). Dotted lines are seismic lines. Also shown are location of wells.

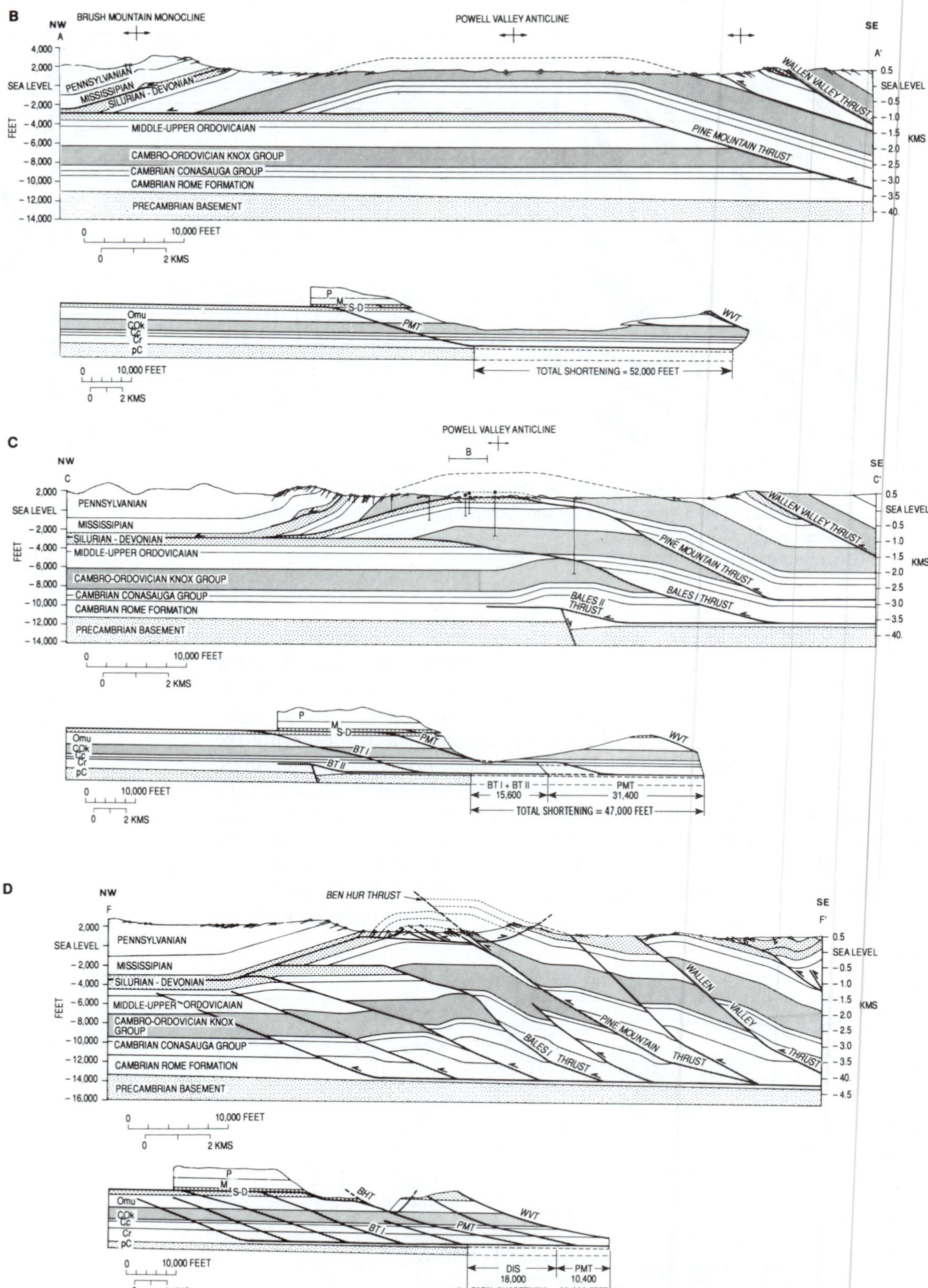

Figure E.8 (*continued*) (*B*) Balanced cross section and restored counterpart through the Wheeler area (*A–A'*). (*C*) Balanced cross section and restored counterpart through the Martin Creek window (*C–C'*). (*D*) Balanced cross section and restoration through the Big Stone Gap area (*F–F'*). [From Mitra (1988). Published with permission of Geological Society of America.]

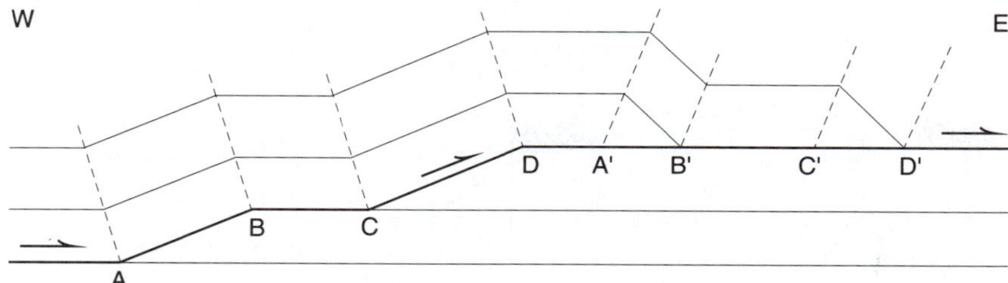

Figure E.9 Cross section drafted in a way that the hanging-wall ramps and flats can be restored such that they match the footwall ramps and flats. [After Marshak and Woodward (1988), *in* Marhsak/Mitra, BASIC METHODS OF STRUCTURAL GEOLOGY, © 1988, p. 313. Reprinted by permission of Prentice Hall, Upper Saddle River, New Jersey.]

information, or drilling information. Plot along the surface profile of the cross section the locations of fault contacts and formation contacts as well as dip information for faults and bedding.

By way of execution, identify the footwall ramps and flats for each fault at depth by using the geometry of the hanging-wall structure as a kind of template. The spacing between dip panels in the hanging-wall rock will often correspond to the lengths of flats and ramps on the footwall. Upon interpreting the flat/ramp geometry for each fault, be sure that the hanging-wall cutoffs across each formation match in length the footwall cutoffs across each formation (Figure E.9). In the words of Marshak and Woodward [(1988), p. 313]:

> . . . before deformation the shape of the base of the hanging wall is identical with the shape of the top of the footwall, and the hanging-wall cutoff of a given contact must be adjacent to the footwall cutoff of the same contact. After deformation, therefore, the shape of the base of the hanging wall that we observe in outcrop must have its counterpart along the top of the footwall somewhere at depth. Likewise, footwall shapes observed in outcrop may be representative of the shape of the hanging wall that has been eroded.

By way of testing the interpretation, restore the cross section by straightening out each of the beds to conform to the original regional configuration (see Figure E.8*B*,*C*,*D*). Take the fault traces, and the surface topography too, along for the ride. The original geographic locations where the thrusts broke up-section toward the surface will become apparent, as will the degree of regional shortening. Unsightly gaps and mismatches will appear in the restoration if the section being restored was misbalanced.

F. PREPARING SUBSURFACE CONTOUR MAPS

Subsurface exploration in the search for petroleum, metals, groundwater, and other natural resources provides an important source of data for descriptive analysis. Drilling and seismics produce data on structure and lithology in the deep third dimension. Depths at which specific rock formations are encountered become the basis for constructing **structure contour maps**. Structure contour maps describe the structural form of rock bodies at depth. Knowing the surface elevation where the drill hole is "collared," and knowing the depth to the top of a particular formation of interest, it is straightforward to determine the elevation of the top of the formation of interest in the subsurface. Elevations determined in this way can then be contoured in the same way that topographic maps are contoured (Figure F.1). Frequently such maps are used to describe the structure of sedimen-

Figure F.1 Contouring of a structure contour map on the basis of raw data giving elevations of the top of a marker bed in the subsurface. [Modified from Dutton (1982). Published with permission of American Association of Petroleum Geologists.]

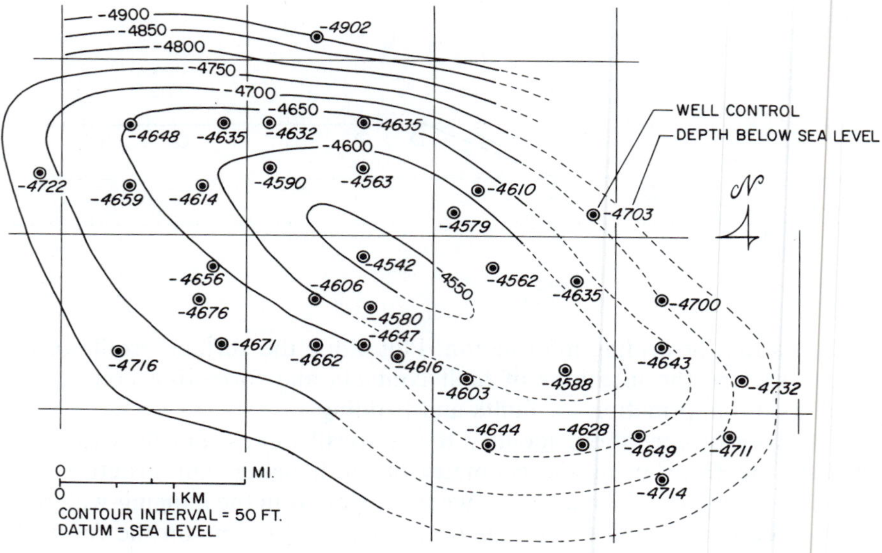

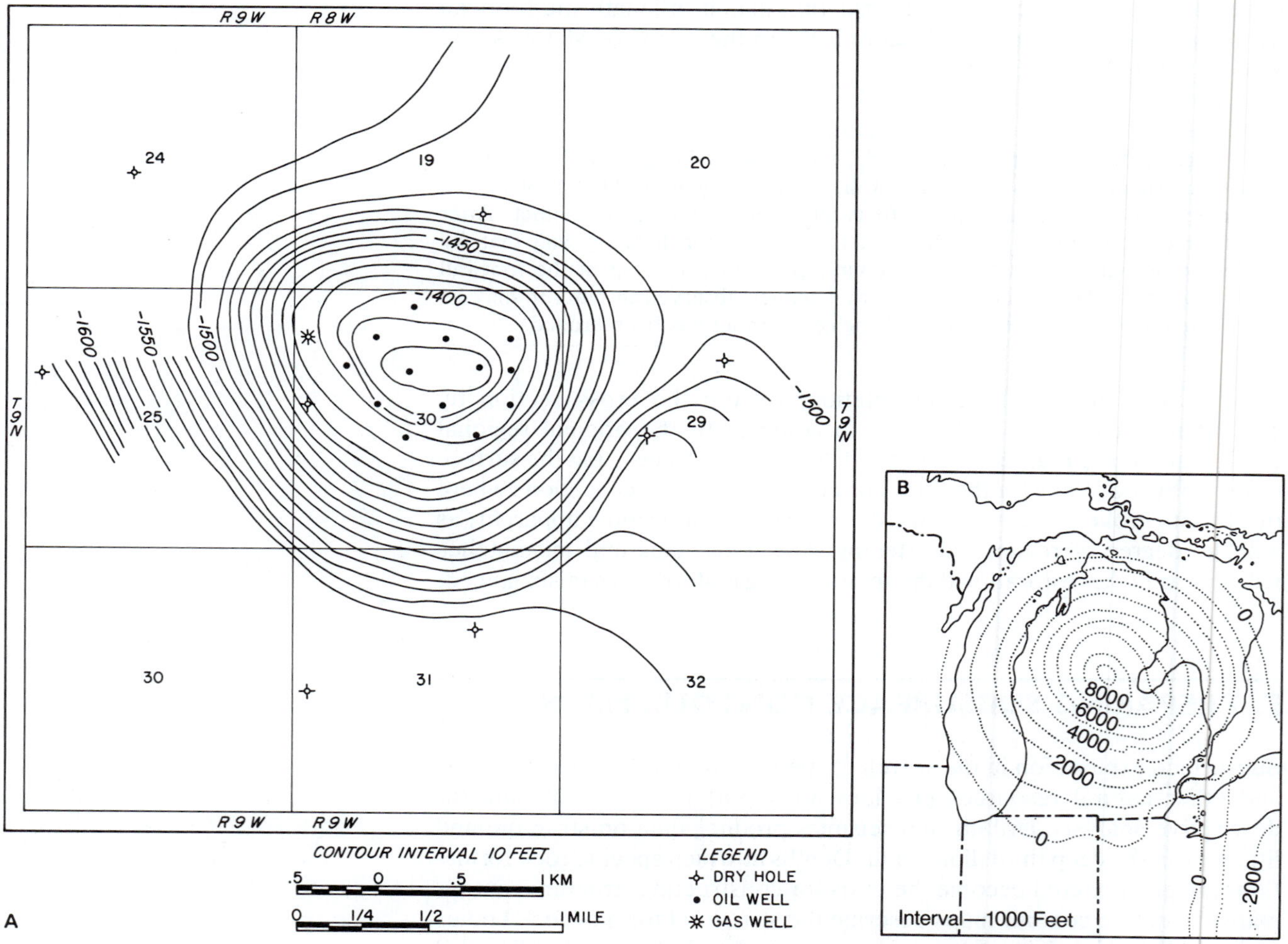

Figure F.2 (*A*) Structure contour map of the Wilfred pool dome, Indiana. [From Dana (1980). Published with permission of American Association of Petroleum Geologists.] (*B*) Structure contour map of the Michigan basin. Contours represent the top of Middle Silurian strata. [From *The Evolution of North America* by P. B. King, fig. 17C, p. 30. Published with permission of Princeton University Press, Princeton, NJ, copyright © 1959.]

tary formations. Domal and basinal patterns are marked by concentrically arranged, closed contours (Figure F.2). More complex structures can be shown equally effectively. Anticlines and synclines, and arches and troughs (Figure F.3) might display combinations of crescent-shaped, straight-lined, and closed contour patterns. **Homoclines**, which are simple tilted structures where bedding dips uniformly in a single direction, are distinguished by subparallel contour lines that steadily decrease (or increase) in elevation value across the map (Figure F.3). Faults are marked by offset contour lines (Figure F.3). Where a rock formation gradually steepens, the contour lines come closer and closer together. The contour lines actually merge into one where the top of the rock formation dips vertically.

Isopach maps are contour maps that describe formation thicknesses. Thicknesses are compiled on the basis of geologic mapping, underground mapping and mining, and subsurface drilling. Values of thicknesses are posted on a base map and contoured. The resulting patterns describe the

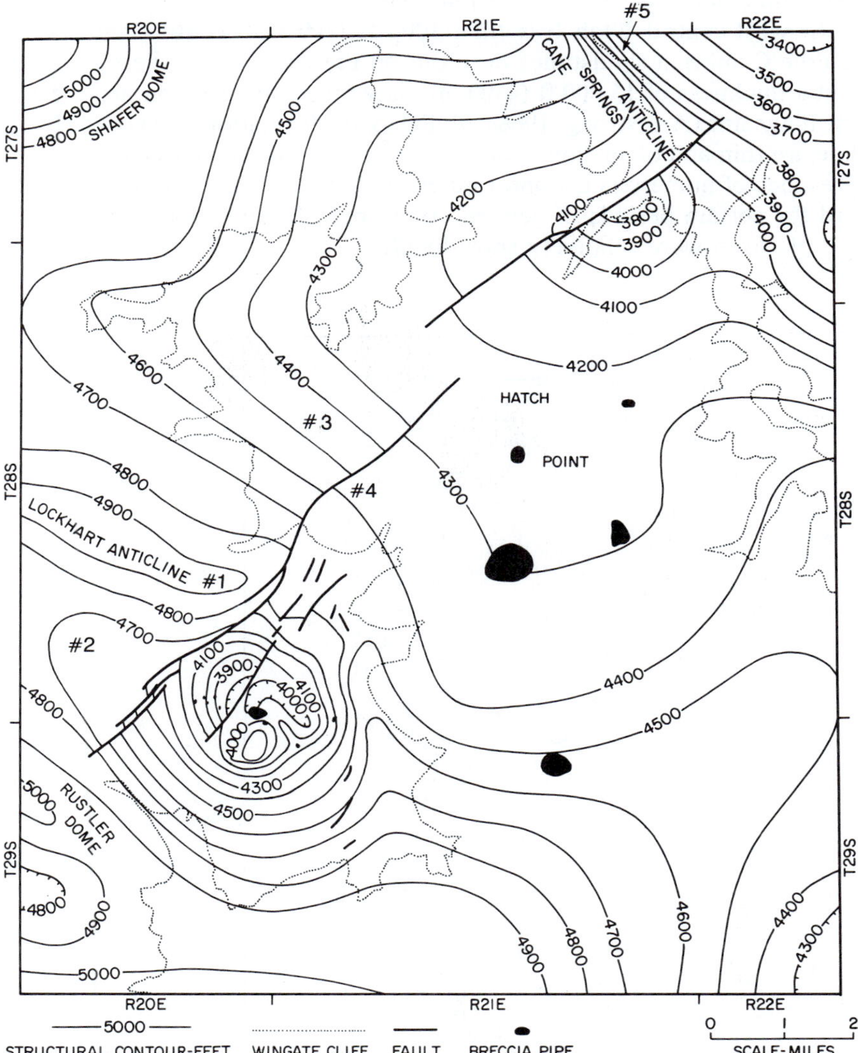

Figure F.3 Structure contour map of a part of the Canyonlands region, Utah, depicting examples of representations of (1) an anticline, (2) a syncline, (3) a homocline, (4) a fault, and (5) steeply dipping strata. [From Huntoon and Richter (1979). Published with permission of Four Corners Geological Society.]

variations in thickness of a particular formation and/or series of formations. The isopach map shown in Figure F.4 shows the variations in thickness of windblown Jurassic sandstone (Nugget and Navajo) in the Rocky Mountain region. The thickness variations indicate the shape and scope of the ancient dune field. I find the isopach map shown in Figure F.5 to be especially interesting, having grown up in Pittsburgh, the son of a father who, as a mining engineer, worked the Pittsburgh Coal Seam. The map provides a picture of the thickest accumulations of organic mass in the ancient Pennsylvanian swamps.

A good example of the regional structural insight afforded by isopach maps is found in maps of sedimentary assemblages of youngest Precambrian and Paleozoic ages in the hinge-line region of the western United States (Figure F.6*A*). The hinge line is generally regarded as the belt in western Utah and eastern Nevada that coincides with a remarkable change in thickness of late Precambrian and Paleozoic sedimentary rocks (Figure F.6*B*). The location of the hinge line coincides closely with the boundary between crust that was thinned during continental rifting in late Precambrian time and crust to the east that was not thinned.

In west-central Utah less than 1000 ft (300 m) of late Cambrian to middle Cambrian sandstones crop out on the east side of the hinge line. Yet a thickness of 15,000 ft (4500 m) marks the same section just west of the hinge line (Armstrong, 1968*a,b*). Paleozoic formations east of the hinge line are thin and of shallow-marine, intertidal, and nonmarine origin. The presence of numerous unconformities, and the lack of a sedimentary record for certain periods of geologic time, reveal that transgressions onto the platform were relatively short-lived.

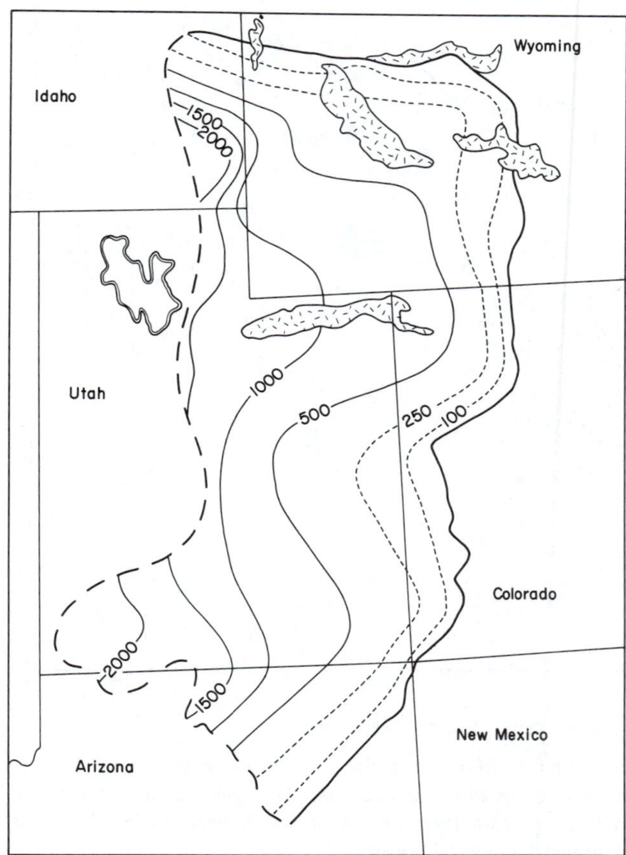

Figure F.4 Isopach map showing variations in thickness (in feet) of the Navajo and Nugget Sandstones (Jurassic). [Redrawn from Jordan (1965).]

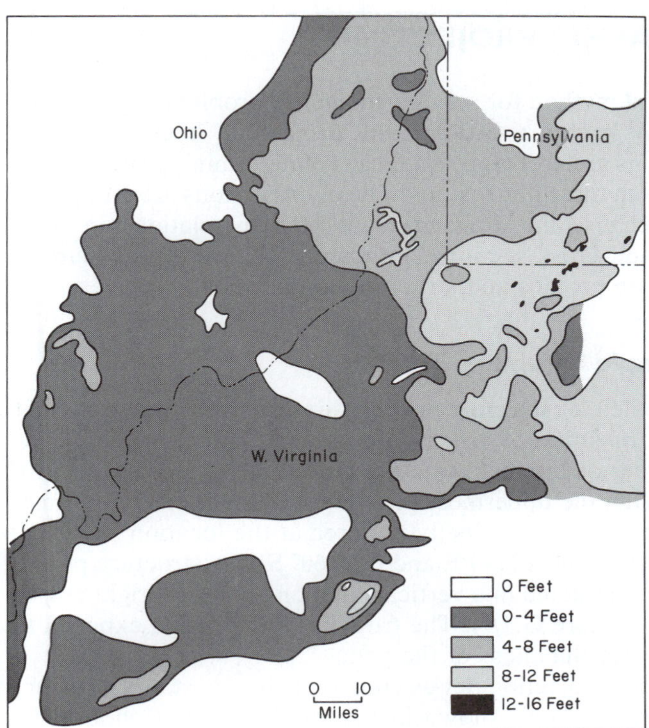

Figure F.5 Isopach map showing variations in thickness of the Pittsburgh Coal Seam. [From Hoover et al. (1969). Published with permission of Geological Society of America and the authors.]

By way of contrast, Paleozoic formations west of the hinge line are thick sequences of deeper-water marine sediments. All periods are presented, and long time periods were marked by apparent continuous sedimentation. The influence of the hinge line on sedimentation patterns is clearly and sensitively revealed in isopach maps of parts or all of the total younger Precambrian and Paleozoic sedimentary record in that region (Figure F.6*C*).

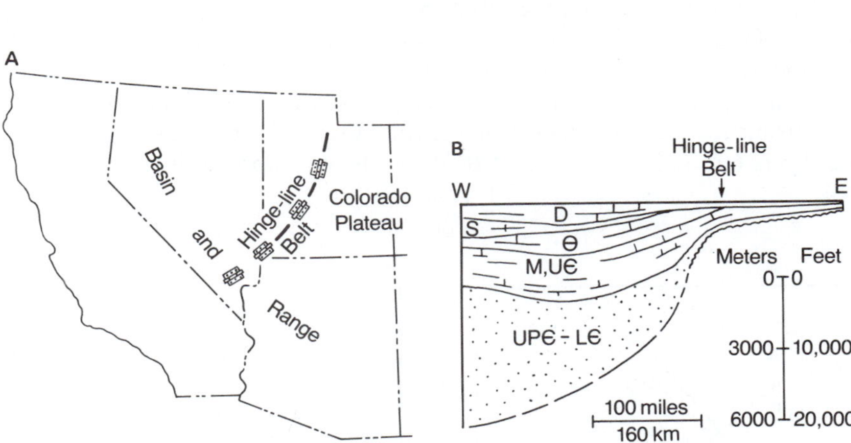

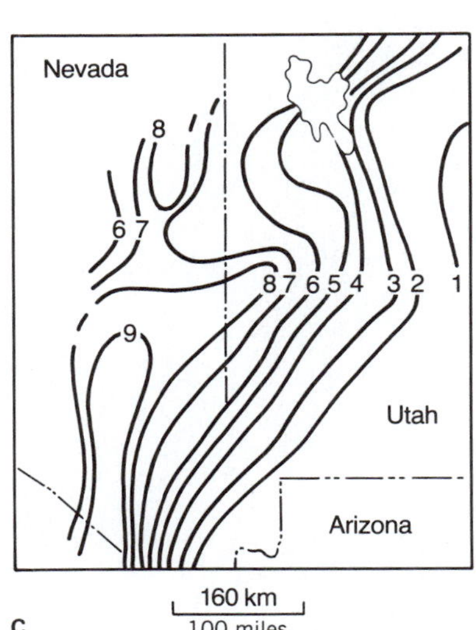

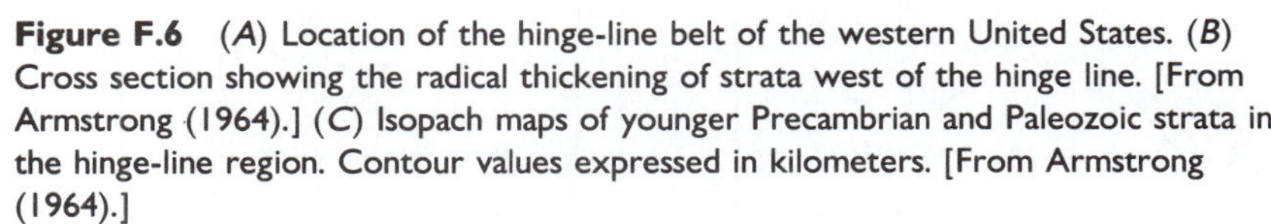
Figure F.6 (*A*) Location of the hinge-line belt of the western United States. (*B*) Cross section showing the radical thickening of strata west of the hinge line. [From Armstrong (1964).] (*C*) Isopach maps of younger Precambrian and Paleozoic strata in the hinge-line region. Contour values expressed in kilometers. [From Armstrong (1964).]

G. ORTHOGRAPHIC PROJECTION

A traditionally useful method for solving geometric problems is a kind of descriptive geometry known as **orthographic projection**. In essence, line-drawing constructions are prepared as a means of determining angular and spatial relationships in three dimensions. The constructions are difficult to visualize at first because the drawings convert map relationships into mixtures of maps and cross sections. Fundamental to the procedure is constructing structure profiles and structure contour lines.

Constructing Simple Structure Profiles

Cross-sectional profiles, or **structure profiles**, are generally drawn at right angles to the trend or strike of structural features. These profiles show the **traces** of plunging or dipping structures, as they would appear in a vertical "cut" through the uppermost part of the earth. Consider a limestone bed that crops out in a perfectly flat area at the location shown in Figure G.1*A*. The bed strikes N40°E and dips 60° SE. A structure profile view of the bed is constructed in a vertical cut along *A* - *A'* at right angles to the line of strike (Figure G.1*B*). The true dip of the bed is exposed to full view, as is the true thickness of the bed.

Using orthographic projection, let us construct step by step a structure profile for the limestone layer shown in Figure G.1*A*. First choose the orientation and location of the **profile line** along which the structure section is to be constructed (Figure G.2*A*). Points *A* and *B* are identified along the profile line such that *A* is on the lower contact of the limestone and *B* is on the upper contact at a location directly along the dip direction from *A*. Project points *A* and *B* from the interior of the map toward the edge of the map (or off the map onto another sheet of paper), where there is more available working space (Figure G.2*B*). Project each reference point by the same distance and in a direction strictly parallel to the strike of the limestone bed.

Draw a line through the **projected points** *A'* and *B'* (Figure G.2*B*). This line represents the topographic **surface profile** for the location of the profile line where the limestone crops out. In this special case the surface profile is perfectly horizontal. In the general case, the surface profile would be marked by some **topographic relief**. Showing such relief in profile would be part of the construction process. Topographic control would be afforded by topographic contour lines on the base map.

Using a protractor, construct the angle of dip of the limestone bed in the subsurface, beneath the surface profile (Figure G.2*C*). For this example the dip is 60° SE. From *A'* and *B'* draw the 60°-dipping lines that corre-

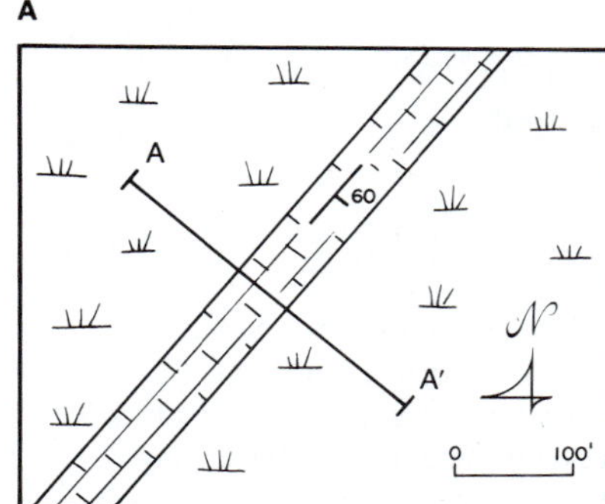

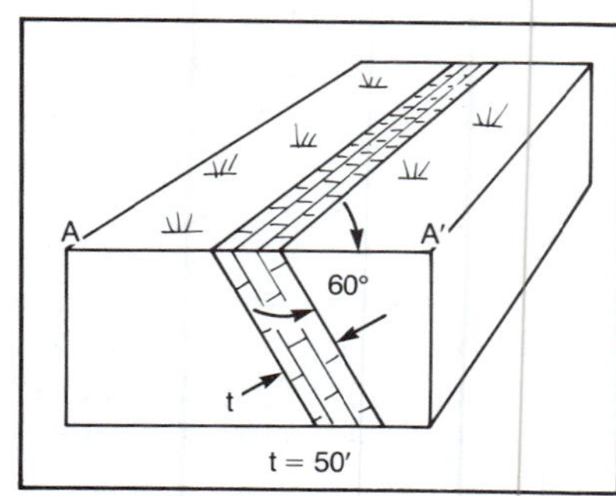

Figure G.1 (*A*) Plan view of a limestone bed that strikes northeast and dips southeast. (*B*) Structure profile view of the dipping bed. The profile view is the front face of the block diagram. It is oriented at right angles to the line of strike.

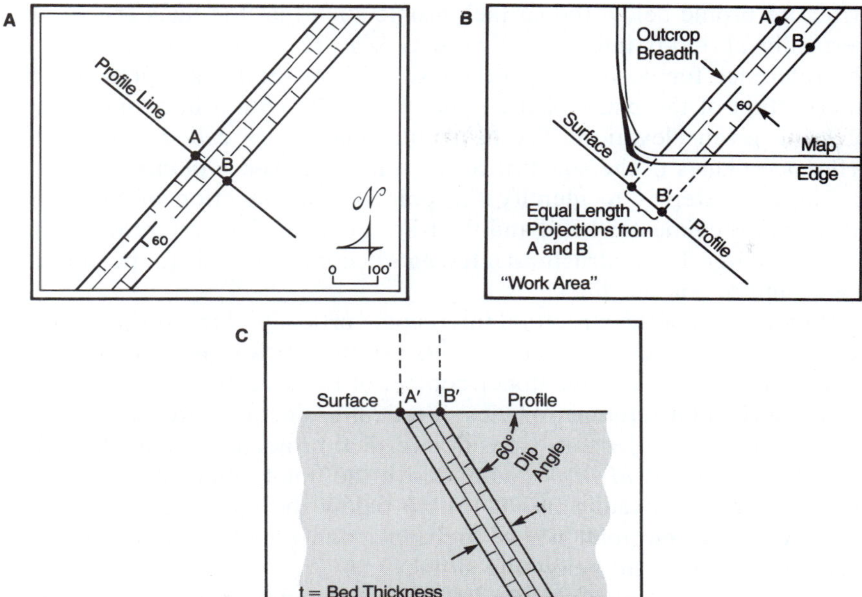

Figure G.2 Steps in constructing a structure profile. (*A*) Draw the profile line (*AB*) for which the structure profile is to be constructed. (*B*) Step the profile line to the edge of the map, or off the map sheet onto another piece of paper, where there is working space. (*C*) Plot the true dip of the bed and draw the subsurface expression of the top and bottom of the bed.

spond to the lower and upper contacts of the limestone bed. This completes the structure profile (Figure G.2*C*). It represents an approximation of the structural form and the **attitude** (i.e., orientation) of the limestone bed at depth. It is a ''projection'' of the form and the dip of the limestone unit based on surface exposures.

Determining the Thickness of a Bed

The thickness of a bed can be measured directly in structure profile view, provided the profile is constructed at right angles to strike. The map scale is used as the guide in determining thickness. Thickness is measured perpendicular to the upper and lower contacts for the bed in question. Figure G.2*C* shows how thickness is measured in structure profile view. Note that the measured thickness of the bed is not the same as **outcrop breadth** (see Figure G.2*B*).

Representing Dipping Planes by Structure Contour Lines

In problems of applied geology, it is commonly necessary to predict the location of rock layers, contacts, and structures at depth. For example, if the surface outcrop of the limestone bed in the preceding example were found to be mineralized, the location of the limestone in the subsurface could be of economic interest. It would then be beneficial to prepare a structure contour map of the limestone bed, using the upper contact of the limestone bed as a datum. Structure contour lines would connect points of equal elevation on the upper surface of the limestone. Since the contour lines, by definition, connect points of equal elevation, they are lines of strike. Each contour line represents a strike line on the limestone bed at some specified elevation.

If the designated contour interval for the structure contour map is, for example, 100 ft (30 m), the pattern of the corresponding contour lines can be found by a series of orthographic construction steps. First, construct a structure profile for the limestone bed, and add horizontal lines to the structure profile below the surface profile such that the lines are spaced vertically at 100-ft intervals (Figure G.3*A*). The map scale is used as the reference for positioning the lines. Each of the lines represents the intersection of the plane of the structure profile with a horizontal plane of some given elevation. The horizontal planes are called topographic reference planes in the subsurface, or simply **reference planes**.

The next step is to identify the points of intersection of the upper surface of the limestone bed and the trace of each of the reference planes (Figure G.3*B*). These **structural intercepts** become the basis for positioning the structure contour lines.

Project each of the structural intercepts vertically to the surface profile (Figure G.3*C*). These projected points are the **vertical projections** of the structural intercepts of the upper contact of the limestone bed with each of the horizontal reference planes. Vertical projections are fundamental to orthographic projection. Consider vertical projection points *M* and *O* as examples. Point *M* lies directly above the point where the top of the limestone bed is exactly 200 ft (61 m) below the surface. Point *O* lies directly above the point where the upper contact of the limestone lies exactly 400 ft (122 m) below the surface.

Finally we shift our construction from structure profile to map view (Figure G.3*D*), and we project **strike lines** from *M* and *O* into the interior

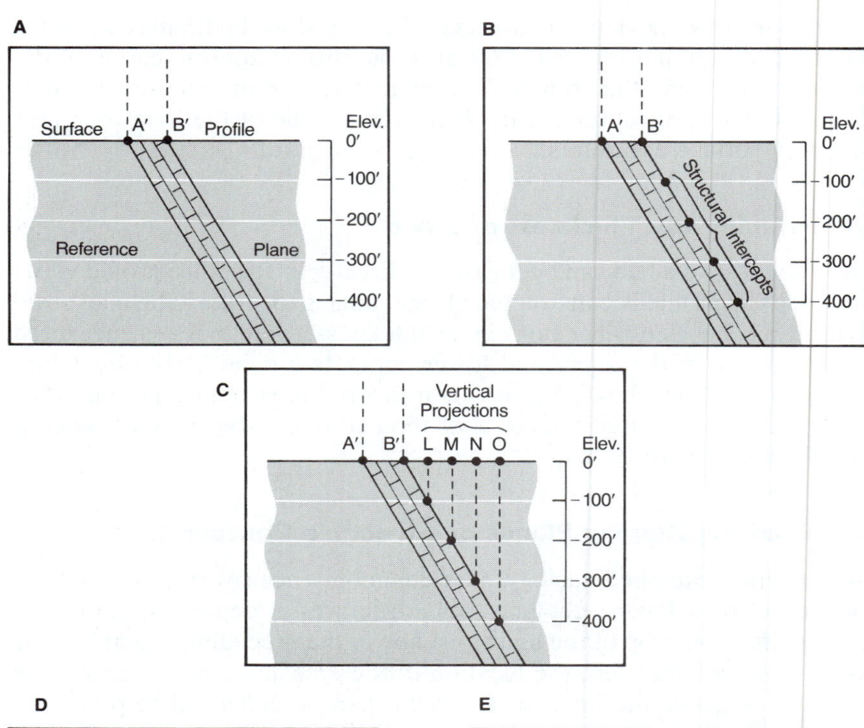

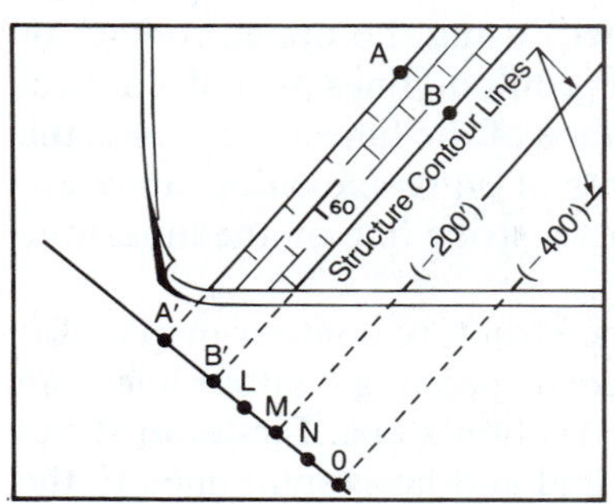

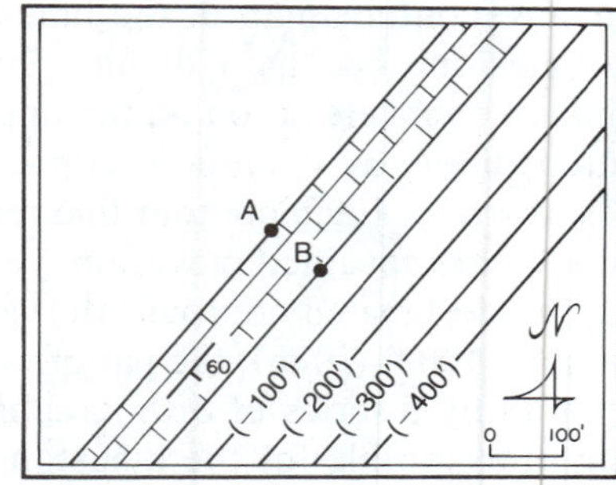

Figure G.3 Construction of structure contour lines representing the upper surface of a dipping bed. (*A*) Starting information: the structure profile view of a dipping bed. (*B*) Using the structure profile view, identify the structural intercepts of the top of the bed with the elevation reference planes shown in the subsurface. (*C*) Plot the vertical projections of each of the structural intercepts. (*D*) Project the vertical projections parallel to the line of strike. (*E*) The finished structure contour map.

of the map. These lines are **structure contour lines**, one of value −200 ft (i.e., 200 ft below the surface), the other of value −400 ft (−122 m). Since point *M is* the vertical projection of a point on the top of the limestone bed at elevation −200 ft, every point on the strike line through *M* must also lie 200 ft below the surface. The structure contour map is completed by drawing strike lines through all the vertical projections (Figure G.3*E*).

The ability to construct structure profiles and structure contour maps on the basis of surface or mine map patterns provides the means to solve a number of practical structural geologic problems.

Measuring Apparent Dip

Apparent dip is the inclination of the trace of a plane in a direction other than the true dip direction. Using the structure contour map displayed in Figure G.3*E*, we can solve for apparent dip of the limestone bed in any direction, for example, along a north–south line. First draw a north-trending profile line from one structure-contour line to another (Figure G.4*A*). The profile line in Figure G.4*A* connects a point on the −400-ft contour line with a point on the −100-ft (−30-m) contour line. Project the end points of this profile line to the edge of the map, or off it, and draw the surface profile (Figure G.4*B*). The end points of the surface profile are vertical projections from the upper surface of the limestone bed at elevations corresponding to the values of the structure contour lines.

Begin to fashion the structure profile by constructing the horizontal reference planes that correspond to the contour lines on which the reference points of the profile line rest (Figure G.4*C*). Then project lines vertically down from the reference points on the surface profile to the corresponding structural intercepts of the limestone bed and the horizontal reference planes. A line connecting the structural intercepts at the −100- and −400-ft levels represents the upper contact of the limestone bed. Its angle of inclination, as measured from the horizontal, is the apparent dip. Its value, as measured with a protractor, is 47° (Figure G.4*C*).

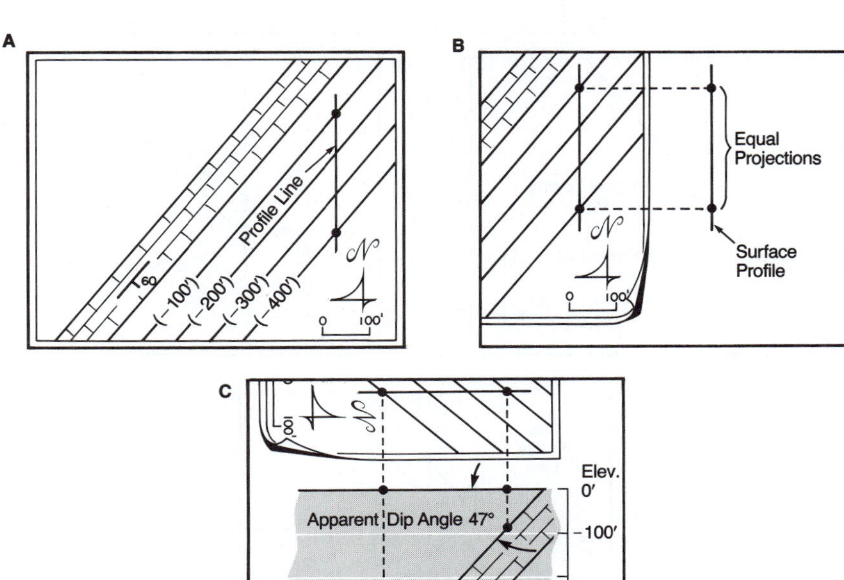

Figure G.4 Orthographic construction method for determining the apparent dip of a bed. (*A*) Designate the location and trend of the profile line along which apparent dip is to be determined. (*B*) Step the profile line to a location where working space is available. (*C*) Construct horizontal reference planes for the −100-ft (−30-m) and −400-ft (−122-m) levels. Project reference points on profile line to corresponding reference planes. Connect these structural intercepts to display apparent dip.

The standard orthographic solution to the apparent dip problem is a shortcut to the orthographic method just described. Here is how to do it. On a sheet of paper designate a **control point** that lies on the upper (or lower) contact of the limestone bed (Figure G.5*A*). Through it draw a strike line (N40°E) representing the strike attitude of the bed. At right angles to the strike line, construct a surface profile and draw the structural profile view of the dipping bed (Figure G.5*B*). Add to this profile view a horizontal reference plane that is positioned some arbitrary but known distance beneath the surface, for example, −100 ft (Figure G.5*C*). Plot the vertical projection of the structural intercept of the reference plane and the dipping bed (Figure G.5*D*), and project a structure contour line across the map from the vertical projection point.

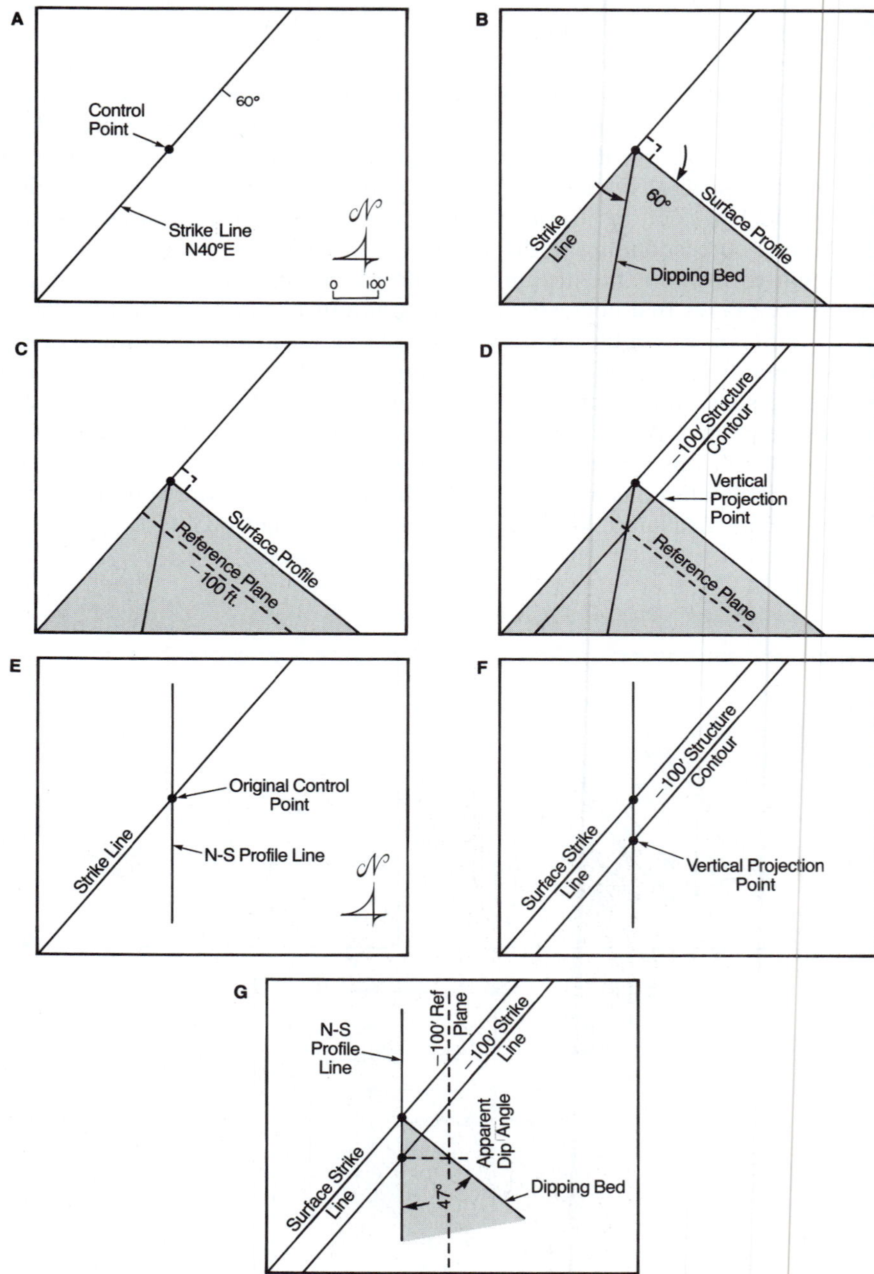

Figure G.5 "Shortcut" method for determining apparent dip. (*A*) Designate control point at convenient location on lower contact of limestone. (*B*) Construct structural profile view of dipping bed. (*C*) Plot the structural intercept representing the vertical projection of the top of the bed at the −100-ft (−30-m) level. (*D*) Identify the vertical projection of the structural intercept. (*E*) Through the control point established in *A*, draw the north–south profile line along which apparent dip is to be determined. (*F*) Identify the location where the north–south profile line intersects the vertical projection of the dipping plane at the −100-ft level. (*G*) Construct a structure profile on the basis of the known elevation of the surface control point *A* and the −100-ft structural intercept. Measure the apparent dip with a protractor.

To find apparent dip as viewed in a north-trending vertical exposure, draw a north–south-trending profile line northward from the original control point (Figure G.5*E*). Where this profile line crosses the strike line of value −100 ft, there lies the vertical projection of the structural intercept of the upper contact of the limestone bed and the −100-ft structure contour (Figure G.5*F*). The apparent dip of the limestone is found by constructing a structure profile along the north-trending surface profile (Figure G.5*G*). The value of the apparent dip is 47°S. Apparent-dip constructions of this type make it clear that true dip of bedding or any other planar structure can be viewed only in profiles oriented *perpendicular* to strike. Exposures of structure profiles oriented *parallel* to strike reveal 0° apparent dip. Sections oriented obliquely to strike disclose intermediate values of dip between 0° and the true dip.

Constructing the Line of Intersection of Two Planes

Determining the trend and plunge of the line of intersection of two planes is fundamental to a number of geometric and geologic problems. Let us solve for the trend and plunge of the line of intersection of a dike and a limestone bed (Figure G.6*A*). Map relationships show that the dike strikes N68°E and dips 45°NW; and the limestone bed strikes N39°W and dips 35°NE. We assume that the dike and the sandstone bed are perfectly planar and that the land surface is perfectly flat. For simplicity, no shifting of the limestone bed due to dike emplacement is shown in Figure G.6*A*.

Let the intersection of the northwest margin of the dike with the upper contact of the limestone bed be a control point for the constructions that follow (Figure G.6*B*). The control point is one of the two intersection points that we need to define the trend and plunge of the line of intersection of the two planes. The second control point will be found in the subsurface, where the top of the limestone bed and the northwest margin of the dike intersect at some known depth.

To find the second control point, first construct structure profile views of both the limestone bed and the dike (Figure G.6*C*). For each structure profile, construct a horizontal reference plane at some specified elevation [e.g., −1000 ft (−300 m)] below each surface profile (Figure G.6*C*).

Identify the structural intercept of the upper surface of the limestone bed and the horizontal reference plane, and then define its vertical projection to the surface profile (Figure G.6*D*). In the same manner, plot the vertical projection of the structural intercept of the dike and the horizontal reference plane.

Construct a strike line through the vertical projection of the limestone/reference plane structural intercept. This line is a map view of the vertical projection of the intersection of the top of the limestone bed with the −1000-ft reference plane (Figure G.6*E*). It is a structure contour on the limestone bed at elevation −1000 ft. In the same fashion, establish a −1000-ft contour line for the dike.

The intersection of the −1000-ft structure contour lines for the dike and the top of the limestone bed, respectively, is the vertical projection of the intersection of the dike and the limestone bed at elevation −1000 ft (Figure G.6*F*). Connect this intersection point with the original control point to define the trend of the intersection of the two planes. Its value is N27°E. To determine the plunge, construct a structure profile of the line of intersection of the dike and the limestone bed (Figure G.6*G*). The plunge measures 34°NE.

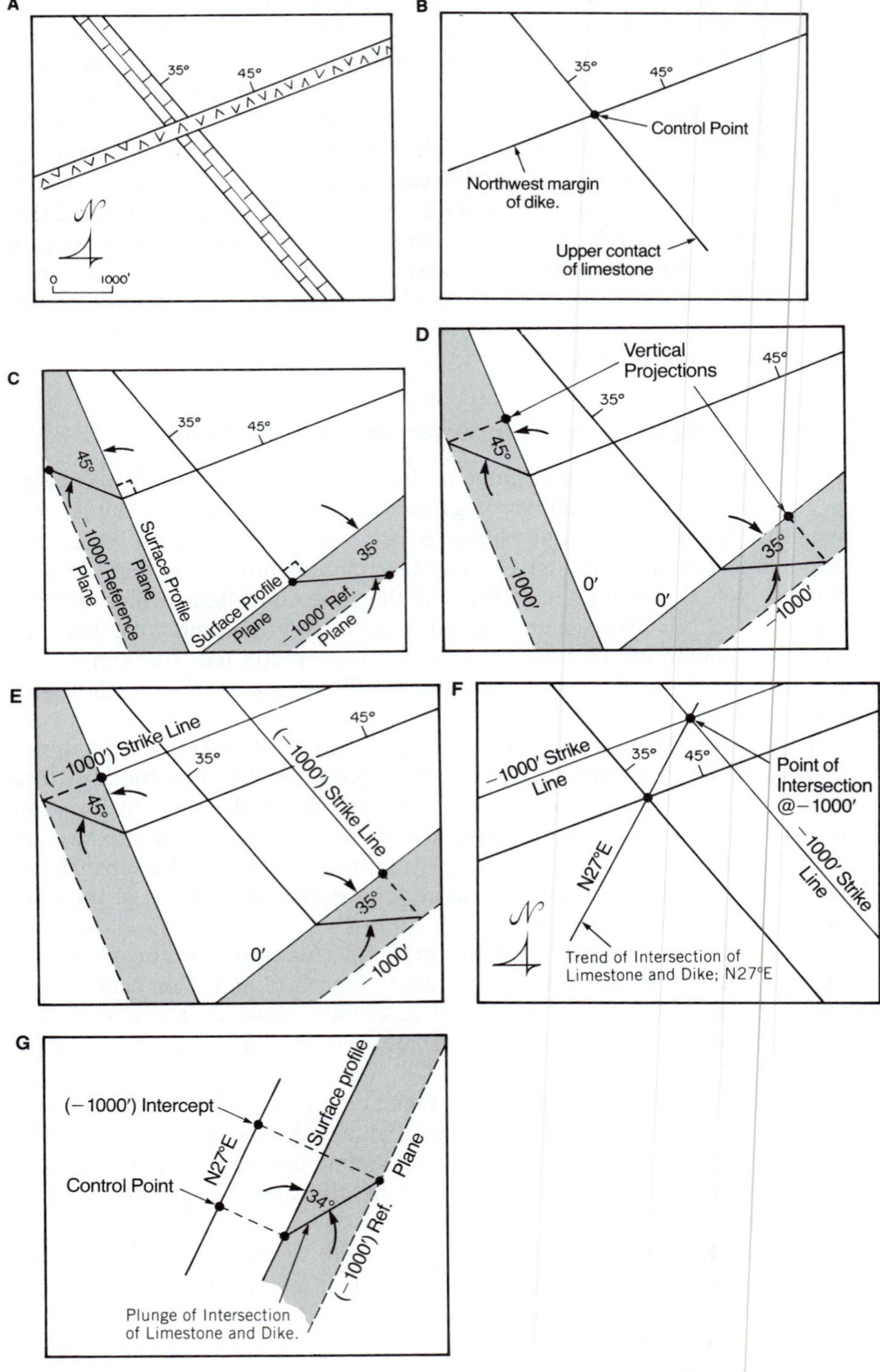

Figure G.6 Orthographic construction for determining the trend and plunge of the intersection of two planes. (*A*) Map of limestone bed that is intruded by a dike. (*B*) Simplified map of the same limestone bed, showing only the elements that are required to solve the problem at hand. (*C*) Draw two structure profiles—one for the dike, another for the limestone bed. Construct each profile, as always, at right angles to the line of strike. (*D*) Identify the vertical projection of the dike with the −1000-ft (−300-m) elevation reference plane. Also identify the vertical projection of the top of the limestone bed with the −1000-ft elevation reference plane. (*E*) Identify the vertical projection of the intersection of the top of the limestone bed with the northwestern margin of the dike at the −1000-ft level. (*F*) Connect the point of intersection of the northwest margin of the dike and the top of the limestone bed at the surface with the vertical projection of the same intersection at the −1000-ft elevation level. This line is the trend of the line of intersection of the dike and the limestone bed. (*G*) Construct a structure profile parallel to the trend of the line of intersection of the dike and the limestone bed. The plunge of the line of intersection is determined by drawing a line in profile view that connects the point of intersection of the dike and limestone bed at the surface level with the point of intersection of the dike and limestone bed at the −1000-ft level.

The Three-Point Problem

In some studies, it may be appropriate to assume that a specific structural feature, like an unconformity, is planar within a given area. If such an assumption is reasonable, the average strike and dip of the unconformity over a relatively large area may be determined through an orthographic construction known as a **three-point problem** (Figure G.7). Solving a three-point problem requires elevation control for at least three points *that lie on a common plane,* in this case the surface of unconformity. The elevation control may be derived from topographic maps used in conjunction with geologic maps and/or subsurface drilling information. The three control points used in three-point constructions should define a triangular array of relatively widely spaced points, which mark different elevations.

Consider points *A, B,* and *C*, at elevations of 3400, 2700, and 2400 ft (1030, 818, and 727 m), respectively (Figure G.7*A*). Each point lies on the angular unconformity whose attitude is sought. Point *B* is intermediate in elevation between points *A* and *C*. The trace of the unconformity as seen in a cross section passing through points *A* and *C* is constructed in vertical profile (Figure G.7*B*). It is inclined at an apparent dip of 28° toward *C*. Somewhere along its trace is a point whose elevation is the same as *B*. The location of this point (*D*) is found by constructing a horizontal reference plane whose elevation is the same as that of *B* (Figure G.7*C*). The intersection of the trace of the unconformity with the reference plane is projected vertically to *D'*, which represents the vertical projection of the point on the unconformity whose elevation is 2700 ft. By connecting *B* and *D'* (i.e., points of equal elevation on the angular unconformity), the line of strike is defined (Figure G.7*D*). Its trend is N44°E.

To determine the dip of the unconformity, it is necessary to construct a structural profile at right angles to the line of strike, and to project the trace of the unconformity into this plane. Such a profile is shown in Figure G.7*E*, with elevations projected into it from control points *A, B,* and *C*. The inclination of the unconformity in this special profile is a measure of true dip, namely 33°SE.

Three-point constructions are not restricted to determining the strike and dip of unconformities. The three-point method can be used effectively to establish the strike and dip of any extensive planar structure, like a bed, formation, fault, or intrusive contact.

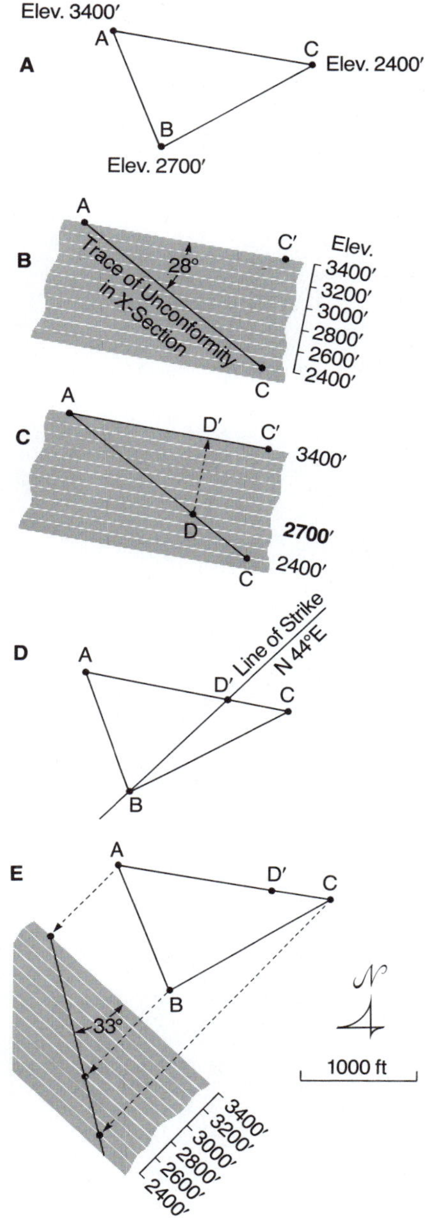

Figure G.7 Three-point problem. (*A*) Starting data: locations and elevations of three points on a common plane, in this case a planar unconformity. (*B*) Trace of unconformity as seen in cross section through points *A* and *C*. (*C*) Identification of point *D* on the trace of unconformity when elevation is the same as point *B*. Identification of vertical projection of point *D* to the line of section *AC*. (*D*) Line of strike connects points *B* and *D'* (*E*) Determination of true dip of unconformity.

H. STEREOGRAPHIC PROJECTION

Stereographic projection is a powerful method for solving geometric problems in structural geology (Bucher, 1944; Phillips, 1971). Stereographic projection differs from orthographic projection in a fundamental way: orthographic projection preserves *spatial relations* among structures, but stereographic projection displays *geometries* and *orientations* of lines and planes without regard to spatial relations.

The use of stereographic projection is preferable to orthographic projection in solving many geometric problems, simply because of ease of operations. Solving for apparent dip, the trend and plunge of the intersection of two planes, and angles between lines and planes in space can be carried out rapidly and accurately using stereographic projection. Orthographic projection, in contrast, requires the slow, careful construction of line drawings. But, orthographic projection remains the only effective way to solve geometric problems when topographic relief, map relationships, and depth to structures in the subsurface are integral to the solution of structural problems. In practice, we combine orthographic and stereographic projection techniques in ways that are practical, efficient, and complementary.

Geometry of Projection

Think of stereographic projection as a procedure comparable to using a three-dimensional protractor. A two-dimensional protractor is simple to use. Using a protractor we can plot trends of lines, measure angles between lines, construct **normals** (i.e., perpendiculars) to lines, and rotate lines by specified angles. Stereographic projection permits the same kinds of operation, but in three-dimensional space. Moreover, both lines and planes can be plotted and analyzed. Equipped with a three-dimensional protractor, we can do the following: plot orientations of lines; plot orientations of planes; determine the orientation of the intersection of two planes; determine the angle between two lines; determine the angle between two planes; measure the angle between a line and a plane; and rotate lines and planes in space about vertical, horizontal, or inclined axes.

All the preceding operations would be simple if it were possible to assemble real lines and planes in space, like Tinkertoys, and measure their geometric properties directly. Using stereographic projection techniques, we figuratively assemble lines and planes within a reference sphere.

The line or plane to be stereographically represented can be thought of as passing through the center of a reference sphere and intersecting its lower hemisphere (Figure H.1*A*). Planes intersect the lower hemisphere in the form of **great circles**; lines intersect the lower hemisphere in **points**. Stereographic projection of lines and planes to points and great circles constitutes a systematic reduction of three-dimensional geometry to two dimensions. The ''flattening'' to two dimensions is achieved by projecting the lower hemisphere intersections to an **equatorial plane** of reference that passes through the center of the sphere (Figure H.1*B*). This is the plane of stereographic projection. The lower hemisphere intersections are projected as rays *upward* through the horizontal reference plane to the *zenith* of the sphere. Where the rays of projection pass through the horizontal reference plane, point or great-circle intersections are produced, and these are **stereograms** or **stereographic projections** of lines and planes. Details of the projection geometry are presented in Phillips (1971).

Steep-plunging lines stereographically project to locations close to the center of the horizontal plane of projection; shallow-plunging lines project to locations near the perimeter of the plane of projection (Figure H.2*A*). Steeply dipping planes stereographically project as great circles that pass near the center of the plane of projection; gently dipping planes project as great circles passing close to the perimeter of the horizontal plane of projection (Figure H.2*B*). The distance that a great circle or point departs from the center of the plane of projection is a measure of the degree of inclination of the plane or line that has been stereographically plotted.

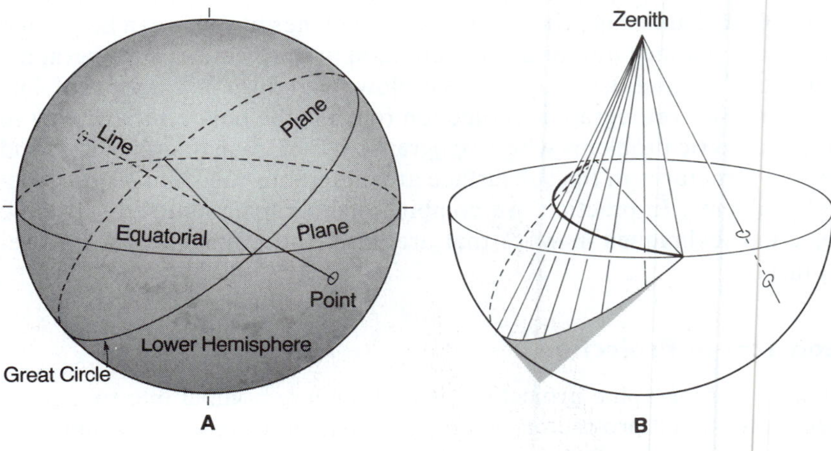

Figure H.1 The inherent three-dimensional geometry of stereographic projection. (*A*) Projection of a plane and a line through the center of a reference sphere. The plane intersects the lower hemisphere of the reference sphere as a great circle. The line intersects the lower hemisphere as a single point. (*B*) Projection of intersection points from the lower hemisphere of the reference sphere to the zenith of the projection.

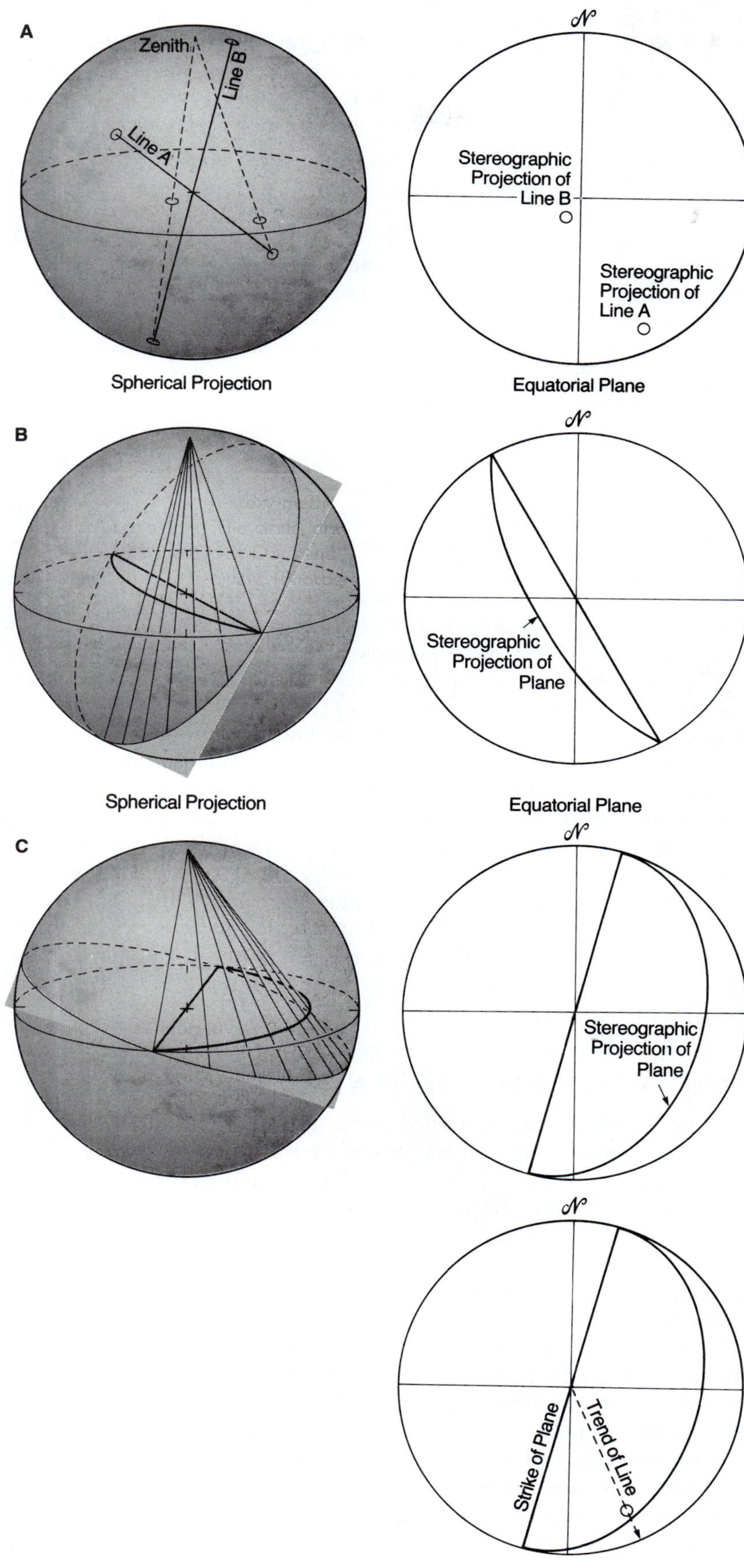

Figure H.2 The distance that a great circle or point lies from the center of the equatorial plane of projection is a measure of the inclination of a plane or line. (*A*) Shallow plunging lines project close to the perimeter of the equatorial plane; steeply plunging lines project close to the center. (*B*) Great circles that represent the orientation of steeply dipping planes pass close to the center of projection; shallow dipping planes are represented by great circles that pass close to the perimeter of the equatorial plane. (*C*) Stereographic representation of the strike of a plane and the trend of a line.

The trend of the line connecting the end points of the great circle corresponds to the strike of the plane that the great circle stereographically portrays (Figure H.2*C*). And the trend of a line connecting the center of the stereographic projection to a point representing a stereographically plotted line is the same as the trend of the line in space (Figure H.2*C*).

Stereographic Net, or Stereonet

In actual practice, stereographic projection of lines and plane is carried out through the use of a **stereographic net,** or **stereonet** for short (Figure H.3). A stereonet displays a network of great-circle and small-circle projections that occupy the equatorial plane of projection of the reference sphere. Both the great circles and small circles are spaced at 2° intervals; every fifth one is darkened so that 10° intervals can be readily counted. The great circles represent a family of planes of common strike whose dips range from 0° to 90°. The planes intersect in a horizontal line represented by the north–south line of the net. The small circles may be thought of as the paths along which lines would move when rotated about a horizontal axis oriented parallel to the ordinate of the net. The combination of small and great circles constitutes an orientation framework for stereographically plotting lines and planes.

There are two different kinds of stereonet: Wulff nets and Schmidt nets (see Figure H.3). Constructions are carried out the same way on each (Phillips, 1971, p. 61). Structural geologists find the Schmidt net (see Figure H.3*A*) to be the most versatile, for reasons to be explained later.

To prepare the Schmidt net for use, tape or glue it to a heavy backing, such as cardboard or Masonite. Insert a thumbtack through the backing and through the exact center of the net, taping the tack to the underside of the net so it cannot fall free (Figure H.4). A sheet of tracing paper is placed on the net so that the paper, punctured by the thumbtack, can rotate about the tack. A small square of clear tape applied to the back of the tracing paper, covering the area where the thumb tack will go through, prevents the paper from ripping during plotting. All construction work is carried out on the tracing paper, which is oriented with respect to north by marking a **north index** at a point corresponding to the top of the north–south line. This geographically orients the overlay for the constructions to be carried out.

Plotting the Trend and Plunge of a Line

Before any problems can be solved stereographically, it is necessary to learn how to represent the orientations of lines and planes on a stereonet. Lines are easiest to plot. Consider a line that plunges 26° N40°E. To represent the trend and plunge of this line stereographically, first plot the trend, in degrees, on the outer perimeter of the stereographic net. To do this, measure east from north by 40° (Figure H.5*A*). This can be accomplished simply by using the stereographic net as you would a protractor, counting clockwise from the north index on the tracing paper along the periphery of the net to 40°. The 40° azimuth corresponds to the 40° small-circle intercept on the perimeter of the net. Mark the point at 40° with a **trend-index mark** (*t*) (Figure H.5*A*).

To plot the 26° plunge, first rotate the overlay clockwise until *t* comes to rest on the right end of the east–west line of the net (Figure H.5*B*). This is one of two lines (the N–S axis being the other) where inclination can be directly measured and plotted. The plunge is measured by counting 26° from the perimeter of the net along the east–west line toward the

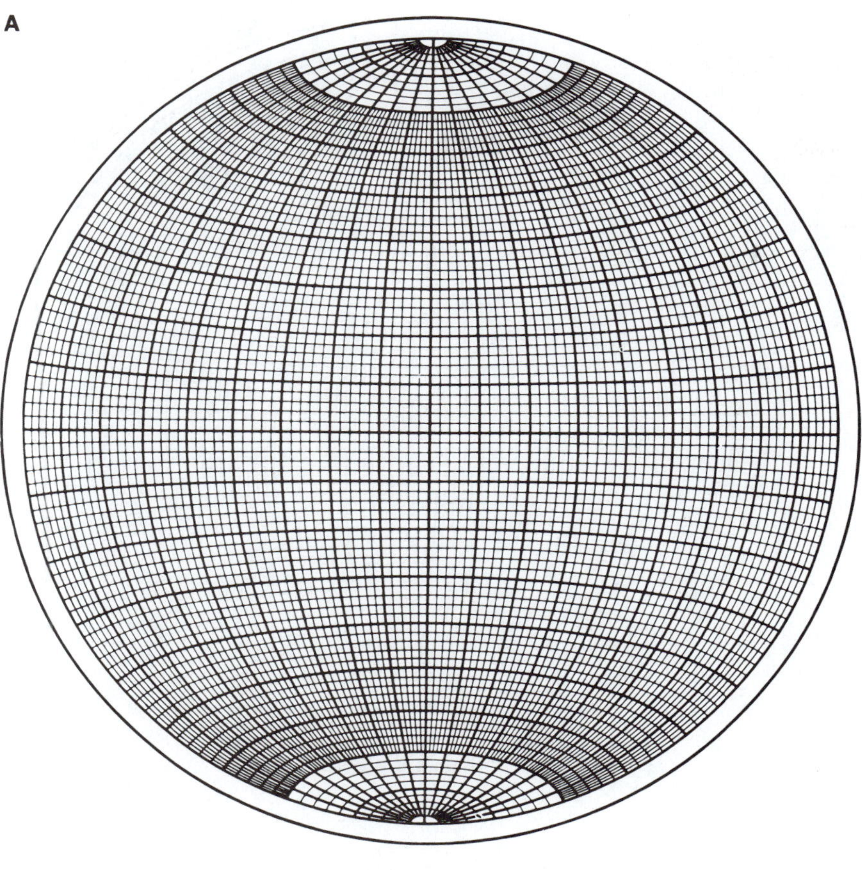

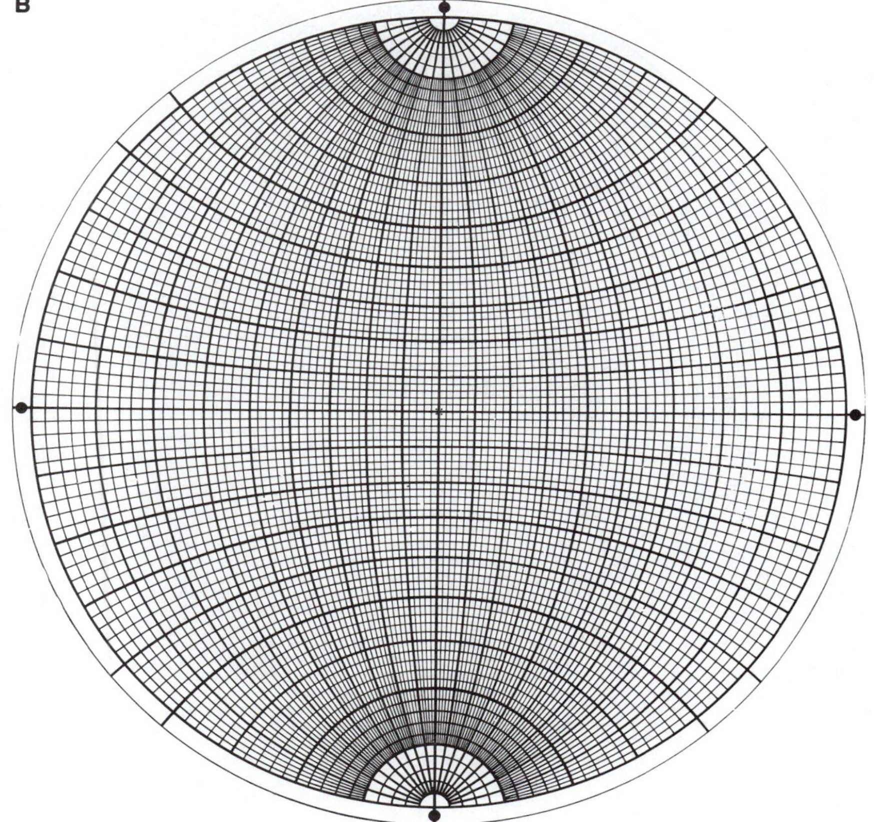

Figure H.3 Types of stereographic net. (*A*) The Schmidt net, an equal-area net. (*B*) The Wulff net, an equal-angle net.

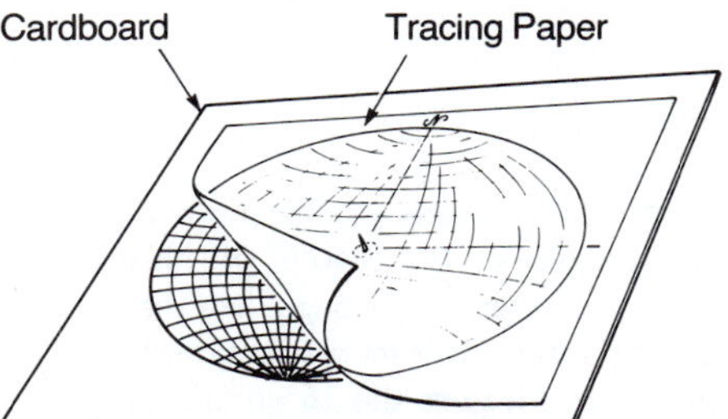

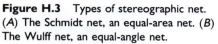

Figure H.4 Stereographic net ready for action. (From *Structural Geology of Folded Rocks* by E. T. H. Whitten. Originally published by Rand-McNally and Company, Skokie, Ill., copyright © 1966. Published with permission of John Wiley & Sons, Inc., New York.)

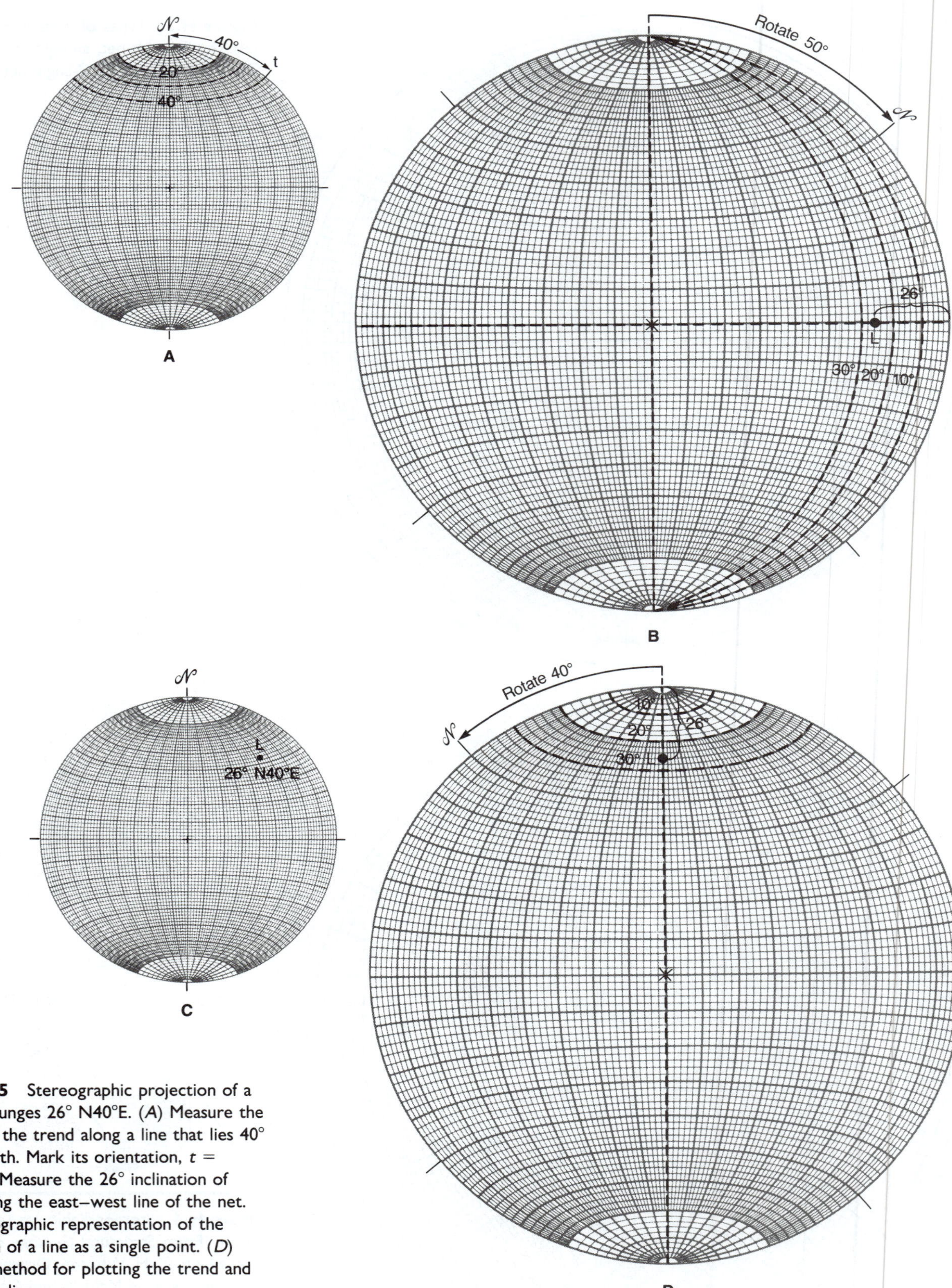

Figure H.5 Stereographic projection of a line that plunges 26° N40°E. (*A*) Measure the azimuth of the trend along a line that lies 40° east of north. Mark its orientation, t = trend. (*B*) Measure the 26° inclination of plunge along the east–west line of the net. (*C*) Stereographic representation of the orientation of a line as a single point. (*D*) Shortcut method for plotting the trend and plunge of a line.

center of the net. Point *L* represents the 26° plunging line (Figure H.5*B*). By rotating the tracing paper counterclockwise such that the north index again becomes aligned with the top of the north–south line ("home position"), point *L* can be viewed in its proper orientation framework (Figure H.5*C*). As a general check, it can be seen that point *L* lies in the northeast quadrant, corresponding to a northeast trend. Furthermore, it falls relatively close to the perimeter, reflecting a rather shallow plunge.

One shortcut is available. Plotting the N40°E trend can be achieved simply by rotating the tracing paper counterclockwise such that the north index comes to rest on 40°W (Figure H.5*D*). This automatically orients the N40°E trend line along the north–south line of the net, along which the 26° plunge can be directly measured.

As a second example, let us plot stereographically the orientation of a line plunging 75° S65°W. First define the trend by measuring 65° west of south (Figure H.6*A*). Point *t* represents the trend of this line. Next rotate *t* until it coincides with the left end of the east–west line (Figure H.6*B*). Measure the value of the plunging line by counting inward 75° from the perimeter of the net (*L* marks the 75°-plunging line). Rotate the tracing paper back to home position and view *L* in its proper orientation (Figure H.6*C*). Note that *L* plots close to the center of the net because of its steep plunge.

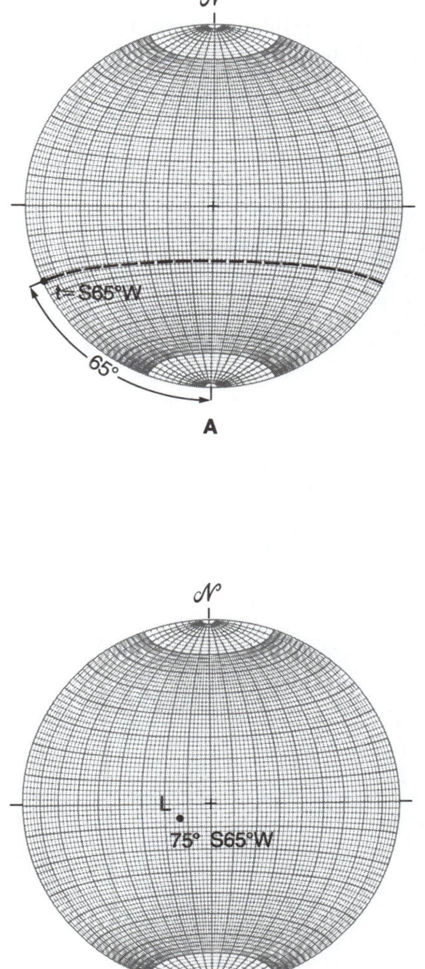

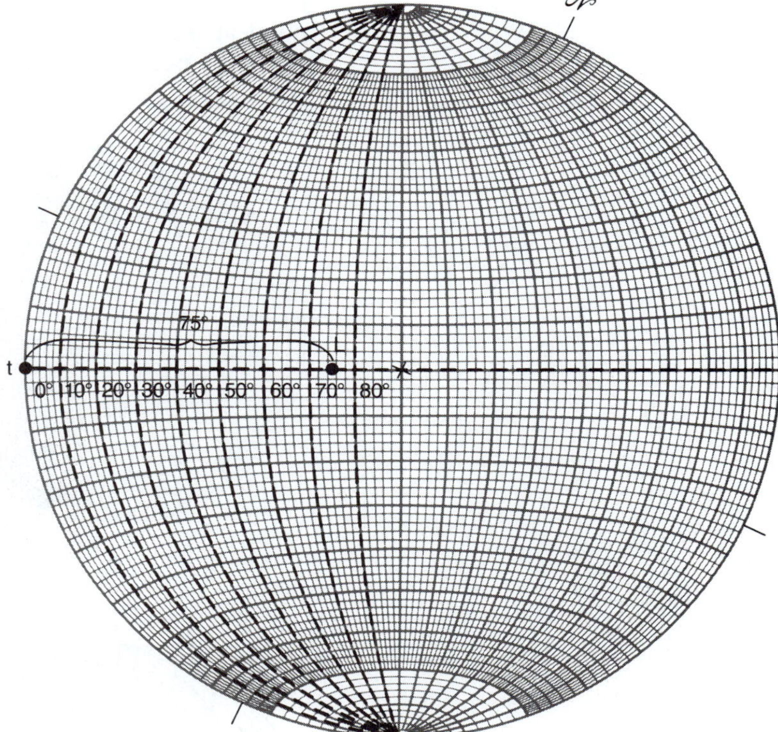

Figure H.6 Stereographic projection of a line that plunges 75° S65°W. (*A*) Measure the azimuth of the trend of the line. (*B*) Rotate the trend line to the east–west line of the net. Measure the plunge inclination along the east–west line, moving from the edge of the net toward the center. (*C*) Stereographic result as seen when overlay is in home position.

Plotting the Strike and Dip of a Plane

Let us now stereographically plot the orientation of a plane. Consider a plane striking N40°W and dipping 30°SW. The strike line of this plane is found by rotating the tracing paper clockwise until the north index comes to rest on 40°E (Figure H.7*A*,*B*). The strike can be visually shown by tracing the north–south line of the net (Figure H.7*B*), although this is not strictly necessary. When the strike line of the plane is rotated into coincidence with the north–south line of the stereonet, the dip of the plane can be plotted. Count inward 30° from the perimeter of the net along the east–west line, which is the line of dip when the strike line of the plane coincides with the north–south line of the net (Figure H.7*C*). Then trace the great circle that coincides with the 30° SW dip. By rotating the north index back to home position (Figure H.7*D*), the strike line and the great circle become aligned in an orientation that corresponds stereographically to a plane striking N40°W and dipping 30°SW.

There is another way to represent the orientation of a plane stereographically. The orientation of any plane in space can be described uniquely by the orientation of a line perpendicular to the plane. If the trend and plunge of a normal (**pole**) to a plane is known, the orientation of the plane itself is also established. The pole to a vertical plane is horizontal, and it stereographically plots as a point on the perimeter of the stereonet. The pole to a horizontal plane is vertical, and it plots stereographically as a point at the very center of the stereonet. The pole to an inclined plane plots as a point somewhere in the interior of the net, but not at its center.

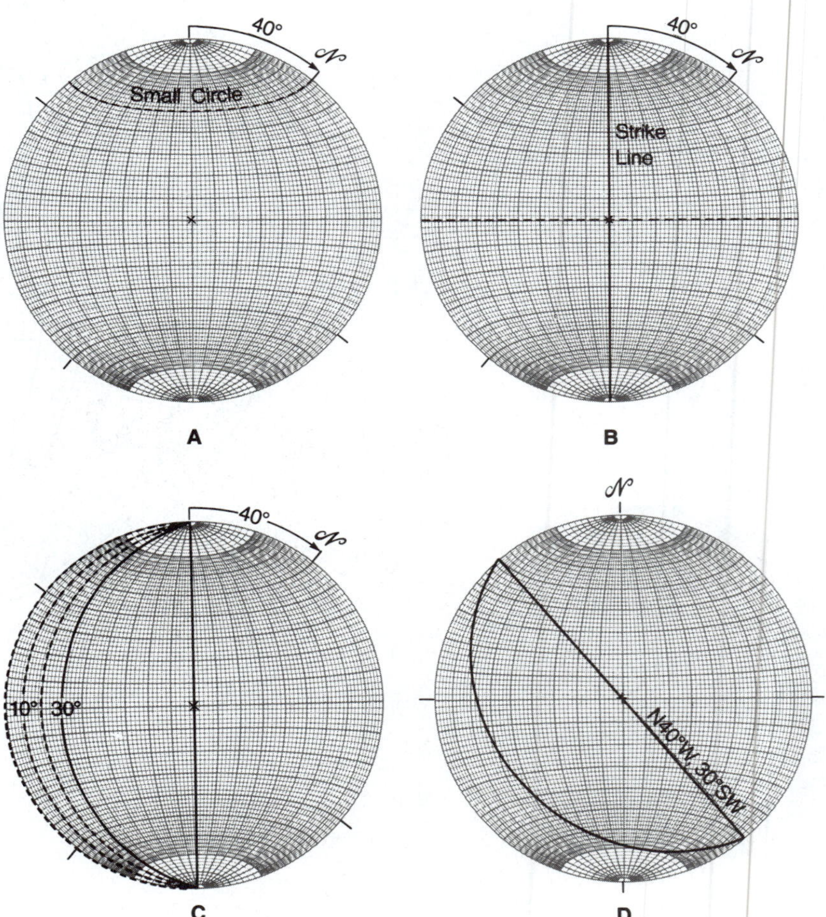

Figure H.7 Stereographic projection of a plane that strikes N40°W and dips 30°SW. (*A*) Measure the azimuth of strike, 40° west of north. (*B*) Draw the strike line. (*C*) Measure the 30° dip inclination along the east–west line of the net, moving from the edge of the net toward its center. (*D*) The stereographic representation of a plane is the combination of a strike line and a great circle.

(To visualize this, place a pencil between your fingers, pointing it down and away from your palm. The pencil represents the pole to the plane of your hand. When your hand is horizontal, the pencil points straight down. When your hand is held vertically, the pencil points horizontally.) When large numbers of planes must be plotted stereographically on a single projection, it is far more practical to plot each plane as a pole, not as the combination of a strike line and a great circle. The resulting diagram is cleaner and more conducive to interpretation of preferred orientations.

The procedure for plotting poles to planes stereographically is reasonably straightforward. Consider the pole to a plane that strikes N80°E and dips 20°SE. Figure H.8*A* shows the orientation of the plane plotted stereographically as the combination of a strike line and a great circle. By definition, the pole to this plane is oriented 90° to the plane, measured in a vertical plane perpendicular to strike. To plot this pole stereographically, rotate the strike line of the plane so that it coincides with the north–south line of the stereonet (Figure H.8*B*). In this orientation, the line of true dip in the plane lies on the east–west line of the net. From the point representing the line of true dip of the plane, measure 90° along the east–west line (Figure H.8*C*). The position of the pole is 20° beyond the center point of the net and is plotted as point *P*. Rotating the north index back to home position results in placement of the pole in its proper orientation (Figure H.8*D*).

In practice, plotting a pole to a plane need not include plotting the plane as a great circle. Rather, the strike line of the plane is rotated so

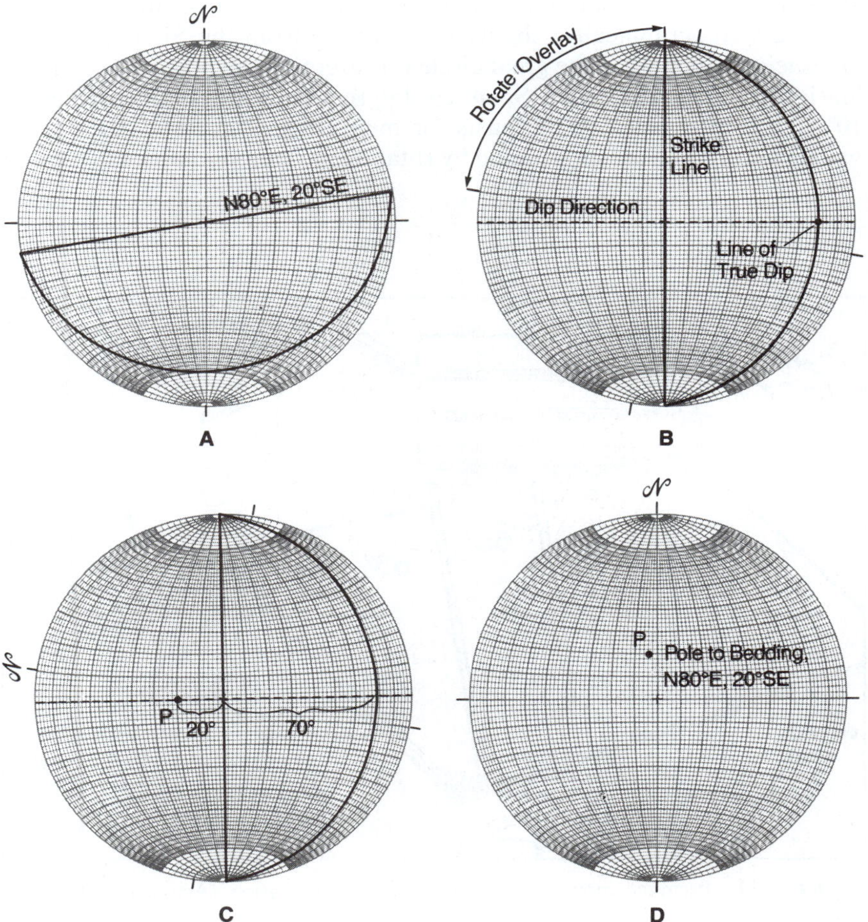

Figure H.8 Stereographic projection of the pole to a plane. (*A*) Stereographic representation of a plane that strikes N80°E and dips 20°SE. (*B*) Rotate strike line into parallelism with the north–south line of the net. (*C*) Identify the pole to the plane by measuring 90° along the east–west line of the net from the point that represents the inclination of true dip. (*D*) Final portrayal of the stereographic representation of the plane as a pole.

Figure H.9 Slickenlines on a fault surface, a structural geologic example of lines in a plane. (Photograph by G. H. Davis.)

that it coincides with the north–south line of the stereographic net (as in Figure H.8*C*); then the pole to the plane is found by measuring along the east–west line *outward* from the center of the net, into the quadrant opposite the dip direction of the plane.

Plotting the Orientation of a Line in a Plane

Many structural relationships involve the presence of a line in a plane, like slickenlines on a fault surface (Figure H.9). Stereographically, the point representing the trend and plunge of a line in a plane must lie on the great circle representing the strike and dip of the plane. Consider the geometry of a fault that strikes N10°E and dips 44°NW, containing slickenlines that plunge 40° N50°W. If the stereographic orientations of these two elements are plotted independently, it is found that the trend and plunge of the slickenlines are represented by a point that falls on the great circle corresponding to the strike and dip of the fault (Figure H.10).

Another way to describe the orientation of a line in a plane is to measure the **rake** (or **pitch**) of the line. Rake is the angle between a line and the strike line of the plane in which it is found (Figure H.11). If the orientation of a line, as measured in the field, is expressed in the field notebook as a rake angle within a plane of known orientation, its stereographic portrayal can be plotted readily. Consider, for example, a line that rakes 65°SE in a plane whose orientation is N40°W, 45°SW. The orientation of the plane is plotted stereographically as a great circle (Figure H.12*A*). Since the line lies in this plane, the point representing the trend and plunge of the line must lie on the great circle representing the plane. The rake angle of 65°SE is measured from the S40°E end of the strike line. To show the line stereographically, simply measure 65° from the SE quadrant of the tracing paper along the great circle representing the plane in which the line is found (Figure H.12*B*). Small-circle/great-circle intercepts, spaced at 10° and 2° intervals, are the basis for measuring. The trend-and-plunge values for this line are interpreted by rotating the stereographically plotted

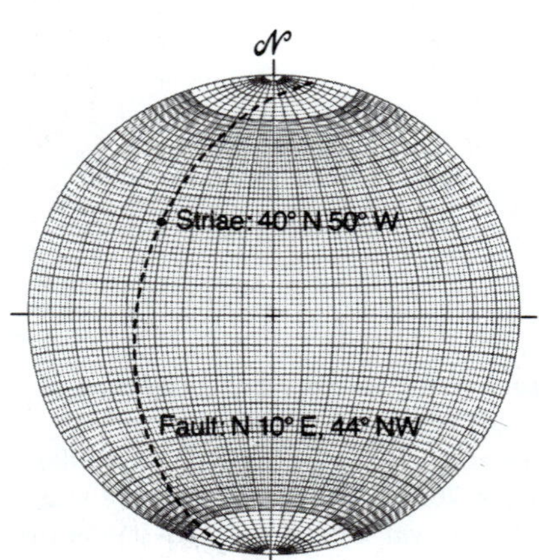

Figure H.10 If a line lies in a plane, the stereographic projection of the line as a point must fall on the great circle that stereographically represents the orientation of the plane.

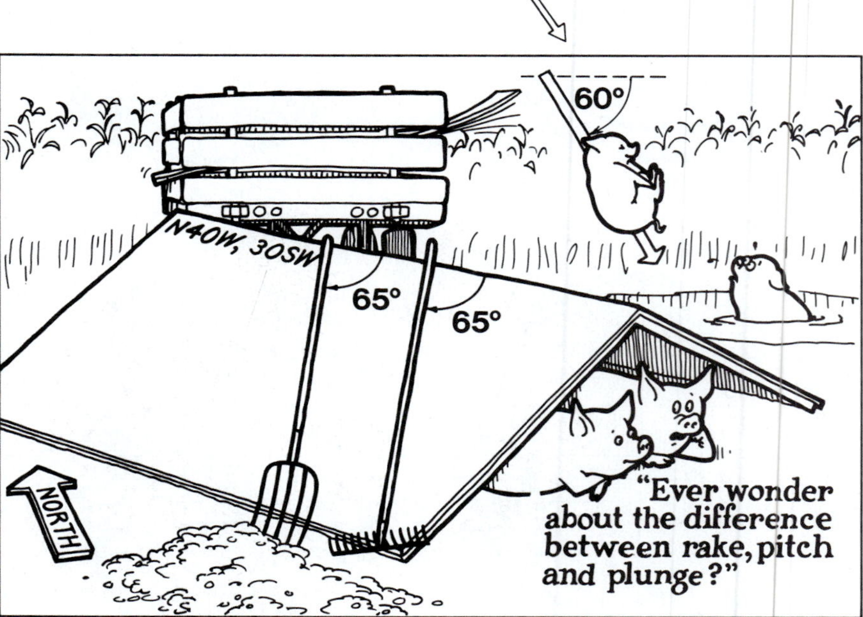

Figure H.11 Barnyard conversation about tools and recreation. (Artwork by D. A. Fisher.)

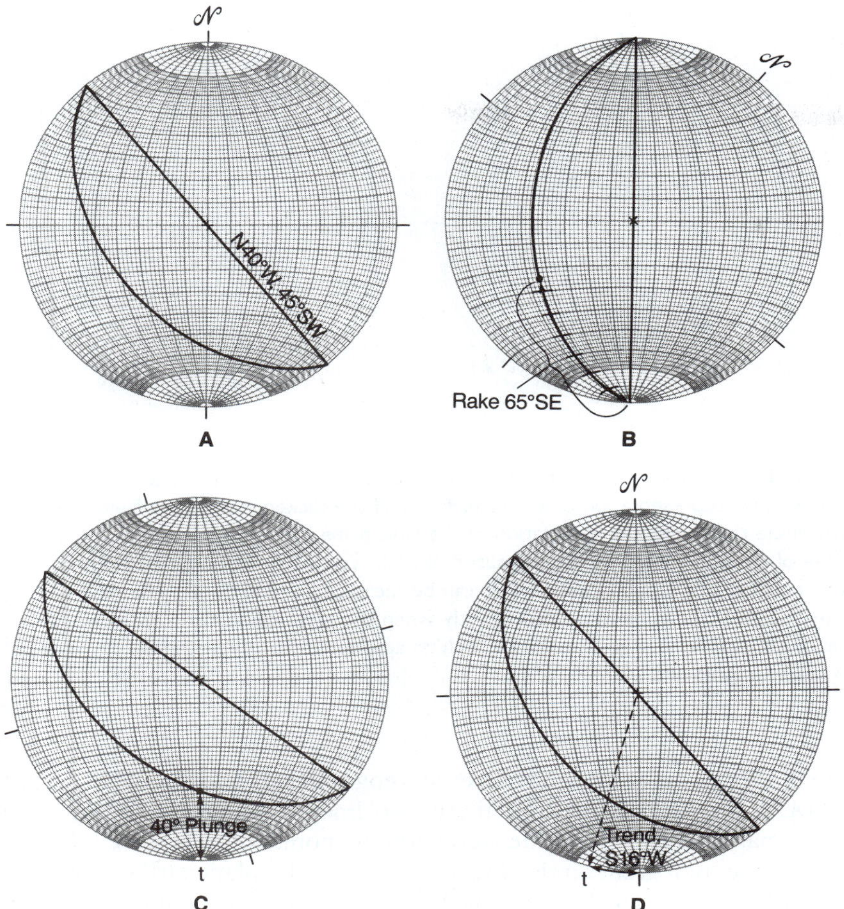

Figure H.12 Stereographic meaning of rake. (*A*) Stereographic representation of a plane that strikes N40°W and dips 45°SW. (*B*) Counting the 65° rake angle that describes the orientation of a line in the plane. (*C*) Measurement of the plunge of the line contained in the dipping plane. (*D*) Measurement of the trend of the line contained in the dipping plane.

point to the east–west or north–south line of the projection, and measuring the inclination of the point from the horizontal, in this case 40° (Figure H.12*C*). While the tracing paper is in the same position, the trend index *t* of the point can be marked on the perimeter of the net. Rotating the overlay to home position, the trend can be interpreted, in this case S16°W (Figure H.12*D*).

Converting trend and plunge to rake is a reasonably smooth operation as well. The fault surface stereographically represented in Figure H.13*A* is positioned stereographically in such a way to emphasize the 23° **plunge** value of slickenlines. The trend of the slickenlines is S44°E (Figure H.13*B*). **Rake** of the slickenlines is found by first aligning the great circle representing the fault with the corresponding great circle on the stereographic net (Figure H.13*C*). The rake of 25°SE is measured along the great circle, inward from the perimeter to the point representing the trend and plunge of slickenlines.

Measuring the Angle Between Two Lines

If someone walked up to you on the street and asked you to compute the angle between two lines in space, one plunging 16° N42°E and the other plunging 80° S16°E, how would you do it? The stereographic solution is based on knowledge that two lines define a plane, and that the angle between the two lines is measured in the plane common to both. The orientations of the lines are given as 16°/042° (line 1) and 80°/164°

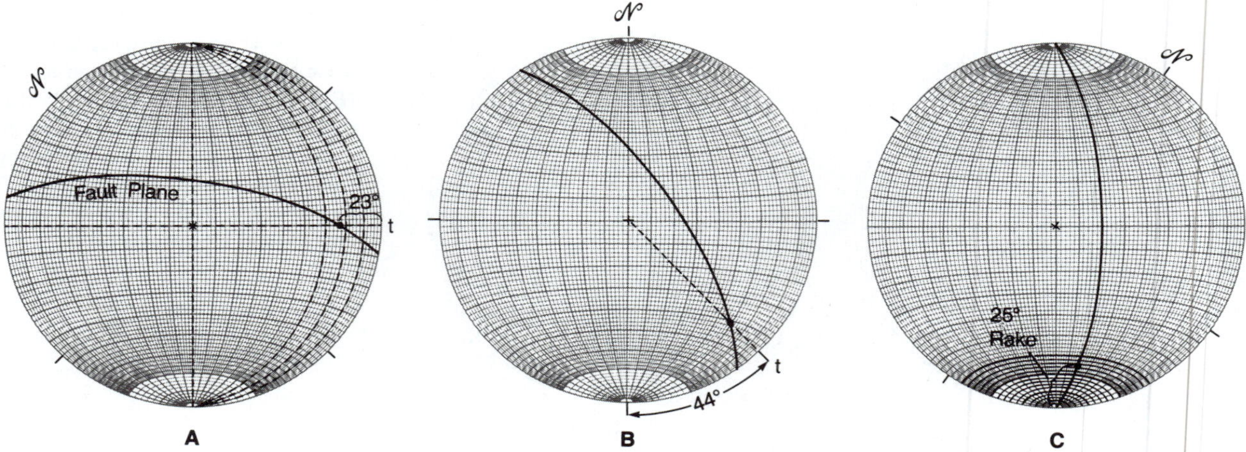

Figure H.13 Converting the trend and plunge of a line to rake in a plane. (*A*) Stereographic representation of a fault surface and the slickenlines on the fault. The great circle portraying the orientation of the fault plane is oriented such that the plunge of the slickenlines can be measured directly. (*B*) With the north index mark in home position, the trend of slickenlines can be measured as S44°E. (*C*) The rake of slickenlines (25°S) can be measured directly when the great circle that represents the fault orientation is rotated to coincide with an appropriately oriented great circle on the stereographic net.

(line 2). We can plot these two lines stereographically, as shown in Figure H.14*A*. The plane defined by these two lines is found by rotating the tracing paper overlay until the stereographic points representing the lines lie on a common great circle (Figure H.14*B*). (The plane strikes 040° and dips 81°SE.) The angle between the lines is measured by counting 2° small-circle intercepts along the great circle representing the common plane (Figure H.14*C*). The acute angle separating these points is 80°; the obtuse angle is 100°. Always carry a stereonet in the street. You never know who you might meet.

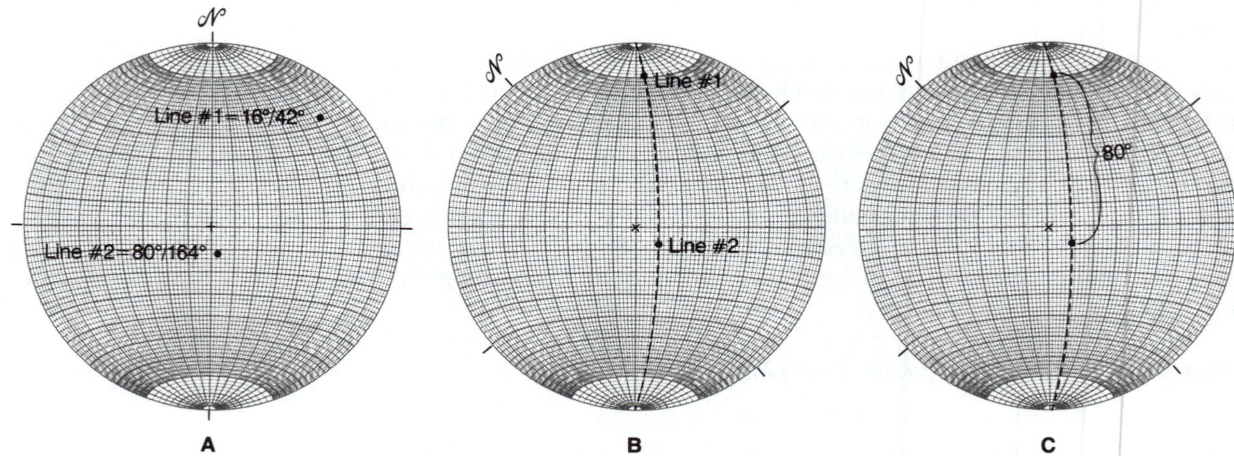

Figure H.14 Stereographic determination of the angle between two lines. (*A*) Plot the orientation of the lines stereographically as points. (*B*) Fit the lines to a common great circle (i.e., to a common plane). (*C*) Measure the acute angle between the two lines, by counting along the common great circle that connects the points.

Measuring the Angle Between Two Planes

The angle between any two planes, like two faults or two joints, is the same as the angle between the poles to the planes. Consequently, if the orientations of two planes are plotted as poles, measuring the angle between the poles reduces the problem to measuring the angle between two lines. And we have seen how this is done.

Figure H.15*A* shows two planes, plane 1 striking 305° and dipping 26°SW, plane 2 striking 010° and dipping 41°NW. The poles to these planes are shown stereographically. To measure the angle between the planes, simply align the two poles on the same great circle (Figure H.15*B*) and measure the acute angle between the poles by counting 2° small-circle intercepts along the great circle. For this example, the angle is 36°.

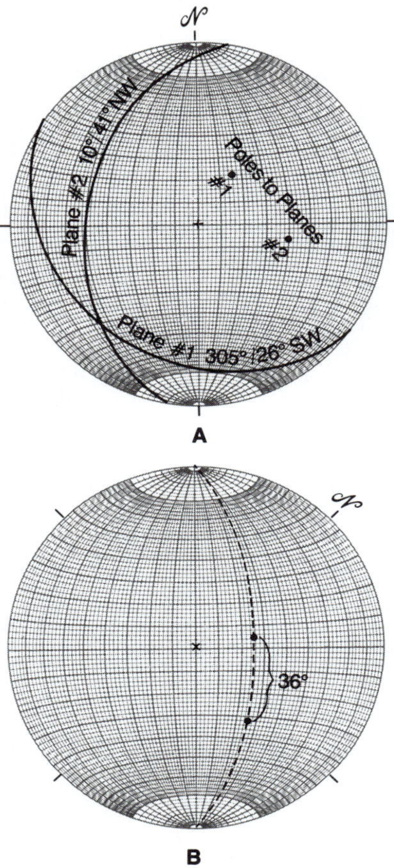

Figure H.15 Stereographic determination of the angle between two planes. (*A*) Plot the orientations of the planes both as great circles and as poles. (*B*) Fit the poles of the planes to a common great circle, and then measure the acute angle between the poles.

Determining the Orientation of the Intersection of Two Planes

The payoff for learning the principles of stereographic projection is derived from the ease with which certain geometric problems can be solved. One of the best examples of the effectiveness of stereographic projection is determining the trend and plunge of the intersections of two planes. Consider two planes, one striking N49°E and dipping 42°SE, the other striking N10°W and dipping 65°NE (Figure H.16*A*). The intersection of the great circles is a point *i* whose orientation is that of the line of the intersection of the two planes. The trend of the line *i* can be determined by drawing (projecting) a straight line from the center of the projection through line *i* to the perimeter, and measuring the orientation of this line with respect to north or south. For this example, the trend is S34°E (Figure H.16*B*). The plunge of the line is found by rotating line *i* to the east–west or north–south lines and measuring its inclination from the perimeter (Figure H.16*C*). The plunge as measured in this example is 42°.

Apparent dip problems are a special case of determining the line of intersection of two planes. Figure H.17*A* shows bedrock in a seacoast exposure. Bedding strikes N24°E and dips 79°SE. What would be the apparent dip for these beds, as observed in a vertical cliff face that trends N40°E?

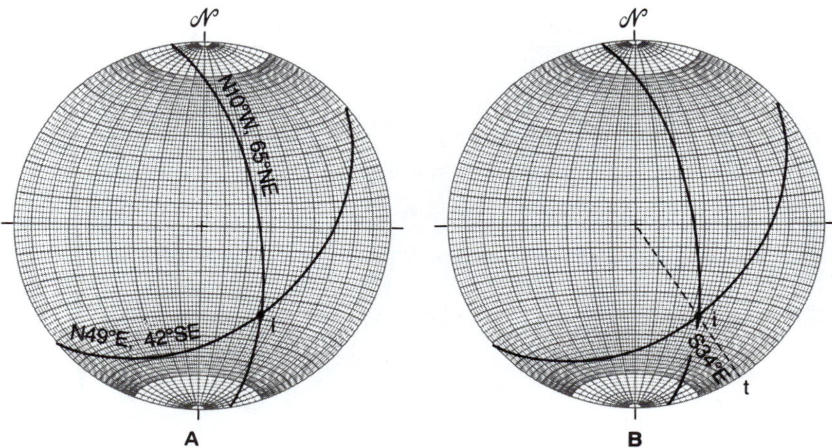

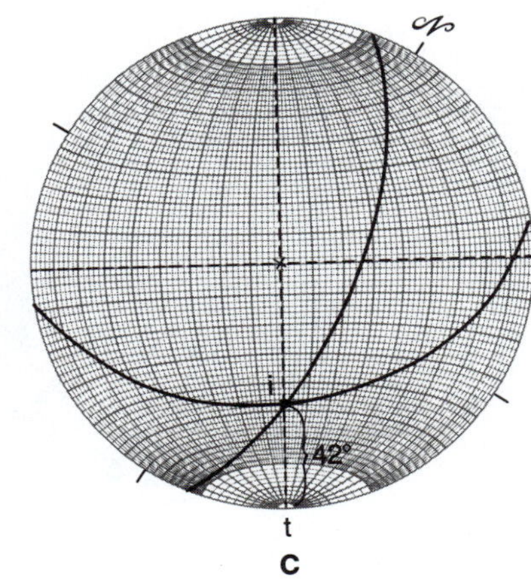

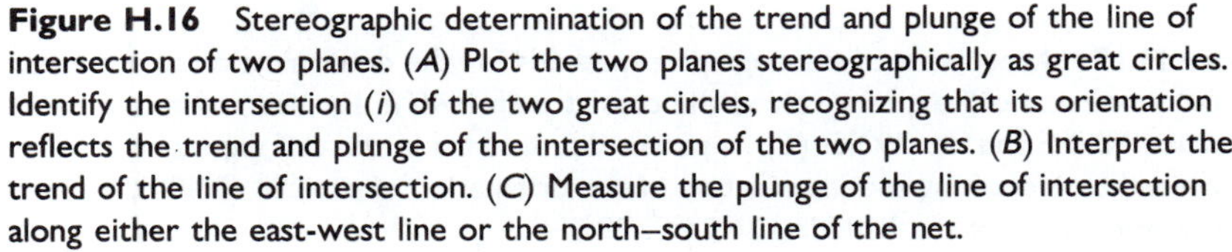

Figure H.16 Stereographic determination of the trend and plunge of the line of intersection of two planes. (*A*) Plot the two planes stereographically as great circles. Identify the intersection (*i*) of the two great circles, recognizing that its orientation reflects the trend and plunge of the intersection of the two planes. (*B*) Interpret the trend of the line of intersection. (*C*) Measure the plunge of the line of intersection along either the east-west line or the north–south line of the net.

To solve, plot the orientation of the bed and the orientation of the vertical face stereographically (Figure H.17*B*). The intersection of the two planes is a point that represents the trend and plunge of the line of intersection of the bed and the vertical cliff face. The plunge value is in fact the apparent dip of the plane, namely 54°NE. Examples like this begin to scratch the surface of the power of the three-dimensional protractor known as the stereographic net.

Stereographic Projection as a Statistical Tool

We are often required in structural analysis to determine the average orientation of a certain structural element, or to determine whether the range of orientations of a particular structure is in some way systematic. One way to identify **preferred orientations** of structures might be to plot lines or poles to planes stereographically on a Wulff net (see Figure H.3*B*) and to evaluate the extent to which the plotted points tend to cluster or to spread in systematic ways. But were we to do this, we would find that the Wulff net, otherwise known as the equal-angle net, has an undesirable built-in bias. Two degree areas bounded by great and small circles on a Wulff net are not of equal size. Those toward the periphery of the net are larger than those toward the center (see Figure H.3*B*). The central part of the Wulff net takes in a greater range of orientations than the peripheral parts. Consequently, even orientations gathered from a table of random numbers and plotted stereographically would show an uneven concentration of points across the face of the circular net. The distribution of points might convey the incorrect impression that most of the lines, plotted as points, reflect relatively steeply plunging orientations. This geometric peculiarity invalidates the equal-angle net as a useful *statistical* device for evaluating preferred orientations. Instead, the Schmidt net, the equal-area net, is used.

The geometry of projection of the Schmidt net (see Figure H.3*A*) is such that 2° areas bounded by great and small circles are the same size across the net (Phillips, 1971). Since 2° great-circle/small-circle areas are the same size across the entire face of a Schmidt net, randomly distributed orientations will appear random, whereas nonrandom concentrations of

Figure H.17 Application of stereographic methods for the determination of apparent dip. (*A*) A vertical cliff exposure along the coast trends N40°E. Inclined strata in the cliff strike N24°E and dip 79°SE. (*B*) Apparent dip of strata in the cliff exposure is determined by identifying the intersection of the orientation of the cliff (N40°E, 90°) and the orientation of the beddding (N24°E, 79°SE).

stereographically plotted points reflect preferred orientations. Contouring the values of the **density distribution** of plotted points provides a measure of the degree of preferred orientation.

Evaluating Preferred Orientations

A stereographic projection of 65 poles to bedding in Cretaceous strata in the Mule Mountains near Bisbee, Arizona, is shown in Figure H.18*A*. The general concentration of points is near the center of the projection, signifying that the beds whose orientations are plotted are rather gently dipping. But what is the specific orientation of the bedding, as expressed in strike and dip? And what is the strength of the preferred orientation? Answering these questions requires contouring the density distribution of the plotted points on the face of the stereonet.

To evaluate density distribution, the equal-area net is subdivided into a gridwork of many overlapping circular areas, each of which corresponds to 1% of the area of the stereographic projection. Density is described in terms of percentage of total data points falling within a given 1% area of the stereographic projection.

Pole-density distribution is thus calculated as follows:

$$\text{Density (\%)} = \frac{\text{Number of points within 1\% area of net}}{\text{Total number of data points}} \times 100$$

To subdivide a 20-cm-diameter stereogram into overlapping 1% circular areas, a square grid is constructed such that the spacing of grid intersections is 1 cm. The square gridwork is overlain by the tracing paper on which the data points were originally stereographically plotted (Figure H.18*B*). The grid intersection points are used as control points for systematically moving a **center counter** whose area is 1% that of the stereogram (Whitten, 1966, pp. 20–26). In this manner the **number** of data points that lie within each of the overlapping 1% areas is counted. The counts are posted on yet another overlay (Figure H.18*C*).

A calculator is used to convert the numbers of data points in each of the 1% areas to percentage of total data points (Figure H.18*C*). It is the **percentage values** of density distribution that are contoured (Figure H.18*D*). Since the 1% areas are overlapping, and because most data points are counted more than once, the density distribution values represent a smoothed portrayal of the raw orientations.

Ideally, the finished contour diagram (Figure H.18*E*) should be marked by some constant contour interval, with the number of contour-line values not exceeding five or six. To emphasize the density distribution, it is common practice to shade the diagrams in such a way that the zones of highest density are darkest and most pronounced. For the example of bedding orientations in the Mule Mountains, the completed diagram discloses a bull's-eye, **unimodal distribution** of points. The eye or center of the contoured pattern corresponds to the preferred orientation of the bedding.

The counting and contouring process is relatively straightforward, except when considering density distribution values close to or on the perimeter of the projection. Consider the point plot of 85 measurements of mineral lineation in gneiss in the Rincon Mountains of southern Arizona (Figure H.19*A*). The lineations are low plunging, as can be judged by their closeness to the perimeter of the projection. The number of data points falling within each 1% area in the interior of the stereographic projection can be counted using a center counter. But in this example, individual 1% area counting circles sprawl beyond the perimeter of the projection, and thus

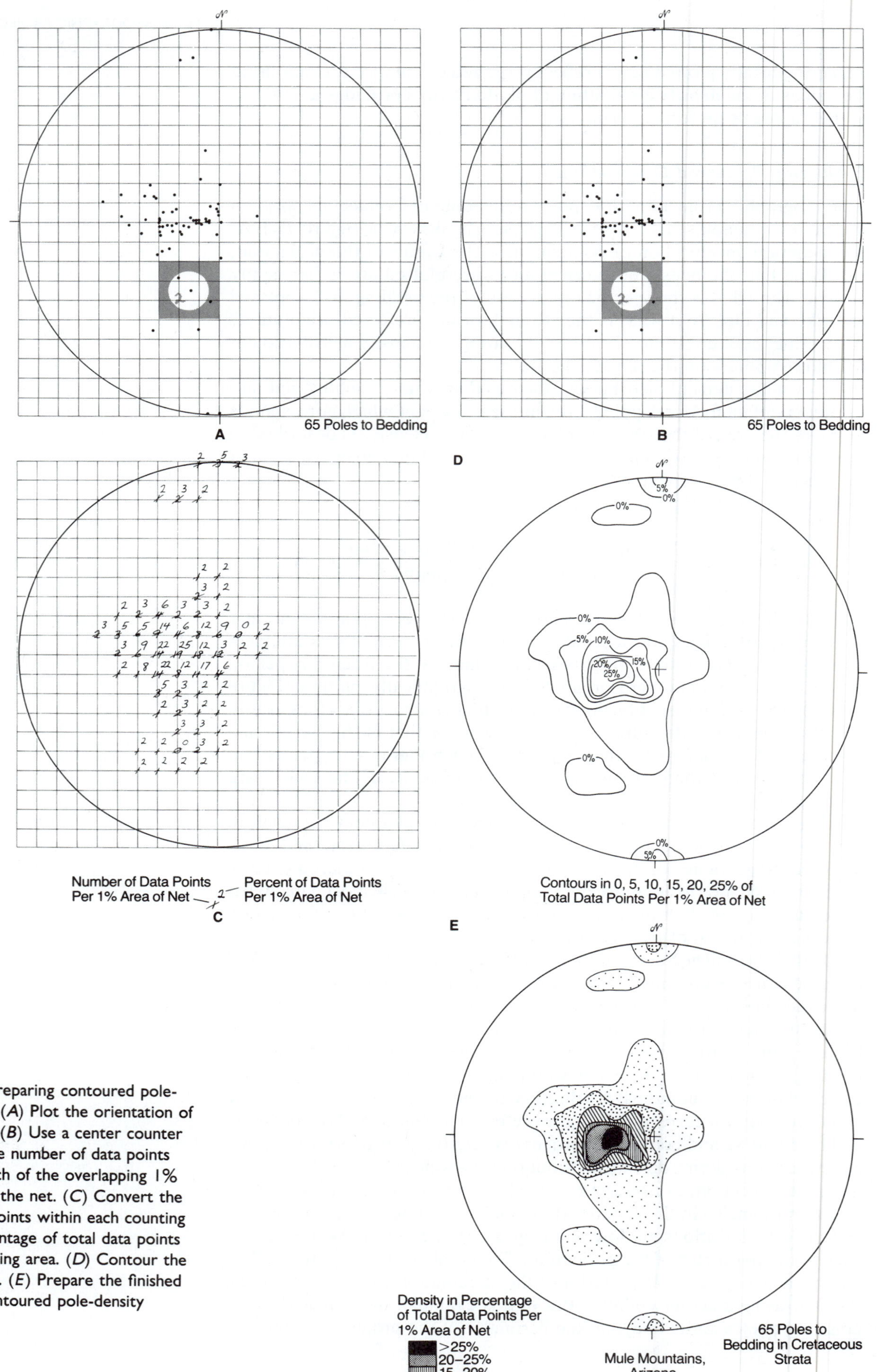

Figure H.18 Preparing contoured pole-density diagrams. (*A*) Plot the orientation of bedding as poles. (*B*) Use a center counter and determine the number of data points that lie within each of the overlapping 1% counting areas of the net. (*C*) Convert the number of data points within each counting area to the percentage of total data points within each counting area. (*D*) Contour the percentage values. (*E*) Prepare the finished version of the contoured pole-density diagram.

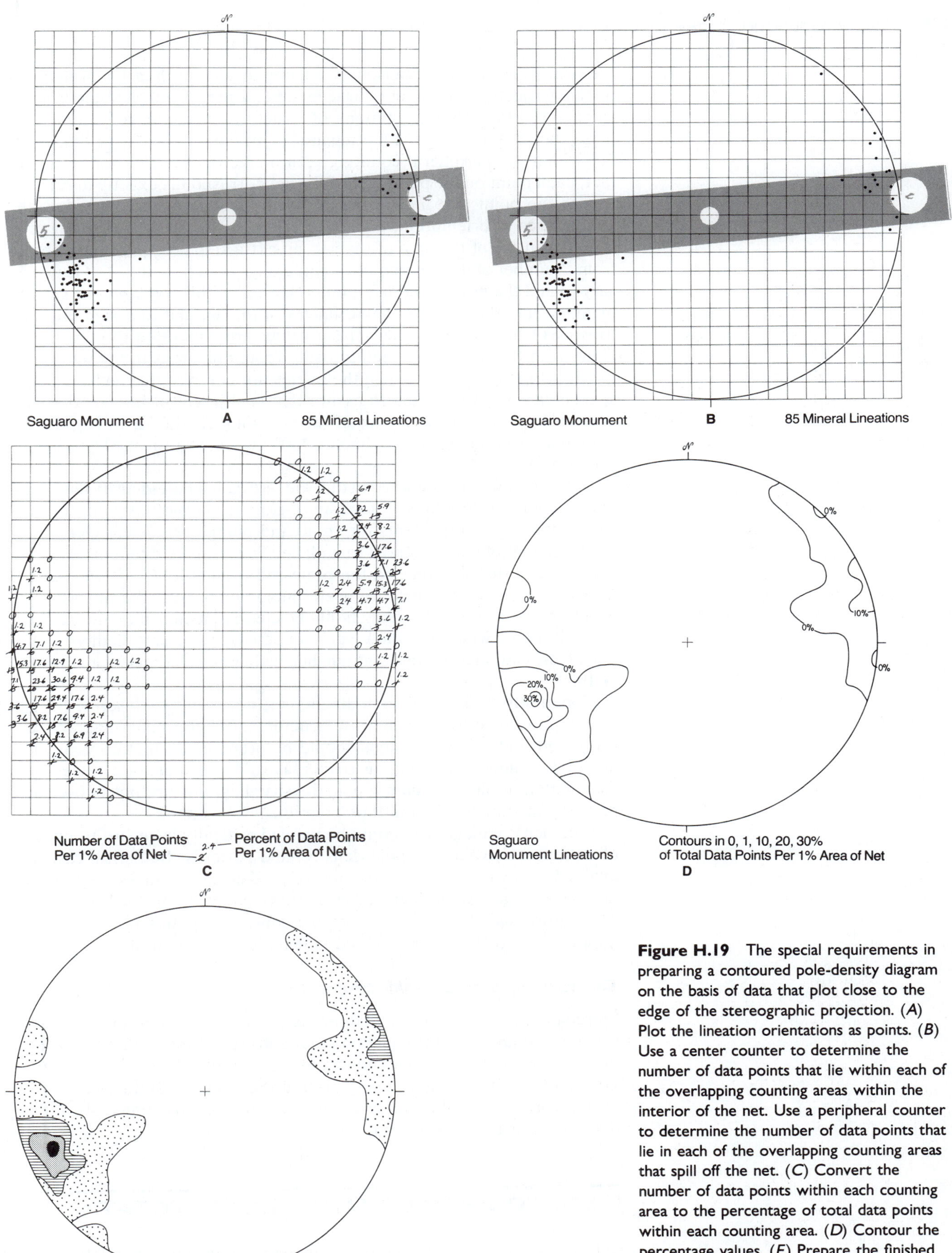

Figure H.19 The special requirements in preparing a contoured pole-density diagram on the basis of data that plot close to the edge of the stereographic projection. (*A*) Plot the lineation orientations as points. (*B*) Use a center counter to determine the number of data points that lie within each of the overlapping counting areas within the interior of the net. Use a peripheral counter to determine the number of data points that lie in each of the overlapping counting areas that spill off the net. (*C*) Convert the number of data points within each counting area to the percentage of total data points within each counting area. (*D*) Contour the percentage values. (*E*) Prepare the finished version.

the points near or at the periphery occupy areas less than 1% of the area of the projection. To count these points, a **peripheral counter** is used (Figure H.19*B*). The orientations of structural elements represented by points lying in peripheral counting areas correspond closely to those that fall within the peripheral counting area diametrically (180°) opposite. In fact, each point on the perimeter of the net corresponds exactly in orientation to the point that lies on the perimeter 180° away. Recognizing this equivalence in orientation, and given that each peripheral counting area is less than 1% of the area of the net, an extra step is required to convert from "number of data points" to "percentage of data points." It is necessary to add the number of data points that fall within *each pair* of supplementary partial circles on the perimeter of the net and to assign this sum to *each* of the partial circles. These values are converted to percentage, along with the values assigned to center counting areas in the interior of the net (Figure H.19*C*). When the values are contoured (Figure H.19*D*), special care must be taken to assure that each point of intersection of a contour line with the perimeter of the net is matched by a corresponding intersection point on the perimeter 180° away. The finished product is shown in Figure H.19*E*.

Contour diagrams prepared in this way are called **pole-density diagrams**, where "pole" refers loosely to a stereographically plotted point, regardless of whether it represents the trend and plunge of a linear element or the trend and plunge of a pole to a plane. Pole-density diagrams describe the range in distribution and the preferred orientation(s), if any, of the measured structures. They are useful in summarizing large quantities of geometric data. Each diagram should be clearly labeled according to area of study (e.g., where the data were collected), structural element, number of measurements, and contour-line values) (see Figures H.18*E* and H.19*E*).

Commonly an array of such diagrams is necessary to describe the full orientation range for a system of structures. Where many diagrams are presented, individual diagrams should be prepared so that contour lines are of a common interval, in percentage. When there are fewer than 50 orientation measurements, pole-density diagrams are not particularly meaningful statistically and thus point diagrams suffice.

Pole-density diagrams should be viewed critically in regard to number of data points, values of pole density, and patterns. Statistical tests are available to evaluate the significance of pole-density values in light of the number of data points (Kamb, 1959). Many types of pattern are possible. The nature and the symmetry of pole-density diagrams have important implications for the geometry and kinematics of structural systems.

Software for Stereographic Projection

Stereographic projection, including the production of contour diagrams, is made fast and easy through computer software, such as that generously made available free for noncommercial use by Rick Allmendinger, Cornell University. Students and faculty around the country, and in many parts of the world, are using such programs to plot, contour, and analyze their data much faster than by hand.

I. STEREOGRAPHIC EVALUATION OF ROTATION

Certain stereographic techniques are indispensable in the kinematic analysis of rotational operations. Stereographic techniques can be used to pic-

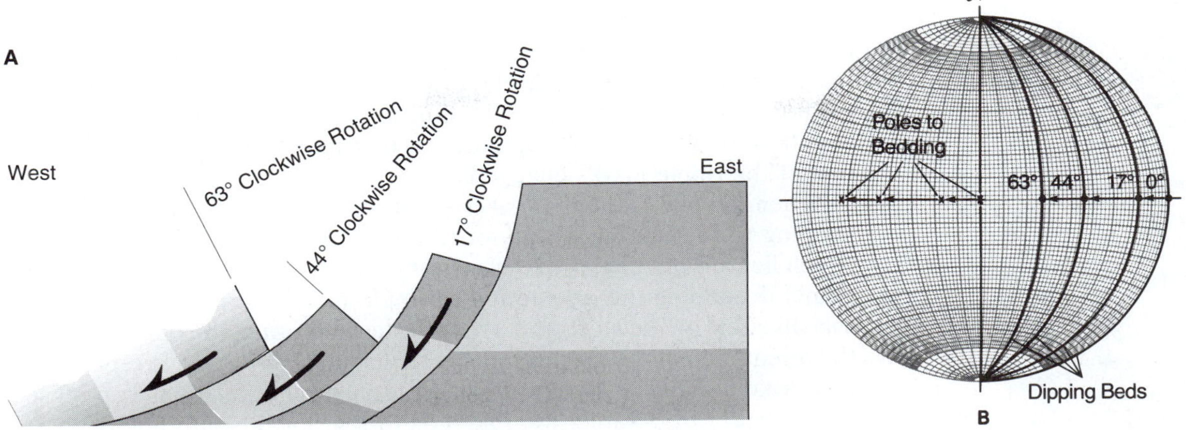

Figure I.1 (*A*) Rotation of strata by listric faulting, portrayed at different stages. (*B*) Stereographic representation of the rotation of strata.

ture the rotation of lines and planes in geometric space. A stereonet is used to plot the path or **locus** of points representing lines and planes at various stages of rotation. The plotting procedure is based on facts regarding the orientation of the axis of rotation, sense of rotation, and the magnitude of rotation in degrees. Let us consider some examples.

Example of Rotational Faulting

Figure I.1*A* shows strata in various stages of rotation during listric normal faulting. Orientation of strata at each stage is portrayed stereographically in Figure I.1B, both through the use of great circles representing strike and dip of bedding and through poles to bedding. Poles provide the most convenient representation. The track of poles, taken together, is the locus of points representing progressive rotation of the strata by faulting. Note that the sequence of poles to bedding is aligned along a small circle perpendicular to the axis of rotation.

Example of Rotation of Layers by Folding

Progressive folding of a layer of rock is portrayed in Figure I.2*A*. In Figure I.2*B*, the strike and dip of bedding on the limbs of the fold are shown stereographically at each stage of deformation. The poles to bedding, taken together, represent the locus of points describing the rotation of bedding during folding. The locus is in the form of a great circle that lies 90° from the axis of rotation, just like the great-circle distribution of poles to bedding in the example of rotation during listric faulting (see Figure I.1*B*).

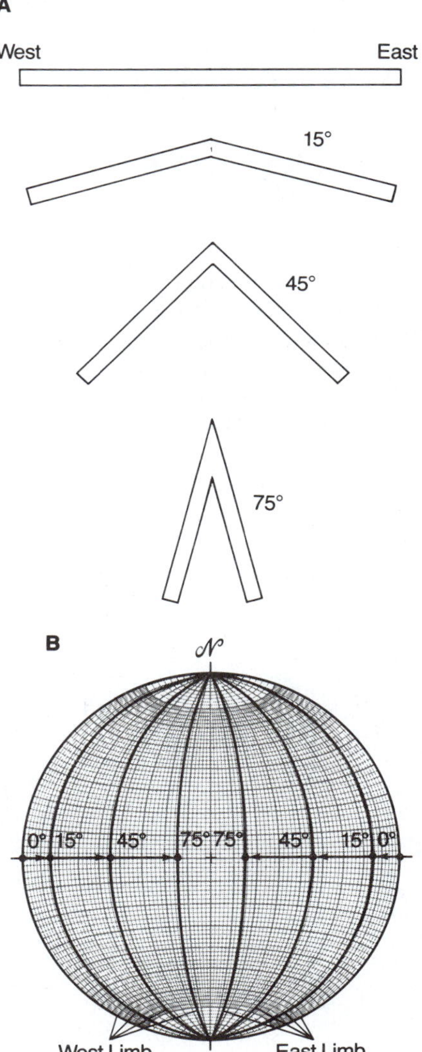

Figure I.2 (*A*) Rotation of strata during progressive folding. (*B*) Stereographic representation of the rotation of bedding during the folding.

Stereographic Unfolding of Layers with Ripple Marks

The general case of rotation of strata by folding is shown in Figure I.3, where layered strata are flexed about an inclined axis. The axis of rotation plunges 30° S45°E. The locus of poles to folded bedding describes a great circle lying 90° from the axis.

The ability to stereographically rotate lines in space is essential in structural studies. The geological lines we focus on now are the crests and troughs of ripple marks (Figure I.4*A*). Consider a horizontal layer of

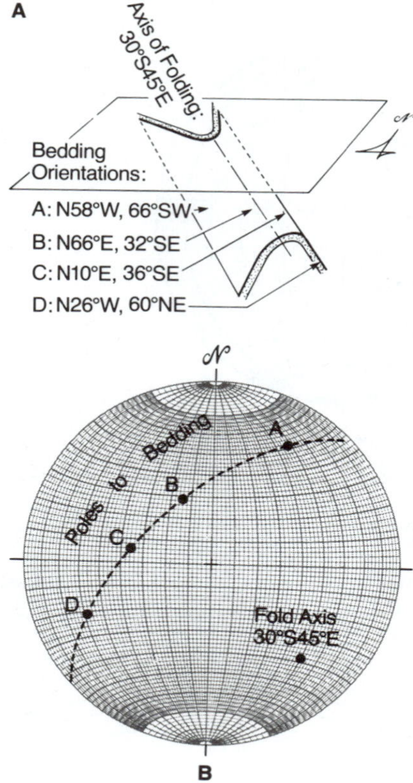

Figure I.3 (*A*) Rotation of bedding by folding about an inclined axis. (*B*) Stereographic portrayal of poles to the rotated, folded bedding, measured at locations *A–D*. Note that the poles lie on a great circle whose pole is the fold axis.

sandstone containing parallel-aligned current ripple marks (Figure I.4*B*). The ripple marks lie in the plane of bedding, and thus a point that stereographically portrays the trend and plunge of the ripple marks must fall on the great circle representing the orientation of bedding (Figure I.4*C*). If the horizontal sandstone layer is subjected to 48° of southeastward tilting about a horizontal, N70°E-trending axis (Figure I.4*D*), how will the orientation of the ripple marks change?

When the bed is tilted 48° about a horizontal axis of rotation, poles to bedding at the various stages of tilting trace out part of a great circle, which lies at right angles to the axis of rotation (Figure I.4*E*). In contrast, points describing the orientation of ripple marks during tilting trace out a small circle on the stereonet. At each stage of tilting, the point describing the orientation of the ripple marks remains on the great circle representing the strike and dip of bedding. These stereographic relations can be seen more clearly by rotating the fold axis to the north–south line of the net and observing that poles to bedding all lie on the x axis of the net, and noting the correspondence between the movement path of ripple mark orientations and one of the small-circle traces embossed on the underlying net (Figure I.4*F*). The angle between the ripple mark orientation and the axis of rotation, at each stage of tilting, remains constant, matching the angle between the trend of the ripple marks and the trend of the axis of rotation before the tilting commenced (see Figure I.4*B*).

Suppose as field geologists we encounter an outcrop of tilted strata revealing both bedding and ripple marks (Figure I.5*A*). How can we determine the original orientation of the ripple marks? We begin by measuring the strike and dip of bedding (N80°W, 80°SW) and the trend and plunge of ripple marks (39° N88°W). Alternatively, we measure the strike and dip of bedding (N80°W, 80°SW) and the rake of the ripple marks in the plane of bedding (39°W). When these data are plotted stereographically (Figure I.5*B*), the point representing the trend and plunge of the ripple marks is seen to lie on the great circle that describes the strike and dip of bedding. The original orientation of the ripple marks is found by rotating bedding to its inferred original horizontal orientation. To do this an axis of rotation must be chosen.

For this problem, the line of strike of bedding is the right choice. Its orientation is N80°W. To rotate stereographically about any horizontal axis, the point(s) representing the selected axis of rotation must be brought into alignment with north (or south) on the perimeter of the stereonet. This is achieved simply by rotating the tracing overlay clockwise by 80° (Figure I.5*C*). Thus positioned, the axis of rotation lies as the central axis to the small-circle paths of rotation. By rotating the pole to bedding and the point representing trend and plunge of ripple marks about this axis, and in the proper sense, the bed containing the ripple marks is, in effect, lifted from its steeply inclined orientation to horizontal (Figure I.5*C*). The points traverse small circles from the interior of the net to the perimeter. Once accomplished, the original trend of the ripple marks can be measured: namely, S61°W.

Multiple Rotations

A final example of rotational kinematics is required to present the general operation(s) that can be adapted to all rotational problems, regardless of complexity. The goal is to be able to carry out successive rotational operations about axes of different orientations. The order in which rotational operations are performed vitally influences the final orientation of the rotated body. Multiple rotations are simply a series of single rotations

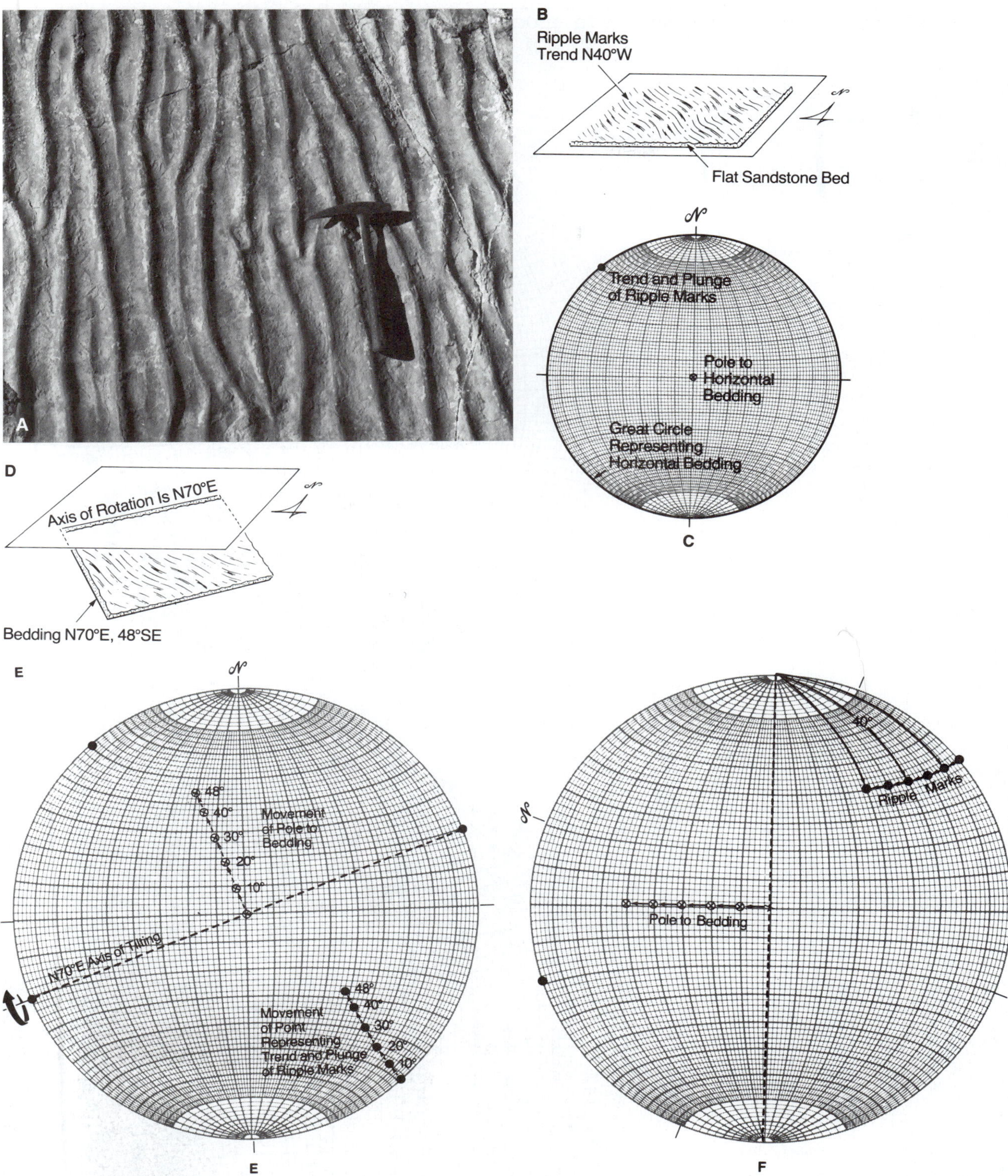

Figure I.4 (*A*) Geologic lines in a plane, namely, crests and troughs of current ripple marks in the Dakota Sandstone. (Photograph by J. R. Stacy. Courtesy of United States Geological Survey.) (*B*) Ripple marks trending N40°W in horizontal sandstone bed. (*C*) Stereographic portrayal of the orientations of the sandstone bed and the ripple marks before tilting. (*D*) Bedding and ripple marks after tilting. (*E*) Stereographic portrayal of the tilting of the sandstone bed and the consequent change in orientation of the ripple marks. (*F*) Stereographic view showing how the movement path of the pole to bedding follows the great-circle trace along the east–west line of the net, whereas the movement path of the point representing the orientation of the ripple marks follows a small-circle path.

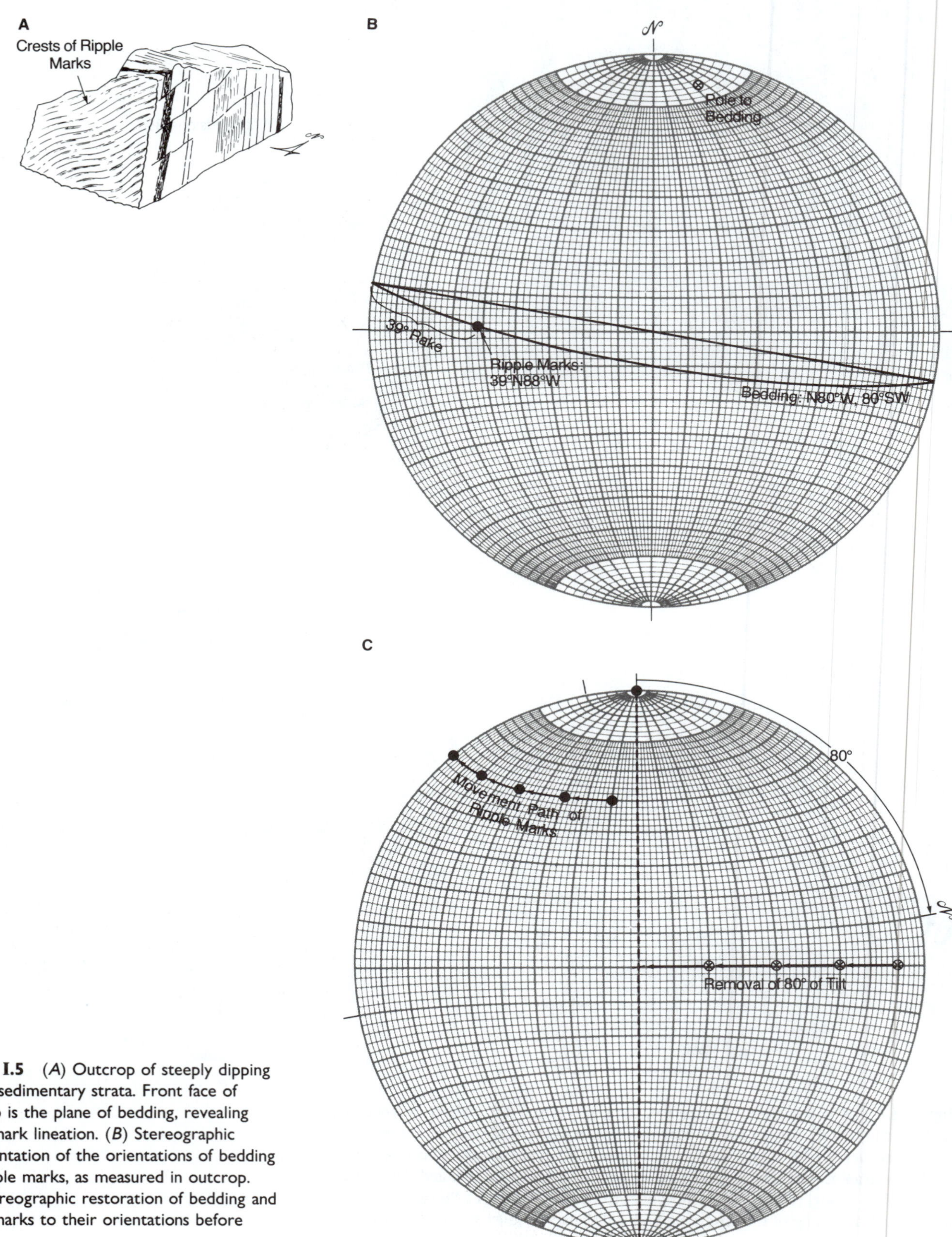

Figure I.5 (*A*) Outcrop of steeply dipping faulted sedimentary strata. Front face of outcrop is the plane of bedding, revealing ripple mark lineation. (*B*) Stereographic representation of the orientations of bedding and ripple marks, as measured in outcrop. (*C*) Stereographic restoration of bedding and ripple marks to their orientations before tilting.

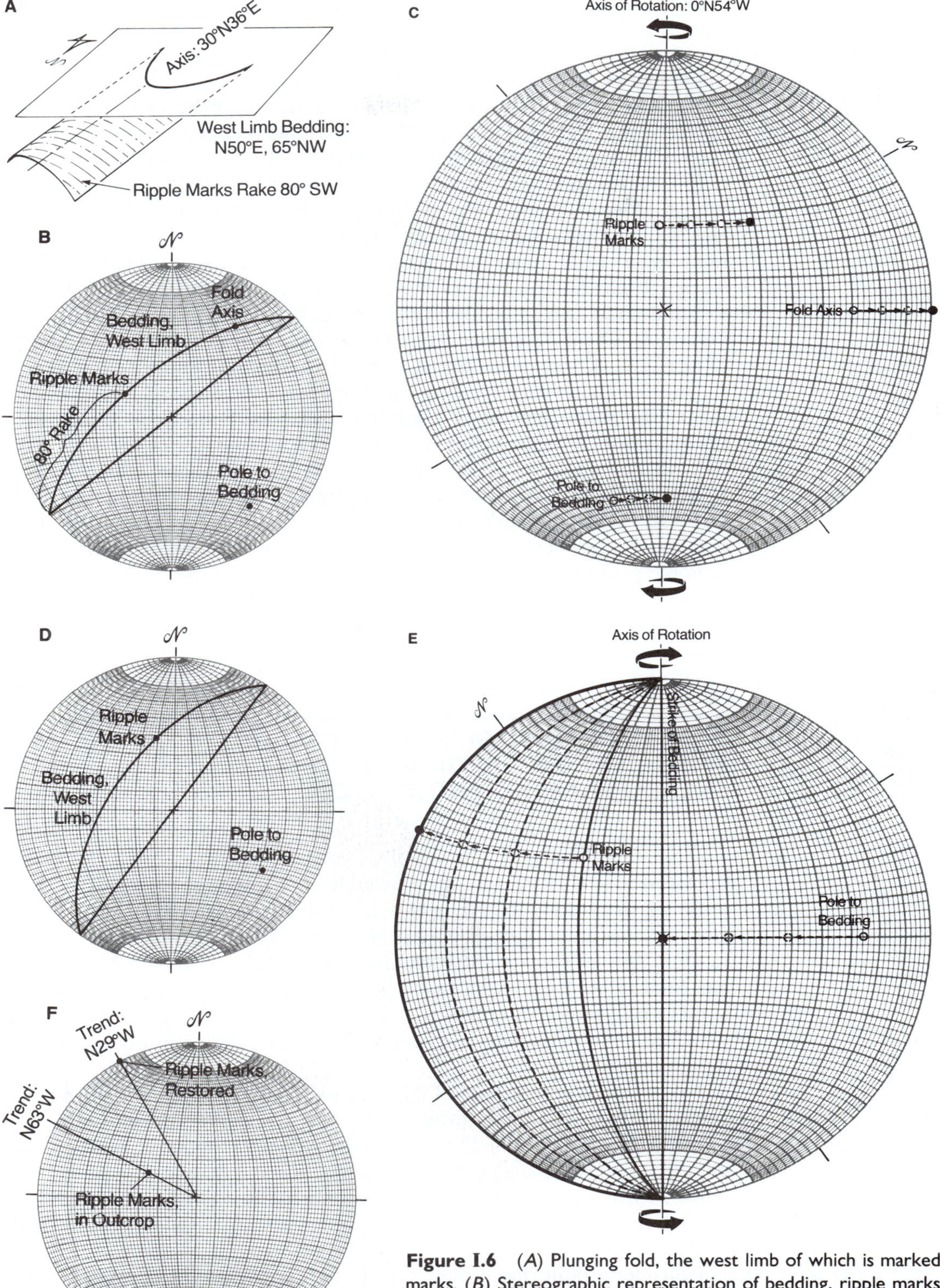

Figure I.6 (*A*) Plunging fold, the west limb of which is marked by abundant ripple marks. (*B*) Stereographic representation of bedding, ripple marks, and the axis of folding. (*C*) Rotation of fold axis to horizontal, and rotation of the pole to bedding and the ripple marks by a like amount. (*D*) Orientation of west-limb bedding and the ripple marks following rotation of the fold axis to horizontal. (*E*) Rotation of bedding to horizontal about an axis parallel to the strike of bedding. (*F*) Measurement of the trend of ripple marks in the restored configuration.

applied in sequence. If we can do one, we should be able to do more than one.

Suppose we are interested in determining the original orientation of ripple marks that lie in a steeply inclined limb of a plunging fold (Figure I.6*A*). The fold axis plunges 30° N36°E. Bedding on the west limb of this fold strikes N50°E and dips 65°NW. The ripple marks rake 80°SW. These geometric data are plotted stereographically in Figure I.6*B*. The first step in the restoration is to rotate the fold axis to horizontal, and at the same time to rotate the pole to bedding and the ripple marks by the same amount. The rotation axis chosen to achieve this plunges 0° N54°W, at right angles to the trend of the fold axis. The rotation axis is brought to the north position on the perimeter of the net (Figure I.6*C*) and is lifted to horizontal, traversing 30° along the east–west line of the net to the perimeter. The pole to bedding and the point defining the trend and plunge of ripple marks move 30° in the same sense but along small-circle paths.

Once this initial step has been accomplished, the great circle perpendicular to the pole to bedding is plotted (Figure I.6*D*). It passes through the rotated position of the ripple mark axes. The strike line of the bedding is then aligned along the north–south line of the net (Figure I.6*E*), serving as the axis of rotation about which the final operation is completed. Bedding and the ripple marks are lifted to horizontal, traversing small-circle paths to the perimeter of the net (Figure I.6*E*). Returning the tracing paper to home position (Figure I.6*F*), the original trend of the ripple marks can be directly measured. Its value is N29°W, substantially different from N63°W, the trend of the ripple marks as they appear in outcrop.

J. DETERMINING SLIP ON A FAULT THROUGH ORTHOGRAPHIC AND STEREOGRAPHIC PROJECTION

The Challenge

Sometimes it is necessary to determine the direction, magnitude, and sense of slip from geologic map relationships, using the combination of orthographic and stereographic projection. There are a number of ways to do it, but all require the identification of a real or geometric line that has been offset by faulting. A classic method in evaluating slip on faults is one made famous in *Structural Geology* by Billings (1954, 1972). The "Billings" fault problem has become a rite of passage in undergraduate structural problems.

Not many reference lines of this type exist in nature. On the other hand, **geometric lines** exist in deformed rock systems. Billings drew attention to the kind of geometric line that is formed by the intersection of two geologic planes, like a dike and a bed. If such a line is cut and displaced by a fault, a basis for evaluating translation is readily available. The concept is relatively simple (Figure J.1). The actual solution requires 3-D visualization.

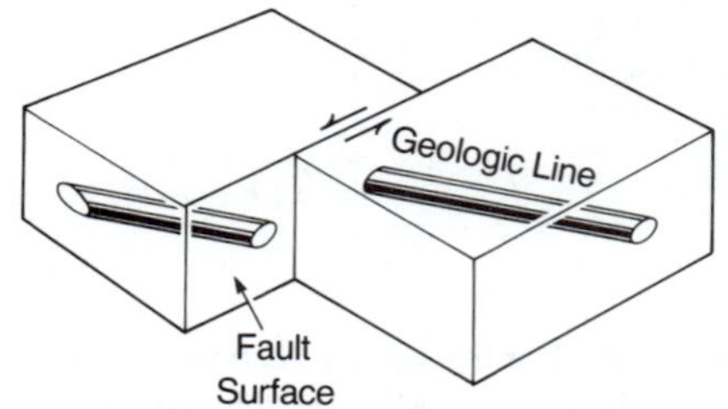

Figure J.1 Schematic portrayal of the offset of a once-continuous line by faulting. Reconstruction of the line permits the direction, sense, and magnitude of the translation vector due to faulting to be calculated.

Consider a fault that strikes N70°E and dips 75°5E (Figure J.2*A*). Assume that it is exposed in a perfectly flat area whose elevation is 2000 ft (609 m). A dike and a distinctive limestone bed crop out on both the north and south sides of the fault. The dike strikes Nl0°E and dips 45°SE. The limestone bed strikes N60°W and dips 50°SW. What is the displacement vector describing the movement of the south block of the fault relative to the north block?

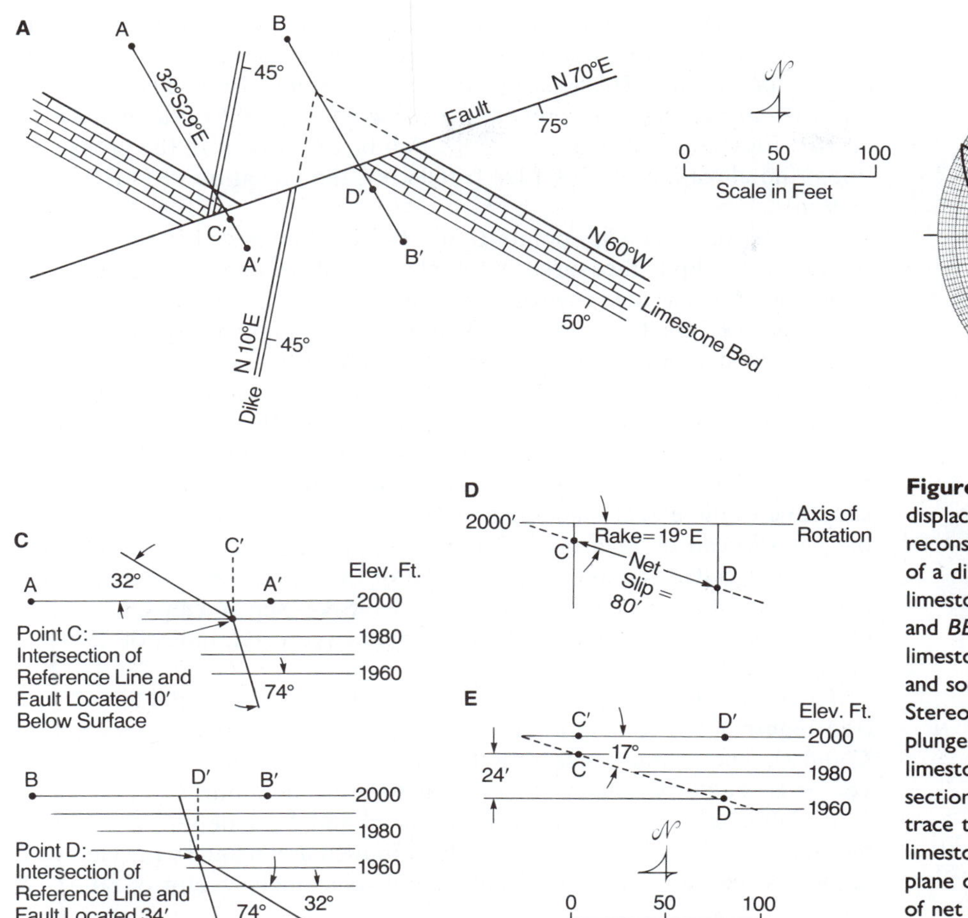

Figure J.2 Determination of the displacement vector for a fault by reconstructing the faulted line of intersection of a dike and a limestone bed. (*A*) Faulted limestone bed and faulted dike. Lines *AA′* and *BB′* are the vertical projections of the limestone–dike intersections on the north and south sides of the fault, respectively. (*B*) Stereographic determination of the trend and plunge of the line of intersection of the limestone bed and the dike. (*C*) Cross sections showing the relation of the fault trace to the line of intersection of the limestone bed and dike. (*D*) View in the plane of the fault showing net slip and rake of net slip. (*E*) Cross section showing plunge of the displacement vector.

The intersection of the dike and the limestone bed provides the geometric line of reference for establishing slip on this fault. The trend and plunge of the reference line is determined stereographically by plotting the intersection of the great circles representing the dike and the limestone bed, respectively (Figure J.2*B*). The orientation of the line is 32° S29°E. The vertical projection of the line of intersection of the dike and the limestone bed on the north block of the fault is plotted as line *A–A′* (see Figure J.2*A*). This line passes through the surface intersection of the trace of the dike and the trace of the limestone bed as seen on the surface in outcrop. The counterpart of this line on the south block of the fault is *B-B′*.

To find the displacement vector for fault translation, it is necessary to determine where the two geometric lines of intersection, one on each side of the fault, pierce the fault surface. This step is ordinarily the most difficult to visualize and to construct graphically. To see the relationship, construct vertical cross sections that pass through the vertical projections (*A–A′* and *B–B′*) of the intersection of the dike and the limestone bed (Figure J.2*C*). The 32° plunge of the line of intersection of the dike with the limestone bed is plotted on each of the cross sections. The **apparent dip** of the fault is also plotted, and its value is determined stereographically. In this example, the apparent dip of 74° happens to be merely 1° less than true dip. Where the line of intersection of the dike and the limestone bed meets the inclined trace of the fault, there lies the reference point. For the north block of the fault, *C* marks the intersection point and point *C′*

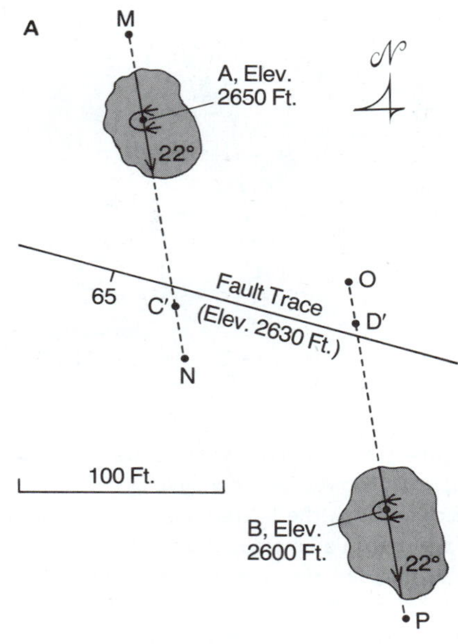

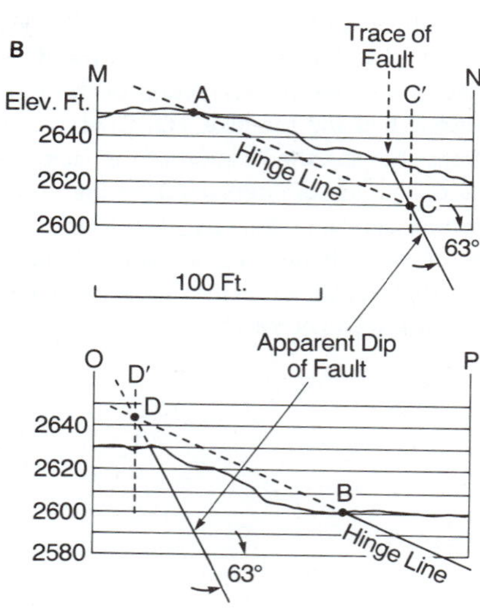

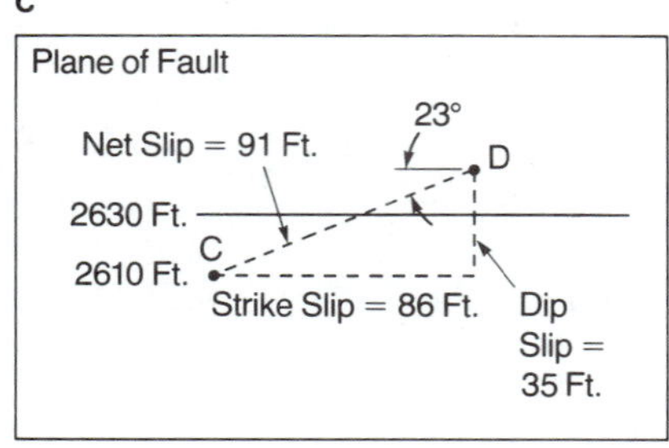

Figure J.3 Determination of the translation along a fault on the basis of the offset of the hinge line of a fold. (*A*) Map relationships showing the fault offset of the trace of an overturned syncline. (*B*) Structure profiles showing where hinge lines intersect fault surface. (*C*) Slip components as seen in the fault plane.

marks its vertical projection (Figure J.2*C*). For the south block of the fault, the intersection point is *D*, and the vertical projection is *D'* (Figure J.2*C*). *Points C and D originally occupied the same position before faulting; they are equivalent points that can be connected by a displacement vector.* Elevations of these reference points are evident in the cross sections: the elevation of *C* is 1990 ft (606 m); the elevation of *D* is 1966 ft (599 m).

To determine the slip, points *C* and *D* must be viewed in the plane of the fault, for slip is measured in the plane of movement. To achieve this view, the fault surface is figuratively rotated to the surface about its strike line (Figure J.2*D*). The fault plane is thus laid out in the plane of the paper so that the **net slip** can be directly measured. The axis of rotation, the strike line, is the intersection of the fault with the horizontal surface of the terrane. Its elevation is 2000 ft. The trace of this axis of rotation is shown in Figure J.2*D*. Reference points *C* and *D* are plotted along lines perpendicular to the axis of rotation, and at down-dip distances corresponding to the elevations of *C* and *D*. Once *C* and *D* have been plotted, the net slip can be measured. The magnitude of the net slip is 80 ft (24 m). This is called the **net slip**. The rake of the net slip on the fault plane is 19°E. Net slip can be subdivided into two **strike–slip** and **dip-slip** components (Figure J.2*D*), which in this example have magnitudes of 75 ft (23 m) and 25 ft (7.6 m), respectively.

The trend of the displacement vector is the azimuth of the line *C'–D'* (see Figure J.2*A*) connecting the vertical projections of reference points *C* and *D*. The azimuth measures N78°E. The plunge of the displacement vector is found by constructing a structure profile along *C'–D'* (Figure J.2*E*), showing the relative positions of *C* and *D* at depth. The plunge measures 17°. Thus, the southeast block moved down and to the east with respect to the northwest block, along a displacement vector of 80 ft, 17° N 78°E.

Using the Faulted Hinge of a Fold to Determine Slip

The faulted hinge line of a fold also can be used as a basis for determining fault slip. The hinge of a folded layer is like the crease in a folded piece of paper. In the mapped relationship shown in Figure J.3*A*, the hinge of a syncline plunges 22°S10°E. The fold hinge is offset by a fault that strikes N75°W and dips 65°SW. Displaced parts of the faulted fold hinge crop out on both sides of the fault (locations *A* and *B*, Figure J.3*A*). Before faulting, these hinges were part of a straight, continuous hinge line in a sandstone bed. Slip on the fault is deduced by projecting each of the offset hinge segments into the plane of the fault. Once accomplished, the distance and direction between these projected lines can be measured.

The orthographic solution involves constructing vertical structural profiles oriented parallel to the trend of the fold axis (N10°W)(Figure J.3*B*). On the footwall side, the hinge line projects downward at 22° toward the projected position of the fault in the subsurface. The projection of the hinge line pierces the fault at point *C*, elevation 2610 ft (791 m). On the hanging-wall side, the hinge line rises 22° skyward from its outcrop elevation to point *D* (elevation 2645 ft; 801 m) where it pierces the upward projection of the fault plane.

As in the previous example, the **net slip** on the fault is oblique, neither parallel to the strike of the fault or down the dip of the fault. The net slip can be divided into two components: the **strike–slip** and the **dip-slip** (Figure J.3*C*). The actual strike–slip and dip-slip components of translation are measured in the plane of faulting (Figure J.3*C*). To accomplish this, points

C and *D* are plotted in the plane of the fault at their respective elevations and locations. The strike–slip, dip-slip, and net-slip components of translation are 86 ft, 35 ft, and 91 ft (26 m, 11 m, and 28 m), respectively (Figure J.3*C*). The rake of the displacement vector on the fault plane is 23°W. This corresponds to a net-slip orientation of 21° N86°W, as computed stereographically.

K. STEREOGRAPHIC RELATIONSHIP OF FAULTS TO PRINCIPAL STRESS DIRECTIONS

Interpreting Fault Orientations from Stress Directions

The likely orientations of faults that should form in a given stress field can be evaluated stereographically, provided the angle of internal friction is known. In interpreting likely fault orientations, we assume that rocks are homogeneous and isotropic, and that the Coulomb law of failure holds.

Let us consider what kinds of faults should develop in a stress field where σ_1 = 0° N20°E, σ_2 = 0° N70°W, and σ_3 is vertical. Assume θ = 30°. We know that thrust-slip faults should form simply on the basis of the vertical orientation of σ_3. The probable orientations of the thrusts can be predicted by applying what we have already learned about the formation of conjugate faults. *Conjugate faults intersect in* σ_2. *And the trace of each fault, when viewed in the* σ_1/σ_3 *plane, is oriented at an angle of* $\theta = (90 - \phi)/2$ *with respect to* σ_1. (See Chapter 6.)

To give these relationships geometric reality, we stereographically plot points that portray the orientations of the principal stress directions (Figure K.1*A*). We then define the σ_1/σ_3 principal plane by fitting σ_1 and σ_3 to a common great circle (Figure K.1*B*). Counting 30° along the great circle from σ_1, we locate two reference points (*1* and *2*) that represent the intersection of the fault planes with the σ_1/σ_3 plane. By fitting σ_2 and reference point 1 to a common great circle, one of the thrust-slip faults is defined (Figure K.1*C*). And by fitting reference point 2 and σ_2 to a common great circle, the second thrust-slip fault of the conjugate set is established. The actual orientations of the faults can then be determined by normal stereographic procedures (Figure K.1*D*). Comparable solutions for defining the orientations of strike–slip and normal-slip faults within appropriate stress fields are shown in Figures K.2 and K.3, respectively.

If one or more of the principal stress directions is inclined, the solution is more difficult to visualize, but the stereographic operations are the same. Consider the general situation wherein none of the principal stress directions are vertical or horizontal; instead all are inclined (Figure K.4). As before, we define the σ_1/σ_3 plane by fitting σ_1 and σ_3 to a common great circle, and then we set off reference points from σ_1 along the great circle at distances corresponding to the value of θ. The orientations of the fault planes are each defined by one of the reference points and σ_2 (see Figure K.4).

Using Faults to Interpret Stress

In the course of field-oriented structural analysis, it is common to work through this problem backward, as an "inverse problem." Suppose we want to interpret principal stress directions on the basis of the orientations of two conjugate faults. The faults are first plotted stereographically as great circles, as in Figure K.5*A*. The intersection of the two great circles

Figure K.1 Stereographic representation of the relation of faults to the principal stress directions. (*A*) Stereographic portrayal of principal stress directions. (*B*) Identification of the σ_1/σ_3 plane, and the plotting of reference points (1 and 2) at $\theta = 30°$ from σ_1 on the σ_1/σ_3 great circle. (*C*) The orientation of one of the thrust faults is represented by a great circle that passes through reference point 1 and σ_2. The orientation of the second thrust fault is defined by the great circle that passes through reference point 2 and σ_2. (*D*) Portrayal of actual orientations of the faults and the slickenline striae that occur along them.

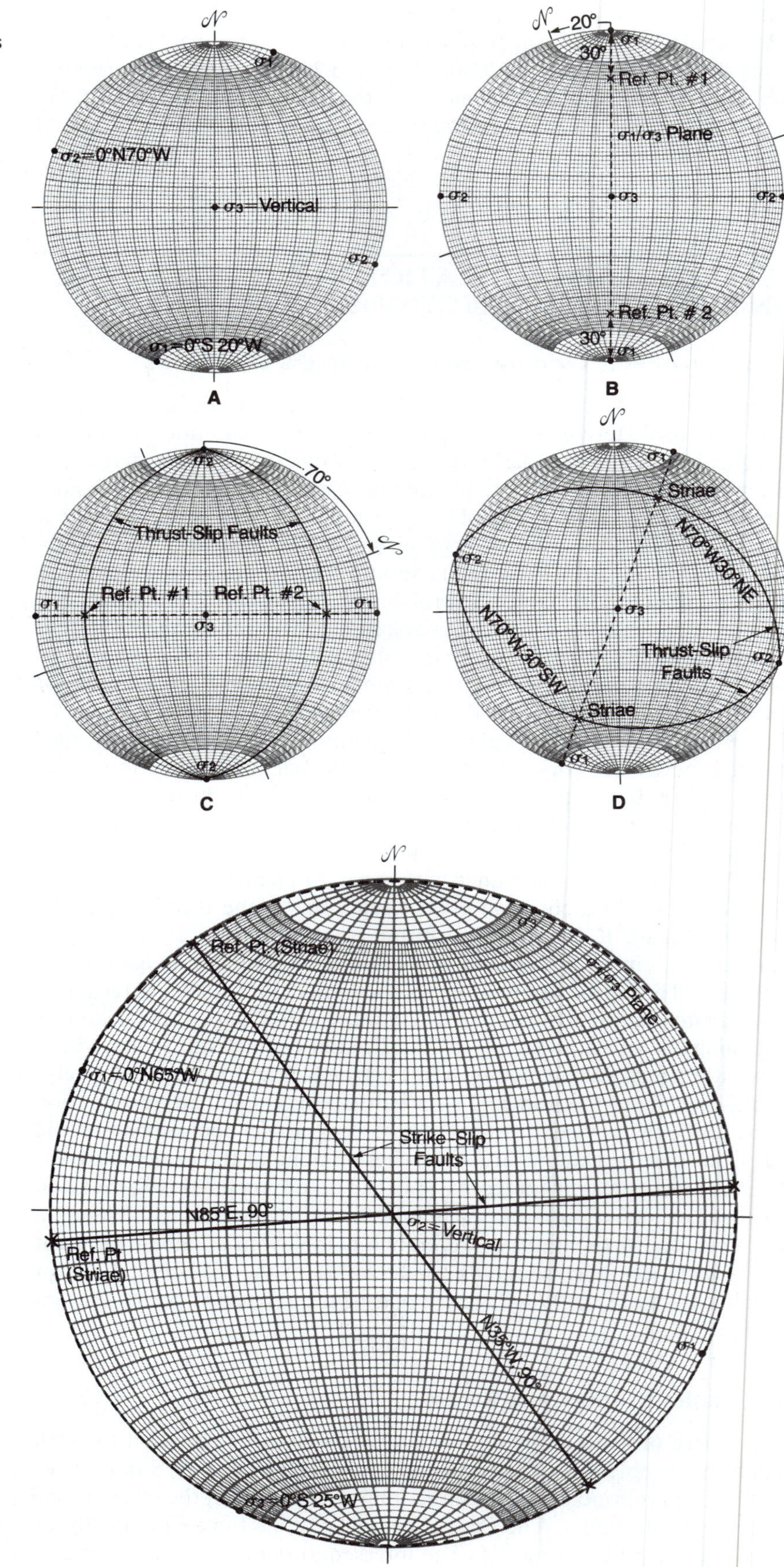

Figure K.2 Stereographic representation of the relation of conjugate strike–slip faults to the principal stress directions.

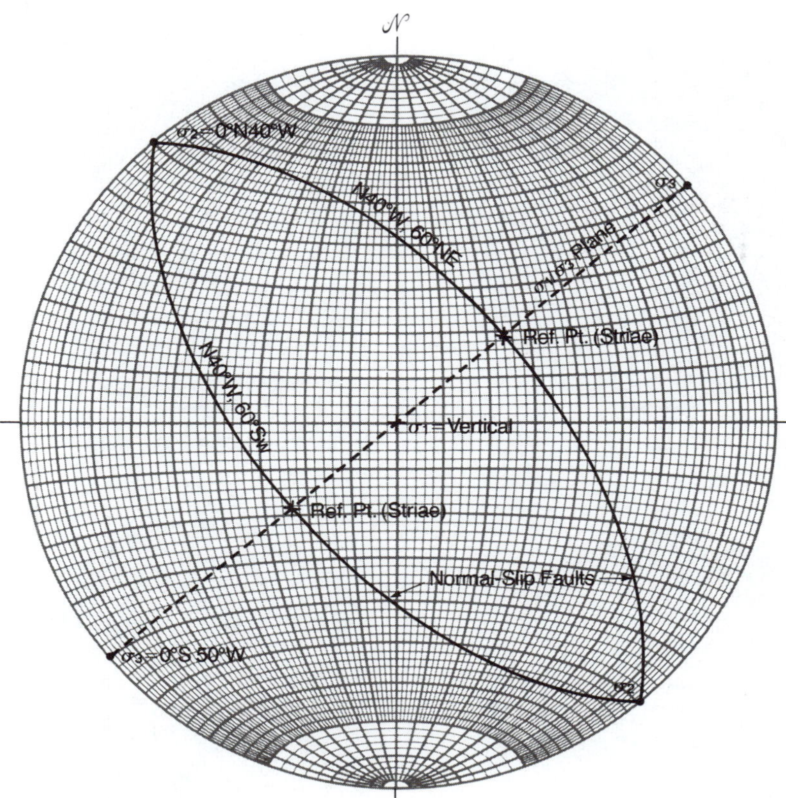

Figure K.3 Stereographic representation of the relation of conjugate normal-slip faults to the principal stress directions.

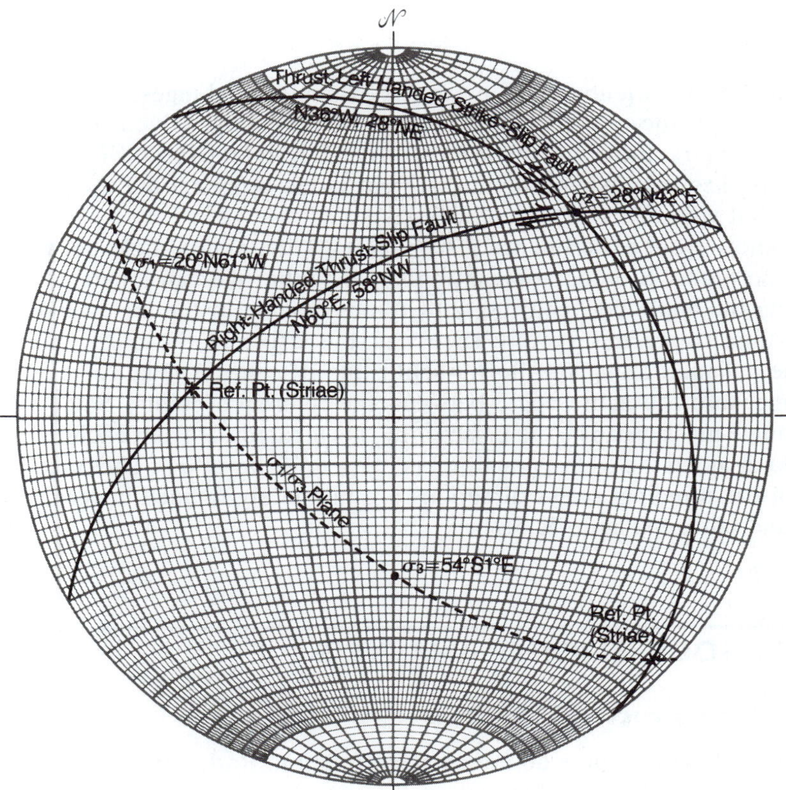

Figure K.4 Stereographic determination of the orientations of conjugate faults that would develop in a stress system characterized by inclined principal stress directions.

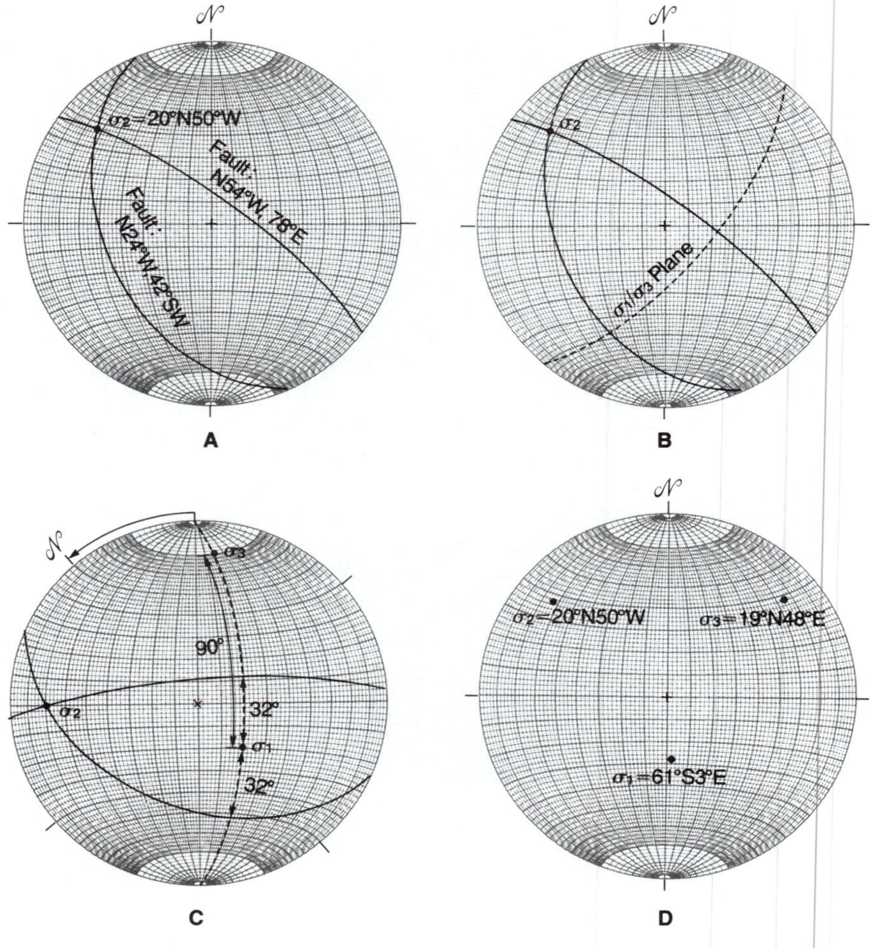

Figure K.5 Interpretation of principal stress directions on the basis of the known orientations of conjugate faults. (*A*) Stereographic representation of the faults as great circles. The faults intersect in σ_2. (*B*) Identification of the σ_1/σ_3 plane, the great circle whose pole is σ_2. (*C*) Determination of the stereographic locations of σ_1 and σ_3. (*D*) Portrayal of the orientations of σ_1, σ_2, and σ_3, as solved.

is taken to be the orientation of σ_2; in this example, σ_2 plunges 20° N50°W. We know that the σ_1/σ_3 plane is perpendicular to σ_2. Given the trend and plunge of σ_2, the σ_1/σ_3 plane must strike N40°E and dip 70°SE (Figure K.5*B*). σ_1 lies somewhere along the great circle representing the σ_1/σ_3 planes, but where?

We know from experiments that σ_1 bisects the acute angle between conjugate faults. Stereographically this means that we can locate σ_1 by bisecting the angle between the fault traces that lie in the σ_1/σ_3 plane. To do this, rotate the overlay such that the strike of the σ_1/σ_3 plane becomes aligned along the north–south line of the net (Figure K.5*C*). Then measure the acute angle between the reference points. Since the acute angle measures 64°, the bisector (σ_1) is located 32° from each reference point along the σ_1/σ_3 plane; σ_3 is perpendicular to σ_1. The orientations of the principal stress directions are ultimately defined in terms of trend and plunge (Figure K.5*D*).

L. METHODS IN JOINT ANALYSIS

Establishing Structural Domains

Joints and shear fractures do not exist in isolation. Rather they are members of enormously large families of fractures with literally millions of members. Every regional rock assemblage, like a granitic batholith or a plateau of sedimentary rock, is pervaded by jointing. It is impossible to try to explain the origin of every joint or shear fracture in an outcrop, let alone every joint or shear fracture within a regional rock assemblage. Instead, we try to explain the origin of dominant **sets** of joints and shear fractures that can be identified through statistical analysis of the orientations and physical properties of the joints and shear fractures within a given system (Figure L.1).

To begin to discover order among millions upon millions of joints and shear fractures, it is essential to subdivide regional rock assemblages into **structural domains**, each of which may be thought of as containing its own fractures system. Strictly speaking, two or more **sets** comprise a **system**. But we find it preferable, from an operational point of view, to regard the term **fracture system** as the entire family of joints and shear fractures within a given structural domain, regardless of whether all the individual surfaces can be neatly packaged into sets of like orientation.

Structural domains are designated on the basis of geographic boundaries, lithologic contacts, structural subdivisions, ages of rock formations, and combinations of these and other factors. The criteria vary according to the scope of the investigation (see Chapter 5).

Figure L.1 (*A*) Three sets of barrels containing Spanish olives. (Photograph by B. J. Young.) (*B*) Two sets of joints (one conspicuous, one less so) cut the Moab Member of the Entrada Sandstone (Jurassic) at Arches National Park, Utah. Spacing of joints is 25 m to 40 m. (Photograph by R. Dyer.)

Once structural domains have been assigned, the work begins. The substance of detailed joint and shear fracture analysis lies in evaluating the geologic characteristics of the fracture system that occupies each structural domain. There is no single conventional procedure for doing this, even though efforts have been made to standardize nomenclature and methods. Most studies are based on structural analysis of joints and shear fractures at selected **stations** within each structural domain. Less commonly, and usually for very specific applied purposes, **fracture-pattern maps** are generated.

The Mapping of Joints and Shear Fractures

In regions where joints and shear fractures are distinctly expressed in the weathered landscape, the mapping of the fractures is best achieved photogeologically. Aerial images often display remarkable portrayals of fracture patterns. With patience and care, the intricacies of joint and shear fracture patterns can be accurately reproduced in inked tracings, such as those rendered by Cruikshank and others (1991) for the Arches National Park area, Utah (Figure L.2). A masterful example of regional photogeologic mapping of jointing is found in Kelley's (1955) fracture-pattern map of the Colorado Plateau. He very carefully mapped the traces of joint and fault patterns for each of the major Laramide uplifts to determine whether a geometric relation between overall fold geometry and joint geometry could be discovered. Conventional large-scale aerial photography (either black and white or color) serves nicely as a base for most studies. But for regional analysis of fracture systems, it is advantageous to use some of the extraordinary space-satellite imagery that is available. Advances in remote-sensing techniques that have accompanied the proliferation of regional imagery make photogeologic analysis of jointing even more attractive.

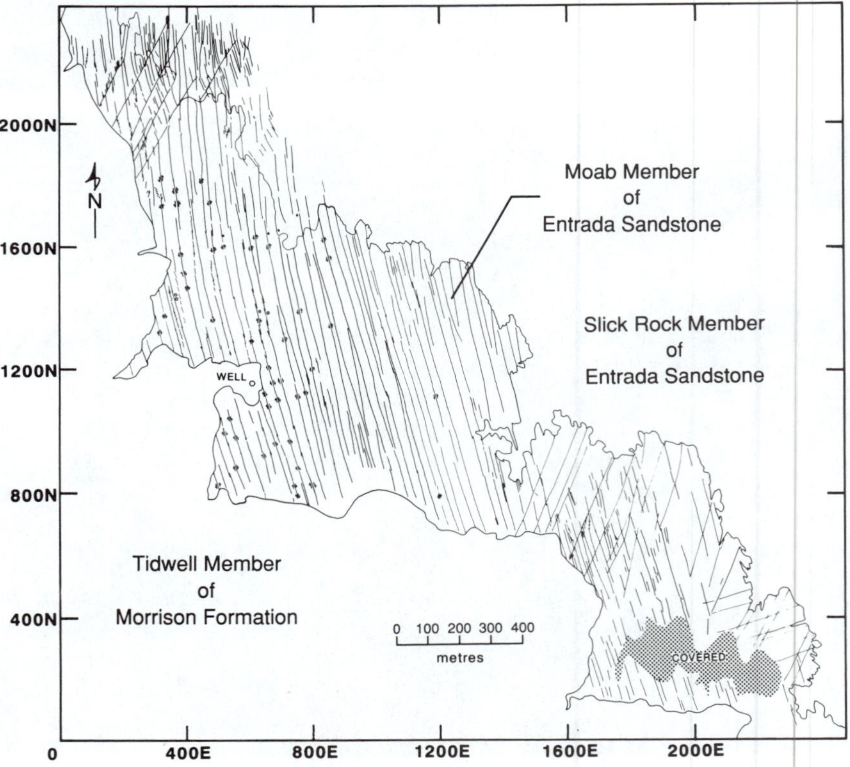

Figure L.2 Ink tracing of jointing in the Garden area, Arches National Park region, Utah. The traces of systematic fractures include joints, faulted joints, and jointed faults. There are three sets of joints in the area, trending about N10°W, N30°E, and N60°E. (Reprinted with permission from *Journal of Structural Geology*, v. 13, Cruikshank, K. M., Zhao, G., and Johnson, A. M., Analysis of minor fractures associated with joints and faulted rocks, 1991, Elsevier Science Ltd., Pergamon Imprint, Oxford, England.)

There are shortcomings associated with using the photogeologic approach, exclusively, in the mapping of joints and shear fractures. Low-dipping fractures are automatically screened from view, thus biasing the data toward moderate to steep-dipping fractures. Furthermore, it is impossible to measure from photographs the dip and dip direction of joints and shear fractures and to gather information regarding the types of fractures that exist in each structural domain.

Where jointing and shear fracturing are not well expressed in the landscape, or are too subtle or fine to be resolved on aerial images, the "mapping" of fractures is reduced exclusively to systematically measuring the orientations of representative joints and shear fractures in the field. Under these circumstances, aerial photos do not afford any special leverage in analysis, and thus either a topographic map or an aerial photograph can be used as a base map. Generally, little attempt is made to show the actual physical traces of the fractures, for they are generally far too numerous and much too short to portray at reasonable map scales. Instead, the orientations of the dominant systematic sets of joints and shear fractures are portrayed through standard joint symbology, plotted as close as possible to the locations where the fractures are measured. The symbols are drafted in dense overlapping clusters that literally fill the map to overflowing. Under ideal circumstances such fracture-pattern maps show at a glance the chief types of joint and shear fracture and their respective orientations. Under less favorable circumstances, the maps are a collage of confusion, doing little to clarify the basic elements of the fracture pattern.

It is common practice to measure eye-catching joints and shear fractures as a part of the normal geologic mapping process. Representative structures are plotted according to type and orientation. The few and scattered data that are typically posted on geologic maps do not constitute a basis for structural analysis. Instead, they simply provide a preliminary forecast of the dominant joint and shear fracture sets that might exist in the area of investigation.

Map Symbology

Map symbology for representing joint and shear fracture data on fracture-pattern maps and normal geologic maps is presented in Figure L.3. Symbols distinguish ordinary featureless joints, plumose joints, veins, crystal fiber veins, en echelon veins and joints, and slickenlined shear fractures. The symbol for plumose joints includes portrayal of the direction of convergence of the plumose markings. The strike symbol for ordinary veins can be color-coded to distinguish veins according to mineralogy and/or alteration assemblages. The symbol for crystal-fiber veins allows the trend of the fibers and their curvature, if any, to be shown. The symbol for en echelon veins and fractures includes a portrayal of the trend of the line of bearing. Symbols for striated shear fractures distinguish whether the lineation is a crystal fiber lineation or slickenlines of another origin. The symbols, taken as a whole, convey useful geometric and kinematic information.

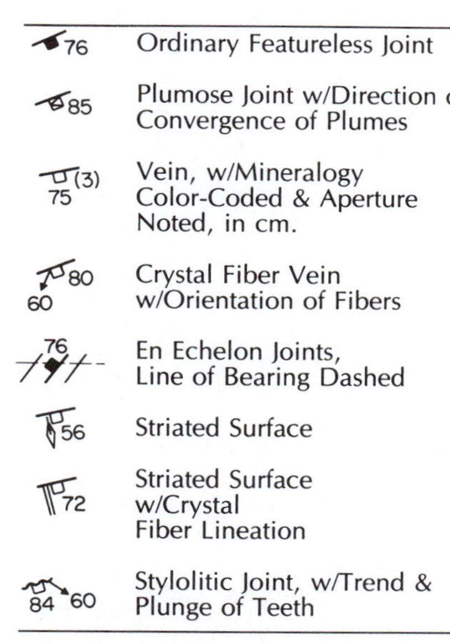

Figure L.3 Map symbology for joints, veins, shear fractures, and stylolites.

Choosing Stations for Structural Analysis

It is impossible to examine all joints and shear fractures that are contained in a given structural domain. Thus, standard practice is to evaluate jointing and shear fracturing through detailed structural analysis at selected **stations**. The strategy is to learn the nature of the overall fracture system through systematic examination of representative **subareas** within the domain.

A **sampling station** established for fracture analysis is very small compared to the size of the structural domain within which it lies. It is a site of well-exposed jointed bedrock where joints and shear fractures are classified and measured. In most studies the stations are simply outcrop areas of varying size and shape. In a restricted sense, stations can be designated as circular or square **inventory areas** of specified dimension, or as relatively short **sample lines** of specified traverse length and direction.

Measuring Orientations of Joints and Shear Fractures

In practice, there are two basic approaches that are used in collecting orientation data at sample stations. One, which we might call the **selection method**, involves selecting only certain joints and shear fractures for measurement and study. The second approach, which we might call the **inventory method**, requires measuring and classifying every single joint and shear fracture at a station site.

The basis of the selection method is to restrict analysis to joints and shear fractures that are continuous and through-going, and are conspicuously associated with other fractures of similar appearance and orientation. It is neither very easy nor very objective to decide which fracture surface should be measured at each station, and which should be left alone. But in spite of this difficulty, many workers have found the selection method to be practical and useful. Parker (1942) restricted his joint-orientation measurements in the Appalachian Plateau region of New York to those fractures that appeared straight and continuous. Hodgson (1961), in analyzing the fracture pattern of part of the Colorado Plateau, measured only smooth planar continuous joints, especially those that displayed plumose markings. Rehrig and Heidrick (1972, 1976) measured joints and veins that occurred in sets of three or more, ignoring all others.

In contrast to the selection method, the inventory method requires measuring all the joints and shear fractures that occupy a sampling station. Measurements are taken within an inventory area or along a sample line. In sampling along a line it is common practice to stretch out a tape measure to some prescribed length (normally less than 20 m), and to classify and measure the orientation of every joint that is intercepted by the tape.

A novel inventory approach is one we refer to as the **circle-inventory method** (Titley, 1976). First a circle of some known and predetermined diameter (normally less than 3 m) is traced out on a bedrock surface of perfectly exposed fractured rock (Figure L.4). The circle is drawn with a piece of (carpenter's) chalk attached to a string of suitable length. Then the orientation and trace length of each fracture within the circle are measured. Data are recorded in the manner shown in Table L.1. To avoid measuring the same fracture surface twice, it is helpful to trace out the full length of each joint or shear fracture with chalk after it is measured. Outcrops are generally available in which the three-dimensional expression of each fracture surface is clear, allowing orientations to be measured in terms of strike and dip. But where outcrops are so smooth and flat that only the straight-line traces of the fractures are evident in the bedrock surface, trends alone can be measured.

The time and tedium of the circle-inventory operation depends largely on the size of the sampling circle and the abundance of the joints and shear fractures that must be measured. Some initial trial-and-error planning may permit selection of an optimum circle diameter. When analyzing fractures in sedimentary and volcanic rocks, it is useful to select a circle radius

Figure L.4 Circle-inventory method for sampling the orientations of joints and joint-related structures. (*A*) The circle, drawn on bedrock with chalk. (*B*) With joints measured and traced, Matt Davis and "Katie" let their handiwork be admired. Katie still has her chalk in her paw. (Photographs by G. H. Davis.)

TABLE L.1
Circle-inventory method for evaluating fracture density

Station #34, Tucson Mountains
Lower Cretaceous Sandstone
Radius of Inventory Circle = 19 cm; Area = 1134 cm²

Trend of Fracture	*Length of Fracture (cm)*
N79E	35
N46W	36
N04W	30
N32E	25
NS	22
EW	20
N16E	20
N42W	39
N41W	13
N44W	17
N43W	11
N30W	28
N41E	18
N82W	13
N64W	13
NS	6
N12W	20
N23E	14
N11E	7
	387 cm = Cumulative Fracture Length

$$\text{Fracture Density} = \frac{387\text{ cm}}{1134\text{ cm}^2} = \boxed{.34\text{ cm}^{-1}}$$

whose magnitude is some function of layer thickness. In fact, setting the radius equal to layer thickness works out reasonably well in thin- to medium-bedded rocks. However, if the host rock for the fractures is very thick bedded or massive, for example 15 to 150 m (50 to 500 ft) or more, it becomes necessary to adapt the circle-inventory method to a photogeologic approach—unless there is plenty of chalk, string, and time available. The photogeologic adaptation requires drawing an inventory circle of appropriate diameter onto an aerial photograph, or a transparent overlay of the photograph, then measuring the trends and trace lengths of fractures directly from the photograph. North arrow and scale provide the control. Subsequent field investigations can focus on measuring dip magnitudes of fracture sets and classifying the fractures according to their physical and kinematic characteristics.

Measuring Density of Joints and Shear Fractures

The abundance of joints and shear fractures at a given station is described through the evaluation of **fracture density**. Fracture density can be measured and described in a number of ways: average spacing of fractures; number of fractures in a given area; total cumulative length of fractures in a specified area; surface area of all fractures within a given volume of rock. The measure of fracture density used in conjunction with the circle-inventory method is the summed length of all fractures within an inventory circle, divided by the area of the circle:

$$\rho_f = \frac{L}{\pi r^2}$$

where

ρ_f = fracture density

L = cumulative length of all fractures

r = radius of inventory circle.

An example of the computation is shown in Table L.1. The fracture density is expressed in units of length/area (e.g., ft/ft^2, cm/cm^2, m/m^2, km/km^2). In practice, the values of fracture density are converted to the reciprocal form (ft^{-1}, cm^{-1}, m^{-1}, and km^{-1}). Thus a fracture density of 0.57 ft/ft^2 would be expressed as 0.57 ft^{-1}. Comparative analysis of fracture density, however expressed, is especially useful in applied studies. Petroleum geologists Harris, Taylor, and Walper (1960) compared joint density to degree of curvature of the flanks of two oil-producing domes in Wyoming. They were interested in evaluating the relationship between fracture-induced permeability and degree of folding. As part of their study they found it necessary to "normalize" the natural variations in joint density that are due to differences in rock type and bedding thickness.

Haynes and Titley (1980) compared quantitative differences in fracture density in veined, mineralized rocks of the Sierrita porphyry copper deposit south of Tucson. Their purpose was to explore for centers of intrusion and/or mineralization, using joint-density variations as a guide. Wheeler and Dickson (1980) sought to evaluate whether systematic changes in joint density can disclose the locations of known or hidden faults. They prepared contour maps showing variations in joint density in a part of the Central Appalachians, and compared these maps with the known fault distribution as revealed on geologic maps of the same region.

Recording the Data

It is important to keep a systematic record of the descriptive characteristics of joints and shear fractures that are examined at each sampling station. Notebook entries should include site location; rock type; orientation of bedding or foliation (if present); bed or layer thickness (if applicable); offset (if detectable) along joint or shear fracture; geometry and orientation of plumose and rib structure; nature and locations of origins; orientation of joints and shear fractures; geometry of joint intersections and terminations; and measurements pertinent to computing fracture density. Block diagrams that schematically portray the array of joints and joint-related structures are also a good idea (Figure L.5).

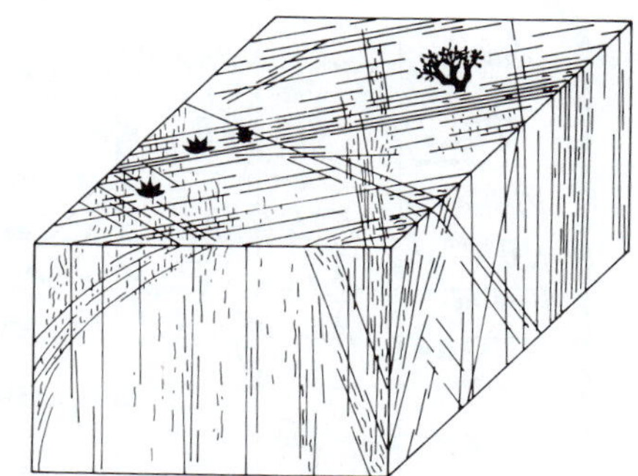

Figure L.5 Block diagram portrayal of the nature and orientation of dikes, joints, and joint-related structures in Laramide granitic plutons in southern Arizona. (By permission, T. L. Heidrick and S. R. Titley, "Fracture and Dike Patterns in Laramide Plutons and Their Structural and Tectonic Implications," fig. 4.1, in *Advances in Geology of the Porphyry Copper Deposits: Southwest North America*, S. R. Titley, editor. University of Arizona Press, Tucson, copyright © 1982.)

Nick Nickelson uses a method for recording fracture and fault data at a given station that he learned from Ernst Cloos. Rather than recording the orientation data in columns, he fills a page of his notebook with strike-and-dip symbology, based on his measurements (Figure L.6). No attempt is made to portray the relative locations within the station outcrop where each reading was taken. However, the bearing of the strike of each feature, or the trend of slickenlines, is constructed accurately using a protractor. In this way, the "map" at a glance provides a visual portrayal of preferred orientations and systematic relations, if indeed they exist. The method takes just a little longer than lining up readings in columns, but the results are much, much more conducive to clear thinking about the meaning of the structures. This method of course works for all minor structures in outcrops. As Nick points out: "You will *never* record a structure improperly because when you complete your data collection, you orient your notebook north and compare your symbols with the outcrop, seeing if you have shown proper strike, dip, rake, or plunge directions." (Nick Nickelson, personal communication, 1993.)

Systematically recorded fracture data provide a basis for answering the typical questions that are raised in the structural analysis of joints and shear fractures. Are certain fractures associated with specific rock types? How does fracture density vary according to lithology and layer thickness? Does fracture density and/or orientation change according to structural location? Are rocks of different ages characterized by different fracture patterns? Can favored directions of vein mineralization be recognized? The data that are collected and recorded, station by station, provide a playground for comparative analysis of a statistical nature.

Preparing Fracture-Orientation Diagrams

Orientation data collected during the course of fracture analysis may be summarized in **pole diagrams** and **pole-density diagrams**. Pole diagrams (and contoured pole-density diagrams) are three-dimensional stereographic displays of strike-and-dip data (Figure L.7*A*). Preferred orientations of fractures emerge from pole diagrams as dense clusterings of poles to joints. The orientation of the center of distribution represents the average orientation of the set.

Where three-dimensional control on the attitude of joints and shear fractures is not attainable, either due to the nature of the bedrock surfaces or because the fracture orientations are gathered from aerial photographs, it is appropriate to present the orientation data on **rose diagrams** or **strike histograms**, that is, on two-dimensional plots (Figure L.7*B*,*C*). In preparing rose diagrams and strike histograms, the trend and/or strike data are first organized into **class intervals** of 5° or 10°, encompassing the orientation range from west through north to east. The number and percentage of

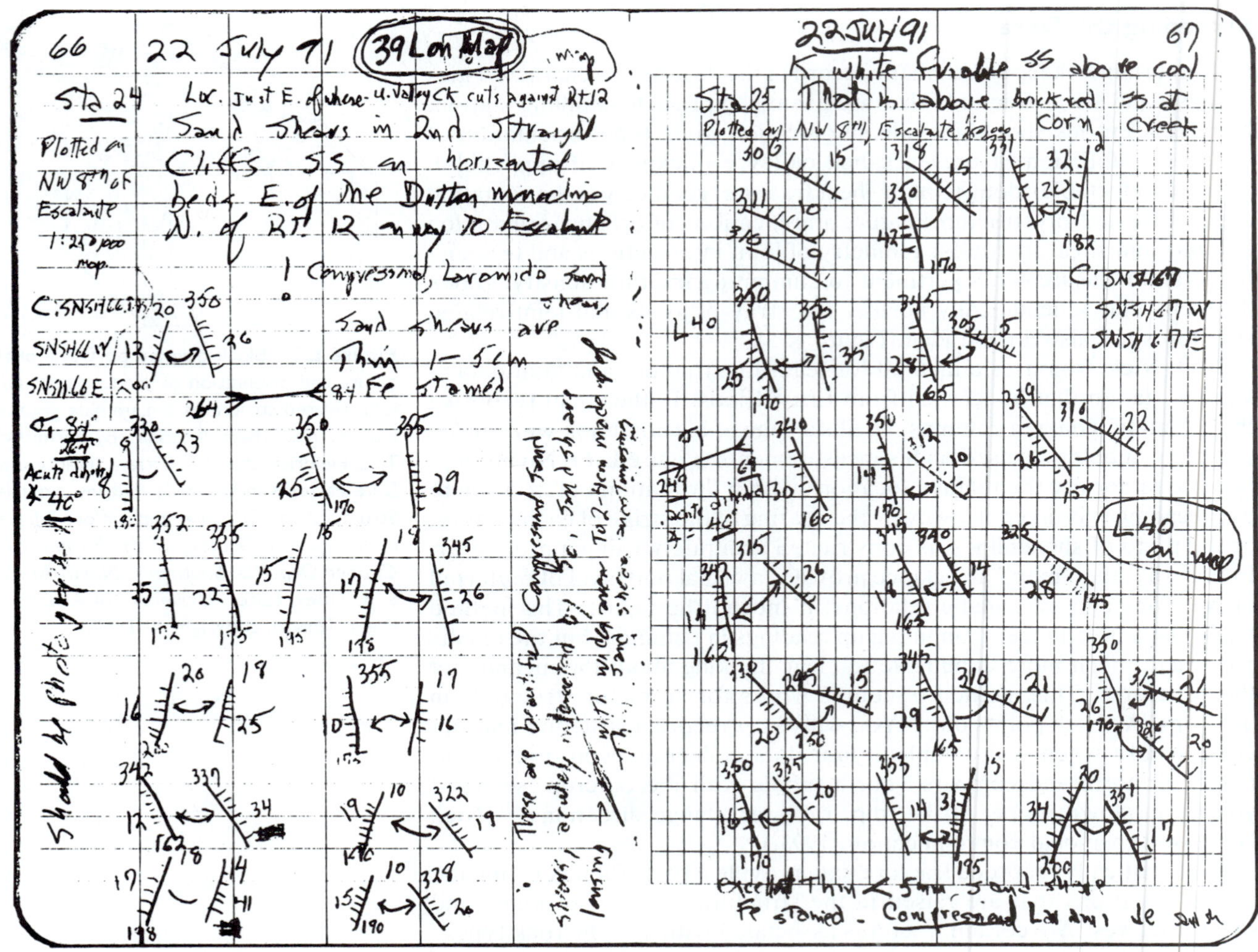

Figure L.6 The Ernst Cloos method for notebook-recording of structural data, including joint data. These data reflect orientations of deformation bands in sandstone.

readings that fall within each class interval are then tallied. Data thus arranged are plotted in one of two ways. For rose diagrams, class intervals are distinguished by rays subtending arcs of 5° or 10° that extend outward from a common point (Figure L.8). A family of concentric circles provides scaled control for the number (or percentage) of fracture-orientation readings that occupy each class interval. For strike histograms, class intervals are plotted along the y-axis of an x-y plot; numbers (or percentages) of readings are plotted along the x axis (Figure L.8). Trends of dominant fracture sets coincide with high-frequency peaks.

Rehrig and Heidrick (1972) effectively used rose diagrams in presenting the orientations of steeply dipping Late Cretaceous to Eocene dikes, veins, and elongate plutons in the southern Basin and Range of Arizona and New Mexico (see Figure 5.70). Their pole-density diagrams display the orientations of 12,412 mineralized joints, veins, dikes, and faults in Late Cretaceous-Eocene plutons in the American Southwest. The data were subdivided according to three domains: productive plutons, ore-related plutons, and nonproductive plutons. The data in each diagram were contoured according to statistical probability of random distribution.

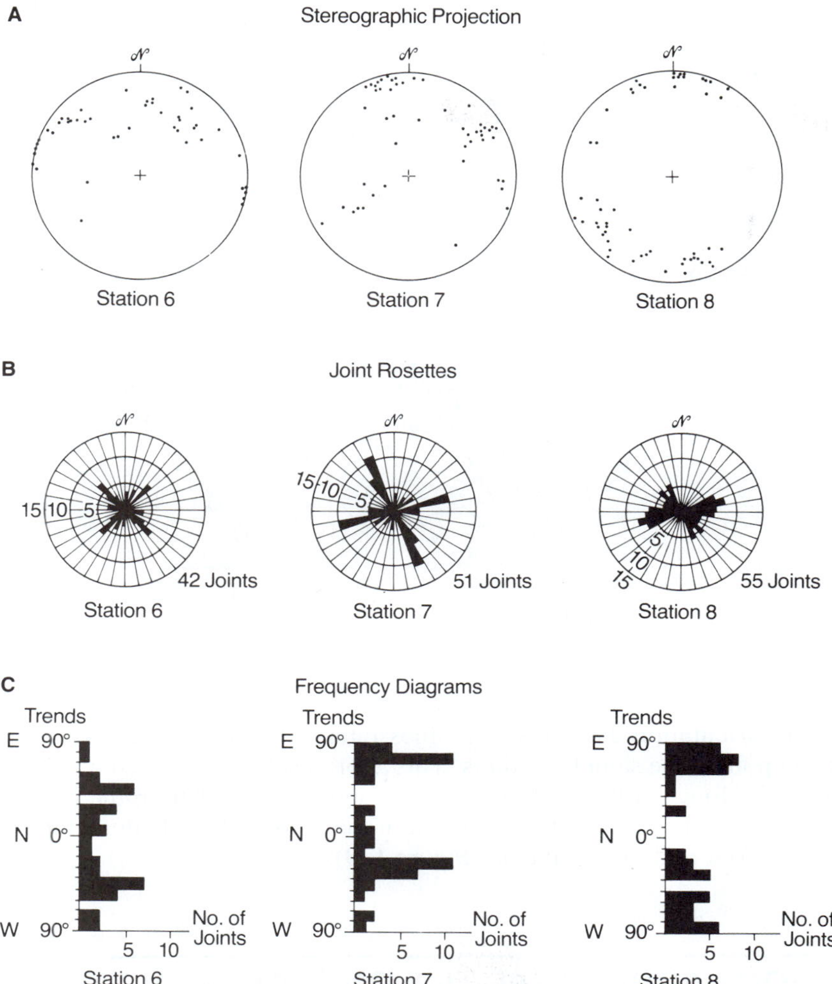

Figure L.7 Three kinds of plots for displaying the orientations of joints: the data reflect orientations of joints in Cretaceous sandstones in the Tucson Mountains, Arizona. (*A*) Stereographic pole diagrams. (*B*) Joint rosettes (rose diagrams). (*C*) Frequency diagrams. (Data collected and plotted by R. Chavez.)

Preferred orientations of sets of joints or shear fractures are commonly estimated by eye from fracture-orientation diagrams. This is easy to do where preferred orientations are obvious, but is difficult and subjective where orientations are diffuse. Fortunately, fracture data and fracture-density diagrams lend themselves to rigorous statistical analysis. Joint enthusiasts are not without computer packages that integrate orientation data and statistical analysis of the data.

It is beneficial to distinguish the orientations of different types of fractures, based upon physical characteristics, or joints in different rock units. Separate orientation diagrams can be made for each type. Depending on the purpose of study, individual fracture-orientation diagrams can be prepared for each station, for a given array of stations, for an entire structural domain, and/or for an entire region.

Although fracture-orientation diagrams are normally presented as discrete illustrations in a technical report, it is sometimes quite effective to post the diagrams on a structural geologic map of the region within which the fracture analysis was carried out. The diagrams are positioned as close as possible to the locations of sampling stations. To provide greater visual impression of the dominant fracture trends, **joint and shear fracture trajectories** can be constructed whose orientations are everywhere parallel to the average trends of high-angle fracture sets recognized in each of the

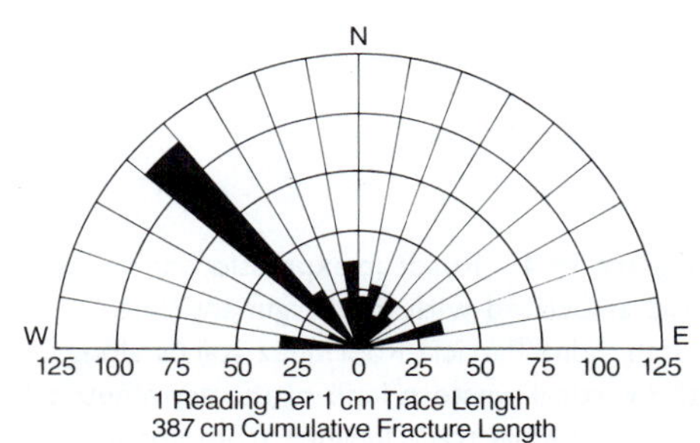

Figure L.8 Rose diagram of joint orientation data presented in Table L.1.

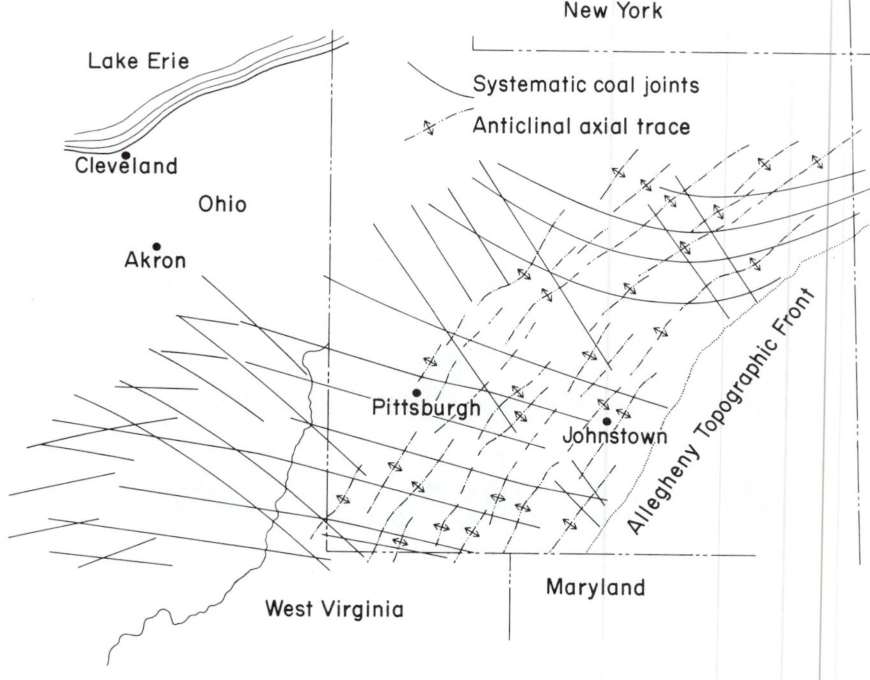

Figure L.9 Use of joint trajectories in picturing the dominant joint orientations in coal in the Appalachian Plateau. [From Nickelsen and Hough (1967). Published with permission of Geological Society of America and the authors.]

fracture-orientation diagrams. The lines intersect at the locations of the control points, the sample stations. This approach has been used by Nickelson and Hough (1967) and by Engelder and Geiser (1980) in the classic joint analysis that has been carried out in the Appalachian Plateau region of New York and Pennsylvania (Figure L.9).

M. SOME ADDITIONAL TIPS ON FOLD ANALYSIS

Direct Measurement of Axes and Axial Surfaces

It might be useful to review from start to finish the steps involved in measuring the orientations of axes and axial surfaces of folds (Figure M.1). Where folds are so small that they are completely exposed in single outcrops, the orientations of fold axes and axial surfaces are usually easy to visualize and can be measured directly with a compass. Conditions of

Figure M.1 Roadcut expression of minor fold associated with the Virgin anticline, Utah. Edna Patricia Rodriquez stands along the vertical western limb of the asymmetrical syncline. Axial trace inclined from upper right to lower left. (Photograph by G. H. Davis.)

measurement are ideal when the hinge of one or more of the folded layers has weathered out as a pencil-like or rodlike linear form. The hinge is taken to be a line whose orientation is parallel to the axis of the fold. Its orientation is measured and described in terms of trend and plunge and is entered in the field notebook under the "fold axis" orientation. To help visualize the hinge line when it is not weathered out in full three-dimensional relief, it is useful to align a pencil parallel to the orientation of the hinge line. The pencil provides a tangible, physical guide in shooting the trend and plunge.

Measuring the orientation of the axial surface of a fold in outcrop requires a degree of physical dexterity. Since axial surfaces are geometric elements, there is no real physical surface on which to directly measure the axial surface orientation unless, by chance or by nature's design, there are cleavage surfaces or joint surfaces that are exactly parallel to the axial surface. In the absence of a natural physical surface on which to take a direct measurement, the strike-and-dip orientation is measured on a clipboard or field notebook held in one hand, parallel to the axial surface. To be sure that the clipboard or notebook is aligned properly, the trace of the axial surface must be visible on at least two surfaces of the outcrop. Using as reference two (or more) axial traces common to a single axial surface, the clipboard (or notebook) can be aligned in the unique attitude that satisfies the trend of each. Axial surface orientation is measured and then posted in the field notebook in terms of strike and dip.

Map Determination of Axial Surface Orientation

There is a useful way to determine the orientation of the axial surface of a large fold, using a topographically controlled geologic map as a guide as well as some stereographic projection. This method does not depend on measuring the orientation of the bisecting surface as an approximation to the axial surface. The procedure is to plot stereographically the trend and plunge of two or more axial traces of the fold as points, and to fit these points (representing the orientations of lines) to a common great circle. Two lines define a plane. The great circle, constructed in this manner, describes the orientation of the plane.

An example of the use of this method is shown in Figure M.2. The data base is a simplified geologic map of an overturned anticline. As revealed by the topographic contour lines, the fold is exposed in a plateau-like terrain cut by a steep-walled canyon (Figure M.2*A*). The topography affords excellent exposures of the axial trace of the anticline, both in map and cross-sectional views. On the plateau surface the axial trace of the fold is seen to trend N5°E along a reasonably straight line that passes, as it should, through hinge points of successively folded layers. The plunge of the axial trace, as calculated trigonometrically (or graphically) from the elevation control available, is 6° NE. The axial trace of the fold exposed to view in the northeast wall of the canyon trends S47°E. Its plunge is 32°SE (Figure M.2*A*). The two axial trace orientations are lines, one plunging 6° N5°E, and the other plunging 32° S47°E. When these two lines are plotted stereographically and fitted to a common great circle, the overall attitude of the axial surface (axial plane) is found to be N2°W, 42°NE (Figure M.2*B*).

Plotting Fold-Orientation Data on Geologic Maps

Fold-orientation data, whether measured directly in the field or computed stereographically from representative strike-and-dip orientations, should

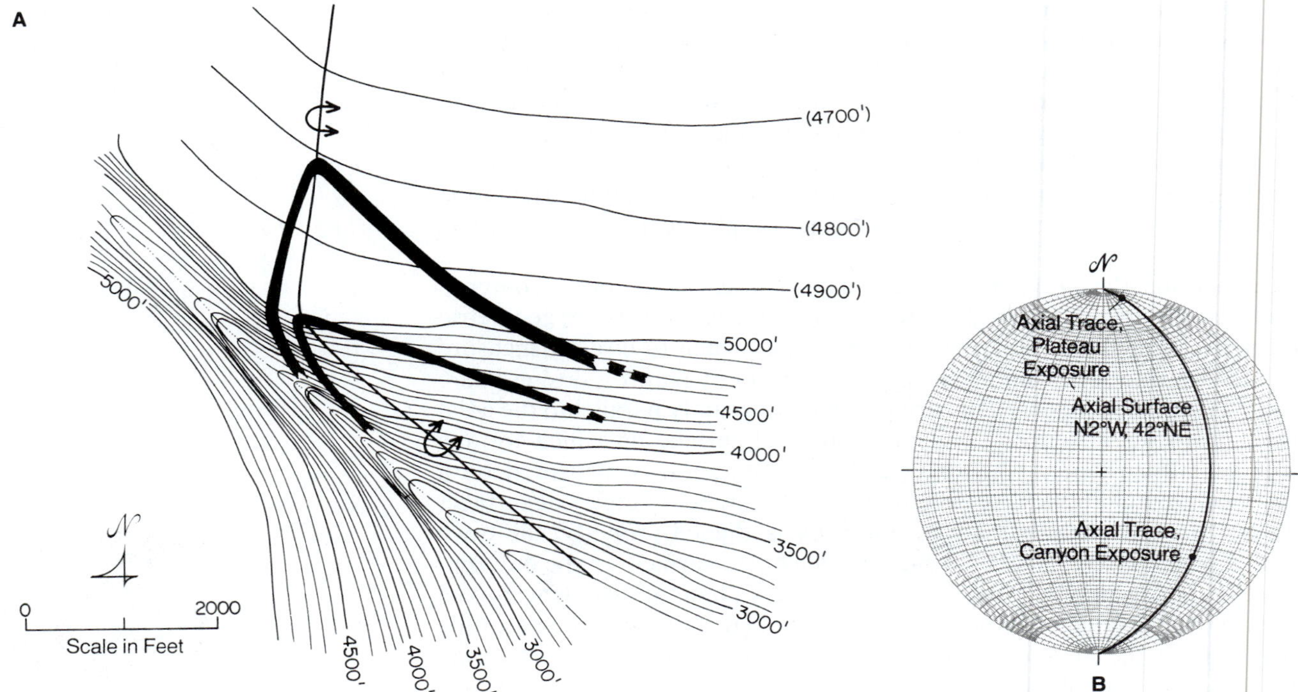

Figure M.2 Determination of the orientation of the axial surface of a fold on the basis of axial trace orientations. (*A*) Geologic map expression of folded beds (thick black bands). Two nearly straight axial traces can be identified, one crossing the high, flat tableland, the other exposed along the canyon wall. Orientations of these axial traces can be measured from the map relationships. (*B*) The axial surface orientation is determined stereographically by fitting the two axial trace orientations to a common great circle.

become part of the geologic map record. How many of the orientation data are plotted depends largely on the size and scale of the final, rendered map. Orientation data that do not become part of the final geologic map are not lost. Rather, they constitute an integral part of stereographic displays of the preferred orientations of bedding, axes, and axial surfaces, displays that support the map relationships in a complementary way.

Formation contacts and marker beds that are plotted in the normal course of geologic mapping serve to outline the general form of map-scale folds. Strike-and-dip data representing the orientations of bedding and/or foliation help to disclose the fold configurations. The main fold-orientation information, however, is represented by symbology for axial trace, axis, and axial surface attitude. The axial trace of each fold is drawn not as a ruled straight line but a smooth, straight to curved line that passes through the hinge points of the folded formations and marker beds (Figure M.3). The trend and plunge of the fold axis and the strike and dip of the axial surface are plotted at one or more locations along each axial trace. Plotting axis and axial surface data at multiple locations along a given axial trace is necessary to describe the changes in fold orientation and configuration across the area of investigation.

The conventional symbols used in representing map-scale folds on geologic maps are listed in Table M.1 The symbols are used to identify whether a given fold is an anticline or syncline, or an antiform or synform. Overturned folds receive special attention. Like the standard symbols for anticlines and synclines, the symbols for overturned anticlines and

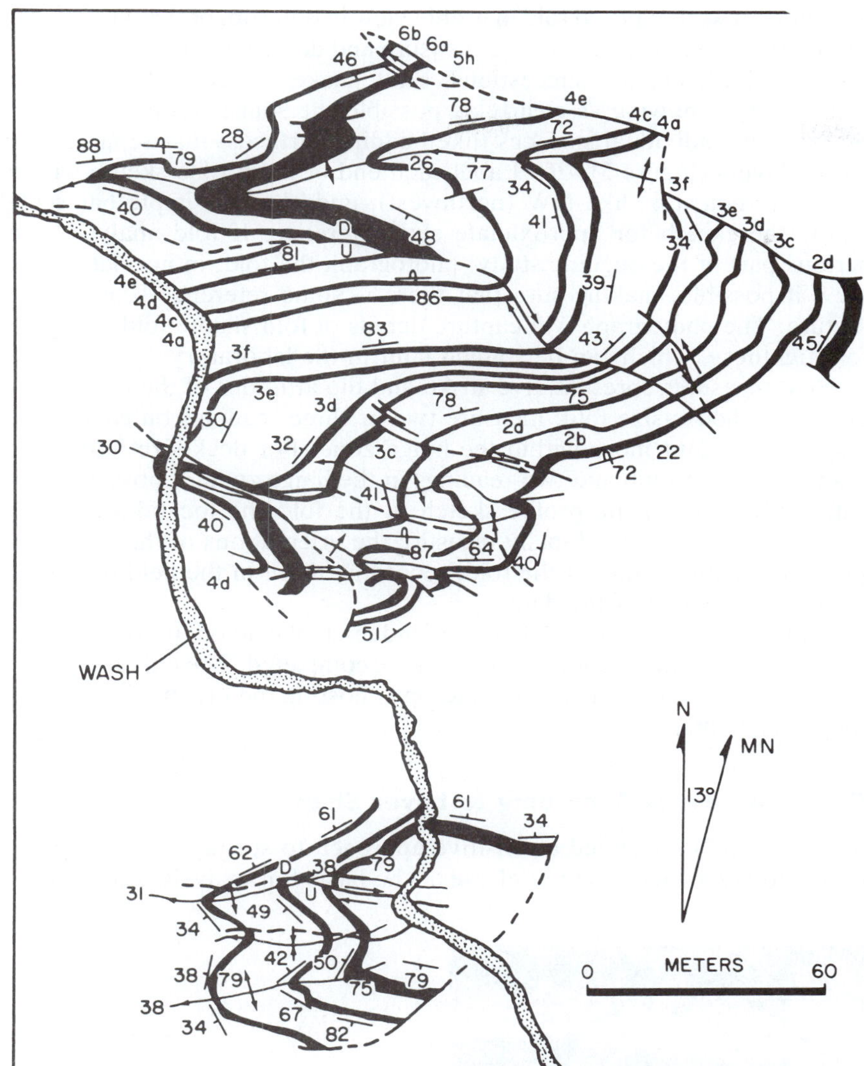

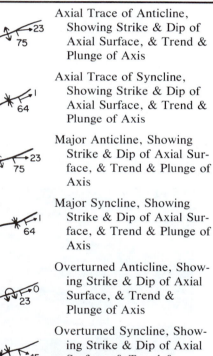

Figure M.3 Structural geologic map of folded strata at the Agua Verde locality near Tucson, Arizona. Alphanumeric symbols like "3f" serve to identify marker beds. (Map based on mapping by undergraduate student, D. Gossett.)

synclines represent the dip directions of the limbs with small arrows. For overturned folds, these arrows point in a common direction, like the dip of the limbs of the overturned folds. The arrowheads point toward the U-shaped closure on the symbol of an overturned syncline. They point away from the U-shaped closure on the symbol for an overturned syncline.

Representative minor folds are plotted on geologic maps using special symbols (Table M.1). Each symbol portrays the strike and dip of the axial surface, the trend and plunge of the fold axis, and the generalized form of the folded surface as it is viewed down plunge. Symmetrical folds are shown as M-shaped forms. Asymmetrical folds are shown as Z-shaped and S-shaped forms. It is visually useful to orient the M's, S's, and Z's on the map in such a way that the axial trace of the minor fold bisects the form of the fold. If the scale of mapping permits, strike-and-dip orientations of the limbs of the minor folds should be posted as well.

Suggestions for Describing Folds in Outcrop

Analysis of the orientations of shapes of folded surfaces demands care in systematically collecting and recording the pertinent data. Suppose we

TABLE M.1
Geologic map symbology for folds and fold elements

Symbol	Description
23, 75	Axial Trace of Anticline, Showing Strike & Dip of Axial Surface, & Trend & Plunge of Axis
1, 64	Axial Trace of Syncline, Showing Strike & Dip of Axial Surface, & Trend & Plunge of Axis
23, 75	Major Anticline, Showing Strike & Dip of Axial Surface, & Trend & Plunge of Axis
1, 64	Major Syncline, Showing Strike & Dip of Axial Surface, & Trend & Plunge of Axis
0, 23	Overturned Anticline, Showing Strike & Dip of Axial Surface, & Trend & Plunge of Axis
47, 45	Overturned Syncline, Showing Strike & Dip of Axial Surface, & Trend & Plunge of Axis
30, 68	Dextral Minor Fold, Showing Strike & Dip of Axial Surface & Trend & Plunge of Axis
3, 85	Sinistral Minor Fold, Showing Strike & Dip of Axial Surface & Trend & Plunge of Axis
21, 72	Symmetrical Minor Fold, Showing Strike & Dip of Axial Surface & Trend & Plunge of Axis

encounter a well-exposed fold in profile view in outcrop, of the type shown in Figure M.4*A*. What should we measure and describe before leaving the outcrop? Here are our suggestions. First, make a sketch of the fold in profile view, showing as clearly as possible the shape of one or two or more of the individual surfaces (like bedding surfaces) that separate the folded layers (Figure M.4*B*). Label each end of the profile sketch with a direction indicator, like NW (northwest), and place a simple bar scale below the sketch for approximate size reference. If fold analysis is a serious part of the ongoing study, photograph the fold, in normal profile view if possible, making sure that there is some reference scale in the picture. The photograph will capture details of fold shape, fold size, and fold tightness, which can be studied later in the laboratory

Next measure representative strike-and-dip attitudes of the folded layers. For a house-size fold, measure two or three readings on each limb, and three or four more within the hinge zone. For desk-size folds, one reading on each limb and two readings in the hinge zone, if possible, will suffice. Record on the profile sketch of the fold the locations of each reading (Figure M.4*B*). Finally measure the orientations of the axial surface and the hinge line of the fold, recording these in the field notebook along with the rest of the data.

With this information in hand, the fundamental field operations bearing on shape, size, and orientation analysis are completed. The only remaining step before heading to the next outcrop is posting the orientation data on the geologic map.

Classifying Folds According to Layer Shape

Ramsay (1967) presented a sensitive approach to sorting out the shapes of folded layers according to classes. The basis for analysis is a geologic

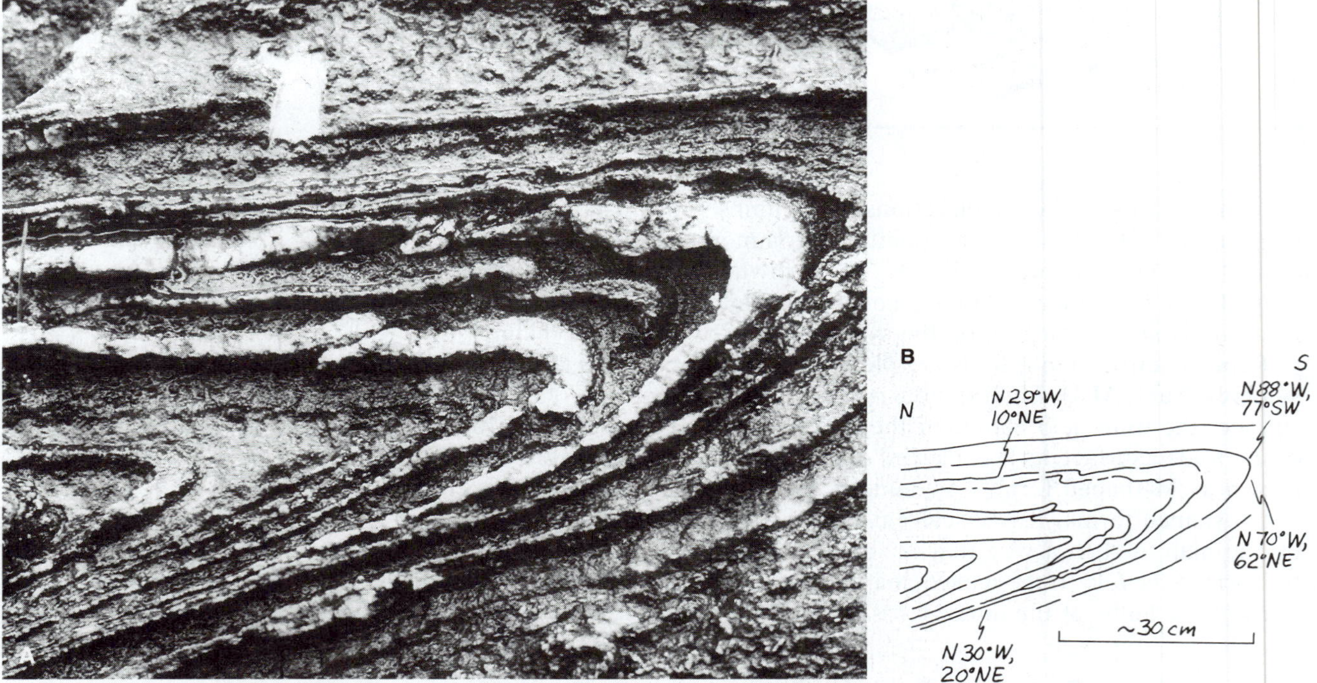

Figure M.4 (*A*) Profile view of fold in highly deformed sequence of marble and fine-grained quartzite. (Photograph by G. H. Davis.) (*B*) Notebook sketch of the fold, including orientation measurements.

cross section or a profile-view photograph (Figure M.5*A*). The method requires comparing layer thickness measured at a number of locations on the limbs of a fold with layer thickness measured in the hinge. The thickness that is measured is **true thickness**. Limb thickness (t_α) is identified according to **limb inclination** (α) as measured when the fold is positioned as a perfectly upright antiform (Figure M.5*B*). Limb thickness is taken to be the spacing between tangents drawn through control points of equal inclination value on the upper and lower surfaces of the folded layer (Figure M.5*B*). The thicknesses thus measured are recorded and become the basis for constructing a graph allowing fold class to be precisely assigned.

The graph that reveals layer shape is a plot of **relative thickness (t')** versus limb inclination (α) (Figure M.5*C*). Relative thickness (t') is the ratio of thickness (t_α) measured at a location on the fold limb to thickness (t_o) measured in the hinge of the fold:

$$t' = \frac{t_\alpha}{t_o},$$

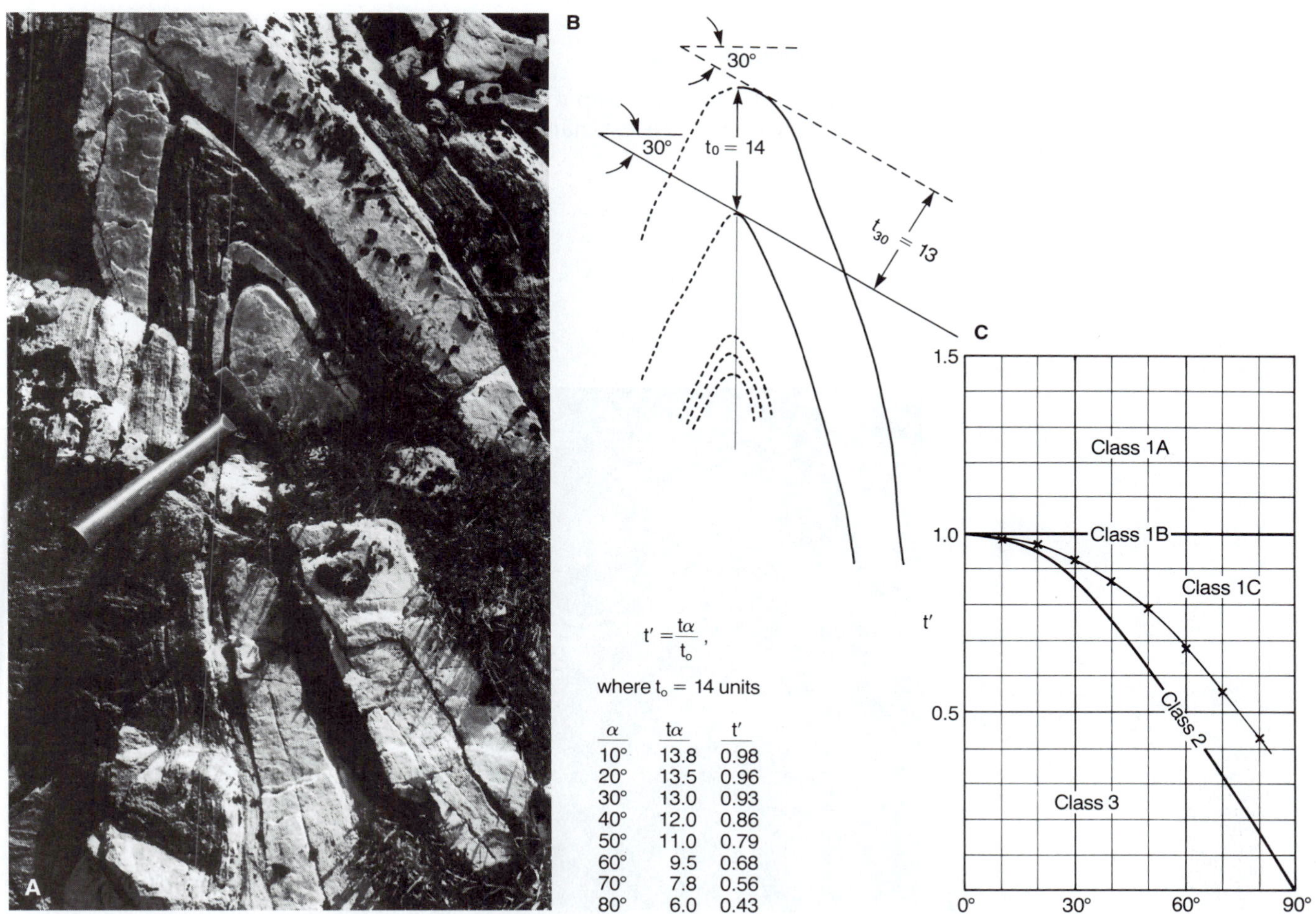

α	tα	t'
10°	13.8	0.98
20°	13.5	0.96
30°	13.0	0.93
40°	12.0	0.86
50°	11.0	0.79
60°	9.5	0.68
70°	7.8	0.56
80°	6.0	0.43

Figure M.5 (*A*) Normal profile view of a tightly folded sequence of layers. (*B*) Construction steps used in determining layer thickness at layer inclinations of 0° and 30°. (*C*) The standard graphical expression of the variations in layer thickness that characterize each of the major classes of folds. Note that the fold shown in *A* plots as a class 1C fold. (From *Folding and Fracturing of Rocks* by J. G. Ramsay. Published with permission of McGraw-Hill Book Company, New York, copyright © 1967.)

where

t' = relative thickness

t_α = limb thickness

limb inclination = α

t_o = layer thickness in hinge

$\alpha = 0°$

Using their measured values of t_α and t_o, t' is calculated and recorded in the manner shown in Figure M.5*B*. Then each set of (t') values is plotted to form the graphic construction shown in Figure M.5*C*. The plotted points for a given fold layer are connected by a smooth curve. The locus of the curve within the graph can then be compared with the subareas and lines occupied by the fundamental classes of folds. Graphs of this type show that the full range of layer shapes of folds is very expansive and that concentric folds (class IB) and similar folds (class 2) are indeed very special cases.

A Special Touch

As an elegant follow-up to this layer shape analysis, Richard Lisle (1992) shows us how to estimate strain in flattened buckle folds. His simple,

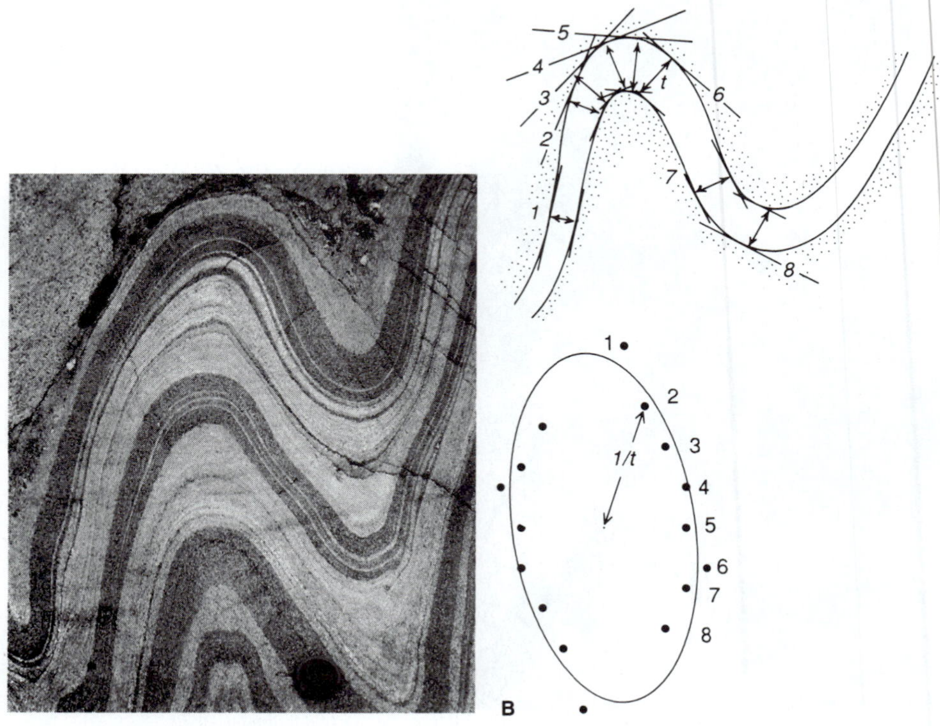

Figure M.6 (*A*) Photograph of folds in Precambrian banded gneiss. (*B*) Measurement of orthogonal thicknesses of folded layer. Note that measurements are made perpendicular to tangents 1–8 . (*C*) Construction of strain ellipse reflecting flattening. Simply plot inverse thicknesses from a common point in the direction of the appropriate line of tangent. (Reprinted with permission from *Journal of Structural Geology*, v. 13, Lisle, R. J., Strain estimation from flattened buckle folds, 1992, Elsevier Science Ltd., Pergamon Imprint, Oxford, England.)

direct method assumes that a given class IC, II, or III fold had a class IB (parallel) shape before flattening. To determine the amount of flattening in a fold such as shown in Figure M.6*A*, measure the orthogonal thicknesses (t) perpendicular to tangents drawn to folded layering (Figure M.6*B*), and then plot inverse thicknesses ($1/t$) from a common point, each in the direction of the line of tangency (Figure M.6*C*). A strain ellipse emerges that discloses strain due to flattening!

N. STUDYING SHEAR ZONES IN THE FIELD

When we encounter a shear zone in the field, there are numerous things to describe, measure, and think about. We need to measure and record the orientation of the shear zone, as well as the foliation and lineation within the zone. If foliation and lineation vary in orientation across the zone, multiple measurements (strikes and dips) will help us evaluate whether the variation is systematic across the shear zone. We describe the character of the shear zone, including its length, width, enclosed rock types, style of deformation, and amount of separation on any offset markers. We need to carefully describe what exactly is defining the foliation and lineation (quartz ribbons, compositional banding), so that we can evaluate their strain significance.

If we look at an appropriately oriented rock face, we may recognize features indicative of the sense of shear (Chapter 9). The proper rock face to examine for sense-of-shear indicators is a face cut parallel to the stretching lineation and perpendicular to the shear zone or foliation. Such indicators can be very subtle and may require us to really put our nose to the outcrops and look carefully. If we find a sense-of-shear indicator, we describe it and draw a sketch, being particularly careful to properly record angular relations between the feature, foliation, and other fabrics; these angles may have strain significance. Record your interpretation of the sense of shear (dextral or top to the right, sinistral or top to the left, or some other clear terminology, such as "top to the west" or "northeast side up", and *be sure to record which way you are looking* so that you can relate your relative sense of shear to geographic directions. A shear zone that is dextral when viewed one way will be sinistral when viewed in the opposite (180°) direction.

An easy trap to fall into is to walk up to an outcrop, see a single sense of shear indicator, record the sense of shear, and walk away without really looking at the entire outcrop. Instead, take time to see how many sense of shear indicators are present and whether they are all consistent. Record if indicators are lacking in what otherwise should be good rock types and appropriate amounts of strain for indicators. Indicate if you found a few indicators that appear to register a conflicting sense of shear opposite to the vast majority of indicators. Finally, make as few assumptions as possible.

Foliation will not generally represent the S_1S_2 plane of the finite strain ellipse. Shear sense indicators in a shear zone formed by *coaxial* deformation should give no consistent sense of shear. If formed by simple shear, foliation will be inclined at 45° to the shear zone at the onset of the simple shear. It will progressively lean over in the direction of shear with increasing strain and nearly coincide with the shear zone only in very strongly deformed rocks. The stretching lineation, like the foliation, may not strictly parallel to the line of transport, but will approach parallelism at very high shear strains.

REFERENCES

Allis, R. G., 1981, Continental underthrusting beneath the southern Alps of New Zealand: Geology, v. 9, p. 303–307.

Alvarez, W., Engelder, T., and Lowrie, W., 1976, Formation of spaced cleavage and folds in brittle limestone by dissolution: Geology, v. 4, p. 698–701.

Alvarez, W., Engelder, T., and Geiser, P. A., 1978, Classification of solution cleavage in pelagic limestones: Geology, v. 6, p. 263–266.

Anderson, E. M., 1951, The dynamics of faulting and dyke formation with applications to Britain: Oliver & Boyd, Edinburgh, 206 p.

Anderson, R. E., 1971, Thin-skin distension in Tertiary rocks of southeastern Nevada: Geological Society of America Bulletin, v. 82, p. 43–58.

Anderson, R. E., 1973, Large magnitude Late Tertiary strike slip faults, north of Lake Mead, Nevada: United States Geological Survey Professional Paper 794, 18 p.

Anderson, R. N., and Noltimier, H. C., 1973, A model for the horst and graben structure of midocean ridge crests based upon spreading velocity and basalt delivery to the oceanic crust: Geophysical Journal of the Royal Astronomical Society, v. 34, p. 137–147.

Angelier, J., 1979, Determination of the mean principal directions of stresses for a given fault population: Tectonophysics, v. 56, p. T17–T26.

Angelier, J., 1990, Inversion of field data in fault tectonics to obtain the regional stress. III. A new rapid direct inversion method by analytical means: International Journal of Geophysics, v. 103, p. 363–376.

Angelier, J., 1994, Fault slip analysis and palaeostress reconstruction, *in* Hancock, P. L. (ed.), Continental deformation: Pergamon Press, Oxford, p. 53–100.

Angelier, J., Colletta, B., and Anderson, R. E., 1985, Neogene paleostress changes in the Basin and Range: a case study at Hoover Dam, Nevada–Arizona: Geological Society of America Bulletin, v. 96, p. 347–361.

Armstrong, R. L., 1964, Geochronology and geology of the eastern Great Basin in Nevada and Utah: Ph.D. dissertation, Yale University, New Haven, Connecticut, 202 p.

Armstrong, R. L., 1968a, Sevier orogenic belt in Nevada and Utah: Geological Society of America Bulletin, v. 79, p. 429–458.

Armstrong, R. L., 1968b, The Cordilleran miogeosyncline in Nevada and Utah: Utah Geological and Mineralogical Survey Bulletin 78, 58 p.

Armstrong, R. L., 1972, Low-angle (denudation) faults, hinterland of the Sevier orogenic belt, eastern Nevada and western Utah: Geological Society of America Bulletin, v. 83, p. 1729–1754.

Ashby, M. F., 1972, A first report on deformation–mechanism maps: Acta Metallurgica et Materialia, v. 20, p. 887–897.

Ashby, M. F., and Verrall, R. A., 1973, Diffusion accommodated flow and superplasticity: Acta Metallurgia et Materialia., v. 21, p. 149–163.

Atkinson, B. K., 1987, Introduction to fracture mechanics and its geophysical applications, *in* Atkinson, B. K. (ed.), Fracture Mechanics of Rock: Academic Press, London, p. 1–26.

Atwater, T. M., 1970, Implications of plate tectonics for the Cenozoic tectonic evolution of western North America: Geological Society of America Bulletin, v. 81, p. 3513–3536.

Aydin, A., 1978, Small faults formed as deformation bands in sandstone: Pure and Applied Geophysics, v. 116, p. 913–930.

Aydin, A., and DeGraff, J. M., 1988, Evolution of polygonal fracture patterns in lava flows: Science, v. 239, p. 471–476.

Aydin, A., and Johnson, A. M., 1978, Development of faults as zones of deformation bands and as slip surfaces in sandstone: Pure and Applied Geophysics, v. 116, p. 931–942.

Aydin, A., and Johnson, A. M., 1983, Analysis of faulting in porous sandstone: Journal of Structural Geology, v. 5, p. 19–31.

Bahat, D., and Engelder, T., 1984, Surface morphology on cross-fold joints of the Appalachian Plateau, New York and Pennsylvania: Tectonophysics, v. 104, p. 299–313.

Bailey, E. B., 1938, Eddies in mountain structure: Geological Society of London Quarterly Journal, v. 94, p. 607–625.

Baker, A. A., 1936, Geology of the Monument Valley–Navajo Mountain Region, San Juan County, Utah: United States Geological Survey Bulletin 865, 106 pp.

Balk, R., 1937, Structural behaviour of igneous rocks with special reference to interpretations by Hans Cloos and collaborators: Geological Society of America Memoir 5, 177 p.

Balk, R., 1949, Structure of Grand Saline salt dome, Van Zandt County, Texas: American Association of Petroleum Geologists Bulletin, v. 33, p. 1791–1829.

Ballard, R. D., and van Andel, Tj. H., 1977, Morphology and tectonics of the inner rift valley at lat. 36°50′N on the mid-Atlantic ridge: Geological Society of America Bulletin, v. 88, p. 507–530.

Bally, A. W., Gordy, P. L., and Stewart, G. A., 1966, Structure, seismic data, and orogenic evolution of southern Canadian Rockies: Canadian Association of Petroleum Geologists Bulletin, v. 14, p. 337–381.

Bally, A. W., and Snelson, S., 1980, Realms of subsidence, *in* Miall, A. D. (ed.), Facts and principles of world petroleum occurrence: Canadian Society of Petroleum Geologists Memoir 6, p. 9–94.

Barber, A. J., 1982, Interpretations of the tectonic environment of southwest Japan: Proceedings of the Geological Association, v. 93, p. 131–145.

Barnes, C. W., 1974, Interference and gravity tectonics in the Gray Mountain area, Arizona, *in* Karlstrom, T. N. V., Swann, G. A., and Eastwood, R. L. (eds.), Geology of northern Arizona, Part II—area studies and field guide: Northern Arizona University, Flagstaff, Arizona, p. 442–453.

Barnes, C. W., and Marshall, D. R., 1974, A dynamic model for monoclinal uplift and gravity gliding: Geological Society of America Abstracts with Programs, v. 6, p. 424.

Barnett, J. A. M., Mortimer, J., Rippon, J. H., Walsh, J. L., and Watterson, J., 1987, Displacement geometry in the volume containing a single normal fault: American Association of Petroleum Geologists Bulletin, v. 71, p. 925–938.

Batiza, R., 1978, Geology, petrology, and geochemistry of Isla Tortuga, a recently formed tholeiitic island in the Gulf of California: Geological Society of America Bulletin, v. 89, p. 1309–1324.

Beach, A., 1980, Retrogressive metamorphic processes in shear zones with special reference to the Lewisian complex, *in* Carreras, J., Cobbold, P. R., Ramsay, J. G., and White, S. H. (eds.), Shear zones in rocks: Journal of Structural Geology, v. 2, p. 257–263.

Bebout, D. G., Loucks, R. G., and Gregory, A. R., 1978, Frio sandstone reservoirs in the deep subsurface along the Texas Gulf Coast: University of Texas Bureau of Economic Geology Report of Investigation 91, 93 p.

Becker, C. L., 1949, Progress and power: Alfred A. Knopf, Inc., New York, 116 p.

Behrmann, J. H., 1987, A precautionary note on shear bands as kinematic indicators: Journal of Structural Geology, v. 9, p. 659–666.

Benioff, H., 1954, Orogenesis and deep crustal structure: additional evidence from seismology: Geological Society of America Bulletin, v. 65, p. 385–400.

Berg, R. R., 1962, Mountain flank thrusting in Rocky Mountain foreland, Wyoming and Colorado: American Association of Petroleum Geologists Bulletin, v. 46, p. 2019–2032.

Berthé, D., Choukroune, P., and Jegouza, P., 1979, Orthogneiss, mylonite and noncoaxial deformation of granites: the example of the South Amorican shear zone: Journal of Structural Geology, v. 1, p. 31–42.

Berthé, D., and Brun, J. P., 1980, Evolution of folds during progressive shear in the South Amorican shear zone, France: Journal of Structural Geology, v. 2, p. 127–133.

Beutner, E. C., 1978, Slaty cleavage and related strain in Martinsburg Slate, Delaware Water Gap, New Jersey: American Journal of Science, v. 278, p. 1–23.

Bevis, M., and Isacks, B., 1984, Hypocentral trend surface analysis: probing the geometry of Benioff zones: Journal of Geophysical Research, v. 89, no. B7, p. 6153–6170.

Bevis, M., Schutz, B., Taylor, F. W., and Recy, J., 1993, Direct observations of convergence and back-arc spreading at the Tonga Arc (1988–1992): Geological Society of America Abstracts with Programs, v. 25, p. 242.

Bijlaard, P. P., 1946, On the elastic stability of thin plates supported by a continuous medium: Royal Dutch Academy of Science Proceedings, v. 49, p. 1189–1199.

Billings, M. P., 1954, Structural geology (2nd edition): Prentice Hall, Englewood Cliffs, New Jersey, 514 p.

Billings, M. P., 1972, Structural geology (3rd edition): Prentice-Hall, Englewood Cliffs, New Jersey, 606 p.

Biot, M. A., 1957, Folding instability of a layered viscoelastic medium under compression: Royal Society of London Proceedings, Series A, v. 242, p. 211–228.

Biot, M. A., 1959, On the instability of folding deformation of a layered viscoelastic medium under compression: Journal of Applied Mechanics, v. 26, p. 393–400.

Biot, M. A., Ode, H., and Roever, W. L., 1961, Experimental verification of the folding of stratified viscoelastic media: Geological Society of America Bulletin, v. 72, p. 1621–1630.

Bishop, R. S., 1977, Shale diapir emplacement in south Texas: Laward and Sherriff examples: Transactions of the Gulf Coast Association of Geological Societies, v. 27, p. 20–31.

Bjorn, L. I., 1970, Natural stress-values obtained in different parts of the Fennoscandian rock masses: Proceedings of the Second Congress of the International Society for Rock Mechanics, "Jaroslav Cerni" Institute for Development of Water Resources, Belgrade, Yugoslavia, p. 209–212.

Blackwood, J. R. (ed.), 1969, College talks: Oxford University Press, New York, 177 p.

Blake, D. R., and Roy, C. J., 1949, Unusual stylolites: American Journal of Science, v. 247, p. 779–790.

Bobyarchick, A. R., 1986, The eigenvalues of steady flow in Mohr space: Tectonophysics, v. 122, p. 35–51.

Bohannon, R. G., Grow, J. A., Miller, J. J., and Blank, R. H., 1993, Seismic stratigraphy and tectonic development of Virgin River depression and associated basins, southeastern Nevada and northwestern Arizona: Geological Society of America Bulletin, v. 105, p. 501–520.

Bolton, T., 1989, Geologic maps; their solution and interpretation: Cambridge University Press, Cambridge, United Kingdom, 144 p.

Bond, G. C., and Kominz, M. A., 1988, Evolution of thought on passive continental margins from the origin of geosynclinal theory (1860) to the present: Geological Society of America Bulletin, v. 100, p. 1909–1933.

Bott, M. H. P., 1967, Solution of the linear inverse problem in magnetic interpretation with application to oceanic magnetic anomalies: Geophysical Journal of the Royal Astronomical Society, v. 13, p. 313–323.

Bott, M. H. P., 1982, The interior of the Earth, its structure, constitution and evolution (2nd edition): Edward Arnold, London, 403 p.

Bott, M. H. P., 1982, The mechanism of continental splitting, *in* Hales, A. L. (ed.), Geodynamics final symposium: Tectonophysics, v. 81, p. 301–309.

Boyer, S. E., and Elliott, D., 1982, Thrust systems: American Association of Petroleum Geologists Bulletin, v. 66, p. 1196–1230.

Brace, W. F., Paulding, B. W., Jr., and Scholz, C., 1966, Dilatancy in the fracture of crystalline rocks, Journal of Geophysical Research, v. 71, p. 3939–3953.

Bronowski, J., 1973, The ascent of man: Little, Brown, & Company, Boston, 448 p.

Brown, W. G., 1975, Casper Mountain area (Wyoming)—structural model of Laramide deformation (abstract): American Association of Petroleum Geologists Bulletin, v. 59, p. 906.

Bruce, C. H., 1972, Pressured shale and related sediment deformation: Gulf Coast Association of Geological Societies Transactions, v. 22, p. 23–31.

Brun, J.-P., Soukoutis, D., and Van Den Driessche, J., 1994, Analogue modeling of detachment fault systems and core complexes: Geology, v. 22, p. 319–322.

Bucher, W. H., 1944, The stereographic projection, a handy tool for the practical geologist: Journal of Geology, v. 52, p. 191–212.

Bullard, E. C., Everett, J. E., and Smith, A. G., 1965, The fit of the continents around the Atlantic: Philosophical Transactions of the Royal Society of London, v. 258A, p. 41–51.

Burchfiel, B. C., Chen, Z., Hodges, K. V., Liu, Y., Royden, L. H., Deng, C., and Xu, J., 1992, The South Tibetan detachment system, Himalayan orogen: extension contemporaneous with and parallel to shortening in a collisional mountain belt: Geological Society of America Special Paper 268, 41 p.

Burchfiel, B. C., and Stewart, J. H., 1966, "Pull-apart" origin of the central segment of Death Valley, California: Geological Society of America Bulletin, v. 77, p. 439–441.

Burkhard, M., 1993, Calcite twins, their geometry, appearance and significance as stress-strain markers and indicators of tectonic regime: a review, *in* Casey, M., Dietrich, D., Ford, M., Watkinson, J., and Hudleston, P. J. (eds.), The geometry of naturally deformed rocks: Journal of Structural Geology, v. 15, p. 351–368.

Butler, R. W. H., 1992, Evolution of Alpine fold-thrust complexes: a linked kinematic approach, *in* Mitra, S. and Fisher, G. W. (eds), Structural geology of fold and thrust belts: The Johns Hopkins University Press, Baltimore, p. 29–44.

Butler, R. W. H., Grasso, M., and LaManna, F., 1992, Origin and deformation of the Neogene-Recent Maghrebian foredeep at the Gela Nappe, Southeast Sicily: Journal of the Geological Society, v. 149, p. 547–556.

Byerlee, J. D., 1967, Frictional characteristics of granite under high confining pressure: Journal of Geophysical Research, v. 72, p. 3639–3648.

Byerlee, J. D., 1978, Friction of rocks: Pure and Applied Geophysics, v. 116, p. 615–626.

Capuano, R. M., 1993, Evidence of fluid flow in microfractures in geopressured shales: American Association of Petroleum Geologists Bulletin, v. 77, no. 8, p. 1303–1314.

Carey, S. W., 1953, The rheid concept in geotectonics: Geological Society of Australia Journal, v. 1, p. 67–117.

Carey, S. W., 1958, Note on the columnar jointing of Tasmanian dolerite: Dolerite: a symposium, Tasmanian University, p. 229–230.

Carey, S. W., 1962, Folding: Journal of Alberta Association of Petroleum: Geologists, v. 10, p. 95–144.

Carreras, J., Estrada, A., and White, S., 1977, The effects of folding on the c-axis fabrics of a quartz mylonite: Tectonophysics, v. 39, p. 3–24.

Carter, N. L., Friedman, M., Logan, J. M., and Stearns, D. W., 1981, Mechanical behavior of crustal rocks; the Handin volume: Geophysical Monograph, v. 24, 326 p.

Carter, N. L., and Hansen, F. D., 1983, Creep of rock salt: Tectonophysics, v. 92, p. 275–333.

Chapman, D. S., and Pollack, H. N., 1977, Regional geotherms and lithospheric thickness: Geology, v. 5, p. 265–268.

Chapple, W. M., 1978, Mechanics of thin-skinned fold-and-thrust belts: Geological Society of America, v. 89, p. 1189–1198.

Chase, C., 1978, Plate kinematics: the Americas, east Africa, and the rest of the world: Earth and Planetary Sciences Letters, v. 37, p. 355–368.

Chester, J. S., Logan, J. M., and Spang, J. H., 1991, Influence of layering and boundary conditions on fault-bend and fault-propagation folding: Geological Society of America Bulletin, v. 103, p. 1059–1072.

Cloos, E., 1946, Lineation: A critical review and annotated bibliography: Geological Society of America Memoir 18, 122 p.

Cloos, E., 1947, Oolite deformation in the South Mountain fold, Maryland: Geological Society of America Bulletin, v. 58, p. 843–918.

Cloos, E., 1955, Experimental analysis of fracture patterns: Geological Society of America Bulletin, v. 66, p. 241–256.

Cloos, E., 1968, Experimental analysis of Gulf Coast fracture patterns: American Association of Petroleum Geologists Bulletin, v. 52, p. 420–444.

Cloos, H., 1922, Uber Ausbau und Anwendung der granittektonischen Methode: Preussischen Geologischen Landesanstalt, v. 89, p. 1–18.

Cloos, H., 1936, Einfuhrung in die Geologie, ein Lehrbuch der inneren Dynamik: Borntraeger, Berlin, 503 p.

Cobbold, P. R., 1976, Fold shapes as functions of progressive strain: a discussion on natural strain and geological structure: Royal Society of London Philosophical Transactions, Series A , v. 283, p. 129–138.

Cobbold, P. R., Gapais, D., Means, W. D., and Treagus, S. H. (eds.), 1987, Shear criteria in rocks: Journal of Structural Geology, v. 9, 778 p.

Cobbold, P. R., and Quinquis, H., 1980, Development of sheath folds in shear regimes: Journal of Structural Geology, v. 2, p. 119–126.

Compton, R. R., 1985, Geology in the field: John Wiley & Sons, New York, 398 p.

Coney, P. J., 1978, Mesozoic-Cenozoic Cordilleran plate tectonics, *in* Smith, R. B., and Eaton, G. P. (eds.), Cenozoic tectonics and regional geophysics of the western Cordillera: Geological Society of America Memoir 152, p. 33–50.

Coney, P. J., 1981, Accretionary tectonics in western North America, *in* Dickinson, W. R., and Payne, W. D. (eds.), Relations of tectonics to ore deposits in the southern Cordillera: Arizona Geological Society Digest, v. 15, p. 23–37.

Coney, P. J., and Reynolds, S. J., 1977, Cordilleran Benioff zones: Nature, v. 270, p. 403–406.

Coney, P. J., Siberling, N. J., Jones, D. L., and Richter, D. H., 1981, Structural relations along the leading edge of Wrangellia terrane in the Clearwater Mountains, Alaska: United States Geological Survey Circular 823-B, p. 56–58.

Coney, P. J., and Harms, T. A., 1984, Cordilleran metamorphic core complexes: Cenozoic extensional relics of Mesozoic compression: Geology, v. 12, p. 550–554.

Conybeare, C. E. B., and Crook, K. A. W., 1968, Manual of sedimentary structures: Commonwealth of Australia, Department of National Development, Bureau of Mineral Resources, Geology and Geophysics Bulletin, v. 102, 327 p.

Cook, F. A., Brown, L. D., and Oliver, J. E., 1980, The southern Appalachians and the growth of continents: Scientific American, v. 243, p. 156–168.

Cooper, J. R., and Silver, L. T., 1964, Geology and ore deposits of the Dragoon quadrangle, Cochise County, Arizona: United States Geological Survey Professional Paper 416, 196 p.

Cooper, M. A., and Williams, G. D. (eds.), 1989, Inversion tectonics: Geological Society of London Special Publication No. 44, Blackwell Scientific Publications, London, 375 p.

Cosgrove, J. W., 1980, The tectonic implications of some small scale structures in the Mona Complex of Holy Isle, North Wales: Journal of Structural Geology, v. 2, p. 383–396.

Coulomb, C. A., 1773, Sur une application des regles de maximus et minimis a quelques problemes de statique relatifs a l'architecture: Academie Royale des Sciences, Memoires de Mathematique et de Physique par divers Savants, v. 7, p. 343–382.

Cox, A. V., Dalrymple, G. B., and Doell, R. R., 1967, Reversals of the Earth's magnetic field: Scientific American, v. 216, p. 44–45.

Cox, A., and Hart, R., 1986, Plate tectonics; how it works: Blackwell Scientific Publications, London, 392 p.

Craddock, J. P., and Pearson, A., 1994, Non-coaxial horizontal shortening strains preserved in twinned amygdule calcite, DSDP Hole 433, Suiko Seamount, Northwest Pacific Plate: Journal of Structural Geology, v. 16, no. 5, p. 719–724.

Crespi, J. M., 1982, The relationship of cleavage in carbonate rocks to folding and faulting in Agua Verde Wash, southeastern Arizona: Implications of volume loss: M.S. thesis, 123 pp.

Crittenden, M. D., Jr., Coney, P. J., and Davis, G. H. (eds.), 1980, Cordilleran metamorphic core complexes: Geological Society of America Memoir 153, 490 p.

Cross, T. A., and Pilger, R. H., 1982, Controls of subduction geometry, location of magmatic arcs, and tectonics of arc and back-arc regions: Geological Society of America, v. 93, p. 545–562.

Crowell, J. C., 1959, Problems of fault nomenclature: American Association of Petroleum Geologists Bulletin, v. 43, p. 2653–2674.

Crowell, J. C., 1974, Origin of late Cenozoic basins in southern California, *in* Dickinson, W. R. (ed.), Tectonics and sedimentation: Society of Economic Paleontologists and Mineralogists Special Publication 22, p. 190–204.

Crowell, J. C., and Ramirez, V. R., 1979, Late Cenozoic faults in southeastern California, *in* Crowell, J. C., and Sylvester, A. G. (eds.), Tectonics of the juncture between the San Andreas fault system and the Salton trough, southeastern California: Guidebook for Geological Society of America meeting, San Diego, 1979, p. 27–39.

Cruikshank, K. M., and Johnson, A. M., 1993, High-amplitude folding of linear-viscous multilayers: Journal of Structural Geology, v. 15, p. 79–94.

Cruikshank, K. M., Zhao, G., and Johnson, A. M., 1991, Analysis of minor fractures associated with joints and faulted joints: Journal of Structural Geology, v. 13, p. 865–886.

Currie, J. B., Patnode, A. W., and Trump, R. P., 1962, Development of folds in sedimentary strata: Geological Society of America Bulletin, v. 73, p. 655–674.

Dahlen, F. A., Suppe, J., and Davis, D. M., 1984, Mechanics of fold-and-thrust belts and accretionary wedges (continued): Cohesive Coulomb theory: Journal of Geophysical Research, v. 89, p. 10087–10101.

Dahlstrom, D. C. A., 1960, The upper detachement in concentric folding: Bulletin of Canadian Petroleum Geology, v. 17, p. 326–346.

Dahlstrom, D. C. A., 1969, Balanced cross sections: Canadian Journal of Earth Sciences, v. 6, p. 743–757.

Dahlstrom, D. C. A., 1970, Structural geology in the eastern margin of the Canadian Rocky Mountains: Bulletin Canadian Petroleum Geologists, v. 18, p. 332–406.

Dale, T. N., 1923, The commercial granites of New England: United States Geological Survey Bulletin 738, 488 p.

Dana, S. W., 1980, Analysis of gravity anomaly over coral-reef oil fields: Wilfred pool, Sullivan County, Indiana: American Association of Petroleum Geologists Bulletin, v. 64, p. 400–413.

Davis, D. M., and Engelder, T., 1985, The role of salt in fold-and-thrust belts: Tectonophysics, v. 119, p. 67–88.

Davis, D. M., Suppe, J., and Dahlen, F. A., 1983, Mechanics of fold-and-thrust belts and accretionary wedges: Journal of Geophysical Research, v. 88, p. 1153–1172.

Davis, G. A., Anderson, J. L., Frost, E. G., and Shackelford, T. J., 1980, Mylonitization and detachment faulting in the Whipple-Buckskin-Rawhide Mountains terrane, southeastern California and western Arizona, *in* Crittenden, M. D., Jr., Coney, P. J., and Davis, G. H. (eds.), Cordilleran metamorphic core complexes: Geological Society of America Memoir 153, p. 79–129.

Davis, G. H., 1972, Deformational history of the Caribou strata-bound sulfide deposit, Bathurst, New Brunswick, Canada: Economic Geology, v. 67, p. 634–655.

Davis, G. H., 1978, The monocline fold pattern of the Colorado Plateau, *in* Matthews, V. (ed.), Laramide folding associated with basement block faulting in the western United States: Geological Society of America Memoir 151, p. 215–233.

Davis, G. H., 1979, Laramide folding and faulting in southeastern Arizona: American Journal of Science, v. 279, p. 543–569.

Davis, G. H., 1980, Structural characteristics of metamorphic core complexes, *in* Crittenden, M. D., Jr., Coney, P. J., and Davis, G. H. (eds.), Cordilleran metamorphic core complexes: Geological Society of America Memoir 153, p. 35–77.

Davis, G. H., 1981, Regional strain analysis of the superposed deformations in southeastern Arizona and the eastern Great Basin, *in* Dickinson, W. R., and Payne, W. D. (eds.), Relation of tectonics to ore deposits in the southern Cordillera: Arizona Geological Society Digest, v. 14, p. 155–172.

Davis, G. H., 1983, Shear-zone model for the origin of metamorphic core complexes: Geology, v. 11, p. 348–351.

Davis, G. H., 1987, Saguaro National Monument, Arizona: outstanding display of structural characteristics of metamorphic core complexes, *in* Hill, M. L. (ed.), Centennial Field Guide Volume 1, Cordilleran Section of the Geological Society of America: Geological Society of America, Boulder, p. 35–40.

Davis, G. H., Eliopulos, G. J., Frost, E. G., Goodmundson, R. C., Knapp, R. B., Liming, R. B., Swan, M. M., and Wynn, J. C., 1974, Recumbent folds—focus of an investigative workshop in tectonics: Journal of Geological Education, v. 22, p. 204–208.

Davis, G. H., and Hardy, J. J., Jr., 1981, The Eagle Pass detachment, southeastern Arizona: Product of mid-Miocene listric(?) normal faulting in the southern Basin and Range: Geological Society of America Bulletin, Part 1, v. 92, p. 749–762.

Davis, G. H., Showalter, S. R., Benson, G. S., McCalmont, L. S., and Cropp, F. W., 1981, Guide to the geology of the Salt River Canyon region, Arizona: Arizona Geological Society Digest, v. 13, p. 48–97.

Davis, G. H., Gardulski, A. F., and Lister, G. S., 1987, Shear zone origin of quartzite mylonite and mylonitic pegmatite in the Coyote mountains: Journal of Structural Geology, v. 9, p. 289–298.

Davison, I., 1994, Linked fault systems; extensional, strike-slip and contractional, *in* Hancock, P. L. (ed.), Continental Deformation: Pergamon Press, Oxford, p. 121–142.

Dean, S. L., Kulander, B. R., and Skinner, J. M., 1988, Structural chronology of the Alleghenian orogeny in southeastern West Virginia: Geological Society of America Bulletin, v. 100, p. 299–310.

DeGraff, J. M., and Aydin, A., 1987, Surface morphology of columnar joints and its significance to mechanics and direction of joint growth: Geological Society of America, v. 99, p. 605–617.

DeGraff, J. M., Long, P. E., and Aydin, A., 1989, Use of joint-growth directions and rock textures to infer thermal regimes during solidification of basaltic lava flows: Journal of Volcanology and Geothermal Research, v. 38, p. 309–324.

Dennis, A. J., and Secor, D. T., 1987, A model for the development of crenulations in shear zones with applications from the Southern Appalachian Piedmont: Journal of Structural Geology, v. 9, p. 809–817.

Dennis, J. G., 1972, Structural geology: Ronald Press, New York, 532 p.

DeSitter, L. U., 1964, Structural geology (2nd edition): McGraw-Hill Book Company, New York, 551 p.

Dewey, J. F., 1972, Plate tectonics: Scientific American, v. 226, p. 56–72.

Dewey, J. F., and Bird, J. M., 1970, Mountain belts and the new global tectonics: Journal of Geophysical Research, v. 75, p. 2625–2647.

Dewey, J. R., and Bird, J. M., 1971, Origin and emplacement of the ophiolite suite: Appalachian ophiolites in Newfoundland: Journal of Geophysical Research, v. 76, p. 3179–3206.

Dickinson, W. R., 1973, Reconstruction of past arc-trench systems from petrotectonic assemblages in the island arcs of the western Pacific, *in* Coleman, P. J. (ed.), The western Pacific: island arcs, marginal seas, geochemistry: University of Western Australia Press, Nedlands, West Australia, p. 569–601.

Dickinson, W. R., 1974, Plate tectonics and sedimentation, *in* Dickinson, W. R. (ed.), Tectonics and sedimentation: Society of Economic Paleontologists and Mineralogists Special Publication 22, p. 1–27.

Dickinson, W. R., 1977, Tectono-stratigraphic evolution of subduction controlled sedimentary assemblages, *in* Talwani, M., and Pitman, W. C. (eds.), Island arcs, deep-sea trenches, and back-arc basins: American Geophysical Union Maurice Ewing Series 1, p. 33–40.

Dickinson, W. R., 1980, Plate tectonics and key petrologic associations, *in* Strangway, D. W. (ed.), The continental crust and its mineral deposits: Geological Association of Canada Special Paper 20, p. 341–360.

Dickinson, W. R., and Seely, D. S., 1979, Structure and stratigraphy of forearc regions: American Association of Petroleum Geologists Bulletin, v. 63, p. 2–31.

Dickinson, W. R., and Payne, W. D. (eds.), 1981, Relation

of tectonics to ore deposits in the southern Cordillera: Arizona Geological Society Digest, v. 14, 288 p.

Dietrich, D., 1986, Calcite fabrics around folds as indicators of deformation history: Journal of Structural Geology, v. 8, p. 655–668.

Dietz, R. S., 1961, Continent and ocean basin evolution by spreading of sea floor: Nature, v. 190, p. 854–857.

Dietz, R. S., 1968, Shatter cones in cryptoexplosion structures, *in* French, B. M., and Short, N. M. (eds.), Shock metamorphism of natural materials: Mono Book Corporation, Baltimore, p. 267–285.

Dietz, R. S., 1972, Shatter cones (shock fractures) in astroblemes: 24th International Geological Congress (Montreal, Canada): Harpell's, Quebec, p. 112–118.

Dietz, R. S., and Holden, J. I. C., 1970, Reconstruction of Pangaea: Breakup and dispersion of continents, Permian to present: Journal of Geophysical Research, v. 75, p. 4939–4955.

Dokka, R. K., 1993, The Eastern California shear zone and its role in the creation of young extensional zones in the Mojave Desert region, *in* Craig, S. D. (ed.), Structure, tectonics, and mineralization of the Walker Lane: Geological Society of Nevada, Walker Lane Symposium, p. 161–187.

Donath, F. A., 1961, Experimental study of shear failure in anisotropic rocks: Geological Society of America Bulletin, v. 72, p. 985–989.

Donath, F. A., 1962, Analysis of Basin-Range structure, south-central Oregon: Geological Society of America Bulletin, v. 73, p. 1–16.

Donath, F. A., 1970a, Rock deformation apparatus and experiments for dynamic structural geology: Journal of Geological Education, v. 18, p. 1–12.

Donath, F. A., 1970b, Some information squeezed out of rock: American Scientist, v. 58, p. 54–72.

Donath, F. A., and Parker, R. B., 1964, Folds and folding: Geological Society of America Bulletin, v. 75, p. 45–62.

Duncan, R. A., 1981, Hotspots in the southern oceans: an absolute frame of reference for motion of the Gondwana continents: Tectonophysics, v. 74, p. 29–42.

Durney, D. W., and Ramsay, J. G., 1973, Incremental strains measured by syntectonic crystal growths, *in* De Jong, K. A., and Scholten, R. (eds.), Gravity and tectonics: John Wiley & Sons, New York, p. 67–96.

Dutton, S. P., 1982, Pennsylvanian fan-delta and carbonate deposition, Mobeetie field, Texas Panhandle: American Association of Petroleum Geologists Bulletin, v. 66, p. 389–407.

Eaton, G. P., 1980, Geophysical and geological characteristics of the crust of the Basin and Range province, *in* Continental tectonics: National Academy of Sciences, Washington, D. C., p. 96–113.

Elasser, W. M., 1968, The mechanics of continental drift in Gondwanaland revisited—new evidence for continental drift: Proceedings of the American Philosophical Society, v. 112, p. 344–353.

Elliot, D., 1983, The construction of balanced cross-sections: Journal of Structural Geology, v. 5, p. 101.

Elliott, D., and Johnson, M. R. W., 1980, Structural evolution in the northern part of the Moine thrust belt, northwest Scotland: Royal Society of Edinburgh Transactions, v. 71, p. 69–96.

Ellis, P. G., and McClay, K. R., 1988, Listric extensional fault systems; results of analogue model experiments: Basin Research, v. 1, p. 55–70.

Engbretson, D. C., Cox, A., and Gordon, R. G., 1985, Relative motions between oceanic and continental plates in the Pacific basin: Geological Society of America Special Paper 206, 59 p.

Engelder, T., 1984, Loading paths to joint propagation during a tectonic cycle: an example from the Appalachian Plateau, U.S.A.: Eos, American Geophysical Union Transactions, v. 65, p. 1118.

Engelder, T., 1985, Loading paths to joint propagation during a tectonic cycle: an example from the Appalachian Plateau, U.S.A.: Journal of Structural Geology, v. 7, p. 459–476.

Engelder, T., 1987, Joints and shear fractures in rock, *in* Atkinson, B. K. (ed.), Fracture mechanics of rock: Academic Press, London, p. 27–69.

Engelder, T., and Engelder, R., 1977, Fossil distortion and decollement tectonics of the Appalachian Plateau: Geology, v. 5, p. 457–460.

Engelder, T., and Geiser, P., 1979, The relationship between pencil cleavage and lateral shortening within the Devonian section of the Appalachian Plateau, New York: Geology, v. 7, p. 460–464.

Engelder, T., and Geiser, P., 1980, Relationship between jointing sequence and paleostress fields on the Appalachian Plateau of New York: Geological Society of America Abstracts with Programs, v. 12, p. 33.

Engelder, T., and Geiser, P., 1980, On the use of regional joints sets as trajectories of paleostress fields during the development of the Appalachian Plateau, New York: Journal of Geophysical Research, Series B, v. 85, p. 6319–6341.

Engelder, T., and Marshak, S., 1985, Disjunctive cleavage formed at shallow depths in sedimentary rocks: Journal of Structural Geology, v. 7, p. 327–343.

Engelder, T., and Marshak, S., 1988, Analysis of data from rock-deformation experiments, *in* Marshak, S., and Mitra, G. (eds.), Basic methods of structural geology: Prentice-Hall, Englewood Cliffs, New Jersey, p. 193–212.

Engelder, T., and Oertel, G., 1985, The correlation between undercompaction and tectonic jointing within the Devonian Catskill Delta: Geology, v. 13, p. 863–866.

England, P. C., and Houseman, G. A., 1986, Finite strain calculations of continental deformation; 2, comparison with the India–Asia collision zone: Journal of Geophysical Research, v. 91, p. 3664–3676.

England, P., and Wortel, R., 1980, Some consequences of the subduction of young slabs: Earth and Planetary Science Letters, v. 47, p. 403–415.

Ernst, W. G., 1975, Metamorphism and plate tectonic regimes: benchmark papers in geology: Halsted Press, New York, 440 p.

Escher, A., Escher, J. C., and Watterson, J., 1975, The reorientation of the Kangamiut dike swarm, West Greenland: Canadian Journal of Earth Sciences, v. 12, no. 2, p. 158–173.

Eskola, P. E., 1949, The problem of mantled gneiss domes: Geological Society of London Quarterly Journal, v. 104, p. 461–476.

Etchecopar, A., and Malavieille, J., 1987, Computer models of pressure shadows; a method for strain measurement and shear-sense determination, *in* Cobbold, P. R., Gapais, D., Means, W. D., and Treagus, S. H. (eds.), Shear criteria in rocks: Journal of Structural Geology, v. 9, p. 667–677.

Etchecopar, A., and Vasseur, G., 1987, A 3-D kinematic model of fabric development in polycrystalline aggregates: comparisons with experimental and natural samples: Journal of Structural Geology, v. 9, p. 705–717.

Farnetani, C. G., and Richards, M. A., 1994, Numerical investigations of the mantle plume initiation model for flood basalt events: Journal of Geophysical Research, v. 99, no. B7, p. 13,813–13,833.

Feininger, T., 1978, The extraordinary striated outcrop at Saqsaywaman, Peru: Geological Society of America Bulletin, v. 89, p. 494–503.

Fink, J., 1980, Surface folding and viscosity of rhyolite flows: Geology, v. 8, p. 250–254.

Fletcher, R. C., and Pollard, D. D., 1981, Anticrack model for pressure solution surfaces: Geology, v. 9, p. 419–425.

Fleuty, M. J., 1964, The description of folds: Geological Association Proceedings, v. 75, p. 461–492.

Fleuty, M. J., 1975, Slickensides and slickenlines: Geological Magazine, v. 112, p. 319–322.

Flinn, D., 1962, On folding during three-dimensional progressive deformation: Geological Society of London Quarterly Journal, v. 118, p. 385–433.

Folk, R. L., 1965, Henry Clifton Sorby (1826–1908), the founder of petrography: Journal of Geological Education, v. 13, p. 43–47.

Forsyth, D. W., and Uyeda, S., 1975, On the relative importance of the driving forces of plate motion: Geophysical Journal of the Royal Astronomical Society, v. 43, p. 163–200.

Fossen, H., 1992, The role of extensional tectonics in the Caledonides of South Norway: Journal of Structural Geology, v. 14, p. 1033–1046.

Fox, P. J., and Gallow, D. G., 1984, Tectonic model for ridge-transform-ridge plate boundaries: implications for the structure of oceanic lithosphere: Tectonophysics, v. 104, p. 205–242.

Friedman, M., Handin, J., Logan, J. M., Min, K. D., and Stearns, D. W., 1976, Experimental folding of rocks under confining pressure. Part III. Faulted drape folds in multilithologic layered specimens: Geological Society of America Bulletin, v. 87, p. 1049–1066.

Fry, N., 1979, Random point distributions and strain measurement in rocks: Tectonophysics, v. 60, p. 69–105.

Gash, S. P. J., 1971, A study of surface features relating to brittle and semi-brittle fracture: Tectonophysics, v. 12, no. 5, p. 349–391.

Geiser, P., Kligfield, R., and Geiser, J., 1986, Using a microcomputer for interactive section construction and balancing: American Association of Petroleum Geologists Bulletin, v. 70, p. 594–595.

Ghosh, S. K., 1968, Experiments of buckling of multilayers which permit interlayer gliding: Tectonophysics, v. 6, p. 207–249.

Ghosh, S. K., 1993, Structural geology: fundamentals and modern developments: Pergamon Press, Oxford, 598 p.

Gibbons, W., 1994, Suspect terranes, *in* Hancock, P. L. (ed.), Continental deformation: Pergamon Press, Oxford, p. 305–319.

Goodwin, A. M., 1976, Giant impacting and the development of continental crust, *in* Windley, B. F. (ed.), The early history of the Earth: John Wiley & Sons Ltd., Chichester, England, p. 77–95.

Gray, D. R., 1977a, Morphologic classification of crenulation cleavages: Journal of Geology, v. 85, p. 229–235.

Gray, D. R., 1977b, Some parameters which affect the morphology of crenulation cleavages: Journal of Geology, v. 85, p. 763–780.

Gray, D. R., 1979, Microstructure of crenulation cleavages: An indication of cleavage origin: American Journal of Science, v. 279, p. 97–128.

Gray, D. R., 1981, Compound tectonic fabrics in singly folded rocks from southwest Virginia, U.S.A.: Tectonophysics, v. 78, p. 229–248.

Gray, D. R., and Durney, D. W., 1979, Investigations on the mechanical significance of crenulation cleavage: Tectonophysics, v. 58, p. 35–79.

Griesbach, C. L., 1891, Geology of central Himalayas: Geological Survey of India, Memoir 23.

Grieve, R. A. F., and Pesonen, L. J., 1992, The terrestrial impact cratering record: Tectonophysics, v. 216, p. 1–30.

Griffith, A. A., 1924, Theory of rupture: Proceedings of the First International Congress on Applied Mechanics, Delft, the Netherlands, p. 55–63.

Griggs, D. T., Turner, F. J., and Heard, H. C., 1960, Deformation of rocks at 500° to 800°C, *in* Griggs, D. T., and Handin, J. (eds.), Rock deformation: Geological Society of America Memoir 79, p. 39–104.

Groshong, R. H., Jr., 1975, "Slip" cleavage caused by pressure solution in a buckle fold: Geology, v. 3, p. 411–413.

Groshong, R. H., Jr., 1988, Low-temperature deformation mechanisms and their interpretation: Geological Society of America Bulletin, v. 100, p. 1329–1360.

Guilbert, J. M., and Park, C. F., Jr., 1986, The geology of ore deposits: W. H. Freeman, New York, 985 p.

Gustafson, J. K., Burrell, H. C., and Garretty, M. D., 1950, Geology of the Broken Hill ore deposit, New South Wales,

Australia: Geological Society of America Bulletin, v. 61, p. 1369–1437.

Hafner, W., 1951, Stress distribution and faulting: Geological Society of America Bulletin, v. 62, p. 373–398.

Halbouty, M. T., 1969, Hidden and subtle traps in Gulf Coast: American Association of Petroleum Geologists Bulletin, v. 53, p. 3–29.

Hamblin, W. K., 1965, Origin of "reverse drag" on the downthrown side of normal faults: Geological Society of America Bulletin, v. 76, p. 1145–1164.

Hamilton, W., 1964, Geologic map of the Big Maria Mountains NE Quadrangle, Riverside County, California, and Yuma County, Arizona: U.S. Geological Survey Geologic Quadrangle Map GQ–350.

Hamilton, W., 1978, Mesozoic tectonics of the western United States, *in* Howell, D. G., and McDougall, K. A. (eds.), Mesozoic paleogeography of the western United States: Society of Economic Paleontologists and Mineralogists, Pacific Coast Paleogeography Symposium, p. 33–70.

Hamilton, W., 1979, Tectonics of the Indonesian region: United States Geological Survey Professional Paper 1078, 345 p.

Hamilton, Warren, 1987, Mesozoic geology and tectonics of the Big Maria Mountains region, southeastern California, *in* Dickinson, W. R., and Klute, M. A. (eds.), Mesozoic rocks of southern Arizona and adjacent areas: Arizona Geological Society Digest, v. 18, p. 33–47.

Hancock, P. L., 1985, Brittle microtectonics: principles and practice: Journal of Structural Geology, v. 7, p. 437–458.

Hancock, P. L. (ed), 1994, Continental deformation: Pergamon Press, Oxford, 421 p.

Hancock, P. L., and Barka, A. A., 1987, Kinematic indicators on active normal faults in western Turkey: Journal of Structural Geology, v. 9, p. 573–584.

Handin, J., 1966, Strength and ductility, *in* Clark, S. P. Jr., ed., Handbook of physical constants: Geological Society of America Memoir 97, p. 223–289.

Handin, J., 1969, On the Coulomb-Mohr failure criterion: Journal of Geophysical Research, v. 74, p. 5343–5348.

Handin, J., and Hager, R. V., 1957, Experimental deformation of sedimentary rocks under confining pressure: tests at room temperature on dry samples: American Association of Petroleum Geologists Bulletin, v. 41, p. 1–50.

Handin, J., Hager, R. V., Jr., Friedman, M., and Feather, J. N., 1963, Experimental deformation of sedimentary rocks under confining pressure: pore pressure tests: American Association of Petroleum Geologists Bulletin, v. 47, p. 717–755.

Hanmer, S., and Passchier, C., 1991, Shear-sense indicators: a review: Geological Survey of Canada Paper 90-17, 72 p.

Hanmer, S., Bowring, S., Van Breemen, O., and Parrish, R., 1992, Great Slave Lake shear zone, NW Canada: mylonitic record of Early Proterozoic continental convergence, collision and indentation: Journal of Structural Geology, v. 14, p. 757–774.

Harding, T. P., 1973, Newport-Inglewood trend, California—an example of wrenching style of deformation: American Association of Petroleum Geologists Bulletin, v. 57, p. 97–116.

Harding, T. P., 1993, Evidence of fluid flow in microfractures: American Association of Petroleum Geologists Bulletin, v. 77, p. 1303–1314.

Harris, J. F., Taylor, G. L., and Walper J. L., 1960, Relation of deformation features in sedimentary rocks to regional and local structure: American Association of Petroleum Geologists Bulletin, v. 44, p. 1853–1873.

Harris, L. D., 1979, Similarities between the thick-skinned Blue Ridge anticlinorium and thin-skinned Powell Valley anticline: Geological Society of America Bulletin, Part 1, v. 90, p. 525–539.

Haynes, F. M., and Titley, S. R., 1980, The evolution of fracture-related permeability within the Ruby Star granodiorite, Sierrita porphyry copper deposit, Pima County, Arizona: Economic Geology, v. 75, p. 673–683.

Hayward, A. B., and Graham, R. H., 1989, Some geometrical characteristics of inversion, *in* Cooper, M. A. and Williams, G. D. (eds.), Inversion tectonics: Geological Society of London Special Publication No. 44, Blackwell Scientific Publications, London, p. 17–39.

Heard, H. C., 1963, Effect of large changes in strain rate in the experimental deformation of Yule marble: Journal of Geology, v. 71, p. 162–195.

Heidrick, T. L., and Titley, S. R., 1982, Fracture and dike patterns in Laramide plutons and their structural and tectonic implications: American Southwest, *in* Titley, S. R. (ed.), Advances in geology of the porphyry copper deposits: Southwest North America: University of Arizona Press, Tucson, Arizona, p. 73–91.

Heim, A., 1921, Geologie der Schweiz, v. 2/1: Tauchnitz, Leipzig.

Helgeson, D. E., and Aydin, A., 1991, Characteristics of joint propagation across layer interfaces in sedimentary rocks: Journal of Structural Geology, v. 13, p. 897–911.

Hendry, H. E., and Stauffer, M. R., 1977, Penecontemporaneous folds in cross-bedding: inversion of facing criteria and mimicry of tectonic folds: Geological Society of America Bulletin, v. 88, p. 809–812.

Hess, H. H., 1962, History of ocean basins, *in* Engel, A. E. J., James, H. L., and Leonard, B. F. (eds.), Petrologic studies: a volume in honor of A. F. Buddington: Geological Society of America, Boulder, Colorado, p. 599–620.

Hey, R. N., Dennebier, F. K., and Morgan, W. J., 1980, Propagating rifts on mid-ocean ridges: Journal of Geophysical Research, v. 85, p. 3647–3658.

Hey, R. N., Menard, H. W., Atwater, T. M., and Caress, D. W., 1988, Changes in direction of seafloor spreading revisited: Journal of Geophysical Research, v. 93, p. 2803–2811.

Higgins, M. W., 1971, Cataclastic rocks: United States Geological Survey Professional Paper 687, 97 p.

Hill, M. L., 1959, Dual classification of faults: Geological Society of America Bulletin, v. 43, p. 217–221.

Hills, E. S., 1972, Elements of structural geology (2nd edition): John Wiley & Sons, New York, 502 p.

Hirth, G., and Tullis, J., 1992, Dislocation creep regimes in quartz aggregates: Journal of Structural Geology, v. 14, p. 145–160.

Hobbs, B. E., Means, W. D., and Williams, P. F., 1976, An outline of structural geology: John Wiley & Sons, New York, 571 p.

Hodgson, R. A., 1961, Classification of structures on joint surfaces: American Journal of Science, v. 259, p. 493–502.

Holmes, A., 1964, Principles of physical geology: Ronald Press, New York, 1288 p.

Hoover, J. R., Malone, R., Eddy, G., and Donaldson, A., 1969, Regional position, trend, and geometry of coals and sandstones of the Monongahela Group and Waynesburg formation in the Central Appalachians, *in* Donaldson, A. C. (ed.), Some Appalachian coals and carbonates: models of ancient shallow-water deposition: West Virginia Geological and Economic Survey, Morgantown, West Virginia, p. 157–192.

Hubbert, M. K, 1937, Theory of scale models as applied to the study of geologic structures: Geological Society of America Bulletin, v. 48, p. 1459–1519.

Hubbert, M. K., 1951, Mechanical basis for certain familiar geologic structures: Geological Society of America Bulletin, v. 62, p. 355–372.

Hubbert, M. K., and Rubey, W. W., 1959, Role of fluid pressure in mechanics of overthrust faulting. Part 1: Geological Society of America Bulletin, v. 70, p. 115–166.

Hudleston, P. J., 1973, Fold morphology and some geometrical implications of theories of fold development: Tectonophysics, v. 16, p. 1–46.

Huiqi, L., McClay, K. R., and Powell, C. M., 1992, Physical models of thrust wedges, *in* McClay, K. R. (ed.), Thrust tectonics: Chapman and Hall, New York, p. 71–81.

Hull, D., and Bacon, D. S., 1984, Introduction to dislocations: Pergamon Press, New York, 251 p.

Humphreys, E. D., and Weldon, R. J., 1994, Deformation across the western United States: a local estimate of Pacific-North America transform deformation: Journal of Geophysical Research, v. 99, p. 19,975–20,010.

Hunt, C. B., Averitt, P., and Miller, R. L., 1953, Geology and geography of the Henry Mountain region, Utah: United States: Geological Survey Professional Paper 228, 234 p.

Huntoon, P. W., and Richter, H. R., 1979, Breccia pipes in the vicinity of Lockhart Basin, Canyonlands area, Utah, *in* Baars, D. L. (ed.), Permianland: Four Corners Geological Society Guidebook, 9th Field Conference, p. 47–53.

Ingersol, R. V., 1988, Tectonics of sedimentary basins: Geological Society of America Bulletin, v. 100, p. 1704–1719.

Ingraffea, A. R., 1987, Theory of crack initiation and propagation in rock, *in* Atkinson, B. K. (ed.), Fracture mechanics of rock: Academic Press, London, p. 71–110.

Isacks, B., and Molnar, P., 1969, Mantle earthquake mechanisms and the sinking of the lithosphere: Nature, v. 223, p. 1121–1124.

Isacks, B., Oliver, J., and Sykes, L. R., 1968, Seismology and the new global plate tectonics: Journal of Geophysical Research, v. 73, p. 5855–5899.

Jackson, J. A., Gagnepain, J., Houseman, G., King, G. C. P., Papadimitriou, P., Soufleris, C., and Virieux, J., 1982, Seismicity, normal faulting, and the geomorphological development of the Gulf of Corinth (Greece): the Corinth earthquakes of February and March, 1981: Earth and Planetary Science Letters, v. 57, p. 377–397.

Jacobs, L. L., 1993, Quest for the African dinosaurs: ancient roots of the modern world: Villard Books, New York, 314 p.

Jaeger, J. C., and Cook, N. G. W., 1976, Fundamentals of rock mechanics: Halsted Press, New York, 585 p.

Jagnow, D. H., 1979, Cavern development in the Guadalupe Mountains: Adobe Press, Albuquerque, New Mexico, 55 p.

Jessell, M. W., 1988a, Simulation of fabric development in recrystallizing aggregates; I, Description of the model: Journal of Structural Geology, v. 10, p. 771–778.

Jessell, M. W., 1988b, Simulation of fabric development in recrystallizing aggregates; II, Example model runs: Journal of Structural Geology, v. 10, p. 779–793.

Johnson, A. M., 1970, Physical processes in geology: Freeman, Cooper, and Company, San Francisco, 577 p.

Johnson, A. M., 1977, Styles of folding: mechanics and mechanisms of folding of natural elastic materials: Elsevier Scientific Publishing Company, Amsterdam, 406 p.

Jordan, T. H., 1988, Structure and formation of the continental tectosphere, *in* Menzies, M., and others (eds.), Oceanic and continental lithosphere; similarities and differences: Journal of Petrology, p. 11–37.

Jordan, W. M., 1965, Regional environmental study of the Early Mesozoic Nugget and Navajo Sandstones: Ph.D. dissertation, University of Wisconsin, Madison, Wisconsin, 206 p.

Kamb, W. B., 1959, Theory of preferred orientation developed by crystallization under stress: Journal of Geology, v. 67, p. 153–170.

Karig, D. E., 1971, Origin and development of marginal basins in the western Pacific: Journal of Geophysical Research, v. 76, p. 2542–2561.

Karlstrom, K. E., and Bowring, S. A., 1991, Styles and timing of Early Proterozoic deformation in Arizona: constraints on tectonic models, *in* Karlstrom, K. E. (ed.), Proterozoic geology and ore deposits of Arizona: Arizona Geological Society Digest, v. 19, p. 1–10.

Karato, S. I., and Wu, P., 1993, Rheology of the upper mantle: a synthesis: Science, v. 260, p. 771–778.

Kearey, P., and Vine, F. J., 1990, Global tectonics: Blackwell Scientific Publications, Boston, 302 p.

Kelley, V. C., 1955, Monoclines of the Colorado Plateau: Geological Society of America Bulletin, v. 66, p. 789–804.

Kerrich, R., 1978, An historical review and synthesis of research on pressure solution: Zentbl. Miner. Geol. Palaeontol., v. 1, p. 512–550.

King, P. B., 1959, The evolution of North America: Princeton University Press, Princeton, New Jersey, 189 p.

Knipe, R. J., 1989, Deformation mechanisms; recognition from natural tectonites: Journal of Structural Geology, v. 11, p. 127–146.

Knopf, E. B., and Ingerson, E., 1938, Structural petrology: Geological Society of America Memoir 6, 270 p.

Krantz, R. W., 1986, The odd-axis model: orthorhombic fault patterns and three-dimensional strain fields: PhD dissertation, The University of Arizona, Tucson, 97 p.

Krantz, R. W., 1988, Multiple fault sets and three-dimensional strain: theory and application: Journal of Structural Geology, v. 10, p. 225–237.

Kranz, R. L., 1983, Microcracks in rocks; a review, *in* Friedman, M., and Toksoez, M. N. (eds.), Continental tectonics; structure, kinematics and dynamics: Tectonophysics, v. 100, p. 449–480.

Kronenberg, A. K., Segall, P., and Wolf, G. H., 1990, Hydrolytic weakening and penetrative deformation within a natural shear zone, *in* Duba, A. G., Durham, W. B., Handin, J. W., and Wang, H. F. (eds.), The brittle-ductile transition in rocks: Geophysical Monograph, v. 56, p. 21–36.

Kuenen, P. H., and DeSitter, L. U., 1938, Experimental investigation into the mechanism of folding: Leidse Geological Mededlingen, v. 9, p. 217–239.

Lapworth, C., 1885, The highland controversy in British geology; its causes, course and consequences: Nature, v. 32, p. 558–559.

Laubscher, H. P., 1977, Fold development in the Jura: Tectonophysics, v. 37, p. 337–362.

Law, R. D., Casey, M., and Knipe, R. J., 1986, Kinematic and tectonic significance of microstructures and crystallographic fabrics within quartz mylonites from the Assynt and Eriboll regions of the Moine thrust zone, NW Scotland: Royal Society of Edinburgh Transactions, v. 77, p. 99–125.

Lawn, B. R., and and Wilshaw, T. R., 1975, Fracture of brittle solids: Cambridge University Press, Cambridge, 204 p.

Le Pichon, X., 1968, Sea floor spreading and continental drift: Journal of Geophysical Research, v. 73, p. 3661–3697.

Le Pichon, X., and Sibuet, J., 1981, Passive margins: a model of formation: Journal of Geophysical Research, v. 86, p. 3708–3720.

Lisle, R. J., 1992, Strain estimation from flattened buckle folds: Journal of Structural Geology, v. 14, p. 369–371.

Lisle, R. J., 1994, Palaeostrain analysis, *in* Hancock, P. L. (ed.), Continental deformation: Pergamon Press, New York, p. 28–42.

Lister, G. S., and Hobbs, B. E., 1980, The simulation of fabric development during plastic deformation and its application to quartzite: the influence of deformation history: Journal of Structural Geology, v. 2, p. 355–370.

Lister, G. S., and Snoke, A. W., 1984, S-C mylonites: Journal of Structural Geology, v. 6, p. 617–638.

Lister, G. S., and Williams, P. F., 1983, The partitioning of deformation in flowing rock masses: Etheridge, M. A., and Cox, S. F. (eds.), Deformation processes in tectonics: Tectonophysics, v. 92, p. 1–33.

Lloyd, G. E., and Knipe, R. J., 1992, Deformation mechanisms accommodating faulting of quartzite under upper crustal conditions: Journal of Structural Geology, v. 14, p. 127–143.

Lockner, D., and Byerlee, J. D., 1977a, Acoustic emission and fault formation in rocks: Series on Rock and Soil Mechanics, v. 2, no. 3, p. 99–107.

Lockner, D., and Byerlee, J. D., 1977b, Hydrofracture in Weber Sandstone at high confining pressure and differential stress: Journal of Geophysical Research, v. 82, p. 2018–2026.

Lockner, D., and Byerlee, J. D., 1977c, Acoustic emission and creep in rock at high confining pressure and differential stress: Seismological Society of America Bulletin, v. 67, p. 247–258.

Lowell, J. D., 1968, Geology of the Kalamazoo ore body, San Manuel district, Arizona: Economic Geology, v. 63, p. 645–654.

Lowell, J. D., and Genik, G. J., 1972, Sea-floor spreading and structural evolution of southern Red Sea: American Association of Petroleum Geologists Bulletin, v. 56, p. 247–259.

Lowell, J. D., Genik, G. J., Nelson, T. H., and Tucker, P. M., 1975, Petroleum and plate tectonics of the southern Red Sea, *in* Fisher, A. G., and Judson, S (eds.), Petroleum and global tectonics: Princeton University Press, Princeton, New Jersey, p. 129–153.

Lowman, P. D., Jr., 1976, A satellite view of diapiric Archean granites in western Australia: Journal of Geology, v. 84, p. 237–238.

Lowman, P. D., Jr., 1981, A global tectonic activity map with orbital photographic supplement: NASA Technical Memorandum 82073, 117 p.

Lowry, H. F., 1969, in College Talks, Blackwood, J. R. (ed.): Oxford University Press, New York, 177 p.

Lundin, E. R., 1989, Thrusting of the Claron Formation, the Bryce Canyon region, Utah: Geological Society of America Bulletin, v. 101, p. 1038–1050.

Macdonald, G. A., 1967, Forms and structures of extrusive basaltic rocks in Hess, H. H., and Poldervaart, A. (eds.), Basalts—the Poldervaart treatise on rocks of basalt composition: Wiley-Interscience Publishers, New York, v. 1, p. 1–61.

MacDonald, K. C., and Fox, P. J., 1983, Overlapping spreading centres: new accretion geometry on the East Pacific Rise: Nature, v. 302, p. 55–58.

MacDonald, K. C., Scheirer, D. S., and Carbotte, S. M., 1991, Mid-ocean ridges: discontinuities, segments and giant cracks: Science, v. 253, p. 986–994.

Mackin, J. H., 1950, The down-structure method of viewing geologic maps: Journal of Geology, v. 58, p. 55–72.

Malavieille, J., 1987, Kinematics of compressional and extensional ductile shearing deformation in a metamorphic core complex of the northeastern Basin and Range: Journal of Structural Geology, v. 9, p. 541–554.

Malcolm, W., 1912, Gold fields of Nova Scotia: Canadian Geological Survey Memoir 20-E, 331 p.

Mann, P., Hempton, M. R., Bradley, D. C., and Burke, K., 1983, Development of pull-apart basins: Journal of Geology, v. 91, p. 529–554.

Manton, O. O., 1965, The orientation and origin of shatter-cones in the Vredefort Ring: New York Academy of Science Annals, v. 123, p. 1017–1049.

Marlow, P. C., and Etheridge, M. A., 1977, Development of a layered crenulation cleavage in mica schists of the Kanmantoo Group near Macclesfield, South Australia: Geological Society of America Bulletin, v. 88, p. 873–882.

Marshak, S., 1986, Structure and tectonics of the Hudson Valley fold-thrust belt, eastern New York State: Geological Society of America Bulletin, v. 97, p. 354–368.

Marshak, S., and Engelder, T., 1985, Development of cleavage in a fold-thrust belt in eastern New York: Journal of Structural Geology, v. 7, p. 345–359.

Marshak, S., and Tabor, J. R., 1989, Structure of the Kingston orocline in the Appalachian fold-thrust belt, New York: Geological Society of America Bulletin, v. 101, no. 5, p. 683–701.

Marshak, S., and Woodward, N., 1988, Introduction to cross-section balancing, *in* Marshak, S., and Mitra, G. (eds.), Basic methods of structural geology: Prentice-Hall, Englewood Cliffs, New Jersey, 303–326.

Martin, R. G., 1978, Northern and eastern Gulf of Mexico continental margin, stratigraphic and structural framework, *in* Bouma A. H., Moore, G. T., and Coleman, J. M. (eds.), Framework, facies, and oil-trapping characteristics of the upper continental margin: American Association of Petroleum Geologists Studies in Geology 7, p. 21–42.

Mattauer, M., 1975, Sur le mechanisme de formation de la schistosité dans l'Himalaya: Earth and Planetary Science Letters, v. 8, p. 144–154.

McClay, K. R., 1989, Analogue models of inversion tectonics, *in* Cooper, M. A., and Williams, G. D. (eds.), Inversion tectonics: Geological Society of London Special Publication No. 44, Blackwell Scientific Publications, London, p. 41–59.

McClay, K. R., 1992a, Glossary of thrust tectonic terms, *in* McClay, K. R. (ed.), Thrust tectonics: Chapman and Hall, London, p. 419–433.

McClay, K. R., (ed.), 1992b, Thrust tectonics: Chapman and Hall, London, 447 p.

McClay, K. R., and Coward, M. P., 1981, The Moine thrust zone: an overview, *in* McClay, K. R., and Price, N. J. (eds.), Thrust and nappe tectonics: Geological Society of London Special Publication 9, Blackwell Scientific Publications, p. 241–260.

McClay, K. R., and Ellis, P. G., 1987, Geometries of extensional fault systems developed in model experiments: Geology, v. 15, p. 341–344.

McClay, K. R., and Price, N. J. (eds.), 1981, Thrust and nappe tectonics: Geological Society of London, Special Publication 9, Blackwell Scientific Publications, 539 p.

McClay, K. R., and Scott, A. D., 1991a, Experimental models of hangingwall deformation in ramp-flat listric extensional fault systems, *in* Cobbold, P. R. (ed.), Experimental and numerical modelling of continental deformation: Tectonophysics, v. 188, p. 85–96.

McClay, K. R., Waltham, D. A., Scott, A. D., and Abousetta, A., 1991b, Physical and seismic modelling of listric normal fault geometries, *in* Roberts, A. M., Yielding, G., and Freeman, B. (eds.), 1991, The geometry of normal faults: Geological Society of London Special Publication No. 56, p. 231–239.

McGill, G. E., and Stromquist, A. W., 1979, The grabens of Canyonlands National Park, Utah: geometry, mechanics, and kinematics: Journal of Geophysical Research, v. 4, p. 4547–4563.

McKenzie, D. P., and Bickle, M. J., 1988, The volume and composition of melt generated by extension of the lithosphere: Journal of Petrology, v. 29, p. 625–629.

McKenzie, D. P., and Morgan, W. J., 1969, The evolution of triple junctions: Nature, v. 224, p. 125–133.

McKinstry, H. E., 1961, Mining geology: Prentice-Hall, Englewood Cliffs, New Jersey, 680 p.

Means, W. D., 1976, Stress and strain: Springer-Verlag, New York, 339 p.

Means, W. D., 1980, High temperature simple shearing fabrics; a new experimental approach: Journal of Structural Geology, v. 2, p. 197–202.

Means, W. D., 1987, A newly recognized type of slickenside striation: Journal of Structural Geology, v. 9, p. 585–590.

Means, W. D., 1990, Review paper: Kinematics, stress, deformation, and material behavior: Journal of Structural Geology, v. 12, no. 8, p. 953–971.

Means, W. D., Hobbs, B. E., Lister, G. S., and Williams, P. F., 1980, Vorticity and non-coaxiality in progressive deformations: Journal of Structural Geology, v. 2, no. 3, p. 371–378.

Melosh, H. J., and Raefsky, A., 1980, The dynamical origin of subduction zone topography: Geophysical Journal of the Royal Astronomical Society, v. 60, p. 333–354.

Menges, C. M., 1981, The Sonoita Creek basin: implications for late Cenozoic evolution of basins and ranges in southeastern Arizona: M.S. thesis, University of Arizona, Tucson, 239 p.

Merle, O., and Guillier, B., 1989, The building of the Central Swiss Alps: an experimental approach: Tectonophysics, v. 165, p. 41–56.

Michener, J. A., 1959, Hawaii: Random House, New York, 937 p.

Middleton, G. V., and Wilcock, P. R., 1994, Mechanics in the earth and environmental sciences: Cambridge University Press, Cambridge, England, 459 p.

Milnes, A. G., 1979, Albert Heim's general theory of natural deformation (1878): Geology, v. 7, p. 99–103.

Mitra, S., 1987, Regional variations in deformation mechanisms and structural styles in the central Appalachian orogenic belt: Geological Society of America Bulletin, v. 98, p. 569–590.

Mitra, S., 1988, Three-dimensional geometry and kinematic evolution of the Pine Mountain thrust system, southern Appalachians: Geological Society of America Bulletin, v. 100, p. 72–95.

Mitra, S., 1992, Balanced structural interpretations in fold and thrust belts, *in* Mitra, S., and Fisher, G. W. (eds.), Structural geology of fold and thrust belts: Johns Hopkins University Press, Baltimore, p. 53–77.

Miyashiro, A., 1973, Metamorphism and metamorphic belts: William Clowes & Sons, Ltd., London, 492 p.

Miyashiro, A., 1974, Volcanic rock series in island arcs and active continental margins: American Journal of Science, v. 274, p. 321–355.

Mohr, O. C., 1990, Welche Umstande bedingen die Elastizitatsgrenze und den Bruch eines Materials: Zeitschrift der Vereines Deutscher Ingenieure, v. 44, p. 1524–1530 and 1572–1577.

Molnar, P., 1988, Continental tectonics in the aftermath of plate tectonics: Nature, v. 335, p. 131–137.

Molnar, P., and England, P., 1990, Temperatures, heat flux and frictional stress near major thrust faults: Journal of Geophysical Research, Series B: Solid Earth and Planets, v. 95, p. 4833–4856.

Molnar, P., and Tapponier, P., 1975, Cenozoic tectonics of Asia: effect of a continental collision: Science, v. 189, p. 419–425.

Morgan, W. J., 1968, Rises, trenches, great faults, and crustal blocks: Journal of Geophysical Research, v. 73, p. 1959–1982.

Morgan, W. J., 1971, Convection plumes in the lower mantle: Nature, v. 230, p. 42–43.

Morgan, W. J., 1972, Convection plumes and plate motions: American Association of Petroleum Geologists Bulletin, v. 56, p. 203–213.

Morgan, W. J., 1981, Hot spot tracks and the opening of the Atlantic and Indian oceans, *in* Emiliani, C. (ed.), The oceanic lithosphere: John Wiley & Sons, New York, p. 443–487.

Mosher, S., 1981, Pressure solution deformation of the Purgatory Conglomerate from Rhode Island: Journal of Geology, v. 89, p. 35–55.

Murrell, S. A. F., 1990, Brittle-to-ductile transitions in polycrystalline nonmetallic materials, *in* Barber, D. J., and Meredith, P. G. (eds.), Deformation processes in minerals, ceramics and rocks: Unwin Hyman, London, p. 109–134.

Nettleton, L. L., 1934, Fluid mechanics of salt domes: American Association of Petroleum Geologists Bulletin, v. 18, p. 1175–1204.

Neuhauser, K. R., 1988, Sevier-age ramp-style thrust faults at Cedar Mountain, northwestern San Rafael Swell (Colorado Plateau), Emery County, Utah: Geology, v. 16, p. 299–302.

Nickelsen, R. P., 1972, Attributes of rock cleavage in some mudstones and limestones of the Valley and Ridge province, Pennsylvania: Pennsylvania Academy of Science Proceedings, v. 46, p. 107–112.

Nickelsen, R. P., and Hough, V. D., 1967, Jointing in the Appalachian Plateau of Pennsylvania: Geological Society of America Bulletin, v. 78, p. 609–630.

Nicolas, A., 1987, Principles of rock deformation: petrology and structural geology: D. Reidel Publishing Company, Dordrecht, Netherlands, 208 p.

Nicolas, A., and Le Pichon, X., 1980, Thrusting of young lithosphere in subduction zones with special reference to structures in ophiolitic peridotites: Earth and Planetary Science Letters, v. 46, p. 397–406.

Nicolas, A., and Poirier, J. P., 1976, Crystalline plasticity and solid state flow in metamorphic rocks: selected topics in geological sciences: John Wiley & Sons, London, 444 p.

Norton, D., and Knapp, R., 1977, Transport phenomena in hydrothermal systems: the nature of porosity: American Journal of Science, v. 277, p. 913–936.

Nur, A., and Simmons, G., 1970, The origin of small cracks in igneous rocks: International Journal of Rock Mechanics and Mining Science, v. 7, p. 307–312.

O'Driscoll, E. S., 1962, Experimental patterns in superposed similar folding: Journal of Alberta Society of Petroleum Geologists, v. 10, p. 145–167.

O'Driscoll, E. S., 1964a, Cross fold deformation by simple shear: Economic Geology, v. 59, p. 1061–1093.

O'Driscoll, E. S., 1964b, Interference patterns from inclined shear fold systems: Canadian Petroleum Geologists Bulletin, v. 12, p. 279–310.

Oertel, G., 1965, The mechanism of faulting in clay experiments: Tectonophysics, v. 2, p. 343–393.

Oertel, G., 1970, Deformation of a slaty, lapillar tuff in the Lake District, England: Geological Society of America Bulletin, v. 81, p. 1173–1188.

Olson, J., and Pollard, D. D., 1988, Inferring stress states from detailed joint geometry, *in* Cundall, P. A., Sterling, R. L., and Starfield, A. M., (eds.), Key questions in rock mechanics: Proceedings of the 29th U. S. Symposium on Rock Mechanics, A. A. Balkoma, Rotterdam, p. 159–167.

Olson, J., and Pollard, D. D., 1989, Inferring paleostresses from natural fracture patterns: a new method: Geology, v. 17, p. 345–348.

Park, R. G., 1988, Geological structures and moving plates: Blackie, Glasgow, 337 p.

Park, C. F., and MacDiarmid, R. A., 1964, Ore deposits: W. H. Freeman, San Francisco, 475 p.

Parker, J. M., 1942, Regional systematic jointing in slightly deformed sedimentary rocks: Geological Society of America Bulletin, v. 53, p. 381–408.

Parker, T. J., and McDowell, A. N., 1955, Model studies of salt-dome tectonics: American Association of Petroleum Geologists Bulletin, v. 39, p. 2384–2470.

Parsons, B., and Sclater, J. G., 1977, An analysis of the variation of ocean floor bathymetry and heat flow with age: Journal of Geophysical Research, v. 82, p. 803–827.

Passchier, C. W., 1987, Stable positions of rigid objects in non-coaxial flow; a study in vorticity analysis, *in* Cobbold, P. R., Gapais, D., Means, W. D., and Treagus, S. H. (eds.), Shear criteria in rocks: Journal of Structural Geology, v. 9, p. 679–690.

Passchier, C. W., 1994, Mixing in flow perturbations: a model for development of mantled porphyroclasts in mylonites: Journal of Structural Geology, v. 16, p. 733–741.

Passchier, C. W., and Simpson, C., 1986, Porphyroclast systems as kinematic indicators: Journal of Structural Geology, v. 8, p. 831–843.

Passchier, C. W., Myers, J. S., and Kröner, A., 1990, Field geology of high-grade gneiss terrains: Springer-Verlag, New York, 150 p.

Passchier, C. W., and Sokoutis, D., 1993, Experimental modelling of mantle porphyroclasts: Journal of Structural Geology, v. 15, p. 895–909.

Paterson, M. S., and Weiss, L. E., 1966, Experimental deformation and folding of phyllite: Geological Society of America Bulletin, v. 77, p. 343–374.

Paterson, W. A., 1978, A technique for approximating values for the formation factor parameters of "m" and "a" and formation water resistivities of shaly formations: The Log Analyst, v. 19, p. 12–22.

Paterson, S. R., and Tobisch, O. T., 1992, Rates of processes in magmatic arcs: implications for the timing and nature of pluton emplacement and wall rock deformation: Journal of Structural Geology, v. 14, p. 291–300.

Pavlis, T. L., 1986, The role of strain heating in the evolution of megathrusts: Journal of Geophysical Research, v. 91, p. 6522–6534.

Peach, B. N., Horne, J., Gunn, W., Clough, C. T., Hinxman, L. W., and Teall, J. J. H., 1907, The geological structure of the Northwest Highlands of Scotland: Memoir of the Geological Survey of the United Kingdom.

Peacock, D. C. P., and Sanderson, D. J., 1992, Effects of layering and anisotropy on fault geometry: Journal of the Geological Society of London, v. 149, Part 5, p. 793–802.

Peacock, S., 1993, The importance of blueschist-eclogite dehydration reactions on subducting oceanic crust: Geological Society of America Bulletin, v. 105, p. 684–694.

Peltzer, G., Tapponnier, P., and Cobbold, P., 1982, An experimental study of intracontinental deformation in Asia: Eos, American Geophysical Union Transactions, 63, p. 1096–1097.

Phillips, F. C., 1971, The use of stereographic projection in structural geology: Edward Arnold, London, 90 p.

Phillips, J., 1844, Orientation movements in the parts of stratified rocks: British Association for the Advancement of Science Report 1843, p. 60–61.

Pitman, W. C., III, and Talwani, M., 1972, Sea-floor spreading in the North Atlantic: Geological Society of America Bulletin, v. 83, p. 619–646.

Plafker, G., 1965, Tectonic deformation associated with the 1964 Alaska earthquake: Science, v. 148, p. 1675–1687.

Plafker, G., 1976, Tectonic aspects of the Guatemala earthquake of 4 February 1976: Science, v. 193, p. 1201–1208.

Platt, J. P., 1986, Dynamics of orogenic wedges and the uplift of high-pressure metamorphic rocks: Geological Society of America Bulletin, v. 97, p. 1037–1053.

Platt, J. P., and Vissers, R. L. M., 1980, Extensional structures in anisotropic rocks: Journal of Structural Geology, v. 2, p. 397–410.

Pollard, D. D., and Aydin, A., 1988, Progress in understanding jointing over the past century: Geological Society of America Bulletin, v. 100, p. 1181–1204.

Pollard, D. D., and Segall, P., 1987, Theoretical displacements and stresses near fractures in rock; with applications to faults, joints, veins, dikes, and solution surfaces, *in* Atkinson, B. K. (ed.), Fracture mechanics of rock: Academic Press, London, p. 277–349.

Ponce De Leon, M. I., and Choukroune, P., 1980, Shear zones in the Iberian arc: Journal of structural geology, v. 2, p. 63–68.

Powell, C. McA., 1979, A morphological classification of rock cleavage, *in* Bell, T. H., and Vernon, R. H. (eds.), Microstructural processes during deformation and metamorphism: Tectonophysics, v. 58, p. 21–34.

Powell, J. W., 1873, Geological structure of a district of country lying to the north of the Grand Canyon of the Colorado: American Journal of Science, v. 5, p. 456–465.

Powell, J. W., 1875, Exploration of the Colorado River of the west and its tributaries: United States Government Printing Office, Washington, D. C., 291 p.

Price, N. J., 1959, Mechanics of jointing in rocks: Geological Magazine, v. 96, p. 149–167.

Price, N. J., and Cosgrove, J. W., 1990, Analysis of geological structures: Cambridge University Press, Cambridge, England, 502 p.

Price, R. A., and Mountjoy, E. W., 1970, Geologic structure of the Canadian Rocky Mountains between Bow and Athabasca rivers—progress report, *in* Wheeler, J. O. (ed.): Geological Association of Canada Special Paper 6, p. 7–25.

Price, R. A., Mountjoy, E. W., and Cook, G. G., 1978, Geologic map of Mount Goodsir (west half), British Columbia: Geological Survey of Canada, map 1477A, 1:50,000.

Proffett, J. M., Jr., 1977, Cenozoic geology of the Yerington district, Nevada, and its implications for the nature and origin of Basin and Range faulting: Geological Society of America Bulletin, v. 88, p. 247–266.

Quinquis, H., Audren, C., Brun, J. P., and Cobbold, P. R., 1978, Intense progressive shear in Ile de Groix blueschists and compatibility with subduction or obduction: Nature, v. 273, p. 43–45.

Ragan, D. M., 1969, Introduction to concepts of two-dimensional strain and their application with the use of card-deck models: Journal of Geological Education, v. 17, p. 135–141.

Ragan, D. M., 1973, Structural geology, an introduction to geometrical techniques (2nd edition): John Wiley & Sons, New York, 208 p.

Ragan, D. M., 1985, Structural geology, an introduction to geometrical techniques (3rd edition): John Wiley & Sons, New York, 393 p.

Ramberg, H., 1955, Natural and experimental boudinage and pinch-and-swell structures: Journal of Geology, v. 63, p. 512–526.

Ramberg, H., 1959, Evolution of ptygmatic folding: Norsk Geologisk Tidsskrift, v. 39, p. 99–151.

Ramberg, H., 1962, Contact strain and folding instability of a multilayered body under compression: Geologische Rundschau, v. 51, p. 405–439.

Ramberg, H., 1963, Evolution of drag folds: Geological Magazine, v. 100, p. 97–106.

Ramberg, H., 1967, Gravity, deformation and the Earth's crust as studied by centrifuged models: Academic Press, New York, 214 p.

Ramberg, H., 1973, Model studies in gravity-controlled tectonics by the centrifuge technique, *in* De Jong, K. A., and Scholten, R. (eds.), Gravity and tectonics: John Wiley & Sons, New York, p. 49–66.

Ramberg, H., 1975, Particle paths, displacement and progressive strain applicable to rocks: Tectonophysics, v. 28, p. 1–37.

Ramsay, J. G., 1958, Superposed folding at Loch Monar, Inverness-shire and Ross-shire: Quarterly Journal of the Geological Society of London, v. 113, p. 271–308.

Ramsay, J. G., 1967, Folding and fracturing of rocks: McGraw-Hill Book Company, New York, 560 p.

Ramsay, J. G., 1969, The measurement of strain and displacement in orogenic belts, *in* Kent, P. E., Satterthwaite, G. E., and Spencer, A. M. (eds.), Time and place in orogeny: Geological Society of London Special Publication 3, p. 43–79.

Ramsay, J. G., 1980a, Shear zone geometry: a review: Journal of Structural Geology, v. 2, p. 83–99.

Ramsay, J. G., 1980b, The crack-seal mechanism of rock deformation: Nature, v. 284, p. 135–139.

Ramsay, J. G., and Graham, R. H., 1970, Strain variation in shear belts: Canadian Journal of Earth Sciences, v. 7, p. 786–813.

Ramsay, J. G., and Huber, M. I., 1983, The techniques of modern structural geology, v. 1: Strain analysis: Academic Press, London, 307 p.

Ramsay, J. G., and Huber, M. I., 1987, The techniques of modern structural geology, v. 2: Folds and fractures: Academic Press, London, 381 p.

Rawnsley, K. D., Rives, T., Petit, J.-P., Hencher, S. R., and Lumsden, A. C., 1992, Joint development in perturbed stress fields near faults, *in* Burg, J.-P., Mainprice, D., and Petit, J.-P. (eds.), Mechanical instabilities in rocks and tectonics; a selection of papers: Journal of Structural Geology, v. 14, p. 939–951.

Reches, Z., 1976, Analysis of joints in two monoclines in Israel: Geological Society of America Bulletin, v. 87, p. 1654–1662.

Reches, Z., 1978a, Analysis of faulting in three-dimensional strain field: Tectonophysics, v. 47, p. 109–129.

Reches, Z., 1983, Faulting of rocks in three-dimensional strain fields: II. Theoretical analysis: Tectonophysics, v. 95, p. 133–156.

Reches, Z., and Dieterich, J. H., 1983, Faulting of rocks in three-dimensional strain fields: 1. Failure of rocks in polyaxial, servo-control experiments: Tectonophysics, v. 95, p. 111–132.

Reches, Z., and Johnson, A. M., 1978, Development of monoclines. Part II. Theoretical analysis of monoclines, *in* Matthews, V. (ed.), Laramide folding associated with basement block faulting in the western United States: Geological Society of America Memoir 151, p. 273–311.

Redfern, R., 1986, The making of a continent: American Geological Institute, Royal Smeets, 242 p.

Rehrig, W. A., and Heidrick, T. L., 1972, Regional fracturing in Laramide stocks of Arizona and its relationship to porphyry copper mineralization: Economic Geology, v. 67, p. 198–213.

Rehrig, W. A., and Heidrick, T. L., 1976, Regional tectonic stress during the Laramide and late Tertiary intrusive periods, Basin and Range province, Arizona: Arizona Geological Digest, v. 10, p. 205–228.

Reinhardt, J., 1992, Low-pressure, high-temperature metamorphism in a compressional tectonic setting: Mary Kathleen Fold Belt, northeastern Australia: Geological Magazine, v. 129, p. 41–57.

Reks, I. J., and Gray, D. R., 1982, Pencil structure and strain in weakly deformed mudstone and siltstone: Journal of Structural Geology, v. 4, p. 161–176.

Reynolds, S. J., 1988, Geologic map of Arizona: Arizona Geological Survey Map 26, scale 1:1,000,000.

Reynolds, S. J., and Lister, G. S., 1987, Structural aspects of fluid-rock interactions in detachment zones: Geology, v. 15, p. 362–366.

Reynolds, S. J., and Lister, G. S., 1990, Folding of mylonitic zones in Cordilleran metamorphic core complexes: evidence from near the mylonitic front: Geology, v. 18, p. 216–219.

Reynolds, S. J., Richard, S. M., Haxel, G. B., Tosdal, R. M., and Laubach, S. E., 1988, Geologic setting of Mesozoic and Cenozoic metamorphism in Arizona, *in* Ernst, W. G. (ed.), Metamorphism and crustal evolution of the western United States (Rubey Volume VII): Prentice Hall, Englewood Cliffs, New Jersey, p. 466–501.

Rich, J. L., 1934, Mechanics of low-angle overthrust faulting as illustrated by Cumberland thrust block, Virginia, Kentucky, and Tennessee: American Association of Petroleum Geologists Bulletin, v. 18, p. 1584–1596.

Richardson, R. M., 1992, Ridge forces, absolute plate motions, and the intraplate stress field, *in* Zoback, M. L. (ed.), The World Stress Map Project: Journal of Geophysical Research, v. 97, p. 11,739–11,748.

Richter, D., 1976, Allgemeine Geologie: Walter de Gruyter, Berlin, 366 p.

Ridgeway, J., 1920, Preparation of illustrations for the reports of the United States Geological Survey, with brief descriptions of processes of reproduction: United States Geological Survey, Washington, D. C., 101 p.

Riedel, W., 1929, Zur Mechanik geologischer Brucherscheinungen. Ein Beitrag zum Problem der "Fiederspalten": Centralblatt für Mineralogie, Geologie, und Paleontologie, Part B, p. 354–368.

Rigby, J. K., 1953, Some transverse stylolites: Journal of Sedimentary Petrology, v. 23, p. 265–271.

Rispoli, R., 1981, The stress fields about strike-slip faults inferred from stylolites and tension gashes: Tectonophysics, v. 75, p. 29–36.

Roberts, J. C., 1961, Feather fractures and the mechanics of rock jointing: American Journal of Science, v. 259, p. 481–492.

Rowan, M. G., and Kligfield, R., 1992, Kinematics of large-scale asymmetric buckle folds in overthrust shear: an example from the Helvetic nappes, *in* McClay, K. R., Thrust tectonics: Chapman and Hall, New York, p. 165–174.

Roy, A. B., 1978, Evolution of slaty cleavage in relation to diagenesis and metamorphism: a study from the Hunsruckschiefer: Geological Society of America Bulletin, v. 89, p. 1775–1785.

Royse, F., Jr., Warner, M. A., and Reese, D. L., 1975, Thrust belt structural geometry and related stratigraphic problems, Wyoming-Idaho-northern Utah, *in* Bolyard, D. W. (ed.), Deep drilling frontiers of the central Rocky Mountains Symposium: Rocky Mountain Association of Geologists, p. 41–54.

Rummel, F., 1987, Fracture mechanics approach to hydraulic fracturing stress measurements, *in* Atkinson, B. K., Fracture mechanics of rock: Academic Press, London, p. 217–239.

Rutter, E. H., 1976, The kinetics of rock deformation by pressure solution: Philosophical Transactions of the Royal Society of London, Series A, v. 283, p. 203–219.

Rutter, E. H., Maddock, R. H., Hall, S. H., and White, S. H., 1986, Comparative microstructures of natural and experimentally produced clay-bearing fault gouges: Pure and Applied Geophysics, v. 124, p. 3–30.

Ryan, H. F., and Scholl, D. W., 1989, The evolution of forearc structures along an oblique convergent margin, central Aleutian arc: Tectonics, v. 8, p. 497–516.

Ryan, M. P., and Sammis, C. G., 1978, Cyclic fracture mechanisms in cooling basalt: Geological Society of America Bulletin, v. 89, p. 1295–1308.

Rykkelid, E., and Andresen, A., 1994, Late Caledonian extension in the Ofoten area, northern Norway: Tectonophysics, v. 231, p. 157–169.

Salisbury, M. H., and Keen, C. E., 1993, Listric faults imaged in oceanic crust: Geology, v. 21, p. 117–120.

Sammis, C. G., Biegel, R., and King, G., 1986, A self-similar model for the kinematics of gouge deformation: American Geophysical Union Transactions, v. 67, p. 1187.

Sammis, C. G., King, G., and Biegel, R., 1987, The kinematics of gouge formation: Pure and Applied Geophysics, v. 125, p. 777–812.

Sander, B., 1930, Gefügekunde der Gesteine: Springer-Verlag, Vienna, 352 p.

Sander, B., 1970, The study of fabrics of geological bodies: Pergamon Press, New York, 641 p.

Sandiford, M., 1989, Horizontal structures in granulite terrains: a record of mountain building or mountain collapse?: Geology, v. 17, p. 449–452.

Sanford, A. R., 1959, Analytical and experimental study of simple geologic structures: Geological Society of America Bulletin, v. 70, p. 19–52.

Schmidt, C. J., James, C., and Shearer, J. N., 1981, Estimate of displacement in major zone of tear-faulting in fold and thrust belt, Southwest Montana: American Association of Petroleum Geologists Bulletin, v. 65, p. 986–987.

Scholz, C. H., 1968a, Experimental study of the fracturing process in brittle rock: Journal of Geophysical Research, v. 73, p. 1447–1454.

Scholz, C. H., 1968d, Microfracturing and the inelastic deformation of rock in compression: Journal of Geophysical Research, v. 73, p. 1417–1432.

Scholz, C. H., 1968e, The frequency-magnitude relation of microfracturing in rock and its relation to earthquakes: Seismological Society of America Bulletin, v. 58, p. 399–415.

Schreiber, J. F., Jr., 1974, Field descriptions of sedimentary rocks, *in* Davis, G. H. (ed.), Geology field camp manual: University of Arizona, Tucson, Arizona, p. 97–110.

Schulz, B., 1990, Prograde-retrograde P-T-t deformation path of Austroalpine micaschists during Variscan continental collision: Journal of Metamorphic Petrology, v. 8, p. 629–643.

Schweitzer, J., and Simpson, C., 1986, Cleavage development in dolomite of the Elbrook Formation, southwest Virginia: Geological Society of America Bulletin, v. 97, p. 754–764.

Scotese, C. R., 1991, Jurassic and Cretaceous plate tectonic reconstructions: Palaeogeography, palaeoclimatology, palaeoecology, v. 87, p. 493–501.

Secor, D. T., Jr., 1965, Role of fluid pressure in jointing: American Journal of Science, v. 263, p. 633–646.

Serway, R. A., 1990, Physics for Scientists and Engineers (3rd edition): Saunders, Philadelphia: 336 p.

Sharp, R. P., and Carey, D. L., 1976, Sliding stones, Racetrack Playa, California: Geological Society of America Bulletin, v. 87, p. 1704–1717.

Sharpe, D., 1847, On slaty cleavage: Geological Society of London Quarterly Journal, v. 3, p. 74–105.

Sherwin, J. A., and Chapple, W. M., 1968, Wavelengths of single layer folds: a comparison between theory and observation: American Journal of Science, v. 266, p. 167–179.

Shoemaker, E. M., 1979, Synopsis of the geology of Meteor Crater, *in* Shoemaker, E. M., and Kieffer, S. W. (eds.),

Geology of Meteor Crater, Arizona: Arizona State University Laboratory for Meteorite Studies, no. 17, p. 1–11.

Sibson, R. H., 1975, Fault rocks and fault mechanisms: Journal of the Geological Society of London, v. 133, p. 191–213.

Sibson, R. H., 1980, Transient discontinuities in ductile shear zones: Journal of Structural Geology, v. 1, p. 165–171.

Siddans, A. W. B., 1972, Slaty cleavage, a review of research since 1815: Earth Science Reviews, v. 8, p. 205–232.

Sieh, K. E., 1978, Prehistoric large earthquakes produced by slip on the San Andreas fault at Pallett Creek, California: Journal of Geophysical Research, v. 83, p. 3907–3939.

Simpson, C., 1986a, Microstructural evidence for northeastward movement on the Vincent-Orocopia-Chocolate Mountains thrust system, Geological Society of America, Cordilleran Section Abstracts with Programs, v. 18, p. 185.

Simpson, C., 1986b, Determination of movement sense in mylonites: Journal of Geological Education, v. 34, p. 246–261.

Simpson, C., 1988, Analysis of two-dimensional finite strain, *in* Marshak, S., and Mitra, G. (eds.), Basic methods of structural geology: Prentice-Hall, Englewood Cliffs, New Jersey, p. 333–359.

Simpson, C., and DePaor, D. G., 1993, Strain and kinematic analysis in general shear zones: Journal of Structural Geology, v. 15, p. 1–20.

Simpson, C., and Schmid, S. M., 1983, An evaluation of criteria to deduce the sense of movement in sheared rocks: Geological Society of America Bulletin, v. 94, p. 1281–1288.

Smiley, T. L., 1964, On understanding geochronological time: Arizona Geological Society Digest, v. 7, p. 1–12.

Smith, R. B., and Braile, L. W., 1993, Topographic signature, space-time evolution, and physical properties of the Yellowstone–Snake River volcanic system: the Yellowstone hotspot, *in* Snoke, A. W., Steidtmann, J. R., and Roberts, S. M. (eds.), Geology of Wyoming: Geological Survey of Wyoming Memoir No. 5, p. 694–754.

Smith, R. B., and Braile, L. W., 1994, The Yellowstone hotspot: Journal of Volcanology and Geothermal Research, v. 61, p. 121–187.

Smith, R. L., 1960a, Ash flows: Geological Society of America Bulletin, v. 71, p. 795–842.

Smith, R. L., 1960b, Zones and zonal variations in welded ash flows: United States Geological Survey Professional Paper 354-F, p. 149–159.

Smithson, S. B., Brewer, J., Kaufman, S., Oliver, J., and Hurich, C., 1978, Nature of Wind River thrust, Wyoming, from COCORP deep reflection data and from gravity data: Geology, v. 6, p. 648–652.

Sorby, H. C., 1853, On the origin of slaty cleavage: Edinburgh New Philosophical Journal, v. 55, p. 137–148.

Sorby, H. C., 1856, On slaty cleavage as exhibited in the Devonian limestones of Devonshire: Philosophical Magazine, v. 11, p. 20–37.

Sorensen, K., 1983, Growth and dynamics of the Nordre Stromfjord shear zone: Journal of Geophysical Research, v. 88, no. B4, p. 3419–3437.

Spencer, E. W., 1993, Geologic maps; a practical guide to the interpretation and preparation of geologic maps for geologists, geographers, engineers and planners: Macmillan Publishing Company, New York, 147 p.

Spencer, J., 1984, Role of tectonic denudation in warping and uplift of low-angle normal faults: Geology, v. 12, p. 95–98.

Spencer, E. W., 1969, Introduction to the structure of the Earth: McGraw-Hill, New York, 597 p.

Spray, J. G., 1989, Friction phenomena in rock: an introduction: Journal of Structural Geology, v. 11, p. 783–785.

Spry, A. H., 1969, Metamorphic textures: Pergamon Press, Oxford, 350 p.

Stearns, D. W., 1978, Faulting and forced folding in the Rocky Mountain foreland, *in* Matthews, V. (ed.), Laramide folding associated with basement block faulting in the western United States: Geological Society of America Memoir 151, p. 1–37.

Stein, R., and Eckstrom, G., 1992, Seismicity and geometry of a 110-km-long blind thrust fault: Journal of Geophysical Research, v. 97, no. B4, p. 4865–4883.

Stock, P., 1992, A strain model for antithetic fabric rotation in shear band structure: Journal of Structural Geology, v. 14, p. 1267–1275.

Stockdale, P. B., 1922, Stylolites: their nature and origin: Indiana University Studies, Bloomington, v. 9, 97 p.

Suppe, J., 1980a, A retrodeformable cross section of northern Taiwan: Geological Society of China Proceedings, no. 23, p. 46–55.

Suppe, J., 1980b, Imbricated structure of western foothills belt, south-central Taiwan: Petroleum Geology of Taiwan, no. 17, p. 1–16.

Suppe, J., 1983, Geometry and kinematics of fault-bend folding: American Journal of Science, v. 283, p. 684–721.

Suppe, J., 1985, Principles of structural geology: Prentice-Hall, Englewood Cliffs, New Jersey, 537 p.

Swan, M. M., 1976, The Stockton Pass fault: an element of the Texas lineament: M.S. thesis, University of Arizona, Tucson, 119 p.

Sykes, L. R., 1967, Mechanism of earthquakes and nature of faulting on the mid-oceanic ridges: Journal of Geophysical Research, v. 72, p. 2131–2153.

Sylvester, A. G., 1984, Palmdale Bulge, *in* Hester, R. L., and Hallinger, D. E. (eds.), San Andreas Fault Cajon Pass to Wrightwood: American Association of Petroleum Geologists Pacific Section Guidebook 55, p. 1.

Syme Gash, P. J., 1971, A study of surface features relating to brittle and semi-brittle fractures: Tectonophysics, v. 12, p. 349–391.

Tapponnier, P., Peltzer, G., Le Dain, A. Y., Armijo, R., and Cobbold, P., 1982, Propagating extrusion tectonics in Asia; new insights from simple experiments with plasticine: Geology, v. 10, p. 611–616.

Titley, S. R. (ed.), 1982, Advances in geology of the porphyry copper deposits: Southwest North America: University of Arizona Press, Tucson, 560 p.

Trudgill, B., and Cartwright, J., 1994, Relay-ramp forms and normal-fault linkages: Geological Society of America Bulletin, v. 106, p. 1143–1157.

Tullis, J., 1990a, Experimental constraints on flow laws for the crust: Eos, American Geophysical Union Transactions, v. 71, p. 1579.

Tullis, J., 1990b, Experimental studies of deformation mechanisms and microstructure in quartzo-feldspathic rocks, *in* Barber, D. J., and Meredith, P. G. (eds.), Deformation processes in minerals, ceramics and rocks: Unwin Hyman, London, p. 190–227.

Tullis, J., Dell'Angelo, L., and Yund, R. A., 1990, Ductile shear zones from brittle precursors in feldspathic rocks; the role of dynamic recrystallization, *in* Duba, A. G., Durham, W. B., Handin, J. W., and Wang, H. F. (eds.), The brittle-ductile transition in rocks: Geophysical Monograph, v. 56, p. 67–82.

Tullis, T. E., and Tullis, J., 1986, Experimental rock deformation techniques, *in* Hobbs, B. E., and Heard, H. C. (eds.), Mineral and rock deformation; laboratory studies; the Paterson volume: Geophysical Monograph, Monash University, Clayton, Victoria, Australia, v. 36, p. 297–324.

Tullis, T. E., and Wood, D. S., 1975, Correlation of finite strain from both reduction bodies and preferred orientation of mica in slate from Wales: Geological Society of America Bulletin, v. 86, p. 632–638.

Turner, F. J., and Weiss, L. E., 1963, Structural analysis of metamorphic tectonites: McGraw-Hill Book Company, New York, 560 p.

Twiss, R. J., and Moores, E. M., 1992, Structural geology: W. H. Freeman & Company, New York, 532 p.

Urai, J. L., Means, W. D., and Lister, G. S., 1986, Dynamic recrystallization of minerals, in Hobbs, B. E., and Heard, H. C. (eds.), Mineral and rock deformation: laboratory studies; the Paterson volume: Geophysical Monograph, v. 36, p. 161–199.

Urai, J. L., Spiers, C. J., Zwart, H. J., and Lister, G. S., 1986, Weakening of rock salt by water during long-term creep: Nature, v. 324, p. 554–557.

Urai, J. L., Williams, P. F., and van Roermund, H. L. M., 1991, Kinematics of crystal growth in syntectonic fibrous veins: Journal of Structural Geology, v. 13, p. 823–836.

U.S. Geological Survey, 1988, Crater Lake National Park and vicinity, Oregon: 1:62,500-scale topographic map with text and block diagrams, U.S. Geological Survey, Denver.

Uyeda, S., 1978, The new view of the Earth: W. H. Freeman & Company, San Francisco, 217 p.

Van den Driessche, J., and Brun, J.-P., 1987, Rolling structures at large shear strain, *in* Cobbold, P. R., Gapais, D., Means, W. D., and Treagus, S. H. (eds.), Shear criteria in rocks: Journal of Structural Geology, v. 9, p. 691–704.

Van Hise, R., 1896, Principles of North American pre-Cambrian geology: United States Geological Survey 16th Annual Report, Part 1, p. 581–844.

Vendeville, B., Cobbold, P. R., Davy, P., Brun, J. P., and Choukroune, P., 1987, Physical models of extensional tectonics at various scales, *in* Coward, M. P., Dewey, J. F., and Hancock, P. L. (eds.), Continental extensional tectonics: Geological Society of London Special Publication, v. 28: Blackwell Scientific Publications, London, p. 95–107.

Vendeville, B. C., and Jackson, M. P. A., 1992, The rise of diapirs during thin-skinned extension: Marine and Petroleum Geology, v. 9, p. 331–353.

Vine, F. J., and Matthews, D. H., 1963, Magnetic anomalies over oceanic ridges: Nature, v. 199, p. 947–949.

Wartolowska, J., 1972, An example of the processes of tectonic stylolitization: Bulletin de l'Académie Polonaise des Sciences, Series des Sciences de la Terre, v. 20, p. 197–204.

Wadi, D. N., 1953, Geology of India: Macmillan and Co., London, 531 p.

Wegener, A., 1915, Die Entstehung der Kontinente und Ozeane: Vieweg, Braunschweig.

Wegener, A., 1936, Die Entstehung der Kontinente und Ozeana, Feunfte unveraenderte Auflage (Die Wissenschaft Bd. 66—Wilhelm Westphal), xii: Braunschweig, Friedr., Vieweg & Sohn, 242 p.

Weijmers, R., 1993, Progressive deformation of single layers under constantly oriented boundary stress: Journal of Structural Geology, v. 15, p. 911–922.

Weiss, L. E., 1972, The minor structures of deformed rocks: a photographic atlas: Springer-Verlag, New York, 431 p.

Weiss, L. E., 1980, Nucleation and growth of kink bands: Tectonophysics, v. 65, p. 1–38.

Wernicke, B., 1981, Low-angle normal faults in the Basin and Range province: nappe tectonics in an extending orogen: Nature, v. 291, p. 645–648.

Wernicke, B. P., 1985, Uniform-sense normal simple shear of the continental lithosphere: Canadian Journal of Earth Sciences, v. 22, p. 108–125.

Wernicke, B., and Burchfiel, B. C., 1982, Modes of extension tectonics: Journal of Structural Geology, v. 4, p. 105–115.

Wernicke, B., Axen, G. L., and Snow, J. K., 1988, Basin and Range extensional tectonics at the latitude of Las Vegas, Nevada: Geological Society of America Bulletin, v. 100, p. 1738–1757.

Wheeler, R. L., and Dickson, J. M., 1980, Intensity of systematic joints, methods, and application: Geology, v. 8, p. 230–233.

White, S. H., Bretan, P. G., and Rutter, E. H., 1986, Fault-zone reactivation; kinematics and mechanisms: Royal Society of London Philosophical Transactions, Series A: Mathematical and Physical Sciences, v. 317, p. 81–92.

White, S. H., Burrows, S. E., Carreras, J., Shaw, N. D., and Humphreys, F. J., 1980, On mylonites in ductile shear zones, *in* Carreras, J., Cobbold, P. R., Ramsay, J. G., and White, S. H. (eds.), Shear zones in rocks: Journal of Structural Geology, v. 2, p. 175–187.

White, S. H., and Knipe, R. J., 1978, Transformation and

reaction-enhanced ductility in rocks: Geological Society of London Journal, v. 135, Part 5, p. 513–516.

Whitten, E. T. H., 1966, Structural geology of folded rocks: Rand-McNally, Skokie, Illinois, 663 p.

Wilcox, R. E., Harding, T. P., and Seely, D. R., 1973, Basic wrench tectonics: American Association of Petroleum Geologists Bulletin, v. 57, p. 74–96.

Williams, G. D., Powell, C. M., and Cooper, M. A., 1989, Geometry and kinematics of inversion tectonics, *in* Cooper, M. A., and Williams, G. D. (eds.), Inversion tectonics: Geological Society of London Special Publication No. 44, Blackwell Scientific Publications, London, England, p. 3–15.

Williams, H., 1942, The geology of Crater Lake National Park, Oregon: Carnegie Institute Publication 540, 162 p.

Williams, P. F., 1976, Relationships betwen axial plane foliation and strain: Tectonophysics, v. 30, p. 181–196.

Willis, B., 1894, The mechanics of Appalachian structure: United States Geological Survey 13th Annual Report, Part 2, p. 213–281.

Wilson, C. J. L., 1981, Experimental folding and fabric development in multilayered ice: Tectonophysics, v. 78, p. 139–159.

Wilson, C. J. L., 1986, Deformation induced recrystallization of ice: the application of in situ experiments: Geophysical monograph 36, p. 213–232.

Wilson, G., 1961, The tectonic significance of small-scale structures and their importance to the geologist in the field: Annales de la Societé Geologique de Belgique, v. 84, p. 424–548.

Wilson, G., 1982, Introduction to small-scale geologic structures: George Allen & Unwin (Publishers) Ltd., London, 128 p.

Wilson, J. T., 1965, A new class of faults and their bearing on continental drift: Nature, v. 207, p. 343–347.

Wise, D. U., 1964, Microjointing in basement, Middle Rocky Mountains of Montana and Wyoming: Geological Society of America Bulletin, v. 75, p. 287–306.

Woodcock, N. H., and Fischer, M., 1986, Strike-slip duplexes: Journal of Structural Geology, v. 8, p. 725–735.

Woodcock, N. J., and Schubert, C., 1994, Continental strike slip tectonics, *in* Hancock, P. L. (ed.), Continental deformation: Pergamon Press, New York, p. 251–263.

Woodward, N. B., Boyer, S. E., and Suppe, J., 1985, An outline of balanced cross sections: Studies in Geology (Knoxville), v. 11, 170 p.

Woodworth, J. B., 1896, On the fracture system of joints, with remarks on certain great fractures: Boston Society of Natural History Proceedings, v. 27, p. 163–184.

Worrall, D. M., 1977, Structural development of Round Mountain area, Uinta County, Wyoming, *in* Heisey, E. L. (ed.), Rocky Mountain thrust belt, geology and resources: Wyoming Geological Association Guidebook 29, p. 537–541.

Worrall, D. M., and Snelson, S., 1989, Evolution of the northern Gulf of Mexico, with emphasis on Cenozoic growth faulting and the role of salt, *in* Bally, A. W., and Palmer, A. R. (eds.), The geology of North America: An overview, DNAG, v. 4: Geological Society of America, Boulder, Colorado, p. 97–138.

Wright, L. A., and Troxel, B., 1973, Shallow fault interpretation of Basin and Range structure, southwestern Great Basin, *in* De Jong, K. A., and Scholten, R. (eds.), Gravity and tectonics: John Wiley & Sons, New York, p. 397–407.

Wright, T. O., and Platt, L. B., 1982, Pressure dissolution and cleavage in the Martinsburg Shale: American Journal of Science, v. 282, p. 122–135.

Yeats, R. S., Huftile, G. J., Grigsby, F. B., 1988, Oak Ridge fault, Ventura fold belt, and the Sisar decollement, Ventura basin, California: Geology, v. 16, p. 1112–1116.

Zen, E., White, W. S., Hadley, J. B., and Thompson, J. B., Jr. (eds.), 1968, Studies of Appalachian geology, northern and marine, Interscience Publishers, New York and London, 475 p.

Zhang, Y., Hobbs, B. E., and Jessell, M. W., 1993, Crystallographic preferred orientation development in a buckled single layer: a computer simulation: Journal of Structural Geology, v. 15, p. 265–276.

Zhao, G., and Johnson, A., 1992, Sequence of deformations recorded in joints and faults, Arches National Park, Utah: Journal of Structural Geology, v. 14, p. 225–236.

Zingg, T., 1935, Beitrag zur Schotteranalyze: Schweizer Mineralogische und Petrographische Mitteilungen, v. 15, p. 39–140.

Zoback, M. L., 1992, First and second order patterns of stress in the lithosphere; the World Stress Map Project, *in* Zoback, M. L. (ed.), The World Stress Map Project: Journal of Geophysical Research, Series B, Solid Earth and Planets, v. 97, p. 11,703–11,728.

AUTHOR INDEX

SUBJECT INDEX